Physics of Continuous Matter

Second Edition

Physics of Continuous Matter

Second Edition

Exotic and Everyday Phenomena in the Macroscopic World

B Lautrup
The Niels Bohr International Academy, The Niels Bohr Institute
Copenhagen, Denmark

CRC Press
Taylor & Francis Group
Boca Raton London New York

CRC Press is an imprint of the
Taylor & Francis Group, an **informa** business
A TAYLOR & FRANCIS BOOK

The front cover shows a section of the wake shed by flags that flap in a flowing soap film. The flow field is visualized using optical interference, which reveals the slight thickness variations in the thin film. Credits: Teis Schnipper (Technical University of Denmark) and Jun Zhang (New York University).

CRC Press
Taylor & Francis Group
6000 Broken Sound Parkway NW, Suite 300
Boca Raton, FL 33487-2742

First issued in paperback 2019

ISBN-13: 978-0-4200-7700-1 (hbk)
ISBN-13: 978-0-367-86511-5 (pbk)

Visit the Taylor & Francis Web site at
http://www.taylorandfrancis.com

and the CRC Press Web site at
http://www.crcpress.com

Contents

Preface

This second edition of *Physics of Continuous Matter* is, as the first, primarily an introduction to the basic ideas of continuum physics and their application to the wealth of macroscopic phenomena. The equations of continuum mechanics were developed between 1750 and 1850, and are so simple that they can be derived from Newtonian particle mechanics in a couple of pages. Unfortunately, they are generally unsolvable by analytic methods, although they have during the last half century yielded to direct numerical computation. Over the years many approximate methods have been developed that give insight into the rich physics hidden in the basic equations, and these methods take, as earlier, center stage in this book.

In this edition the main "story line" has been kept largely intact. Everywhere the physics arguments and their mathematical presentation have been rethought and to some extent rewritten to improve clarity and consistency. Some structural changes could not be avoided. An originally introductory chapter on Cartesian vector and tensor algebra has been moved into an appendix, while two chapters on gravity and stars have been dropped. The original chapter on ideal flow has been split into two chapters on, respectively, incompressible and compressible inviscid fluids. Conversely, the former chapter on lubrication is now joined with the chapter on creeping flow. A new chapter has been added on elasticity of slender rods, as well as two new chapters on, respectively, energy and entropy. After a single introductory chapter, this second edition proceeds immediately to the basics of continuum physics.

The book is aimed squarely at third-year undergraduate or older students. Although I originally taught second-year students from the first edition, much space and time were wasted on mathematical concepts that the students naturally would encounter in a second-year course on electromagnetism. Now these concepts are merely recapitulated in appendices that may be sampled as the need arises. The necessary mathematical tools are developed along with the physics on a "need-to-know" basis in order to avoid lengthy and boring mathematical preliminaries seemingly without purpose. Mathematical rigor is, as before, only used when it is necessary for a clear understanding of the physics.

The important thing to learn from this book is how to reason about physics, both qualitatively and—especially—quantitatively. Numerical simulations may be fine for obtaining solutions to practical problems, but are of very little aid in obtaining real understanding. Physicists must learn to think in terms of fundamental principles and generic methods. Solving one problem after another of a similar kind seems unnecessary and wasteful. This does not mean that the physicist should not be able to reach a practical result through calculation, but the physical principles behind equations and the conditions underlying approximations must never be lost from sight.

A structured text is, in my opinion, an important prerequisite for learning new material. Beyond parts, chapters, sections, subsections, and paragraphs, further structuring has been introduced in this edition by marking certain section and subsection headings with "Application" or "Case". These categories may be seen as "worked examples" and can mostly be included or left out according to the reader's preferences. Headers marked with an asterisk indicate that the subject matter is harder or falls outside the main line of presentation, and may be left out in a first reading without breaking the continuity of the study. The book is divided into six parts: Fluids at rest, Solids at rest, Fluids in motion, Balance and Conservation, Selected topics, and Appendices. The part headings now include a short description of the chapters contained in the part with some comments on what a minimal curriculum could contain.

Each chapter ends with a collection of problems that has been somewhat expanded in this edition. Most problems are meant to illustrate and further develop the physical and mathematical concepts introduced in the chapter text. Problems marked with an asterisk are generally harder than the unmarked ones. Answers to all odd-numbered problems are found in the back of the book. On request, a complete collection of solutions to all problems is available to course teachers, although my experience shows that such collections rapidly tend to leak into the hands of the students. Secrets are, as demonstrated many times, hard to keep in the age of the Internet. Anyway, students are strongly urged to try their hands on the problems before turning to the answers.

Illustrations have, as before, been included everywhere. Not only are there now more margin drawings to aid the understanding of the text, but also more photographs to lighten the presentation. Images resulting from simulations have been redone and improved. As in the first edition, micro-biographies of the major players in the historic development of continuum physics have been provided in the margin, and some have now been illustrated with portraits. The reader should again (and again) be warned that the history of continuum physics is much more complicated than can be learned here. Nevertheless, I believe that it is important for the understanding of this subject at least to be able to place the major contributors properly in time, space, and physics context.

The field of continuum physics is so highly developed that it cannot be given justice in a single textbook. Nearly every chapter in this book represents a separate subfield, each covered by a number of specialized textbooks and monographs, some of which are referred to in the list of references. There are numerous, more advanced topics that could—or perhaps should—have earned a place in this book, for example viscoelastic, anisotropic, and artificial materials; elasticity of membranes; non-linear elasticity; nonlinear surface waves; magneto- and plasma-hydrodynamics; turbulence modeling; finite volume numerics; and so on. What is and is not included reflects the intended level, my personal predilections and, of course, the limitations imposed by the size of the book[1]. But I hope that students of this book will have acquired the basic tools necessary for entering into topics not covered here.

My colleagues through many years John Renner Hansen, Poul Olesen, Poul Henrik Damgaard, and Mogens Høgh Jensen are thanked for their kind and generous support. Besides the persons mentioned in the first edition, special thanks go to Anders Andersen, Hassan Aref, Tomas Bohr, Predrag Cvitanovic, Joachim Mathiesen, and Niels Kjær Nielsen, and to all the students around the world who have sweated over it. Finally, I thank my wife Birthe Østerlund for her unwavering support during the ardors of the preparation of the second edition over the past two years.

The book is written for adults with a serious intention to learn physics. I have selected for readers what I think are the interesting topics in continuum physics, and presented these as pedagogically as I can without trying to cover everything. I sincerely hope that my own joy in understanding and explaining the physics shines through everywhere.

Benny Lautrup
Copenhagen, Denmark

[1]Extra material may be found at the book's homepage, which also contains the programs used for working out the figures, examples, cases, and applications. See `http://www.lautrup.nbi.dk/continuousmatter2`.

1 Continuous matter

The everyday experience of the smoothness of matter is an illusion. Since the beginning of the twentieth century it has been known with certainty that the material world is composed of microscopic atoms and molecules, responsible for the macroscopic properties of ordinary matter. Long before the actual discovery of molecules, chemists had inferred that something like molecules had to exist, even if they did not know how big they were. Molecules *are* small—so small that their existence may be safely disregarded in all our daily doings. Although everybody possessing a powerful microscope will note the irregular Brownian motion of small particles in a liquid, it took quite some mental effort to move from the everyday manipulation of objects and recognize that this is a sign that molecules are really there.

Continuum physics deals with the systematic description of matter at length scales that are large compared to the molecular scale. Most macroscopic length scales occurring in practice are actually huge in molecular units, typically in the hundreds of millions. This enormous ratio of scales isolates theories of macroscopic phenomena from the details of the microscopic molecular world. A general meta-law of physics claims that the physical laws valid at one length scale are not very sensitive to the details of what happens at much smaller scales. Without this meta-law, physics would in fact be impossible, because we never know what lies below our currently deepest level of understanding.

The microscopic world impinges upon the macroscopic almost only through material constants, such as coefficients of elasticity and viscosity, characterizing the interactions between macroscopic amounts of matter. It is, of course, an important task for the physics of materials to derive the values of these constants, but this task lies outside the realm of continuum physics. It is nevertheless sometimes instructive to make simple models of the underlying atomic or molecular structure in order to obtain an understanding of the origin of macroscopic phenomena and of the limits to the continuum description.

This chapter paints in broad outline the transition from molecules to continuous matter, or mathematically speaking from point particles to fields. It is emphasized that the macroscopic continuum description must necessarily be statistical in nature, but that random statistical fluctuations are strongly suppressed by the enormity of the number of molecules in any macroscopic material object. The modern fields of nanophysics and biophysics straddle the border between the continuum and particle descriptions of matter, resulting in numerous new phenomena outside the scope of classical continuum physics. These topics will not be covered here. The central theme of this book is the recasting of Newton's laws for point particles into a systematic theory of continuous matter, and the application of this theory to the wealth of exotic and everyday phenomena in the macroscopic material world.

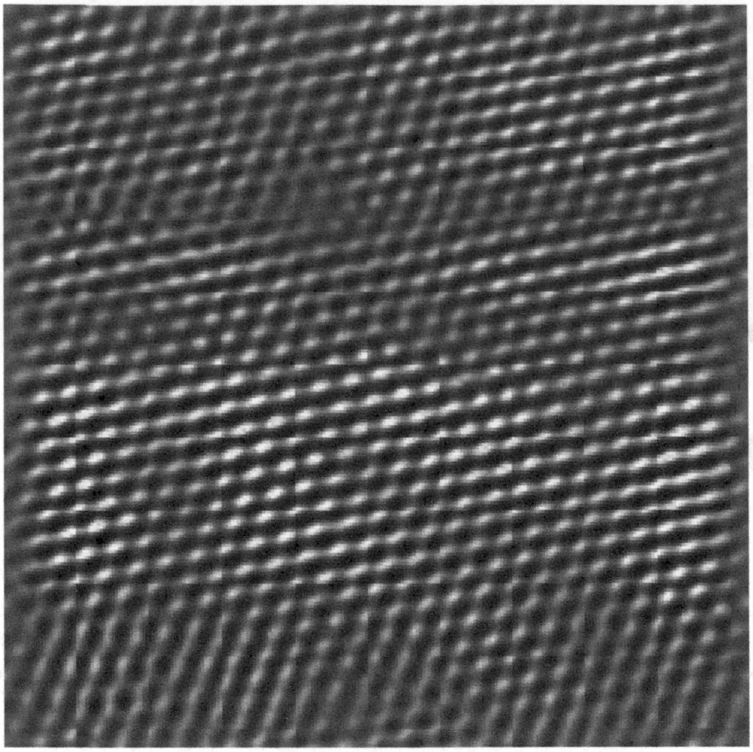

Figure 1.1. How continuous matter really looks at the atomic scale. Noise-filtered image of freshly cleaved Mica obtained by atomic force microscopy, approximately 225 Angstrom on a side. This granularity of matter is ignored in continuum physics. (Source: Mark J. Waner, PhD dissertation, Michigan State University, 1998. With permission.)

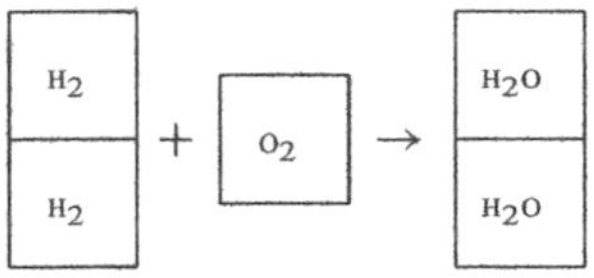

The quantitative meaning of a chemical formula. The boxes represent fixed amounts, for example moles.

1.1 Molecules

Chemical reactions such as $2H_2 + O_2 \rightarrow 2H_2O$ are characterized by simple integer coefficients. Two measures of hydrogen plus one measure of oxygen yield two measures of water without anything left over of the original ingredients. What are these measures? For gases at the same temperature and pressure, a measure is simply a fixed volume, for example a liter, so that two liters of hydrogen plus one liter of oxygen yield two liters of water vapor, assuming that the water vapor without condensing can be brought to the same temperature and pressure as the gases had before the reaction. In 1811, Count Avogadro of Italy proposed that the simple integer coefficients in chemical reactions between gases could be explained by the rule that equal volumes of gases contain equal numbers of molecules (at the same temperature and pressure).

Lorenzo Romano Amadeo Carlo Avogadro (1776–1856). Italian philosopher, lawyer, chemist, and physicist. Count of Quaregna and Cerratto. Formulated that equal volumes of gas contain equal numbers of molecules. Also argued that simple gases consist of diatomic molecules. (Source: Wikimedia Commons.)

The various measures do not weigh the same. A liter of oxygen is roughly 16 times heavier than a liter of hydrogen at the same temperature and pressure. The mass of any amount of water vapor must—of course—be the sum of the masses of its ingredients, hydrogen and oxygen. The reaction formula tells us that two liters of water vapor weigh roughly $(2 \times 1) + (1 \times 16) = 18$ times a liter of hydrogen. Such considerations led to the introduction of the concept of relative molecular mass (or weight) in the ratio 1:16:9 for molecular hydrogen, molecular oxygen, and water. Today, most people would prefer to write these proportions as 2:32:18, reflecting the familiar molecular masses of H_2, O_2, and H_2O, respectively. In practice, relative molecular masses deviate slightly from integer values, but for the sake of argument we shall disregard that here.

Mole and molar mass

In the beginning there was no way of fixing an absolute scale for molecular mass. To define a scale that was practical for the chemist at work in his laboratory, the *molar mass* of atomic hydrogen (H) was arbitrarily set to be 1 gram. The ratios of molecular masses obtained from chemical reactions would then determine the molar mass of any other substance. Thus the molar mass of hydrogen gas (H_2) is 2 grams and that of oxygen gas (O_2) is 32 grams, whereas water (H_2O) has a molar mass of 18 grams because the chemical reaction tells us that $(2\times 2)+(1\times 32) = 2\times 18$ grams. This system could be extended to all chemical reactions allowing the determination of molar mass for any substance participating in such processes.

An amount of a substance with mass equal to its molar mass is called a *mole* and the symbol used for the unit is mol. Thus 1 gram of atomic hydrogen, 2 grams of molecular hydrogen, 32 grams of molecular oxygen, or 18 grams of water all make up one mole. The chemical reaction formula $2H_2 + O_2 \rightarrow 2H_2O$ simply expresses that 2 moles hydrogen gas plus 1 mole oxygen gas produces 2 moles water. According to Avogadro's hypothesis, *the number of molecules in a mole of any substance is the same*, appropriately called Avogadro's number by Perrin and denoted N_A. In 1908 Perrin carried out the first modern determination of its value from Brownian motion experiments. Perrin's experiments relying on Einstein's recent (1905) theory of Brownian motion were not only seen as a confirmation of this theory but also as the most direct evidence for the reality of atoms and molecules.

Jean-Baptiste Perrin (1870–1942). French physicist. Received the Nobel Prize for his work on Brownian motion in 1926. He founded several French scientific institutions, among them the now-famous *Centre National de la Recherche Scientifique (CNRS)*.

Today, Avogadro's number is defined to be the number of atoms in exactly 12 grams of the fundamental carbon isotope ^{12}C, which therefore has molar mass equal to exactly 12 g mol^{-1}. Avogadro's number is determined empirically, and the accepted 2006 value [1] is $N_A = 6.02214179(30) \times 10^{23}$ molecules per mole, with the parenthesis indicating the absolute error on the last digits.

Unit of mass: The definition of Avogadro's number depends on the definition of the unit of mass, the kilogram, which is (still) defined by a prototype from 1889 stored by the *International Bureau of Weights and Measures* near Paris, France. Copies of this prototype and balances for weighing them can be made to a precision of one part in 10^9. Maybe already in 2011 a new definition of the kilogram will replace this ancient one [MMQ&06], for example by defining the kilogram to be the total mass of an exact number of ^{12}C atoms. Avogadro's number will then also become an exact number without error.

Molecular separation length

Consider a sample of a pure substance with volume V and mass M. If the molar mass of the substance is denoted $M_{\rm mol}$, the number of moles in the sample is $n = M/M_{\rm mol}$, and the number of molecules $N = nN_A$. The volume per molecule is V/N, and a cube with this volume would have sides of length

$$L_{\rm mol} = \left(\frac{V}{N}\right)^{1/3} = \left(\frac{M_{\rm mol}}{\rho N_A}\right)^{1/3}, \tag{1.1}$$

where $\rho = M/V$ is the mass density. This *molecular separation length* sets the scale at which the molecular granularity of matter dominates the physics, and any conceivable continuum description of bulk matter must utterly fail.

For liquids and solids where the molecules touch each other, this length is roughly the size of a molecule. For solid iron we get $L_{\rm mol} \approx 0.23$ nm, and for liquid water $L_{\rm mol} \approx 0.31$ nm. Since by Avogadro's hypothesis equal gas volumes contain an equal number of molecules, the molecular separation length for *any* (ideal) gas at normal temperature and pressure ($p = 1$ atm and $T = 20°C$) becomes $L_{\rm mol} \approx 3.4$ nm. There is a lot of vacuum in a volume of gas, in fact about 1000 times the true volume of the molecules at normal temperature and pressure.

$*$ **Mixtures**[1]**:** The above expression for L_{mol} may also be used for a mixture of pure substances, provided M_{mol} is taken to be a suitable average over the molar masses M_i^{mol} of the i-th pure component (consisting of only one kind of molecules). For a mixture sample of mass M containing the mass M_i of each component, the total mass becomes the sum $M = \sum_i M_i$. The number of moles of the i-th component is $n_i = M_i/M_i^{\text{mol}}$ and the total number of moles in the sample is $n = \sum_i n_i$. Characterizing the composition of the mixture by the *molar fraction* $X_i = n_i/n$ of each component, the average molar mass, $M_{\text{mol}} = M/n$, becomes

$$M_{\text{mol}} = \sum_i X_i M_i^{\text{mol}}, \tag{1.2}$$

where $\sum_i X_i = 1$. If we instead describe the composition by the *mass fraction* $Y_i = M_i/M = X_i M_i^{\text{mol}}/M_{\text{mol}}$ of each component, and use that $\sum_i X_i = 1$, the average molar mass is determined by the reciprocal sum,

$$\frac{1}{M_{\text{mol}}} = \sum_i \frac{Y_i}{M_i^{\text{mol}}}, \tag{1.3}$$

where $\sum_i Y_i = 1$.

Dry air is a molar mixture of 78.08% nitrogen, 20.95% oxygen and 0.93% argon with an average molar mass of $M_{\text{mol}} = 28.95 \text{ g mol}^{-1}$. By mass the mixture is 75.56% nitrogen, 23.15% oxygen, and 1.29% argon, and has of course the same average molar mass.

Intermolecular forces and states of matter

Johannes Diederik van der Waals (1837–1923). Dutch physicist. Developed an equation of state for gases, now carrying his name. Received the Nobel Prize in 1910 for his work on liquids and gases.

Apart from the omnipresent gravitational interaction between all bodies, material interactions are entirely electromagnetic in nature, from the fury of a tornado to the gentlest kiss. A detailed understanding of the so-called van der Waals forces acting between neutral atoms and molecules falls outside the scope of this book. Generally, however, the forces are strongly repulsive if the molecules are forced closer than their natural sizes allow, and moderately attractive when they are moved apart. This tug of war between repulsion and attraction determines an equilibrium distance between them that is comparable to the molecular size. In Figure 1.2 is shown an example of such a potential, the famous Lennard-Jones potential (see Problem 1.2).

When huge numbers of molecules are put together to make up a body, they may arrange themselves in a number of different ways to minimize their total energy. The total energy receives negative contributions from the intermolecular potential energy, which attempts to bind the molecules to each other near equilibrium, and positive contributions from the kinetic energy in their thermal motion, which tends to make them fly apart. The three classic states of neutral matter—solid, liquid, and gas—depend, broadly speaking, on the competition between negative binding energy and positive thermal energy.

Solids: In *solid matter* the binding is so strong that thermal motion cannot overcome it. The molecules remain bound to each other by largely elastic forces, and constantly undergo small-amplitude thermal motion around their equilibrium positions. If increasing external forces are applied, solids will begin to *deform* elastically, until they eventually become plastic or even fracture. A solid body retains its shape independently of the shape of a container large enough to hold it, apart from small deformations, for example due to gravity.

Liquids: In *liquid matter* the binding is weaker than in solid matter, although it is still hard for a molecule on its own to leave the company of the others through an open liquid surface. The molecules stay in contact but are not locked to their neighbors. Molecular conglomerates may form and stay loosely connected for a while, as for example chains of water molecules.

[1]The asterisk indicates that this part of the text can be skipped in a first reading.

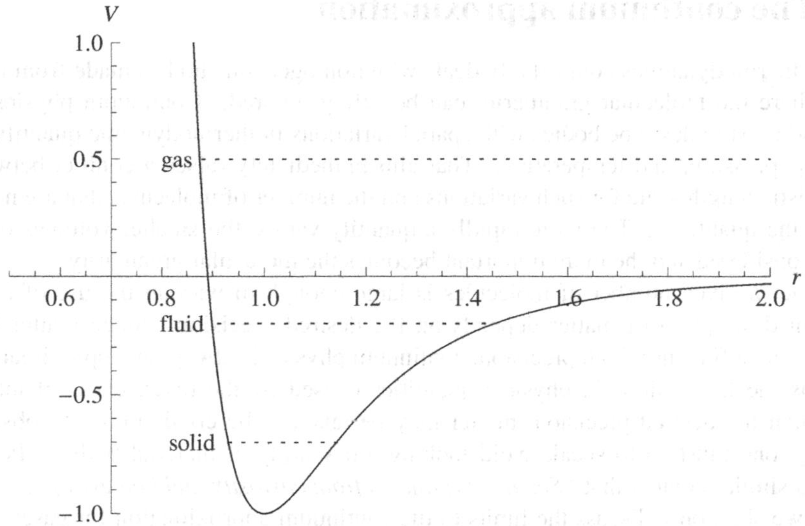

Figure 1.2. Sketch of the intermolecular potential energy $V(r)$ between two roughly spherical neutral molecules as a function of the distance r between their centers. It is attractive at moderate range and strongly repulsive at close distance. The equilibrium between attraction and repulsion is found at the minimum of the potential. Here the units are arbitrarily chosen such that the equilibrium distance becomes $r = 1$ and the minimum potential $V(1) = -1$. The horizontal dashed lines suggest the thermal energy levels for solid, liquid, and gaseous matter.

Under the influence of external forces, for example gravity, a liquid will undergo bulk motion, called *flow*, a process that may be viewed as a kind of continual fracturing. A liquid will not expand to fill an empty container completely, but will under the influence of external forces eventually adapt to its shape wherever it touches it.

Gases: In *gaseous matter* the molecules are bound so weakly that the thermal motion easily overcomes it, and they essentially move around freely between collisions. A gas will always expand to fill a closed empty container completely. Under the influence of external forces, for example a piston pushed into the container, a gas will quickly flow to adapt to the changed container shape.

Granular matter

Molecules and atoms represent without discussion the basic granularity of all macroscopic matter, but that does not mean that all macroscopic matter can be viewed as continuous. Barley grain, quartz sand, living beings, buildings, and numerous materials used in industry are examples of matter having a discrete, granular substructure that influences the material properties even at considerably larger length scales. The main difference between granular matter and molecular matter lies in the friction that exists between the granular elements. Although grain may flow like water in great quantities, the friction between the grains can make them stick to each other, forming a kind of solid that may block narrow passages. A heap of grain does not flatten out like a puddle of water, and neither do the wind-blown ripples of dry sand in the desert. You may also build a castle of wet sand, the main part of which remains standing even when dried out.

In this book we shall not consider granular matter as such, although at sufficiently large length scales, granular matter may in many respects behave like continuous matter, whether the grains are tiny as the quartz crystals in sand or enormous as the galaxies in the universe.

1.2 The continuum approximation

Classical thermodynamics comfortably deals with homogeneous bodies made from moles of matter where the molecular granularity can be safely ignored. Continuum physics on the other hand aims to describe bodies with spatial variations in thermodynamic quantities, such as density, pressure, and temperature. That aim immediately raises a conflict between the characteristic length scale for such variations and the number of molecules that are necessary to define the quantities. The more rapidly a quantity varies, the smaller volumes of matter must be considered and the more important becomes the molecular granularity.

Bernard le Bovier de Fontenelle (1657–1757). French intellectual, poet, author, and philosopher. Member of both the French Academy of Letters and of Sciences. Made popular accounts of the theories of René Descartes, whom he admired greatly.

Whether a given number of molecules is large enough to warrant the use of a smooth continuum description of matter depends on the desired precision. Since matter is never continuous at sufficiently high precision, continuum physics is always an approximation. But as long as the fluctuations in physical quantities caused by the discreteness of matter are smaller than the desired precision, matter may be taken to be continuous. To observe the continuity, one must so to speak avoid looking too sharply at material bodies. Fontenelle stated in a similar context that *"Science originates from curiosity and bad eyesight"*.

Here we shall only discuss the limits to the continuum approximation for gases. Similar limits exist for solids and liquids but are more difficult to estimate. The gas results may nevertheless be used as an upper limit to the fluctuations (see Section 23.3 on page 398).

Density fluctuations

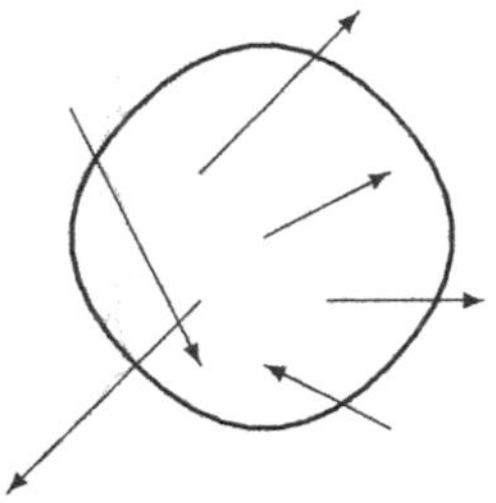

In a gas the molecules move rapidly in and out of a small volume with typical velocities comparable to the speed of sound.

Consider a fixed small volume V of a pure gas with molecules of mass m. If at a given time t the number of molecules in this volume is N, the mass density at this time becomes

$$\rho = \frac{Nm}{V}. \tag{1.4}$$

Due to rapid random motion of the gas molecules, the number N will be different at a later time $t + \Delta t$. Provided the time interval Δt is much larger than the time interval τ between molecular collisions, the molecules in the volume V will at the later time be an essentially random sample of molecules taken from a much larger region around V. The probability that any particular molecule from this larger region ends up in the volume V will be tiny. From general statistical considerations it follows that the root-mean-square size of the fluctuations in the number of molecules is given by $\overline{\Delta} N = \sqrt{N}$ (see Figure 1.3, and Problem 1.1). Since the density is linear in N, the relative fluctuation in density becomes

$$\frac{\overline{\Delta}\rho}{\rho} = \frac{\overline{\Delta} N}{N} = \frac{1}{\sqrt{N}}. \tag{1.5}$$

In classical macroscopic thermodynamics where typically $N \sim N_A$, the relative fluctuation becomes of magnitude 10^{-12} and can safely be ignored.

In continuum physics this is not so. If we want the relative density fluctuation to be smaller than a given value $\overline{\Delta}\rho/\rho \lesssim \epsilon$, we must require $N \gtrsim \epsilon^{-2}$. The smallest acceptable number of molecules, ϵ^{-2}, occupies a volume $\epsilon^{-2} L_{\text{mol}}^3$, where L_{mol} is the molecular separation length (1.1). A cubic cell with this volume has side length

$$\boxed{L_{\text{micro}} = \epsilon^{-2/3} L_{\text{mol}}.} \tag{1.6}$$

Thus, to secure a relative precision $\epsilon = 10^{-3}$ for the density, the microscopic cell should contain 10^6 molecules and have side length $L_{\text{micro}} = 100 L_{\text{mol}}$. For an ideal gas under normal conditions we find $L_{\text{micro}} \approx 0.34\ \mu\text{m}$.

The micro scale diverges for $\epsilon \to 0$, substantiating the claim that *it is impossible to maintain a continuum description to arbitrarily high precision.*

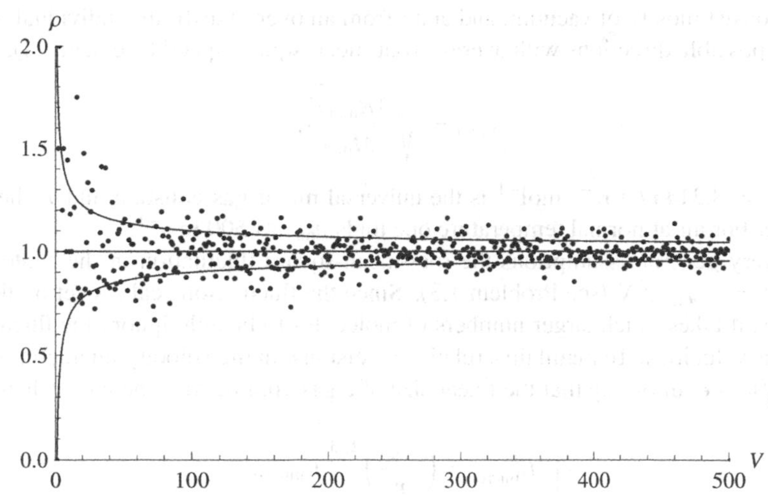

Figure 1.3. Simulation of density fluctuations as a function of volume size. A three-dimensional "universe" consisting of $20 \times 20 \times 20 = 8{,}000$ cells of unit volume is randomly filled with as many unit mass "molecules". On average each of the 8,000 cells should contain a single molecule, corresponding to a density of $\rho = 1$. A "volume" consisting of V cells will not contain precisely V molecules, and thus has a density that deviates from unity. The plot shows the density of a random collection of V cells as a function of V. The drawn curves, $\rho = 1 \pm 1/\sqrt{V}$, indicate the expected fluctuations.

Macroscopic smoothness

As a criterion for a smooth continuum description we demand that the relative variation in density between neighboring cells should be less than the desired precision ϵ. The change in density between the centers of neighboring cells along some direction x is of magnitude $\Delta\rho \approx L_{\text{micro}} \, |\partial\rho/\partial x|$. Demanding the relative variation to be smaller than the precision, $\Delta\rho/\rho \lesssim \epsilon$, we obtain a constraint on the magnitude of the density derivative,

$$\left|\frac{\partial\rho}{\partial x}\right| \lesssim \frac{\rho}{L_{\text{macro}}}, \tag{1.7}$$

where

$$\boxed{L_{\text{macro}} = \epsilon^{-1} L_{\text{micro}}.} \tag{1.8}$$

As long as the above condition is fulfilled, the density may be considered to vary smoothly, because the density changes over the micro-scale are imperceptible. Any significant change in density must take place over distances larger than L_{macro}. With $\epsilon = 10^{-3}$ we find $L_{\text{macro}} \approx 1000 L_{\text{micro}} \approx 0.34$ mm for an ideal gas under normal conditions.

The thickness of interfaces between macroscopic bodies is typically on the order of L_{mol} and thus much smaller than L_{macro}. Consequently, these regions of space fall outside the smooth continuum description. In continuum physics, interfaces appear instead as surface discontinuities in the otherwise smooth macroscopic description of matter.

Velocity fluctuations

Everyday gas speeds are small compared to the molecular velocities—unless one is traveling by jet aircraft or cracking a whip. What we normally mean by wind is the bulk drift of air, not the rapid molecular motions. So even if the individual molecules move very fast in random directions, the center of mass of a collection of N molecules in a small volume V will normally move with much slower velocity, which for large N approximates the drift speed v.

Gases consist mostly of vacuum, and apart from an overall drift, the individual molecules move in all possible directions with average root-mean-square speed (see page 24),

$$v_{\text{mol}} = \sqrt{\frac{3R_{\text{mol}}T}{M_{\text{mol}}}}, \tag{1.9}$$

where $R_{\text{mol}} = 8.31447\ \text{J K}^{-1}\ \text{mol}^{-1}$ is the universal molar gas constant and T the absolute temperature. For air at normal temperature one finds $v_{\text{mol}} \approx 500\ \text{m s}^{-1}$.

Under very general assumptions the root-mean-square fluctuation in the center-of-mass speed is $\Delta v = v_{\text{mol}}/\sqrt{N}$ (see Problem 1.3). Since the fluctuation scale is set by the molecular velocity, it takes much larger numbers of molecules to be able ignore the fluctuations in everyday gas velocities. To maintain a relative precision ϵ in the velocity fluctuations we must require $\Delta v/v \lesssim \epsilon$, implying that the linear size of a gas volume must be larger than

$$\boxed{L'_{\text{micro}} = \left(\frac{v_{\text{mol}}}{v}\right)^{2/3} L_{\text{micro}}.} \tag{1.10}$$

The velocity fluctuations of a gentle steady wind, say $v \approx 0.5\ \text{m s}^{-1}$, can be ignored with precision $\epsilon \approx 10^{-3}$ for volumes of linear size larger than $L'_{\text{micro}} = 100L_{\text{micro}} \approx 34\ \mu\text{m}$. The smoothness scale should similarly be $L'_{\text{macro}} = \epsilon^{-1}L'_{\text{micro}}$, which in this case becomes $L'_{\text{macro}} \approx 34$ mm. A hurricane wind, $v \approx 50\ \text{m s}^{-1}$, only requires volumes of linear size $L'_{\text{micro}} = 4.6L_{\text{micro}} \approx 1.6\ \mu\text{m}$ to yield the desired precision, but in this case fluctuations due to turbulence will anyway completely swamp the molecular fluctuations.

* Mean free path

Another condition for obtaining a valid continuum description is that molecules should interact with each other to "iron out" strong differences in velocities. If there were no interactions, a molecule with a given velocity would keep on moving with that velocity forever. In solids and liquids where the molecules are closely packed, these interactions take place over a molecular separation length.

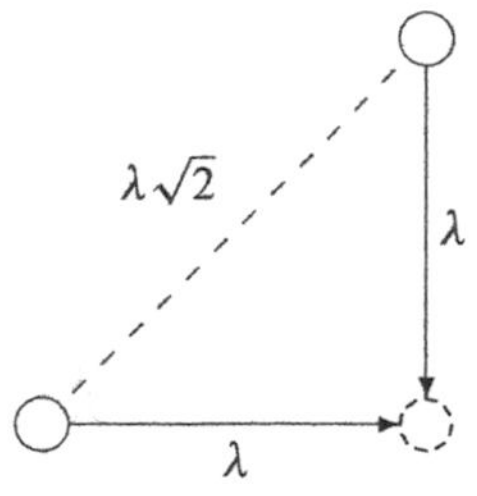

Two molecules on collision course with equal orthogonal velocities. Each moves the distance λ between collisions, and thus $\lambda\sqrt{2}$ relative to each other.

In a gas every molecule traces out a straight path through the vacuum until it collides with another molecule. The *mean free path* λ is defined as the average distance traveled by a single molecule. The other molecule is, however, not a "sitting duck" but travels itself on average also a distance λ before it is hit by the first. Since the movement directions are arbitrary, the second molecule "sees" the first coming toward it along a straight line with a speed that on average is orthogonal to the first and thus $\sqrt{2}$ times the root-mean-square speed v_{mol}. Denoting the molecular diameter by d, the collision will only happen if the center of the second molecule is within a "striking distance" d from the path of the first, that is, inside a cylinder of radius d. There is on average one molecule in the molecular volume L_{mol}^3, so the mean free path is defined such that the average volume swept out by the moving spheres equals the molecular volume, or $\pi d^2\sqrt{2}\lambda = L_{\text{mol}}^3$. The mean free path thus becomes

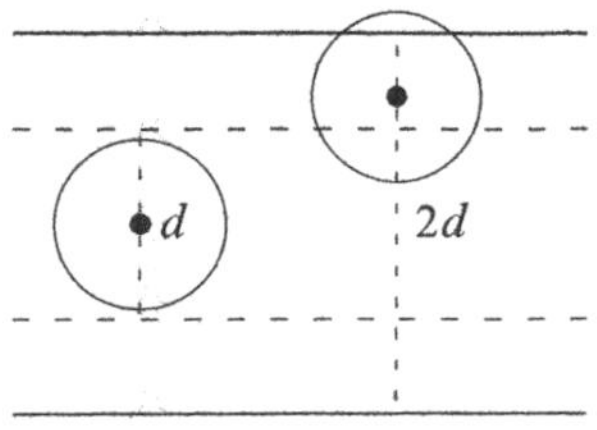

A sphere of diameter d will collide with any other sphere of the same diameter with its center inside a cylinder of diameter $2d$.

$$\lambda = \frac{L_{\text{mol}}^3}{\sqrt{2}\pi d^2} = \frac{M_{\text{mol}}}{\sqrt{2}\pi d^2 \rho N_A}. \tag{1.11}$$

Since $\lambda/L_{\text{mol}} \sim \rho^{-2/3}$, the mean free path will in sufficiently dilute gases always become larger than the micro scale and should instead be used to define the smallest linear scale for the continuum description.

Air consists mainly of nitrogen and argon molecules with an average molar mass $M_{\text{mol}} \approx 29\ \text{g mol}^{-1}$ and average diameter $d \approx 0.37$ nm [2]. At normal temperature and pressure the mean free path becomes $\lambda \approx 65$ nm, which is five times smaller than the microscopic length scale for the density, $L_{\text{micro}} \approx 340$ nm (for $\epsilon = 10^{-3}$). The mean collision time may be estimated as $\tau \approx \lambda/v_{\text{mol}}$, and becomes for air $\tau \approx 0.13$ ns.

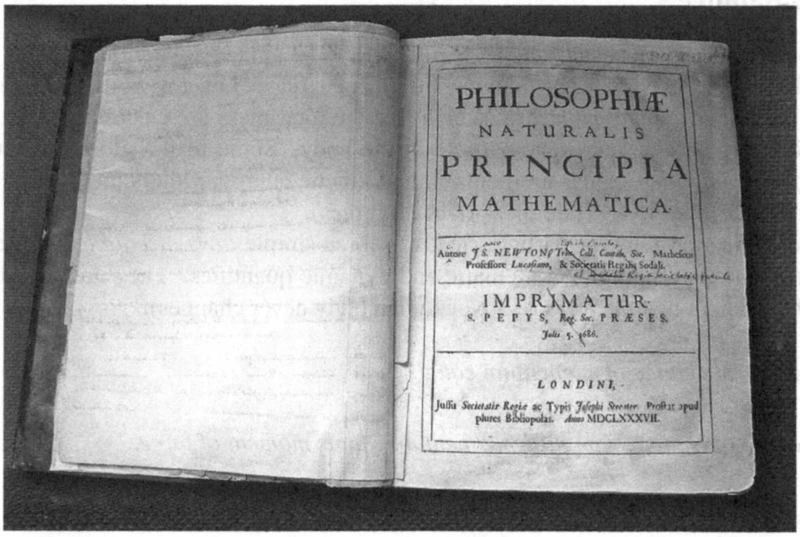

Figure 1.4. Newton's own copy of *Principia* with handwritten corrections for the second edition. (Source: Andrew Dunn (2004). Wikimedia Commons.)

1.3 Newtonian mechanics

In Newtonian mechanics the elementary material object is a point particle ("molecule") with a fixed mass. Newton postulated three laws for such particles, which formed the basis for rational mechanics in the following centuries and became a role model for all other natural sciences. It is not the intention here to enter into a discussion of the consistency of these laws or other objections that can be raised, but just to present them in a short form supposed to be suitable for any reader of this book. Here they are:

Sir Isaac Newton (1643–1727). English physicist and mathematician. Founded classical mechanics on three famous laws in his books *Philosophiae Naturalis Principia Mathematica* (1687). Newton developed calculus to solve the equations of motion, and formulated theories of optics and of chemistry. He still stands as perhaps the greatest scientific genius of all time. (Source: Wikimedia Commons.)

1. *There exist (inertial) reference frames in which a particle moves with constant velocity along a straight line when it is not acted upon by any forces.*
2. *The mass times the acceleration of a particle equals the sum of all forces acting on it.*
3. *If a particle acts on another with a certain force, the other particle acts back on the first with an equal and opposite force.*

In Appendix A you will find a little refresher course in Newtonian mechanics.

Newton's Second Law is *the* fundamental equation of motion. Mathematically, this law is expressed as a second-order differential equation in time. Since the force acting on any given particle can depend on the positions and velocities of the particle itself and of other particles as well as on external parameters, the dynamics of a collection of particles becomes a web of coupled ordinary second-order differential equations in time. Even if macroscopic bodies *are* huge collections of atoms and molecules, it is completely out of the question to solve the resulting web of differential equations. In addition, there is the problem that molecular interactions are quantum mechanical in nature, so that Newtonian mechanics does not apply at the atomic level. This knowledge is, however, relatively new and has as mentioned earlier some difficulty in making itself apparent at the macroscopic level. So although quantum mechanics definitely rules the world of atoms, its special character is rarely amplified to macroscopic proportions, except in low-temperature phenomena such as superconductivity and superfluidity.

Laws of balance

In Newtonian particle mechanics, a "body" is taken to be a fixed collection of point particles with unchangeable masses, each obeying the Second Law. For any body one may define various *global mechanical quantities*, which like the total mass are calculated as sums over contributions from each and every particle in the body. Some of the global quantities are *kinematic*: momentum, angular momentum, and kinetic energy. Others are *dynamic*: force, moment of force, and power (rate of work of all forces).

Newton's Second Law for particles leads to three simple *laws of balance*—often called conservation laws—relating the kinematic and dynamic quantities. They are (in addition to the trivial statement that the mass of a Newtonian body never changes):

- *The rate of change of momentum equals force,*
- *The rate of change of angular momentum equals moment of force,*
- *The rate of change of kinetic energy equals power.*

Even if these laws are insufficient to determine the dynamics of a multi-particle body, they represent seven individual constraints (two vectors and one scalar) on the motion of any collection of point particles, regardless of how complex it is.

In continuum mechanics any volume of matter may be considered to be a body. As the dynamics unfolds, matter is allowed to be exchanged between the environment and the body, but we shall see that *the laws of balance can be directly carried over to continuum mechanics when exchange of matter between a body and its environment is properly taken into account.* Combined with simplifying assumptions about the macroscopic behavior, for example symmetry, the laws of balance for continuous matter also turn out to be quite useful for obtaining quick solutions to a variety of problems.

Material particles

In continuum physics we generally speak about *material particles* as the elementary constituents of continuous matter, obeying Newton's equations. A material particle will always contain a large number of molecules but may in the continuum description be thought of as infinitesimal or point-like. From the preceding analysis we know that material particles cannot be truly infinitesimal, but represent *the smallest bodies that may consistently be considered part of the continuum description within the required precision.* Continuum physics does not "on its own" go below the level of the material particles. Even if the mass density may be determined by adding together the masses of all the molecules in a material particle and dividing the sum by the volume of the particle, this procedure falls, strictly speaking, outside continuum physics.

Although we normally think of material particles as being identical in different types of matter, it is sometimes necessary to go beyond the continuum approximation and look at their differences. In solids, we may with some reservation think of *solid particles* as containing a fixed collection of molecules, whereas in liquids and especially in gases we should not forget that the molecules making up a *fluid particle* at a given instant will shortly be replaced by other molecules. If the molecular composition of the matter in the environment of a material particle has a slow spatial variation, this incessant molecular game of "musical chairs" may slowly change the composition of the material inside the particle. Such *diffusion processes* driven by spatial variations in material properties lie at the very root of fluid mechanics. Even a spatial velocity variation will drive momentum diffusion, causing internal (viscous) friction in the fluid.

1.4 Reference frames

Physics is a quantitative discipline using mathematics to relate measurable quantities expressed in terms of real numbers. In formulating the laws of nature, undefined mathematical primitives—for example the points, lines and circles of Euclidean geometry—are not particularly useful, and such concepts have for this reason been eliminated and replaced by numerical representations everywhere in physics. This step necessitates a specification of the practical procedures by which these numbers are obtained in an experiment; for example, which units are being used.

Behind every law of nature and every formula in physics, there is a framework of procedural descriptions, a *reference frame*, supplying an operational meaning to all physical quantities. Part of the art of doing physics lies in comprehending this, often tacitly understood, infrastructure to the mathematical formalism. The reference frame always involves physical objects—balances to measure mass, clocks to measure time and rulers to measure length—that are not directly a part of the mathematical formalism. Precisely because they *are* physical objects, they can at least in principle be handed over or copied, and thereby shared among experimenters. This is in fact still done for the unit of mass (see page 3).

Agreement on a common reference frame for all measurable quantities is a necessary condition for doing science. The system of units, the *Système Internationale (SI)*, is today fixed by international agreement; but even if our common frame of reference for units is thus defined by social convention, physics is nevertheless objective. In principle our frames of reference could be shared with all other beings in the universe, or alternatively given a precise translation to theirs.

Time

Time is the number you read on your *clock*. There is no better definition. Clocks are physical objects that may be shared, compared, copied, and synchronized to create an objective meaning of *time*. Most clocks, whether they are grandfather clocks with a swinging pendulum or oscillating quartz crystals, are based on periodic physical systems that return to the same state again and again. Time intervals are simply measured by counting periods. There are also aperiodic clocks, for example hour glasses, and clocks based on radioactive elements. It is especially the latter that allow time to be measured on geological time scales. On extreme cosmological time scales the very concept of time becomes increasingly more theory laden; see for example [RZ09].

Like all macroscopic physical systems, clocks are subject to small fluctuations in the way they run. Some clocks are considered better than others because they keep time more stably with respect to copies of themselves as well as with clocks built on other principles. Grandfather clocks are much less stable than mechanical maritime chronometers that in turn are less stable than modern quartz clocks. The international frame of reference for time is always based on the most stable clocks currently available.

Unit of time: The unit of time, the *second*, was formerly defined as 1/86,400 of a mean solar day. But the Earth's rotation is not that stable, and since 1966 the second has been defined by international agreement as the duration of 9,192,631,770 oscillations of the microwave radiation absorbed in a certain hyperfine transition in cesium-133, a metal that can be found anywhere on Earth [1]. A beam of cesium-133 atoms is used to stabilize a quartz oscillator at the right frequency by a resonance method, so what we call an atomic clock is really an atomically stabilized quartz clock. The intrinsic precision in this time standard has been continually improving and is now about 4×10^{-16} corresponding to about 1 second in 80 million years [LHJ07].

In the extreme mathematical limit, time may be taken to be a real number, and in Newtonian physics its value is assumed to be universally knowable.

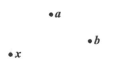

Figure 1.5. Independently of how different their reference frames, two observers who agree on the unique reality of any point in space can in principle determine the coordinate transformation relating them by listing their respective coordinates for each and every point in space. (Source: Fragment of "The Creation of Adam", by Michelangelo Buonarroti (1511). Wikimedia Commons.)

Space

It is a mysterious and so far unexplained fact that physical *space* has three dimensions, which means that it takes exactly three real numbers to locate a point in space. These numbers are called the *coordinates* of the point, and the reference frame for coordinates is called the *coordinate system.* It must contain all the operational specifications for locating a point given the coordinates, and conversely obtaining the coordinates given the location. In this way we have relegated all philosophical questions regarding the *real* nature of points and of space to the operational procedures contained in the reference frame.

$\bullet a$

$\bullet b$

$\bullet x$

Points may be visualized as dots on a piece of paper. Each point is labeled by its position in the chosen coordinate system (not visualized here).

Earth coordinates: On Earth everybody navigates by means of a geographical coordinate system agreed upon by international convention, in which a point is characterized by latitude δ, longitude λ, and elevation h. Latitude and longitude are angles fixed by the Earth's rotation axis and the position of the former Royal Observatory in Greenwich near London (UK). Elevation is defined as the signed height above the average sea level. The modern Global Positioning System (GPS) uses instead "fixed points in the sky" in the form of at least 24 satellites, and the geographical coordinates of any point on Earth as well as the absolute time in this point is determined from radio signals received from four or more of these satellites.

The triplet of coordinates that locates a point in a particular coordinate system is called its *position* in that coordinate system, and usually marked with a single symbol printed in boldface[2], for example $\boldsymbol{x} = (x_1, x_2, x_3)$. Given the position $\boldsymbol{x} = (x_1, x_2, x_3)$ of a point in one coordinate system, we now require that the position $\boldsymbol{x}' = (x_1', x_2', x_3')$ of the exact same point in another coordinate system must be calculable from the first:

$\bullet \boldsymbol{a} \leftrightarrow \boldsymbol{a}'$

$\bullet \boldsymbol{b} \leftrightarrow \boldsymbol{b}'$

$\bullet \boldsymbol{x} \leftrightarrow \boldsymbol{x}'$

In different coordinate systems the same points have different coordinates, connected by a transformation $\boldsymbol{x}' = \boldsymbol{f}(\boldsymbol{x})$.

$$x_1' = f_1(x_1, x_2, x_3), \qquad x_2' = f_2(x_1, x_2, x_3), \qquad x_3' = f_3(x_1, x_2, x_3). \tag{1.12}$$

More compactly this may be written $\boldsymbol{x}' = \boldsymbol{f}(\boldsymbol{x})$.

The postulate that there should exist a unique, bijective transformation connecting any given pair of coordinate systems reflects that *physical reality is unique* (see Figure 1.5), and

[2]The printed boldface notation is hard to reproduce in calculations with pencil on paper, so other graphical means are commonly used, for example a bar ($\overline{x}$), an arrow ($\vec{x}$), or underlining ($\underline{x}$).

that *different coordinate systems are just different ways of representing the same physical space in terms of real numbers.*

What's in a symbol?: There is nothing sacred about the symbols used for the coordinates. Mostly the coordinates are given suggestive or conventional symbolic names in particular coordinate systems, for example δ, λ, and h for the Earth coordinates above. In physics the familiar Cartesian coordinates are often denoted x, y, and z, while in more formal arguments one may retain the index notation, x_1, x_2, and x_3. Cylindrical coordinates are denoted r, ϕ, and z, and spherical coordinates r, θ, and ϕ. These three coordinate systems are in fact the only ones to be used in this book.

Length

From the earliest times humans have measured the *length* of a road between two points, say $\boldsymbol{a}$ and $\boldsymbol{b}$, by counting the number of steps it takes to walk along this road. This definition of length depends, however, on how you are built. In order to communicate to others the length of a road, the count of steps must be accompanied by a clear definition of the length of a step in terms of an agreed-upon unit of length.

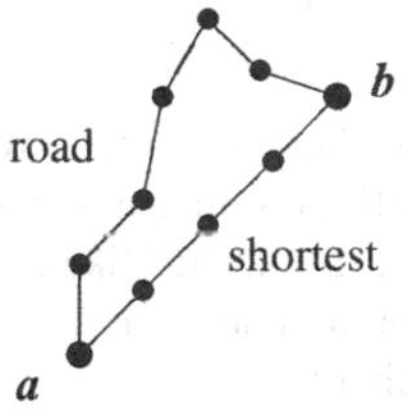

The length of the road between the positions $\boldsymbol{a}$ and $\boldsymbol{b}$ is measured by counting steps along the road. Different roads have different lengths, but normally there is a unique shortest road.

Unit of length: Originally the units of length—inch, foot, span, and fathom—were directly related to the human body, but increasing precision in technology demanded better-defined units. In 1793 the meter was introduced as a ten millionth of the distance from equator to pole on Earth, and until far into the twentieth century a unique "normal meter" was stored in Paris, France. In 1960 the meter became defined as a certain number of wavelengths of a certain spectral line in krypton-86, an isotope of a noble gas that can be found anywhere on Earth. Since 1983 the meter has been defined by international convention to be the distance traveled by light in exactly 1/299,792,458 of a second [1]. The problem of measuring lengths has thus been transferred to the problem of measuring time, which makes sense because the precision of the time current standard is at least a thousand times better than any proper length standard.

This method for determining the length of a path may be refined to any desired practical precision by using very short steps. In the extreme mathematical limit, the steps become infinitesimally small, and the road becomes a continuous path.

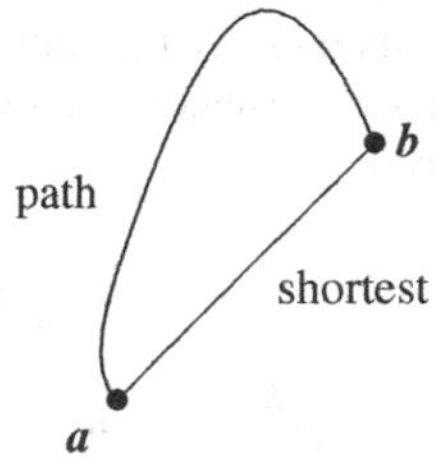

In the mathematical limit the shortest continuous path connecting $\boldsymbol{a}$ and $\boldsymbol{b}$ is called the geodesic. Normally, there is only one geodesic between any two points.

Distance

The shortest path between two points is called a *geodesic* and represents the "straightest line" between the points. In Euclidean space, a geodesic is indeed what we would intuitively call a straight line. On the spherical surface of the Earth geodesics are great circles, and airplanes and ships travel along them for good reason. The *distance* between two points is defined to be the length of a geodesic connecting them. Since the points are completely defined by their coordinates $\boldsymbol{a}$ and $\boldsymbol{b}$ in the chosen coordinate system, the distance must be a real positive function $d(\boldsymbol{a}, \boldsymbol{b})$ of the two sets of coordinates.

It is clear that the distance between two points must be the same in all coordinate systems, because it can, in principle, be determined by laying out rulers or counting steps between two points without any reference to coordinate systems. *Distance is a property of space rather than a property of the coordinate system.* The actual distance function $d'(\boldsymbol{a}', \boldsymbol{b}')$ in a new coordinate system may be different from the old, $d(\boldsymbol{a}, \boldsymbol{b})$, but the numerical values have to be the same,

$$d'(\boldsymbol{a}', \boldsymbol{b}') = d(\boldsymbol{a}, \boldsymbol{b}), \tag{1.13}$$

where $\boldsymbol{a}' = \boldsymbol{f}(\boldsymbol{a})$ and $\boldsymbol{b}' = \boldsymbol{f}(\boldsymbol{b})$ are calculated by the coordinate transformation (1.12). Knowing the distance function $d(\boldsymbol{a}, \boldsymbol{b})$ in one coordinate system, it may be calculated in any other coordinate system by means of the appropriate coordinate transformation.

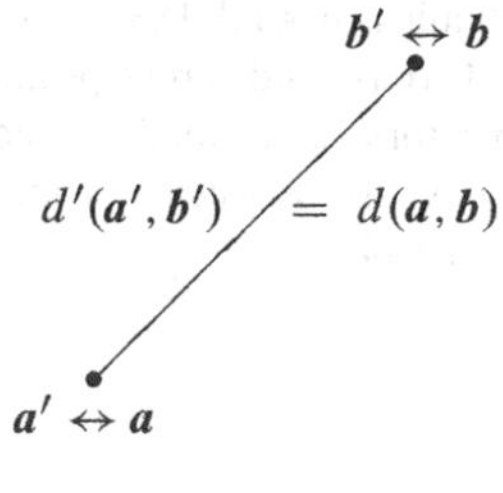

The distance is invariant under coordinate transformations.

1.5 Cartesian coordinate systems

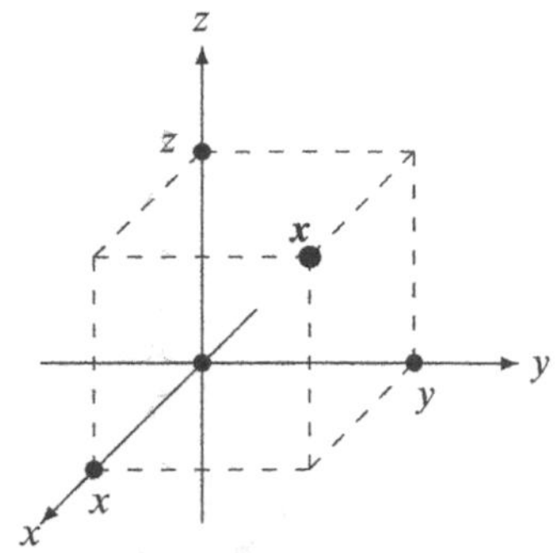

A Cartesian coordinate system, here with coordinates labeled x, y, and z. Looks just like the ones you got to know (and love) in high school.

The space in which we live is nearly *flat* everywhere. Its geometry is Euclidean, meaning that Euclid's axioms and the theorems deduced from them are valid everywhere. After Einstein we know, however, that space is not perfectly flat. In the field of gravity from a massive body, space necessarily curves, but normally only very little. At the surface of the Earth the radius of curvature of space due to Earth's gravity is comparable to the distance to the Sun. The kind of physics that is the subject of this book may always be assumed to take place in perfectly flat Euclidean space.

The Cartesian distance function

In Euclidean geometry one derives Pythagoras' theorem, which relates the lengths of the sides of any right-angled triangle. The simplicity of Pythagoras' theorem favors the use of right-angled *Cartesian coordinate systems* in which the distance between two points in space is the square root of the sum of the squares of their coordinate differences.

$$d(\boldsymbol{a}, \boldsymbol{b}) = \sqrt{(a_1 - b_1)^2 + (a_2 - b_2)^2 + (a_3 - b_3)^2}. \tag{1.14}$$

René Descartes (1596–1650). French scientist and philosopher, father of analytic geometry. Developed a theory of mechanical philosophy, later to be superseded by Newton's work. Confronted with doubts about reality, he saw thought as the only argument for existence: "I think, therefore I am".

Cartesian coordinates provide by far the most compact description of the geometry of flat space, and we shall in this book systematically describe physics "through the eyes" of Cartesian coordinates—even when we employ cylindrical or spherical coordinates.

Cartesian vectors

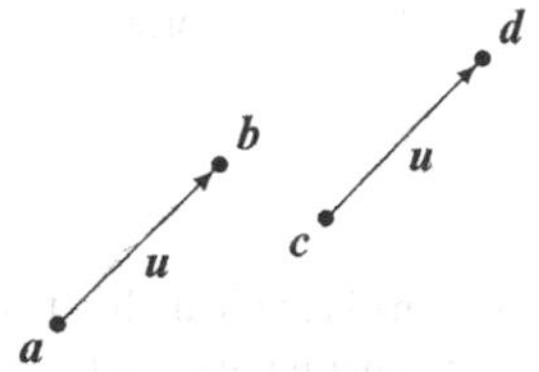

The vector $\boldsymbol{u}$ connects the point $\boldsymbol{a}$ with the point $\boldsymbol{b}$. The same vector also connects the points $\boldsymbol{c}$ and $\boldsymbol{d}$.

Since the distance between two points only depends on coordinate differences, triplets of coordinate differences naturally play a major role in Cartesian coordinate systems. Such triplets are called *vectors* and are marked with boldface in the same way as positions, for example $\boldsymbol{u} = (u_1, u_2, u_3) = (b_1 - a_1, b_2 - a_2, b_3 - a_3)$, which is constructed from the coordinate differences of the positions $\boldsymbol{b}$ and $\boldsymbol{a}$. Geometrically, this vector may be visualized as an arrow connecting $\boldsymbol{a}$ to $\boldsymbol{b}$. As there is no "memory" in a vector about the absolute positions of the points, the same vector $\boldsymbol{u}$ will *carry*[3] you from any given position to another, provided the positions have the same coordinate difference. A position triplet $\boldsymbol{x}$ connects the origin of the coordinate system $\boldsymbol{0} = (0, 0, 0)$ to the particular point in space located by $\boldsymbol{x}$, and positions may for this reason also be viewed as vectors (in Cartesian coordinates[4]). The mathematical concept of a vector is usually attributed to W. R. Hamilton.

William Rowan Hamilton (1805–1865). Irish mathematical physicist. Created the Hamiltonian formulation of classical mechanics around 1833. Introduced dot- and cross-products of vectors and applied vector analysis to physical problems around 1845.

The structure of the Cartesian distance function makes it natural to define linear operations on vectors, such as multiplication with a constant, $k\boldsymbol{u} = (ku_1, ku_2, ku_3)$; addition, $\boldsymbol{u} + \boldsymbol{v} = (u_1 + v_1, u_2 + v_2, u_3 + v_3)$; and subtraction, $\boldsymbol{u} - \boldsymbol{v} = (u_1 - v_1, u_2 - v_2, u_3 - v_3)$. Mathematically, this makes the set of all vectors a three-dimensional vector space. Other operations, such as the dot-product, $\boldsymbol{u} \cdot \boldsymbol{v} = u_1 v_1 + u_2 v_2 + u_3 v_3$, and the cross-product, $\boldsymbol{u} \times \boldsymbol{v} = (u_2 v_3 - u_3 v_2, u_3 v_1 - u_1 v_3, u_1 v_2 - u_2 v_1)$, may also be defined and have simple geometric interpretations.

In Appendix B Cartesian vector and tensor algebra is set up and analyzed in full detail. Most readers should already have met and worked with geometric vectors and will only need to consult this appendix in rare cases, for example to get a proper understanding of tensors and coordinate transformations. Coordinate transformations are central to analytic geometry and lead to the characterization of geometric quantities—scalars, vectors, and tensors—by the way they transform rather than in often ill-defined geometric terms.

[3]The word "vector" is Latin for "one who carries", derived from the verb "vehere", meaning to carry, and also known from "vehicle". In epidemiology a "vector" is a carrier of disease.

[4]In non-Euclidean spaces or in curvilinear coordinate systems, the vector concept is only meaningful for infinitesimal coordinate differences in the infinitesimal neighborhood of any point (the local Euclidean tangent space).

1.6 Fields

In the extreme mathematical limit, material particles are taken to be truly infinitesimal and all physical properties of the particles as well as the forces acting on them are described by smooth—or at least piecewise smooth—functions of space and time. Continuum physics is therefore a theory of *fields*. Mathematically, a field f is simply a real-valued function, $f(\boldsymbol{x}, t) = f(x_1, x_2, x_3, t)$ or $f(\boldsymbol{x}, t) = f(x, y, z, t)$, of the spatial coordinates $\boldsymbol{x}$ and time t, *representing the value of a physical quantity in this point of space at this time*[5].

English	*field*
German	*feld*
Dutch	*veld*
Danish	*felt*
Swedish	*fält*
French	*champ*
Italian	*campo*
Spanish	*campo*
Russian	*polje*

The use of the word "field" in physics to denote a function of the spacetime coordinates has an unclear origin. The original meaning of phrases such as "gravitational field", "electric field", or "magnetic field" was presumably to denote regions of gravitational, electric, or magnetic influences in the otherwise empty space around a body. The meaning was later shifted to the mathematical representation of the strength (and direction) of such influences in every point of space.

The fields of continuum physics

We have already met the mass density field $\rho(\boldsymbol{x}, t)$. Knowing this field, the mass dM of a material particle occupying the volume dV near the point $\boldsymbol{x}$ at time t can be calculated as

$$dM = \rho(\boldsymbol{x}, t)\, dV. \tag{1.15}$$

We shall mostly suppress the explicit space and time variables and just write $dM = \rho\, dV$.

Sometimes a collection of functions is also called a field and the individual real-valued members are called its *components*. The most fundamental field of fluid mechanics, the velocity field $\boldsymbol{v} = (v_1, v_2, v_3)$ or $\boldsymbol{v} = (v_x, v_y, v_z)$, has three components, one for each of the Cartesian coordinate directions. The velocity field $\boldsymbol{v}(\boldsymbol{x}, t)$ determines the momentum,

$$d\mathcal{P} = \boldsymbol{v}(\boldsymbol{x}, t)\, dM, \tag{1.16}$$

of a material particle of mass dM near $\boldsymbol{x}$ at time t. The velocity field will be of major importance in formulating the dynamics of continuous systems.

Besides fields characterizing the state of the material, such as mass density and velocity, it is convenient to define fields that characterize the forces acting on and within the material. The gravitational acceleration field $\boldsymbol{g}(\boldsymbol{x}, t)$ penetrates all bodies from afar and acts on a material particle of mass dM with a force

$$d\mathcal{F} = \boldsymbol{g}(\boldsymbol{x}, t)\, dM. \tag{1.17}$$

Using that $dM = \rho\, dV$ we may also write gravity as a *body force* (or *volume force*) of the form, $d\mathcal{F} = \boldsymbol{f}\, dV$, with a density of force $\boldsymbol{f} = \rho \boldsymbol{g}$. It has infinite range, and the same is true for electromagnetism, but that also ends the list. No other forces in nature seem to have infinite range.

Some force fields are only meaningful for regions of space where matter is actually present, as for example the density ρ or the pressure field p, which acts across the imagined contact surfaces that separate neighboring volumes of a fluid at rest. Pressure is, however, not the only *contact force*. Fluids in motion, solids and more general materials, have more complicated contact forces that can only be fully described by the nine-component stress field, $\boldsymbol{\sigma} = \{\sigma_{ij}\}$, which is a (3×3) matrix field with rows and columns labeled by Cartesian coordinates: $i, j = 1, 2, 3$ or $i, j = x, y, z$.

Mass density, velocity, gravity, pressure, and stress are the usual fields of continuum mechanics and will all be properly introduced in the chapters to come. Other fields are thermodynamic, like the temperature T, the specific internal energy U, or the specific entropy S. Some describe different states of matter, for example the electric charge density ρ_e and current density $\boldsymbol{j}_e$ together with the electric and magnetic field strengths, $\boldsymbol{E}$ and $\boldsymbol{B}$. Like gravity $\boldsymbol{g}$, these force fields are thought to exist in regions of space completely devoid of matter.

[5] In mathematics the tendency is to use coordinates labeled by integers. In physics we shall—as mentioned before—mostly label the coordinate axes x, y, and z, and use these labels as vector indices $\boldsymbol{u} = (u_x, u_y, u_z)$. The general position is denoted $\boldsymbol{x} = (x, y, z)$ and becomes admittedly a bit inconsistent because of the strange relations, $x_x = x$, $x_y = y$ and $x_z = z$. In many physics texts the general position is instead denoted $\boldsymbol{r} = (x, y, z)$, but that comes with its own esthetic problems in more formal analysis. There seems to be no easy way to get the best of both worlds.

There are also fields that refer to *material properties*, for example the coefficient of shear elasticity μ of a solid and the coefficient of shear viscosity η of a fluid. Such fields are often nearly constant within homogeneous bodies, that is, independent of space and time, and are mostly treated as *material constants* rather than true fields.

Field equations

Like all physical variables, fields evolve with time according to dynamical laws, called *field equations*. In continuum mechanics, the central equation of motion descends directly from Newton's Second Law applied to every material particle. Mass conservation, which is all but trivial and most often tacitly incorporated in particle mechanics, turns in continuum theory into an equation of motion for the mass density. Still other field equations such as Maxwell's equations for the electromagnetic fields have completely different and non-mechanical origins, although they do couple to the mechanical equations of motion via the Lorentz force.

Mathematically, field equations are *partial differential equations* in both space and time. This makes continuum mechanics considerably more difficult than particle mechanics where the equations of motion are ordinary differential equations in time. On the other hand, this greater degree of mathematical complexity also leads to a plethora of new and sometimes quite unexpected phenomena. Mathematically, field equations in three-dimensional space would be quite cumbersome to deal with, were it not for an efficient extension of vector methods to what is now called called *vector calculus* or *field calculus*. It is introduced along the way in the chapters to come and presented with some rigor in Appendix C.

In some field theories, for example Maxwell's electromagnetism, the field equations are *linear* in the fields, but that is not the case in continuum mechanics. The *non-linearity* of the field equations of continuum mechanics is caused by the velocity field, which behaves like a "wind" that carries other fields (and itself) along in the motion. This adds a further layer of mathematical difficulty to this subject, making it very different from linear theories—and much richer. The non-linearity leads to dynamic instabilities and gives rise to the chaotic and as yet not fully understood phenomenon of *turbulence*, well known from our daily dealings with water and air.

Local versus global descriptions

In modern textbooks on continuum physics there has been a tendency to avoid introducing the physical concept of a material particle. Instead these presentations rely on the Newtonian global laws of balance to deduce the local continuum description—in the form of partial differential equations—by purely mathematical means. Although quite elegant and apparently free of physical interpretation problems, such an approach unfortunately obscures the conditions under which the local laws may be assumed to be valid.

In this book the concept of a material particle has been carefully introduced in the proper physical context set by the desired precision of the continuum description. The advantage of such an approach is that the local description of continuous systems in terms of partial differential equations in space and time may be interpreted as representing the Newtonian laws applied to individual material particles. Furthermore, this "materialistic" approach allows us to set physical limits to the validity of the partial differential equations involving local quantities such as the density and the velocity. Such quantities cannot be assumed to be physically meaningful at distance scales smaller than the microscopic length scale L_{micro}. Furthermore, to maintain a continuum description, major spatial changes in these quantities should not take place in regions smaller than the macroscopic length scale L_{macro}, a condition that limits the magnitude of the spatial derivatives found in partial differential equations. Through the dynamical equations expressed as partial differential equations, the spatial limits imposed by precision also set limits on the magnitude of the partial time derivatives.

Mathematically, the local and global equations are equivalent, and must both be presented in any textbook, including this one. The local equations allow us to find exact solutions, either analytically or numerically, while the global equations are well suited for getting approximative solutions and making estimates.

Physical reality of force fields

Whereas the mass density and the pressure only have physical meaning in regions actually containing matter (but may be defined to be zero in vacuum), the gravitational field is assumed to exist and take non-vanishing values even in the vacuum. It specifies the force that would be exerted on a unit mass particle at a given point, but the field is assumed to be there even if no particles are present. In non-relativistic Newtonian physics, the gravitational field has no independent physical meaning and may be completely eliminated and replaced by non-local forces acting between material bodies. The truc physical objects appear to be the material bodies, and the gravitational field is just a mathematical convenience for calculating the gravitational forces exerted by and on these bodies according to Newton's law of gravity. There are no independent dynamical equations that tell us how the Newtonian field of gravity changes with time. When material bodies move around or change shape, their fields of gravity adapt instantaneously everywhere in space to reflect these changes.

In relativistic mechanics, on the other hand, fields take on a completely different meaning. The reason is that instantaneous action-at-a-distance cannot take place. If matter is moved, the current view is that it will take some time before the field of gravity adjusts to the new positions, because no signal can travel faster than light. As we understand it today, gravity is mediated by a field that emanates from massive bodies and in the manner of light takes time to travel through a distance. If the Sun were suddenly to blink out of existence, it would take eight long minutes before daylight was switched off and the Earth set free in space. Due to relativity, force fields must also travel independently, obey their own equations of motion, and carry physical properties such as energy and momentum. Electromagnetic waves bringing radio and TV signals to us are examples of force fields thus liberated from their origin. Gravitational waves have not yet been observed directly. Indirectly they have been observed in the spin-down of binary neutron star systems, which cannot be fully understood unless gravitational radiation is taken into account [WT05].

Even if we do not deal with relativistic theories of the continuum, and therefore may consider the gravitational field to be merely a mathematical convenience, it may nevertheless be wise, at least in the back of our minds, to think of the field of gravity as having an independent physical existence. Then we shall havc no philosophical problem endowing it with physical properties such as energy, even in matter-free regions of space.

Is matter *really* discrete or continuous?

Although continuum physics is always an approximation to the underlying discrete molecular level, this is not the end of the story. At a deeper level it turns out that matter is best described by another continuum formalism, relativistic quantum field theory, in which the discrete particles—electrons, protons, neutrons, nuclei, atoms, and everything else—arise as quantum excitations of the fields. Relativistic quantum field theory without gravitation emerged in the middle of the twentieth century as *the* basic description of the subatomic world, but in spite of its enormous success it is still not clear how to include gravity.

Just as the continuity of macroscopic matter is an illusion, the quantum field continuum may itself one day be replaced by even more fundamental discrete or continuous descriptions of space, time, and matter. It is by no means evident that there could not be a fundamental length in nature setting an ultimate lower limit to distance and time, and theories of this kind have in fact been proposed [Whe89]. It appears that we do not know, and perhaps will never know, whether matter at its deepest level is truly continuous or truly discrete.

Problems

1.1 Consider a small volume V of a much larger volume of gas, such that the probability for any molecule to be found in V is exceedingly small. If the average number of molecules in V is known to be N, the probability of finding precisely n of the molecules in V is given by the Poisson distribution,

$$\Pr(n|N) = \frac{N^n}{n!}e^{-N}.$$

Show that

(a) It is normalized.
(b) The mean value is $\langle n \rangle = N$.
(c) The variance is $\Delta N^2 = \langle (n-N)^2 \rangle = N$, that is $\Delta N = \sqrt{N}$.

1.2 The Lennard–Jones potential is often used to describe the interaction energy between two neutral atoms. It is given by the conventional formula

$$V(r) = 4\epsilon\left(\frac{\sigma^{12}}{r^{12}} - \frac{\sigma^6}{r^6}\right),$$

where r is the distance between centers of the atoms. The parameters ϵ and σ have dimensions of energy and length.

(a) Determine the equilibrium distance $r = a$ where the potential is minimal, and its minimal value.
(b) Determine the leading behavior of the potential around minimum.
(c) Calculate the frequency of radial harmonic vibrations around equilibrium with one atom held fixed.
(d) As an example, take argon, which has molar mass $M_{\text{mol}} = 40\ \text{g mol}^{-1}$, molar energy $\epsilon = 1\ \text{kJ mol}^{-1}$, and equilibrium distance $a = 2.87$ Å.

1.3 Consider a collection of N identical molecules (a "material particle") taken from a large volume of gas. Let the instantaneous molecular velocities be $\boldsymbol{v}_n$ for $n = 1, 2 \cdots, N$. Collisions with other molecules in the gas at large will randomly change the velocity of each of the selected molecules, but if there is no overall drift in the gas, the velocity of individual molecules should average out to zero, $\langle \boldsymbol{v}_n \rangle = \boldsymbol{0}$, the velocities of different molecules should be uncorrelated, $\langle \boldsymbol{v}_n \boldsymbol{v}_m \rangle = \boldsymbol{0}$ for $n \neq m$, and the average of the square of the velocity should be the same for all molecules, $\langle \boldsymbol{v}_n^2 \rangle = v_0^2$.

(a) Show that the root-mean-square average of the center-of-mass velocity of the collection equals $v_0/\sqrt{N}$.

1.4 Any distance function must satisfy the axioms

$$d(\boldsymbol{a}, \boldsymbol{a}) = 0,$$
$$d(\boldsymbol{a}, \boldsymbol{b}) = d(\boldsymbol{b}, \boldsymbol{a}), \qquad \text{(symmetry)}$$
$$d(\boldsymbol{a}, \boldsymbol{b}) \leq d(\boldsymbol{a}, \boldsymbol{c}) + d(\boldsymbol{c}, \boldsymbol{b}). \qquad \text{(triangle inequality)}$$

Show that the Cartesian distance function (1.14) satisfies these axioms.

Part I

Fluids at rest

Fluids at rest

In the following chapters we only consider fluids without any macroscopic motion. Microscopically, matter is never truly at rest because of random molecular motion (heat).

The essential chapters for a minimal curriculum are Chapter 2 and part of Chapter 3. Chapters 4 and 5 may be included in part or whole depending on the level and length of the course.

List of chapters

2. **Pressure:** The intuitive concept of pressure in static fluids is brought on a formal footing by means of the pressure field. The basic equations of hydrostatics are developed and applied to the sea, the atmosphere, and the outermost convective layer of the Sun.

3. **Buoyancy and stability:** Archimedes' principle is derived from the balance of forces and expanded to include balance of moments. The stability of floating objects is analyzed.

4. **Hydrostatic shapes:** The flattening of the Earth as well as the tides of Earth are analyzed. The tides that Jupiter generates on its moon Io are calculated.

5. **Surface tension:** The physical origin and formal definition of surface tension are presented. Soap bubbles are described and the importance of Marangoni forces is emphasized. The Young–Laplace law is derived and the Rayleigh–Plateau instability analyzed. Capillary effects, menisci, and drop shapes are discussed.

2

Pressure

If the Sun did not shine, if no heat were generated inside the Earth, and no energy radiated into space, all the winds in the air and the currents in the sea would die away, and the air and water on the planet would in the end come to rest in equilibrium with gravity. In the absence of external driving forces or time-dependent boundary conditions, and in the presence of dissipative contact forces, any fluid must eventually reach a state of *hydrostatic equilibrium*, where nothing moves anymore and all fields become constant in time. This state must be the first approximation to the sea, the atmosphere, the interior of a planet, or a star.

In mechanical equilibrium of a continuous system there is everywhere a balance between short-range *contact forces* and long-range *body forces*, such as gravity. Contact interactions between material bodies or even between parts of the same body take place across *contact surfaces*. A contact force acting on a tiny patch of a surface can in principle take any direction relative to the surface, and may be resolved into a normal and a tangential component. If the normal component has the same orientation as the surface normal, it is called a *tension force*, and if opposite a *pressure force*. The component acting tangentially to the surface is called a *shear force* or a *traction force*. Fluids in motion, and solids at rest or in motion, are able to sustain shear forces, whereas fluids at rest cannot. Should shear forces arise in a fluid at rest, it will begin to *flow* until it again reaches mechanical equilibrium without shear forces.

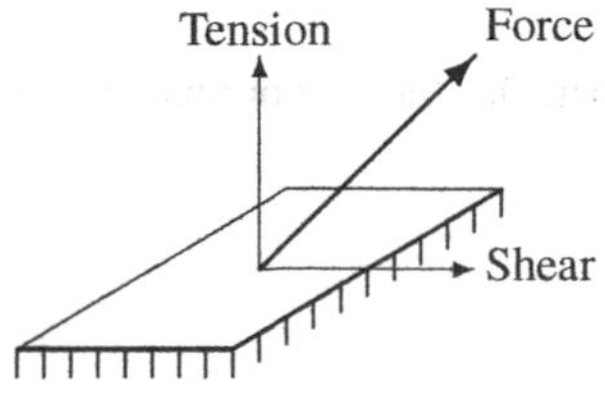

The force acting on the material underneath a small patch of a surface can always be resolved into a perpendicular pressure force and a tangential shear force. The pressure is positive if the force is directed toward the patch, and negative if it (as here) is directed away from it.

In this chapter we shall first investigate the basic properties of pressure, and afterward develop the mathematical formalism that permits us to analyze hydrostatic equilibrium in the sea and the atmosphere. Along the way we shall recapitulate some basic rules of thermodynamics. In the following chapters we shall continue to study the implications of hydrostatic equilibrium for balloons and ships, and the shapes of large fluid bodies subject only to gravity and small fluid bodies subject mainly to surface tension.

2.1 What is pressure?

Pressure is defined as *normal force per unit of area*. The SI–unit for pressure is accordingly newton per square meter, but was in 1971 given the name *pascal* and the special symbol $\text{Pa} = \text{N m}^{-2}$. Earlier units for pressure were the *bar* ($1 \text{ bar} = 10^5 \text{ Pa}$) and the *standard atmosphere* ($1 \text{ atm} = 101,325 \text{ Pa}$), which is close to the average air pressure at sea level. Modern television weather forecasters are now abandoning the older units and tend to quote air pressure in hectopascals rather than in millibars, even if they are exactly the same ($1 \text{ hPa} = 100 \text{ Pa} = 10^{-3} \text{ bar} = 1 \text{ millibar}$).

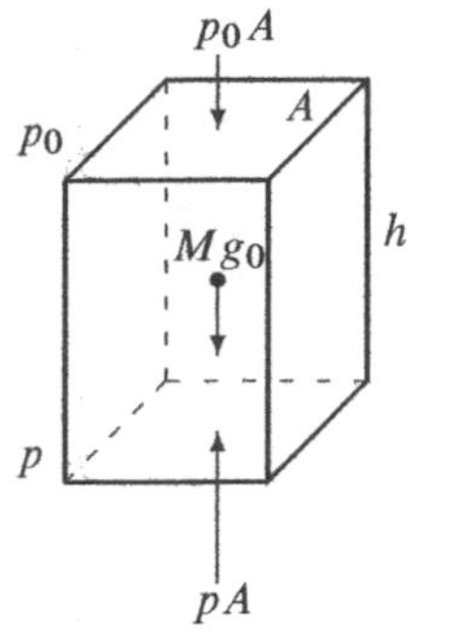

A column of sea water. The pressure difference between bottom and top must carry the weight of the water in the box. Notice that the sum of the forces vanishes.

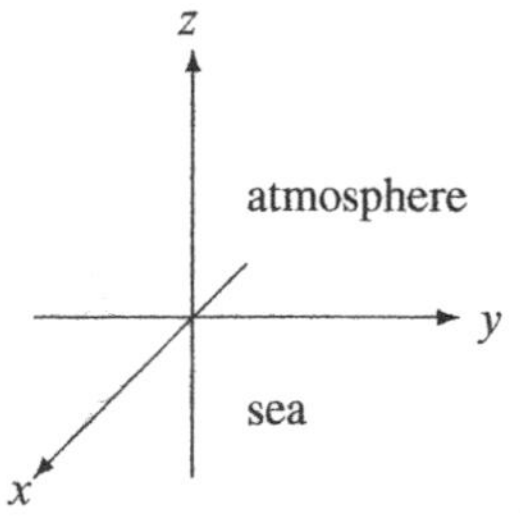

The flat-Earth coordinate system.

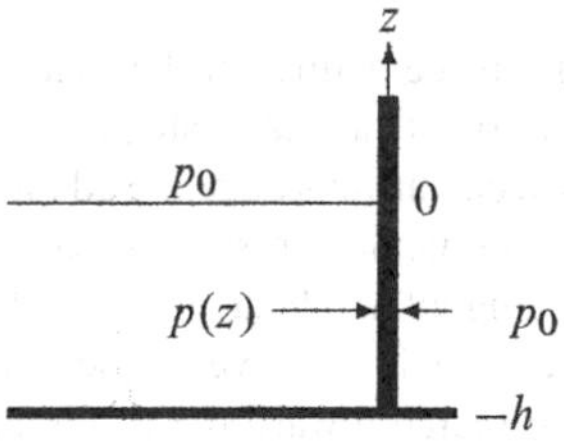

Water stemmed up behind a sluice gate. The pressure varies linearly with depth $-z$.

Evangelista Torricelli (1608–1647). Italian physicist. Constructed the first mercury barometer in 1643 and noted that the barometric pressure varied from day to day. Served as companion and secretary for Galileo in the last months of Galileo's life. Died from typhoid fever at the age of 39.

Case: The incompressible sea

Before presenting a formal definition of the pressure field, we use simple arguments to calculate it in the sea. In the first approximation, water is incompressible and has everywhere the same mass density ρ_0. A vertical box of water in the sea with cross-sectional area A and height h has volume Ah, so that the total mass in the box is $M = \rho_0 A h$ and its weight Mg_0 where g_0 is standard gravity. The pressure p_0 at the top acts on the box with a downward force p_0A whereas the pressure p at the bottom acts with an upward force pA. In mechanical equilibrium, the upward force pA must equal the sum of the downward forces,

$$pA = Mg_0 + p_0A. \tag{2.1}$$

If this equation is not fulfilled, the non-vanishing total vertical force on the column of water will begin to move it upward or downward.

Dividing by the area A, the pressure at the bottom of the box becomes

$$p = p_0 + \rho_0 g_0 h. \tag{2.2}$$

To get an expression for the pressure field we introduce a *flat-Earth coordinate system* with vertical z-axis and the surface of the sea at $z = 0$. Taking a box with the top at $z = 0$ and the bottom at $z = -h$, we may write the pressure as a function of the coordinates, that is, as a field,

$$\boxed{p = p_0 - \rho_0 g_0 z,} \tag{2.3}$$

where p_0 is the surface pressure (equal to the atmospheric pressure). The pressure field is independent of the horizontal coordinates, x and y, because the flat-Earth sea has the same properties everywhere at the same depth.

With $\rho_0 \approx 10^3$ kg m^{-3} and $g_0 \approx 10$ m s^{-2}, the scale of the pressure increase per unit of depth in the sea becomes $\rho_0 g_0 \approx 10^4$ Pa m^{-1} $= 1$ bar/10 m. Thus, at the deepest point in the sea—the Challenger Deep of the Mariana Trench in the Western Pacific Ocean—with $z \approx -11$ km, the pressure is a about 1,100 bar. The assumption of constant density is reasonably well justified even to this depth, because the density of water is only about 5% higher there than at the surface (see page 34).

Example 2.1 [Sluice gate force]: Water is stemmed up to height h behind a sluice gate of width L. On the water surface and on the outer side of the gate there is atmospheric pressure p_0. What is the total horizontal force acting on the gate when $h = 20$ m and $L = 30$ m?

Using that the pressure is the same in all directions in a fluid at rest (see the following section), we find by integrating the pressure excess, $p(z) - p_0$, on the inside of the gate,

$$\mathcal{F} = \int_{-h}^{0} (p(z) - p_0)\, L dz = -\rho_0 g_0 L \int_{-h}^{0} z\, dz = \frac{1}{2}\rho_0 g_0 L h^2. \tag{2.4}$$

This result should have been anticipated, because the pressure rises linearly with depth, such that the total force is simply the product of the area of the sluice gate Lh with the average pressure excess $\langle p - p_0 \rangle = \frac{1}{2}\rho_0 g_0 h$ acting on the gate. With the given numbers the force becomes $\mathcal{F} \approx 6 \times 10^7$ N, corresponding to the weight of 6,000 metric tons of water.

Example 2.2 [Torricelli's mercury barometer]: The mercury barometer is a glass tube, sealed at one end, filled with mercury and placed vertically with the open end in a little dish containing mercury. If the tube is long enough (about 1 m), the mercury will sink down until the height of the mercury column stabilizes at about 76 cm, depending on the ambient air pressure p. Above the mercury column there is nearly vacuum with a bit of mercury vapor exerting only a small pressure $p_0 \approx 0$. With $\rho_0 = 13.534$ g cm^{-3} and $p = 1$ atm we get indeed $h = (p - p_0)/\rho_0 g_0 = 76.34$ cm from Equation (2.2). Conversely, from the variations in the actual height the actual atmospheric pressure can be determined. Until fairly recently a mercury-based sphygomanometer with a U-shaped glass tube of about 30 cm height open in one side was always used to measure resting blood pressure (see Problem 2.1).

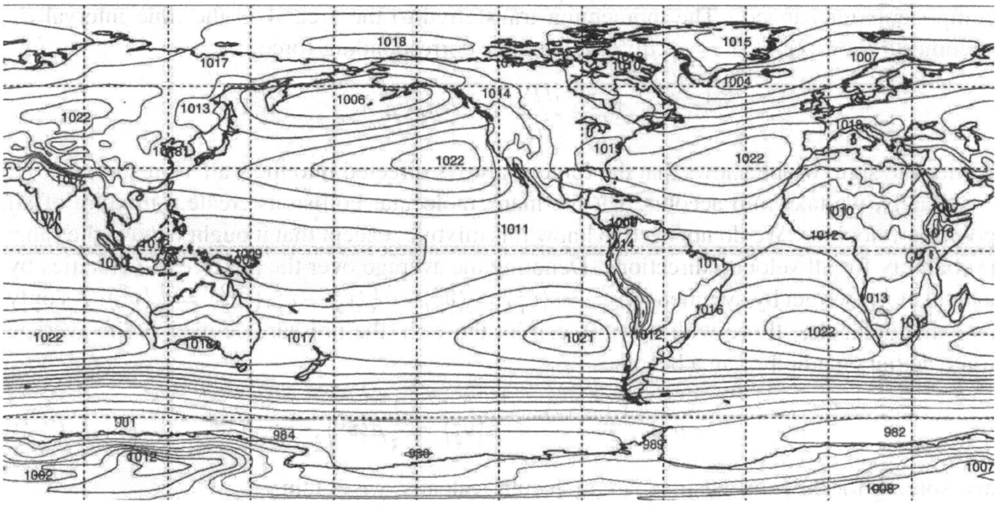

Figure 2.1. The mean air pressure at sea level is not constant, but varies by a few percent around the standard atmosphere: 1 atm = 101, 325 Pa = 1.01325 bar. This picture shows the world-wide sea level air pressure averaged over a 22-year period (1979–2001). The contours are isobars and the numbers are millibars. (Source: European Center for Medium-Range Weather Forecasts (ECMWF). With permission.)

Case: Incompressible atmosphere?

Since air is highly compressible, it makes little sense to use the linear pressure field (2.3) for the atmosphere, except just above the sea. Anyway, if continued beyond this region the pressure would reach zero at a height $z = h_0$, where

$$h_0 = \frac{p_0}{\rho_0 g_0}. \tag{2.5}$$

With surface pressure $p_0 = 1$ atm and air density $\rho_0 = 1.18$ kg m^{-3} at 25°C we get $h_0 = 8.72$ km, which is a tiny bit lower than the height of Mount Everest (8.848 km). This is, of course, meaningless since climbers have reached the summit of that mountain without oxygen masks. Nevertheless, this height sets the correct scale for major changes in the atmospheric properties. As seen in Figure 2.2 on page 36, the linear decrease of air pressure with height is quite a decent approximation for the first 2 km.

Microscopic origin of pressure in gases

In a liquid the molecules directly touch each other and the container walls, and liquid pressure may be seen as a result of these contacts. A gas consists mostly of vacuum with the molecules moving freely along straight lines between collisions with each other, and the gas pressure on a solid wall arises from the incessant random molecular bombardment of the wall.

We shall under very general and reasonable assumptions obtain the pressure from the average rate of molecular momentum transfer to a small flat area A of the wall orthogonal to the z-axis. Let us begin with the (evidently unreasonable) assumption that all molecules have the same velocity vector $\boldsymbol{v} = (v_x, v_y, v_z)$ with $v_z < 0$, such that they all eventually will strike the wall at $z = 0$. The molecules that collectively hit the area A in a small time interval dt must all lie within a "striking distance" $dz = -v_z dt$ from the wall, so that the total mass of these molecules becomes $dM = \rho A dz = \rho A(-v_z)dt$, where ρ is the mass density of the gas. Next we assume that all molecules are reflected perfectly from the wall, thereby changing the z-component of the velocity from v_z to $-v_z$ while leaving the other

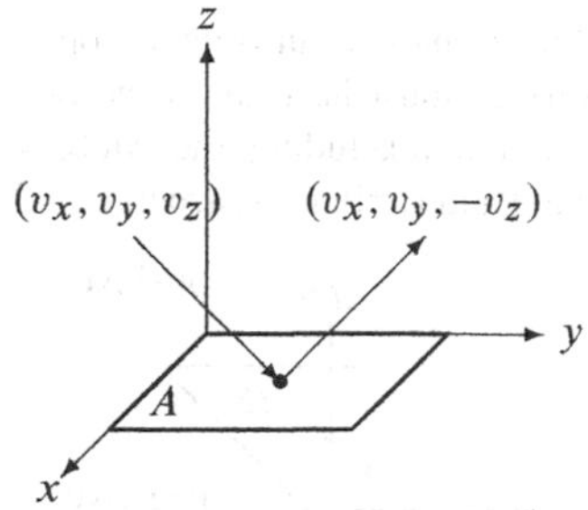

A molecule of mass m that is reflected from the wall changes the sign of its z-component but leaves the other components unchanged, such that the momentum transferred to the wall is $2mv_z$.

components unchanged. The momentum transferred to the area A in the time interval dt becomes $d\mathcal{P}_z = 2v_z dM = -2\rho v_z^2 A dt$, and the corresponding force

$$\mathcal{F}_z = \frac{d\mathcal{P}_z}{dt} = -2\rho v_z^2 A. \tag{2.6}$$

Notice the sign, which shows that the force is always directed into the wall, as it should.

Finally, we take into account that the many molecular collisions create a mixture of all possible velocities. We do not need to know this mixture, except that it ought to have the same probability for all velocity directions. Denoting the average over the mixture of velocities by a bracket $\langle\cdot\rangle$, we get by symmetry $\langle v_x^2\rangle = \langle v_y^2\rangle = \langle v_z^2\rangle = \frac{1}{3}\langle v_x^2 + v_y^2 + v_z^2\rangle = \frac{1}{3}\langle \boldsymbol{v}^2\rangle$. As only half the molecules, those with $v_z < 0$, can hit the wall, the pressure defined as the average force acting on a unit of area becomes

$$p \equiv \frac{1}{2}\frac{\langle -\mathcal{F}_z\rangle}{A} = \rho\langle v_z^2\rangle = \frac{1}{3}\rho\langle \boldsymbol{v}^2\rangle, \tag{2.7}$$

and solving for the root-mean-square molecular velocity, we obtain

$$\boxed{v_{\text{mol}} \equiv \sqrt{\langle \boldsymbol{v}^2\rangle} = \sqrt{\frac{3p}{\rho}}.} \tag{2.8}$$

We shall see later that this velocity—as one might suspect—is of the same order of magnitude as the velocity of sound. At a pressure of $p = 1$ atm and temperature $T = 18\,°\text{C} = 291\text{K}$, the density of air is $\rho \approx 1.2$ kg m^{-3}, and the molecular velocity calculated from the above expression becomes $v \approx 500$ m s^{-1}.

2.2 The pressure field

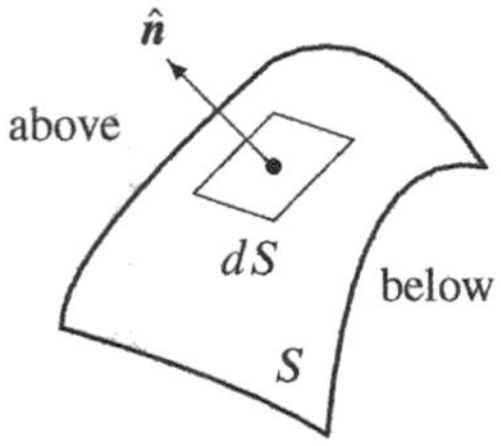

All normals to an oriented open surface must have the same orientation (excluding the Möbius band and similar constructs).

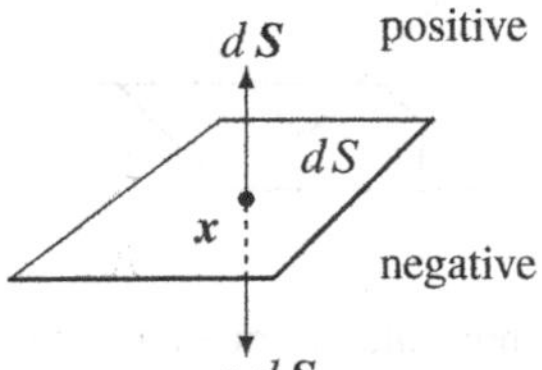

The force on a vector surface element under positive pressure is directed against the normal toward the material on the negative side.

Pressure is a contact force acting on surfaces. A surface need not be a real interface where material properties change dramatically but may just be an imaginary surface separating two parts of the same body from each other. The surface can be divided into a huge number of tiny flat vector surface elements, each being the product of its area dS and the unit vector $\hat{\boldsymbol{n}}$ in the direction of the normal to the surface,

$$d\boldsymbol{S} = \hat{\boldsymbol{n}}\,dS. \tag{2.9}$$

There is nothing intrinsic in a surface that defines the orientation of the normal, that is, whether the normal is really $\hat{\boldsymbol{n}}$ and not $-\hat{\boldsymbol{n}}$. A choice must, however, be made, and having done that, one may call the side of the surface element into which the normal points *positive* and the other of course *negative* (or alternatively *above* and *below*). Neighboring surface elements are required to be oriented toward the same side. By universal convention the normal of a closed surface is chosen to be directed out of the enclosed volume, so that the enclosed volume always lies at the negative side (more details can be found in Appendix C).

A fluid at rest cannot sustain shear forces, so all contact forces on a surface must act along the normal at every point of the surface. The force exerted by the material at the *positive* side of the surface element $d\boldsymbol{S}$ (near $\boldsymbol{x}$) on the material at the *negative* side is written as

$$\boxed{d\boldsymbol{\mathcal{F}} = -p(\boldsymbol{x})\,d\boldsymbol{S},} \tag{2.10}$$

where $p(\boldsymbol{x})$ is the pressure field. Convention dictates that a positive pressure exerts a force directed *toward* the material on the negative side of the surface element, and this explains the minus sign. A negative pressure that pulls at a surface is sometimes called a *tension*.

The total pressure force acting on any oriented surface S is obtained by adding (that is, integrating) all the little vector contributions from each surface element,

$$\mathcal{F} = -\int_S p \, d\boldsymbol{S}. \tag{2.11}$$

Here we have again suppressed the explicit dependence on $\boldsymbol{x}$. This is the force that acts on the cork in the champagne bottle, moves the pistons in the cylinders of your car engine, breaks a dam, and fires a bullet from a gun. It is also this force that lifts fish, ships, and balloons.

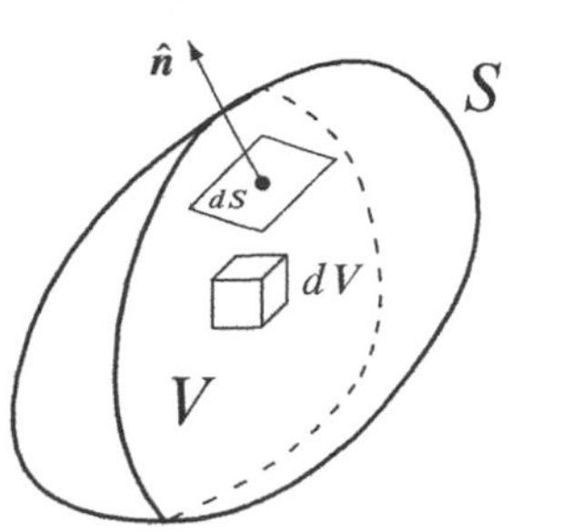

A volume V defined by the closed surface S has by convention all normals oriented toward the outside.

Total pressure force on a material particle

A material particle in the fluid is like any other body subject to pressure forces from all sides, but being infinitesimal it is possible to derive a general expression for the resultant force. Let us choose a material particle in the shape of a small rectangular box with sides dx, dy, and dz, and thus a volume $dV = dxdydz$. Since the pressure is slightly different on opposite sides of the box, the z-component of total pressure force (2.11) becomes

$$d\mathcal{F}_z = (p(x,y,z) - p(x,y,z+dz))dxdy \approx -\frac{\partial p(x,y,z)}{\partial z} dxdydz.$$

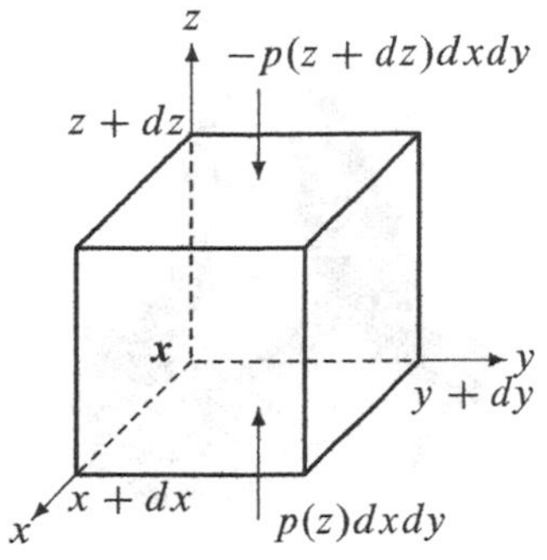

Pressure forces in the z-direction on a material particle in the shape of a rectangular box with sides dx, dy, and dz. The x and y coordinates are suppressed in the pressure.

Here we have used that the normals to the closed surface of the box all point out of the box, so that $dS_z = \pm dxdy$ on the faces of the box orthogonal to the z-axis and $dS_z = 0$ on the other faces. Including similar results for the other coordinate directions we obtain the total pressure force

$$\boxed{d\mathcal{F} = -\left(\frac{\partial p}{\partial x}, \frac{\partial p}{\partial y}, \frac{\partial p}{\partial z}\right) dV = -\nabla p \, dV.} \tag{2.12}$$

Here we have introduced the *gradient operator* or *nabla*, $\nabla = (\partial/\partial x, \partial/\partial y, \partial/\partial z)$. If you need more details, please consult Appendix C where all these things are explained.

The resultant of all pressure forces acting on a tiny material particle is apparently equivalent to a volume force with a density of force equal to the negative gradient of the pressure, $d\mathcal{F}/dV = -\nabla p$. It is, however, not a true long-range volume force like gravity or electromagnetism but an effective expression for the total contact force on the material particle. We shall in Chapter 6 see that a similar result holds when shear contact forces are at play.

Johann Carl Friedrich Gauss (1777–1855). German mathematician of great genius. Contributed to number theory, algebra, non-Euclidean geometry, and complex analysis. In physics he developed the magnetometer. The older (cgs) unit of magnetic strength is named after him. (Source: Wikimedia Commons.)

Gauss' theorem

The total pressure force on a body of volume V can thus be obtained in two ways: either by adding (integrating) all the pressure forces that act on its surface S, or by adding (integrating) all the forces on its constituent material particles. Reversing the sign on both sides, we must in other words have

$$\oint_S p \, d\boldsymbol{S} = \int_V \nabla p \, dV. \tag{2.13}$$

The circle through the surface integral is only there to remind us that the surface is closed. This is one version of *Gauss' theorem*, a purely mathematical relation between the integral of an arbitrary function $p(\boldsymbol{x})$ over a closed surface S and the integral of its gradient $\nabla p(\boldsymbol{x})$ over the enclosed volume V.

Usually *Gauss' theorem* is formulated as a relation between the surface integral over a vector field $\boldsymbol{U}(\boldsymbol{x})$ and the volume integral over its *divergence* $\nabla \cdot \boldsymbol{U} = \nabla_x U_x + \nabla_y U_y + \nabla_z U_z = \partial U_x/\partial x + \partial U_y/\partial y + \partial U_z/\partial z$,

$$\boxed{\oint_S \boldsymbol{U} \cdot d\boldsymbol{S} = \int_V \nabla \cdot \boldsymbol{U}\, dV.} \tag{2.14}$$

The following argument shows that the two formulations are equivalent. Starting with the x-component of (2.13) with $p \to U_x$, we obtain $\oint_S U_x dS_x = \int_V \nabla_x U_x\, dV$. Adding the similar contributions from the two other directions we obtain (2.14). Conversely, we may choose $\boldsymbol{U} = (p, 0, 0)$ in (2.14) to obtain the x-component of (2.13), and similarly for the other directions. A more rigorous mathematical proof of Gauss' theorem is found in Appendix C.

Same pressure in all directions?

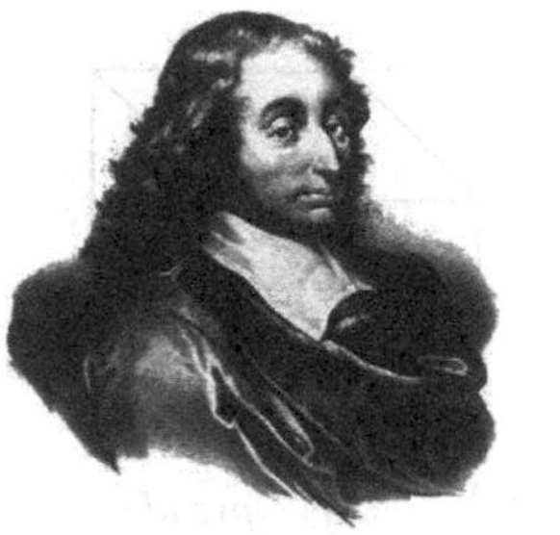

Blaise Pascal (1623–1662). French mathematician and physicist. Founded probability theory. Wrote the first book on the systematic theory of hydrostatics. Constructed what may be viewed as the first digital calculator. He spent his later years with religious thinking in the Cistercian abbey of Port-Royal. More than one property of pressure goes under the name of Pascal's law. (Source: Wikimedia Commons.)

One might think that the pressure could depend on the orientation of the surface element $d\boldsymbol{S}$, but that is actually not the case. Newton's Third Law guarantees that the pressure force exerted *by* the material below *on* the material above is $-d\boldsymbol{\mathcal{F}}$. Since the surface element also changes sign we can write the definition (2.10) as $-d\boldsymbol{\mathcal{F}} = -p(\boldsymbol{x})(-d\boldsymbol{S})$, which shows that the pressure $p(\boldsymbol{x})$ is the same, whether you calculate the force from above or below. This conclusion is part of a broader theorem, called *Pascal's law*, which says that *the pressure in a fluid at rest is the same in all directions*. Pressure is, in other words, a true field that only depends on space (and possibly also time).

The simple reason for pressure being the same in all directions in hydrostatic equilibrium is that the pressure acts on the surface of a body whereas a body force like gravity acts on the volume. If we let the body shrink, the contribution from the body force will vanish faster than the contribution from the surface force because the volume vanishes faster than the surface area. In the limit of vanishing body size, only the surface force is left, but it must then itself vanish in hydrostatic equilibrium where the total force on all parts of a body has to vanish.

* **Proof of Pascal's law:** Consider a material particle in the shape of a right tetrahedron oriented along the coordinate axes, as shown in the margin figure. Each of the three triangles making up the sides of the tetrahedron in the coordinate planes is in fact the projection of the front face $d\boldsymbol{S} = (dS_x, dS_y, dS_z)$ onto that face. By elementary geometry the areas of the three projected triangles are dS_x, dS_y and dS_z. Assume now that the pressure is actually different on the four faces of the tetrahedron: p on the front face, and p_x, p_y, and p_z on the faces in the coordinate planes. The total pressure force acting on the body is then

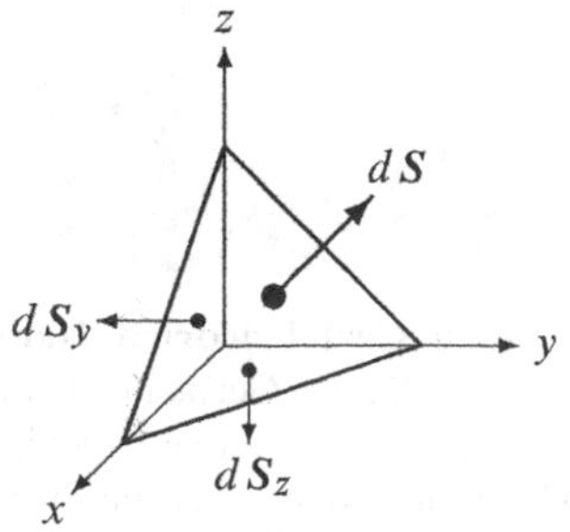

A right tetrahedron with three sides in the coordinate planes. The vector normals to the sides are all pointing out of the volume (dS_x is hidden from view).

$$\begin{aligned} d\boldsymbol{\mathcal{F}}_S &= -p\,d\boldsymbol{S} + p_x dS_x \hat{\boldsymbol{e}}_x + p_y dS_y \hat{\boldsymbol{e}}_y + p_z dS_z \hat{\boldsymbol{e}}_z \\ &= ((p_x - p)dS_x, (p_y - p)dS_y, (p_z - p)dS_z). \end{aligned}$$

For a fluid at rest the sum of the total surface force $d\boldsymbol{\mathcal{F}}_S$ and a volume force $d\boldsymbol{\mathcal{F}}_V = \boldsymbol{f}\, dV$ (for example, gravity) must vanish:

$$d\boldsymbol{\mathcal{F}}_S + d\boldsymbol{\mathcal{F}}_V = \boldsymbol{0}. \tag{2.15}$$

The idea is now to show that for sufficiently small tetrahedrons the volume force can be neglected. To convince yourself of this, consider a geometrically congruent tetrahedron with all lengths scaled down by a factor $\lambda < 1$. Since the volume scales as the third power of λ whereas the surface areas only scale as the second power, the above equation for the scaled tetrahedron becomes $\lambda^2 d\boldsymbol{\mathcal{F}}_S + \lambda^3 d\boldsymbol{\mathcal{F}}_V = \boldsymbol{0}$ or equivalently

$$d\boldsymbol{\mathcal{F}}_S + \lambda d\boldsymbol{\mathcal{F}}_V = \boldsymbol{0}. \tag{2.16}$$

In the limit of $\lambda \to 0$, it follows that the total contact force must vanish, that is, $d\boldsymbol{\mathcal{F}}_S = 0$, and that is only possible for $p_x = p_y = p_z = p$. The pressure must indeed be the same in all directions.

2.3 Hydrostatics

Two classes of forces are at play in standard continuum physics. The first class consists of *body forces* that act throughout the volume of a body. The only known body forces are gravity and electromagnetism, but in standard hydrostatics we shall only discuss gravity. The second class consists of *contact forces* acting only on the surface of a body. Pressure is the only contact force that is possible in fluids at rest, but there are others in fluids in motion and in solids. The full set of possible contact forces will be introduced in Chapter 6.

Global hydrostatic equilibrium

The total force acting on a fluid volume V enclosed by the surface S receives two contributions: one from gravity and one from pressure. The gravitational acceleration field $\boldsymbol{g} = \boldsymbol{g}(\boldsymbol{x})$ is a body force that acts on every material particle with a force $d\boldsymbol{\mathcal{F}} = \boldsymbol{g}\, dM$, where $dM = \rho\, dV$ is the mass of the particle. Pressure acts as described in the preceding section with a force $d\boldsymbol{\mathcal{F}} = -p\, d\boldsymbol{S}$ on every surface element of the body. Explicitly, these two contributions to the total force on the fluid body are

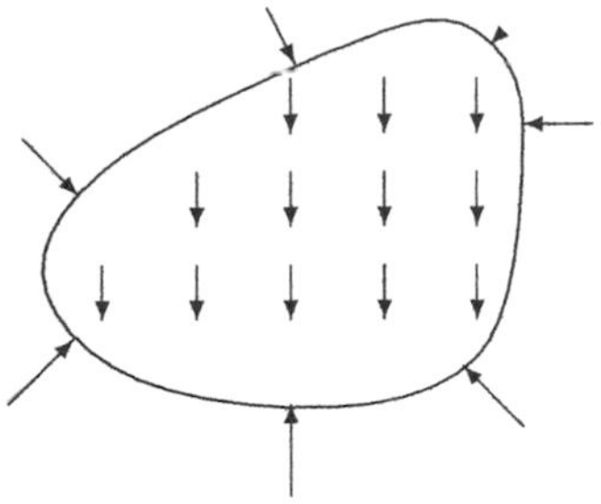

A body field like gravity acts on a body over its entire volume, while a contact field like pressure only acts on its surface.

$$\boldsymbol{\mathcal{F}}_G = \int_V \rho \boldsymbol{g}\, dV, \qquad \boldsymbol{\mathcal{F}}_B = -\oint_S p\, d\boldsymbol{S}. \tag{2.17}$$

The force of gravity is called the *weight* of the fluid body, and the force due to pressure is called the *buoyancy* force.

In mechanical equilibrium the total force must vanish,

$$\boxed{\boldsymbol{\mathcal{F}} = \boldsymbol{\mathcal{F}}_G + \boldsymbol{\mathcal{F}}_B = \int_V \rho \boldsymbol{g}\, dV - \oint_S p\, d\boldsymbol{S} = \boldsymbol{0}} \tag{2.18}$$

for *each and every* volume V of fluid. This is the *equation of global hydrostatic equilibrium*. It states that *the weight of the fluid in the volume must exactly balance its buoyancy*. In Section 2.1 we used intuitively that the weight of a vertical box of sea water must be in balance with the pressure forces on the bottom and the top of the box, and could in this way derive the pressure in the sea. If the balance is not perfect, as for example if a small volume of fluid is heated or cooled relative to its surroundings, the fluid *must* start to move, either upward if the buoyancy force wins over the weight or downward if it loses.

The problem with the global equilibrium equation is that we have to know the force density $\rho \boldsymbol{g}$ and the pressure p in order to calculate the integrals. Sometimes symmetry considerations can get us a long way (see the example below), but in general we need a local form of the equation of hydrostatic equilibrium valid at each point $\boldsymbol{x}$ in any geometry, to be able to determine the hydrostatic pressure anywhere in a fluid.

Example 2.3 [The incompressible sea]: In constant gravity and using flat-Earth coordinates, we have $\boldsymbol{g}(\boldsymbol{x}) = \boldsymbol{g}_0 = (0, 0, -g_0)$. For symmetry reasons the sea on the flat Earth ought to have the same properties for all x and y, such that the pressure can only depend on z. Taking a rectangular vertical box with height h and cross-sectional area A, we find

$$\boldsymbol{\mathcal{F}}_G = -\rho_0 g_0 A h \hat{\boldsymbol{e}}_z, \qquad \boldsymbol{\mathcal{F}}_B = \left(-p(0)A + p(-h)A\right)\hat{\boldsymbol{e}}_z, \tag{2.19}$$

because the pressure contributions from the sides of the box cancel. Global hydrostatic balance immediately leads back to Equation (2.2).

Local hydrostatic equilibrium

The total force on a material particle is of course also the sum of its weight $d\mathcal{F}_G = \boldsymbol{g}\, dM$ and its effective buoyancy $d\mathcal{F}_B = -\nabla p\, dV$ from (2.12),

$$d\mathcal{F} = d\mathcal{F}_G + d\mathcal{F}_B = (\rho\boldsymbol{g} - \nabla p)\, dV. \tag{2.20}$$

The expression in the parenthesis is called the *effective force density*,

$$\boxed{\boldsymbol{f}^* = \rho\boldsymbol{g} - \nabla p.} \tag{2.21}$$

It must again be emphasized that the effective force density is the sum of the true long-range force (gravity) and the resultant of all the short-range pressure forces acting on its surface.

In hydrostatic equilibrium the total force $d\mathcal{F} = \boldsymbol{f}^*\, dV$ must vanish for every material particle, implying that the effective force density must vanish everywhere, $\boldsymbol{f}^* = \boldsymbol{0}$, or

$$\boxed{\nabla p = \rho\,\boldsymbol{g}.} \tag{2.22}$$

This is the *local equation of hydrostatic equilibrium.* It is a differential equation for the pressure, valid everywhere in a fluid at rest, and it encapsulates in an elegant way all the physics of hydrostatics. Gauss' theorem (2.13) allows us to convert the local equation back into the global one (2.18), showing that there is complete mathematical equivalence between the local and global formulations of hydrostatic equilibrium.

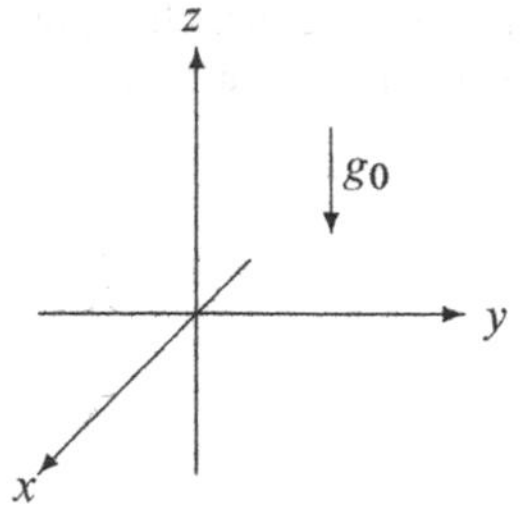

The flat-Earth coordinate system with constant gravity pointing against the direction of the z-axis.

Case: Hydrostatics in constant gravity

In a flat-Earth coordinate system, the constant field of gravity is $\boldsymbol{g}(\boldsymbol{x}) = (0, 0, -g_0)$ for all $\boldsymbol{x}$. Written out explicitly in coordinates, the equation of local equilibrium (2.22) becomes three individual equations:

$$\frac{\partial p}{\partial x} = 0, \qquad \frac{\partial p}{\partial y} = 0, \qquad \frac{\partial p}{\partial z} = -\rho\, g_0. \tag{2.23}$$

The two first equations express that the pressure does not depend on x and y but only on z, which confirms the previous argument based on symmetry. It also shows that, independently of the shape of a fluid container, the pressure will always be the same at a given depth (in constant gravity). For the special case of constant density, $\rho(z) = \rho_0$, the last equation may immediately be integrated to yield the previous result (2.3) for the pressure field in the incompressible sea.

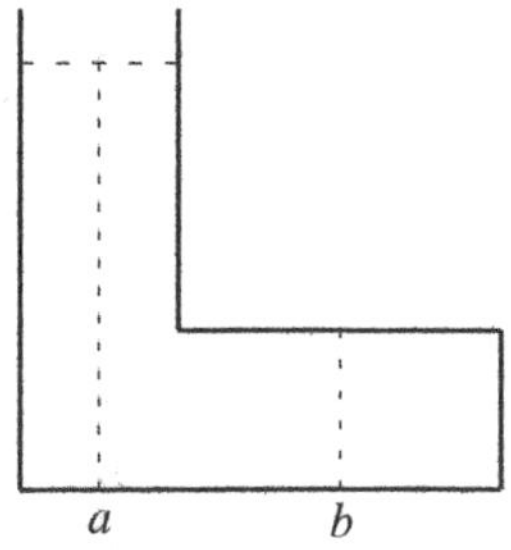

A schematic "boot". The pressure at a horizontal bottom under the open shaft of the boot (a) has to carry the full weight of the liquid above, but what about the pressure in the toe of the boot (b) where the water column is much shorter?

Paradoxes of hydrostatics

It is easily understood why the pressure can only depend on z in an open liquid container, because it must everywhere carry the weight of the liquid column above. But what if the liquid column does not reach all the way to the surface, as in a "boot" filled with water (see the margin figure)? Since there is only a short column of water in the "toe", it raises the question of where the constant pressure comes from. What is it "up against"? The only possibility is that the material of the boot must supply the necessary downward forces to compensate for the missing weight of the water column. These forces arise as a response to the elastic extension of the material of the boot caused by the pressure. If the material in the toe cannot deliver a sufficient response, it will break and a fountain of water will erupt from the toe.

Another "paradox" goes back to Pascal. Wine is often stored for aging in large wooden casks containing cubic meters of the precious liquid. Such casks may only be able to withstand perhaps 30% pressure difference between liquid and atmosphere. Even if it would seem natural to mount a pipeline to transport the wine to and from the casks, such a system comes with its own problems. If, for example, the pipe entrance lies 5 m above the casks (that often are placed in cellars), and if the unwary employee fills the cask and the whole pipe with wine, half an atmosphere extra pressure will be added to the cask, which may as a consequence develop a leak or even burst. Paradoxically, this does not depend on the diameter of the pipe.

Finally, what happens if someone presents you with a fluid density $\rho(x, y, z)$ that actually depends on the horizontal coordinates? Such a situation could, for example, arise when you remove a separating wall in a container with petrol on one side and water on the other. The hydrostatic equations (2.23) cannot be fulfilled because they imply that $-g_0\partial\rho/\partial x = \partial^2 p/\partial x\partial z = 0$, contradicting the assumption that ρ varies with x. The conclusion must be that the liquid cannot be static but must flow until the dependence on the horizontal coordinates has gone. In the end, all the petrol will lie in a horizontal layer on top of the water. Non-horizontal separating surfaces between immiscible fluids at rest can simply not exist.

2.4 Equation of state

The local equation of hydrostatic equilibrium (2.22) with an externally given gravitational field is not enough in itself, but needs some relation between pressure and density. In the preceding sections we mostly thought of fluids with constant density and could then integrate the hydrostatic equation and determine the pressure, but in general that is not possible.

Ordinary thermodynamics provides us with a relationship between density ρ, pressure p, and absolute temperature T, called the *equation of state*,

$$F(\rho, p, T) = 0. \tag{2.24}$$

It is valid for any macroscopic amount of homogeneous isotropic fluid in thermodynamic equilibrium. The actual form of the equation of state for a particular substance is derived from the molecular properties, a subject that falls outside the scope of this book.

In continuum physics where conditions can change from point to point, we shall always assume that each material particle is in thermodynamic equilibrium with its environment, such that the equation of state holds for the local values of density, pressure, and temperature fields in every point of space:

$$F\left(\rho(\boldsymbol{x}), p(\boldsymbol{x}), T(\boldsymbol{x})\right) = 0. \tag{2.25}$$

In full generality the function F could also depend explicitly on $\boldsymbol{x}$, for example if the molecular composition depends on the position, but we shall not consider that possibility.

Benoit Paul Émile Clapeyron (1799–1864). French engineer and physicist. Formulated the ideal gas law from previous work by Boyle, Mariotte, Charles, Gay-Lussac, and others. Established what is now called Clapeyron's formula for the latent heat in the change of state of a pure substance. (Source: Wikimedia Commons.)

The ideal gas law

The *ideal gas law* is the oldest and by far the most famous equation of state, normally credited to Clapeyron (1834) although there were other contributors, as nearly always in the history of physics. It is usually presented in terms of the volume V of a mass M of the gas and the number $n = M/M_{\rm mol}$ of moles in this volume,

$$p\,V = n R_{\rm mol}\,T, \tag{2.26}$$

where $R_{\rm mol} = 8.31447\ \mathrm{J\,K^{-1}\,mol^{-1}}$ is the *universal molar gas constant*[1].

[1] In physics and chemistry the universal molar gas constant is nearly always denoted R. We shall however use the more systematic notation $R_{\rm mol}$ and leave the single-letter symbol R free for other uses, for example the specific gas constant $R = R_{\rm mol}/M_{\rm mol}$. It occurs naturally in the local form of the ideal gas law (2.27), and simplifies the formalism. Since this notation is slightly atypical, we shall always make appropriate remarks when it is used.

	$10^3 M_{mol}$	$10^{-3} R$
H_2	2.0	4.157
He	4.0	2.079
Ne	20.2	0.412
N_2	28.0	0.297
O_2	32.0	0.260
Ar	39.9	0.208
CO_2	44.0	0.189
Air	29.0	0.287
	$\frac{\text{kg}}{\text{mol}}$	$\frac{\text{J}}{\text{K kg}}$

The molar mass, M_{mol}, and the specific gas constant, $R = R_{mol}/M_{mol}$, for a few gases (in SI units).

Ludwig Boltzmann (1844–1906). Austrian theoretical physicist. Made fundamental contributions to statistical mechanics and the understanding of the relations between the macroscopic and microscopic descriptions of nature. Believed firmly in the reality of atoms of molecules, even if most of his contemporaries did not. His recurrent depressions eventually led him to suicide. (Source: Wikimedia Commons.)

Since $\rho = M/V = nM_{mol}/V$ we may write the ideal gas law in the local form,

$$p = R\rho T, \qquad R = \frac{R_{mol}}{M_{mol}}. \tag{2.27}$$

In the margin the *specific gas constant* R is tabled for a few gases. Various extensions of the ideal gas law take into account the excluded molecular volume as well as the strong molecular repulsion (see Problem 2.7). The ideal gas law is not only valid for pure gases but also for a mixture of pure gases provided one uses the average molar mass (see page 4). Real air with $M_{mol} = 28.95 \text{ g mol}^{-1}$ is quite well described by the ideal gas law at normal temperatures, although in precise calculations it may be necessary to include various corrections (see for example [Kaye and Laby 1995]).

* Microscopic origin of the ideal gas law

Even if the derivation of the equation of state belongs to statistical mechanics, it is nevertheless of interest to see how the ideal gas law arises in the simplest possible molecular picture of a gas at rest, already discussed on page 23. The gas consists of particles of mass m that are free except for random collisions with the walls and with each other, so that the velocity vector, $\boldsymbol{v} = (v_x, v_y, v_z)$, of a typical particle incessantly fluctuates around $\boldsymbol{v} = \boldsymbol{0}$.

In statistical mechanics one derives (under certain conditions that we shall not discuss here) the *equipartition theorem*, which states that each independent kinematic variable on average carries the thermal energy $\frac{1}{2}k_B T$, where $k_B = 1.38065 \times 10^{-23}\ \text{J K}^{-1}$ is Boltzmann's constant. Accordingly, the average kinetic energy of a molecule of mass m becomes

$$\frac{1}{2}m\left\langle \boldsymbol{v}^2\right\rangle = \frac{1}{2}m\left\langle v_x^2\right\rangle + \frac{1}{2}m\left\langle v_y^2\right\rangle + \frac{1}{2}m\left\langle v_z^2\right\rangle = \frac{3}{2}k_B T, \tag{2.28}$$

and inserting this into (2.7) we obtain

$$p = \frac{1}{3}\rho\left\langle \boldsymbol{v}^2\right\rangle = \rho\frac{k_B T}{m}. \tag{2.29}$$

Multiplying the numerator and denominator of the fraction on the right by Avogadro's number N_A and using that the molar mass is $M_{mol} = N_A m$, we obtain the ideal gas law with the molar gas constant $R_{mol} = N_A k_B$, as it should be.

Case: Isothermal atmosphere

Everybody knows that the atmospheric temperature varies with height; but if we nevertheless assume that it has constant temperature, $T(\boldsymbol{x}) = T_0$, and combine the equation of hydrostatic equilibrium (2.23) with the ideal gas law (2.27), we obtain

$$\frac{dp}{dz} = -\rho g_0 = -\frac{g_0 p}{R T_0}. \tag{2.30}$$

This is an ordinary differential equation for the pressure, and using the initial condition $p = p_0$ for $z = 0$, we find the solution

$$p = p_0 e^{-z/h_0}, \qquad h_0 = \frac{R T_0}{g_0} = \frac{p_0}{\rho_0 g_0}. \tag{2.31}$$

In the last step we have again used the ideal gas law at $z = 0$ to show that the expression for h_0 is identical to the incompressible atmospheric scale height (2.5).

In the isothermal atmosphere the pressure thus decreases exponentially with height on a characteristic length scale again set by $h_0 \approx 8{,}728$ m calculated for 1 atm and 25°C. The pressure at the top of Mount Everest ($z = 8{,}848$ m) is now finite and predicted to be 368 hPa. As seen in Figure 2.2, the isothermal approximation is better than the incompressible one and fits the data to a height of about 4 km. In Section 2.6 we shall obtain a third estimate, the homentropic atmosphere, lying between the two.

Barotropic equation of state

The equation of state is in general not a simple relation between density and pressure, which may be plugged into the equation of local hydrostatic equilibrium, but also involves temperature. To solve the general problem we need a further dynamic equation for the temperature, called the heat equation, which is derived from energy balance (see Chapter 22). Only if the temperature is somehow given in advance can we use the equation of state as it stands and solve the equilibrium equation.

Sometimes there exists a so-called *barotropic* relationship between density and pressure,

$$F(\rho(\boldsymbol{x}), p(\boldsymbol{x})) = 0, \tag{2.32}$$

which does not depend on the local temperature $T(\boldsymbol{x})$. This is in fact not as far-fetched as it might seem at first. The condition of constant density $\rho(\boldsymbol{x}) = \rho_0$ that we used in the preceding section to calculate the pressure in the sea is a trivial example of such a relationship in which the density is independent of both pressure and temperature. A less trivial example is obtained if the walls containing a fluid at rest are held at a fixed temperature T_0. The omnipresent heat conduction will eventually cause all of the fluid to attain this temperature, $T(\boldsymbol{x}) = T_0$; and in this state of *isothermal equilibrium* the equation of state (2.25) simplifies to $F(\rho(\boldsymbol{x}), p(\boldsymbol{x}), T_0) = 0$, which is indeed a barotropic relationship.

Polytropic relation: In the following we shall often meet a barotropic relation of the so-called *polytropic* form,

$$p = C\rho^\gamma, \tag{2.33}$$

where C is independent of ρ and γ is a constant, called the *polytropic index.*

A polytropic relation arises naturally in an ideal gas when heat conduction can be completely disregarded, so that all processes take place without heat transfer. In thermodynamics such processes are called *adiabatic*, and the polytropic index is called the *adiabatic index.* Gases with diatomic molecules, like nitrogen and oxygen, have $\gamma \approx 7/5$, whereas monatomic noble gases have $\gamma \approx 5/3$ and multiatomic gases, like water vapor or carbon dioxide, have $\gamma \approx 4/3$ (see the margin table). Since the local entropy stays constant during reversible, adiabatic processes, they are also said to be *isentropic* (see Appendix E).

	γ	$10^{-3} c_p$
H_2	1.41	14.30
He	1.63	5.38
Ne	1.64	1.05
N_2	1.40	1.04
O_2	1.40	0.91
Ar	1.67	0.52
CO_2	1.30	0.82
Air	1.40	1.00
Units		$\mathrm{J\,K^{-1}kg^{-1}}$

Table of the adiabatic index and the specific heat at constant pressure ($c_p = \gamma R/(\gamma - 1)$) for a few nearly ideal gases.

Pressure potential

In elementary physics it is shown that the gravitational field is conservative and can be obtained from the gradient of the *gravitational potential* $\Phi(\boldsymbol{x})$:

$$\boldsymbol{g}(\boldsymbol{x}) = -\nabla\Phi(\boldsymbol{x}). \tag{2.34}$$

The function $\Phi(\boldsymbol{x})$ represents the *potential energy per unit of mass*, so that a body of mass m at the point $\boldsymbol{x}$ has a gravitational potential energy $m\Phi(\boldsymbol{x})$. For constant gravity in flat-Earth coordinates we have $\boldsymbol{g}(\boldsymbol{x}) = (0, 0, -g_0)$, and the potential is simply $\Phi = g_0 z$.

In terms of the potential, the equilibrium equation (2.22) may now be written as

$$\nabla\Phi + \frac{\nabla p}{\rho} = 0. \tag{2.35}$$

If the density is constant, $\rho(\boldsymbol{x}) = \rho_0$, it follows that

$$\Phi^* = \Phi + \frac{p}{\rho_0} \tag{2.36}$$

is also a constant, independent of $\boldsymbol{x}$. In flat-Earth gravity, where $\Phi = g_0 z$, the constancy of $\Phi^* = g_0 z + p/\rho_0$ leads directly to the linear pressure field (2.3) in the incompressible sea.

More generally, it is always possible to integrate the hydrostatic equation for any barotropic fluid with $\rho = \rho(p)$. To do this, we define the so-called *pressure potential*,

$$w(p) = \int \frac{dp}{\rho(p)}, \tag{2.37}$$

which becomes $w = p/\rho_0$ for constant density. It then follows from the chain rule for differentiation that

$$\nabla w(p) = \frac{dw}{dp}\nabla p = \frac{1}{\rho}\nabla p,$$

so that hydrostatic equilibrium (2.35) may be written as $\nabla\Phi^* = \mathbf{0}$, where

$$\Phi^* = \Phi + w(p). \tag{2.38}$$

The requirement $\Phi^*(\boldsymbol{x}) = \text{const}$ is completely equivalent to the equation of hydrostatic equilibrium for barotropic fluids. There is no agreement in the literature about a name for Φ^*, but we have chosen to call it the *effective potential* because it combines the true gravitational potential with the pressure potential, and because the effective force density (2.21) in general is given by $\boldsymbol{f}^* = \rho\boldsymbol{g}^*$ with $\boldsymbol{g}^* = -\nabla\Phi^*$.

Example 2.4 [Isothermal gas]: Under isothermal conditions, the pressure potential of an ideal gas is calculated by means of the ideal gas law (2.27)

$$w = \int RT_0\frac{dp}{p} = RT_0 \log p. \tag{2.39}$$

In flat-Earth gravity, the constancy of $\Phi^* = g_0 z + RT_0 \log p$ immediately leads to the exponentially decreasing pressure (2.31) in the isothermal atmosphere.

Example 2.5 [Polytropic fluid]: For fluids obeying a polytropic relation (2.33), the pressure potential becomes

$$w = \int C\gamma\rho^{\gamma-1}\frac{d\rho}{\rho} = C\frac{\gamma}{\gamma-1}\rho^{\gamma-1} = \frac{\gamma}{\gamma-1}\frac{p}{\rho}. \tag{2.40}$$

When the fluid is an ideal gas with $p = R\rho T$, this takes the even simpler form

$$w = \frac{\gamma}{\gamma-1}RT = c_p T, \qquad c_p = \frac{\gamma}{\gamma-1}R, \tag{2.41}$$

where c_p is the specific heat at constant pressure (see margin table on the preceding page). In Section 2.6 we shall use this expression to model the atmosphere.

2.5 Bulk modulus

The archetypal thermodynamics experiment is carried out on a fixed amount $M = \rho V$ of a fluid placed in a cylindrical container with a moveable piston. When you slightly increase the force on the piston, the volume of the chamber decreases, so that $dV < 0$. The pressure in the fluid must necessarily always increase, $dp > 0$; for if the pressure instead always diminished, an arbitrarily small extra force would send the piston right down to the bottom of the chamber leaving zero volume for the gas. Since a larger volume diminishes proportionally more for a given pressure increase, we define the *bulk modulus* as the pressure increase dp per *fractional decrease* in volume, $-dV/V$, or

$$K = \frac{dp}{-dV/V} = \frac{dp}{d\rho/\rho} = \rho\frac{dp}{d\rho}. \tag{2.42}$$

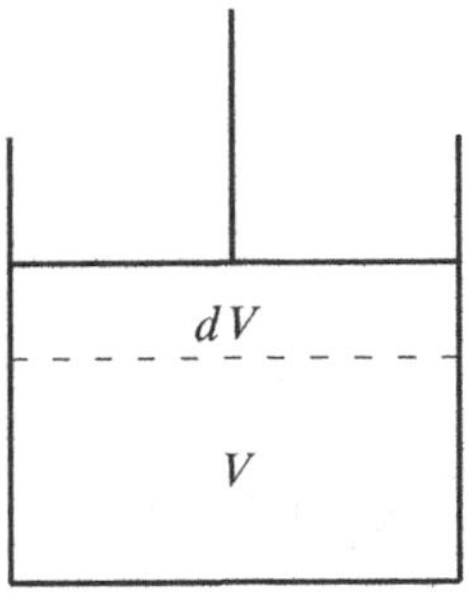

The archetypal thought-experiment in thermodynamics: A cylindrical chamber with a moveable piston.

In the second step we have used the constancy of the mass $M = \rho V$ of the fluid in the volume to derive that $dM = \rho dV + V d\rho = 0$, from which we get $-dV/V = d\rho/\rho$.

The above definition makes immediate sense for a barotropic fluid, where $p = p(\rho)$ is a function of density. For general fluid states it is necessary to specify the conditions under which the bulk modulus is defined, for example whether the temperature is held constant (isothermal) or whether there is no heat transfer (adiabatic or isentropic). Thus, the equation of state (2.27) for an ideal gas implies that the isothermal bulk modulus is

$$K_T = \left(\rho\frac{\partial p}{\partial \rho}\right)_T = p, \tag{2.43}$$

Liquid	T[°C]	K[GPa]
Mercury	20	24.93
Glycerol	0	3.94
Water	20	2.18
Benzene	25	1.04
Ethanol	20	0.89
Methanol	20	0.82
Hexane	20	0.60

Isothermal bulk modulus for common liquids [Lide 1996].

where the index—as commonly done in thermodynamics—indicates that the temperature T is held constant. Similarly, for an isentropic ideal gas obeying the polytropic relation (2.33) that the *isentropic bulk modulus* becomes

$$K_S = \left(\rho\frac{dp}{d\rho}\right)_S = \gamma p, \tag{2.44}$$

where the index indicates that the entropy S is held constant. It larger than the isothermal bulk modulus (2.43) by a factor of $\gamma > 1$, because adiabatic compression also increases the temperature of the gas, which further increases the pressure.

The definition of the bulk modulus (and the above equation) shows that it is measured in the same units as pressure, for example pascals, bars, or atmospheres. The bulk modulus is actually a measure of *incompressibility*, because the larger it is, the greater is the pressure increase that is needed to obtain a given fractional increase in density. The inverse bulk modulus $\beta = 1/K$ may be taken as a measure of *compressibility*.

Example 2.6 [Water compression]: For sea water the bulk modulus is $K \approx 23.2$ kbar $=$ 2.32 GPa. As long as the pressure change is much smaller than the bulk modulus, $dp \ll K$, we may calculate the relative change in density from (2.42) to be $d\rho/\rho \approx dp/K$. In the deepest abyss of the sea, the pressure is about 1,100 bar, implying that the relative density change is $d\rho/\rho \approx 1/20 \approx 5\%$.

Case: Polytropic water

The bulk modulus of sea-water is nearly constant, although it does increase slowly with pressure, nearly doubling between 1 and 3,000 bar [Kaye and Laby 1995]. Disregarding any temperature dependence, we model the data by a linear function,

$$K = K_0 + \gamma(p - p_0), \tag{2.45}$$

with $K_0 = 23.2$ kbar, $p_0 = 1$ bar, and $\gamma = 6$.

Inserting this expression into the definition (2.42) it becomes

$$\frac{dp}{K_0 + \gamma(p - p_0)} = \frac{d\rho}{\rho}.$$

Integrating both sides of this equation, and fixing the integration constant by setting the density to $\rho = \rho_0$ at $p = p_0$, we finally arrive at the following barotropic equation of state for water, valid well beyond the depth of the deepest abyss:

$$p = p_0 + \frac{K_0}{\gamma}\left[\left(\frac{\rho}{\rho_0}\right)^{\gamma} - 1\right]. \tag{2.46}$$

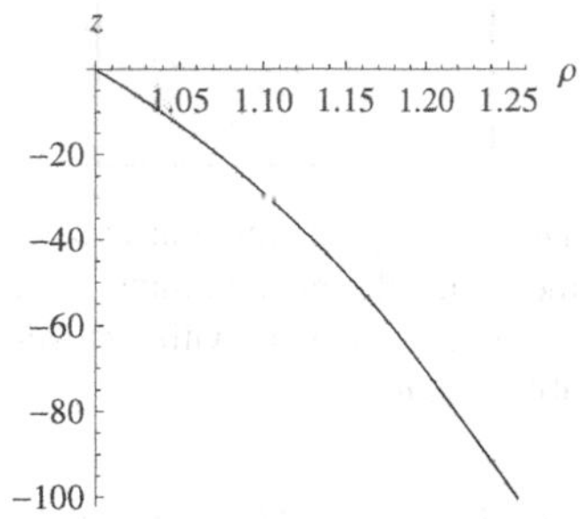

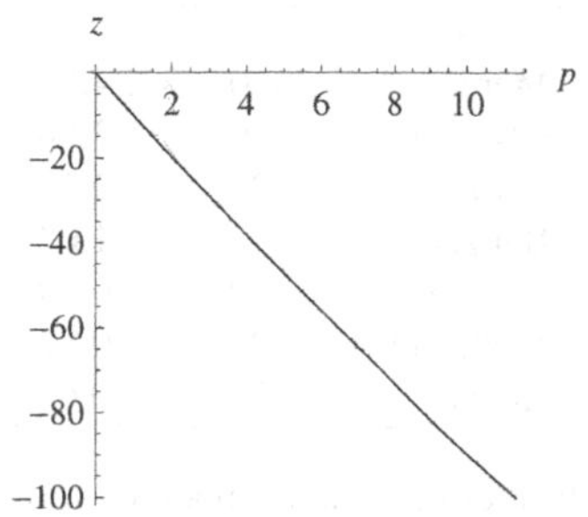

Polytropic water as a function of depth z in units of km. **Top:** density ρ in units of g cm^{-3} and **Bottom:** pressure p in units of kbar.

Clearly, this is a variant of the polytropic relation (2.33).

Differentiating with respect to z we get

$$\frac{dp}{dz} = \frac{K_0}{\rho_0^{\gamma}}\rho^{\gamma-1}\frac{d\rho}{dz}.$$

In flat-Earth equilibrium we have $dp/dz = -\rho g_0$, and a dividing both sides by ρ, the differential equation may immediately be integrated to yield

$$-g_0 z + \frac{K_0}{(\gamma-1)\rho_0} = \frac{K_0}{(\gamma-1)\rho_0^{\gamma}}\rho^{\gamma-1},$$

where the integration constant on the left-hand side has been determined by demanding $\rho = \rho_0$ at $z = 0$. Solving for the density we get

$$\rho(z) = \rho_0\left(1 - \frac{z}{h_1}\right)^{1/(\gamma-1)}, \qquad h_1 = \frac{K_0}{(\gamma-1)\rho_0 g_0}. \tag{2.47}$$

With the given parameters the length scale becomes $h_1 = 47.3$ km. The pressure as a function of depth z is obtained by inserting this into the equation of state (2.46). Since the resulting exponent, $\gamma/(\gamma-1) = 1.2$, is close to unity, the pressure increases nearly linearly with depth, as seen in the margin figures. At the deepest point of the sea, $z = -10.924$ km, the density 4.3% higher than the surface density. The pressure at the deepest point is increased by about 2.2%, corresponding to 24 bar above the pressure that would have been obtained if the density had been constant.

2.6 Application: Earth's homentropic atmosphere

Anyone who has flown in a modern passenger jet and listened to the pilot can testify that the atmospheric temperature falls with height. So the atmosphere is *not* in isothermal equilibrium, and this is perhaps not so surprising, since the "container walls" of the atmosphere, the ground and outer space, have different temperatures. There must be a heat flow through the atmosphere between the ground and outer space, maintained by the inflow of solar radiation and the outflow of geothermal energy. But air is a bad conductor of heat, so although heat conduction does play a role, it is not directly the cause of the temperature drop in the atmosphere.

Of much greater importance are the indirect effects of solar heating, the *convection* that creates air currents, winds, and local turbulence, continually mixing different layers of the atmosphere. The lower part of the atmosphere, the *troposphere*, is quite unruly and vertical mixing happens at time scales that are much shorter than those required to reach thermal equilibrium. There is in fact no true hydrostatic equilibrium state for the real atmosphere. Even if we disregard large-scale winds and weather systems, horizontal and vertical mixing always takes place at small scales, and a realistic model of the atmosphere must take this into account.

Vertical mixing

Let us imagine that we take a small amount of air and exchange it with another amount with the same mass, taken from a different height with a different volume and pressure. In order to fill out the correct volume, one air mass would have to contract and the other expand. If this is done quickly, there will be no time for heat exchange with the surrounding air, and one air mass will consequently be heated up by compression and the other cooled down by expansion. If the atmosphere initially were in isothermal equilibrium, the temperature of the swapped air masses would not be the same as the temperature of the surrounding air in the new positions, and the atmosphere would be brought out of equilibrium.

If, however, the surrounding air initially had a temperature distribution, such that the swapped air masses after the expansion and compression would arrive at precisely the correct temperatures of their new surroundings, a kind of "equilibrium" could be established, in which the omnipresent vertical mixing had essentially no effect. Intuitively, it is reasonable to expect that the end result of fast vertical mixing and slow heat conduction might be precisely such a state. It should, however, not be forgotten that this state is not a true equilibrium state but rather a dynamically balanced state depending on the incessant small-scale turbulent motion in the atmosphere.

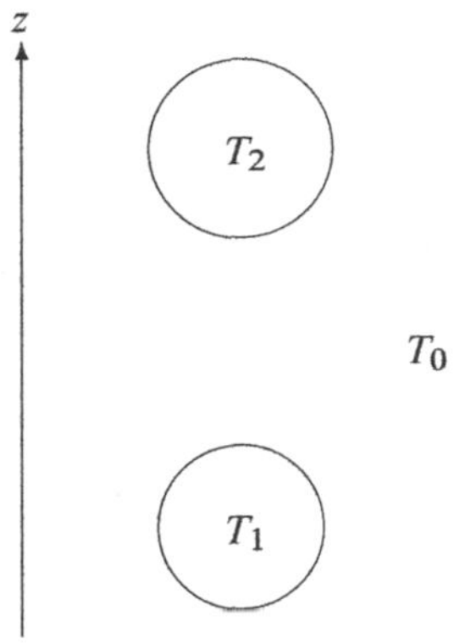

Swapping air masses from different heights. If the air has temperature T_0 before the swap, the compressed air would be warmer, $T_1 > T_0$, and the expanded air colder, $T_2 < T_0$.

Homentropic equation of state

A process that takes place without exchange of heat between a system and its environment is said to be *adiabatic*. If furthermore the process is *reversible*, it will conserve the entropy and is called *isentropic*. In that case it follows from the thermodynamics of ideal gases (see Appendix E) that an isentropic compression or expansion of a fixed amount M of an ideal gas that changes the volume from V_0 to V and the pressure from p_0 to p, will obey the rule

$$pV^\gamma = p_0 V_0^\gamma. \tag{2.48}$$

Here γ is the *adiabatic index*, which for a gas like air with diatomic molecules is approximately $\gamma \approx 7/5 = 1.4$.

Expressed in terms of the density $\rho = M/V$, an isentropic process that locally changes density ρ_0 and pressure p_0 to ρ and p must obey the local *polytropic* relation:

$$\boxed{p\rho^{-\gamma} = p_0\rho_0^{-\gamma}.} \tag{2.49}$$

In the dynamically balanced state of the atmosphere described above, this rule would not only be valid for a local process, but also apply to the swapping of two air masses from different heights, for example from the ground $z = 0$ to any value of z. The lower part of the atmosphere, the troposphere, is at least approximatively in such a *homentropic state* in which this law is valid everywhere, with the right-hand side being a constant expressed in terms of the ground values of pressure and density.

The pressure potential was calculated in Example 2.5 with the result

$$\boxed{w = c_p T, \qquad c_p = \frac{\gamma}{\gamma - 1} R.} \tag{2.50}$$

The constant $R = R_{\rm mol}/M_{\rm mol}$ is the specific gas constant, and c_p is the *specific heat* at constant gas pressure (see Appendix E for details). For dry air with $R \approx 0.287\ \mathrm{J\ K^{-1}\ g^{-1}}$ and $\gamma = 7/5$, its value is $c_p \approx 1\ \mathrm{J\ K^{-1}\ g^{-1}}$, which is about 4 times smaller than for liquid water.

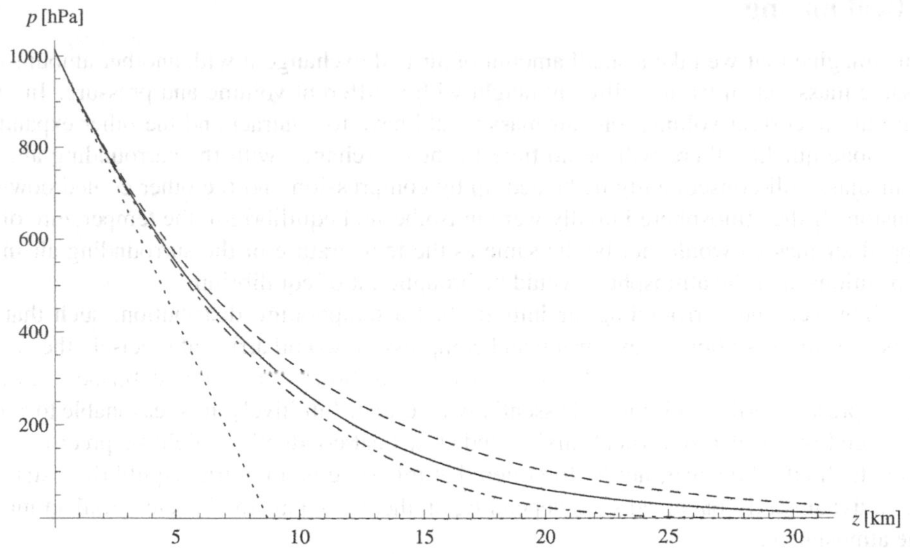

Figure 2.2. The pressure as a function of altitude (for $T_0 = 25°C$) in three different atmospheric models analyzed in this chapter: constant density (dotted), isothermal (dashed), and homentropic (dot-dashed). The solid curve is the Standard Atmosphere (1976) model (see page 37).

Homentropic solution

It was shown before on page 32 that hydrostatic equilibrium implies that $\Phi^* = \Phi + w(p) = g_0 z + c_p T$ is a constant, independent of $\boldsymbol{x}$. Defining the value of this constant to be $\Phi^* = c_p T_0$, we find the temperature

$$T(z) = T_0 - \frac{g_0}{c_p} z, \tag{2.51}$$

where T_0 is the sea-level temperature. It is convenient to rewrite this expression as

$$T = T_0 \left(1 - \frac{z}{h_2}\right), \qquad h_2 = \frac{c_p T_0}{g_0} = \frac{\gamma}{\gamma - 1} h_0, \tag{2.52}$$

where h_0 is the scale height (2.31) for the isothermal atmosphere. For $\gamma = 7/5$ and $T_0 = 25°C$, we find the isentropic scale height $h_2 \approx 31$ km.

The ideal gas law combined with the homentropic law (2.49) implies that $T\rho^{1-\gamma}$ and $p^{1-\gamma}T^{\gamma}$ are also constant, so that the density and pressure become

$$\rho = \rho_0 \left(1 - \frac{z}{h_2}\right)^{1/(\gamma-1)}, \qquad p = p_0 \left(1 - \frac{z}{h_2}\right)^{\gamma/(\gamma-1)}, \tag{2.53}$$

where ρ_0 and p_0 are the density and pressure at sea level, respectively. Temperature, density, and pressure all vanish at $z = h_2$, which is of course unphysical.

In Figure 2.2 the pressure in the various atmospheric models has been plotted together with the 1976 Standard Atmosphere model (see below). Even if the homentropic model gives the best fit, it fails at altitudes above 10 km. The real atmosphere is more complicated than any of the models considered here.

The atmospheric temperature lapse rate

The negative vertical temperature gradient $-dT/dz = g_0/c_p$ is called the *atmospheric temperature lapse rate*. Its value is 9.8 K km^{-1} for dry air. At the top of Mount Everest ($z = 8.848$ km), the temperature is for $T_0 = 25$°C predicted to be −61°C, the density 0.50 kg m^{-3}, and the pressure 306 hPa.

Water vapor is always present in the atmosphere and will condense to clouds in rising currents of air. The latent heat released during condensation heats up the air, so that the temperature lapse rate becomes smaller than in dry air, perhaps more like 6–7 K km^{-1}, which corresponds to choosing an effective adiabatic index of $\gamma \approx 1.23$. Using this value with $T_0 = 25$°C, the scale height increases to 46 km, and we now predict the pressure at the top of Mount Everest to be 329 hPa, and the temperature −33°C, which is not so far from the measured average summit temperature of about −25°C.

Clouds may eventually precipitate out as rain, and when the now drier air afterward descends again, for example on the lee side of a mountain, the air will heat up at a higher rate than it cooled during its ascent on the windward side and become quite hot, a phenomenon called *föhn* in the Alps.

Atmospheric stability

In the real atmosphere the temperature lapse rate will generally not be constant throughout the troposphere. Sunlight will often heat bubbles of air relative to the surrounding atmosphere because of local topographic variations. Since the pressure must be the same in the heated bubble and the atmosphere, the ideal gas law tells us that a higher temperature is always accompanied by lower mass density. The bubble will thus weigh less than the volume of air it displaces, and must according to Archimedes principle begin to rise (see the following chapter). As it rises it will always be in pressure balance with the environment, but the temperature will generally not have time to equilibrate because of the low heat conductivity of air.

If the temperature of the bubble drops faster with falling pressure than that of the atmosphere, it will soon stop rising. If, on the other hand, the bubble temperature drops slower with pressure than that of its environment, the bubble temperature will increase relative to the environment, and the bubble will continue to rise faster and faster. This will, for example, happen if a bubble of warm humid air rises in a relatively dry atmosphere. To begin with, it rises slowly because of the small initial temperature excess in the bubble; but when the humidity begins to condense, the released heat will make the bubble's temperature drop much slower. Strong vertical currents may be generated in this way, for example in thunderstorms.

* The Standard Atmosphere (1976)

A number of standard models for the atmosphere have been created in the last half of the twentieth century to aid pilots and flight engineers to know in advance the temperature, pressure, and density of the air at the increasing altitudes reached by their aircraft. The US Standard Atmosphere (1976)—which is essentially the same as the International Standard Atmosphere (1976)—was created to provide a numerically trustworthy model valid up to about 100 km. With minor corrections this has become the accepted model of the dry atmosphere for suborbital altitudes.

The input to the model is an empirical vertical temperature distribution $T(z)$. Using the ideal gas law (2.27), the mass density may be eliminated from the hydrostatic equation (2.23) to arrive at the differential equation

$$\frac{dp}{dz} = -\frac{g_0 p}{RT}, \tag{2.54}$$

where R is the specific gas constant for air. Since T is assumed to be known, this equation

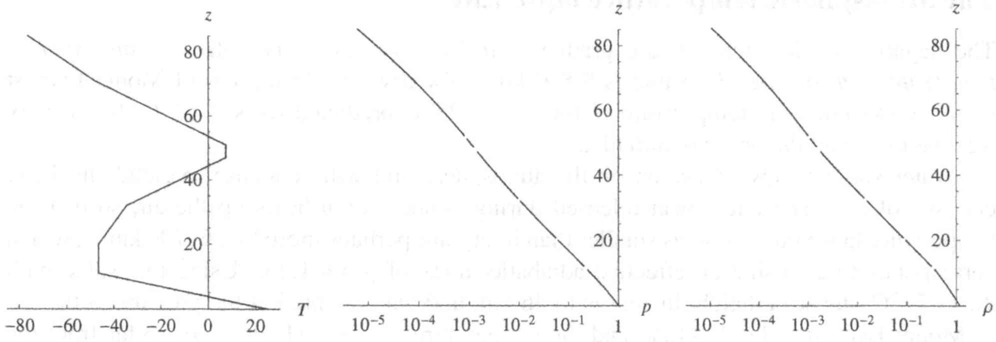

Figure 2.3. Plots of temperature, pressure, and density in the Standard Atmosphere (1976) for sea-level temperature $T_0 = 25°\text{C}$. The vertical height is measured in kilometers, the temperature in Celsius, the pressure in bars, and the density in kilograms per cubic meter.

Layer	z	$\frac{dT}{dz}$
Mesopause	85	
High mesosphere	71	-2.0
Low mesosphere	51	-2.8
Stratopause	47	0.0
High stratosphere	32	+2.8
Low stratosphere	20	+1.0
Tropopause	11	0.0
Troposphere	0	-6.5

The temperature gradient in the layered model of the US Standard Atmosphere (1976). The height z is measured in km and the gradient dT/dz in K m^{-1}.

can be immediately integrated to yield

$$p(z) = p_0 \exp\left(-\lambda_0 \int_0^z \frac{dz'}{T(z')}\right), \qquad \lambda_0 = \frac{g_0}{R}, \tag{2.55}$$

where p_0 is the pressure at $z = 0$. The exponential coefficient $\lambda_0 = 34.16$ K km^{-1} is called the *atmospheric constant*.

In the margin table is shown the temperature gradient for the Standard Atmosphere (1976). It is given as a piecewise constant function, implying that the temperature itself is piecewise linear. This way of presenting the model is convenient for analytic or numeric calculations. The result for the pressure is shown as the fully drawn curve in Figure 2.2. Plots of temperature, pressure, and density are shown in Figure 2.3 for $T_0 = 25°\text{C}$.

∗ 2.7 Application: The Sun's convective envelope

Stars, like the Sun, are self-gravitating, gaseous, and almost perfectly spherical bodies that generate heat by thermonuclear processes in a fairly small region close to the center. The heat is transported to the surface by radiation, conduction, and convection and eventually released into space as radiation. Like planets, stars also have a fairly complex structure with several layers differing in chemical composition and other physical properties.

Our Sun consists—like the rest of the universe—of roughly 75% hydrogen and 25% helium (by mass), plus small amounts of other elements (usually called "metals"). The radius of its photosphere is $a = 700{,}000$ km, its mass $M_0 = 2 \times 10^{30}$ kg, and its total luminosity (power) 3.8×10^{26} W. The gases making up the Sun are almost completely ionized, so that they form a *plasma* consisting of positively charged hydrogen ions (H^+), doubly charged helium ions (He^{2+}), and negatively charged electrons (e^-). The mean molar mass of the plasma is $M_{\text{mol}} = 0.59$ g mol^{-1}. It is smaller than unity because the nearly massless electrons make up about 52% of all the particles in the plasma.

In this section we shall ignore heat production and only model the convective outer layer as a homentropic ideal gas, in other words modeling a Sun that does not shine! A much more enlightening and comprehensive analysis of the structure of stellar interiors can, for example, be enjoyed in [Hansen and Kawaler 1994]. The generally accepted Standard Model for the Sun is found in [CDDA&96].

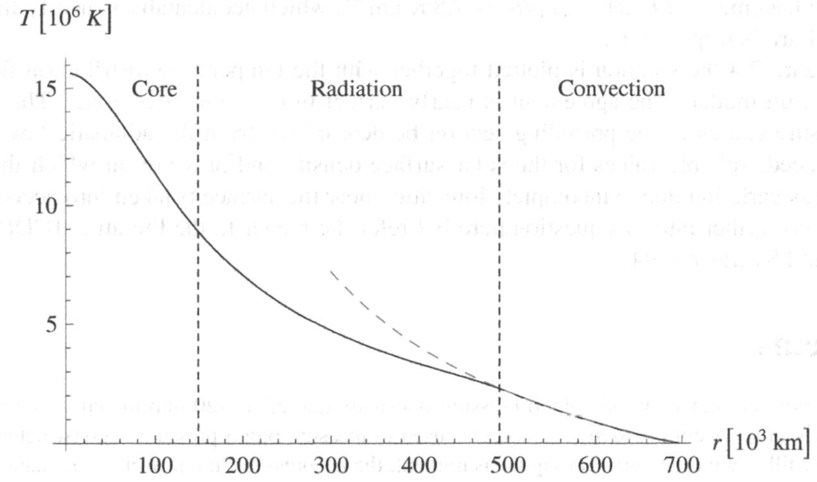

Figure 2.4. The temperature distribution in the Sun as a function of the distance from the center. The vertical lines are boundaries between the three major layers of the Sun. The fully drawn curve is extracted from the Standard Sun model [CDDA&96] and [3]. The dashed curve is the approximative homentropic solution (2.57) for the convective envelope.

The convective envelope

In the preceding section we argued for the case of Earth's troposphere that—provided the time scale for vertical convective mixing is fast compared to heat conduction—a *homentropic* dynamical equilibrium state will be established in the ideal gas for which $p\rho^{-\gamma}$ is constant. We shall now apply this argument to the convective envelope of the Sun, which has a thickness of about 200,000 km. In the deeper layers, heat transport is dominated by radiative or material conduction, not by convection.

The density in the convective zone is so low that the zone contains only about 3% of the Sun's mass, so that we may approximate the gravitational potential in this region by that of a point particle with the Sun's mass,

$$\Phi = -G\frac{M_0}{r}, \tag{2.56}$$

where $G = 6.674 \times 10^{-11}\ \mathrm{N\,m^2\,kg^{-2}}$ is the universal gravitational constant [1]. The pressure potential for a homentropic gas is as before given by (2.50). Since the solar plasma consists of monatomic "molecules", the adiabatic index is $\gamma = 5/3$, leading to a specific heat capacity at constant pressure $c_p = 35.1\ \mathrm{J\,K^{-1}\,g^{-1}}$. This is 35 times that of the Earth's atmosphere, mainly because of the low average molar mass of the solar plasma.

Hydrostatic equilibrium (2.38) requires the effective potential $\Phi^* = \Phi + c_p T$ to be constant. Fixing this constant by means of the surface temperature T_0, we obtain the expression

$$T(r) = T_0 + T_1\left(\frac{a}{r} - 1\right), \quad \text{where } T_1 = \frac{GM_0}{c_p\, a}. \tag{2.57}$$

With the Sun's parameters we find $T_1 = 5.4 \times 10^6$K, a value that also sets the scale for the Sun's central temperature. The precise surface of the Sun is not too well defined, but normally chosen to be that spherical surface at radius $r = a$ where the plasma turns into neutral gas and becomes transparent. The transition to complete transparency actually takes place over a few hundred kilometers. The surface temperature is about $T_0 \approx 6{,}000$ K, which is about 1,000 times smaller than T_1 and can mostly be ignored. At the surface of the Sun the temperature lapse rate becomes $-dT/dr = T_1/a = 7.8\,\mathrm{K\,km^{-1}}$, which accidentally is nearly the same as in the Earth's troposphere.

In Figure 2.4 the solution is plotted together with the temperature distribution from the Standard Sun model. The agreement is nearly perfect in the convective zone. The density and pressure can as in the preceding section be determined from the adiabatic law. To do that one needs reliable values for the solar surface density and pressure, in which the mean molar mass variation due to incomplete ionization near the surface is taken into account. We shall not go further into this question here but refer the reader to the literature [CDDA&96, Hansen and Kawaler 1994].

Problems

2.1 The normal human systolic blood pressure is usually quoted as 120 mm mercury (above atmospheric pressure). A clinical *sphygomanometer* used to measure blood pressure is constructed from a U-tube half filled with mercury. During measurement, the air pressure in one arm of the manometer is supplied by an inflated blood-constricting cuff around the upper arm or the wrist, whereas the other arm of the manometer is exposed to atmospheric pressure. **(a)** How long should the arms of the manometer be when it must accommodate a measurement range of $\pm 100\%$ around normal?

2.2 Consider a canal with a dock gate that is 12 m wide and has water depth 9 m on one side and 6 m on the other side. Calculate

(a) The pressure in the water on both sides of the gate at a height z over the bottom.
(b) The total force on the gate.
(c) The total moment of force around the bottom of the gate.
(d) The height over the bottom at which the total force acts.

2.3 An underwater lamp is covered by a hemispherical glass with a radius of $a = 15$ cm and is placed with its center at a depth of $h = 3$ m on the side of the pool. **(a)** Calculate the total horizontal force from the water on the lamp when there is air at normal pressure inside.

2.4 Using a manometer, the pressure in an open container filled with liquid is found to be 1.6 atm at a height of 6 m over the bottom, and 2.8 atm at a height of 3 m. **(a)** Determine the density of the liquid and **(b)** the height of the liquid surface.

2.5 An open jar contains two non-mixable liquids with densities $\rho_1 > \rho_2$. The heavy layer has thickness h_1 and the light layer on top of it has thickness h_2. **(a)** An open glass tube is now lowered vertically into the liquids toward the bottom of the jar. Describe how high the liquids rise in the tube (disregarding capillary effects). **(b)** The open tube is already placed in the container with its opening close to the bottom when the heavy fluid is poured in, followed by the light. How high will the heavy fluid rise in the tube?

2.6 Show that a mixture of ideal gases (see page 4) also obeys the equation of state (2.27).

2.7 The equation of state due to van der Waals is

$$\left(P + \frac{n^2 a}{V^2}\right)(V - nb) = nRT,$$

where a and b are constants. It describes gases and their condensation into liquids. **(a)** Calculate the isothermal bulk modulus. **(b)** Under which conditions can it become negative, and what does that mean?

2.8 Calculate the pressure and density in the flat-Earth sea, assuming constant bulk modulus. **(a)** Show that both quantities are singular at a certain depth and calculate this depth.

* **2.9** Calculate the isentropic scale height for the Mars atmosphere.

3

Buoyancy and stability

Fishes, whales, submarines, balloons, and airships all owe their ability to float to *buoyancy*, the lifting power of water and air. The understanding of the physics of buoyancy goes back as far as antiquity and probably sprung from the interest in ships and shipbuilding in classic Greece. The basic principle is due to Archimedes. His famous law states that the buoyancy force on a body is equal and oppositely directed to the weight of the fluid that the body displaces. Before his time it was thought that the shape of a body determined whether it would sink or float.

"Buoy", pronounced "booe", probably of Germanic origin. A tethered floating object used to mark a location in the sea.

The shape of a floating body and its mass distribution do, however, determine whether it will float stably or capsize. Stability of floating bodies is of vital importance to shipbuilding, and to anyone who has ever tried to stand up in a small rowboat. Newtonian mechanics not only allows us to derive Archimedes' principle for the equilibrium of floating bodies, but also to characterize the deviations from equilibrium and calculate the restoring forces. Even if a body floating in or on water is in hydrostatic equilibrium, it will not be in complete mechanical balance in every orientation, because the center of mass of the body and the center of mass of the displaced water, also called the center of buoyancy, do not in general coincide.

Archimedes of Syracuse (287–212 BC). Greek mathematician, physicist, and engineer. Discovered the formulae for area and volume of cylinders and spheres, and invented rudimentary infinitesimal calculus. Formulated the Law of the Lever, and wrote two volumes on hydrostatics titled *On Floating Bodies*, containing his Law of Buoyancy. Killed by a Roman soldier. (Source: Photograph of Fields Medal courtesy Stefan Zachow. Wikimedia Commons.)

The mismatch between the centers of mass and buoyancy for a floating body creates a moment of force, which tends to rotate the body toward a stable equilibrium. For submerged bodies, submarines, fishes, and balloons, the stable equilibrium will always have the center of gravity situated directly below the center of buoyancy. But for bodies floating stably on the surface, ships, ducks, and dumplings, the center of gravity is mostly found directly *above* the center of buoyancy. It is remarkable that such a configuration can be stable. The explanation is that when the stably floating body is tilted away from equilibrium, the center of buoyancy moves instantly to reflect the new volume of displaced water. Provided the center of gravity does not lie too far above the center of buoyancy, this change in the displaced water creates a moment of force that counteracts the tilt.

3.1 Archimedes' principle

Mechanical equilibrium takes a slightly different form than global hydrostatic equilibrium (2.18) on page 27 when a body of another material is immersed in a fluid. If its material is incompressible, the body retains its shape and displaces an amount of fluid with exactly the same volume. If the body is compressible, like a rubber ball, the volume of displaced fluid will be smaller. The body may even take in fluid, like a sponge or the piece of bread you dunk into your coffee, but we shall disregard this possibility in the following.

Weight and buoyancy

A body that is partially immersed with a piece inside and another outside the fluid may formally be viewed as a body that is fully immersed in a fluid with properties that vary from place to place. This also covers the case where part of the body is in vacuum, which may be thought of as a fluid with vanishing density and pressure.

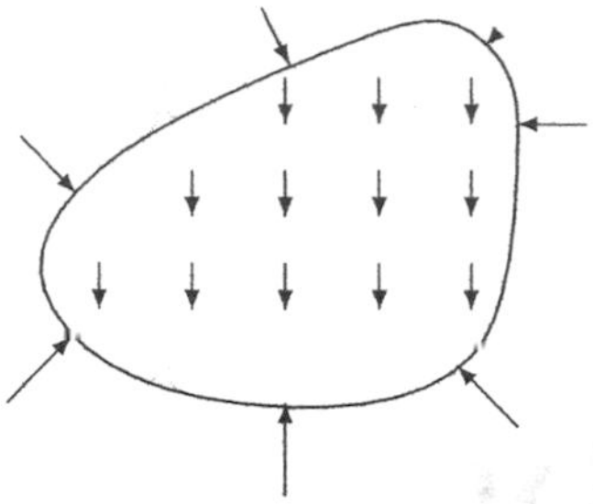

Gravity pulls on a body over its entire volume while pressure only acts on the surface.

Let the actual, perhaps compressed, volume of the immersed body be V with surface S. In the field of gravity, an unrestrained body with mass density ρ_{body} is subject to two forces: its weight,

$$\mathcal{F}_G = \int_V \rho_{\text{body}}\, g\, dV, \tag{3.1}$$

and the buoyancy due to pressure acting on its surface,

$$\mathcal{F}_B = -\oint_S p\, dS. \tag{3.2}$$

If the total force $\mathcal{F} = \mathcal{F}_G + \mathcal{F}_B$ does not vanish, an unrestrained body will accelerate in the direction of $\mathcal{F}$ according to Newton's Second Law. Therefore, in mechanical equilibrium, weight and buoyancy must precisely cancel each other at all times to guarantee that the body will remain in place.

Assuming that the body does not itself contribute to the field of gravity, the local balance of forces in the fluid, expressed by Equation (2.22) on page 28, will be the same as before the body was placed in the fluid. In particular the pressure in the fluid cannot depend on whether the volume V contains material that is different from the fluid itself. The pressure acting on the surface of the immersed body must for this reason be identical to the pressure on a body of fluid of the same shape, but then the global equilibrium condition (2.18) on page 27 for any volume of fluid tells us that $\mathcal{F}_G^{\text{fluid}} + \mathcal{F}_B = \mathbf{0}$, or

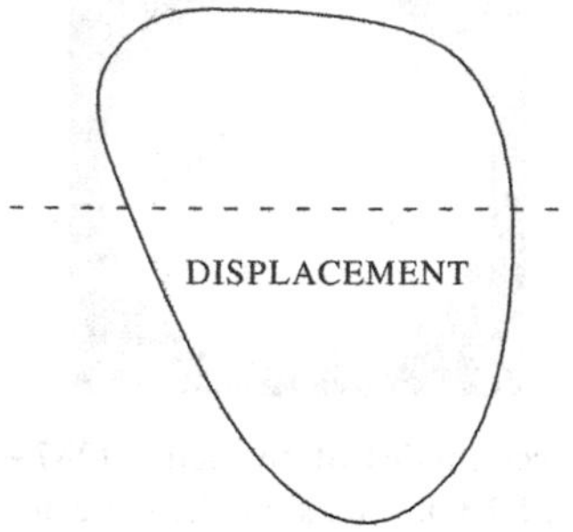

A body partially submerged in a liquid. The *displacement* is the amount of water that has been displaced by the body below the waterline. In equilibrium the weight of the displaced fluid equals the weight of the body.

$$\mathcal{F}_B = -\mathcal{F}_G^{\text{fluid}} = -\int_V \rho_{\text{fluid}}\, g\, dV. \tag{3.3}$$

This theorem is indeed Archimedes' principle:

- *The force of buoyancy is equal and opposite to the weight of the displaced fluid.*

The total force on the body may now be written

$$\mathcal{F} = \mathcal{F}_G + \mathcal{F}_B = \int_V (\rho_{\text{body}} - \rho_{\text{fluid}}) g\, dV, \tag{3.4}$$

explicitly confirming that when the body is made from the same fluid as its surroundings—so that $\rho_{\text{body}} = \rho_{\text{fluid}}$—the resultant force vanishes automatically. In general, however, the distributions of mass in the body and in the displaced fluid will be different.

Karl Friedrich Hieronymus Freiherr von Münchhausen (1720–1797). German soldier, hunter, nobleman, and delightful story-teller. In one of his stories, he lifts himself out of a deep lake by pulling at his bootstraps.

Münchhausen effect: Archimedes' principle is valid even if the gravitational field varies across the body, but fails if the body is so large that its own gravitational field cannot be neglected, such as would be the case if an Earth-sized body fell into Jupiter's atmosphere. The additional gravitational compression of the fluid near the surface of the body increases the fluid density and thus size of the buoyancy force. In semblance with Baron von Münchhausen's adventure, the body in effect lifts itself by its bootstraps (see Problems 3.8 and 3.9).

Constant field of gravity

In a constant gravitational field, $g(x) = g_0$, everything simplifies. The weight of the body and the buoyancy force become instead

$$\mathcal{F}_G = M_{\text{body}}\, g_0, \qquad \mathcal{F}_B = -M_{\text{fluid}}\, g_0. \tag{3.5}$$

Since the total force is the sum of these contributions, one might say that buoyancy acts as if the displacement were filled with fluid of negative mass $-M_{\text{fluid}}$. In effect the buoyancy force acts as a kind of antigravity.

The total force on an unrestrained object is now

$$\mathcal{F} = \mathcal{F}_G + \mathcal{F}_B = (M_{\text{body}} - M_{\text{fluid}})g_0. \tag{3.6}$$

If the body mass is smaller than the mass of the displaced fluid, the total force is directed upward (i.e., against gravity) and the body will begin to rise, and conversely if the force is directed downward it will sink. Alternatively, if the body is kept in place, the restraints must deliver a force $-\mathcal{F}$ to prevent the object from moving.

In constant gravity, a body can only hover motionlessly inside a fluid (or on its surface) if its mass equals the mass of the displaced fluid:

$$M_{\text{body}} = M_{\text{fluid}}. \tag{3.7}$$

A fish achieves this balance by adjusting the amount of water it displaces (M_{fluid}) through contraction and expansion of its body by means of an internal gas-filled bladder. A submarine, in contrast, adjusts its mass (M_{body}) by pumping water in and out of ballast tanks while keeping its shape unchanged.

Bermuda Triangle Mystery: It has been proposed that the mysterious disappearance of ships near Bermuda could be due to a sudden release of methane from the vast deposits of methane hydrates known to exist on the continental shelves. What effectively could happen is the same as when you shake a bottle of soda. Suddenly the water is filled with tiny gas bubbles with a density near that of air. This lowers the average density of the frothing water to maybe only a fraction of normal water, such that the mass of the ship's displacement falls well below the normal value. The ship is no more in buoyant equilibrium and drops like a stone, until it reaches normal density water or hits the bottom where it will usually remain forever because it becomes filled with water on the way down. Even if this sounds like a physically plausible explanation for the sudden disappearance of surface vessels, there is no consensus that this is what really happened in the Bermuda Triangle, nor in fact whether there is a mystery at all [4].

The physical phenomenon is real enough. It is, for example, well known to whitewater sailors that "holes" can form in which highly aerated water decreases the buoyancy, even to the point where it cannot carry their craft. You can yourself do an experiment in your kitchen using a partially filled soda bottle with a piece of wood floating on the surface. When you tap the bottle hard, carbon dioxide bubbles are released, and the "ship" sinks. In this case the "ship" will however quickly reappear on the surface because it does not take in water.

Example 3.1 [Gulf Stream surface height]: The warm Gulf Stream originates in the Gulf of Mexico and crosses the Atlantic to Europe after having followed the North American coast to Newfoundland. It has a width of about 100 km and a depth of about $d = 1{,}000$ m. Its temperature is about $\Delta T = 10°\text{C}$ above the surrounding ocean, of course warmer close to its origin and close to the surface. Since the fractional expansion of seawater (also called the expansivity) is about $\alpha = 2 \times 10^{-4}\ \text{K}^{-1}$, and taking into account that the temperature excess varies through the water column, we estimate the increase in surface height from half the depth: $h = \alpha\Delta T \times d/2$. Inserting the numbers we find $h \approx 1$ m. This agrees quite well with known values and with seasonal variations in surface height of about $\Delta h \approx 15$ cm [KSH99].

3.2 The gentle art of ballooning

Joseph Michel Montgolfier (1740–1810). Experimented (together with his younger brother **Jacques Étienne (1745–1799)**) with hot-air balloons. On November 21, 1783, the first humans flew in such a balloon for a distance of 9 km at a height of 100 m above Paris. Only one of the brothers ever flew, and then only once!

Jacques Alexandre César Charles (1746–1823). French physicist. The first to use hydrogen balloons for manned flight, and made on December 1, 1783, an ascent to about 3 km. Discovered Charles' law, a forerunner of the ideal gas law, stating that the ratio of volume to absolute temperature (V/T) is constant for a given pressure.

The first balloon flights are credited to the Montgolfier brothers who on November 21, 1783, flew an untethered manned hot-air balloon, and to Jacques Charles who on December 1 that same year flew an untethered manned hydrogen gas balloon (see Figure 3.1). In the beginning there was an intense rivalry between the advocates of Montgolfier and Charles type balloons, respectively called *la Montgolfière* and *la Charlière*, which presented different advantages and dangers to the courageous fliers. Hot-air balloons were easier to make although prone to catch fire, while hydrogen balloons had greater lifting power but could suddenly explode. By 1800 the hydrogen balloon had won the day, culminating in the huge (and dangerous) hydrogen airships of the 1930s. Helium balloons are much safer, but also much more expensive to fill. In the last half of the twentieth century, hot-air balloons again came into vogue, especially for sports, because of the availability of modern strong lightweight materials (nylon) and suitable gas fuel (propane).

The balloon equation

Let M denote the mass of the balloon at height z above the ground. This includes the gondola, the balloon skin, the payload (passengers), but not the gas (be it hot air, hydrogen or helium). The mass of the balloon can diminish if the balloon captain decides to throw out stuff from the gondola to increase its maximal height, also called the *ceiling*, and often sand bags are carried as ballast for this purpose. The condition (3.7) for the balloon to float stably at some height z above the ground now takes the form[1]

$$\boxed{M + \rho' V = \rho V,} \tag{3.8}$$

where ρ' the average density of the gas, ρ is the average density of the displaced air, and V the volume of the gas at height z. If the left-hand side of this equation is smaller or larger than the right-hand side, the balloon will rise or fall.

Hot-air balloons

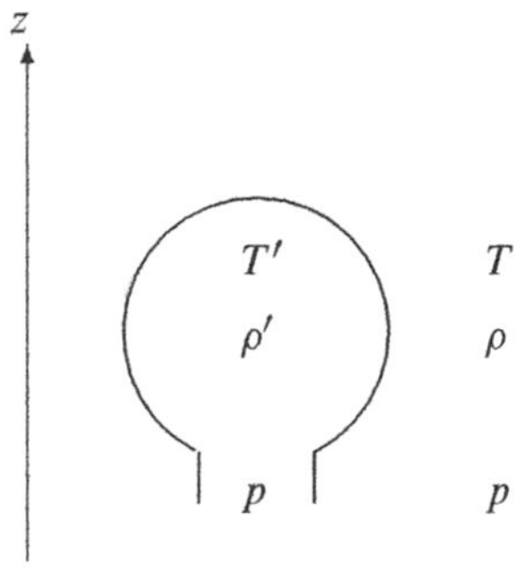

A hot-air balloon has higher temperature $T' > T$ and lower density $\rho' < \rho$ but essentially the same pressure as the surrounding atmosphere because it is open below.

A hot-air balloon is open at the bottom so that the inside pressure is always the same as the atmospheric pressure outside. The air in the balloon is warmer ($T' > T$) than the outside temperature and the density is correspondingly lower ($\rho' < \rho$). Using the ideal gas law (2.27) and the equality of the inside and outside pressures, we obtain $\rho' T' = \rho T$ so that the inside density is $\rho' = \rho T/T'$. Up to a height of about 10 km one can use the expressions (2.52) and (2.53) for the homentropic temperature and density of the atmosphere.

Example 3.2 [La Montgolfière]: The first Montgolfier balloon used for human flight on November 21, 1783, was about 15 m in diameter with an oval shape and had a constant volume $V \approx 60{,}000$ foot$^3 \approx 1700$ m^3. It carried two persons, rose to a ceiling of $z \approx 1{,}000$ m and flew about 9 km in 25 minutes. The machine is reported to have weighed 1,600 lbs $\approx$ 725 kg, and adding the two passengers and their stuff the total mass to lift could have been $M \approx 850$ kg. The November air being fairly cold and dense, we guess that its density must have been $\rho = 1.1$ kg m^{-3} at 1,000 m above Paris, and from (3.8) we then conclude that the hot air density must have been $\rho' = (\rho V - M)/V = 0.60$ kg m^{-3}. Taking $T = 0°C = 273$ K it follows from the ideal gas law that the hot air temperature was $T' = T\rho/\rho' = 501$ K $= 228°C$. This is uncomfortably close to the famous ignition temperature of paper, 451°F = 233°C. The balloon's material did actually get scorched by the burning straw used to heat the air in flight, but the fire was quickly extinguished with water brought along for this eventuality.

[1]On the right-hand side we have left out the tiny buoyancy $\rho V_0 = M\rho/\rho_0$ due to the displaced air volume V_0 of the balloon's solid material with average density $\rho_0 = M/V_0$. Typically $\rho/\rho_0 \lesssim 10^{-3}$.

Figure 3.1. Contemporary pictures of the first balloons. The first ascents were witnessed by huge crowds. Benjamin Franklin, scientist and one of the founding fathers of the United States, also witnessed the first ascents and became deeply interested in the future possibilities of this invention, but did not live to see the first American hot-air balloon flight in 1793. **Left:** : The Montgolfier hot-air balloon. The first ascent was piloted by Jean-Francois Pilatre de Rozier and the Marquis d'Arlandes. In 1785 Rozier attempted to cross the English Channel with a combination of a hydrogen and hot-air balloon, but the balloon suddenly deflated and he died in the crash. (Source: Wikimedia Commons.) **Right:** The Charles hydrogen balloon. The first ascent was piloted by Charles himself and Nicolas-Louis Robert. At the end of the first flight Robert jumped out, and the balloon with Charles rapidly shot up to about 3,000 m. Charles actually never flew again. (Source: Wikimedia Commons.)

Gas balloons

A modern large hydrogen or helium balloon typically begins its ascent being only partially filled, assuming an inverted tear-drop shape. During the ascent the gas expands because of the fall in ambient air pressure, and eventually the balloon becomes nearly spherical and stops expanding (or bursts) because the "skin" of the balloon cannot stretch further. To avoid bursting, the balloon can be fitted with a safety valve, or simply be open at the bottom. Since the density of the surrounding air falls with height, causing the buoyancy to drop, the balloon will eventually reach a ceiling where it would hover permanently if it did not lose gas. In the end no balloon stays aloft forever[2].

Example 3.3 [La Charlière]: The hydrogen balloon used by Charles for the ascent on December 1, 1783, carried two passengers to a height of 600 m. The balloon was made from rubberized silk and nearly spherical with a diameter of 27 ft, giving it a nearly constant volume of $V = 292\ \text{m}^3$. It was open at the bottom to make the pressure the same inside and outside. Taking also the temperatures to be the same inside and outside, it follows from the ideal gas law that the ratio of densities equals the ratio of molar masses, $\rho'/\rho = M_{\text{hydrogen}}/M_{\text{air}} \approx 1/15$, independently of height. Assuming a density $\rho = 1.1\ \text{kg m}^{-3}$ at the ceiling, we obtain from (3.8) $M = (\rho - \rho')V = 300$ kg. Assuming furthermore a skin thickness of 1.5 mm and a skin density 0.4 g cm^{-3}, the skin mass becomes 128 kg, leaving only 172 kg for the gondola and the two passengers. Assuming they each weight 60 kg, the mass of the gondola can have been only about 50 kg, subject of course to the uncertainty in the assumptions that we have made.

[2]Curiously, no animals seem to have developed balloons for floating in the atmosphere, although both the physics and chemistry of gas ballooning appears to be within reach of biological evolution.

3.3 Stability of floating bodies

Although a body may be in buoyant equilibrium, such that the total force composed of gravity and buoyancy vanishes, $\mathcal{F} = \mathcal{F}_G + \mathcal{F}_B = \mathbf{0}$, it may still not be in complete mechanical equilibrium. The total moment of all the forces acting on the body must also vanish; otherwise an unrestrained body will necessarily start to rotate. In this section we shall discuss the mechanical stability of floating bodies, whether they float on the surface, like ships and ducks, or float completely submerged, like submarines and fish. To find the stable configurations of a floating body, we shall first derive a useful corollary to Archimedes' principle concerning the moment of force due to buoyancy.

Moments of gravity and buoyancy

The total moment is like the total force a sum of two terms,

$$\mathcal{M} = \mathcal{M}_G + \mathcal{M}_B, \tag{3.9}$$

with one contribution from gravity,

$$\mathcal{M}_G = \int_V \boldsymbol{x} \times \rho_{\text{body}} \boldsymbol{g} \, dV, \tag{3.10}$$

and the other from pressure, called the moment of buoyancy,

$$\mathcal{M}_B = \oint_S \boldsymbol{x} \times (-p \, d\boldsymbol{S}). \tag{3.11}$$

If the total force vanishes, $\mathcal{F} = \mathbf{0}$, the total moment will be independent of the origin of the coordinate system, as may be easily shown.

Assuming again that the presence of the body does not change the local hydrostatic balance in the fluid, the moment of buoyancy will be independent of the nature of the material inside V. If the actual body is replaced by an identical volume of the ambient fluid, this fluid volume must be in total mechanical equilibrium, such that both the total force as well as the total moment acting on it have to vanish. Using that $\mathcal{M}_G^{\text{fluid}} + \mathcal{M}_B = \mathbf{0}$, we get

$$\mathcal{M}_B = -\mathcal{M}_G^{\text{fluid}} = -\int_V \boldsymbol{x} \times \rho_{\text{fluid}} \boldsymbol{g} \, dV. \tag{3.12}$$

We have in other words shown that

- *The moment of buoyancy is equal and opposite to the moment of the weight of the displaced fluid.*

This result is a natural corollary to Archimedes' principle, and of great help in calculating the buoyancy moment. A formal proof of this theorem, starting from the local equation of hydrostatic equilibrium, is found in Problem 3.4.

Constant gravity and mechanical equilibrium

In the remainder of this chapter we assume that gravity is constant, $\boldsymbol{g}(\boldsymbol{x}) = \boldsymbol{g}_0$, and that the body is in buoyant equilibrium so that it displaces exactly its own mass of fluid, $M_{\text{fluid}} = M_{\text{body}} = M$. The density distributions in the body and the displaced fluid will in general be different, $\rho_{\text{body}}(\boldsymbol{x}) \neq \rho_{\text{fluid}}(\boldsymbol{x})$ for nearly all points.

The moment of gravity (3.10) may be expressed in terms of the center of mass $\boldsymbol{x}_G$ of the body, here called the *center of gravity*:

$$\mathcal{M}_G = \boldsymbol{x}_G \times M\boldsymbol{g}_0, \qquad \boldsymbol{x}_G = \frac{1}{M}\int_V \boldsymbol{x}\,\rho_{\text{body}}\,dV. \tag{3.13}$$

Similarly the moment of buoyancy (3.12) may be written,

$$\mathcal{M}_B = -\boldsymbol{x}_B \times M\boldsymbol{g}_0, \qquad \boldsymbol{x}_B = \frac{1}{M}\int_V \boldsymbol{x}\,\rho_{\text{fluid}}\,dV, \tag{3.14}$$

where $\boldsymbol{x}_B$ is the moment of gravity of the displaced fluid, also called the *center of buoyancy*. Although each of these moments depends on the choice of origin of the coordinate system, the total moment,

$$\boxed{\mathcal{M} = (\boldsymbol{x}_G - \boldsymbol{x}_B) \times M\boldsymbol{g}_0,} \tag{3.15}$$

will be independent. This is also evident from the appearance of the difference of the two center positions. A shift of the origin of the coordinate system will affect the centers of gravity and buoyancy in the same way and therefore cancel out.

As long as the total moment is non-vanishing, the unrestrained body is not in complete mechanical equilibrium, but will start to rotate toward an orientation with vanishing moment. Except for the trivial case where the centers of gravity and buoyancy coincide, the above equation tells us that the total moment can only vanish if the centers lie on the same vertical line, $\boldsymbol{x}_G - \boldsymbol{x}_B \propto \boldsymbol{g}_0$. Evidently, there are two possible orientations satisfying this condition: one where the center of gravity lies below the center of buoyancy, and another where the center of gravity is above. At least one of these must be stable, for otherwise the body would never come to rest.

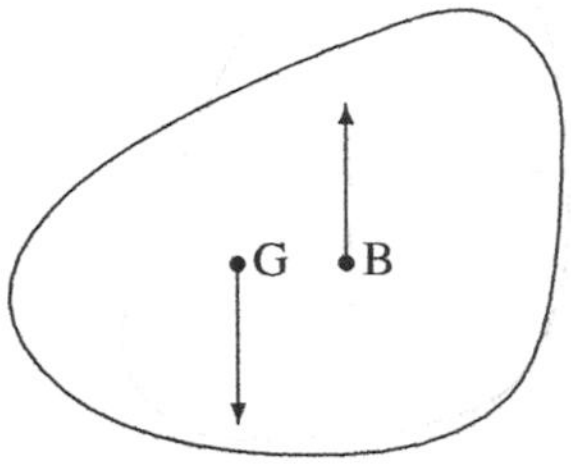

Fully submerged body in buoyant equilibrium with non-vanishing total moment (which here sticks out of the paper). The moment will for a fully submerged body always tend to rotate it (here anticlockwise) such that the center of gravity is brought below the center of buoyancy.

Fully submerged body

In a fully submerged rigid body, for example a submarine, both centers are always in the same place relative to the body, barring possible shifts in the cargo. If the center of gravity does not lie directly below the center of buoyancy, but is displaced horizontally, for example by rotating the body, the direction of the moment will always tend to turn the body so that the center of gravity is lowered with respect to the center of buoyancy. The only stable equilibrium orientation of the body is where the center of gravity lies vertically below the center of buoyancy. Any small perturbation away from this orientation will soon be corrected and the body brought back to the equilibrium orientation, assuming of course that dissipative forces (friction) can seep off the energy of the perturbation, for otherwise it will oscillate. A similar argument shows that the other equilibrium orientation with the center of gravity above the center of buoyancy is unstable and will flip the body over, if perturbed the tiniest amount.

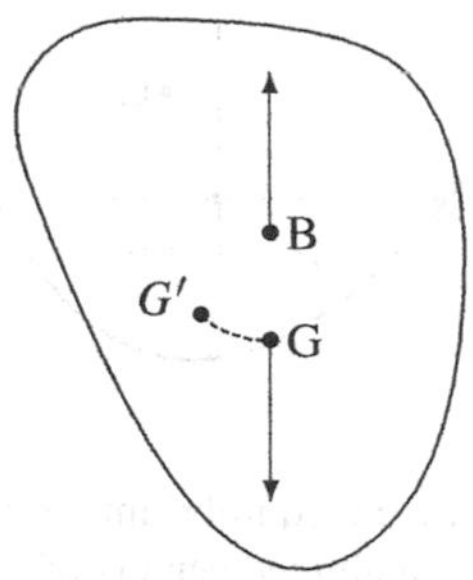

A fully submerged rigid body (a submarine) in stable equilibrium must have the center of gravity situated directly below the center of buoyancy. If G is moved to G', for example by rotating the body, a restoring moment is created that sticks out of the plane of the paper, as shown in the upper figure.

Now we understand better why the gondola hangs below an airship or balloon. If the gondola were on top, its higher average density would raise the center of gravity above the center of buoyancy and thereby destabilize the craft. Similarly, a fish goes belly-up when it dies because various gases fill the swim bladder, which enlarges into the belly and reverses the positions of the centers of gravity and buoyancy. It also loses buoyant equilibrium and floats to the surface, and stays there until it becomes completely waterlogged and sinks to the bottom. Interestingly, a submarine always has a conning tower on top to serve as a bridge when sailing on the surface, but its weight will be offset by the weight of heavy machinery at the bottom, so that the boat remains fully stable when submerged. Surfacing, the center of buoyancy of the submarine is obviously lowered, but as we shall see below this need not destabilize the boat even if it comes to lie below the center of gravity.

Body floating on the surface

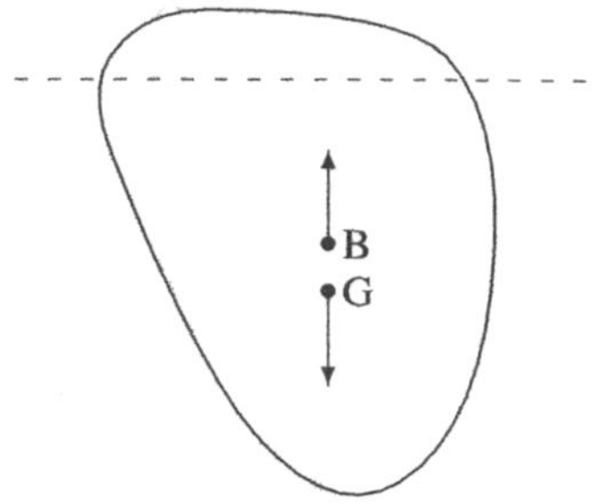

A floating body may have a stable equilibrium with the center of gravity directly *below* the center of buoyancy.

At the surface of a liquid, a body such as a ship or an iceberg will according to Archimedes' principle always arrange itself so that the mass of displaced liquid exactly equals the mass of the body. Here we assume that there is vacuum or a very light fluid such as air above the liquid. The center of gravity is always in the same place relative to the body if the cargo is fixed (see however Figure 3.2), but the center of buoyancy depends now on the orientation of the body, because the volume of displaced fluid changes place and shape, while keeping its mass constant, when the body orientation changes.

Stability can—as always—only occur when the two centers lie on the same vertical line, but there may be more than one stable orientation. A sphere made of homogeneous wood floating on water is stable in all orientations. None of them are in fact truly stable, because it takes no force to move from one to the other (disregarding friction). This is however a marginal case.

A floating body may, like a submerged body, possess a stable orientation with the center of gravity directly *below* the center of buoyancy. A heavy keel is, for example, used to lower the center of gravity of a sailing ship so much that this orientation becomes the only stable equilibrium. In that case it becomes virtually impossible to capsize the ship, even in a very strong wind. The stable orientation for most floating objects, such as ships, will in general have the center of gravity situated directly *above* the center of buoyancy. This happens always when an object of constant mass density floats on top of a liquid of constant mass density, for example an iceberg on water. Since ice, like water, is homogeneous, the part of the iceberg that lies below the waterline must have its center of buoyancy in exactly the same place as its center of gravity. The part of the iceberg lying above the water cannot influence the center of buoyancy whereas it always will shift the center of gravity upward.

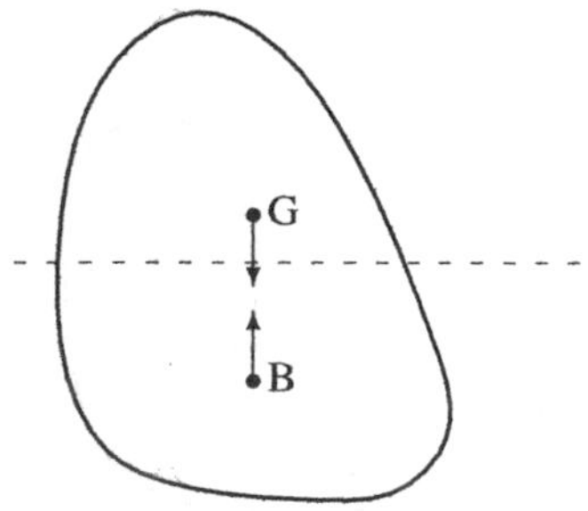

A floating body generally has a stable equilibrium with the center of gravity directly *above* the center of buoyancy.

How can that situation ever be stable? Will the moment of force not be of the wrong sign if the ship is perturbed? Why don't ducks and tall ships capsize spontaneously? The qualitative answer is that when the body is rotated away from such an equilibrium orientation, the volume of displaced water will change place and shape in such a way as to shift the center of buoyancy back to the other side of the center of gravity, reversing thereby the direction of the moment of force to restore the equilibrium. We shall now make this argument quantitative.

3.4 Ship stability

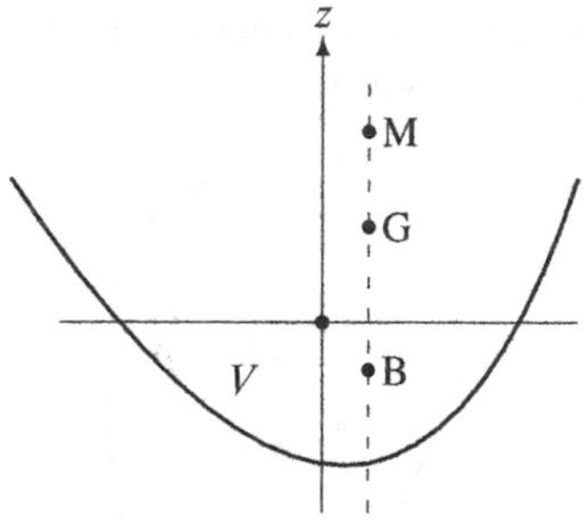

Ship in an equilibrium orientation with aligned centers of gravity (G) and buoyancy (B). The metacenter (M) lies in this case above the center of gravity, so that the ship is stable against small perturbations. The horizontal line at $z = 0$ indicates the surface of the water. In buoyant equilibrium, the ship always displaces the volume V, independently of its orientation.

Sitting comfortably in a small rowboat, it is fairly obvious that the center of gravity lies above the center of buoyancy, and that the situation is stable with respect to small movements of the body. But many a fisherman has learned that suddenly standing up may compromise the stability and throw him out among the fishes. There is, as we shall see, a strict limit to how high the center of gravity may be above the center of buoyancy. If this limit is violated, the boat becomes unstable and capsizes. As a practical aid to the captain, the limit is indicated by the position of the so-called *metacenter*, a fictive point usually placed on the vertical line through the equilibrium positions of the centers of buoyancy and gravity (the "mast"). The stability condition then requires the center of gravity to lie below the metacenter (see the margin figure).

Initially, we shall assume that the ship is in complete mechanical equilibrium with vanishing total force and vanishing total moment of force. The aim is now to calculate the moment of force that arises when the ship is brought slightly out of equilibrium. If the moment tends to turn the ship back into equilibrium, the initial orientation will be stable, otherwise it is unstable.

Figure 3.2. The Flying Enterprise (1952). A body can float stably in many orientations, depending on the position of its center of gravity. In this case the list to port was caused by a shift in the cargo which moved the center of gravity to the port side. The ship and its lonely captain Carlsen became famous because he stayed on board during the storm that eventually sent it to the bottom. (Source: *Politiken*, Denmark. With permission.)

Center of roll

Most ships are mirror symmetric in a plane, but we shall be more general and consider a "ship" of arbitrary shape. In a flat-Earth coordinate system with vertical z-axis, the waterline is naturally taken to lie at $z = 0$. In the waterline the ship covers a horizontal region A of arbitrary shape. The geometric center or *area centroid* of this region is defined by the average of the position,

$$(x_0, y_0) = \frac{1}{A}\int_A (x, y)\, dA, \tag{3.16}$$

where $dA = dx\, dy$ is the area element. Without loss of generality we may always place the coordinate system such that $x_0 = y_0 = 0$. In a ship that is mirror symmetric in a vertical plane, the area center will also lie in this plane.

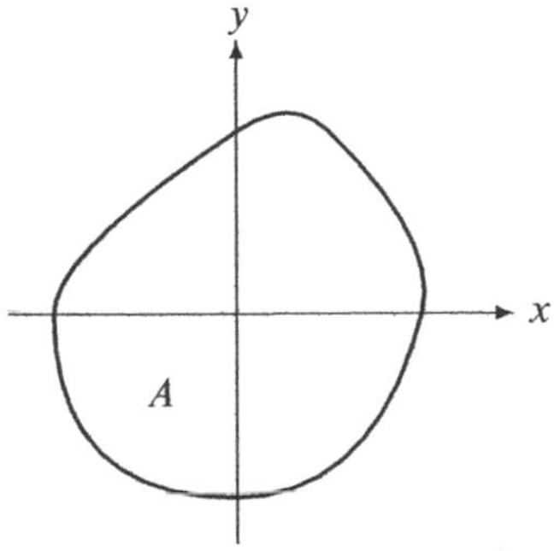

The area A of the ship in the waterline may be of quite arbitrary shape.

To discover the physical significance of the centroid of the waterline area, the ship is tilted (or "heeled" as it would be in maritime language) through a tiny positive angle α around the x-axis, such that the equilibrium waterline area A comes to lie in the plane $z = \alpha y$. The net change ΔV in the volume of the displaced water is to lowest order in α given by the difference in volumes of the two wedge-shaped regions between new and the old waterline. Since the displaced water is removed from the wedge at $y > 0$ and added to the wedge for $y < 0$, the volume change becomes

$$\Delta V = -\int_A z\, dA = -\alpha \int_A y\, dA = 0. \tag{3.17}$$

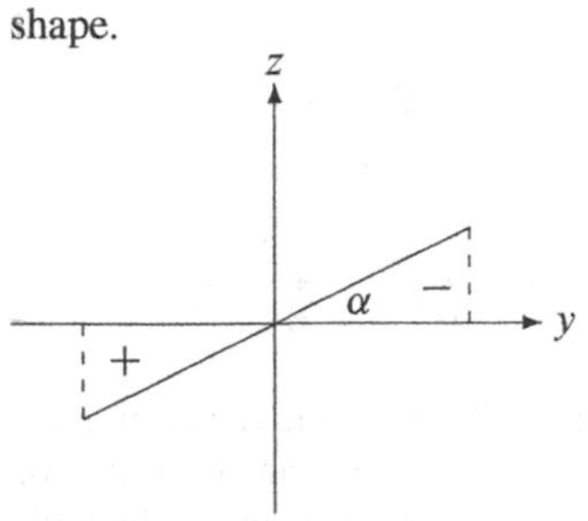

Tilt around the x-axis. The change in displacement consists of moving the water from the wedge to the right into the wedge to the left.

In the last step we have used that the origin of the coordinate system coincides with the centroid of the waterline area (i.e., $y_0 = 0$). There will be corrections to this result of order α^2 due to the actual shape of the hull just above and below the waterline, but they are disregarded here. To leading order the two wedges have the same volume.

Figure 3.3. The Queen Mary 2 set sail on its maiden voyage on January 2, 2004. It was at that time the world's largest ocean liner with a length of 345 m, a height of 72 m from keel to funnel, and a width of 41 m. Having a draft of only 10 m, its superstructure rises an impressive 62 m over the waterline. The low average density of the superstructure, including 2,620 passengers and 1,253 crew, combined with the high average density of the 117 megawatt engines and other heavy facilities close to the bottom of the ship nevertheless allow the stability condition (3.28) to be fulfilled. (Source: Wikimedia Commons.)

Since the direction of the x-axis is quite arbitrary, the conclusion is that the ship may be heeled around any line going through the centroid of the waterline area without any first-order change in volume of displaced water. This guarantees that the ship will remain in buoyant equilibrium after the tilt. The centroid of the waterline area may thus be called the ship's *center of roll*.

The metacenter

Taking water to have constant density, the center of buoyancy is simply the geometric average of the position over the displacement volume V (below the waterline),

$$(x_B, y_B, z_B) = \frac{1}{V}\int_V (x, y, z)\, dV. \tag{3.18}$$

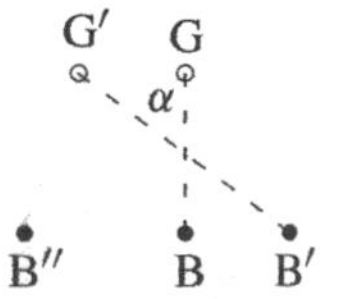

The tilt rotates the center of gravity from G to G′, and the center of buoyancy from B to B′. In addition, the change in displaced water shifts the center of buoyancy back to B″. In stable equilibrium this point must for $\alpha > 0$ lie to the left of the new center of gravity G′.

In equilibrium, the horizontal positions of the centers of buoyancy and gravity must be equal $x_B = x_G$ and $y_B = y_G$. The vertical position z_B of the center of buoyancy will normally be different from the vertical position of the center of gravity z_G, which depends on the actual mass distribution of the ship, determined by its structure and load.

The tilt around the x-axis changes the positions of the centers of gravity and buoyancy. The center of gravity $\boldsymbol{x}_G = (x_G, y_G, z_G)$ is supposed to be fixed with respect to the ship (see however Figure 3.2) and is to first order in α shifted horizontally by a simple rotation through the infinitesimal angle α,

$$\delta y_G = -\alpha z_G. \tag{3.19}$$

There will also be a vertical shift, $\delta z_G = \alpha y_G$, but that is of no importance to the stability because gravity is vertical and therefore the shift creates no moment.

The center of buoyancy is also shifted by the tilt, at first by the same rule as the center of gravity, but because the displacement also changes there will be another contribution Δy_B to the total shift, so that we may write

$$\delta y_B = -\alpha z_B + \Delta y_B. \tag{3.20}$$

As discussed above, the change in the shape of the displacement amounts to moving the water from the wedge at the right ($y > 0$) to the wedge at the left ($y < 0$). The ensuing change in the horizontal position of the center of buoyancy may according to (3.18) be calculated by averaging the position change $y - y_B$ over the volume of the two wedges,

$$\Delta y_B = -\frac{1}{V}\int_A (y - y_B)\, z\, dA = -\frac{\alpha}{V}\int_A y^2\, dA = -\alpha\frac{I}{V},$$

where

$$\boxed{I = \int_A y^2\, dA} \tag{3.21}$$

is the second-order moment of the waterline area A around the x-axis. The movement of displaced water will also create a shift in the x-direction, $\Delta x_B = -\alpha J/V$ where $J = \int_A xy\, dA$, which does not destabilize the ship.

The total horizontal shift in the center of buoyancy may thus be written

$$\delta y_B = -\alpha\left(z_B + \frac{I}{V}\right). \tag{3.22}$$

This shows that the complicated shift in the position of the center of buoyancy can be written as a simple rotation of a point M that is *fixed* with respect to the ship with z-coordinate,

$$\boxed{z_M = z_B + \frac{I}{V}.} \tag{3.23}$$

This point, called the *metacenter*, is usually placed on the straight line that goes through the centers of gravity and buoyancy, such that $x_M = x_G = x_B$ and $y_M = y_G = y_B$. The calculation shows that when the ship is heeled through a small angle, the center of buoyancy will always move so that it stays vertically below the metacenter.

The metacenter is a purely geometric quantity (for a liquid with constant density), depending only on the displacement volume V, the center of buoyancy x_B, and the second-order moment of the shape of the ship in the waterline. The simplest waterline shapes are

Rectangular waterline area: If the ship has a rectangular waterline area with sides $2a$ and $2b$, the roll center coincides with the center of the rectangle, and the second moment around the x-axis becomes

$$I = \int_{-a}^{a} dx \int_{-b}^{b} dy\, y^2 = \frac{4}{3}ab^3. \tag{3.24}$$

If $a > b$, this is the smallest moment around any tilt axis because $ab^3 < a^3b$.

Elliptic waterline area: If the ship has an elliptical waterline area with axes $2a$ and $2b$, the roll center coincides with the center of the ellipse, and the second moment around the x-axis becomes

$$I = \int_{-a}^{a} dx \int_{-b\sqrt{1-x^2/a^2}}^{b\sqrt{1-x^2/a^2}} y^2\, dy = \frac{4}{3}ab^3 \int_0^1 (1-t^2)^{3/2}\, dt = \frac{\pi}{4}ab^3. \tag{3.25}$$

Notice that this is about half the value for the rectangle.

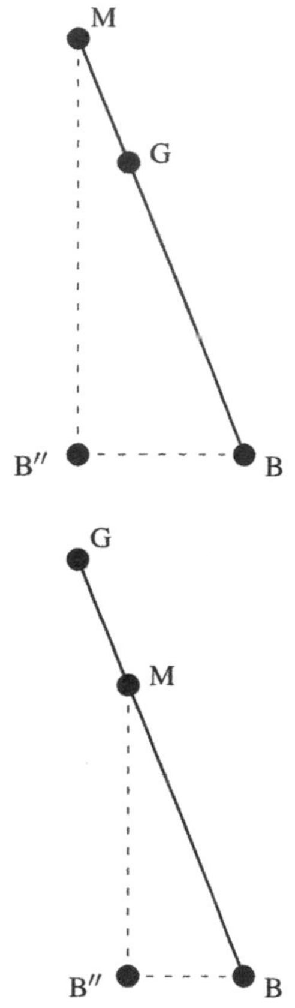

The metacenter M lies always directly above the actual center of buoyancy B″. **Top:** The ship is stable because the metacenter lies above the center of gravity. **Bottom:** The ship is unstable because the metacenter lies below the center of gravity.

Stability condition

The tilt generates a *restoring moment* around the x-axis, which may be calculated from (3.15),

$$\mathcal{M}_x = -(y_G - y_B)Mg_0. \tag{3.26}$$

Since we have $y_G = y_B$ in the original mechanical equilibrium, the difference in coordinates after the tilt may be written, $y_G - y_B = \delta y_G - \delta y_B$, where δy_G and δy_B are the small horizontal shifts of order α in the centers of gravity and buoyancy, calculated above. The shift $\Delta x_B = -\alpha J/V$ will create a moment $M_y = -\Delta x_B M g_0$, which tends to pitch the ship along x but does not affect its stability.

In terms of the height of the metacenter z_M, the restoring moment becomes

$$\boxed{\mathcal{M}_x = \alpha(z_G - z_M)Mg_0.} \tag{3.27}$$

For the ship to be stable, the restoring moment must counteract the tilt and thus have opposite sign of the tilt angle α. Consequently, the stability condition becomes

$$z_G < z_M. \tag{3.28}$$

Evidently, *the ship is only stable when the center of gravity lies below the metacenter.* For an alternative derivation of the stability condition, see Problem 3.13.

Example 3.4 [Elliptical rowboat]: An elliptical rowboat with vertical sides has major axis $2a = 2$ m and minor axis $2b = 1$ m. The smallest moment of the elliptical area is $I = (\pi/4)ab^3 \approx 0.1\ \text{m}^4$. If your mass is 75 kg and the boat's is 50 kg, the displacement will be $V = 0.125\ \text{m}^3$, and the draft $d \approx V/4ab = 6.25$ cm, ignoring the usually curved shape of the boat's hull. The coordinate of the center of buoyancy becomes $z_B = -3.2$ cm and the metacenter $z_M = 75$ cm. Getting up from your seat may indeed raise the center of gravity so much that it gets close to the metacenter and the boat begins to roll violently. Depending on your weight and mass distribution, the boat may even become unstable and turn over.

Metacentric height and righting arm

The orientation of the coordinate system with respect to the ship's hull was not specified in the analysis, which is therefore valid for a tilt around any direction. For a ship to be fully stable, the stability condition must be fulfilled for all possible tilt axes. Since the displacement V is the same for all choices of tilt axis, the second moment of the area on the right-hand side of (3.23) should be chosen to be the *smallest* one. Often it is quite obvious which moment is the smallest. Many modern ships are extremely long with the same cross-section along most of their length and with mirror symmetry through a vertical plane. For such ships the smallest moment is clearly obtained with the tilt axis parallel to the longitudinal axis of the ship.

The restoring moment (3.27) is proportional to the vertical distance, $z_M - z_G$, between the metacenter and the center of gravity, also called the *metacentric height*. The closer the center of gravity comes to the metacenter, the smaller the restoring moment will be, and the longer the period of rolling oscillations will be. The actual roll period depends also on the true moment of inertia of the ship around the tilt axis (see Problem 3.11). Whereas the metacenter is a purely geometric quantity that depends only on the ship's actual draft, the center of gravity depends on the way the ship is actually loaded. A good captain should always know the positions of the center of gravity and the metacenter of his ship before he sails, or else he may capsize when casting off.

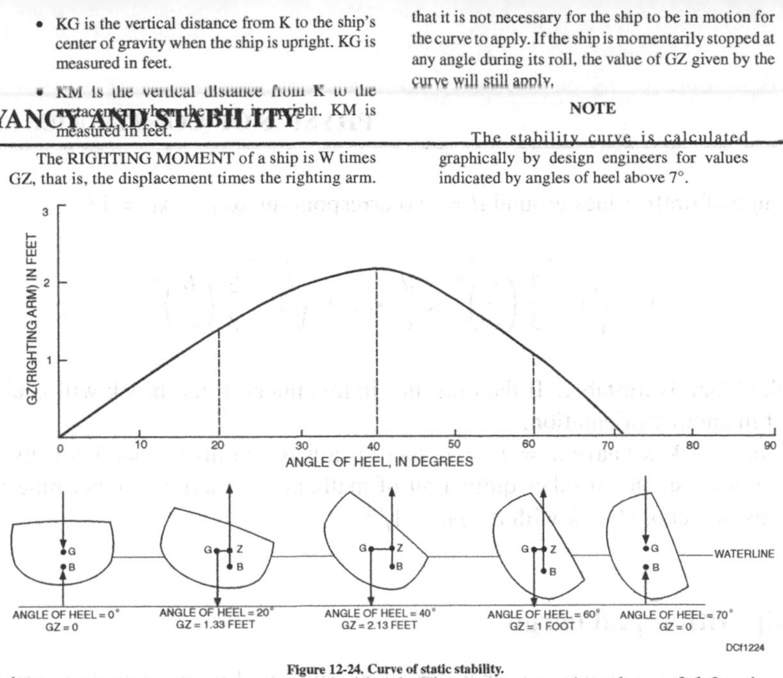
• KG is the vertical distance from K to the ship's center of gravity when the ship is upright. KG is measured in feet.
• KM is the vertical distance from K to the metacenter when the ship is upright. KM is measured in feet.
The RIGHTING MOMENT of a ship is W times GZ, that is, the displacement times the righting arm.
that it is not necessary for the ship to be in motion for the curve to apply. If the ship is momentarily stopped at any angle during its roll, the value of GZ given by the curve will still apply.
NOTE
The stability curve is calculated graphically by design engineers for values indicated by angles of heel above 7°.
GZ(RIGHTING ARM) IN FEET
ANGLE OF HEEL, IN DEGREES
WATERLINE
ANGLE OF HEEL = 0° GZ = 0
ANGLE OF HEEL = 20° GZ = 1.33 FEET
ANGLE OF HEEL = 40° GZ = 2.13 FEET
ANGLE OF HEEL = 60° GZ = 1 FOOT
ANGLE OF HEEL = 70° GZ = 0
DCf1224
Figure 12-24. Curve of static stability.
12-12

Figure 3.4. Stability curve for large angles of heel. The metacenter is only useful for tiny heel angles where all changes are linear in the angle. For larger angles one uses instead the 'righting arm' which is the distance between the center of gravity and the vertical line through the actual center of buoyancy. Instability sets in when the righting arm reaches zero, in the above plot for about 72° heel. (Source: John Pike, Global Security. With permission.)

The metacenter is only useful for tiny heel angles where all changes are linear in the angle. For larger angles one uses instead the *righting arm* which is the horizontal distance $|y_G - y_B|$ between the center of gravity and the vertical line through the actual center of buoyancy. The restoring moment is the product of the righting arm and the weight of the ship, and instability sets in when the righting arm reaches zero for some non-vanishing angle of heel (see Figure 3.4).

Case: Floating block

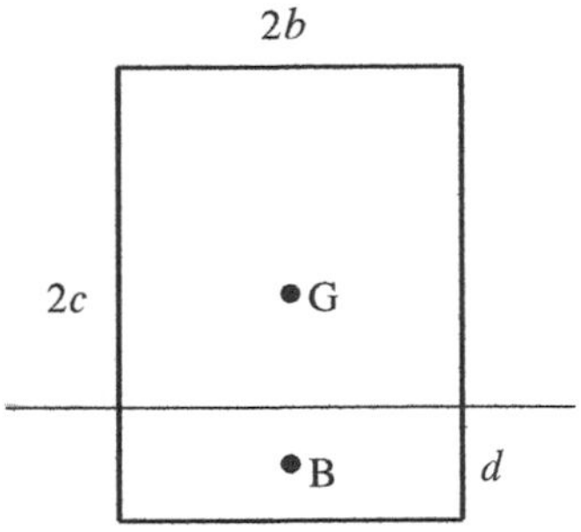
2b
2c
G
B
d

Floating block with height h, draft d, width $2b$, and length $2a$ into the paper.

The simplest non-trivial case in which we may apply the stability criterion is that of a rectangular block of dimensions $2a$, $2b$, and $2c$ in the three coordinate directions. Without loss of generality we may assume that $a > b$. The center of the waterline area coincides with the roll center and the origin of the coordinate system with the waterline at $z = 0$. The block is assumed to be made from a uniform material with constant density ρ_1 and floats in a liquid of constant density ρ_0.

In hydrostatic equilibrium we must have $M = 4abd\rho_0 = 8abc\rho_1$, or

$$\frac{\rho_1}{\rho_0} = \frac{d}{2c}. \tag{3.29}$$

The position of the center of gravity is $z_G = c - d$ and the center of buoyancy $z_B = -d/2$. Using (3.24) and $V = 4abd$, the position of the metacenter is

$$z_M = -\frac{d}{2} + \frac{b^2}{3d}. \tag{3.30}$$

Rearranging the stability condition, $z_M > z_G$, it may be written as

$$\left(\frac{d}{c} - 1\right)^2 > 1 - \frac{2b^2}{3c^2}. \tag{3.31}$$

When the block dimensions obey $a > b$ and $b/c > \sqrt{3/2} = 1.2247\ldots$, the right-hand side becomes negative and the inequality is always fulfilled. On the other hand, if $b/c < \sqrt{3/2}$

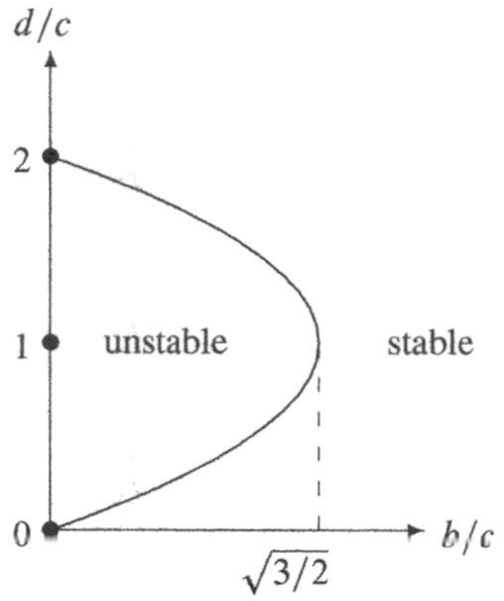

Stability diagram for the floating block.

there is a range of draft values around $d = c$ (corresponding to $\rho_1/\rho_0 = 1/2$),

$$1 - \sqrt{1 - \frac{2}{3}\left(\frac{b}{c}\right)^2} < \frac{d}{c} < 1 + \sqrt{1 - \frac{2}{3}\left(\frac{b}{c}\right)^2}, \tag{3.32}$$

for which the block is unstable. If the draft lies in this interval, the block will keel over and come to rest in another orientation.

For a cubic block we have $a = b = c$, there is always a range around density $\rho_1/\rho_0 = 1/2$ that cannot be stable. It takes quite a bit of mathematical labor to determine the stable configurations of a cubic block with $\rho_1/\rho_0 = 1/2$.

Case: Ship with liquid cargo

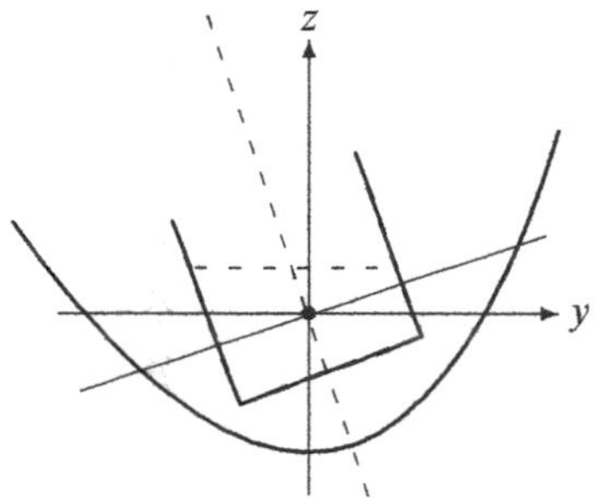

Tilted ship with an open container filled with liquid.

Many ships carry liquid cargos, oil or water, or nearly liquid cargoes such as grain. When the tanks are not completely filled, this kind of cargo may strongly influence the stability of the ship. The main effect of an open liquid surface inside the ship is the movement of real liquid, which shifts the center of mass in the same direction as the movement of displaced water shifts the center of buoyancy. This competition between the two centers may directly lead to instability because the movement of the real liquid inside the ship nearly cancels the stabilizing movement of the displaced water. On top of that, the fairly slow sloshing of the real liquid compared to the instantaneous movement of the displaced water can lead to dangerous oscillations that may also capsize the ship.

For the case of a single open tank, the calculation of the restoring moment must now include the liquid cargo. Disregarding sloshing, a similar analysis as before shows that there will be a horizontal change in the center of gravity from the movement of a wedge of real liquid of density ρ_1,

$$\Delta y_G = -\alpha \frac{\rho_1 I_1}{M} = -\alpha \frac{\rho_1}{\rho_0} \frac{I_1}{V}, \tag{3.33}$$

where I_1 is the second moment of the open liquid surface. The metacenter position now becomes

$$\boxed{z_M = z_B + \frac{I}{V} - \frac{\rho_1}{\rho_0} \frac{I_1}{V}.} \tag{3.34}$$

The effect of the moving liquid is to lower the metacentric height (or shorten the righting arm) with possible destabilization as a result. The unavoidable inertial sloshing of the liquid may further compromise the stability. The destabilizing effect of a liquid cargo is often counteracted by dividing the hold into a number of smaller compartments by means of bulkheads along the ship's principal roll axis.

Car ferry instability: In heavy weather or due to accidents, a car ferry may inadvertently get a layer of water on the car deck. Since the car deck of a roll-on/roll-off ferry normally spans the whole vessel, we have $\rho_1 = \rho_0$ and $I_1 \approx I$, implying that $z_M \approx z_B < z_G$, nearly independent of the thickness h of the layer of water (as long as h is not too small). The inequality $z_M < z_G$ spells rapid disaster, as several accidents with car ferries have shown. Waterproof longitudinal bulkheads on the car deck would stabilize the ro-ro ferry, but are usually avoided because they would hamper efficient loading and unloading of the cars.

* Principal roll axes

It has already been pointed out that the metacenter for absolute stability is determined by the smallest second moment of the waterline area. To determine that we instead tilt the ship around an axis, $\boldsymbol{n} = (\cos\phi, \sin\phi, 0)$, forming an angle ϕ with the x-axis. Since this configuration is obtained by a simple rotation through ϕ around the z-axis, the transverse coordinate to be used in calculating the second moment becomes $\tilde{y} = y\cos\phi - x\sin\phi$ (see Equation (B.29b)), and we find the moment,

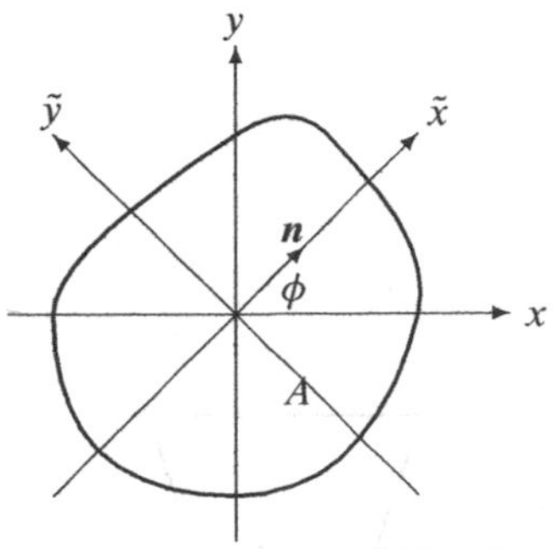

Tilt axis $\boldsymbol{n}$ forming an angle ϕ with the x-axis.

$$\tilde{I} = \int_A \tilde{y}^2\, dA = I_{yy}\cos^2\phi + I_{xx}\sin^2\phi - 2I_{xy}\sin\phi\cos\phi, \tag{3.35}$$

where I_{xx}, I_{yy} and I_{xy} are the elements of the symmetric matrix,

$$\mathbf{I} = \begin{pmatrix} I_{xx} & I_{xy} \\ I_{yx} & I_{yy} \end{pmatrix} = \int_A \begin{pmatrix} x^2 & xy \\ xy & y^2 \end{pmatrix} dA, \tag{3.36}$$

which is actually a two-dimensional tensor.

The extrema of $\tilde{I}(\phi)$ are easily found by differentiation with respect to ϕ. They are

$$\phi_1 = \frac{1}{2}\arctan\frac{2I_{xy}}{I_{xx} - I_{yy}}, \qquad \phi_2 = \phi_1 + \frac{\pi}{2}, \tag{3.37}$$

with the respective area moments $I_1 = \tilde{I}(\phi_1)$ and $I_2 = \tilde{I}(\phi_2)$. The two angles determine orthogonal *principal directions*, 1 and 2, in the ship's waterline area. The principal direction with the smallest second-order moment around the area centroid has the lowest metacentric height. If I_1 is the smallest moment and the actual roll axis forms an angle ϕ with the 1-axis, we can calculate the moment I for any other axis of roll forming an angle ϕ with the x-axis from

$$I = I_1\cos^2\phi + I_2\sin^2\phi. \tag{3.38}$$

Since $I_1 < I_2$, it follows trivially that if the ship is stable for a tilt around the first principal axis, it will be stable for a tilt around any axis.

Problems

3.1 A stone weighs $\mathcal{F}_1 = 1{,}000$ N in vacuum and $\mathcal{F}_0 = 600$ N when submerged in water of density ρ_0. **(a)** Calculate the volume V and **(b)** average density ρ_1 of the stone.

3.2 A hydrometer is an instrument used to measure the density of a liquid. A certain hydrometer with mass $M = 4$ g consists of a roughly spherical bulb and a long thin cylindrical stem of radius $a = 2$ mm. The sphere is weighed down so that the apparatus will float stably with the stem pointing vertically upward and crossing the fluid surface at some point. **(a)** How much deeper will it float in alcohol with mass density $\rho_1 = 0.78$ g cm^{-3} than in oil with mass density $\rho_2 = 0.82$ g cm^{-3}? You may disregard the tiny density of air.

3.3 A cylindrical wooden stick with density $\rho_1 = 0.65$ g cm^{-3} floats in water (density $\rho_0 = 1$ g cm^{-3}). The stick is loaded down with a lead weight with density $\rho_2 = 11$ g cm^{-3} at one end such that it floats in a vertical orientation with a fraction $f = 1/10$ of its length out of the water. **(a)** What is the ratio M_1/M_2 between the masses of the wooden stick and the lead weight? **(b)** As a function of the density of the wood, how large a fraction of the stick can be out of the water in hydrostatic equilibrium (disregarding questions of stability)?

3.4 Prove without assuming constant gravity that the hydrostatic moment of buoyancy equals (minus) the moment of gravity of the displaced fluid (corollary to Archimedes' principle).

3.5 Assuming constant gravity, show that for a body not in buoyant equilibrium (that is for which the total force $\mathcal{F}$ does not vanish), there is always a well-defined point of attack $\boldsymbol{x}_0$ such that the total moment of gravitational plus buoyant forces is given by $\mathcal{M} = \boldsymbol{x}_0 \times \mathcal{F}$.

3.6 A right rotation cone has half opening angle α and height h. It is made from a homogeneous material of density ρ_1 and floats in a liquid of density $\rho_0 > \rho_1$. **(a)** Determine the stability condition on the mass ratio ρ_1/ρ_0 when the cone floats vertically with the peak downward. **(b)** Determine the stability condition on the mass ratio when the cone floats vertically with the peak upward. **(c)** What is the smallest opening angle that permits simultaneous stability in both directions?

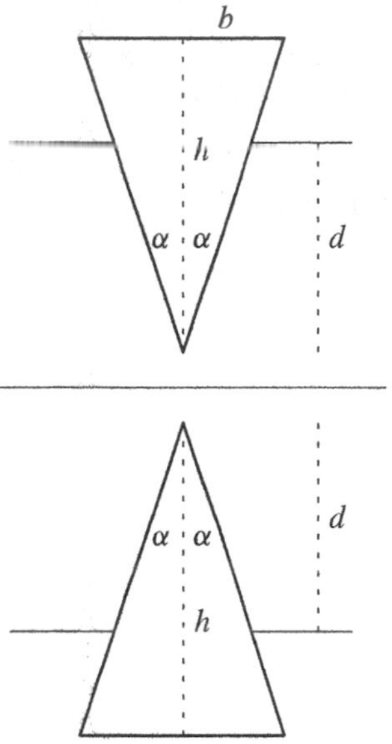

Top: Triangular ship of length L (into the paper) floating with its peak vertically downward. **Bottom:** The peak vertically upward. Note that in this case the draft is $h - d$.

* **3.7** A ship of length L has a longitudinally invariant cross-section in the shape of an isosceles triangle with half opening angle α and height h. It is made from homogeneous material of density ρ_1 and floats in a liquid of density $\rho_0 > \rho_1$. **(a)** Determine the stability condition on the mass ratio ρ_1/ρ_0 when the ship floats vertically with the peak downward. **(b)** Determine the stability condition on the mass ratio when the ship floats vertically with the peak upward. **(c)** What is the smallest opening angle that permits simultaneous stability in both directions?

3.8 Two identical homogenous spheres of mass M and radius a are situated a distance $D \gg a$ apart in a barotropic fluid. Due to their field of gravity, the fluid will be denser near the spheres. There is no other gravitational field present, the fluid density is ρ_0, and the pressure is p_0 in the absence of the spheres. One may assume that the pressure corrections due to the spheres are small everywhere in comparison with p_0. **(a)** Show that the spheres will repel each other and calculate the magnitude of the force to leading order in a/D. **(b)** Compare with the gravitational attraction between the spheres. **(c)** Under which conditions will the total force between the spheres vanish?

* **3.9** A barotropic compressible fluid is in hydrostatic equilibrium with pressure $p(z)$ and density $\rho(z)$ in a constant external gravitational field with potential $\Phi = g_0 z$. A finite body having a "small" gravitational field $\Delta\Phi(\boldsymbol{x})$ is submerged in the fluid. **(a)** Show that the change in hydrostatic pressure to lowest order of approximation is $\Delta p(\boldsymbol{x}) = -\rho(z)\Delta\Phi(\boldsymbol{x})$. **(b)** Show that for a spherically symmetric body of radius a and mass M, the extra surface pressure is $\Delta p = g_1 a \rho(z)$, where $g_1 = GM/a^2$ is the magnitude of surface gravity. **(c)** Show that the buoyancy force is increased.

3.10 Let $\mathbf{I}$ be a symmetric (2×2) matrix. Show that the extrema of the corresponding quadratic form $\boldsymbol{n} \cdot \mathbf{I} \cdot \boldsymbol{n} = I_{xx} n_x^2 + 2 I_{xy} n_x n_y + I_{yy} n_y^2$ where $n_x^2 + n_y^2 = 1$ are determined by the eigenvectors of $\mathbf{I}$ satisfying $\mathbf{I} \cdot \boldsymbol{n} = \lambda \boldsymbol{n}$.

3.11 Show that in a stable orientation the angular frequency of small oscillations around a principal tilt axis of a ship is

$$\omega = \sqrt{\frac{M g_0 (z_M - z_G)}{J}},$$

where J is the moment of inertia of the ship around this axis.

* **3.12** A ship has a waterline area that is a regular polygon with $n \geq 3$ edges. Show that the area moment tensor (3.36) must have $I_{xx} = I_{yy}$ and $I_{xy} = 0$.

3.13 **(a)** Show that the work (or potential energy) necessary to tilt a ship through an angle α to second order in α is

$$W = -\frac{1}{2}\alpha^2 M_0 g_0 (z_G - z_B) + \frac{1}{2}\rho_0 g_0 \int_{A_0} z^2 \, dA,$$

where $z = \alpha y$, and **(b)** show that this leads to the stability condition (3.28).

4 Hydrostatic shapes

It is primarily the interplay between gravity and contact forces that shapes the macroscopic world around us. The seas, the atmosphere, planets and stars all owe their shape to gravity, and even our own bodies bear witness to the strength of gravity at the surface of our massive planet. The sea is obviously horizontal at short distances, but bends below the horizon at larger distances following the planet's curvature. The Earth as a whole is spherical and so is the sea but that is only the leading approximation. The Moon's gravity tugs at the water in the seas and raises tides, and even the massive Earth is itself flattened by the centrifugal forces of its own rotation. What physics principle should we invoke to determine the equilibrium shape of our planet as well as the shape of the shallow oceans covering its surface?

Disregarding surface tension, the answer is based on the principle that in hydrostatic equilibrium, surfaces of constant gravitational potential, of constant pressure, and of constant density must all coincide. If an interface between a fluid and vacuum crosses an equipotential surface (as shown in the margin figure), there will arise a tangential component of gravity along the interface, and that can only be balanced by shear contact forces, which a fluid at rest is unable to supply. An iceberg rising out of the sea does not obey this principle because it is solid, not fluid, and therefore able to sustain internal shear forces. Neither is the principle valid for fluids in motion. Waves in the sea are in fact "waterbergs" that normally move along the surface, but under special circumstances they may actually stay in one place, as for example in a river flowing over a big stone.

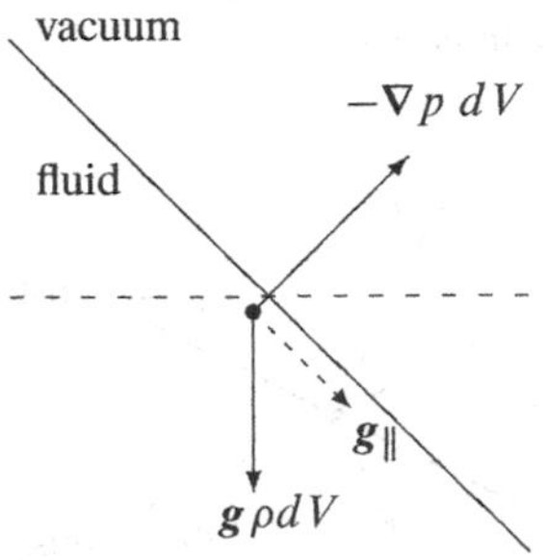

Interface (fully drawn) between a fluid and vacuum that intersects a horizontal equipotential surface (dashed) at an angle. The pressure must be constant (zero) at the interface, implying that the pressure gradient is orthogonal to it. It is thus unable to oppose vertical gravity as it must in hydrostatic equilibrium. The leftover tangential component of gravity $g_\parallel$ will set the fluid in motion along the interface.

In this chapter the influence of gravity on the static shape of large bodies of fluid is analyzed, the primary goal being the calculation of the flattened shape of the rotating planet. We shall also get an understanding of the shape of the tides, although they are not in hydrostatic equilibrium. Surface tension can be ignored for these large bodies of fluid but is important for tiny amounts of fluid and will be discussed in Chapter 5.

4.1 Fluid interfaces in hydrostatic equilibrium

The intuitive argument about the impossibility of creating a hydrostatic "waterberg" must in fact follow from the equations of hydrostatic equilibrium. We shall now demonstrate that an equilibrium interface between two fluids with different local densities ρ_1 and ρ_2 will always coincide with an equipotential surface. Since the gravitational field is the same on both sides of the interface, hydrostatic balance $\nabla p = \rho g$ and the jump in density implies that there is a jump in the pressure gradient across the interface, so that on one side it is $(\nabla p)_1 = \rho_1 g$ and

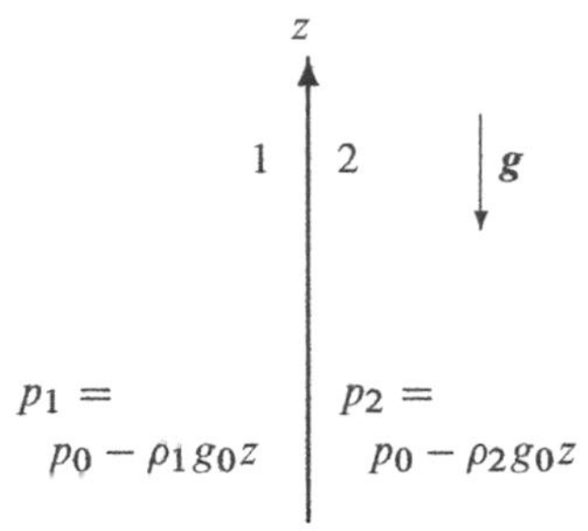

An impossible vertical interface between two fluids with different densities at rest in constant gravity. Even if the hydrostatic pressures on the two sides are the same for $z = 0$, they will be different everywhere else along the interface.

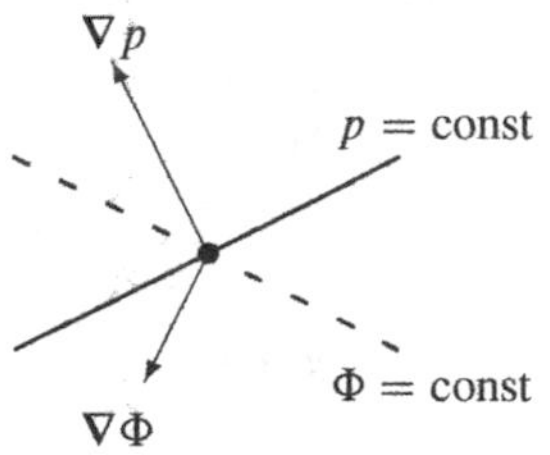

If isobars and equipotential surfaces cross, hydrostatic balance $\nabla p + \rho\nabla\Phi = \mathbf{0}$ becomes impossible.

on the other $(\nabla p)_2 = \rho_2 \boldsymbol{g}$. Assuming for the sake of argument that gravity is not orthogonal to the interface, it follows that there will be a jump in the tangential pressure gradient. If the pressures are equal at one point, they must therefore be different a small distance away along the interface. Newton's Third Law, however, requires the pressure to be the same on both sides of the interface (as long as there is no surface tension), so this contradiction can only be avoided if gravity is everywhere orthogonal to the interface, implying that the interface is an equipotential surface.

If, on the other hand, the fluid densities are exactly the same on both sides of the interface but the fluids themselves are different, for example water colored red and blue, the interface is not forced to follow an equipotential surface. This is, however, an unusual and highly unstable situation. The slightest deviation from equality in density on the two sides will call gravity in to make the interface horizontal. Stable vertical interfaces between fluids at rest are simply not seen, except perhaps in extremely viscous fluids where the time to reach equilibrium can be very long. An often cited example of vertical fluid interfaces points to old cathedral window glasses that are sometimes observed to be thicker at the bottom than at the top. That this could be due to slow glass flow over centuries appears, however, to be wrong [Zan98].

Isobars and equipotential surfaces

Surfaces of constant pressure, satisfying $p(\boldsymbol{x}) = p_0$, are called *isobars*. Through every point of space runs one and only one isobar, namely the one corresponding to the pressure at that point. The gradient of the pressure is everywhere normal to the local isobar surface, in the same way as gravity, $\boldsymbol{g} = -\nabla\Phi$, is everywhere normal to the local equipotential surface defined by $\Phi(\boldsymbol{x}) = \Phi_0$. Local hydrostatic equilibrium, $\nabla p = \rho \boldsymbol{g} = -\rho\nabla\Phi$, tells us that the normal to an isobar is everywhere parallel with the normal to an equipotential surface. This can only be the case if *isobars coincide with equipotential surfaces in hydrostatic equilibrium.* If an isobar crosses an equipotential surface anywhere at a finite angle, the two normals can not be parallel (see the margin figure).

Since the curl of a gradient trivially vanishes, it follows from hydrostatic equilibrium that

$$\mathbf{0} = \nabla \times \nabla p = \nabla \times (\rho \boldsymbol{g}) = \nabla\rho \times \boldsymbol{g} + \rho\nabla \times \boldsymbol{g} = -\nabla\rho \times \nabla\Phi,$$

where we in the last step have put $\boldsymbol{g} = -\nabla\Phi$. This implies that $\nabla\rho \sim \nabla\Phi$, so the surfaces of constant density (called *isopycnics*) must also coincide with the equipotential surfaces in hydrostatic equilibrium, and therefore with the isobars. Note that the coincidence of isobars and isopycnics in hydrostatic equilibrium is valid in general, not only for barotropic fluids where pressure and density are simply related.

4.2 The centrifugal force

Newton's Second Law of motion is only valid in *inertial* reference frames, where free particles move on straight lines with constant velocity. In classical mechanics a particle is commonly said to be free if it is not subject to forces caused by identifiable sources, for example the mass or electric charge of a material body. In reference frames that are accelerated or rotating relative to inertial frames, one may formally write the Second Law in its usual form, but the price to pay is the inclusion of certain force-like terms that do not have any obvious sources, but only depend on the motion of the reference frame. Such terms are called *fictitious forces*, although they are by no means pure fiction, as one becomes painfully aware when standing in a bus that suddenly stops. A more reasonable name might be *inertial forces*, since they arise as a consequence of the inertia of material bodies. In this chapter we only discuss the role of the centrifugal force for hydrostatics in rotating reference frames while in Chapter 18 we shall analyze the role of another fictitious force, the Coriolis force, which is of great importance for hydrodynamics in rotating reference frames such as the Earth.

Antigravity of rotation

Suppose you—in an inertial frame of reference—tie a stone of mass m to an unstretchable string of length s and swing it steadily in a circle with constant angular velocity Ω. By Newton's Second Law, the inward acceleration $\Omega^2 s$ of stone's circular motion requires an inward *centripetal force* $m\Omega^2 s$ from the string. But seen from the rotating frame where the stone is at rest, the inward string force appears to be unbalanced, so to uphold the Second Law we must postulate an outward fictitious *centrifugal force* $m\Omega^2 s$.

The centrifugal force acts everywhere in a uniformly rotating coordinate system. It is directed perpendicularly outward from the axis of rotation and of magnitude $m\,\Omega^2\,s$, where s is the shortest distance to the axis. In a Cartesian coordinate system, it is most convenient to let the z-axis coincide with the rotation axis, so that the shortest vector from the axis to the point $\boldsymbol{x} = (x, y, z)$ is $\boldsymbol{s} = (x, y, 0)$. Since the centrifugal force is proportional to the mass of the particle, it mimics a gravitational field,

$$\boldsymbol{g} = \Omega^2 \boldsymbol{s} = \Omega^2 (x, y, 0). \tag{4.1}$$

This fictitious gravitational field is the negative gradient of an equally fictitious potential

$$\Phi = -\frac{1}{2}\Omega^2 s^2 = -\frac{1}{2}\Omega^2 (x^2 + y^2). \tag{4.2}$$

The centrifugal field being directed away from the axis of rotation acts as a kind of *antigravity* field that will try to split things apart and throw objects off carousels and rotating planets. The antigravity field of rotation is, however, cylindrical in shape rather than spherical and has consequently the greatest influence at the equator of a planet, and none at the poles.

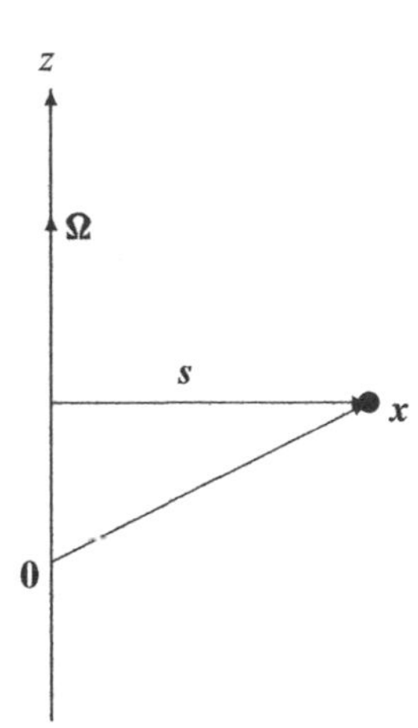

The geometry of a rotating system is characterized by a rotation vector $\boldsymbol{\Omega}$ directed along the axis of rotation, with magnitude equal to the angular velocity. The vector $\boldsymbol{s}$ is orthogonal to the axis, and its length is the shortest distance from the axis to the point $\boldsymbol{x}$. Normally, as here, the z-axis is chosen to coincide with the rotation axis.

Case: Newton's bucket

A bucket of water set in rotation is an example going back to Newton himself. Internal friction (viscosity) in the water will after some time bring it to rest relative to the rotating bucket, and the whole thing will end up rotating as a solid body, although—as everybody has noticed—the surface will not be horizontal but bends upward toward the rim. In a rotating coordinate system the total acceleration field becomes $\boldsymbol{g} = (\Omega^2 x, \Omega^2 y, -g_0)$, including both real gravity and the fictitious centrifugal acceleration, and the total potential becomes

$$\Phi = g_0 z - \frac{1}{2}\Omega^2 \left(x^2 + y^2\right). \tag{4.3}$$

Assuming that the air has constant pressure at the air-water interface, the potential must also be constant, $\Phi = \Phi_0$, making the shape of the surface a paraboloid of revolution,

$$z = z_0 + \frac{\Omega^2}{2g_0}\left(x^2 + y^2\right), \tag{4.4}$$

where $z_0 = \Phi_0/g_0$ is the water height at the center. In a bucket of diameter 20 cm rotating once per second, the water stands 2 cm higher at the rim than in the center.

Since water is incompressible we know from (2.36) that the pressure is

$$p = p_0 - \rho_0(\Phi - \Phi_0) = p_0 + \rho_0 g_0 (z_0 - z) + \frac{1}{2}\rho_0 \Omega^2 \left(x^2 + y^2\right), \tag{4.5}$$

where p_0 is the air pressure. At a fixed height z, the pressure in the water grows toward the rim, reflecting everywhere the change in height of the water column above.

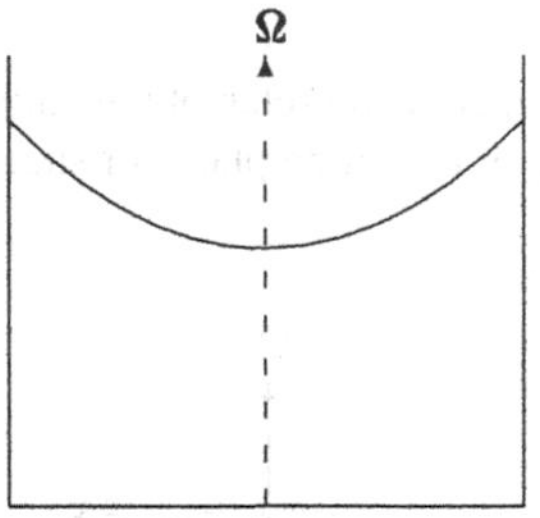

The water surface in a rotating bucket takes a parabolic shape because of the centrifugal force.

Example 4.1 [Ultracentrifuge]: An ultracentrifuge of radius 10 cm contains water (without an open surface) and rotates at $\Omega = 60{,}000$ rpm $\approx 6{,}300\ \mathrm{s}^{-1}$. The centrifugal acceleration at the rim becomes 400,000 times standard gravity and the pressure close to 2,000 atm, which is double the pressure at the bottom of the deepest abyss in the sea. At such pressures, the water density is increased by 10%.

4.3 The figure of Earth

On a rotating planet or star, centrifugal forces will add a cylindrical component of "antigravity" to the gravitational acceleration field. Its strength may be characterized by the ratio of centrifugal acceleration to surface gravity at the equator. Since the surface gravity is $g_0 \approx GM/a^2$, where M is the mass of the planet and a its equatorial radius, we find

$$q = \frac{\Omega^2 a}{g_0} \approx \frac{\Omega^2 a^3}{GM}. \tag{4.6}$$

With the parameters of the Earth ($a \approx 6371$ km, $g_0 \approx 9.8$ m s^{-1}, $\Omega \approx 73\times 10^{-6}$ s^{-1}), we get $q \approx 3.5\times 10^{-3} = 0.35\%$, confirming that centrifugal forces are quite unimportant under most circumstances. However, if our planet rotated as a rigid sphere with a period of 85 min, we would find $q = 1$, and people and everything else at the equator could (and would) levitate!

Example 4.2 [Rotating neutron star (pulsar)]: Neutron stars belong to the most extreme objects in the universe. The heaviest neutron stars have radius $a = 10$ km; mass twice the Sun's: $M = 4 \times 10^{30}$ kg; and rotate with a period down to 1.4 ms, corresponding to $\Omega = 4500$ s^{-1}. The surface gravity becomes $g_0 = 2.7 \times 10^{12}$ m s^{-2} and the centrifugal acceleration at the equator $\Omega^2 a = 2 \times 10^{11}$ m s^{-2}, making the "levitation parameter" $q = \Omega^2 a/g_0 = 0.076$. This is 22 times larger than Earth's, but still small compared to unity.

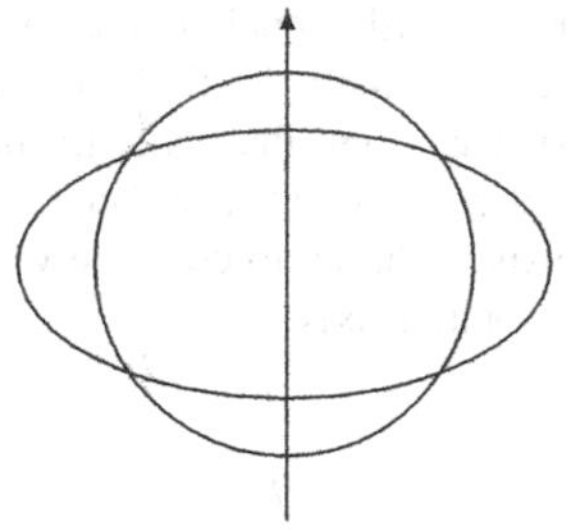

Exaggerated sketch of the flattening of a rotating planet or star.

On a perfectly spherical Earth, the tiny centrifugal "antigravity" nevertheless makes the road from pole to equator slightly downhill, effectively creating a centrifugal "valley" at the equator. A first guess would be that the depth of this valley is 0.35% of Earth's radius, which comes to about 22 km. If such a centrifugal valley suddenly came to exist on a spherical Earth, all the water would like huge tides run toward the equator (ignoring the Coriolis force; see Chapter 18). Since there *is* land at the equator, we may conclude that the massive Earth itself must over time have flowed into the centrifugal valley and filled it up. The measured difference between the equatorial and polar radii is in fact 21.4 km, in agreement with the estimate above. Coincidentally, this is roughly the same as the difference between the highest mountain top and the deepest ocean trench.

The theoretical calculation of the "figure of Earth" has—like the tides of Earth—been a topic that has attracted the best minds of past centuries (see for example [Lamb 1993]). We shall here consider the simplest possible model, which nevertheless gives a good approximation of the shape of a slowly rotating planet.

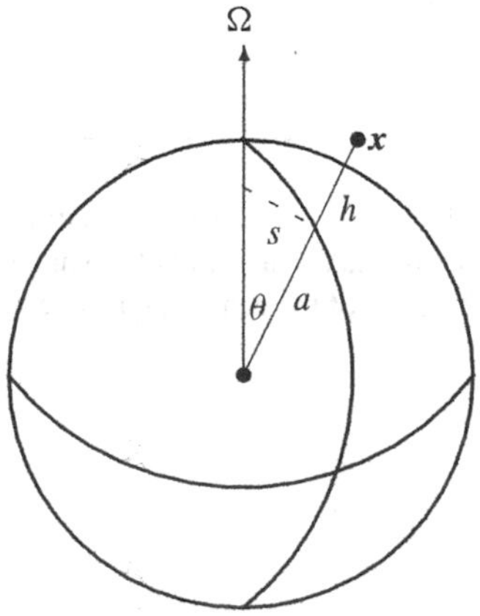

Centrifugal deformation of a spherical planet with radius a. The point $\boldsymbol{x}$ sits on the actual surface, which lies a radial distance $h = h(\theta)$ above the sphere.

Centrifugal deformation of a spherical planet

The centrifugal deformation of an originally spherical planet with radius a can only depend on the distance $s = a \sin\theta$ to the axis, where θ is the polar angle. If $h(\theta)$ denotes the radial shift of the surface due to rotation, as shown in the margin figure, the total surface potential becomes (to lowest order in the deviation from a perfect sphere)

$$\Phi = g_0 h(\theta) - \frac{1}{2}\Omega^2 a^2 \sin^2\theta + \Delta\Phi(\theta). \tag{4.7}$$

The first term is the gravitational potential and the second the centrifugal potential at the surface. The last term, $\Delta\Phi(\theta)$, is the gravitational potential due to the mass moved around by the deformation itself. It is not immediately clear whether this "self-potential" can be ignored, for although the amount of shifted mass is small compared to the mass of the planet, it lies so close to the surface that its influence could be significant. We shall soon see that it would be very wrong to ignore it.

If we anyway ignore the self-potential, setting $\Delta\Phi = 0$, and require the total potential to be constant, we obtain the following expression for the surface height above the sphere:

$$h = \frac{1}{2} a q \sin^2\theta + \text{const}, \tag{4.8}$$

with a so far unknown constant. But the total mass of the planet cannot change because of its rotation, and assuming that the shifted material has constant density ρ_1, its total volume must also be unchanged. Since the surface element of the sphere is $dS = 2\pi a \sin\theta\, a d\theta$, the condition that the change in volume should vanish becomes for small deformations, $|h| \ll a$,

$$\oint h(\theta)\, dS \propto \int_0^{\pi} h(\theta) \sin\theta\, d\theta = 0. \tag{4.9}$$

This fixes the constant, and one may easily verify that the condition is fulfilled by

$$\boxed{h = h_0 \left(\sin^2\theta - \frac{2}{3}\right) = -h_0 \left(\cos^2\theta - \frac{1}{3}\right),} \tag{4.10}$$

with $h_0 = \frac{1}{2} a q$. The expression clearly shows that the deformation is positive at the equator ($\theta = 90°$) and negative at the poles $\theta = 0°, 180°$, as we foresaw qualitatively at the beginning of this section. Unfortunately, the radial difference between the equator and the poles, the *deformation range* $h_0 \approx 11$ km, is only half our earlier "guesstimate".

The poor result warns us that the gravitational potential of the shifted material must play an important role. We shall prove below that if the general shape of the deformation takes the form (4.10), the self-potential becomes (to lowest order)

$$\Delta\Phi = -\frac{3}{5}\frac{\rho_1}{\rho_0} g_0 h(\theta), \tag{4.11}$$

where ρ_1 is the density of the shifted material and $\rho_0 = M/\frac{4}{3}\pi a^3$ is the average density of the planet. Inserting this into (4.7), the deformation range becomes [Lamb 1993, p. 361]

$$\boxed{h_0 = \frac{1}{2} a q \left(1 - \frac{3}{5}\frac{\rho_1}{\rho_0}\right)^{-1}.} \tag{4.12}$$

For Earth the average density of the mantle material is $\rho_1 \approx 4.5\ \text{g cm}^{-3}$ whereas the average density of the Earth as a whole is $\rho_0 \approx 5.5\ \text{g cm}^{-3}$. With these densities the number in parenthesis becomes 0.51, so that one gets $h_0 = 21.9$ km, in good agreement with the the actual value of 21.4 km. This result must, however, be viewed as somewhat fortuitous, because the self-potential should be calculated from the actual redistribution of matter throughout the Earth, including its varying compressibility, and not just from density averages.

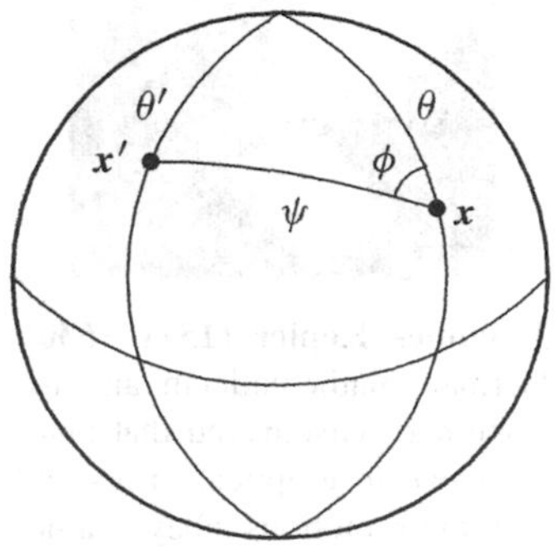

Spherical triangle formed by the various angles.

$*$ **Proof:** The extra gravitational potential due to the shifted material is calculated by adding the potentials from all its material particles,

$$\Delta\Phi(\boldsymbol{x}) = -G \int \frac{dM'}{|\boldsymbol{x} - \boldsymbol{x}'|} = -G\rho_1 \oint_{|\boldsymbol{x}'|=a} \frac{h(\theta')}{|\boldsymbol{x} - \boldsymbol{x}'|} dS'. \tag{4.13}$$

In the last step we have used that the shifted material is a thin layer of thickness h, so that the (signed) mass element becomes $dM' \approx \rho_1 h(\theta') dS'$, where dS' is the surface element of the sphere, $|\boldsymbol{x}'| = a$. To first order in h, we only need to evaluate the integral on the sphere itself for $|\boldsymbol{x}| = a$. Writing $(\boldsymbol{x} - \boldsymbol{x}')^2 = a^2 + a^2 - 2a^2\cos\psi = 2a^2(1 - \cos\psi)$ where ψ is the angle between $\boldsymbol{x}$ and $\boldsymbol{x}'$, and using the spherical cosine relation $\cos\theta' = \cos\theta\cos\psi + \sin\theta\sin\psi\cos\phi$ and the area element $dS' = a^2\sin\psi\, d\phi\, d\psi$ (see the margin figure), we get

$$\begin{aligned}
\Delta\Phi(\theta) &= G\rho_1 a h_0 \int_0^{\pi}\int_0^{2\pi} \frac{\cos^2\theta' - \frac{1}{3}}{\sqrt{2(1-\cos\psi)}}\sin\psi\, d\phi\, d\psi \\
&= 2\pi G\rho_1 a h_0 \int_0^{\pi} \frac{\cos^2\theta\cos^2\psi + \frac{1}{2}\sin^2\theta\sin^2\psi - \frac{1}{3}}{\sqrt{2(1-\cos\psi)}}\sin\psi\, d\psi \\
&= 2\pi G\rho_1 a h_0 \int_{-1}^{1} \frac{\cos^2\theta\, u^2 + \frac{1}{2}\sin^2\theta(1-u^2) - \frac{1}{3}}{\sqrt{2(1-u)}}\, du \\
&= 4\pi G\rho_1 a h_0 \int_0^1 \left(\cos^2\theta\left(1 - 4v^2 + 4v^4\right) + \frac{1}{2}\sin^2\theta\left(4v^2 - 4v^4\right) - \frac{1}{3}\right) dv \\
&= \frac{4}{5}\pi G\rho_1 a h_0\left(\cos^2\theta - \frac{1}{3}\right).
\end{aligned}$$

The last integrals are carried out by substituting $\cos\psi = u$ and $u = 1 - 2v^2$. Since $g_0 = GM/a^2 = \frac{4}{3}\pi G\rho_0 a$, we arrive at (4.11). The remarkable insight that a mass distribution with angular dependence $\cos^2\theta - \frac{1}{3}$ yields a potential with the same angular dependence is actually a general theorem valid for all Legendre polynomials, of which the second is in fact $P_2(\cos\theta) = \frac{1}{2}(3\cos^2\theta - 1)$ (see Problem 4.4).

Galileo Galilei (1564–1642). Italian natural philosopher, astronomer, mathematician, and craftsman. Considered the father of the modern scientific method. Carried out gravity experiments with falling objects and inclined planes. Built telescopes and was the first to see the mountains of the Moon, the large moons of Jupiter, and that the Milky Way is made from stars. (Source: Wikimedia Commons.)

4.4 The Earth, the Moon, and the tides

Kepler thought that the Moon would influence the waters of Earth and raise tides, but Galilei found this notion of Kepler's completely crazy and compared it to common superstition[1]. After Newton we know that the Moon's gravity acts on everything on Earth, also on the water in the sea, and attempts to pull it out of shape, thereby creating the tides. But since high tides occur roughly at the same time at antipodal points of the Earth, and therefore twice a day, the explanation is not simply that the Moon pulls the sea toward itself.

The rotation of the Earth, combined with the orbiting motion of both the Earth and the Moon around their common center of mass, makes the tides an essentially dynamic phenomenon that cannot be treated correctly by the methods of hydrostatics used in this chapter. A number of basic features can nevertheless be exposed by simple approximations (and careful wording of explanations). The best natural scientists and mathematicians of the eighteenth and nineteenth centuries worked on the tides and the dynamics of tidal waves. Fairly complete presentations may be found in Lamb's classic book [Lamb 1993], or the more modern account in [Melchior 1978].

Johannes Kepler (1571–1630). German mathematician and astronomer. Discovered that planets move in elliptical orbits and that their motions obey mathematical laws. (Source: Wikimedia Commons.)

The tidal field

The Moon is not quite spherical, but so small and far away that we may approximate its gravitational potential across the Earth with that of a point mass m, situated at the Moon's position $\boldsymbol{x}_0$,

$$\Phi = -\frac{Gm}{|\boldsymbol{x} - \boldsymbol{x}_0|}. \tag{4.14}$$

Choosing a coordinate system with the origin at the center of the Earth and the z-axis in the direction of the Moon, we have $\boldsymbol{x}_0 = (0, 0, D)$, where $D = |\boldsymbol{x}_0|$ is the Moon's distance. This

[1]Galileo wrote about Kepler: "But among all the great men who have philosophized about this remarkable effect, I am more astonished at Kepler than at any other. Despite his open and acute mind, and though he has at his fingertips the motions attributed to the earth, he has nevertheless lent his ear and his assent to the moon's dominion over the waters, and to occult properties, and to such puerilities." [Cohen 1985, p. 145].

coordinate system is in fact not inertial, but orbits around the common center of mass of the Earth-Moon system while in the same time rotating so that the z-axis always points toward the Moon. The center of mass lies on the Earth-Moon line a distance $d = Dm/(m + M) =$ 4,671 km from the center of the Earth (of mass M), about 1,700 km below the surface.

The Moon is approximately 60 Earth radii away, so its potential across the Earth may conveniently be expanded in powers of $a/D \approx 1/60$. As we shall see, it is necessary to go to second order. First we write $|\boldsymbol{x} - \boldsymbol{x}_0| = \sqrt{x^2 + y^2 + (z - D)^2} = \sqrt{D^2 - 2zD + r^2}$, where $r = \sqrt{x^2 + y^2 + z^2}$, and then we expand to second order in z/D and r/D,

$$\begin{aligned}\frac{1}{|\boldsymbol{x} - \boldsymbol{x}_0|} &= \frac{1}{\sqrt{D^2 - 2zD + r^2}} = \frac{1}{D}\left(1 - \frac{2z}{D} + \frac{r^2}{D^2}\right)^{-1/2} \\ &\approx \frac{1}{D}\left(1 - \frac{1}{2}\left(-\frac{2z}{D} + \frac{r^2}{D^2}\right) + \frac{3}{8}\left(-\frac{2z}{D}\right)^2\right) \\ &= \frac{1}{D}\left(1 + \frac{z}{D} + \frac{3z^2 - r^2}{2D^2}\right).\end{aligned}$$

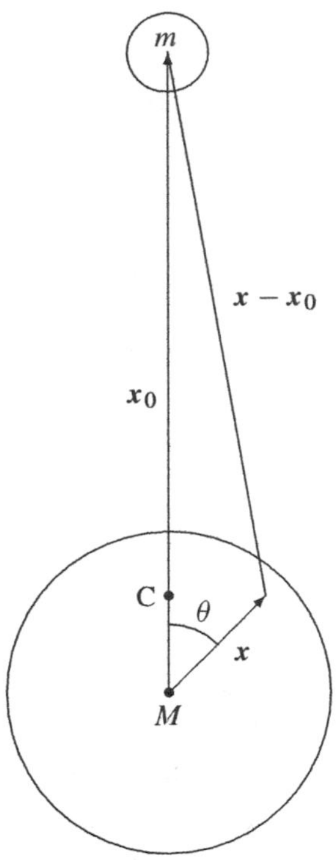

Geometry of the Earth and the Moon (not to scale). The point C is the center of mass of the system.

The first term in this expression leads to a constant potential $-Gm/D$, which may of course be ignored. The second term corresponds to a constant gravitational field in the direction toward the Moon $g_z = Gm/D^2 \approx 33\ \mu\text{m s}^{-2}$. This field provides the centripetal force necessary to keep the Earth in its circular orbit around the common center of mass. Anyway, in a constant gravitational field everything in and on the planet falls in the same way, as Galilei taught. Consequently, spaceship Earth is completely unaware of the two leading terms in the Moon's potential, so these terms cannot raise the tides. Galilei was right to leading non-trivial order, and that's actually not so bad!

Tides arise from the variation in the Moon's gravitational field across the Earth, to leading order given by the third term in the expansion of the potential. Introducing the angle θ between the direction to the Moon and the observation point on Earth, we have $z = r\cos\theta$, and the Moon's tidal potential becomes

$$\boxed{\Phi = -\frac{1}{2}\left(3\cos^2\theta - 1\right)\frac{r^2 Gm}{D^3}.} \tag{4.15}$$

This expansion may of course be continued indefinitely to higher powers of r/D. The coefficients $P_n(\cos\theta)$ are Legendre polynomials; here $P_2(\cos\theta) = \frac{1}{2}(3\cos^2\theta - 1)$.

The Moon's tidal field is calculated from the gradient of the potential. In spherical coordinates (see Appendix D) the gradient becomes

$$g_r = -\frac{\partial\Phi}{\partial r} = \left(3\cos^2\theta - 1\right)\frac{rGm}{D^3}, \tag{4.16a}$$

$$g_\theta = -\frac{1}{r}\frac{\partial\Phi}{\partial\theta} = -\frac{3}{2}\sin 2\theta\,\frac{rGm}{D^3}, \tag{4.16b}$$

$$g_\phi = -\frac{1}{r\sin\theta}\frac{\partial\Phi}{\partial\phi} = 0. \tag{4.16c}$$

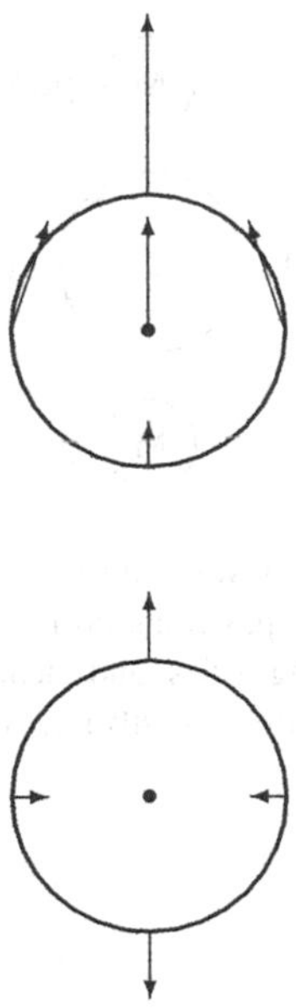
Top: How the Moon's gravity varies over the Earth (exaggerated). **Bottom:** The Moon's gravity with the unobservable constant acceleration canceled. This figure explains why the tides have a semi-diurnal period.

The radial component g_r is largest for $\theta = 0°$ and $180°$, and smallest for $\theta = 90°$. The tangential component g_θ is largest in magnitude for $\theta = 45°$ and $135°$. At the surface of the Earth the scale of the vertical part of the Moon's tidal field is $3aGm/D^3 \approx 1.65\ \mu\text{m s}^{-2}$.

Concluding, we repeat that tide-generating forces arise from variations in the Moon's gravitational field across the Earth. The field is generally not radial but has a tangential component about half the size of the vertical which is what drives the motion of the water in the tides. From the sign and shape of the potential as a function of angle, we see that the Moon lowers the gravitational potential just below its position ($\theta = 0°$), and at the antipodal point on the opposite side of the Earth ($\theta = 180°$). Sometimes these "poles" are called the Moon and anti-Moon positions. Similarly, for $\theta = 90°$ the potential is raised by half the amount it was lowered at the "poles". These changes in the gravitational potential act as if the Moon had created circular cup-shaped valleys at the "poles" and a ring-shaped hill around the "equator".

Tidal range

The lunar tides are, as mentioned, unavoidably time dependent, and it is not possible to understand the details without going into the complex dynamics of the tidal waves set in motion by the tidal fields of the Moon (and Sun) acting on the rotating Earth. Time-independent, static tides would only be possible if the Earth's rotation could be slowed down such that the Moon always stood vertically above the same place. That being out of the question (see, however, Section 4.5), we can nevertheless calculate the shape such static tides would have had. That exercise gives us important information about the scale of the real tides.

At a small (signed) height h above the spherical surface of the Earth, $r = a$, the Earth's own gravitational potential is $g_0 h$, where g_0 is surface gravity. The total instantaneous gravitational potential near the Earth's surface thus becomes, in the leading approximation,

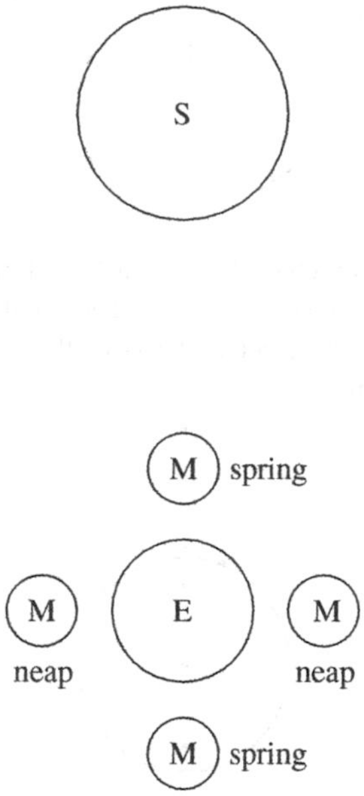

Earth and Sun with the Moon in extreme positions corresponding to spring tides and neap tides. The figure is (vastly) not to scale.

$$\Phi = g_0 h - \frac{1}{2}\left(3\cos^2\theta - 1\right)\left(\frac{a}{D}\right)^2 \frac{Gm}{D}, \tag{4.17}$$

and the shape of the equipotential surface corresponding to $\Phi = 0$ becomes

$$h = \frac{1}{2}\left(3\cos^2\theta - 1\right)\left(\frac{a}{D}\right)^2 \frac{Gm}{g_0 D}. \tag{4.18}$$

Integrating this expression over the surface of the sphere, we find as in (4.9) that the total signed volume of the deformation vanishes—as it should.

The maximal height of the null potential surface is found at $\theta = 0$ and the minimal at $\theta = 90°$. The difference, which we shall call the *basic tidal range*, becomes

$$\boxed{H_0 = \frac{3}{2}\left(\frac{a}{D}\right)^2 \frac{Gm}{g_0 D} = \frac{3}{2} a \frac{m}{M}\left(\frac{a}{D}\right)^3,} \tag{4.19}$$

where the last form is obtained by substituting $g_0 = GM/a^2$. Inserting the values for the Earth and Moon, we get $H_0 \approx 54$ cm.

In spite of the Sun being so far away, its mass is so large that the tidal range due the Sun turns out to be just half as large as that of the Moon, about 25 cm. When the Moon is new or full, the Sun and the Moon are nearly aligned and reinforce each other's tidal fields, creating large *spring tides*. When the Moon is half, the angle between Moon and Sun is 90°, giving rise to small *neap tides*. Assuming that the Moon and the Sun both lie in the equatorial plane of the Earth (which they don't), the extreme tidal ranges become 54 ± 25 cm, or 79 cm at springs and 29 cm at neaps, varying by almost a factor of 3. The self-potential (4.11) can at most increase the tidal range by 12% due to the low density of water.

Although we shall not discuss the dynamics of the ocean tides, the driving forces are as mentioned provided by the tidal fields of the Moon and the Sun. For the tides to reach full height, water must move in from huge areas of the Earth, as is evident from the shallow shape of the potential. Where that is not possible, for example in lakes and enclosed seas, the observed tidal range becomes much smaller than in the open oceans. Local geography may also influence tides. In bays and river mouths, funneling can cause tides to build up to huge values. Spring tide ranges up to 16 m have been observed in the Bay of Fundy in Canada.

The solid parts of the Earth will also react to the Moon's field with elastic deformation (rather than flow) with a range up toward the basic tidal range. This deformation has been

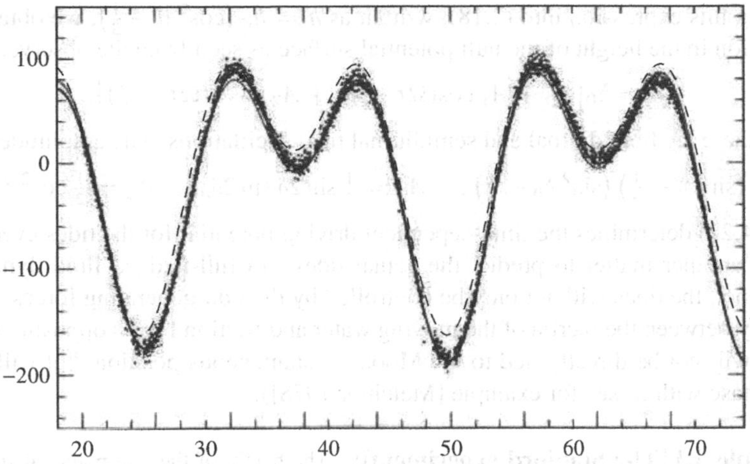

Figure 4.1. Variation in the vertical gravitational acceleration over a period of 56 hours in units of 10^{-8} m s^{-2} $\approx 10^{-9} g_0$ measured in Stanford, California, on December 8–9, 1996. The semidiurnal as well as diurnal tidal variations are prominently visible as dips in the curves. The dashed curve is obtained from (4.21) with parameters given in Example 4.3. The barely visible fully drawn curves are drawn by the authors from more sophisticated tidal models. (Source: Reproduced from [PCC99]. With permission.)

indirectly measured [PCC99] to a precision of a few percent in the daily 0.1 ppm variations in the strength of gravity (see Figure 4.1). There are also tidal effects in the atmosphere, but they are normally dominated by other atmospheric motions.

* Tidal cycles

The rotation of the Earth cannot be neglected. It is, after all, the cyclic variation in the water level seen at the coasts of seas and large lakes that makes the tides observable. Since the axis of rotation of the Earth is neither aligned with the direction to the Moon nor orthogonal to it, the local tidal forces acquire a diurnal cycle superimposed on the basic semidiurnal one. This is clearly seen in the data plotted in Figure 4.1, although in this case it looks more like a small semidiurnal cycle superposed on a larger diurnal.

For a fixed position on the surface of the Earth, the dominant variation in the lunar zenith angle θ is due to Earth's diurnal rotation. In addition, there are many other sources of periodic variations in the lunar angle [Melchior 1978], which we shall ignore here[2]. Now, let the fixed observer position at the surface of the Earth have (easterly) longitude ϕ and (northerly) latitude δ. The lunar angle θ at time t is then calculated from the spherical triangle formed by the north pole, the lunar position, and the observer's position,

$$\cos\theta = \sin\delta\sin\delta_0 + \cos\delta\cos\delta_0\cos(\Omega t + \phi), \qquad (4.20)$$

where $\Omega = 2\pi/T$ and T is the length of the lunar day (about 24 hours and 48 minutes). The "constant" δ_0 is the latitude of the lunar position which actually changes a bit from day to day due to the Moon's orbital motion. The origin of time has been chosen such that the Moon at $t = 0$ is directly above the Greenwich meridian $\phi = 0$.

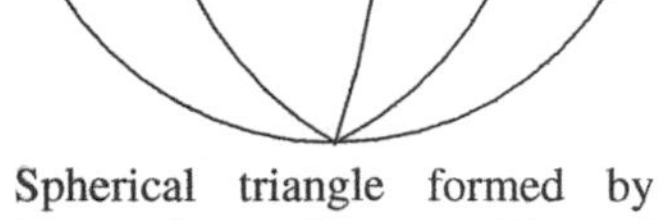

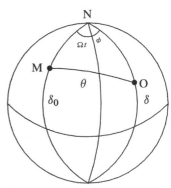

Spherical triangle formed by Moon (M), observer (O), and north pole (N).

[2]The most prominent such source is the lunar orbital period of a little less than a month, which increases the diurnal period of the Moon to nearly 25 hours. Over a lunar cycle the elliptical orbit of the Moon changes the Earth-Moon distance and thereby influences the tidal range. Furthermore, the orbital plane of the Moon inclines about 5° with respect to the ecliptic (the orbital plane of the Earth around the Sun), and precesses with this inclination around the ecliptic in a little less than 19 years. The Earth's equator is itself inclined about 23° to the ecliptic and precesses around it in about 25,700 years. Due to lunar orbit precession, the angle between the equatorial plane of the Earth and the plane of the lunar orbit will range over $23 \pm 5°$, that is, between 18° and 28°, in about 9 years.

Inserting this expression into (4.18), written as $h = h_0 \left(\cos^2\theta - \frac{1}{3}\right)$, we obtain the temporal variation in the height of the null potential surface as seen from the observer's place,

$$h = h_0\big[A_0 + A_1 \cos(\Omega t + \phi) + A_2 \cos 2(\Omega t + \phi)\big]. \tag{4.21}$$

Evidently, there are both diurnal and semidiurnal tidal oscillations. The amplitudes are

$$A_0 = \tfrac{3}{2}\left(\sin^2\delta - \tfrac{1}{3}\right)\left(\sin^2\delta_0 - \tfrac{1}{3}\right), \quad A_1 = \tfrac{1}{2}\sin 2\delta \sin 2\delta_0, \quad A_2 = \tfrac{1}{2}\cos^2\delta\cos^2\delta_0.$$

Although (4.21) determines the time-dependent driving potential for the tides everywhere on Earth, it is another matter to predict the actual tides. In full-fledged fluid dynamics on a rotating planet, the tides will not only be controlled by the tide-generating forces, but also by the interplay between the inertia of the moving water and friction forces opposing the motion. High tides will not be directly tied to the Moon's instantaneous position, but will in general be out of phase with it (see for example [Melchior 1978]).

Example 4.3 [The Stanford experiment]: The height of the null potential surface (4.18) represents in fact the instantaneous radial tidal field (4.16a), because $g_r = 2g_0h/a$. This field was measured at Stanford University in 1999 [PCC99], and the data is displayed in Figure 4.1. A decent fit (dashed curve) can be obtained from the expression in (4.21), given the known latitude of Stanford, $\delta = 37.4°$, and reasonable guesses for the other (unpublished) parameters, $\delta_0 = 20°$ and $h_0 = 90$ cm. Presumably the larger value of h_0, compared to the basic range $H_0 = 54$ cm, is caused partly by the solar tidal field and partly by the effect of the self-potential due to the tidal motion of the solid ground, although these details are not given in [PCC99].

* 4.5 Application: The tides of Io

Suppose for the sake of argument that Earth was brought to rotate synchronously with the Moon's period of 27.3 days. Ignoring the Sun and all other objects, the Earth and the Moon would move in unison along ideally circular orbits around their common center of mass. In a co-rotating frame of reference with origin in the center of mass, both the Earth and the Moon would be static, and we would be allowed to use hydrostatics to calculate the actual shape of the static "tides" on Earth created by the Moon's gravity in combination with the centrifugal field in the rotating frame. Although it would be considerably larger than the ordinary moving tides, this permanent deformation would, however, be far more difficult to detect for the unsophisticated observer.

Even if this is a speculative example, there are other cases of *bound rotation*. First of all, the Moon itself is in bound rotation, always turning the same side toward the Earth. More interestingly, the Galilean moons of Jupiter are similarly in bound rotation in their orbits around the huge planet, and small variations in orbital radius during a revolution lead to quite large changes in the otherwise permanent tides. Io, the moon closest to Jupiter, is constantly being exercised by these forces, moving its surface up and down by about 100 m. This makes it the most volcanic object in the solar system because of frictional heat developed in the process. Io is in fact the only known astronomical object that has a surface with color and structure reminiscent of hot, bubbling pizza [5].

Permanent deformation

For simplicity we shall in the following analysis stick with the preceding notation for a system consisting of a "planet" of mass M in bound rotation with a "moon" of mass m orbiting each other with a nearly constant distance D. The center of mass of this system lies a distance $d = Dm/(M + m)$ from the center of the planet. In a frame of reference rotating around the axis through the center of mass, the moon and the planet are both at rest, and the planet does not rotate. It does not matter if the moon rotates because we are interested only in the static deformation of the planet.

Parameter	Unit	Static Earth	Moon	Io
M	kg	5.97×10^{24}	7.35×10^{22}	8.93×10^{22}
m	kg	7.35×10^{22}	5.97×10^{24}	1.89×10^{27}
D	m	$384. \times 10^{6}$	$384. \times 10^{6}$	$423. \times 10^{6}$
ΔD	m	42.5×10^{6}	42.5×10^{6}	3.4×10^{6}
a	m	6.37×10^{6}	1.74×10^{6}	1.82×10^{6}
T	s	3.36×10^{6}	3.36×10^{6}	153×10^{3}
ρ_0	kg m^{-3}	5.51×10^{3}	3.35×10^{3}	3.53×10^{3}
ρ_1	kg m^{-3}	4.5×10^{3}	$3. \times 10^{3}$	$2. \times 10^{3}$
h_c	m	28.8	14.4	2.18×10^{3}
h_g	m	1.1	42.6	6.55×10^{3}
$h_{\max}$	m	10.3	33.2	5.09×10^{3}
$\Delta h_{\max}$	m	0.2	9.4	114

Table 4.1. Tides of three objects in bound rotation (hypothetical static Earth, Moon, and Io). The average density ρ_1 of shifted material is mainly guessed.

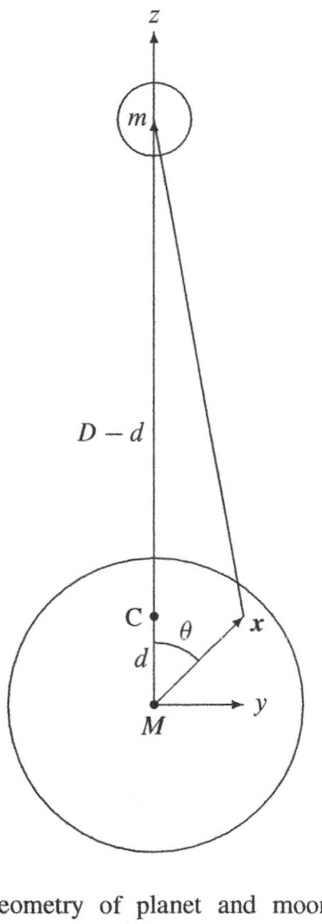

Geometry of planet and moon in a co-rotating frame of reference with rotation axis through the center of mass of the system (C). The coordinate system has origin at the center of the planet, z-axis through the center of the moon, and y-axis in the orbital plane. The x-axis is parallel with the axis of rotation and points out of the paper.

Let us as before choose a Cartesian coordinate system with origin at the center of the planet, the z-axis towards the moon and the x-axis parallel with the rotation axis. The sum of the centrifugal and gravitational potentials then becomes

$$\Phi = -\frac{1}{2}\Omega^2(y^2 + (z-d)^2) - z\frac{Gm}{D^2} - \frac{1}{2}\left(2z^2 - x^2 - y^2\right)\frac{Gm}{D^3}, \tag{4.22}$$

where we have kept the linear term in the gravitational potential. This demonstrates that the total linear term in the potential, $z(\Omega^2 d - Gm/D^2)$, vanishes because of Kepler's Third Law of planetary motion, $\Omega^2 D^3 = G(m+M)$, as it must.

By the methods used in the preceding sections, we may write down the height of the null potential surface above the planet, expressed in spherical coordinates,

$$h = h_c\left(\cos^2\theta + \sin^2\theta\sin^2\phi - \frac{2}{3}\right) + h_g\left(\cos^2\theta - \frac{1}{3}\right), \tag{4.23}$$

with centrifugal and gravitational ranges,

$$h_c = \frac{1}{2}\frac{\Omega^2 a^4}{GM}\left(1 - \frac{3}{5}\frac{\rho_1}{\rho_0}\right)^{-1}, \qquad h_g = \frac{3}{2}a\frac{m}{M}\left(\frac{a}{D}\right)^3\left(1 - \frac{3}{5}\frac{\rho_1}{\rho_0}\right)^{-1}, \tag{4.24}$$

including the amplifying factor due to the self-potential (4.11). Since the planet is not rotating in this coordinate system, this is the hydrostatic shape that its surface in the end will actually assume. The coincidence of the moon's orbital motion and the planet's own rotation implies via Kepler's Third Law that the ratio of gravitational to centrifugal ranges is

$$\frac{h_c}{h_g} = \frac{\Omega^2 D^3}{3Gm} = \frac{1}{3}\left(1 + \frac{M}{m}\right). \tag{4.25}$$

If $M \gg m$, as is the case for the hypothetical static Earth in bound rotation, the centrifugal tides are much greater than the gravitational. Conversely, for the moons of Jupiter we have $M \ll m$, and the centrifugal tides will only be one third of the gravitational.

As the moon circles the planet, the orbital radius D changes by ΔD. From the above results we estimate that the change in the maximal height, $h_{\max} = \frac{1}{3}h_c + \frac{2}{3}h_g$, becomes

$$\Delta h_{\max} = \frac{\partial h_{\max}}{\partial D}\Delta D = -2h_g\frac{\Delta D}{D}. \tag{4.26}$$

The planet literally "breathes" in tune with the moon's periodic motion relative to the stars (disregarding perihelion shifts). In Table 4.1 the tidal parameters have been calculated for selected objects. Notice the impressive variation (114 m) in the tidal range of Io.

Problems

4.1 Estimate the change in sea level if the air pressure locally rises by $\Delta p = 20$ hPa and stays there (disregarding surface tension).

4.2 The tides raises the water above the average level at the Moon and anti-Moon positions. **(a)** How much water is found in each of these tidal bulges?

4.3 Calculate **(a)** the mean height $\langle h \rangle$ and **(b)** the tidal range at the observer's position in the quasi-static approximation (4.21).

4.4 Show that $\nabla^2[f(r)(3\cos^2\theta - 1)] = g(r)(3\cos^2\theta - 1)$ and determine $g(r)$ as a function of $f(r)$ and its derivatives.

5

Surface tension

At the interface between two materials, physical properties change rapidly over distances comparable to the molecular separation scale. Since a molecule at the interface is exposed to a different environment than inside the material, it will also have a different binding energy. In the continuum limit where the transition layer becomes a mathematical surface separating one material from the other, the difference in molecular binding energy manifests itself as a macroscopic *surface energy density*. And where energy is found, forces are not far away.

Molecules sitting at a free liquid surface against vacuum or gas have weaker binding than molecules in the bulk. The missing (negative) binding energy can therefore be viewed as a positive energy added to the surface itself. Since a larger area of the surface contains larger surface energy, external forces must perform positive work against internal surface forces to increase the total area of the surface. Mathematically, these internal surface forces are represented by *surface tension*, defined as the normal force per unit of length. This is quite analogous to bulk tension (negative pressure), defined as the normal force per unit of area.

In homogeneous matter such as water, surface tension does not depend on how much the surface is already stretched. Certain impurities, called *surfactants*, have dramatic influence on surface tension because they agglomerate on the surface and form an elastic skin that resists stretching with increasing force. Best known among these are soaps and detergents from which one can blow such beautiful bubbles. A lipid bilayer also surrounds every living cell and separates the internal biochemistry from the environment.

Although surface tension is present at all interfaces, it is most important for small fluid bodies (at the human scale). At length scales where surface tension does come into play, the shape of an interface bears little relation to the gravitational equipotential surfaces that dominate the static shapes of large-scale systems analyzed in the preceding chapter. In this chapter, surface tension is introduced and only applied to hydrostatic systems. The physics of surface tension has a long history and excellent books exist on this subject, for example the recent [de Gennes et al. 2002].

5.1 Basic physics of surface tension

Although surface tension may be taken as a primary phenomenological concept in continuum physics, it is nevertheless instructive first to make a simple molecular model that captures the essential features of the phenomenon and even allows us to make an estimate of its magnitude. Afterward we shall define the concept without recourse to molecules.

Molecular origin of surface energy

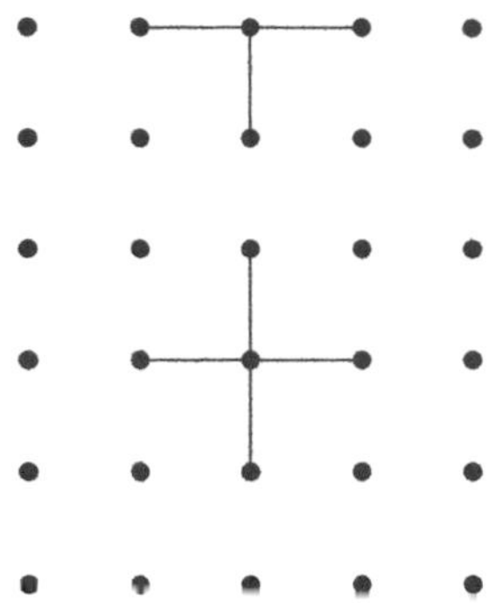

Cross-section of a primitive three-dimensional model of a material interfacing to vacuum. A molecule at the surface has only five bonds compared to the six that link a molecule in the interior. The missing bond energy at the surface gives rise to a positive surface energy density, which is the cause of surface tension.

Even if surface tension is most important for liquids, we shall here consider a simple regular "solid" surrounded by vacuum. The molecules are placed in a cubic grid (see the margin figure) with grid length equal to the molecular separation length $L_{\rm mol}$. Each molecule in the interior has six bonds to its neighbors whereas a surface molecule has only five. If the total binding energy of a molecule in the bulk is ϵ, a surface molecule will only be bound by $\frac{5}{6}\epsilon$. The missing binding energy corresponds to having added an extra positive energy $\frac{1}{6}\epsilon$ for each surface molecule.

The binding energy may be estimated as $\epsilon \approx h\, m$, where h is the specific heat of evaporation (evaporation enthalpy per unit of mass) and $m = M_{\rm mol}/N_A = \rho L_{\rm mol}^3$ is the mass of a single molecule. Dividing the molecular surface energy $\frac{1}{6}\epsilon$ by the molecular area scale $L_{\rm mol}^2$, we arrive at the following estimate of the surface energy density[1]:

$$\alpha \approx \frac{\frac{1}{6}\epsilon}{L_{\rm mol}^2} \approx \frac{hm}{6L_{\rm mol}^2} \approx \frac{1}{6}\, h\, \rho\, L_{\rm mol}. \tag{5.1}$$

The appearance of the molecular separation scale is quite understandable because $\rho L_{\rm mol}$ represents the effective surface mass density of a layer of thickness $L_{\rm mol}$. One might expect that the smallness of the molecular scale would make the surface energy density insignificant in practice, but the small thickness is offset by the fairly large values of ρ and h in normal liquids, for example water.

In the margin table a few measured and estimated values are shown. For water the mass density is $\rho \approx 10^3$ kg m^{-3} and the specific evaporation heat $h \approx 2.4 \times 10^6$ J kg^{-1}, leading to the estimate $\alpha \approx 0.126$ J m^{-2}, which should be compared to the measured value of 0.072 J m^{-2} at 25°C. Although not impressive, these estimates are not too bad, considering the crudeness of the model.

	α	Est
Water	72	126
Methanol	22	63
Ethanol	22	56
Bromine	41	44
Mercury	337	222
Unit	mJ m^{-2}	

Surface energy (or surface tension) for some liquids at 25°C in units of millijoule per square meter. The estimates are obtained using (5.1). (Data from [Lide 1996].)

Work and surface energy density

Increasing the area of an interface with surface energy density α by a tiny amount dA requires an amount of work equal to the surface energy contained in the extra piece of interface,

$$\boxed{dW = \alpha\, dA.} \tag{5.2}$$

Except for the sign, this is quite analogous to the mechanical work $dW = -p\, dV$ performed against bulk pressure when the volume of the system is expanded by dV. But where a volume expansion under positive pressure performs work *on* the environment, increasing the area against a positive surface energy density requires work *from* the environment.

The surface energy density associated with a liquid or solid interface against vacuum (or gas) is always positive because of the missing negative binding energy of the surface molecules. The positivity of the surface energy density guarantees that such interfaces seek toward the minimal area, consistent with the other forces that may be at play, for example gravity. Small raindrops and air bubbles are for this reason nearly spherical. Larger falling raindrops are also shaped by viscous friction and air currents, giving them a more complicated shape (see Figure 5.1).

Interfaces between solids and liquids or between liquids and liquids are not required to have positive interfacial energy density. The sign depends on the strength of the cohesive forces holding molecules of a material together compared to the strength of the adhesive forces between the opposing molecules of the interfacing materials. The interface between a solid and liquid is normally not deformable, and a negative interfacial energy density has

[1] There is no agreed-upon symbol for surface tension, which is variously denoted α, γ, σ, S, Υ, and T.

no dramatic effect. If on the other hand the interfacial energy density between two liquids is negative, a large amount of energy can be released by maximizing the area of the interface. Folding the interface like crumpled paper, the two fluids are mixed thoroughly instead of being kept separate. Fluids that readily mix with each other, such as alcohol and water, may be viewed as having negative interfacial energy density, although the concept is not particularly well-defined in this case. Immiscible fluids like oil and water must on the other hand have positive interfacial energy density, which makes them seek toward minimal interface area with maximal smoothness.

Since an interface has no macroscopic thickness, it may be viewed as being locally flat everywhere, implying that the surface energy density cannot depend on the curvature of the interface, but only on the properties of the interfacing materials. If these are homogeneous and isotropic—as they often are—the value of the energy density will be the same everywhere on the interface. In the following we shall focus on hydrostatics of homogeneous materials and assume that surface tension takes the same value everywhere on a static interface (with the exception of soap films).

Force and surface tension

The resistance against extension of a free surface shows that the surface has an internal *surface tension*, which we shall now see equals the surface energy density α. Suppose we wish to stretch the surface along a straight line of length L by a uniform amount ds (see the margin figure). Since the area is increased by $dA = Lds$, it takes an amount of work $dW = \alpha L ds$, from which we conclude that the force we apply orthogonally to the line is $\mathcal{F} = dW/ds = \alpha L$. In equilibrium, the applied normal force per unit of length, $\mathcal{F}/L = \alpha$, defines the internal surface tension α. Surface tension is thus identical to surface energy density, and is of course measured in the same unit, $\text{N m}^{-1} = \text{J m}^{-2}$.

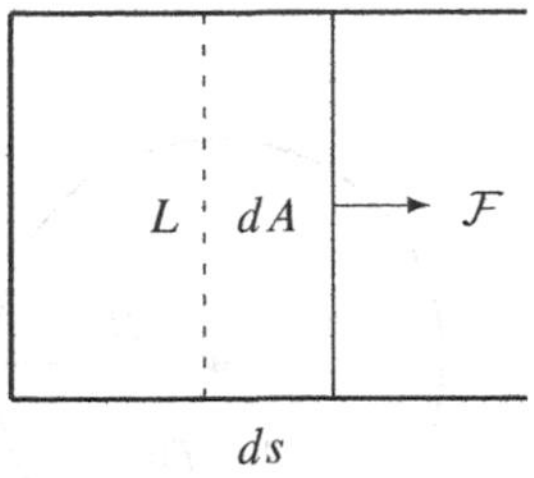

The basic surface tension experiment is a two-dimensional version of the three-dimensional experiment with a piston in a container. Using a soap film and metal wire as both "container" and moveable "piston", it can easily be carried out in the home.

Formally, surface tension is defined in much the same way as pressure was defined in Equation (2.10). Let an oriented open surface be divided into two parts by an oriented curve, such that the surface has a uniquely defined left- and right-hand side with respect to the curve. If $\hat{\boldsymbol{n}}$ denotes the normal to the surface, then

$$\boxed{d\mathcal{F} = \alpha\, d\boldsymbol{\ell} \times \hat{\boldsymbol{n}}} \tag{5.3}$$

is the force that the right-hand side of the surface exerts on the left-hand side through the curve element $d\boldsymbol{\ell}$ (see the margin figure). The total force is obtained by integrating this expression along the curve. It follows immediately that to shift a piece of the curve of length L by the amount ds orthogonally to the curve the work is given by (5.2) with $dA = Lds$.

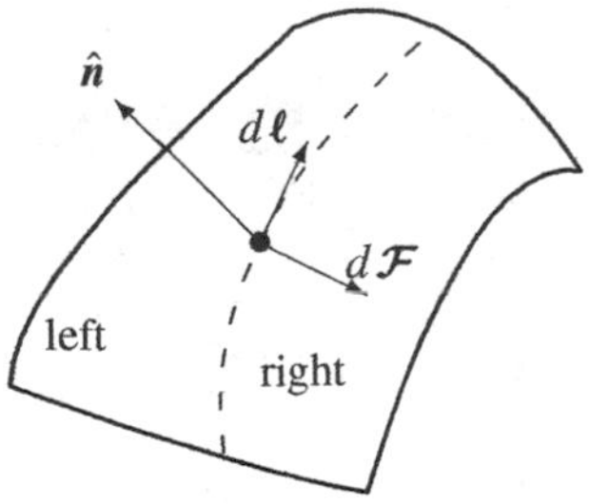

A piece of an oriented surface divided by an oriented curve (dashed). The cross-product of the curve element $d\boldsymbol{\ell}$ and the surface normal $\hat{\boldsymbol{n}}$ determines the direction of the force with which the piece of the surface on the right acts on the piece on the left across the curve element.

At an interface between homogeneous materials, surface tension does not depend on how much the interface has already been stretched, and this makes the interface quite different from an elastic membrane that like the skin of a balloon, resists stretching with increasing force because elastic tension increases as the deformation grows (Hooke's law; see Chapter 8). Soap films and biological membranes are exceptions and behave like elastic membranes because of their peculiar surfactant molecules. That is in fact what accounts for their great stability (see Section 5.2).

Pressure excess in a sphere

Everybody knows that the pressure inside a toy balloon is higher than outside, and that the excess pressure is connected with the tension in the taut balloon skin. Here we shall calculate the surface pressure discontinuity for a droplet of a homogeneous liquid hovering weightlessly, for example in a spacecraft. In the absence of all external forces, surface tension will attempt to make the droplet spherical because that shape has the smallest area for a given volume.

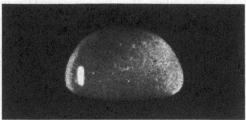

Figure 5.1. Steadily falling raindrops are subject to the combined effects of surface tension, gravity, friction, and air currents. **Left:** Small raindrops are dominated by surface tension and therefore nearly spherical. Here the drop radius is slightly larger than the capillary length of 2.7 mm and the drop is slightly oval. **Right:** Larger raindrops assume a typical "hamburger" shape. (Source: Photographs by Choji Magono, Yokohama National University. Reproduced from [McD54].)

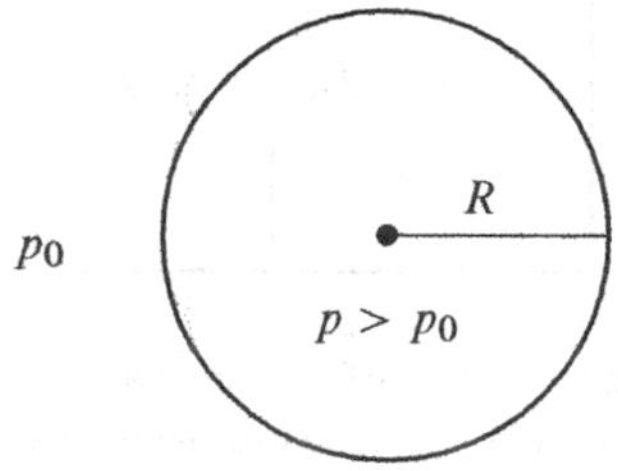

Surface tension makes the pressure higher inside a spherical droplet or bubble.

Surface tension will also attempt to contract the ball but is stopped by the build-up of an extra positive pressure Δp inside the liquid. If we increase the radius R by an amount dR, we must perform the extra work $dW = \alpha\, dA - \Delta p\, dV$, where dA is the change in surface area and dV is the change in volume. Since $dA = d(4\pi R^2) = 8\pi R\, dR$ and $dV = d\left(\frac{4}{3}\pi R^3\right) = 4\pi R^2\, dR$, the work becomes $dW = (\alpha 8\pi R - \Delta p\, 4\pi R^2)\, dR$. In equilibrium, where the contraction forces must balance the pressure forces, there should be nothing to gain or lose, that is, $dW = 0$, from which it follows that

$$\Delta p = \frac{2\alpha}{R}. \tag{5.4}$$

It should be emphasized that the above argument holds as well for a spherical air bubble in the same liquid: the pressure excess inside a bubble is exactly the same as in a droplet of the same radius.

Typical meteorological weather symbols. (Source: Wikimedia Commons.)

Raindrops: A spherical raindrop of diameter 1 mm has an excess pressure of only about 300 Pa, which is tiny compared to atmospheric pressure (10^5 Pa). In clouds, raindrops are tiny with diameter about 1 μm, implying a pressure excess a thousand times larger, about 3 atm. Falling raindrops are deformed by the interaction with the air and never take the familiar teardrop shape with a pointed top and a rounded bottom, so often used in drawings—and even by most meteorology offices (see the margin figure). Small raindrops with radius up to about 2 mm are nearly perfect spheres, but for larger radius they become increasingly flattened (see Figure 5.1). Eventually, when the radius exceeds 4.5 mm, a raindrop will break up into smaller drops due to the interaction with the air.

Capillary length

When can we disregard the influence of gravity on the shape of a raindrop? Consider a spherical raindrop of radius R falling steadily through the air. Since the air drag force must be balanced by the weight of the drop, the condition must be that the variation in hydrostatic pressure inside the drop should be negligible compared to the pressure excess due to surface tension. We must in other words require that $\rho_0 g_0 2R \ll 2\alpha/R$, where ρ_0 is the density of water (minus the tiny density of air). Solving for R, the inequality becomes $R \ll L_c$, where

$$L_c = \sqrt{\frac{\alpha}{\rho_0 g_0}} \tag{5.5}$$

is the so-called *capillary length* (or *capillary constant*). For gravity to be negligible, the radius of the drop should be much smaller than the capillary length. The capillary length equals 2.7 mm for the air–water interface at 25°C.

Although derived for a sphere, the capillary length is also the characteristic length scale at which static pending drops and captive bubbles begin to deviate from being spherical (see Section 5.8). To understand how the capillary length, calculated from the density of water, can also be meaningful for an air bubble in water, one must remember that the surrounding water imposes a changing hydrostatic pressure over the bubble surface, giving rise to its buoyancy. More generally the density parameter in (5.5) should be taken to be $|\Delta\rho|$, where $\Delta\rho$ is the density difference between the fluid in the sphere and the fluid that surrounds it. When the two densities are nearly equal, the capillary length can become very large.

Marangoni forces

Variations in surface tension create both normal and shear *Marangoni forces* in the surface. Such variations can arise from inhomogeneous material properties, or from temperature variations. Surface tension generally decreases with rising temperature, which is one of the reasons we use hot water for washing ourselves and our clothes. Unless balanced by other forces, Marangoni forces cannot be sustained in a liquid at rest but will unavoidably set it into motion. In static soap bubbles, gravity is in fact balanced by variations in surface tension caused by the elastic reaction to stretching of the bubble "skin". Apart from a discussion of this particular case in the following section, we shall not investigate the Marangoni effect and its consequences further here (see however [6]).

Carlo Guiseppe Matteo Marangoni (1840–1925). Italian physicist. Studied the effect of spatial variations in surface tension for his doctoral dissertation in 1865.

Soap-powered boat: The Marangoni effect has been exploited in toys, for example the soap-powered boat. The "boat" is made from a small piece of wood, polystyrene, or cardboard cut out in a shape like the one in the margin figure. The "boat" is driven around on the water surface by a small piece of soap mounted at the rear end. The dissolving soap lowers the surface tension of the water behind the boat, and the unbalance with the larger surface tension ahead of the boat pulls the boat forward. The fun stops of course when the dissolving soap has reached the same concentration all over the water surface. A piece of camphor is also effective as a motor for the little boat but is harder to come by.

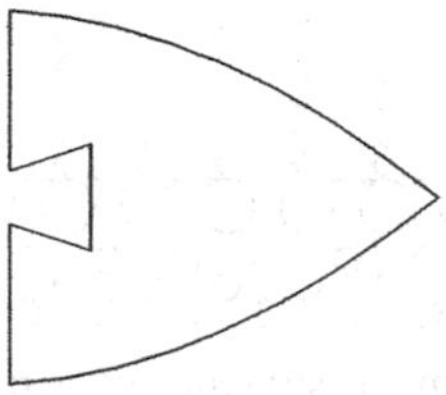

Shape of soap-powered boat. A pea-sized piece of soap or camphor is deposited in the small cutout to the left.

5.2 Soap bubbles

Children and adults alike love to play with soap bubbles, and have done so for centuries [Boys 1959]. If you start to think a bit about what is really going on with these fantastic objects, many questions arise. Why do we need soap for making bubbles? Why can't we make bubbles out of pure water? How big can they become? How long can they last? Freely floating soap bubbles are fun because their weight (minus buoyancy) is so tiny that they remain floating in the slightest updraft (Figure 5.2). The skin of a soap bubble is so strong that a bubble can be supported by a hand. Bubbles are not the only soap films that can be created. Using wire frames, it is possible to make all kinds of geometric shapes (see Figure 5.3) that have attracted the lasting interest of mathematicians [CW06].

The nature of bubble liquid

An air-filled bubble created from a thin liquid film always has a huge surface area compared to the area of a spherical volume containing the same amount of liquid as the film. Any external "agent"—gravity, drafts of air, encounters with other objects, evaporation, and draining—will stretch the skin in some places and maybe compress it in others. If the surface tension is constant, as it is for homogeneous liquids like pure water, the same force will keep on stretching the skin until it becomes so thin that the bubble bursts, and instantly turns into one or more spherical droplets. Large air-filled bubbles with a skin made from pure water are not seen because they are inherently unstable.

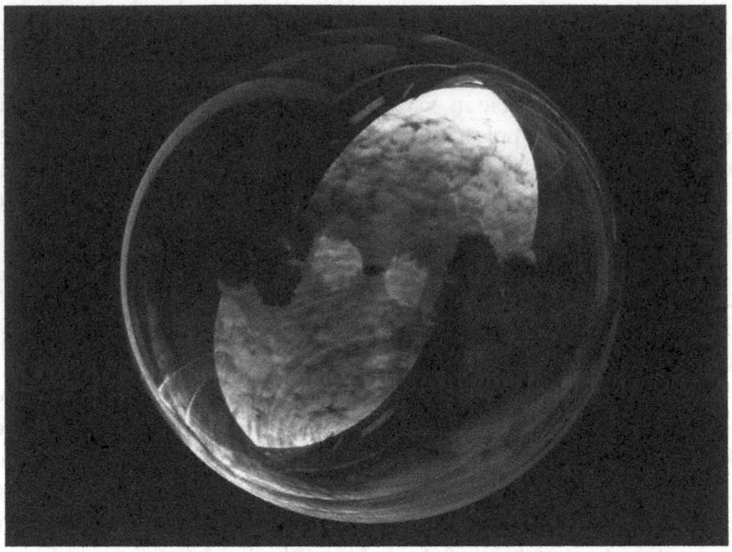

Figure 5.2. Free-floating, nearly perfectly spherical soap bubble with reflection of clouds (Source: Courtesy Mila Zinkova via Wikipedia.)

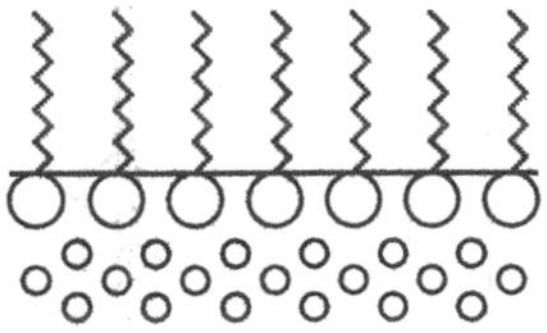

Schematic structure of a monolayer of soap molecules on a horizontal water surface. The hydrophobic tails stick up into the air with the hydrophilic heads buried in water (small circles).

A suitable *surfactant* [de Gennes et al. 2002, Chap. 7], for example soap, can stabilize the skin. Soap molecules are salts of fatty acids with hydrophilic (electrically charged) heads and hydrophobic (neutral) tails, making soap "love" both watery and fatty substances. This is the main reason why we use soap as detergent for washing; another is that the surface tension of soapy water is only about one third of that of pure water. Almost all surface impurities will in fact diminish the surface tension of water because they reduce the fairly strong native attraction between water molecules. On the surface of soapy water, the soap molecules tend to orient themselves with their hydrophobic tails sticking out of the surface and their heads buried in the water. In a thin film of soapy water, the tails of the soap molecules stick out on both sides and form a bilayer with water in between.

The stabilizing effect of soap is primarily due to the coherence of the bilayer in conjunction with the lower surface tension of soapy water. When the soap film is stretched locally, the soap molecules are pulled apart and the surface tension increases because the intermolecular forces increase with distance (see Figure 1.2). Long before the structure of soap molecules was known, such a mechanism was suspected to lie behind bubble stability [Boys 1959]. Viscosity can also stabilize bubbles by slowing drainage, and this is why glycerin is often included in soap bubble recipes. The soap molecules crowding at the surface furthermore prolong the lifetime of a soap bubble by diminishing evaporation.

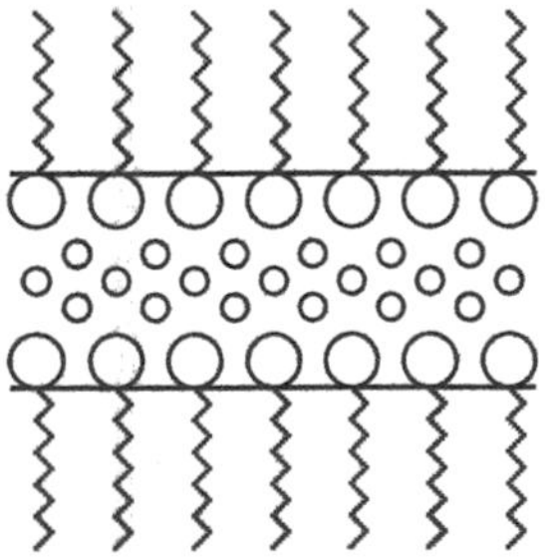

Schematic structure of a bilayer of soap molecules. The hydrophobic tails stick out on both sides of the membrane while the hydrophilic heads are buried in the thin layer of water.

Mass of a bubble

A spherical air-filled soap bubble of radius R has two surfaces, one from air to soapy water and one from soapy water to air. Each gives rise to a pressure jump of $2\alpha/R$, where α is the monolayer surface tension. The total pressure inside a soap bubble is thus $\Delta p = 4\alpha/R$ larger than outside (although α is only about a third of that of pure water). Taking $\alpha \approx 0.025\ \mathrm{N\ m^{-1}}$ and a nominal radius $R \approx 10$ cm we find $\Delta p \approx 1$ Pa. The relative pressure increase is tiny, $\Delta p/p_0 \approx 10^{-5}$ where $p_0 \approx 1$ bar is the ambient air pressure. The ideal gas law (2.27) tells us that the relative density increase is the same, $\Delta\rho/\rho_0 = \Delta p/p_0$ where ρ_0 is the ambient air density. Whereas the true mass of the air in the nominal bubble is $M_0 = \frac{4}{3}\pi R^3 \rho_0 \approx 5$ g, the effective mass of the air reduced by buoyancy is only $\Delta M = \frac{4}{3}\pi R^3 \Delta\rho \approx 50\ \mu$g.

The play in spectral colors often seen in soap bubbles tells us that the thickness of the bubble skin must be comparable to the wavelength of light — about half a micrometer — which makes it about 1,000 times thicker than the size of the hydrophilic head of a soap molecule. Taking the skin thickness of the nominal bubble to be constant, $\tau = 1\ \mu\text{m}$, the mass of the water in the skin becomes $M_1 = 4\pi R^2 \tau \rho_1 \approx 0.13$ g, where ρ_1 is the density of water. This is much larger than the effective mass of the air ΔM, and one may easily verify that the ratio $\Delta M / M_1 \approx 4 \times 10^{-4}$ is independent of the radius and thus the same for all soap bubbles with the nominal skin thickness.

Concluding, we have shown that buoyancy almost completely cancels the weight of the air in the bubble, so that the effective mass of the whole bubble always equals the mass of its skin, $M \approx M_1$. This tiny mass is the reason that even a gentle air current can make a bubble float away. In Example 17.3 on page 297, we shall estimate the terminal fall velocity of fairly large spherical bubbles like this to be about $0.3\ \text{m s}^{-1}$, independent of the radius.

* The need for Marangoni forces

The weight of the bubble skin is the reason that surface tension cannot be perfectly constant in equilibrium but must increase when the skin is stretched. Suppose for definiteness that a bubble is falling with steady velocity through the air, only subject to the force of gravity and to drag-forces from the surrounding air. The external forces must of course cancel each other globally to secure steady motion, but will not cancel perfectly everywhere on the skin of the bubble. Whereas gravity acts in the same way all over the skin, drag acts most strongly on the lower part of the bubble. Consequently, internal forces are necessary to secure that all parts of the bubble remain at rest with respect to each other.

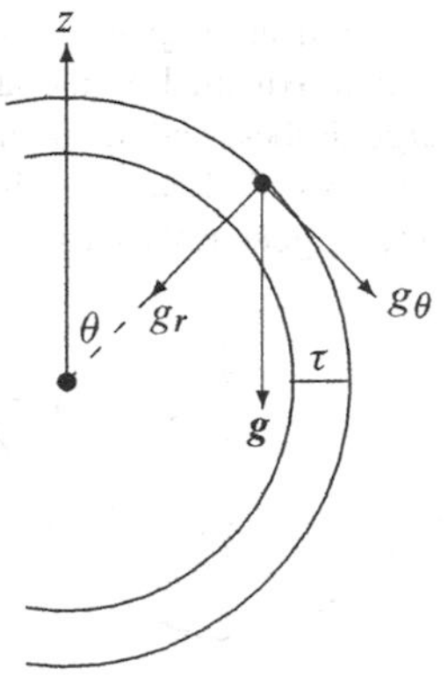

Gravity acting on the skin of the bubble, resolved in radial and tangential components, $g_r = -g_0 \cos\theta$ and $g_\theta = g_0 \sin\theta$.

Here we shall disregard drag and supporting forces and only discuss how internal forces locally can balance gravity on the upper part of the bubble. Consider a little circle on the sphere with fixed polar angle θ to the vertical. The mass of the skin in a small circular band $d\theta$ is $dM = \rho_1 \tau \times 2\pi R \sin\theta \times R d\theta$. The tangential component of its weight can only be balanced by a corresponding variation $d\alpha$ in surface tension of soapy water over the interval $d\theta$. Taking into account that the skin has two such surfaces, the tangential force balance becomes $g_0 \sin\theta dM + 2\pi R \sin\theta\, 2d\alpha = 0$, or

$$\frac{d\alpha}{d\theta} = -\tfrac{1}{2}\rho_1 \tau R g_0 \sin\theta. \tag{5.6}$$

Assuming R to be nearly constant, the solution is $\alpha(\theta) = \alpha + \rho_1 \tau R g_0 \cos^2 \tfrac{1}{2}\theta$, where α is the surface tension of the unstretched surface at $\theta = 180°$. The ratio of the two terms in $\alpha(\theta)$ becomes a dimensionless number, called the *Bond number*,

$$\boxed{\mathsf{Bo} = \frac{\rho_1 \tau R g_0}{\alpha} = \frac{\tau R}{L_c^2},} \tag{5.7}$$

where $L_c = \sqrt{\alpha / \rho_1 g_0} \approx 1.6$ mm is the capillary length of soapy water. The Bond number may also be viewed as the ratio between the weight of the bubble skin and the surface tension force at the equator (see Problem 5.1).

For the nominal bubble with $\tau = 1\ \mu\text{m}$, $R = 10$ cm, and $\alpha = 0.025\ \text{N m}^{-1}$, we find $\mathsf{Bo} = 0.04$. The relative variation in surface tension over the bubble is thus about 4%, indicating that the skin must be stretched by a comparable amount to deliver the necessary elastic Marangoni forces. The radial component of the skin's weight also demands a local change in the bubble radius of the same magnitude to secure a constant pressure difference between the air inside and outside the bubble (see Problem 5.2). For $R = 2.5$ m, the Bond number becomes unity, and the bubble undergoes major deformation, and perhaps breakup. You should probably never expect to encounter free-floating soap bubbles larger than a couple of meters[2].

[2]The largest free-floating bubble ever made had a volume of $2.98\ \text{m}^3$ [7].

5.3 Pressure discontinuity

The discontinuity in pressure across a spherical interface (5.4) is a special case of a general law discovered by Thomas Young (1805) and Pierre-Simon Laplace (1806). The law expresses the pressure discontinuity at a given point of an interface as the product of the local value of surface tension and twice the mean curvature of the interface. To derive this law and appreciate its power, we need first to study some aspects of the local geometry of curves and surfaces.

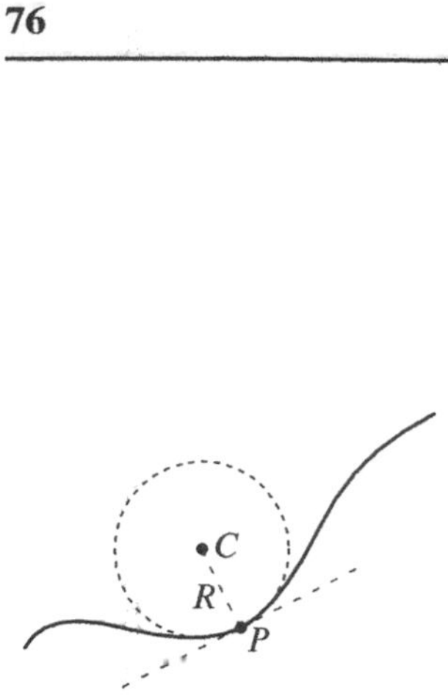

The local geometric properties of a smooth planar curve (fully drawn) near a given point are are approximated by the tangent (large dashes) and the osculating circle (small dashes) having the same tangent as the curve.

Local curvature of a planar curve

In the neighborhood of a given point P on a smooth planar curve, the behavior of the curve can be described geometrically by a series of increasingly precise approximations. In the first non-trivial approximation, the curve is viewed as locally straight and approximated by its *tangent* in P. In the second approximation, the curve is viewed as locally circular and represented by its *osculating circle* in P with the same tangent as the curve. The radius R of the osculating circle is called the *radius of curvature* of the curve in the point P and the center C, which must necessarily lie on the normal to the tangent, is called the *center of curvature*. The curve's *curvature* κ in P is defined as the inverse radius of curvature, $\kappa = 1/R$. As the point P moves along the smooth curve, both the direction of the tangent and the radius of curvature will change.

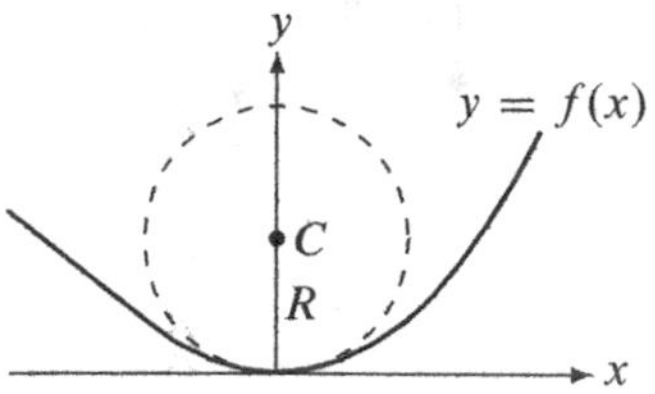

The curve $y = f(x)$ coincides near the origin with the osculating circle to second order in x.

To obtain a quantitative relation between the radius of curvature and the local properties of the curve, we introduce a local coordinate system with origin in the point P and the x-axis coinciding with the tangent in P. In this coordinate system the curve is described by a function $y = f(x)$ with $f(0) = f'(0) = 0$, so that to second order in x we have $y \approx \frac{1}{2}x^2 f''(0)$. In these coordinates, the equation for the osculating circle is $x^2 + (y - R)^2 = R^2$ and for small x and y we find to leading order that $y \approx x^2/2R$. The osculating circle coincides with the curve $f(x)$ to second order when its curvature equals the second-order derivative in P,

$$\kappa = \frac{1}{R} = f''(0). \tag{5.8}$$

In Problem 5.6 an expression is derived for the curvature at any point along a curve where the tangent is not necessarily parallel with the x-axis, but for now the above result suffices.

Notice that with this choice of coordinate system the radius of curvature is positive if the center of curvature lies above the curve, and negative if it lies below. The choice of sign is, however, pure convention, and there is no intrinsic geometric significance associated with the sign of the radius of curvature in a particular point. What is significant are the changes of sign that may occur when comparing curvatures in different points.

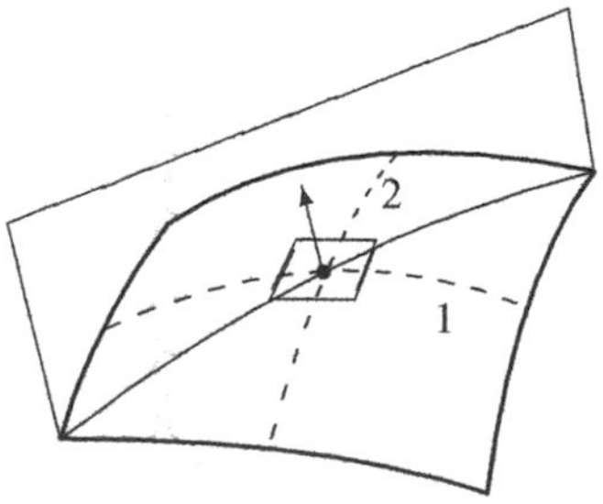

A plane containing the normal in a point of a surface intersects the surface in a planar curve. The extreme values of the signed radius of curvature define the principal directions (dashed). The small rectangle has sides parallel with the principal directions.

Local curvature of a surface

Suppose you cut a smooth surface with a plane that contains the normal to a given point P on the surface. This plane intersects the surface in a smooth planar curve (a so-called normal section) with a radius of curvature R in the point P. The radius of curvature is given a conventional sign depending on which side of the surface the center of curvature is situated. As the intersection plane is rotated around the normal, the center of curvature moves up and down along the normal between the two extremes of the signed radius of curvature. The extreme radii, R_1 and R_2, are called the *principal radii of curvature* and the corresponding *principal directions* are, as we shall now prove, orthogonal.

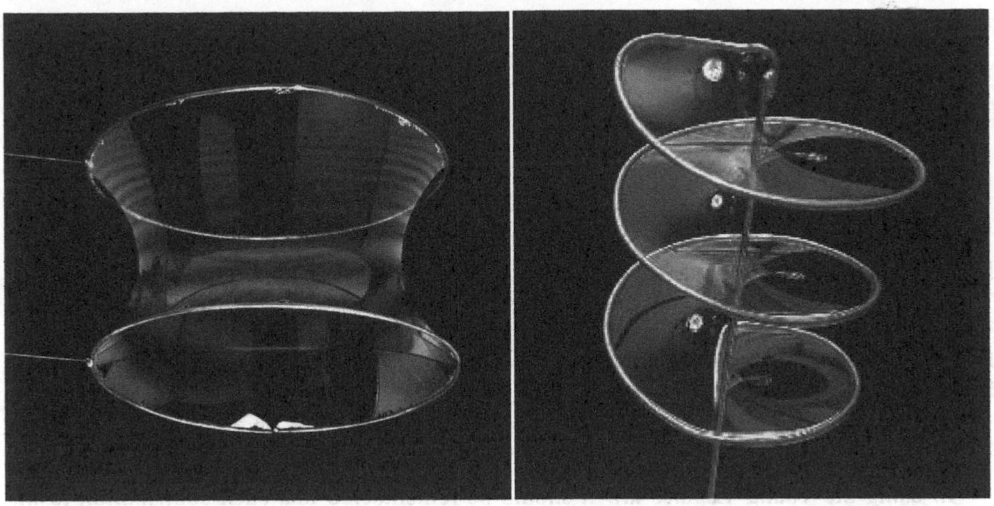

Figure 5.3. Soap films with zero mean curvature. **Left:** Catenoid film (Source: San Francisco Exploratorium.) **Right:** Helical film. (Source: Giorgio Carboni. With permission.)

Proof: We introduce a local coordinate system with origin in P and the z-axis along the normal to the surface in P. To second order in x and y, the most general smooth surface is

$$z = \tfrac{1}{2}ax^2 + \tfrac{1}{2}by^2 + cxy, \qquad (5.9)$$

where a, b, and c are constants. Introducing polar coordinates $x = r\cos\phi$ and $y = r\sin\phi$, and using (5.8), we find the curvature of the normal section in the direction ϕ,

$$\frac{1}{R(\phi)} = \left.\frac{\partial^2 z}{\partial r^2}\right|_{r=0} = a\cos^2\phi + b\sin^2\phi + 2c\sin\phi\cos\phi. \qquad (5.10)$$

Differentiating with respect to ϕ we find that the extremal values are determined by $\tan 2\phi = 2c/(a-b)$. Assuming $a \neq b$, this equation has two orthogonal solutions for $\phi = \phi_0 = \frac{1}{2}\arctan(2c/(a-b))$ and $\phi = \phi_0 + 90°$, and this proves our claim.

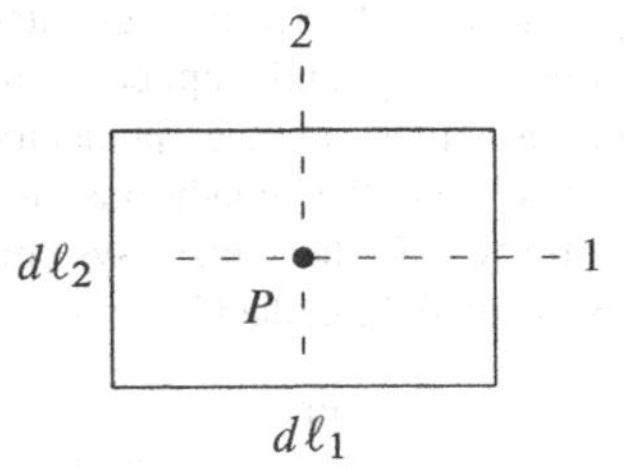

Tiny rectangle aligned with the principal directions, seen from "above". The sides are $d\ell_1$ and $d\ell_2$.

The Young–Laplace law

Consider now a small rectangle with sides $d\ell_1$ and $d\ell_2$ aligned with the principal directions in a given point of the surface, and let to begin with the principal radii of curvature, R_1 and R_2, be both positive. Surface tension pulls at this rectangle from all four sides, but the curvature of the surface makes each of these forces not quite parallel with the tangent plane. In the 1-direction, surface tension acts with two nearly opposite forces of magnitude $\alpha d\ell_2$, each forming a tiny angle of magnitude $\frac{1}{2}d\ell_1/R_1$ with the tangent to the surface in the 1-direction. Projecting both of these forces on the normal, we calculate the total force in the direction of the center of curvature C_1 as $d\mathcal{F} = 2\times\alpha d\ell_2 \times \frac{1}{2}d\ell_1/R_1 = \alpha dA/R_1$ where $dA = d\ell_1\, d\ell_2$ is the area of the rectangle. Dividing by dA we obtain the excess in pressure $\Delta p = \alpha/R_1$ on the side of the surface containing the center of curvature. Finally, adding the contribution from the 2-direction we arrive at the *Young–Laplace* law for the pressure discontinuity due to surface tension:

$$\boxed{\Delta p = \alpha\left(\frac{1}{R_1} + \frac{1}{R_2}\right).} \qquad (5.11)$$

For the sphere with $R_1 = R_2 = R$, we recover immediately the preceding result (5.4).

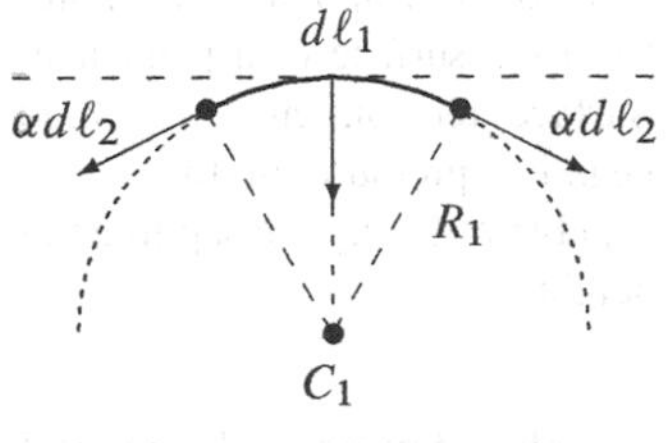

The rectangular piece of the surface with sides $d\ell_1$ and $d\ell_2$ is subject to two tension forces along the 1-direction resulting in a normal force pointing toward the center of curvature C_1. The tension forces in the 2-direction contribute analogously to the normal force.

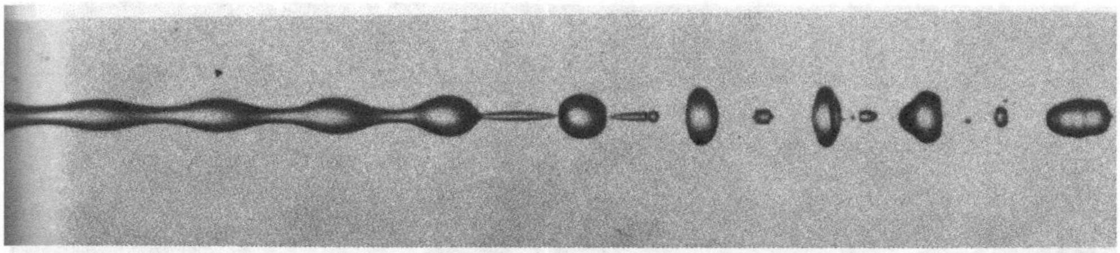

Figure 5.4. Breakup of a water jet emerging from a vertical tube (on the left) with diameter 4 mm. The picture has been turned 90° for graphical reasons. The jet is perturbed by the sound from a loudspeaker with a spatial wavelength corresponding to 4.6 diameters, which is somewhat larger than the Rayleigh–Plateau critical value of π diameters, but close to the wavelength of fastest breakup. (Source: Reproduced from [RJ71]. With permission.)

Pierre Simon marquis de Laplace (1749–1827). French mathematician, astronomer, and physicist. Developed gravitational theory and applied it to perturbations in the planetary orbits and the conditions for stability of the solar system. (Source: Wikimedia Commons.)

In using the Young–Laplace law it should be remembered that *each contribution to the pressure discontinuity is positive on the side of the surface containing the center of curvature, otherwise negative.* The Young–Laplace law is in fact valid if R_1 and R_2 are the radii of curvature of *any two orthogonal normal sections*, not necessarily along the principal directions, because it follows from (5.10) that

$$\frac{1}{R(\phi)} + \frac{1}{R(\phi + 90°)} = a + b, \tag{5.12}$$

which is independent of ϕ. The half sum of the reciprocal radii of curvature, $\frac{1}{2}(R_1^{-1} + R_2^{-1})$, is called the *mean curvature* of the surface in a given point. Soap films enclosing no air have the same pressure on both sides, $\Delta p = 0$, implying that the mean curvature must vanish everywhere. That does not mean that such films are planar. On the contrary, using suitable wire frames, spectacular shapes can be created (see Figure 5.3).

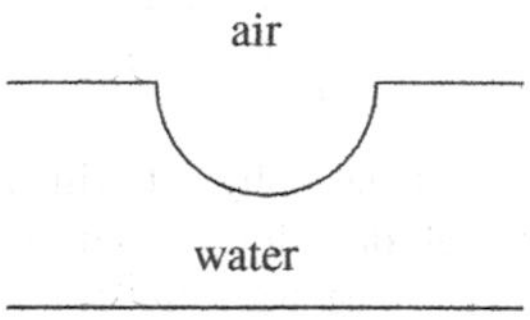

Sketch of the spherical meniscus formed by evaporation of water from the surface of a plant leaf. Surface tension creates a high negative pressure in the water, capable of lifting the sap to great heights.

Example 5.1 [How sap rises in plants]: Plants evaporate water through tiny pores on the surface of their leaves. This creates a hollow air-to-water surface with radii of curvature comparable to the radius a of the pore. Both centers of curvature lie outside the water, leading to a *negative* pressure excess in the water. For a pore of diameter, $2a \approx 1\ \mu$m, the excess pressure inside the water will be about $\Delta p = -2\alpha/a \approx -3$ atm. This pressure is capable of lifting sap through a height of 30 m, but in practice, the lifting height is considerably smaller because of resistance in the xylem conduits of the plant through which the sap moves. Taller plants and trees need correspondingly smaller pore sizes to generate sufficient negative pressures, even down to -100 atm! Recent research has confirmed this astonishing picture [Tyr03, HZ08].

5.4 The Rayleigh–Plateau instability

Joseph Antoine Ferdinand Plateau (1801–1883). Belgian experimental physicist. Studied soap bubbles extensively. In 1829 he stared into the Sun for 25 seconds to study the aftereffects in the eye, an experiment which he believed led to his total loss of vision about 14 years later.

The spontaneous breakup of the jet of water emerging from a thin pipe (see Figure 5.4) is a well-known phenomenon, first studied by Savart in 1833. In spite of being blind, Plateau found experimentally in 1873 that the breakup begins when the water jet has become longer than its circumference. The observation was explained theoretically by Lord Rayleigh in 1878 as caused by the pressure discontinuity due to surface tension. Here we shall verify Plateau's conclusion in a simple calculation without carrying out a full stability analysis involving the dynamic equations of fluid mechanics (see the extensive review of liquid jets [EV08]).

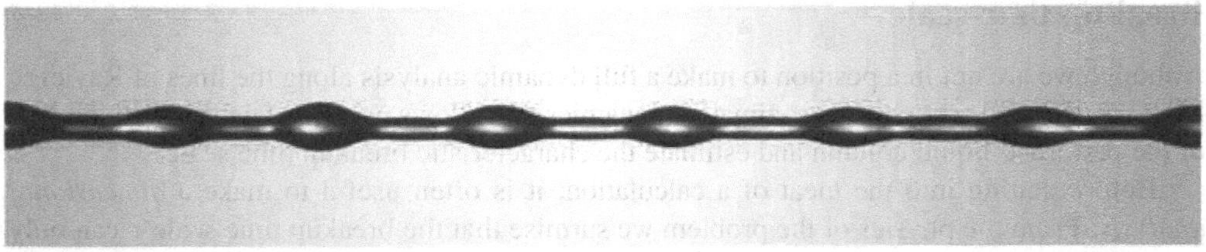

Figure 5.5. The Rayleigh–Plateau instability makes it quite difficult to coat a fiber by pulling it out of a liquid bath, even if the liquid does wet a horizontal sheet of the fiber material. Spiders use this instability to deposit globules of glue with regular spacing on their snare lines. In the picture a film of silicone oil deposited on a cylindrical wire breaks up in a regular pattern of beads due to the instability. In this case the radius of the wire (0.1 mm) is so small that gravity has no influence. (Source: Reproduced from [LSQ06]. With permission.)

Critical wavelength

Consider an infinitely long cylindrical column of incompressible liquid at rest in the absence of gravity with its axis along the z-axis. Due to the cylindrical form, surface tension will make the pressure larger inside by $\Delta p = \alpha/a$, where a is its radius. This uniform pressure excess will attempt to squeeze the column uniformly toward a smaller radius, but if the column is infinitely long it will be impossible to move the liquid out of the way, and that allows us to ignore this problem.

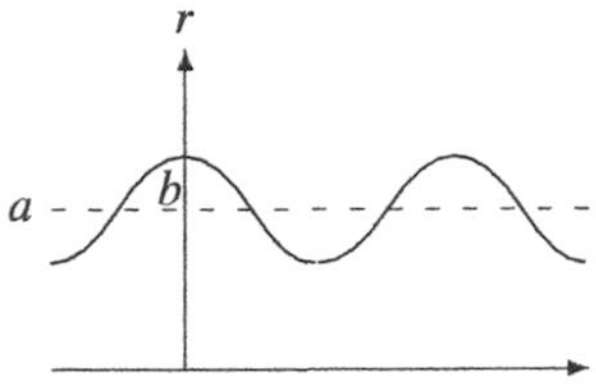

A harmonic perturbation $r(z) = a + b\cos kz$ of the radius of the fluid cylinder. The amplitude b of the perturbation is vastly exaggerated.

What cannot be ignored, however, are local radial perturbations $r = r(z)$ that make the column bulge in some places and contract in others. The amount of extra liquid to move out of the way in response to pressure variations will then be finite. Any sufficiently small radial perturbation can always be resolved into a sum over perturbations of definite wavelengths (Fourier expansion), and to lowest order of approximation it is therefore sufficient to consider a harmonic perturbation with general wavelength λ and amplitude $b \ll a$ of the form

$$r(z) = a + b\cos kz, \tag{5.13}$$

where $k = 2\pi/\lambda$ is the wavenumber. For tiny $b \ll a$ where the cylinder is barely perturbed, the two principal radii of curvature at any z become $R_1 \approx r(z) = a + b\cos kz$ and $1/R_2 \approx r''(z) = -bk^2\cos kz$, the latter obtained using (5.8). To first order in b, the pressure excess thus becomes

$$\Delta p = \alpha\left(\frac{1}{R_1} - \frac{1}{R_2}\right) \approx \frac{\alpha}{a} + \frac{\alpha b}{a^2}\left(k^2a^2 - 1\right)\cos kz, \tag{5.14}$$

where we have chosen the explicit sign of R_2 such that both radii curvature contribute to the pressure excess in a bulge. The parenthesis in the final expression shows that there are now two radically different cases, depending on whether ka is larger or smaller than unity. Since $k = 2\pi/\lambda$, the so-called *critical value* of the wavelength corresponding to $ka = 1$ equals the circumference of the jet, $\lambda_c = 2\pi a$.

John William Strutt, 3rd Baron Rayleigh (1842–1919). Discovered and isolated the rare gas argon for which he got the Nobel Prize (1904). Published the influential book *The Theory of Sound* on vibrations in solids and fluids in 1877–1878. (Source: Wikimedia Commons.)

For $ka > 1$ or $\lambda < \lambda_c$, the pressure is higher in the bulges ($\cos kz > 0$) than in the constrictions ($\cos kz < 0$). Consequently, the liquid will be driven away from the bulges and into the constrictions, thereby diminishing the amplitude b of the perturbation and thus the pressure difference. Since the liquid seeks back toward the unperturbed state, the column is *stable* in this regime. Conversely, for $ka < 1$ or $\lambda > \lambda_c$, the pressure is higher in the constrictions than in the bulges, and liquid will be driven away from the constrictions and into the bulges, thereby increasing the amplitude of the perturbation. As seen in Figures 5.4 and 5.5, the instability leads in the end to breakup of the column into individual drops, although this cannot be demonstrated by analyzing only infinitesimal perturbations as we have done here.

* Breakup time scale

Although we are not in a position to make a full dynamic analysis along the lines of Rayleigh [Ray78], we shall set up an approximative dynamics that allows us to model the time evolution of the perturbed liquid column and estimate the characteristic breakup time scale.

Before getting into the meat of a calculation, it is often useful to make a *dimensional analysis*. From the physics of the problem we surmise that the breakup time scale τ can only be a function of the radius a, the density ρ, and the surface tension α, that is, $\tau = f(a, \rho, \alpha)$. Since the dimension of density is kg m^{-3} and the dimension of surface tension is kg s^{-2}, the ratio ρ/α has dimension s^2 m^{-3}. To obtain a time we must multiply with the third power of the radius a, and then take the square root, to get the unique form

$$\tau \sim \sqrt{\frac{a^3\rho}{\alpha}}. \tag{5.15}$$

The only thing that is missing is the constant in front, which cannot be obtained by dimensional analysis, but requires a serious dynamical calculation.

Our dynamical approximation consists of picking a half wavelength section of the liquid column between a minimum and maximum of the perturbation, and enclose it between solid walls (see the margin figure). The perturbed liquid column can be viewed as consisting of such sections. To first order in b, the volume of the liquid in half the bulge is

$$\int_0^{\lambda/4} \pi\left(r(z)^2 - a^2\right) dz \approx 2\pi a b \int_0^{\lambda/4} \cos kz \, dz = ab\lambda.$$

Since the liquid is incompressible, the whole liquid column section must simultaneously be shifted along the axis through a distance z to build up the bulge. Equating the volume of shifted liquid with the volume in the bulge, $\pi a^2 z = ab\lambda$, we get $z = b\lambda/\pi a$.

The shift in the column section is driven by the pressure difference between minimum and maximum,

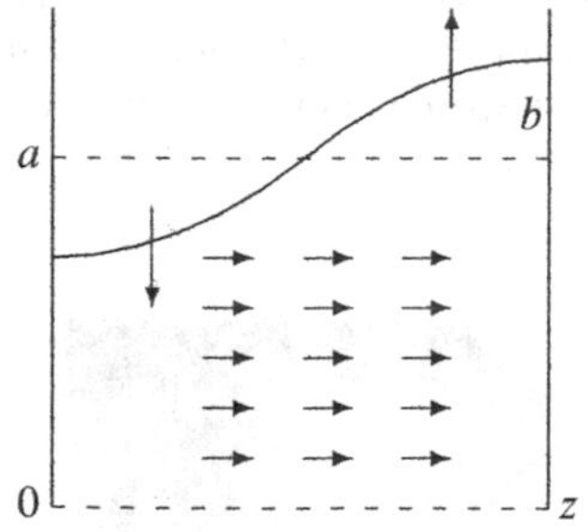

An increase in the amplitude b of the perturbation leads to a "horizontal" flow toward the right along the column axis.

$$\delta p = \Delta p|_{kz=\pi} - \Delta p|_{kz=0} = 2\frac{\alpha b}{a^2}\left(1 - k^2a^2\right). \tag{5.16}$$

Newton's Second Law for the column section becomes $M d^2z/dt^2 = \mathcal{F}_z$ where $M = \frac{1}{2}\pi a^2 \lambda \rho$ is the mass of the column and $\mathcal{F}_z = \pi a^2 \delta p$ is the driving force. Putting it all together we arrive at the following equation of motion for b:

$$\frac{d^2b}{dt^2} = \frac{b}{\tau^2} \qquad \text{with} \qquad \frac{1}{\tau^2} = \frac{\alpha}{\pi a \rho} k^2\left(1 - k^2a^2\right). \tag{5.17}$$

This confirms that in the unstable regime, $ka < 1$, the amplitude grows exponentially, $b = b_0 \exp(t/\tau)$, from an initial amplitude b_0. In the stable regime, $ka > 1$, it oscillates instead harmonically.

Normally the column is perturbed by a superposition of many wavelengths, but the one that corresponds to the smallest value of τ will completely dominate the sum at large times because of the exponential growth in the unstable regime. As a function of the wave number k, the right-hand side of $1/\tau^2$ is maximal for $k^2a^2 = 1/2$, and this shows that the minimal breakup time scale and its corresponding wavelength are, respectively

$$\tau_{\min} = \sqrt{\frac{4\pi a^3 \rho}{\alpha}}, \qquad \lambda_{\min} = 2\pi\sqrt{2}\, a = \sqrt{2}\, \lambda_c. \tag{5.18}$$

At the fastest time scale the jet is chopped into pieces of size $\lambda_{\min}$, which is $\pi\sqrt{2} \approx 4.44$ times its diameter (Rayleigh obtained 4.508). This is why the jet in Figure 5.4 is perturbed harmonically with a wavelength of 4.6 diameters. In Figure 5.4 the jet diameter is 4 mm, implying $\lambda_{\min} = 18$ mm and $\tau_{\min} = 37$ ms. In Problem 5.5 it is calculated that the broken liquid column turns into spherical drops with diameter nearly double the column diameter.

5.5 Contact angle

The two-dimensional interface between two homogeneous fluids makes contact with a solid container wall along a one-dimensional *contact line*. For the typical case of a three-phase contact between solid, liquid, and gas the *contact angle* χ is defined as the angle between the solid and the interface (inside the liquid). Water and air against clean glass meet in a small acute contact angle, $\chi \approx 0°$, whereas mercury and air meet glass at an obtuse contact angle of $\chi \approx 140°$. Due to its small contact angle, water is very efficient in *wetting* many surfaces, whereas mercury has a tendency to contract into pearls.

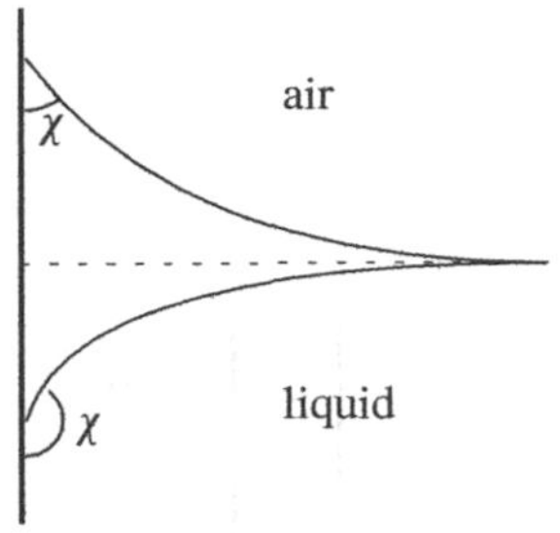

An air/liquid interface meeting a solid wall. The upper curve makes an acute contact angle, like water, whereas the lower curve makes an obtuse contact angle, like mercury.

It should be emphasized that the contact angle is extremely sensitive to surface properties, fluid composition, and additives. This is especially true for the water-air-glass contact angle. In the household we regularly use surfactants that diminish both the surface tension and contact angle, thereby better enabling dishwater to wet greasy surfaces on which it otherwise would tend to pearl. Oppositely, after washing our cars we apply a wax that makes rainwater pearl and prevents it from wetting the surface, thereby diminishing rust and corrosion.

Young's law

The contact angle is a material constant that depends on the properties of all three materials coming together. Whereas material adhesion can maintain a tension normal to the solid wall, the tangential tension has to vanish at the contact line, or it will start to move. This yields an equilibrium relation due to Thomas Young (1805)[3],

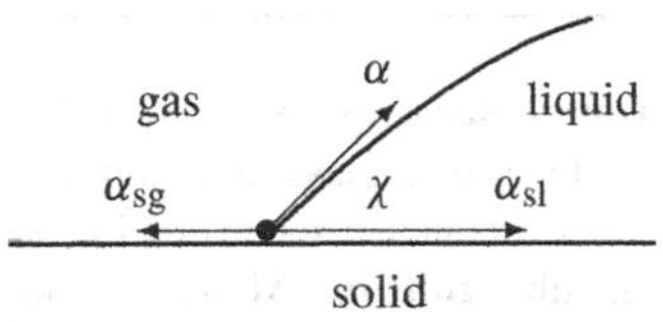

Liquid and gas meeting at a solid wall in a line orthogonal to the paper. The tangential component of surface tension along the wall must vanish.

$$\alpha_{sg} = \alpha_{sl} + \alpha \cos \chi, \tag{5.19}$$

where α_{sg} and α_{sl} denote the solid/gas and solid/liquid surface tensions. Given the three surface tensions, we may directly calculate the contact angle,

$$\cos \chi = \frac{\alpha_{sg} - \alpha_{sl}}{\alpha}. \tag{5.20}$$

This expression is, however, not particularly useful because the three surface tensions are not equally well known. As shown in the example below, estimates can sometimes be carried out that lead to reasonable values for the contact angle. It is, however, normally better to view χ as an independent material constant that can be directly measured and thereby provide information about the difference $\alpha_{sg} - \alpha_{sl}$.

The condition for existence of an equilibrium contact angle is evidently that the cosine lies between -1 and $+1$. For $0 < \alpha_{sg} - \alpha_{sl} < \alpha$, the contact angle is acute, $0 < \chi < 90°$, and the contact is said to be *mostly wetting*, while for $-\alpha < \alpha_{sg} - \alpha_{sl} < 0$, the contact angle is obtuse, $90° \leq \chi \leq 180°$, and the contact is said to be *mostly non-wetting* [de Gennes et al. 2002]. If on the other hand $\alpha_{sg} - \alpha_{sl} > \alpha$, the contact line cannot be static, but will be drawn toward the gas while dragging the liquid along, and thereby spreading the liquid all over a horizontal surface. Such a contact is for good reason said to be *completely wetting*. Similarly, if $\alpha_{sg} - \alpha_{sl} < -\alpha$, the contact is *completely non-wetting* and the contact line will recede from the gas on a horizontal surface until all the liquid collects into nearly spherical pearls, a phenomenon called *dewetting*.

A large amount of recent research concerns the dynamics of contact lines under various conditions (see [de Gennes et al. 2002] and [Gen85]). In the following we shall limit the analysis to static contact lines with a well-defined contact angle.

---

[3]Sometimes this law is called the Young–Dupré law, thereby crediting the French mathematician/physicist Athanase Dupré (1808–1869) for relating this law to the thermodynamic work function for adhesion.

Example 5.2 [Contact angle of water on ice]: An estimate of the contact angle of water on ice at 0°C can be obtained from the latent heats of evaporation of liquid water, and from the latent heats of melting and sublimation of ice. For liquid water the molar enthalpy of evaporation is $h_{\text{evap}} = 45.051$ kJ mol^{-1}, whereas for ice the molar enthalpy of melting is $h_{\text{melt}} = 6.010$ kJ mol^{-1} and the molar enthalpy of sublimation $h_{\text{subl}} = 51.059$ kJ mol^{-1} [8]. Notice that $h_{\text{subl}} = h_{\text{melt}} + h_{\text{evap}}$ as one would expect. Using the densities $\rho_{\text{ice}} = 916.72$ kg m^{-3} and $\rho_{\text{water}} = 1000$ kg m^{-3} at 0°C, we find from (5.1) the following values: $\alpha_{sg} = 0.138$ N m^{-1}, $\alpha_{sl} = 0.016$ N m^{-1}, and $\alpha = 0.129$ N m^{-1}. From Equation (5.20) we estimate $\cos\chi \approx 0.95$ or $\chi \approx 19°$, which is quite reasonable since the value is quoted to lie between 12° and 24° [VSM02].

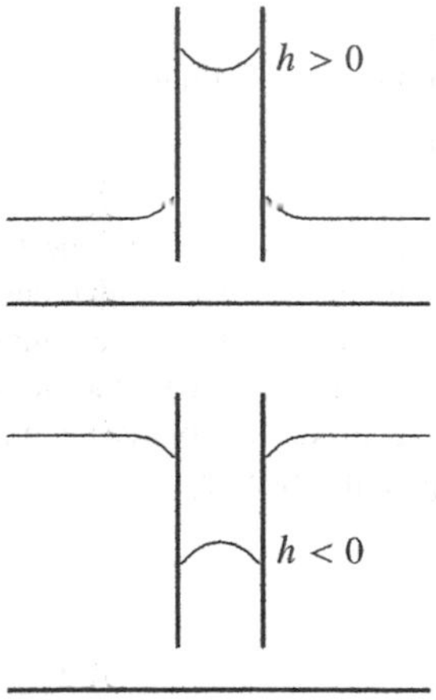

Top: Water rises above the ambient level in a glass tube and displays a concave meniscus inside the tube. **Bottom:** Mercury sinks below the general level in a capillary glass tube and displays a convex meniscus.

James Jurin (1684–1750). English physician. An ardent proponent of Newton's theories. Credited for the discovery of the inverse dependence of the capillary height on the radius of the tube in 1718, although others had noticed it before [de Gennes et al. 2002, p. 49].

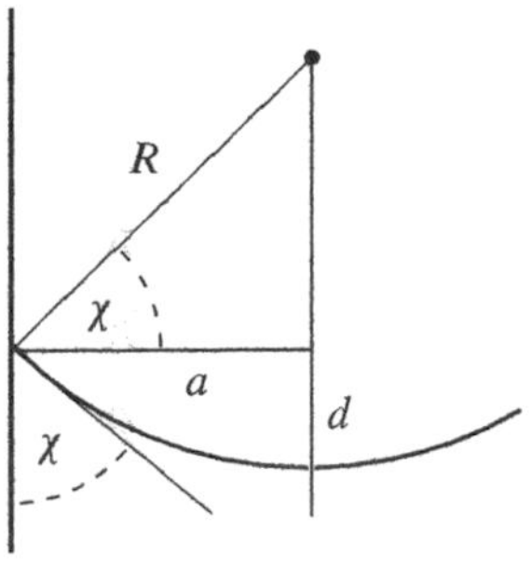

Approximately spherical surface with acute contact angle in a narrow circular tube.

Capillary effect

Water has a well-known ability to rise above the ambient level in a narrow, vertical, cylindrical glass tube that is lowered into the liquid. This is called the *capillary effect* and takes place because the surface tension of glass in contact with air is larger than that of glass in contact with water, $\alpha_{sg} > \alpha_{sl}$. The water rises to height h where the upward force due to the difference in surface tension balances the downward weight of the raised water column (disregarding the volume of the meniscus-shaped surface),

$$(\alpha_{sg} - \alpha_{sl})2\pi a \approx \rho_0 g_0 \pi a^2 h. \tag{5.21}$$

Using the Young law (5.19) we arrive at *Jurin's height*,

$$\boxed{h \approx 2\frac{L_c^2}{a}\cos\chi,} \tag{5.22}$$

where L_c is the capillary length (5.5). The height is positive for acute contact angles ($\chi < 90°$) and negative for obtuse ($\chi > 90°$). It fairly obvious that the approximation of ignoring the shape of the meniscus is equivalent to assuming that the tube radius is much smaller than the capillary length, $a \ll L_c$. In Section 5.7 we shall determine the exact shape of the surface meniscus for any tube radius.

For mostly wetting liquids, the rise may also be seen as caused by the acute contact angle at the water-air-glass contact line, which makes the surface inside the tube concave, such that the center of curvature lies outside the liquid. Surface tension will create a negative pressure just below the liquid surface, and that is what lifts the water column (as in Example 5.1). Mercury with its obtuse contact angle displays instead a convex surface shape, creating a positive pressure jump that makes the liquid drop down to a level where the pressure just below the surface equals the ambient hydrostatic pressure at the same level outside.

When the tube radius a is small compared to the capillary length L_c, gravity has no effect on the shape, and the surface may be assumed to be part of a sphere of radius R. The geometric construction in the margin figure yields the tube radius $a \approx R\cos\chi$ and the central depth $d \approx R(1 - \sin\chi)$. Eliminating R, we obtain

$$\boxed{d \approx a\frac{1-\sin\chi}{\cos\chi}.} \tag{5.23}$$

Both the capillary depth and the capillary height are, as mentioned, modified for larger radius, $a \gtrsim L_c$, where the surface flattens in the middle.

Example 5.3: A certain capillary tube has diameter $2a = 1$ mm. Water with $\chi \approx 0°$ and $L_c \approx 2.7$ mm rises to $h = +30$ mm with a surface depth $d = +0.50$ mm. Mercury with contact angle $\chi \approx 140°$ and $L_c = 1.9$ mm sinks, on the other hand, to $h = -11$ mm and depth $d = -0.23$ mm under the same conditions.

Figure 5.6. Spider walking on water. This is also an effect of surface tension; see [HCB03] for analysis of water striding. (Source: Reproduced from [HCB03]. With permission.)

Case: Walking on water

Many species of animals—mostly insects and spiders— are capable of walking on the surface of water (see Figure 5.6 and the comprehensive review [BH06b]). Compared to the displacement necessary to float, only a very small part of the water-walking animal's body is below the ambient water surface level. The animal's weight is mostly carried by surface tension while buoyancy, otherwise so important for floating objects, is negligible.

The water is deformed by the animal's weight and makes contact with its "feet" along a total (combined) perimeter length L. The animal must somehow "wax its feet" to obtain an obtuse contact angle $\chi > 90°$ so they avoid getting wet. In a given contact point, the angle between the tangent to the "foot" and the undeformed water surface is denoted θ (see the margin figure). For simplicity we shall assume that this angle is the same everywhere along the perimeter curve. Surface tension acts along the tangent to the deformed water surface and simple geometry tells us that the angle of this force with the vertical is $\psi = 90° - \theta + 180° - \chi$. Projecting the surface tension on the vertical, we obtain the following condition for balancing the animal's weight:

$$\alpha L \cos \psi = M g_0. \tag{5.24}$$

The ratio between the weight and surface tension force is called the *Bond number*,

$$\boxed{\mathsf{Bo} = \frac{M g_0}{\alpha L},} \tag{5.25}$$

where M is the mass of the animal. If $\mathsf{Bo} < 1$, surface tension is capable of carrying the weight of the animal with and angle determined by $\cos \psi = \mathsf{Bo}$. If $\mathsf{Bo} > 1$, this is not possible, and the animal literally falls through. The reciprocal number $\mathsf{Je} = 1/\mathsf{Bo}$, which must be larger than unity for anyone who wants to walk on water has—jestingly—been called the *Jesus number* [Vogel 1988].

An animal with mass $M = 10$ mg requires the total length L of the contact perimeter to be larger than 1.3 mm to make $\mathsf{Bo} < 1$. Heavier animals need correspondingly larger feet.

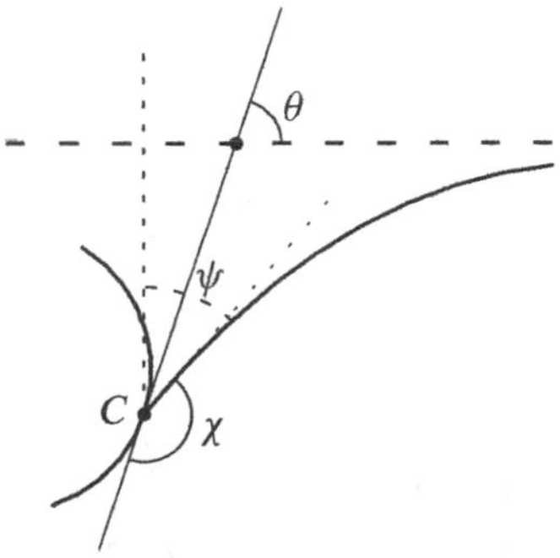

Geometry of water-walking with a contact line through C. The arc on the left represents the "foot", and the curve on the right the deformed water surface. The contact angle is χ and the angle between the tangent and the horizontal undeformed water surface (dashed) is θ. Surface tension acts along the tangent (dotted) to the deformed water surface. Its projections on the vertical (dashed) through the angle ψ must add up to the animal's weight.

5.6 Meniscus at a flat wall

Even if the capillary effect is best known from cylindrical tubes, any fluid interface will take a non-planar shape near a solid wall where the interface rises or falls relative to the ambient horizontal fluid level. The crescent shape of the interface is called a *meniscus*. A vertical flat wall is the simplest possible geometry, and in this case there is no capillary effect because the interface must approach the ambient level far from the wall. Taking the x-axis horizontal and the z-axis vertical, the meniscus shape may be assumed to be independent of y and described by a simple planar curve $z = z(x)$ in the xz-plane.

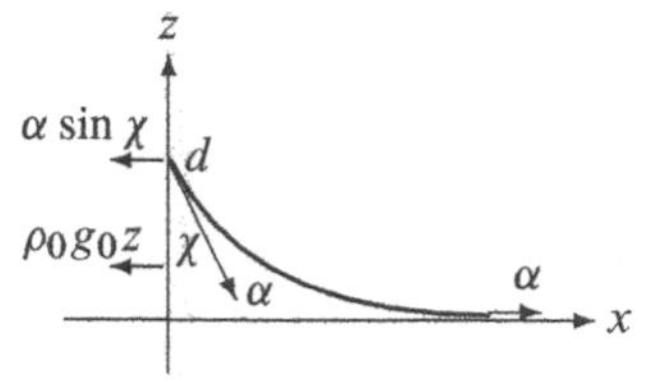

The meniscus starts at the wall with the given contact angle (here acute) and approaches the ambient interface level far from the wall.

The height at the wall

The meniscus height d at the wall can be calculated exactly by balancing the external horizontal forces acting on the fluid (see the margin figure). Far from the wall for $x \to \infty$ the force (per unit of y) is just α, whereas at the wall for $x = 0$ there is the wall's reaction $\alpha \sin \chi$ to the pull from the surface itself plus its reaction to the hydrostatic pressure exerted by the fluid, $\int_0^d \rho_0 g_0 z \, dz = \frac{1}{2} \rho_0 g_0 d^2$. The equation of balance thus becomes

$$\alpha \sin \chi + \frac{1}{2} \rho_0 g_0 d^2 = \alpha,$$

which yields

$$d = L_c \sqrt{2(1 - \sin \chi)} = 2L_c \sin \frac{90^\circ - \chi}{2}. \tag{5.26}$$

Although derived for acute contact angles, the last expression is also valid for obtuse angles. It should be emphasized that this is an exact result. For water with $\chi \approx 0$, we find $d \approx \sqrt{2} L_c \approx 3.9$ mm, whereas for mercury with $\chi = 140^\circ$ we get $d \approx -1.6$ mm.

Geometry of planar curves

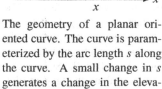

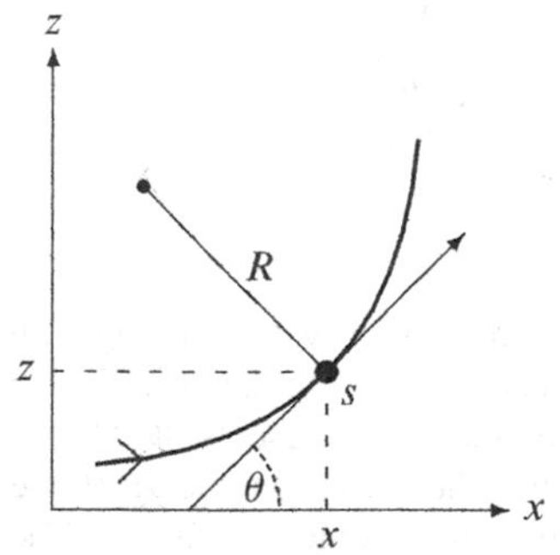

The geometry of a planar oriented curve. The curve is parameterized by the arc length s along the curve. A small change in s generates a change in the elevation angle θ determined by the local radius of curvature. Here the radius of curvature is positive.

The best way to handle the geometry of a planar oriented curve is to introduce two auxiliary parameters: the *arc length* s along the curve, and the *elevation angle* θ between the x-axis and the oriented tangent to the curve. Due to the monotonic increase of arc length as a point moves along the oriented curve, it is mostly convenient to parameterize the curve by means of the arc length, $\theta = \theta(s)$, $x = x(s)$, and $z = z(s)$.

A small increment $s \to s + ds$ in the arc length moves the point (x, z) an amount (dx, dz) along the tangent, and we obtain immediately from the definition of θ, and the fact that $ds = R \, d\theta$,

$$\frac{dx}{ds} = \cos \theta, \qquad \frac{dz}{ds} = \sin \theta, \qquad \frac{d\theta}{ds} = \frac{1}{R}. \tag{5.27}$$

If the elevation angle θ is an increasing function of the arc length s, the radius of curvature R must be taken to be positive, otherwise negative. One should be aware that this sign convention may not agree with the physical sign convention for the Young–Laplace law (5.11). Depending on the arrangement of the fluids with respect to the interface, it may—as we shall now see—be necessary to introduce an explicit sign to get the physics right.

The meniscus equation

To obtain an expression for the geometric radius of curvature, we employ hydrostatic balance at every point of the meniscus. For simplicity we assume that there is air above the liquid and that the air pressure on the interface is constant, $p = p_0$. Then the pressure in the liquid just below the surface is $p = p_0 + \Delta p$, where Δp is given by the Young–Laplace law (5.11). In terms of the local geometric radius of curvature R, we have for an acute angle of contact $R_1 = -R$ because the center of curvature of the concave meniscus lies outside the liquid. The other principal radius of curvature is infinite, $R_2 = \infty$, because the meniscus is flat in the y-direction.

The pressure just below the surface, $p = p_0 - \alpha/R$, is always smaller than atmospheric. Since surface tension does not change the equation of hydrostatic equilibrium, the pressure in the liquid will—as in the sea (page 22)—be constant at any horizontal level z, and given by $p = p_0 - \rho_0 g_0 z$ where p_0 is the ambient pressure. Just below the surface $z = z(s)$ we have

$$p_0 - \rho_0 g_0 z = p_0 - \frac{\alpha}{R}, \tag{5.28}$$

and using this equation to eliminate $1/R$ we obtain from (5.27)

$$\frac{d\theta}{ds} = \frac{z}{L_c^2}, \tag{5.29}$$

where we have also introduced the capillary length $L_c = \sqrt{\alpha/\rho_0 g_0}$. Differentiating the above equation once more with respect to s and making use of (5.27), we obtain

$$\boxed{L_c^2 \frac{d^2\theta}{ds^2} = \sin\theta.} \tag{5.30}$$

This second-order ordinary differential equation for the planar meniscus is expressed entirely in the elevation angle $\theta = \theta(s)$.

Pendulum connection: The meniscus equation (5.30) is nothing but the equation for an *inverted mathematical pendulum*. A mathematical pendulum of length ℓ and unit mass obeys Newton's Second Law in a constant field of gravity g (see the margin figure),

$$\ell \frac{d^2\phi}{dt^2} = -g \sin\phi.$$

Taking $\phi = 180° - \theta$ and $t = (s/L_c)\sqrt{\ell/g}$ we arrive at the meniscus equation. Although it is amusing to translate the solutions to the meniscus equation into pendulum motions, we shall not exploit this connection further here.

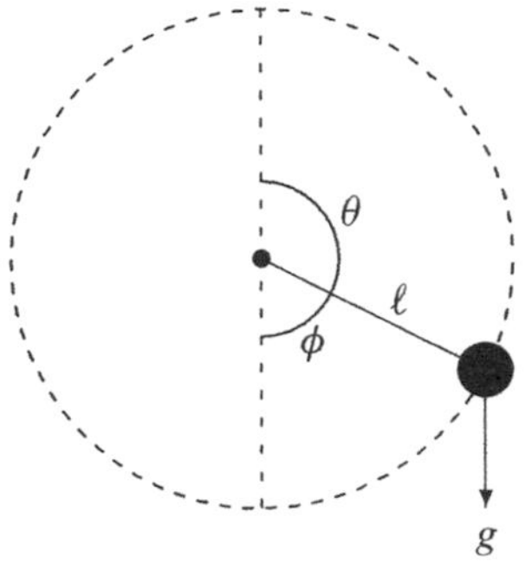

A mathematical pendulum. The angle ϕ is measured from stable equilibrium $\phi = 0$. The elevation angle is $\theta = 180° - \phi$.

Analytic solution

Since L_c is the only length scale in the problem we may without loss of generality take $L_c = 1$, so that all lengths are measured in units of L_c. Multiplying the meniscus equation (5.30) with $d\theta/ds$, its first integral becomes

$$\frac{1}{2}\left(\frac{d\theta}{ds}\right)^2 = 1 - \cos\theta. \tag{5.31}$$

The constant on the right-hand side has been determined from the conditions that $\theta \to 0$ and $d\theta/ds \to 0$ for $s \to \infty$. From this equation we get

$$\frac{d\theta}{ds} = \sqrt{2(1-\cos\theta)} = -2\sin\frac{\theta}{2}, \tag{5.32}$$

where we have used that the contact angle is acute so that $d\theta/ds > 0$ and $\theta < 0$ (see the margin figure).

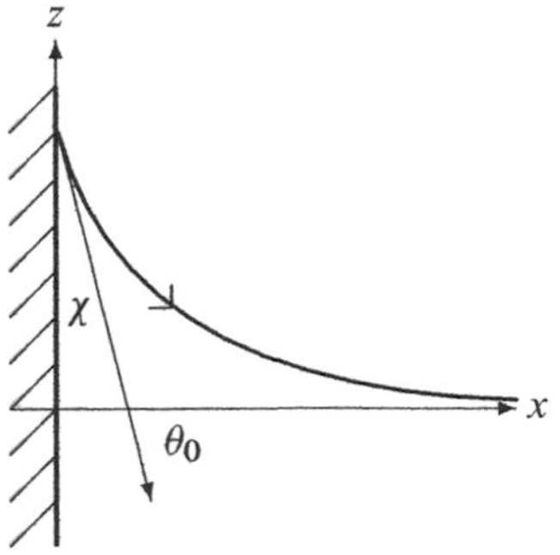

Geometry of single plate meniscus for acute contact angle $\chi < 90°$. The elevation angle θ is always negative with initial value $\theta_0 = \chi - 90°$ at the wall for $s = 0$. The other initial values are $x = 0$, $z = d$ for $s = 0$.

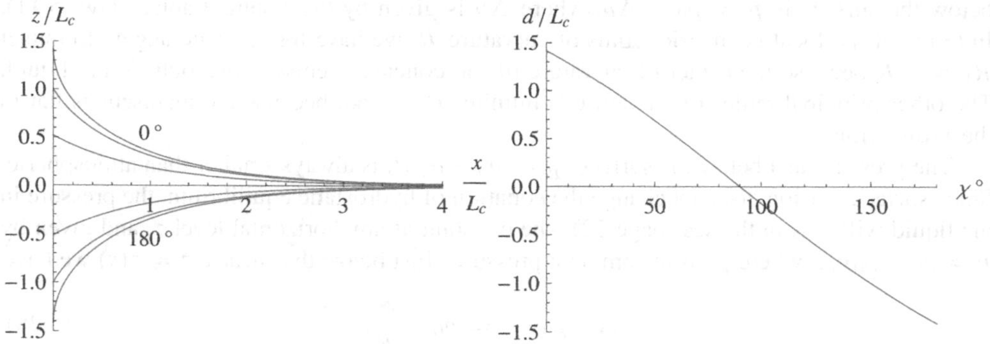

Figure 5.7. **Left:** Flat-wall menisci for all contact angles in steps of 30°. Close to the wall the menisci are linear, and far from the wall they approach the x-axis exponentially. **Right:** Height of the meniscus at the wall as a function of the contact angle χ.

Combining this equation and (5.27) we obtain, using well-known trigonometric relations,

$$\frac{dx}{d\theta} = \sin\frac{\theta}{2} - \frac{1}{2\sin\frac{\theta}{2}}, \qquad \frac{dz}{d\theta} = -\cos\frac{\theta}{2}. \tag{5.33}$$

Finally integrating these equations we get the meniscus shape in parametric form,

$$\boxed{x = 2\cos\frac{\theta_0}{2} + \log\left|\tan\frac{\theta_0}{4}\right| - 2\cos\frac{\theta}{2} - \log\left|\tan\frac{\theta}{4}\right|, \qquad z = -2\sin\frac{\theta}{2}.} \tag{5.34}$$

The integration constant is chosen such that $x = 0$ for $\theta = \theta_0 = \chi - 90°$ and $z \to 0$ for $\theta \to 0$. The solution is valid for both acute and obtuse contact angles, and is plotted in Figure 5.7L for a selection of contact angles.

At the wall the meniscus height becomes (see Figure 5.7R)

$$d = z(\theta_0) = 2\sin\frac{90° - \chi}{2}, \tag{5.35}$$

which is identical to the earlier result (5.26). Far from the wall for $\theta \to 0$, we have $x \approx -\log|\theta/4|$ and $z \approx -\theta$, so that

$$z \approx \pm 4e^{-x}. \tag{5.36}$$

The meniscus decays exponentially with a sign depending on whether the contact angle is acute or obtuse.

5.7 Meniscus in a cylindrical tube

Many static interfaces, for example the meniscus in a cylindrical tube but also droplets and bubbles are invariant under rotation around an axis, allowing us to establish a fairly simple formalism for the shape of the interface, similar to the two-dimensional formalism of the preceding section.

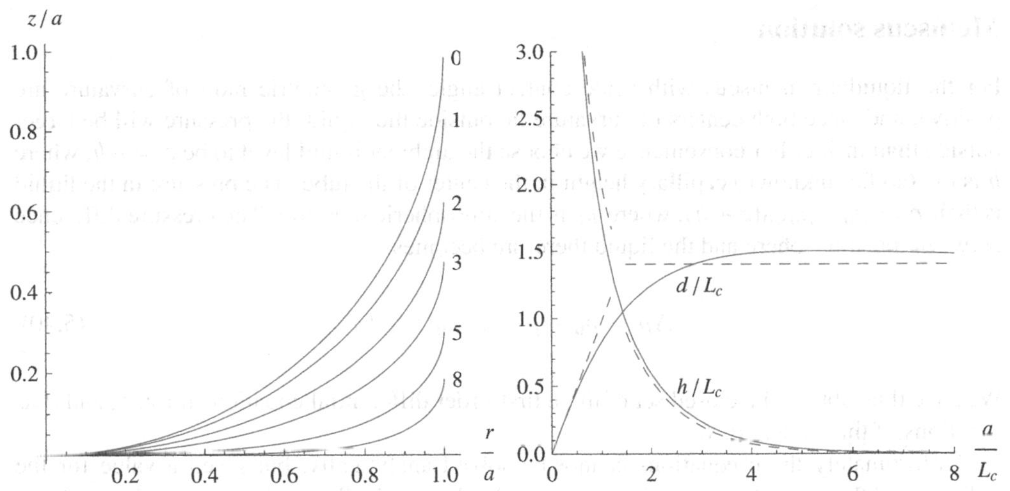

Figure 5.8. Meniscus and capillary effect in "water" with $\chi = 1°$ for a cylindrical tube of radius a. **Left:** Meniscus shape $z(r)/a$ plotted as a function of r/a for $a/L_c = 0, 1, 2, 3, 5, 8$. Note how the shape is nearly spherical for $a/L_c \lesssim 1$ and only weakly dependent on a. **Right:** Computed capillary height h and depth d as functions of a (solid lines). The dashed curves are obtained from the meniscus estimates (5.42).

Geometry of axial surfaces

In cylindrical coordinates an axially surface (invariant under rotations around the axis) is described by a planar curve in the rz-plane. Using again the arc length s along the curve and the angle of elevation θ for its slope, we find as in the planar case,

$$\frac{dr}{ds} = \cos\theta, \qquad \frac{dz}{ds} = \sin\theta. \tag{5.37}$$

The first principal radius of curvature may be directly taken from the two-dimensional case, whereas it takes some work to show that the second center of curvature lies on the z-axis (see Problem 5.4), such that the radial distance becomes $r = R_2 \sin\theta$, so that

$$\frac{d\theta}{ds} = \frac{1}{R_1}, \qquad \frac{\sin\theta}{r} = \frac{1}{R_2}. \tag{5.38}$$

Adding the two last equations and using the Young–Laplace expression (5.11), we get

$$\frac{d\theta}{ds} + \frac{\sin\theta}{r} = \frac{1}{R_1} + \frac{1}{R_2} = \frac{\Delta p}{\alpha}. \tag{5.39}$$

As before, one should keep in mind that the sign convention for the geometric radii of curvature may not agree with the physical sign convention for the Young–Laplace law (5.11), and that one must be careful to get the sign of Δp right.

A special class of solutions is furnished by open soap film surfaces with vanishing mean curvature, $\Delta p = 0$ (see Figure 5.3). Their shapes are entirely dependent on the boundary conditions provided by the the stiff wire frames used to create and maintain them. There is essentially only one axially symmetric shape with vanishing mean curvature, namely the catenoid in Figure 5.3L (see Problem 5.9).

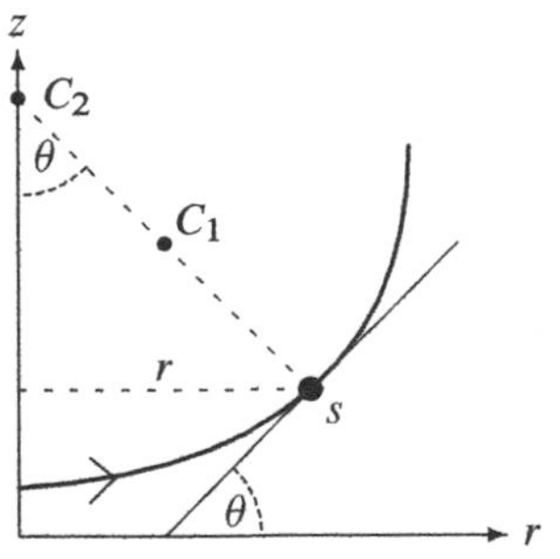

The oriented curve is parameterized by the arc length s. A small change in s generates a change in the elevation angle θ determined by the local radius of curvature R_1 with center C_1. The second radius of curvature R_2 lies on the z-axis with center C_2.

Meniscus solution

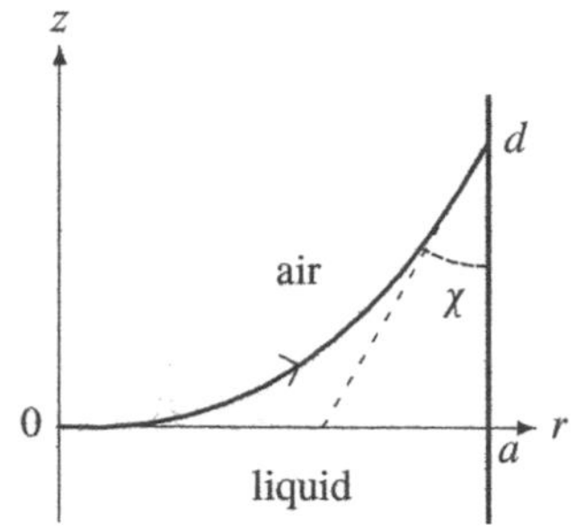

Meniscus with acute angle of contact. Both centers of curvature lie outside the liquid.

For the liquid/air meniscus with acute contact angle, the geometric radii of curvature are positive, and since both centers of curvature lie outside the liquid, the pressure will be larger outside than inside. For convenience we choose the ambient liquid level to be $z = -h$, where h is the (so far unknown) capillary height in the center of the tube. The pressure in the liquid is then $p = p_0 - \rho_0 g_0 (z + h)$, where p_0 is the atmospheric pressure. The pressure difference between the atmosphere and the liquid therefore becomes

$$\Delta p = p_0 - p = \rho_0 g_0 (z + h). \tag{5.40}$$

We have thus obtained a closed set of three first-order differential equations for r, z, and θ as functions of the arc length s.

Unfortunately these equations cannot be solved analytically, but given a value for the radius a and the contact angle χ, they may be solved numerically as a boundary value problem in the interval $0 \leq s \leq s_0$ with the boundary conditions

$$r = 0, \qquad z = 0, \qquad \theta = 0, \qquad (s \to 0); \tag{5.41a}$$

$$r = a, \qquad z = d, \qquad \theta = 90° - \chi, \qquad (s \to s_0), \tag{5.41b}$$

where d, h, and s_0 are unknown. Integrating, one obtains not only the functions $r(s)$, $z(s)$, and $\theta(s)$, but also the values of the unknown parameters h, d, and s_0. The numeric solutions for the menisci in water with $\chi \approx 0$ are displayed in Figure 5.8L for a selection of values of the radius a.

The corresponding capillary height h and depth d are shown as functions of a in Figure 5.8R, and are compared with the small and large a approximations already obtained in (5.22), (5.23), (5.36), and (5.26),

$$h \approx 2\frac{L_c^2}{a}\cos\chi, \qquad d \approx a\frac{1 - \sin\chi}{\cos\chi}, \qquad \text{for } a \ll L_c; \tag{5.42a}$$

$$h \approx \pm 4 L_c e^{-a/L_c}, \qquad d \approx 2L_c \sin\frac{90° - \chi}{2}, \qquad \text{for } a \gg L_c. \tag{5.42b}$$

We have here replaced z by h and x by a in Equation (5.36) to get an estimate of the central height h. As seen from Figure 5.8R, these approximations fit the numeric solutions quite well in their respective regions.

* 5.8 Application: Sessile drops and captive bubbles

Everyone is familiar with drops of liquid sitting stably (i.e., sessile) on top of a horizontal surface. Small droplets tend to be spherical in shape, but gravity flattens larger drops into puddles when their size exceeds the capillary length. On a wetting surface where the contact angle is tiny, water droplets are always quite flat and tend to spread out into thin layers. Mercury, on the other hand, with its obtuse contact angle of 140° against many solid surfaces, is seen to form small, nearly spherical droplets. Larger mercury drops tend to form puddles that likewise can be hard to keep in place. The "quickness" of mercury (also known as quicksilver) is mainly due its high density, which makes gravity easily overcome surface adhesion, if for example the horizontal surface is tilted a tiny bit.

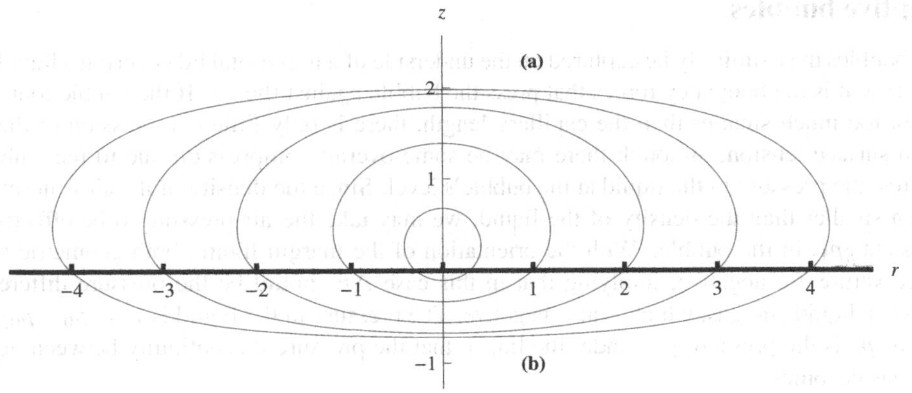

Figure 5.9. **(a)** Shapes of sessile liquid droplets in air, and **(b)** captive air bubbles in the same liquid. The contact angle is obtuse in both cases, $\chi = 140°$ ("mercury"). For the complementary contact angle $\chi = 40°$, the picture is simply turned upside down with the flat shapes on the top and the rounded ones on the bottom. All lengths in the figure are measured in units of the capillary length L_c.

Sessile drops

The coordinate system is chosen so that the plate surface is at $z = 0$ (see the margin figure). Both centers of curvature lie inside the liquid so that the pressure is everywhere higher in the liquid than in the atmosphere. The pressure in the liquid is hydrostatic $p = p_1 - \rho_0 g_0 z$, where p_1 is the pressure at $z = 0$. Since the geometric radii of curvature are positive with the indicated orientation of the curve, the pressure jump is in this case the difference between liquid and atmospheric pressure:

$$\Delta p = p - p_0 = p_1 - p_0 - \rho_0 g_0 z = \rho_0 g_0 (h - z). \tag{5.43}$$

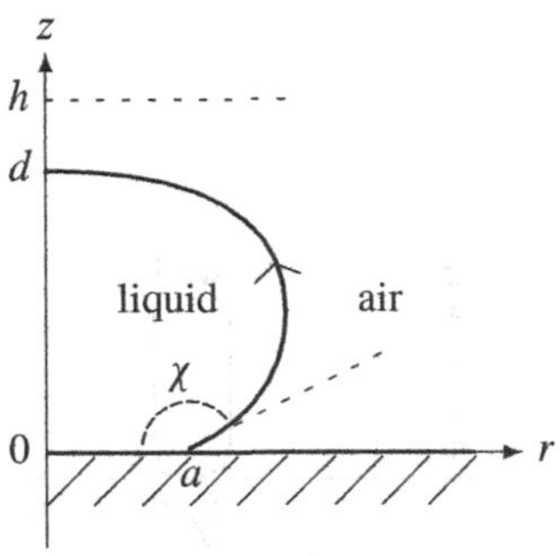

Geometry of sessile drop on a horizontal plate. The contact angle χ is here obtuse (as for mercury), the circular contact area (the droplet's "footprint") has radius a, and the central height of the droplet is d.

Here we have defined the height h by setting $p_1 - p_0 = \rho_0 g_0 h$. Since the radii of curvature are always positive, we must have $h > d$ such that the pressure on the top of the droplet at $z = d$ is always higher than atmospheric, as one would expect.

The solution of the three coupled first-order differential equations (5.37) and (5.39) now proceeds as in the preceding section. The boundary conditions are in this case

$$r = a, \quad z = 0, \quad \theta = 180° - \chi \quad (s = 0); \tag{5.44a}$$

$$r = 0, \quad z = d, \quad \theta = 180° \quad (s = s_0). \tag{5.44b}$$

Given the "footprint" radius a and the contact angle χ, the droplet shapes and the unknown parameters d, h, and s_0 can be determined. The droplet shapes are shown in Figure 5.9 for a choice of footprint radii and an obtuse contact radius $\chi = 140°$, corresponding to mercury on glass.

For $a \ll L_c$ the droplet is spherical while for $a \gg L_c$ it becomes a flat puddle (see Figure 5.9a). Estimating in the same way as before (see Problem 5.10), we find

$$d \approx a\frac{1 - \cos\chi}{\sin\chi} \quad \text{for } a \ll L_c, \qquad d \approx 2L_c \sin\frac{\chi}{2} \quad \text{for } a \gg L_c. \tag{5.45}$$

Mercury with $\chi = 140°$ has maximal puddle height $d \approx 1.88L_c$, in close agreement with Figure 5.9a. Water on a wetting surface with $\chi \approx 1°$ has maximal puddle height $d \approx L_c\chi \approx 50\ \mu$m. A bucket of water thrown on the floor really makes a mess.

Captive bubbles

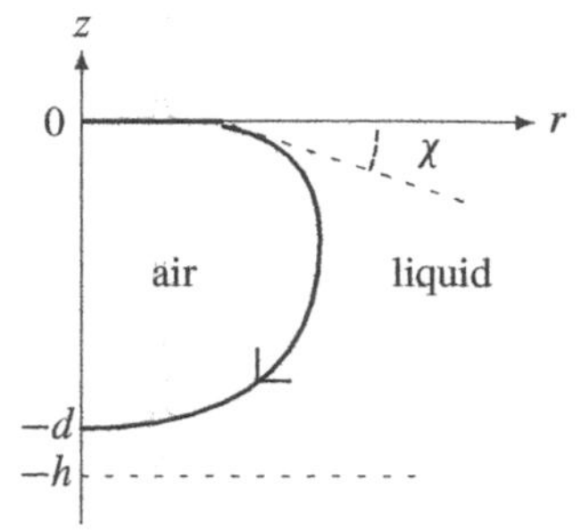

Captive air bubble under a horizontal lid in a liquid with acute angle of contact, for example in water.

Air bubbles may similarly be captured on the underside of a horizontal lid covering a liquid. In this case it is the buoyancy forces that press the bubble against the lid. If the bubble footprint is not too much smaller than the capillary length, there is only little compression of the air from surface tension, although there may be some overall compression due to the ambient hydrostatic pressure in the liquid at the bubble's level. Since the density of the air is normally much smaller than the density of the liquid, we may take the air pressure to be effectively constant, p_1, in the bubble. With the orientation of the margin figure, both geometric radii of curvature are negative, implying that in this case Δp should be the pressure difference between liquid and gas, which is also negative. The pressure in the liquid is $p = p_0 - \rho_0 g_0 z$, where p_0 is the pressure just under the lid, so that the pressure discontinuity between liquid and gas becomes

$$\Delta p = p - p_1 = p_0 - p_1 - \rho_0 g_0 z = -\rho_0 g_0 (z + h). \tag{5.46}$$

Here we have introduced an artificial "height" h by setting $p_1 - p_0 = \rho_0 g_0 h$. This defines the level $z = -h$ where the pressure jump would vanish. Since the pressure discontinuity between liquid and gas is negative, we must have $h > d$ where d is the height of the bubble.

The transformations $(r, z, \theta) \to (r, -z, -\theta)$ and $\Delta p \to -\Delta p$ leave (5.37) and (5.39) invariant. The bubble shapes may for this reason be obtained from the drop shapes provided one replaces the contact angle by its complement, $\chi \to 180° - \chi$. The bubble shapes are also shown in Figure 5.9 below the horizontal line.

* 5.9 Application: Pendant drops and tethered bubbles

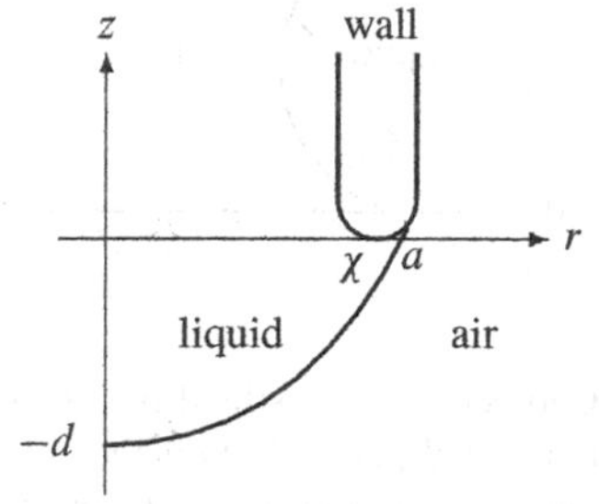

A liquid drop hanging from a tube with strongly exaggerated wall thickness. Any true contact angle can be accommodated by the 180° turn at the end of the tube's material. The apparent contact angle χ between the liquid surface and the horizontal can in principle take any value.

Whereas sessile drops in principle can have unlimited size, pendant (hanging) liquid drops will detach and fall if they become too large. Likewise, air bubbles tethered to a solid object in a liquid will detach and rise if they grow too large. Here we shall only discuss pendant drops because the shapes of tethered bubbles are complementary to the shapes of pendant drops (like captive bubbles are complementary to sessile; see the preceding section). Drops can hang from all kinds of objects. Under a flat, horizontal plate, drops may, for example, form from condensation. In that case the constraint is that the liquid has to have the correct contact angle with the plate.

A drop may also hang from the tip of a vertical capillary tube, for example a pipette fitted with a rubber bulb, through which liquid can be slowly injected to feed it. If the feeding rate is slow enough, the drop will grow through a sequence of equilibrium shapes with continuously increasing volume until it becomes unstable and detaches. The advantage of this arrangement is that the rounded end of the tube can accommodate any contact angle (see the margin figure). The constraint is in this case that all drop shapes must have the same footprint radius. We shall for simplicity also assume that the tube wall has vanishing thickness and that the three-phase contact line is fixed at the tip of the tube. The actual dynamics of the detachment of the drop is highly interesting, but cannot be studied within the hydrostatic framework of this chapter (see however [Egg97, WPB99]).

Tate's Law

The naive reason that a pendant droplet can be in stable equilibrium is that surface tension α pulls so strongly upward at the rim of the tube that it overcomes the weight of the drop. Denoting the capillary tube radius by a and the angle between the liquid surface and the horizontal by χ, this condition becomes

$$M g_0 < 2\pi a \alpha \sin \chi, \tag{5.47}$$

where M is the total mass of the drop.

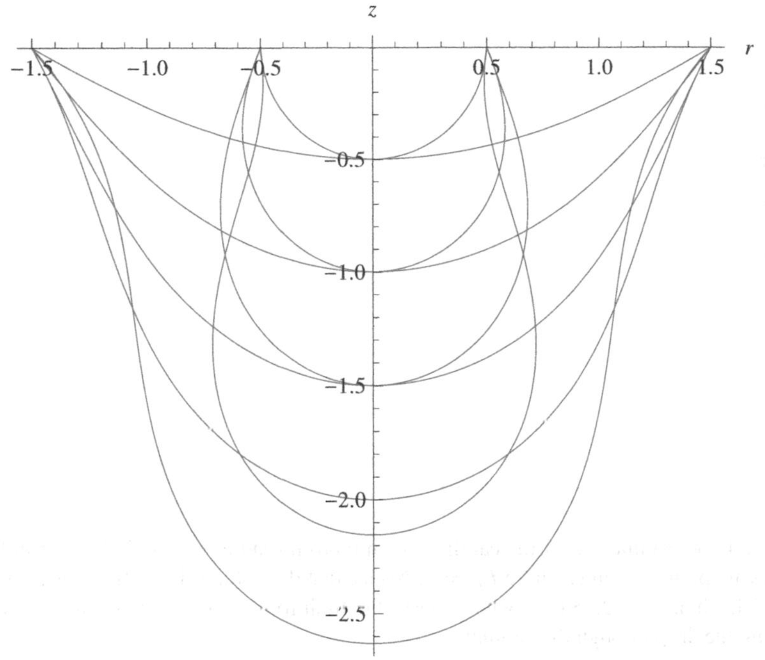

Figure 5.10. Two sets of static drop shapes hanging from the mouth of capillary tube with radius $a = 0.5$ and $a = 1.5$, respectively (in units where $L_c = 1$). For each set there is no static solution with larger volume than shown here, which indicates the limit to the size of the static shapes that can be created by slowly feeding the drops through the tubes.

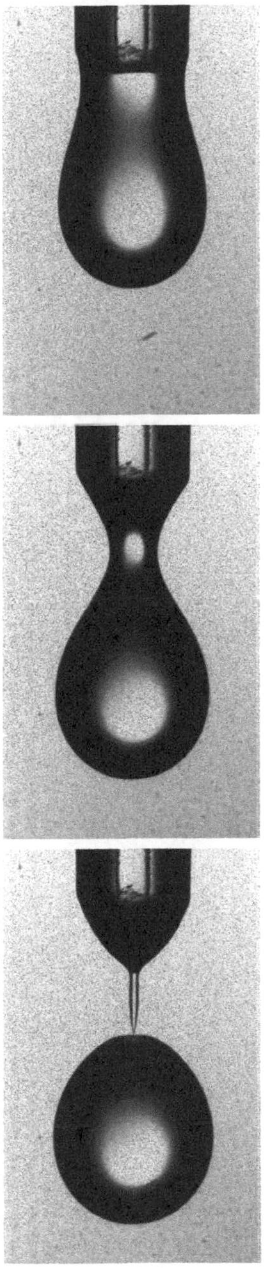

Pendant drop that is slowly fed with liquid through a tube. **Top:** Typical shape similar to the static ones in Figure 5.10. **Middle:** Stretching the neck. **Bottom:** The moment of detachment. (Source: Courtesy Anders Andersen, Technical University of Denmark.)

For the capillary, the right-hand side is maximal for $\chi = 90°$, so that the drop should detach when the mass satisfies

$$M g_0 = 2\pi a \alpha. \tag{5.48}$$

This is Tate's law from 1864 [Tat64]. It provides a simple method for measurement of surface tension because the mass of a drop can be determined by weighing the liquid accumulated by many drops. There are of course corrections to this simple expression, the main one being that only part of the liquid in a pendant drop actually detaches (see [YXB05]).

The mass of a drop can be determined from the balance of forces at $z = 0$,

$$M g_0 = 2\pi a \alpha \sin\chi - \pi a^2 \Delta p_0. \tag{5.49}$$

On the left we have the weight of the drop and on the right the total force due to surface tension minus the force due to the pressure excess Δp_0 in the liquid at $z = 0$. In Problem 5.12 it is shown that this is an exact relation, derivable from the differential equations. Notice that the inequality (5.47) corresponds to $\Delta p_0 > 0$, implying that the pressure should be above atmospheric at $z = 0$ for the inequality to hold. Tate's law implies that the drop detaches when $\chi = 90°$ and the pressure excess vanishes, $\Delta p_0 = 0$.

Numeric solution

The pressure excess in the liquid takes the same form as before,

$$\Delta p = \rho_0 g_0 (h - z), \tag{5.50}$$

where $z = h$ is the height at which the liquid pressure equals the atmospheric. The solution of the three coupled first-order differential equations, (5.37) through (5.39), proceeds as

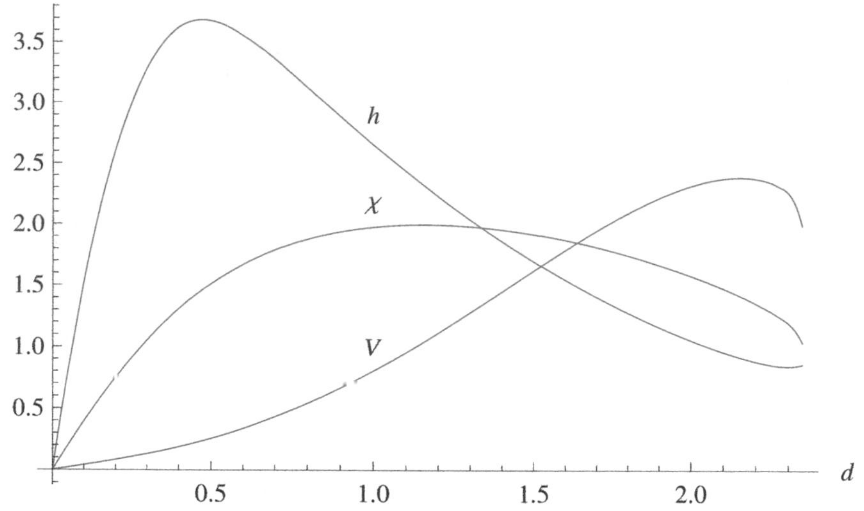

Figure 5.11. Drop parameters for the capillary with footprint radius $a = 0.5$ pictured in Figure 5.10. All quantities are plotted in units where $L_c = 1$. Notice that the volume $V = M/\rho_0$ has a maximum as a function of depth at $d = 2.154\ldots$, which marks the limit to the static shapes that can be created by slowly feeding the drop through the capillary.

previously with the boundary conditions

$$r = 0, \qquad z = -d, \qquad \theta = 0 \qquad (s = 0); \tag{5.51a}$$

$$r = a, \qquad z = 0, \qquad \theta = \chi \qquad (s = s_0). \tag{5.51b}$$

Since there is no constraint on the angle χ, one may specify the depth d and the radius a, leaving h, χ, and s_0 as unknowns.

In Figure 5.10 two sets of solutions are shown for a capillary tubes with radii $a = 0.5$ and $a = 1.5$ in units where $L_c = 1$. The solutions are parameterized by the depth d. The shapes with $d = 2.154\ldots$ and $d = 2.631\ldots$ are limiting cases, and although there are static solutions with larger values of d, they have a smaller volume $V = M/\rho_0$ and cannot be reached by feeding the drop through the tube. They must for this reason be unstable even if they are static. In Figure 5.11 the parameters for the $a = 0.5$ set of pendant drop shapes are shown as a function of depth.

At $z = 0$ we have $\Delta p = \rho_0 g_0 h$, implying that we must have $h > 0$ to obtain the naive inequality (5.47). This is in fact the case for all the shapes shown in Figure 5.10 for $a = 0.5$, whereas the largest shape for $a = 1.5$ has $h = -0.27$, such that part of the drop's weight is carried by the negative pressure excess $z = 0$. Negative pressure excess, which by the Young–Laplace law is the same as negative mean curvature, can only occur if the first principal radius of curvature R_1 becomes negative at $z = 0$, which it is also seen to be the case in the figure.

Problems

5.1 Show that the Bond number (5.7) can be written as the ratio of the weight of the bubble and the force due to surface tension acting on its equator.

5.2 The inside and outside air pressures as well as their difference $\Delta p_0 = p_{\text{inside}} - p_{\text{outside}}$ may be assumed to be constant across a soap bubble skin.

(a) Show that instead of (5.6) we get the coupled equations

$$\frac{d\alpha}{d\theta} = -\tfrac{1}{2}\rho_1\tau R g_0 \sin\theta, \qquad \Delta p_0 = \frac{4\alpha}{R} + \rho_1\tau g_0 \cos\theta.$$

(b) Show that the solutions are

$$\alpha = \frac{\alpha_0}{(1 - K\cos\theta)^2}, \qquad R = \frac{R_0}{(1 - K\cos\theta)^3},$$

where $K = \rho_1\tau g_0/\Delta p_0 = \mathsf{Bo}_0/4$. Evidently we must require $K < 1$ for the bubble not to break.

5.3 A column of incompressible fluid is subject to an infinitesimal radial fluctuation of the form $r = a - b\cos(2\pi z/\lambda)$ where $b \ll a$ is the magnitude of the perturbation and $\lambda \gg b$ is the wavelength. **(a)** Calculate the volume V of one wavelength of the column. **(b)** Calculate the area A of one wavelength of the column in the leading approximation in b/λ. **(c)** Show that the Rayleigh–Plateau stability condition is equivalent to the requirement that the area A should grow with b when the volume V is held fixed. Instability occurs when the area diminishes.

5.4 Determine the local radii of curvature for an axially symmetric surface (Section 5.7) by expanding the shape $z = f(r)$ with $r = \sqrt{x^2 + y^2}$ to second order around $x = x_0$, $y = 0$ and $z = z_0$.

5.5 Calculate the size of the spherical drops that arise when a liquid column breaks up (assume that the fastest breakup yields one drop for each wavelength).

5.6 Show that for a general planar curve, $y = y(x)$ or $x = x(y)$, the radius of curvature is

$$\frac{1}{R} = \frac{y''(x)}{\left[1 + y'(x)^2\right]^{\frac{3}{2}}} = -\frac{x''(y)}{\left[1 + x'(y)^2\right]^{\frac{3}{2}}}.$$

5.7 **(a)** Calculate the force per unit of length that the planar meniscus exerts on the vertical wall. **(b)** Do objects floating on the surface attract or repel each other?

5.8 Show that for a general cylindrical curve, $z = z(r)$ or $r = r(z)$, we have

$$\frac{1}{R_1} = \frac{z''(r)}{\left[1 + z'(r)^2\right]^{\frac{3}{2}}} = -\frac{r''(z)}{\left[1 + r'(z)^2\right]^{\frac{3}{2}}}, \qquad \frac{1}{R_2} = \frac{z'(r)}{r\sqrt{1 + z'(r)^2}} = \frac{1}{r(z)\sqrt{1 + r'(z)^2}}.$$

5.9 Show that the shape of the zero mean curvature soap film in Figure 5.3 is given by

$$r = a\cosh\frac{z}{a}$$

with suitable choice of a and origin of z. Hint: Use the result of Problem 5.8.

5.10 Show that Equation (5.45) can be derived from a simple model, just as in Section 5.6.

5.11 Assume that the pendant drop is spherical. Use Tate's law to obtain an approximate expression for its radius R when it detaches and falls.

5.12 **(a)** Show that the partial volume of a pendant droplet of arc-length s (with $s = 0$ at the bottom of the drop) is

$$V(s) = \left(2\pi r \sin\theta - \pi r^2 \frac{\Delta p}{\alpha}\right) L_c^2, \tag{5.52}$$

where $\Delta p = \rho_0 g_0 (h - z)$. Hint: Differentiate with respect to s to prove $dV = \pi r^2 dz$.
(b) Show that the above expression leads to Equation (5.49).

Part II

Solids at rest

Solids at rest

In the following chapters we only consider solids permanently at rest without any macroscopic local or global motion. The essential chapters for a minimal curriculum are 6, 7, and 8. Chapter 9 can be included in part or whole. Chapter 10 is a bit harder. Readers who want to focus on fluid mechanics alone need only Chapter 6 and a bit of Chapter 7.

List of chapters

6. **Stress:** Friction is discussed as an example of shear stress. The stress tensor field is formally introduced. The equations of mechanical equilibrium in all kinds of matter are formulated. The symmetry of the stress tensor is discussed. Although only applied to elastic solids, the formalism for stress developed here is valid for all kinds of matter, also fluids.

7. **Strain:** The displacement field is introduced and analyzed in the limit of infinitesimal strain gradients. The strain tensor is introduced and its geometrical meaning exposed. Work and energy are analyzed, and large deformation briefly discussed.

8. **Elasticity:** Hooke's law is discussed and generalized to linear homogeneous materials. Young's modulus and Poisson's ratio are introduced. Fundamental examples of static uniform deformation are discussed. Expressions for the elastic energy are obtained.

9. **Basic elastostatics:** The most important elementary applications of linear elasticity are analyzed: gravitational settling, bending and twisting beams, spheres and tubes under pressure.

10. **Slender rods:** Slender elastic rods are analyzed and presented with characteristic examples, such as cantilevers, bridges, and coiled springs.

11. **Computational elastostatics:** The basic theory behind numeric computation of linear elastostatic solutions is presented, and a typical example is analyzed and computed.

6

Stress

In fluids at rest, pressure is the only contact force. For solids at rest or in motion, and for viscous fluids in motion, this simple picture is no longer valid. Besides pressure-like forces acting along the normal to a contact surface, there may also be shearing forces acting in any tangential direction to the surface. The relevant local quantity analogous to pressure is the shear force per unit of area, called the *shear stress*. Friction, so well-known from everyday life, is always a result of shear stress acting in the contact areas between material bodies. The integrity of a solid body is largely secured by internal shear stresses.

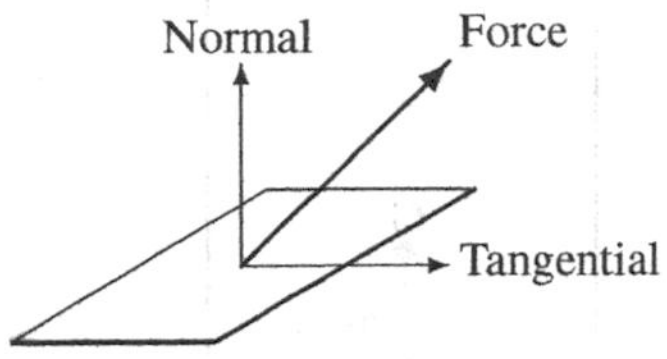

The force on a small piece of a surface can be resolved into a normal pressure force and a tangential shear force.

The two major classes of materials, fluids and solids react differently to external stresses. Whereas fluids respond by *flowing*, solids respond by *deforming*. Elastic deformation usually grows linearly with stress, although every material in the end becomes plastic and deforms permanently, and may eventually rupture. We shall as far as possible maintain a general view of the physics of continuous systems, applicable to all types of materials, also those that do not fall into the major classes. The number of artificial and even exotic materials is rapidly growing because of the needs for special material properties in technology.

In this chapter the emphasis is on the theoretical formalism for contact forces, independent of whether they occur in solids, fluids, or intermediate forms. The vector notation used up to this point is not adequate to the task, because contact forces not only depend on the spatial position but also on the orientation of the surface on which they act. A collection of nine stress components, called the *stress tensor*, is used to describe the full range of contact forces that may come into play. The concept of stress and its properties is a centerpiece of the continuum mechanics, and therefore of this book.

6.1 Friction

Friction between solid bodies is a shear force known to us all. We hardly think of friction forces, even though all day long we are served by them and do service to them. Friction is the reason that the objects we hold are not slippery as a piece of soap in the bathtub, but instead allow us to grab, drag, rub, and scrub. Most of the work we do is done against friction, from stirring the coffee to making fire by rubbing two pieces of wood against each other.

The law of static friction goes back to Amontons (1699) and the law of dynamic friction to Coulomb (1779). The full story of friction between solid bodies is complicated, and in spite of our everyday familiarity with friction, there is still no universally accepted microscopic explanation of the phenomenon [GM01, Kes01].

Guillaume Amontons (1663–1705). French physicist. Made improvements to the barometer, hygrometer, and thermometer.

Charles-Augustin de Coulomb (1736–1806). French physicist, best known from the law of electrostatics and the unit of electric charge that carries his name.

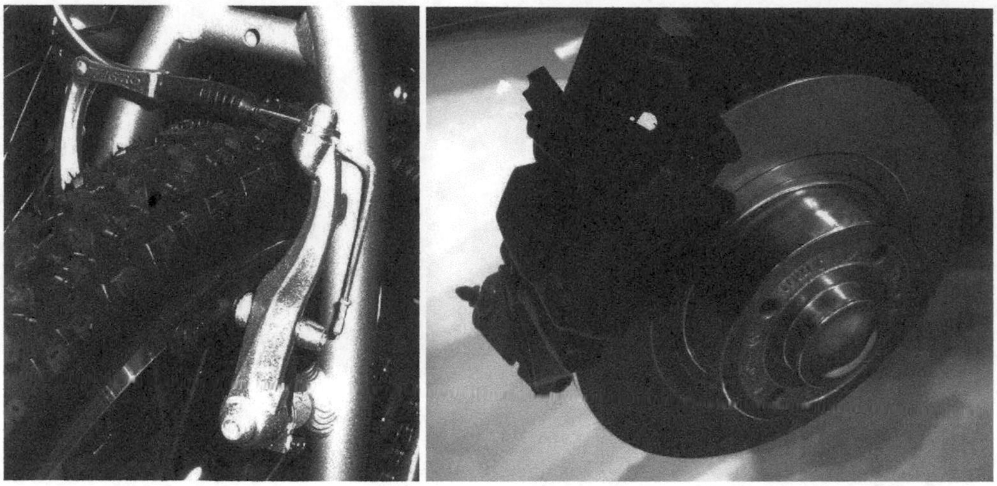

Figure 6.1. The construction of a friction brake depends on how much energy it has to dissipate. **Left:** In bicycles, the pads are pressed against the metallic rim of the wheel by suitably levered finger forces. (Source: Wikimedia Commons.) **Right:** In cars, disc brakes require hydraulics to press the pads against the metallic discs. (Source: Courtesy David Monniaux. Wikimedia Commons.)

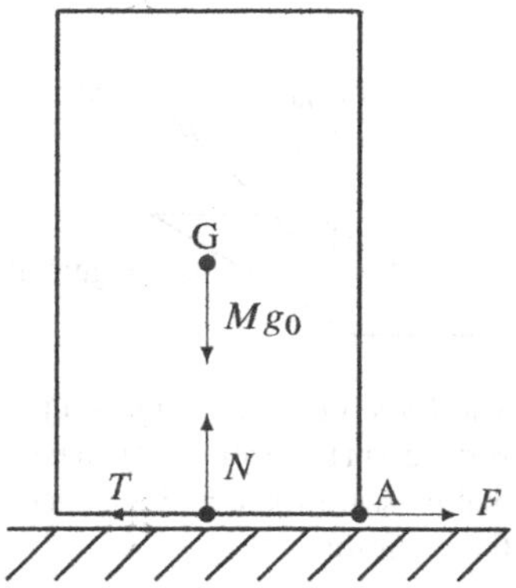

Balance of forces on a crate at rest on a horizontal floor. The point of attack A is here chosen at floor level to avoid creating a moment of force that could turn over the crate.

Static friction

Consider a heavy crate at rest, standing on a horizontal floor. The bottom of the crate acts on the floor with a force equal to its weight Mg_0, and the floor acts back on the bottom of the crate with an equal and opposite normal reaction force of magnitude $N = Mg_0$. If you try to drag the crate along the floor by applying a horizontal force, F, you may discover that the crate is so heavy that you are not able to budge it, implying that the force you apply must be fully balanced by an opposite tangential or shearing friction force (also called traction), $T = F$, between the floor and the crate.

Empirically, such *static friction* can take any magnitude up to a certain maximum, which is proportional to the normal load,

$$T < \mu_0 N. \tag{6.1}$$

The dimensionless constant of proportionality μ_0 is called the *coefficient of static friction.* In our daily doings it may take a quite sizable value, say 0.5 or greater, making dragging comparable to lifting. Its value depends on what materials are in contact and on the roughness of the contact surfaces (see the margin table on the following page).

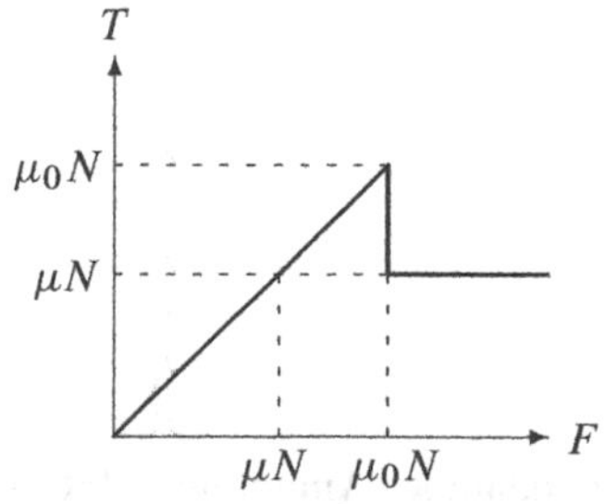

Sketch of tangential reaction (traction) T as a function of applied force F. Up to $F = \mu_0 N$, the traction adjusts itself to the applied force, $T = F$. At $F = \mu_0 N$, the tangential reaction drops abruptly to a lower value, $T = \mu N$, and stays there regardless of the applied force.

Dynamic friction

If you are able to pull with sufficient strength, the crate suddenly starts to move, but friction will still be present and you will have to do real work to move the crate any distance. Empirically, this *dynamic* (or sliding) friction is proportional to the normal load,

$$T = \mu N, \tag{6.2}$$

with a coefficient of dynamic friction, μ, that is always smaller than the corresponding coefficient of static friction, $\mu < \mu_0$. This is why you have to heave strongly to get the crate set into motion, whereas afterward a smaller force, $F = \mu N$, suffices to keep it going at constant speed. Notice that whereas static friction requires no work, the rate of work, $P = FU$, performed against dynamic friction grows linearly with velocity U, and this work is irretrievably lost to heat.

Static and dynamic friction are both independent of the size of the contact area, so that a crate on legs is as hard to drag as one without. The best way to diminish the force necessary to drag the crate is to place it on wheels.

	μ_0	μ
Glass/glass	0.9	0.4
Rubber/asphalt	0.9	0.7
Steel/steel	0.7	0.6
Metal/metal	0.6	0.4
Wood/wood	0.4	0.3
Steel/ice	0.1	0.05
Steel/teflon	0.05	0.05

Typical friction coefficients for various combinations of materials.

Why do cars have wheels?: The contact between a rolling wheel and the road surface is always static, although the actual contact area changes as the wheel rolls. Static friction performs no work, so friction losses are only found in the bearing of the wheels and other parts of the transmission (and some in the deformation of the tires). Being lubricated, these bearings are much less lossy than dynamic friction would be (see Chapter 14).

When a moving car brakes, energy is dissipated by sliding friction in the brake assembly (see Figure 6.1R) where the normal load is controlled by the force applied to the brake pedals. A larger weight generates proportionally larger static friction between wheel and road, so a car's minimal braking distance will in the first approximation be independent of how heavily it is loaded, and that is lucky for the flow of traffic. When braking a car, it is best to avoid blocking the wheels so that the car skids because dynamic (sliding) friction is smaller than static (rolling) friction (see Problem 6.2). More importantly, a skidding car has no steering control, so modern cars are equipped with anti-skid brake systems that automatically adjust brake pressure to avoid skidding.

6.2 Stress fields

The above arguments indicate that it is necessary to introduce one or more fields describing the distribution of normal and tangential forces acting on a surface. In analogy with pressure, such *stress fields* represent the local force per unit of area, whether it is normal or tangential, and are measured in the same units as pressure, namely the pascal, $\text{Pa} = \text{N m}^{-2}$. A stress that acts along the normal to a surface is called a *tension stress* whereas a stress that acts against the normal is called a *pressure stress*. A stress that acts along a tangential direction to a surface is called a *shear stress*.

External and internal stresses

The stresses acting in the interface between a body and its environment are *external*. As for pressure we shall also speak about *internal stresses* acting across any imagined surface in the body. Internal stresses abound in the macroscopic world around us. Whenever we come into contact with the environment (and when do we not?), internal stresses are set up in the materials we touch, as well as in our own bodies. The precise distribution of stress in a body depends not only on the external forces applied to the body but also on the types of material the body is made from and on other macroscopic quantities such as temperature.

There is no guarantee that the normal stress is the same for any choice of surface through a given point. The proof of Pascal's law (page 26)—that the pressure is the same in all directions of the surface normal—assumed that the material was a fluid at rest. For fluids in motion and for solids in general, this proof does not hold, but there is a generalized version of the theorem.

In the absence of external forces there is usually no internal stress in a material, although fast cooling may freeze stresses permanently into certain materials, for example glass, and provoke an almost explosive release of stored energy when triggered by a sudden impact.

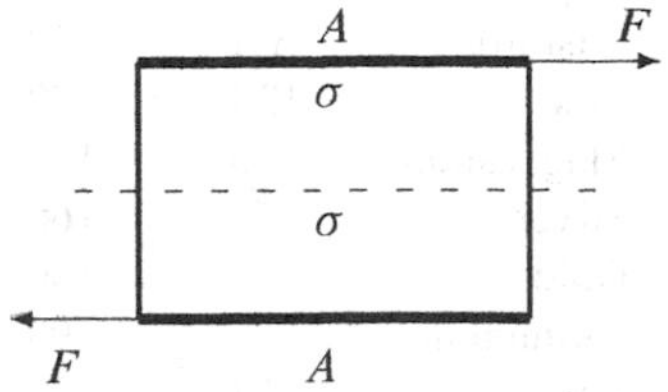

Clamped slab of homogeneous material. The shear force F at the upper clamp is balanced by an oppositely directed fixation force F on the lower clamp. The shear stress $\sigma = F/A$ is the same on all inner planes parallel with the clamps (dashed).

Estimating stress

In many situations it is quite straightforward to estimate average stresses in a body from a knowledge of the external forces that act on the body. Consider, for example, a slab of homogeneous solid material bounded by two stiff flat clamps of area A, firmly glued to it. A tangential force F applied to one clamp with the other held fixed will deform the slab a bit in the direction of the applied force, so that there is an average shear stress $\sigma = F/A$ acting on the upper surface of the slab in the direction of the force.

The fixed clamp will by Newton's Third Law act back on the slab with a force of the same magnitude but in the opposite direction. If we make an imaginary cut through the slab parallel with the clamps, then the upper part of the slab must likewise act like a clamp on the lower with the shear force F, so that the average internal shear stress acting in the cut must again be $\sigma = F/A$. Had a normal external force also been applied to the clamps, we would have gone through the same type of argument to convince ourselves that the average normal stress would be the same everywhere in the cut.

For bodies with a more complicated geometry and non-uniform external load, internal stresses are not so easily calculated, although their average magnitudes may often be estimated, especially for symmetric bodies. In analogy with friction one may assume that variations in shear and normal stresses are roughly of the same order of magnitude for unexceptional material, external load, and body geometry.

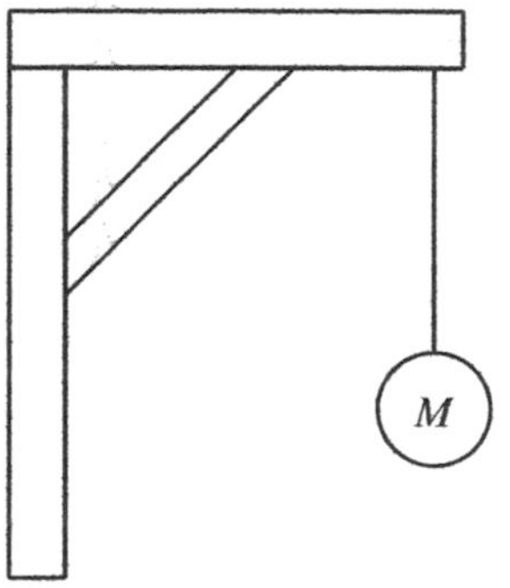

The classic gallows construction.

Example 6.1 [Classic gallows]: The classic gallows is constructed from a vertical pole, a horizontal beam, and a diagonal strut. A body of mass $M = 70$ kg hangs at the extreme end of the horizontal beam of cross-section $A = 100$ cm^2. The body's weight must be balanced by a shear stress in the beam of average magnitude $\sigma \approx Mg_0/A \approx 70\,000$ Pa ≈ 0.7 bar. The actual distribution of shear stress will generally vary over the cross-section of the beam and the position of the chosen cross-section, but its average magnitude should be of the estimated value. There will also be a pressure force that has to vary over the cross-section to balance the moment of force that the body exerts on the beam (see Section 9.3).

Example 6.2 [Water mains]: The half-inch water mains in your house have an inner pipe radius $a \approx 0.6$ cm. Tapping water at a high rate, the internal friction in the water (viscosity) creates shear stresses opposing the flow in the water and on the inner pipe wall. Assuming that the pressure drops by $\Delta p \approx 0.1$ bar $= 10^4$ Pa over a length of $L \approx 10$ m of the pipe, we may calculate the shear stress imposed on the water by the inner surface of the pipe, because the pressure drop between the ends of the pipe is the only other external force acting on the water. Setting the force due to the pressure difference equal to the total shear force on the inner surface, we get

$$\pi a^2 \Delta p = 2\pi a L \sigma, \tag{6.3}$$

from which it follows that $\sigma = \Delta p\, a/2L \approx 3$ Pa. This stress is of the same magnitude as the pressure drop $\Delta p \times 2a/L$ over a stretch of pipe of length equal to the diameter.

Tensile strength

	Yield	Tensile
Tungsten		3450
Titanium	275	420
Steel	250	415
Nickel	58	317
Copper	70	220
Cast iron	*NA*	172
Zinc	124	172
Magnesium	90	172
Silver	55	160
Gold		130
Aluminium	35	90
Tin	12	22
Lead		17
Unit	MPa	MPa

Typical yield stress and tensile strength for common metals. The values may vary widely for different specimens, depending on purity, heat treatment, and other factors.

When external forces grow large, a solid body will deform strongly, become plastic, or even fracture and break apart. The maximal tension a material can sustain without failure is called the *tensile strength* of the material. Similarly, the *yield stress* is defined as the stress beyond which otherwise elastic solids begin to undergo plastic deformation from which the material does not automatically recover when released. Such materials are also called *ductile*. Copper is an example of a ductile metal that can be drawn or stretched without breaking. Brittle materials, such as concrete or cast iron, do not have a yield point but fail suddenly when the tension reaches a certain strength.

For metals the tensile strength typically lies in the region of several hundred megapascals (i.e., several kilobars) as shown in the margin table. Modern composite carbon fibers can have tensile strengths up to several gigapascals, whereas "ropes" made from single-wall carbon nanotubes are reported to have tensile strengths of up to 50 gigapascals or even larger [YFAR00].

Example 6.3 [Steel rod]: Plain steel has a tensile strength of 415 MPa. A quick estimate shows that a steel rod with a diameter of 2 cm breaks if loaded with more than 13,000 kg. Adopting a safety factor of 10, one should not load it with more than 1,300 kg, which also brings it well below the yield point for steel.

6.3 The nine components of stress

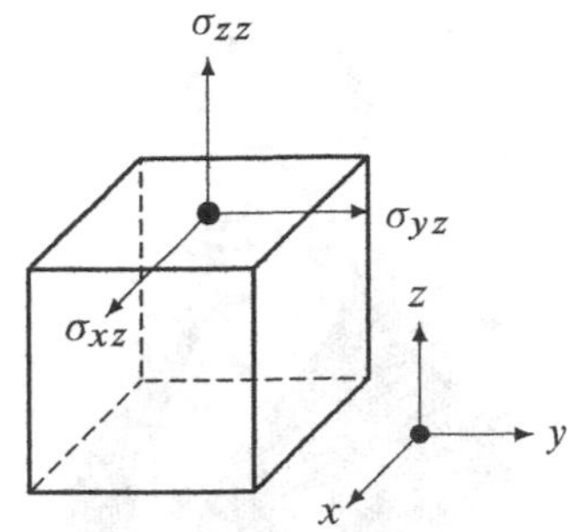

Components of stress acting on a surface element dS_z in the xy-plane.

Shear forces are more complicated than normal forces, because there is an infinity of tangential directions on a surface. In a coordinate system where a force $d\mathcal{F}_x$ is applied along the x-direction to a material surface element dS_z with its normal in the z-direction, the shear stress will be denoted $\sigma_{xz} = d\mathcal{F}_x/dS_z$, instead of just σ. Similarly, if the shear force is applied in the y-direction, the stress would be denoted $\sigma_{yz} = d\mathcal{F}_y/dS_z$; and if a normal force is applied along the z-direction, the normal stress is consistently denoted $\sigma_{zz} = d\mathcal{F}_z/dS_z$. The sign is chosen such that a positive value of σ_{zz} corresponds to a pull or *tension*.

Cauchy's stress hypothesis

Continuing this argument for surface elements with normal along x or y, it appears to be necessary to use at least nine numbers to indicate the state of stress in a given point of a material in a Cartesian coordinate system. *Cauchy's stress hypothesis* (to be proved below) asserts that these nine numbers are in fact all that is needed to determine the force $d\mathcal{F} = (d\mathcal{F}_x, d\mathcal{F}_y, d\mathcal{F}_z)$ on an arbitrary surface element, $dS = (dS_x, dS_y, dS_z)$, according to the rule

$$\begin{aligned} d\mathcal{F}_x &= \sigma_{xx}dS_x + \sigma_{xy}dS_y + \sigma_{xz}dS_z, \\ d\mathcal{F}_y &= \sigma_{yx}dS_x + \sigma_{yy}dS_y + \sigma_{yz}dS_z, \\ d\mathcal{F}_z &= \sigma_{zx}dS_x + \sigma_{zy}dS_y + \sigma_{zz}dS_z. \end{aligned} \tag{6.4}$$

This may also be written more compactly with index notation,

$$d\mathcal{F}_i = \sum_j \sigma_{ij} dS_j, \tag{6.5}$$

where the indices range over the coordinate labels, here x, y, and z. Each coefficient $\sigma_{ij} = \sigma_{ij}(\boldsymbol{x}, t)$ depends on the position (and time), and is thus a field in the normal sense of the word.

∗ Proof of Cauchy's stress hypothesis: As in the proof of Pascal's law (page 26) we take again a surface element in the shape of a tiny triangle with area vector $dS = (dS_x, dS_y, dS_z)$. This triangle and its projections on the coordinate planes form together a little body in the shape of a right tetrahedron. Let the external (vector) forces acting on the three triangular faces of the tetrahedron be denoted, respectively, $d\mathcal{F}_x$, $d\mathcal{F}_y$, and $d\mathcal{F}_z$. Adding the external force $d\mathcal{F}$ acting on the skew face and a possible volume force $\boldsymbol{f}\,dV$, Newton's Second Law for the small tetrahedron becomes

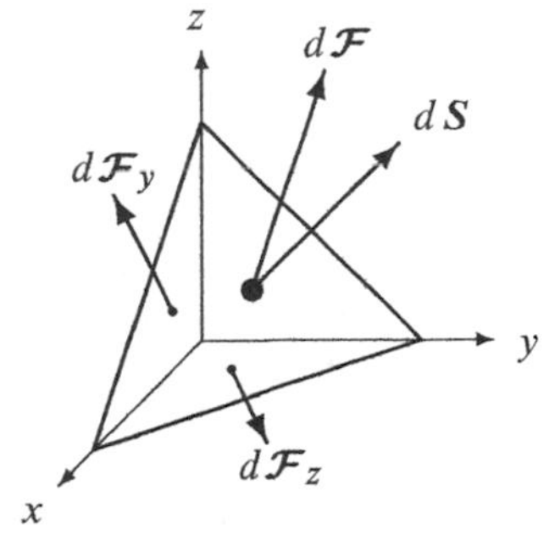

The tiny triangle and its projections form a tetrahedron. The (vector) force acting on the triangle in the xy-plane is called $d\mathcal{F}_z$, and the force acting on the triangle in the zx-plane is $d\mathcal{F}_y$. The force $d\mathcal{F}_x$ acting on the triangle in the yz-plane is hidden from view. The force $d\mathcal{F}$ acts on the skew triangle.

$$\boldsymbol{w}\,dM = \boldsymbol{f}\,dV + d\boldsymbol{\mathcal{F}}_x + d\boldsymbol{\mathcal{F}}_y + d\boldsymbol{\mathcal{F}}_z + d\boldsymbol{\mathcal{F}}, \tag{6.6}$$

where $\boldsymbol{w}$ is the center-of-mass acceleration of the tetrahedron and $dM = \rho\,dV$ its mass.

The volume of the tetrahedron scales like the third power of its linear size, whereas the surface area only scales like the second power. Making the tetrahedron progressively smaller, the body force term and the acceleration term will vanish faster than the surface terms. In the limit of a truly infinitesimal tetrahedron, only the surface terms survive, so that we must have $d\boldsymbol{\mathcal{F}}_x + d\boldsymbol{\mathcal{F}}_y + d\boldsymbol{\mathcal{F}}_z + d\boldsymbol{\mathcal{F}} = \boldsymbol{0}$. Taking into account that the area projections dS_x, dS_y, and dS_z point *into* the tetrahedron, we define the stress vectors along the coordinate directions as $\boldsymbol{\sigma}_x = -d\boldsymbol{\mathcal{F}}_x/dS_x$, $\boldsymbol{\sigma}_y = -d\boldsymbol{\mathcal{F}}_y/dS_y$, and $\boldsymbol{\sigma}_z = -d\boldsymbol{\mathcal{F}}_z/dS_z$. Consequently,

$$d\boldsymbol{\mathcal{F}} = \boldsymbol{\sigma}_x dS_x + \boldsymbol{\sigma}_y dS_y + \boldsymbol{\sigma}_z dS_z. \tag{6.7}$$

This shows that the force on an arbitrary surface element may be written as a linear combination of three basic stress vectors along the coordinate axes. Introducing the nine coordinates of the three stress vectors, $\boldsymbol{\sigma}_x = (\sigma_{xx}, \sigma_{yx}, \sigma_{zx})$, $\boldsymbol{\sigma}_y = (\sigma_{xy}, \sigma_{yy}, \sigma_{zy})$, and $\boldsymbol{\sigma}_z = (\sigma_{xz}, \sigma_{yz}, \sigma_{zz})$, we arrive at Cauchy's hypothesis (6.4).

The stress tensor

Augustin-Louis, Baron Cauchy (1789–1857). French mathematician who produced an astounding 789 papers. Contributed to the foundations of elasticity, hydrodynamics, partial differential equations, number theory, and complex functions. (Source: Wikimedia Commons.)

Together the nine fields, $\{\sigma_{ij}\}$, make up a single geometric object, called the *stress tensor field*[1], first introduced by Cauchy in 1822. It must be emphasized that this collection of nine fields cannot be viewed geometrically as consisting of either nine scalar or three vector fields, but must be considered together as one geometric object, a *tensor field* $\{\sigma_{ij}(\boldsymbol{x},t)\}$. There is unfortunately no simple, intuitive way of visualizing the most general stress tensor field. Vector and tensor calculus is developed in some detail in Appendix B.

It is often convenient to write the stress tensor as a matrix,

$$\boldsymbol{\sigma} = \{\sigma_{ij}\} = \begin{pmatrix} \sigma_{xx} & \sigma_{xy} & \sigma_{xz} \\ \sigma_{yx} & \sigma_{yy} & \sigma_{yz} \\ \sigma_{zx} & \sigma_{zy} & \sigma_{zz} \end{pmatrix}, \tag{6.8}$$

so that the force may be written compactly as a matrix equation,

$$\boxed{d\boldsymbol{\mathcal{F}} = \boldsymbol{\sigma} \cdot d\boldsymbol{S}.} \tag{6.9}$$

Notice how a dot is used to denote the product of a matrix and a vector. This is consistent with the rule for dot-products of vectors.

Writing $d\boldsymbol{S} = \hat{\boldsymbol{n}}\, dS$ where $\hat{\boldsymbol{n}}$ is the normal to the surface, the force per unit of area becomes $d\boldsymbol{\mathcal{F}}/dS = \boldsymbol{\sigma} \cdot \hat{\boldsymbol{n}}$, often called the *stress vector*. Although it is a vector, it is *not a vector field* in the strict sense of the word, because it also depends on the normal to the surface on which it acts. The following example demonstrates this point.

Example 6.4 [A tensor field]: A tensor field of the form

$$\{\sigma_{ij}\} = \{x_i x_j\} = \begin{pmatrix} x^2 & xy & xz \\ yx & y^2 & yz \\ zx & zy & z^2 \end{pmatrix} \tag{6.10}$$

is a tensor product (see Equation (B.8) on page 597) and thus by construction a true tensor. The "stress vector" acting on a surface with normal in the direction of the x-axis becomes

$$\boldsymbol{\sigma}_x = \boldsymbol{\sigma} \cdot \hat{\boldsymbol{e}}_x = \begin{pmatrix} x \\ y \\ z \end{pmatrix} x. \tag{6.11}$$

It is not a true vector because of the factor x on the right-hand side.

Total force

Including a true body force, $d\boldsymbol{\mathcal{F}} = \boldsymbol{f}\, dV$, for example gravity $\boldsymbol{f} = \rho \boldsymbol{g}$, the total force on a body of volume V with surface S becomes

$$\boxed{\boldsymbol{\mathcal{F}} = \int_V \boldsymbol{f}\, dV + \oint_S \boldsymbol{\sigma} \cdot d\boldsymbol{S},} \tag{6.12}$$

which in index notation takes the form

$$\mathcal{F}_i = \int_V f_i\, dV + \oint_S \sum_j \sigma_{ij}\, dS_j. \tag{6.13}$$

The circle through the surface integral is only there to remind us that the surface is closed.

[1]Some texts use the symbols $\{\tau_{ij}\}$ or $\{T_{ij}\}$ for the stress tensor field.

The surface integral can be converted into a volume integral by means of Gauss' theorem, Equation (2.13) on page 25, applied to each component of the stress tensor,

$$\mathcal{F}_i = \int_V f_i \, dV + \int_V \sum_j \nabla_j \sigma_{ij} \, dV \equiv \int_V f_i^* \, dV. \tag{6.14}$$

In the last step we have—in complete analogy with the hydrostatic definition of the effective force density (2.21)—introduced the *effective force density*,

$$\boxed{f_i^* = f_i + \sum_j \nabla_j \sigma_{ij}.} \tag{6.15}$$

In matrix notation the second term may be written compactly as a tensor divergence,

$$\boldsymbol{f}^* = \boldsymbol{f} + \nabla \cdot \boldsymbol{\sigma}^\top, \tag{6.16}$$

where $\boldsymbol{\sigma}^\top$ is the *transposed* matrix, defined as $\sigma_{ji}^\top = \sigma_{ij}$.

The effective force density is not just a formal quantity because the total force on a material particle is indeed $d\boldsymbol{\mathcal{F}} = \boldsymbol{f}^* \, dV$ (see the margin figure). That we can write the total force as a sum over contributions of this kind confirms that matter may be viewed as a collection of material particles. It must as before be emphasized that the effective force density $\boldsymbol{f}^*$ is not a long-range volume force, but a local expression that for a tiny material particle equals the sum of the true long-range force (say gravity) and all the short-range contact forces acting on its surface.

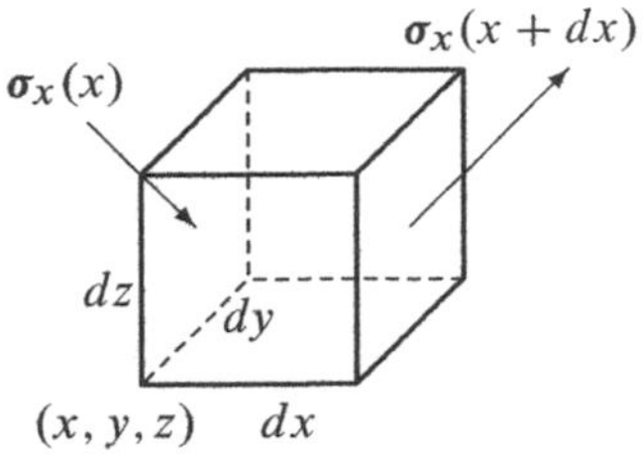

The total contact force on a small box-shaped material particle is calculated from the differences of the stress vectors acting on the sides. Suppressing the dependence on y and z, the resultant contact force on dS_x becomes $d\boldsymbol{\mathcal{F}} = (\boldsymbol{\sigma}_x(x + dx) - \boldsymbol{\sigma}_x(x)) dS_x \approx \nabla_x \boldsymbol{\sigma}_x \, dV$. Adding the similar contributions from dS_y and dS_z, one arrives at the total surface force, $d\boldsymbol{\mathcal{F}} = (\nabla_x \boldsymbol{\sigma}_x + \nabla_y \boldsymbol{\sigma}_y + \nabla_z \boldsymbol{\sigma}_z) \, dV = \nabla \cdot \boldsymbol{\sigma}^\top \, dV$.

Mechanical pressure

In hydrostatic equilibrium, where the only contact force is pressure and the force element is $d\boldsymbol{\mathcal{F}} = -p \, d\boldsymbol{S}$, a comparison with (6.9) shows that the stress tensor must be

$$\boldsymbol{\sigma} = -p \, \mathbf{1} \qquad \text{or equivalently} \qquad \sigma_{ij} = -p \, \delta_{ij}, \tag{6.17}$$

$$\mathbf{1} = \{\delta_{ij}\} = \begin{pmatrix} 1 & 0 & 0 \\ 0 & 1 & 0 \\ 0 & 0 & 1 \end{pmatrix}$$

The unit matrix (or tensor).

where $\mathbf{1}$ is the (3×3) unit matrix. In index notation the unit matrix is represented by the Kronecker delta tensor, δ_{ij}, which equals 1 for $i = j$ and 0 otherwise (see Appendix B on page 595).

Generally, however, the stress tensor will also have off-diagonal non-vanishing components, and its diagonal components will be different from each other. A diagonal component behaves like a (negative) pressure, and often the negative diagonal components are called the "pressures" along the coordinate directions,

$$p_x = -\sigma_{xx}, \qquad p_y = -\sigma_{yy}, \qquad p_z = -\sigma_{zz}. \tag{6.18}$$

Here it should be remembered that the triplet (p_x, p_y, p_z) of diagonal elements of the stress tensor does not behave like a Cartesian vector (see Section B.6). It is in fact not a well-defined geometric object at all.

There is no unique way of defining *the* pressure in general materials, so wherever pressure is used, it must be accompanied by a suitable definition. Sometimes the pressure will be identified with the *mechanical pressure*, defined as the average of the three pressures along the axes,

$$p = \tfrac{1}{3}\left(p_x + p_y + p_z\right) = -\tfrac{1}{3}\left(\sigma_{xx} + \sigma_{yy} + \sigma_{zz}\right). \tag{6.19}$$

This makes sense because the sum over the diagonal elements of a matrix, the trace $\operatorname{Tr} \boldsymbol{\sigma} \equiv \sum_i \sigma_{ii} = \sigma_{xx} + \sigma_{yy} + \sigma_{zz}$, is invariant under Cartesian coordinate transformations (see Section B.5). Defining pressure in this way ensures that it is a scalar field, taking the same value in all coordinate systems. There is in fact no other scalar linear combination of the nine stress components with that property.

Example 6.5: For the stress tensor given in Example 6.4, the pressures along the coordinate axes become $p_x = -x^2$, $p_y = -y^2$, and $p_z = -z^2$. Evidently, they do not form a vector, but the mechanical pressure,

$$p = -\tfrac{1}{3}\left(x^2 + y^2 + z^2\right), \tag{6.20}$$

is clearly a scalar, invariant under rotations of the Cartesian coordinate system.

6.4 Mechanical equilibrium

In mechanical equilibrium, the total force on any body must vanish, for if it does not, the body must begin to move. Accordingly, the general condition for mechanical equilibrium is that $\boldsymbol{\mathcal{F}} = \mathbf{0}$ for all volumes V. Applied to material particles, it implies that the force on each and every material particle must vanish, $\boldsymbol{f}^* = \mathbf{0}$. We thus arrive at *Cauchy's equilibrium equation*,

$$\boldsymbol{f} + \nabla \cdot \boldsymbol{\sigma}^\top = \mathbf{0}, \tag{6.21}$$

which in tensor notation becomes

$$\boxed{f_i + \sum_j \nabla_j \sigma_{ij} = 0,} \tag{6.22}$$

In spite of their simplicity *these partial differential equations govern mechanical equilibrium in all kinds of continuous matter*, be it solid, fluid, or whatever. For a fluid at rest with pressure being the only stress component, we have $\boldsymbol{\sigma} = -p\mathbf{1}$, and recover with $\boldsymbol{f} = \rho \boldsymbol{g}$ the equation of hydrostatic equilibrium, $\rho\,\boldsymbol{g} - \nabla p = \mathbf{0}$.

It is instructive to write out the three coupled partial differential equations in full:

$$\begin{aligned} f_x + \nabla_x \sigma_{xx} + \nabla_y \sigma_{xy} + \nabla_z \sigma_{xz} &= 0, \\ f_y + \nabla_x \sigma_{yx} + \nabla_y \sigma_{yy} + \nabla_z \sigma_{yz} &= 0, \\ f_z + \nabla_x \sigma_{zx} + \nabla_y \sigma_{zy} + \nabla_z \sigma_{zz} &= 0. \end{aligned} \tag{6.23}$$

These three equations are in themselves not sufficient to determine the stress distribution in continuous matter, but must be supplemented by *constitutive equations* connecting the stress tensor with the variables describing the state of the matter. For fluids at rest, the equation of state serves this purpose by relating pressure to mass density (and temperature), as we saw in Chapter 2. In solids at rest the constitutive equations are more complicated and relate stress to strain, a tensor that describes the state of deformation (Chapter 7).

Symmetry

There is one additional condition (also going back to Cauchy) that is normally imposed on the stress tensor, namely that it must be *symmetric*,

$$\boxed{\sigma_{ij} = \sigma_{ji}, \qquad \text{or} \qquad \boldsymbol{\sigma}^\top = \boldsymbol{\sigma}.} \tag{6.24}$$

Symmetry only affects the shear stress components,

$$\sigma_{xy} = \sigma_{yx}, \qquad \sigma_{yz} = \sigma_{zy}, \qquad \sigma_{zx} = \sigma_{xz}, \tag{6.25}$$

and thereby reduces the number of independent stress components from nine to six. Here we shall now present a simple, though not entirely correct, argument for symmetry. In the following section we analyze the properties of asymmetric stress tensors and the deeper reasons for imposing symmetry.

Consider a material particle in the shape of a tiny rectangular box with sides a, b, and c. The shear force acting in the y-direction on a face in the x-plane is $\sigma_{yx}bc$ whereas the shear force acting in the x-direction on a face in the y-plane is $\sigma_{xy}ac$. On opposite faces the contact forces have opposite sign in mechanical equilibrium (their difference is as we have seen of order abc). Since the total force vanishes, the total moment of force on the box may be calculated around any point we wish. Using the lower left corner, we get (since the moments of the diagonal elements of the stress tensor cancel automatically)

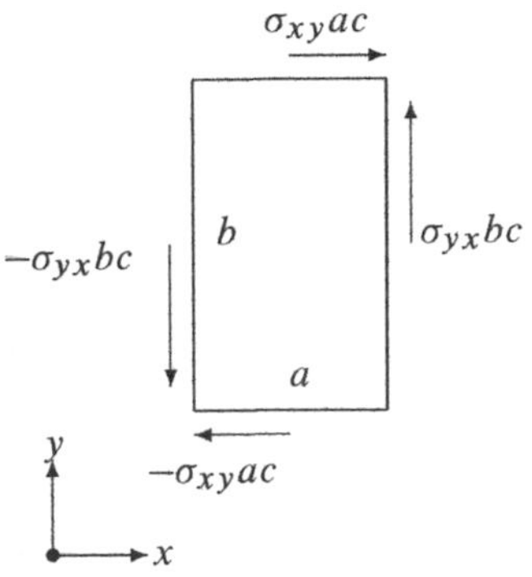

An asymmetric stress tensor will produce a non-vanishing moment of force on a small rectangular box (the z-direction not shown).

$$\mathcal{M}_z = a\,\sigma_{yx}bc - b\,\sigma_{xy}ac = (\sigma_{yx} - \sigma_{xy})\,abc.$$

This shows that if the stress tensor is asymmetric, $\sigma_{xy} \neq \sigma_{yx}$, there will be a resultant moment on the box. In mechanical equilibrium this cannot be allowed, since such a moment would begin to rotate the box, and consequently the stress tensor must be symmetric. Conversely, when the stress tensor is symmetric, mechanical equilibrium of the forces alone guarantees that all local moments of force will also vanish.

Being a symmetric matrix, the stress tensor may be diagonalized. The eigenvectors of a symmetric stress tensor define the principal directions of stress and the eigenvalues define the principal tensions or stresses. In the *principal basis* there are no off-diagonal elements, that is, no shear stresses, only pressure-like stresses. The principal basis is generally different from point to point in space.

Example 6.6: The stress tensor in Example 6.4 has one radial eigenvector $\hat{\boldsymbol{e}}_r = \boldsymbol{x}/r$ with eigenvalue $r^2 = x^2 + y^2 + z^2$ and two degenerate eigenvectors with eigenvalue 0 orthogonal to $\hat{\boldsymbol{e}}_r$, for example, the spherical unit vectors $\hat{\boldsymbol{e}}_\theta$ and $\hat{\boldsymbol{e}}_\phi$.

Boundary conditions

Cauchy's equation of mechanical equilibrium (6.22) constitutes a set of coupled partial differential equations valid inside a body's volume, and such equations require boundary conditions to be imposed on the surface of the body. The stress tensor is a local physical quantity, or rather collection of quantities, and may, like pressure in hydrostatics, be assumed to be continuous in regions where material properties change continuously. Across real boundaries, surfaces or interfaces, where material properties may change abruptly, Newton's Third Law demands (in the absence of surface tension) that the two sides act on each other with equal and opposite forces. Since the surface elements of the two sides are opposite to each other, it follows that the stress vector, $\boldsymbol{\sigma} \cdot \hat{\boldsymbol{n}} = \{\sum_j \sigma_{ij} n_j\}$, must be continuous across a surface with normal $\hat{\boldsymbol{n}}$ (in the absence of surface tension). This is best expressed by the vanishing of the surface discontinuity of the stress vector,

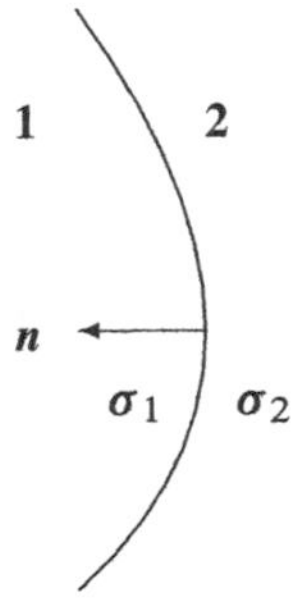

Contact surface separating body 1 from body 2. Newton's Third Law requires continuity of the stress vector $\boldsymbol{\sigma} \cdot \boldsymbol{n}$ across the boundary, that is, $\boldsymbol{\sigma}_1 \cdot \boldsymbol{n} = \boldsymbol{\sigma}_2 \cdot \boldsymbol{n}$, or $\Delta\boldsymbol{\sigma} \cdot \boldsymbol{n} = \boldsymbol{0}$.

$$\boxed{[\boldsymbol{\sigma} \cdot \hat{\boldsymbol{n}}] = \boldsymbol{0},} \tag{6.26}$$

where the bracket $[\cdot]$ indicates the difference between the two sides. This does *not* mean that all the components of the stress tensor should be continuous. Since it is a vector condition, it imposes continuity on three linear combinations of stress components, but leaves three other combinations free to jump discontinuously (for the symmetric stress tensor). Surprisingly it does not follow that the mechanical pressure (6.19) is continuous. In full-fledged continuum mechanics, the mechanical pressure loses the appealing intuitive meaning it acquired in hydrostatics.

Example 6.7 [What may jump and what may not]: Consider a plane interface in the yz-plane with normal along x. The stress components σ_{xx}, σ_{yx}, and σ_{zx} must then be continuous because they specify the stress vector on such a surface. Symmetry implies that σ_{xy} and σ_{xz} are likewise continuous. The remaining three independent components σ_{yy}, σ_{zz}, and $\sigma_{yz} = \sigma_{zy}$ are allowed to jump at the interface. In particular, the mechanical pressure, $p = -(\sigma_{xx} + \sigma_{yy} + \sigma_{zz})/3$, may (and will usually) be discontinuous.

* 6.5 Asymmetric stress tensors

The stress tensor was introduced at the beginning of this chapter as a quantity that furnished a complete description of the contact forces that may act on any surface element. There is nothing in the definition of the stress tensor that requires it to be symmetric, so we cannot exclude the possibility that it could be asymmetric. In normal elastic materials where stress only depends on deformation, it will automatically be symmetric, as we shall see in Chapter 8.

Total moment of force

The basic principle of continuum physics is that matter may be viewed as a collection of material particles obeying Newton's laws. We have seen that the total force (6.14) is the integral over individual forces $d\mathcal{F} = \boldsymbol{f}^* \, dV$ acting on each and every material particle, so in keeping with this principle, the total moment of force on a body with volume V must be

$$\mathcal{M} = \int_V \boldsymbol{x} \times d\mathcal{F} = \int_V \boldsymbol{x} \times \boldsymbol{f}^* \, dV. \tag{6.27}$$

In Chapter 20 we shall see that this moment is indeed the time derivative of the total angular momentum of the body, as it ought to be. Since mechanical equilibrium requires $\boldsymbol{f}^* = \boldsymbol{0}$ everywhere, it follows that the total moment of force vanishes automatically in mechanical equilibrium, whether the stress tensor is symmetric or not.

How can it then be that we (on the preceding page) were able to demonstrate the symmetry of the stress tensor by demanding that the moment of force on a material particle should vanish? To find the answer we first write the total moment in index notation,

$$\mathcal{M} = \sum_i \hat{\boldsymbol{e}}_i \mathcal{M}_i = \int_V \sum_{ijk} \epsilon_{ijk} \hat{\boldsymbol{e}}_i x_j \left(f_k + \sum_l \nabla_l \sigma_{kl} \right) dV, \tag{6.28}$$

where ϵ_{ijk} is the totally antisymmetric Levi-Civita tensor, defined in Equation (B.25) on page 600. Rewriting the stress term by means of the trivial identity,

$$x_j \sum_l \nabla_l \sigma_{kl} = \sum_l \nabla_l (x_j \sigma_{kl}) - \sigma_{kj}, \tag{6.29}$$

and applying Gauss theorem to the divergence, the total moment can be rewritten as the sum of the moment of external body and surface forces, plus an extra term,

$$\mathcal{M} = \int_V \boldsymbol{x} \times \boldsymbol{f} \, dV + \oint_S \boldsymbol{x} \times \boldsymbol{\sigma} \cdot d\boldsymbol{S} + \int_V \sum_{ijk} \hat{\boldsymbol{e}}_i \epsilon_{ijk} \sigma_{jk} \, dV. \tag{6.30}$$

The last contribution to the total moment may be viewed as a kind of internal moment associated with the material of the body. It vanishes for all volumes if and only if the stress tensor is symmetric.

In our discussion of symmetry above, we only calculated the moment of the external surface stresses acting on a material particle, that is, the second term in (6.30). But the external moment does not have to vanish in equilibrium if the stress tensor is asymmetric, only the sum of the external and internal moments must vanish. So the argument fails in general, and is vacuous if the stress tensor is already symmetric. We are, however, left with an unanswered question about the nature of the extra internal moment distribution due to the asymmetry.

Non-classical continuum theories

The stress tensor allows us to calculate the force on any open surface in a body. But open surfaces do not define physical bodies; only closed surfaces do. This gives rise to an ambiguity in the stress tensor that makes it possible to convert any asymmetric stress tensor into a symmetric tensor without changing the forces on physical bodies (see [MPP72, Appendix A] and [Landau and Lifshitz 1986, p. 7]).

Even if it is thus formally possible to construct an equivalent symmetric stress tensor from an asymmetric one, the relation between them is, however, non-local. This is an embarrassment because such non-locality is in disagreement with the basic principle in continuum physics that forces and moments should be calculable from the local state of matter, that is, from the properties of the material particles.

Manifestly asymmetric stress tensors have been used in various generalizations of classical continuum theory, generally called *Cosserat theory*. In such theories a material particle is endowed with geometric properties in addition to its position, for example angles describing its spatial orientation. These theories have turned out to be useful for the continuum treatment of modern complex materials that the classical theory is incapable of handling. We shall not go further into these extensions of classical continuum physics; see, for example, [Narasimhan 1993, p. 493] for a short review.

Eugène Maurice Pierre Cosserat (1866–1931). French mathematician. Worked in mathematics and astronomy, and created in 1909 together with his brother, an engineer, the now-famous extension to the classical continuum theory of elasticity.

Problems

6.1 A crate is dragged over a horizontal floor with sliding friction coefficient μ. Determine the angle α with the vertical of the total reaction force.

6.2 A car of mass m moves with speed v. **(a)** Estimate the minimal braking distance without skidding and the corresponding braking time. **(b)** Do the same if it skids from the beginning to the end. For numerics, use $v = 100\ \text{km}\,\text{h}^{-1}$. The static coefficient of friction between rubber and the surface of a road may be taken to be $\mu_0 = 0.9$, whereas for sliding friction it is $\mu = 0.7$. **(c)** Why is it good for traffic that the braking distance is independent of the mass?

6.3 A strong man pulls a jumbo airplane slowly but steadily, exerting a force of $\mathcal{F} = 2000$ N on a rope. The plane has $N = 32$ wheels, each touching the ground in a square area $A = 40 \times 40\ \text{cm}^2$. **(a)** Estimate the shear stress between the rubber and the tarmac. **(b)** Estimate the shear stress between the tarmac and his feet, each with area $A = 5 \times 25\ \text{cm}^2$.

6.4 **(a)** Estimate the maximal height h of a mountain made from rock with density about 3,000 kg m^{-3} when the maximal stress the material can tolerate before it deforms permanently is 300 MPa. **(b)** How high could it be on Mars where the surface gravity is 3.7 m s^{-2}? **(c)** Are the results reasonable?

6.5 A certain stress tensor has all components equal, that is, $\sigma_{ij} = \tau$ for all i, j. Find its eigenvalues and eigenvectors.

6.6 Show that if the stress tensor is diagonal in all coordinate systems, then it can only contain pressure.

6.7 A stress tensor and a rotation matrix are given by

$$\sigma = \begin{pmatrix} 15 & -10 & 0 \\ -10 & 5 & 0 \\ 0 & 0 & 20 \end{pmatrix}, \qquad \mathsf{A} = \begin{pmatrix} 3/5 & 0 & -4/5 \\ 0 & 1 & 0 \\ 4/5 & 0 & 3/5 \end{pmatrix}.$$

Calculate the stress tensor in the rotated coordinate system $\boldsymbol{x}' = \mathsf{A} \cdot \boldsymbol{x}$.

6.8 **(a)** Show that the average of a unit vector $\boldsymbol{n}$ over all directions obeys

$$\langle n_i n_j \rangle = \frac{1}{3}\delta_{ij}.$$

(b) Use this to show that the average of the normal stress acting on an arbitrary surface element equals (minus) the mechanical pressure (6.19).

* **6.9 [Stick-slip mechanism]** A body of mass m stands still on a horizontal floor. The coefficients of static and kinetic friction between body and floor are μ_0 and μ. An elastic string with string constant k is attached to the body in a point close to the floor. The string can only exert a force on the body when it is stretched beyond its relaxed length. When the free end of the string is pulled horizontally with constant velocity v, intuition tells us that the body will have a tendency to move in fits and starts.

(a) Calculate the amount s that the string is stretched, just before the body begins to move.
(b) Write the equation of motion for the body when it is just set into motion, for example in terms of the distance x that the point of attachment of the string has moved and the time t elapsed since the motion began.
(c) Show that the solution to this equation is

$$x = \frac{v}{\omega}(\omega t - \sin \omega t) + (1 - r)s(1 - \cos \omega t),$$

where $\omega = \sqrt{k/m}$, $r = \mu/\mu_0$.
(d) Assuming that the string stays stretched, calculate at what time $t = t_0$ the body stops again?
(e) Find the condition for the string to be stretched during the whole motion.
(f) How long will the body stay in rest, before moving again?

6.10 One may define three *invariants*, that is, scalar functions, of the stress tensor in any point. The first is the trace $I_1 = \sum_i \sigma_{ii}$, the second $I_2 = \frac{1}{2}\sum_{ij}(\sigma_{ii}\sigma_{jj} - \sigma_{ij}\sigma_{ji})$ that has no special name, and the determinant $I_3 = \det \boldsymbol{\sigma}$. **(a)** Show that the characteristic equation for the matrix $\boldsymbol{\sigma}$ can be expressed in terms of the invariants. **(b)** Can you find an invariant for an asymmetric stress tensor that vanishes if symmetry is imposed?

* **6.11** A space elevator (fictionalized by Arthur C. Clarke in *Fountains of Paradise* (1978)) can be created if it becomes technically feasible to lower a line down to Earth from a geostationary satellite at height h. Assume that the line is unstretchable and has constant cross-section A and constant density ρ. **(a)** Calculate the maximal tension $\sigma = -\mathcal{F}/A$ (force per unit of area) in the line. **(b)** Determine the numerical value of the ratio of tension to density, σ/ρ, and compare with the tensile strength (breaking tension) of a known material.

7 Strain

All materials deform when subjected to external forces but different materials react in different ways. Elastic materials bounce back again to the original configuration when the forces cease to act. Others are plastic and retain their shape after deformation. Viscoelastic materials behave like elastic solids under rapid deformation, but creep like viscous liquid over longer periods of time. Elasticity is itself an idealization limited to a certain range of forces. If the external forces become excessive, all materials become plastic and undergo permanent deformation or may even fracture.

When a body is deformed, its material is displaced away from its original position. Small deformations are mathematically much easier to handle than large deformations where parts of a body become greatly and non-uniformly displaced relative to other parts, as for example when you crumple a piece of paper. A rectilinear coordinate system embedded in the original body and deformed along with the material of the body becomes a curvilinear coordinate system after the deformation. It can therefore come as no surprise that the general theory of finite deformation is mathematically at the same level of difficulty as general curvilinear coordinate systems. Luckily our buildings and machines are rarely subjected to such violent treatment, and in most cases the deformation may be assumed to be tiny.

Although displacement is naturally described by a vector field, the description of deformation inevitably leads to the introduction of a new tensor quantity, the *strain tensor*, which characterizes the state of *local deformation* or *strain* in a material. It can come as no surprise that material strain causes tension or stress—as do strained relations among people. In this chapter we shall focus exclusively on the description of strain, and postpone the discussion of the stress-strain relationship for elastic materials to Chapter 8.

7.1 Displacement

The prime example of deformation is a *uniform scaling* in which the coordinates of all material particles in a body are multiplied with the scale factor κ. A material particle originally situated in the point $\boldsymbol{X}$ is thus displaced to the point

$$\boldsymbol{x} = \kappa \boldsymbol{X}. \tag{7.1}$$

It is emphasized that *both* $\boldsymbol{X}$ *and* $\boldsymbol{x}$ *refer to the same coordinate system*. Uniform scaling with $\kappa > 1$ is also called uniform *dilatation*, whereas scaling with $0 < \kappa < 1$ is called uniform *compression*. Negative scaling with $\kappa < 0$ is physically impossible.

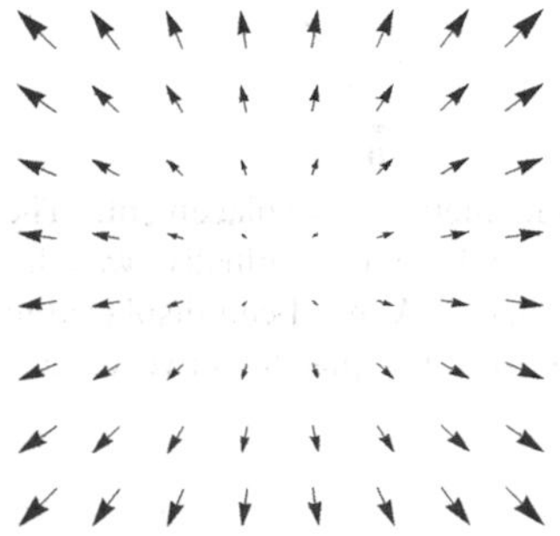

Uniform dilatation. The arrows indicate how material particles are displaced.

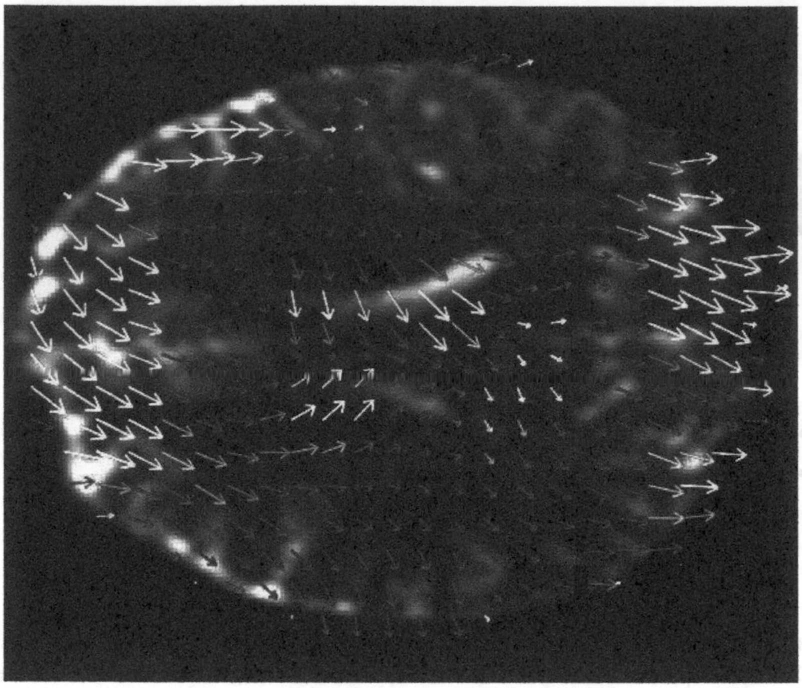

Figure 7.1. In statistical brain image analysis it is necessary to align brain images from different individuals so that structurally similar regions are brought to overlap. This process, called registration, may be viewed as a deformation of one brain into a standard brain. In the above image the small arrows picture the displacement field connecting two brain images. (Source: Courtesy Hauke Bartsch.)

The only point that does not change place during uniform scaling is the origin of the coordinate system. Although it superficially looks as if the origin of the coordinate system plays a special role, this is not really the case. All relative positions of material particles scale in the same way, because $\boldsymbol{x} - \boldsymbol{y} = \kappa(\boldsymbol{X} - \boldsymbol{Y})$, independent of the origin of the coordinate system. There is no special center for a uniform scaling, either geometrically or physically. The origin of the coordinate system is simply an *anchor point* for the mathematical description of scaling.

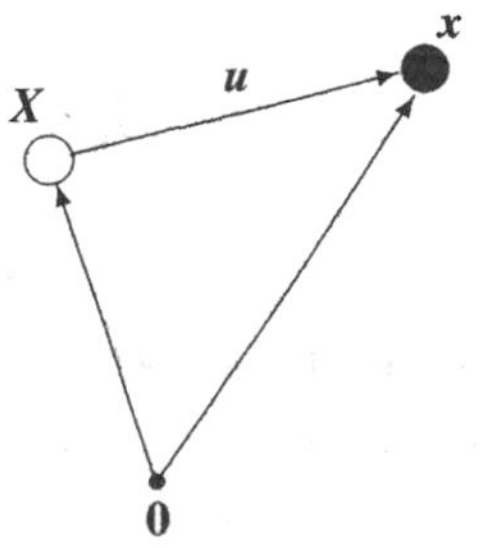

Geometry of displacement. The particle that originally was located at $\boldsymbol{X}$ has been displaced to $\boldsymbol{x}$ by the displacement vector $\boldsymbol{u}$.

Linear displacements

Under a displacement, the center-of-mass of a material particle is moved from its original position $\boldsymbol{X}$ to its actual position $\boldsymbol{x}$. The *displacement vector* is always defined as the difference between the *actual* and the *original* coordinates,

$$\boldsymbol{u} = \boldsymbol{x} - \boldsymbol{X}. \tag{7.2}$$

For the case of uniform scaling, the displacement vector becomes

$$\boldsymbol{u} = (\kappa - 1)\,\boldsymbol{X} = \left(1 - \frac{1}{\kappa}\right)\boldsymbol{x}. \tag{7.3}$$

Mathematically we are—as shown above—completely free to express the displacement as a function of either the original position $\boldsymbol{X}$ or the actual position $\boldsymbol{x}$ of the material particle. For scaling, the displacement is in both cases a linear function of the coordinates.

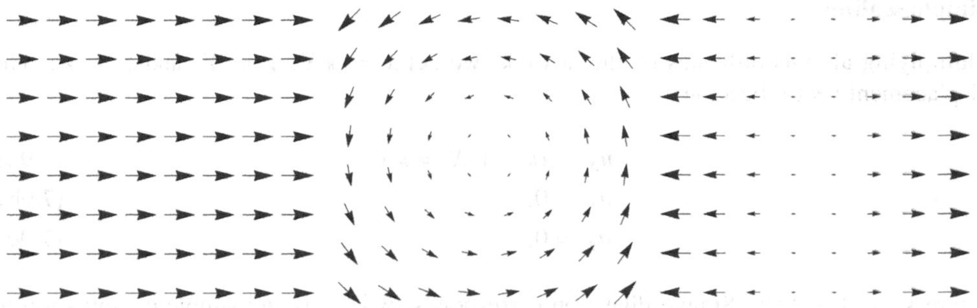

Figure 7.2. Arrow plots of the displacement fields for simple translation, rotation, and dilatation.

More generally, a linear displacement (and its inverse) takes the form

$$x = \mathsf{A} \cdot X + b, \qquad X = \mathsf{A}^{-1} \cdot (x - b). \tag{7.4}$$

where A is a non-singular constant matrix and b is a constant vector. As for scaling, the displacement vector may be expressed as a function of either the original or the actual positions,

$$u = (\mathsf{A} - 1) \cdot X + b = \left(1 - \mathsf{A}^{-1}\right) \cdot x + \mathsf{A}^{-1} \cdot b. \tag{7.5}$$

There is strong similarity between the general linear displacements and the transformations of Cartesian coordinates (seen Appendix B), but the class of linear displacements is larger, because the matrix A is not restricted to be orthogonal.

The general linear displacement may, like Cartesian coordinate transformations, be resolved into simpler types, namely translation along a coordinate axis, rotation by a fixed angle around a coordinate axis, and scaling by a fixed factor along a coordinate axis (see Figure 7.2). The physically impossible reflections in a coordinate axis are excluded. We shall not prove here that the general linear displacement may be resolved in this way but instead rely on geometric intuition.

Simple translation

A rigid body translation of the material through a distance b along the x-axis is described by $x = X + b$, $y = Y$, and $z = Z$. The displacement vector becomes

$$u_x = b, \qquad u_y = 0, \qquad u_z = 0. \tag{7.6}$$

The geometric relationships in a body are evidently unchanged under any translation, so this is not a deformation.

Simple rotation

A rigid body rotation through the angle ϕ around the z-axis takes the form

$$x = X\cos\phi - Y\sin\phi, \qquad X = x\cos\phi + y\sin\phi; \tag{7.7a}$$

$$y = X\sin\phi + Y\cos\phi, \qquad Y = -x\sin\phi + y\cos\phi; \tag{7.7b}$$

$$z = Z, \qquad Z = z. \tag{7.7c}$$

The corresponding displacement vector components are

$$u_x = -X(1-\cos\phi) - Y\sin\phi \qquad = x(1-\cos\phi) - y\sin\phi, \tag{7.8a}$$

$$u_y = X\sin\phi - Y(1-\cos\phi) \qquad = x\sin\phi - y(1-\cos\phi), \tag{7.8b}$$

$$u_z = 0 \qquad = 0. \tag{7.8c}$$

Since all distances in the body are unchanged, this is not a deformation.

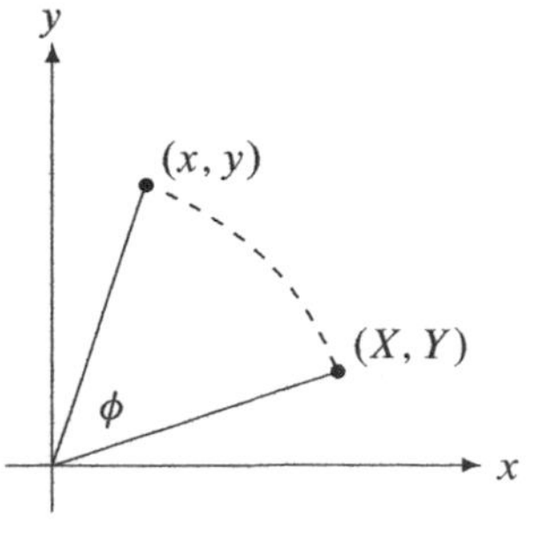

A rigid body rotation through an angle ϕ moves the material particle at (X, Y) to (x, y).

Simple scaling

Multiplying all x-coordinates by the factor κ, we get $x = \kappa X$, $y = Y$, and $z = Z$. The displacement vector becomes

$$u_x = (\kappa - 1)X = kx, \tag{7.9a}$$
$$u_y = 0, \tag{7.9b}$$
$$u_z = 0, \tag{7.9c}$$

where $k = 1 - 1/\kappa$. Simple dilatation corresponds to $k > 0$ and simple compression to $k < 0$. Uniform scaling (7.1) is a combination of three such scalings by the same factor along the three coordinate axes. Scaling is a true deformation.

7.2 The displacement field

In this book we have systematically adopted a "materialistic" attitude toward the description of continuous matter. Field values, such as the density $\rho(\boldsymbol{x})$ and gravity $\boldsymbol{g}(\boldsymbol{x})$, represent the physical properties in the immediate neighborhood of the point $\boldsymbol{x}$. When a macroscopic material body is deformed, all its material particles are in general simultaneously displaced. In keeping with the materialistic attitude, we let the field $\boldsymbol{X}(\boldsymbol{x})$ denote the original position of the material particle now situated at $\boldsymbol{x}$, so that the *displacement field* becomes a function of the actual position,

$$\boxed{\boldsymbol{u}(\boldsymbol{x}) = \boldsymbol{x} - \boldsymbol{X}(\boldsymbol{x}).} \tag{7.10}$$

This (material) representation of displacement is also called the *Euler representation*.

Mathematically, we could—as we did for the linear displacements—solve the equation, $\boldsymbol{X}(\boldsymbol{x}) = \boldsymbol{X}$ for $\boldsymbol{x}$, to obtain the actual position in terms of the original, $\boldsymbol{x} = \boldsymbol{x}(\boldsymbol{X})$. In that case, the displacement, $\boldsymbol{u} = \boldsymbol{x}(\boldsymbol{X}) - \boldsymbol{X}$, becomes a function of the original position. Although it seems physically awkward, this *Lagrange representation* of displacement is conceptually convenient in many situations and has played a great role in the long history of continuum physics. Here we shall mainly deal with slowly varying displacement fields, for which there is essentially no difference between the Euler and the Lagrange representations. A bit of the general theory of arbitrary displacements is presented in Section 7.5.

Joseph Louis Lagrange (1736–1813). Italian-born mathematician and astronomer who spent the last half of his life in France. Contributed in particular to the systematic mathematical analysis of Newtonian mechanics, in particular with his treatise *Mécanique Analytique* from 1788. Lagrangian mechanics is still taught at every university.

Local deformation

A general displacement field also includes all kinds of ordinary rigid body translations and rotations, and it would be wrong to classify all displacement fields as deformations. Sailing a submarine at the surface of the water will only translate or rotate it horizontally, not deform it, whereas taking it to the bottom of the sea will also compress it. *A true deformation must involve changes in geometric relationships*, that is, lengths and angles, in the body.

Large-scale deformation can be very complex. Think of all the loops and knots that weavers make from a roll of yarn. We should for this reason not expect to find a simple formalism for global deformation. Weaving, knitting, folding, winding, writhing, wringing, and squashing may bring particles that were originally far apart into close proximity. Even the wildest weave consists, however, locally of small pieces of straight yarn that have only been translated, rotated, stretched or contracted, but not folded, spindled, or mutilated. We may therefore expect to find a much simpler description of deformation for very small pieces of matter.

Displacement of an infinitesimal "needle"

Consider a tiny elongated piece of matter, a "needle" or *material vector* $\boldsymbol{a}$, now actually connecting the points $\boldsymbol{x}$ and $\boldsymbol{x}+\boldsymbol{a}$ in the displaced material. Before the displacement this needle connected the points $\boldsymbol{X} = \boldsymbol{X}(\boldsymbol{x})$ and $\boldsymbol{X}+\boldsymbol{a}_0 = \boldsymbol{X}(\boldsymbol{x}+\boldsymbol{a})$. Subtracting these equations and using that $\boldsymbol{X}(\boldsymbol{x}) = \boldsymbol{x} - \boldsymbol{u}(\boldsymbol{x})$, we find

$$\boldsymbol{a}_0 = \boldsymbol{X}(\boldsymbol{x}+\boldsymbol{a}) - \boldsymbol{X}(\boldsymbol{x}) = \boldsymbol{a} - \boldsymbol{u}(\boldsymbol{x}+\boldsymbol{a}) + \boldsymbol{u}(\boldsymbol{x}). \tag{7.11}$$

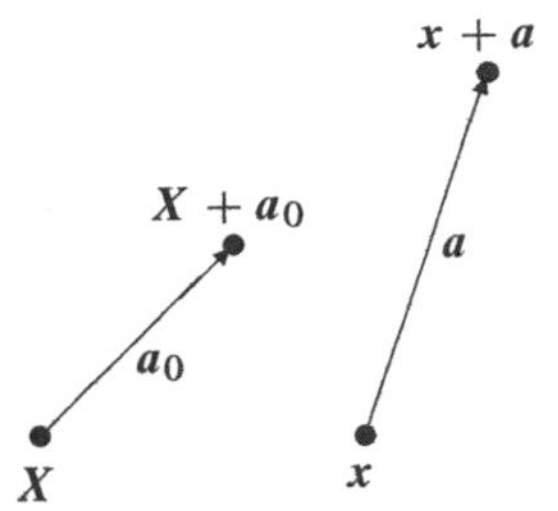

Displacement of a tiny material needle from $\boldsymbol{a}_0$ to $\boldsymbol{a}$. It may be translated, rotated, and scaled. Only the latter corresponds to a true deformation.

Expanding to the displacement field to first order in $\boldsymbol{a}$, we get

$$\begin{aligned}\boldsymbol{u}(\boldsymbol{x}+\boldsymbol{a}) &= \boldsymbol{u}(\boldsymbol{x}) + a_x\frac{\partial \boldsymbol{u}(\boldsymbol{x})}{\partial x} + a_y\frac{\partial \boldsymbol{u}(\boldsymbol{x})}{\partial y} + a_z\frac{\partial \boldsymbol{u}(\boldsymbol{x})}{\partial z} + \mathcal{O}\left(a^2\right) \\ &= \boldsymbol{u}(\boldsymbol{x}) + (\boldsymbol{a}\cdot\nabla)\,\boldsymbol{u}(\boldsymbol{x}) + \mathcal{O}\left(a^2\right).\end{aligned} \tag{7.12}$$

This shows that the displacement changes an infinitesimal needle vector by

$$\boxed{\delta\boldsymbol{a} \equiv \boldsymbol{a} - \boldsymbol{a}_0 = (\boldsymbol{a}\cdot\nabla)\,\boldsymbol{u}(\boldsymbol{x}).} \tag{7.13}$$

In index notation this may be written as

$$\delta a_i = \sum_j a_j\,\nabla_j u_i. \tag{7.14}$$

The coefficients of the linear transformation of $\boldsymbol{a}$ are computed from the set of derivatives of the displacement field, $\{\nabla_j u_i\}$, also called the *displacement gradients*. For a general linear displacement (7.5), we find $\nabla_j u_i = \delta_{ij} - (\mathsf{A}^{-1})_{ij}$.

In matrix (dyadic) notation, (7.14) becomes

$$\delta\boldsymbol{a} = (\boldsymbol{a}\cdot\nabla)\,\boldsymbol{u} = \boldsymbol{a}\cdot(\nabla\boldsymbol{u}) = (\nabla\boldsymbol{u})^{\top}\cdot\boldsymbol{a}, \tag{7.15}$$

where $(\nabla\boldsymbol{u})^{\top}$ is the transposed matrix of displacement gradients.

Example 7.1 [Simple rotation]: The displacement gradient matrix of a simple rotation (7.7) is

$$\{\nabla_j u_i\} = \begin{pmatrix} 1-\cos\phi & -\sin\phi & 0 \\ \sin\phi & -1+\cos\phi & 0 \\ 0 & 0 & 0 \end{pmatrix}, \tag{7.16}$$

where the index i enumerates the rows and j the columns.

Slowly varying displacement field

Displacements and coordinates have dimension of length, so that *the displacement gradients are dimensionless*, that is, pure numbers. This makes it meaningful to speak of small displacement gradients in an absolute way. A displacement field is said to be *slowly varying* if all the displacement gradients are small everywhere,

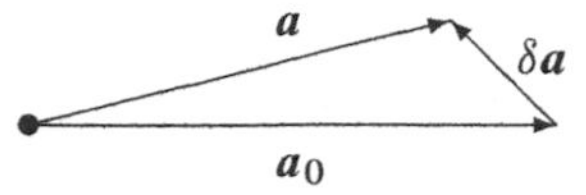

The change in a needle vector is small compared to its length when the displacement gradients are small.

$$\left|\nabla_j u_i(\boldsymbol{x})\right| \ll 1, \tag{7.17}$$

for all i, j, and $\boldsymbol{x}$. If we define the norm of a matrix as $|\mathsf{A}| = \sum_{ij}\left|A_{ij}\right|^2$, this may also be written $|\nabla\boldsymbol{u}| \ll 1$. Except for Section 7.5 where a few aspects of the theory of finite deformations are presented, we shall from now on assume that the displacement field is slowly varying, so that the change in a material needle is much smaller than its length, $|\delta\boldsymbol{a}| \ll |\boldsymbol{a}|$.

Small displacement gradients do not automatically guarantee that the displacement field itself is small compared to the size of the body because the displacement could include a rigid body translation to the other end of the universe, and that would not affect its gradient. But relative to a fixed anchor point in the body, a slowly varying field will always be small compared to the size L of the body and thus fulfill

$$|\boldsymbol{u}(\boldsymbol{x})| \ll L. \tag{7.18}$$

A displacement field satisfying this condition everywhere will in general also be slowly varying, though there are notable exceptions. If you, for example, make a small crease in your shirt when you iron it, the displacement gradients will be almost infinitely large in the crease although none of the shirt's material is greatly displaced compared to its size.

Example 7.2 [Small rotations]: For small rotation angle $\phi \ll 1$, the displacement field of a simple rotation (7.8) becomes $\boldsymbol{u} = (-y, x, 0)\ \phi$, and the gradient matrix (7.16) becomes

$$\{\nabla_j u_i\} = \begin{pmatrix} 0 & -\phi & 0 \\ \phi & 0 & 0 \\ 0 & 0 & 0 \end{pmatrix} \tag{7.19}$$

to lowest order in ϕ. The change in $\boldsymbol{a}$ becomes $\delta a_x = -\phi a_y$ and $\delta a_y = \phi a_x$. This may also be written as a cross-product, $\delta \boldsymbol{a} = \boldsymbol{\phi} \times \boldsymbol{a}$, where $\boldsymbol{\phi} = \phi\, \hat{\boldsymbol{e}}_z$.

Cauchy's strain tensor

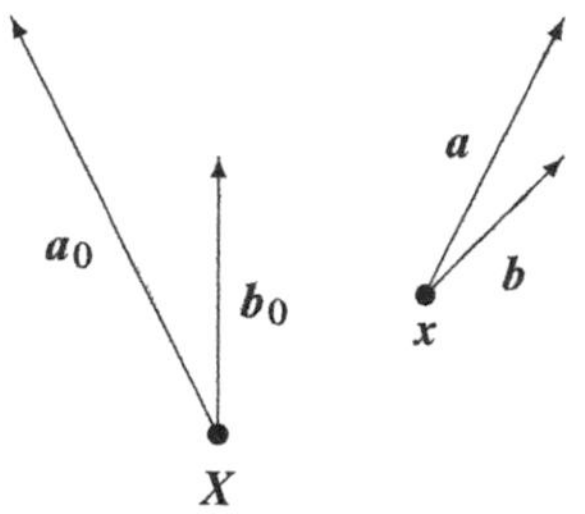

Displacement of a pair of infinitesimal material needles may affect their lengths as well as the angle between them.

The scalar product of two needles $\boldsymbol{a} \cdot \boldsymbol{b}$ is unchanged by translation and rotation, so it ought to be a useful indicator for a change in geometry. Using (7.13), we calculate the change in the scalar product $\delta(\boldsymbol{a} \cdot \boldsymbol{b}) \equiv \boldsymbol{a} \cdot \boldsymbol{b} - \boldsymbol{a}_0 \cdot \boldsymbol{b}_0$ to first order in the small displacement gradients,

$$\begin{aligned} \delta(\boldsymbol{a} \cdot \boldsymbol{b}) &= \delta \boldsymbol{a} \cdot \boldsymbol{b} + \boldsymbol{a} \cdot \delta \boldsymbol{b} \\ &= (\boldsymbol{a} \cdot \nabla)\boldsymbol{u} \cdot \boldsymbol{b} + (\boldsymbol{b} \cdot \nabla)\boldsymbol{u} \cdot \boldsymbol{a} \\ &= \sum_{ij} (\nabla_i u_j + \nabla_j u_i)\, a_i b_j \end{aligned}$$

or

$$\delta(\boldsymbol{a} \cdot \boldsymbol{b}) = 2 \sum_{ij} u_{ij} a_i b_j = 2\, \boldsymbol{a} \cdot \mathbf{u} \cdot \boldsymbol{b}, \tag{7.20}$$

where $\mathbf{u} = \{u_{ij}\}$ is the symmetrized displacement gradient tensor,

$$\boxed{u_{ij} = \tfrac{1}{2}(\nabla_i u_j + \nabla_j u_i).} \tag{7.21}$$

This tensor is called *Cauchy's (infinitesimal) strain tensor*, or just the *strain tensor* when that is unambiguous. It is sometimes convenient to write this relation in matrix form

$$\mathbf{u} = \tfrac{1}{2}\left(\nabla \boldsymbol{u} + \nabla \boldsymbol{u}^\top\right), \tag{7.22}$$

where $(\nabla \boldsymbol{u})_{ij} = \nabla_i u_j$ and $(\nabla \boldsymbol{u}^\top)_{ij} = \nabla_j u_i$ is its transposed.

The *strain tensor contains all the information about the local geometric changes caused by the displacement* and is accordingly a good measure of local deformation. All bodily translations and rotations have been automatically taken out, and any displacement that is a combination of translations and rotations must consequently yield a vanishing strain tensor. It should, however, be emphasized that *Cauchy's expression is only valid for small displacement gradients.* When that is not the case, a more complicated expression must be used, involving the square of the displacement gradients (see Section 7.5). The relative error committed by using Cauchy's approximation rather than the true strain tensor is therefore of the same magnitude as the displacement gradients themselves.

It is instructive and useful for practical calculations to write out all the components of the strain tensor explicitly, once and for all. The six independent components are

$$u_{xx} = \nabla_x u_x, \qquad u_{yz} = u_{zy} = \tfrac{1}{2}(\nabla_y u_z + \nabla_z u_y), \tag{7.23a}$$

$$u_{yy} = \nabla_y u_y, \qquad u_{zx} = u_{xz} = \tfrac{1}{2}(\nabla_z u_x + \nabla_x u_z), \tag{7.23b}$$

$$u_{zz} = \nabla_z u_z, \qquad u_{xy} = u_{yx} = \tfrac{1}{2}(\nabla_x u_y + \nabla_y u_x). \tag{7.23c}$$

Had we not assumed that the displacement field was slowly varying, there would as mentioned above also have been quadratic terms in the displacement gradients, and the strain tensor might take large values. But with our assumption of small displacement gradients (7.17), the strain tensor field is likewise small, $|u_{ij}(\boldsymbol{x})| \ll 1$ for all i, j, and $\boldsymbol{x}$. Contrariwise, a small strain tensor does not imply that the displacement gradients are small.

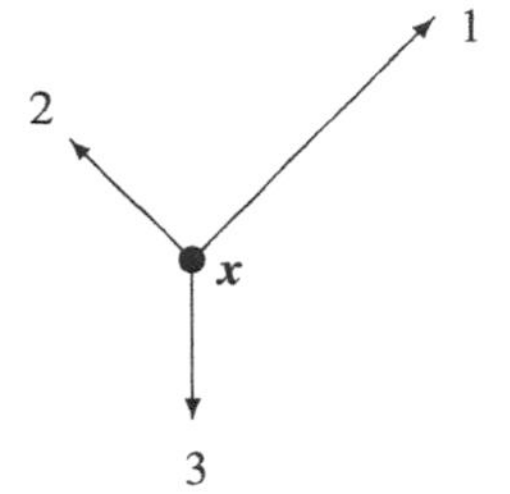

Principal strain basis in a point $\boldsymbol{x}$. The deformation consists entirely of scale changes along the principal axes, often shown by the lengths of the basis vectors.

Example 7.3 [Simple linear displacements]: The matrix of displacement gradients of a simple translation, $\boldsymbol{u}(\boldsymbol{x}) = (b, 0, 0)$, vanishes trivially, and so does the strain tensor. This confirms that a translation is not a deformation. For small angles of rotation, $|\phi| \ll 1$, the displacement gradient matrix (7.19) is antisymmetric. Cauchy's symmetric strain tensor therefore vanishes, confirming that a small rotation is not a deformation. For a simple scaling $\boldsymbol{u} = k(x, 0, 0)$, the displacement gradient matrix is symmetric and therefore equals the strain tensor

$$\{u_{ij}\} = k \begin{pmatrix} 1 & 0 & 0 \\ 0 & 0 & 0 \\ 0 & 0 & 0 \end{pmatrix}. \tag{7.24}$$

Evidently this is a true deformation.

* Principal axes of strain

According to its definition (7.21), the strain tensor is born *symmetric* under exchange of its indices

$$\boxed{u_{ij} = u_{ji}.} \tag{7.25}$$

It differs in this respect from the stress tensor (see page 104), for which symmetry is not self-evident but must be viewed as a constitutive equation.

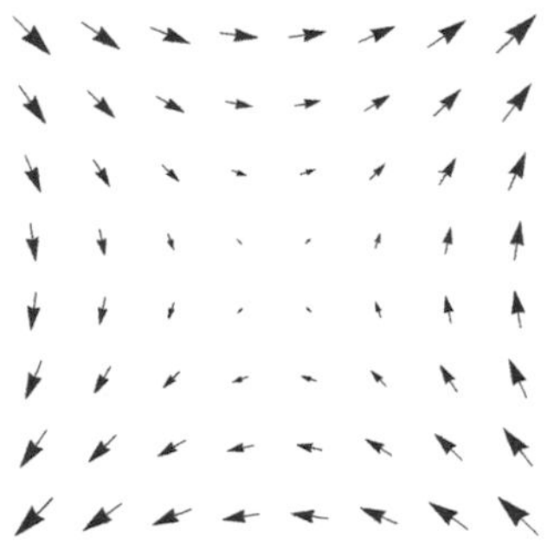

Arrow plot of the two-dimensional Lagrangian linear displacement field $\boldsymbol{u} = (y, x, 0)$ in the square $-1 < x < 1$ and $-1 < y < 1$. The material is dilated along one diagonal and contracted along the other. These are the principal directions of strain everywhere (see Problem 7.5).

A symmetric tensor can always be *diagonalized*. The eigenvectors of the strain tensor at a given point are called the *principal axes of strain*, and form a *principal basis* at every point of the body. In the principal basis for any given point, the strain tensor is diagonal, and the angles between the principal axes are unchanged under the displacement. The signs and magnitudes of the eigenvalues determine how much the material is being stretched or contracted along the principal axes. It should, however, be remembered that the principal basis of the strain tensor varies from point to point in space, and defines three orthogonal unit vector fields and three scalar eigenvalue fields (see Figure 7.3).

A symmetric tensor has six independent component whereas the displacement field has only three independent components. Every strain tensor must consequently satisfy consistency or compatibility conditions that remove three degrees of freedom. These conditions are formulated in Problem 7.11 and will not be further discussed here.

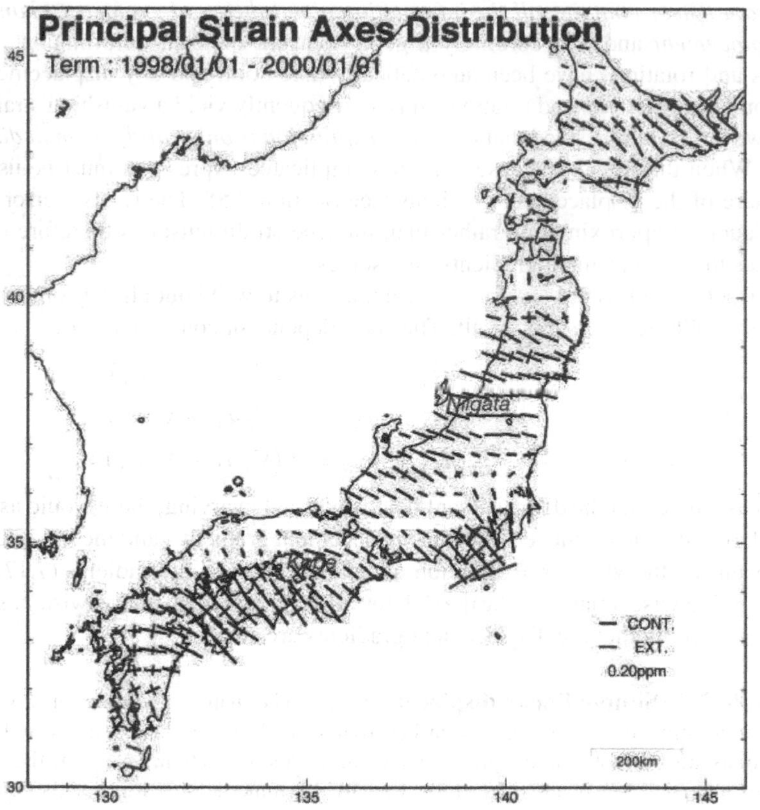

Figure 7.3. Principal strain axis distribution for ground displacements in Japan, determined by GPS over two years. Only the two horizontal axes are shown with lengths proportional to the magnitude of the eigenvalues. The black axes (running mainly southeast-to-northwest) indicate contraction and the gray extension. (Source: Copyright 2003-2004 Geospatial Information Authority of Japan. Reproduced with permission.)

7.3 Geometrical meaning of the strain tensor

The strain tensor contains all the relevant information about local changes in geometric relationships, such as lengths of material needles and the angles between them. Other local geometric quantities, for example curve, surface, and volume elements, are also changed under a deformation.

Lengths and angles

It is useful for the following discussion to define the *projection* u_{ab} of a tensor u_{ij} on the directions of two arbitrary vectors $\boldsymbol{a}$ and $\boldsymbol{b}$,

$$u_{ab} = \hat{\boldsymbol{a}} \cdot \mathbf{u} \cdot \hat{\boldsymbol{b}} = \frac{\boldsymbol{a} \cdot \mathbf{u} \cdot \boldsymbol{b}}{|\boldsymbol{a}|\ |\boldsymbol{b}|}. \tag{7.26}$$

Then we may simply write

$$\delta(\boldsymbol{a} \cdot \boldsymbol{b}) = 2\, |\boldsymbol{a}|\ |\boldsymbol{b}|\, u_{ab} \tag{7.27}$$

for the change in a scalar product (7.20).

The change in the length of a needle, $\delta\,|a| \equiv |a| - |a_0|$, is obtained by setting $b = a$ in (7.27), and using that $2u_{aa}\,|a|^2 = \delta(a \cdot a) = \delta(|a|^2) = 2\,|a|\,\delta\,|a|$, we get

$$\boxed{\frac{\delta\,|a|}{|a|} = u_{aa}.} \tag{7.28}$$

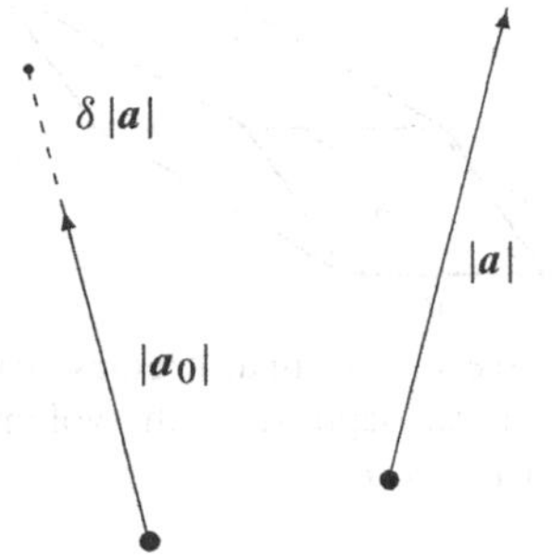

The length of the original needle is $|a_0| = |a| - \delta\,|a|$, where $\delta\,|a| = u_{aa}\,|a|$ is determined by the diagonal projection of the strain tensor.

The diagonal strain projection u_{aa} thus equals the *fractional change of lengths* in the direction a. Obtaining this relation is part of the reason behind the conventional factor 2 in the definition (7.21) of the strain tensor. Another reason is given in Problem 7.8.

Introducing the angle ϕ between two needles, we have $a \cdot b = |a|\,|b|\cos\phi$, and thus

$$\delta(a \cdot b) = \delta\,|a|\,|b|\cos\phi + |a|\,\delta\,|b|\cos\phi - |a|\,|b|\sin\phi\,\delta\phi.$$

Solving for $\delta\phi$ and using(7.27) and (7.28), we get

$$\delta\phi \equiv \phi - \phi_0 = \frac{(u_{aa} + u_{bb})\cos\phi - 2u_{ab}}{\sin\phi}. \tag{7.29}$$

For actually orthogonal vectors, such as the coordinate axes, we have $\phi = 90°$, and the change in angle simplifies to

$$\boxed{\delta\phi = -2u_{ab}.} \tag{7.30}$$

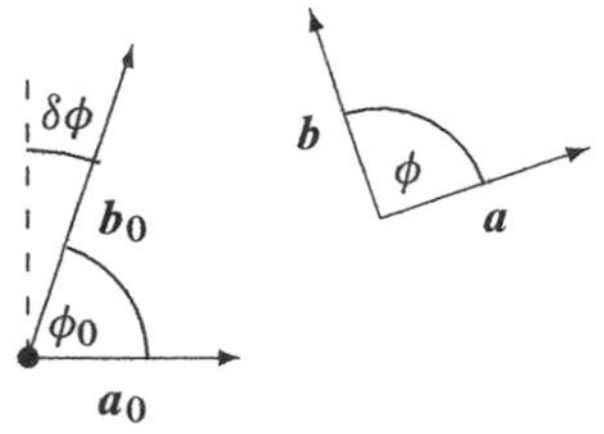

The original angle between two vectors is $\phi_0 = \phi - \delta\phi$, where in this case the actual angle is $\phi = 90°$. The Euler representation makes the drawing a bit awkward because it is based on the actual geometry and not the original.

The off-diagonal projections of the strain tensor thus determine the change in angle between actually orthogonal needles.

Infinitesimal elements

Curve, surface, and volume integrals appear everywhere in continuum physics, and the mathematics of these integrals is discussed in Appendix C. When material is displaced, the infinitesimal elements of the integrals also change, and the formalism of displacement with small gradients introduced in this chapter allows us to calculate how they transform.

Curve element

A curve element is nothing but a small needle. Under a displacement, the curve element changes from $d\boldsymbol{\ell}_0$ to $d\boldsymbol{\ell}$, given by the needle expression (7.13),

$$\boxed{\delta(d\boldsymbol{\ell}) \equiv d\boldsymbol{\ell} - d\boldsymbol{\ell}_0 = d\boldsymbol{\ell} \cdot \nabla \boldsymbol{u} = \nabla \boldsymbol{u}^{\mathsf{T}} \cdot d\boldsymbol{\ell}.} \tag{7.31}$$

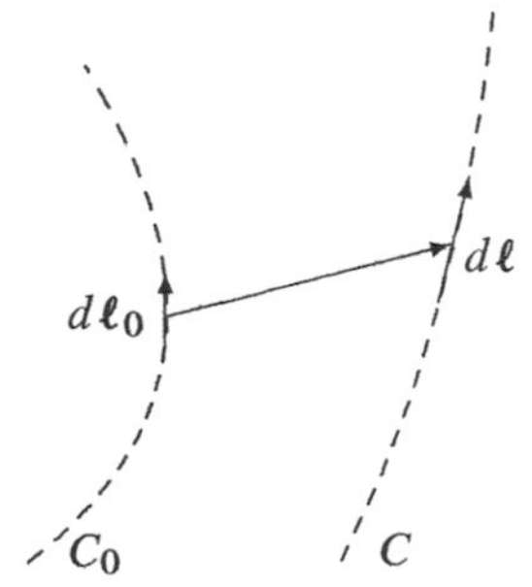

A line element is stretched and rotated by the displacement that changes the curve from C_0 to C.

Here we have as before used a compact matrix notation for the displacement gradient tensor, $(\nabla \boldsymbol{u})_{ij} = \nabla_i u_j$. The transposed displacement gradient matrix (with rows and columns interchanged) then becomes $(\nabla \boldsymbol{u})^{\mathsf{T}}_{ij} = \nabla_j u_i$.

Example 7.4: Let $\boldsymbol{g}(\boldsymbol{x})$ be some vector field, and consider the integral $\int_C \boldsymbol{g} \cdot d\boldsymbol{\ell}$ along a curve C. Assuming that the end points are not displaced, the change in the integral becomes

$$\begin{aligned}\delta \int_C \boldsymbol{g} \cdot d\boldsymbol{\ell} &= \int_C \delta\boldsymbol{g} \cdot d\boldsymbol{\ell} + \int_C \boldsymbol{g} \cdot \delta(d\boldsymbol{\ell}) \\ &= \int_C (\boldsymbol{u} \cdot \nabla)\boldsymbol{g} \cdot d\boldsymbol{\ell} + \int_C \boldsymbol{g} \cdot (\nabla \boldsymbol{u})^{\mathsf{T}} \cdot d\boldsymbol{\ell}.\end{aligned} \tag{7.32}$$

The first term stems from the change in the field $\boldsymbol{g}$ under the displacement, $\delta\boldsymbol{g} = \boldsymbol{g}(\boldsymbol{x}) - \boldsymbol{g}(\boldsymbol{x} - \boldsymbol{u}) \approx (\boldsymbol{u} \cdot \nabla)\boldsymbol{g}$. This requires the displacement to be small compared to the length scales for changes in the $\boldsymbol{g}$-field. The second term represents the change in the curve elements, and only requires the displacement gradients to be small.

Volume element

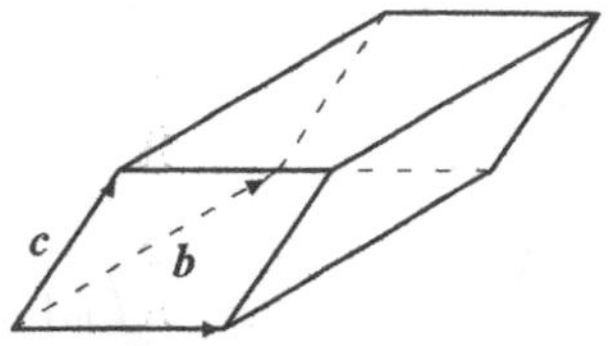

Three infinitesimal needles span a parallelepiped with volume $dV a \times b \cdot c$.

Consider a tiny material volume, $dV = \boldsymbol{a} \times \boldsymbol{b} \cdot \boldsymbol{c}$, of a parallelepiped spanned by three linearly independent infinitesimal needles. Under the displacement the volume changes by

$$\begin{aligned}\delta(dV) &= \delta(\boldsymbol{a} \times \boldsymbol{b} \cdot \boldsymbol{c}) \\ &= \delta\boldsymbol{a} \times \boldsymbol{b} \cdot \boldsymbol{c} + \boldsymbol{a} \times \delta\boldsymbol{b} \cdot \boldsymbol{c} + \boldsymbol{a} \times \boldsymbol{b} \cdot \delta\boldsymbol{c} \\ &= (\boldsymbol{a} \cdot \nabla)\boldsymbol{u} \times \boldsymbol{b} \cdot \boldsymbol{c} + (\boldsymbol{b} \cdot \nabla)\boldsymbol{a} \times \boldsymbol{u} \cdot \boldsymbol{c} + (\boldsymbol{c} \cdot \nabla)\boldsymbol{a} \times \boldsymbol{b} \cdot \boldsymbol{u} \\ &= \left[\boldsymbol{b} \times \boldsymbol{c}\,(\boldsymbol{a} \cdot \nabla) + \boldsymbol{c} \times \boldsymbol{a}\,(\boldsymbol{b} \cdot \nabla) + \boldsymbol{a} \times \boldsymbol{b}\,(\boldsymbol{c} \cdot \nabla)\right] \cdot \boldsymbol{u}.\end{aligned}$$

In the last step we used that $\boldsymbol{u} \times \boldsymbol{b} \cdot \boldsymbol{c} = \boldsymbol{b} \times \boldsymbol{c} \cdot \boldsymbol{u}$ and $\boldsymbol{a} \times \boldsymbol{u} \cdot \boldsymbol{c} = \boldsymbol{c} \times \boldsymbol{a} \cdot \boldsymbol{u}$ to pull the only $\boldsymbol{x}$-dependent factor $\boldsymbol{u}(\boldsymbol{x})$ out to the right.

We now use an identity between four arbitrary vectors,

$$\boldsymbol{a} \times \boldsymbol{b}\,(\boldsymbol{c} \cdot \boldsymbol{d}) + \boldsymbol{b} \times \boldsymbol{c}\,(\boldsymbol{a} \cdot \boldsymbol{d}) + \boldsymbol{c} \times \boldsymbol{a}\,(\boldsymbol{b} \cdot \boldsymbol{d}) = (\boldsymbol{a} \times \boldsymbol{b} \cdot \boldsymbol{c})\,\boldsymbol{d}. \tag{7.33}$$

It expresses the simple fact that in a three-dimensional space, four vectors will always be linearly dependent. It may be verified by dotting from left with $\boldsymbol{a}$, $\boldsymbol{b}$, and $\boldsymbol{c}$ (see also Problem B.10 on page 608). Replacing $\boldsymbol{d}$ by ∇, it follows immediately that

$$\boxed{\delta(dV) = \nabla \cdot \boldsymbol{u}\, dV.} \tag{7.34}$$

We have thus shown that for small displacement gradients, the divergence of the displacement field, $\nabla \cdot \boldsymbol{u} = \sum_i \nabla_i u_i$, determines the fractional change $\delta(dV)/dV$ in the local volume. There are other ways of deriving this relation (see Problem 7.12).

Example 7.5 [Simple linear displacements]: A translation $\boldsymbol{u} = \boldsymbol{b}$ does not change the density because $\nabla \cdot \boldsymbol{u} = 0$. Likewise for an infinitesimal rotation around the z-axis through a small angle ϕ, the displacement field $\boldsymbol{u} = (-y, x, 0)\phi$ has vanishing divergence, so that the density is unchanged. A uniform scaling $\boldsymbol{u} = k\boldsymbol{x}$ has $\nabla \cdot \boldsymbol{u} = 3k$. If $k > 0$, the volume increases (and the density diminishes) with a relative contribution k from each dimension.

The change in volume of a material particle induces a change in the local density of the displaced matter. Using that the mass of the particle is unchanged during the displacement,

$$\delta(dM) = \delta(\rho dV) = \rho\,\delta(dV) + \delta\rho\, dV = 0,$$

and making use of (7.34), the change in density becomes

$$\boxed{\delta\rho = -\rho\,\nabla \cdot \boldsymbol{u}.} \tag{7.35}$$

Thus, if the divergence vanishes, there is no change in volume or density.

Surface element

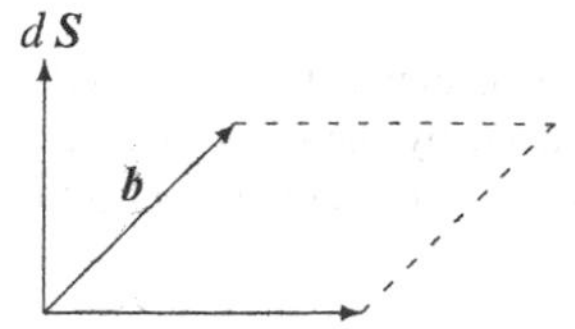

Two infinitesimal needles span a parallelogram with area $d\boldsymbol{S} = \boldsymbol{a} \times \boldsymbol{b}$.

A tiny material surface element, $d\boldsymbol{S} = \boldsymbol{a} \times \boldsymbol{b}$, also changes under a displacement. Using that the volume element equals $dV = \boldsymbol{c} \cdot d\boldsymbol{S}$ we find from (7.34) that

$$\boldsymbol{c} \cdot \delta(d\boldsymbol{S}) = \delta(\boldsymbol{c} \cdot d\boldsymbol{S}) - \delta\boldsymbol{c} \cdot d\boldsymbol{S} = \nabla \cdot \boldsymbol{u}(\boldsymbol{c} \cdot d\boldsymbol{S}) - (\boldsymbol{c} \cdot \nabla\boldsymbol{u}) \cdot d\boldsymbol{S}.$$

Since $\boldsymbol{c}$ is an arbitrary vector, it can be "divided out", and we get (using matrix notation)

$$\boxed{\delta(d\boldsymbol{S}) = (\nabla \cdot \boldsymbol{u}\,\mathbf{1} - \nabla\boldsymbol{u}) \cdot d\boldsymbol{S}.} \tag{7.36}$$

Both the magnitude and direction of the vector surface element are changed by the displacement, but the rule is quite different from that of the vector curve element (7.31).

7.4 Work and energy

Deforming a body takes work, a fact known to everyone who has ever kneaded clay or dough. In these cases, the work you perform seems to get lost inside the material; but in other cases, as for example when you squeeze an elastic rubber ball, the material appears to store the work and release it again when you relinquish your grip. Many ball games like ping-pong or tennis rely entirely on the elasticity of the ball. No material is, however, perfectly elastic. Some work is always lost to internal friction. A hard steel ball may jump many times on a hard floor, but eventually it loses all its energy and comes to rest, partly due to air resistance, partly due to losses in the ball and, perhaps more importantly, in the floor. But even when your work seems to disappear into the dough, this is not really the case. The energy you have put into the dough has in the end been converted into heat that, however, cannot easily be recovered. We shall analyze the interplay of mechanics and heat in Chapter 22.

In continuum physics it can be quite subtle to derive the correct energy relations. The simplest way to proceed is to *follow the work.* This is quite analogous to the admonition, "follow the money", often used with success to uncover economic or political fraud.

Virtual displacement work

A volume of an arbitrary material that is not in mechanical equilibrium, will left to itself seek toward equilibrium. If the effective force $d\mathcal{F} = \boldsymbol{f}^*\, dV$ acting on a material particle of volume dV does not vanish everywhere, the particle is (literally) forced to move until the effective force vanishes. If we wish to keep all material particles in their non-equilibrium positions, we must act on the body with an external volume distribution of so-called *virtual forces*, $\boldsymbol{f}' = -\boldsymbol{f}^*$, to compensate for the effective internal forces. Even if such forces may be impossible to realize in practice, they will—in a thought experiment—freeze all the material particles in their positions for as long as we wish.

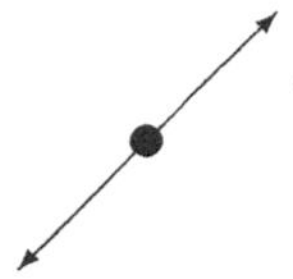

Every material particle can be kept in place by acting on it with an additional external force that balances the already existing effective body force on the particle.

Imagine now that in this frozen state all the material particles of the body are displaced infinitesimally by $\delta\boldsymbol{u}(\boldsymbol{x})$. To prevent the constant external forces on the surface of the body from performing work, we shall choose to keep the surface S of its volume V unchanged, so that the infinitesimal displacement must vanish at the surface, $\delta\boldsymbol{u}(\boldsymbol{x}) = \boldsymbol{0}$ for $\boldsymbol{x} \in S$. The work of the virtual forces under this displacement then becomes

$$\delta W = \int_V \boldsymbol{f}' \cdot \delta\boldsymbol{u}\, dV = -\int_V \boldsymbol{f}^* \cdot \delta\boldsymbol{u}\, dV. \tag{7.37}$$

Inserting $\boldsymbol{f}^* = \boldsymbol{f} + \nabla\cdot\boldsymbol{\sigma}^\top$ we find for the stress term (using index notation for clarity),

$$\sum_{ij} (\nabla_j \sigma_{ij})\, \delta u_i = \sum_{ij} \nabla_j(\sigma_{ij}\delta u_i) - \sum_{ij} \sigma_{ij} \nabla_j \delta u_i.$$

Integrating over V, the first term is converted into a surface integral that vanishes because the displacement field $\delta\boldsymbol{u}$ vanishes at the surface. We now introduce a convenient notation for the trace of the product of two matrices $\mathbf{A}\!:\!\mathbf{B} = \mathrm{Tr}(\mathbf{A}\cdot\mathbf{B}) = \sum_{ij} A_{ij} B_{ji}$. The work of the virtual forces may then be written as

$$\delta W = -\int_V \boldsymbol{f} \cdot \delta\boldsymbol{u}\, dV + \int_V \boldsymbol{\sigma} : \nabla\delta\boldsymbol{u}\, dV, \tag{7.38}$$

where $\boldsymbol{f}$ is the true force density due to long-range forces and $\boldsymbol{\sigma} : \nabla\delta\boldsymbol{u} = \sum_{ij} \sigma_{ij} \nabla_j \delta u_i$.

The first term represents the part of the displacement work that is spent by the virtual forces *against* the true body forces, for example gravity. In that case the work contributes to the gravitational energy of the body. The second term represents the part of the work of the virtual forces that is spent against the *internal stresses* in the body,

$$\delta W_{\text{deform}} = \int_V \boldsymbol{\sigma} : \nabla \delta \boldsymbol{u} \, dV. \tag{7.39}$$

If the stress tensor is symmetric (which it normally is), $\sigma_{ij} = \sigma_{ji}$, the integrand may be written as

$$\boldsymbol{\sigma} : \nabla \delta \boldsymbol{u} = \sum_{ij} \sigma_{ij} \nabla_j \delta u_i = \sum_{ij} \sigma_{ij} \delta u_{ij} = \boldsymbol{\sigma} : \delta \mathbf{u},$$

where $\delta u_{ij} = \frac{1}{2}(\nabla_i \delta u_j + \nabla_j \delta u_i)$ is the infinitesimal change in the strain tensor. Evidently the work (7.39) is associated with deformation of the material and contributes to the *deformation energy* of the body. In Chapter 8 we shall derive an explicit expression for the deformation energy.

Example 7.6 [Thermodynamic work]: Suppose the stresses are only due to pressure, $\sigma_{ij} = -p\,\delta_{ij}$. Then the deformation work becomes

$$\delta W_{\text{deform}} = -\int_V p \, \nabla \cdot \delta \boldsymbol{u} \, dV. \tag{7.40}$$

Since $\delta(dV) = (\nabla \cdot \delta \boldsymbol{u})\, dV$ is the change in volume of a material particle, we see that the deformation work is identical to the thermodynamic work $-p\,\delta(dV)$ summed over all material particles.

∗ 7.5 Large deformations

Ronald Samuel Rivlin (1915–2005). British-born mathematician and physicist. Contributed to the understanding of non-linear materials during the 1940s and 1950s. Discovered exact non-linear solutions for isotropic materials.

When the condition (7.17) for slowly varying displacement is not fulfilled, we can no longer use the simple Cauchy strain tensor (7.21). The local description of large deformation is essentially equivalent to the formalism of general curvilinear coordinate systems, but because space is Euclidean the description is not quite as complicated as that of truly non-Euclidean spaces [Green and Zerna 1992]. Although many aspects of the theory of large deformations were developed in the nineteenth century, the subject was not fully established until the mid-twentieth century through Rivlin's work on non-linear materials. Here we shall only touch briefly on the most general aspects of large deformation theory, which is mathematically a rather challenging subject [Green and Adkins 1960, Doghri 2000].

The Euler representation

When there are no restrictions on the magnitude of the displacement field or the displacement gradients, the transformation $\boldsymbol{x} \to \boldsymbol{X}(\boldsymbol{x})$ becomes a completely general non-singular differentiable point-to-point map between two regions of space representing the actual body and its original situation. There is then no particular reason to split off the displacement, except to make contact with the description of small deformations in the preceding part of this chapter. The local properties of the displacement field will still be of importance, because the map is nearly linear in the neighborhood of any point in the body.

Consider again an infinitesimal material "needle" (in the actual body) described by the vector $d\boldsymbol{x}$. It originated in a material needle with coordinates

$$dX_i = \sum_j \frac{\partial X_i}{\partial x_j} dx_j. \tag{7.41}$$

The scalar product of the infinitesimal vectors $d\boldsymbol{X}$ and $d\boldsymbol{Y}$ then becomes

$$d\boldsymbol{X} \cdot d\boldsymbol{Y} = \sum_{ij} g_{ij}(\boldsymbol{x}) dx_i dx_j, \tag{7.42}$$

where the tensor field

$$g_{ij} = \sum_k \frac{\partial X_k}{\partial x_i} \frac{\partial X_k}{\partial x_j} \tag{7.43}$$

is called the *Eulerian deformation tensor*. It contains all information about geometric changes taking place under the displacement. Writing

$$g_{ij} = \delta_{ij} - 2u_{ij}, \tag{7.44}$$

the change in the scalar product can be written in the same way as the expression (7.20) for slowly varying displacement,

$$d\boldsymbol{x} \cdot d\boldsymbol{y} - d\boldsymbol{X} \cdot d\boldsymbol{Y} = 2\sum_{ij} u_{ij} dx_i dy_j. \tag{7.45}$$

Finally we insert $\boldsymbol{X}(\boldsymbol{x}) = \boldsymbol{x} - \boldsymbol{u}(\boldsymbol{x})$ and arrive at the strain tensor in the Euler representation,

$$\boxed{u_{ij} = \frac{1}{2}\left(\frac{\partial u_j}{\partial x_i} + \frac{\partial u_i}{\partial x_j} - \sum_i \frac{\partial u_k}{\partial x_i}\frac{\partial u_k}{\partial x_j}\right).} \tag{7.46}$$

It only differs from the infinitesimal Cauchy strain tensor (7.21) by the last non-linear term. This tensor was first introduced by Emilio Almansi in 1911 and Georg Hamel in 1912 (see [Chandrasekharaiah and Debnath 1994]).

Emilio Almansi (1869–1948). Italian mathematical physicist. Worked on non-linear elasticity theory, electrostatics, and celestial mechanics.

Georg Hamel (1877–1954). German mathematician. Solved one of the famous Hilbert problems in his doctoral thesis under Hilbert (1901).

Example 7.7 [Uniform scaling]: For uniform scaling $\boldsymbol{x} = \kappa \boldsymbol{X}$, we have $\boldsymbol{X} = \kappa^{-1}\boldsymbol{x}$, so that $\partial X_i/\partial x_j = \kappa^{-1}\delta_{ij}$. Consequently the displacement gradient tensor becomes

$$\frac{\partial u_i}{\partial x_j} = \frac{\partial x_i}{\partial x_j} - \frac{\partial X_i}{\partial x_j} = \left(1-\kappa^{-1}\right)\delta_{ij}, \tag{7.47}$$

from which we get the Euler–Almansi strain tensor,

$$u_{ij} = \frac{1}{2}\left[\left(1-\kappa^{-1}\right)\delta_{ij} + \left(1-\kappa^{-1}\right)\delta_{ij} - \left(1-\kappa^{-1}\right)^2\delta_{ij}\right] = \frac{1}{2}\left(1-\kappa^{-2}\right)\delta_{ij}, \tag{7.48}$$

valid for any value of κ. For $\kappa = 1 + k$ with $|k| \ll 1$, this expression becomes $u_{ij} = k\delta_{ij}$ to leading order.

The Lagrange representation

Even if the Lagrange representation of large deformation goes against the "materialistic" attitude of this book—in which the properties of material particles are functions of their actual positions—it is sometimes useful, for example in numerical computations. The basic relation between the two representations is simply that the Lagrangian displacement field by $\boldsymbol{U}(\boldsymbol{X})$ must take the same value as the Eulerian one at corresponding points, so that

$$\boldsymbol{x} = \boldsymbol{X} + \boldsymbol{U}(\boldsymbol{X}) \qquad \text{with } \boldsymbol{U}(\boldsymbol{X}) = \boldsymbol{u}(\boldsymbol{x}). \tag{7.49}$$

Given the Euler displacement field, $\boldsymbol{u}(\boldsymbol{x})$, this non-linear equation may be solved for $\boldsymbol{U}(\boldsymbol{X})$.

The local analysis now proceeds as before via the infinitesimal line element,

$$d\boldsymbol{x} = \sum_j \frac{\partial \boldsymbol{x}_i}{\partial X_j} dX_j. \tag{7.50}$$

The scalar product becomes

$$d\boldsymbol{x} \cdot d\boldsymbol{y} = \sum_{ij} G_{ij} dX_i dY_j, \qquad G_{ij} = \sum_k \frac{\partial x_k}{\partial X_i} \frac{\partial x_k}{\partial X_j}, \tag{7.51}$$

where G_{ij} is the *Lagrangian deformation tensor.* Writing

$$G_{ij} = \delta_{ij} + 2U_{ij} \tag{7.52}$$

it follows from (7.49) that the *Lagrangian strain tensor* is

$$\boxed{U_{ij} = \frac{1}{2}\left(\frac{\partial U_j}{\partial X_i} + \frac{\partial U_i}{\partial X_j} + \sum_k \frac{\partial U_k}{\partial X_i}\frac{\partial U_k}{\partial X_j}\right).} \tag{7.53}$$

George Green (1793–1841). English, largely self-taught mathematician and mathematical physicist. Contributed to hydrodynamics, electricity and magnetism, and partial differential equations.

It is called the *Lagrange–Green strain tensor.*

Example 7.8 [Uniform scaling]: For a uniform scaling $\boldsymbol{x} = \kappa \boldsymbol{X}$, the Lagrangian displacement field becomes

$$\boldsymbol{U}(\boldsymbol{X}) = \boldsymbol{x} - \boldsymbol{X} = (\kappa - 1)\boldsymbol{X}. \tag{7.54}$$

The Lagrange–Green strain tensor becomes

$$U_{ij} = \tfrac{1}{2}\left[(\kappa-1)\delta_{ij} + (\kappa-1)\delta_{ij} + (\kappa-1)^2\delta_{ij}\right] = \tfrac{1}{2}(\kappa^2-1)\delta_{ij} \tag{7.55}$$

for any value of κ. It vanishes for $\kappa = \pm 1$, that is, for no displacement and for a (physically impossible) pure reflection in the origin.

The scalar product of two infinitesimal needles then becomes

$$d\boldsymbol{x} \cdot d\boldsymbol{y} = \kappa^2 d\boldsymbol{X} \cdot d\boldsymbol{Y} \tag{7.56}$$

and just reflects that all vectors are scaled by the same amount κ.

Problems

7.1 Prove that Equation (7.7) is the correct transformation for a simple rotation.

7.2 Calculate displacement gradients and the strain tensor for the transformation

$$u_x = \alpha(5x - y + 3z),$$
$$u_y = \alpha(x + 8y),$$
$$u_z = \alpha(-3x + 4y + 5z),$$

where α is small.

7.3 A displacement field is given by

$$u_x = \alpha(x + 2y) + \beta x^2,$$
$$u_y = \alpha(y + 2z) + \beta y^2,$$
$$u_z = \alpha(z + 2x) + \beta z^2,$$

where α and βL are small (with L being the size of the body).
(a) Calculate the divergence and curl of this field. **(b)** Calculate Cauchy's strain tensor.

7.4 Calculate the strain tensor for the displacement field $\boldsymbol{u} = (Ax + Cy, Cx - By, 0)$ where A, B, C are small constants. Under what condition will the volume be unchanged?

7.5 Calculate the strain tensor for $\boldsymbol{u} = \alpha(y, x, 0)$ where $0 < \alpha \ll 1$. Determine the principal directions of strain and the change in length scales along these directions.

7.6 **(a)** Calculate the displacement gradients and the strain tensor for the displacement field $\boldsymbol{u} = \alpha(y^2, xy, 0)$ with $|\alpha| \ll 1/L$, where L is the size of the body. **(b)** Calculate the principal directions of strain and the scaling factors.

7.7 Show that the change in a scalar product under a deformation is derivable from changes in length, that is, from the diagonal projections u_{aa} of the strain tensor.

7.8 **(a)** Show that the general displacement rule for an infinitesimal needle (7.13) may be written

$$\boldsymbol{a}' = \boldsymbol{a} + \boldsymbol{\phi} \times \boldsymbol{a} + \mathbf{u} \cdot \boldsymbol{a},$$

where $\boldsymbol{\phi} = \frac{1}{2}\nabla \times \boldsymbol{u}$ and $\mathbf{u} = \{u_{ij}\}$ is Cauchy's strain tensor (7.21). **(b)** What does the second term mean?

7.9 Show that the most general solution, for which Cauchy's strain tensor (7.21) vanishes, is

$$u_x = A + Dy + Ez,$$
$$u_y = B - Dx + Fz,$$
$$u_z = C - Ex - Fy,$$

where A, B, C are arbitrary constants and D, E, F are small.

7.10 A deformable material undergoes two successive displacements, $\boldsymbol{x}' = \boldsymbol{x} + \boldsymbol{u}(\boldsymbol{x})$ and $\boldsymbol{x}'' = \boldsymbol{x}' + \boldsymbol{u}'(\boldsymbol{x}')$, both having small strain. Calculate the final strain tensor for the total deformation u''_{ij} relative to the original reference state.

7.11 Show that Cauchy's strain tensor (7.21) satisfies the relation (going back to Saint-Venant)

$$\nabla_i \nabla_j u_{kl} + \nabla_k \nabla_l u_{ij} = \nabla_i \nabla_l u_{kj} + \nabla_k \nabla_j u_{il}.$$

Conversely, it can be shown (but not here) that if this relation is fulfilled for a symmetric tensor field u_{ij}, then there is a displacement field such that the strain tensor is given by Cauchy's strain tensor.

7.12 **(a)** Show that a small volume change can be represented by a surface integral

$$\delta V = \oint_S \boldsymbol{u} \cdot d\boldsymbol{S}.$$

(b) Use this result to derive (7.34).

8

Hooke's law

When you bend a wooden stick, the reaction grows notably stronger the further you go—until it perhaps breaks with a snap. If you release the bending force before it breaks, the stick straightens out again and you can bend it again and again without it changing its reaction or its shape. That is what we call elasticity.

In elementary mechanics the elasticity of a spring is expressed by *Hooke's law*, which says that the amount a spring is stretched or compressed beyond its relaxed length is proportional to the force acting on it. In continuous elastic materials, Hooke's law implies that strain is a linear function of stress. Some materials that we usually think of as highly elastic, for example rubber, do not obey Hooke's law except under very small deformations. When stresses grow large, most materials deform more than predicted by Hooke's law and in the end reach the elasticity limit where they become plastic or break.

The elastic properties of continuous materials are determined by the underlying molecular level but the relation is complicated, to say the least. Luckily, there are broad classes of materials that may be described by a few material parameters that can be determined empirically. The number of such parameters depends on the how complex the internal structure of the material is. We shall almost exclusively concentrate on structureless, isotropic elastic materials, described by just two material parameters: Young's modulus and Poisson's ratio.

In this chapter, the emphasis will be on matters of principle. We shall derive the basic equations of linear elasticity, but only solve them in the simplest possible cases. In Chapter 9 we shall solve these equations in a number of cases of more practical interest.

Robert Hooke (1635–1703). English biologist, physicist, and architect (no verified contemporary portrait exists). In physics he worked on gravitation, elasticity, built telescopes, and discovered diffraction of light. His famous law of elasticity goes back to 1660. First stated in 1676 as a Latin anagram *ceiiinosssttuv*, he revealed it in 1678 to stand for *ut tensio sic vis*, meaning "as is the extension, so is the force".

8.1 Young's modulus and Poisson's ratio

Massless elastic springs obeying Hooke's law are a mainstay of elementary mechanics. If a spring of relaxed length L is anchored at one end and pulled by some external "agent" at the other with a force $\mathcal{F}$, its length is increased to $L + x$. Hooke's law states that there is proportionality between force and extension,

$$\mathcal{F} = kx. \tag{8.1}$$

The constant of proportionality, k, is called the *spring constant*. Newton's Third Law guarantees of course that the spring must act back on the external "agent" with a force $-kx$.

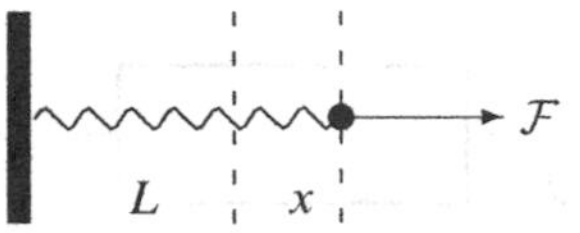

A spring of relaxed length L anchored at the left and pulled toward the right by an external force $\mathcal{F}$ will be stretched by $x = \mathcal{F}/k$.

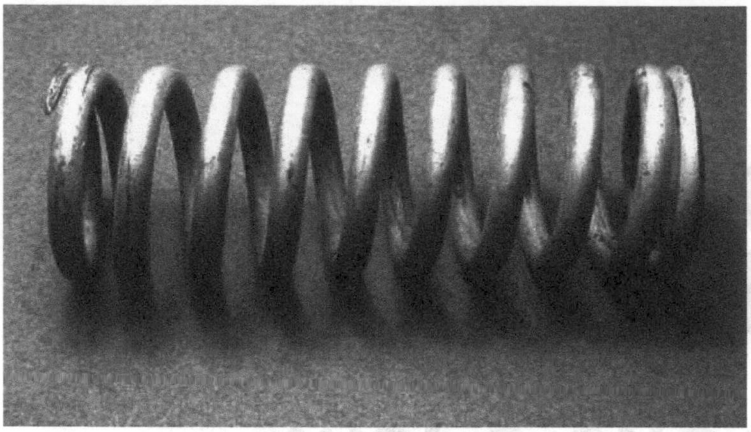

Figure 8.1. Springs come in many shapes. Here a relaxed coiled spring that responds elastically under compression and stretching. (Source: Courtesy Jean-Jacques Milan. Wikimedia Commons.)

Thomas Young (1773–1829). English physician, physicist, and Egyptologist. He observed the interference of light and was the first to propose that light waves are transverse vibrations, explaining thereby the origin of polarization. He contributed much to the translation of the Rosetta stone. (Source: Wikimedia Commons.)

Young's modulus

Real springs, such as the one pictured in Figure 8.1, are physical bodies with mass, shape, and internal structure. Almost any solid body, anchored at one end and pulled at the other, will react like a spring when the force is not too strong. Basically, this reflects that interatomic forces are approximately elastic when the atoms are only displaced slightly away from their positions (Problem 8.1). Many elastic bodies that we handle daily, for example rubber bands, piano wire, sticks, or water hoses, are long rod-like objects with constant cross-section, typically made from homogeneous and *isotropic* material without any particular internal structure. Their uniform composition and simple form make such rods convenient models for real, material springs.

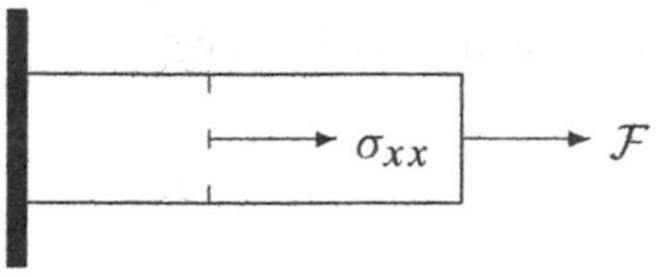

The same tension must act on any cross-section of the rod-like spring.

The force $\mathcal{F} = kx$ necessary to extend the length L of a rod-like material spring by a small amount x must be proportional to the area, A, of the spring's cross-section. For if we bundle N such springs loosely together to make a thicker spring of area NA, the total force will have to be Nkx in order to get the same change of length. This shows that the relevant quantity to speak about is not the force itself, but rather the (average) normal stress, $\sigma_{xx} = N\mathcal{F}/NA = kx/A$, which is independent of the number of sub-springs, and thus independent of the area A of the cross-section. Since the same force $\mathcal{F}$ must act on any cross-section of the rod, the stress must be the same at each point along the spring. Likewise, for a smaller piece of the spring of length $L' < L$, the uniformity implies that it will be stretched proportionally less such that $x'/L' = x/L$. This indicates that the relevant parameter is not the absolute change of length x but rather the relative longitudinal extension or strain $u_{xx} = x/L$, which is independent of the length L of the spring. Consequently, the quantity

$$E = \frac{\sigma_{xx}}{u_{xx}} = \frac{\mathcal{F}/A}{x/L} = k\frac{L}{A} \tag{8.2}$$

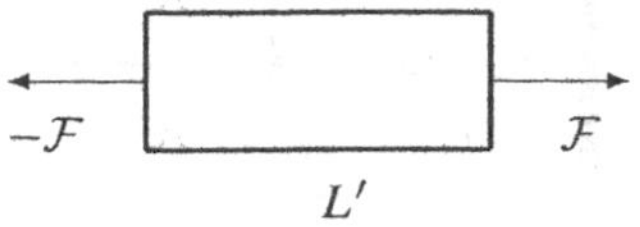

The force acts in opposite directions at the terminal cross-sections of a smaller slice of the spring. The extension is proportionally smaller.

must be independent of the length L of the spring, the area A of its cross-section, and the extension x (for $|x| \ll L$). It is a material parameter, called the *modulus of extension* or *Young's modulus* (1807). Given Young's modulus we may calculate the actual spring constant,

$$\boxed{k = E\frac{A}{L},} \tag{8.3}$$

for any spring, made from this particular material, of length L and cross-section A .

Young's modulus characterizes the behavior of the material of the spring when stretched in one direction. The relation (8.2) also tells us that a unidirectional tension σ_{xx} creates a relative extension,

$$u_{xx} = \frac{\sigma_{xx}}{E}, \tag{8.4}$$

of the material. Evidently, Hooke's law leads to a local linear relation between stress and strain, and materials with this property are generally called *linearly elastic*.

Young's modulus is by way of its definition (8.2) measured in the same units as pressure, and typical values for metals are, like the bulk modulus (2.42), of the order of 10^{11} Pa $=$ 100 GPa $=$ 1 Mbar. In the same way as the bulk modulus is a measure of the incompressibility of a material, Young's modulus is a measure of the *instretchability*. The larger it is, the harder it is to stretch the material. In order to obtain a large strain $u_{xx} \approx 100\%$, one would have to apply stresses of magnitude $\sigma_{xx} \approx E$. Such strains are, of course, not permitted in the theory of small deformations, but Young's modulus nevertheless sets the scale. The fact that the yield stress for metals is roughly a thousand times smaller than Young's modulus shows that for metals the elastic strain can never become larger than 10^{-3}. This in turn justifies the assumption of small displacement gradients underlying the use of Cauchy's strain tensor.

Material	E	ν
Wolfram	411	0.28
Nickel (hard)	219	0.31
Iron (soft)	211	0.29
Plain steel	205	0.29
Cast iron	152	0.27
Copper	130	0.34
Titanium	116	0.32
Brass	100	0.35
Silver	83	0.37
Glass (flint)	80	0.27
Gold	78	0.44
Quartz	73	0.17
Aluminium	70	0.35
Magnesium	45	0.29
Lead	16	0.44
Units	GPa	

Young's modulus and Poisson's ratio for various isotropic materials [Kaye and Laby 1995]. These values are typically a factor 1,000 larger than the tensile strength. Single-wall carbon nanotubes have been reported with a Young's modulus of up to 1,500 GPa [YFAR00].

Example 8.1 [Rope pulling contest]: At company outings, employees often play the game of pulling in teams at each end of a rope. Before the inevitable terminal instability sets in, there is often a prolonged period where the two teams pull with almost equal force $\mathcal{F}$. If the teams each consist of ten persons, all pulling with about their average weight of 70 kg, the total force becomes $\mathcal{F} = 7{,}000$ N. For a rope diameter of 5 cm, the stress becomes quite considerable, $\sigma_{xx} \approx 3.6$ MPa. If Young's modulus is taken to be $E = 360$ MPa, the rope will stretch by $u_{xx} \approx 1\%$.

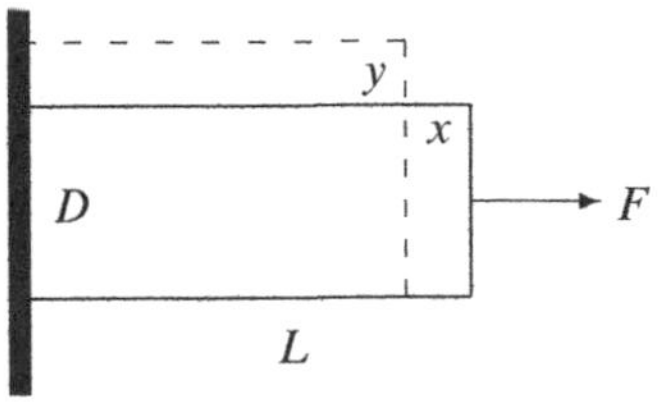

A rod-like spring normally contracts in transverse directions when pulled at the ends.

Poisson's ratio

Normal materials will always contract in directions transverse to the direction of extension. If the transverse size, the "diameter" D of a rod-like spring changes by y, the transverse strain becomes on the order of $u_{yy} = y/D$, and will in general be negative for a positive stretching force $\mathcal{F}$. In linearly elastic materials, the transverse strain must also be proportional to $\mathcal{F}$, so that the ratio u_{yy}/u_{xx} will be independent of $\mathcal{F}$. The negative of this ratio,

$$\nu = -\frac{u_{yy}}{u_{xx}}, \tag{8.5}$$

is called *Poisson's ratio* (1829)[1]. It is also a material parameter characterizing isotropic materials, and as we shall see below there are no others. Poisson's ratio is dimensionless, and must as we shall see lie between -1 and $+0.5$, although it is always positive for natural materials. Typical values for metals lie around 0.30 (see the margin table).

Whereas longitudinal extension can be understood as a consequence of elastic atomic bonds being stretched, it is harder to understand why materials should contract transversally. The reason is, however, that in an isotropic material there are atomic bonds in all directions, and when bonds that are not purely longitudinal are stretched, they create a transverse tension that can only be relieved by transverse contraction of the material (see the model material below). It is, however, possible to construct artificial materials that expand when stretched (see page 131).

Simeon Denis Poisson (1781–1840). French mathematician. Contributed to electromagnetism, celestial mechanics, and probability theory. (Source: Wikimedia Commons.)

[1]Poisson's ratio is also sometimes denoted σ, but that clashes too much with the symbol for the stress tensor. Later we shall in the context of fluid mechanics also use ν for the kinematic viscosity, a choice that does not clash seriously with the use here.

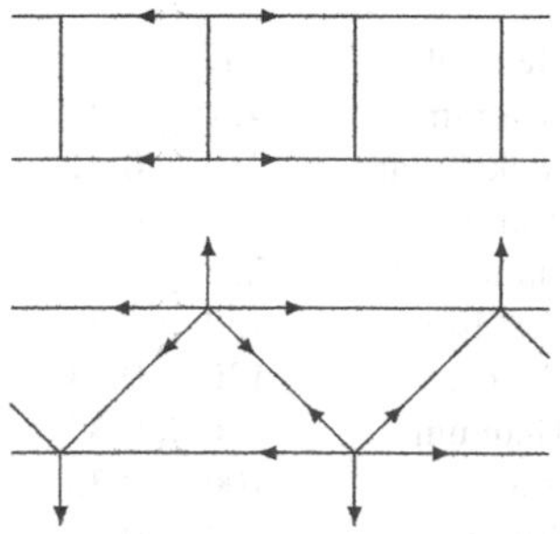

Ladder with purely transverse rungs (top) and with skew rungs (bottom). The forces acting on the sides balance the transverse contraction forces.

Model material: A ladder constructed from ideal springs (see the margin figure) with rungs orthogonal to the sides will not experience a transverse contraction when stretched. If, on the other hand, some of the rungs are skew (making the ladder unusable), they will be stretched along with the ladder. But that will necessarily generate forces that tend to contract the ladder transversally and these forces either have to be balanced by external forces or relieved by actual contraction of the ladder.

The stresses and strains of stretching

Consider a stretched rod-like object laid out along the x-direction of the coordinate system. The only non-vanishing stress component is a constant tension or pull $P = \mathcal{F}/A$ along x, so that the complete symmetric stress tensor becomes

$$\sigma_{xx} = P, \qquad \sigma_{yy} = \sigma_{zz} = 0, \qquad \sigma_{xy} = \sigma_{yz} = \sigma_{zx} = 0. \tag{8.6}$$

From (8.4) and (8.5) we obtain the corresponding diagonal strain components, and adding the vanishing shear strains, the complete strain tensor becomes

$$u_{xx} = \frac{P}{E}, \qquad u_{yy} = u_{zz} = -\nu\frac{P}{E}, \qquad u_{xy} = u_{yz} = u_{zx} = 0. \tag{8.7}$$

In Section 8.3 we shall see that this is actually a feasible deformation that can be represented by a suitable displacement field.

8.2 Hooke's law in isotropic matter

Hooke's law is a linear relation between force and extension, and continuous materials with a linear relation between stress and strain implement the local version of Hooke's law. Since there are six independent strain components and six independent stress components, a general linear relation between them can become quite involved. In isotropic matter where there are no internal material directions defined, the local version of Hooke's law takes the simplest form possible. In the following we shall mostly consider *isotropic homogeneous* matter with the same material properties everywhere. Toward the end of this section we shall, however, briefly touch upon anisotropic materials, such as crystals.

The Lamé coefficients

Gabriel Lamé (1795–1870). French mathematician, engineer, and physicist. Worked on curvilinear coordinates, number theory, and mathematical physics. (Source: Wikimedia Commons.)

The absence of internal directions in isotropic matter tells us that there are only two tensors available to construct a linear relation between the tensors σ_{ij} and u_{ij}. One is the strain tensor u_{ij} itself; the other is the Kronecker delta δ_{ij} multiplied with the trace of the strain tensor $\sum_k u_{kk}$. The trace is in fact the only scalar quantity that can be formed from a linear combination of the strain tensor components. Consequently, the most general strictly linear tensor relation between stress and strain is of the form (Cauchy 1822, Lamé 1852),

$$\boxed{\sigma_{ij} = 2\mu\, u_{ij} + \lambda\, \delta_{ij} \sum_k u_{kk}.} \tag{8.8}$$

The coefficients λ and μ are material constants, called *elastic moduli* or *Lamé coefficients.* Whereas λ has no special name, μ is called the *shear modulus* or the *modulus of rigidity*, because it controls the magnitude of shear (off-diagonal) stresses. Since the strain tensor is dimensionless, the Lamé coefficients are, like the stress tensor itself, measured in units of pressure. We shall see below that the Lamé coefficients are directly related to Young's modulus and Poisson's ratio.

Written out in full detail, the local version of Hooke's law takes the form

$$\sigma_{xx} = (2\mu + \lambda)u_{xx} + \lambda(u_{yy} + u_{zz}), \qquad \sigma_{yz} = \sigma_{zy} = 2\mu u_{yz}, \tag{8.9a}$$

$$\sigma_{yy} = (2\mu + \lambda)u_{yy} + \lambda(u_{zz} + u_{xx}), \qquad \sigma_{zx} = \sigma_{xz} = 2\mu u_{zx}, \tag{8.9b}$$

$$\sigma_{zz} = (2\mu + \lambda)u_{zz} + \lambda(u_{xx} + u_{yy}), \qquad \sigma_{xy} = \sigma_{yx} = 2\mu u_{xy}. \tag{8.9c}$$

For $\mu = 0$, the shear stresses vanish and the stress tensor becomes proportional to the unit matrix, $\sigma_{ij} = \lambda\delta_{ij}\sum_k u_{kk}$. This shows that an isotropic elastic material with vanishing shear modulus is in some respects similar to a fluid at rest—although it is definitely not a fluid. It is sometimes convenient to use this observation to verify the result of a calculation by comparison with a similar calculation for a fluid at rest.

In deriving the local version of Hooke's law we have tacitly assumed that all stresses vanish when the strain tensor vanishes. In a relaxed (undeformed) isotropic material at rest with constant temperature, the pressure must also be constant, so that the above relationship should be understood as the extra stress caused by the deformation itself. If the temperature varies across the undeformed body, thermal expansion will give rise to thermal stresses that also must be taken into account.

Finally, it should also be mentioned that the stress and strain tensors in the Euler representation are both viewed as functions of the actual position of a material particle. In the Lagrange representation they are instead viewed as functions of the position of a material particle in the undeformed body. The difference is, however, negligible for slowly varying displacement fields (small displacement gradients), and will be ignored here.

Form invariance of natural laws: The arguments leading to the most general isotropic relation (8.8) depend strongly on our understanding of tensors as geometric objects in their own right. As soon as we have cast a law of nature in the form of a scalar, vector, or tensor relation, its validity in all Cartesian coordinate systems is immediately guaranteed. The form invariance of the natural laws under transformations that relate different observers has since Einstein been an important guiding principle in the development of modern theoretical physics.

Young's modulus and Poisson's ratio in relation to the Lamé coefficients

Young's modulus and Poisson's ratio must be functions of the two Lamé coefficients. To derive the relations between these material parameters, we insert the stresses (8.6) and strains (8.7) for the simple stretching of a material spring into the general relations (8.9), and get

$$P = (2\mu + \lambda)\frac{P}{E} - 2\lambda\nu\frac{P}{E}, \qquad 0 = -(2\mu + \lambda)\nu\frac{P}{E} + \lambda\left(-\nu\frac{P}{E} + \frac{P}{E}\right).$$

Solving these equations for E and ν, we obtain

$$\boxed{E = \mu\frac{3\lambda + 2\mu}{\lambda + \mu}, \qquad \nu = \frac{\lambda}{2(\lambda + \mu)}.} \tag{8.10}$$

Conversely, we may also express the Lamé coefficients in terms of Young's modulus and Poisson's ratio:

$$\boxed{\lambda = \frac{E\nu}{(1 - 2\nu)(1 + \nu)}, \qquad \mu = \frac{E}{2(1 + \nu)}.} \tag{8.11}$$

In practice it is Young's modulus and Poisson's ratio that are found in tables, and the above relations immediately allow us to calculate the Lamé coefficients.

Mechanical pressure and bulk modulus

The trace of the stress tensor (8.8) becomes

$$\sum_i \sigma_{ii} = (2\mu + 3\lambda) \sum_i u_{ii} \tag{8.12}$$

because the trace of the Kronecker delta is $\sum_i \delta_{ii} = 3$. Since the stress tensor in Hooke's law represents the change in stress due to the deformation, we find the change in the mechanical pressure (6.19) caused by the deformation

$$\Delta p = -\frac{1}{3} \sum_i \sigma_{ii} = -\left(\lambda + \frac{2}{3}\mu\right) \sum_i u_{ii}. \tag{8.13}$$

The trace of the strain tensor was previously shown in (7.35) to be proportional to the relative local change in density, $\sum_i u_{ii} = \nabla \cdot \boldsymbol{u} = -\Delta\rho/\rho$, and using the definition of the bulk modulus (2.42) $K = \rho dp/d\rho \approx \rho\, \Delta p/\Delta\rho$, one finds the (isothermal) bulk modulus:

$$\boxed{K = \lambda + \frac{2}{3}\mu = \frac{E}{3(1-2\nu)}.} \tag{8.14}$$

The bulk modulus equals Young's modulus for $\nu = 1/3$, which is in fact a typical value for ν in many materials. The Lamé coefficients and the bulk modulus are all proportional to Young's modulus and thus of the same order of magnitude.

Inverting Hooke's law

Hooke's law (8.8) may be inverted so that strain is instead expressed as a linear function of stress. Solving for u_{ij} and making use of (8.12), we get

$$u_{ij} = \frac{\sigma_{ij}}{2\mu} - \frac{\lambda}{2\mu(3\lambda + 2\mu)} \delta_{ij} \sum_k \sigma_{kk}. \tag{8.15}$$

Introducing Young's modulus and Poisson's ratio from (8.10), this takes the simpler form

$$\boxed{u_{ij} = \frac{1+\nu}{E}\sigma_{ij} - \frac{\nu}{E}\delta_{ij} \sum_k \sigma_{kk}.} \tag{8.16}$$

Explicitly, we find for the six independent components

$$u_{xx} = \frac{\sigma_{xx} - \nu(\sigma_{yy} + \sigma_{zz})}{E}, \qquad u_{yz} = u_{zy} = \frac{1+\nu}{E}\sigma_{yz}, \tag{8.17a}$$

$$u_{yy} = \frac{\sigma_{yy} - \nu(\sigma_{zz} + \sigma_{xx})}{E}, \qquad u_{zx} = u_{xz} = \frac{1+\nu}{E}\sigma_{zx}, \tag{8.17b}$$

$$u_{zz} = \frac{\sigma_{zz} - \nu(\sigma_{xx} + \sigma_{yy})}{E}, \qquad u_{xy} = u_{yx} = \frac{1+\nu}{E}\sigma_{xy}. \tag{8.17c}$$

Evidently, if the only stress is $\sigma_{xx} = P$, we obtain immediately from the correct relations (8.7) for simple stretching.

Positivity constraints

The bulk modulus $K = \lambda + \frac{2}{3}\mu$ cannot be negative, because a material with negative K would expand when put under pressure (see however [WL05]). Imagine what would happen to such a strange material if placed in a closed vessel surrounded by normal material, for example air. Increasing the air pressure a tiny bit would make the strange material expand, causing a further pressure increase followed by expansion until the whole thing blows up. Likewise, materials with negative shear modulus, μ, would mimosa-like pull away from a shearing force instead of yielding to it. Formally, it may be shown (see Section 8.4) that the conditions $3\lambda + 2\mu > 0$ and $\mu > 0$ follow from the requirement that the elastic energy density should be bounded from below. Although λ, in principle, may assume negative values, natural materials always have $\lambda > 0$.

Young's modulus cannot be negative because of these constraints, and this confirms that a rope always stretches when pulled at the ends. If there were materials with the ability to contract when pulled, they would also behave magically. As you begin climbing up a rope made from such material, it pulls you further up. If such materials were ever created, they would spontaneously contract into nothingness at the first possible occasion or at least into a state with a positive value of Young's modulus.

Poisson's ratio $\nu = \lambda/2(\lambda + \mu)$ reaches its maximum $\nu \to 1/2$ for $\lambda \to \infty$. The limit $\nu \to 1/2$ corresponds to $\mu \to 0$, and in that limit there are no shear stresses in the material, which as mentioned in some respects behaves like a fluid at rest. Since the bulk modulus $K = \lambda + \frac{2}{3}\mu$ is positive, we have $\lambda > -\frac{2}{3}\mu$, and since Poisson's ratio is a monotonically increasing function of λ, the minimal value is obtained for $\lambda = -\frac{2}{3}\mu$ corresponding to $\nu = -1$.

Although natural materials shrink in the transverse directions when stretched, and thus have $\nu > 0$, there may actually exist so-called *auxetic* materials that expand transversally when stretched without violating the laws of physics. Composite materials and foams with isotropic elastic properties and negative Poisson's ratio have in fact been made [Lak92, Bau03, HCG&08].

Limits to Hooke's law

Hooke's law for isotropic materials, expressed by the linear relationship between stress and strain, is only valid for stresses up to a certain value, called the *proportionality limit*. Beyond the proportionality limit, nonlinearities set in, and the present formalism becomes invalid. Eventually one reaches a point, called the *elasticity limit*, where the material ceases to be elastic and undergoes permanent deformation without much further increase of stress but instead accompanied by loss of energy to heat. In the end, the material may even fracture. Hooke's law is, however, a very good approximation for most metals under normal conditions where stresses are tiny compared to the elastic moduli.

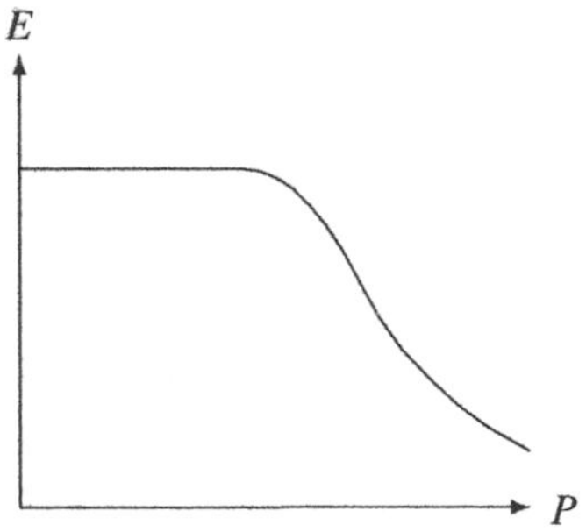

Sketch of how Young's modulus might vary as a function of increased tension. Beyond the proportionality limit, its effective value becomes generally smaller.

* Anisotropic materials

Anisotropic (also called *aeolotropic*) materials have different properties in different directions. In the most general case, Hooke's law takes the form

$$\boxed{\sigma_{ij} = \sum_{kl} E_{ijkl} u_{kl},} \tag{8.18}$$

where the coefficients E_{ijkl} form a tensor of rank 4, called the *elasticity tensor*.

For isotropic materials, one may verify from (8.8) that the elasticity tensor takes the form,

$$E_{ijkl} = \lambda\delta_{ij}\delta_{kl} + \mu\left(\delta_{ik}\delta_{jl} + \delta_{jk}\delta_{il}\right). \tag{8.19}$$

In general materials, there are more than two material constants, the actual number depending on the intrinsic complexity of the internal structure of the material.

Requiring only that the stress tensor be symmetric, the elasticity tensor connects the six independent components of stress with the six independent components of strain and can in principle contain $(6 \times 6) = 36$ independent parameters. If one further demands that an elastic energy function should exist (see page 135), this (6×6) array of coefficients must itself be symmetric under the exchange $ij \leftrightarrow kl$, that is, $E_{ijkl} = E_{klij}$, and can thus only contain $(6 \times 7)/2 = 21$ independent parameters. The orientation of a completely asymmetric material relative to the coordinate system takes three parameters (Euler angles), so altogether there may be up to 18 truly independent material constants characterizing the intrinsic elastic properties of such a material, a number actually realized by triclinic crystals [Landau and Lifshitz 1986, Green and Adkins 1960].

Example 8.2 [Cubic symmetry]: In crystals with cubic symmetry, for example diamond, the atoms or molecules are arranged in a lattice of cubic cells with identical structure. In the natural coordinate system along the lattice axes, the material has fourfold rotation symmetry (90°) around the each of the three coordinate axes and mirror symmetry in each of the the three coordinate planes. There are thus $4 \times 3 \times 3 = 36$ elements in the symmetry group (and not an infinity as in isotropic materials). These symmetries demand that the set of equations expressing Hooke's law remains unchanged under change of sign of any of the coordinate axes and under interchange of any two coordinate axes, and this limits their form severely. The diagonal components take the same form as for full isotropy, whereas the off-diagonal components also take the same form but with a different material constant κ for pure shear,

$$\sigma_{xx} = (2\mu + \lambda)u_{xx} + \lambda(u_{yy} + u_{zz}), \qquad \sigma_{yz} = \sigma_{zy} = 2\kappa u_{yz}, \tag{8.20a}$$

$$\sigma_{yy} = (2\mu + \lambda)u_{yy} + \lambda(u_{zz} + u_{xx}), \qquad \sigma_{zx} = \sigma_{xz} = 2\kappa u_{zx}, \tag{8.20b}$$

$$\sigma_{zz} = (2\mu + \lambda)u_{zz} + \lambda(u_{xx} + u_{yy}), \qquad \sigma_{xy} = \sigma_{yx} = 2\kappa u_{xy}. \tag{8.20c}$$

Cubic symmetry thus has three material constants. Notice that this form is only valid in a coordinate system with axes coinciding with the lattice axes.

8.3 Static uniform deformation

To see how Hooke's law works for continuous systems, we now turn to the extremely simple case of a *static uniform deformation* in which the strain tensor u_{ij} takes the same value everywhere in a body at all times. Hooke's law (8.8) then ensures that the stress tensor is likewise constant throughout the body, so that all its derivatives vanish, $\nabla_k \sigma_{ij} = 0$. From the condition (6.22) for mechanical equilibrium, $f_i + \sum_j \nabla_j \sigma_{ij} = 0$, it follows that $f_i = 0$ so that uniform deformation necessarily excludes body forces. Conversely, in the presence of body forces such as gravity or electromagnetism, there must always be non-uniform deformation of an isotropic material, in fact a quite reasonable conclusion.

Furthermore, at the boundary of a uniformly deformed body, the stress vector $\boldsymbol{\sigma} \cdot \boldsymbol{n}$ is required to be continuous, and this puts strong restrictions on the form of the external forces that may act on the surface of the body. Uniform deformation is for this reason only possible under very special circumstances, but when it applies, the displacement field is nearly trivial.

Uniform compression

In a fluid at rest with a constant pressure P, the stress tensor is $\sigma_{ij} = -P\delta_{ij}$ everywhere. If a solid body made from isotropic material is immersed into this fluid, the natural guess is that the pressure will also be P inside the body. Inserting $\sigma_{ij} = -P\delta_{ij}$ into (8.16) and using that $\sum_k \sigma_{kk} = -3P$, the strain may be written as

$$u_{ij} = -\frac{P}{3K}\delta_{ij}. \tag{8.21}$$

Since $u_{xx} = \nabla_x u_x = -P/3K$, we may immediately integrate this equation (and the similar ones for u_{yy} and u_{zz}) and obtain a particular solution to the displacement field,

$$u_x = -\frac{P}{3K}x, \qquad u_y = -\frac{P}{3K}y, \qquad u_z = -\frac{P}{3K}z. \tag{8.22}$$

The most general solution is obtained by adding an arbitrary small rigid body displacement to this solution.

Note that we arrived at this result by making an educated *guess* for the form of the stress tensor inside the body. It could in principle be wrong but is in fact correct due to a uniqueness theorem to be derived in Section 8.4. The theorem guarantees, in analogy with the uniqueness theorems of electrostatics, that provided the equations of mechanical equilibrium and the boundary conditions are fulfilled by the guess (which they are here), there is essentially only one solution to any *elastostatic* problem. The only liberty left is the arbitrary small rigid body displacement that may always be added to the solution.

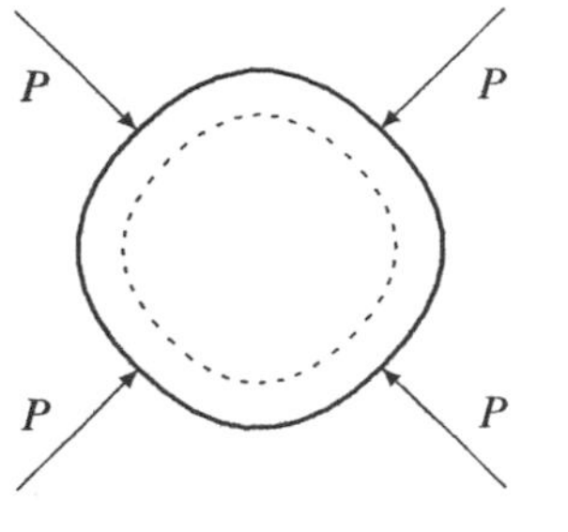

A body made from isotropic, homogenous material subject to a uniform external pressure will be uniformly compressed.

Uniform stretching

At the beginning of this chapter we investigated the reaction of a rod-like material body stretched along its main axis, say the x-direction, by means of a tension $\sigma_{xx} = P$ acting uniformly over its constant cross section. If there are no other external forces acting on the body, the natural guess is that the only non-vanishing component of the stress tensor is $\sigma_{xx} = P$ throughout the body, because that gives no surface stresses on the sides of the rod. Inserting this into (8.15), we obtain as before the strains (8.7). The corresponding displacement field is again found by integrating $\nabla_x u_x = u_{xx}$ etc., and we find the particular solution

$$u_x = \frac{P}{E}x, \qquad u_y = -\nu\frac{P}{E}y, \qquad u_z = -\nu\frac{P}{E}z. \tag{8.23}$$

It describes a simple dilatation along the x-axis and a compression toward the x-axis in the yz-plane.

If the rod-like spring is clamped on the sides by a hard material, the boundary conditions are instead $u_y = u_z = 0$ on the sides. In that case the only non-vanishing constant strain is u_{xx}, and the solution is obtained along the same lines as above (see Problem 8.2).

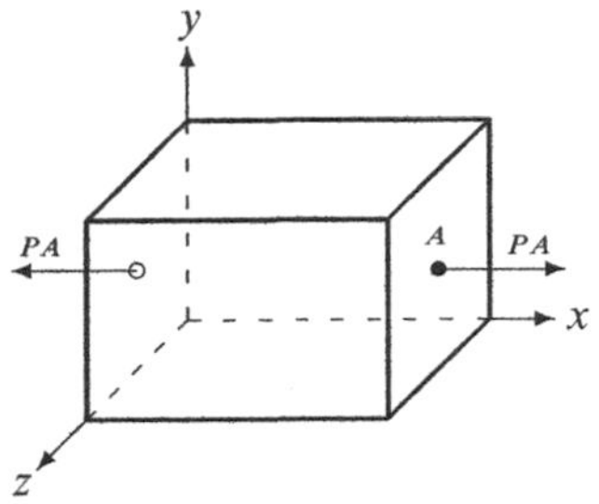

Uniformly rectangular stretched rod under constant longitudinal tension P. The shape of the cross-section can be arbitrary, but must be constant throughout the rod, with area A.

Uniform shear

Finally we return to the example from page 99 of a clamped rectangular slab of homogeneous, isotropic material subject to shear stress along one side (here the x-direction). As we argued there, the shear stress $\sigma_{xy} = P$ must be constant throughout the material. What we did not appreciate at that point was that the symmetry of the stress tensor demands that also $\sigma_{yx} = P$ everywhere in the material. As a consequence, there must also be shearing forces acting on the ends of the slab whereas the remaining sides are free (see the margin figure).

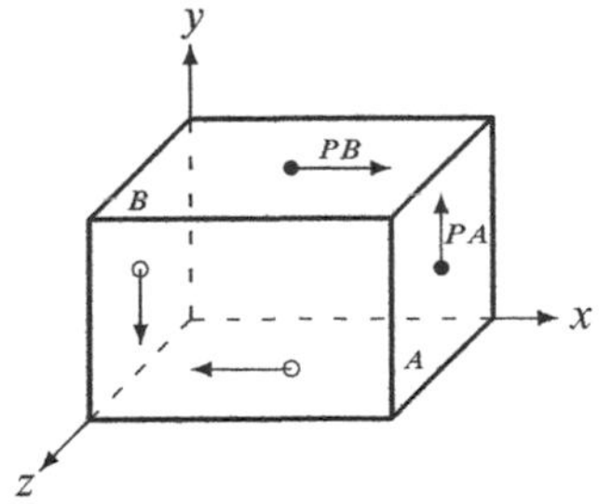

Clamped rectangular slab under constant shear stress $\sigma_{xy} = \sigma_{yx} = P$. The upper clamp is acted upon by an external force $\mathcal{F}_x = PB$, where B is the area of the clamp, while the force on the lower is $\mathcal{F}_x = -PB$. The symmetry of the stress tensor demands a clamp force $\mathcal{F}_y = PA$ on the right-hand side and a force $\mathcal{F}_y = -PA$ on the left-hand side, where A is the area of that clamp.

Assuming that there are no other stresses, the only strain component becomes $u_{xy} = P/2\mu$, and using that $2u_{xy} = \nabla_x u_y + \nabla_y u_x$, we find a particular solution

$$u_x = \frac{P}{\mu} y, \qquad u_y = u_z = 0. \tag{8.24}$$

In these coordinates the displacement in the x-direction vanishes for $y = 0$ and grows linearly with y. Notice that each infinitesimal "needle" besides the deformation is also rotated by a small angle $\phi = \frac{1}{2}(\nabla_x u_y - \nabla_y u_x) = -P/2\mu$ around the z-axis (see Problem 7.8).

* Finite uniform deformation in the Euler representation

In Section 7.5 we discussed the extension of the strain tensor formalism to strongly varying displacement fields. We found that in the Euler representation where the displacement field is a function of the actual coordinates $\boldsymbol{x}$, the correct definition of the strain tensor is non-linear,

$$u_{ij} = \frac{1}{2}\left(\nabla_i u_j + \nabla_j u_i - \sum_k \nabla_i u_k \nabla_j u_k\right). \tag{8.25}$$

General arguments again ensure that a linear relationship[2] between stress and strain for a homogeneous, isotropic material is given by (8.8). We shall for simplicity assume that the elastic coefficients are constant.

Under these assumptions it follows as before that a constant stress tensor implies a constant strain tensor. In the three cases discussed above, the strain tensors are calculated from the stress tensors in the same way as before. Differences first arise in the calculation of the Eulerian displacement fields. For uniform compression, the strain tensor is again $u_{ij} = -(P/3K)\delta_{ij}$, and since the displacement field must be a uniform scaling we find the solution,

$$u_x = Ax, \qquad u_y = Ay, \qquad u_z = Az, \qquad \text{with } A = 1 - \sqrt{1 + \frac{2P}{3K}}. \tag{8.26}$$

For $|P| \ll K$ this turns into the solution (8.22). Notice that the solution breaks down for a dilatation with $P < -\frac{3}{2}K$. The reason for the singular behavior for $P = -\frac{3}{2}K$ is that it yields $\boldsymbol{u} = \boldsymbol{x}$, implying that all material particles originally were found at the same position, $\boldsymbol{X} = \boldsymbol{x} - \boldsymbol{u} = \boldsymbol{0}$. Similar behavior is found for uniform stretching and uniform shear.

8.4 Elastic energy

The work performed by the external force in extending a spring further by the amount dx is $dW = \mathcal{F}\,dx = kx\,dx$. Integrating this expression, we obtain the total work $W = \frac{1}{2}kx^2$, which is identified with the well-known expression for the elastic energy, $\mathcal{E} = \frac{1}{2}kx^2$, stored in a stretched or compressed spring. Calculated per unit of volume $V = AL$ for a rod-like spring, we find the density of elastic energy in the material,

$$\varepsilon = \frac{\mathcal{E}}{V} = \frac{kx^2}{2V} = \frac{1}{2}Eu_{xx}^2 = \frac{\sigma_{xx}^2}{2E}, \tag{8.27}$$

where $u_{xx} = x/L$ and $\sigma_{xx} = \mathcal{F}/A$. Poisson's ratio ν does not appear because the transverse contraction given by Poisson's ratio can play no role in building up the elastic energy of a stretched spring when there are no forces acting on the sides of the spring.

[2]If linearity is not required, the most general stress tensor for an isotropic material will be a linear combination of the three matrices $\mathbf{1}$, $\mathbf{u}$, and $\mathbf{u}^2$ with coefficients that may depend on the three scalar invariants of the strain tensor $\mathbf{u}$. Such materials are called *hyperelastic* [Doghri 2000] and may, for example, be used to describe the elasticity of rubber.

Deformation work

In the general case, strains and stresses vary across the body, and the calculation becomes more complicated. To determine the general expression for the elastic energy density we use the expression derived in Section 7.4 for the virtual work performed under an infinitesimal change δu_{ij} of the strain field,

$$\delta W_{\text{deform}} = \int_V \boldsymbol{\sigma} : \delta \mathbf{u} \, dV, \tag{8.28}$$

where $\boldsymbol{\sigma} : \delta \mathbf{u} = \sum_{ij} \sigma_{ij} \delta u_{ij}$. The volume V is the actual volume of the body, not the volume of the undeformed body, but the difference between integrating over the actual volume and the undeformed volume is negligible for a slowly varying displacement field.

To build up a complete strain field, one must divide the process into infinitesimal steps and add the work for each step. Here one should remember that the stress tensor depends on the strain tensor, $\boldsymbol{\sigma} = \boldsymbol{\sigma}(\mathbf{u})$ or $\sigma_{ij} = \sigma_{ij}(\{u_{kl}\})$, and therefore also changes along the road (in the six-dimensional space of possible symmetric deformation tensors). For the final result to be independent of the road along which the strain is built up, we must require that the cross-derivatives are equal,

$$\boxed{\frac{\partial \sigma_{ij}}{\partial u_{kl}} = \frac{\partial \sigma_{kl}}{\partial u_{ij}}.} \tag{8.29}$$

When this condition is fulfilled, the work spent in building up the deformation may be interpreted as an energy stored in the deformation of the body.

Denoting the deformation energy density $\varepsilon = \varepsilon(\mathbf{u})$, the total deformation energy becomes

$$\mathcal{E} = \int_V \varepsilon \, dV, \tag{8.30}$$

and the stress-strain relation is derived from the energy density,

$$\sigma_{ij} = \frac{\partial \varepsilon}{\partial u_{ij}}. \tag{8.31}$$

Evidently, this stress tensor satisfies the condition (8.29).

Energy in general linear materials

If the stress tensor is a general linear function of the strain tensor as in (8.18),

$$\sigma_{ij} = \sum_{kl} E_{ijkl} u_{kl}, \tag{8.32}$$

the above condition translates into the symmetry of the elasticity tensor,

$$E_{ijkl} = E_{klij}. \tag{8.33}$$

When this condition is fulfilled, the energy of a strain in its own stress field may now be built up by adding infinitesimal strain changes, starting from zero stress. On average, each infinitesimal strain change only meets half the final stress, so that the total energy density becomes

$$\boxed{\varepsilon = \tfrac{1}{2}\boldsymbol{\sigma} : \mathbf{u} = \frac{1}{2} \sum_{ij} \sigma_{ij} u_{ji} = \frac{1}{2} \sum_{ijkl} E_{ijkl} u_{ij} u_{kl}.} \tag{8.34}$$

Not unsurprisingly, it is a quadratic polynomial in the strain tensor components.

Isotropic materials

For isotropic materials with σ_{ij} given by Hooke's law (8.8), the energy density simplifies to

$$\varepsilon = \mu \sum_{ij} u_{ij}^2 + \frac{1}{2}\lambda \left(\sum_i u_{ii}\right)^2 = \mu \mathrm{Tr}\left(\mathbf{u}^2\right) + \frac{1}{2}\lambda \mathrm{Tr}(\mathbf{u})^2. \tag{8.35}$$

The energy density must be bounded from below, for if it were not, elastic materials would be unstable, and an unlimited amount of energy could be obtained by increasing the state of deformation. From the boundedness, it follows immediately that the shear modulus must be positive, $\mu > 0$, because otherwise we might make one of the off-diagonal components of the strain tensor, say u_{xy}, grow without limit and extract unlimited energy. The condition on λ is more subtle because the diagonal components of the strain tensor are involved in both terms. For a uniform deformation with $u_{ij} = k\delta_{ij}$, the energy density becomes $\varepsilon = \frac{3}{2}(3\lambda + 2\mu)k^2$, implying that $3\lambda + 2\mu > 0$, that is, that the bulk modulus (8.14) is positive. In Problem 8.6 it is shown that there are no stronger conditions.

The energy density may also be expressed as a function of the stress tensor,

$$\varepsilon = \frac{1+\nu}{2E} \sum_{ij} \sigma_{ij}^2 - \frac{\nu}{2E}\left(\sum_i \sigma_{ii}\right)^2 = \frac{1+\nu}{2E}\mathrm{Tr}\left(\boldsymbol{\sigma}^2\right) - \frac{\nu}{2E}\mathrm{Tr}(\boldsymbol{\sigma})^2. \tag{8.36}$$

Under uniform stretching where the only non-vanishing stress is σ_{xx}, we have $\mathrm{Tr}\left(\boldsymbol{\sigma}^2\right) = \mathrm{Tr}(\boldsymbol{\sigma})^2 = \sigma_{xx}^2$, and all the dependence on Poisson's ratio ν cancels out, bringing us back to the energy density in a rod-like spring (8.27).

Total energy in an external field

In an external gravitational field, a small displacement $\boldsymbol{u}(\boldsymbol{x})$ will lead to a change in gravitational potential energy of a material particle (defined in Equation (2.34) on page 31),

$$d\mathcal{E} = \left(\Phi(\boldsymbol{x}) - \Phi(\boldsymbol{X})\right) dM = \left(\Phi(\boldsymbol{x}) - \Phi(\boldsymbol{x} - \boldsymbol{u}(\boldsymbol{x}))\right) dM \approx -\rho \boldsymbol{g} \cdot \boldsymbol{u}\, dV. \tag{8.37}$$

In the last step we expanded the potential to first order in $\boldsymbol{u}$ and used $\boldsymbol{g} = -\nabla\Phi$. Since the elastic energy is of second order in the displacement gradients, we should for consistency also expand the potential to second order; but for normal bodies, the second-order contribution to the gravitational energy density is, however, much smaller than the elastic energy density (see Problem 8.7).

Adding the gravitational contribution, the total energy of a linearly elastic body in a conservative external force field $\boldsymbol{f} = \rho \boldsymbol{g}$ is, accordingly,

$$\mathcal{E} = -\int_V \boldsymbol{f} \cdot \boldsymbol{u}\, dV + \int_V \tfrac{1}{2}\boldsymbol{\sigma}:\mathbf{u}\, dV, \tag{8.38}$$

where as before $\boldsymbol{\sigma}:\mathbf{u} = \sum_{ij} \sigma_{ij} u_{ji}$.

* Uniqueness of elastostatics solutions

To prove the uniqueness of the solutions to the mechanical equilibrium equations (6.22) for linearly elastic materials, we assume that we have found two solutions $\boldsymbol{u}^{(1)}$ and $\boldsymbol{u}^{(2)}$ that both satisfy the equilibrium equations and the boundary conditions. Due to the linearity, the difference between the solutions $\boldsymbol{u} = \boldsymbol{u}^{(1)} - \boldsymbol{u}^{(2)}$ generates a difference in strain $u_{ij} = \frac{1}{2}(\nabla_i u_j + \nabla_j u_i)$ and a difference in stress $\sigma_{ij} = \sum_{kl} E_{ijkl} u_{kl}$. Under the assumption that the body forces are identical for the two fields (as is the case in constant gravity), the change in stress satisfies the equation, $\sum_j \nabla_j \sigma_{ij} = 0$, and by means of Gauss' theorem we obtain

$$0 = \int_V \sum_{ij} u_i \nabla_j \sigma_{ij} \, dV = \oint_S \sum_{ij} u_i \sigma_{ij} \, dS_j - \int_V \sum_{ij} \sigma_{ij} \nabla_j u_i \, dV.$$

Here the surface integral vanishes because of the boundary conditions, which either specify the same displacements for the two solutions at the surface, that is, $u_i = 0$, or the same stress vectors, that is, $\sum_j \sigma_{ij} dS_j = 0$. Using the symmetry of the stress tensor, we get

$$0 = \sum_{ij} \sigma_{ij} \nabla_j u_i = \sum_{ij} \sigma_{ij} u_{ij} = \sum_{ijkl} E_{ijkl} u_{ij} u_{kl}.$$

The integrand is of the same form as the energy density (8.34), which is always assumed to be positive definite. Consequently, the integral can only vanish if the strain tensor for the difference field vanishes everywhere in the body, that is, $u_{ij} = 0$.

We have thus shown that given the boundary conditions, there is essentially only one solution to the equations of mechanical equilibrium in linear elastic materials. Although the two displacement fields may, in principle, differ by a small rigid body displacement, they will give rise to identical deformations everywhere in the body. Therefore, if we have somehow guessed a solution satisfying the equations of mechanical equilibrium and the boundary conditions, it will necessarily be the right one.

Problems

8.1 Two particles interact with a potential energy $V(r)$, depending only on their mutual distance r. Show that the force between them obeys Hooke's law near an equilibrium position.

8.2 A beam with constant cross-section is fixed such that its sides cannot move. One end of the beam is also held in place while the other end is pulled with a uniform tension P. Determine the strains, stresses, and the displacement field in the beam.

8.3 A rectangular beam with its axis along the x-axis is fixed on the two sides orthogonal to the y-axis but left free on the two sides orthogonal to the z-axis. The beam is held fixed at one end and pulled with a uniform tension P at the other. Determine the strains, stresses, and the displacement field in the beam.

8.4 Show that the field of uniform compression (8.22) is a superposition of uniform stretching fields (8.23) in the three coordinate directions.

*** 8.5 [Two-dimensional elasticity]** In two dimensions (x, y),

(a) Show that Cauchy's strain tensor always fulfills the condition (see Problem 7.11),

$$\nabla_y^2 u_{xx} + \nabla_x^2 u_{yy} = 2\nabla_x \nabla_y u_{xy}. \tag{8.39}$$

(b) Assume that the only non-vanishing components of the stress tensor are σ_{xx}, σ_{yy}, and $\sigma_{xy} = \sigma_{yx}$. Formulate Cauchy's equilibrium equations for this stress tensor in the absence of volume forces.

(c) Show that the solution to the equilibrium equations takes the form,

$$\sigma_{xx} = \nabla_y^2 \phi, \qquad \sigma_{yy} = \nabla_x^2 \phi, \qquad \sigma_{xy} = -\nabla_x \nabla_y \phi,$$

where ϕ is an arbitrary function of x and y.

(d) Calculate the strain tensor in terms of ϕ in an isotropic elastic medium, and show that ϕ must satisfy the biharmonic equation

$$\nabla_x^4\phi + \nabla_y^4\phi + 2\nabla_x^2\nabla_y^2\phi = 0.$$

(e) Determine a solution of the displacement field (u_x, u_y, u_z), when $\phi = xy^2$ and Young's modulus is set to $E = 1$. Hint: Begin by integrating the diagonal elements of the strain tensor, and afterward add extra terms to u_x to get the correct off-diagonal elements.

8.6 **(a)** Show that one may write the energy density (8.34) in the following form

$$\varepsilon = \tfrac{1}{2}[\lambda - 2\mu(3\alpha^2 - 2\alpha)]\left(\sum_i u_{ii}\right)^2 + \mu\sum_{ij}\left(u_{ij} - \alpha\sum_k u_{kk}\delta_{ij}\right)^2,$$

where α is arbitrary. **(b)** Use this to argue that $3\lambda + 2\mu > 0$ and that this is the strictest condition on λ.

8.7 Estimate the ratio of the second-order gravitational energy density (8.37) to the elastic energy density for a body of size L, and show that it is normally tiny.

8.8 **(a)** Calculate the non-linear Euler field for a linearly elastic beam subject to finite uniform stretching $\sigma_{xx} = P$. **(b)** What is the range of validity of the solution?

8.9 **(a)** Calculate the non-linear Euler field for a linearly elastic slab subject to finite uniform shear $\sigma_{xy} = P$. **(b)** What is the range of validity of the solution?

9 Basic elastostatics

Flagpoles, bridges, houses, and towers are built from elastic materials, and are designed to stay in one place with at most small excursions away from equilibrium due to wind and water currents. Ships, airplanes, and space shuttles are designed to move around, and their structural integrity depends crucially on the elastic properties of the materials from which they are made. Almost all human constructions and natural structures depend on elasticity for stability and ability to withstand external stresses [Vogel 1988, Vogel 1998].

It can come as no surprise that the theory of static elastic deformation, *elastostatics*, is a huge engineering subject. Engineers must know the internal stresses in their constructions in order to predict risk of failure and set safety limits, and that is only possible if the elastic properties of the building materials are known, and if they are able to solve the equations of elastostatics, or at least get decent approximations of them. Today computers aid engineers in getting precise numeric solutions to these equations and allow them to build critical structures, such as submarines, supertankers, airplanes, and space vehicles, in which over-dimensioning of safety limits is deleterious to fuel consumption as well as to construction costs.

In this fairly long chapter we shall recapitulate the equations of elastostatics for bodies made from isotropic materials and apply them to the classic highly symmetric body geometries for which analytic solutions can be obtained. With the present availability of numeric simulation tools, there is no reason to spend a lot of time on more complex examples. In Chapter 10 we shall apply the results to one-dimensional bodies (rods), and in Chapter 11 we shall outline the principles for obtaining numeric solutions. Many specialized textbooks cover elastostatics to greater depth than here, for example [Bower 2010], [da Silva 2006], [Landau and Lifshitz 1986], and [Sedov 1975].

9.1 Equations of elastostatics

Basic elastostatics concerns deformations in bodies with highly symmetric non-exceptional shapes, subject to simple combinations of external forces. The bodies are assumed to be made from an isotropic and homogeneous linear elastic material, but even with all these assumptions it still takes some work to find the analytic solutions, when it can be done at all. One of the simplest cases—the gravitational settling of a circularly cylindrical body standing on a hard horizontal floor—cannot be solved analytically, although a decent analytic approximation can be found (Section 9.2).

The fundamental equations of elastostatics are obtained from a combination of the results of the three preceding chapters,

$$f_i + \sum_j \nabla_j \sigma_{ij} = 0, \quad \text{mechanical equilibrium (6.22)} \tag{9.1a}$$

$$\sigma_{ij} = 2\mu\, u_{ij} + \lambda \delta_{ij} \sum_k u_{kk}, \quad \text{Hooke's law (8.8)} \tag{9.1b}$$

$$u_{ij} = \tfrac{1}{2}\left(\nabla_i u_j + \nabla_j u_i\right), \quad \text{Cauchy's strain tensor (7.21)} \tag{9.1c}$$

We shall only use these equations with time-independent external gravity where the body force, if present, is given by $\boldsymbol{f} = \rho \boldsymbol{g}$. The body force could also be of electromagnetic origin, for example caused by an inhomogeneous electric field acting on a polarizable dielectric.

Inserting the two last equations into the first we get in index notation,

$$\sum_j \nabla_j \sigma_{ij} = 2\mu \sum_j \nabla_j u_{ij} + \lambda \nabla_i \sum_j u_{jj} = \mu \sum_j \nabla_j^2 u_i + (\lambda + \mu) \nabla_i \sum_j \nabla_j u_j.$$

Rewriting this in vector notation, the equilibrium equation finally takes the form

$$\boxed{\boldsymbol{f} + \mu \nabla^2 \boldsymbol{u} + (\lambda + \mu) \boldsymbol{\nabla} \boldsymbol{\nabla} \cdot \boldsymbol{u} = \boldsymbol{0},} \tag{9.2}$$

called *Navier's equation of equilibrium* or the *Navier–Cauchy equilibrium equation*. For the analytic calculations in this chapter it actually turns out to be more enlightening to use the three basic equations (9.1) rather than this equation.

Claude Louis Marie Henri Navier (1785–1836). French engineer; worked on applied mechanics, elasticity, fluid mechanics, and suspension bridges. Formulated the first version of the elastic equilibrium equation in 1821, a year before Cauchy gave its final form. (Source: Wikimedia Commons.)

The displacement gradients are always assumed to be tiny, allowing us to ignore all nonlinear terms in the above equations, as well as to view the displacement field as a function of the original coordinates of the undeformed material rather than the actual coordinates. Effectively we use the Lagrangian representation with the spatial variable denoted $\boldsymbol{x}$ rather than $\boldsymbol{X}$. Furthermore, the linearity of the equilibrium equations allows us to *superpose* solutions. Thus, for example, if you both compress and stretch a body uniformly, the displacement field for the combined operation will simply be the sum of the respective displacement fields.

The boundary conditions are often implicit in the mere posing of an elastostatics problem. Typically, a part of the body surface is "glued" to a hard surface where the displacement has to vanish, and where the environment automatically provides the external reaction forces necessary to balance the surface stresses. On the remaining part of the body surface, explicit external forces implement the "user control" of the deformation. In regions where the external forces vanish, the body surface is said to be *free*. For the body to remain at rest, the total external force and the total external moment of force must always vanish.

Estimates

Confronted with partial differential equations, it is always useful to get a rough idea of the size of a particular solution. It should be emphasized that such estimates just aim to find the right orders of magnitude of the fields, and that there may be special circumstances in a particular problem that invalidate them. If that is the case, or if precision is needed, there is no way around analytic or numeric calculation.

Imagine, for example, that a body made from elastic material is subjected to surface stresses of a typical magnitude P and that the deformation of the body due to gravity can be ignored. A rough guess on order of magnitude of the stresses in the body is then also $|\sigma_{ij}| \sim P$. The elastic moduli λ, μ, E, and K are all of the same magnitude, so that the deformation is on the order of $|u_{ij}| \sim P/E$, here disregarding Poisson's ratio, which is anyway of order unity. Since the deformation is calculated from gradients of the displacement field, the variation in displacement across a body of typical size L may be estimated to be on the order of $|\Delta u_i| \sim L\,|u_{ij}| \sim LP/E$.

Example 9.1 [Deformation of the legs of a chair]: Standing with your full weight of $M \approx 70$ kg on the seat of a chair supported by four wooden legs with total cross-sectional area $A \approx 30\ \text{cm}^2$ and $L \approx 50$ cm long, you exert a stress $P \approx Mg_0/A \approx 230$ kPa $= 2.3$ bar on the legs. Taking $E \approx 10^9$ Pa, the deformation will be about $|u_{ij}| \sim P/E \approx 2.3 \times 10^{-4}$ and the maximal displacement about $|\Delta u_i| \sim PL/E \approx 0.1$ mm. The squashing of the legs of the chair due to your weight is barely visible.

On the other hand, if gravity is dominant, mechanical equilibrium (9.1a) allows us to estimate the variation in stress over the vertical size L of the body to be $|\Delta\sigma_{ij}| \approx \rho g L$. The corresponding variation in strain becomes $|\Delta u_{ij}| \sim L\rho g/E$ for non-exceptional materials. Since u_{ij} is dimensionless, it is convenient to define the *gravitational deformation scale*,

$$D \sim \frac{E}{\rho g}, \tag{9.3}$$

so that $|\Delta u_{ij}| \sim L/D$. The length D characterizes the scale for major gravitational deformation (of order unity), and small deformations require $L \ll D$. Finally, the gravitationally induced variation in the displacement over a vertical distance L is estimated to be of magnitude $|\Delta u_i| \sim L\,|\Delta u_{ij}| \sim L^2/D \ll L$.

Example 9.2 [Gravitational settling of a tall building]: How much does a tall building settle under its own weight when it is built? Let the height of the building be $L = 413$ m, its ground area $A = 63 \times 63\ \text{m}^2$, and its average mass density one tenth of water, $\rho = 100\ \text{kg m}^{-3}$, including walls, columns, floors, office equipment, and people. The weight of it all is carried by steel columns taking up about $f = 1\%$ of its ground area. It now follows that the total mass of the building is $M = \rho AL = 1.6 \times 10^8$ kg and the stress in the supports at ground level is $P = Mg_0/fA \approx 400$ bar. Taking Young's modulus to be that of steel, $E = 2 \times 10^{11}$ Pa, the deformation range becomes $|\Delta u_{ij}| \approx P/E \approx \rho g_0 L/fE$, so that the deformation scale becomes huge, $D \approx fE/\rho g_0 \approx 2{,}000$ km. The strain range is about 2×10^{-4}, about the same as for the chair, and the top of the building settles by merely $L^2/D \approx 8$ cm.

Saint-Venant's principle

Suppose an elastic body that is already in mechanical equilibrium is loaded with an additional static force distribution that only acts inside a compact region which is small compared with the general size of the body. The total force as well as the total moment of this distribution must of course vanish, for otherwise the body will not remain in equilibrium. How far will the deformation due to these forces reach? Loosely formulated, Saint-Venant's principle claims that *the deformation due to a localized external force distribution with vanishing total force and total moment of force will not reach much beyond the linear size of the region of force application.* This is illustrated numerically in Figure 9.1 where a pressure distribution with zero total force and total moment is applied to one end of a circular cylinder. One sees that the deformation barely reaches one diameter into the cylinder.

Adhémar Jean Claude Barré de Saint-Venant (1797–1886). French engineer. Worked on mechanics, elasticity, hydrostatics, and hydrodynamics. Rederived the Navier–Stokes equations in 1843, avoiding Navier's molecular approach, but did not get credited for these equations with his name. In 1853 he formulated the principle for which he is most remembered.

Together with the superposition principle, Saint-Venant's principle is very useful for dealing with the structural statics problems encountered by engineers. It guarantees that the details of how forces are applied locally have essentially no influence on the deformation farther away, as long as the total force and total moment applied in the local region are unchanged. Thus, for example, the difference in floor load when standing with legs together or apart has no consequences for the deformation of the floor except near the place you stand, because your weight is the same and the moment of force vanishes in both positions. Shifting your weight from one foot to the other will, on the other hand, change the moment of force that you exert on the floor, a moment that must be balanced by an opposite moment from the possibly distant floor supports.

As intuitively right as it may appear, Saint-Venant's principle has been difficult to prove in full generality for bodies of all shapes, although mathematical proofs can be found in

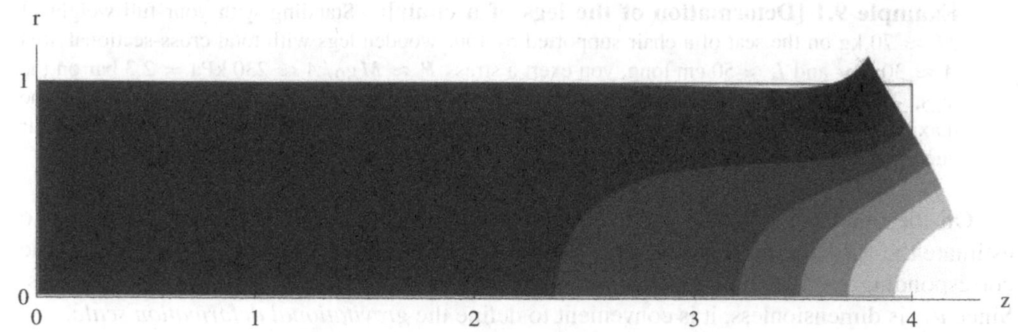

Figure 9.1. Numerical demonstration of Saint-Venant's principle. A circular cylinder with radius $a = 1$ and length $L = 4$ is placed with its z-axis along the bottom (only half shown here). You should rotate the figure around the z-axis to see the three-dimensional image. The originally relaxed cylinder is loaded at the right end with a radially varying pressure distribution $p_z = 2r^2 - 1$, for which both the total force and total moment of force on the cylinder vanish. Young's modulus is taken to be 1000, and Poisson's ratio 1/3. The gray-levels indicate the pressure p_z along z in the cylinder, with black being positive (compression) and white negative (extension). The pressure distribution is seen to extend about one diameter ($2a$) from the right end, in accordance with Saint-Venant's principle. The surface deformation is strongly exaggerated.

certain regular geometries [Davis and Selvadurai 1994, IV02, BT08]. Somewhat contrived counterexamples are found in other regular geometries [vM45, Soutas-Little 1999]. The discussion has lasted for 150 years and does not seem to want to end. In the following we shall along with most engineers use the principle to good measure without further worrying about proofs.

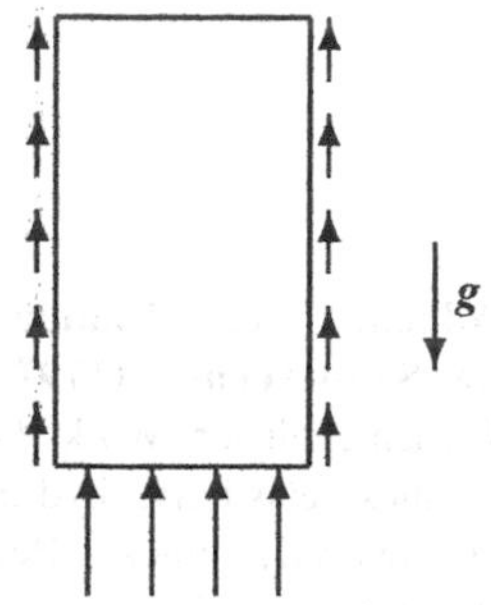

Shear stresses may aid in carrying the weight of a vertical column of elastic material.

9.2 Standing up to gravity

Solid objects, be they mountains, bridges, houses, or coffee cups, standing on a horizontal surface are deformed by gravity, and deform in turn, by their weight, the supporting surface. Intuition tells us that gravity makes such objects settle toward the ground and squashes their material so that it bulges out horizontally, unless prevented by constraining walls. In a fluid at rest, each horizontal surface element has to carry the weight of the column of fluid above it, and this determines the pressure in the fluid. In a solid at rest, this is more or less also the case, except that shear elastic stresses in the material are able to distribute part or all of the vertical load from the column in the horizontal directions.

Uniform settling

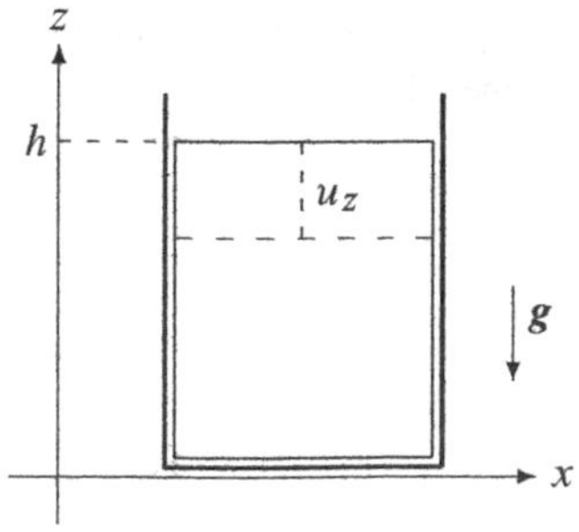

Elastic "sea" of material undergoing a downward displacement because of gravity. The container has fixed, slippery walls.

An infinitely extended slab of homogeneous and isotropic elastic material placed on a horizontal surface is a kind of "elastic sea", which like the fluid sea may be assumed to have the same properties everywhere in a horizontal plane. In a flat-Earth coordinate system, where gravity is given by $\boldsymbol{g} = (0, 0, -g_0)$, we expect a uniformly vertical displacement, which only depends on the z-coordinate,

$$\boldsymbol{u} = (0, 0, u_z(z)) = u_z(z)\,\hat{\boldsymbol{e}}_z. \tag{9.4}$$

In order to realize this "elastic sea" in a finite system, it must be surrounded by fixed, vertical, and slippery walls. The vertical walls forbid horizontal but allow vertical displacement, and at the bottom, $z = 0$, we place a horizontal supporting surface that forbids vertical displacement. At the top, $z = h$, the elastic material is left free to move without any external forces acting on it.

The only non-vanishing strain is $u_{zz} = \nabla_z u_z$. From the explicit form of Hooke's law (8.9), we obtain the non-vanishing stresses

$$\sigma_{xx} = \sigma_{yy} = \lambda u_{zz}, \qquad \sigma_{zz} = (\lambda + 2\mu) u_{zz}, \tag{9.5}$$

and Cauchy's equilibrium equation (9.1a) simplifies in this case to

$$\nabla_z \sigma_{zz} = \rho_0 g_0, \tag{9.6}$$

where ρ_0 is the constant mass density of the undeformed material. Using the boundary condition $\sigma_{zz} = 0$ at $z = h$, this equation may immediately be integrated to

$$\sigma_{zz} = -\rho_0 g_0 (h - z). \tag{9.7}$$

The vertical pressure $p_z = -\sigma_{zz} = \rho_0 g_0 (h - z)$ is positive and rises linearly with depth $h - z$, just as in the fluid sea. It balances everywhere the full weight of the material above, but this was expected since there are no shear stresses to distribute the vertical load. The horizontal pressures $p_x = p_y = p_z \lambda/(\lambda + 2\mu)$ are also positive but smaller than the vertical, because both λ and μ are positive in normal materials. The horizontal pressures are eventually balanced by the stiffness of the fixed vertical walls surrounding the elastic sea.

The strain

$$u_{zz} = \nabla_z u_z = \frac{\sigma_{zz}}{\lambda + 2\mu} = -\frac{\rho_0 g_0}{\lambda + 2\mu}(h - z) \tag{9.8}$$

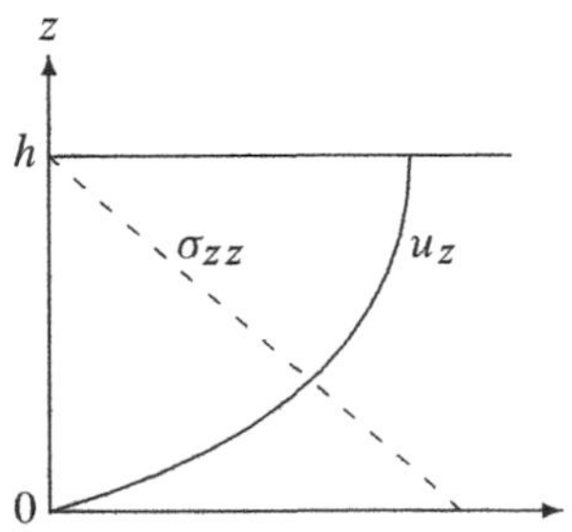

Sketch of the displacement (solid curve) and stress (dashed) for the elastic "elastic sea in a box".

is negative, corresponding to a compression. The characteristic length scale for major deformation is in this case chosen to be

$$D = \frac{\lambda + 2\mu}{\rho_0 g_0} = \frac{1 - \nu}{(1 + \nu)(1 - 2\nu)} \cdot \frac{E}{\rho_0 g_0}. \tag{9.9}$$

Integrating the strain with the boundary condition $u_z = 0$ for $z = 0$, we finally obtain

$$\boxed{u_z = -\frac{h^2 - (h - z)^2}{2D}.} \tag{9.10}$$

The displacement is always negative, largest in magnitude at the top, $z = h$, and vanishes at the bottom. At the top it varies quadratically with height h, as expected from the estimate in the preceding section.

Shear-free settling

What happens if we remove the vertical container walls around the elastic sea? A fluid would of course spill out all over the place, but an elastic material is only expected to settle a bit more while bulging horizontally out where the walls were before. A cylindrical piece of jelly placed on a flat plate is perhaps the best image to have in mind. In spite of the basic simplicity of the problem, there is no analytic solution to this problem. The numerical solution is shown in Figure 9.2.

But if one cannot find the right solution to a problem, it is common practice in physics (as well as in politics) to redefine the problem to fit a solution that one *can* get! What we can obtain is a simple solution with no shear stresses (in the chosen coordinates), but the price we pay is that the vertical displacement will not vanish everywhere at the bottom of the container, as it ought to.

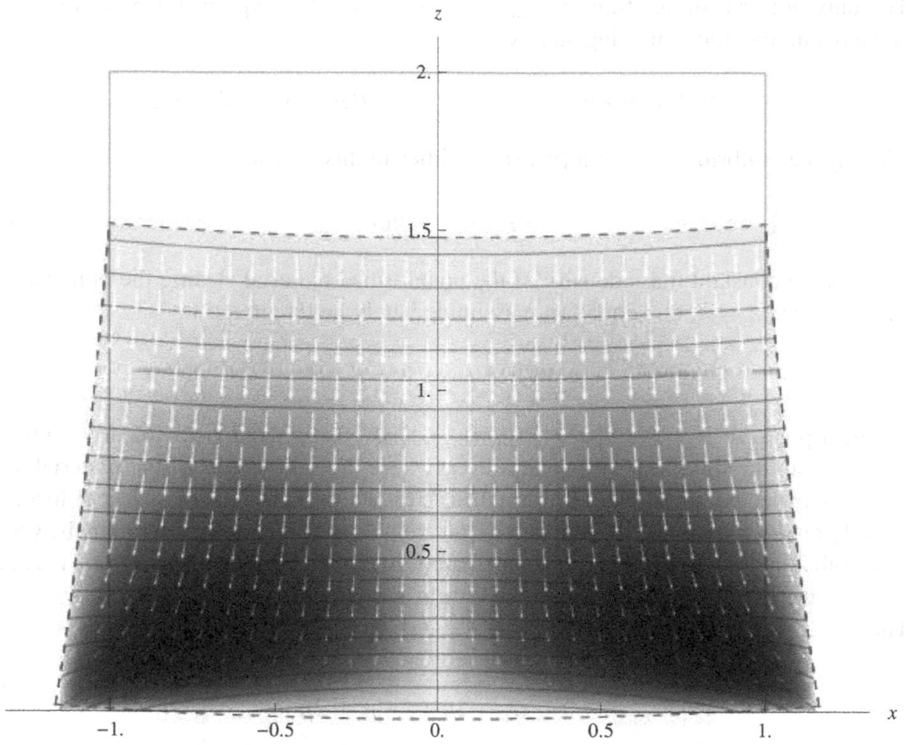

Figure 9.2. Numeric solution to the gravitational settling of a cylindrical block of fairly soft material ("jelly"). The square outline marks the undeformed shape with equal radius $a = 1$ and height $h = 2$, Poisson's ratio $\nu = 1/3$, and deformation scale $D = E/\rho_0 g_0 = 4$. The image is a cut through the xz-plane and should be rotated around the vertical z-axis. The white arrows are proportional to the displacement field. The gray-level background indicates shear stress, $\sigma_{rz} = \sigma_{zr}$, with black being high. The fully drawn, gray, and nearly horizontal lines are isobars for the vertical pressure, showing that it is higher in the middle than at the edges at a given horizontal level. The dashed lines indicate the shape of the deformed block in the shear-free approximation, normalized to vanishing average vertical displacement at $z = 0$. The agreement between model and data is quite impressive.

The equilibrium equation (9.1a) with all shear stresses set to zero, that is, $\sigma_{xy} = \sigma_{yz} = \sigma_{zx} = 0$, now simplifies to

$$\nabla_x \sigma_{xx} = 0, \qquad \nabla_y \sigma_{yy} = 0, \qquad \nabla_z \sigma_{zz} = \rho_0 g_0. \tag{9.11}$$

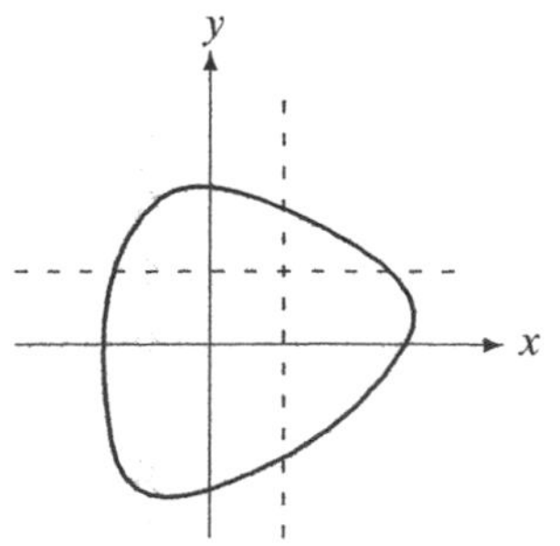

Horizontal cross-section of elastic block with vertical sides. Straight lines running parallel with the axes of the coordinate system must cross the outer perimeter in at least two places.

The first equation says that σ_{xx} does not depend on x, or in other words that σ_{xx} is constant on straight lines parallel with the x-axis. But such lines must always cross the vertical sides, where the x-component of the stress vector, $\sum_j \sigma_{xj} n_j = \sigma_{xx} n_x$, has to vanish, and consequently we must have $\sigma_{xx} = 0$ everywhere. In the same way, it follows that $\sigma_{yy} = 0$ everywhere. Finally, the third equation tells us that σ_{zz} is linear in z, and using the condition that $\sigma_{zz} = 0$ for $z = h$, we find

$$\sigma_{zz} = -\rho_0 g_0 (h - z). \tag{9.12}$$

This shows that every column of material carries the weight of the material above it. This result was again to be expected because there are no shear stresses to redistribute the vertical load.

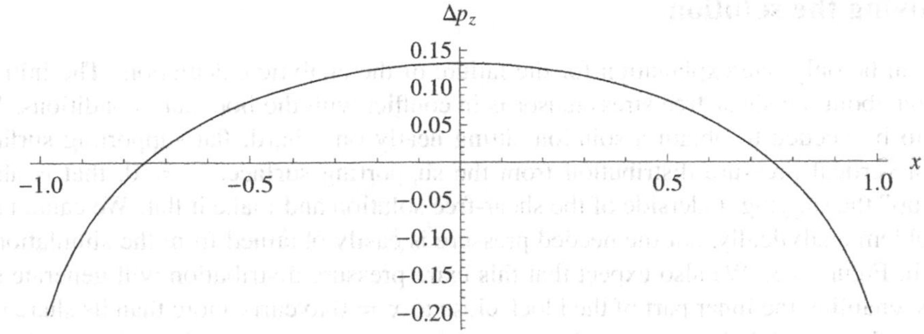

Figure 9.3. The extra vertical pressure distribution at the bottom of a cylindric block, obtained from the numeric solution shown in Figure 9.2. The x-coordinate refers to the undeformed body. The total force given by the integral $\mathcal{F}_z = \int_0^1 \Delta p_z(r) 2\pi r\, dr$ vanishes as it should, and the total moment of force vanishes for symmetry reasons.

From the inverse Hooke's law (8.15), the non-vanishing strain components are

$$u_{xx} = u_{yy} = \nu \frac{\rho_0 g_0}{E}(h - z), \qquad u_{zz} = -\frac{\rho_0 g_0}{E}(h - z), \tag{9.13}$$

where E is Young's modulus and ν Poisson's ratio. The natural deformation scale is in this case

$$D = \frac{E}{\rho_0 g_0}. \tag{9.14}$$

Using that $u_{xx} = \nabla_x u_x$ and $u_{yy} = \nabla_y u_y$, the horizontal displacements may readily be integrated with boundary conditions $u_x = u_y = 0$ at $x = y = 0$. The integration of $u_{zz} = \nabla_z u_z$ is also straightforward and leads naively to the same result as for the elastic sea (9.10), up to an arbitrary function of x and y. This function is determined by the vanishing of the shear stresses u_{xz} and u_{yz}. One may easily verify that the following solution has the desired properties

$$\boxed{\begin{aligned} u_x &= \nu \frac{(h - z)x}{D}, \\ u_y &= \nu \frac{(h - z)y}{D}, \\ u_z &= \frac{h^2 - (h - z)^2}{2D} + \nu \frac{x^2 + y^2}{2D} + K, \end{aligned}} \tag{9.15}$$

where K is an arbitrary constant. In any plane parallel with the xy-plane, the horizontal displacement represents a uniform expansion with a z-dependent scale factor that vanishes for $z = h$ and is maximal at $z = 0$.

The trouble with this solution is that we cannot impose the bottom boundary condition, $u_z = 0$ for $z = 0$, for any choice of K. For $K = 0$, the vertical displacement only vanishes at the center $x = y = 0$. Instead of describing the deformation of a block of material resting on a hard and flat horizontal surface, we have obtained a solution that only rests in a single point. For the case of a cylindric block of radius a, the numerical simulation is compared to the shear-free solution in Figure 9.2. In this case we have chosen $K = -\nu a^2/4D$, such that the integral over the bottom displacement vanishes, $\int_A u_z\, dA = 0$. The block now "rests" on a circle with radius $a/\sqrt{2}$ instead of a point, but also sinks a bit below the surface.

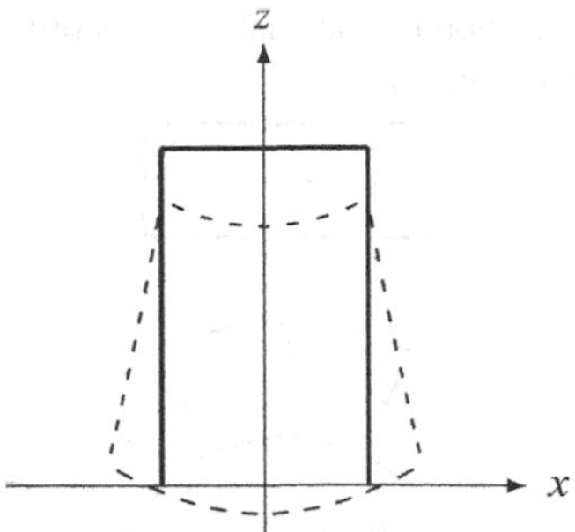

Shear-free model for the gravitational settling of a block of elastic material ("jelly on a plate") in the plane $y = 0$. The model is not capable of fulfilling the boundary condition $u_z = 0$ at $z = 0$ and describes a block that only rests on a circle.

Improving the solution

There can be only one explanation for the failure of the analytic calculation: The initial assumption about the shear-free stress tensor is in conflict with the boundary conditions. What seems to be needed to obtain a solution sitting neatly on a hard, flat supporting surface is an extra vertical pressure distribution from the supporting surface, $z = 0$, that is able to "shore up" the sagging underside of the shear-free solution and make it flat. We cannot solve this problem analytically, but the needed pressure is easily obtained from the simulation and shown in Figure 9.3. We also expect that this extra pressure distribution will generate shear stresses, enabling the inner part of the block close to $x = 0$ to carry more than its share of the weight of the material above it, and the outer part near $x = a$ to carry less. This is also seen in the numerical solution in Figure 9.2.

For a tall block with height larger than the diameter, Saint Venant's principle comes to the rescue. The extra pressure exerts vanishing total force on the block because the bottom pressure must carry the weight of the block in both cases, and the total moment of the extra pressure also vanishes for symmetry reasons. The shear-free solution will for this reason be a good approximation to the settling of a tall block, except in a region near the ground comparable to the radius of the block. The simulation in figure 9.2 clearly shows that Saint-Venant's principle holds, even for a cylindrical block with height equal to its diameter.

Bending a beam by wrenching it at the ends.

9.3 Bending a beam

Sticks, rods, girders, struts, masts, towers, planks, poles, and pipes are all examples of a generic object, which we shall call a *beam*. Geometrically, a beam consists of a bundle of straight parallel lines or *rays*, covering the same cross-section in any plane orthogonal to the lines. Physically, we shall assume that the beam is made from homogeneous and isotropic elastic material.

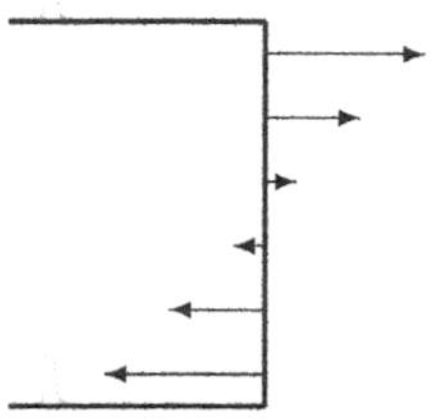

A bending couple may be created by applying normal stresses only to a terminal.

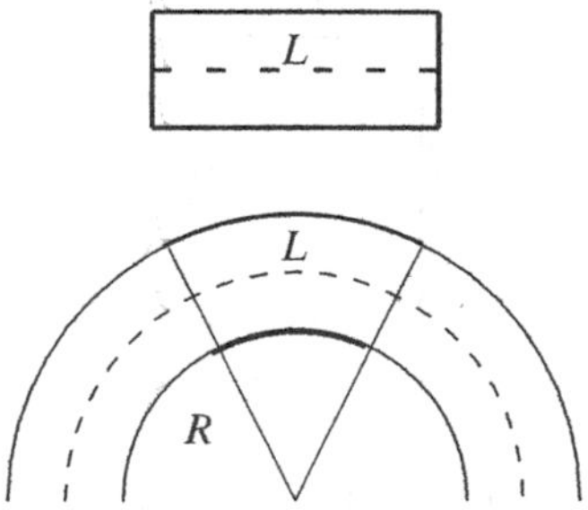

A uniformly bent beam is a part of a circular ring.

Uniform pure bending

There are many ways to bend a beam. A cantilever is a beam that is fixed at one end and bent like a horizontal flagpole or a fishing rod. A beam may also be supported at the ends and weighed down in the middle like a bridge, but the cleanest way to bend a beam is probably to grab it close to the ends and wrench it like a pencil so that it adopts a uniformly curved shape. Ideally, in *pure bending*, body forces should be absent and external stresses should only be applied to the terminal cross-sections. On average these stresses should neither stretch nor compress the beam, but only provide external couples (moments of force) at the terminals. It should be noted that such couples do not require shear stresses, but may be created by normal stresses alone that vary in strength over the terminal cross-sections. If you try, you will realize that it is in fact rather hard to bend a pencil in this way. Bending a rubber eraser by pressing it between two fingers is somewhat easier, but tends to add longitudinal compression as well.

The bending of the beam is also assumed to be *uniform*, such that the physical conditions, stresses and strains, will be the same everywhere along the beam. This is only possible if the originally straight beam of length L is deformed to become a section of a circular ring of radius R with every ray becoming part of a perfect circle. In that case, it is sufficient to consider just a tiny slice of the beam in order to understand uniform bending for a beam of any length. We shall see later that non-uniform bending can also be handled by piecing together little slices with varying radius of curvature. Furthermore, by appealing to Saint-Venant's principle and linearity, we may even calculate the properties of a beam subject to different types of terminal loads by judicious superposition of displacement fields.

Choice of coordinate system

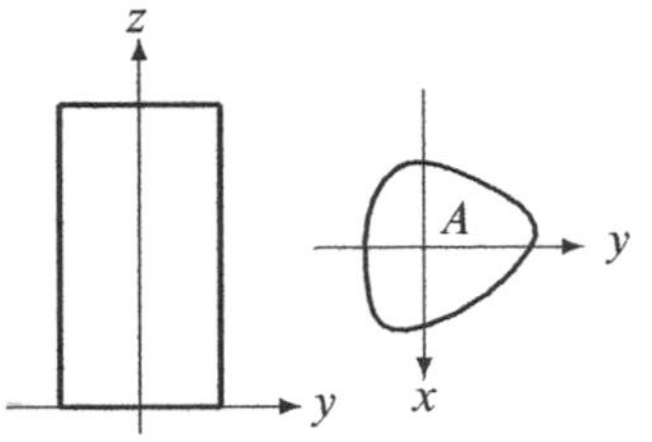

The unbent beam is aligned with the z-axis. Its cross-section, A, in the xy-plane is the same for all z.

In a Cartesian coordinate system, we align the undeformed beam with the z-axis, and put the terminal cross-sections at $z = 0$ and $z = L$. The length L of the beam may be chosen as small as we please. The cross-section A in the xy-plane may be of arbitrary shape, but we may—without loss of generality and for reasons to become clear below—position the coordinate system in the xy-plane with its origin coinciding with the *area centroid* (see Equation (3.16) on page 49), such that the area integral over the coordinates vanishes,

$$\int_A x\, dA = \int_A y\, dA = 0. \tag{9.16}$$

Finally, we require that the central ray after bending becomes part of a circle in the yz-plane with radius R and its center on the y-axis at $y = R$. The radius R is obviously the length scale for major deformation, and must be assumed large compared to the transverse dimensions of the beam.

Shear-free bending

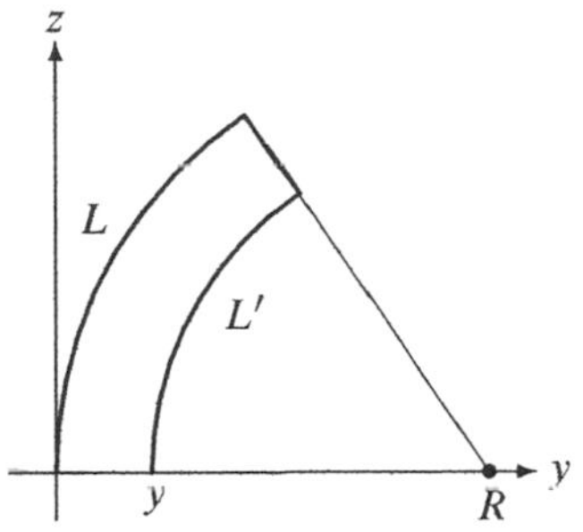

The length of the arc at x must satisfy $L'/(R-y) = L/R$.

What precisely happens in the beam when it is bent depends on the way the actual stresses are distributed on its terminals, although by Saint-Venant's principle the details should only matter near the terminals (see Figure 9.1). In the simplest case we may view the beam as a loose bundle of thin elastic strings that do not interact with each other, but are stretched or compressed individually according to their position in the beam without generating shear stresses. Let us fix the central string so that it does not change its length L when bent into a circle of radius R. A simple geometric construction (see the margin figure) then shows that a nearby ray in position x will change its length to L', satisfying $(R-y)/L' = R/L$. Thus according to (7.28) on page 117, the beam experiences a longitudinal strain,

$$u_{zz} = \frac{\delta L}{L} = \frac{L' - L}{L} = -\frac{y}{R}. \tag{9.17}$$

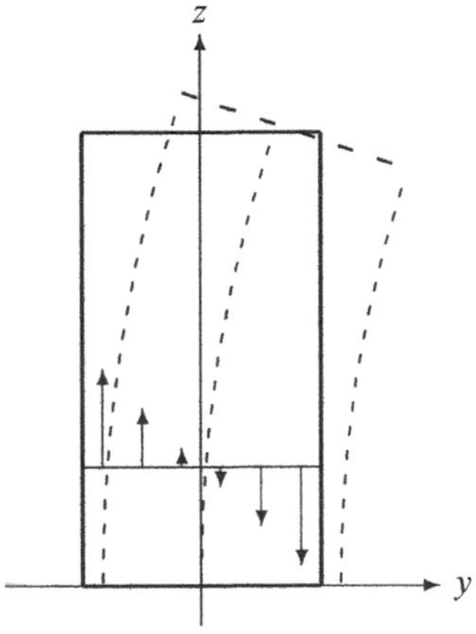

Sketch of the bending of a beam. The arrows show the strain u_{zz}.

For negative y the material of the beam is being stretched, while for positive y it is being compressed. For the strain to be small, we must have $|y| \ll R$ everywhere in the beam.

Under the assumption that the bending is done without shear and that there are no forces acting on the sides of the beam, it follows as in the preceding section that $\sigma_{xx} = \sigma_{yy} = 0$. The only non-vanishing stress is $\sigma_{zz} = E u_{zz}$, and the non-vanishing strains are as before found from the inverted Hooke's law (8.15),

$$u_{xx} = u_{yy} = -\nu u_{zz} = \nu \frac{y}{R}, \tag{9.18}$$

where ν is Poisson's ratio. This shows that the material is being stretched horizontally and compressed vertically for $y > 0$ and conversely for $y < 0$.

Using $u_{ii} = \nabla_i u_i$, and requiring that the central ray is only bent, not stretched, a particular solution is found to be

$$\boxed{\begin{aligned} u_x &= \nu \frac{xy}{R}, \\ u_y &= \frac{z^2}{2R} + \nu \frac{y^2 - x^2}{2R}, \\ u_z &= -\frac{yz}{R}. \end{aligned}} \tag{9.19}$$

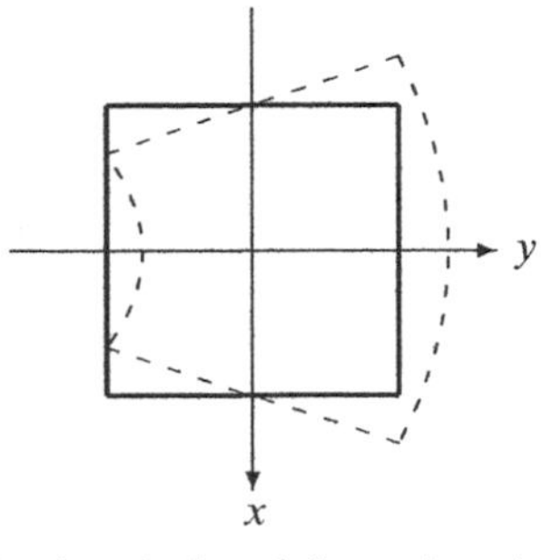

Sketch of the deformation in the xy-plane of a beam with quadratic cross-section. This deformation may easily be observed by bending a rubber eraser.

The second term in u_y is, like in the preceding section, forced upon us by the requirement of no shear stresses (and strains). For displacement gradients to be small, all dimensions of the beam have to be small compared to R. Note that the beam's actual dimensions do not appear in the displacement field, which is therefore a generic solution for pure bending of any beam. For a simple quadratic beam, the deformation is sketched in the margin.

Total force and total moment of force

The only non-vanishing stress component is as mentioned above,

$$\sigma_{zz} = E u_{zz} = -\frac{E}{R} y. \tag{9.20}$$

It is a tension for negative y, and we consequently expect the material of the beam to first break down at the most distant point of the cross-section opposite the direction of bending, as common experience also tells us.

The total force acting on a cross-section vanishes,

$$\mathcal{F}_z = \int_A \sigma_{zz}\, dS_z = -\frac{E}{R}\int_A y\, dA = 0, \tag{9.21}$$

because the origin of the coordinate system is chosen to coincide with the centroid of the beam cross-section (9.16).

The moments of the longitudinal stress in any cross-section are

$$\mathcal{M}_x = \int_A y\, \sigma_{zz}\, dS_z = -\frac{E}{R}\int_A y^2\, dA, \tag{9.22a}$$

$$\mathcal{M}_y = -\int_A x\, \sigma_{zz}\, dS_z = \frac{E}{R}\int_A xy\, dA, \tag{9.22b}$$

$$\mathcal{M}_z = 0. \tag{9.22c}$$

The component $\mathcal{M}_x$ orthogonal to the bending plane is called the *bending moment*. The integral

$$I = \int_A y^2\, dA \tag{9.23}$$

is the *area moment* that we previously introduced on page 51 in connection with ship stability.

The component $\mathcal{M}_y$ vanishes only if

$$\int_A xy\, dA = 0. \tag{9.24}$$

This will, for example, be the case if the beam has circular cross-section, is mirror symmetric under reflection in either axis, or more generally if the axes coincide with the *principal directions* of the cross-section. There are always two orthogonal principal directions for any shape of cross-section (page 55).

The Euler–Bernoulli law

Liberating ourselves from the coordinate system, we can express the magnitude of the bending moment $\mathcal{M}_b = -\mathcal{M}_x$ in terms of the unsigned radius of curvature R (or curvature $\kappa = 1/R$),

$$\boxed{\mathcal{M}_b = \frac{EI}{R} = EI\kappa.} \tag{9.25}$$

This is the *Euler–Bernoulli law*. The product EI is called the *flexural rigidity* or *bending stiffness* of the beam. The larger it is, the larger is the moment required to bend it with a given radius of curvature. The unit of flexural rigidity is Pa m^4 = N m^2.

The Euler–Bernoulli law is also valid if the beam is subject to *constant* normal or shear stresses in the cross-section, because such forces do not contribute to the total moment around the centroid of the cross-section. In many engineering applications the Euler–Bernoulli law, combined with the superposition principle and Saint-Venant's principle, is enough to determine how much a beam is deformed by external loads.

Rectangular beam: A rectangular beam with sides $2a$ and $2b$ along x and y has moment of inertia

$$I = \int_{-a}^{a} dx \int_{-b}^{b} y^2 dy = \frac{4}{3}ab^3. \tag{9.26}$$

It grows more rapidly with the width of the beam in the direction of bending (y) than orthogonally to it (x). This agrees with the common experience that to obtain a given bending radius R, it is much easier to bend a beam in the direction where it is thinnest.

Elliptic beam: An elliptical beam with major axes $2a$ and $2b$ along x and y has moment of inertia

$$I = \int_{-a}^{a} dx \int_{-b\sqrt{1-x^2/a^2}}^{b\sqrt{1-x^2/a^2}} y^2\, dy = \frac{4}{3}ab^3 \int_0^1 (1-t^2)^{3/2}\, dt = \frac{\pi}{4}ab^3, \tag{9.27}$$

which is only a little more than half of the rectangular result. An elliptical spring would thus bend about twice the amount of a flat spring of similar dimensions for the same applied moment of force. For a circular beam with radius a, the moment of inertia $I = \frac{\pi}{4}a^4$ is the same in all directions.

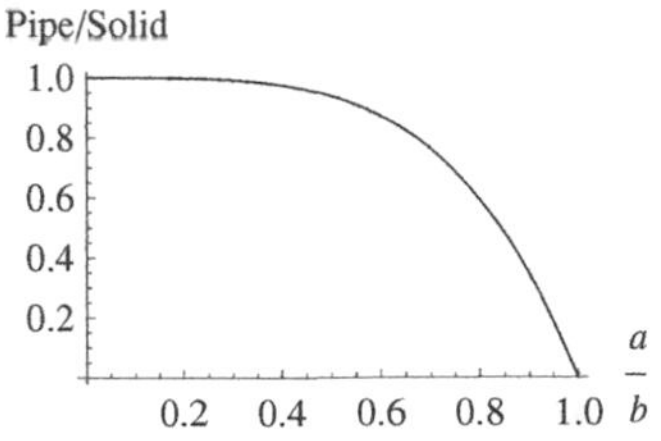

Moment of inertia of a pipe relative to a solid rod of the same outer radius plotted as a function of the ratio of inner and outer radii.

Circular pipe: For a circular pipe with inner radius $r = a$ and outer radius $r = b$, we find

$$I = \int_{a \le r \le b} y^2\, dA = \int_{r \le b} y^2\, dA - \int_{r \le a} y^2\, dA = \frac{\pi}{4}\left(b^4 - a^4\right), \tag{9.28}$$

which (of course) is the difference between the moments of inertia of two circular beams. Due to the fourth power, the moment of inertia and thereby the flexural rigidity is not very dependent on the inner radius (see the margin figure).

Example 9.3 [Flexural rigidity of a water pipe]: A 1-inch water pipe has inner and outer diameters $2a = 2.54$ cm and $2b = 3.36$ cm, implying $I = 4.29\ \text{cm}^4$. Taking $E = 200$ GPa we get $EI = 8{,}400\ \text{N m}^2$, which is 2/3 of the flexural rigidity of a massive rod with the same outer radius. It is hard to bend this pipe. If you hang at the end of a $L = 1$ m long horizontal 1-inch pipe with your full weight of $M = 75$ kg the terminal bending moment becomes LMg_0, and the Euler-Bernoulli law yields the radius of curvature 11 m. Assuming pure bending, the pipe is only bent by about $\phi = L/R = 5°$, corresponding to a terminal deflection of $y = 4$ cm.

Example 9.4 [Flexural rigidity of microtubules]: Microtubules are used for many purposes in the living cell but in particular they provide structural rigidity to the cytoskeleton. They are hollow cylinders made from polymerized protein units arranged in a helical pattern with nearly always 13 such units in each turn. Over the years, measurements have yielded a large scatter in the flexural rigidities from $1–40 \times 10^{-24}\ \text{N m}^2$ [HCFJ07]. This is now understood as arising from the strong anisotropy in the elastic properties along and around the tubules, preventing pure shear-free bending from being established [PLJ&06].

Yield radius of curvature

As discussed on page 100, the *yield stress* σ_{yield} is defined to be the stress at which a metal becomes ductile and retains its deformed shape. Using (9.20) with $y = -a$ we obtain the maximal stress Ea/R in a bent circular beam with radius a, so that for the beam to remain elastic, we find the following limits on the radius of curvature and bending moment:

$$R \gtrsim a\frac{E}{\sigma_{\text{yield}}}, \qquad \mathcal{M}_b \lesssim \frac{\pi}{4}a^3\sigma_{\text{yield}}. \tag{9.29}$$

The smallest value of R is naturally called the *yield radius of curvature* and the largest value of the bending moment $\mathcal{M}$ the *yield moment*. Interestingly it is independent of Young's modulus.

Example 9.5 [Paper-clips]: Anyone who has played with a paper-clip during a dull board meeting knows that—although its purpose of keeping papers together requires it to be elastic—it does not take much force to bend it into all kinds of interesting shapes. Paper-clips come in many sizes. They are usually made from galvanized steel wire with diameters ranging from 0.5 mm to 2 mm. For steel the typical ratio of Young's modulus to yield stress is about 1,000, such that the yield radius for a small paper-clip is about 25 cm and for a large 4 times that. The corresponding yield moment for the small clip becomes about 3 N mm, perfect for the kind of light finger manipulation that should not attract much attention. For the large clip with 4 times the radius of the small, the yield moment becomes 64 times larger, and attempting to bend the clip into a new shape will hardly go unnoticed by the chairman.

Bending energy

A bent beam contains an elastic energy equal to the work performed by the external forces while bending it. Since the only non-vanishing stress is $\sigma_{zz} = -Ey/R$, it follows immediately from Equation (8.35) on page 136 that the energy density in the beam is $\epsilon = \sigma_{zz}^2/2E = Ey^2/2R^2$. Integrating over the beam cross-section we get the total bending energy per unit of beam length,

$$\frac{d\mathcal{E}_b}{d\ell} = \frac{EI}{2R^2} = \frac{\mathcal{M}_b^2}{2EI}, \tag{9.30}$$

where I is the area moment (9.23). It is of course a constant for pure bending where all cross-sections are equivalent, but can also be used for deformations with varying bending moment. We shall return to this in the next chapter.

Extension versus bending

If a beam of length L and cross-sectional area $A \approx a^2$ is subjected to a longitudinal terminal force $\mathcal{F}_z$, its central ray will according to Equation (8.23) on page 133 stretch by $u_z \approx L\mathcal{F}_z/AE$. If it instead is subjected to a transverse terminal force $\mathcal{F}_y$, the central ray will according to (9.19) deflect from a straight line by about $u_y \approx L^2/2R \sim L^3\mathcal{F}_y/EI$, because $\mathcal{M}_b \sim L\mathcal{F}_y$ at $z \approx 0$. Since $I \sim Aa^2$, the ratio of longitudinal to transverse displacement becomes

$$\left|\frac{u_z}{u_y}\right| \sim \frac{a^2}{L^2}\left|\frac{\mathcal{F}_z}{\mathcal{F}_y}\right|. \tag{9.31}$$

This shows that for thin beams with $a \ll L$, and comparable longitudinal and transverse forces, $|\mathcal{F}_z/\mathcal{F}_y| \approx 1$, the longitudinal displacement is always negligible relative to the transverse.

9.4 Twisting a shaft

The drive shaft in older cars and in trucks connects the gear box to the differential and transmits engine power to the rear wheels. In characterizing engine performance, maximum torque is often quoted because it creates the largest shear force between wheels and road and therefore maximal acceleration, barring wheel-spin. Although the shaft is made from steel, it will nevertheless undergo a tiny deformation in the form of a *torsion* or *twist*.

Pure torsion

Let the shaft be a beam with circular cross-section of radius a and axis coinciding with the z-axis[1]. The deformation is said to be a *pure torsion* if the shaft's material is rotated by a constant amount τ per unit of length, such that a given cross-section at the position z is rotated by an angle τz relative to the cross-section at $z = 0$. The constant τ that measures the rotation angle per unit of length of the beam is called the *torsion*. Its inverse, $1/\tau$, is the length of a beam that undergoes a pure torsion through 1 radian. It could be called the *torsion length*, although this is not commonly used. It is analogous to the radius of curvature $R = 1/\kappa$ for pure bending, which is the length of a beam that is bent through 1 radian.

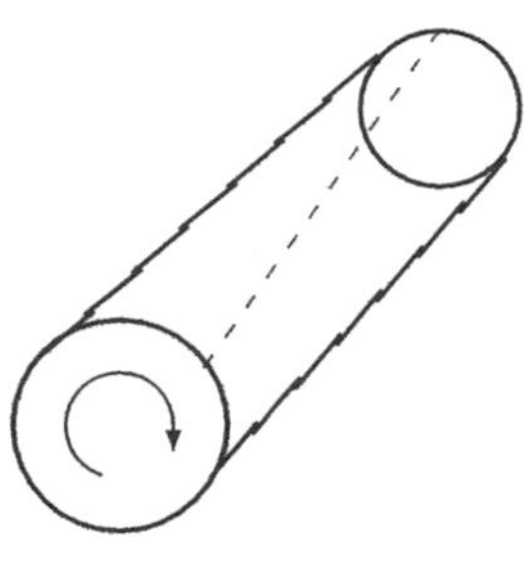

Pure torsion consists of rotating every cross-section by a fixed amount per unit of length.

The uniform nature of pure torsion allows us to consider just a small slice of the shaft of length L, which is only twisted through a tiny angle, $\tau L \ll 1$. Since the physical conditions are the same in all such slices, we can later put them together to make a shaft of any length. To lowest order in the vector angle $\boldsymbol{\phi} = \tau z \hat{\boldsymbol{e}}_z$, the displacement field in the slice becomes

$$\boldsymbol{u} = \boldsymbol{\phi} \times \boldsymbol{x} = \tau z \hat{\boldsymbol{e}}_z \times \boldsymbol{x} = \tau z(-y, x, 0). \tag{9.32}$$

Not surprisingly, it is purely tangential and is always much smaller than the radius of the shaft, $|\boldsymbol{u}| = \tau\,|z|\,|\boldsymbol{x}| \leq \tau L a \ll a$, because $\tau L \ll 1$.

Strains and stresses

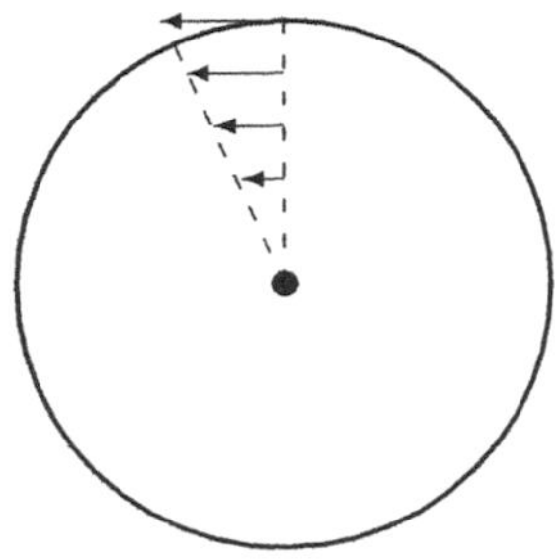

The displacement field for a rotation through a tiny angle τz (exaggerated here) is purely tangential and grows linearly with the radial distance.

The displacement gradient tensor becomes

$$\{\nabla_j u_i\} = \begin{pmatrix} 0 & -\tau z & -\tau y \\ \tau z & 0 & \tau x \\ 0 & 0 & 0 \end{pmatrix}, \tag{9.33}$$

and for this matrix to be small, we must also require $\tau a \ll 1$, or in other words that the twist must be small over a length of the shaft comparable to its radius. The only non-vanishing strains are

$$u_{xz} = u_{zx} = -\tfrac{1}{2}\tau y, \qquad u_{yz} = u_{zy} = \tfrac{1}{2}\tau x, \tag{9.34}$$

and the corresponding stresses are obtained from the isotropic Hooke's law (8.8) on page 128,

$$\sigma_{xz} = \sigma_{zx} = -\mu\tau y, \qquad \sigma_{yz} = \sigma_{zy} = \mu\tau x. \tag{9.35}$$

Inserting these stresses into the equilibrium equation (9.1a), it is seen that it is trivially fulfilled. In vector form these equations may be written as

$$\boldsymbol{\sigma} \cdot \hat{\boldsymbol{e}}_r = \boldsymbol{0}, \qquad \boldsymbol{\sigma} \cdot \hat{\boldsymbol{e}}_z = \mu\tau \hat{\boldsymbol{e}}_z \times \boldsymbol{x}, \tag{9.36}$$

where $\hat{\boldsymbol{e}}_r = (x, y, 0)/r$ is the radial unit vector (and $r = \sqrt{x^2 + y^2}$). The first of these equations shows that on the cylindrical surface of the shaft, the stress vector vanishes, as it should when there are no external forces acting there.

In order to realize a pure torsion, the correct stress distribution (9.36) must be applied to the ends of the shaft. Applying a different stress distribution, for example by grabbing both ends of the shaft with monkey wrenches and twisting, leads to a different solution near the end, but the pure torsion solution should, according to Saint-Venant's principle, still be valid everywhere, except within one diameter from the ends.

[1] The solution for circular cross-section was first obtained by Coulomb in 1787, whereas the more complicated case of beams with non-circular cross-section was analyzed by Saint-Venant in 1855 (see [Landau and Lifshitz 1986, p. 59] or [Sokolnikoff 1956, p. 109]). There is in fact no explicit solution in the general non-circular case.

The Coulomb–Saint-Venant law

In any cross-section we may calculate the total moment of force around the shaft axis, in this context called the *torque*. On a surface element $d\boldsymbol{S}$, the moment is $d\boldsymbol{\mathcal{M}} = \boldsymbol{x} \times d\boldsymbol{\mathcal{F}} = \boldsymbol{x} \times \boldsymbol{\sigma} \cdot d\boldsymbol{S}$. Since the cross-section lies in the xy-plane, the z-component of the torque becomes

$$\mathcal{M}_z = \int_A (x\sigma_{yz} - y\sigma_{xz})\, dA = \mu\tau \int_A (x^2 + y^2) dx dy. \tag{9.37}$$

Liberating ourselves from the coordinate system, the torque $M_t = M_z$ can always be written analogously to the Euler–Bernoulli law (9.25),

$$\boxed{\mathcal{M}_t = \mu J\, \tau,} \tag{9.38}$$

where J for a circular pipe with inner and outer radii, a and b, becomes

$$J = \int_{a<r<b} (x^2 + y^2) dx dy = \frac{\pi}{2} \left(b^4 - a^4\right). \tag{9.39}$$

For non-circular cross-sections the result may be found in [Landau and Lifshitz 1986, p. 62].

The quantity μJ is called the *torsional rigidity* of the beam. For a pipe with circular cross-section we have $J = 2I$, implying that $EI = (1+\nu)\mu J$, where ν is Poisson's ratio. Knowing the torsional rigidity and the torque $\mathcal{M}_t$, one may calculate the torsion, $\tau = \mathcal{M}_z/\mu J$ and conversely. Like the flexural rigidity, the torsional rigidity grows rapidly with the radius of the beam, and pipes with walls that are not too thin have also nearly the same torsional rigidity as a massive rod.

Transmitted power

If the shaft rotates with constant angular velocity Ω, the material at the point (x, y, z) will have velocity $\boldsymbol{v}(x, y) = \Omega\hat{\boldsymbol{e}}_z \times \boldsymbol{x} = \Omega(-y, x, 0)$. The shear stresses acting on an element of the cross-section, $d\boldsymbol{S} = \hat{\boldsymbol{e}}_z dx dy$, will transmit a *power* (i.e., work per unit of time) of $dP = \boldsymbol{v} \cdot d\boldsymbol{\mathcal{F}} = \boldsymbol{v} \cdot \boldsymbol{\sigma} \cdot d\boldsymbol{S}$ to the shaft. Integrating over the cross-section the total power becomes

$$P = \int_A \boldsymbol{v} \cdot \boldsymbol{\sigma} \cdot d\boldsymbol{S} = \int_A \Omega(x\sigma_{yz} - y\sigma_{xz}) dx dy = \Omega\mathcal{M}_z = \boldsymbol{\Omega} \cdot \boldsymbol{\mathcal{M}}. \tag{9.40}$$

As the derivation shows, this relation does not depend on the actual stress distribution but is generally valid for the instantaneous power delivered by the torque $\boldsymbol{\mathcal{M}}$ acting on a body rotating with angular velocity vector $\boldsymbol{\Omega}$.

Example 9.6 [Car engine]: The typical torque delivered by a family car engine can be of the order of 100 N m. If the shaft rotates with 3,000 rpm, corresponding to an angular velocity of $\Omega \approx 314\ \mathrm{s}^{-1}$, the transmitted power is about 31.4 kW, or 42 horsepower. For a solid drive shaft made of steel with shear modulus $\mu = 80$ GPa and radius $a = 2$ cm, the torsional rigidity becomes $C = \frac{1}{2}\pi a^2 \mu \approx 2 \times 10^4\ \mathrm{N\,m^2}$. In direct drive without gearing, the torsion becomes $\tau \approx 0.005\ \mathrm{m}^{-1} = 0.3\,^\circ\ \mathrm{m}^{-1}$. For a car with rear-wheel drive, the length of the drive shaft may be about 2 m, and the total twist amounts to about $0.6\,^\circ$. The maximal shear stress in the material is $\sigma = \mu\tau a \approx 8 \times 10^6$ Pa $= 80$ bar at the rim of the shaft.

Torsion energy

A twisted beam contains elastic energy, just like a bent beam. Since there are only four stress components that are non-vanishing, we find from the energy density (8.35) on page 136 that the torsional energy per unit of beam length is

$$\frac{d\mathcal{E}_t}{d\ell} = \frac{1}{2}\mu J \tau^2 = \frac{\mathcal{M}_t^2}{2\mu J}. \tag{9.41}$$

It is of course constant for pure torsion where all cross-sections are equivalent, but may also be used for deformations with varying torque.

* 9.5 Application: Radial deformation of a spherical body

A spherically shaped vessel withstands external pressure better than any other shape and has for that reason been used for extreme deep-sea exploration. Intuitively this is clear from symmetry alone. Since the water pressure is the same in all directions, the spherical shape ought to be the one that best withstands collapse, because the collapse—so to speak—can find no place to begin. Nevertheless, in films of deep-sea diving, you see the rivets beginning to pop on the inside of the vessel. Why is that?

Uniform radial displacement

Spheres can be deformed in an infinity of different ways, but we shall in the following analysis only consider radial displacement fields of the form

$$\boldsymbol{u} = u_r(r)\,\hat{\boldsymbol{e}}_r, \tag{9.42}$$

where $\hat{\boldsymbol{e}}_r = \boldsymbol{x}/r$ is the radial unit vector in coordinates with origin in the center of the sphere. Note that we keep the redundant index r to remind us that this is the radial component. At this point we could introduce full-fledged spherical coordinates (see Appendix D), but the simplicity of the radial field allows us to open a bag of tricks that makes it unnecessary.

Equilibrium equation

Using that the radial unit vector is the gradient or the radial coordinate, $\hat{\boldsymbol{e}}_r = \boldsymbol{\nabla} r$, we can write the uniform radial displacement as the gradient of a helper field $\psi(r)$,

$$\boldsymbol{u} = u_r \hat{\boldsymbol{e}}_r = \frac{d\psi}{dr}\boldsymbol{\nabla} r = \boldsymbol{\nabla}\psi(r), \qquad \psi(r) = \int u_r(r)\,dr. \tag{9.43}$$

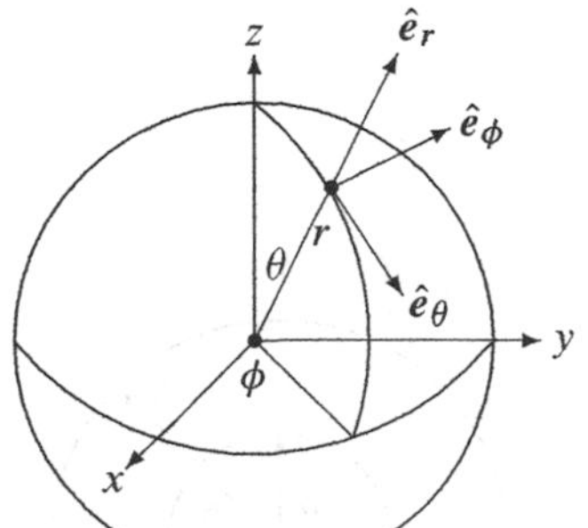

Spherical coordinates with radial basis vector $\hat{\boldsymbol{e}}_r$ and tangential basis vectors $\hat{\boldsymbol{e}}_\theta$ and $\hat{\boldsymbol{e}}_\phi$.

A short calculation shows that the Laplacian of the displacement field becomes itself a gradient,

$$\boldsymbol{\nabla}^2\boldsymbol{u} = \boldsymbol{\nabla}^2\boldsymbol{\nabla}\psi = \boldsymbol{\nabla}\boldsymbol{\nabla}^2\psi = \boldsymbol{\nabla}\boldsymbol{\nabla}\cdot\boldsymbol{\nabla}\psi = \boldsymbol{\nabla}\boldsymbol{\nabla}\cdot\boldsymbol{u},$$

so that the Navier–Cauchy equation (9.2) takes the much simpler form

$$(2\mu + \lambda)\boldsymbol{\nabla}\boldsymbol{\nabla}\cdot\boldsymbol{u} = -\boldsymbol{f}. \tag{9.44}$$

Since $\boldsymbol{\nabla}\cdot\boldsymbol{u}$ can only depend on r for symmetry reasons, it follows that the only body force consistent with the radial assumption must itself be radial, $\boldsymbol{f} = f_r(r)\hat{\boldsymbol{e}}_r$.

Gauss theorem leads to the following identity for the volume integral over the sphere,

$$4\pi r^2 u_r = \oint_S \boldsymbol{u}\cdot d\boldsymbol{S} = \int_V \nabla\cdot\boldsymbol{u}\, dV = \int_0^r \nabla\cdot\boldsymbol{u}\, 4\pi r^2 dr. \tag{9.45}$$

Differentiating both sides with respect to r, we get

$$\nabla\cdot\boldsymbol{u} = \frac{1}{r^2}\frac{d(r^2 u_r)}{dr} = \frac{du_r}{dr} + 2\frac{u_r}{r}, \tag{9.46}$$

so that the Navier–Cauchy equation reduces to a second-order ordinary differential equation,

$$\boxed{(2\mu+\lambda)\frac{d}{dr}\left(\frac{1}{r^2}\frac{d(r^2u_r)}{dr}\right) = -f_r.} \tag{9.47}$$

The only radial body force that naturally could come into play is of course gravity in a solid planet, for example the Moon (see Problem 9.7).

Strains and stresses

Using dyadic notation, the displacement gradients are simple to evaluate:

$$\nabla\boldsymbol{u} = \nabla\left(u_r\frac{\boldsymbol{x}}{r}\right) = \nabla\left(\frac{u_r}{r}\right)\boldsymbol{x} + \frac{u_r}{r}\mathbf{1} = \frac{du_r}{dr}\hat{\boldsymbol{e}}_r\hat{\boldsymbol{e}}_r + \frac{u_r}{r}(\mathbf{1} - \hat{\boldsymbol{e}}_r\hat{\boldsymbol{e}}_r). \tag{9.48}$$

It is obviously symmetric, and thus equal to the strain tensor, $\mathbf{u} = \nabla\boldsymbol{u}$. The only non-vanishing components of the strain tensor are its projections on the radial and tangential directions,

$$u_{rr} = \hat{\boldsymbol{e}}_r\cdot\mathbf{u}\cdot\hat{\boldsymbol{e}}_r = \frac{du_r}{dr}, \qquad u_{tt} = \hat{\boldsymbol{e}}_t\cdot\mathbf{u}\cdot\hat{\boldsymbol{e}}_t = \frac{u_r}{r}, \tag{9.49}$$

where $\hat{\boldsymbol{e}}_t$ is any tangential unit vector orthogonal to $\hat{\boldsymbol{e}}_r$.

Finally we obtain the non-vanishing components of the stress tensor ,

$$\sigma_{rr} = 2\mu u_{rr} + \lambda(u_{rr} + 2u_{tt}), \qquad \sigma_{tt} = 2\mu u_{tt} + \lambda(u_{rr} + 2u_{tt}), \tag{9.50}$$

where we have used that the divergence (9.46) may be written as $\nabla\cdot\boldsymbol{u} = u_{rr} + 2u_{tt}$. We now recognize that the factor 2 in front of u_{tt} stems from the two orthogonal tangential directions, say $\hat{\boldsymbol{e}}_{t1}$ and $\hat{\boldsymbol{e}}_{t2}$, in the local basis.

Spherical shell under external pressure

In the absence of gravity, $f_r = 0$, the most general solution is easily found from (9.47),

$$u_r = Ar + \frac{B}{r^2}, \tag{9.51}$$

where A and B are integration constants. The strains and stresses become

$$u_{rr} = A - \frac{2B}{r^3}, \qquad u_{tt} = A + \frac{B}{r^3}, \tag{9.52}$$

$$\sigma_{rr} = (2\mu+3\lambda)A - 4\mu\frac{B}{r^3}, \qquad \sigma_{tt} = (2\mu+3\lambda)A + 2\mu\frac{B}{r^3}. \tag{9.53}$$

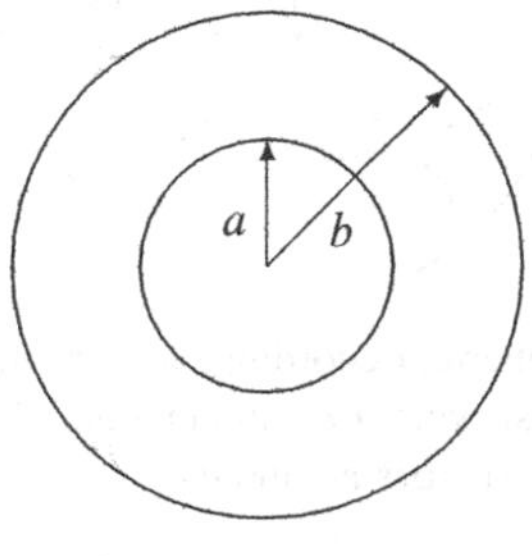

A spherical shell with inner radius a and outer radius b.

Let the spherical shell have inner and outer radii a and b. The boundary conditions are now $\sigma_{rr} = 0$ for $r = a$ and $\sigma_{rr} = -P$ for $r = b > a$ (corresponding to a positive external pressure P and vacuum inside), leading to two linear equations for A and B. The solution is

$$A = -\frac{b^3}{b^3-a^3}\frac{P}{2\mu+3\lambda}, \qquad B = -\frac{a^3b^3}{b^3-a^3}\frac{P}{4\mu}. \tag{9.54}$$

Notice that both integration constants are negative.

Displacement: Expressed in terms of Young's modulus E and Poisson's ratio ν, the displacement becomes

$$\boxed{u_r = -\frac{b^3}{b^3-a^3}\left((1-2\nu)r + (1+\nu)\frac{a^3}{2r^2}\right)\frac{P}{E},} \tag{9.55}$$

which is always negative, as we would expect under a compression.

From the displacement we may calculate the change in thickness $d = b - a$,

$$\delta d = u_r(b) - u_r(a) = -\frac{b^3}{b^2+ab+a^2}\left(1-2\nu-(1+\nu)\frac{a(a+b)}{2b^2}\right)\frac{P}{E}. \tag{9.56}$$

Surprisingly, for a range of values of b/a including $b/a = 1$, the expression in parenthesis is negative, such that $\delta d > 0$ and the shell actually *thickens* when compressed (see Problem 9.6). A similar result holds for a cylindrical tube (Problem 9.8). This is probably why you see rivets jumping out of the hull in films of submarines going too deep, because they are literally being pulled out by the increase of the wall thickness caused by compression.

Strains: The strains become

$$u_{rr} = -\frac{b^3}{b^3-a^3}\left(1-2\nu-(1+\nu)\frac{a^3}{r^3}\right)\frac{P}{E}, \tag{9.57a}$$

$$u_{tt} = -\frac{b^3}{b^3-a^3}\left(1-2\nu+(1+\nu)\frac{a^3}{2r^3}\right)\frac{P}{E}. \tag{9.57b}$$

Since $-1 < \nu < 1/2$, the tangential strain u_{tt} is always negative, corresponding to a tangential compression of the material. The radial strain u_{rr} is always positive at the inside of the shell at $r = a$, and can even be positive at the outside $r = b$. This is of course related to the thickening of the shell. Although the material is everywhere displaced radially inward, it will always expand at the inside for $r = a$, but may expand or contract at the outside.

Stresses: The stresses become

$$\sigma_{rr} = -\frac{b^3}{b^3-a^3}\left(1-\frac{a^3}{r^3}\right)P, \tag{9.58a}$$

$$\sigma_{tt} = -\frac{b^3}{b^3-a^3}\left(1+\frac{a^3}{2r^3}\right)P. \tag{9.58b}$$

Again there is surprise: The stresses are independent of the elastic properties of the material.

Thin shell approximation

For a thin shell with thickness $d = b - a \ll a$ we put $r = a + s\,d$ with $0 \leq s \leq 1$ and expand to leading order in the thickness,

$$u_r \approx -(1-\nu)\frac{a^2}{2d}\frac{P}{E}, \qquad \delta d \approx \nu a\frac{P}{E}, \tag{9.59a}$$

$$u_{rr} \approx \nu\frac{a}{d}\frac{P}{E}, \qquad u_{tt} \approx -\frac{1}{2}(1-\nu)\frac{a}{d}\frac{P}{E}, \tag{9.59b}$$

$$\sigma_{rr} \approx -sP, \qquad \sigma_{tt} \approx -\frac{1}{2}\frac{a}{d}P. \tag{9.59c}$$

Again we note that the shell thickens, and that there is radial expansion and tangential compression for $0 < \nu < 1/2$. The tangential stress is greater than the applied pressure by a large factor $a/2d$.

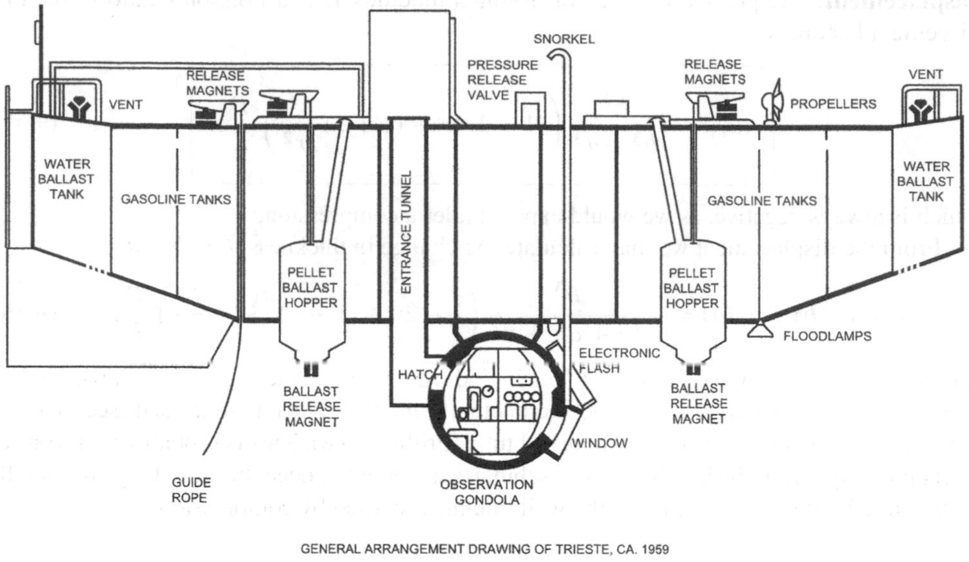

Figure 9.4. Sketch of the structure of the bathyscaphe Trieste that reached the bottom of the nearly 11,000-m deep Challenger Deep in the Mariana Trench on January 23, 1960. The physical properties of the spherical crew cabin are discussed in Example 9.8. (Source: Wikimedia Commons. Courtesy Ralph Sutherland.)

Example 9.7 [The bathysphere]: The first deep-sea vessel, the bathysphere, was spherical and tethered to a surface vessel. It reached a record depth of 923 m in 1932 with two pilots in its tiny chamber. It was made from cast steel with inner diameter $2a = 4.75$ ft $= 145$ cm and thickness $d = 1$ in. $= 2.54$ cm. Since $d/a \approx 0.035$ it is appropriate to use the thin shell formulas. Taking $P = 100$ bar, $E = 200$ GPa and $\nu = 1/3$, we find $u_r \approx -0.35$ mm, $\delta d \approx 12$ μm, $u_{rr} = -u_{tt} \approx 5 \times 10^{-4}$, and an impressive tangential stress, $\sigma_{tt} \approx -1400$ bar. This is not far from the tensile strength of steel, which is about 2,000–4,000 bar, although the compressive strength may be a factor 10 larger than that.

Example 9.8 [The bathyscaphe]: The tethering of the bathysphere to a surface vessel limited the depth that could be reached by this construction. Instead, a self-contained dirigible deep-sea vessel, called the bathyscaphe by its inventor Auguste Picard, was constructed in the late 1940s. A second version, named the *Trieste*, was built and launched in 1953 (see Figure 9.4). Most of the vessel was cylindrical and with open flotation tanks containing gasoline (for buoyancy) and other equipment that could tolerate the enormous pressure. The crew chamber was as before a sphere, this time with a diameter of $2a = 200$ cm and thickness $d = 5$ in. $= 12.7$ cm. Since $d/a = 0.127$, we use the thin shell formulas and expect approximation errors on the order of 13%. With the same material parameters as above we find that $u_r \approx -1.4$ mm, $\delta d \approx 0.18$ mm, and $\sigma_{tt} \approx -4{,}300$ bar . The sphere was carefully constructed by the German Krupp Steel Works to avoid potentially deadly weaknesses in its hull and observation ports. It reached the ultimate depth on Earth, the nearly 11-km Challenger Deep of the Mariana Trench, on January 23, 1960, a feat yet to be repeated.

* 9.6 Application: Radial deformation of a cylindrical body

Cylindrical pipes (or tubes) carrying fluids under pressure are found everywhere, in living organisms and in machines, not forgetting the short moments of intense pressure in the barrel of a gun or cannon. How much does a pipe expand under pressure, and how is the deformation distributed? What are the stresses in the material, and where will it tend to break down?

Uniform radial displacement

The ideal pipe is a right circular cylinder with inner radius a, outer radius b, and length L, made from homogeneous and isotropic elastic material. When subjected to a uniform internal pressure, the pipe is expected to expand radially and perhaps also contract longitudinally. Initially, we shall prevent the contraction by clamping the ends of the pipe such that its length remains unchanged while it exerts a negative pressure on the clamps.

Under these assumptions, the only freedom for the pipe is to expand radially and uniformly such that the displacement fields takes the form

$$u = u_r(r)\,\hat{e}_r, \tag{9.60}$$

where $u_r(r)$ is only a function of the axial distance $r = \sqrt{x^2 + y^2}$. The three unit vectors (see the margin figure),

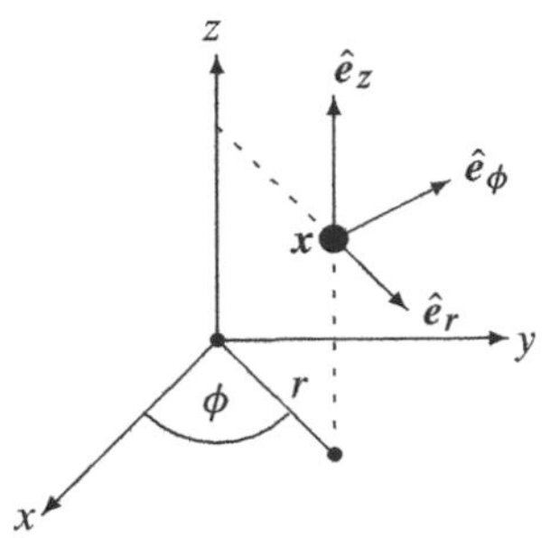

Cylindrical coordinates and basis vectors (see Appendix D).

$$\hat{e}_r = \frac{(x, y, 0)}{r}, \qquad \hat{e}_\phi = \frac{(-y, x, 0)}{r}, \qquad \hat{e}_z = (0, 0, 1), \tag{9.61}$$

form a local basis for any cylindrical geometry (Appendix D). The following analysis proceeds roughly along the same path as in the preceding section.

Equilibrium equation

Since $\nabla r = \hat{e}_r$, it follows from the radial assumption (9.60) that the displacement field may be written as the gradient of another field,

$$u = u_r \hat{e}_r = \frac{d\psi}{dr}\nabla r = \nabla\psi(r), \qquad \psi(r) = \int u_r(r)\,dr. \tag{9.62}$$

The Laplacian of the displacement field can now be written as a gradient, $\nabla^2 u = \nabla^2\nabla\psi = \nabla\nabla^2\psi = \nabla\nabla\cdot u$, so that the Navier–Cauchy equation (9.2) takes the much simpler form

$$(2\mu + \lambda)\,\nabla\nabla\cdot u = -f. \tag{9.63}$$

Cylindrical symmetry demands that $\nabla\cdot u$ can only depend on r, and thus that the body force density, if present, must also be radial, $f = f_r(r)\,\hat{e}_r$.

Using again that the divergence only depends on r, and Gauss' theorem, we obtain the following identity for the volume integral over a piece of the cylinder of length L:

$$2\pi r L u_r = \oint_S u\cdot dS = \int_V \nabla\cdot u\,dV = \int_0^r \nabla\cdot u\,2\pi r L\,dr. \tag{9.64}$$

Differentiating both sides with respect to r, we get

$$\nabla\cdot u = \frac{1}{r}\frac{d(ru_r)}{dr} = \frac{du_r}{dr} + \frac{u_r}{r}, \tag{9.65}$$

and finally we arrive at the ordinary second-order differential equation in r,

$$\boxed{(\lambda + 2\mu)\,\frac{d}{dr}\left(\frac{1}{r}\frac{d(ru_r)}{dr}\right) = -f_r.} \tag{9.66}$$

In most cases, the radial body force vanishes, $f_r = 0$. The only natural candidate is the centrifugal force on a cylinder rotating around its axis (see Problem 9.9).

Strains and stresses

The displacement gradients are most easily calculated in Cartesian coordinates, where

$$u_x = x\frac{u_r}{r}, \qquad u_y = y\frac{u_r}{r}. \tag{9.67}$$

Using that $\nabla_x r = x/r$ and $\nabla_y r = y/r$, the non-vanishing displacement gradients become

$$\begin{aligned}
\nabla_x u_x &= \frac{u_r}{r} + \frac{x^2}{r}\frac{d(u_r/r)}{dr} = \frac{x^2}{r^2}\frac{du_r}{dr} + \frac{y^2}{r^2}\frac{u_r}{r},\\
\nabla_y u_y &= \frac{u_r}{r} + \frac{y^2}{r}\frac{d(u_r/r)}{dr} = \frac{y^2}{r^2}\frac{du_r}{dr} + \frac{x^2}{r^2}\frac{u_r}{r},\\
\nabla_x u_y = \nabla_y u_x &= \frac{xy}{r}\frac{d(u_r/r)}{dr} = \frac{xy}{r^2}\frac{du_r}{dr} - \frac{xy}{r^2}\frac{u_r}{r}.
\end{aligned}$$

Expressed in dyadic notation (see page 597), the displacement gradients may be compactly written as

$$\boldsymbol{\nabla u} = \frac{du_r}{dr}\hat{e}_r\hat{e}_r + \frac{u_r}{r}\hat{e}_\phi\hat{e}_\phi. \tag{9.68}$$

The right-hand side is a symmetric matrix and thus identical to Cauchy's strain tensor $\mathbf{u}$. We note that the trace of this matrix equals $\boldsymbol{\nabla}\cdot\boldsymbol{u}$, as it should.

The only non-vanishing components of the strain tensor are

$$u_{rr} = \frac{du_r}{dr}, \qquad u_{\phi\phi} = \frac{u_r}{r}. \tag{9.69}$$

The non-vanishing stress tensor components are found from Hooke's law (8.8) by projecting on the basis vectors

$$\sigma_{rr} = 2\mu u_{rr} + \lambda\left(u_{rr} + u_{\phi\phi}\right), \tag{9.70a}$$

$$\sigma_{\phi\phi} = 2\mu u_{\phi\phi} + \lambda\left(u_{rr} + u_{\phi\phi}\right), \tag{9.70b}$$

$$\sigma_{zz} = \lambda\left(u_{rr} + u_{\phi\phi}\right). \tag{9.70c}$$

Here we have used that the trace of the strain tensor is independent of the basis, so that $\sum_k u_{kk} = u_{xx} + u_{yy} = u_{rr} + u_{\phi\phi}$. The longitudinal stress, σ_{zz}, appears as a consequence of the clamping of the ends of the cylinder.

Clamped pipe under internal pressure

In the simplest case there are no body forces, $f_r = 0$, and we find immediately the solution,

$$u_r = Ar + \frac{B}{r}, \tag{9.71}$$

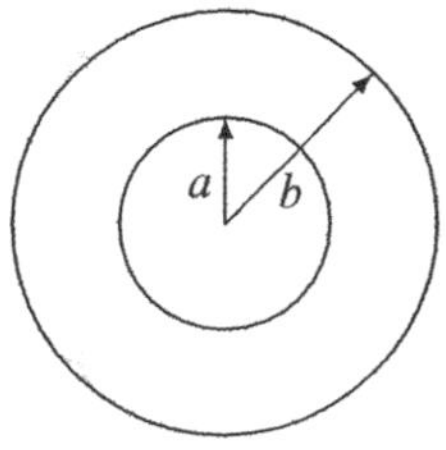

Cylindrical pipe with inner radius a and outer radius b.

where A and B are integration constants to be determined by the boundary conditions. The non-vanishing strains and stresses become

$$u_{rr} = A - \frac{B}{r^2}, \qquad u_{\phi\phi} = A + \frac{B}{r^2}, \tag{9.72}$$

$$\sigma_{rr} = 2A(\lambda+\mu) - \frac{2\mu B}{r^2}, \qquad \sigma_{\phi\phi} = 2A(\lambda+\mu) + \frac{2\mu B}{r^2}, \qquad \sigma_{zz} = 2A\lambda. \tag{9.73}$$

The boundary conditions are $\sigma_{rr} = -P$ at the inside surface $r = a$ and $\sigma_{rr} = 0$ at the outside surface $r = b$. The minus sign may be a bit surprising, but remember that the normal to the inside surface of the pipe is in the direction of $-\hat{e}_r$, so that the stress vector $\sigma_{rr}(-\hat{e}_r) = P\hat{e}_r$ points in the positive radial direction, as it should.

The boundary conditions are solved for A and B, and we find

$$A = \frac{a^2}{b^2 - a^2}\frac{P}{2(\lambda + \mu)}, \qquad B = \frac{a^2 b^2}{b^2 - a^2}\frac{P}{2\mu}. \tag{9.74}$$

Note that both are positive.

Displacement field: Expressed in terms of Young's modulus E and Poisson's ratio ν, the radial displacement field becomes

$$u_r = (1+\nu)\frac{a^2}{b^2 - a^2}\left((1-2\nu)r + \frac{b^2}{r}\right)\frac{P}{E}. \tag{9.75}$$

Since $\nu \leq 1/2$, the radial displacement field is always positive and monotonically decreasing for $a \leq r \leq b$. It reaches its maximum at the inner surface, $r = a$, confirming the intuition that the pressure in the pipe should push the innermost material farthest away from its original position.

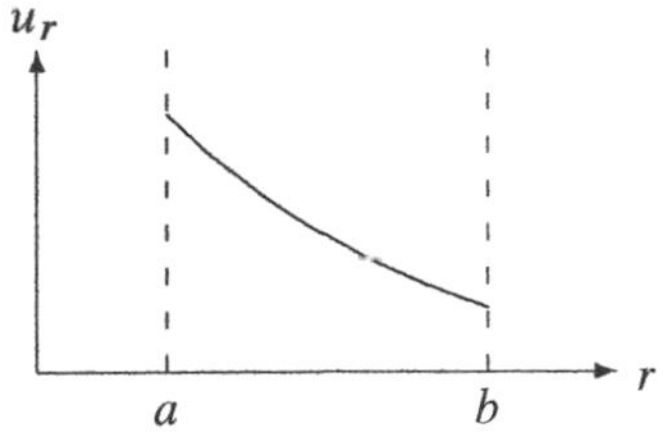

Sketch of the displacement field.

Strain tensor: The non-vanishing strain tensor components become

$$u_{rr} = (1+\nu)\frac{a^2}{b^2 - a^2}\left(1 - 2\nu - \frac{b^2}{r^2}\right)\frac{P}{E}, \tag{9.76a}$$

$$u_{\phi\phi} = (1+\nu)\frac{a^2}{b^2 - a^2}\left(1 - 2\nu + \frac{b^2}{r^2}\right)\frac{P}{E}. \tag{9.76b}$$

For normal materials with $0 < \nu \leq 1/2$, the radial strain u_{rr} is negative, corresponding to a compression of the material, whereas the tangential strain $u_{\phi\phi}$ is always positive, corresponding to an extension. There is no longitudinal strain because of the clamping of the ends of the pipe.

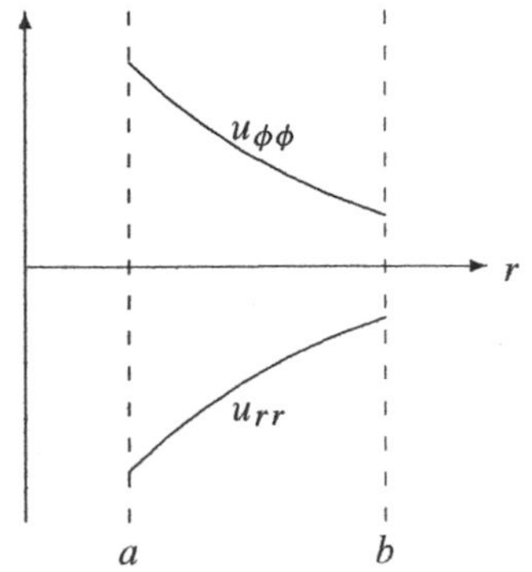

Sketch of strain components.

The scale of the strain is again set by the ratio P/E. For normal materials under normal pressures, for example an iron pipe with $E \approx 1$ Mbar subject to a water pressure of a few bars, the strain is only on the order of parts per million, whereas the strains in the walls of your garden hose or the arteries in your body are much larger. When the walls become thin, that is, for $d = b - a \ll a$, the strains grow stronger because of the denominator $b^2 - a^2 \approx 2da$, and actually diverge toward infinity in the limit. This is in complete agreement with our understanding that the walls of a pipe need to be of a certain thickness to withstand the internal pressure.

Stress tensor: The non-vanishing stress tensor components become

$$\sigma_{rr} = -\frac{a^2}{b^2 - a^2}\left(\frac{b^2}{r^2} - 1\right)P, \tag{9.77a}$$

$$\sigma_{\phi\phi} = \frac{a^2}{b^2 - a^2}\left(\frac{b^2}{r^2} + 1\right)P, \tag{9.77b}$$

$$\sigma_{zz} = 2\nu\frac{a^2}{b^2 - a^2}P. \tag{9.77c}$$

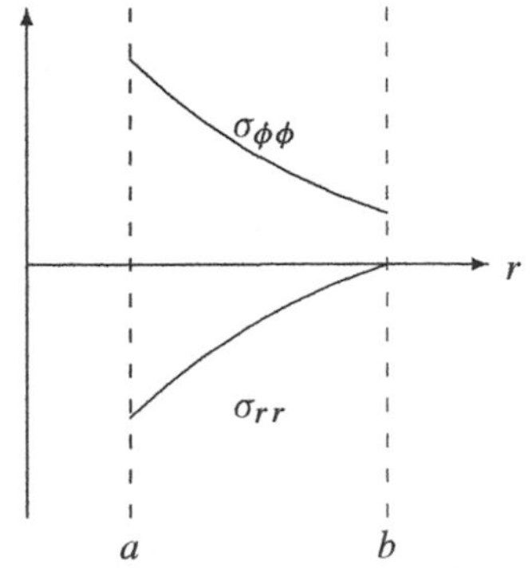

Sketch of stress components.

None of the stresses depend on Young's modulus E, and only the longitudinal stress depends on Poisson's ratio ν.

The radial pressure $p_r = -\sigma_{rr}$ can never become larger than P because we may write

$$\frac{p_r}{P} = \frac{b^2 - r^2}{b^2 - a^2} \frac{a^2}{r^2}, \tag{9.78}$$

which is the product of two factors, both smaller than unity for $a < r < b$. The tangential pressure $p_\phi = -\sigma_{\phi\phi}$ and the longitudinal pressure $p_z = -\sigma_{zz}$ are both negative (tensions), and can become large for thin-walled pipes. The average pressure

$$p = \frac{1}{3}\left(p_r + p_\phi + p_z\right) = -\frac{2}{3}(1+\nu)\frac{a^2}{b^2 - a^2} P \tag{9.79}$$

is also negative and like the longitudinal pressure is constant throughout the material. The average pressure does not vanish at $r = b$, and this confirms the suspicion voiced on page 103 that the pressure behaves differently in a solid with shear stresses than the pressure in a fluid at rest, where it has to be continuous across boundaries in the absence of surface tension.

Blowup: A pipe under pressure blows up if the material is extended beyond a certain limit. Compression does not matter, except for very large pressures. The point where the pipe breaks is primarily determined by the point of maximal local tension. As we have seen, this occurs at the inside of the pipe for $r = a$, where

$$\sigma_{\phi\phi} = \frac{b^2 + a^2}{b^2 - a^2} P. \tag{9.80}$$

When this tension exceeds the tensile strength in a brittle material, a crack will develop where the material has a small weakness, and the pipe blows up from the inside!

Example 9.9 [One-inch water pipe]: A standard American 1-inch iron water pipe has $2a =$ 0.957 in. and $2b = 1.315$ in. Taking the yield strength of iron to be 200 MPa = 2,000 bar, the blowup pressure becomes $P = 615$ bar. Even with a safety factor 10, normal water pressures can never blow up such a pipe, as long as corrosion has not thinned the wall too much.

Example 9.10 [Frost bursting]: Broken water pipes in winter are a common phenomenon. The reason is that water expands by about 9% when freezing at 0°C. After freezing it contracts slowly if the temperature continues to drop. The bulk modulus of solid ice is $K = 8.8$ GPa = 88,000 bar. If the water is prevented from expanding along the pipe, for example by being blocked by already frozen regions, it will in principle be able to develop a radial pressure of about 9% of the bulk modulus, or 8,000 bar, which is four times larger than the yield strength of iron. No wonder that pipes burst! The calculation is, however, only an estimate, because other phases of ice exist at high pressures.

Unclamped pipe

In older houses where central heating pipes have been clamped too tight by wall fixtures, major noise problems can arise because no normal fixtures can withstand the large pressures that arise when the water temperature changes and the pipes expand and contract longitudinally. In practice, pipes should always be thought of as being unclamped.

The constancy of the longitudinal tension (9.77c) permits us to solve the case of an unclamped pipe by superposing the above solution with the displacement field for uniform stretching (8.23) on page 133. In the cylindrical basis, the field of uniform stretching becomes (after interchanging x and z)

$$u_r = -\nu r \frac{Q}{E}, \qquad u_z = z\frac{Q}{E}, \tag{9.81}$$

where Q is the tension applied to the ends.

Choosing Q equal to the longitudinal tension (9.77c) in the clamped pipe,

$$Q = 2\nu \frac{a^2}{b^2 - a^2} P, \tag{9.82}$$

and subtracting the stretching field from the clamped pipe field (9.75), we find for the unclamped pipe that

$$u_r = \frac{a^2}{b^2 - a^2}\left((1-\nu)r + (1+\nu)\frac{b^2}{r}\right)\frac{P}{E}, \tag{9.83a}$$

$$u_z = -2\nu \frac{a^2}{b^2 - a^2} z \frac{P}{E}. \tag{9.83b}$$

The strains are likewise obtained from the clamped strains (9.76a) by subtracting the strains for uniform stretching, and we get

$$u_{rr} = \frac{a^2}{b^2 - a^2}\left(1 - \nu - (1+\nu)\frac{b^2}{r^2}\right)\frac{P}{E}, \tag{9.84a}$$

$$u_{\phi\phi} = \frac{a^2}{b^2 - a^2}\left(1 - \nu + (1+\nu)\frac{b^2}{r^2}\right)\frac{P}{E}, \tag{9.84b}$$

$$u_{zz} = -2\nu \frac{a^2}{b^2 - a^2}\frac{P}{E}. \tag{9.84c}$$

The superposition principle guarantees that the radial and tangential stresses are the same as before and given by (9.77), while the longitudinal stress now vanishes, $\sigma_{zz} = 0$.

Thin wall approximation

Most pipes have thin walls relative to their radius. Let us introduce the wall thickness, $d = b - a \ll a$, and put the radial distance, $r = a + sd$ with $0 \leq s \leq 1$. In the leading approximation we get the displacement field for the unclamped pipe:

$$u_r \approx a \frac{a}{d}\frac{P}{E}, \qquad u_z \approx -z\nu \frac{a}{d}\frac{P}{E}. \tag{9.85}$$

The corresponding strains become

$$u_{rr} \approx -\nu \frac{a}{d}\frac{P}{E}, \qquad u_{\phi\phi} \approx \frac{a}{d}\frac{P}{E}, \qquad u_{zz} \approx -\nu \frac{a}{d}\frac{P}{E}. \tag{9.86a}$$

The strains all diverge for $d \to 0$, and the condition for small strains is now $P/E \ll d/a$. Finally, we get the non-vanishing stresses

$$\sigma_{rr} \approx -(1-s)P, \qquad \sigma_{\phi\phi} \approx \frac{a}{d} P. \tag{9.87a}$$

The radial pressure $p_r = -\sigma_{rr}$ varies between 0 and P as it should when s ranges from 0 to d. It is always positive and of order P, whereas the tangential tension $\sigma_{\phi\phi}$ diverges for $d \to 0$. Blowups always happen because the tangential tension becomes too large.

Problems

9.1 Show that Navier's equation of equilibrium (9.2) may be written as

$$\nabla^2 \boldsymbol{u} + \frac{1}{1-2\nu}\nabla\nabla\cdot\boldsymbol{u} = -\frac{1}{\mu}\boldsymbol{f},$$

where ν is Poisson's ratio.

9.2 A body made from isotropic elastic material is subjected to a body force in the z-direction, $f_z = kxy$. Show that the displacement field,

$$u_x = Ax^2yz, \qquad u_y = Bxy^2z, \qquad u_z = Cxyz^2,$$

satisfies the equations of mechanical equilibrium for suitable values of A, B, and C.

* **9.3** A certain gun has a steel barrel of length of $L = 1$ m, a bore diameter of $2a = 1$ cm. The charge of gunpowder has length $x_0 = 3$ cm and density $\rho_0 = 1$ g cm^{-3}. The bullet in front of the charge has mass $m = 5$ g. The expansion of the ideal gases left by the explosion of the charge at $t = 0$ is assumed to be isentropic with index $\gamma = 7/5$. **(a)** Determine the velocity $\dot{x}$ as a function of x for a bullet starting at rest from $x = x_0$. **(b)** Calculate the pressure just after the explosion. **(c)** Calculate the pressure when the bullet leaves the muzzle with a velocity of $U = 800$ m s^{-1}. **(d)** Calculate the initial and final temperatures when the average molar mass of the gases is $M_{\text{mol}} = 30$ g mol^{-1}. **(e)** Calculate the maximal strains in the steel on the inside of the barrel when it has thickness $d = b - a = 5$ mm. **(f)** Calculate the stresses and compare with the tensile strength of the steel. Will the barrel blow up?

* **9.4** **(a)** Show that the most general solution to the uniform shear-free bending of a beam, originally placed along the z-axis, is

$$u_x = a_x - \phi_z y + \phi_y z - \alpha\nu x + \tfrac{1}{2}\beta_x\left(z^2 - \nu(x^2 - y^2)\right) - \beta_y \nu xy,$$

$$u_y = a_y + \phi_z x - \phi_x z - \alpha\nu y + \tfrac{1}{2}\beta_y\left(z^2 - \nu(y^2 - x^2)\right) - \beta_x \nu xy,$$

$$u_z = a_z - \phi_y x + \phi_x y + \alpha z - \beta_x xz - \beta_y yz,$$

where the coefficients are all constants. **(b)** Interpret the coefficients. Hint: Use that u_{zz} must be linear in x and y.

9.5 Show that a shift in the y-coordinate, $y \to y - \alpha$, in the shear-free bending field (9.19), corresponds to adding in a uniform stretching deformation (plus a simple translation).

9.6 Determine the range of values a/b for which a spherical shell actually thickens when $0 < \nu < \frac{1}{2}$. Hint: See (9.56).

9.7 **(a)** Calculate how a massive sphere of radius a and constant mass density ρ_0 is deformed elastically under its own gravity. **(b)** Calculate the surface displacement, and **(c)** the central radial strain, and **(d)** stress for the Earth. Use $a = 6371$ km, $G = 6.6 \times 10^{-11}$ N m^2 kg^{-2}, $E = 500$ GPa, $\nu = 0.3$, and $\rho_0 = 5.5$ g cm^{-3}.

9.8 Calculate the displacement, strain, and stress for a clamped evacuated pipe subject to an external pressure P. Can the pipe wall actually thicken during compression?

9.9 A massive cylinder with radius a and constant density ρ_0 rotates around its axis with constant angular frequency Ω. **(a)** Find the centrifugal force density in cylindrical coordinates, rotating with the cylinder. **(b)** Calculate the displacement for the case where the ends of the cylinder are clamped to prevent change in length and the sides of the cylinder are free. **(c)** Show that the tangential strain always corresponds to an expansion, whereas the radial strain corresponds to an expansion close to the center and a compression close to the rim. **(d)** Where will the breakdown happen?

10 Slender rods

The regular object geometries analyzed in the preceding chapter were all three-dimensional, meaning that their sizes in different directions were comparable. Some of the recurring problems in elasticity concern bodies with widely different sizes in different directions. A rod is much thinner than it is long and thus effectively one-dimensional. A plate is much thinner than it is wide, making it effectively two-dimensional. In the mathematical limit a rod becomes a curve described by a vector function of one parameter, while a plate becomes a surface described by a vector function of two parameters.

Mathematics is, however, not physics. The mechanical properties of a rod depend on the shape and size of its cross-section and the material from which it is made. The Euler–Bernoulli law for bending and the Coulomb–Saint-Venant law for twisting provide the connection between the mathematical description of a rod's curvature and torsion and the physical forces and moments at play. As long as the radius of curvature and the torsion length of a rod are much larger than the effective diameter of its cross-sections, all the components of the strain tensor will be small. This does, however, not guarantee that the *deflection* of a rod from its initial shape will be small in comparison with the radius of curvature and the torsion length. Thus, for example, the deflection of a longbow is always comparable to its radius of curvature.

The present chapter opens with a discussion of bending with small deflection and no torsion, resulting in a differential equation that can be solved analytically in nearly all practical situations. The chapter continues with an analysis of the famous buckling instability that is encountered when a straight rod is compressed longitudinally and suddenly, spontaneously deviates from the straight shape. Finite deflection without torsion is also tractable and allows us, for example, to calculate the shape of a relaxed stringed bow. Finally, the combination of bending and twisting of rods is analyzed and applied to the case of a coiled spring.

10.1 Small deflections without torsion

Pure bending is an ideal that is rarely met in practice where initially straight beams can be bent and twisted by a multitude of forces. Some forces act locally, like the supports that carry a bridge; others are distributed all over the beam, like the weight of the beam itself. In this section we shall only consider a straight beam or rod, that is bent—but not twisted—by a tiny amount. The beam is, as before, initially placed along the z-axis between $z = 0$ and $z = L$, and bent in the direction of the y-axis by external forces acting only in the yz-plane and external moments only in the direction of the x-axis.

If the beam cross-section is not circular, the principal axes of the cross-section must be aligned with the x- and y-axes, for otherwise an internal moment will arise along the y-axis (see page 148), which complicates matters. The deformed rod is described by the displacement, $y = y(z)$, of its centroid, also called the *deflection* of the rod. We shall in this section assume that the deflection varies slowly along the rod, or so that its derivative is small everywhere, that is,

$$\left|\frac{dy}{dz}\right| \ll 1. \tag{10.1}$$

Provided there is a point on the rod that is not deflected, the maximal deflection will always be small compared to the length of the rod, because $|y| \lesssim |dy/dz|_{\max} L \ll L$. Similarly, we have $|dy/dz| \lesssim |d^2y/dz^2|_{\max} L \simeq L/R_{\min}$, where $R_{\min}$ is the minimal radius of curvature. This indicates that the minimal radius of curvature of the rod should be much larger than its length, $R_{\min} \gg L$.

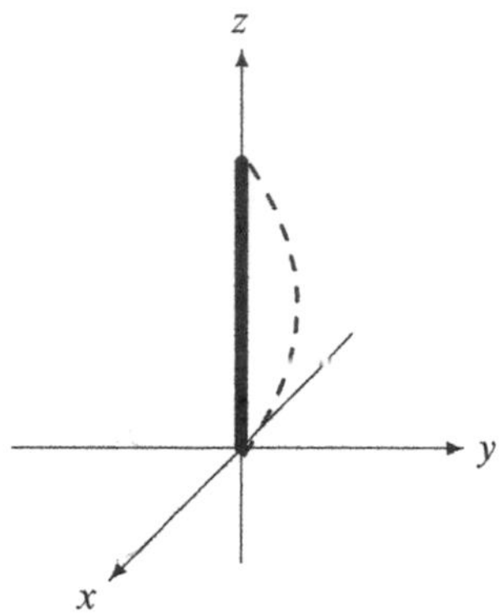

Initial position of the undeformed beam and a possible planar deflection in the y-direction (dashed).

Local balance of forces and moments

Consider now the cross-section of the bent rod at z. By the assumption of planar bending, the internal stresses in the material make the part of the rod lying above this cross-section act on the part below with a total transverse force $\mathcal{F}_y(z)$ and a total longitudinal force $\mathcal{F}_z(z)$, as well as a total moment of force $\mathcal{M}_x(z)$, calculated around the centroid of the cross-section. Besides these internal forces, there are external forces and moments of force acting on the rod. Some of these act locally in a point, like the pillars of a bridge or the weight of a car on the bridge; others are distributed along the rod, such as the weight of the bridge's material.

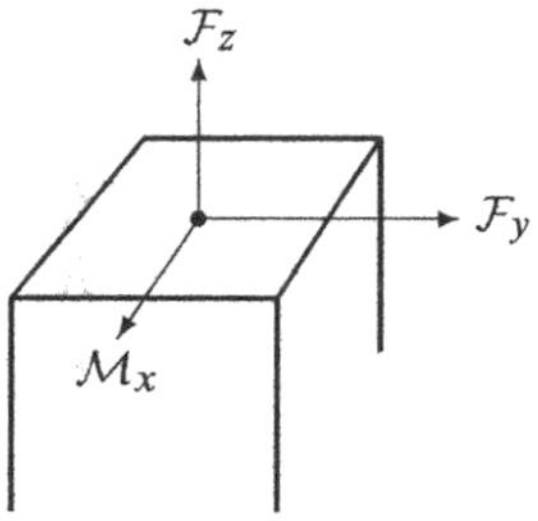

Forces and moments due to internal stresses in a cross-section, here drawn rectangular.

Everywhere between the points of attack of external point forces and moments, it is fairly simple to set up the local balance of forces and moments that secure mechanical equilibrium. If $K_y(z)dz$ denotes the transverse resultant of the distributed external forces acting on a small piece dz of the rod, the y-component of the total force on this small piece must vanish, leading to (see the lowest margin figure)

$$\mathcal{F}_y(z + dz) - \mathcal{F}_y(z) + K_y(z)\,dz = 0,$$

and similarly for the longitudinal distributed force K_z. Dividing by dz we get

$$\boxed{\frac{d\mathcal{F}_y}{dz} = -K_y, \qquad \frac{d\mathcal{F}_z}{dz} = -K_z.} \tag{10.2}$$

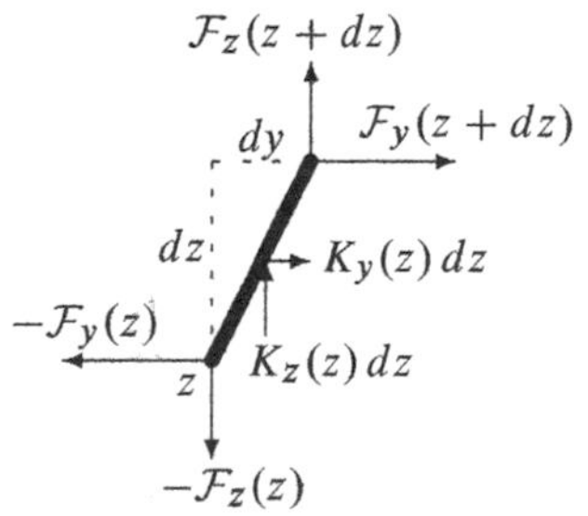

The forces on a small piece of the bent rod. The x-axis comes out of the paper.

The total moment of force around the the centroid of the cross-section at z must also vanish,

$$\mathcal{M}_x(z + dz) - \mathcal{M}_x(z) + \mathcal{F}_z(z)dy - \mathcal{F}_y(z)dz = 0,$$

and dividing by dz, this equation becomes

$$\boxed{\frac{d\mathcal{M}_x}{dz} = \mathcal{F}_y - \mathcal{F}_z\frac{dy}{dz}.} \tag{10.3}$$

By the assumption, $|dy/dz| \ll 1$, the last term on the right may be disregarded, unless the longitudinal force $\mathcal{F}_z$ is much larger than the transverse force $\mathcal{F}_y$. Everything else being equal, it is the transverse force $\mathcal{F}_y$ that is important for bending the rod (see also Equation (9.31) on page 150), so for now we drop the longitudinal term on the right, effectively setting $\mathcal{F}_z = 0$.

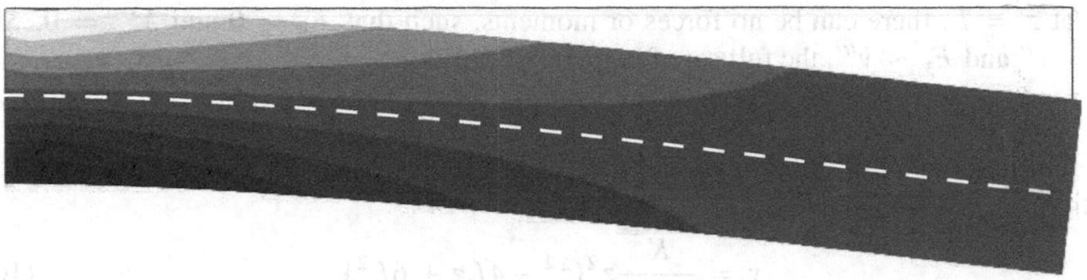

Figure 10.1. Simulation of cantilever bent by its own weight. The undeformed shape is outlined in dark. In arbitrary units (for example SI), the cantilever has length $L = 6$ and a quadratic cross-section with side lengths 1, yielding an area moment $I = 1/12$. Young's modulus is $E = 10^4$ and the weight per unit of length $K = 3$. The contours indicate the values of longitudinal stress, σ_{zz}, and clearly show that the rod's material is stretched on the upper side and compressed at the lower. The dashed white line, which follows the central ray perfectly, is the slender rod prediction (10.8).

Bending moment

The hypothesis due to Saint-Venant is now that the local bending moment may be obtained from the Euler–Bernoulli law (9.25) with the local curvature given by $\kappa = d^2y/dz^2$, an expression that is correct to order $|dy/dz|^2$. Combining it with force balance (10.2) and moment balance (10.3) (without the longitudinal term), we get the rod equations,

$$\mathcal{M}_x = -EI\frac{d^2y}{dz^2}, \qquad \mathcal{F}_y = \frac{d\mathcal{M}_x}{dz}, \qquad K_y = -\frac{d\mathcal{F}_y}{dz}. \tag{10.4}$$

Given the transverse distributed force K_y together with suitable boundary conditions, these equations can be solved for the transverse deflections $y(z)$ of a rod. Note that they are valid even if the cross-section, the area moment or the material properties change along the rod.

Case: Horizontal uniform rod

For simplicity we now assume that the rod has constant cross-section A, constant flexural rigidity EI, and constant mass density ρ. Taking the z-axis to be horizontal and the y-axis pointing downward along the direction of constant gravity g_0, the transverse force distribution becomes $K_y \equiv K = \rho A g_0$ and $K_z = 0$. Combining the preceding equations we obtain an amazing fourth-order ordinary differential equation for the deflection,

$$EI\frac{d^4y}{dz^4} = K. \tag{10.5}$$

The solution to this equation is a fourth-order polynomial in z with four unknown coefficients,

$$y = a + bz + cz^2 + dz^3 + \frac{K}{24EI}z^4. \tag{10.6}$$

Evidently we need four boundary conditions to determine a solution. We shall now discuss how this works out for a few well-known constructions.

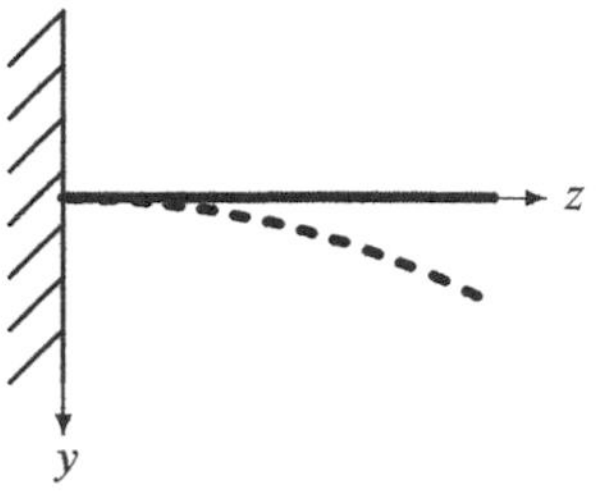

The cantilever is clamped to the wall at one end, but free to move at the other end.

Cantilever bent by its own weight: A cantilever is a horizontal rod that is clamped to a wall at $z = 0$ but free to move at $z = L$. Cantilevers are found in many constructions, for example flagpoles, jumping boards, and cranes. At $z = 0$, the clamped state demands that there is no deflection and that the rod is horizontal, such that $y = 0$ and $y' = 0$. In the free state at $z = L$, there can be no forces or moments, such that $\mathcal{F}_y = 0$ and $M_x = 0$. Since $M_x \sim y''$ and $F_y \sim y'''$, the full set of boundary conditions are

$$y(0) = 0, \qquad y'(0) = 0, \qquad y''(L) = 0, \qquad y'''(L) = 0, \tag{10.7}$$

with a prime denoting differentiation with respect to z. One may directly verify that the solution is,

$$y = \frac{K}{24EI} z^2(z^2 - 4Lz + 6L^2). \tag{10.8}$$

The free end bends down by $y(L) = KL^4/8EI$. As seen in Figure 10.1, the slender-rod approximation is quite good even for a not very slender rod.

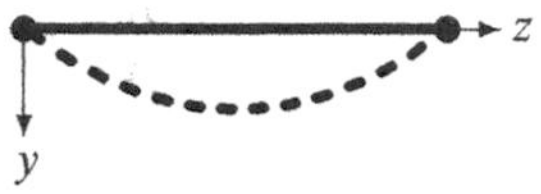

Bridge supported by hinges at the ends.

Bridge with hinged supports: A bridge is supported at its ends by pylons with hinges that fix the ends but allow them to rotate. In this case the boundary conditions are that there is no displacement and no moments at the ends:

$$y(0) = 0, \qquad y''(0) = 0, \qquad y(L) = 0, \qquad y''(L) = 0. \tag{10.9}$$

The solution is

$$y = \frac{K}{24EI} z(L - z)(L^2 + Lz - z^2). \tag{10.10}$$

In the middle, the bridge bends down by $y(L/2) = 5KL^4/384EI$.

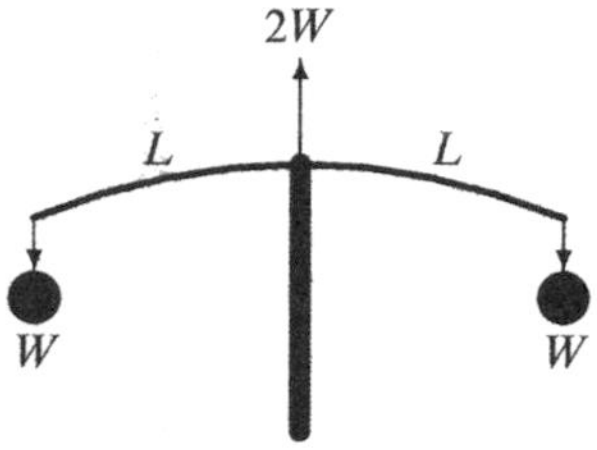

Yoke with two equal weights W supported in the middle. Each arm of the yoke has length L and may be viewed as an end-loaded cantilever.

Light yoke with heavy loads: Yokes have been used since time immemorial for carrying fairly heavy loads across the shoulders. We idealize the yoke in the form of a straight rod with weight much smaller than the weight of the loads at the extreme ends. These loads must have nearly equal weight W, or the yoke will tip. Disregarding the weight of the yoke itself ($K = 0$), we can view each half of the yoke as a cantilever loaded at one end, leading to the boundary conditions

$$y(0) = 0, \qquad y'(0) = 0, \qquad y''(L) = 0, \qquad y'''(L) = -\frac{W}{EI}. \tag{10.11}$$

The last condition was obtained from the rod equations (10.4) applied to the end point, $W = \mathcal{F}_y(L) = -EIy'''(L)$. The solution is

$$y(z) = \frac{W}{6EI} z^2(3L - |z|) \qquad \text{for } -L \le z \le L. \tag{10.12}$$

Notice that the third derivative jumps from $+W/EI$ to $-W/EI$ when passing through $z = 0$. Something like this actually has to happen. For if there were no jump, the fourth-order curve would be continuous at $z = 0$, and that is not possible because that would imply no external forces and thus belie the upward supporting point force $2W$ acting at $z = 0$.

10.2 Buckling instability

A walking stick must be chosen with care. Too sturdy, and it will be heavy and unyielding; too slender, and it may buckle or even collapse under your weight when you lean on it. Stability against buckling and sudden collapse is of course of great technologically importance, considering all the struts, columns, and girders that are found in human buildings and machines (see Figure 10.2). Luckily, however, buckling does not happen until the compressive forces on the beam terminals exceed a certain threshold. First determined by Euler, this threshold load can be used to estimate the point of failure of a column and to set safety limits. Sometimes the buckled rod itself is of technological interest, for example the longbow, which may be viewed as a rod brought beyond the buckling threshold and captured in its bent shape by a taut string between its ends.

Figure 10.2. Impact buckling of a steel support column in Cortland St./WTC Station (New York City) after the collapse of the World Trade Center on September 11, 2001. (Source: Metropolitan Transportation Authority, New York City Transit.)

Euler's threshold for buckling

In the preceding section we found that the longitudinal force on the rod, $\mathcal{F}_z$, is only important for small deflections when it is much larger than the transverse force $\mathcal{F}_y$. In the absence of distributed forces $K_x = K_y = 0$, it follows from (10.2) that $\mathcal{F}_y$ and $\mathcal{F}_z$ are constants along the rod. For simplicity we assume that $\mathcal{F}_y = 0$, so that the rod is only subject to a longitudinal compression force, $\mathcal{F}_z = -\mathcal{F}$, imposed on the upper terminal of the rod (see the margin figure). Integrating local moment balance (10.3) once, we then obtain

$$\boxed{EI\frac{d^2y}{dz^2} = -\mathcal{F}\,y.} \tag{10.13}$$

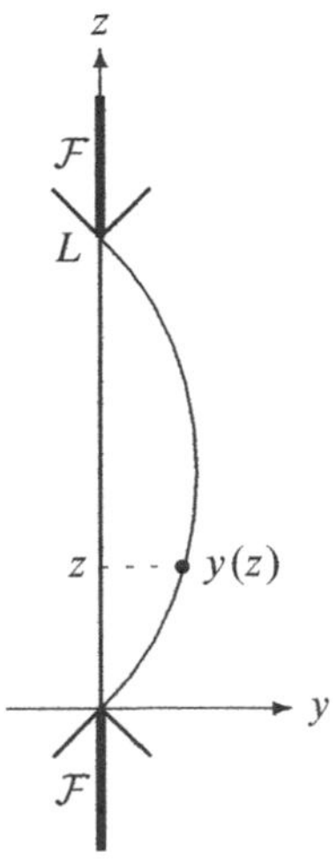

The centroid line of a deformed beam with compressional external forces acting along the z-axis.

This relation simply expresses the vanishing of the total moment on the part of the rod below the cross-section at z, because the external moment is $-\mathcal{F}y$ and the internal moment $\mathcal{M}_x = -EId^2y/dz^2$.

The above is nothing but the standard harmonic equation with wavenumber $k = \sqrt{\mathcal{F}/EI}$, and its general solution is $y = A\sin kz + B\cos kz$, where A and B are constants. Applying the boundary conditions that $y(0) = y(L) = 0$, the buckling solution becomes[1]

$$y = A\sin kz, \qquad k = \frac{n\pi}{L}, \tag{10.14}$$

where n is an arbitrary integer. Since $k = \sqrt{\mathcal{F}/EI}$, we arrive at the following expression for the force:

$$\mathcal{F} = k^2EI = n^2\frac{\pi^2EI}{L^2}, \qquad \text{for } n = 0, 1, 2, \ldots. \tag{10.15}$$

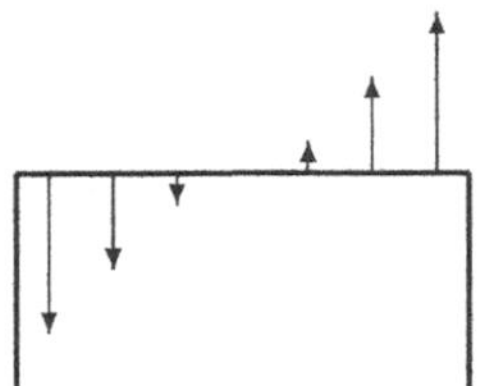

Sketch of the stress forces that create the bending couple on the piece of the rod below z. The material is expanded away from the center of curvature and compressed toward it (on the left).

Strangely, buckling solutions only exist for certain values of the applied force, corresponding to integer values of n. How should this be understood?

[1] The buckling of a stalk with one free end is analyzed in problem 10.1; see [Doghri 2000] for other examples.

Everyday experience tells us that we can normally lean quite heavily on a walking stick without danger of it buckling. A small force cannot bend the beam but only compress it longitudinally, an effect we have not taken into account in the above calculation. As the applied force $\mathcal{F}$ increases, it will eventually reach the threshold value corresponding to $n = 1$ above, called the *Euler threshold*,

$$\boxed{\mathcal{F}_E = \frac{\pi^2 E I}{L^2}.} \tag{10.16}$$

At this point the longitudinal compression mode becomes unstable and the first buckling solution takes over at the slightest provocation. To prove that this is indeed what takes place requires a stability analysis, which we shall carry out below. In practice only the lowest mode is seen, unless a strong force is rapidly applied, in which case the rod may become permanently deformed, or even crumple and collapse. The buckled column in Figure 10.2 appears approximatively to be a solution with $n = 2$, permanently frozen into the steel because the yield stress was surpassed in the violent event.

Example 10.1 [Wooden walking stick]: A wooden walking stick has length $L = 1$ m and circular cross-section of diameter $2a = 2$ cm. Taking Young's modulus $E = 10^{10}$ Pa and using (9.27), the buckling threshold becomes $\mathcal{F}_E = 775$ N, corresponding to the weight of 79 kg. If you weigh more, it would be prudent to choose a slightly thicker stick. Since the area moment I grows as the fourth power of the radius, increasing the diameter to 1 inch (2.54 cm) raises the Euler threshold to the weight of 206 kg, which should be sufficient for most people.

Stability analysis

Suppose the rod initially is already compressed with a longitudinal terminal force $\mathcal{F}$ and now has the length L. Let us now perturb the straight rod by bending it ever so slightly such that the centroid falls on a chosen curve $y = y(z)$. To do that we need to impose extra (virtual) forces on the rod, and we shall now show that the work of these forces is

$$\boxed{W = \frac{1}{2} E I \int_0^L y''(z)^2 \, dz - \frac{1}{2} \mathcal{F} \int_0^L y'(z)^2 \, dz.} \tag{10.17}$$

The first term is easy, because it represents the pure bending energy of the perturbation obtained from (9.30) with the bending moment $\mathcal{M}_x = -EIy''$. This energy must of course be provided by the work of the virtual forces. The second term arises from the increase in length of the rod due to the perturbation, which releases a bit of the compression energy initially present, and therefore diminishes the amount of work that the virtual forces have to do. Since the perturbation is infinitesimal, we can calculate the (negative) work as $-\mathcal{F}\Delta L$, where ΔL is the increase in length of the perturbed rod. Using that the line element in the yz-plane is

$$d\ell = \sqrt{dy^2 + dz^2} = dz\sqrt{1 + y'(z)^2} \approx dz + \frac{1}{2} y'(z)^2 dz, \tag{10.18}$$

the total increase in length becomes $\Delta L = \int_0^L (d\ell - dz)$, which leads to the second term.

As long as the virtual work W is positive, the undeformed beam is stable when left on its own, because there are no external "agents" around to perform the necessary work. This is evidently the case for $\mathcal{F} = 0$. If, on the other hand, the virtual work W is negative for some choice of $y(z)$, the undeformed beam is unstable and will spontaneously deform without the need of work from any external "agent". Since the second term in (10.17) is always negative for positive $\mathcal{F}$, the rod will always become unstable for a sufficiently large value of $\mathcal{F}$. The

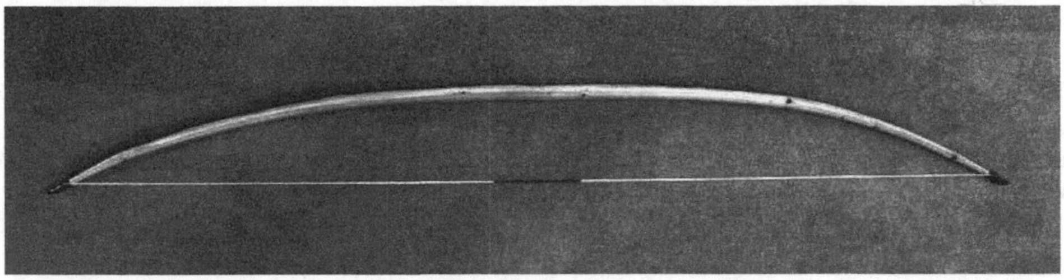

Figure 10.3. English wooden bow. Being visibly thinner toward the ends, this bow does not correspond perfectly to any of the ideal shapes calculated in the text. (Source: Wikimedia Commons.)

lowest possible value, $\mathcal{F} = \mathcal{F}_c$, where this can happen, is called the *critical load*. At this point it takes no work to begin to deform the beam.

The shape of the critical perturbation is determined by that choice of $y(z)$ which yields the smallest value of W for a given $\mathcal{F}$. It can be determined by variation of the perturbation in W, and leads—not unsurprisingly—to the Euler perturbation $y = A \sin kz$ with $k = n\pi/L$. Inserting this solution into (10.17), the integrals are now trivial and we get

$$\boxed{W = \frac{n^2\pi^2 A^2}{4L}(n^2\mathcal{F}_E - \mathcal{F}),} \tag{10.19}$$

where $\mathcal{F}_E$ is the Euler threshold (10.16). For $F < F_E$, the work is positive for all $n \geq 1$, so that the undeformed beam is stable against any such perturbation. The smallest value of $\mathcal{F}$ for which the work can become zero corresponds to $n = 1$, and this shows that the Euler threshold is the critical load. At this point, the amplitude of the perturbation can grow without requiring us to perform any work on the system. The approximation of small perturbations will, however, soon become invalid, but experience tells us that the rod finds a new equilibrium in which it is bent by a finite amount.

10.3 Large deflections without torsion

A stringed bow (see Figure 10.3) may be viewed as a straight rod that has been brought beyond the buckling threshold and is kept in mechanical equilibrium by the tension in the unstretchable bowstring. In this case, the deflection of the rod is not small compared to the dimension of the bow, but the strains in the material are still small as long as the radius of curvature of the bow is much larger than the transverse dimensions of the beam. In this section we shall develop the formalism for large deflections of the central ray, without torsion and with all bending taking place in a plane. For simplicity we assume that there is no compression or shear in the rod. In the following section we turn toward the general theory of rods with torsion.

Planar deflection

The balance of forces in the yz-plane (10.2) and the bending moment along the x-axis (10.3) are also valid for large planar deflections, where the length and radius of curvature of the rod are comparable. To describe the curved planar rod we use the formalism already introduced for bubble shapes (page 84) in which the curve is described by the curve length s from one end and the elevation angle, $\theta = \theta(s)$, here chosen relative to the z-axis. From the planar geometry we immediately get the relations (see the margin figure),

$$\frac{dz}{ds} = \cos\theta, \qquad \frac{dy}{ds} = \sin\theta, \qquad \frac{d\theta}{ds} = -\frac{\mathcal{M}_x}{EI}, \tag{10.20}$$

where we have also used the Euler–Bernoulli law (9.25) to eliminate the radius of curvature, taking into account that a positive moment along x generates a negative curvature.

Differentiating once more with respect to s, and making use of (10.3), it follows that

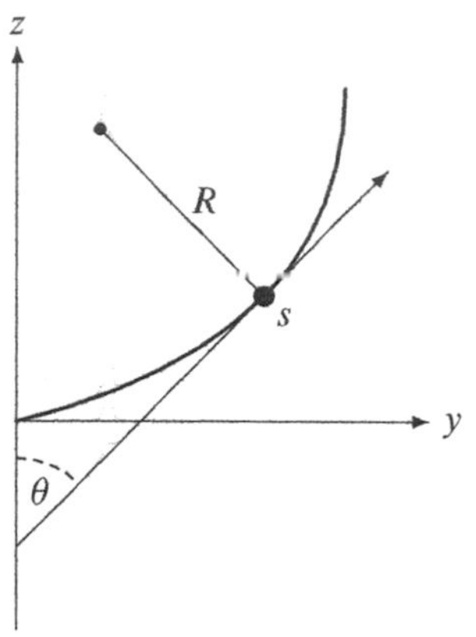

The geometry of planar deflection. The curve is parameterized by the arc length s along the curve. A small change in s generates a change in the elevation angle θ determined by the local radius of curvature. With this choice of θ, the radius of curvature is negative because θ diminishes when s increases.

$$EI\frac{d^2\theta}{ds^2} = -\mathcal{F}_y\cos\theta + \mathcal{F}_z\sin\theta. \tag{10.21}$$

If there are no distributed forces, both $\mathcal{F}_y$ and $\mathcal{F}_z$ are constants. In that case, this equation is a variant of the equation for a mathematical pendulum (see Problem 10.2). Multiplying by $d\theta/ds$, this equation can immediately be integrated to yield

$$\frac{1}{2}EI\left(\frac{d\theta}{ds}\right)^2 = -\mathcal{F}_y\sin\theta - \mathcal{F}_z\cos\theta + C, \tag{10.22}$$

where C is an integration constant, determined by the boundary conditions. This equation can always be solved by quadrature.

Case: Shape of an ideal stringed bow

The stringed bow (see the margin figure) is kept in mechanical equilibrium by the string force, $\mathcal{F}_z = -\mathcal{F}$, whereas $\mathcal{F}_y = 0$. The ends are hinged such that $d\theta/ds \sim \mathcal{M}_x = 0$ for $z = 0$ and $z = L$. Denoting the opening angle, $\theta = \pm\alpha$, at the ends, this fixes the constant in the integrated equation (10.22), which becomes

$$\left(\frac{d\theta}{ds}\right)^2 = 2k^2(\cos\theta - \cos\alpha), \tag{10.23}$$

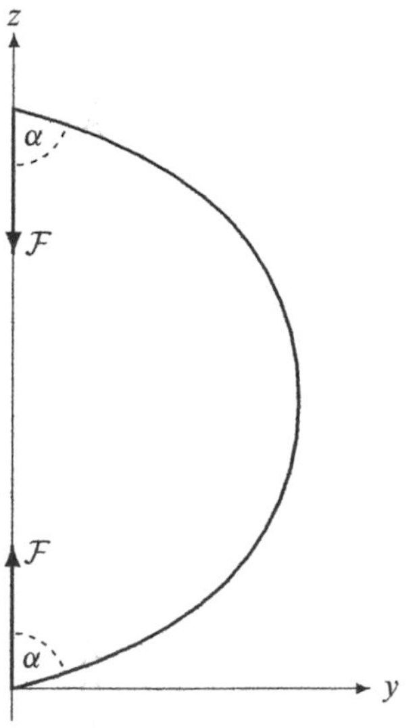

The geometry of a stringed bow with opening angle α. It is kept in mechanical equilibrium by a force $\mathcal{F}_z = -\mathcal{F}$ (and $\mathcal{F}_y = 0$).

where as before $k = \sqrt{\mathcal{F}/EI}$. Since the angle decreases with s, so that $d\theta/ds < 0$, we find

$$s = \frac{1}{k}\int_\theta^\alpha \frac{d\theta'}{\sqrt{2(\cos\theta' - \cos\alpha)}}, \tag{10.24}$$

which is an elliptic integral that is easy to evaluate numerically. For $\theta = -\alpha$, the left-hand side becomes equal to the length of the bow L, and this equation provides a relation between $\sqrt{\mathcal{F}/\mathcal{F}_E} = Lk/\pi$ and α.

Finally, we may calculate $dy/d\theta$ and $dz/d\theta$ and integrate to obtain the Cartesian coordinates as functions of the elevation angle θ:

$$y = \frac{1}{k}\sqrt{2(\cos\theta - \cos\alpha)}, \qquad z = \frac{1}{k}\int_\theta^\alpha \frac{\cos\theta'}{\sqrt{2(\cos\theta' - \cos\alpha)}}\,d\theta'. \tag{10.25}$$

Together these expressions define the shape of the bow parameterized by θ. In Figure 10.4 some shapes are plotted for various opening angles.

Example 10.2 [Wooden longbow]: A certain longbow is constructed from a circular wooden rod of length $L = 150$ cm and diameter $2a = 15$ mm. The bow is stringed with an opening angle of $\alpha = 20\,°$. The maximal string distance from the bow becomes $d \approx 16$ cm and the stringed height $h \approx 145$ cm. The moment of inertia becomes $I \approx 2.5 \times 10^{-9}$ m^4 and taking $E = 10$ GPa, the Euler threshold becomes $\mathcal{F}_E \approx 109.0$ N, corresponding to a weight of 11 kg, which is quite manageable for most people. Numeric integration yields $\mathcal{F} = 110.6$ N, which is very close to the Euler threshold.

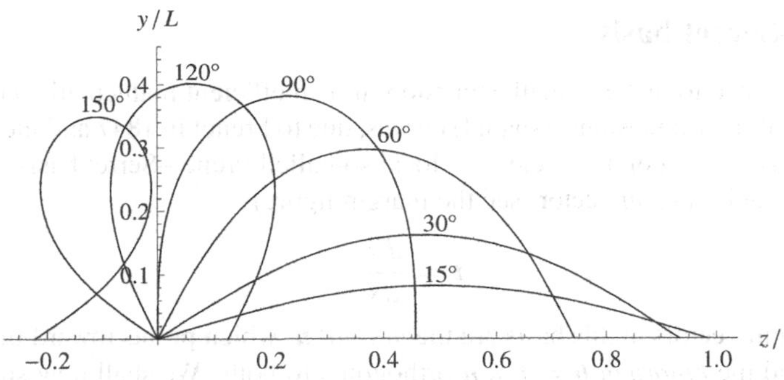

Figure 10.4. Ideal bow shapes for various opening angles . At $\alpha \approx 130°$, the ends of the "bow" cross.

10.4 Mixed bending and twisting

The analysis in the preceding sections can be generalized to arbitrary three-dimensional deformations of rods, including both bending and twisting. We shall again for simplicity assume that the rod is made from a homogeneous, isotropic material, but now we also assume that it has constant circular cross-section with radius a and area $A = \pi a^2$. The radius is as before assumed to be so much smaller than the length L of the rod that stretching and shearing can be disregarded, leaving only room for bending and twisting. Since the area moment for twisting is twice the area moment for bending $J = 2I = \frac{\pi}{2}a^4$, we obtain the relation $EI = (1+\nu)\mu J$ between the flexural and torsional rigidities.

The three-dimensional theory of slender rods goes back to Kirchhoff (1859). The presentation in this section owes much to [Landau and Lifshitz 1986]. A modern presentation, including rods with non-circular cross-sections, can be found in [Bower 2010, ch. 10].

Gustav Robert Kirchhoff (1824–1887). German physicist. Developed the famous laws of electrical circuit theory. Founded spectrum analysis (with Bunsen) and applied it to sunlight, using the dark absorption lines to determine its composition. Published a much-used, four-volume "Lectures on Mathematical Physics". (Source: Wikimedia Commons.)

Local balance of forces and moments

Let the actual shape of the rod's central ray be given by the vector function $\boldsymbol{x} = \boldsymbol{x}(s)$ of the curve length s (the so-called *natural parametrization*). In each circular cross-section at s there are internal stresses that integrate up to a total internal force $\boldsymbol{F}(s)$ and moment of force $\mathcal{M}(s)$. We may as before without loss of generality assume that point-like external forces and moments only act on the end terminals of the rod, because forces or moments acting somewhere between the terminals can be handled by dividing the rod into pieces and imposing suitable continuity conditions where they join. Distributed external forces, for example gravity or viscous drag, act with a vector force $\boldsymbol{K}(s)\,ds$ on any infinitesimal piece of the curve between s and $s + ds$. We assume that there are no distributed moments of force.

The balance of forces on a small piece of the rod at rest (see the margin figure) then becomes $\mathcal{F}(s + ds) - \mathcal{F}(s) + \boldsymbol{K}(s)\,ds = 0$, or after division with ds,

$$\boxed{\frac{d\mathcal{F}}{ds} = -\boldsymbol{K}.} \tag{10.26}$$

Similarly, the balance of moments around the center of the cross-section at s takes the form $\mathcal{M}(s + ds) - \mathcal{M}(s) + d\boldsymbol{x} \times \mathcal{F}(s + ds) = \boldsymbol{0}$, and to first order in ds, we find

$$\boxed{\frac{d\mathcal{M}}{ds} = -\frac{d\boldsymbol{x}}{ds} \times \mathcal{F}.} \tag{10.27}$$

These are the basic equations that determine the rod shape from the external forces and moments, together with the Euler–Bernoulli and Coulomb–Saint-Venant constitutive equations relating the internal moments to curvature and torsion.

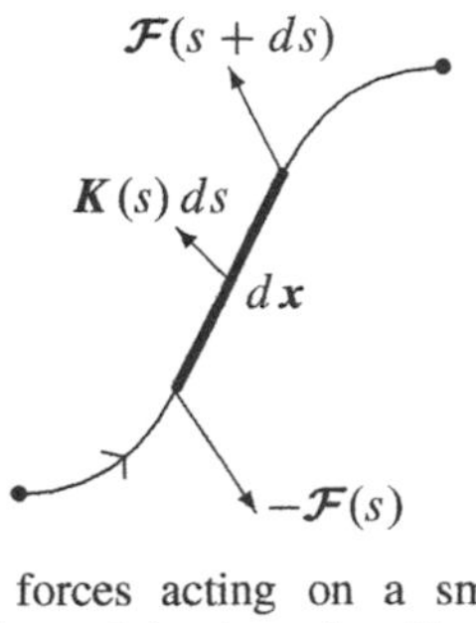

The forces acting on a small, nearly straight piece $d\boldsymbol{x}$ of length $ds = |d\boldsymbol{x}|$ of the deformed rod. At the end terminals the missing internal forces and moments must be supplied by external agents.

The Frenet–Serret basis

Jean Frédéric Frenet (1816–1900). French mathematician. Only published a few papers and a book. Directed the astronomical observatory in Lyon, and made meteorological observations.

Joseph Alfred Serret (1819–1885). French mathematician. Worked on differential geometry, number theory, and calculus.

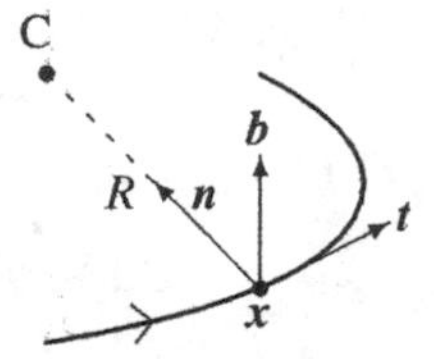

The Frenet–Serret basis consists of the tangent vector $\boldsymbol{t}$, the normal $\boldsymbol{n}$, and the binormal $\boldsymbol{b}$. The normal points toward the center of curvature C. As the point $\boldsymbol{x}$ moves along the curve, the basis turns around $\boldsymbol{b}$ and twists around $\boldsymbol{t}$ to reflect the geometric curvature and torsion of the oriented curve.

At this point we need to make a small excursion into an efficient mathematical description of the geometry of three-dimensional (spatial) curves, due to Frenet in 1847 and independently to Serret in 1851. At every point of a curve, a local so-called Frenet–Serret basis is established, starting with the unit *tangent* vector (see the margin figure)

$$\boldsymbol{t} = \frac{d\boldsymbol{x}}{ds}. \tag{10.28}$$

The two other unit vectors in this basis are the *normal* $\boldsymbol{n}$, which points toward the local center of curvature, and the *binormal* $\boldsymbol{b} = \boldsymbol{t} \times \boldsymbol{n}$ orthogonal to both. We shall now show that these vectors satisfy the relations,

$$\frac{d\boldsymbol{t}}{ds} = \kappa\,\boldsymbol{n}, \qquad \frac{d\boldsymbol{b}}{ds} = -\tau\,\boldsymbol{n}, \qquad \frac{d\boldsymbol{n}}{ds} = \tau\,\boldsymbol{b} - \kappa\,\boldsymbol{t}, \tag{10.29}$$

where $\kappa = \kappa(s)$ is the *local curvature* and $\tau = \tau(s)$ the *local torsion* of the curve.

Proof: As the point s moves a distance ds along the curve with local curvature $\kappa = 1/R$, the tangent vector rotates toward the center of curvature through an angle $d\phi = ds/R$, such that the change in the tangent vector is $d\boldsymbol{t} = \boldsymbol{n}\,d\phi = \kappa\boldsymbol{n}\,ds$. The change $d\boldsymbol{b}$ in the binormal is orthogonal to $\boldsymbol{b}$ because $\boldsymbol{b}$ is a unit vector, but since $d\boldsymbol{b} = d\boldsymbol{t} \times \boldsymbol{n} + \boldsymbol{t} \times d\boldsymbol{n} = \boldsymbol{t} \times d\boldsymbol{n}$, it is also orthogonal to $\boldsymbol{t}$. Consequently, $d\boldsymbol{b}$ rotates around $\boldsymbol{t}$ in the direction $-\boldsymbol{n}$, such that $d\boldsymbol{b} = -\boldsymbol{n}d\psi$, where $d\psi = \tau ds$ is the angle of twist over the distance ds. The last equation follows from $\boldsymbol{n} = \boldsymbol{b} \times \boldsymbol{t}$.

Whereas the curvature κ by definition is never negative, the torsion τ can take both signs. If the curvature vanishes everywhere, $\kappa(s) = 0$, the tangent is constant and the curve becomes straight. In that case torsion can be anything we want and it loses its geometric meaning. If the torsion vanishes everywhere, $\tau(s) = 0$, it follows that $\boldsymbol{b}$ is a constant, implying that the curve lies in a plane orthogonal to $\boldsymbol{b}$. This case was discussed in the preceding section. Generally, it may be shown (see Problem 10.8) that the curvature $\kappa(s)$ and the torsion $\tau(s)$ are sufficient to determine the shape $\boldsymbol{x}(s)$ of the curve, given suitable boundary conditions.

Moments of bending and twisting

Since bending corresponds to a rotation around the binormal and twisting to a rotation around the tangent, we shall assume that the internal moment can only have components along these directions,

$$\boxed{\boldsymbol{\mathcal{M}} = \mathcal{M}_b\,\boldsymbol{b} + \mathcal{M}_t\,\boldsymbol{t}.} \tag{10.30}$$

This is of course the same as saying that the normal component vanishes, $\mathcal{M}_n = \boldsymbol{\mathcal{M}} \cdot \boldsymbol{n} = 0$. For a rod with non-circular cross-section, $\mathcal{M}_n$ would generally be non-vanishing, even for pure bending.

From moment balance (10.27), we obtain

$$\frac{d\mathcal{M}_b}{ds}\boldsymbol{b} + \frac{d\mathcal{M}_t}{ds}\boldsymbol{t} + (\kappa\mathcal{M}_t - \tau\mathcal{M}_b)\boldsymbol{n} = -\boldsymbol{t} \times \boldsymbol{\mathcal{F}},$$

and projecting this equation on the local basis, we get

$$\frac{d\mathcal{M}_b}{ds} = -\mathcal{F}_n, \qquad \frac{d\mathcal{M}_t}{ds} = 0, \qquad \kappa\mathcal{M}_t - \tau\mathcal{M}_b = \mathcal{F}_b. \tag{10.31}$$

Evidently $\mathcal{M}_t$ must be constant along rod with circular cross-section, as long as there are only internal bending and twisting moments. If $\mathcal{M}_t$ is non-zero, the only way that the rod can be twisted is by applying external torsion moments $-\mathcal{M}_t$ and $\mathcal{M}_t$ on the start and end terminals to compensate for the missing internal moments.

Constitutive equations for small deflections

Common experience with highly bendable elastic beams—thin metal wires, electrical cables, garden hoses—tells us that unrestricted twisting and bending can lead to highly contorted shapes. Conversion of twist energy into bending energy may lead to torsional buckling where a part of the rod loops back and writhes around itself [GPL05]. The theory of large deflections of three-dimensional rod shapes is a difficult subject and has been under intense study by physicists and mathematicians since Kirchoff opened the ball.

Paperclips, discussed in Example 9.5, and coiled springs like the ones shown in Figure 10.5, are examples of slender rods given permanently bent and twisted equilibrium shapes by forces strong enough to overcome the yield stress of the material. The elasticity of such objects under further small deformations around the relaxed equilibrium shapes is often what makes them practically useful: for paperclips to hold sheets of paper together, and for coiled springs to dampen the influence of bumps in the road on the passenger compartments of vehicles, or to slam a mousetrap shut.

Let $\kappa(s)$ and $\tau(s)$ be the geometric curvature and torsion of the relaxed rod with no external load. Under the influence of external forces and moments, the rod deforms into a new equilibrium state, but we assume that the external load is so small that the deflection $\Delta\boldsymbol{x}(s)$ is everywhere small compared to the radius of curvature $1/\kappa$ and the torsion length $1/\tau$. The constitutive equations are (without proof) assumed to be given by the Euler–Bernoulli law (9.25) and the Coulomb–Saint-Venant law (9.38):

$$\mathcal{M}_b = EI\Delta\kappa, \qquad \mathcal{M}_t = \mu J\Delta\tau, \tag{10.32}$$

where $\Delta\kappa(s)$ and $\Delta\tau$ are the (small) changes in curvature and torsion, caused by the external load. The preceding analysis tells us that $\Delta\tau$ must be constant, even if the relaxed state has a frozen-in geometric torsion that varies with s.

10.5 Application: The helical spring

Here we shall only analyze the deformation of a slender circular rod for which the central ray has been permanently shaped into a perfect helix. The helix is uniquely defined by having constant geometric curvature and torsion. It has a beautiful non-trivial regularity that may well be the reason for the fascination it evokes. Everybody has probably made a, not quite perfect, permanent helix by winding a copper wire around a cardboard cylinder, and then removing the cylinder. Helices are found ubiquitously in natural objects—from carbon nanotubes, DNA, and bacteria, to horns and vines of large animals and plants—as well as in artificial structures from screws to staircases [CGM06].

Geometry of a perfect helix

A perfect helix is an infinitely long curve that winds around an imaginary cylinder with constant radius a and constant elevation angle α (see the margin figure). In cylindrical coordinates r, ϕ, and z (see Appendix D), the natural parametrization of the helix becomes

$$r = a, \qquad \phi = s\frac{\cos\alpha}{a}, \qquad z = s\sin\alpha, \qquad (-\infty < s < \infty). \tag{10.33}$$

In Cartesian coordinates this may be written more compactly as

$$\boldsymbol{x}(s) = a\hat{\boldsymbol{e}}_r + s\sin\alpha\hat{\boldsymbol{e}}_z, \tag{10.34}$$

where $\hat{\boldsymbol{e}}_r = (\cos\phi, \sin\phi, 0)$, $\hat{\boldsymbol{e}}_\phi = (-\sin\phi, \cos\phi, 0)$, and $\hat{\boldsymbol{e}}_z = (0, 0, 1)$ are the usual basis vectors for cylindrical coordinates.

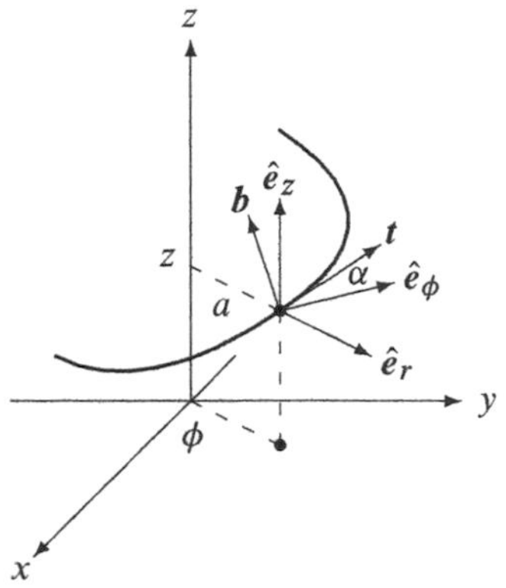

A piece of a right-handed helix with constant radius a and constant elevation angle α, spiraling up along the z-axis. The tangent and binormal of the Frenet–Serret basis are shown.

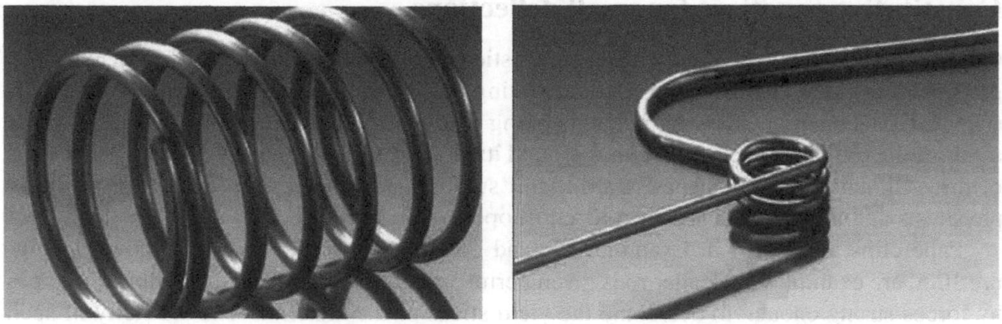

Figure 10.5. Two kinds of helical springs. On the left, an extension/compression spring, and on the right a torsion spring. (Source: Spring Solutions Pty Ltd, Australia.)

Using $d\hat{e}_r/d\phi = \hat{e}_\phi$ and $d\hat{e}_\phi/d\phi = -\hat{e}_r$ we find the Frenet–Serret basis

$$t = \cos\alpha\,\hat{e}_\phi + \sin\alpha\,\hat{e}_z, \qquad b = -\sin\alpha\,\hat{e}_\phi + \cos\alpha\,\hat{e}_z, \qquad n = -\hat{e}_r, \tag{10.35}$$

with curvature and torsion

$$\kappa = \frac{\cos^2\alpha}{a}, \qquad \tau = \frac{\cos\alpha\sin\alpha}{a}. \tag{10.36}$$

Both of these are constant along the helix. The helix is in fact the only curve for which they are both constant (see Problem 10.7).

In practice, we would rather describe a finite helical spring of length L by the total turning angle ψ and the height h. These parameters are found by setting $s = L$ in (10.33),

$$\psi = \frac{L\cos\alpha}{a}, \qquad h = L\sin\alpha. \tag{10.37}$$

The finite spring thus covers the intervals $0 \le s \le L$, $0 \le \phi \le \psi$, and $0 \le z \le h$. In these variables the curvature and torsion become

$$\kappa = \frac{\psi}{L}\sqrt{1 - \frac{h^2}{L^2}}, \qquad \tau = \frac{\psi h}{L^2}, \tag{10.38}$$

where we have used $\cos\alpha = \sqrt{1 - \sin^2\alpha} = \sqrt{1 - h^2/L^2}$.

Small, perfectly helical deflections

The relaxed helix is now deformed by suitable terminal loads to be discussed below. We shall insist that these loads are chosen such that the deformed helix is also a perfect helix with constant parameters $a + \Delta a$ and $\alpha + \Delta\alpha$. Expressed in terms of the change in turning angle $\Delta\psi$ and the change in height Δh, we find from (10.38) the following first-order changes in curvature and torsion:

$$\Delta\kappa = \frac{\Delta\psi}{L}\cos\alpha - \frac{\Delta h}{aL}\sin\alpha, \qquad \Delta\tau = \frac{\Delta\psi}{L}\sin\alpha + \frac{\Delta h}{aL}\cos\alpha. \tag{10.39}$$

Inserting these into the constitutive equations, the bending and twisting moments become

$$\mathcal{M}_b = EI\Delta\kappa, \qquad \mathcal{M}_t = \mu J\Delta\tau, \tag{10.40}$$

expressed in terms of $\Delta\psi$ and Δh.

Projecting $\mathcal{M} = \mathcal{M}_b \boldsymbol{b} + \mathcal{M}_t \boldsymbol{t}$ on the cylindrical directions $\hat{\boldsymbol{e}}_z$ and $\hat{\boldsymbol{e}}_\phi$, we get

$$\mathcal{M}_z = \mathcal{M}_b \cos\alpha + \mathcal{M}_t \sin\alpha, \qquad \mathcal{M}_\phi = -\mathcal{M}_b \sin\alpha + \mathcal{M}_t \cos\alpha. \tag{10.41}$$

In terms of the changes in turning angle and height, we have

$$\mathcal{M}_z = A\frac{\Delta\psi}{L} - C\frac{\Delta h}{aL}, \qquad \mathcal{M}_\phi = B\frac{\Delta h}{aL} - C\frac{\Delta\psi}{L}, \tag{10.42}$$

where

$$A = EI\cos^2\alpha + \mu J \sin^2\alpha, \tag{10.43a}$$
$$B = EI\sin^2\alpha + \mu J \cos^2\alpha, \tag{10.43b}$$
$$C = (EI - \mu J)\cos\alpha \sin\alpha. \tag{10.43c}$$

All three constants are positive definite because $EI - \mu J = \nu\mu J$ for a circular rod.

Implementation

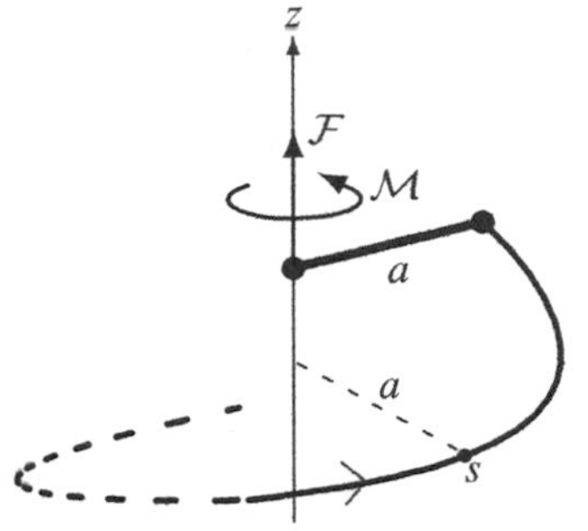

A possible implementation of the load on the end terminal of a helical spring. The external force $\mathcal{F}$ exerts a pull along the z-axis and the external moment $\mathcal{M}$ wrenches the spring around the z-axis.

Since $\mathcal{M}_z$ and $\mathcal{M}_\phi$ are both independent of s, they determine the external moments that must be applied to the end terminal of the rod at $s = L$. We assume that the start terminal at $s = 0$ is clamped in such a way that it can provide all the necessary reaction forces. The symmetry of the helix indicates that the only possible external load that may lead to a perfectly helical deformation consists of a force $\mathcal{F} = \mathcal{F}\hat{\boldsymbol{e}}_z$ and a moment $\mathcal{M} = \mathcal{M}\hat{\boldsymbol{e}}_z$, both acting along the z-axis.

One way of implementing such loads is shown in the margin figure, where a stiff lever transmits both the central external force $\mathcal{F}$ and the central external moment $\mathcal{M}$ to the end terminal. The external force creates a terminal moment $\mathcal{M}_\phi = a\mathcal{F}$, but it appears that the external moment creates a terminal force $\mathcal{F}_\phi = \mathcal{M}/a$. That cannot be right because the total force $\mathcal{F} = \mathcal{F}_z\hat{\boldsymbol{e}}_z + \mathcal{F}_\phi\hat{\boldsymbol{e}}_\phi$ has to be constant along the rod according to (10.2) (when there is no distributed load). We shall appeal to Saint-Venant's principle and assume that some diameters away from the rod's terminals we obtain the perfect helix with $\mathcal{M}_\phi = a\mathcal{F}$ and $\mathcal{M}_z = \mathcal{M}$.

Finally, putting it all together, we obtain expressions of the form

$$\boxed{\mathcal{F} = k_F\,\Delta h - m\,\Delta\psi, \qquad \mathcal{M} = k_M\,\Delta\psi - m\,\Delta h,} \tag{10.44}$$

with spring constants k_F for extension, k_M for torsion, and m for "cross-over":

$$k_F = \frac{B}{a^2 L}, \qquad k_M = \frac{A}{L}, \qquad m = \frac{C}{aL}. \tag{10.45}$$

Note that for small α, the force is dominated by the torsional rigidity whereas the moment is dominated by the flexural rigidity. Paradoxically, the elasticity of a coiled compression spring, like the shock absorber in your car, depends mainly on twisting the rod, whereas the elasticity of a torsion spring, like the one in a mousetrap, functions primarily by bending the rod.

Locked compression spring: A helical spring is fixated so that its end terminals cannot turn with respect to each other. Using the expressions for circular rods with rod radius b and $\Delta\psi = 0$, we get

$$k_F = \frac{Eb^4}{8Na^3}\frac{1+\nu\sin^2\alpha}{1+\nu}\cos\alpha, \qquad m = -\frac{Eb^4}{8Na^2}\frac{\nu}{1+\nu}\cos^2\alpha\sin\alpha, \tag{10.46}$$

where $N = \psi/2\pi$ is the number of turns of the spring.

Example 10.3 [Car suspension]: The four suspension springs in a certain car have diameter $2a = 10$ cm, rod diameter $2b = 1$ cm, number of turns $N = 5$, and elevation angle $\alpha = 15°$. The length is $L = 163$ cm and the relaxed height $h = 42$ cm. Taking $E = 200$ GPa, and $\nu = 1/3$, we find the spring constant $k_F = 18.6$ kN/m and the cross-over constant $m = -75$ N. If the passenger compartment (including passengers) has total mass 1000 kg and is suspended by four such springs, we find the compression $\Delta h = -13$ cm and the reaction moment $\mathcal{M} = -10$ Nm.

Problems

10.1 Calculate the buckling threshold for a rod that is clamped at one end so that it can neither move nor turn around that point. The other end is free to move and subject to only a longitudinal compressional force but no moment.

10.2 Show that (10.21) can always be cast as the equation for the mathematical pendulum when the forces are constants.

* **10.3** **(a)** Estimate the shortening of the stringed bow due to compression, and **(b)** show that it is negligible for a slender rod.

10.4 Calculate the general shape of a clamped stalk bent by a terminal force $\mathcal{F}$.

* **10.5** **(a)** Show by variation of $y(z)$ that the equation for the minimum of the work functional (10.17) is

$$EI\frac{d^4y}{dz^4} = -\mathcal{F}\frac{d^2y}{dz^2},$$

which is a version of the Euler–Bernoulli equation. Hint: Assume that the variation $\delta y(z)$ and its derivatives vanish at the integration limits. **(b)** Show that this leads to (10.13) when $\mathcal{F}$ is constant.

10.6 Show that the Frenet–Serret basis rotates as a solid body with rotation vector per unit of curve length, $\boldsymbol{\Omega} = \kappa\,\boldsymbol{b} + \tau\,\boldsymbol{t}$.

* **10.7** Prove that if both curvature κ and torsion τ are constants (independent of s), the curve becomes a *perfect helix* (10.33).

10.8 Show that the following fourth-order ordinary differential equation is satisfied by any curve $\boldsymbol{x}(s)$,

$$\frac{d}{ds}\left(\frac{1}{\tau}\frac{d}{ds}\left(\frac{1}{\kappa}\frac{d^2\boldsymbol{x}}{ds^2}\right)\right) + \frac{\tau}{\kappa}\frac{d^2\boldsymbol{x}}{ds^2} + \frac{d}{ds}\left(\frac{\kappa}{\tau}\frac{d\boldsymbol{x}}{ds}\right) = \boldsymbol{0},$$

where $\kappa = \kappa(s)$ and $\tau = \tau(s)$ are the curvature and torsion functions, respectively.

11 Computational elastostatics

Historically, almost all of the insights into elasticity were obtained from analytic calculations carried out by some of the best scientists of the time using the most advanced methods available to them, sometimes even inventing new mathematical concepts and methods along the way. Textbooks on the theory of elasticity are often hard to read because of their demands on the reader for command of mathematics.

In the last half of the twentieth century, the development of the digital computer changed the character of this field completely. Faced with a problem in elastostatics, modern engineers quickly turn to numerical computation. The demand for prompt solutions to design problems has over the years evolved these numerical methods into a fine art, and numerous commercial and public domain programs are now available.

In this short chapter we shall illustrate how it is possible to solve a concrete problem numerically and provide sufficient detail to implement a computer program. It is not the intention here to expose the wealth of tricks of the trade but just present the basic reasoning behind the numerical approach and the various steps that must be carried out. First, one must decide on the field equations and boundary conditions that are valid for the problem at hand. Second, the infinity of points in continuous space must be replaced by a finite set of points or volumes, often organized in a regular grid, and the fundamental equations must be approximated on this set. Third, a method must be chosen for an iterative approach toward the desired solution, and convergence criteria must be established that enable one to monitor the progress of the computation and calculate error estimates.

11.1 Theory of the numeric method

As we do not, from the outset, know the solution to the problem we wish to solve by numerical means, we must begin by making an educated guess about the initial displacement field. Unless we are incredibly lucky, this guess will fail to satisfy the mechanical equilibrium equations and the boundary conditions, resulting in both a non-vanishing effective body force and a violation of Newton's Third Law on the boundary. The idea is now to create an iterative procedure that through a sequence of tiny changes in displacement, will proceed from the chosen initial state toward the equilibrium state.

Basic equations

Given an arbitrary (infinitesimal) displacement field $\boldsymbol{u}(\boldsymbol{x})$, the fundamental equations underlying computational elastostatics are

$$u_{ij} = \tfrac{1}{2}(\nabla_i u_j + \nabla_j u_i) \qquad \text{Cauchy's strain tensor (7.21)}, \tag{11.1a}$$

$$\sigma_{ij} = \sum_{kl} E_{ijkl} u_{kl} \qquad \text{general Hooke's law (8.18)}, \tag{11.1b}$$

$$f_i^* = f_i + \sum_j \nabla_j \sigma_{ij} \qquad \text{effective body force (6.15)}. \tag{11.1c}$$

In numeric computation there is no reason to limit the formalism to isotropic materials, so we have used the most general form of Hooke's law. If somebody presents you with a displacement field for a body, these equations represent a "machine" or "program" that through purely local operations (i.e., differentiation) allows you to calculate the effective body force acting in each and every point of the body. In mechanical equilibrium the effective body force $\boldsymbol{f}^*$ must of course vanish everywhere in the volume V of the body, but since we do not know the equilibrium displacement field—otherwise the numerical computation would be unnecessary—$\boldsymbol{f}^*$ will in general not vanish while we are seeking the solution.

The solution must also satisfy boundary conditions on the surface S of the body. We shall assume that part of the surface, S_0, is permanently "glued" to undeformable and unmoveable external bodies such that the displacement has to vanish here. At the remainder of the body surface, $S_P = S - S_0$, the surface stress is under "user control" and fixed to be $\boldsymbol{P}(\boldsymbol{x})$. The solution must in other words satisfy

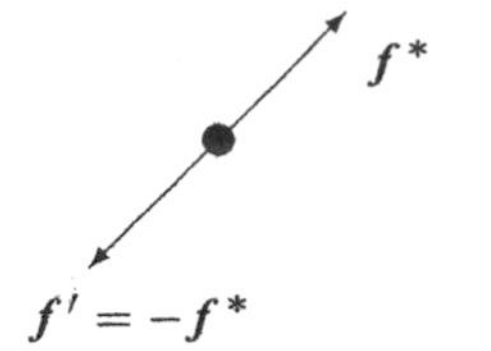

Every material particle can be kept in place by means of a virtual body force, $\boldsymbol{f}' = -\boldsymbol{f}^*$, which balances the effective internal body force on the particle.

$$\boldsymbol{f}^*(\boldsymbol{x}) = \boldsymbol{0} \qquad \text{for } \boldsymbol{x} \in V, \tag{11.2a}$$

$$\boldsymbol{u}(\boldsymbol{x}) = \boldsymbol{0} \qquad \text{for } \boldsymbol{x} \in S_0, \tag{11.2b}$$

$$\boldsymbol{\sigma}(\boldsymbol{x}) \cdot \boldsymbol{n}(\boldsymbol{x}) = \boldsymbol{P}(\boldsymbol{x}) \qquad \text{for } \boldsymbol{x} \in S_P, \tag{11.2c}$$

where $\boldsymbol{n}(\boldsymbol{x})$ is the normal to the surface in $\boldsymbol{x}$. Where the external surface stresses vanish, $\boldsymbol{P}(\boldsymbol{x}) = \boldsymbol{0}$, the body surface is said to be *free*.

Virtual work and potential energy

Starting with an educated guess for the displacement field $\boldsymbol{u}(\boldsymbol{x})$, which vanishes on S_0, there is as mentioned no guarantee that the effective force $\boldsymbol{f}^*$ vanishes nor that the boundary condition $\boldsymbol{\sigma} \cdot \boldsymbol{n} = \boldsymbol{P}$ is fulfilled. To prevent the material particles from beginning to move, we must as in Section 7.4 add a *virtual body force* $\boldsymbol{f}' = -\boldsymbol{f}^*$ to the material particles as well as a *virtual surface stress vector* $\boldsymbol{P}' = \boldsymbol{\sigma} \cdot \boldsymbol{n} - \boldsymbol{P}$ to secure that Newton's Third Law is fulfilled in each point of S_P. If we now change the displacement field by a tiny amount $\delta\boldsymbol{u}(\boldsymbol{x})$, which also vanishes at S_0, the work of the virtual forces becomes

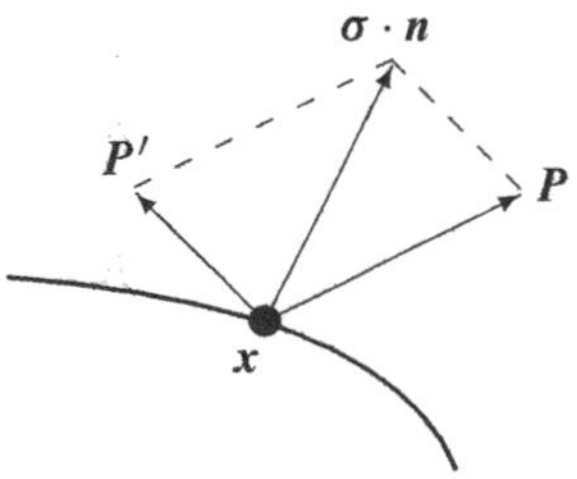

At every point of the surface S_P, we must add a virtual stress $\boldsymbol{P}' = \boldsymbol{\sigma} \cdot \boldsymbol{n} - \boldsymbol{P}$ to secure continuity of the stress vector, demanded by Newton's Third Law.

$$\boxed{\delta W = -\int_V \boldsymbol{f}^* \cdot \delta\boldsymbol{u}\, dV + \int_{S_P} \delta\boldsymbol{u} \cdot (\boldsymbol{\sigma} \cdot \boldsymbol{n} - \boldsymbol{P})\, dS.} \tag{11.3}$$

Carrying out a partial integration as in Section 7.4, and using that $\delta\boldsymbol{u} = \boldsymbol{0}$ on S_0, we find

$$\delta W = -\int_V \boldsymbol{f} \cdot \delta\boldsymbol{u}\, dV + \int_V \boldsymbol{\sigma} : \delta\mathbf{u} - \int_{S_P} \delta\boldsymbol{u} \cdot \boldsymbol{P}\, dS, \tag{11.4}$$

where as before $\boldsymbol{\sigma} : \delta\mathbf{u} = \sum_{ij} \sigma_{ij} \delta u_{ji}$. The first term in the above equation is the virtual work against the external body forces, the second the virtual work against the internal stresses, and the last the virtual work against the external surface forces.

Since neither f nor P depends on u, the infinitesimal virtual work may immediately be integrated to yield the (potential) energy of any displacement (see for example [Doghri 2000, p. 35] for a more rigorous treatment),

$$\mathcal{E} = -\int_V f \cdot u \, dV + \frac{1}{2}\int_V \sigma : \mathbf{u} \, dV - \int_{S_P} u \cdot P \, dS. \tag{11.5}$$

Here the first term is the energy of the external body forces (gravity), the second the elastic energy residing in the deformation, and the last the energy of the external surface forces. Notice that the last term is present even if the boundary condition (11.2c) is fulfilled. In Section 7.4 we did not get this term because we assumed that the displacement was fixed, $\delta u = 0$, on all of the body surface S.

Relaxing toward equilibrium

The potential energy (11.5) is a positive definite quadratic form in the displacement field and its derivatives, and has consequently a unique minimum. At this minimum, the energy must be stationary $\delta\mathcal{E} = \delta W = 0$ under any variation δu that vanishes on S_0, and it follows immediately from (11.3) that this implies $f^* = \mathbf{0}$ in V and $\sigma \cdot n = P$ on S_P. *The unique minimum of the energy is therefore the desired equilibrium solution.* This result forms the basis for obtaining the numerical solution.

Starting with an arbitrary displacement field satisfying $u = \mathbf{0}$ on S_0, the iterative *relaxation procedure* consists of designing a sequence of small displacement steps δu that all drain energy away from the body. In each iteration the total energy of the body will decrease until it reaches the unique minimum. Thus, in the end, the relaxation procedure will arrive at the desired equilibrium state.

Gradient descent

A common relaxation procedure is to choose the displacement change in a step to be

$$\delta u = \epsilon f^* \text{ in } V, \qquad \delta u = -\epsilon(\sigma \cdot n - P) \text{ on } S_P, \tag{11.6}$$

where ϵ is a positive quantity, called the *step-size*. Relaxing the displacement in this way guarantees that the integrands in both terms of (11.3) are manifestly negative everywhere in the volume V and on the surface S_P, and thus drains energy away from every material particle in the body that is not already in equilibrium. Since the displacement incessantly "walks downhill" *against* the gradient of the total energy (in the space of all allowed displacement fields), it is naturally called *gradient descent*.

Gradient descent is not a foolproof method, even when the energy (as in a linear elastic material) is a quadratic function of the displacement field with a unique minimum. In particular, the step-size ϵ must be chosen judiciously. Too small, and the procedure may never seem to converge; too large, and it may overshoot the minimum and go into oscillations or even diverge. Many fine tricks have been invented to get around these problems and speed up convergence [Press et al. 1992, Braess 2001], for example increasing the step-size when the convergence is too slow and decreasing it when it is too fast. Another rather effective method is *conjugate gradient descent* in which the optimal step-size is calculated in advance by searching for a minimum of the total energy along the initial direction of descent. Having found this minimum, the procedure is repeated with a new direction of steepest descent, which necessarily must be orthogonal to the old one. In the remainder of this chapter we shall just use the straightforward technique of the dedicated downhill skier, always selecting the steepest gradient for the next "step".

11.2 Discretization of space

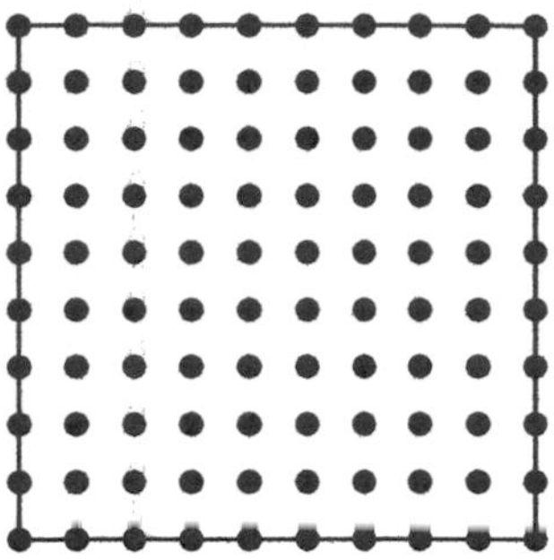

A two-dimensional (10×10) square grid. There are 36 points at the boundary and 64 inside. Small grids have a lot of boundary.

The infinity of points in space cannot be represented in a finite computer. In numerical simulations of the partial differential equations of continuum physics, smooth space is often replaced by a finite collection of points, a grid or lattice on which the various fields "live" (see for example [Anderson 1995, Griebel et al. 1998]). In Cartesian coordinates the most convenient grid for a rectangular volume $a \times b \times c$ is a rectangular lattice with $(N_x + 1) \times (N_y + 1) \times (N_z + 1)$ points that are equally spaced at coordinate intervals $\Delta x = a/N_x$, $\Delta y = b/N_y$ and $\Delta z = c/N_z$. The grid coordinates are numbered by $n_x = 0, 1, \ldots, N_x$; $n_y = 0, 1, \ldots, N_y$; and $n_z = 0, 1, \ldots, N_z$, and the various fields can only exist at the positions $(x, y, z) = (n_x \Delta x, n_y \Delta y, n_z \Delta z)$.

There are many other ways of discretizing space besides using rectangular lattices, for example triangular, hexagonal or even random lattices. The choice of grid depends on the problem itself, as well as on the field equations and the boundary conditions. The coordinates in which the system is most conveniently described may not be Cartesian but curvilinear, and that leads to quite a different discretization. The surface of the body may or may not fit well with the chosen grid, but that problem may be alleviated by making the grid very dense at the cost of computer time and memory. When boundaries are irregular, as they usually are for real bodies, an adaptive grid that can fit itself to the shape of the body may be the best choice. Such a grid may also adapt to put more points where they are needed in regions of rapid variation of the displacement field.

Finite difference operators with first-order errors

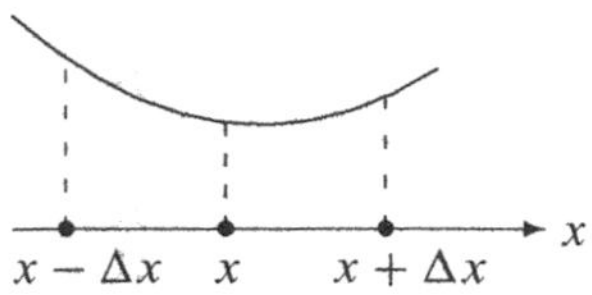

Forward and backward finite differences can be very different, and may as here even have opposite signs.

In a discrete space, coordinate derivatives of fields such as $\nabla_x f(x, y, z)$ must be approximated by finite differences between the field values at the allowed points. Using only the nearest neighbors on the grid, there are two basic ways of forming such differences at a given *internal* point of the lattice, namely forward and backward:

$$\widehat{\nabla}_x^{+} f(x) = \frac{f(x+\Delta x) - f(x)}{\Delta x}, \qquad \widehat{\nabla}_x^{-} f(x) = \frac{f(x) - f(x-\Delta x)}{\Delta x}. \tag{11.7}$$

Here and in the following we suppress for clarity the "sleeping" coordinates y and z and furthermore assume that finite differences in these coordinates are defined analogously.

According to the rules of differential calculus, both of these expressions will in the limit of $\Delta x \to 0$ converge toward $\nabla_x f(x)$. Inserting the Taylor expansion

$$f(x + \Delta x) = f(x) + \Delta x \nabla_x f(x) + \tfrac{1}{2}\Delta x^2 \nabla_x^2 f(x) + \tfrac{1}{6}\Delta x^3 \nabla_x^3 f(x) + \cdots,$$

we find indeed

$$\widehat{\nabla}_x^{\pm} f(x) = \nabla_x f(x) \pm \tfrac{1}{2}\Delta x \nabla_x^2 f(x) + \cdots,$$

with an error (the second term) that is of first order in the interval Δx.

Finite difference operators with second-order errors

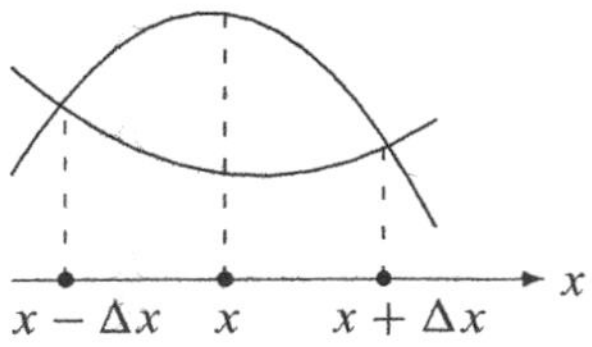

The central difference is insensitive to the value at the center. The two curves shown here have the same symmetric difference but behave quite differently.

It is clear from the above expression that the first-order error may be suppressed by forming the average of forward and backward difference operators, called the *central difference*,

$$\widehat{\nabla}_x f(x) = \frac{1}{2}\left(\widehat{\nabla}_x^{+} + \widehat{\nabla}_x^{-}\right) f(x) = \frac{f(x+\Delta x) - f(x-\Delta x)}{2\Delta x}. \tag{11.8}$$

Expanding the function values to third order, we obtain

$$\widehat{\nabla}_x f(x) = \nabla_x f(x) + \tfrac{1}{6}\Delta x^2 \nabla_x^3 f(x) + \cdots,$$

with errors of second order only. The central difference does not involve the field value at the central point x, so one should be wary of possible "leapfrog" or "flipflop" numeric instabilities in which half the points of the lattice behave differently than the other half.

On a boundary, the central difference cannot be calculated, and we are forced to use one-sided differences. On the left boundary one must use the forward difference and on the right the backward one. In order to consistently avoid $\mathcal{O}(\Delta x)$ errors, one may instead of one-step differences (11.7) use one-sided two-step difference operators (see Problem 11.1),

$$\widehat{\nabla}_x^+ f(x) = \frac{-f(x+2\Delta x) + 4f(x+\Delta x) - 3f(x)}{2\Delta x}, \tag{11.9a}$$

$$\widehat{\nabla}_x^- f(x) = \frac{f(x-2\Delta x) - 4f(x-\Delta x) + 3f(x)}{2\Delta x}. \tag{11.9b}$$

The coefficients are chosen here such that the leading order corrections vanish. Expanding to third order we find indeed that

$$\widehat{\nabla}_x^\pm f(x) = \nabla_x f(x) \mp \tfrac{1}{3}\Delta x^2 \nabla_x^3 f(x) + \cdots,$$

which shows that the one-sided differences represent the derivative at the point x with leading errors of $\mathcal{O}(\Delta x^2)$ only.

Other schemes involving more distant neighbors to suppress even higher-order errors are of course also possible, but here we shall only use the second-order expressions.

Numeric integration

In simulations it will also be necessary to calculate various line, surface, and volume integrals over discretized space. Since the fields are only known at the points of the discrete lattice, the integrals must be replaced by suitably weighted sums over the lattice points.

Consider, for example, a one-dimensional integral over an interval $0 \le x \le a$ along the x-axis. Dividing the interval into a set of N sub-integrals over the lattice spacing $\Delta x = a/N$, we may write the integral as an exact sum of sub-integrals,

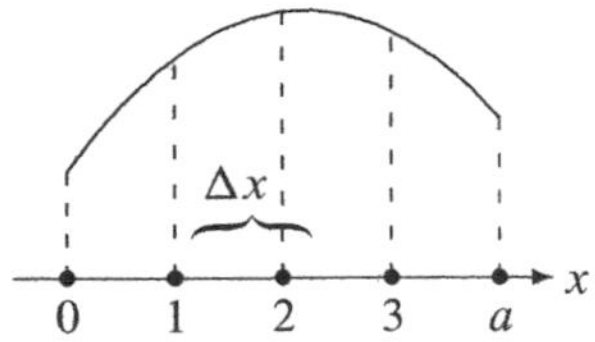

The interval $0 \le x \le a$ has four subintervals of size Δx numbered $n = 0, 1, 2, 3$.

$$\int_0^a f(x)\,dx = \sum_{n=0}^{N-1} \int_{x_n}^{x_{n+1}} f(x)\,dx, \tag{11.10}$$

where $x_n = n\Delta x$. By means of the Taylor expansion at x, we find for a single sub-interval

$$\int_x^{x+\Delta x} f(x')dx' = \Delta x f(x) + \tfrac{1}{2}\Delta x^2 \nabla_x f(x) + \tfrac{1}{6}\Delta x^3 \nabla_x^2 f(x) + \cdots. \tag{11.11}$$

In the lowest order of approximation we replace the first-order derivative by the forward difference operator (11.7a), and get

$$\begin{aligned}\int_{x_n}^{x_{n+1}} f(x')dx') &= \Delta x f(x_n) + \tfrac{1}{2}\Delta x^2 \widehat{\nabla}_x f(x_n) + \mathcal{O}(\Delta x^3)\\ &= \tfrac{1}{2}\Delta x\big(f(x_{n+1}) + f(x_n)\big) + \mathcal{O}(\Delta x^3).\end{aligned}$$

Finally, adding the N contributions, we arrive at

$$\int_0^a f(x)\,dx = \tfrac{1}{2}\big(f(0) + f(a)\big)\,\Delta x + \sum_{n=1}^{N-1} f(x_n)\,\Delta x + \mathcal{O}(\Delta x^2). \tag{11.12}$$

This is the well-known trapezoidal rule [Press et al. 1992, p. 131].

In higher dimensions one may integrate each dimension according to this formula. Again there exist schemes for numerical integration on a regular grid with more complicated weights and correspondingly smaller errors, for example Simpson's famous formula [Press et al. 1992, p. 134], which is correct to $\mathcal{O}(\Delta x^4)$.

11.3 Application: Gravitational settling in two dimensions

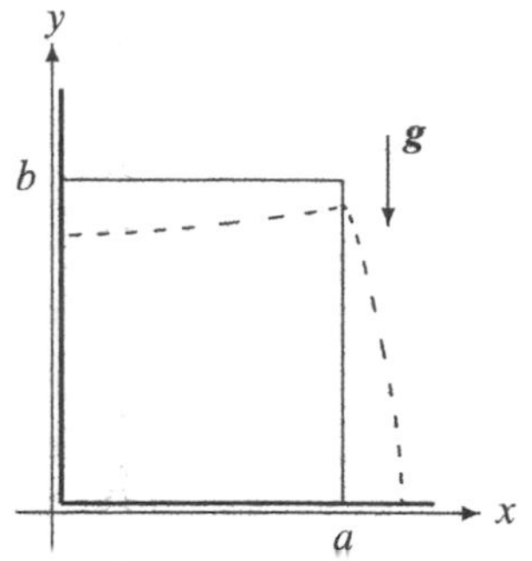

Expected shape of two-dimensional gravitational settling. When the wall at $x = a$ is removed, the elastic material will bulge out, because of its own weight.

One of the simplest non-trivial problems that does not seem to admit an exact analytic solution is the gravitational settling of an infinitely long, horizontal block of elastic material with a rectangular cross-section of dimensions $a \times b$. When one of the vertical sides is removed, the material bulges out (see the margin figure). In this somewhat academic case we follow the conventions normally used in two dimensions and take the y-axis to be vertical. The wall that is removed is situated at $x = a$ whereas the wall at $x = 0$ remains in place. It is reasonable to assume that there can be no displacement in the z-direction direction, that is, $u_z = 0$ everywhere, because it would have to move infinitely much material. It is also reasonable to assume that the displacements u_x and u_y only depend on x and y, but not on z. The problem has become effectively two-dimensional, although there are vestiges of the three-dimensional problem, for example the non-vanishing stress along the z-direction, but that can be ignored.

Equations

The components of the two-dimensional strain tensor are

$$u_{xx} = \nabla_x u_x, \tag{11.13a}$$

$$u_{yy} = \nabla_y u_y, \tag{11.13b}$$

$$u_{xy} = \tfrac{1}{2}(\nabla_x u_y + \nabla_y u_x). \tag{11.13c}$$

The corresponding stresses are found from Hooke's law (8.9) ,

$$\sigma_{xx} = 2\mu u_{xx} + \lambda(u_{xx} + u_{yy}), \tag{11.14a}$$

$$\sigma_{yy} = 2\mu u_{yy} + \lambda(u_{xx} + u_{yy}), \tag{11.14b}$$

$$\sigma_{xy} = \sigma_{yx} = 2\mu\, u_{xy}. \tag{11.14c}$$

Finally, the components of the effective force are

$$f_x^* = \nabla_x \sigma_{xx} + \nabla_y \sigma_{xy}, \tag{11.15a}$$

$$f_y^* = \nabla_x \sigma_{xy} + \nabla_y \sigma_{yy} - \rho_0 g_0. \tag{11.15b}$$

Note that only first-order partial derivatives are used in these equations. This makes it trivial to convert the equations to the discrete lattice by replacing the derivatives by finite difference operators.

Boundary conditions

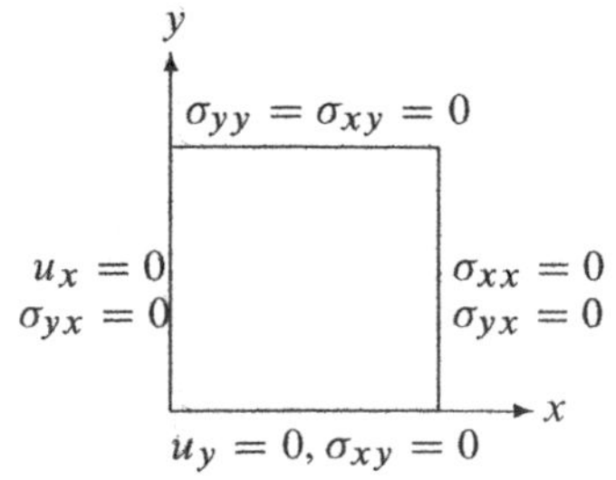

Boundary conditions for the rectangular block.

The boundary consists of the two fixed surfaces at $x = 0$ and $y = 0$ and the free surfaces at $x = a$ and $y = b$. We shall adopt the following boundary conditions:

$\sigma_{xx} = 0$,	$\sigma_{yx} = 0$	free surface at $x = a$,	(11.16a)
$\sigma_{yy} = 0$,	$\sigma_{xy} = 0$	free surface at $y = b$,	(11.16b)
$u_x = 0$,	$\sigma_{yx} = 0$	fixed wall at $x = 0$,	(11.16c)
$u_y = 0$,	$\sigma_{xy} = 0$	fixed wall at $y = 0$.	(11.16d)

Here we have assumed that the fixed surfaces are slippery, so that the shear stress must vanish. That is however not the only choice.

Had we instead chosen the fixed walls to be sticky so that the elastic material was unable to slip along the sides, the tangential displacements at these boundaries would also have to vanish, that is, $u_y = 0$ at $x = 0$ and $u_x = 0$ at $y = 0$. The tangential stress $\sigma_{xy} = \sigma_{yx}$ would, on the other hand, be left free to take any value determined by the field equations.

Shear-free solution

Since the shear stress vanishes at all boundaries, it is tempting to solve the equations by requiring the shear stress also to vanish throughout the block, $\sigma_{xy} = \sigma_{yx} = 0$, as we did for the three-dimensional settling in Section 9.2. One may verify that the following field solves the field equations,

$$u_x = \frac{\nu}{1-\nu}\frac{(b-y)x}{D}, \qquad u_y = -\frac{b^2-(b-y)^2}{2D} + \frac{\nu}{1-\nu}\frac{3x^2-a^2}{6D}, \tag{11.17}$$

where the characteristic deformation scale is

$$D = \frac{4\mu(\lambda+\mu)}{(2\mu+\lambda)\rho_0 g_0} = \frac{E}{(1-\nu^2)\rho_0 g_0}. \tag{11.18}$$

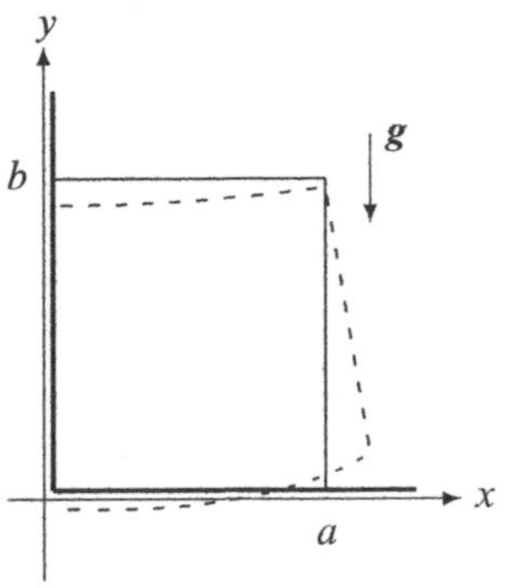

The shear-free solution sinks partly into the bottom of the box. An extra vertical pressure is needed from below in order to fulfill the boundary conditions.

The solution is of the same general form as in the three-dimensional case (9.15) on page 145, but the dependence on Poisson's ratio ν is different because of the different geometry. As before, this solution also fails to meet the boundary conditions at the bottom, here $y = 0$. The arbitrary constant in u_y has been chosen such that $\int_0^a u_y\, dx = 0$ at $x = 0$.

Convergence measures

The approach toward equilibrium may, for example, be monitored by means of the integral over the square of the effective force field, which should converge toward zero, if the algorithm works. We shall choose the monitoring parameter to be

$$\chi = \frac{1}{\rho_0 g_0}\sqrt{\frac{1}{ab}\int_0^a dx \int_0^b dy \left(f_x^{*2} + f_y^{*2}\right)}. \tag{11.19}$$

It is normalized such that $\chi = 1$ in the undeformed state where $u_x = u_y = 0$ and thus $f_x^* = 0$ and $f_y^* = -\rho_0 g_0$. The integrals are calculated using the trapezoidal rule. The iterative process can then be stopped when the value of χ falls below any desired accuracy, say $\chi \lesssim 0.01$.

Another possibility is to calculate the potential energy (11.5) with $\boldsymbol{P} = \boldsymbol{0}$:

$$\mathcal{E} = \int_0^a dx \int_0^b dy \left[\frac{1}{2}(u_{xx}\sigma_{xx} + u_{yy}\sigma_{yy} + 2u_{xy}\sigma_{xy}) + \rho_0 g_0 u_y\right]. \tag{11.20}$$

This quantity should decrease monotonically from $\mathcal{E} = 0$ in the undeformed state toward its (negative) minimum. Since it—like χ—is well-defined in the continuum, its value should be relatively independent of how fine-grained the discretization is, as long as the lattice is large enough. It is, however, harder to determine the relative accuracy attained.

Iteration cycle

Assuming that the discretized displacement field on the lattice (u_x, u_y) satisfies the boundary conditions, we may calculate the strains (u_{xx}, u_{yy}, u_{zz}) from (11.13) by means of the discrete derivatives, and the stresses $(\sigma_{xx}, \sigma_{yy}, \sigma_{xy})$ from Hooke's law (11.14). Stress boundary conditions are then imposed and the effective force field (f_x^*, f_y^*) is calculated from (11.15). At this point the monitoring parameter χ may be checked and if below the desired accuracy, the iteration process is terminated. If not, the corrections

$$\delta u_x = \epsilon f_x^*, \qquad \delta u_y = \epsilon f_y^* \tag{11.21}$$

are added into the displacement field, boundary conditions are imposed on the displacement field, and the cycle repeats.

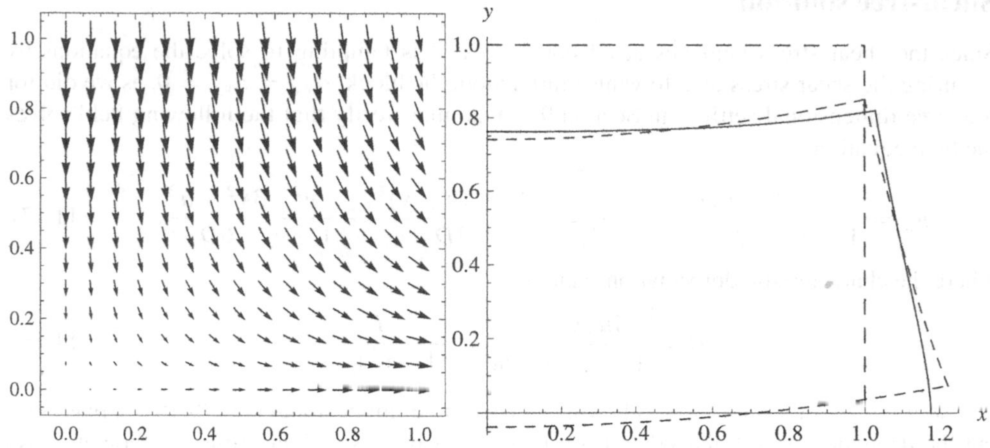

Figure 11.1. Computed deformation of a square two-dimensional block. On the left the equilibrium displacement field is plotted by means of little arrows (not to scale). On the right is plotted the outline of the deformed block. The displacement vanishes as it must at the fixed walls. The protruding material (solid line) has a slightly convex shape rather than the concave shape in the shear-free approximation (dashed lines).

The iteration process may be viewed as a dynamical process that in the course of (computer) time makes the displacement field converge toward its equilibrium configuration. The true dynamics of deformation (see Chapter 24) go on in real time and are quite different. Since dissipation in solids is not included here, the true dynamics are unable to eat away energy and make the system relax toward equilibrium. Releasing the block from the undeformed state, as we do here, would instead create vibrations and sound waves that would reverberate forever throughout the system.

Choice of parameters

Since we are mostly interested in the shape of the deformation, we may choose convenient values for the input parameters. They are the box sides $a = b = 1$, the lattice sizes $N_x = N_y = 20$, Young's modulus $E = 2$, Poisson's ratio $\nu = 1/3$, and the force of gravity $\rho_0 g_0 = 1$. The step-size is chosen of the form

$$\epsilon = \frac{\omega}{E} \frac{\Delta x^2 \Delta y^2}{\Delta x^2 + \Delta y^2}, \tag{11.22}$$

where ω is called the *convergence parameter*. The reason for this choice is that the effective force is proportional to Young's modulus E and (due to the second-order spatial derivatives) to the inverse squares of the grid spacings, say $1/\Delta x^2 + 1/\Delta y^2 = (\Delta x^2 + \Delta y^2)/\Delta x^2 \Delta y^2$. The convergence parameter ω is consequently dimensionless and may be chosen to be of order unity to get fastest convergence. In the present computer simulation, the largest value that could be used before numeric instabilities set in was $\omega = 1$.

Programming hints

The fields are represented by real arrays, containing the field values at the grid points, for example

$$UX[i, j] \Leftrightarrow u_x(i\Delta x, j\Delta y), \qquad UY[i, j] \Leftrightarrow u_y(i\Delta x, j\Delta y), \tag{11.23}$$

and similarly for the strain and stress fields. Allocating separate arrays for strains and stresses may seem excessive and can be avoided, but when lattices are as small as here, it does not matter. Anyway, the days of limited memory are over.

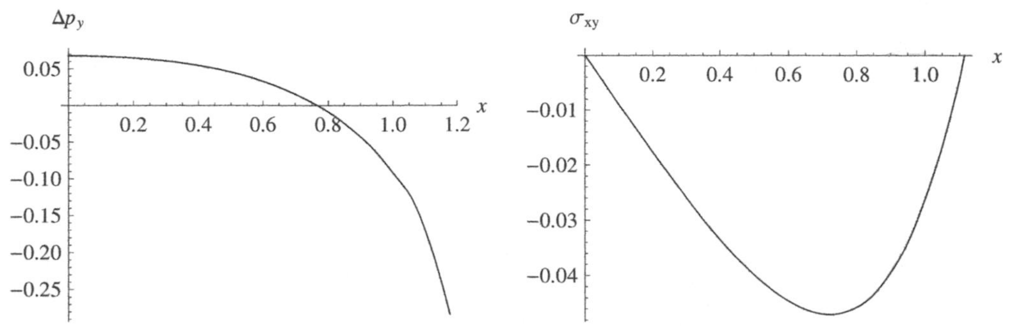

Figure 11.2. The computed vertical pressure excess, $\Delta p_y = p_y - p_0$ with $p_y = -\sigma_{yy}$ and $p_0 = \rho_0 g_0 b$, is plotted on the left for $y = 0$ as a function of the displaced x. On the right, the corresponding shear stress is plotted at $y = 0.5$. The pressure is higher in the central region than the shear-free estimate p_0, and the shear stress is negative (but small) and thus adds to the force exerted by gravity on the central part. The curves have been linearly interpolated between the data points. The small kink in Δp_y just beyond $x = 1$ is an artifact of the coarse lattice.

The iteration cycle is implemented as a loop, containing a sequence of calls to sub-routines that evaluate strains, stresses, effective forces, and impose boundary conditions, followed by a step that evaluates the monitoring parameters and finally updates the displacement arrays before the cycle repeats. The iteration loop is terminated when the accuracy has reached the desired level, or the number of iterations has exceeded a chosen maximum.

Results

After about 2,000 iteration cycles the monitoring parameter χ has fallen from 1 to about 0.01 where it seems to remain without further change. This is most probably due to the brute enforcing of boundary values. The limiting value of χ diminishes with increasing lattice volume $N = N_x N_y$, in accordance with the lessened importance of the boundary, which decreases like $1/\sqrt{N}$ relative to the volume.

The final displacement field and its influence on the outline of the original box is shown in Figure 11.1. One notes how the displacement does not penetrate into the fixed bottom wall as it did in the shear-free approximation. In Figure 11.2 the vertical pressure excess $\Delta p_y = p_y - p_0$, where $p_y = -\sigma_{yy}$ and $p_0 = \rho_0 g_0 b$ is plotted as a function of x at the bottom of the block ($y = 0$). Earlier we argued that there would have to be an extra normal reaction from the bottom in order to push up the sagging solution to the shear-free equations. This is also borne out by the plot of Δp_y, which has roughly the same shape throughout the block. Since the vertical pressure in the central region is now larger than the weight of the column of material above (p_0), we expect that there must be a negative shear stress on the sides of the column to balance the extra vertical pressure, as is also evident from Figure 11.2.

Problems

11.1 Show that the coefficients in the one-sided two-step differences (11.9) are uniquely determined.

Problems

Part III

Fluids in motion

Fluids in motion

The following chapters deal with the basic aspects of macroscopic motion in continuous matter. A minimal curriculum in fluid mechanics requires Chapters 12, 13, 15 and part of 16.

List of chapters

12. **Continuum dynamics:** The velocity field is the basic field for fluids in motion. Mass conservation and Newton's Second Law applied to material particles lead to the fundamental dynamic equations for any kind of matter. In this chapter it is only applied to two simple cases: Big Bang and non-relativistic cosmology.
13. **Nearly ideal flow:** Incompressible frictionless (Euler) fluids satisfy the simplest equations of motion. Benoulli's theorem is derived and applied to a few cases. Vorticity and circulation are introduced and typical potential flow solutions are solved.
14. **Compressible flow:** At velocities close to the speed of sound, all fluids are compressible. A simplified one-dimensional theory of compressible steady flow is presented and applied to two famous rocket engines.
15. **Viscosity:** Real fluids are viscous. Velocity-driven planar incompressible flow is solved in a few simple cases. The Navier–Stokes equations are formulated and the classification of flows by Reynolds number introduced. Viscous attenuation of sound is analyzed.
16. **Plates and pipes:** Pressure-driven planar and pipe flow (Poiseuille) are solved. Phenomenology of turbulent pipe flow is discussed. Cylindrical velocity-driven flow (Couette) is solved.
17. **Creeping flow:** In the limit of vanishing Reynolds number, fluid dynamics becomes linear. The exact solution around a solid ball is analyzed. The phenomenology of drag around spherical and other objects is discussed also for higher Reynolds numbers. Finally, the theory of lubrication is outlined.
18. **Rotating fluids:** Large-scale geophysical flows all take place on our rotating planet and are strongly influenced by the Corioli's force. Geostrophic flow and the shape of the Ekman boundary layer are analyzed. The structure of bathtub vortices is discussed and an urban legend is debunked.
19. **Computational fluid dynamics:** The basic theory of computational fluid dynamics is presented and applied to laminar channel entrance flow.

12 Continuum dynamics

The running water in a brook, a waterfall, the streaming wind, and the rolling sea are all examples of fluids in motion, while the vibrations of a church bell, the swaying of a tree, or the tremor of an earthquake are examples of solids in motion. It is one of the wonders of nature that all this richness is contained in Newton's equations of motion applied to continuous matter. Easy to write down, these continuum equations of motion only have analytic solutions in a number of idealized and highly constrained situations (see Chapter 16). Nevertheless, such solutions offer valuable insights into the dynamics, which is otherwise accessible only through experiment and computer calculations.

The motion of solids is generally less rich than fluid motion, and it is precisely for this reason that solids are used to build structures like houses, bridges, and machines. Solids remember their original shape, fluids forget. Fluids and solids are extremes in the world of continuous matter, and there are many materials with properties in between. Two basic mechanical equations govern the motion of continuous matter. One is the conservation of mass, which states that the only way the mass of a volume of matter can change is through movement of material across its boundary. The other is Newton's Second Law applied to continuous systems, which is equivalent to momentum balance (or momentum conservation, as some prefer to say). Together with suitable expressions for the forces at play in the material, we are led to the equations of motion for the mass density and velocity fields.

In this chapter we shall analyze continuous matter in motion without distinguishing between particular types, although we shall mostly think of fluids. We shall only apply the equations of motion to the whole universe, and show that they lead to a surprisingly sensible cosmology. Later chapters will deal with much more earthly aspects of matter in motion.

12.1 The velocity field

Trying to define a velocity field $\boldsymbol{v}(\boldsymbol{x}, t)$, we are faced with the problem that the molecules, especially in the gaseous state, move rapidly around, even in matter at rest, and this motion must somehow be averaged out. Newtonian particle mechanics (see Appendix A) tells us, however, that *the center of mass of a collection of particles moves as a single particle with mass equal to the sum of the individual masses of all the particles, acted upon by the sum of the individual forces acting on each particle in the collection*. The only meaningful definition of the velocity field $\boldsymbol{v}(\boldsymbol{x}, t)$ is therefore that it *represents the center-of-mass velocity of the collection of molecules in a material particle with center-of-mass position $\boldsymbol{x}$ at time t*.

Figure 12.1. Two spectacular examples of fluids in motion. It is remarkable that both phenomena are controlled by Newton's equations of motion, although they cannot be solved to yield the details shown here. **Left:** A fish-bone pattern created in the laboratory by letting two water jets collide. (Source: John W. M. Bush, MIT. With permission.) **Right:** The Iguazu Falls in Argentina and Brazil (Source: Wikimedia Commons. Courtesy Reinhard Jahn.)

The total momentum of all the molecules in a material particle of mass dM with center of mass in the point $\boldsymbol{x}$ at time t may consequently be written (see Appendix A)

$$\boxed{d\boldsymbol{\mathcal{P}} = \boldsymbol{v}(\boldsymbol{x}, t)\, dM = \boldsymbol{v}\, \rho\, dV,} \tag{12.1}$$

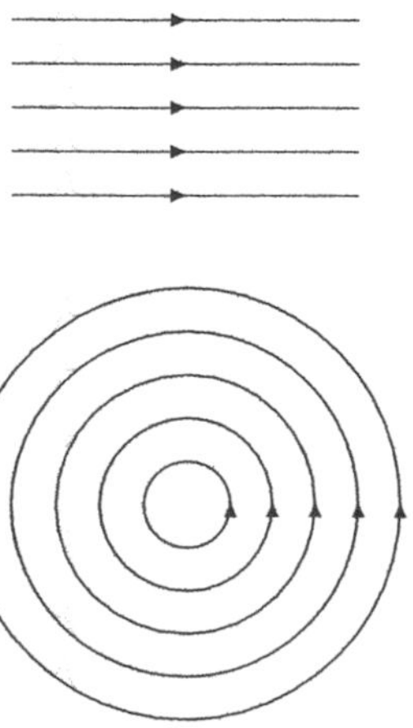

Streamlines for rigid body translation, and for the rigid body rotation (bottom).

where ρ is the mass density field and dV is the volume of the particle. Identifying the velocity field with the center-of-mass velocity of material particles thus permits us directly to apply Newtonian particle mechanics to the material particles themselves without worrying about their shape or internal structure.

Example 12.1: The velocity field of a non-rotating rigid body moving with constant velocity $\boldsymbol{U}$ is $\boldsymbol{v}(\boldsymbol{x}, t) = \boldsymbol{U}$. If instead the body is rotating around the origin of the coordinate system with constant angular velocity vector $\boldsymbol{\Omega}$, the velocity field is $\boldsymbol{v}(\boldsymbol{x}, t) = \boldsymbol{\Omega} \times \boldsymbol{x}$. Both fields are visualized in the margin figure by means of streamlines.

The velocity field is like the mass density, a fluctuating quantity obtained from an average over nearly random individual molecular contributions. In Chapter 1 it was shown that, provided the linear dimension of the material particle is larger than the micro scale L'_{micro}, defined in Equation (1.10) on page 8, the relative fluctuation in center-of-mass velocity will be smaller than the measurement precision.

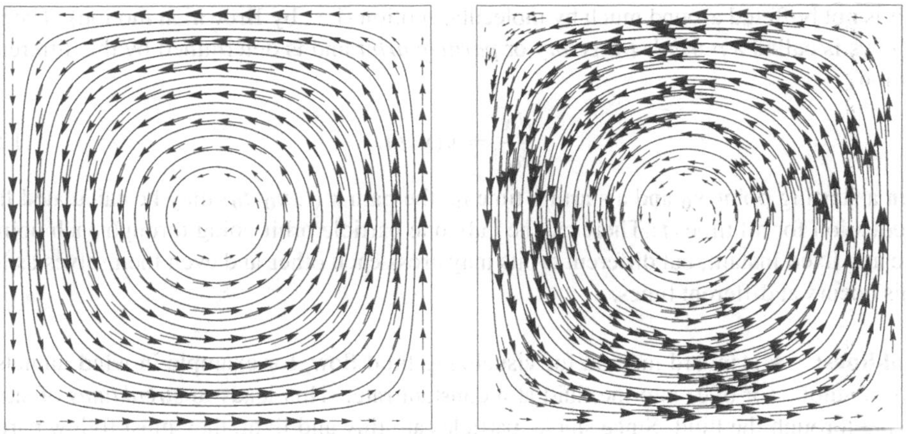

Figure 12.2. Plots of streamlines and little arrows for the stationary incompressible flow $\boldsymbol{v} = (\sin x \cos y, -\cos x \sin y, 0)$ in the region $0 \leq x \leq \pi$ and $0 \leq y \leq \pi$ (see also Example 12.2). The whirl (and its mirror images) are repeated periodically throughout the xy-plane. **Left:** Regular plot on a Cartesian (16 × 16) grid. **Right:** Random plot of the field in 512 points.

Flow visualization

Wind and water currents are normally invisible unless polluted by foreign matter. A gentle breeze in the air can be observed from the motion of dust particles dancing in the sunshine or the undulations of smoke from a cigarette. Even a tornado first becomes visible when water vapor condenses near its center or debris is picked up and thrown around. Modern technology does, on the other hand, permit us indirectly to visualize the velocity fields. Doppler radar is used for visualizing damaging winds in violent storms, and Doppler acoustics is used to visualize blood flow in the heart. On paper or screen, real or computed flows can be visualized in a number of different ways.

Streamlines: *Streamlines* are curves that are everywhere tangent to the velocity field at a fixed time. Such curves are solutions to the ordinary differential equation

$$\frac{d\boldsymbol{x}}{dt} = \boldsymbol{v}(\boldsymbol{x}, t_0), \tag{12.2}$$

where the velocity field is calculated for a fixed value of time t_0. Starting at any point $\boldsymbol{x}_0$ at $t = t_0$ we may use this equation to determine the path $\boldsymbol{x} = \boldsymbol{x}(t, \boldsymbol{x}_0, t_0)$ of the streamline. Because the velocity field is evaluated at a fixed moment in time t_0, there will be only one tangent and thus only one streamline through every point of space. Streamlines depict the velocity field at a single instant in time and can never intersect. Streamlines depend on the motion of the coordinate system: A flow to the right in one coordinate system may move to the left in another that moves with respect to the first.

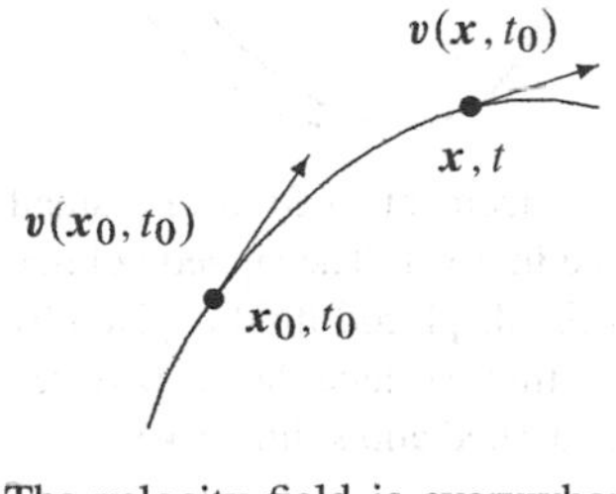

The velocity field is everywhere tangent to the streamline at a given time t_0.

The instantaneous velocity field can also be visualized by means of little arrows attached to a regular grid of points, each of a length and direction proportional to the velocity field in the point (see Figure 12.2). Sometimes it is more illustrative and permits finer flow details to be seen if the arrows are drawn from a random selection of points, because the density of arrows can be higher.

Particle trajectories: Imagine you drop a tiny particle—a speck of dust—into a fluid, and watch how it is carried along with the fluid in its motion. The speck of dust should be neutrally buoyant in the fluid, and so small that its mass plays no role, but on the other hand so large that it is not buffeted around much by molecular motion (i.e., by Brownian motion). The path it follows is called a *particle trajectory* or *particle orbit* and is determined by the differential equation

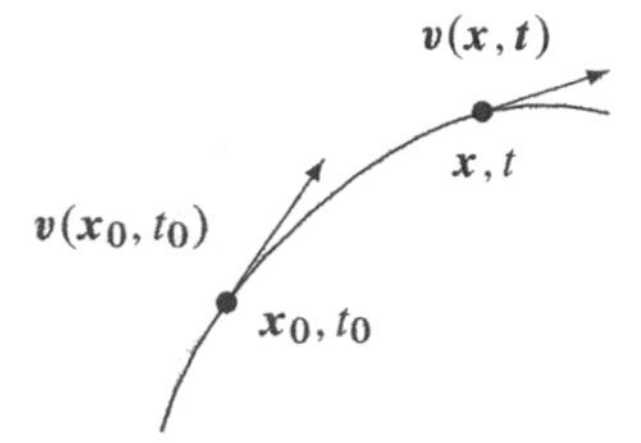

The instantaneous velocity field is everywhere tangent to the particle orbit.

$$\frac{dx}{dt} = v(x,t). \tag{12.3}$$

Given a starting point x_0 and a starting time t_0, the path $x(t, x_0, t_0)$ may be calculated from this equation for all times t. There can be only one particle orbit going through each point in space at a fixed instant, but different orbits may cross each other and even themselves as long as this occurs at different times.

Streaklines: A standard method for visualizing fluid flow, for example in wind tunnels, is to inject smoke (or dye) into the fluid at a constant rate. This leads to *streaklines* of smoke weaving through the fluid. Since smoke particles are tiny and light, they must follow particle orbits. But at a given time t, a streakline is obtained from the particle orbit $x(t, x_0, t_0)$ by varying the start time t_0 while keeping fixed the point x_0 from which the smoke emanates.

For *steady flow* where the velocity field is independent of time, $v(x,t) = v(x)$, the particle orbits evidently coincide with the streamlines, and since the streamlines in this case can only depend on the time difference $t - t_0$, the streaklines will also coincide with them. For *unsteady flow* with time-dependent velocity field, the relationship between the three types of flow lines can be quite hard to visualize (see Problems 12.1, 12.2, and 12.3). One should never forget that streamlines can be quite misleading for unsteady flow.

12.2 Incompressible flow

Most liquids are perceived as incompressible under ordinary circumstances. A fluid is, however, effectively incompressible when flow speeds are much smaller than the velocity of sound. The fluid—so to speak—"prefers to get out of the way" of external forces rather than being compressed (see Chapter 14).

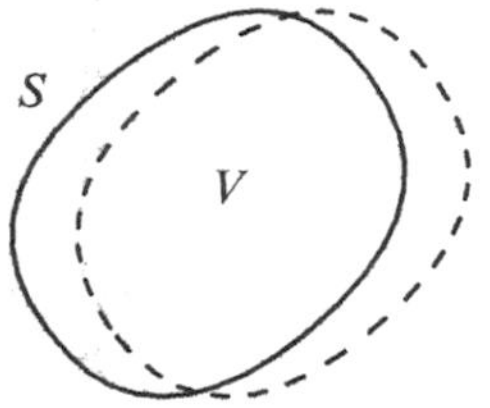

Displacement of fluid in a small time interval. The signed volume of the displaced fluid is given by the thin surface layer between the dashed and solid curves.

All materials must nevertheless in principle be compressible. For in truly incompressible matter the sound velocity would be infinite, and that violates the relativistic injunction against any signal moving faster than the speed of light. Incompressibility is always an approximation and should, strictly speaking, be viewed as a condition on the flow rather than an absolute material property. It is nevertheless such an important condition that we shall devote the major part of the remainder of this book to the study of incompressible flows.

Global and local forms

Consider a *fixed* (time-independent) surface S enclosing a volume V in a fluid flowing with velocity field $v(x,t)$. The (signed) volume of the fluid that leaves (or enters) the volume through any outward-oriented surface element dS during a tiny time interval δt is $v\delta t \cdot dS$. Incompressibility means that the total volume of displaced fluid must vanish, and dividing by the constant δt and integrating $v \cdot dS$ over S we arrive at the *global incompressibility condition*

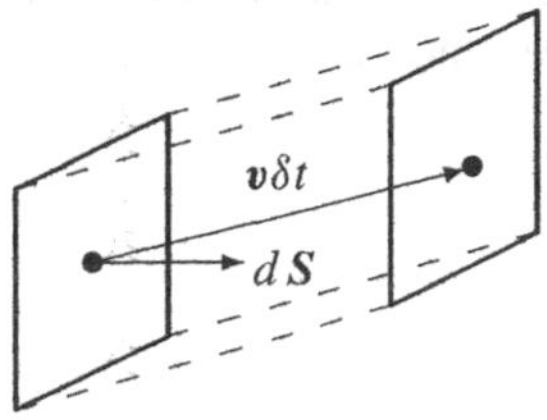

The volume $v\delta t \cdot dS$ passes through a surface element dS in a small time interval δt.

$$\boxed{\oint_S v \cdot dS = 0.} \tag{12.4}$$

Incompressible matter cannot accumulate anywhere, and equal volumes of incompressible fluid must enter and leave through any fixed surface per unit of time. Since the above condition refers to a single instant of time, it does in fact not matter whether the surface is fixed or moves in any way you may desire.

Applying Gauss' theorem (2.14) on page 26 to the surface integral we find $\int_V \nabla\cdot\boldsymbol{v}\, dV = 0$ for any volume V of fluid. Since this must *a fortiori* be true for point-like material particles, we obtain the *local incompressibility condition*

$$\boxed{\nabla \cdot \boldsymbol{v} = 0.} \tag{12.5}$$

The divergence of the velocity field vanishes in incompressible flow. Notice that this local incompressibility condition does not refer to any volume of matter, only to the velocity field itself. A divergence-free field is sometimes called *solenoidal*.

Example 12.2 [Artificial flow pattern]: The flow described by the time-independent velocity field

$$\boldsymbol{v} = (\sin x \cos y, -\cos x \sin y, 0) \tag{12.6}$$

is incompressible because

$$\nabla \cdot \boldsymbol{v} \equiv \frac{\partial v_x}{\partial x} + \frac{\partial v_y}{\partial y} + \frac{\partial v_z}{\partial z} = \cos x \cos y - \cos x \cos y = 0. \tag{12.7}$$

Due to the periodicity in both x and y, the flow forms a regular square array of stationary whirls, one of which is shown in Figure 12.2. There is probably no practical way of generating this particular flow pattern.

Leonardo da Vinci (1452–1519). Italian renaissance artist, architect, scientist, and engineer. A universal genius who made fundamental contributions to almost every field. Also a highly practical man who concerned himself with the basic mechanical principles behind everyday machines, and sometimes also future machines, such as the helicopter. (Source: Wikimedia Commons.)

Case: Leonardo's law

Leonardo da Vinci knew—and used—that the water speed decreases when a canal or river becomes wider and increases when it becomes narrower [Tokaty 1994]. He discovered the simple law that the product of the cross-sectional area of a canal and the average flow velocity through this area is always the same.

Consider, for example, an aqueduct or canal and mark two fixed planar cross-sections A_1 and A_2. Leonardo's law then says that the average water velocities v_1 and v_2 through these cross-sections must obey the relation

$$A_1 v_1 = A_2 v_2. \tag{12.8}$$

The law expresses the rather self-evident fact that the same volume of incompressible water has to pass any cross-section of the canal per unit of time. We use it all the time when dealing with water.

Leonardo's law follows from the global condition of incompressibility (12.4). Together with the sides of the canal, the two cross-sections define a volume to which the condition can be applied. Since no water can flow through the open surface and the sides of the canal, the surface integral only receives contributions from the two cross-sections, and consequently,

$$\oint \boldsymbol{v} \cdot d\boldsymbol{S} = \int_{A_2} \boldsymbol{v} \cdot d\boldsymbol{S} - \int_{A_1} \boldsymbol{v} \cdot d\boldsymbol{S} = 0. \tag{12.9}$$

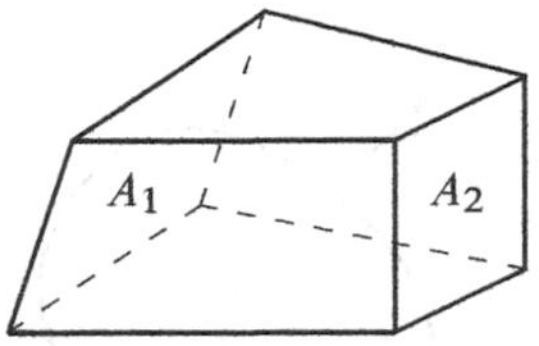

Aqueduct with varying cross-section. The volume of water passing through the cross-section A_1 is the same as the amount of water passing through A_2, or $A_1 v_1 = A_2 v_2$, where v_1 and v_2 are the flow velocities through the areas.

The average flow velocity through a cross-section A of the canal is

$$v = \frac{1}{A} \int_A \boldsymbol{v} \cdot d\boldsymbol{S}. \tag{12.10}$$

It then follows from (12.9) that the product $A\, v$ is the same everywhere along the canal. This modern formulation of Leonardo's law is valid, independent of whether the flow is orderly or turbulent.

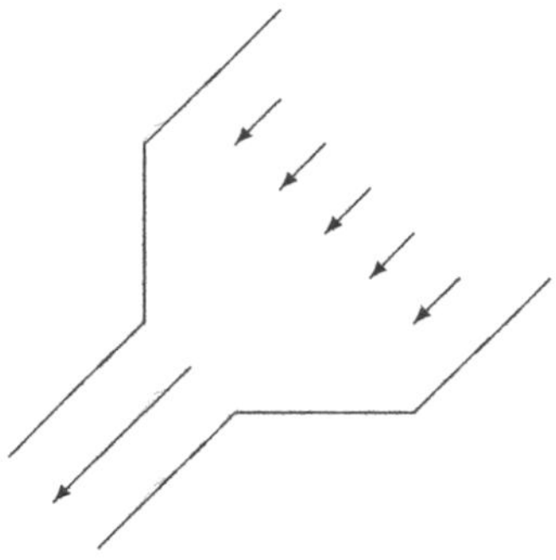

The flow velocity in the thin needle of a syringe is much higher than in the liquid chamber.

Example 12.3 [Hypodermic syringe]: A hypodermic syringe contains a few cubic centimeters of liquid in a small chamber about 1 cm in diameter and a few centimeters long. The liquid is injected through a hollow needle with an inner diameter of about 1 mm in the course of a few seconds. Since the ratio of cross-sections is 100, Leonardo's law tells us that the speed of the liquid in the needle is about 100 times larger than the speed with which the piston of the syringe is pushed, that is, on the order of meters per second. No surprise that it sometimes hurts.

Leondardo's law is definitely *not* valid for compressible flow. When you push the piston of your bicycle pump into its cylindric chamber filled with air, while holding a finger at the exit valve, the average flow velocity will decrease toward the end of the chamber where it has to vanish because no air can pass through there. But the cross-section of the chamber is constant, so the product of cross-section and average velocity *cannot* be constant throughout the chamber. To deal with compressible flow we need a stronger law.

12.3 Mass conservation

In Newtonian mechanics, mass is *conserved*, meaning that it can be neither created nor destroyed, only moved around[1]. The mass of a collection of point particles ("molecules") can only change by addition or removal of particles. Since all matter is made from molecules, this must mean that the only way the mass in a given volume of continuous matter can change is by mass flowing in or out of the volume through its surface. This almost trivial remark leads to the first of the central differential equations of continuum dynamics, the *equation of continuity* first derived by Euler in 1753.

Global and local forms

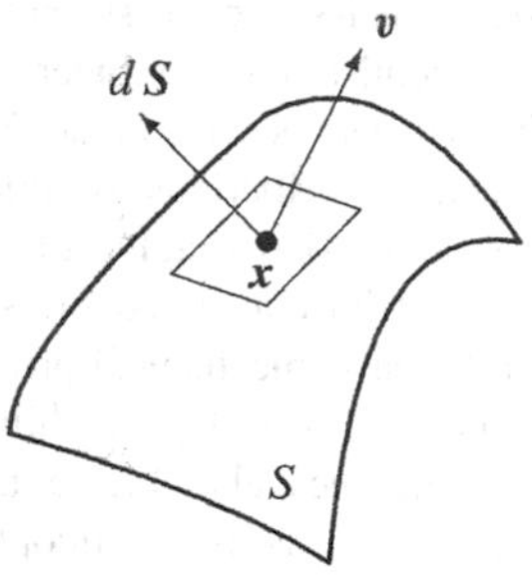

Matter moves with velocity $\boldsymbol{v}(\boldsymbol{x},t)$ through the surface element $d\boldsymbol{S}$ near $\boldsymbol{x}$ at time t.

Again let S be a *fixed* closed surface enclosing the volume V in a fluid of density $\rho(\boldsymbol{x},t)$ flowing with velocity $\boldsymbol{v}(\boldsymbol{x},t)$. The (signed) amount of mass leaving through a surface element $d\boldsymbol{S}$ in the time interval δt is $\rho\,\boldsymbol{v}\delta t \cdot d\boldsymbol{S}$; and since mass is conserved, the rate that mass is gained by the volume must equal the rate at which mass flows *into* the volume through its closed surface S (as always, oriented outward),

$$\boxed{\frac{d}{dt}\int_V \rho\, dV = -\oint_S \rho\,\boldsymbol{v}\cdot d\boldsymbol{S}.} \tag{12.11}$$

This is the *global equation of mass conservation* for an arbitrary *fixed* control volume. It expresses the obvious fact that the mass you gain while you eat equals the mass of the food you pass into your mouth.

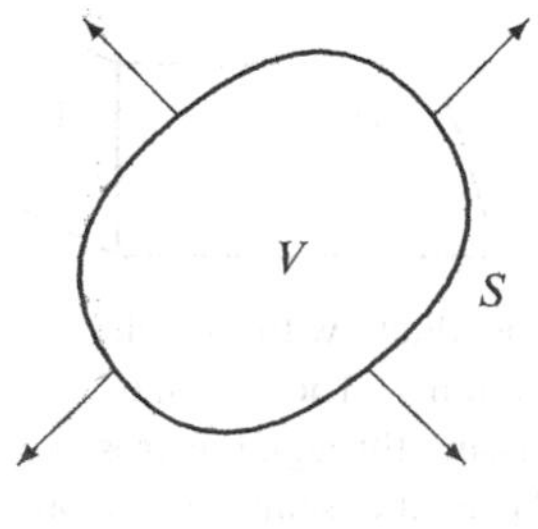

The amount of mass in a fixed volume V can only decrease by a net outflow through its closed surface S.

Terminology: The vector field $\rho\,\boldsymbol{v}$ is called the *current density of mass*, and its integral over any surface (open or closed) is called the *mass flow rate*. In older "dialects" of flow parlance the current density of mass is instead called the *mass flux*. In other dialects it is sometimes called the *mass flux density* and its integral is called the *mass flux*. To confuse things further, it follows from the definition of the velocity field (12.1) that the same quantity $\rho\,\boldsymbol{v}$ also performs in the role of *momentum density*.

[1]Mass conservation does not hold in relativity where Einstein's famous equation, $E = mc^2$, tells us that energy and mass are equivalent physical quantities. Only the total energy of a system, including the rest mass energy of all massive particles, is strictly conserved. Unfortunately (or fortunately?), it is not easy to get access to the rest mass energy, because one needs to annihilate every particle in a body by bringing it together with the corresponding antiparticle. More down to earth, the negative gravitational energy of our planet (about 10^{32} J) diminishes, according to Einstein's formula, its mass by about 10^{15} kg.

For a fixed control volume we may move the time derivative inside the volume integral and use Gauss' theorem (2.14) on the surface integral to obtain

$$\int_V \left(\frac{\partial \rho}{\partial t} + \nabla \cdot (\rho \boldsymbol{v})\right) dV = 0.$$

As this has to be true for any fixed volume V, we conclude that mass conservation requires

$$\boxed{\frac{\partial \rho}{\partial t} + \nabla \cdot (\rho \boldsymbol{v}) = 0} \qquad (12.12)$$

for all points $\boldsymbol{x}$ in space and all times t. This is the *local equation of mass conservation*, also called the *equation of continuity*. Although derived from global mass conservation applied to fixed volumes, it is itself a local relation completely without reference to volumes. Going backward through the preceding argument, we may conversely derive global mass conservation from the continuity equation. In Chapter 20 we shall investigate the global equation of mass conservation and equations of balance for other quantities using arbitrary moving surfaces.

Example 12.4 [Bicycle pump]: A bicycle pump consists of a piston that can be pushed into an air-filled cylindric chamber with constant cross-section A. We shall imagine that we block the exit valve (which otherwise provides the bicycle pump with its normal purpose). Let the piston's distance from the end wall be $x = a(t)$ at time t. If the piston moves slowly enough (compared to the velocity of sound), the density is the same throughout the chamber, $\rho(t) = M/Aa(t)$, where M is the (constant) mass of the air in the chamber. Assuming that only the x-component of the velocity field is non-vanishing, the equation of continuity becomes

$$\frac{\partial \rho(t)}{\partial t} + \frac{\partial(\rho(t) v_x(x,t))}{\partial x} = 0.$$

The unique solution to this differential equation, which vanishes for $x = 0$, is

$$v_x(x,t) = -x\frac{\dot{\rho}(t)}{\rho(t)} = x\frac{\dot{a}(t)}{a(t)}. \qquad (12.13)$$

Whether you pull or push the piston (slowly!), the velocity always varies linearly with the distance from the end of the chamber.

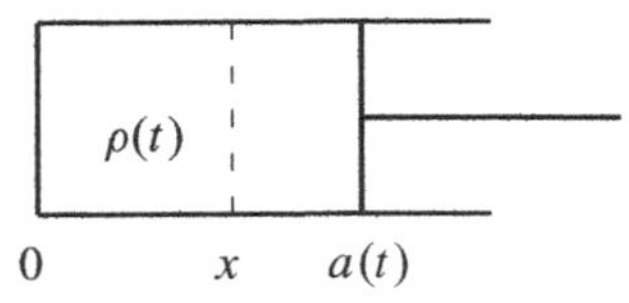

Bicycle pump. The velocity field varies linearly with x.

Material rate of change of density

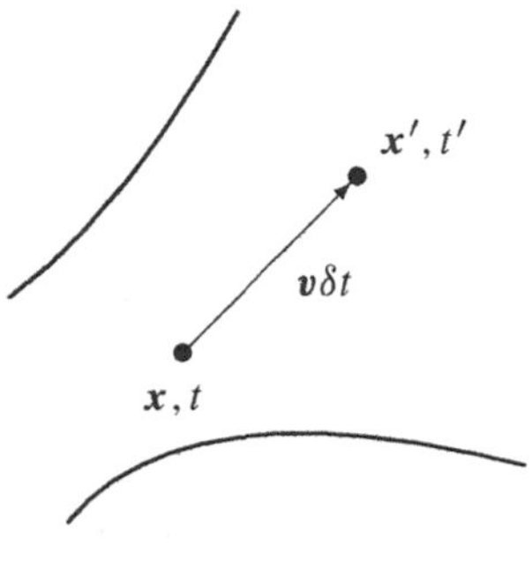

A particle may be swept along with the flow into a region of different density and velocity.

How does the neighborhood look from the point of view of a small object riding along with the motion of the material? A speck of dust being sucked into a vacuum cleaner will find itself in a region of higher air velocity (relative to the ground) and lower pressure and density than outside the machine, even if the flow of air is completely steady with velocity, pressure and density being constant everywhere in time, perhaps because you have stopped moving the head of the cleaner while thinking of such things. The ambient flow thus determines the physical environment of *comoving* material particles, whether they are grains of dust or parts of the fluid itself.

A material particle at the point $\boldsymbol{x}$ at time t riding along with the flow will at time $t' = t + \delta t$ have been displaced to the point $\boldsymbol{x}' = \boldsymbol{x} + \boldsymbol{v}(\boldsymbol{x}, t)\delta t$. Expanding to first order in δt we find the following change in mass density along the particle's motion:

$$\begin{aligned}\delta\rho &= \rho(\boldsymbol{x} + \boldsymbol{v}\delta t, t + \delta t) - \rho(\boldsymbol{x}, t)\\ &\approx v_x\delta t\frac{\partial\rho(\boldsymbol{x},t)}{\partial x} + v_y\delta t\frac{\partial\rho(\boldsymbol{x},t)}{\partial y} + v_z\delta t\frac{\partial\rho(\boldsymbol{x},t)}{\partial z} + \delta t\frac{\partial\rho(\boldsymbol{x},t)}{\partial t}\\ &= \left(\frac{\partial\rho}{\partial t} + (\boldsymbol{v}\cdot\nabla)\rho\right)\delta t.\end{aligned}$$

Dividing by δt we obtain the rate of change of the density around a comoving particle, also called the *material (or comoving) rate of change* of the density,

$$\boxed{\frac{D\rho}{Dt} = \frac{\partial\rho}{\partial t} + (\boldsymbol{v}\cdot\nabla)\rho.} \tag{12.14}$$

Note that this is a linear combination of first-order time and space derivatives of the density.

Using the relation (see Appendix C),

$$\nabla\cdot(\rho\boldsymbol{v}) = (\boldsymbol{v}\cdot\nabla)\rho + \rho\nabla\cdot\boldsymbol{v}, \tag{12.15}$$

we may write the equation of continuity (12.12) as

$$\frac{D\rho}{Dt} = -\rho\,\nabla\cdot\boldsymbol{v}. \tag{12.16}$$

This shows that the local incompressibility condition, $\nabla\cdot\boldsymbol{v} = 0$, is equivalent to saying that the material rate of change of the density vanishes for incompressible flow.

Rolling boulder: Intuitively, this is not so surprising even if the actual density field $\rho(\boldsymbol{x},t)$ might itself depend on both space and time. A rigid body, for example a boulder rolling down a mountainside, is by all counts incompressible, even if its mass density may vary with space and time according to the mineral composition of the rock and the way it rolls. But the density stays—of course—the same in the neighborhood of any particular mineral grain, independent of how the boulder moves.

Material rate of change of volume

In a small time interval δt, all points of the fluid are displaced by $\delta\boldsymbol{u} = \boldsymbol{v}\delta t$. It now follows from Equation (7.34) on page 118 that the volume dV of a material particle changes by $\delta(dV) = \nabla\cdot\delta\boldsymbol{u}\,dV = \nabla\cdot\boldsymbol{v}\delta t\,dV$. The rate of change of the comoving volume element is thus,

$$\frac{D(dV)}{Dt} = \nabla\cdot\boldsymbol{v}\,dV. \tag{12.17}$$

Combining this with the continuity equation (12.16) it follows that the material rate of change of mass $dM = \rho\,dV$ of a material particle vanishes:

$$\frac{D(dM)}{Dt} = \frac{D(\rho\,dV)}{Dt} = \frac{D\rho}{Dt}dV + \rho\frac{D(dV)}{Dt} = 0. \tag{12.18}$$

Again, since there can be no mass flow in or out of a comoving volume of fluid, this should not come as a surprise.

Material time derivative

The material rate of change of the density (12.14) contains the differential operator,

$$\frac{D}{Dt} = \frac{\partial}{\partial t} + \boldsymbol{v}(\boldsymbol{x},t) \cdot \nabla, \tag{12.19}$$

applied to the density field. This operator, called the *local material (or comoving) time derivative*, is a mixed differential operator in time and space that can be applied to any field to calculate the comoving rate of change of that field. The first term $\partial/\partial t$ is the *local rate of change* of the field in a fixed point $\boldsymbol{x}$, whereas the second term, called the *advective rate of change* $\boldsymbol{v} \cdot \nabla$, represents the effect of following the motion of the material in the environment of the point[2].

For the trivial time-independent position vector field $\boldsymbol{x}$, we find

$$\frac{D\boldsymbol{x}}{Dt} = (\boldsymbol{v}(\boldsymbol{x},t) \cdot \nabla)\boldsymbol{x} = \boldsymbol{v}(\boldsymbol{x},t). \tag{12.20}$$

The material time derivative of the position thus equals the velocity field, as one would expect.

* Eulerian displacement field

In the following chapters we shall see that the mass density and velocity fields suffice for the formulation the dynamics of ordinary fluids. Intuitively, a fluid particle tends to quickly "forget" where it came from. In all other kinds of matter, such as elastic and viscoelastic materials, the original position is "remembered" and the actual displacement of a material particle from its *reference position* plays an important role.

Let the field $\boldsymbol{X}(\boldsymbol{x},t)$ denote the reference position of a material particle actually found in the point $\boldsymbol{x}$ at time t. The *Eulerian displacement field* is then defined as the difference between the actual position and the reference position,

$$\boldsymbol{u}(\boldsymbol{x},t) = \boldsymbol{x} - \boldsymbol{X}(\boldsymbol{x},t). \tag{12.21}$$

For definiteness one may think of the reference position as the actual position at $t = 0$, so that $\boldsymbol{X}(\boldsymbol{x},0) = \boldsymbol{x}$ and $\boldsymbol{u}(\boldsymbol{x},0) = \boldsymbol{0}$ for all $\boldsymbol{x}$.

Since the reference position is the same for all points along a given particle trajectory, its material derivative must vanish, $D\boldsymbol{X}/Dt = \boldsymbol{0}$, and using (12.20) the velocity field becomes

$$\boldsymbol{v} = \frac{D\boldsymbol{u}}{Dt} = \frac{\partial \boldsymbol{u}}{\partial t} + (\boldsymbol{v} \cdot \nabla)\boldsymbol{u}. \tag{12.22}$$

The velocity field is simply the material derivative of the Eulerian displacement field. Conversely, given the velocity field this equation may be solved for the displacement field. The solution is in general non-trivial.

---

[2]There is no general agreement in the literature on a symbol for the material time derivative. Some texts use the ordinary differential operator d/dt and others use a notation like $(d/dt)_{\text{system}}$, but it seems as if the notation D/Dt used here is the most common. We shall sometimes shorten it to the more compact operator form $D_t \equiv D/Dt$. Neither is there a universally accepted name for the advective part, even if this word means "carried along with". In other texts it is called the *convective* or *inertial* part.

12.4 Equations of continuum dynamics

Newton's Second Law states that "mass times acceleration equals force" for a point particle of fixed mass. Continuum physics is concerned with volumes of matter of finite extent, and the smallest volumes consistent with the continuum description are the material particles. Comoving material particles are point-like and have constant mass, which makes them good candidates for Newtonian point particles, and we shall always treat them as such although this point of view is not without subtlety.

Material acceleration field

The acceleration of a material particle defines the *material acceleration field*,

$$\boxed{\boldsymbol{w} = \frac{D\boldsymbol{v}}{Dt} = \frac{\partial \boldsymbol{v}}{\partial t} + (\boldsymbol{v} \cdot \boldsymbol{\nabla})\boldsymbol{v}.} \tag{12.23}$$

The first term, the *local acceleration* $\partial \boldsymbol{v}/\partial t$, is most important for rapidly varying small-amplitude velocity fields, such as sound waves in solids or fluids (see Section 14.1). The second term, the *advective acceleration* $(\boldsymbol{v} \cdot \boldsymbol{\nabla})\boldsymbol{v}$, is most important for flows with strong spatial variations in the velocity field. In the limit of *steady flow* where the local acceleration vanishes, $\partial \boldsymbol{v}/\partial t = \boldsymbol{0}$, the advective term will be the only cause of acceleration. We become acutely aware of the advective acceleration in a little boat that drifts down toward the narrows of a steadily flowing river when the steady flow speeds up according to Leonardo's law. The advective acceleration is essentially of the same nature as the "fictitious" or inertial accelerations that arise in moving coordinate systems (see Chapter 18).

Newton's Second Law for continuous matter

Applying Newton's Second Law to any comoving material particle with mass $dM = \rho\, dV$, its equation of motion becomes

$$dM\,\boldsymbol{w} = d\boldsymbol{\mathcal{F}}, \tag{12.24}$$

where $d\boldsymbol{\mathcal{F}}$ is the total force acting on the particle. Dividing by dV we arrive at the seductively simple equation, $\rho \boldsymbol{w} = \boldsymbol{f}^*$, where $\boldsymbol{f}^* = d\boldsymbol{\mathcal{F}}/dV$ is the *effective force density*. It was defined earlier in Equation (6.15) on page 103, and we repeat it here:

$$f_i^* = f_i + \sum_j \nabla_j \sigma_{ij}, \tag{12.25}$$

where f_i is the true body force density and σ_{ij} the stress tensor. Finally, inserting the acceleration field (12.23) we have arrived at *Cauchy's equation* (1827):

$$\boxed{\rho \frac{D\boldsymbol{v}}{Dt} \equiv \rho \left(\frac{\partial \boldsymbol{v}}{\partial t} + (\boldsymbol{v} \cdot \boldsymbol{\nabla})\boldsymbol{v} \right) = \boldsymbol{f}^*.} \tag{12.26}$$

Remarkably, *this equation governs the dynamics of all continuous matter.*

Different types of materials—gases, liquids, solids, and whatnot—are characterized by different expressions for the stress tensor. The last two and a half centuries of continuum mechanics have in a sense "only" been an exploration of the rich ramifications of this equation. The remainder of this book will proceed along the same lines.

* Field equations of motion

It is instructive in the general case to formulate the complete dynamics as a set of *equations of motion* for the mass density ρ, the velocity $\boldsymbol{v}$, and the displacement $\boldsymbol{u}$:

$$\frac{\partial \rho}{\partial t} = -(\boldsymbol{v}\cdot\nabla)\rho - \rho\nabla\cdot\boldsymbol{v}, \tag{12.27a}$$

$$\frac{\partial \boldsymbol{v}}{\partial t} = -(\boldsymbol{v}\cdot\nabla)\boldsymbol{v} + \frac{1}{\rho}\boldsymbol{f}^*, \tag{12.27b}$$

$$\frac{\partial \boldsymbol{u}}{\partial t} = -(\boldsymbol{v}\cdot\nabla)\boldsymbol{u} + \boldsymbol{v}. \tag{12.27c}$$

To close this set of equations, we need a constitutive equation of the form

$$\boldsymbol{f}^* = \boldsymbol{f}^*[\rho, \boldsymbol{u}, \boldsymbol{v}](\boldsymbol{x}, t), \tag{12.28}$$

relating the effective force density to these fields. The brackets indicate that $\boldsymbol{f}^*$ is a local function of the values of the fields as well as their spatial derivatives (in principle to any order).

Knowing all the fields (and the relevant spatial derivatives) at any particular instant of time t, we may calculate the right-hand sides of the equations of motion and thereby the instantaneous rate of change for each field component on the left-hand sides. Each component may then be advanced to time $t + dt$ by multiplying its rate of change by the small time interval dt and adding the results to the respective field value at time t. Basically this is how a numerical solution of the field equations is carried out, although the road is fraught with difficulties (see Chapter 19).

In full generality, all three differential equations (12.27) are needed, but for ordinary Newtonian viscous fluids, the effective force does not depend on the displacement $\boldsymbol{u}$, so that the two first equations of motion close on their own. Finally, it should be noted that if the effective force also depends on non-mechanical fields, for example the temperature or pressure, additional equations of motion are needed for those fields to close the set (see Chapter 22).

* Galilean invariance

Galilei remarked already in 1632 that mechanical experiments yield the same results whether done on a ship at rest or in uniform motion relative to the static sea. A few decades later Newton included this law of inertia in his formulation of mechanics (see page 9). Galilean invariance must therefore be implicit in the continuum field equations.

To verify this explicitly we consider a new frame of reference that moves with constant velocity $\boldsymbol{U}$ relative to the old, and coincides with the old at $t = 0$. The coordinate transformation becomes

$$\boldsymbol{x}' = \boldsymbol{x} - \boldsymbol{U}t. \tag{12.29}$$

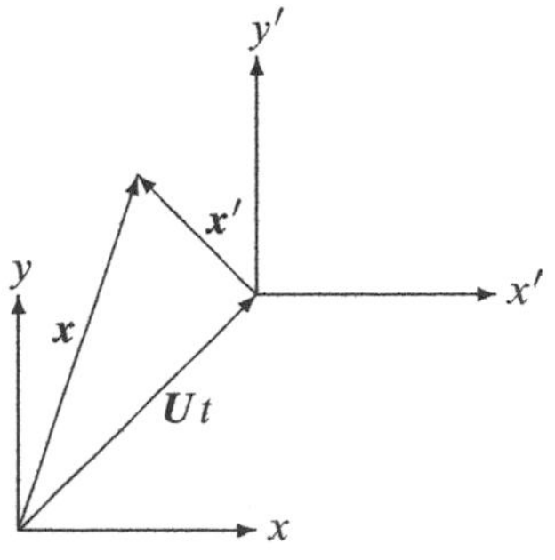

The geometry of the Galilei transformation.

It can now be shown that the equations of motion take the same form for the fields:

$$\rho'(\boldsymbol{x}', t) = \rho(\boldsymbol{x}, t), \tag{12.30a}$$

$$\boldsymbol{v}'(\boldsymbol{x}', t) = \boldsymbol{v}(\boldsymbol{x}, t) - \boldsymbol{U}, \tag{12.30b}$$

$$\boldsymbol{u}'(\boldsymbol{x}', t) = \boldsymbol{u}(\boldsymbol{x}, t) - \boldsymbol{U}t, \tag{12.30c}$$

$$\boldsymbol{f}'^*(\boldsymbol{x}', t) = \boldsymbol{f}^*(\boldsymbol{x}, t). \tag{12.30d}$$

The proof is relatively straightforward (see Problem 12.11). The last equation is actually a requirement that constrains the possible forms of the effective force.

12.5 Application: Big Bang

A cloud of non-interacting particles, grains, or fragments is clearly a poor model for continuous matter, but it is nevertheless of interest to study the equations of motion for the velocity field in this most simple case where all volume and contact forces are absent. It may even be used as a primitive model for the expanding universe with galaxies playing the role of grains. But the lack of interaction violates the continuum conditions discussed in Chapter 1 because there will be no dynamic smoothing of the fields by collisions. Given a certain initial velocity, any grain will like a ghost continue unhindered with the same velocity through the cloud for all time, and thus have infinite mean free path. The model should definitely be taken with a grain of salt.

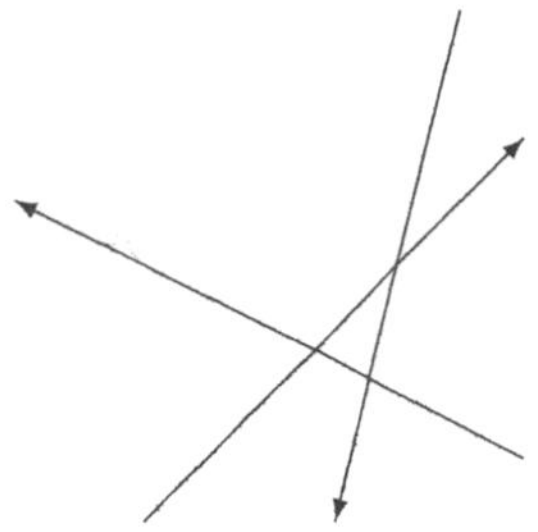

In a cloud of non-interacting particles, and in the absence of gravity, every particle moves forever in a straight line with constant velocity.

Explosion field

Suppose that the cloud is created in an explosion where the fragments have stopped interacting immediately after the event and now move freely away from each other. Since there are neither contact forces nor body forces, we have $\boldsymbol{f}^* = \boldsymbol{0}$, and every material particle is unaccelerated. Consequently, the comoving acceleration (12.23) must vanish everywhere:

$$\frac{\partial \boldsymbol{v}}{\partial t} + (\boldsymbol{v} \cdot \nabla)\boldsymbol{v} = 0. \tag{12.31}$$

In spite of this being the simplest possible dynamical equation, it looks complicated enough, and if presented without any other explanation, we would have some difficulty solving it because of its non-linearity. Underneath we know, however, that it only implements the law of inertia, with all particles moving with constant velocity along straight lines.

Suppose the explosion happened at the point $\boldsymbol{x} = \boldsymbol{0}$ at time $t = 0$. The fragments were almost instantly given random velocities in all directions, and then stopped interacting. After that the fragments will be separated according to their velocities with the fastest fragments being farthest away, so that at time t, the velocity of any fragment found at $\boldsymbol{x}$ must be

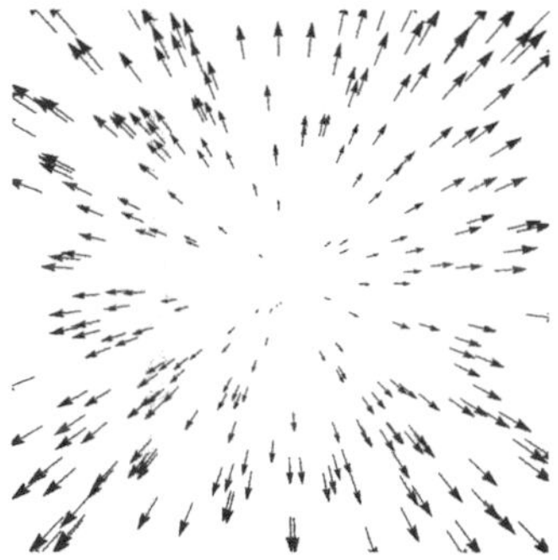

Explosion fragments become separated according to initial velocity, because those fragments that accidentally have the largest initial velocity will ever after be farthest away from the center.

$$\boxed{\boldsymbol{v}(\boldsymbol{x}, t) = \frac{\boldsymbol{x}}{t},} \tag{12.32}$$

independent of the details of the explosion. Clearly, it is the local creation event that smoothes the field. To see that the field indeed satisfies (12.31), we calculate the x-component,

$$\begin{aligned}\frac{\partial v_x}{\partial t} + (\boldsymbol{v} \cdot \nabla)v_x &= \frac{\partial (x/t)}{\partial t} + \frac{x}{t}\frac{\partial (x/t)}{\partial x} + \frac{y}{t}\frac{\partial (x/t)}{\partial y} + \frac{z}{t}\frac{\partial (x/t)}{\partial z} \\ &= -\frac{x}{t^2} + \frac{x}{t^2} \\ &= 0,\end{aligned}$$

and similarly for the y- and z-components.

Hubble's law

The explosion field (12.32) is of the same form as Hubble's law, which states that all galaxies move away from us (and each other) with speeds that are proportional to their distances, in our notation written as

$$\boldsymbol{v} = H_0\,\boldsymbol{x}, \tag{12.33}$$

where H_0 is called the *Hubble constant*. In general relativity, this law is understood as a consequence of a uniform expansion of space itself since the initial Big Bang. In the primitive model used here, with gravity ignored, the age of the universe equals the inverse of the Hubble constant $t_0 = 1/H_0$.

Although first determined in 1929, it has until recently been very difficult for astronomy to settle on a reliable experimental value for the Hubble constant. The latest published present-day value is $H_0 = 73.2$ kilometers per second per megaparsec with an uncertainty of about 4% [SBD&07], or in SI-units $H_0 = 2.37(8) \times 10^{-18}\ \mathrm{s}^{-1}$. The inverse comes to $t_0 = 1/H_0 = 13.4(5)$ billion years, but the problem is that the age of the universe determined from cosmology is model dependent, so the true age can be quite different from t_0. In the last decade it has, however, become clear that the universe appears to be nearly flat [SBD&07], which in turn makes the age of the universe nearly equal to the inverse Hubble constant.

Edwin Powell Hubble (1889–1953). American astronomer. Although originally obtaining a BS degree in astronomy and mathematics, he continued in law but only practiced it for a year before turning back to astronomy. Demonstrated in 1924 that the observed spiral clouds were galaxies, far outside the Milky Way. Discovered the distance-velocity relationship that now carries his name in 1929.

* 12.6 Application: Newtonian cosmology

Continuing our investigation of the expanding universe, we now wish to include the gravitational field in the dynamical equation (12.26) but still allow no contact forces, so that $f^* = \rho g$. The equation of motion then becomes

$$\frac{\partial v}{\partial t} + (v \cdot \nabla)\, v = g \tag{12.34}$$

instead of (12.31). Despite being non-relativistic, this model captures essential elements of cosmology, although a proper understanding does require general relativity [Weinberg 1972]. The following analysis is in particular indebted to [Weinberg 2008].

Cosmic democracy

In days of old, the Earth was thought to be at the center of the universe. Since Copernicus, this thinking has been increasingly replaced by the more "democratic" view that the Earth, the Sun, and the Galaxy are but common members of the universe of no particular distinction. The end of this line of thought is the extreme Copernican view that for cosmological considerations, every place in the universe is as good as any other. As we shall see, this *Cosmological Principle* or *Principle of Cosmic Democracy* is quite useful (for a recent discussion, see [Ell08]).

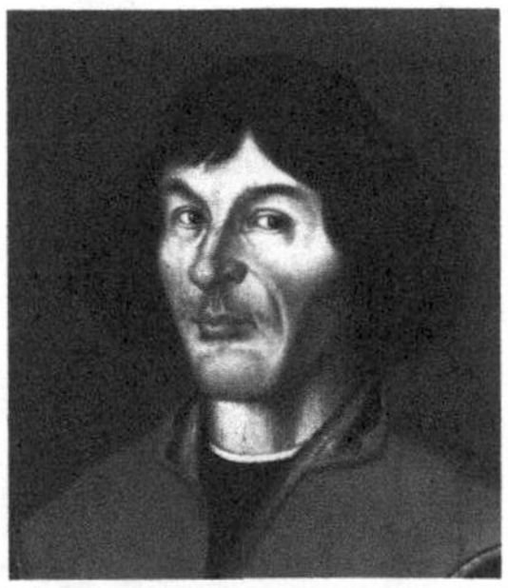

Nicolaus Copernicus (1473–1543)). Polish astronomer. Studied when young both medicine and astronomy. Revolutionized the understanding of the solar system with his anonymous booklet *Little Commentary* from 1514, and especially with his life's work, the 200-page book *De revolutionibus orbium coelestium*, published in Latin in the year of his death (1543). His views of the solar system were widely criticized but supported by both Kepler and Galileo. (Source: Wikimedia Commons.)

Mass density: It immediately follows from this principle that at a particular instant of time t, the (average) mass density cannot depend on where you are,

$$\boxed{\rho(x,t) = \rho(t),} \tag{12.35}$$

and must thus be the same everywhere.

Velocity field: The Hubble expansion of the universe,

$$\boxed{v(x,t) = H(t)\, x,} \tag{12.36}$$

with a time-dependent Hubble "constant", $H(t)$, does not look democratic because it seems to single out the center of the coordinate system. It is, in fact, completely democratic, because

$$v(x,t) - v(y,t) = H(t)(x - y). \tag{12.37}$$

This means that an observer in a galaxy at point y will also see the other galaxies recede from him according to the same Hubble law as ours.

Field of gravity: Can gravity be democratic? In Newtonian cosmology it is not possible to view the universe as a homogeneous whole when it comes to gravity. What is gravity in an infinite universe? Symmetry would seem to argue that it should vanish because there is as much matter pulling from one side as from the opposite, but is that right?

To get around this problem, let us for a while think of the universe as a huge sphere with vacuum outside, and let us put the origin of the coordinate system at the center of this sphere. In elementary physics it is shown that the strength of gravity at a given point $\boldsymbol{x}$ depends only on the amount of mass, $M(r) = \frac{4}{3}\pi r^3 \rho$, inside the sphere with radius $r = |\boldsymbol{x}|$, whereas one may forget the mass outside this radius. In other words, the field of gravity at the point $\boldsymbol{x}$ is

$$\boldsymbol{g}(\boldsymbol{x}, t) = -G\frac{M(r)}{r^2}\hat{\boldsymbol{e}}_r = -\frac{4\pi}{3}G\,\rho(t)\,\boldsymbol{x}. \tag{12.38}$$

Consider now a freely falling galaxy at another point $\boldsymbol{y}$ at time t. Its acceleration equals the local gravitational acceleration field, $\boldsymbol{g}(\boldsymbol{y}, t)$, but *Einstein's Equivalence Principle* tells us that a freely falling observer at $\boldsymbol{y}$ cannot determine the local field of gravity. The apparent acceleration of another freely falling galaxy at the point $\boldsymbol{x}$ therefore becomes

$$\boldsymbol{g}(\boldsymbol{x}, t) - \boldsymbol{g}(\boldsymbol{y}, t) = -\frac{4}{3}\pi G\rho(t)(\boldsymbol{x} - \boldsymbol{y}), \tag{12.39}$$

as if the galaxy at $\boldsymbol{y}$ were also at the center of the universe. An observer at $\boldsymbol{y}$ will—just like us—think that his universe is a huge sphere centered on him, even if we know that this is not really the case. In general relativity, this problem happily goes away.

Cosmological equations

Using $\nabla \cdot \boldsymbol{x} = 3$, we obtain from (12.36) and the equation of continuity (12.12)

$$\boxed{\dot{\rho} = -3H\rho,} \tag{12.40}$$

where as before a dot denotes differentiation with respect to time. Similarly, inserting (12.36) into (12.34) and using $(\boldsymbol{x} \cdot \nabla)\boldsymbol{x} = \boldsymbol{x}$, this equation becomes, after removal of a common factor,

$$\boxed{\dot{H} + H^2 = -\frac{4\pi}{3}G\rho.} \tag{12.41}$$

Newtonian cosmology thus reduces to just two coupled ordinary differential equations for the mass density and the Hubble "constant". Note that the reference to the center of the universe has disappeared completely, and we may now again think of a truly infinite universe with equal rights for all observers.

The cosmic scale factor

The simplest way to solve these equations is by introducing a new quantity with the dimension of length, $a(t)$, called the *cosmic scale factor*, satisfying

$$\frac{\dot{a}}{a} = H. \tag{12.42}$$

From the equation of continuity (12.40) we get

$$\frac{d}{dt}(\rho a^3) = \dot{\rho}a^3 + 3\rho a^2\dot{a} = -3H\rho a^3 + \rho\,3a^2 Ha = 0,$$

and this shows that the mass, $M = \frac{4}{3}\pi\rho(t)\,a(t)^3$, in an expanding sphere of radius $a(t)$ is constant in time.

Eliminating H from (12.41), we obtain the following differential equation for cosmic scale factor:

$$\ddot{a} = -G\frac{M}{a^2}. \tag{12.43}$$

Evidently, it is identical to the equation of motion for a test particle at a distance a from the center, moving radially in the gravitational field of the mass M inside this distance.

Critical density

The "equation of motion" (12.43) implies that the "energy" of a unit-mass test particle is:

$$E = \frac{1}{2}\dot{a}^2 - \frac{GM}{a}, \tag{12.44}$$

It must be conserved in the time evolution of the scale factor, that is, $\dot{E} = 0$, as may easily be verified. Eliminating the mass, and using (12.42) to eliminate $\dot{a}$, it may be written in the form

$$E = \frac{4}{3}\pi G a^2 (\rho_c - \rho), \tag{12.45}$$

where ρ_c is the *critical density*,

$$\boxed{\rho_c = \frac{3H^2}{8\pi G}.} \tag{12.46}$$

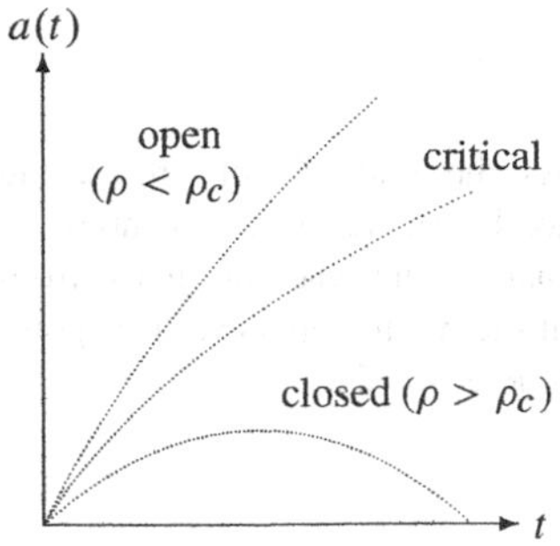

Time evolution of the cosmic scale factor depending on the actual average mass density compared to the critical density.

It can be calculated from present-day observation of the Hubble constant, $H = H_0$, and using the WMAP value $H_0 = 73.2$ km s^{-1} Mpc^{-1} [SBD&07], one finds $\rho_c \approx 1.00(4) \times 10^{-26}$ kg m^{-3}, corresponding to about six protons per cubic meter.

From the particle analogy, we know that the scale factor will "escape" to infinity for $E > 0$, but turn around and "fall back" for $E < 0$. This means that the expansion will continue forever for $\rho < \rho_c$, but eventually must turn around and become a contraction if $\rho > \rho_c$. The two types of universe are called *open* or *closed*, respectively. For the critical value $\rho = \rho_c$, the expansion also continues forever but at the slowest possible pace.

Age of the universe

The energy equation (12.44) may be solved for $\dot{a}$ with the result

$$\dot{a} = \sqrt{2\left(E + \frac{GM}{a}\right)}.$$

Demanding that $a = 0$ for $t = 0$, the solution is given implicitly by

$$t = \int_0^a \frac{dr}{\sqrt{2(E + GM/r)}}, \tag{12.47}$$

which must be the time elapsed since the scale factor was zero, i.e., since the Big Bang.

For the critical case $E = 0$, called the *flat cosmology*, the integral becomes

$$t = \frac{2}{3}\frac{a^{3/2}}{\sqrt{2GM}}. \tag{12.48}$$

Using $H = \dot{a}/a = \sqrt{2GM}a^{-3/2}$, it follows that $tH = \frac{2}{3}$. With the present-day value of the Hubble constant, $H = H_0$, the age of the universe would be only about 9 billion years, disagreeing strongly with the age of globular clusters and other data (reviewed in [Weinberg 2008, p. 59ff]). The inevitable conclusion is that this model cannot be right.

The cosmological constant

The problem is, as mentioned earlier, that the actual age of the universe determined from cosmology is model dependent. In the last decade, observations have convincingly shown that the expansion of the universe is accelerating (see [Weinberg 2008, p. 45ff]), rather than decelerating as gravity alone would demand. The cause of the acceleration is not (yet) known, but the simplest model, which implicitly contains acceleration, goes back to Einstein himself. In 1917 he introduced the so-called *cosmological constant* to allow the construction of a static model of the universe that would remain unchanged for all time. Hubble's discovery of the actual expansion closed for many years the case for the cosmological constant but recent discoveries have completely changed that.

The non-relativistic framework at our disposal in this book does not allow us to carry out a proper analysis of the relation of the cosmological constant to the expansion of the universe, but we can nevertheless taste its flavor in a naive model. Suppose the universe, in addition to ordinary matter, is filled with a ghostly "vacuum material" with a positive mass density, ρ_v, constant all over the universe and independent of time[3]. Since the total "vacuum mass" in a sphere of radius $a(t)$ is $M_v(t) = \frac{4\pi}{3}a(t)^3\rho_v$, the particle energy (12.44) should be replaced by

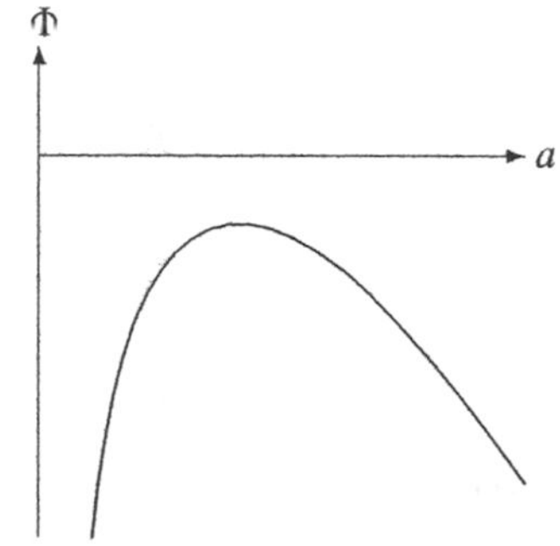

As the scale factor grows, the total gravitational potential of matter and vacuum goes from attractive to repulsive at a point where $\rho = 2\rho_v$.

$$E = \frac{1}{2}\dot{a}^2 - \frac{G(M + M_v)}{a} = \frac{1}{2}\dot{a}^2 - \frac{GM}{a} - \frac{4\pi}{3}Ga^2\rho_v. \tag{12.49}$$

The last two terms make up the modified gravitational potential.

Requiring this energy to be constant, we obtain the equation of motion for a test particle,

$$\ddot{a} = -\frac{GM}{a^2} + \frac{8\pi}{3}G\rho_v a = -\frac{4\pi}{3}G(\rho - 2\rho_v)a, \tag{12.50}$$

where ρ as before is the mass density of ordinary matter, still obeying the continuity equation (12.40). For sufficiently small mass density, $\rho < 2\rho_v$, the gravitational deceleration apparently turns into a *gravitational acceleration*. From the above equations we can see that it happens because the amount of "vacuum material" inside a sphere of radius a grows with time while the amount of ordinary matter stays constant.

Expressing the particle energy in terms of the critical density, it becomes

$$E = \frac{4\pi}{3}Ga^2(\rho_c - (\rho + \rho_v)). \tag{12.51}$$

The critical case, $E = 0$, which for a number of reasons is favored theoretically, now requires

$$\rho + \rho_v = \rho_c. \tag{12.52}$$

The recent WMAP observations are consistent with[4] $\rho \approx 0.26\rho_c$ and thus $\rho_v \approx 0.74\rho_c$, implying that now $\rho \approx 0.35\rho_v$. At the transition from deceleration to acceleration in (12.50), the ordinary mass density was twice the vacuum density, $\rho = 2\rho_v$, which is $2/0.35 \approx 5.7$ times larger than now. Since the mass density varies as a^{-3}, the cosmic scale factor a was about $5.7^{-1/3} = 0.56$ times smaller at the transition than it is now. An integral similar to the one in (12.47), but including the vacuum energy shows that the transition took place when the universe was very close to half as old as it is now (Problem 12.12).

[3]The vacuum mass density ρ_v is related to the cosmological constant Λ by $\rho_v = \Lambda c^2/8\pi G$. It is equivalent to a constant vacuum energy density $\rho_v c^2$, now mostly called "dark energy" to distinguish it from "ordinary matter" for which the density decreases as the universe expands.

[4]It must be emphasized that the mass density ρ includes the mass of the "cold dark matter" that appears to be necessary for the consistency of galactic models. The actual baryonic "luminous matter" that we ourselves and everything around us is made from, is only about 4 % of ρ_c.

Problems

12.1 A velocity field is given by $\boldsymbol{v}(\boldsymbol{x}, t) = (a, bt, 0)$, where a and b are constants. Determine the **(a)** streamlines, **(b)** particle orbits, and **(c)** streaklines. Try to convince yourself that this is indeed how the smoke trail from a chimney would behave. Draw the various curves for $a = b = 1$.

12.2 A velocity field is given by $\boldsymbol{v} = a(\cos\omega t, \sin\omega t, 0)$, where a and ω are constants. The field resembles that of a pad sander, used to smooth wooden surfaces. **(a)** Draw streamlines, **(b)** particle orbits, and **(c)** streaklines.

12.3 The wind suddenly turns from south to west. Draw streamlines, particle orbits, and streaklines before and after the event.

12.4 A water pipe with diameter 1 inch branches into two pipes with diameters 3/4 inch and 1/2 inch. Water is tapped from the largest branch at double the rate as from the other. What is the ratio of velocities in the pipes?

12.5 Calculate the time-derivative of $\rho(\boldsymbol{x}(t), t)$, where $\boldsymbol{x}(t)$ is a particle orbit, and show that it is identical to the comoving derivative.

12.6 Consider an incompressible steady flow in a stream with constant depth, $z = d$, bounded on one side by a straight line $y = 0$, and on the other side by a curve $y = h(x)$, which is slowly varying $|dh/dx| \ll 1$. **(a)** Calculate the average flow velocity in the x-direction (for fixed x). **(b)** Approximately calculate the comoving acceleration in the flow. **(c)** What should the shape of the curve be in order for the comoving acceleration to be independent of x?

12.7 Consider an incompressible steady flow in a circular tube along the x-axis with a slowly changing radius $r = a(x)$, satisfying $|da(x)/dx| \ll 1$. **(a)** Calculate the average flow velocity in the x-direction. **(b)** Approximately calculate the comoving acceleration in the tube, and **(c)** determine what shape of tube will lead to constant comoving acceleration.

12.8 Consider a comoving material particle. Use the Jacobi determinant of the infinitesimal displacement from $\boldsymbol{x}$ to $\boldsymbol{x}' = \boldsymbol{x} + \boldsymbol{v}\delta t$ to derive (12.17).

* **12.9** The right-hand side of (12.27c) is a linear algebraic equation for the velocity field. Show that the solution may be written (in dyadic or matrix notation) as

$$\boldsymbol{v} = \frac{\partial \boldsymbol{u}}{\partial t} \cdot (1 - \nabla \boldsymbol{u})^{-1}.$$

Although this equation in principle could be used to permanently eliminate the velocity field, such a step would introduce unpleasant inverse matrices everywhere.

12.10 Show that the dynamics of continuous matter (12.26) may be recast as a *continuity equation for momentum* (see also Chapter 20),

$$\frac{\partial(\rho \boldsymbol{v})}{\partial t} + \nabla \cdot (\rho \boldsymbol{v} \boldsymbol{v}) = \boldsymbol{f}^*.$$

Hint: Use index notation for the proof.

* **12.11** Prove the form invariance of the equations of motion (12.27) under the Galilean transformation (12.30).

* **12.12** Calculate the lifetime of the critical universe when vacuum energy is included.

Problems

13 Nearly ideal flow

The most important fluids of our daily life, air and water, are lively and easily set into irregular motion. Getting out of a bathtub creates visible turbulence in the soapy water, whereas we have to imagine the unruly air behind us when we jog. Internal friction, or *viscosity*, seems to play only a minor role in these fluids. Other fluids, like honey and grease, are highly viscous, do not become turbulent in daily use, and would certainly be very hard to swim in.

The earliest modern quantitative model of fluid behavior goes back to Euler about 250 years ago and did not include viscosity. Although introduced by Newton, viscosity first entered fluid mechanics in its modern formulation about a century later. Fluids with no viscosity are said to be *inviscid*, but are also called *ideal* or *perfect*. An ideal fluid does not hang on to solid surfaces but is able to slip along container walls with finite velocity, whereas a real fluid has to adjust its velocity field to match the surface speed of the solid objects it encounters.

Leonhard Euler (1707–1783). Swiss mathematician who made fundamental contributions to calculus, geometry, number theory, and to practical ways of solving mathematical problems. His books on differential calculus (1755) and integral calculus (1768–1770) have been especially useful for physics. (Source: Wikimedia Commons.)

Being lively or sluggish is not an absolute property of a fluid but rather a condition of the circumstances under which it flows. Lava may be very sluggish in small amounts but when it streams down a mountainside, it appears quite lively. In Chapter 15 we shall learn that fluid flow may be characterized by a dimensionless number, called the *Reynolds number*, which is large compared to unity for lively flow and small for sluggish flow. Although truly ideal fluids do not really exist, except for a component of superfluid helium close to zero kelvin, viscous fluids may nevertheless flow with such high Reynolds number that they behave as *nearly ideal*. But independent of how large the Reynolds number of the flow, there will always be thin viscous *boundary layers* hugging the walls of fluid conduits.

13.1 Euler equation for incompressible ideal flow

In an ideal homogeneous fluid, the only contact force is pressure, but in distinction to hydrostatics where the pressure is in balance with the body forces (gravity), the effective density of force $\boldsymbol{f}^* = \rho \boldsymbol{g} - \nabla p$ may now be non-vanishing. Inserted into the general dynamic equation (12.26) on page 198 we obtain after dividing by the constant density $\rho(\boldsymbol{x}, t) = \rho_0$

$$\boxed{\frac{\partial \boldsymbol{v}}{\partial t} + (\boldsymbol{v} \cdot \nabla)\, \boldsymbol{v} = \boldsymbol{g} - \frac{\nabla p}{\rho_0}, \qquad \nabla \cdot \boldsymbol{v} = 0.} \tag{13.1}$$

These two *Euler equations* govern the dynamics of incompressible ideal fluid, also called *Euler fluid*. Compressible ideal flow with variable density will be analyzed in Chapter 14.

The role of pressure

The Euler equations constitute a closed set of four equations for the four fields, v_x, v_y, v_z, and p (assuming that $\boldsymbol{g}$ is known). Although it appears that we miss an explicit equation of motion for the pressure, it is in fact determined indirectly by the divergence condition. To see this we calculate the divergence of the first Euler equation, and obtain,

$$\nabla^2 p = \rho_0 \nabla \cdot \boldsymbol{g} - \rho_0 \nabla \cdot ((\boldsymbol{v} \cdot \nabla)\boldsymbol{v}). \tag{13.2}$$

This is nothing but the Poisson equation, so well known from electrostatics and Newtonian gravity. Together with suitable boundary conditions, it can always be solved and determines the pressure field at a given time from the velocity field (and gravity) at the same time.

Non-locality: The solutions to the Poisson equation are, however, *non-local*, meaning that the pressure in a point depends on the velocity field everywhere, in the same way as Coulomb's law expresses the electrostatic potential in a point as the integral over the instantaneous distribution of electric charge in all other points. Thus, *any change in the velocity field is instantly communicated to the rest of the fluid via the pressure.* This is an unavoidable consequence of incompressibility; but like true rigidity, true incompressibility is an ideal that cannot be reached with real materials where the velocity of sound sets an upper limit to the small-amplitude signal propagation speed. The rapid global communication of pressure changes in nearly incompressible fluids is known from everyday experience where the closing of a defect water faucet can result in rather noisy "water-hammer" responses from the house piping. The non-locality of pressure also creates problems in numerical simulations of incompressible flow (Chapter 19).

Boundary conditions

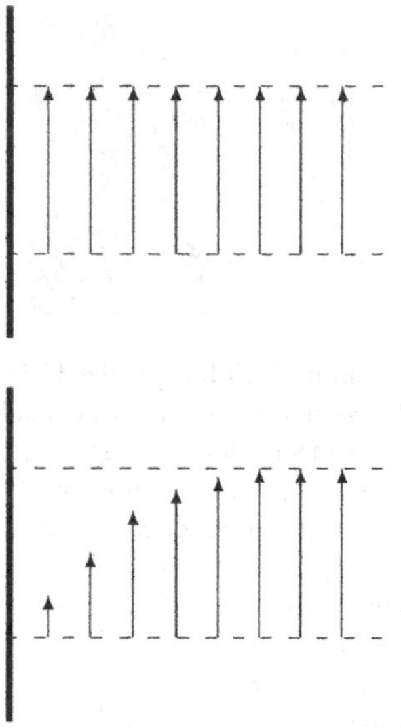

Top: **Top:** Ideal fluid. The tangential component of the velocity field is non-vanishing all the way to the solid wall on the left. **Bottom:** Viscous fluid. The tangential component of the velocity field rises linearly from a wall and joins smoothly with the flow at large.

Partial differential equations need boundary conditions to connect the fields on both sides of a material interface. As in hydrostatics, Newton's Third Law demands that the pressure—in the absence of surface tension—must be continuous across the interface. Furthermore, since fluid cannot accumulate in a material interface, the velocity field must be continuous along the normal to the interface. There is no restriction on the density, which usually jumps at a material interface. The complete set of *jump conditions* valid at a material interface may accordingly be written (with a bracket [·] to indicate the difference between the two sides) as

$$[\boldsymbol{v} \cdot \boldsymbol{n}] = 0, \qquad [p] = 0, \tag{13.3}$$

where $\hat{\boldsymbol{n}}$ is the local normal to the interface. In truly ideal flow, the tangential velocity component is unrestricted. Ideal fluids are thus able to slip along container surfaces with finite tangential velocity, and there may even be internal surfaces of discontinuity with different tangential flow velocities on the two sides.

Slip or no-slip: Tangential flow discontinuities are, however, always unstable to the introduction of even the smallest amount of viscosity. The omnipresent viscosity of real fluids demands that the tangential velocity must also be continuous at a material interface (see Chapter 15), and this requirement unavoidably causes boundary layers to arise near all conduit walls.

What is the use of ideal fluids?

One may rightly ask why one should study ideal fluids at all when they do not really exist. Would it not be better from the outset to focus on viscous fluids? Some authors in fact do that [Batchelor 67], and afterward proceed to discuss nearly ideal flow as a limiting case. Although natural from a theoretical point of view, we humans do live in a world of nearly ideal flows of air and water, and we shall for this reason begin by discussing the many practical uses of the Euler equations while along the way keeping an eye on the unavoidable deviations from ideality due to viscosity.

* 13.2 Application: Collapse of a spherical cavity

The simplest non-trivial solution to the time-dependent Euler equations is the collapse of a spherical cavity. Cavities may arise in liquids subject to rapid strong pressure variations, for example caused by the churning of a ship's propellers (see [Brennen 1995]). The quantitative analysis of the growth and subsequent collapse of spherical cavities goes back to Rayleigh in 1917 [Ray17]. In recent years, interest has focused on the related problem of tiny spherical gas bubbles forced to oscillate rapidly by imposing a radial sound wave on the liquid. Under the right circumstances, a short (picosecond) flash of light will be emitted by the gas at the peak of bubble compression in each cycle. This fascinating phenomenon is called *sonoluminescence* [PP77, TL03, NAT&05].

Here we shall first consider the general problem of spherical cavity dynamics and then specialize to the collapse of an empty cavity.

Ideal incompressible radial flow

A time-dependent purely radial flow is described by a velocity field $\boldsymbol{v} = v(r,t)\,\hat{\boldsymbol{e}}_r$. We do not have to solve the continuity equation to find v because mass conservation guarantees that the mass flux $4\pi r^2 v(r,t)\rho_0$ is the same for all r. The Euler equations now reduce to

$$\frac{\partial v}{\partial t} + v\frac{\partial v}{\partial r} = -\frac{1}{\rho_0}\frac{\partial p}{\partial r}, \qquad v = \frac{C(t)}{r^2}, \tag{13.4}$$

where the "constant" $C(t)$ does not depend on r. Inserting v in the first, we obtain $\partial p/\partial r$ as an explicit function of r, which integrates to

$$p = p_0(t) + \rho_0\left(\frac{\dot{C}(t)}{r} - \frac{C(t)^2}{2r^4}\right), \tag{13.5}$$

with the overdot indicating differentiation with respect to time. Evidently $p_0(t)$ is the pressure at spatial infinity ($r = \infty$) and may, as shown, in principle depend on time.

Let now $R(t)$ be the instantaneous radius of the surface of the cavity, $\dot{R}(t)$ the surface velocity, and $P(t)$ the surface pressure in the liquid. Setting $r = R$, $p = P$, and using that $C = r^2 v = R^2\dot{R}$, we obtain the second-order differential equation with respect to time,

$$\boxed{R\ddot{R} + \frac{3}{2}\dot{R}^2 = \frac{\Delta p}{\rho_0},} \tag{13.6}$$

where $\Delta p = P - p_0$ is the possibly time-dependent pressure difference between the cavity surface and the distant environment. To solve it we rewrite it as

$$\frac{d}{dt}\left(R^3\dot{R}^2\right) = 2\frac{\Delta p}{\rho_0}R^2\dot{R}. \tag{13.7}$$

Assuming Δp is a known function of R (but not of $\dot{R}$ or t), this equation can be solved by integrating both sides with respect to t and using $\dot{R}\,dt = dR$.

This will be the case for an empty cavity as well as for a gas-filled cavity in global thermodynamic equilibrium with uniform pressure, density, and temperature. During the extremely fast final compression of a gas bubble in a sonoluminescence experiment, the assumption of global thermodynamic equilibrium cannot be maintained, and the internal dynamics of the gas in the cavity will have to be taken into account.

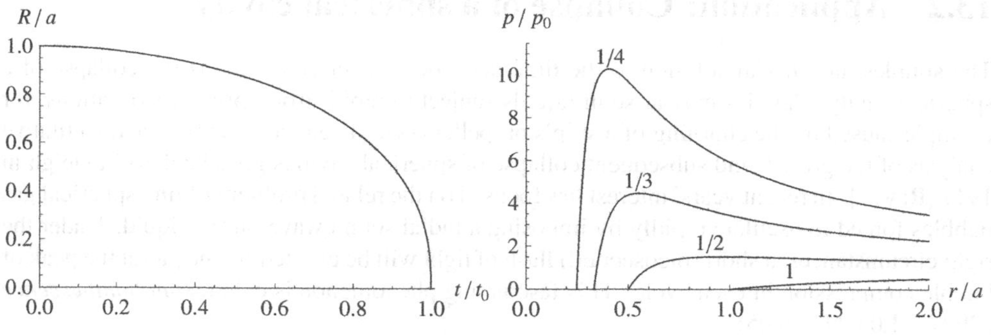

Figure 13.1. Collapse of an empty cavity in dimensionless units. **Left:** The variation of the cavity radius with time. **Right:** The pressure as a function of r for $R/a = 1, 1/2, 1/3, 1/4$. The maximum pressure diverges as the cavity radius approaches zero.

Empty cavity

Consider now an empty cavity with positive constant external pressure $p_0 > 0$ and constant internal pressure $P = 0$, so that $\Delta p = -p_0$ (disregarding surface tension; see Problem 13.1). Denoting the initial radius $R(0) = a$ and requiring it to be maximal, $\dot{R}(0) = 0$, integration of (13.7) yields (as may easily be verified)

$$\dot{R}^2 = \frac{2p_0}{3\rho_0}\left(\frac{a^3}{R^3} - 1\right). \tag{13.8}$$

Collapse corresponds to the negative root, $\dot{R} < 0$, and integration yields t as function of R:

$$t = \int_R^a \frac{dR}{\dot{R}} = \sqrt{\frac{3\rho_0}{2p_0}} \int_R^a \left(\frac{a^3}{r^3} - 1\right)^{-1/2} dr = \frac{a}{c}\sqrt{\frac{3}{2}} \int_{R/a}^1 \frac{dx}{\sqrt{x^{-3} - 1}}. \tag{13.9}$$

In the last step we also introduced $c = \sqrt{p_0/\rho_0}$, which has the dimension of velocity.
The time for the cavity to collapse to $R = 0$ can be evaluated analytically:

$$t_0 = \frac{a}{c}\sqrt{\frac{3\pi}{2}} \frac{\Gamma(5/6)}{\Gamma(1/3)} = 0.915\frac{a}{c}. \tag{13.10}$$

The pressure is finally obtained from (13.5) with $C = R^2\dot{R}$, and may be written as

$$\frac{p}{p_0} = 1 - \frac{R}{r} + \frac{1}{3}\frac{R}{r}\left(1 - \frac{R^3}{r^3}\right)\left(\frac{a^3}{R^3} - 1\right). \tag{13.11}$$

The variation of the cavity radius with time and the liquid pressure with radial distance and time are plotted in Figure 13.1. Since the maximal pressure for $R \to 0$ diverges as $p_{\max} \approx p_0 2^{-8/3} a^3/R^3$, all constant material properties are eventually challenged in the collapse. The assumption of incompressibility ceases to be valid when the maximal pressure approaches the bulk modulus K of the liquid. The above expression for the maximal pressure shows that the compressibility radius where $p = K$ is $R_K = 2^{-8/9} a (p_0/K)^{1/3}$.

Example 13.1 [Empty cavity in water]: An empty cavity in water with $\rho_0 = 1\ \text{g cm}^{-3}$ has initial radius $a = 1$ cm and external pressure $p_0 = 1$ bar. The velocity scale becomes $c = 10\ \text{m s}^{-1}$ and the collapse takes $t_0 = 0.9$ ms. For water we have $K \approx 23$ kbar, so the compressibility radius becomes $R_K = 2^{-8/9} a (p_0/K)^{1/3} \approx 0.2$ mm. This radius is reached only 24 ns before the collapse is complete. Here we have disregarded the slowdown due to the unavoidable presence of water vapor and gas residuals in the cavity.

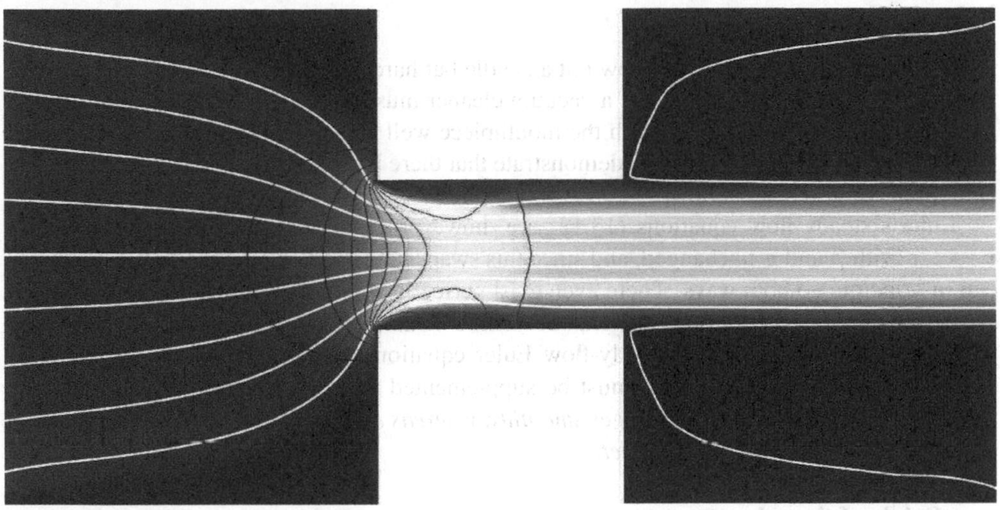

Figure 13.2. Two-dimensional simulation of nearly ideal steady flow through a channel (in the absence of gravity). The Reynolds number is about 500. The incompressible fluid enters at the left of the image under uniform pressure and exits at the right with a lower, also uniform, pressure. The gray-level represents the speed of the fluid, and the streamlines indicate the flow direction as well as the flow speed (by their density). Pressure contours are also shown and generally cross the streamlines. There are clearly visible boundary layers near the walls of the narrow channel, and the uniformly spaced streamlines indicate that there is nearly homogeneous flow in the central region of the channel and in the outlet jet. The boundary layers separate from the channel walls at the outlet and continue along the two edges of the jet, slowly causing it to widen. Near the walls of the exit region there is a weak secondary flow running against the main flow. For decreasing viscosity—or better, for increasing Reynolds number—this general picture will persist, although the boundary layers become progressively thinner until turbulence unavoidably sets in (see Chapter 28).

13.3 Steady incompressible ideal flow

Steady incompressible flow $\boldsymbol{v}(\boldsymbol{x})$ in a static gravitational field $\boldsymbol{g}(\boldsymbol{x})$ obeys the time independent Euler equations,

$$\boxed{(\boldsymbol{v}\cdot\nabla)\,\boldsymbol{v} = \boldsymbol{g} - \frac{\nabla p}{\rho_0}, \qquad \nabla\cdot\boldsymbol{v} = 0.} \tag{13.12}$$

Truly steady flow is like true incompressibility, an idealization only valid to a certain approximation. A river may flow steadily for days and weeks, while over several months seasonal changes in rainfall makes its water level rise and subside. Although strictly speaking such flow is unsteady, it may be called *nearly steady* or *quasi-stationary*.

One might think that these equations apply only to nearly ideal liquids with large bulk modulus, for example water. But as we shall prove later, a *fluid in steady flow is effectively incompressible when the flow velocity is everywhere much smaller than the speed of sound in the fluid.* Even when running, biking, or driving a car, the streaming air around you may be taken to be incompressible, whereas there will be considerable compression of the air at the front of a modern passenger jet. A simple intuitive explanation of this theorem is that below the speed of sound, a local increase in pressure caused by a moving object tends to drive the fluid out of the way rather than compressing it. The formal proof is given in Chapter 14.

Inlet-outlet asymmetry

Everybody knows that it is easy blow out a candle but hard and a bit dangerous to try to suck it out. Likewise, the mouthpiece of a vacuum cleaner must be held close to the floor while a garden leaf-blower can be used with the mouthpiece well above the ground. Such examples, and the simulations in Figure 13.2, demonstrate that there is strong asymmetry between inlets and outlets in nearly ideal flow.

Euler's steady-flow equations (13.12) are, however, invariant under the transformation $\boldsymbol{v} \to -\boldsymbol{v}$ with p and $\boldsymbol{g}$ unchanged, and since this swaps inlets and outlets, the flow asymmetry can in principle go both ways. There even exist perfectly symmetric solutions, for example a "river" of fluid passing through the channel with constant velocity and pressure everywhere (see the margin figure). The steady-flow Euler equations are by themselves incapable of selecting the correct solution but must be supplemented with the rule that *inflow toward a channel from a wider region converges smoothly, whereas outflow from a channel into a wider region continues as a well-defined jet.*

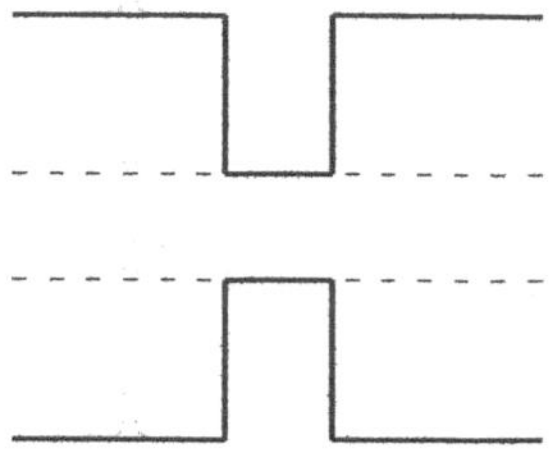

Symmetric ideal flow solution through a narrow channel. The pressure is constant everywhere and the velocity is constant in the "river" and vanishes outside.

Origin of the rule: To understand the rule, we must appeal to the time-dependent Euler equations (13.1). Starting with fluid at rest ($\boldsymbol{v} = \boldsymbol{0}$), the first Euler equation becomes $\partial \boldsymbol{v}/\partial t = -\nabla p$ (in the absence of gravity). All the fluid is thus accelerated in the direction of the negative pressure gradient, that is, from left to right in Figure 13.2. Initially the velocity is nearly zero, and the flow is as far from being ideal as possible. Even the tiniest viscosity will play an important role and generate a creeping flow (see Chapter 17) that converges smoothly toward the inlet and fans out smoothly again after passing the outlet. As the velocity increases, the advective (inertial) acceleration grows faster than the tiny viscous resistance and soon overcomes it. The inlet flow stays smooth while the inertia of the fluid in the narrow channel makes the outlet flow shoot out as a jet that only loses a small amount of its large kinetic energy to viscous friction along its surface. The jet stays intact for a while, although it will in the end spread out and decay further downstream.

Daniel Bernoulli (1700–1782). Dutch-born mathematician who made major contributions to the theory of elasticity, fluid mechanics and the mechanics of musical instruments. Bernoulli pointed out the relation between pressure and velocity in the world's first book on hydrodynamics, *Hydrodynamica*, which he published in 1738. (Source: Wikimedia Commons.)

Bernoulli's theorem

The negative sign of the pressure gradient in Euler's steady-flow equation shows that in the absence of gravity, a flow accelerating in a certain direction must be accompanied by a drop in pressure in the same direction. As a consequence, regions of high flow velocity will generally have a lower pressure than regions of low velocity.

Bernoulli's theorem (or Bernoulli's principle) implements this by stating that in ideal incompressible steady flow, the field,

$$\boxed{H = \tfrac{1}{2}\boldsymbol{v}^2 + \Phi + \frac{p}{\rho_0},} \tag{13.13}$$

is *constant along streamlines*. This field—which we shall call the Bernoulli field—may be viewed as an extension of the effective potential (2.36) on page 32 to fluid in motion. The theorem is extended to compressible fluids in Section 14.2.

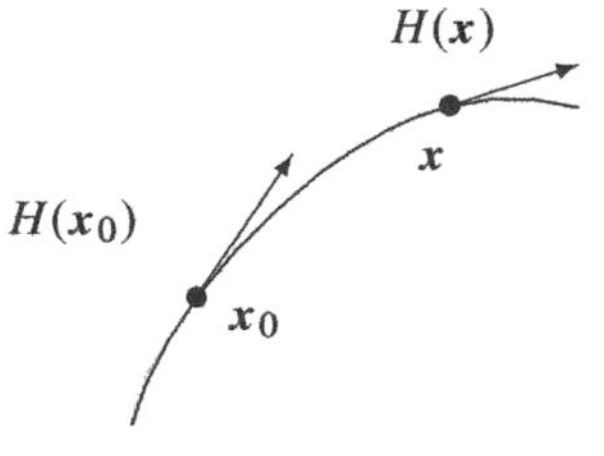

In steady flow, H is constant along a streamline, i.e., $H(\boldsymbol{x}) = H(\boldsymbol{x}_0)$.

Proof: The material rate of change of H along a particle orbit becomes, in general,

$$\begin{aligned}\frac{DH}{Dt} &= \boldsymbol{v}\cdot\frac{D\boldsymbol{v}}{Dt} + \frac{D\Phi}{Dt} + \frac{1}{\rho_0}\frac{Dp}{Dt}\\ &= \boldsymbol{v}\cdot\left(\boldsymbol{g} - \frac{1}{\rho_0}\nabla p\right) + \frac{\partial\Phi}{\partial t} + \boldsymbol{v}\cdot\nabla\Phi + \frac{1}{\rho_0}\left(\frac{\partial p}{\partial t} + \boldsymbol{v}\cdot\nabla p\right)\\ &= \frac{\partial\Phi}{\partial t} + \frac{1}{\rho_0}\frac{\partial p}{\partial t},\end{aligned}$$

where we in the second step used the Euler time-dependent Equation (13.1). In steady flow the gravitational potential and the pressure are both time independent, and the Bernoulli field will indeed be constant along particle orbits and therefore along streamlines.

The first two terms in the Bernoulli field make up the total mechanical (kinetic plus potential) energy of a unit mass particle, also called the *specific mechanical energy*. Bernoulli's theorem states that any change in the specific mechanical energy—whether due to a change in velocity or to a change of position in a gravitational field—must be balanced by an opposite change in the pressure. Since pressure performs work on a moving fluid, it is not surprising that Bernoulli's theorem is equivalent to energy balance in ideal flow (page 353). We shall later see that the Bernoulli function is closely related to the *specific enthalpy*.

Terminology: The importance of Bernoulli's theorem for practical hydrodynamical applications has led to several different terminologies, not always wholly consistent and systematic. Since $\rho_0 H = \frac{1}{2}\rho_0 \boldsymbol{v}^2 + \rho_0 \Phi + p$ has dimension of pressure, it is called the *total pressure* whereas the kinetic energy density $\frac{1}{2}\rho_0 \boldsymbol{v}^2$ is called the *dynamic pressure*, as opposed to the ordinary pressure p also called the *static pressure*. A point where the velocity vanishes is called a *stagnation point*, and the quantity $\rho_0 \Phi + p$ is called the *stagnation pressure*.

Another engineering terminology is only useful in constant gravity: $\boldsymbol{g} = (0, 0, -g_0)$, where $\Phi = g_0 z$. Since $H/g_0 = v^2/2g_0 + z + p/\rho_0 g_0$ has the dimension of length, the expression $v^2/2g_0$ is called the *velocity head*. The vertical height z of a point on the streamline above some fixed reference level $z = 0$ is called the *elevation head* or sometimes the *static head*, while $p/\rho_0 g_0$ is called the *pressure head*. The combination $z + p/\rho_0 g_0$ is called the *hydraulic head* and H/g_0 is called the *total head*.

Using Bernoulli's theorem

Bernoulli's theorem is highly useful because many of the flows that we deal with in our daily lives are nearly steady, nearly ideal and nearly incompressible. Bernoulli's theorem often provides us with a first idea about the behavior of a flow in a given geometry. Thus, for example, the drop in pressure accompanying the higher air speed above a wing relative to the air speed below lies at the root of lift generation for flying animals and machines[1].

Example 13.2 [Life of a flatfish]: Lift is usually thought of as beneficial but that may not always be the case. Some fish hide by burrowing superficially into the sandy bottom of a stream. The fish's curved upper surface forces the passing water to speed up, leading to a pressure drop above the fish that grows with the square of the flow velocity (see Example 13.9). If the stream velocity increases, the fish may be lifted out of the sand, whether it wants to or not. To avoid that may be why flatfish are indeed—flat.

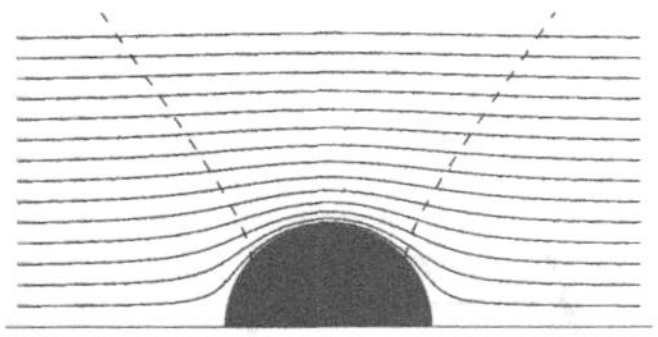

The incompressible flow has to quicken to get over an obstacle (here a half sphere) on the bottom of a stream. By Bernoulli's theorem, there will be a lower pressure above the obstacle, that is, a lift. The pressure is lower than ambient pressure between the dashed lines and higher outside.

In many cases of practical interest, fluid streams through a network of ducts with inlets and outlets that are under external control (see, for example, [Fox and McDonald 1985, White 1999]). Assuming that there are well-defined streamlines connecting inlets with outlets, Bernoulli's theorem is typically used to relate the velocities and pressures at inlets and outlets, even if nothing is known about details of the the internal flow in the system. Viscosity is, however, never completely absent but is almost always negligible well away from the boundaries of the containers and conduits that we use to handle fluids (as seen in Figure 13.2).

Exploiting the constancy of H along a streamline is always an approximation, and in any realistic problem there will be what the engineers call "head loss" due to viscosity, secondary flow, and turbulence. Typically it is also assumed that the flow velocity U is the same all over the cross-section A of an inlet or outlet. Such a pattern is called *plug flow*, and the volume flow rate through the inlet or outlet is simply calculated as $Q = AU$. This is by no means guaranteed to be correct, and engineers sometimes put in a *discharge factor* C to take care of the lower velocities at the sides of a duct, writing $Q = CAU$ where U is now the velocity of a central streamline well away from the sides of the duct (Problem 13.6).

[1] Contrary to popular belief, the difference in air speed is primarily caused by the wing's angle of attack rather than by its shape (see Chapter 29). Otherwise aircraft could not fly inverted!

Giovanni Batista Venturi (1746–1822). Italian physicist, mathematician, philosopher, and priest. Contemporary of Euler and Bernoulli. Collected and published Galileo's writings.

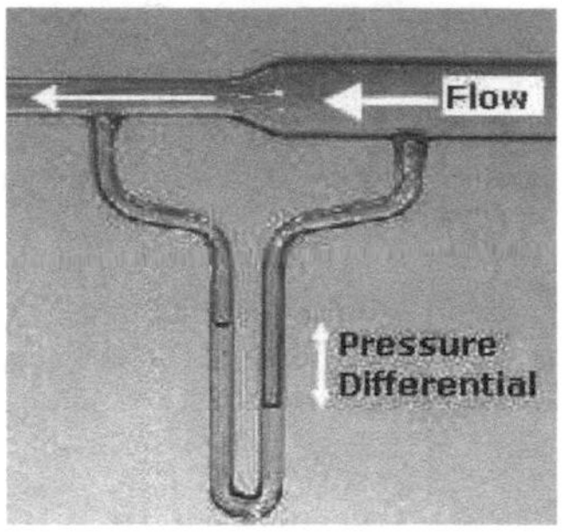

Venturi effect demonstration. Air streams from the right into a narrower tube where it speeds up. The shift in water level in the shunt tube underneath can be used to determine the pressure drop and thus the flow speed. (Source: Wikimedia Commons.)

Case: The Venturi effect

A simple duct with gently varying cross-section carries a constant volume flow rate Q of incompressible fluid. For simplicity we assume the duct is horizontal, such that gravity can be disregarded. For a streamline running through the duct, Bernoulli's theorem tells us that

$$H = \tfrac{1}{2}v^2 + \frac{p}{\rho_0} \tag{13.14}$$

takes the same value everywhere along the streamline. Approximating the velocity with its average $U = Q/A$ over the cross-section A, the pressure becomes

$$p = \rho_0 \left(H - \frac{Q^2}{2A^2} \right). \tag{13.15}$$

This demonstrates the *Venturi effect*: the pressure rises when the duct cross-section increases, and conversely (see the margin figure).

The Venturi effect is used in a number of technological devices, from the carburetor in car engines to perfume atomizers, as well as pumps and flow meters. The Venturi effect has also been used to clarify otherwise puzzling findings in dynamic cerebrovascular ultrasound tests [RR02].

Case: Torricelli's law

A barrel of wine has a little spout close to the bottom. If the plug in the spout is suddenly removed, the hydrostatic pressure accelerates the wine to stream out with considerable speed. Provided the spout is narrow compared to the size of the barrel, a nearly steady flow will soon establish itself. This is a case where Bernoulli's theorem readily yields a quantitative result.

Consider a streamline starting near the top of the barrel and running with the flow down through the middle of the spout. Near the top at a height $z = h$ over the position of the spout, the fluid is almost at rest, that is, $v \approx 0$. The pressure is atmospheric, $p = p_0$, and the gravitational potential may be taken to be $g_0 h$ so that

$$H_{\text{top}} = g_0 h + \frac{p_0}{\rho_0}. \tag{13.16}$$

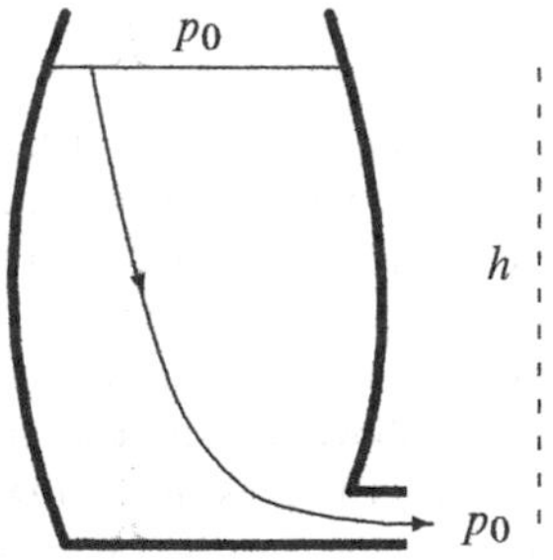

Wine running out of a barrel. The wine emerges with the same speed as it would have obtained by falling freely through the height h of the fluid in the barrel.

Just outside the spout, the fluid has some horizontal velocity v, and the pressure is also atmospheric, $p = p_0$, with no contribution from gravity, because the potential has been chosen to vanish here. Hence,

$$H_{\text{bottom}} = \frac{1}{2}v^2 + \frac{p_0}{\rho_0}. \tag{13.17}$$

Equating the values of H at the top and the bottom we find that

$$\frac{1}{2}v^2 + \frac{p_0}{\rho_0} = g_0 h + \frac{p_0}{\rho_0},$$

which has the solution

$$\boxed{v = \sqrt{2g_0 h}.} \tag{13.18}$$

Surprisingly, this is exactly the same velocity as a drop of wine would have obtained by falling freely from the top of the barrel to the spout. This result is called *Torricelli's law* (1644), and preceding Bernoulli's theorem by more than a century it was in its time a major step forward in the understanding of fluids. We could, in fact, have come to the same conclusion simply by converting the gravitational energy of an amount of water (almost) at rest at the surface into kinetic energy at the spout exit. In the absence of viscosity, there can be no loss of energy along the way.

In a sense the barrel acts as a device for diverting the vertical momentum of the falling liquid into the horizontal direction. All streamlines yield the same result, except the ones running very near to the walls of the barrel and spout where the unavoidable viscosity slows down the flow in the boundary layers. Even if the spout is replaced by a pipe that is not horizontal but turns and twists, the exit velocity from the pipe will equal the free-fall velocity from the fluid surface at the top of the barrel to the actual exit level.

Example 13.3 [Wine barrel]: A large cylindrical wine barrel has diameter 1 m and height 2 m. According to Torricelli's law, the wine will emerge from the spout with the free-fall speed of about 6.3 m s^{-1}. If the spout opening has diameter 5 cm, about 12.3 L (liters) of wine will be spilled on the floor per second. At this rate it would take a bit more than 2 min to empty the barrel, but we shall now see that it actually takes double because the level sinks.

Quasi-stationary emptying

If the barrel has constant cross-section A_0, Leonardo's law (12.8) tells us that the average vertical flow velocity in the barrel is $v_0 = vA/A_0$, where $A \ll A_0$ is the cross-section of the spout and $v = \sqrt{2g_0 z}$ is the average horizontal flow velocity through the spout when the water level is z. Since $dz/dt = -v_0$, we obtain the following differential equation for quasi-stationary emptying of the barrel:

$$\frac{dz}{dt} = -\frac{A}{A_0}\sqrt{2g_0 z}. \tag{13.19}$$

Integrating this equation with initial value $z = h$ for $t = 0$, we obtain the time it takes to empty the barrel,

$$t_0 = \int_h^0 \frac{dt}{dz}\,dz = -\int_h^0 \frac{A_0}{A}\frac{dz}{\sqrt{2g_0 z}} = \frac{A_0}{A}\sqrt{\frac{2h}{g_0}}. \tag{13.20}$$

It equals the free-fall time from height h multiplied by the usually huge ratio of the barrel and spout cross-sections. For the cylindrical wine barrel of the example above, the free-fall time is 0.64 s, but it takes 400 times longer, about 4 min, to empty the barrel.

Henri Pitot (1695–1771). French mathematician, astronomer, and hydraulic engineer. Invented the Pitot tube around 1732 to measure the flow velocity in the river Seine.

Case: The Pitot tube

Fast aircraft often have a sharply pointed nose that on closer inspection is seen to end in a little open tube. On other aircraft the tube may stick orthogonally out from the side and is bent forward into the oncoming airstream. This device is called a *Pitot tube* and is used in many variants to measure flow speeds in gases and liquids. In its original form, the Pitot tube is just an open glass tube bent through a right angle. The tube is lowered into a river streaming steadily with velocity U, with one end turned horizontally toward the current and the other vertically in the air above. The flow will stem water up into the vertical part of the tube, until the static pressure of the water column balances the dynamic pressure from the flow. After the flow has steadied, the water in the tube rises to some height h above the river surface.

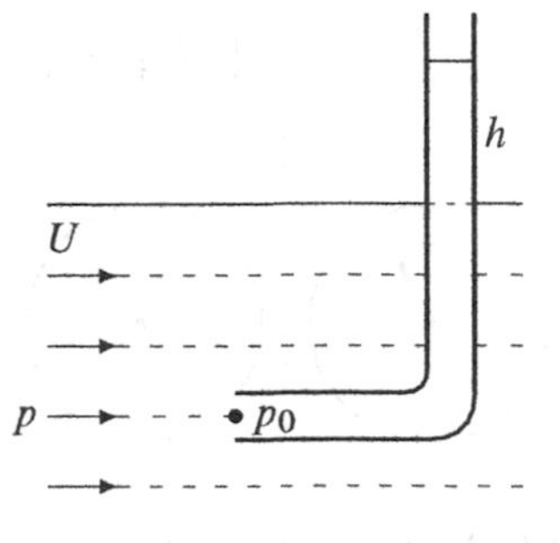

The principle of the Pitot tube. The pressure increase along the stagnating streamline must equal the weight of the raised water column.

In the steady state, the water speed must vanish at the entrance to the horizontal part of the tube, and a horizontal streamline arriving here from afar must come to an end in a so-called *stagnation point*. The gravitational potential is constant everywhere along the horizontal streamline, and can be disregarded, so that Bernoulli's theorem for this streamline reads

$$\frac{p_0}{\rho_0} = \tfrac{1}{2}U^2 + \frac{p}{\rho_0}, \tag{13.21}$$

where p is the pressure at infinity and p_0 is the stagnation pressure.

The excess pressure $\Delta p = p_0 - p = \frac{1}{2}\rho_0 U^2$ must also equal the extra hydrostatic pressure from the water column above the water surface, $\Delta p = \rho_0 g_0 h$, so that

$$U = \sqrt{2g_0 h}. \tag{13.22}$$

Again we find the simple and surprising result that the speed of the water is exactly what it would be after a free fall from the height h. At a typical flow speed of 1 m s^{-1}. the water in the tube is raised 5 cm above the river level.

Even if they do not know Bernoulli's theorem, farmers know better than to leave the barn door open toward the wind in a storm. A gust of wind will not only decrease the pressure above the barn roof because the wind has to move faster to get over the building, but the Pitot effect will also increase the pressure inside, giving the roof a double reason to blow off

Example 13.4 [Water scoop]: Forest fires are often combated by aircraft dropping large amounts of sea or lake water. To avoid landing and take-off, the aircraft collects water by lowering a scoop into the water while flying slowly at very low altitude. If the scoop turns directly forward and the aircraft velocity is U, it can, like the Pitot tube, raise the water to a maximal height, $h = U^2/2g_0$. Even for a speed as low as $U = 120$ km h$^{-1} \approx 33$ m s^{-1}, this comes to $h = 56$ m. In practice, the height of the water tank over the lake surface is much smaller, so that the water should ideally arrive in the tank with nearly maximal speed U. A scoop with an opening area of just $A \approx 300$ cm^2 can deliver water at a rate of $Q = UA \approx 1$ m^3 s^{-1}. Turbulence lowers this somewhat, but typically such an aircraft can collect 6 m^3 in just 12 s.

13.4 Vorticity

The value of the Bernoulli field $H(\boldsymbol{x})$ in a point $\boldsymbol{x}$ is only a function of the streamline going through this point. Different streamlines will in general have different values of H. It is, however, often possible to relate the values of H for bundles of streamlines.

Asymptotically uniform flow

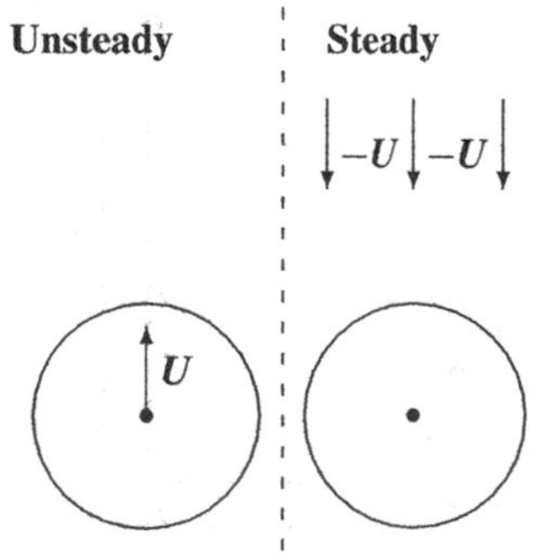

A body moving at constant speed through a fluid is physically equivalent to the same body being at rest in a steady flow that is asymptotically uniform.

A general and frequently occurring case is a body moving with constant velocity $\boldsymbol{U}$ through a fluid initially at rest with pressure p_0 in the absence of gravity. The body creates an unsteady disturbance in the fluid around itself, but far ahead of the body the fluid should remain undisturbed at rest. The *relativity of motion* in Newtonian mechanics tells us that this unsteady flow in the fluid at rest is physically equivalent to a steady flow around the same body at rest in a fluid which far upstream moves with uniform steady velocity $-\boldsymbol{U}$. We tacitly used this equivalence in the water scoop example above. Consequently, for any streamline coming in from far upstream, the Bernoulli field must take the same value at any point along it where the velocity is $\boldsymbol{v}$ and the pressure p,

$$\frac{1}{2}\boldsymbol{v}^2 + \frac{p}{\rho_0} = \frac{1}{2}\boldsymbol{U}^2 + \frac{p_0}{\rho_0}. \tag{13.23}$$

Thus, if one can be sure that all streamlines started out infinitely far upstream, then the Bernoulli field would take the same value everywhere in the fluid. Unfortunately this cannot be guaranteed.

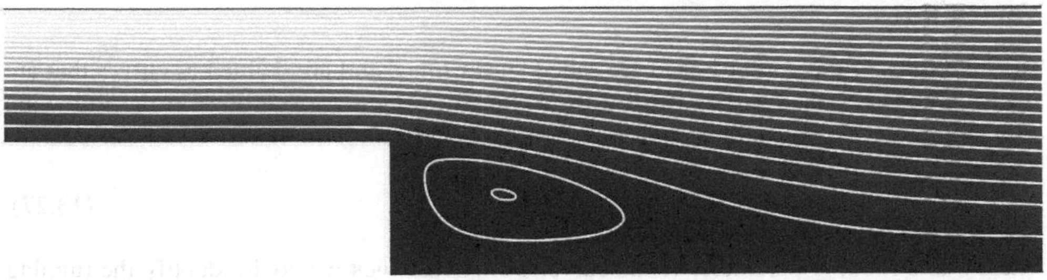

Figure 13.3. Two-dimensional flow over a step. The streamlines coming in from the left do not cover the whole volume of fluid because a whirl has formed behind the step. The gray-level indicates the magnitude of the velocity field. Some viscosity is necessary to make the simulation work, so this is not truly ideal but only nearly ideal flow well away from the solid boundaries. The Reynolds number is 100.

The vorticity field

The simple result that H is spatially constant for asymptotically uniform flow is spoiled by the possibility that there may be circulating streamlines unconnected with the flow at infinity. Common experience indicates that such circulating flow may occur in the wake of the disturbance created by a change in the shape of a container or a moving body (see Figure 13.3). We are thus naturally led to the study of local circulation in a fluid, and we shall see that H *is* in fact globally constant if there is no local circulation anywhere.

Using the Euler equation for steady incompressible non-viscous flow (13.12), the gradient of the Bernoulli field including gravity becomes (in index notation)

$$\nabla_i H = \frac{1}{2}\nabla_i \boldsymbol{v}^2 + \nabla_i \Phi + \frac{1}{\rho_0}\nabla_i p = \boldsymbol{v}\cdot(\nabla_i \boldsymbol{v}) - (\boldsymbol{v}\cdot\nabla)v_i = (\boldsymbol{v}\times(\nabla\times\boldsymbol{v}))_i\,.$$

Defining the *vorticity field* as the curl of the velocity field,

$$\boldsymbol{\omega} = \nabla\times\boldsymbol{v}, \qquad (13.24)$$

we have for steady flow

$$\boxed{\nabla H = \boldsymbol{v}\times\boldsymbol{\omega}\,.} \qquad (13.25)$$

The vorticity field also goes back to Cauchy (1841) and is a quantitative measure of the local circulation in the fluid. In any region where the vorticity field vanishes, we have $\nabla H = 0$, so the Bernoulli field must take the same value everywhere in that region, that is, $H(\boldsymbol{x}) = H_0$. Flow completely free of vorticity is said to be *irrotational* and leads to a particularly simple formalism that we shall present in Section 13.6.

Example 13.5: The curl of the field $\boldsymbol{v} = (x^2, 2xy, 0)$ is $\boldsymbol{\omega} = (0, 0, 2y)$. The curl of $\boldsymbol{v} = (y^2, 2xy, 0)$ vanishes, so this velocity field is irrotational.

Example 13.6 [Rigidly rotating flow]: A trivial example of a flow with vorticity is a steadily rotating rigid body. If the rotation vector of the body is $\boldsymbol{\Omega}$, the velocity field becomes $\boldsymbol{v} = \boldsymbol{\Omega}\times\boldsymbol{x}$, with the vorticity field

$$\boldsymbol{\omega} = \nabla\times(\boldsymbol{\Omega}\times\boldsymbol{x}) = \boldsymbol{\Omega}(\nabla\cdot\boldsymbol{x}) - (\boldsymbol{\Omega}\cdot\nabla)\boldsymbol{x} = 3\boldsymbol{\Omega} - \boldsymbol{\Omega} = 2\boldsymbol{\Omega}\,. \qquad (13.26)$$

The vorticity is in this case constant and equal to twice the rotation vector. In Problem 7.8 on page 123, the slightly puzzling factor of 2 is shown to be related to the sharing of the displacement gradients between rotation and deformation.

Vortex lines

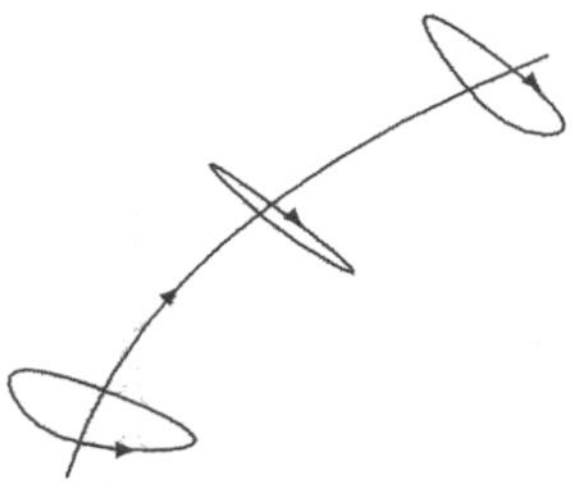

Around a vortex line there is local circulation of fluid.

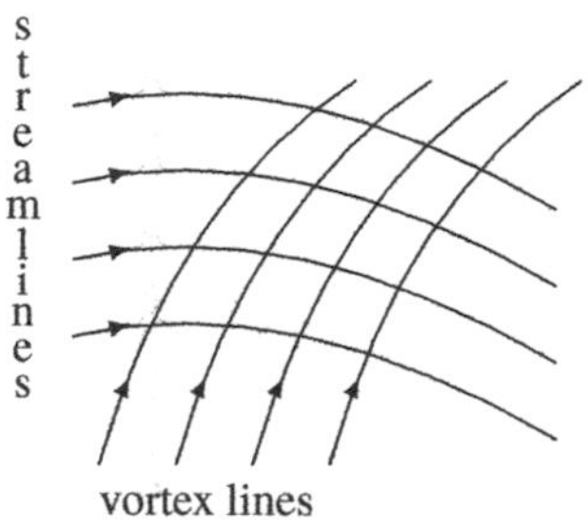

The Bernoulli field is constant on surfaces made from vortex lines and streamlines.

The field lines of the vorticity field are called *vortex lines*, and are defined as curves that are always tangent to the vorticity field. Like streamlines, they are solutions to the following ordinary differential equation at a given fixed time t_0,

$$\frac{d\boldsymbol{x}}{ds} = \boldsymbol{\omega}(\boldsymbol{x}, t_0), \tag{13.27}$$

where s is a running parameter of the curve. For streamlines we could identify the running parameter with time, but this is not the case here, where s has the dimension of time multiplied by length. Also like streamlines, vortex lines cannot cross each other, and since the vorticity field is a "curl", it is rigorously divergence-free, $\nabla \cdot \boldsymbol{\omega} = 0$. This implies that vortex lines cannot emerge or terminate anywhere in a fluid but must keep on going until they reach the boundaries of the flow (which may be at spatial infinity). In well-behaved geometries, vortex lines may form closed curves.

In steady flow, vortex lines are of course independent of the chosen time t_0. Bernoulli's theorem, $(\boldsymbol{v} \cdot \nabla)H = 0$, follows immediately from (13.25) by dotting with $\boldsymbol{v}$. Similarly, by dotting with $\boldsymbol{\omega}$ we obtain $(\boldsymbol{\omega} \cdot \nabla)H = 0$, showing that the Bernoulli field is also constant along vortex lines. Together these results show that *the Bernoulli field is constant on the two-dimensional surfaces formed by combining vortex lines and streamlines.* They are sometimes called *Lamb surfaces*, and sometimes *Bernoulli surfaces*.

Equation of motion for vorticity

The vorticity field is derived from the velocity field, so the equation of motion for the vorticity field must follow from the equation of motion (13.1) for the velocity field. Retracing the steps leading to (13.25), but now including the time dependence of the velocity field, we obtain,

$$\frac{\partial \boldsymbol{v}}{\partial t} = \boldsymbol{v} \times \boldsymbol{\omega} - \nabla H, \tag{13.28}$$

which of course reduces to (13.25) for $\partial \boldsymbol{v}/\partial t = \boldsymbol{0}$. Taking the curl of both sides of this equation, and using that the curl of a gradient vanishes, we arrive at the equation of motion for vorticity:

$$\boxed{\frac{\partial \boldsymbol{\omega}}{\partial t} = \nabla \times (\boldsymbol{v} \times \boldsymbol{\omega}).} \tag{13.29}$$

There is a major lesson to draw from this equation: If the vorticity vanishes identically, $\boldsymbol{\omega}(\boldsymbol{x}, t) = \boldsymbol{0}$, throughout a region V of space at time t, then we have $\partial \boldsymbol{\omega}/\partial t = \boldsymbol{0}$ for all $\boldsymbol{x}$ in V at t. Consequently, the vorticity field will not change in the next instant, and continuing this argument, we conclude that if the vorticity field vanishes in the volume V at time t, it will vanish in this volume forever after (and forever before!). In other words, *vorticity cannot be generated by the flow of an ideal fluid but must be present from the outset.* If you, for example, accelerate a body from rest in a truly ideal fluid, initially also at rest, the flow will forever remain without vorticity because $\boldsymbol{\omega} = \boldsymbol{0}$ at the beginning.

Vorticity from boundary layers: The whirling air that trails a speeding car or an airplane must for this reason somehow be caused by viscous forces, independent of how tiny the viscosity is. As mentioned before, truly ideal fluids do not exist. Even the tiniest viscosity will create viscous boundary layers close to the solid walls of obstacles and the confining walls of ducts. The braking action due to friction in these boundary layers will make the streaming fluid "tumble" and thereby create vorticity. Although this, strictly speaking, invalidates the theorem above, the theorem is nevertheless quite useful in many practical cases because the boundary layers can be assumed to be very thin, as for example in aerodynamics (see Chapter 29).

13.5 Circulation

Vorticity is a local property of a fluid, indicating how much the fluid rotates in the neighborhood of a point. The corresponding global concept is called *circulation* and is formally defined as the integrated projection of the velocity field onto a closed curve C,

$$\Gamma(C,t) = \oint_C \boldsymbol{v}(\boldsymbol{x},t)\cdot d\boldsymbol{\ell}. \tag{13.30}$$

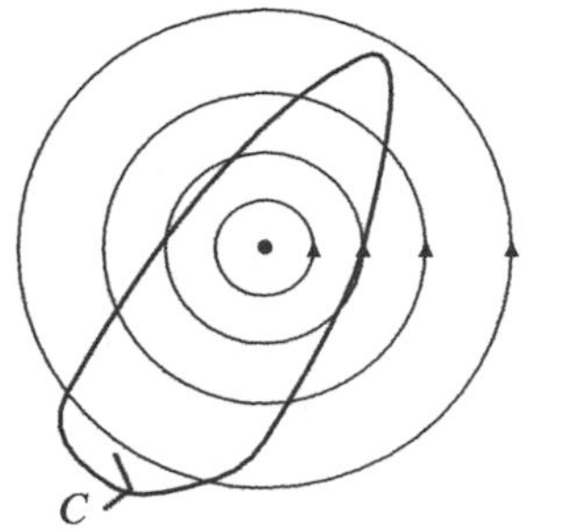

A closed curve C encircling a whirl.

If C encircles a region of whirling fluid, the projection of the velocity field onto the curve will tend to be of the same sign all the way around. Whether it is positive or negative depends on whether the curve runs with the whirling flow or against it. The circulation may be calculated for any curve, not just a streamline encircling a whirl, although that might be the natural thing to do in some situations. Although the integral is done at a fixed time, the curve may even change with time.

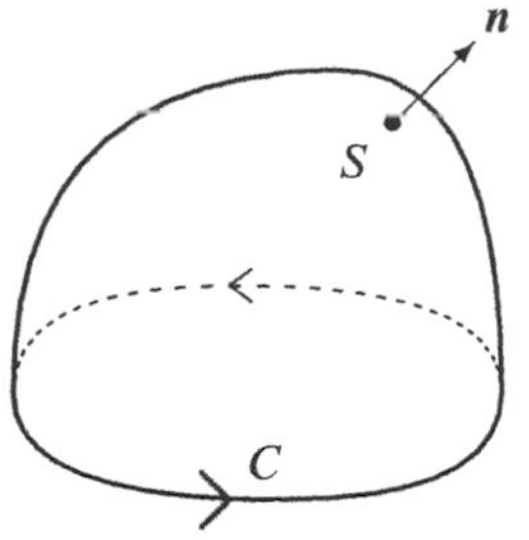

A surface S with perimeter C. The normal to the surface is consistent with the orientation of C (here using a right-hand rule).

Stokes' theorem

The most important theorem about circulation is attributed to Stokes who set it as one of the problems for Smith's Prize Exam at Cambridge in 1854 (although it actually goes back to Kelvin). The theorem states that the circulation of a vector field $\boldsymbol{v}(\boldsymbol{x},t)$ around a closed curve C equals the flux of its curl $\nabla\times\boldsymbol{v}(\boldsymbol{x},t)$ through any surface S bounded by the curve,

$$\boxed{\oint_C \boldsymbol{v}\cdot d\boldsymbol{\ell} = \int_S \nabla\times\boldsymbol{v}\cdot d\boldsymbol{S}.} \tag{13.31}$$

It does not matter which surface S the flux is calculated for, as long as it has C as boundary, lies entirely within the fluid, and is oriented consistently with the orientation of C (see Problem 13.7). Nor does it matter what interpretation you give to the vector field $\boldsymbol{v}$.

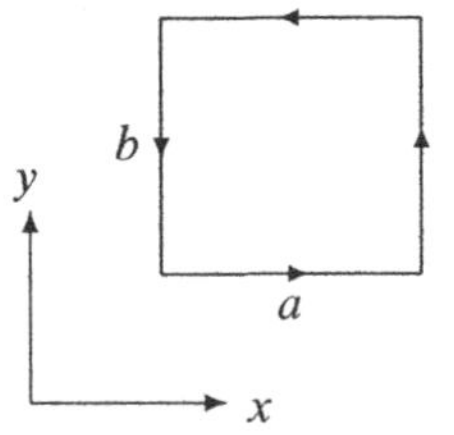

Circulation around a small rectangle of dimensions $a\times b$.

Proof: The relation between global and local quantities is, as before, established by calculating the global quantity for an infinitesimal geometric figure, in this case a tiny rectangle in the xy-plane with sides a and b. To first order in the sides, the circulation around the rectangle becomes (suppressing both z and t)

$$\begin{aligned}\oint_{a\times b}\boldsymbol{v}\cdot d\boldsymbol{\ell} &= \int_x^{x+a} v_x(x',y)\,dx' + \int_y^{y+b} v_y(x+a,y')\,dy' \\ &\quad - \int_x^{x+a} v_x(x',y+b)\,dx' - \int_y^{y+b} v_y(x,y')\,dy' \\ &\approx \int_y^{y+b} a\,\nabla_x v_y(x,y')\,dy' - \int_x^{x+a} b\,\nabla_y v_x(x',y)\,dx' \\ &\approx ab\big(\nabla_x v_y(x,y) - \nabla_y v_x(x,y)\big) = ab\,(\nabla\times\boldsymbol{v})_z.\end{aligned}$$

The last expression is indeed the projection $(\nabla\times\boldsymbol{v})\cdot d\boldsymbol{S}$ of the vorticity field on the small vector surface element of the rectangle, $d\boldsymbol{S} = ab\hat{\boldsymbol{e}}_z$.

Consider now an open surface, not necessarily planar, built up from little rectangles of this kind. Adding together the circulation for each rectangle, the contributions from the inner common edges cancel and one is left only with the circulation around the outer perimeter of the surface, which is Stokes' theorem. The proof also implies that the shape of the surface S does not matter (see Problem 13.7). If you do not like the ruggedness of the perimeter in the rectangular approximation to the surface, a somewhat improved proof is found in Appendix C.

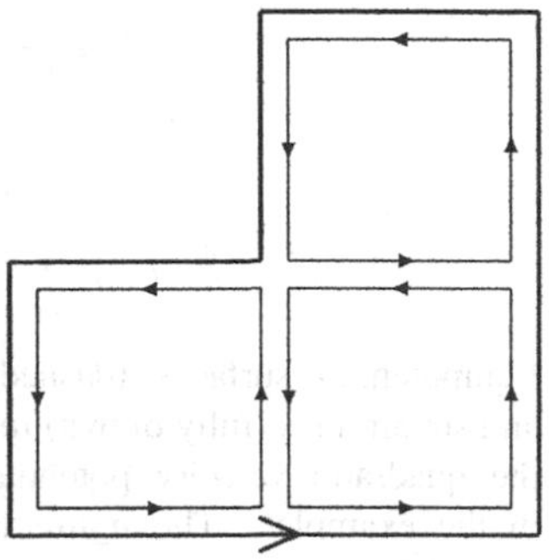
Adding rectangles together, the circulation cancels along the edges where two rectangles meet. This is also valid if the rectangles bend into the other coordinate directions.

13.6 Potential flow

In the absence of viscosity a flow that is irrotational in some region of space will, as we have seen, stay irrotational at all times in that region. Although viscosity will always create vorticity in boundary layers, at high Reynolds numbers these layers usually stay close to the boundaries and leave the flow at large nearly irrotational.

From $\nabla \times \boldsymbol{v} = \mathbf{0}$ it follows that the velocity field is a gradient (see Appendix C for a proof),

$$\boldsymbol{v} = \nabla\Psi. \tag{13.32}$$

The scalar field Ψ is called the *flow potential* or the *velocity potential*[2]. Such *potential flow* obeys a much simpler formalism than flow with vorticity, in particular when it is also incompressible. The results to be derived below for non-viscous incompressible flow can be generalized to compressible flow, although much of the simplicity is lost (see Problem 14.3).

Incompressible potential flow

In an incompressible fluid the vanishing of the divergence of the velocity field implies that the flow potential must satisfy Laplace's equation,

$$\boxed{\nabla^2\Psi = 0.} \tag{13.33}$$

Inserting $\boldsymbol{\omega} = \mathbf{0}$ and $\boldsymbol{v} = \nabla\Psi$ into the rewritten Euler equation (13.28), we immediately find $\nabla(H + \partial\Psi/\partial t) = \mathbf{0}$. This means that $H + \partial\Psi/\partial t$ cannot depend on the spatial coordinates, but only on time. Inserting the Bernoulli function (13.13) and solving for the pressure, we obtain

$$\boxed{p = C - \rho_0\left(\frac{1}{2}\boldsymbol{v}^2 + \Phi + \frac{\partial\Psi}{\partial t}\right),} \tag{13.34}$$

where the "constant" $C = C(t)$ is a function of time only. All the original nonlinearity of the Euler equation has thus been relegated to the expression for the pressure.

Potential flow in incompressible fluid is much simpler than flow with vorticity because a lot is known about the solutions to Laplace's equation. Any solution to Laplace's equation or linear superpositions of such solutions will in fact define a solvable potential flow problem. A particular problem geometry "simply" requires that we find the solution that fulfills the boundary conditions.

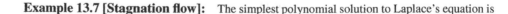

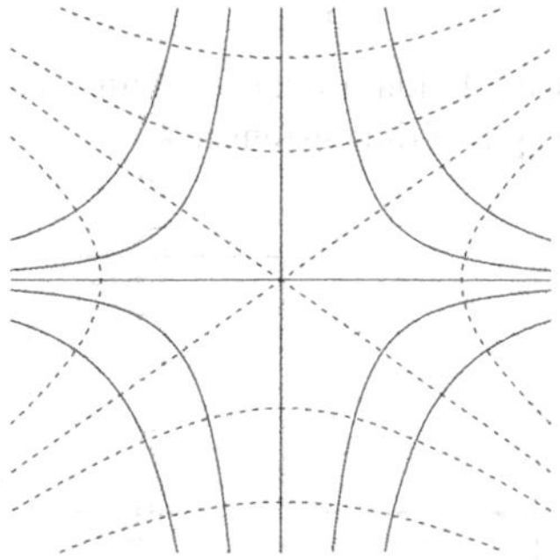

Equipotential surfaces (dashed) and streamlines (fully drawn) for the quadratic velocity potential in the example. The figure is plotted in a plane containing the vertical z-axis and should be rotated around it to give the three-dimensional picture. Notice that the streamlines are orthogonal to the equipotential surfaces.

Example 13.7 [Stagnation flow]: The simplest polynomial solution to Laplace's equation is

$$\Psi = \frac{x^2 + y^2 - 2z^2}{2\tau}, \qquad \boldsymbol{v} = \frac{(x, y, -2z)}{\tau}, \tag{13.35}$$

where τ has the dimension of time. The equipotential surfaces are rotation hyperboloids around the z-axis. The structure of the field is shown in the margin figure. Since $v_z = 0$ for $z = 0$, we may replace the xy-plane with a solid surface (at $z = 0$). For $z \geq 0$ the field therefore describes the ideal flow near the stagnation point at $x = y = z = 0$.

[2]There is no general agreement on the symbol for the velocity potential, although the preferred notation appears to be ϕ. In this book we use Ψ to avoid clashes with the azimuthal coordinate in cylindrical or spherical systems.

* The stream function

Flows that are invariant under translation along or rotation around an axis are effectively two-dimensional and often used in fluid mechanics because of the greater simplicity of the formalism (see for example [Panton 2005, ch. 12]). Here we shall only discuss steady flow, where $\boldsymbol{v} = \boldsymbol{v}(\boldsymbol{x})$ does not depend on time.

Axial invariance in Cartesian coordinates: In Cartesian coordinates, translational invariance along the cylinder axis z reduces the incompressibility condition to

$$\frac{\partial v_x}{\partial x} + \frac{\partial v_y}{\partial y} = 0. \tag{13.36}$$

It is explicitly fulfilled by a *stream function* $\psi(x, y)$ satisfying

$$v_x = \frac{\partial \psi}{\partial y}, \qquad v_y = -\frac{\partial \psi}{\partial x}. \tag{13.37}$$

Such a function always exists, because for given v_x we may calculate $\psi = -\int v_x\,dy$, and

$$v_y = \int \frac{\partial v_y}{\partial y}\,dy = -\int \frac{\partial v_x}{\partial x}\,dy = -\int \frac{\partial^2 \psi}{\partial x \partial y}\,dy = -\frac{\partial \psi}{\partial x}$$

up to an arbitrary function of x that can be absorbed in ψ without disturbing v_x.

The gradient of the stream function is orthogonal to the velocity field,

$$\boldsymbol{v} \cdot \nabla\psi = -\frac{\partial \psi}{\partial y}\frac{\partial \psi}{\partial x} + \frac{\partial \psi}{\partial x}\frac{\partial \psi}{\partial y} = 0. \tag{13.38}$$

Since a streamline is defined by $d\boldsymbol{x}/dt = \boldsymbol{v}(\boldsymbol{x})$, it follows that $d\psi/dt = \boldsymbol{v} \cdot \nabla\psi = 0$. The stream function is therefore constant along any streamline, which is of course the reason for its name. It also means that *in steady, ideal flow one may replace any contour surface,* $\psi = \text{const}$, *with a solid boundary* because the fluid flows parallel to such a surface.

The only component of vorticity is

$$\omega_z = \frac{\partial v_y}{\partial x} - \frac{\partial v_x}{\partial y} = -\frac{\partial^2 \psi}{\partial x^2} - \frac{\partial^2 \psi}{\partial y^2} = -\nabla^2\psi. \tag{13.39}$$

In potential flow where $\omega_z = 0$, the Cartesian stream function must also satisfy Laplace's equation in two dimensions.

Axial invariance in cylindrical coordinates: In cylindrical coordinates (r, ϕ, z) translational invariance along the cylinder axis z reduces the incompressibility condition to (see Equation (D.14) on page 621),

$$\frac{1}{r}\frac{\partial (r v_r)}{\partial r} + \frac{1}{r}\frac{\partial v_\phi}{\partial \phi} = 0. \tag{13.40}$$

This condition is satisfied by,

$$v_r = \frac{1}{r}\frac{\partial \psi}{\partial \phi}, \qquad v_\phi = -\frac{\partial \psi}{\partial r}, \tag{13.41}$$

One may verify that $\boldsymbol{v} \cdot \nabla\psi = 0$, so that ψ is indeed the stream function.

In these coordinates we find from (D.15) the only non-vanishing component of vorticity,

$$\omega_z = \frac{\partial v_\phi}{\partial r} + \frac{v_\phi}{r} - \frac{1}{r}\frac{\partial v_r}{\partial \phi} = -\left(\frac{\partial^2 \psi}{\partial r^2} + \frac{1}{r}\frac{\partial \psi}{\partial r} + \frac{1}{r^2}\frac{\partial^2 \psi}{\partial \phi^2}\right). \tag{13.42}$$

The expression in parentheses is, as before, the Laplacian of the stream function, which must vanish in potential flow. That is not surprising, since we have in fact only made a coordinate transformation from plane Cartesian coordinates (x, y) to plane polar coordinates (r, ϕ).

It must, however, be emphasized that there are other cases where the symmetry of a three-dimensional problem implies that there are effectively only two spatial variables. In such cases the requirement of vanishing vorticity does not necessarily lead to the Laplace equation for ψ. Here follows two examples.

Azimuthal invariance in cylindrical coordinates: In cylindrical coordinates (r, ϕ, z), azimuthal invariance in ϕ reduces the divergence condition to (see Equation (D.14))

$$\frac{1}{r}\frac{\partial(rv_r)}{\partial r} + \frac{\partial v_z}{\partial z} = 0. \tag{13.43}$$

This condition is satisfied by

$$v_r = -\frac{1}{r}\frac{\partial\psi}{\partial z}, \qquad v_z = \frac{1}{r}\frac{\partial\psi}{\partial r}. \tag{13.44}$$

One may readily verify that $\boldsymbol{v} \cdot \boldsymbol{\nabla}\psi = 0$, so that $\psi(r, z)$ is indeed the stream function.

The only component of vorticity (which must vanish in potential flow) is

$$\omega_\phi = \frac{\partial v_r}{\partial z} - \frac{\partial v_z}{\partial r} = -\frac{1}{r}\left(\frac{\partial^2\psi}{\partial r^2} - \frac{1}{r}\frac{\partial\psi}{\partial r} + \frac{\partial^2\psi}{\partial z^2}\right). \tag{13.45}$$

The expression in parentheses is not the Laplacian (D.19).

Azimuthal invariance in spherical coordinates: In spherical coordinates (r, θ, ϕ), azimuthal invariance in ϕ leads to the divergence condition (see Equation (D.35) on page 624)

$$\frac{1}{r^2}\frac{\partial(r^2 v_r)}{\partial r} + \frac{1}{r\sin\theta}\frac{\partial(\sin\theta\, v_\theta)}{\partial\theta} = 0. \tag{13.46}$$

This condition is satisfied by

$$v_r = \frac{1}{r^2\sin\theta}\frac{\partial\psi}{\partial\theta}, \qquad v_\theta = -\frac{1}{r\sin\theta}\frac{\partial\psi}{\partial r}. \tag{13.47}$$

Again one may verify that $\boldsymbol{v} \cdot \boldsymbol{\nabla}\psi = 0$, so that $\psi(r, \theta)$ is indeed the stream function.

The only component of vorticity (which must vanish in potential flow) is

$$\omega_\phi = \frac{\partial v_\theta}{\partial r} + \frac{v_\theta}{r} - \frac{1}{r}\frac{\partial v_r}{\partial\theta} = -\frac{1}{r\sin\theta}\left(\frac{\partial^2\psi}{\partial r^2} + \frac{1}{r^2}\frac{\partial^2\psi}{\partial\theta^2} - \frac{\cot\theta}{r^2}\frac{\partial\psi}{\partial\theta}\right). \tag{13.48}$$

The expression in parentheses is not the Laplacian (D.40).

13.7 Application: Cylinder in uniform crosswind

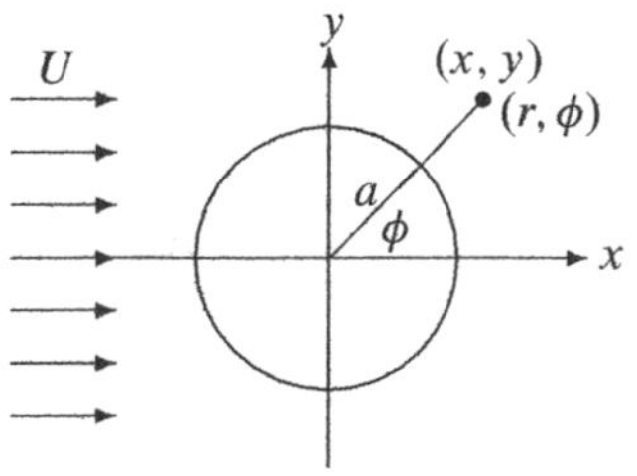

Cylinder of radius a in an asymptotically uniform crosswind U.

A circular cylinder with axis along the z-axis and radius a is placed in an asymptotically uniform "crosswind" U along the x-axis. This does not break the translational invariance along the z-axis so that the velocity potential $\Psi(x, y)$ and the stream function $\psi(x, y)$ are both independent of z. Evidently, the natural choice is cylindrical coordinates (r, ϕ, z).

Velocity potential

Asymptotically, for $r \to \infty$, the velocity potential must approach the field of a constant uniform crosswind, $\Psi \to Ux = Ur\cos\phi$. The linearity of the Laplace equation (13.33) demands that the velocity potential is linear in the asymptotic values,

$$\Psi = U\cos\phi\, f(r), \tag{13.49}$$

where $f(r)$ is a so far unknown function that approaches r for $r \to \infty$.

Inserting Ψ into the cylindrical Laplacian (see Equation (D.19) on page 622),

$$\nabla^2\Psi = \frac{\partial^2\Psi}{\partial r^2} + \frac{1}{r^2}\frac{\partial^2\Psi}{\partial\phi^2} + \frac{\partial^2\Psi}{\partial z^2} + \frac{1}{r}\frac{\partial\Psi}{\partial r}, \tag{13.50}$$

we obtain

$$\frac{d^2 f}{dr^2} + \frac{1}{r}\frac{df}{dr} - \frac{f}{r^2} = 0. \tag{13.51}$$

Since all three terms are of order $1/r^2$, we should look for power law solutions of the form $f \sim r^\alpha$. Inserting this into the equation we find $\alpha = \pm 1$ so that the most general solution is of the form $f = Ar + B/r$, where A and B are arbitrary constants.

The asymptotic condition implies $A = 1$, whereas B is determined by requiring the radial velocity field $v_r = \partial\Psi/\partial r$ to vanish at the surface of the cylinder, or $f'(a) = 0$. This leads to $B = a^2$, so that the solution is

$$\Psi = Ur\cos\phi\left(1 + \frac{a^2}{r^2}\right). \tag{13.52}$$

The stream function is determined in Problem 13.10.

Velocity and pressure fields

Calculating the gradient of $\Psi(r)$ in cylindrical coordinates by means of Equation (D.13) on page 621 we finally obtain the non-vanishing components of the velocity field

$$v_r = \nabla_r\Psi = \frac{\partial\Psi}{\partial r} = U\cos\phi\left(1 - \frac{a^2}{r^2}\right), \tag{13.53a}$$

$$v_\phi = \nabla_\phi\Psi = \frac{1}{r}\frac{\partial\Psi}{\partial\phi} = -U\sin\phi\left(1 + \frac{a^2}{r^2}\right). \tag{13.53b}$$

The radial flow vanishes at the surface of the cylinder as it should, whereas the tangential flow, $v_\phi|_{r=a} = -2U\sin\phi$, only vanishes at the front and rear stagnation points $\phi = 0, \pi$. The flow is visualized in Figure 13.4L.

The pressure is obtained from (13.34). In the absence of gravity and normalized to vanish at infinity, it becomes

$$p = \frac{1}{2}\rho_0\left(U^2 - v^2\right) = \frac{1}{2}\rho_0 U^2\frac{a^2}{r^2}\left(4\cos^2\phi - 2 - \frac{a^2}{r^2}\right). \tag{13.54}$$

On the surface of the cylinder it simplifies to

$$p_a = p|_{r=a} = \frac{1}{2}\rho_0 U^2\left(4\cos^2\phi - 3\right), \tag{13.55}$$

which is negative for $30° < \phi < 150°$.

The up/down invariance of the pressure (under $\phi \to -\phi$) shows that the total force in the y-direction, called *lift*, must vanish—as one would expect from the symmetry. What is more surprising is that due to the forward/backward invariance of the pressure (under $\phi \to \pi - \phi$), the total force along the x-direction, called *drag*, must also vanish (even on the upper half). We shall return to this apparent paradox in Section 13.9.

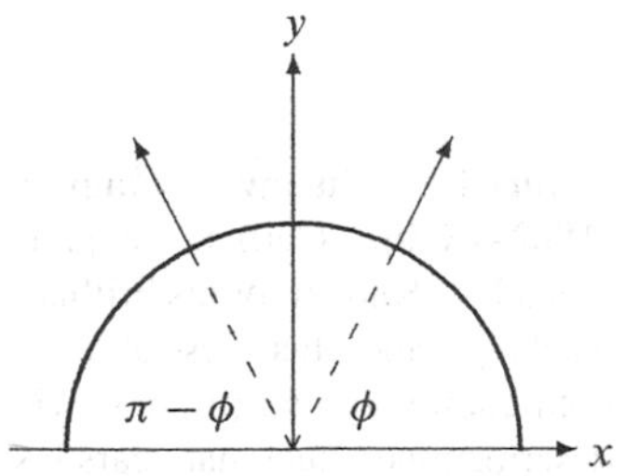

The projection of the pressure force on the x-axis is equal and opposite for ϕ and $\pi - \phi$.

Lift on a half cylinder

The solution also describes the ideal flow over a cylinder that is half buried at the flat bottom because the normal velocity field in the symmetry plane $y = 0$ vanishes, as it must according to the ideal-flow boundary conditions.

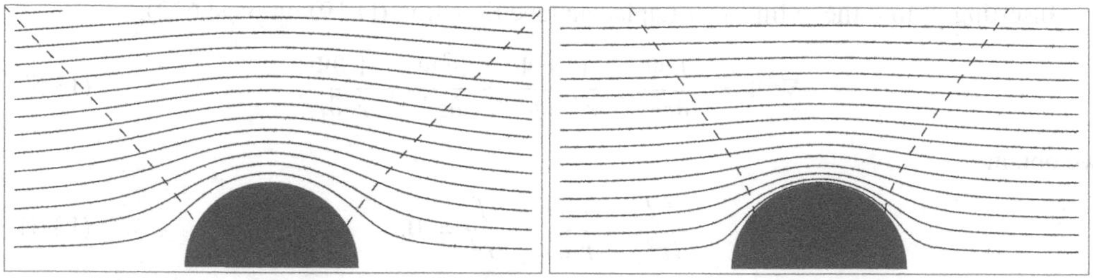

Figure 13.4. Comparison of potential flows around a cylinder (left) and sphere (right). Only the upper half is shown, and the horizontal bottom may in both cases be replaced by a solid surface. The pressure vanishes at infinity and on the dashed lines (and is negative between). The streamlines have been obtained as contours of the stream functions calculated in Problem 13.10. They are forward-backward symmetric and equidistant far from the objects.

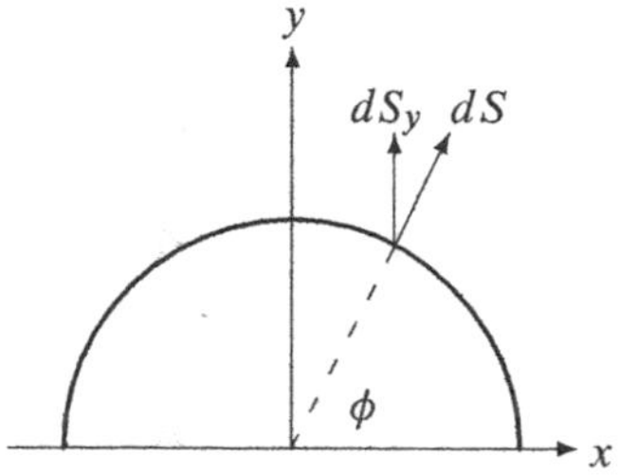

The projection of the surface element $dS = La d\phi$ on the y-axis is $dS_y = \sin\phi dS$.

The vertical lift on a stretch of the half cylinder of length L becomes (see the margin figure)

$$\mathcal{L}_y = -\int_{\phi=0}^{\pi} p_a dS_y = -\int_{\phi=0}^{\pi} p_a \sin\phi La\, d\phi = \frac{5}{3}\rho_0 U^2 La. \tag{13.56}$$

If the average density of the cylinder is $\rho_1 > \rho_0$, the lift-to-weight ratio becomes

$$\frac{\mathcal{L}_y}{Mg_0} = \frac{5}{3\pi}\frac{\rho_0}{\rho_1 - \rho_0}\frac{U^2}{ag_0} \tag{13.57}$$

when buoyancy is also taken into account. Evidently, there is a critical flow speed beyond which the lift becomes larger than the weight, and a half-buried cylindrical body may be lifted out into the stream.

Example 13.8 [Half-buried pipe]: A circular pipeline of diameter $2a = 1$ m carrying oil is half buried at the bottom of the sea. The average density of the filled pipe, including the mass of the pipe, is twice that of water. According to the above formula, the critical velocity becomes $U \approx 3\ \mathrm{m\,s^{-1}}$. Obviously the pipeline should be anchored firmly to the sea bottom if the local current can attain this velocity.

The Magnus effect

Heinrich Gustav Magnus (1802–1870). German experimentalist. Known for his brilliant teaching and his research in both chemistry and physics. He described the effect that bears his name in 1852, although Newton had apparently much earlier understood the underlying cause.

A velocity potential $\Psi = C\phi$, where C is a constant, also satisfies the Laplace equation, $\nabla^2\Psi = 0$, and the corresponding irrotational velocity field is purely azimuthal:

$$v_r = 0, \qquad v_\phi = \frac{C}{r}. \tag{13.58}$$

On page 281 it will be shown that such a field can be created by a steadily rotating cylindrical spindle with radius a, angular velocity Ω, and $C = \Omega a^2$.The circulation (13.30) along any circle with center at the cylinder axis is easily calculated: $\Gamma = 2\pi r \times C/r = 2\pi C$. Since the vorticity vanishes everywhere for $r \geq a$, one might think that Stokes' theorem (13.31) would imply $\Gamma = 0$, but Stokes' theorem cannot be applied here because the surface spanned by the circle unavoidably must cut through the cylinder. The flow is not simply connected.

Since Laplace's equation is linear in Ψ, we may add the circulating field to the cylinder solution (13.53). The pressure is calculated as before, and due to the square of the sum of the two velocity fields, there will be a cross-contribution that is not symmetric under $\phi \to -\phi$:

$$\Delta p = \rho_0 U \sin\phi \left(1 + \frac{a^2}{r^2}\right)\frac{C}{r}. \tag{13.59}$$

This asymmetric pressure is called the *Magnus effect*.

The lift produced by the Magnus effect becomes:

$$\mathcal{L}_y = -\int_{\phi=0}^{2\pi} \Delta p_a dS_y = -\rho_0 U L \Gamma. \tag{13.60}$$

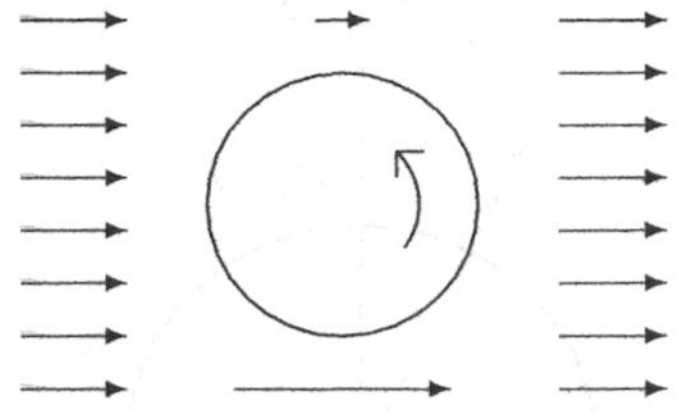

Sketch of the Magnus effect. A rotating cylinder drags air along near its surface (due to even the tiniest viscosity). This increases the wind velocity below the cylinder and decreases it above. By Bernoulli's theorem the pressure will be lower below and higher above, and consequently gives rise to a negative (i.e., downwards) lift. A similar argument explains, for example, why a Ping-Pong ball with topspin dives toward the table.

Intuitively it is easy to understand the Magnus effect, including the sign (see the margin figure). The Magnus effect also produces lift for a rotating sphere in a uniform stream with important consequences for all ballgames (see [DGQC10]). The above equation is of the same form as the famous Kutta-Joukowsky theorem (29.21) on page 524 which relates lift and circulation for aircraft wings.

13.8 Application: Sphere in a uniform stream

A sphere of radius a inserted into an asymptotically uniform flow with velocity U along the z-axis. The natural coordinates are of course spherical (Appendix D), and the symmetry of the problem implies that the velocity potential cannot depend on the azimuthal angle, so that the velocity potential $\Psi(r,\theta)$ and the stream function $\psi(r,\theta)$ only depend on r and θ.

Velocity potential

Asymptotically, for $r \to \infty$, the velocity potential has to approach the uniform flow $\Psi \to Uz = Ur\cos\theta$, and the linearity of the Laplace equation (13.33) requires the velocity potential to be linear in the asymptotic flow,

$$\Psi = U\cos\theta f(r). \tag{13.61}$$

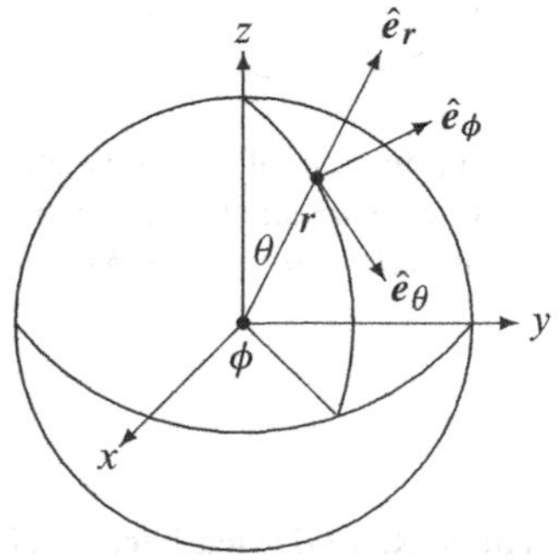

Spherical coordinates and their basis vectors.

Inserted into the spherical Laplacian (D.40), it becomes an ordinary differential equation,

$$\frac{d^2 f}{dr^2} + \frac{2}{r}\frac{df}{dr} - \frac{2}{r^2} f = 0, \tag{13.62}$$

for which the most general solution is $f = Ar + B/r^2$.

The asymptotic condition implies that $A = 1$, and the vanishing of the radial field $v_r = \nabla_r \Psi$ at the surface of the sphere requires $f'(a) = 0$, leading to $B = \frac{1}{2}a^3$. The velocity potential is thus

$$\Psi = Ur\cos\theta\left(1 + \frac{a^3}{2r^3}\right). \tag{13.63}$$

The stream function is calculated in Problem 13.10.

Velocity and pressure fields

The non-vanishing components of the velocity field is calculated from the spherical representation of the gradient (D.32) on page 624,

$$v_r = \nabla_r \Psi = \frac{\partial \Psi}{\partial r} = U\cos\theta\left(1 - \frac{a^3}{r^3}\right), \tag{13.64a}$$

$$v_\theta = \nabla_\theta \Psi = \frac{1}{r}\frac{\partial \Psi}{\partial \theta} = -U\sin\theta\left(1 + \frac{a^3}{2r^3}\right). \tag{13.64b}$$

The flow is visualized in Figure 13.4R.

The pressure is obtained from (13.34) and normalized to vanish at infinity,

$$p = \frac{1}{2}\rho_0 U^2 \frac{a^3}{r^3}\left(3\cos^2\theta - 1 - \frac{1}{4}(1 + 3\cos^2\theta)\frac{a^3}{r^3}\right). \tag{13.65}$$

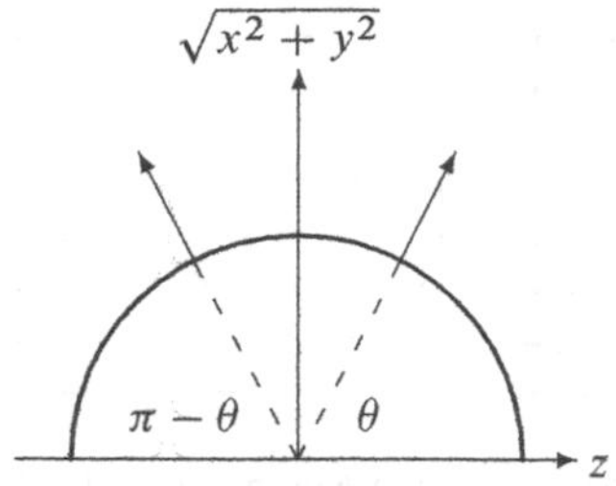

There is no drag because the projection of the pressure force on the z axis is equal and opposite for θ and $\pi - \theta$.

On the surface of the sphere the pressure becomes

$$p_a = p|_{r=a} = \frac{1}{2}\rho_0 U^2 \frac{9\cos^2\theta - 5}{4}. \tag{13.66}$$

It is negative in the interval $42° \lesssim \theta \lesssim 138°$. Again the symmetry $\phi \to 2\pi - \phi$ shows that there is no lift on the sphere, and the symmetry $\theta \to \pi - \theta$ shows that there is no drag on the sphere (and not even on the upper half).

Lift on a half sphere

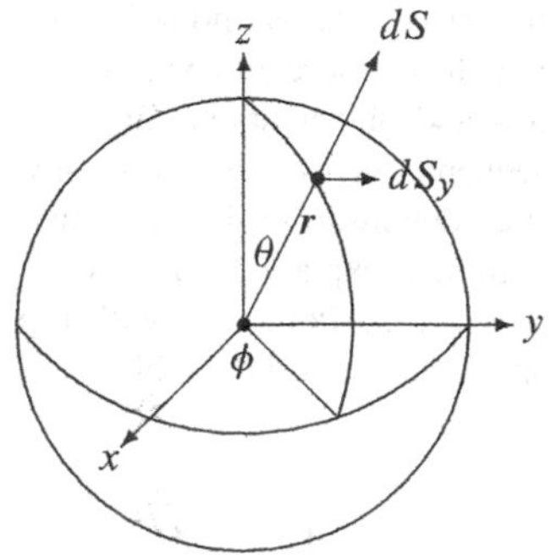

The projection of the area element $dS = a^2 \sin\theta \, d\phi d\theta$ on the y-axis is $dS_y = \sin\theta \sin\phi \, dS$.

The vertical lift on the upper half of the sphere becomes (see the margin figure)

$$\mathcal{L}_y = -\int_{\phi=0}^{\pi} p_a \, dS_y = -a^2 \int_0^{\pi} d\theta \int_0^{\pi} d\phi \, p_a \sin^2\theta \sin\phi = \frac{11\pi}{32}\rho_0 U^2 a^2. \tag{13.67}$$

If the density of the sphere is $\rho_1 > \rho_0$, the lift-to-weight ratio becomes

$$\frac{\mathcal{L}_y}{M g_0} = \frac{33}{128}\frac{\rho_0}{\rho_1 - \rho_0}\frac{U^2}{a g_0}. \tag{13.68}$$

As for the half cylinder, there is a critical stream velocity beyond which the lift becomes greater than the weight.

Example 13.9 [Half-buried fish]: A spherical fish with radius $a = 10$ cm and average density 10% higher than water (with the swim-bladder deflated) lies half buried in the sand at the bottom of a stream. The critical velocity for lift-off is merely $U \approx 0.6$ m s^{-1}, so this fish would do much better by flattening its shape or burrowing deeper.

13.9 d'Alembert's paradox

Jean le Rond d'Alembert (1717–1783). French mathematician. Introduced the concept of partial differential equations and was the first to solve such an equation.

The absence of drag in steady potential flow, which we have explicitly verified for the cylinder and sphere, may be formally shown to be true for any body shape (see page 534). But since everyday experience tells us that a steadily moving object is subject to drag from the fluid that surrounds it, even if the viscosity is vanishingly small, we have exposed a problem called *d'Alembert's paradox*. The paradox has the same origin as the inlet-outlet asymmetry discussed in Section 13.3, namely the symmetry of the steady-flow Euler equations (13.12) under reflection of the velocity field, $\boldsymbol{v}(\boldsymbol{x}) \to -\boldsymbol{v}(\boldsymbol{x})$. We shall see in Chapter 29 that d'Alembert's paradox may in fact be viewed positively as a statement about the smallness of drag compared to lift for streamlined bodies in nearly ideal fluids.

Drag from the trailing wake

Potential flow may be a mathematically correct solution for ideal flow but it misses important aspects of the physics of real flow. Although a tiny viscosity may not give rise to an appreciable friction force between body and fluid, it will generate vorticity close to the surface of the body. The vorticity will then spread into the fluid and produce a *trailing wake* behind the moving body in which the pressure is lower than in front, resulting in a net drag force.

Alternatively, one may understand the drag as caused by the constant loss of kinetic energy to the wake. In potential flow around an object, the fluid does not create a wake but returns to its original state with no loss of kinetic energy, implying that there is no resultant drag. But even if no kinetic energy is lost from the fluid in steady potential flow, there will be kinetic energy in the flow around the body. In the reference frame where the fluid is asymptotically at rest and the sphere moves with velocity $-\boldsymbol{U}$, the total kinetic energy of a sphere and the fluid around it becomes (see Problem 13.11)

$$\mathcal{T} = \frac{1}{2}MU^2 + \int_{r \geq a} \frac{1}{2}\rho_0(\boldsymbol{v} - \boldsymbol{U})^2\, dV = \frac{1}{2}\left(M + \frac{2\pi}{3}a^3\rho_0\right)U^2, \tag{13.69}$$

where M is the mass of the sphere and $\boldsymbol{v}$ is the potential flow velocity field (13.64).

This shows that the *effective mass* of the sphere plus fluid is $M + \frac{1}{2}m$, where $m = \frac{4}{3}\pi a^3\rho_0$ is the mass of the fluid displaced by the sphere. If the sphere moves steadily with constant velocity, the total kinetic energy is also constant, and no external forces need to perform any work. If, however, the sphere is accelerated by means of an external force $\mathcal{F}$, the rate of work of this force must be equal to the rate of loss of kinetic energy, $\mathcal{F}U = \dot{\mathcal{T}} = (M + \frac{1}{2}m)U\dot{U}$. Dividing by U we obtain the "equation of motion" for the sphere:

$$M\dot{U} = \mathcal{F} - \tfrac{1}{2}m\dot{U}. \tag{13.70}$$

The change in the flow pattern around an accelerated sphere thus produces an apparent *dynamic drag force* $\Delta\mathcal{F} = -\frac{1}{2}m\dot{U}$ against the direction of motion. Since m is the liquid mass displaced by the sphere, Archimedes would have loved this result—except for the factor $\frac{1}{2}$.

Problems

$*$ **13.1** **(a)** Show that in the final stages of the collapse of an empty spherical cavity the radius of the cavity is given by

$$R(t) = aC\left(1 - \frac{t}{t_0}\right)^{2/5} \qquad \text{for } t_0 - t \ll t_0,$$

where $C \approx 1.28371$ is a purely numeric constant. Note that the remarkable 2/5 power-dependence on time is also found in strong atmospheric blasts (page 454).

(b) Include surface tension and determine how that changes the formalism for the empty cavity collapse.

(c) Under what conditions can surface tension be ignored?

13.2 A siphon is a popular tool for stealing gasoline from a car. A tube of diameter 1 cm is inserted into the gasoline tank and sucked full of gasoline. The filled tube is quickly lowered into the opening of a 10-liter canister that lies about 20 cm below the level of gasoline in the tank. How long does it take to fill the canister?

13.3 There is a small correction to the flow from the wine barrel (page 214) because the velocity of the flow does not vanish exactly on the top of the barrel. Estimate this correction from the ratio of the barrel cross-section A_0 and the spout cross-section A.

13.4 A wine barrel has two spouts with different cross-sections A_1 and A_2 at the same horizontal level. Show that under steady flow conditions the wine emerges with the same speed from the two spouts.

13.5 An ideal incompressible liquid (water) of density ρ_0 streams through a horizontal tube of radius a. To determine the average flow velocity U_0, a small ring of thickness c is welded into the tube. The ring has outer radius a and inner radius $b = a - c$. A manometer is built into the system in the form of a small bypass partially filled with mercury of density ρ_1. The mercury surface lies a distance d below the tube's inner surface before the liquid is set into motion. The pressure in the liquid may everywhere be assumed to be constant across the tube cross-section.

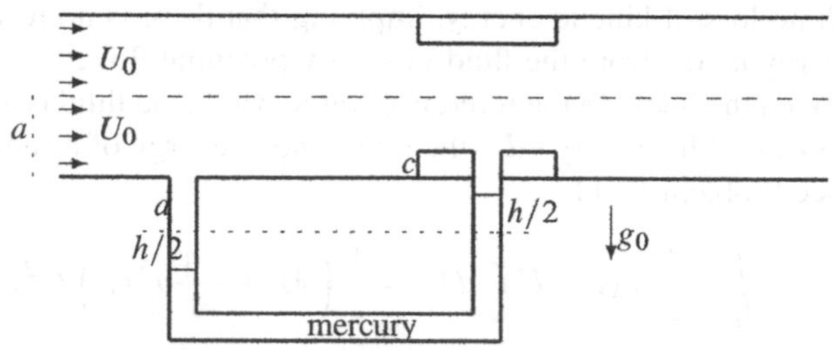

The following numbers may be used: $a = 15$ cm, $c = 1$ cm, $d = 5$ cm, $g_0 = 981$ cm^2 s^{-1}, $\rho_0 = 1$ g cm^{-3}, $\rho_1 = 13.6$ g cm^{-3}, $U_0 = 5$ m s^{-1}.

(a) Calculate the average velocity of the water when it passes the ring.
(b) Determine the pressure difference between the bypass openings.
(c) Calculate the difference h in the mercury levels in the bypass.
(d) Find the maximal velocity U_0 that can be allowed under the condition that the mercury does not enter the mainstream.

13.6 An incompressible non-viscous fluid flows out of a cistern with water level h through a small circular drain with radius a. The flow through the drain is non-uniform with velocity at a radius r from the center given by

$$v(r) = U\left(1 - \frac{r^2}{a^2}\right)^{\kappa},$$

where $U > 0$ is the central velocity and $\kappa > 0$ is a non-uniformity parameter. **(a)** Calculate the total volume flow through the drain. **(b)** Assuming that Bernoulli's theorem is valid for the central streamline, determine the reduction factor in the volume flux due to the non-uniformity.

* **13.7** Show that Stokes' theorem is independent of the shape of the surface.

* **13.8** Show that the vector area of a surface bounded by a closed curve C is given by

$$\int_S d\boldsymbol{S} = \frac{1}{2}\oint_C \boldsymbol{x} \times d\boldsymbol{\ell}.$$

Hint: Multiply with a constant vector $\boldsymbol{\Omega}$ and use Gauss' theorem.

13.9 A cylindrical worm with radius 3 mm lies half buried in sand at the bottom of the sea. Its density is 10% higher than the density of water. Calculate the critical speed at which the worm is lifted out of the water.

* **13.10** Show that the stream function for **(a)** a cylinder in a uniform crosswind and **(b)** a sphere in a uniform stream are, respectively,

$$\psi = Ur\sin\phi\left(1 - \frac{a^2}{r^2}\right), \qquad \psi = \frac{1}{2}Ur^2\sin^2\theta\left(1 - \frac{a^3}{r^3}\right). \tag{13.71}$$

* **13.11** **(a)** Carry out the integral of the effective mass in (13.69). **(b)** Show that the momentum of the fluid is infinite.

14 Compressible flow

Even if air and other gases appear to be quite compressible in our daily doings, we have until now only analyzed incompressible flow and even sometimes applied it to gases. The reason is—as pointed out before—that a gas in steady flow "prefers to get out of the way" rather than become compressed when it encounters an obstacle. Unless one entraps the gas, for example in a balloon or bicycle tire, it will be effectively incompressible in steady flow as long as its velocity relative to obstacles and container walls is well below the speed of sound.

But when steady flow speeds approach the speed of sound in the fluid, compression becomes unavoidable. At such speeds the fluid has, so to speak, not enough time to get out of the way. Normal passenger jets routinely cruise at speeds just below the speed of sound and considerable compression of air must be expected at the front end of the aircraft. High-speed projectiles and fighter jets move at several times the speed of sound while space vehicles and meteorites move at many times the speed of sound. At supersonic speeds the compression at the front of a moving object becomes so strong that a pressure discontinuity or shock is formed that trails the object and is perceived as a sonic boom.

In unsteady flow the effective incompressibility of fluids cannot be taken for granted, even if the velocity is well below sound velocity. Rapid changes in the boundary conditions will generate small-amplitude compression waves, called *sound*, in all fluids. When you clap your hands you create momentarily a small disturbance in the air that propagates to your ear. Likewise, the diaphragm of the loudspeaker in your radio vibrates in tune with the music carried by the radio waves and the electric currents in wires connecting it to the radio, and transfers these vibrations to the air where they continue as sound.

In this chapter we shall begin by investigating compressible flow in ideal fluids, first for harmonic sound waves, and then for sonic and supersonic steady flow through ducts and nozzles. At the end of Chapter 15 we shall present the basic formalism for compressible viscous flow and calculate the viscous attenuation in otherwise harmonic wave propagation.

14.1 Small-amplitude sound waves

Although harmonic compression waves propagate through the fluid at the speed of sound, the amplitude of the velocity oscillations in a sound wave is normally very small compared to the speed of sound. No significant bulk movement of air takes place over longer distances but locally the air oscillates back and forth with small spatial amplitude, and the velocity, density, and pressure fields oscillate along with it.

Wave equation for sound

The Euler equations for ideal compressible flow are obtained from the general dynamic equation (12.26) with stress tensor, $\sigma_{ij} = -p\,\delta_{ij}$, and the continuity equation (12.12),

$$\frac{\partial \boldsymbol{v}}{\partial t} + (\boldsymbol{v}\cdot\nabla)\boldsymbol{v} = \boldsymbol{g} - \frac{1}{\rho}\nabla p, \qquad \frac{\partial \rho}{\partial t} + \nabla\cdot(\rho\boldsymbol{v}) = 0. \tag{14.1}$$

The first expresses the local form of Newton's Second Law and the second local mass conservation. For simplicity we shall in this section assume that there are no volume forces (gravity), $\boldsymbol{g} = \boldsymbol{0}$ (see however Problem 14.5).

Before any sound is produced the fluid is assumed to be in hydrostatic equilibrium with constant density ρ_0 and constant pressure p_0. We now disturb the equilibrium by setting the fluid into motion with a tiny time-dependent velocity field $\boldsymbol{v}(\boldsymbol{x}, t)$. This disturbance generates small changes in the density $\rho = \rho_0 + \Delta\rho$, and the pressure $p = p_0 + \Delta p$. Expanding to first order in the small quantities, $\boldsymbol{v}$, Δp, and $\Delta\rho$, the Euler equations become

$$\frac{\partial \boldsymbol{v}}{\partial t} = -\frac{1}{\rho_0}\nabla\Delta p, \qquad \frac{\partial \Delta\rho}{\partial t} = -\rho_0\nabla\cdot\boldsymbol{v}. \tag{14.2}$$

Differentiating the second equation with respect to time and making use of the first, we get

$$\frac{\partial^2\Delta\rho}{\partial t^2} = \nabla^2\Delta p. \tag{14.3}$$

Assuming now that the fluid obeys a barotropic equation of state, $p = p(\rho)$, we obtain a first-order relation between the pressure and density changes,

$$\Delta p = \left.\frac{dp}{d\rho}\right|_0 \Delta\rho = \frac{K_0}{\rho_0}\Delta\rho, \tag{14.4}$$

where K_0 is the equilibrium bulk modulus defined in Equation (2.42) on page 33.

Inserting Δp in (14.3) we get a *standard wave equation* for the density change,

$$\boxed{\frac{\partial^2\Delta\rho}{\partial t^2} = c_0^2\nabla^2\Delta\rho,} \tag{14.5}$$

where for convenience we have introduced the constant

$$c_0 = \sqrt{\frac{K_0}{\rho_0}}. \tag{14.6}$$

It has the dimension of a velocity and represents, as we shall see, the *speed of sound* of harmonic waves. For water with $K_0 \approx 2.3$ GPa and $\rho_0 \approx 10^3$ kg m^{-3}, the sound speed comes to about $c_0 \approx 1500$ m s^{-1} ≈ 5500 km h^{-1}.

Fluid	T	c_0
Glycerol	25	1920
Sea water	20	1521
Fresh water	20	1482
Lube oil	25	1461
Mercury	25	1449
Ethanol	25	1145
Hydrogen	27	1310
Helium	0	973
Water vapor	100	478
Neon	30	461
Humid air	20	345
Dry air	20	343
Oxygen	30	332
Argon	0	308
Nitrogen	27	363
Unit	°C	m s^{-1}

Empirical sound speeds in various liquids (above) and gases (below). The temperature of the measurement is also listed.

Isentropic sound speed in an ideal gas

Sound vibrations are normally so rapid that temperature equilibrium is never established. This allows us to assume that the oscillations take place adiabatically, that is, without heat exchange. The bulk modulus of an isentropic ideal gas (Equation (2.44) on page 33) is $K_0 = \gamma p_0$, where γ is the adiabatic index, and we obtain the *isentropic sound velocity*

$$\boxed{c_0 = \sqrt{\gamma\frac{p_0}{\rho_0}} = \sqrt{\gamma R T_0}.} \tag{14.7}$$

In the last step we have used the ideal gas law $p_0 = R\rho_0 T_0$,where $R = R_{\text{mol}}/M_{\text{mol}}$ is the specific gas constant (see page 30). The *isothermal sound velocity* is recovered for $\gamma = 1$.

Example 14.1 [Sound speed in the atmosphere]: For dry air at 20 °C with $\gamma = 7/5$ and $M_{\text{mol}} = 29 \text{ g mol}^{-1}$, the sound speed comes to $c_0 = 343 \text{ m s}^{-1} = 1235 \text{ km h}^{-1}$. Since the temperature of the homentropic atmosphere falls linearly with height in the troposphere (see page 36), the speed of sound varies with height z above the ground as

$$c = c_0\sqrt{1 - \frac{z}{h_2}}, \tag{14.8}$$

where c_0 is the sound speed at sea level and $h_2 \approx 30$ km is the homentropic scale height. At the flying altitude of modern jet aircraft, $z \approx 10$ km, the sound speed has dropped to $c \approx 280 \text{ m s}^{-1} \approx 1000 \text{ km h}^{-1}$ (disregarding the effect of humidity, which makes it higher). Beyond the troposphere this expression begins to fail because the homentropic model of the atmosphere fails.

Plane wave solution

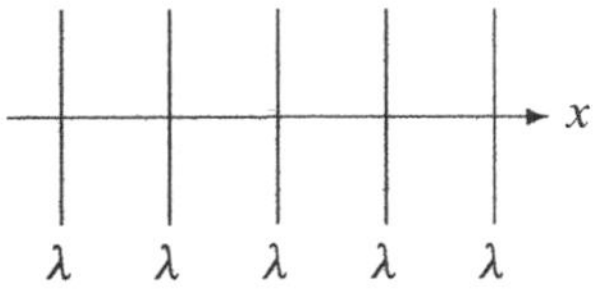

Plane density wave propagating along the x-axis with wavelength λ. The density is constant in all planes orthogonal to the direction of propagation.

An elementary plane density wave moving along the x-axis with wavelength λ, period τ, and amplitude $\rho_1 > 0$ is described by a density correction of the form

$$\Delta\rho = \rho_1 \sin(kx - \omega t), \tag{14.9}$$

where $k = 2\pi/\lambda$ is the wavenumber and $\omega = 2\pi/\tau$ is the circular frequency. Inserting this expression into the wave equation (14.5), we obtain $\omega^2 = c_0^2 k^2$ or $c_0 = \omega/k = \lambda/\tau$. The surfaces of constant density (or pressure) are planes orthogonal to the direction of propagation, corresponding to $kx - \omega t =$ const. Differentiating this equation with respect to time, we see that the planes of constant density move with velocity $dx/dt = \omega/k = c_0$, also called the *phase velocity* of the wave. This shows that c_0 given by (14.6) is indeed the speed of small-amplitude sound in the material.

From the x-component of the Euler equation (14.2) we obtain the only non-vanishing component of the velocity field

$$v_x = v_1 \sin(kx - \omega t), \qquad v_1 = c_0 \frac{\rho_1}{\rho_0}. \tag{14.10}$$

Since $v_y = v_z = 0$, the velocity field of a sound wave in an isotropic fluid is always *longitudinal*, that is, parallel to the direction of wave propagation. The corresponding spatial displacement field u_x, defined by $\partial u_x/\partial t = v_x$ becomes (apart from a constant)

$$u_x = a_1 \cos(kx - \omega t), \qquad a_1 = \frac{v_1}{\omega}, \tag{14.11}$$

where a_1 is the spatial displacement amplitude of the sound wave.

Validity of the approximation

It only remains to verify the approximation of dropping the advective acceleration. The actual ratio between the magnitudes of the advective and local accelerations is

$$\frac{|(\boldsymbol{v} \cdot \nabla)\boldsymbol{v}|}{|\partial \boldsymbol{v}/\partial t|} \approx \frac{k v_1^2}{\omega v_1} = \frac{v_1}{c_0}. \tag{14.12}$$

The condition for the validity of the approximation is thus that *the amplitude of the velocity oscillations should be much smaller than the speed of sound,* $v_1 \ll c_0$. This is equivalent to $\rho_1 \ll \rho_0$, to $p_1 \ll K_0$, and to $a_1 \ll \lambda/2\pi$.

Example 14.2 [Loudspeaker]: A certain loudspeaker transmits sound to air at frequency $\omega/2\pi = 1000 \text{ s}^{-1}$ with diaphragm displacement amplitude of $a_1 = 1$ mm. The velocity amplitude becomes $v_1 = a_1\omega \approx 6 \text{ m s}^{-1}$, and since $v_1/c_0 \approx 1/57$ the approximation of leaving out the advective acceleration is well justified.

* The Jeans instability

In Section 12.6 a question was raised about the meaning of the gravitational field of the whole universe. At that point we decided to view the universe as a huge finite sphere with uniform mass density, and that led us to Newtonian cosmology, which has many features in common with general relativistic cosmology, even if the theories differ at a deeper level. Consider now a region of space that is small at the scale of the universe so that the cosmic gravitational field may be taken to be constant across this region. According to Einstein's equivalence principle such a field of gravity is completely unobservable to a freely falling observer anywhere in this region, and may therefore be disregarded.

A disturbance $\Delta\rho(\boldsymbol{x},t)$ on top of the otherwise uniform mass density ρ_0 generates not only a pressure change $\Delta p = c_0^2\Delta\rho$ but also a local gravitational field $\boldsymbol{g}$ that must be added to the right-hand side of the first of the Euler equations (14.2). It follows from the arguments given above and the Coulomb nature of gravity that the divergence of the gravitational field is proportional to the change in density (with a minus sign due to the attractive nature of gravity),

$$\nabla\cdot\boldsymbol{g} = -4\pi G\Delta\rho, \tag{14.13}$$

where G is the universal gravitational constant. Instead of the simple wave equation (14.5) we now get the modified wave equation[1]

$$\boxed{\frac{\partial^2\Delta\rho}{\partial t^2} = c_0^2\nabla^2\Delta\rho + 4\pi G\rho_0\Delta\rho.} \tag{14.14}$$

Inserting the plane wave (14.9) we obtain the (dispersion) relation

$$\omega^2 = c_0^2k^2 - 4\pi G\rho_0. \tag{14.15}$$

It is only a normal oscillating wave as long as ω^2 is positive, that is, for $c_0k > \sqrt{4\pi G\rho_0}$. For smaller values of the wavenumber the frequency becomes imaginary, signaling an exponentially growing(or decaying) wave amplitude.

Thus, any disturbance containing wavelengths greater than the *Jeans wavelength*,

$$\boxed{\lambda_J = c_0\sqrt{\frac{\pi}{G\rho_0}},} \tag{14.16}$$

is *unstable*. This is the famous *Jeans Instability* from 1902 [Jea02]. For an ideal gas with molar mass M_{mol} and temperature T_0, the isothermal sound velocity is $c_0 = \sqrt{RT_0}$, where $R = R_{\text{mol}}/M_{\text{mol}}$ is the specific gas constant. Taking $\rho_0 = 3.35\times10^{-18}$ kg m^{-3}, corresponding to 1,000 hydrogen molecules per cubic centimeter, $M_{\text{mol}} = 2$ g/mol, and $T_0 = 3$ K, we find $\lambda_J = 1.9\times10^{16}$ m or about 2 light years. The mass of a spherical region of diameter λ_J becomes $M_J = 1.1\times10^{31}$ kg, corresponding to about 6 solar masses, a not unreasonable result in view of the primitive model.

Sir James Hopwood Jeans (1877–1946). British physicist and applied mathematician. Worked on the dynamical theory of gases, electromagnetism, quantum mechanics, astrophysics, and cosmology. (Source: Wikimedia Commons.)

The "Jeans swindle": The arguments leading up to (14.13), which only depends on the local mass density fluctuation $\Delta\rho$ and not on the background density ρ_0, have been subject to much criticism during the 100 years since Jeans published his calculations, and have been stamped with the rather derogatory moniker "Jeans swindle". The result is, however, quite robust and can be given reasonable justification as we did in the beginning of this subsection. A recent mathematical analysis of the several infinities involved in the problem may be found in [Kie03].

[1] It is identical to the *Klein–Gordon equation* with imaginary mass and has been used in the theory of faster-than-light particles (tachyons).

14.2 Steady compressible flow

In steady compressible flow, the velocity, pressure, and density are all independent of time, and the Euler equations take the simpler form,

$$(\boldsymbol{v}\cdot\nabla)\,\boldsymbol{v} = \boldsymbol{g} - \frac{1}{\rho}\nabla p, \qquad \nabla\cdot(\rho\boldsymbol{v}) = 0. \tag{14.17}$$

Here we shall, for simplicity, assume that the fluid obeys a barotropic equation of state, $p = p(\rho)$ or $\rho = \rho(p)$, leaving us with a closed set of five field equations for the five fields, v_x, v_y, v_z, ρ, and p. In this section gravity will mostly be ignored.

Effective incompressibility

First we shall demonstrate the claim made in the preceding section that *in steady flow a fluid is effectively incompressible when the flow speed is everywhere much smaller than the local speed of sound.* The ratio of the local flow speed $\boldsymbol{v}$ (relative to a static solid object or boundary wall) and the local sound speed c is called the (local) *Mach number*,

$$\mathsf{Ma} = \frac{|\boldsymbol{v}|}{c}. \tag{14.18}$$

In terms of the Mach number, the claim is that a steady flow is effectively incompressible when $\mathsf{Ma} \ll 1$ everywhere. Conversely, the flow is truly compressible if the local Mach number somewhere is comparable to unity or larger, $\mathsf{Ma} \gtrsim 1$.

The essential step in the proof is to relate the gradient of pressure to the gradient of density, $\nabla p = (dp/d\rho)\nabla\rho = c^2\nabla\rho$, where $c = \sqrt{dp/d\rho}$ is the local speed of sound. Writing the divergence condition in the form $\nabla\cdot(\rho\boldsymbol{v}) = \rho\nabla\cdot\boldsymbol{v} + (\boldsymbol{v}\cdot\nabla)\rho = 0$ and making use of the Euler equation without gravity, we find the exact result,

$$\nabla\cdot\boldsymbol{v} = -\frac{1}{\rho}(\boldsymbol{v}\cdot\nabla)\rho = -\frac{1}{\rho c^2}(\boldsymbol{v}\cdot\nabla)p = \frac{\boldsymbol{v}\cdot(\boldsymbol{v}\cdot\nabla)\boldsymbol{v}}{c^2}. \tag{14.19}$$

Applying the Schwarz inequality to the numerator (see Problem 14.4), we get

$$|\nabla\cdot\boldsymbol{v}| \le \frac{|\boldsymbol{v}|^2}{c^2}\,|\nabla\boldsymbol{v}| = \mathsf{Ma}^2\,|\nabla\boldsymbol{v}|, \tag{14.20}$$

where $|\nabla\boldsymbol{v}| = \sqrt{\sum_{ij}(\nabla_i v_j)^2}$ is the norm of the velocity gradient matrix.

This relation clearly demonstrates that for $\mathsf{Ma}^2 \ll 1$, the divergence $\nabla\cdot\boldsymbol{v}$ is much smaller than the velocity gradients $\nabla\boldsymbol{v}$, making the incompressibility condition, $\nabla\cdot\boldsymbol{v} = 0$, a good approximation. Typically a flow may be taken to be incompressible when $\mathsf{Ma} \lesssim 0.3$ everywhere, corresponding to $\mathsf{Ma}^2 \lesssim 0.1$.

Ernst Mach (1838–1916). Austrian positivist philosopher and physicist. Made early advances in psycho-physics, the physics of sensations. His rejection of Newton's absolute space and time prepared the way for Einstein's theory of relativity. Proposed the principle that inertia results from the interaction between a body and all other matter in the universe. (Source: Wikimedia Commons.)

Example 14.3 [Mach numbers]: Waving your hands in the air, you generate flow velocities at most of the order of meters per second, corresponding to $\mathsf{Ma} \approx 0.01$. Driving a car at 120 km h^{-1} $\approx$ 33 m s^{-1} corresponds to $\mathsf{Ma} \approx 0.12$. A passenger jet flying at a height of 10 km with velocity about 900 km h^{-1} $\approx$ 250 m s^{-1} has $\mathsf{Ma} \approx 0.9$ because the velocity of sound is only about 1000 km h^{-1} at this height (see Example 14.1). Even if this speed is subsonic, considerable compression of the air must occur, especially at the front of the wings and body of the aircraft. The Concorde and modern fighter aircraft operate at supersonic speeds at Mach 2–3, and the Space Shuttle enters the atmosphere at the hypersonic speed of Mach 25. The strong compression of the air at the frontal parts of such aircraft creates shock waves that appear to us as sonic booms.

Bernoulli's theorem for barotropic fluids

For compressible fluids, Bernoulli's theorem is still valid in a slightly modified form. If the fluid is in a barotropic state with $\rho = \rho(p)$, the Bernoulli field becomes

$$H = \tfrac{1}{2}\boldsymbol{v}^2 + \Phi + w(p), \tag{14.21}$$

where

$$w(p) = \int \frac{dp}{\rho(p)} \tag{14.22}$$

is the *pressure potential*, previously defined on page 32. The proof of the modified Bernoulli theorem is elementary and follows essentially the same lines as on page 212, using $Dw/Dt = (dw/dp)Dp/Dt = \rho^{-1}(\boldsymbol{v} \cdot \nabla)p$.

The most interesting barotropic fluid is an isentropic ideal gas with adiabatic index γ, for which it has been shown on page 35 that the pressure potential is linear in the temperature,

$$w = c_p T, \qquad c_p = \frac{\gamma}{\gamma - 1} R. \tag{14.23}$$

Here c_p is the specific heat of air at constant pressure and $R = R_{\text{mol}}/M_{\text{mol}}$ the specific gas constant. Thus, in the absence of gravity, a drop in velocity along a streamline in isentropic flow is accompanied by a rise in temperature (as well as a rise in both pressure and density).

Isentropic steady flow: There is a conceptual subtlety in understanding isentropic steady flow because of the unavoidable heat conduction that takes place in all real fluids. Since truly steady flow lasts "forever", one might think that there would be ample time for a local temperature change to spread throughout the fluid, regardless of how badly it conducts heat. But remember that steady flow is not static, and fresh fluid is incessantly being compressed or expanded adiabatically, accompanied by local heating and cooling. Provided the flow is sufficiently fast, heat conduction will have little effect. The physics of heat and flow will be discussed in Chapter 22.

Stagnation temperature rise

An object moving through an ideal fluid has at least one stagnation point at the front where the fluid comes to rest relative to the object. There is also at least one stagnation point at the rear of a body, but vortex formation and turbulence will generally disturb the flow so much in this region that the streamlines get tangled and form unsteady whirls. This will often prevent us from using Bernoulli's theorem to relate velocity and pressure at the rear of the body.

At the forward stagnation point the gas is compressed and the temperature will always be higher than in the fluid at large (and similarly at the rear stagnation point if such exists). In the frame of reference where the object is at rest and the fluid asymptotically moves with constant speed and temperature, the flow is steady, and we find from the modified Bernoulli field (14.21) with pressure potential (14.23) in the absence of gravity,

$$\frac{1}{2}\boldsymbol{v}^2 + c_p T = c_p T_0, \tag{14.24}$$

A static airfoil in an air stream coming in horizontally from the left. The pictured streamline (dashed) ends at the forward stagnation point.

where T_0 is the stagnation point temperature and T is the temperature at a point of the streamline where the velocity is $\boldsymbol{v}$ (see the margin figure).

The total temperature rise due to adiabatic compression thus becomes

$$\boxed{\Delta T = T_0 - T = \frac{\boldsymbol{v}^2}{2c_p}.} \tag{14.25}$$

The stagnation temperature rise depends only on the velocity difference between the body and the fluid, and not on the temperature, density, or pressure of the gas. Note that a lower molar mass implies a higher specific heat and thus a smaller stagnation temperature rise.

Example 14.4: A car moving at 100 km h^{-1} has a stagnation temperature rise at the front of merely 0.4 K. For a passenger jet traveling at 900 km h^{-1}, the stagnation temperature rise is a moderate 31 K, whereas a supersonic aircraft traveling at 2,300 km h^{-1} suffers a stagnation point temperature rise of about 200 K. When a re-entry vehicle, such as the Space Shuttle, hits the dense atmosphere with a speed of 3 km s^{-1}, the predicted stagnation point temperature rise would be 4,500 K. At that temperature the air is dissociated and partly ionized, and becomes a glowing plasma with a much lower average molar mass (because of the larger number of low-mass particles in the plasma), and this lowers the stagnation temperature. The plasma also creates a hot shock front a small distance ahead of the exposed surfaces that deflects most of the heat (see Figure 26.6 on page 452), and the surface temperatures do in fact not exceed 2,500 K during reentry. Since such temperatures are nevertheless capable of melting and burning metals, it has been necessary to protect the exposed surfaces of the Space Shuttle with a special heat shield of ceramic tiles. Damage to the thermal protection system led in fact to the disastrous loss of the space shuttle Columbia on February 1, 2003.

Stagnation and sonic properties

It is often convenient to express the ratio of the local temperature to the stagnation point temperature in terms of the local Mach number $\mathsf{Ma} = |\boldsymbol{v}|/c$ where $c = \sqrt{\gamma RT}$ is the local sound velocity. Setting $\boldsymbol{v}^2 = \mathsf{Ma}^2 c^2$ in (14.24), we obtain

$$\frac{T}{T_0} = \left(1 + \tfrac{1}{2}(\gamma - 1)\,\mathsf{Ma}^2\right)^{-1}. \tag{14.26}$$

This relation is valid for any streamline because the stagnation temperature T_0 can be defined by means of (14.24), even if the streamline does not actually end in a stagnation point.

The corresponding pressure and density ratios are obtained from the isentropic relation, $T^\gamma p^{1-\gamma} = T_0^\gamma p_0^{1-\gamma}$, and the ideal gas law, $\rho = p/RT$,

$$\frac{p}{p_0} = \left(\frac{T}{T_0}\right)^{\gamma/(\gamma-1)}, \qquad \frac{\rho}{\rho_0} = \left(\frac{T}{T_0}\right)^{1/(\gamma-1)}. \tag{14.27}$$

In principle the stagnation values T_0, p_0, and ρ_0 can be different for different streamlines. But if for example an object moves through a homogeneous gas that is asymptotically at rest, all stagnation parameters will be true constants independent of the streamline. The flow is then said to be *homentropic*.

A point where the velocity of a steady flow equals the local velocity of sound, $v = c$, is analogously called a *sonic point*. The sonic temperature T_1, pressure p_1, and density ρ_1 are simply related to the stagnation values. Setting $\mathsf{Ma} = 1$ in the expressions above, we get

$$\frac{T_1}{T_0} = \frac{2}{\gamma + 1}, \qquad \frac{p_1}{p_0} = \left(\frac{T_1}{T_0}\right)^{\gamma/(\gamma-1)}, \qquad \frac{\rho_1}{\rho_0} = \left(\frac{T_1}{T_0}\right)^{1/(\gamma-1)}. \tag{14.28}$$

For air with $\gamma = 7/5$ the right hand sides become 0.8333, 0.5282, and 0.6339. The local-to-sonic temperature ratio may now be written

$$\frac{T}{T_1} = \left(1 + \frac{\gamma - 1}{\gamma + 1}\left(\mathsf{Ma}^2 - 1\right)\right)^{-1}, \tag{14.29}$$

from which the corresponding pressure and density ratios may be obtained using expressions analogous to (14.27).

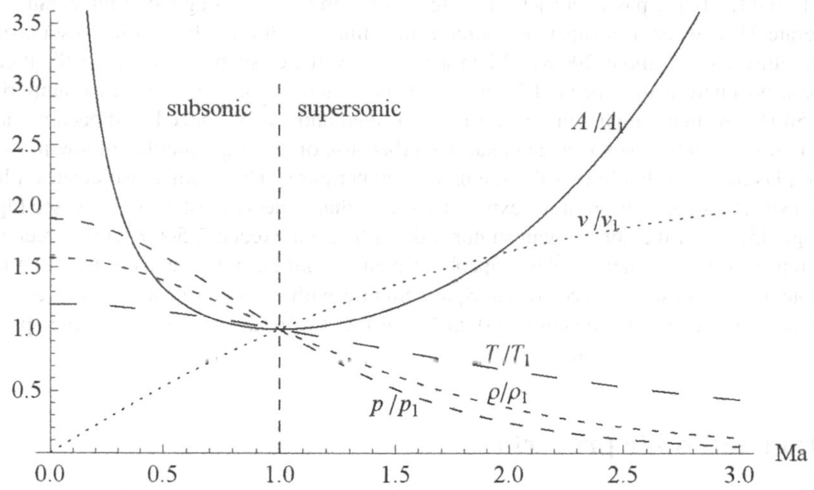

Figure 14.1. Plot of the ratio of local to sonic values as a function of the Mach number in a slowly varying duct for $\gamma = 7/5$. The ratio A/A_1 is a solid line, T/T_1 has large dashes, ρ/ρ_1 medium dashes, p/p_1 small dashes, and v/v_1 is dotted.

Duct with slowly varying cross-section

Consider now an ideal gas flowing through a straight duct with a slowly varying cross-section area $A = A(x)$ orthogonal to the x-axis (see the margin figure). The temperature T, density ρ, pressure p, and normal velocity $v = v_x$ are assumed to be constant over any cross-section, but like the area slowly varying with x. In this quasi-one-dimensional approximation we thus disregard the tiny flow components orthogonal to the x-axis. Since all streamlines have the same parameter values in any cross-section, the flow is homentropic.

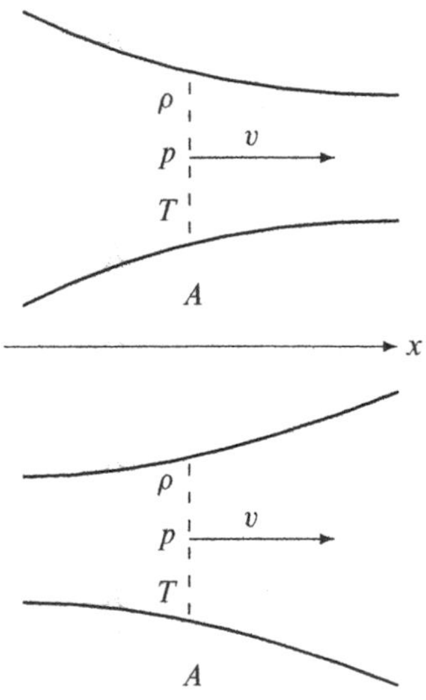

Top: Converging duct. For subsonic flow the velocity increases while the pressure decreases toward the right as in the Venturi effect. **Bottom:** Diverging duct. For supersonic flow the velocity increases while the pressure decreases toward the right.

The constancy of the mass flow rate along the duct, $Q = \rho A v$, provides us with a useful relation between the duct area and the local Mach number. At the sonic point we have the same mass flow as everywhere else, so that $\rho A v = \rho_1 A_1 v_1$. Using $v/c = \mathsf{Ma}$ and $v_1/c_1 = 1$ where c and c_1 are the local and sonic sound velocities, we find

$$\frac{A}{A_1} = \frac{\rho_1 v_1}{\rho v} = \frac{1}{\mathsf{Ma}}\frac{c_1}{c}\frac{\rho_1}{\rho} = \frac{1}{\mathsf{Ma}}\left(\frac{T_1}{T}\right)^{1/2+1/(\gamma-1)},$$

where in the third step we used the ideal gas law. Finally, inserting (14.29), we get

$$\boxed{\frac{A}{A_1} = \frac{1}{\mathsf{Ma}}\left(1 + \frac{\gamma-1}{\gamma+1}\left(\mathsf{Ma}^2 - 1\right)\right)^{1/2+1/(\gamma-1)}}. \tag{14.30}$$

This function, which obviously has minimum $A = A_1$ at the sonic point $\mathsf{Ma} = 1$, is plotted as the solid curve in Figure 14.1, together with the various flow parameters divided by their sonic values. Correspondingly, the current density of mass, $\rho v = Q/A$, can never become larger than the value it takes at the sonic point.

Figure 14.1 is central to the analysis of duct flow. Inspecting the curves we see that subsonic flow ($\mathsf{Ma} < 1$) follows the Venturi principle, such that a *decreasing* duct area implies *increasing* flow velocity and decreasing temperature, pressure, and density (and conversely). But for supersonic flow ($\mathsf{Ma} > 1$) this behavior is reversed, such that an *increasing* duct area now leads to *increasing* velocity and decreasing temperature, pressure, and density (and conversely). This surprising behavior is the key to understanding how supersonic exhaust speeds are obtained in steam turbines, wind tunnels, supersonic aircraft, and rocket engines.

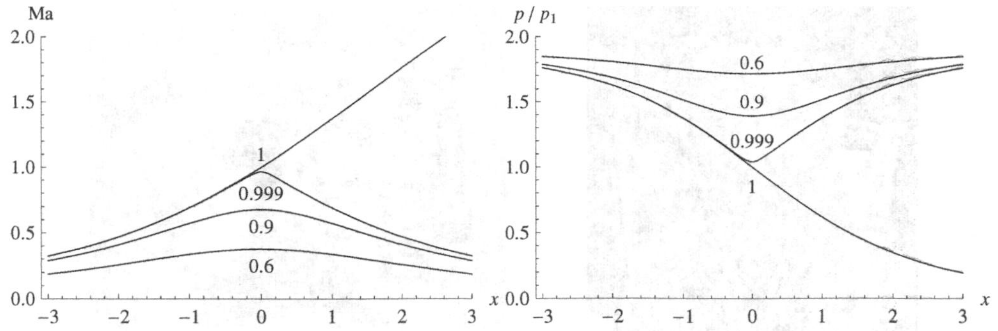

Figure 14.2. Simple symmetric model of a Laval nozzle, $A(x) = A_{\text{throat}} + kx^2$ for $-3 < x < 3$, with $A_{\text{throat}} = 1$ and $k = 0.1$ (and $\gamma = 7/5$). **Left:** Plot of the Mach number $\mathsf{Ma}(x)$ as a function of the duct coordinate x. The different curves are labeled with the ratio A_1/A_{throat}. **Right:** The pressure ratio $p(x)/p_1$ under the same conditions. For $A_1 < A_{\text{throat}}$, the pressure is lowest in the throat (the Venturi effect) but drops to much lower values for $A_1 = A_{\text{throat}}$ when the flow in the diverging part becomes supersonic. In this model the entry and exit Mach numbers are 0.32 and 2.14. The entry pressure is 9.0 times larger than the exit pressure.

Carl Gustav Patrik de Laval (1845–1913). Swedish engineer. Worked on steam turbines and dairy machinery, such as milk-cream separators and milking machines. In 1883 he founded a company that is now called Alfa Laval and is a world leader in heat transfer, material separation, and fluid handling. Discovered in 1888 that a converging-diverging nozzle generates much higher steam speed and thereby higher steam turbine rotation speed. (Source: Wikimedia Commons.)

14.3 Application: The Laval nozzle

In 1888 the Swedish engineer, de Laval, discovered that supersonic steam speeds could be reached in steam turbines by accelerating the steam through a nozzle that first converges and then diverges, like the one shown in the margin figure and modeled in Figure 14.2. This unique design has since been used in all kinds of devices, for example jet and rocket engines, in which one wishes to maximize thrust by accelerating the combustion gases to supersonic speeds. It may even arise naturally in solid-liquid impacts [GPG&10].

The curves plotted in Figure 14.1 indicate that the flow velocity will increase smoothly all the way through a converging-diverging duct, provided it passes through sonic speed precisely at the narrowest point, also called the *throat*. For this to happen, flow conditions must be arranged such that the sonic area exactly equals the physical throat area, $A_1 = A_{\text{throat}}$ (how this is done will be discussed below). Then, as the gas streams through the converging part of the nozzle, the local area A travels down the left-hand, subsonic branch of the area curve while the flow speed U simultaneously increases. Passing the throat at sonic speed, the gas streams through the diverging part while the local area travels up the right-hand supersonic branch and the speed continues to increase. Without the diverging part of the nozzle, the flow could at most reach sonic speed at the exit, but not go beyond. The expansion of the gas in the divergent part is thus essential for obtaining supersonic flow. In fact, the ratio of the nozzle's exit to throat area directly determines the Mach number of the exhaust.

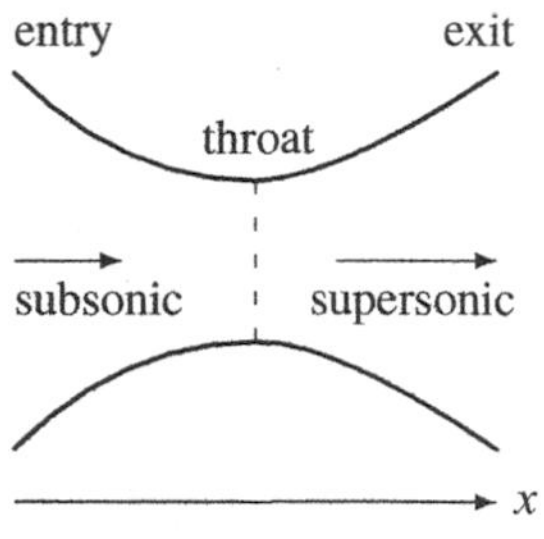

A subsonic flow may become supersonic in a duct with a constriction where the duct changes from converging to diverging. The transition must take place at the narrowest point of the nozzle, called the throat.

Sonic speed is not always reached. Flutes and other musical instruments, including the human vocal tract, have constrictions in the airflow that do not give rise to supersonic flow (which would surely destroy the music). In this case the flow conditions must be such that the sonic area is strictly smaller than the physical throat area, $A_1 < A_{\text{throat}}$. As the gas streams through the converging part of the nozzle, its area travels as before down the left-hand branch of the area curve in Figure 14.1 until it reaches the physical throat area where it turns around and backtracks up along the left-hand branch of the area curve while proceeding through the diverging part of the nozzle. The Mach number never reaches unity and the pressure rises until it passes the throat, after which it falls back again in the exit region.

In Figure 14.2, a model of a Laval nozzle is solved for a few values of A_1/A_{throat}, including the unique *critical* solution, $A_1 = A_{\text{throat}}$, where the flow does become sonic right at the throat, and continues as supersonic afterward. In this case the pressure continues to fall through the exit region.

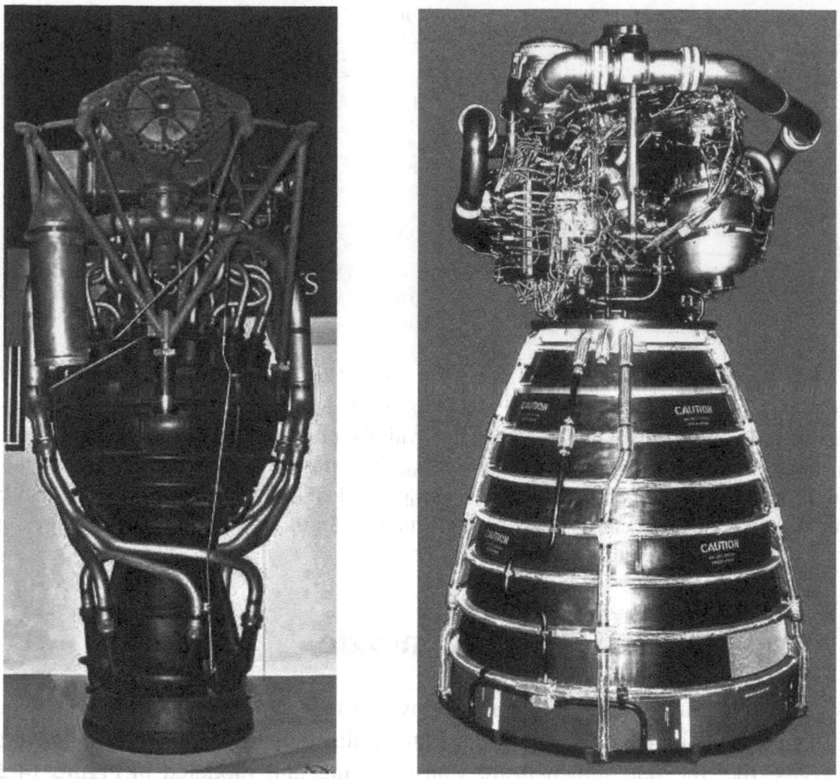

Figure 14.3. **Left:** A4/V2 rocket engine (circa 1943). Developed and used by Germany during the Second World War. After the war it was used to "ignite" the American rocket program. Engine height: 1.7 m. Propellant: ethanol and liquid oxygen. The V2 rocket was powered by one such engine. (Source: Wikimedia Commons.) **Right:** Space Shuttle Main Engine (SSME); developed in USA (circa 1980). Engine height 4.3 m. Propellant: liquid hydrogen and oxygen. The Space Shuttle was powered by three such engines, aided by two solid fuel boosters during lift-off. (Source: RocketDyne Archives. With permission.)

Two rocket engines

Rocket engines, such as those in Figure 14.3, are controlled by the mass flow rate, Q, of propellant that enters the combustion chamber. The propellant is ignited and the resulting combustion gas streams at high temperature T_{entry} into a carefully shaped Laval nozzle, in which the speed becomes supersonic. We shall not discuss the complex transition from subsonic to supersonic flow during startup, but just assume that the engine is now running steadily with the sonic area equal to the throat area, $A_1 = A_{\text{throat}}$. Besides the nozzle geometry, $A = A(x)$, the mass flow rate Q, the entry temperature T_{entry}, we only need to know the average molar mass M_{mol} of the combustion gas and its adiabatic index γ. From these input values, the physical conditions may be calculated everywhere in the engine using the formalism established in the preceding subsection. The main results are shown in table 14.1 for the case of the two important engines pictured in Figure 14.3. We shall now outline the procedure.

The Mach number distribution, $\mathsf{Ma}(x)$, is calculated by solving (14.30) numerically using the known area ratio $A(x)/A_{\text{throat}}$. In particular, the entry Mach number, $\mathsf{Ma}_{\text{entry}} < 1$, may be calculated from the entry-to-throat ratio, $A_{\text{entry}}/A_{\text{throat}}$, and the exit Mach number $\mathsf{Ma}_{\text{exit}} > 1$ from the exit-to-throat ratio, $A_{\text{exit}}/A_{\text{throat}}$. Since the Mach number only depends on the area ratio, engines with congruent geometry perform identically. Scaling up a rocket engine from model to full size is easy—at least in this respect.

Table 14.1. Comparison of A4/V2 and SSME rocket engines.

Input Values			Output Values		
	A4/V2	SSME		A4/V2	SSME
A_{entry}	$0.69\,\text{m}^2$	$0.21\,\text{m}^2$	$\mathsf{Ma}_{\text{entry}}$	0.11	0.15
A_{throat}	$0.13\,\text{m}^2$	$0.054\,\text{m}^2$	$\mathsf{Ma}_{\text{exit}}$	2.47	4.71
A_{exit}	$0.42\,\text{m}^2$	$4.17\,\text{m}^2$	ρ_{entry}	$1.55\,\text{kg}\,\text{m}^{-3}$	$9.68\,\text{kg}\,\text{m}^{-3}$
γ	1.2	1.2	p_{entry}	14.2 bar	203 bar
M_{mol}	$27.3\,\text{g}\,\text{mol}^{-1}$	$14.1\,\text{g}\,\text{mol}^{-1}$	v_{entry}	$117\,\text{m}\,\text{s}^{-1}$	$237\,\text{m}\,\text{s}^{-1}$
T_{entry}	3,000 K	3,600 K	T_{exit}	1,865 K	1,122 K
Q	$125\,\text{kg}\,\text{s}^{-1}$	$494\,\text{kg}\,\text{s}^{-1}$	p_{exit}	0.82 Bar	0.19 bar
			ρ_{exit}	$0.14\,\text{kg}\,\text{m}^{-3}$	$0.028\,\text{kg}\,\text{m}^{-3}$
			v_{exit}	$2{,}040\,\text{m}\,\text{s}^{-1}$	$4{,}160\,\text{m}\,\text{s}^{-1}$
			$\mathcal{R}$(1 bar)	247 kN	1,719 kN
			$\mathcal{R}$(0 bar)	290 kN	2,140 kN

The average molar mass M_{mol} is calculated from the combustion chemistry. For the A4/V2 the fuel is a mixture of 75% ethanol and 25% water, whereas for the SSME it is pure liquid hydrogen. In both engines the oxidizer is liquid oxygen. The engines run "fuel rich" which means that there is fuel left over in the combustion gas after all the oxygen has reacted. For the A4/V2, the exhaust gas becomes a mixture (by mass) of 6% ethanol, 52% carbon dioxide, and 42% water (see Problem 14.7). For the SSME, the exhaust mixture (by mass) is 97% water and 3% hydrogen (see Problem 14.6). The adiabatic index γ, which by the usual rules should be about $4/3 \approx 1.33$, is actually more like 1.2 at these high temperatures because of the excitation of additional molecular vibrational degrees of freedom.

Having determined the Mach number, $\mathsf{Ma}(x)$, the temperature $T(x)$ may now be calculated from (14.28) with $T_1 = T_{\text{throat}}$. The unknown throat temperature T_{throat} is obtained from T_{entry} by setting $\mathsf{Ma} = \mathsf{Ma}_{\text{entry}}$ in this equation. From the temperature we obtain the local sound velocity, $c(x) = \sqrt{\gamma R T(x)/M_{\text{mol}}}$, and the flow speed, $v(x) = \mathsf{Ma}(x)\,c(x)$. The gas density $\rho(x) = Q/v(x)A(x)$ can now be calculated everywhere in the nozzle from the known mass flow. Finally, the pressure in the nozzle is determined by the ideal gas law, $p(x) = R\rho(x)T(x)$ where $R = R_{\text{mol}}/M_{\text{mol}}$ is the specific gas constant. In Table 14.1 the input and output values are shown for the two engines pictured in Figure 14.3.

The total reaction force from the exhaust gas (which is the force that accelerates the rocket) is called the *thrust*. In Chapter 21 (page 357) we shall systematically investigate reaction forces, but here it is fairly simple to write it down,

$$\mathcal{R} = Q v_{\text{exit}} + (p_{\text{exit}} - p_{\text{atm}})A_{\text{exit}}. \tag{14.31}$$

The first term is the rate at which momentum is carried away by the exhaust gases and thereby adding momentum to the rocket itself at the same rate. The second is the force due to the pressure difference between the exhaust gas and the ambient atmosphere. If the design goal is to obtain a particular thrust, this equation can instead be used to determine one other parameter, for example the mass flow rate. The predicted thrust for each of the two rocket engines is also shown in Table 14.1 (for atmospheric pressure and for vacuum). Although the results are estimates, the calculated thrust agrees quite well with the quoted values [9, 10].

Acceleration at lift-off: The initial mass of the V2 rocket was 12,500 kg, and with a thrust of 250 kN which equals twice the initial weight of the rocket, the lift-off acceleration became nearly equal to the acceleration due to gravity. The space shuttle is equipped with three main engines, each delivering 1.7 MN (at the surface of the Earth), and two solid rocket boosters, each delivering 12.5 MN. The total thrust is thus about 30 MN which is 1.5 times the initial weight of 2 million kilograms, so that the Space Shuttle initially accelerates upward with about half the acceleration due to gravity. During ascent the acceleration grows to several times gravity, mainly because fuel is being spent but also because the atmosphere becomes thinner.

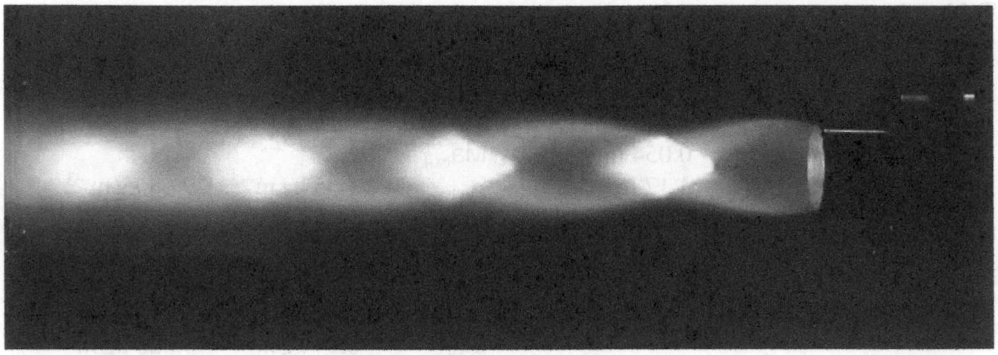

Figure 14.4. Beautiful shock diamonds formed in the exhaust from a small rocket engine with 2.5 kN thrust. (Source: Swiss Propulsion Laboratory. With permission.)

Why the pressure difference?

One may wonder why we allow for a difference between the exit pressure and the ambient pressure in (14.31). In the analysis of incompressible flow, for example Torricelli's law on page 214, we always assumed that the exit and ambient pressures were equal. The justification for this assumption is that any mismatch between exit and ambient pressure will generate an infinite pressure gradient that is instantly communicated to all of the fluid, as discussed in the comment on the non-locality of pressure on page 208.

In subsonic compressible flow such communication is still possible because the local speed of sound is larger than the local speed of the flow. But in the diverging part of a supersonic nozzle there is no way to communicate anything upstream by means of sound waves because the flow speed is everywhere larger than the local speed of sound. The nozzle entry is—so to speak—completely out of touch with what goes on at the exit. As a consequence, a nozzle running supersonic is said to be *choked* because it is not possible to increase the mass flow by lowering the ambient pressure or even applying active suction at the exit. There is, however, nothing against increasing the mass flow simply by increasing the propellant pumping rate. Since $Q = \rho_{\text{entry}} v_{\text{entry}} A_{\text{entry}}$, this will for fixed entry temperature lead to an increase in the entry gas density, the entry pressure and thus in the exit pressure.

Shocks and diamonds

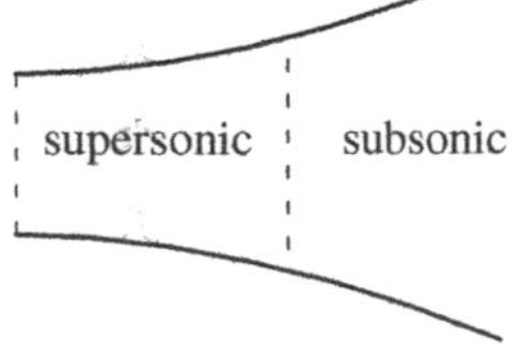

Static shock front in the diverging part of a Laval nozzle when the ambient pressure is higher than the exit pressure, such that the gas is overexpanded.

What actually happens at the exit because of the pressure difference is quite complicated (see for example [Anderson 2004, White 1999, Faber 1995]). If the exit pressure is higher than the ambient pressure, $p_{\text{exit}} > p_{\text{atm}}$, the gas is said to be *underexpanded.* A pattern of standing shock waves, called *shock diamonds* or *Mach diamonds*, will form in the exhaust plume after the nozzle exit. A spectacular case is shown in Figure 14.4, which clearly justifies the name. If, on the other hand, the exit pressure is lower than the ambient pressure, $p_{\text{exit}} < p_{\text{atm}}$, the exhaust gas is said to be *overexpanded.* A static shock front will then form inside the diverging part of the nozzle at a certain distance from the exit (see the margin figure). At the downstream side of the shock front, the supersonic flow drops abruptly to subsonic speed and the pressure, density, and temperature all jump to higher values. As the now subsonic gas proceeds through the remainder of the diverging channel, the velocity will further decrease while the thermodynamic parameters increase in accordance with the subsonic branch of Figure 14.1, until ideally the exit pressure matches the ambient pressure. A shock diamond will also form in the overexpanded case, if the exit pressure following the internal standing shock exceeds the ambient pressure. Shocks will be discussed in some detail in Chapter 26.

Problems

14.1 (a) Show that an ideal small-amplitude pressure wave in a barotropic medium satisfies

$$\frac{\partial}{\partial t}\left(\frac{1}{2}\rho_0 \boldsymbol{v}^2 + \frac{c_0^2 \Delta\rho^2}{2\rho_0}\right) = -\nabla \cdot (\Delta p \boldsymbol{v}).$$

(b) Use this to justify that the energy density in a wave must be

$$\varepsilon = \frac{1}{2}\rho_0 \boldsymbol{v}^2 + \frac{c_0^2 \Delta\rho^2}{2\rho_0}.$$

It contains both kinetic and compressional energy.
(c) Show that the average over a period of a harmonic wave is

$$\langle \varepsilon \rangle = \frac{1}{2}\rho_0 v_1^2,$$

where v_1 is the velocity amplitude.

14.2 A Pitot tube is mounted at the front of an aircraft moving with speed $U = 850$ km h^{-1} relative to the atmosphere, which may be assumed to be an ideal gas with $\gamma = 1.4$, temperature $T = -50°$C, and pressure $p = 400$ hPa. Calculate the relative temperature and pressure increases in the Pitot tube.

14.3 Show that for unsteady, compressible potential flow in a barotropic fluid with $\rho = \rho(p)$, the equations of motion may be chosen to be

$$\frac{\partial \Psi}{\partial t} + \frac{1}{2}\boldsymbol{v}^2 + \Phi + w(p) = 0, \qquad \frac{\partial \rho}{\partial t} + (\boldsymbol{v} \cdot \nabla)\rho = -\rho \nabla^2 \Psi,$$

where $\boldsymbol{v} = \nabla\Psi$ and $w(p) = \int dp/\rho(p)$.

* **14.4** Use the Schwarz inequality

$$\left|\sum_n A_n B_n\right|^2 \leq \sum_n A_n^2 \sum_m B_m^2$$

to derive Equation (14.20).

* **14.5** Consider a non-viscous barotropic fluid in an external time-independent gravitational field $\boldsymbol{g}(\boldsymbol{x})$ with $\nabla \cdot \boldsymbol{g} = 0$. Let $\rho_0(\boldsymbol{x})$ and $p_0(\boldsymbol{x})$ be density and pressure in hydrostatic equilibrium. **(a)** Show that the wave equation for small-amplitude pressure oscillations around hydrostatic equilibrium becomes

$$\frac{1}{c_0^2}\frac{\partial^2 \Delta p}{\partial t^2} = \nabla^2 \Delta p - (\boldsymbol{g} \cdot \nabla)\frac{\Delta p}{c_0^2},$$

where $c_0(\boldsymbol{x})$ is the local sound velocity in hydrostatic equilibrium. **(b)** Estimate under which conditions the extra term can be disregarded in standard gravity for an atmospheric sound wave of wavelength λ.

14.6 The SSME rocket engine propellant consists of 14% hydrogen and 86% oxygen (by mass). The combustion reaction is the well-known

$$2H_2 + O_2 \rightarrow 2H_2O.$$

The molar masses may be taken to be $M_{h2} = 2$, $M_{h2o} = 18$, $M_{o2} = 32$ (all in units of g mol^{-1}). Hint: Use the formalism of page 4.

(a) Calculate the average molar mass of the propellant and the molar fractions of the mixture.
(b) Calculate the molar fractions of hydrogen, oxygen, and water in the exhaust gas.
(c) Calculate the average molar mass of the exhaust gas and the mass fractions of the mixture.

* **14.7** The A4/V2 rocket engine propellant is a mixture (by mass) of about 33% pure ethanol, 11% water, and 56% oxygen. The combustion reaction is

$$C_2H_5OH + 3O_2 \rightarrow 2CO_2 + 3H_2O.$$

The molar masses may be taken to be $M_{h2o} = 18$, $M_{eth} = 46$, $M_{o2} = 32$, $M_{co2} = 44$ (all in units of g mol^{-1}). Hint: Use the formalism of page 4.

(a) Calculate the average molar mass of the propellant and the molar fractions of the mixture.
(b) Calculate the molar fractions of ethanol, oxygen, carbon dioxide, and water in the exhaust gas.
(c) Calculate the average molar mass of the exhaust gas and the mass fractions of the mixture.

15 Viscosity

All fluids are viscous, except for a component of liquid helium close to absolute zero temperature. Air, water, and oil all put up resistance to flow; and part of the money we spend on transport by plane, ship, or car goes to overcome fluid friction. All the energy in the fuel eventually contributes a small amount to heating the atmosphere and the sea.

It is primarily the interplay between the mechanical inertia of a moving fluid and its viscosity, that gives rise to all the interesting and beautiful phenomena, the whirling and the swirling that we are so familiar with. If a volume of fluid is set into motion, inertia would dictate that it continue in its original motion, were it not checked by the action of internal shear stresses. Viscosity acts as a brake on the free flow of a fluid and will eventually make it come to rest in mechanical equilibrium, unless external driving forces continually supply energy to keep it moving. In an Aristotelian sense the "natural" state of a fluid is thus at rest with pressure being the only stress component. Disturbing a fluid at rest slightly, setting it into motion with spatially varying velocity field, will to first order of approximation generate stresses that depend linearly on the spatial derivatives of the velocity field. Fluids with a linear relationship between stress and velocity gradients are said to be *Newtonian*, and the coefficients in this relationship are material constants that characterize the strength of viscosity.

In this chapter the formalism for Newtonian viscosity will be set up, culminating in the formulation of the Navier–Stokes equation for incompressible fluids. The slightly more complicated generalization to compressible fluids is presented at the end of the chapter. Superficially simple, the Navier–Stokes equation is a nonlinear differential equation for the velocity field that nevertheless continues to be a formidable challenge to engineers, physicists, and mathematicians. Understanding its power is the central theme for the remainder of this book.

15.1 Shear viscosity

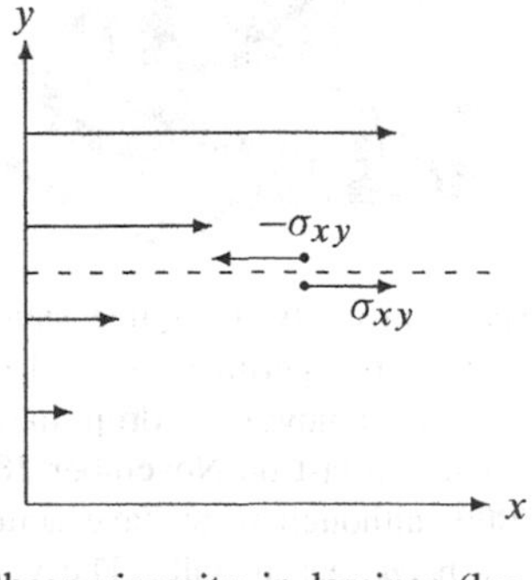

Shear viscosity in laminar (layered) flow. The fluid above the dashed line moves slightly faster than below and exerts a positive shear stress σ_{xy} on the fluid below. By Newton's Third Law the fluid below will exert an opposite stress $-\sigma_{xy}$ on the fluid above.

Consider a fluid flowing steadily along the x-direction with a velocity field $v_x(y)$ that is independent of x but may vary with y. Such a field could, for example, be created by enclosing a fluid between moving plates, and is an elementary example of *laminar* or layered flow. If the velocity field has no y-dependence, so that the fluid is in uniform motion along the x-axis, there should not be any internal stresses. If, on the other hand, the velocity grows with y, so that its gradient is positive $dv_x(y)/dy > 0$, we expect that the fluid immediately *above* a plane $y =$ const will drag along the fluid immediately *below* because of fluid friction and thus exert a *positive* shear stress, $\sigma_{xy}(y) > 0$, on this plane.

Table 15.1. Table of density and dynamic and kinematic viscosity for common substances (at the indicated temperature and at atmospheric pressure). Some of the values are only estimates. Note that air has greater kinematic viscosity than water, and hydrogen greater than olive oil. Glass is usually viewed as a solid, but there are claims (not very well substantiated [Zan98]) that it flows very slowly like a liquid over long periods of time even at normal temperatures.

	T	ρ	η	ν
Hydrogen	20	0.084	8.80×10^{-6}	1.05×10^{-4}
Air	20	1.18	1.82×10^{-5}	1.54×10^{-5}
Water	20	1.00×10^{3}	1.00×10^{-3}	1.00×10^{-6}
Ethanol	25	0.79×10^{3}	1.08×10^{-3}	1.37×10^{-6}
Mercury	25	1.35×10^{4}	1.53×10^{-3}	1.13×10^{-7}
Whole blood	37	1.06×10^{3}	2.7×10^{-3}	2.5×10^{-6}
Olive oil	25	0.90×10^{3}	6.7×10^{-2}	7.4×10^{-5}
Castor oil	25	0.95×10^{3}	0.7	7.4×10^{-4}
Glycerol	20	1.26×10^{3}	1.41	1.12×10^{-3}
Honey(est)	25	1.4×10^{3}	1.4×10^{1}	$1. \times 10^{-2}$
Pitch	20	1.1×10^{3}	2.3×10^{8}	$2. \times 10^{8}$
Glass (est)	20	2.5×10^{3}	10^{18}–10^{21}	10^{15}–10^{18}
Unit	°C	kg m^{-3}	Pa s	m^{2}s^{-1}

It also seems reasonable to expect that a larger velocity gradient will evoke stronger stress. In *Newton's law of viscosity* the shear stress is simply made proportional to the gradient[1],

$$\sigma_{xy}(y) = \eta \frac{dv_x(y)}{dy}. \tag{15.1}$$

The University of Queensland pitch drop experiment, started in 1927. Until now, eight drops have fallen, the last on November 28, 2000, although none have actually been seen to fall. The viscosity is determined to be about 2×10^8 Pa s [EDP84]. (Source: Wikimedia Commons. Courtesy John Mainstone, University of Queensland, Australia.)

The constant of proportionality, η, is called the coefficient of *shear viscosity*, the *dynamic viscosity*, or simply the *viscosity*. It is a measure of how strongly the moving layers of fluid are coupled by friction, and a material constant of the same nature as the shear modulus for elastic materials. We shall see later (page 255) that in compressible fluids there is also a bulk coefficient of viscosity corresponding to the elastic bulk modulus, but that turns out to be rather unimportant in ordinary applications.

The viscosities of naturally occurring fluids range over many orders of magnitude (see Table 15.1 and the margin figure). Since dv_x/dy has dimension of inverse time, the unit for viscosity η is Pa s (pascal seconds). Although this unit is sometimes called Poiseuille, there is in fact no special name for it in the standard (SI) system of units[2].

Molecular origin of viscosity in gases

In gases where molecules are far apart, internal stresses are caused by the incessant molecular bombardment of a boundary surface, transferring momentum in both directions across it. In liquids where molecules are in closer contact, internal stress is caused partly by molecular motion as in gases, and partly by intermolecular forces. The resultant stress in a liquid is a quite complicated combination of the two effects, and we shall for this reason limit the following discussion to the molecular origin of shear stress in gases.

Gas molecules move nearly randomly in all directions at speeds much higher than the velocity field $\boldsymbol{v}(\boldsymbol{x}, t)$, which represents the average non-random component of the molecular motion. In steady laminar planar flow with velocity $v_x(y)$ and positive velocity gradient

[1] We use the letter η rather than the often-used μ for viscosity to avoid a conflict with the shear elastic modulus.
[2] In the older cgs-system, it used to be called poise = 0.1 Pa s.

$dv_x(y)/dy$, a molecule of mass m crossing a surface element dS_y from above will carry an average momentum in the x-direction that is a little larger than $mv_x(y)$. Similarly, a molecule crossing dS_y from below will carry a little less than $mv_x(y)$. Since the same number of molecules pass from above and below, the result will be a net transfer of momentum from the fluid above to the fluid below.

To make a quantitative estimate, let the typical distance between molecular collisions in the gas be λ and the typical time between collisions τ. Disregarding all factors of order unity, a layer of thickness λ above an area element dS_y carries an excess of momentum in the x-direction,

$$d\mathcal{P}_x \approx (v_x(y+\lambda) - v_x(y))\rho\lambda dS_y \approx \rho\lambda^2 \frac{dv_x(y)}{dy} dS_y.$$

The shear stress may be estimated from the transfer of this momentum per unit of time and area, $\sigma_{xy} \approx d\mathcal{P}_x/\tau dS_y$, and indeed takes the form of Newton's law of viscosity (15.1) with a rough estimate of the shear viscosity,

$$\eta \approx \rho \frac{\lambda^2}{\tau} \approx \rho\lambda v_{\text{mol}}. \tag{15.2}$$

In the last expression we used that $\lambda/\tau \approx v_{\text{mol}}$, where $v_{\text{mol}} = \sqrt{3p/\rho} = \sqrt{3RT}$ is the root-mean-square molecular velocity. Over the years the estimate has been refined by means of the kinetic theory of gases, leading to about half the above value [Loeb 1961]. In practice one uses the resulting expression to determine the otherwise rather ill-defined molecular diameter from the measured viscosity (see the margin table).

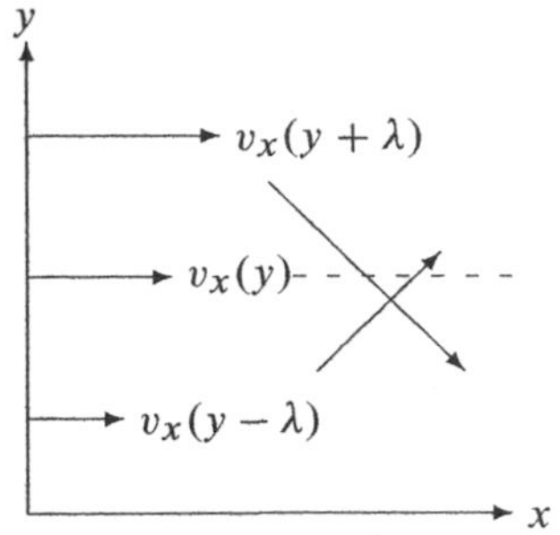

Layers of gas moving with different velocities give rise to shear forces because they exchange molecules with different average velocities.

Gas	η	λ	d
H_2	8.8	122	0.271
He	19.6	193	0.216
Ne	31.3	137	0.256
N_2	17.6	65	0.371
O_2	20.4	71	0.356
Ar	22.3	69	0.360
CO_2	14.7	44	0.454
Air	18.2	67	0.368
Unit	Pa s	nm	nm

Viscosity, mean free path and effective molecular diameter (in nanometers), determined from measured viscosities [2] at 20°C and 1 atm.

Temperature dependence of viscosity

The viscosity of any material depends on temperature. Common experience from the kitchen and industry tells us that most liquids become "thinner" when heated, indicating that the viscosity falls with temperature. Gases on the other hand become more viscous at higher temperatures, simply because the molecules move faster and thus transport more momentum across a surface per unit of time.

For an ideal gas it follows from Equation (1.11) on page 8 that the expression $\lambda\rho$ is a combinations of constants, so that the viscosity becomes $\eta \sim v_{\text{mol}} \sim \sqrt{T}$. Thus, if the viscosity is η_0 at temperature T_0, it may be estimated as

$$\eta = \eta_0 \sqrt{\frac{T}{T_0}} \tag{15.3}$$

at temperature T. Notice that the viscosity is independent of the pressure. Empirically, the viscosity grows slightly faster with temperature because of molecular attraction.

Kinematic viscosity

The viscosity estimate (15.2) seems to point to another measure of viscosity, called the *kinematic viscosity*[3],

$$\nu = \frac{\eta}{\rho}. \tag{15.4}$$

Since the estimate, $\nu \approx \lambda^2/\tau$, does not depend on the unit of mass, this parameter is measured in purely kinematic units[4] of $m^2\,s^{-1}$ (see Table 15.1). In an ideal gas we have $\rho \propto p/T$, so that the kinematic viscosity will depend on both temperature and pressure, $\nu \propto T^{3/2}/p$. For isentropic processes in ideal gases it always decreases with increasing temperature (see Problem 15.1).

[3]The conflicting use of ν for both the kinematic viscosity and Poisson's ratio is pervasive in the literature.
[4]In the older cgs system, the corresponding unit was called stokes $= cm^2\,s^{-1} = 10^{-4}\,m^2\,s^{-1}$.

It is (as we shall see) the kinematic viscosity ν that appears in the Navier–Stokes equation for the velocity field, rather than the dynamic viscosity η. Normally we would think of air as less viscous than water, and hydrogen as less viscous than olive oil, but under suitable conditions it is really the other way around. If a flow is driven by inflow of fluid with a given velocity, air behaves in fact as if it were 10 to 20 times more viscous than water. If instead driven by external forces, air is much easier to set into motion than water because its density is a thousand times smaller, and that is what fools our intuition.

15.2 Velocity-driven planar flow

Before turning to the derivation of the Navier–Stokes equations for viscous flow, we shall explore the concept of shear viscosity a bit further for the simple case of planar flow. Let us, as before, assume that the flow is laminar and planar with the only non-vanishing velocity component being $v_x = v_x(y, t)$, now also allowing for time dependence. It is rather clear that there can be no advective acceleration in such a field, and formally we also find $(\boldsymbol{v} \cdot \boldsymbol{\nabla})v_x = v_x \nabla_x v_x = 0$. In the absence of volume and pressure forces, the Newtonian shear stress (15.1) will be the only non-vanishing component of the stress tensor, and Cauchy's dynamical equation (12.26) on page 198 reduces to

$$\rho \frac{\partial v_x}{\partial t} = f_x^* = \nabla_y \sigma_{xy} = \eta \frac{\partial^2 v_x}{\partial y^2}.$$

Dividing by the density (which is assumed to be constant) we get

$$\boxed{\frac{\partial v_x}{\partial t} = \nu \frac{\partial^2 v_x}{\partial y^2},} \tag{15.5}$$

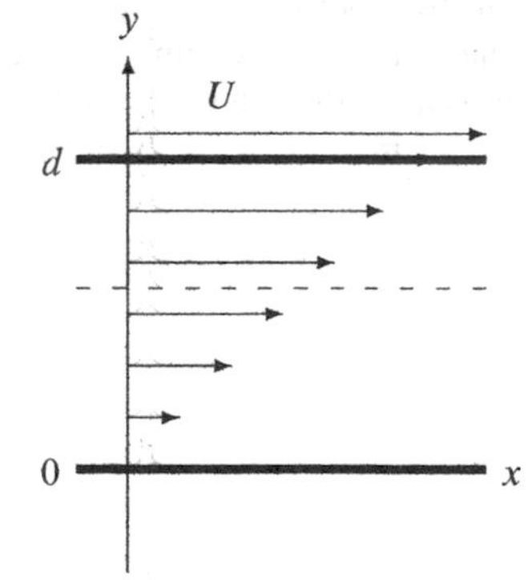

A Newtonian fluid with spatially uniform properties between moving parallel plates. The velocity field varies linearly between the plates and satisfies the no-slip boundary condition that the fluid is at rest relative to both plates. The stress must be the same on any plane in the fluid parallel with the plates (dashed).

where ν is the kinematic viscosity (15.4). This is a simplified version of the Navier–Stokes equation, particularly well suited for the discussion of the basic physics of shear viscosity.

Case: Steady planar flow

In steady flow the left-hand side of (15.5) vanishes, and from the vanishing of the right-hand side it follows that the general solution must be linear, $v_x = A + By$, with arbitrary integration constants A and B. We shall imagine that the flow is maintained between (in principle, infinitely extended) solid plates, one at rest at $y = 0$ and the other moving with constant velocity U at $y = d$. Where the fluid makes contact with the plates, we require it to assume the same speed as the plates; in other words, $v_x(0) = 0$ and $v_x(d) = U$ (this *no-slip* boundary condition will be discussed in more detail later). Solving these conditions we find $A = 0$ and $B = U/d$ such that the field between the plates becomes

$$v_x(y) = \frac{y}{d} U, \tag{15.6}$$

independent of the viscosity. From this expression we obtain the shear stress,

$$\sigma_{xy} = \eta \frac{dv_x}{dy} = \eta \frac{U}{d}. \tag{15.7}$$

It is independent of y, as one might have expected, because in stationary flow the balance of forces (and planar symmetry) requires the stress on any plane parallel with the plates to be the same.

Case: Viscous friction

A thin layer of viscous fluid is often used to lubricate the interface between solid objects. From the above solution to steady planar flow we may calculate the friction force, or *drag*, exerted on the body by the layer of viscous lubricant (see also Section 17.4). Let the would-be contact area between the body and the surface on which it slides be A, and let the thickness of the fluid layer be d everywhere. If the layer is thin, $d \ll \sqrt{A}$, we may disregard edge effects and simply multiply the planar stress (15.7) by the contact area to get the drag force,

$$\mathcal{D} \approx \eta \frac{A}{d} U. \tag{15.8}$$

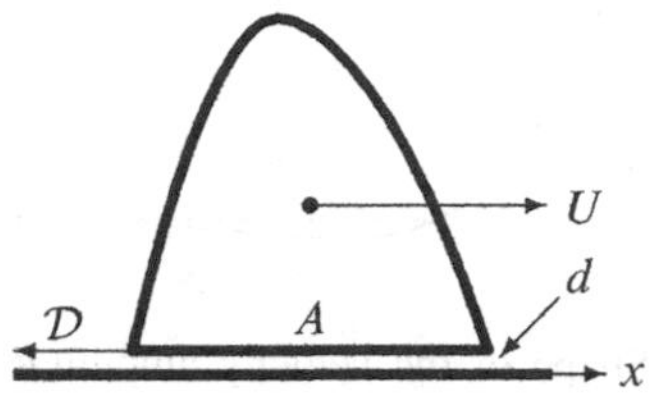

A solid object sliding on a plane lubricated surface with velocity U is subject to a viscous drag $\mathcal{D}$ opposite to the velocity.

The velocity-dependent viscous drag is quite different from the constant drag experienced in solid friction (see Section 6.1 on page 97). The decrease in drag with falling velocity makes the object seem to want to slide "forever", and this is what makes ice sports such as skiing, skating, sledging, and curling interesting. It is scary to brake a car on ice or to aquaplane, because the decreasing deceleration as the speed drops makes the car appear to run away from you. In these cases, a thin layer of liquid water acts as the lubricant.

The quasi-steady horizontal equation of motion for an object of mass M, not subject to forces other than viscous drag opposite the direction of the velocity, becomes

$$M\frac{dU}{dt} = -\eta \frac{A}{d} U. \tag{15.9}$$

Assuming that the thickness of the lubricant layer stays constant (and that is by no means evident) the solution to (15.9) is

$$U = U_0 e^{-t/\tau}, \qquad \tau = \frac{Md}{\eta A}, \tag{15.10}$$

where U_0 is the initial velocity and τ is the characteristic exponential decay time for the velocity. Integrating this expression we obtain the total stopping distance

$$L = \int_0^\infty U\,dt = U_0\tau = \frac{U_0 M d}{\eta A}. \tag{15.11}$$

Although it formally takes infinite time for the sliding object to come to a full stop, it does so in a finite distance! The stopping length grows with the mass of the object, which is quite unlike solid friction, where the stopping length is independent of the mass.

Example 15.1 [Curling]: In the ice sport of *curling*, a "stone" with mass $M \approx 20$ kg is set into motion with the aim of bringing it to a full stop at the far end of an ice rink of length $L \approx 40$ m. The area of the highly polished contact surface toward the ice is $A \approx 700\ \text{cm}^2$ and the initial velocity about $U_0 \approx 3\ \text{m s}^{-1}$. From (15.11) we obtain the thickness of the fluid layer $d \approx 43\ \mu\text{m}$, which does not seem unreasonable, and neither does the decay time $\tau \approx 13$ s. The players' intense sweeping of the ice in front of the moving stone presumably serves to smooth out tiny irregularities in the surface, which could otherwise slow down the stone.

Case: Shear wave

Consider an infinitely extended plate in the xz-plane immersed in an infinite sea of fluid. Let the plate oscillate in its own plane with circular frequency $\omega = 2\pi/\tau$, so that its instantaneous velocity in the x-direction is $U(t) = U_0 \cos\omega t$. The motion of the plate is transferred to the neighboring fluid because of the no-slip condition and then spreads into the fluid at large. To calculate the velocity field is known as *Stokes' Second Problem.*

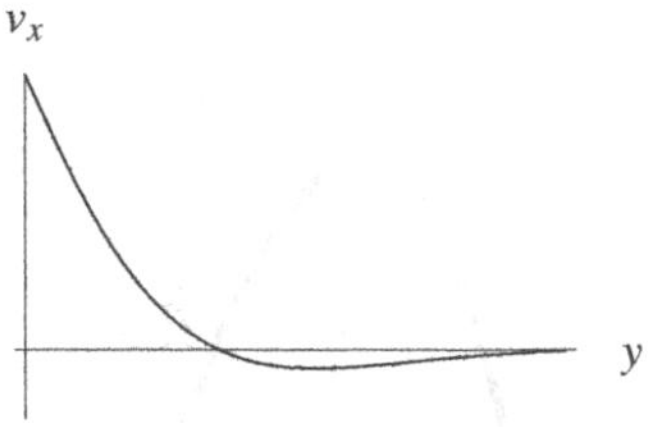

Shape of the velocity amplitude of a shear wave at $t = 0$.

By direct insertion it may be verified that the following field satisfies (15.5) as well as the no-slip boundary condition $v_x = U(t)$ for $y = 0$,

$$v_x(y,t) = U_0 e^{-ky} \cos(\omega t - ky), \qquad k = \sqrt{\frac{\omega}{2\nu}}. \tag{15.12}$$

Evidently, this is a damped wave spreading from the oscillating plate into the fluid. Since the velocity oscillations take place in the x-direction whereas the wave propagates in the y-direction, it is a *transverse* or *shear* wave. The wavenumber k determines the wavelength $\lambda = 2\pi/k = 2\sqrt{\pi\nu\tau}$, as well as the decay length of the exponential, also called the *penetration depth* $d = 1/k = \lambda/2\pi$. The wave is strongly damped and penetrates only a fraction of a wavelength into the fluid. It is really not much of a wave.

A shear wave of frequency 1,000 Hz penetrates only 71 μm in air and 18 μm in water at normal temperature and pressure.

Case: Free momentum diffusion

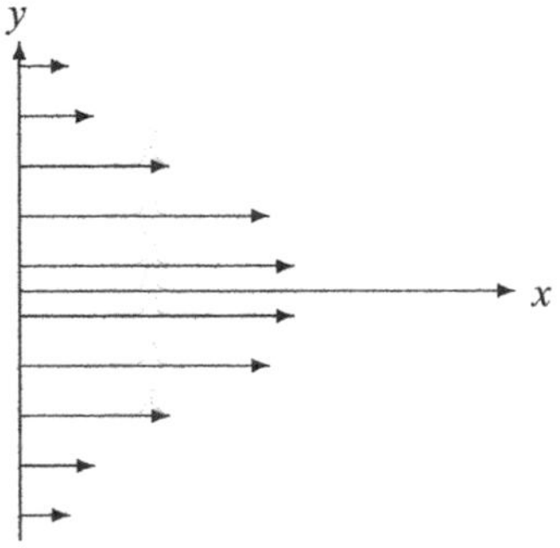

Velocity distribution for a planar Gaussian "river" in an "ocean" of fluid.

The dynamic equation (15.5) is a typical *diffusion equation* with diffusion constant equal to the kinematic viscosity, ν, also called the *momentum diffusivity*. In general, such an equation leads to a spreading of the distribution of the diffused quantity, which in this case is the velocity v_x, or perhaps better, the momentum density ρv_x.

The prototypical example of a flow with momentum diffusion is a "Gaussian river", which starts out at $t = 0$ with shape $v_x = U \exp\left(-y^2/a^2\right)$, where a is a measure of the initial width of the Gaussian. By direct insertion into the planar equation of motion (15.5), it may be verified that the solution at time t is

$$v_x(y,t) = U \frac{a}{\sqrt{a^2 + 4\nu t}} \exp\left(-\frac{y^2}{a^2 + 4\nu t}\right). \tag{15.13}$$

This river spreads with time but stays Gaussian, so that at time t the width parameter has become $\sqrt{a^2 + 4\nu t}$. Although momentum diffuses away from the center of the river, the total momentum must remain constant because there are no external forces acting on the fluid. Kinetic energy is on the other hand dissipated and ends up as heat. The apparent paradox that the kinetic energy can vanish while momentum stays constant is resolved in Problem 15.3.

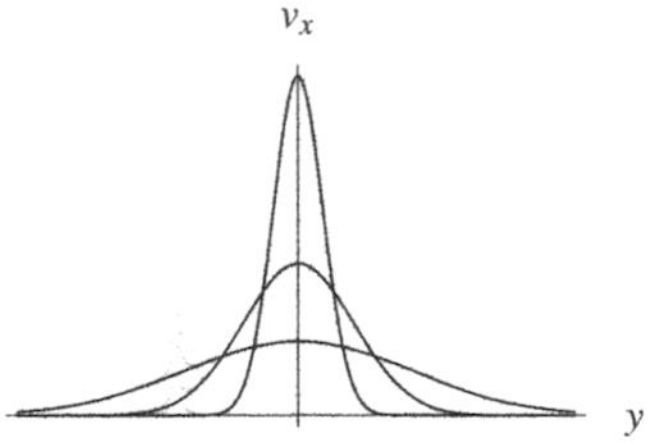

A Gaussian "river" widens and slows down in the course of time because of momentum diffusion, but retains its Gaussian shape.

At sufficiently late times, $t \gg a^2/4\nu$, the width of the Gaussian becomes $\sqrt{4\nu t}$ independent of the original width a. This is, in fact, a general feature of *any* bounded "river" flow. In the course of time it becomes proportional to $\exp(-y^2/4\nu t)$, as shown in Problem 15.5. The exponential factor drops rapidly to zero for values of y larger than

$$\boxed{\delta_{\text{front}} = 2\sqrt{\nu t},} \tag{15.14}$$

which we shall identify with the *momentum diffusion front*. Conversely, a velocity disturbance of size a may be characterized by the time, $t \simeq y^2/4\nu$, it takes for it to spread by momentum diffusion through a distance $y \gg a$.

Case: Growth of a boundary layer

A viscous fluid is initially at rest with respect to an infinite solid plate in the xz-plane ($y = 0$). At time $t = 0$ the plate is suddenly set into motion with velocity U along the x-direction[5]. The motion of the plate is transferred to the neighboring fluid because of the no-slip condition and then spreads into the fluid at large by momentum diffusion. The aim is to calculate the spreading of this *boundary layer* with time, also called *Stokes' First Problem*.

[5]Alternatively, the fluid may be set into motion while the plate remains at rest. This method conforms to the general treatment of boundary layers in Chapter 28.

Assuming that the flow is planar with $v_x = v_x(y,t)$ and $v_y = v_z = 0$, we may use the diffusion equation (15.5) to find the solution. The linearity of this equation guarantees that the velocity everywhere must be proportional to U, and since $\sqrt{\nu t}$ is the only quantity we can construct with dimension of length the velocity field may conveniently be represented as

$$v_x(y,t) = Uf\left(\frac{y}{\sqrt{\nu t}}\right). \tag{15.15}$$

Here the shape function $f(s)$ must satisfy the boundary conditions $f(0) = 1$ and $f(\infty) = 0$.

Inserting the above flow into the planar flow equation (15.5) we are led to an ordinary second-order differential equation for $f(s)$,

$$\boxed{f''(s) + \tfrac{1}{2}sf'(s) = 0.} \tag{15.16}$$

This is also a first-order differential equation for $f'(s)$ with the solution $f'(s) \sim \exp(-s^2/4)$. Integrating and applying the boundary conditions, the result becomes (see the margin figure)

$$f(s) = \frac{1}{\sqrt{\pi}}\int_s^\infty e^{-\frac{1}{4}u^2}\,du. \tag{15.17}$$

Asymptotically for $s \gg 1$, this function approaches unity with a Gaussian tail, $f(s) \sim \exp(-s^2/4) = \exp(-y^2/4\nu t)$, typical of momentum diffusion. Notice that the final expression is independent of what numerical factor is chosen in the argument of f in (15.15).

What is meant by the *boundary layer thickness* depends on the application because the velocity field does not vanish at any finite distance from the plate. An often-used practical choice is the distance $y = \delta_{99}$, where the velocity field only amounts to 1% of the plate velocity (and thus has dropped 99%). Since $f(s) = 0.01$ for $s = 3.64\ldots$, we have (see the margin figure)

$$\delta_{99} = 3.64\sqrt{\nu t}. \tag{15.18}$$

When no misunderstanding is possible, this is usually what is meant by the thickness δ.

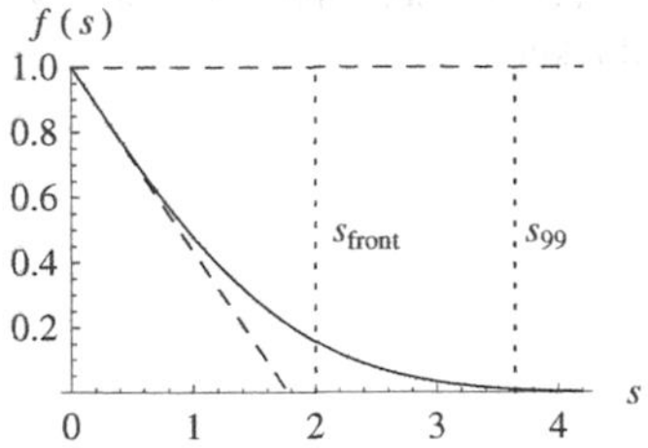

The Stokes layer shape function $f(s)$. The sloping dashed line is tangent at $s = 0$ with inclination $f'(0) = -1/\sqrt{\pi}$. The dotted lines indicate the front and 99% thicknesses.

Infinite diffusion speed: How can it be that the fluid is slowed down everywhere for $t > 0$ when the action of viscosity only set in $t = 0$? The short answer is that we have assumed the fluid to be incompressible, and this—fundamentally untenable—assumption will in itself entail infinite signal speeds. At a deeper level, a diffusion equation like (15.5) is the continuum limit of the dynamics of random molecular motion in the fluid, and although high molecular speeds are strongly damped at speeds beyond $v_{\text{mol}} = \sqrt{3R_{\text{mol}}T/M_{\text{mol}}}$, they may in principle occur.

Origin of vorticity: The vorticity field has only one non-vanishing component,

$$\omega_z = -\frac{\partial v_x}{\partial y} = -\frac{Uf'(s)}{\sqrt{\nu t}} = \frac{U}{\sqrt{\pi\nu t}}e^{-y^2/4\nu t}. \tag{15.19}$$

Before the fluid was set into motion, the flow was everywhere irrotational. Afterward there is vorticity everywhere in the fluid. Where did that come from?

Consider a nearly infinite rectangle with support of length L on the plate. By Stokes' theorem (13.31) on page 219 the total flux of vorticity through the rectangle equals the circulation around its perimeter, $\Gamma = \int \boldsymbol{\omega}\cdot d\boldsymbol{S} = \oint \boldsymbol{v}\cdot d\boldsymbol{\ell}$. The fluid velocity always equals U on the plate, vanishes at infinity, and is orthogonal to the sides of the rectangle, so that we obtain $\Gamma = UL$. Since Γ is constant in time, vorticity cannot be not generated inside the Stokes layer during its growth, only redistributed. Instead, it must arise at the plate surface during the (nearly) instantaneous acceleration of the fluid, and afterward diffuse away from the plate and into the fluid at large without changing the total flux of vorticity (see also Problem 15.6).

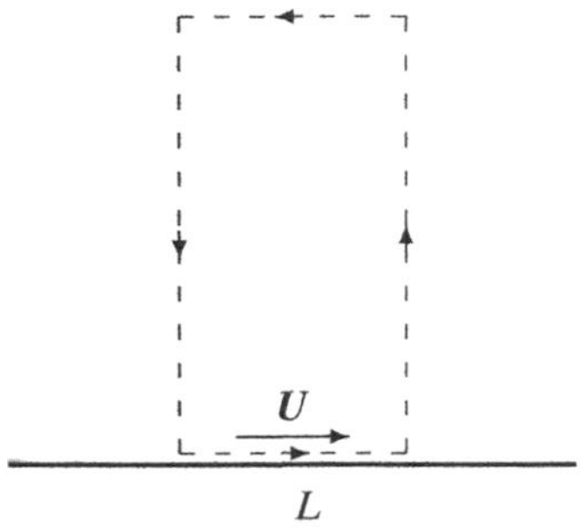

The circulation around an infinitely tall rectangle with side L against the moving wall is $\Gamma = \oint \boldsymbol{v}\cdot d\boldsymbol{\ell} = UL$.

15.3 Dynamics of incompressible Newtonian fluids

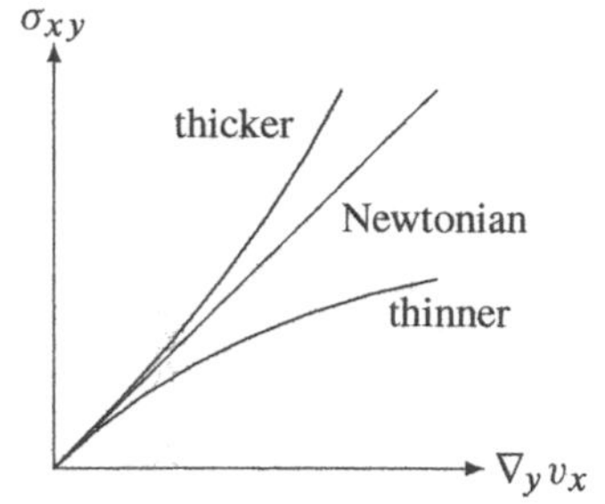

In Newtonian fluids the shear stress σ_{xy} increases linearly with the strain-rate $\nabla_y v_x$, whereas non-Newtonian fluids mostly become thinner and only a few thicker.

Numerous everyday fluids obey Newton's law of viscosity (15.1), for example water, air, oil, alcohol, and antifreeze. A number of common fluids are only approximatively Newtonian, for example paint and blood, and others are strongly non-Newtonian, for example tomato ketchup, jelly, and putty. There also exist *viscoelastic* materials that—depending on frequency—are both elastic and viscous. They are sometimes used in toys that can be slowly deformed like clay but also bounce like rubber balls when dropped on the floor.

In Newtonian fluids the shear stress σ_{xy} is directly proportional to the velocity gradient $\nabla_y v_x$—also called the *shear strain rate*—with proportionality constant equal to the constant shear viscosity η. Most non-Newtonian fluids become "thinner" as the shear strain rate increases, meaning that the shear stress grows slower than linearly. Even the most Newtonian of fluids, water, is thinner at shear strain rates above $10^{12}\ \mathrm{s}^{-1}$. Only few fluids (for example some starches stirred in water) appear to thicken with increasing strain rate. The science of the general flow properties of materials is called *rheology*.

We shall in this section establish the general dynamics for incompressible, isotropic, and homogeneous Newtonian fluids, and postpone the analysis of the slightly more complicated compressible fluids to Section 15.5.

Isotropic viscous stress

Newton's law of viscosity (15.1) is a linear relation between the shear stress and the velocity gradient, only valid in a particular flow geometry. As for Hooke's law for elasticity (page 128) we want a definition of viscous stress that takes the same form in any flow geometry and in any Cartesian coordinate system, leaving us free to choose our own reference frame.

Most fluids are not only Newtonian but also *isotropic*. Liquid crystals are anisotropic, but so special that we shall not consider them here. In an isotropic fluid at rest there are no internal directions at all, and the stress tensor is determined by the pressure, $\sigma_{ij} = -p\,\delta_{ij}$. When such a fluid is set in motion, the velocity field $\boldsymbol{v}(\boldsymbol{x}, t)$ defines a direction in every point of space, but the velocity in a particular point cannot itself provoke stress in the fluid. It is the variation in velocity from point to point that causes stress. Viscous stress must in other words be determined by the velocity gradient tensor, $\nabla_i v_j$. In an incompressible fluid, the trace of this tensor vanishes, $\sum_i \nabla_i v_i = \nabla \cdot \boldsymbol{v} = 0$, so the most general symmetric tensor one can construct from the velocity gradients is of the form

$$\boxed{\sigma_{ij} = -p\,\delta_{ij} + \eta\left(\nabla_i v_j + \nabla_j v_i\right).} \tag{15.20}$$

This is the natural generalization of Newton's law of viscosity (15.1) for incompressible flow to arbitrary Cartesian coordinate systems. Inserting the field of a steady planar flow $\boldsymbol{v} = (v_x(y), 0, 0)$, the only non-vanishing shear stress components are $\sigma_{xy} = \sigma_{yx} = \eta \nabla_y v_x(y)$, demonstrating that the coefficient η is indeed the shear viscosity introduced before.

The Navier–Stokes equations

The right-hand side of Cauchy's general equation of motion (12.26) on page 198 equals the effective density of force $f_i^* = f_i + \sum_j \nabla_j \sigma_{ij}$. Inserting the stress tensor (15.20) and using again $\nabla \cdot \boldsymbol{v} = 0$, we find

$$\sum_j \nabla_j \sigma_{ij} = -\nabla_i p + \eta\left(\sum_j \nabla_i \nabla_j v_j + \sum_j \nabla_j^2 v_i\right) = -\nabla_i p + \eta \nabla^2 v_i.$$

Here we have also assumed that the fluid is *homogeneous* such that the shear viscosity, like the density, does not depend on $\boldsymbol{x}$.

Inserting this expression into Cauchy's equation of motion and converting to ordinary vector notation we finally obtain the fundamental equations for incompressible, isotropic and homogenous fluids, due to Navier (1822) and Stokes (1845),

$$\frac{\partial \boldsymbol{v}}{\partial t} + (\boldsymbol{v} \cdot \nabla)\boldsymbol{v} = \boldsymbol{g} - \frac{1}{\rho_0}\nabla p + \nu \nabla^2 \boldsymbol{v}, \qquad \nabla \cdot \boldsymbol{v} = 0, \tag{15.21}$$

where ρ_0 is the constant density, $\nu = \eta/\rho_0$ is the constant kinematic viscosity, and $\boldsymbol{g} = \boldsymbol{f}/\rho_0$ is the acceleration field of the volume forces (normally due to gravity). Given the acceleration field $\boldsymbol{g}$, we now have four equations for the four fields, v_x, v_y, v_z, and p. Note, however, that whereas the three components of the velocity field are truly dynamic fields, for which the time derivatives are specified, this is not the case for the pressure, which is only determined indirectly through the divergence condition, as discussed in Section 13.1 on page 207.

Relatively simple to look at, the Navier–Stokes equations contain all the complexity of real fluid flow, including that of Niagara Falls! It is therefore clear that one cannot in general expect to find simple solutions. Exact solutions are only found in strongly restricted geometries and under simplifying assumptions concerning the nature of the flow, as in the planar laminar flow examples in the preceding section and the examples to be studied in Chapter 16.

George Gabriel Stokes (1819–1903). British mathematician and physicist. Contributed to the development of field calculus, fluid dynamics, optics, and heat conduction. (Source: Wikimedia Commons.)

Millenium Prize Problem: Among the seven Millenium Prize Problems set out by the Clay Mathematics Institute of Cambridge, Massachusetts, one concerns the existence of smooth, non-singular solutions to the Navier–Stokes equations (even for the simpler case of incompressible flow). The prize money of a million dollars illustrates how little we know and how much we would like to know about the general features of these equations that appear to defy the standard analytic methods for solving partial differential equations.

Boundary conditions

The Navier–Stokes equation for the velocity field is of first order in time, and needs initial values for the fields and their spatial derivatives in order to calculate the field values at later times. But what about physical boundaries, the solid containers of fluids, or even internal boundaries between different fluids? How do the fields behave there? Let us discuss the various fields that we have met one by one.

Density: The density is easy to dispose of, since it is allowed to be discontinuous and jump at a boundary between two materials, so this provides us with no condition at all.

Velocity: The normal component of the velocity field, $v_n = \boldsymbol{v} \cdot \boldsymbol{n}$, must always be continuous across an interface between incompressible materials. If this were not the case, the materials on the two sides of the interface would not move in unison. The tangential velocity component $\boldsymbol{v}_t = \boldsymbol{n} \times (\boldsymbol{v} \times \boldsymbol{n})$ must also be continuous but for a different reason: Although a nearly discontinuous tangential velocity variation may be created close to a wall, for example by hitting the fluid container with a hammer, it cannot remain for long but is quickly smoothed out by viscous momentum diffusion. A viscous fluid never slips along material boundaries but adjusts its velocity to match the velocity of the material on the other side without any discontinuity. This is the previously mentioned *no-slip condition.*

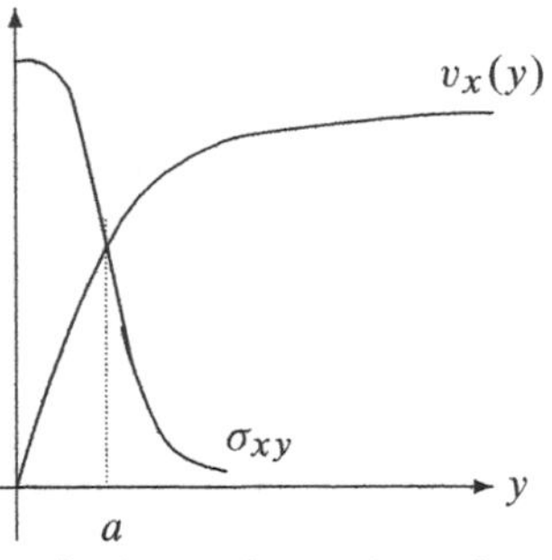

Sketch of strongly varying velocity and shear stress in a region of size a near a boundary. For $a \to 0$ the velocity develops an abrupt jump and the stress becomes infinite. From the Navier–Stokes equation, it follows that the strong decrease in shear stress away from the discontinuity leads to rapid momentum diffusion and smoothing out of the discontinuity.

The complete velocity field may thus always be assumed continuous across any material interface (with a bracket $[\cdot]$ to indicate the difference between the two sides),

$$[\boldsymbol{v}] = \boldsymbol{0}. \tag{15.22}$$

Only if the continuum approximation breaks down can cavitation and shear slippage occur. A notable exception to this rule is an open interface between a liquid and vacuum, where the velocity field will always be discontinuous. Vacuum is not a cohesive material!

Stress: Newton's Third Law only requires the stress vector $\boldsymbol{\sigma} \cdot \boldsymbol{n}$ to be continuous across any material interface (in the absence of surface tension),

$$[\boldsymbol{\sigma} \cdot \hat{\boldsymbol{n}}] = \boldsymbol{0}. \tag{15.23}$$

This only concerns three of the six components of the symmetric stress tensor. The other three are free to jump at the interface (see Example 6.7 on page 106). At an interface between a fluid and vacuum the material stress vector must vanish, $\boldsymbol{\sigma} \cdot \hat{\boldsymbol{n}} = 0$, because the vacuum cannot exert any force on any material object. It is in fact the absence of molecular forces from the vacuum that allows a jump in velocity to exist at an open liquid surface.

Pressure: Pressure is not necessarily continuous across a physical interface between two materials—even in the absence of surface tension (see page 105). For general fluids at rest and ideal fluids in motion, pressure is the only stress component, and must necessarily be continuous because of the continuity of the stress vector. We shall now see that the pressure must also be continuous for an incompressible viscous fluid moving along a solid wall, so that the pressure in the fluid *near* the wall is identical to the pressure acting *on* the wall. In all other cases the pressure will in general be discontinuous.

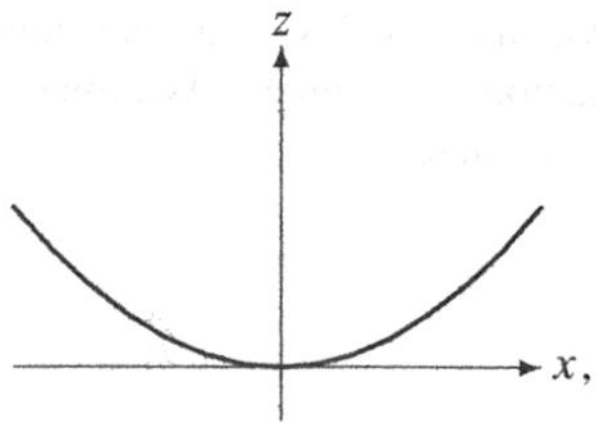

Local coordinate system at a particular point of the interface.

∗ **Proof:** The solid wall may without loss of generality be assumed to be at rest. Let us select any point on the wall and choose a local coordinate system with origin in this point and with the z-axis along the normal to the wall. Near the origin of this coordinate system, the wall may be described by a quadratic function, $z = Ax^2 + By^2 + 2Cxy$. Expanding to first order in the coordinates, the velocity field becomes $\boldsymbol{v}(x, y, z) = x\nabla_x \boldsymbol{v}|_0 + y\nabla_y \boldsymbol{v}|_0$. As the velocity field has to vanish everywhere along the wall it follows that all the tangential derivatives must vanish at the origin, $\nabla_x \boldsymbol{v} = \nabla_y \boldsymbol{v} = \boldsymbol{0}$ for $x = y = z = 0$. From the divergence condition, $\boldsymbol{\nabla} \cdot \boldsymbol{v} = 0$, we find that the normal derivative of the normal component, $\nabla_z v_z = -\nabla_x v_x - \nabla_y v_y = 0$, must also vanish at the origin. Finally, using the stress tensor (15.20) we get $\sigma_{zz} = -p + 2\eta\nabla_z v_z = -p$ at the origin. The continuity of the normal component, σ_{zz}, at the wall consequently guarantees the continuity of the pressure.

∗ Viscous dissipation

When you gently and steadily stir a pot of soup, the fluid will after some time settle down into a nearly steady flow. The fact that you still have to perform work while you stir steadily shows that there must be viscous friction forces at play in the soup. The friction forces between the sides of the pot and the soup cannot perform any work because the fluid is at rest there, due to the no-slip condition. All the work you perform must for this reason be spent against the *internal* friction forces in the soup, the shear stresses acting between the moving layers of the fluid. If you stop stirring, the soup quickly comes to rest and its kinetic energy is *dissipated* into heat. We shall return to dissipation in Chapters 20 and 22.

To calculate the dissipative rate of work, we turn back to the discussion of deformation work resulting in Equation (7.39) on page 120. Since a fluid particle is displaced by $\delta\boldsymbol{u} = \boldsymbol{v}\,\delta t$ in a small time interval δt, fluid motion may be seen as a continuous sequence of infinitesimal deformations with strain tensor $\delta u_{ij} = \frac{1}{2}\left(\nabla_i \delta u_j + \nabla_j \delta u_i\right) = \frac{1}{2}\left(\nabla_i v_j + \nabla_j v_i\right)\delta t$. The symmetrized velocity gradients $v_{ij} \equiv \delta u_{ij}/\delta t = \frac{1}{2}\left(\nabla_i v_j + \nabla_j v_i\right)$ may thus be understood as the *rate of deformation* or *rate of strain* of the fluid material. The rate of work $\dot{W} = \delta W/\delta t$ performed against the internal stresses is, consequently,

$$\dot{W}_{\text{int}} = \int_V \sum_{ij} \sigma_{ij} \nabla_j v_i \, dV = \int_V \sum_{ij} \sigma_{ij} v_{ij} \, dV = \int_V 2\eta \sum_{ij} v_{ij}^2 \, dV. \tag{15.24}$$

In the last step we have inserted the Newtonian stress tensor (15.20) and used that $\sum_i v_{ii} = \boldsymbol{\nabla} \cdot \boldsymbol{v} = 0$. Evidently, the rate of work against internal shear stresses is always positive. It always costs work to keep things moving against friction forces.

15.4 Classification of flows

The most interesting phenomena in fluid dynamics arise from the competition between inertia and viscosity, represented in the Navier–Stokes equation (15.21) by the advective acceleration $(\boldsymbol{v}\cdot\nabla)\boldsymbol{v}$ and the viscous diffusion term $\nu\nabla^2\boldsymbol{v}$. Inertia attempts to continue the motion of a fluid once it is started, whereas viscosity acts as a brake. If inertia is dominant we may leave out the viscous term, arriving again at Euler's equations (13.1) describing lively, *non-viscous* or *ideal* flow, analyzed in Chapter 13. If on the other hand, viscosity is dominant, we may drop the advective term and obtain the basic equations for sluggish *creeping* flow to be analyzed in Chapter 17.

The Reynolds number

As a measure of how much an actual flow is lively or sluggish, one may make a rough estimate, called the *Reynolds number*, for the ratio of the advective to the viscous terms. To get a simple expression we assume that the velocity is of typical size $|\boldsymbol{v}| \approx U$ and that it changes by a similar amount over a region of size L. The order of magnitude of the first-order spatial derivatives of the velocity will then be of magnitude $|\nabla\boldsymbol{v}| \approx U/L$, and the second-order derivatives will be $\left|\nabla^2\boldsymbol{v}\right| \approx U/L^2$. Consequently, the Reynolds number becomes

$$\mathsf{Re} \approx \frac{|(\boldsymbol{v}\cdot\nabla)\boldsymbol{v}|}{|\nu\nabla^2\boldsymbol{v}|} \approx \frac{U^2/L}{\nu U/L^2} = \frac{UL}{\nu}. \tag{15.25}$$

Osborne Reynolds (1842–1912). British engineer and physicist. Contributed to fluid mechanics in general, and to the understanding of lubrication, turbulence, and tidal motion in particular. (Source: Wikimedia Commons.)

The Reynolds number may also be understood as the ratio $\mathsf{Re} \approx t_{\text{diff}}/t_{\text{flow}}$ of the typical diffusion time scale $t_{\text{diff}} \sim L^2/\nu$ through the distance L and the typical flow time scale $t_{\text{flow}} \sim L/U$ through the same distance.

In Table 15.2 the Reynolds number is estimated for a number of flows, covering many orders of magnitude. For small values of the Reynolds number, $\mathsf{Re} \ll 1$, advection plays only little role and the flow just oozes along, while for large values, $\mathsf{Re} \gg 1$, viscosity does not have much influence and the flow tends to be lively. The streamline pattern of creeping flow is always orderly and layered, also called *laminar*, well known from the kitchen when mixing cocoa into dough to make a chocolate cake (although dough is hardly Newtonian!). The laminar flow pattern continues quite far beyond $\mathsf{Re} \simeq 1$, but depending on the flow geometry and other circumstances, there will be a Reynolds number, typically in the region of thousands, where *turbulence* sets in with its characteristic tumbling and chaotic behavior.

Example 15.2 [Bathtub turbulence]: Getting out of a bathtub you create flows with speeds of say $U \approx 1\ \text{m s}^{-1}$ over a distance of $L \approx 1$ m. The Reynolds number becomes $\mathsf{Re} \approx 10^6$ and you are definitely creating visible turbulence in the water. Similarly, when jogging you create air flows with $U \approx 3\ \text{m s}^{-1}$ and $L \approx 1$ m, leading to a Reynolds number around 2×10^5, and you know that you must leave all kinds of little invisible turbulent eddies in the air behind you. The fact that the Reynolds number is smaller in air than in water despite the higher velocity is a consequence of the kinematic viscosity being larger for air than for water.

Example 15.3 [Curling]: For planar flow between two plates (Section 15.1), the velocity scale is set by the velocity difference U between the plates whereas the length scale is set by the distance d between the plates. In the curling Example 15.1 on page 247 we found $U \approx 3\ \text{m s}^{-1}$ and $d \approx 43\ \mu\text{m}$, leading to a Reynolds number $\mathsf{Re} = Ud/\nu \approx 140$. Although not truly creeping flow, it is definitely laminar and not turbulent.

Example 15.4 [Water pipe]: A typical 1/2-inch water pipe has diameter $d \approx 1.25$ cm and that sets the length scale. If the volume flux of water is $Q = 100\ \text{cm}^3\ \text{s}^{-1}$, the average water speed becomes $U = Q/\pi a^2 \approx 0.8\ \text{m s}^{-1}$ and we get a Reynolds number $\mathsf{Re} = Ud/\nu \approx 10^4$, which brings the flow well into the turbulent regime. For olive oil under otherwise identical conditions we get $\mathsf{Re} \approx 0.15$, and the flow would be creeping.

Table 15.2. Table of Reynolds numbers for some moving objects calculated on the basis of typical values of lengths and speeds. Viscosities are taken from Table 15.1. It is perhaps surprising that a submarine operates at a Reynolds number that is larger than that of a passenger jet at cruising speed, but this is mainly due to the kinematic viscosity of air being 15 times larger than that of water.

System	Fluid	Size	Velocity	Reynolds Number
Ship (Queen Mary 2; Figure 3.3)	Water	345	15	5.2×10^9
Submarine (Ohio class; nuclear)	Water	170	12	2.0×10^9
Jet airplane (Boeing 747-400)	Air	71	250	1.2×10^9
Blue whale	Water	33	10	3.3×10^8
Car	Air	5	30	9.7×10^6
Swimming human	Water	2	1	2.0×10^6
Jogging human	Air	1	3	1.9×10^5
Herring	Water	0.3	1	3.0×10^5
Golf ball	Air	0.043	40	1.1×10^5
Ping-Pong ball	Air	0.040	10	2.6×10^4
Fly	Air	0.01	1	6.5×10^2
Flea	Air	0.001	3	1.9×10^2
Gnat	Air	0.001	0.1	6.5
Bacterium	Water	10^{-6}	10^{-5}	10^{-5}
Unit		m	m s^{-1}	

Hydrodynamic similarity

What does it mean if two flows have the same Reynolds number? A stone of size $L = 1$ m sitting in a steady water flow with velocity $U = 2$ m s^{-1} has the same Reynolds number as another stone of size $L = 2$ m in a steady water flow with velocity $U = 1$ m s^{-1}. It even has the same Reynolds number as a stone of size $L = 4$ m in a steady airflow with velocity $U = 8$ m s^{-1}, because the kinematic viscosity of air is about 15 times larger than of water (at normal temperature and pressure). We shall now see that provided the stones are geometrically similar, that is, have congruent geometrical shapes, flows with the same Reynolds numbers are also *hydrodynamically similar* and only differ by their overall length and velocity scales, so that their flow patterns visualized by streamlines will look identical.

In the absence of volume forces, steady incompressible flow is determined by (15.21) with $\boldsymbol{g} = \boldsymbol{0}$ and $\partial \boldsymbol{v}/\partial t = \boldsymbol{0}$, or

$$(\boldsymbol{v} \cdot \nabla)\boldsymbol{v} = -\frac{1}{\rho_0}\nabla p + \nu\nabla^2\boldsymbol{v}. \tag{15.26}$$

Let us rescale all the variables by means of the overall scales ρ_0, U, and L, writing

$$\boldsymbol{v} = U\tilde{\boldsymbol{v}}, \quad \boldsymbol{x} = L\tilde{\boldsymbol{x}}, \quad p = \rho_0 U^2 \tilde{p}, \quad \nabla = \frac{1}{L}\tilde{\nabla}, \tag{15.27}$$

where the "tilded" symbols are all dimensionless. Inserting these variables, the steady flow equation takes the form

$$(\tilde{\boldsymbol{v}} \cdot \tilde{\nabla})\tilde{\boldsymbol{v}} = -\tilde{\nabla}\tilde{p} + \frac{1}{\mathsf{Re}}\tilde{\nabla}^2\tilde{\boldsymbol{v}}. \tag{15.28}$$

The only parameter appearing in this equation is the Reynolds number, which may be interpreted as the inverse of the dimensionless kinematic viscosity. The pressure is as mentioned (page 207) not an independent dynamic variable and its scale is here fixed by the velocity scale, $P = \rho_0 U^2$. If the flow instead is driven by external pressure differences of magnitude P rather than by velocity, the equivalent flow velocity scale is given by $U = \sqrt{P/\rho_0}$.

In congruent flow geometries the no-slip boundary conditions will also be the same, so that any solution of the dimensionless equation can be scaled back to a solution of the original equation by means of (15.27). The three different flows around stones mentioned at the beginning of this subsection may thus all be obtained from the same dimensionless solution if the stones are geometrically similar and the Reynolds numbers identical.

Even if the flows are similar in air and water, the forces exerted on the stones will, however, not be the same. The shear stress magnitude may be estimated as $\sigma \approx \eta |\nabla v| \sim \eta U/L$. The viscous drag on an object of size L will then be of magnitude $\mathcal{D} \approx \sigma L^2 \sim \eta U L = \eta \nu \mathsf{Re}$. The Reynolds numbers are assumed to be the same in the two cases, making the ratio of the viscous drag on the stone in air to that in water about $\mathcal{D}_{\text{air}}/\mathcal{D}_{\text{water}} \approx (\eta\nu)_{\text{air}}/(\eta\nu)_{\text{water}} \approx 0.27$.

Example 15.5 [Flight of the robofly]: The similarity of flows in congruent geometries can be exploited to study the flow around tiny insects by means of enlarged, slower moving models immersed in another fluid. It is, for example, hard to study the air flow around the wing of a hovering fruit fly, when the wing flaps $f = 50$ times per second. For a wing size of $L \approx 4$ mm flapping through 180°, the average velocity becomes $U \approx \pi L f \approx 1.3$ m s^{-1} and the corresponding Reynolds number $\mathsf{Re} \approx UL/\nu \approx 160$. The same Reynolds number can be obtained from a 19 cm plastic wing of the same shape, flapping once every 6 s in mineral oil with kinematic viscosity $\nu = 1.15$ cm^2 s^{-1}, allowing for easy recording of the flow around the wing [BD01] .

Example 15.6 [High-pressure wind tunnels]: In the early days of flight, wind tunnels were extensively used for empirical studies of lift and drag on scaled-down models of wings and aircraft. Unfortunately, the smaller geometrical sizes of the models reduced the attainable Reynolds number below that of real aircraft in flight. A solution to the problem was obtained by operating wind tunnels at much higher than atmospheric pressure. Since the dynamic viscosity η is independent of pressure (page 245), the Reynolds number $\mathsf{Re} = \rho_0 UL/\eta$ scales with the air density and thus with pressure (at a given temperature). The famous *Variable Density Tunnel (VDT)* built in 1922 by the US National Advisory Committee for Aeronautics (NACA) operated on a pressure of 20 atm and was capable of attaining full-scale Reynolds numbers for models only 1/20th of the size of real aircraft [Anderson 1997, p. 301]. The results obtained from the VDT had great influence on aircraft design in the following 20 years.

Flows in different geometries can only be compared in a coarse sense, even if they have the same Reynolds number. A running man has the same Reynolds number as a swimming herring, and a flying gnat the same Reynolds number as a man swimming in castor oil (which cannot be recommended). In both cases the flow geometries are quite different, leading to different streamline patterns. Here the Reynolds number can only be used to indicate the character of the flow, which tends to be turbulent around the running man and the swimming herring, whereas it is laminar around the flying gnat and the man recklessly swimming in castor oil.

* 15.5 Dynamics of compressible Newtonian fluids

In deriving the Navier–Stokes equations for incompressible Newtonian fluids (15.21), we explicitly used the vanishing of the divergence, $\nabla \cdot v = 0$, to limit the most general form (15.20) of a symmetric stress tensor that is linear in the velocity gradients. But when flow velocities become a finite fraction of the velocity of sound, it is—as discussed before—no longer possible to maintain the simplifying assumption of even effective incompressibility. For truly compressible fluids the divergence is non-vanishing, and we have to give up the simple divergence condition and replace it by the continuity equation (12.12). In the same time it also opens the possibility for a slightly more general stress tensor.

Shear and bulk viscosity

The velocity divergence, $\nabla \cdot \boldsymbol{v}$, is the only scalar field that can be constructed by linear combination of the velocity gradients, implying that the only term we can add to the stress tensor (15.20) must be proportional to $\delta_{ij} \nabla \cdot \boldsymbol{v}$. Conventionally the proportionality constant is written $\zeta - \frac{2}{3}\eta$, where η is the shear viscosity, and ζ is a new material parameter, called *bulk viscosity* or the *expansion viscosity*. The complete stress tensor for a compressible isotropic Newtonian fluid in motion thus takes the form (Stokes, 1845)

$$\sigma_{ij} = -p\,\delta_{ij} + \eta\left(\nabla_i v_j + \nabla_j v_i - \frac{2}{3}\nabla \cdot \boldsymbol{v}\,\delta_{ij}\right) + \zeta \nabla \cdot \boldsymbol{v}\,\delta_{ij}. \tag{15.29}$$

Viewing this tensor as a first order expansion in the velocity field, it follows that p must be identified with the thermodynamic pressure, $p = p(\rho, T)$, of the fluid at rest.

Notice that the choice of proportionality constant makes the middle term traceless, so that the the mechanical pressure becomes $p_{\text{mech}} \equiv -\frac{1}{3}\sum_i \sigma_{ii} = p - \zeta \nabla \cdot \boldsymbol{v}$. A viscous fluid in motion thus creates an extra *dynamic pressure*, $-\zeta \nabla \cdot \boldsymbol{v}$, which is negative in regions where the fluid expands ($\nabla \cdot \boldsymbol{v} > 0$) and positive where it contracts ($\nabla \cdot \boldsymbol{v} < 0$). Bulk viscosity is hard to measure, because one must set up physical conditions such that expansion and contraction become important, for example by means of high-frequency sound waves. In the following section we shall analyze viscous attenuation of sound in fluids, and see that it depends on both the shear and the bulk modulus. The measurement of attenuation of sound is quite complicated and yields a rather frequency-dependent bulk viscosity, although it may generally be assumed to be of the same overall magnitude as the coefficient of shear viscosity (see [DG09]).

The Navier–Stokes equations

Inserting the stress tensor (15.29) into Cauchy's equation of motion (12.26) on page 198, we obtain most general form of the *Navier–Stokes equation* (for constant η and ζ),

$$\rho\left(\frac{\partial \boldsymbol{v}}{\partial t} + (\boldsymbol{v} \cdot \nabla)\boldsymbol{v}\right) = \boldsymbol{f} - \nabla p + \eta \nabla^2 \boldsymbol{v} + \left(\zeta + \tfrac{1}{3}\eta\right) \nabla(\nabla \cdot \boldsymbol{v}). \tag{15.30}$$

If the viscosities have a spatial variation, for example due to temperature, the expression becomes more complicated. Together with the equation of continuity (12.12),

$$\frac{\partial \rho}{\partial t} + \nabla \cdot (\rho \boldsymbol{v}) = 0, \tag{15.31}$$

we have obtained four dynamic equations for the four fields v_x, v_y, v_z, and ρ, while the pressure is determined by the thermodynamic equation of state, $p = p(\rho, T)$. For isothermal or isentropic flow the temperature is given algebraically, whereas in the general case we also need a differential *heat equation* to specify the dynamics of the temperature field (Chapter 22).

Boundary conditions

The principal results of the discussion of boundary conditions for incompressible fluids on page 251 remain valid for compressible fluids, namely the continuity of the velocity field and the stress vector across a material interface,

$$[\boldsymbol{v}] = \boldsymbol{0}, \qquad [\boldsymbol{\sigma} \cdot \hat{\boldsymbol{n}}] = \boldsymbol{0}. \tag{15.32}$$

The proof of the continuity of pressure at a solid wall can, however, not be carried through to this case.

Shock fronts: In compressible fluids, nearly singular shock fronts may arise, across which the flow parameters change rapidly (to be analyzed in detail in Section 26.3). In a shock front, the material is the same on both sides of the front, but low-density high-speed fluid on one side is compressed into high-density low-speed fluid on the other. The normal velocity field will thus be discontinuous across a shock from, although mass conservation requires the normal mass flux to be continuous, $[\rho \boldsymbol{v} \cdot \boldsymbol{n}] = 0$. A shock front is, however, not truly singular. Viscosity softens the would-be discontinuity and replaces it by a very steep transition over a finite distance, but for high Reynolds number the front becomes in fact so thin that its thickness approaches the limit to the validity of the continuum approximation.

* Viscous dissipation

The rate of work against internal stresses is slightly more complicated in compressible fluids. Defining the shear strain rate,

$$v_{ij} = \frac{1}{2}\left(\nabla_i v_j + \nabla_j v_i - \frac{2}{3}\nabla \cdot \boldsymbol{v}\,\delta_{ij}\right), \tag{15.33}$$

we obtain from (7.39) on page 120 with $\delta \boldsymbol{u} = \boldsymbol{v}\,\delta t$ the following total rate of work against internal stresses,

$$\dot{W}_{\text{int}} = \int_V \sum_{ij} \sigma_{ij} \nabla_j v_i \, dV = \int_V \left(-p\nabla \cdot \boldsymbol{v} + 2\eta \sum_{ij} v_{ij}^2 + \zeta(\nabla \cdot \boldsymbol{v})^2\right) dV. \tag{15.34}$$

This expression reduces of course to the incompressible expression (15.24) for $\nabla \cdot \boldsymbol{v} = 0$. The first term in the integrand represents the familiar thermodynamic rate of work on the fluid because $\delta(dV) = \nabla \cdot \delta \boldsymbol{u}\, dV = \nabla \cdot \boldsymbol{v}\,\delta t\, dV$ according to Equation (7.34) on page 118. As expected, it is positive during compression ($\nabla \cdot \boldsymbol{v} < 0$) and negative during expansion, and may in principle be recovered completely under under quasistatic, adiabatic conditions. The last two terms are both positive and represent the work done against internal *viscous* stresses. They express the inevitable viscous dissipation of kinetic energy into heat.

* 15.6 Application: Viscous attenuation of sound

It has previously (on page 247) been shown that free shear waves do not propagate through more than about one wavelength from their origin in any type of fluid. In nearly ideal fluids such as air and water, free pressure waves are capable of propagating over many wavelengths. Viscous dissipation (and many other effects) will nevertheless slowly sap their strength, and in the end all of the kinetic energy of the waves will be converted into heat.

In this section we shall calculate the rate of attenuation for damped small-amplitude solutions to the Navier–Stokes equations. The attenuation may equally well be calculated from the general expression for the dissipative work (15.34); see Problem 15.8.

Wave equation

As in the discussion of unattenuated pressure waves in Section 14.1 on page 229 we assume to begin with that a barotropic fluid is in hydrostatic equilibrium, $\boldsymbol{v} = \boldsymbol{0}$, without gravity, $\boldsymbol{g} = \boldsymbol{0}$, so that its density $\rho = \rho_0$ and pressure $p = p(\rho_0)$ are constant throughout space. Consider now a disturbance in the form of a small-amplitude motion of the fluid, described by a velocity field $\boldsymbol{v}$ that is so tiny that the non-linear advective term $(\boldsymbol{v} \cdot \nabla)\boldsymbol{v}$ can be completely disregarded. This disturbance will be accompanied by tiny density corrections, $\Delta\rho = \rho - \rho_0$, and pressure corrections $\Delta p = p - p_0$, which we assume to be of first order in the velocity.

Dropping all higher-order terms, the linearized Navier–Stokes equations become

$$\rho_0 \frac{\partial \boldsymbol{v}}{\partial t} = -\nabla \Delta p + \eta \nabla^2 \boldsymbol{v} + \left(\zeta + \tfrac{1}{3}\eta\right) \nabla(\nabla \cdot \boldsymbol{v}), \tag{15.35a}$$

$$\frac{\partial \Delta\rho}{\partial t} = -\rho_0 \nabla \cdot \boldsymbol{v}, \tag{15.35b}$$

$$\Delta p = c_0^2 \Delta\rho, \tag{15.35c}$$

where c_0 is the speed of sound (14.6). Differentiating the second equation with respect to time and making use of the other two, we obtain to first order in the disturbance,

$$\frac{\partial^2 \Delta\rho}{\partial t^2} = c_0^2 \nabla^2 \Delta\rho + \frac{\zeta + \frac{4}{3}\eta}{\rho_0} \nabla^2 \frac{\partial \Delta\rho}{\partial t}. \tag{15.36}$$

If the last term on the right-hand side were absent, this would be a standard wave equation of the form (14.5) describing free compression waves with phase velocity c_0. It is the last term that causes viscous attenuation, and it defines a characteristic frequency,

$$\omega_0 = \frac{\rho_0 c_0^2}{\zeta + \frac{4}{3}\eta}. \tag{15.37}$$

The quantity $\zeta + \frac{4}{3}\eta$, which naturally appears here, is called the *longitudinal viscosity*.

Taking $\zeta \approx \eta$, the frequency scale becomes of the order of $3 \times 10^9\ \mathrm{s}^{-1}$ in air at normal temperature and pressure, and about $10^{12}\ \mathrm{s}^{-1}$ in water. In view of the huge values of the viscous frequency scale ω_0, the last term in (15.36) will be small for frequencies that are much lower, $\omega \ll \omega_0$. This covers normal sound, including ultrasound in the megahertz region.

The viscous amplitude attenuation coefficient

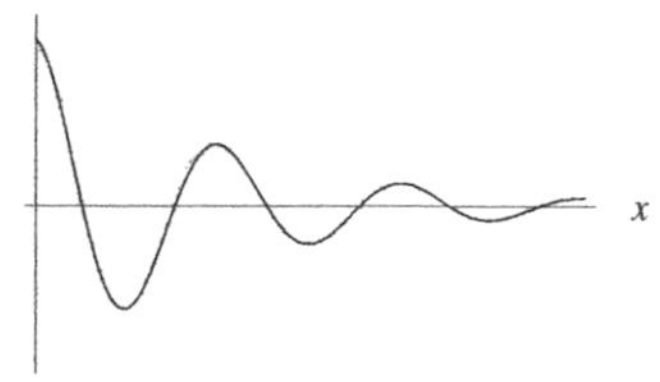

Damped wave.

Let us assume that a wave is created by an infinitely extended plane, a "loudspeaker", situated at $x = 0$ and oscillating in the x-direction with a small amplitude at a definite circular frequency ω. The fluid near the plate has to follow the plate and will be alternately compressed and expanded, thereby generating a damped compression wave of the form

$$\Delta\rho = \rho_1 e^{-\kappa x} \cos(kx - \omega t), \tag{15.38}$$

where k is the wavenumber, and κ is the *viscous amplitude attenuation coefficient*, which determines the length scale $1/\kappa$ for major attenuation, also called the *viscous attenuation length*. Inserting this wave into (15.36), we get to first order in κ/k and ω/ω_0

$$-\frac{\omega^2}{c_0^2} \cos\phi = -k^2 \cos\phi + 2\kappa k \sin\phi - k^2 \frac{\omega}{\omega_0} \sin\phi,$$

where $\phi = kx - \omega t$. This equation can only be fulfilled for $k = \omega/c_0$ and

$$\boxed{\kappa = \frac{k\omega}{2\omega_0} = \frac{\omega^2}{2\omega_0 c_0}.} \tag{15.39}$$

The viscous amplitude attenuation coefficient thus grows quadratically with the frequency, causing high-frequency sound to be attenuated strongly by viscosity.

Example 15.7 [Viscous attenuation in air and water]: In air at normal temperature and pressure, the viscous attenuation length $1/\kappa$ determined by this expression is huge, about 58 km, at a frequency of 1 kHz. At 10 kHz it is 100 times shorter, about 580 m. At 10 MHz the calculated attenuation length in air is only 0.6 mm. In water with higher density and sound velocity,the attenuation length is about 73 cm at 10 MHz. Living tissue is mostly water, and diagnostic imaging typically uses ultrasound between 1 and 15 MHz, which is why the ultrasound emitter is pressed so very close to the skin (and the skin is lubricated with a watery gel).

The drastic reduction in attenuation length $1/\kappa$ with increased frequency is also what makes measurements of the attenuation coefficient much easier at high frequencies. From the viscous attenuation coefficient one may in principle extract the value of the bulk viscosity, but this is complicated by several other fundamental mechanisms that also attenuate sound, such as thermal conductivity, and excitation of molecular rotations and vibrations.

Viscous attenuation in the real atmosphere: In the real atmosphere, many other effects contribute to the attenuation of sound. First, sound is mostly emitted from point sources rather than from infinitely extended vibrating planes, and that introduces a quadratic drop in amplitude with distance. Isentropic compression and heat conduction add other contributions, and other factors like humidity, dust, impurities, and turbulence also contribute, in fact much more than viscosity at the relatively low frequencies that human activities generate (see for example [Faber 1995] for a discussion of the basic physics of sound waves in real gases).

Problems

15.1 **(a)** Calculate the temperature dependence of the kinematic viscosity for isentropic processes in ideal gases. **(b)** What is the exponent of the temperature for monatomic, diatomic, and multiatomic gases?

15.2 A car with mass $M = 1{,}000$ kg moving at velocity $U_0 = 100$ km h^{-1} suddenly hits a patch of ice and begins to slide. The total contact area between the wheels and the ice is $A = 3{,}200$ cm^2, and it is observed to slide to a full stop in about $L = 300$ m. **(a)** Calculate the thickness of the water layer and discuss whether it is a reasonable value. **(b)** What is the time scale for stopping the car?

15.3 Consider the general case of planar momentum diffusion (page 248). Assume that the flow of the incompressible "river" along x vanishes fast at infinity along y, as in the Gaussian case. **(a)** Show that the total volume flux (per unit of length along z) is independent of time. **(b)** Show that the total momentum (per unit of length along both x and z) is likewise constant. **(c)** Show that the kinetic energy (per unit of length along both x and z) always decreases with time. **(d)** For the Gaussian river, calculate the kinetic energy per unit of length in the x and z directions as a function of time. How does it behave for $t \to \infty$?

15.4 Estimate the Reynolds number for **(a)** an ocean current, **(b)** a waterfall, **(c)** a weather cyclone, **(d)** a hurricane, **(e)** a tornado, **(f)** lava running down a mountainside, and **(g)** plate tectonic motion.

* **15.5** **(a)** Let $v_x(y, 0)$ be a velocity field with at most polynomial behavior at infinity. Show that

$$v_x(y,t) = \frac{1}{2\sqrt{\pi \nu t}} \int_{-\infty}^{\infty} \exp\left(-\frac{(y-y')^2}{4\nu t}\right) v_x(y',0)\, dy'$$

satisfies the momentum diffusion equation (15.5) and converges toward $v_x(y, 0)$ for $t \to 0$. **(b)** Show that any initial velocity distribution, which vanishes for $y > a$ at $t = 0$, is Gaussian for $|y| \to \infty$ and $t > 0$.

15.6 Consider a fluid at rest everywhere above a plate that is also at rest for $t < 0$. At $t = 0$ the plate is set into motion with velocity $U(t)$ for $t > 0$. **(a)** Show that the fluid moves with velocity

$$v_x(y,t) = \int_0^t \left(1 - f\left(\frac{y}{\sqrt{\nu(t-t')}}\right)\right) \dot{U}(t')\, dt',$$

where $\dot{U}(t) = dU(t)/dt$ and $f(s)$ given in (15.17). Hint: Show that it satisfies the equation of motion and the boundary conditions. **(b)** Calculate the flux of vorticity through a strip of fluid of length L along the plate (see page 249).

* **15.7** Consider an interface between two incompressible viscous fluids with a viscosity difference $\Delta\eta$ between them. Introduce a local coordinate system in a point of the interface with the z-axis along the normal, so that the interface itself is described by the parabolic form $z = Ax^2 + By^2 + 2Cxy$ in the neighborhood of the origin $x = y = z = 0$.

Show that

(a) $\nabla_x \boldsymbol{v}$ and $\nabla_y \boldsymbol{v}$ are continuous at the origin.
(b) σ_{xz}, σ_{yz} and σ_{zz} are continuous at the origin.
(c) $\nabla_z v_z$ is continuous in the origin.
(d) $\Delta p = 2\Delta\eta\nabla_z v_z$.
(e) $\Delta\nabla_z v_x = \sigma_{xz}\Delta(1/\eta)$ and similarly for $\nabla_z v_y$.
(f) $\Delta\sigma_{xy} = \Delta\eta(\nabla_x v_y + \nabla_y v_x)$.
(g) $\Delta\sigma_{xx} = 2\Delta\eta(\nabla_x v_x + \nabla_z v_z)$ and similarly for $\Delta\sigma_{yy}$.

* **15.8** **(a)** Calculate the average rate of dissipation per unit of volume of an ordinary plane sound wave using (15.34). **(b)** Calculate the attention length for the kinetic energy.

16

Channels and pipes

Even the most common of fluid flows, the water coming out of the kitchen tap or the draught of air around you, are so full of little eddies that their description seems totally beyond exact analysis. Fluid mechanics is the story of coarse approximation, gross oversimplification, and if accurate results are required, numeric computation.

Analytic solutions of the Navier–Stokes equations are few and hard to come by, even for steady flow. They are always found in geometries characterized by a high degree of symmetry, for example planar, cylindrical, or spherical. But symmetry is only a guideline. Even if the geometry of a problem is perfectly symmetric, there is no guarantee that the fluid selects a maximally symmetric solution as its flow pattern. At low Reynolds number, one expects that the maximally symmetric solution should be stable against disturbances but at sufficiently large Reynolds number this will not be the case. The simple laminar flow pattern of the maximally symmetric steady flow is then broken spontaneously or by little irregularities in the geometry, and replaced by time-dependent flow with less than maximal symmetry, or no symmetry at all.

In this chapter we shall study steady, incompressible, viscous flow in the simplest of geometries: planar and cylindrical. The exact solutions presented here are all effectively one-dimensional and lead to ordinary differential equations that are easy to solve. Although the solutions are all of infinite extent they nevertheless provide valuable insight into the behavior of viscous fluids for moderate Reynolds number. For completeness we also include discussions of turbulent pipe flow and of secondary flow between rotating cylinders.

16.1 Steady, incompressible, viscous flow

Most of the fluids encountered in daily life, air, water, gasoline, and oil, are effectively incompressible as long as flow velocities are well below the speed of sound, and often they flow steadily through the channels and pipes that we use to guide them. In looking for exact solutions for viscous flow we shall make the simplifying assumptions that the flow is incompressible and steady, satisfying the Navier–Stokes equations (15.21),

$$(\boldsymbol{v} \cdot \nabla)\boldsymbol{v} = \boldsymbol{g} - \frac{1}{\rho_0}\nabla p + \nu\nabla^2\boldsymbol{v}, \qquad \nabla \cdot \boldsymbol{v} = 0, \tag{16.1}$$

where ρ_0 is the constant density of the fluid and ν the constant kinematic viscosity.

In the following we solve these equations by making assumptions about the form of the fields, based on the symmetries of the particular geometry at hand. While symmetry assumptions ease the labor of solving the differential equations by restricting the fields to be functions of a single variable, they also limit the possible boundary conditions that may be imposed on the fields. In the absence of gravity the field equations are, for example, always solved by the extremely "symmetric" solution, $\boldsymbol{v} = \boldsymbol{0}$ and $p = 0$, but this assumption prevents us from selecting any non-zero values for the velocity and pressure on the inlets and outlets. Symmetry assumptions also make the solution blind to the symmetry breakdown that may happen in the world of real fluids, so it should never be forgotten that although symmetric solutions can be beautiful, they may also be completely irrelevant!

16.2 Pressure-driven channel flow

In Section 15.2 we analyzed velocity-driven planar flow between infinitely extended parallel plates with one of the plates moving at constant velocity with respect to the other. Here we shall solve the case where the plates are fixed and fluid is driven between them by a pressure gradient. For simplicity we assume that there is no gravity; gravity-driven flow will be analyzed in the following section.

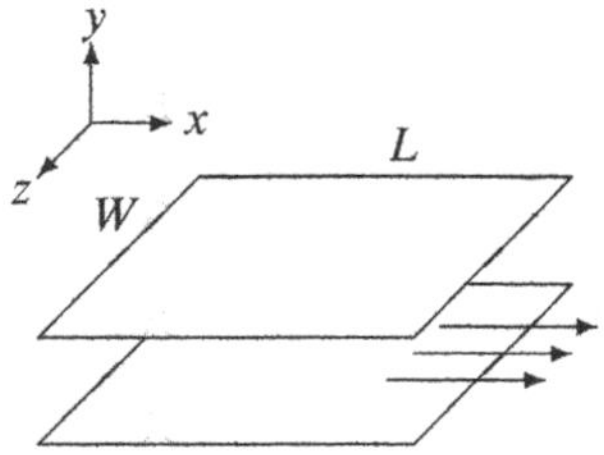

Channel flow between parallel plates of width W and length L.

Maximally symmetric flow

The coordinate system is chosen with the x-axis pointing along the direction of flow and the y-axis orthogonal to the plates. A velocity field respecting the planar symmetry is of the form

$$\boldsymbol{v} = (v_x(y), 0, 0) = v_x(y)\, \hat{\boldsymbol{e}}_x. \tag{16.2}$$

An infinitely extended flow like this is of course unphysical, but should nevertheless offer an approximation to the real flow between plates of finite extent, provided the dimensions of the plates are sufficiently large compared to their mutual distance.

With the assumed form of the flow, the incompressibility condition is automatically fulfilled, $\boldsymbol{\nabla} \cdot \boldsymbol{v} = \nabla_x v_x(y) = 0$. The advective acceleration vanishes likewise, $(\boldsymbol{v} \cdot \boldsymbol{\nabla})\boldsymbol{v} = v_x(y)\nabla_x v_x(y)\hat{\boldsymbol{e}}_x = \boldsymbol{0}$. The Navier–Stokes equation (16.1) then takes the form

$$\boldsymbol{\nabla} p = \eta\, \hat{\boldsymbol{e}}_x \nabla^2 v_x(y).$$

From the y and z components of this equation we get $\nabla_y p = \nabla_z p = 0$, implying that the pressure cannot depend on y and z, or in other words, $p = p(x)$, and the x-component of the above equation becomes

$$\frac{dp(x)}{dx} = \eta \frac{d^2 v_x(y)}{dy^2}. \tag{16.3}$$

The left-hand side depends only on x and the right-hand side only on y, and that is only possible if both sides take the same constant value independent of both x and y. Denoting the common value $-G$, we may immediately solve each of the equations $dp/dx = -G = \eta d^2 v_x/dy^2$ with the result

$$p = p_0 - Gx, \qquad v_x = -\frac{G}{2\eta} y^2 + Ay + B, \tag{16.4}$$

where p_0, A, and B are integration constants. The only freedom left in the planar flow problem lies in the integration constants, and they will be fixed by the boundary conditions of the specific flow configuration.

Specific solution

Let the plates be positioned a distance $d = 2a$ apart, for example at $y = -a$ and $y = a$. Applying the no-slip boundary conditions $v_x(-a) = v_x(a) = 0$ to the general solution, we obtain $A = 0$ and $B = Ga^2/2\eta$, and the velocity field becomes

$$v_x = \frac{G}{2\eta}(a^2 - y^2). \tag{16.5}$$

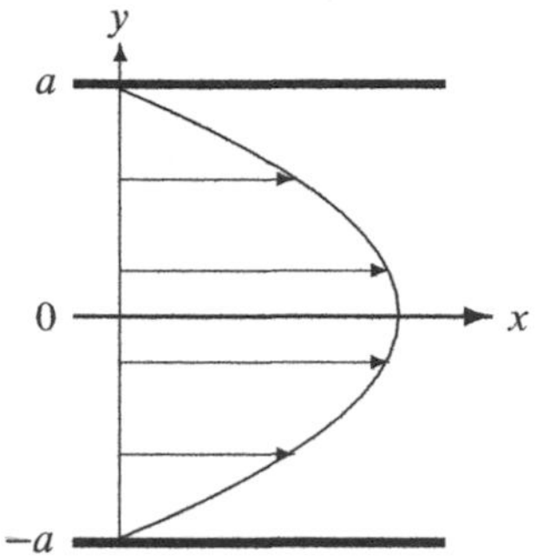

Characteristic parabolic steady velocity profile between the infinitely extended plates with distance $d = 2a$.

It has a characteristic parabolic shape with the maximal velocity $v_x^{\max} = Ga^2/2\eta$ in the middle of the gap, $y = 0$.

Driving pressure and discharge rate

For a real system with finite plates of length L in the direction of flow and width W with a tiny mutual distance, $d \ll L, W$, the flow should be reasonably well described by this solution, except near the edges of the plates (see Figure 16.1).

Two global quantities are immediately measurable for a real system with finite plates. One is the pressure drop, $\Delta p = p(0) - p(L)$, along the length of the plate L. It is called the *driving pressure*, and may be determined by means of a suitable manometer. The other is the total volume of ΔV of (incompressible) fluid discharged during a time T. From these we may calculate the pressure gradient G and the total *volumetric discharge rate* Q,

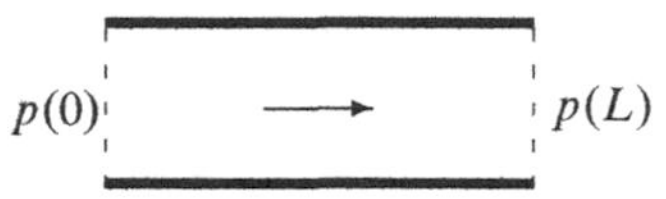

Pressure-driven steady flow between static plates. When $p(0) > p(L)$, the flow goes to the right.

$$G = \frac{\Delta p}{L}, \qquad Q = \frac{\Delta V}{T}. \tag{16.6}$$

Apart from edge effects, both G and Q ought to be constants, independent of the chosen length L or the collection time T.

The solution (16.5) relates these quantities. Integrating the velocity field over the area, $A = Wd$, orthogonal to the flow, we find

$$Q = \int_A \boldsymbol{v} \cdot d\boldsymbol{S} = \int_{-a}^{a} v_x(y)\, W dy = \frac{2GWa^3}{3\eta}. \tag{16.7}$$

From the discharge rate we may calculate the average velocity of the flow,

$$U = \frac{Q}{A} = \frac{Ga^2}{3\eta}. \tag{16.8}$$

It is not surprising that for fixed pressure gradient the average velocity grows with the plate distance, $d = 2a$, because the friction from the walls becomes less and less important with increasing plate distance. The maximal flow velocity is 50% higher than the average, $v_x^{\max} = Ga^2/2\eta = \frac{3}{2}U$.

Reynolds number

The Reynolds number (15.25) on page 253 has been defined in general as the ratio between advective and viscous terms in the Navier–Stokes equation, but since the advective acceleration $(\boldsymbol{v} \cdot \nabla)\boldsymbol{v}$ always vanishes in laminar planar flow, the Reynolds number must, strictly speaking, be zero. How can that be?

The apparent paradox is resolved when we consider what happens when the driving pressure is increased. Laminar flow is then replaced by irregular, time-dependent, even turbulent flow with non-vanishing advective acceleration. The Reynolds number should be understood as a dimensionless characterization of the ratio of advective to viscous forces in terms of the speed and geometry of the general flow. Since the velocity changes from zero at the walls to 50% more than the average in the middle between them, the Reynolds number for pressure-driven flow between parallel plates a distance d apart is conventionally defined as

$$\mathsf{Re} = \frac{Ud}{\nu}. \tag{16.9}$$

This definition makes sense independent of whether the incompressible flow is laminar or turbulent, because the discharge rate Q, and thus the average velocity $U = Q/A$, is the same everywhere along the channel.

Example 16.1: Oil with $\eta = 2 \times 10^{-2}$ Pa s and $\rho_0 = 800$ kg m^{-3} between plates $d = 1$ cm apart is driven by a pressure drop $\Delta p = 10^3$ Pa over a distance $L = 1$ m. The average velocity becomes in this case $U \approx 0.4$ m s^{-1} corresponding to a Reynolds number of $\mathsf{Re} \approx 167$.

Pressure-driven planar flow is known to be stable toward infinitesimal two-dimensional perturbations (in the xy-plane) for $\mathsf{Re} < 5{,}772$ [Lin 1955, Orz71, Graebel 2007], but empirically the flow becomes turbulent already for $\mathsf{Re} \gtrsim 1{,}000$–$2{,}000$, accompanied by breakdown of the two-dimensionality of the flow. The phenomenology of turbulent pressure-driven planar flow is quite similar to that of turbulent pressure-driven pipe flow to be discussed in Section 16.5.

Drag and power

The no-slip condition requires the fluid to be at rest at the bounding walls. It nevertheless exerts a shear stress on the walls, which may be calculated from the planar solution at $y = -a$:

$$\sigma_{\text{wall}} = \sigma_{xy}\big|_{y=-a} = \eta\, \nabla_y v_x\big|_{y=-a} = Ga. \tag{16.10}$$

This *wall shear stress* may be viewed as analogous to static friction between solid bodies.

The total shear force on the walls, called the *drag*, is obtained by multiplying the wall shear stress by the total wall area of both plates,

$$\mathcal{D} = \sigma_{\text{wall}}\, 2LW = GLWd = \Delta p\, A, \tag{16.11}$$

where $A = Wd$ is the area of the opening through which the fluid flows. That the wall drag should equal the total pressure force on the fluid could have been foreseen because the momentum of the fluid between the plates is constant in steady flow, implying that the total force on the fluid, $\mathcal{F}_x = -\mathcal{D} + \Delta p A$, must vanish.

The rate of work, or *power*, of the external pressure forces is similarly obtained:

$$P = \int_A \boldsymbol{v} \cdot \Delta p\, d\boldsymbol{S} = \Delta p \int_A \boldsymbol{v} \cdot d\boldsymbol{S} = \Delta p Q = \frac{2G^2a^3}{3\eta} WL. \tag{16.12}$$

Since the fluid is at rest at the bounding walls, the wall shear stresses perform no work. The work performed on the fluid by the driving pressure is not lost but dissipated into heat by the viscous friction. If the walls are insulating this heat is washed along with the flow and leaves the channel through the exit (see page 382).

Entry length

The steady flow profile (16.5) is actually not established right at the inlet but some distance downstream from it, as shown in Figure 16.1. Since the pressure gradient G varies in the entry region, it is best to control the inflow by the average flow velocity $U = Q/A$ rather than by the inlet pressure. We shall accordingly take the inlet flow to be plug flow with homogeneous

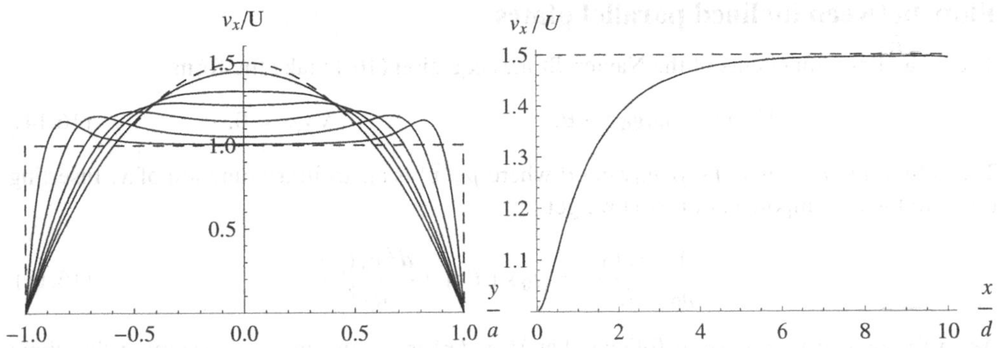

Figure 16.1. Data extracted from the simulation of channel entrance flow at $\mathsf{Re} = 100$ pictured in the margin. **Left:** The velocity profiles for selected values, $x/d = 0, 0.1, 0.3, 0.6, 1, 1.5, 3, \infty$ (with dashed extremes). **Right:** The central flow velocity (at $y = 0$) as a function of the downstream distance x. At $x/d = 5.0$, the curve is only 1% below the asymptotic value $v_x/U = 1.5$.

Simulation (with a commercial program) of channel entrance flow at $\mathsf{Re} = 100$. The flow enters from above with uniform velocity and settles into the parabolic solution (16.5) downstream from the entry. The background gray-level indicates the magnitude of the velocity $|\boldsymbol{v}|$, the white curves are the streamlines, and the crossing gray curves are pressure contours. In the simulation the channel is about three times longer than shown here.

velocity $v_x = U$ and $v_y = 0$ across the inlet. As the fluid moves downstream from the inlet, the no-slip condition at the wall creates a shear wall stress, which by momentum diffusion slows down the fluid in a boundary layer on both sides of the channel. The boundary layers widen as the fluid moves down the channel until they meet. Mass conservation requires the discharge rate and therefore the average velocity to be the same throughout the pipe. To keep the average speed constant the fluid must speed up near the middle of the channel and and in the end reach the parabolic steady flow pattern.

We have earlier estimated (page 248) that the momentum diffusion front in a time interval t is characterized by a transverse spread of size $y = \delta_{\text{front}} = 2\sqrt{\nu t}$. The layers reach the middle of the channel from both sides when $2\sqrt{\nu t} \approx a$, or $t \approx a^2/4\nu$. In this time interval the fluid has on average moved downstream through the distance $L_{\text{entry}} \approx Ut \approx Ua^2/4\nu$. Expressed in terms of the Reynolds number (16.9) this crude estimate of the *entry length* may be written

$$\boxed{\frac{L_{\text{entry}}}{d} \approx k\,\mathsf{Re},} \tag{16.13}$$

where k is a fairly small constant, here estimated to be about $k \approx 1/16 \approx 0.06$. Its actual value depends on how closely one wants the flow to approach the steady flow profile. The simulation data obtained in the margin figure and plotted in Figure 16.1 show how the velocity profile asymptotically approaches the parabola, and that the central flow velocity reaches 99% of the asymptotic value $v_x/U = \frac{3}{2}$ at $x/d = 5.0$, corresponding to $k_{99} = 0.050$.

In Chapter 19 we shall simulate this flow with our own methods and determine the entry length numerically for a wide range of Reynolds numbers. For large Reynolds numbers the entry length is indeed given by the above expression with the more reliable value $k_{99} = 0.041$. A theoretical calculation along the lines of Langhaar's 1942 analysis of pipe flow [Lan42] yields $k_{99} = 0.039$.

16.3 Gravity-driven planar flow

Gravity may also drive the flow between parallel plates if they are inclined an angle θ to the horizon. We choose again a coordinate system with the x-axis in the direction of flow and the y-axis orthogonal to the plates. Assuming constant gravity, the gravitational field is also inclined an angle θ to the negative y-axis.

Flow between inclined parallel plates

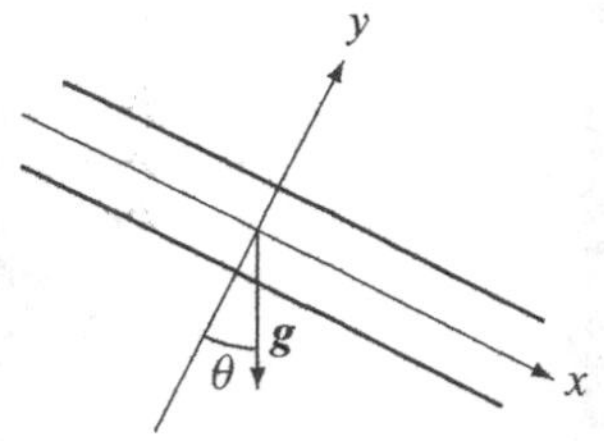

Parallel plates inclined an angle θ to the horizon.

The y- and z-components of the Navier–Stokes equation (16.1) take the form

$$\nabla_y p = -\rho_0 g_0 \cos\theta, \qquad \nabla_z p = 0. \tag{16.14}$$

The solution is $p = p_0(x) - \rho_0 g_0 y \cos\theta$ where $p_0(x)$ is an arbitrary function of x. Inserting this into the x-component of (16.1) we get

$$\frac{1}{\rho_0}\frac{dp_0(x)}{dx} = g_0 \sin\theta + \nu \frac{d^2 v_x(y)}{dy^2}. \tag{16.15}$$

As in the preceding section it follows that $\nabla_x p_0(x)$ is a constant, and assuming that there is no pressure difference between the ends of the plates at $x = 0$ and $x = L$, it follows that $p_0(x)$ must also be constant, $p_0(x) = p_0$. Applying the no-slip boundary conditions, $v_x(0) = v_x(d) = 0$, the complete solution becomes

$$v_x = \frac{g_0 \sin\theta}{2\nu}\left(a^2 - y^2\right), \qquad p = p_0 - \rho_0 g_0 y \cos\theta, \tag{16.16}$$

where p_0 is the constant pressure in the central plane $y = 0$. The pressure is for all θ simply the hydrostatic pressure in a constant field of gravity of strength $g_0 \cos\theta$.

Inclined flow with an open surface

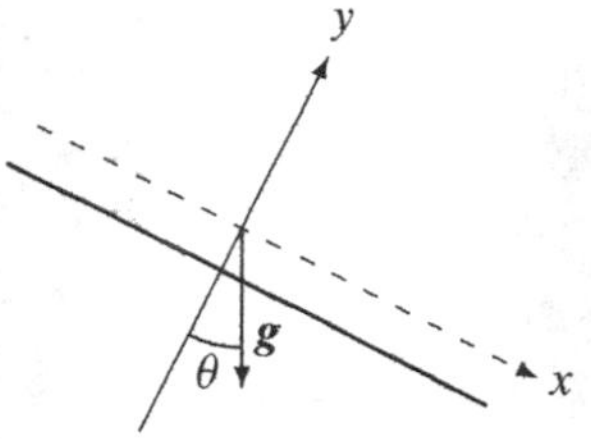

Flow with an open surface on a plate inclined an angle θ to the horizon.

A liquid layer of constant thickness a flowing down an inclined plate with an open surface is another example of purely gravity-driven flow. On the inclined plate, $y = -a$, the no-slip condition again demands $v_x(-a) = 0$, whereas on the open surface, $y = 0$, the pressure must be constant and the shear stress must vanish, $\sigma_{xy} = \eta dv_x/dy = 0$. Since these boundary conditions are fulfilled by (16.16), the velocity and pressure are given by the same expressions as above.

The velocity profile now has its maximum at the open surface. The average velocity is

$$U = \frac{1}{a}\int_{-a}^{0} v_x(y)\, dy = \frac{g_0 a^2 \sin\theta}{3\nu}. \tag{16.17}$$

The volumetric discharge rate in a swath of width W orthogonal to the flow is $Q = UaW$.

The Reynolds number is naturally defined in terms of the layer thickness,

$$\mathsf{Re} = \frac{Ua}{\nu}. \tag{16.18}$$

Introducing the convenient velocity parameter

$$U_0 = \left(\tfrac{1}{3}\nu g_0 \sin\theta\right)^{1/3}, \tag{16.19}$$

we may express all the variables in the convenient form,

$$U = \mathsf{Re}^{2/3} U_0, \qquad a = \mathsf{Re}^{1/3}\frac{\nu}{U_0}, \qquad W = \frac{Q}{\nu\,\mathsf{Re}}. \tag{16.20}$$

Given the inclination angle θ, the viscosity ν, and the discharge rate Q, the Reynolds number Re is the only variable. The length L of the plate is irrelevant as long as $L \gg a$.

Example 16.2 [Liquid film]: A water film of thickness $a = 0.1$ mm flowing down an inclined plate at $\theta = 30°$ has average velocity $U = 16\ \text{mm s}^{-1}$ and Reynolds number $\mathsf{Re} = 1.6$. A 10 mm-layer of glycerol flowing down the same slope has much larger average velocity $15\ \text{cm s}^{-1}$ but a similar Reynolds number of 1.3.

∗ Stability of open-surface flow

Most people are familiar with the unstable nature of a layer of liquid flowing down an inclined surface. Until the middle of the twentieth century a shimmering, unstable curtain of water running down store front windows was used for humidification, cooling, and smell reduction, for example in cheese shops.

Here we shall make a simple estimate that nevertheless captures the correct form of the stability criterion. The basic idea is to calculate the effective surface tension α created by the liquid layer and determine the condition under which it is positive. As discussed in Section 5.1, a negative surface tension will make the interface crumple up, and we shall assume that the same will be the case for the sheet of steadily flowing liquid. The effective surface tension will be determined from the energy per unit of area necessary to expand the liquid sheet sideways, that is, orthogonally to the direction of flow while keeping the volumetric flow rate fixed.

In steady flow the potential energy gained by the downhill flow of the liquid is completely dissipated into heat by viscous friction, and need not be discussed further. Apart from that, the mechanical energy of an area $A = L \times W$ of the liquid layer is the sum of the kinetic energy of the liquid and its potential energy:

$$\mathcal{E} = \rho_0 \int_{-a}^{0} \left(\frac{1}{2}v_x^2 + g_0(y+a)\cos\theta\right) LW dy = \left(\frac{3}{5}U^2 + \frac{1}{2}g_0 a\cos\theta\right)\rho_0 LWa.$$

Notice that the potential energy is calculated relative to the plate at $y = -a$ rather than the surface of the liquid layer, because the layer becomes thinner when we expand it sideways. Expressing all the variables in terms of the Reynolds number and constants, we get

$$\mathcal{E} = \frac{3}{10}\rho_0 QLU_0(5\cot\theta + 2\mathsf{Re})\mathsf{Re}^{-1/3}. \tag{16.21}$$

The energy depends on the width W only through the Reynolds number.

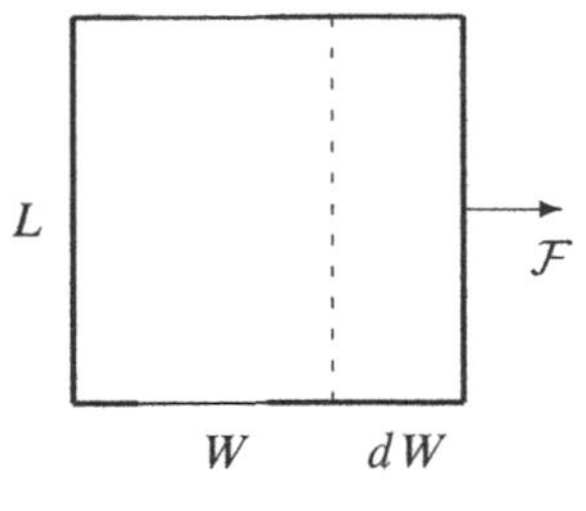

Increasing the flow area by $L\,dW$ changes its energy by $d\mathcal{E} = \alpha\, L\,dW$, where α is the effective surface tension.

To determine the effective surface tension we must calculate the change in energy under a change of width from W to $W + dW$ with the volume flow rate Q held constant. Differentiating through Re we obtain after some algebra the effective surface tension,

$$\alpha = \frac{1}{L}\frac{\partial\mathcal{E}}{\partial W} = -\frac{Q}{\nu W^2 L}\frac{\partial\mathcal{E}}{\partial\mathsf{Re}} = \frac{1}{10}\eta\, U_0(5\cot\theta - 4\mathsf{Re})\mathsf{Re}^{2/3}. \tag{16.22}$$

The condition for α to be positive becomes [Ben66]

$$\boxed{\mathsf{Re}\tan\theta < \frac{5}{4}.} \tag{16.23}$$

This shows that a nearly horizontal open-surface flow with $\theta \ll 1$ and $\mathsf{Re} \lesssim 5/4\theta$ is always stable, while a vertical open-surface flow with $\theta = 90°$ is never stable.

Example 16.3 [Liquid film, continued]: For the water film discussed in the preceding example we find $\mathsf{Re}\tan\theta = 0.94 < 1.25$, and the film will be stable according to the above condition. The effective surface tension becomes tiny and positive $\alpha = 3.5 \times 10^{-6}\ \mathrm{J\,m^{-2}}$. The glycerol flowing down the same slope has much larger average velocity but a similar Reynolds number 1.3, and is also stable because $\mathsf{Re}\tan\theta = 0.75 < 1.25$. The effective surface tension becomes in this case much larger than for the water layer, $0.07\ \mathrm{J\,m^{-2}}$.

16.4 Laminar pipe flow

Jean-Louis-Marie Poiseuille (1799–1869). French physician who studied blood circulation and performed experiments on flow in tubes. Presumably the first to have measured blood-pressure by means of a mercury manometer.

Gotthilf Heinrich Ludwig Hagen (1797–1884). German hydraulic engineer, specialized in waterworks, harbors, and dikes.

Pipes carrying effectively incompressible fluids are ubiquitous, in industry, in the home, and in our own bodies. Household water and almost all other fluids are transported under pressure in long pipes with circular cross-sections. The question of how much fluid a given pressure can drive through a circular tube in a given time interval is one of the most basic problems in fluid mechanics. It was first addressed quantitatively by Poiseuille around 1838 (published in 1841), and unbeknown to the physics community at that time, independently by Hagen in 1839. Today *Poiseuille flow* is used to denote any pressure-driven laminar flow, for example also the channel flow discussed in Section 16.2. To make the following analysis self-contained, it repeats in many respects the analysis of planar flow.

Maximally symmetric flow

An infinitely long circular cylindrical tube is invariant both under translations along its axis and rotations around it. In a coordinate system with the z-axis coincident with the cylinder axis, one velocity field that respects this symmetry is

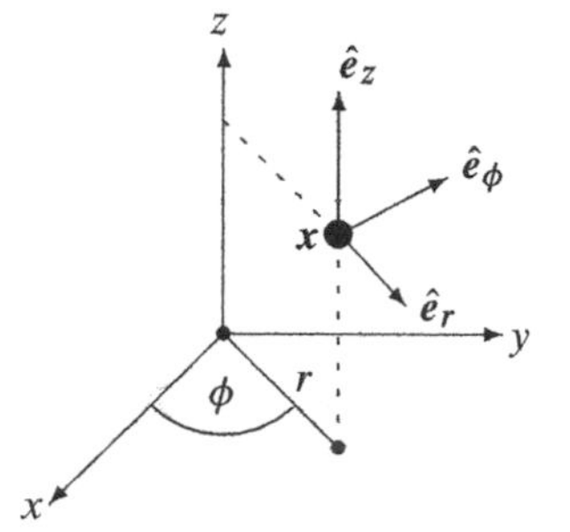

Cylindrical coordinates and basis vectors (see Appendix D).

$$\boldsymbol{v} = (0, 0, v_z(r)) = v_z(r)\,\hat{\boldsymbol{e}}_z, \tag{16.24}$$

where $r = \sqrt{x^2 + y^2}$ is the radial distance from the cylinder axis. In section 16.6 we shall investigate another maximally symmetric flow in this geometry.

The analysis follows basically along the same lines as for planar flow in Section 16.2. To make the calculation proceed smoothly we first use that the gradient operator in cylindrical coordinates takes the form (see Appendix D)

$$\nabla = \hat{\boldsymbol{e}}_r \frac{\partial}{\partial r} + \hat{\boldsymbol{e}}_\phi \frac{1}{r}\frac{\partial}{\partial \phi} + \hat{\boldsymbol{e}}_z \frac{\partial}{\partial z}. \tag{16.25}$$

Applying this operator to the assumed field (16.24), we find $\nabla \cdot \boldsymbol{v} = \hat{\boldsymbol{e}}_r \cdot \partial_r v_z(r)\hat{\boldsymbol{e}}_z = 0$, and $(\boldsymbol{v} \cdot \nabla)\boldsymbol{v} = v_z \partial_z v_z(r)\hat{\boldsymbol{e}}_z = 0$. In the absence of gravity, $\boldsymbol{g} = \boldsymbol{0}$, the Navier–Stokes equation (16.1) now simplifies to

$$\nabla p = \eta\,\hat{\boldsymbol{e}}_z \nabla^2 v_z(r) = \eta\,\hat{\boldsymbol{e}}_z \left(\frac{d^2 v_z}{dr^2} + \frac{1}{r}\frac{dv_z}{dr}\right). \tag{16.26}$$

In the last step we used the well-known expression for the Laplacian in cylindrical coordinates acting on a scalar field (Equation (D.19) on page 622).

From the x- and y-components of this equation we get $\nabla_x p = \nabla_y p = 0$, and consequently the pressure can only depend on z, that is, $p = p(z)$. Since the left-hand side depends only on z whereas the right-hand side depends only on r, neither side can depend on r or z. Denoting the common constant value by $-G$, we obtain the ordinary differential equations,

$$\frac{dp}{dz} = -G, \qquad \eta \frac{1}{r}\frac{d}{dr}\left(r\frac{dv_z}{dr}\right) = -G, \tag{16.27}$$

where we have made a simple rewriting of the last equation. Integrating we find

$$p = p_0 - Gz, \qquad v_z = -\frac{G}{4\eta} r^2 + A \log r + B, \tag{16.28}$$

where p_0, A, and B are integration constants.

Poiseuille solution

Consider now a pipe with inner radius a and diameter $d = 2a$. The fluid velocity cannot be infinite at $r = 0$, implying $A = 0$ in the general solution (16.28). The no-slip boundary condition requires that $v_z(a) = 0$, and this fixes the last integration constant to $B = Ga^2/4\eta$, so that the velocity profile becomes

$$v_z = \frac{G}{4\eta}(a^2 - r^2). \tag{16.29}$$

As for pressure-driven channel flow (16.5), it is parabolic and reaches, as one would expect, its maximal value $v_z^{\max} = Ga^2/4\eta$ at the center of the pipe.

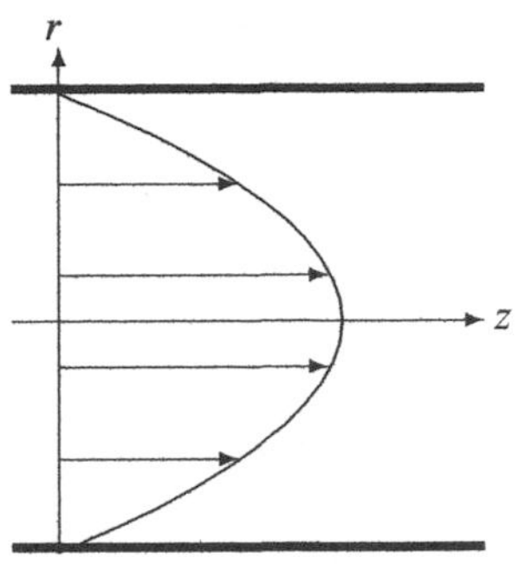

Velocity profile for laminar flow through a circular pipe.

Driving pressure and discharge rate

Two global quantities are—as for channel flow—immediately measurable for a pipe of finite length L. One is the pressure drop between entry and exit, $\Delta p = p(0) - p(L)$, and the other is the total volume of ΔV of (incompressible) fluid discharged during a time T. From these we may calculate the pressure gradient G and the total *volumetric discharge rate* Q,

$$G = \frac{\Delta p}{L}, \qquad\qquad Q = \frac{\Delta V}{T}. \tag{16.30}$$

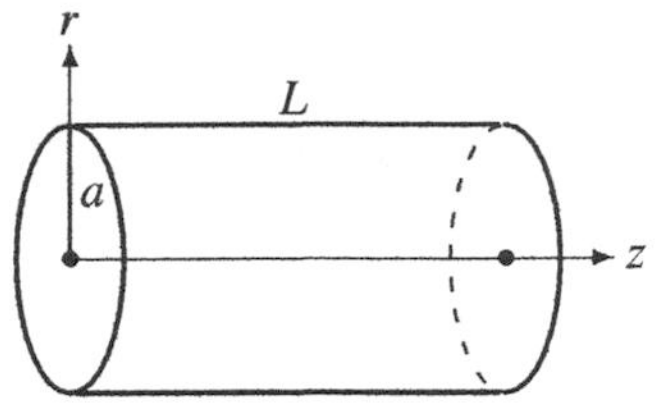

Section of a circular pipe of inner radius a and length L. The pressure is higher at $z = 0$ than at $z = L$ and the pressure drop $\Delta p = p(0) - p(L)$ drives the fluid through the pipe.

Apart from edge effects, both G and Q should be constant when the pipe is sufficiently long and the discharge is collected over a sufficiently long time. We shall return to this point in the discussion of entry length below.

The volumetric discharge rate may immediately be calculated by integrating the velocity field over the cross-section, $A = \pi a^2$, of the pipe,

$$Q = \int_A \boldsymbol{v} \cdot d\boldsymbol{S} = \int_0^a v_z(r)\, 2\pi r\, dr = \frac{G}{4\eta}\pi \int_0^a (a^2 - r^2) 2r dr.$$

Carrying out the integral, this becomes the famous *Hagen–Poiseuille law*, relating the driving pressure gradient and the discharge rate,

$$Q = \frac{\pi a^4}{8\eta} G. \tag{16.31}$$

As could have been expected, the discharge rate grows linearly with the pressure gradient, and inversely with viscosity. The dramatic fourth-power growth with radius could of course have been deduced from dimensional arguments since it is the only missing factor. It expresses the fairly common observation that much more water flows through a wider pipe for a given driving pressure because the shear wall stress becomes less able to hold back the fluid.

The velocity of the flow averaged over the cross-section of the pipe may be calculated from the rate of discharge

$$U = \frac{Q}{\pi a^2} = \frac{Ga^2}{8\eta}. \tag{16.32}$$

The maximal velocity at the center of the pipe is twice the average velocity, $v_z^{\max} = 2U$. The central flow must be faster than for channel flow, because in a pipe there is less room in the central region to compensate for the slow-down of fluid at the pipe wall.

Reynolds number

As for channel flow there is a formal problem in defining the Reynolds number because the advective acceleration formally vanishes in Poiseuille flow. The apparent paradox is, as before, resolved when we consider what happens when the driving pressure is increased. Laminar flow is then replaced by irregular, time-dependent, even turbulent flow with non-vanishing advective acceleration. The Reynolds number should be understood as a dimensionless characterization of the ratio of advective to viscous forces in terms of the speed and geometry of the general flow and becomes

$$\mathsf{Re} = \frac{Ud}{\nu}. \tag{16.33}$$

The choice of the pipe diameter $d = 2a$ as length scale is purely a matter of convention. Osborne Reynolds himself actually used the radius in 1883 when he first discussed what we now call the Reynolds number[1].

Empirically pipe flow remains laminar until turbulence sets in at a Reynolds number between 2,000 and 4,000, with 2,300 as a "nominal" value for smooth pipes. At that point the otherwise linear relationship between volume discharge Q and pressure gradient G becomes nonlinear. For a smooth pipe under *very* carefully controlled conditions, the transition to turbulence can in fact be delayed until a Reynolds number on the order of 100,000. Above that value the flow is so sensitive to disturbances that it becomes practically impossible to avoid turbulence, even if the Poiseuille solution is believed to be formally stable toward infinitesimal perturbations (in the rz-plane) for all Reynolds numbers [BHM03].

Example 16.4 [Aortic flow]: Human blood is not a particularly Newtonian fluid, but its viscosity may approximatively be taken to be $\eta = 2.7 \times 10^{-3}$ Pa s and its density near that of water. The stroke volume of the resting heart is about 70 cm^3, and with a resting heart rate of 60 beats per second, the aortic blood flow rate (averaged over a heartbeat) is $Q \approx 70$ cm^3 s^{-1}. Since the aortic root diameter is $2a \approx 25$ mm, the average blood velocity becomes $U \approx 14$ cm s^{-1} and the Reynolds number $\mathsf{Re} \approx 1{,}300$, well below the turbulent region. The pressure gradient becomes $G \approx 20$ Pa m^{-1}, showing that the pressure drop in the large arteries is small compared to systolic blood pressure, $p \approx 120$ mmHg $\approx 16{,}000$ Pa.

Example 16.5 [Hypodermic syringe]: A hypodermic syringe has a cylindrical chamber with a diameter of about $2b = 1$ cm and a hollow needle with an internal diameter of about $2a = 0.5$ mm. During an injection, about 5 cm^3 of the liquid (here assumed to be water) is gently pressed through the needle in a time of $\Delta t = 10$ s, such that the volume rate is about $Q = 0.5$ cm^3 s^{-1}. The average fluid velocity in the needle becomes $U \approx 2.5$ m s^{-1}, corresponding to a Reynolds number $\mathsf{Re} \approx 1{,}300$, which is in the laminar region below the onset of turbulence. One is thus justified in using the Poiseuille solution for the flow through the needle. The pressure gradient necessary to drive this flow is found from (16.32) and becomes rather large, $G \approx 3.3$ bar m^{-1}. For a needle of length $L \approx 5$ cm, the pressure drop becomes $\Delta p \approx 0.16$ bar $= 16{,}000$ Pa, about the same as the systolic blood pressure. The pressure drop in the fluid chamber can be completely ignored, because the chamber's diameter is 20 times that of the needle.

Example 16.6 [Water pipe]: Household water supply has to reach the highest floor in apartment buildings with pressure "to spare". Pressures must therefore be of the order of bars when water is not tapped. The typical discharge rate from a kitchen faucet is around $Q \approx 100$ cm^3 s^{-1}, leading to an average velocity in a half-inch pipe of about $U \approx 0.8$ m s^{-1}. The Reynolds number becomes about $\mathsf{Re} \approx 10{,}000$, which is well inside the turbulent regime. The pressure gradient calculated from the Hagen–Poiseuille law, $G \approx 160$ Pa m^{-1}, is for this reason untrustworthy.

[1] Reynolds' original paper [Rey83] is still very enjoyable. Its explicit description of the experimental equipment and procedure makes it easy to reproduce in any high-school laboratory.

Case: Ostwald viscometer

In a constant gravitational field the effective vertical pressure gradient is $G = \rho_0 g_0$, and using the Hagen-Poiseuille law (16.31), we obtain the kinematic viscosity,

$$\nu = \frac{\eta}{\rho_0} = \frac{\pi a^4 G}{8\rho_0 Q} = \frac{\pi a^4 g_0}{8Q}. \qquad (16.34)$$

The *Ostwald viscometer* (see the margin figure) is made entirely from glass with no moving parts. The volume flow rate $Q = V/T$ is determined by measuring the passage time T for a known volume V of a liquid through a narrow vertical section of pipe.

The viscosity determined in this way is only meaningful for laminar flow, which requires the Reynolds number to be smaller than about 2,300. From (16.33) we obtain the Reynolds number for the viscometer,

$$\mathsf{Re} = \frac{g_0 a^3}{4\nu^2}, \qquad (16.35)$$

and this shows that for water-like liquids with $\nu \approx 10^{-6}\ \mathrm{m^2\ s^{-1}}$ the tube radius must be smaller than 1 mm in order to avoid turbulence. An Ostwald viscometer is usually designed for determination of the viscosity in a fairly narrow range where the passage time is neither too short nor too long.

Commercial Ostwald viscometer. The viscometer is filled with liquid to the mark above the lower bulb on the left. The upper bulb is then filled to the upper mark by suction, and one measures the time it takes the liquid surface to pass the lower mark. (Source: Poulten Selfe & Lee Ltd., UK. With permission.)

Drag and power

The fluid only interacts with the pipe at the inner surface $r = a$ and attempts to drag it along with a total force or drag $\mathcal{D}$ along z. From the Poiseuille velocity profile (16.29) we obtain the shear stress that the fluid exerts on the pipe wall,

$$\sigma_{\text{wall}} = -\sigma_{zr}\big|_{r=a} = -\eta\nabla_r v_z(r)\big|_{r=a} = \tfrac{1}{2}Ga. \qquad (16.36)$$

Multiplying by the area $2\pi a L$ of the inner pipe surface and eliminating the pressure gradient by (16.32), the drag becomes

$$\boxed{\mathcal{D} = 8\pi\eta U L.} \qquad (16.37)$$

The linear growth of drag with velocity is characteristic of creeping flow at low Reynolds number ($\mathsf{Re} \lesssim 1$; see Chapter 17). At larger Reynolds number the nonlinear effects of the advective acceleration ought to show up, but due to the vanishing advective acceleration of laminar pipe flow the nonlinearity first appears when turbulence sets in for $\mathsf{Re} \gtrsim 2{,}000$.

Newton's Third Law tells us that the pipe wall acts on the fluid with a force $-\mathcal{D}$. The only other force acting on the volume of fluid is the external pressure force, $\Delta p A$ where $A = \pi a^2$. In steady flow the amount of momentum of the fluid in the pipe is constant, and consequently the total force on the volume must vanish, $\mathcal{F}_z = -\mathcal{D} + \Delta p A = 0$ or $\mathcal{D} = \Delta p A$, which is easily verified. The shear stress acting on the pipe wall performs no work because the no-slip condition requires the velocity field to vanish at the wall. Only the driving pressure does work on the fluid, and since it is constant across the pipe cross-section, its rate of work or power becomes

$$P = \int_A \boldsymbol{v} \cdot \Delta p \, d\boldsymbol{S} = \Delta p\, Q = \mathcal{D}U = 8\pi\eta U^2 L. \qquad (16.38)$$

In steady flow, the kinetic energy of the fluid in the pipe is constant, so the power of the driving pressure must equal the rate of kinetic energy loss due to internal friction in the fluid (dissipation). A direct calculation of the rate of dissipation confirms this claim (see Equation (15.24) on page 252 and Problem 16.21).

Wilhelm Ostwald (1853–1932). German scientist. Received the Nobel prize in chemistry in 1909. Considered the father of modern physical chemistry. Invented the Ostwald process for synthesizing nitrates still used in the manufacture of explosives.

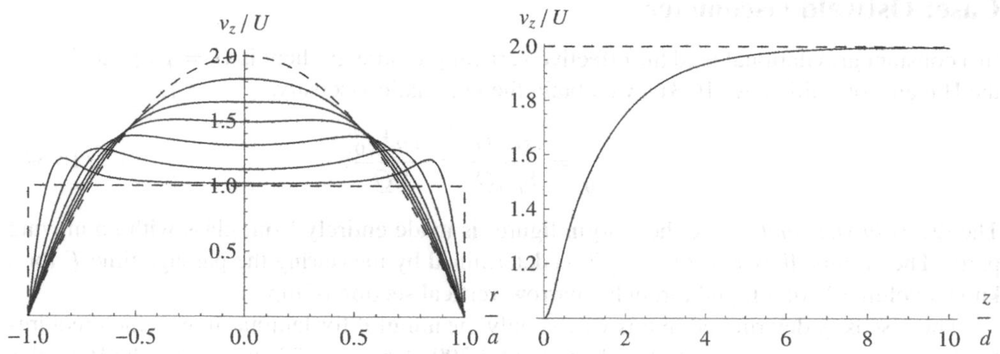

Figure 16.2. Simulated pipe entry flow at $\mathsf{Re} = 100$. **Left:** The fluid enters at $z = 0$ as plug flow (dashed) and the profiles are plotted for $z/d = 0.1, 0.3, 0.6, 1, 1.5, 2.5$ as they converge toward the parabolic shape (dashed). **Right:** The central flow velocity v_z at $r = 0$ as a function of z. At $z \approx 6$, the curve has reached 99% of its asymptotic value.

Entry length

The Poiseuille velocity profile is not established immediately at the entry to a pipe. As the fluid moves downstream from the inlet, the no-slip condition at the inner pipe wall creates a shear wall stress, which by momentum diffusion slows down the fluid in a boundary layer close to the wall. The boundary layer widens as the fluid moves down the pipe, and to keep the average speed constant, the fluid must speed up near the middle of the pipe, for after some distance to reach the parabolic Poiseuille flow pattern (see Figure 16.2).

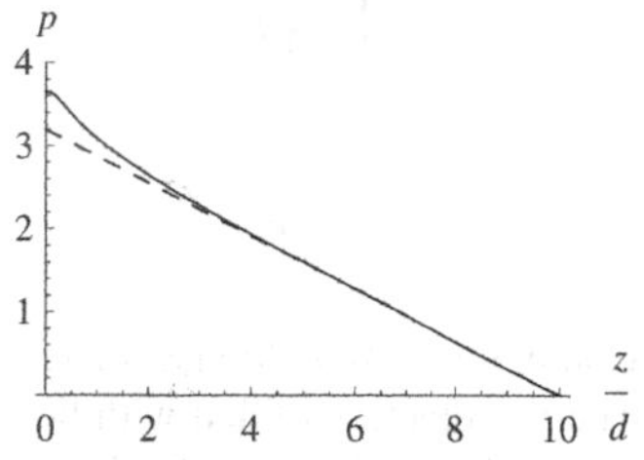

The driving pressure as a function of downstream distance for the simulation in Figure 16.2. It starts out a bit higher than predicted by the linear expression (16.28) for perfect Poiseuille flow (dashed line).

The estimate of the entry length follows the same road as for planar flow (page 265). Momentum diffusion is characterized by a transverse spread of size $2\sqrt{\nu t}$ during a time interval t. The diffusion front reaches the middle of the channel from all sides when $2\sqrt{\nu t} \approx a$, or $t \approx a^2/4\nu$. In this time interval the fluid has on average moved downstream through the distance $z \approx Ut \approx Ua^2/4\nu$, which we take as an estimate for the entry length, L_{entry}. Dividing by $d = 2a$ the entry length relative to the pipe diameter is found to scale linearly with the Reynolds number,

$$\frac{L_{\text{entry}}}{d} \approx k\mathsf{Re}, \tag{16.39}$$

with a crude estimate of the proportionality constant, $k \approx 1/16 \approx 0.063$. It depends, however, on how close the terminal flow velocity $2U$ is approached. From the simulation in Figure 16.2R one sees that the central flow reaches 99% for $z/d = 5.8$, implying that $k_{99} = 0.058$. Langhaar's boundary layer calculation from 1942 yielded $k_{99} = 0.057$ [Lan42]. The pressure gradient is nearly constant over the most of the pipe, except right at the entry where it is about 10% higher (see the margin figure).

In the creeping regime $\mathsf{Re} \lesssim 1$, the entry length becomes constant, about equal to 60% of a diameter (as in planar flow; see Figure 19.3 on page 334). It has—over the years—been quite difficult to settle on an accepted value for the entry length for low Reynolds number (see for example [FM71]). An expression valid up to the onset of turbulence is [White 1991, p. 293]

$$\frac{L_{\text{entry}}}{d} \approx \frac{0.6}{1 + 0.035\mathsf{Re}} + 0.056\mathsf{Re}. \tag{16.40}$$

At the beginning of the transition to turbulence, say $\mathsf{Re} \approx 2{,}300$, this formula yields a 99% entry length of about 130 diameters. In the fully developed turbulent regime, $\mathsf{Re} \gtrsim 4{,}000$, the average velocity distribution is nearly plug-like, and the entry length becomes much smaller, $L_{\text{entry}}/d \approx 4.4\,\mathsf{Re}^{1/6}$, which only comes to about 18 diameters at $\mathsf{Re} \approx 4{,}000$.

Case: Laminar drain

Returning to Torricelli's law for the draining of a cistern (page 214), we are now able to take into account the slowing down of the flow due to viscosity. For simplicity we imagine that the liquid level in the cistern is kept constant by replenishing the cistern with the fluid that exits from the pipe. This allows us to use energy balance to calculate the average velocity U with which the liquid streams through the exit pipe[2].

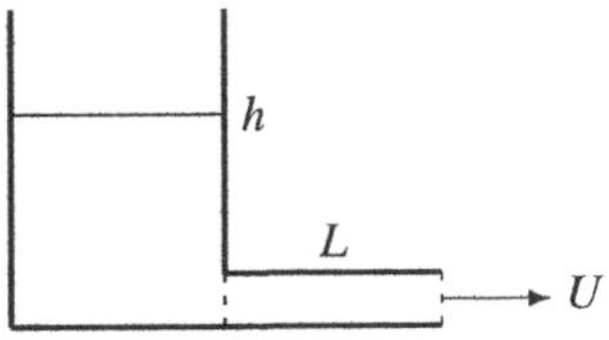

Viscous liquid streaming from a cistern through a long pipe.

The rate at which external work is performed against gravity in moving the fluid from the exit level $z = 0$ back to the cistern level $z = h$ is

$$\dot{W} = \rho_0 g_0 h Q, \tag{16.41}$$

where $Q = UA$ is the volume flow rate through the pipe. We shall assume that the cistern cross-section is so large that the rate of inflow of kinetic energy as well as dissipation in the cistern can be ignored. The external work performed against gravity must either provide kinetic energy to the fluid discharged at the pipe exit or be dissipated into heat,

$$\dot{W} = \dot{\mathcal{K}} + P, \tag{16.42}$$

where $\dot{\mathcal{K}}$ is the rate of discharge of kinetic energy and P is the rate of dissipation.

When the pipe is much longer than the entry length, $L \gg L_{\text{entry}}$, the liquid will always leave the pipe with the Poiseuille velocity profile, $v_z = 2U(1 - r^2/a^2)$. The rate of discharge of kinetic energy through the exit is obtained by adding the contributions from each infinitesimal amount of mass, $d\dot{M} = \rho_0 \boldsymbol{v} \cdot d\boldsymbol{S}$, that leaves the pipe per unit of time through a surface element $d\boldsymbol{S}$,

$$\dot{\mathcal{K}} = \int_A \tfrac{1}{2} v^2 \, d\dot{M} = \int_0^a \frac{1}{2} \rho_0 v_z(r)^2 \cdot v_z(r)\, 2\pi r\, dr = \rho_0 U^3 \pi a^2 = \rho_0 U^2 Q\,.$$

This is actually twice the rate $\frac{1}{2}\rho_0 U^2 Q$ that one would find in plug flow with the same velocity all across the exit, but such a difference must be expected when the velocity field varies over the exit.

Using the expression for Poiseuille dissipation (16.38), we find the ratio of kinetic energy loss to dissipation,

$$\frac{\dot{\mathcal{K}}}{P} = \frac{Ua^2}{8\nu L} = \mathsf{Re}\frac{d}{32L} \approx \frac{L_{\text{entry}}}{2L},$$

where we in the last step have introduced the entry length (16.39) with $k \approx 1/16$. Evidently we may for $L \gg L_{\text{entry}}$ disregard the kinetic energy loss relative to dissipation. Energy balance then simplifies to $\dot{W} \approx P$, and solving for the velocity we finally obtain the average exit pipe velocity (for $L \gg L_{\text{entry}}$),

$$\boxed{U \approx \frac{g_0 a^2}{8\nu}\frac{h}{L}.} \tag{16.43}$$

A bit surprisingly, the velocity grows linearly with h rather than with $\sqrt{h}$, as it would according to Torricelli's law for inviscid flow. But when the height h is increased, the average velocity will grow, and with it the Reynolds number as well as the entry length also grow. Eventually, the entry length will become larger than the length of the pipe, the exit flow will change from Poiseuille to plug flow and come into agreement with Torricelli's law. On top of that, the Reynolds number will sooner or later cross the threshold to turbulence, and a whole new game commences.

[2] In Chapter 20 the laws of balance in continuum mechanics will be systematically presented.

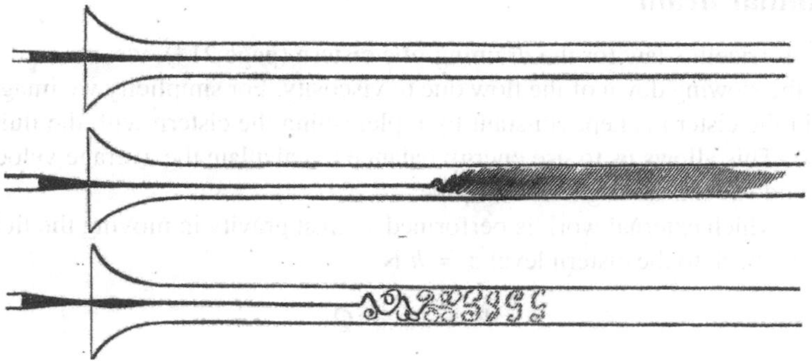

Figure 16.3. Original drawings of pipe flow from Reynolds' 1883 paper [Rey83]. Water enters from the left through the flaring trumpet-like mouthpiece, and dye is continually injected at its center. The top drawing shows how colored and uncolored water remain separated in laminar flow. In the middle is shown what happens when the Reynolds number is so large that turbulence sets in and the color is thoroughly mixed with the water (starting somewhat downstream from the entry). In the bottom drawing the eddies of turbulent flow are revealed by a flash of light. (Source: Reproduced from [Rey83]. With permission.)

16.5 Phenomenology of turbulent pipe flow

In 1883, Osborne Reynolds was the first systematically to investigate the nature of turbulent pipe flow [Rey83]. In spite of their age, the drawings from the original paper are still one of the best illustrations of the two modes of pipe flow. Empirically, laminar flow is only possible up to a certain value of the Reynolds number, beyond which turbulence sets in. Before turbulence is fully developed there is, however, a transition regime characterized by intermittent behavior. After turbulence has become fully established the flow goes for increasing Reynolds number through at least two further distinct stages where different physical processes dominate the character of the flow. Although to this day there is no universally accepted theory of turbulent pipe flow, there exist efficient semiempirical methods.

In fully developed turbulence, the true velocity field varies rapidly in time as well as from place to place throughout the pipe. If dye is injected into the fluid, it quickly spreads through the whole volume, which is quite different from laminar flow where it makes orderly streaks (see Figure 16.3). Since the flowlines of incompressible fluid can neither begin nor end but must form closed loops or go to infinity, turbulent flow will be full of eddies at all scales. In Chapter 31 we shall investigate the eddy distribution. Here we shall only focus on the phenomenological description of turbulent pipe flow.

Drag factor

The incompressibility of the fluid guarantees that the volumetric discharge rate Q through any cross-section of the pipe must always be constant. The average velocity and the Reynolds number are as usual defined from the discharge rate,

$$U = \frac{Q}{A}, \qquad \mathsf{Re} = \frac{2aU}{\nu}, \tag{16.44}$$

where $A = \pi a^2$. Independent of the character of the flow, Q is experimentally easy to determine by measuring the volume of fluid discharged from the pipe over a sufficiently long time interval.

Intuitively, one would expect that the mess of turbulence causes increased resistance to the flow and therefore requires a larger driving pressure Δp than the corresponding laminar flow to obtain the same rate of discharge Q under steady conditions. The total drag on the

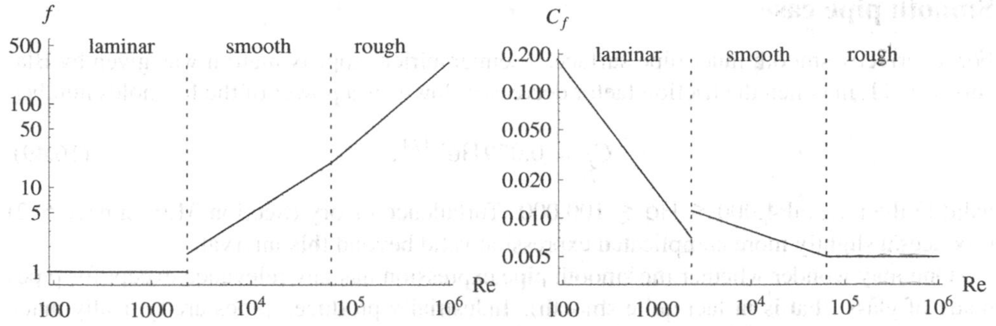

Figure 16.4. Sketch of the drag factor (left) and the Fanning friction factor (right) as functions of the Reynolds number for a pipe with roughness scale, $k/2a = 10^{-3}$, where k is the average height of the irregularities in the pipe wall. The transition from laminar to smooth-pipe turbulent flow nominally takes place at $\mathsf{Re} \approx 2{,}300$ but actually covers the region $2{,}000 \lesssim \mathsf{Re} \lesssim 4{,}000$. The transition from smooth-pipe turbulence to rough-pipe turbulence happens in this case for $\mathsf{Re} \approx 63{,}000$ and is quite gentle.

pipe, $\mathcal{D} = \Delta p A$, is accordingly also expected to be larger than the laminar expression (16.37). As the Reynolds number is the only dimensionless quantity that may be constructed from the average velocity (Problem 16.8), we may in general write the drag as the laminar expression times a dimensionless *drag factor* $f(\mathsf{Re})$, depending only on the Reynolds number,

$$\boxed{\mathcal{D} = 8\pi\eta U L f(\mathsf{Re}).} \tag{16.45}$$

The drag factor is anchored with the value $f(\mathsf{Re}) = 1$ in laminar flow, and grows monotonically with the Reynolds number throughout the turbulent region (see Figure 16.4L). In the transition region between laminar and turbulent flow ($2{,}000 \lesssim \mathsf{Re} \lesssim 4{,}000$), the drag factor is not too well defined because of intermittent shifts between laminar and turbulent flow.

Friction factor

In the turbulent regime, it is more convenient to use another dimensionless measure of turbulent drag, based on the average shear wall stress:

$$\sigma_{\text{wall}} = \frac{\mathcal{D}}{2\pi a L} = 4\eta \frac{U}{a} f(\mathsf{Re}). \tag{16.46}$$

The *Fanning friction factor*[3] is defined as the ratio between the average shear wall stress and the dynamical pressure, $\frac{1}{2}\rho_0 U^2$, where U is the average flow velocity,

$$\boxed{C_f = \frac{\sigma_{\text{wall}}}{\frac{1}{2}\rho_0 U^2} = \frac{16}{\mathsf{Re}} f(\mathsf{Re}).} \tag{16.47}$$

In terms of the friction factor the drag may be written as

$$\mathcal{D} = \pi a L \rho_0 U^2 C_f(\mathsf{Re}). \tag{16.48}$$

In this representation all the dependence on viscosity is hidden in the friction factor.

In Figure 16.4R the friction factor is plotted as a function of the Reynolds number. The friction factor tends to become constant for infinite Reynolds number, except for perfectly smooth pipes. It is thus anchored in the region of extremely turbulent flow, although its limiting value $C_f(\infty)$ depends on the roughness of the inner pipe surface.

John Thomas Fanning (1837–1911). American civil engineer. Chief engineer and consultant for building waterworks for cities and large-scale waterpower stations.

Henri Philibert Gaspard Darcy (1803–1858). French engineer. Pioneered the understanding of fluid flow through porous media and established Darcy's law, which is used in hydrogeology.

[3]In the technical literature one often meets the *Darcy friction factor*, which equals $4C_f$ (see problem 16.10).

Smooth pipe case

For a perfectly smooth inner pipe surface, a semiempirical approximation was given by Blasius in 1911, in which the friction factor decreases slowly as a power of the Reynolds number,

$$C_f \approx 0.079\,\mathsf{Re}^{-1/4}, \tag{16.49}$$

valid in the interval 4,000 $\lesssim$ Re $\lesssim$ 100,000. Turbulence theory (Section 31.6 on page 582) produces a slightly more complicated expression valid beyond this interval.

One may wonder whether the smooth pipe expression has any relevance except for pipes made of glass (that is in fact quite smooth). Industrially produced pipes are typically made from rubber, plastic, iron, copper, or concrete with different degrees of inner surface roughness. But it turns out that any not too irregular surface will tend to behave approximately like a perfectly smooth surface in a range of Reynolds numbers above the onset of turbulence. The reason is that when turbulence sets in, the no-slip condition still requires the fluid velocity to decrease toward zero as one approaches the solid wall. This leads to the formation of viscous wall layers that screen the bulk of the flow from the influence of the surface roughness, thus making the surface appear to be smooth, even if it strictly speaking is not. The wall layers, however, become thinner and the screening less effective for very high Reynolds numbers. Above a certain roughness-dependent value, typically for Re $\gtrsim$ 100,000, the smooth pipe approximation generally loses its usefulness.

Example 16.7 [Water pipe]: A quite normal use of household water was shown in Example 16.6 to lead to a Reynolds number of about Re $\approx$ 10,000, which is well inside the turbulent regime, though not in the rough pipe region. From the smooth pipe formula (16.49) we find the friction factor $f(10{,}000) = 5$. The pressure gradient must therefore be 5 times larger than the Hagen–Poiseuille value (which was 160 Pa m^{-1}), that is, $G \approx 800$ Pa m^{-1}. Drag and dissipation are similarly augmented. Turbulence makes the pipes hiss or "sing" when you tap water at full speed, though most of the noise probably comes from the narrow passages of the faucet where the water speed, Reynolds number, pressure gradient, and drag are largest.

Rough pipe limit

Material	k
Glass	3×10^{-7}
Steel	4.5×10^{-5}
Rubber	1×10^{-5}
Cast iron	2.6×10^{-4}
Concrete	1.5×10^{-3}
Unit	m

Characteristic roughness constants for various materials (from [White 1991]). Note that machining of a surface (sawing, planing, milling, grinding, and polishing) will greatly influence its roughness.

At sufficiently high Reynolds number the irregular turbulent flow is caused by the fluid literally slamming into the small protrusions of the rough pipe surface, rather than by viscous friction. In this limit the drag becomes independent of the actual viscosity, even if it in the end must owe its very existence to viscosity! Correspondingly, the friction coefficient becomes a constant, $C_f(\infty)$, that depends on the character of the rough surface.

The surface roughness may be characterized by the average height k of the surface irregularities, a parameter that depends on the material of the pipe and how it is produced. A semi-empirical expression for the limiting friction coefficient is also given by a power law,

$$C_f(\infty) \approx 0.028 \left(\frac{k}{2a}\right)^{1/4}, \tag{16.50}$$

where k is the typical height variation due to the surface roughness. Turbulence theory again yields a more complicated expression that agrees reasonably with the above power law in the interval $10^{-4} \lesssim k/2a \lesssim 10^{-2}$. Deviations reach 30% at about an order of magnitude beyond this interval in both directions.

The smooth pipe expression crosses the rough pipe expression for Re $\approx 63 \times 2a/k$. The transition from smooth to rough pipe turbulence is fairly soft, certainly not as abrupt as the transition from laminar to turbulent flow (see Figure 16.4). Diagrams covering the whole region and semi-empirical expressions, useful for engineering design purposes, may be found in the technical literature.

Case: Turbulent drain

The calculation of the turbulent drain velocity proceeds in much the same way as for laminar viscous flow (page 273), except that we do not know the actual velocity distribution in the pipe. Although there will also be boundary layers in turbulent pipe flow, we shall for simplicity assume that the bulk of the turbulent flow proceeds through the pipe as a *turbulent plug* with roughly constant average speed across the pipe.

The rate of loss of kinetic energy from the pipe exit is then $\dot{\mathcal{K}} = \frac{1}{2}\rho_0 U^2 Q$ whereas the work to replenish the cistern is $\dot{W} = \rho_0 g_0 h Q$. Writing the dissipative loss as $P = \mathcal{D}U$ with $\mathcal{D}$ given by (16.45), energy conservation (16.42) takes the form

$$2g_0 h = U^2 + \frac{16\nu UL}{a^2} f(\mathsf{Re}) = U^2\left(1 + 2C_f(\mathsf{Re})\frac{L}{a}\right), \qquad (16.51)$$

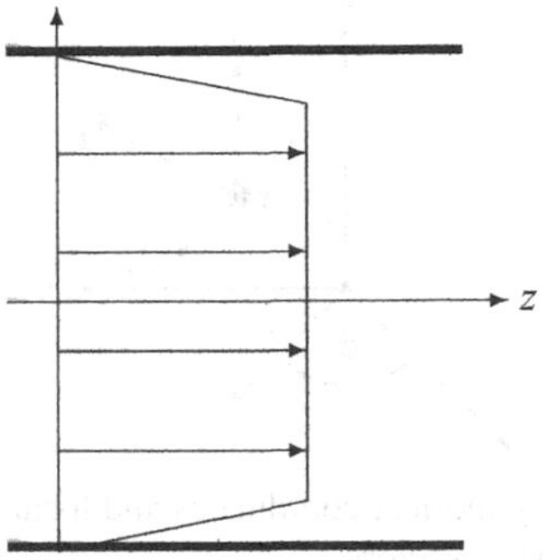

Assumed velocity profile for fully developed turbulent flow through a circular pipe. Apart from thin boundary layers (which we ignore), the velocity field is approximately constant across the pipe.

with $\mathsf{Re} = 2aU/\nu$. The first term on the right-hand side corresponds to Torricelli's law, $U = \sqrt{2g_0 h}$. In the general case this equation has to be solved numerically.

For large Reynolds number we may use the rough pipe form of the friction factor (16.50) to obtain the solution,

$$U = \sqrt{\frac{2g_0 h}{1 + 2C_f(\infty)L/a}}. \qquad (16.52)$$

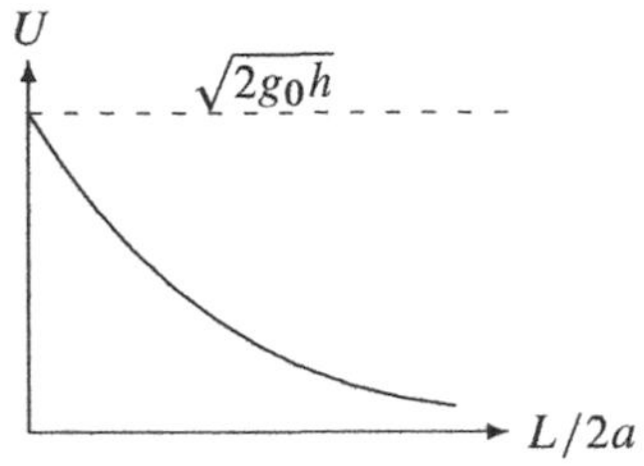

Turbulent drain speed as a function of pipe length. For sufficiently long pipes, the flow becomes laminar.

Torricelli's result is only obtained for sufficiently short pipes where the denominator becomes unity. For sufficiently long pipes the terminal velocity decreases as $1/\sqrt{L}$ and this will eventually bring the Reynolds number down into the smooth pipe region and finally into laminar one where it according to (16.43) decreases as $1/L$.

Example 16.8 [Barrel of wine, continued]: For the barrel of wine (Example 13.3 on page 215), which empties through a wooden spout of length $L = 20$ cm and diameter $2a = 5$ cm, the Reynolds number is about 300,000, well into the rough pipe region. Disregarding entry corrections and assuming a wood roughness of $k \approx 0.5$ mm, corresponding to $k/2a = 0.01$, we find $C_f(\infty) \approx 9 \times 10^{-3}$ and $2C_f(\infty)L/a \approx 0.14$. The Toricelli value for the exit velocity is thus only reduced by about 7% by turbulence.

16.6 Laminar cylindric flow

A layer of fluid lubricating the space between rotating coaxial cylinders is used in nearly all machines. Since the fluid must hang on to the cylindrical surfaces, they will exert a drag on each other that is similar to the drag created by the fluid between moving planar surfaces (page 247). The flow solution is attributed to Couette, but as always in science the true history is tortuous [DMS05]. Today, *Couette flow* is used to denote any kind of motion-driven laminar flow, including laminar channel flow between moving parallel plates.

Maurice Marie Alfred Couette (1858–1943). French physicist from the provincial university of Angers. Published only seven papers, all from 1888 to 1900. Invented what is now called the Couette viscometer.

Maximally symmetric flow

Infinitely extended coaxial cylinders are both translationally symmetric along the rotation axis and rotationally symmetric around it. In cylindrical coordinates a maximally symmetric circulating flow takes the form

$$\boldsymbol{v} = v_\phi(r)\,\hat{\boldsymbol{e}}_\phi, \qquad (16.53)$$

where $\hat{\boldsymbol{e}}_\phi$ is the tangential unit vector and r the distance from the axis (see Appendix D). The field lines are concentric circles.

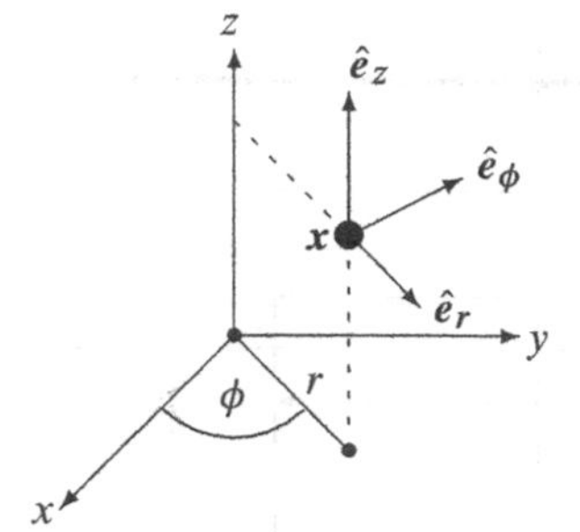

Cylindrical coordinates and local basis vectors

Again we turn to the nabla operator in cylindrical coordinates (16.25) to calculate the tensor gradient of the velocity field. Since there is no z-dependence we find by means of the relation $\partial_\phi \hat{e}_\phi = -\hat{e}_r$,

$$\nabla v = \left(\hat{e}_r \frac{\partial}{\partial r} + \hat{e}_\phi \frac{1}{r}\frac{\partial}{\partial \phi}\right)(v_\phi(r)\hat{e}_\phi) = \hat{e}_r\hat{e}_\phi \frac{dv_\phi}{dr} - \hat{e}_\phi\hat{e}_r \frac{v_\phi}{r}. \tag{16.54}$$

The trace of this tensor yields the divergence that vanishes, $\mathrm{Tr}[\nabla v] = \nabla \cdot v = 0$, because ∇v only has off-diagonal components in the cylindrical basis. This is in agreement with the elementary observation that streamlines neither diverge nor converge in this flow. Dotting the tensor gradient from the left with v, the advective acceleration becomes

$$(v \cdot \nabla)v = v \cdot (\nabla v) = -\hat{e}_r \frac{v_\phi^2}{r}.$$

One should not be surprised: The centripetal acceleration in a circular motion with velocity v_ϕ is indeed directed radially inwards and of size v_ϕ^2/r. Finally, dotting the tensor gradient from the left with ∇, we obtain the Laplacian (see also Equation (D.20) on page 622):

$$\nabla^2 v = \left(\hat{e}_r \frac{\partial}{\partial r} + \hat{e}_\phi \frac{1}{r}\frac{\partial}{\partial \phi}\right)\left(\hat{e}_r\hat{e}_\phi \frac{dv_\phi}{dr} - \hat{e}_\phi\hat{e}_r \frac{v_\phi}{r}\right) = \hat{e}_\phi \frac{d^2 v_\phi}{dr^2} + \hat{e}_\phi \frac{1}{r}\frac{dv_\phi}{dr} - \hat{e}_\phi \frac{v_\phi}{r^2}.$$

In the absence of gravity, the Navier–Stokes equation (16.1) then becomes

$$-\rho_0 \hat{e}_r \frac{v_\phi^2}{r} = -\nabla p + \eta\, \hat{e}_\phi \frac{d}{dr}\left(\frac{1}{r}\frac{d(rv_\phi)}{dr}\right), \tag{16.55}$$

where we in the last term have rewritten the Laplacian in a convenient way.

Projecting it on the three cylindrical basis vectors, $\hat{e}_r$, $\hat{e}_\phi$, and $\hat{e}_z$, we obtain

$$-\rho_0 \frac{v_\phi^2}{r} = -\frac{\partial p}{\partial r}, \tag{16.56a}$$

$$0 = -\frac{1}{r}\frac{\partial p}{\partial \phi} + \eta \frac{d}{dr}\left(\frac{1}{r}\frac{d(rv_\phi)}{dr}\right), \tag{16.56b}$$

$$0 = -\frac{\partial p}{\partial z}. \tag{16.56c}$$

The first equation expresses that the radial pressure gradient must deliver the centripetal force required by the circular motion of the fluid. The last equation shows that the pressure is independent of z, and from the second equation we see by differentiation with respect to ϕ that $\partial^2 p/\partial \phi^2 = 0$. This means that p can at most be linear in ϕ, that is, of the form $p = p_0(r) + p_1(r)\phi$. But here we must require $p_1 = 0$, for otherwise the pressure would have different values for $\phi = 0$ and $\phi = 2\pi$, and that is impossible. The pressure does not depend on ϕ but only on r, and thus it disappears completely from (16.56b).

With the pressure out of the way, the integration of (16.56b) has become almost trivial with the general result,

$$v_\phi = Ar + \frac{B}{r}, \tag{16.57}$$

where A and B are integration constants. Inserting this into (16.56a) and integrating over r, we find the pressure

$$\frac{p}{\rho_0} = C + \frac{1}{2}A^2 r^2 - \frac{1}{2}\frac{B^2}{r^2} + 2AB \log r, \tag{16.58}$$

where C is a third integration constant.

The Couette solution

Suppose the fluid is contained between two long coaxial material cylinders with radii a and $b > a$. We shall for simplicity only study the case where the outer cylinder is held fixed and the inner cylinder rotates like a spindle with constant angular velocity Ω. The boundary conditions, $v_\phi(a) = a\Omega$ and $v_\phi(b) = 0$, then determine A and B, and the velocity profile becomes

$$v_\phi = \frac{\Omega a^2}{r} \frac{b^2 - r^2}{b^2 - a^2}. \tag{16.59}$$

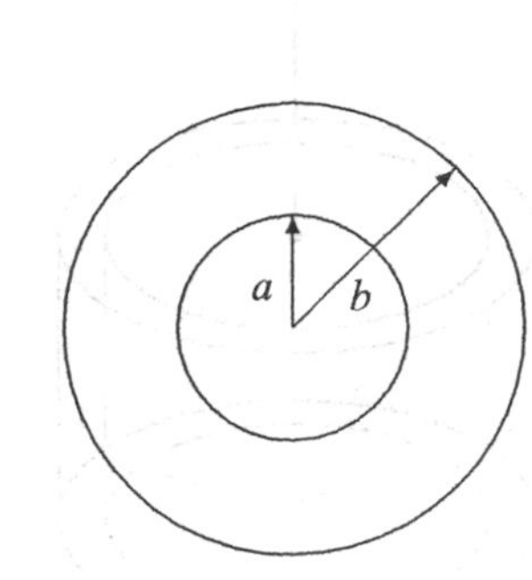

Geometry of Couette flow.

The velocity field decreases monotonically from its value $a\Omega$ at the inner cylinder to zero at the outer. If the outer cylinder also rotates, the expression becomes a bit more complicated (see problem 16.16).

The pressure is found from (16.58),

$$p^* = p_0 + \frac{1}{2}\rho_0 \left(\frac{\Omega a^2}{b^2 - a^2}\right)^2 \left(r^2 - \frac{b^4}{r^2} + 4b^2 \log \frac{b}{r}\right), \tag{16.60}$$

where p_0 is its value at the outer cylinder. The pressure increases monotonically toward p_0.

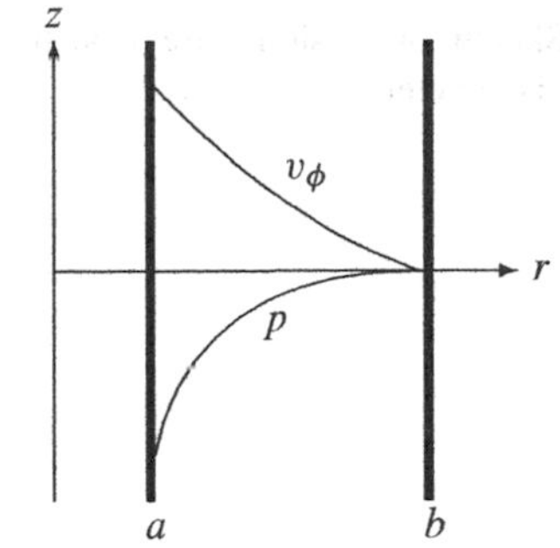

Sketch of the velocity and pressure between cylinders in Couette flow with the outer cylinder held fixed.

Stress, torque, and power

The velocity gradient (16.54) shows that the only non-vanishing shear stress component is

$$\sigma_{\phi r} = \eta(\hat{e}_\phi \cdot (\nabla v) \cdot \hat{e}_r + \hat{e}_r \cdot (\nabla v) \cdot \hat{e}_\phi) = \eta \left(\frac{dv_\phi}{dr} - \frac{v_\phi}{r}\right), \tag{16.61}$$

which upon insertion of the Couette solution (16.59) becomes

$$\sigma_{\phi r} = -2\eta\Omega \frac{a^2 b^2}{b^2 - a^2} \frac{1}{r^2}. \tag{16.62}$$

It represents the viscous friction between the layers of circulating fluid, and the sign is negative because the fluid *outside* the radius r acts as a brake on the motion of the fluid *inside*.

In order to maintain the steady rotation of the inner cylinder (of length L) it is necessary to act on it with a moment of force or torque (and with an opposite moment on the outer cylinder). More generally, multiplying the shear stress $\sigma_{\phi r}$ with the moment arm r and the area $2\pi r L$ of the cylinder at r, we obtain the torque with which the fluid *inside* r acts on the fluid *outside*,

$$\mathcal{M}_z = r\,(-\sigma_{\phi r}) 2\pi r L = 4\pi\eta\Omega L \frac{a^2 b^2}{b^2 - a^2}. \tag{16.63}$$

We could have foreseen that $\mathcal{M}_z$ would be independent of r from angular momentum conservation. In steady flow, the total angular momentum of a layer of fluid contained between any two cylindrical surfaces is constant and there is no transport of angular momentum through the cylindrical surfaces because they are parallel with the velocity. Consequently the total moment of force has to vanish, implying that the moment acting on the inside of the layer must be equal and opposite to the moment that is acting on the outside of the layer.

The rate of work that must be done to keep the inner cylinder rotating is obtained by multiplying the stress on the fluid, $-\sigma_{\phi r}$ at $r = a$, by the velocity $a\Omega$ and the area of the cylinder,

$$P = (-\sigma_{\phi r})\, a\Omega\, 2\pi a L = \mathcal{M}_z \Omega = 4\pi\eta\Omega^2 L \frac{a^2 b^2}{b^2 - a^2}. \tag{16.64}$$

Since the kinetic energy of the fluid is constant in steady flow and since no fluid enters or leaves the system, this must equal the rate of energy dissipation in the fluid.

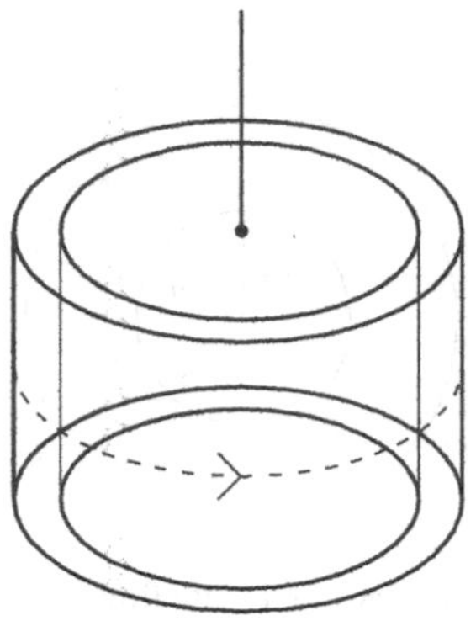

Sketch of torsion wire Couette viscometer.

Couette viscometer: The viscosity of a fluid could directly be determined from the extra motor power necessary to drive the inner cylinder at constant angular velocity. It turns out that pure Couette flow is quite unstable in this configuration. A more stable arrangement (see page 283) consists of hanging the inner cylinder by a torsion wire and rotating the outer cylinder at constant angular velocity. When the chamber is filled with viscous fluid, there will arise a torque on the inner cylinder that is counteracted by the torsion in the wire so that at equilibrium a certain deflection angle is obtained. The torsional rigidity (page 152) of the wire may be determined from the period of free oscillations of the inner cylinder when the chamber is empty. Measuring the equilibrium deflection angle in a fluid, one derives a value for the torque, and thus the viscosity (Problem 16.17).

Case: Unloaded journal bearing

In a lubricated *journal bearing* a liquid, say oil, is trapped in a tiny gap between a rotating shaft (or journal) and its bearing (or bushing). Normally such systems carry a load that brings the shaft off-center (see Section 17.5) but here we shall assume that the journal and its bushing are concentric cylinders, for example kept apart by ball-bearings, such that we may apply the Couette formalism in the limit of very small distance $d = b - a \ll a$ between the cylinders.

To lowest order in d and the distance $s = r - a$ from the shaft, we find that

$$v_\phi = \Omega\, a \left(1 - \frac{s}{d}\right). \tag{16.65}$$

The velocity field is linear in s, just as for planar Couette flow. This is not particularly surprising since the tiny gap between the cylinders looks very much like the gap between parallel plates moving at relative speed Ωa, leading to the Reynolds number:

$$\mathsf{Re} = \frac{\Omega a d}{\nu}. \tag{16.66}$$

We expect turbulence to arise for $\mathsf{Re} \gtrsim 2{,}000$, as in the planar case.

The pressure is found from (16.60) by expanding the logarithm to third order in s and d, and it becomes, in the leading approximation,

$$p = p_0 - \frac{1}{3} a d \Omega^2 \left(1 - \frac{s}{d}\right)^3. \tag{16.67}$$

The pressure variation across the gap is proportional to the size d of the gap and thus normally tiny, unless the quadratic growth with angular velocity overwhelms it. For a bearing of length L, the power becomes (to lowest significant order in d)

$$P = \frac{2\pi\eta\Omega^2 a^3 L}{d}. \tag{16.68}$$

It is inversely proportional to the thickness of the layer of fluid and grows, like the pressure variation, quadratically with the angular velocity. This is why dissipation may cause even a lubricated bearing to burst into flames at sufficiently high rotation speed.

Example 16.9 [Burning bushing]: Consider a vertical wooden shaft in a wooden bushing with diameter $2a = 10$ cm, such as could have been used in an old water mill. Let the gap be $d = 1$ mm and let the lubricant be heavy grease with $\eta \approx 10$ Pa s, corresponding to $\nu \approx 0.01\ \mathrm{m^2\,s^{-1}}$. The power dissipated per unit of lubricant volume is

$$\frac{P}{2\pi a L d} = \frac{\eta\Omega^2 a^2}{d^2}. \tag{16.69}$$

For a modest speed of one rotation per second, $\Omega \approx 2\pi\ s^{-1}$ with a Reynolds number of $\mathsf{Re} \approx 0.03$, this comes to about $1\ \mathrm{J\,cm^{-3}\,s^{-1}}$. If the mill rotates 10 times faster because of torrential rains, this number becomes instead $100\ \mathrm{J\,cm^{-3}\,s^{-1}}$ with a Reynolds number of 0.3. Since wood is a bad heat conductor, the grease may ignite in a matter of minutes, even taking into account that its viscosity decreases with temperature.

Case: Spindle-driven vortex

Suppose a long cylindrical spindle of radius a is inserted into an infinite sea of fluid with vacuum above, and that the spindle is rotated steadily with constant angular velocity Ω. Provided the spindle does not rotate so fast that it creates turbulence, the flow will eventually become steady with an azimuthal velocity field that may be found from the Couette solution with $b \to \infty$,

$$v_\phi = \frac{\Omega a^2}{r}, \qquad p = -\frac{1}{2}\rho_0 v_\phi^2 = -\frac{1}{2}\rho_0 \frac{\Omega^2 a^4}{r^2}. \tag{16.70}$$

The pressure has been chosen to vanish at infinity.

In a gravitational field g_0 pointing downward along the cylinder axis, we must add the hydrostatic pressure, $-\rho_0 g_0 z$, to get the true pressure

$$p = -\frac{1}{2}\rho_0 v_\phi^2 - \rho_0 g_0 z. \tag{16.71}$$

For a nearly ideal liquid the true pressure at an open surface must vanish, just as in hydrostatics. Consequently the shape of the surface is given by

$$z \equiv h(r) = -\frac{v_\phi^2}{2g_0} = -\frac{\Omega^2 a^4}{2g_0 r^2}. \tag{16.72}$$

Thus, at the spindle, $r = a$, the liquid stands $\Omega^2 a^2/2g_0$ below the asymptotic level.

The rule about continuity of pressure across any surface is only strictly valid in hydrostatics (in the absence of surface tension) while in hydrodynamics it is instead the full stress vector that has to vanish at the open surface. For the spindle-driven vortex, the only non-vanishing stresses are

$$\sigma_{\phi r} = -2\eta \frac{\Omega a^2}{r^2}, \qquad \sigma_{rr} = \sigma_{zz} = \sigma_{\phi\phi} = -p. \tag{16.73}$$

At the open surface (16.72), the diagonal stresses all vanish as they should (because the pressure vanishes) whereas the shear stress is always non-zero. This error arises because we illegitimately have reused the Couette solution obtained for an infinitely long cylinder to discuss the open surface case, which is only semi-infinite. Relative to the dynamic pressure, the error becomes $\sigma_{\phi r}/0.5\rho_0 v_\phi^2 = 4/\mathsf{Re}$, where $\mathsf{Re} = \Omega a^2/\nu$, and that is insignificant for $\mathsf{Re} \to \infty$. The structure of driven as well as free vortices will be analyzed in Chapter 27.

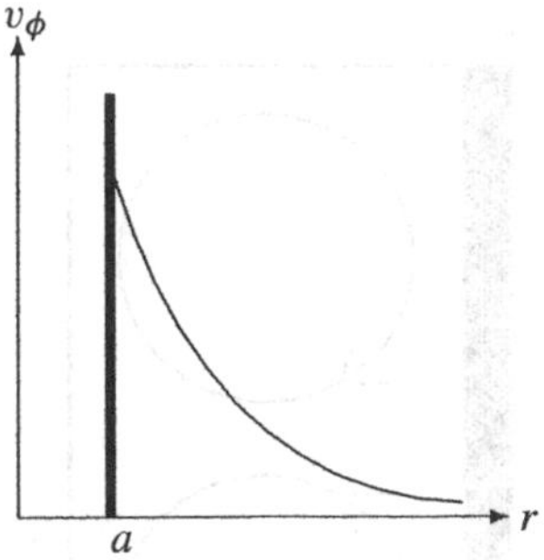

Characteristic $1/r$ velocity profile of spindle-driven vortex.

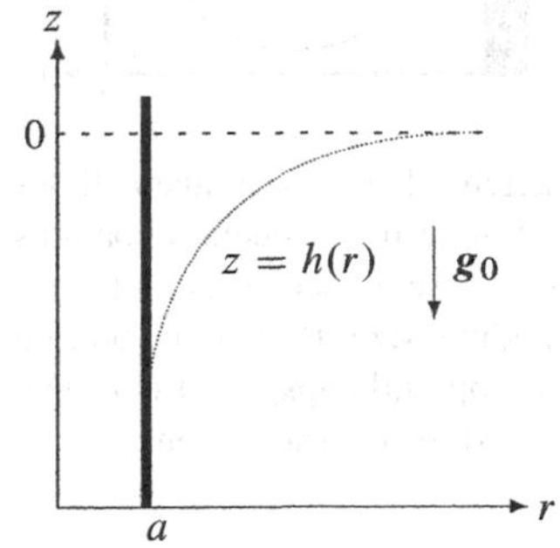

Open liquid surface of a spindle-driven vortex in constant gravity varies as $1/r^2$.

> **Example 16.10:** A spindle with radius $a = 1$ cm making ten turns per second, corresponding to $\Omega = 63$ s^{-1}, would make a depression in the liquid surface, 2 cm, at the spindle. The Reynolds number becomes $\mathsf{Re} \approx 6{,}300$, so that the relative error in ignoring the shear stress at the surface is about 6×10^{-4}.

* 16.7 Secondary flow and Taylor vortices

Real machinery cannot be infinite in any direction. Suppose the cylinders are capped with plates fixed to the outer non-rotating cylinder so that only the inner cylinder rotates. The no-slip condition forces the rotating fluid to slow down and come to rest not only at the outer cylinder, but also at both the end caps, implying that the assumption about a simple circulating flow (16.57) with its z-independent azimuthal velocity $v_\phi(r)$ cannot be right everywhere. In fact, it cannot be right anywhere! Clean Couette flow is, like clean Poiseuille flow, an idealization that can only be approximately realized far from the ends of the apparatus. *Secondary flow* with non-vanishing radial and longitudinal velocity components, v_r and v_z, will have to arise near the end caps to satisfy the no-slip boundary conditions.

Direction of secondary flow

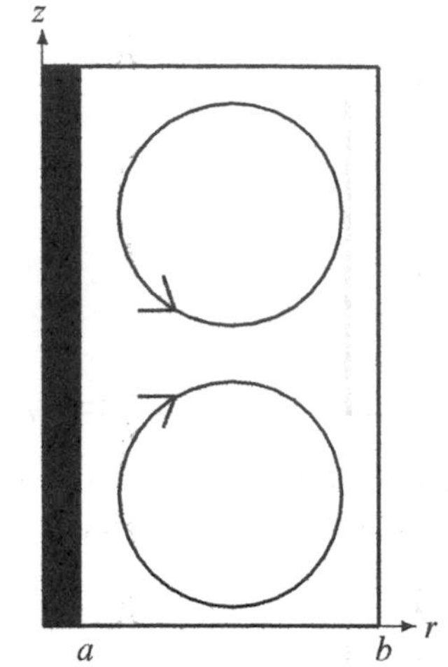

Sketch of the secondary flows that arise in a Couette apparatus as a consequence of its finite longitudinal size. Both at the bottom and top end caps, fluid is driven toward the central region.

We shall begin by assuming that the length of the apparatus is comparable to the gap between the cylinders, say $L \approx 2d$, and that the gap is completely filled with fluid without open surfaces. In the bulk of the fluid, the Couette solution is still a good approximation, and the main job of the pressure is to provide a centripetal force for the circulating fluid, as shown by the radial gradient (16.56a).

The longitudinal pressure gradient $\nabla_z p$ determines how rapidly the pressure can vary in the z-direction, and from the z-component of the full Navier–Stokes equation (16.1) we obtain (still disregarding gravity)

$$\nabla_z p = -\rho_0 (\boldsymbol{v} \cdot \nabla) v_z + \eta \nabla^2 v_z .$$

The right-hand side is expected to be small throughout the fluid because the longitudinal velocity v_z must nearly vanish for the bulk flow in the middle of the apparatus and also has to vanish at the end caps (which the fluid cannot penetrate). The vertical variation in pressure near the bottom is therefore not able to seriously challenge the large values of the centripetally dominated bulk pressure. The pressure is for this reason rather "stiff" as we approach the end caps and remains more or less equal to the pressure in the bulk with only small corrections.

Near the end caps, the still-intact bulk pressure will have an inward directed radial gradient, $-\nabla_r p$, capable of delivering the centripetal force to keep the rapidly rotating fluid in the bulk in circular motion. But the no-slip condition requires the azimuthal velocity to vanish at the end cap, $v_\phi \to 0$ for $z \to 0$, so that the actually required centripetal acceleration $-v_\phi^2/r$ is much smaller there than in the bulk. As a result, the combination of the stiff pressure gradient and the smaller centripetal acceleration gives rise to a net force directed toward the axis, and that force will in turn drive a radial flow inward along the end caps. Since the incompressible moving fluid has to go somewhere, it tends to form two oppositely rotating ring-shaped (toroidal) vortices that encircle the axis.

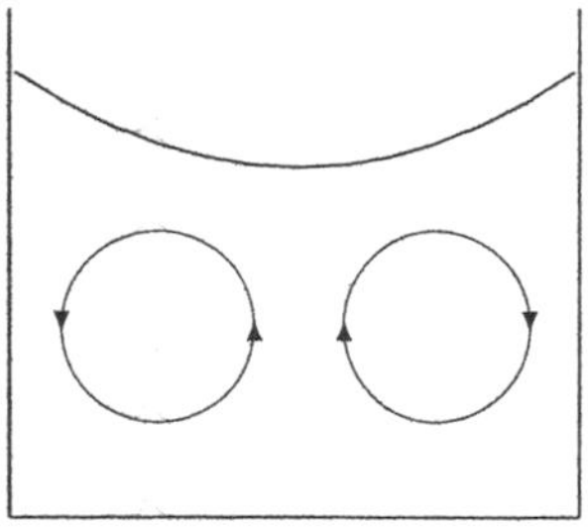

The secondary flow responsible for collecting tea leaves at the center of the bottom of a stirred cup. The parabolic shape of the surface of the rotating liquid is caused by the combination of gravity pointing downward and the centrifugal force pointing outward.

If the axis is vertical and the liquid has an open surface, only a single vortex needs to be formed. It is this kind of secondary flow that drives the tea leaves along the bottom toward the center of a cup of tea after stirring it. As the above argument shows, the flow is independent of the direction of stirring. You cannot, for example, make the tea leaves move out again by "unstirring" your tea.

Rayleigh's stability criterion

In the above discussion, the cylinder length was assumed to be comparable to the gap size. For a long cylinder, the preceding argument seems to indicate that a secondary flow with two highly elongated toroidal "vortices" should appear to satisfy the boundary conditions on the end caps, or just one if there is an open surface. Such elongated vortices are, however, unstable and tend to break up in a number of smaller and more circular vortices.

The instability can be understood by an argument analogous to the discussion of atmospheric stability on page 37. For simplicity we shall, as Rayleigh did, assume that there is no viscosity. Suppose now that a unit mass fluid particle in the bulk of the rotating fluid is shifted outward to a larger value of r by a tiny amount. During the move, the fluid particle will conserve its angular momentum and will in general arrive in its new position with a speed that is different from the ambient speed of the fluid there. The angular momentum of a unit mass particle in the ambient fluid is $r v_\phi$, and depending on how the fluid rotates, it may grow or decrease with increasing distance. If it is constant, such that $d(r v_\phi)/dr = 0$, then the moved particle fits snugly into its new home with just the right velocity. This is a marginal case, realized by the (inviscid) spindle-driven vortex, which has $r v_\phi = \Omega a^2$.

In general, however, the angular momentum $r v_\phi$ of a unit mass particle in the ambient fluid will not be a constant. If it locally grows in magnitude with distance, then a fluid particle displaced to a slightly larger radius will arrive with its original angular momentum and thus have a smaller velocity than its new surroundings. The inward pressure gradient in the ambient fluid will therefore be larger than required to keep the particle in a circular orbit with lower velocity, and the particle will be forced back to where it came from. This is *Rayleigh's stability criterion*:

$$\frac{d\,|rv_\phi|}{dr} > 0. \tag{16.74}$$

The stability criterion will trivially be fulfilled if the inner cylinder is held fixed and the outer rotated, because then the azimuthal velocity $|v_\phi|$ and thus $|rv_\phi|$ must necessarily grow with the radial distance. Such a situation is perfectly stable. Any radial fluctuation in the flow is quickly "ironed out" and the flow remains well-ordered and laminar until turbulence sets in.

Taylor vortices

Conversely, if the above inequality is not fulfilled everywhere, the flow will be unstable and reconfigure itself at the first opportunity. For the Couette solution (16.59) we have

$$\frac{d\,|rv_\phi|}{dr} = -\frac{2a^2\Omega r}{b^2 - a^2}, \tag{16.75}$$

which is evidently negative, and more negative the faster the inner cylinder rotates and the narrower the gap. Viscosity, which has been ignored so far, will delay the onset of instability, but it requires a more careful analysis to determine the precise criterion (see for example [Lin 1955, Acheson 1990, Chandrasekhar 1981, Drazin and Reid 1981]).

For sufficiently large angular driving velocity Ω, the Couette flow in a small gap between long rotating inner and fixed outer cylinders becomes unstable and a string of ring-shaped vortices will arise in the gap. They are called *Taylor vortices* in honor of G. I. Taylor, who first analyzed their properties, both experimentally and theoretically. The Taylor vortices are all mutually counter-rotating and match up with the sense of flow dictated by the boundary conditions on the ends (for pictures see [Ball 1997]).

The formation of secondary flow and the instability leading to toroidal (ring-shaped) Taylor vortices imply a breakdown of the longitudinal symmetry (along z) of the cylinder. The azimuthal symmetry (in ϕ) is to begin with maintained but also breaks down with increasing speed of rotation. The regular Taylor vortices are then replaced by wavy vortices that undulate up and down (in z) while they encircle the axis. Increasing the speed further will increase the number of undulations until the flow finally, after an infinity of such transitions, becomes chaotic and turbulent.

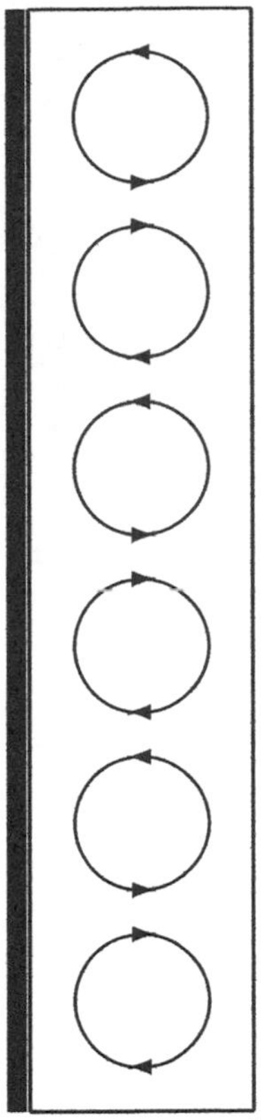

Taylor vortices are toroidal rings of fluid encircling the axis of rotation. In this case, both end caps are fixed and the secondary flow must be directed inward at both top and bottom. That is only possible if an even number of counter-rotating ring vortices are formed.

Problems

16.1 Consider pressure-driven laminar flow between infinitely extended plates moving with constant velocity U with respect to each other in the same direction as the pressure gradient. Determine under which conditions the maximal flow velocity will be found between the plates.

16.2 Consider a pressure-driven laminar flow between infinitely extended plates moving with constant velocity U with respect to each other in a direction orthogonal to the pressure gradient. Determine the planar flow field.

16.3 Why is it impossible to make a pressure-driven planar fluid sheet with an open surface?

16.4 A thin fluid sheet (water) of density ρ_1, viscosity η_1, and thickness d_1 runs steadily down a plate inclined at an angle θ to the horizon. Another plate with the same inclination is placed at a distance $d_1 + d_2$ from the first, and there is another fluid (air) in the gap with density ρ_2, viscosity η_2, and thickness d_2. Assuming laminar flow, calculate the **(a)** pressure and **(b)** velocity fields in the water and in the air between the plates.

16.5 Calculate **(a)** drag and **(b)** dissipation for pressure-driven planar flow expressed in terms of the average velocity U.

16.6 Calculate **(a)** drag and **(b)** dissipation for gravity-driven planar flow with an open surface on an inclined plate in terms of the average velocity.

16.7 Consider laminar gravity-driven flow through a vertical pipe. **(a)** Determine the radius of the pipe as a function of the Reynolds number when the fluid is water. **(b)** Determine the largest radius and **(c)** the entry length just before onset of turbulence.

16.8 **(a)** Show that in pipe flow the only dimensionless combination of the parameters Q, a, ρ_0, η is the Reynolds number (up to trivial transformations). **(b)** Can you make another dimensionless parameter by using $G = \Delta p/L$ instead of Q?

16.9 Pipe resistance is defined as the ratio of driving pressure to mass discharge,

$$R = \frac{\Delta p}{\rho_0 Q}.$$

(a) Show that pipe resistance is additive for pipes connected in series; and **(b)** reciprocally additive for pipes connected in parallel.

16.10 The Darcy friction factor is defined as the ratio of the pressure drop along a diameter and the stagnation pressure,

$$C_{\text{Darcy}} = \frac{Gd}{\frac{1}{2}\rho_0 U^2}.$$

Show that it equals $4C_f$.

16.11 Show that for the Poiseuille profile,

$$\int_0^a v_z(r)^n \, 2\pi r \, dr = \frac{(2U)^n \pi a^2}{n+1},$$

where U is the average velocity.

16.12 **(a)** Show that for a cylindrical tube with hollow annular cross-section defined by $a < r < b$, the solution is given by the general Poiseuille solution (16.28) with

$$A = \frac{G}{4\eta} \frac{b^2 - a^2}{\log b - \log a}, \qquad B = \frac{G}{4\eta} \frac{a^2 \log b - b^2 \log a}{\log b - \log a}.$$

(b) Determine the drag on the tube per unit of length when a fluid is pressed through the tube under a pressure gradient G.

16.13 A pipe has elliptic cross-section with major and minor semi-axes a and b. **(a)** Show that laminar pressure-driven flow of an incompressible fluid with pressure gradient G is given by

$$v_z = \frac{G}{2\eta} \frac{a^2 b^2}{a^2 + b^2} \left(1 - \frac{x^2}{a^2} - \frac{y^2}{b^2}\right), \qquad p = p_0 - Gz.$$

(b) Show that the fields of the circular pipe and the parallel plates are contained in this expression. **(c)** Calculate the volume flow rate (Hint: Use elliptic coordinates $x = ar\cos\theta$ and $y = br\sin\theta$ with area element $dA = abrd\theta dr$). **(d)** Calculate the average flow velocity. **(e)** Calculate the drag over a length L.

16.14 An infinitely long pipe has an annular cross-section in the interval $a < r < b$. The inner wall $r = a$ moves with velocity U whereas the outer wall $r = b$ is at rest. The space between the walls contains a fluid with viscosity η and there is no pressure gradient along the pipe. **(a)** Determine the flow v_z in the pipe. **(b)** Calculate the volume discharge rate Q. Show that $Q \to 0$ for $a \to b$? **(c)** Calculate the drag on a stretch L of the outer wall.

16.15 A permeable circular cylindrical pipe carrying fluid under pressure leaks fluid through its surface into the surrounding ocean at a constant rate. In the steady state, determine the maximally symmetric form of the radial flow pattern outside the pipe.

16.16 Show that the coefficients of the general circulating flow may be written as

$$A = \frac{\Omega_b b^2 - \Omega_a a^2}{b^2 - a^2}, \qquad B = \frac{a^2 b^2 (\Omega_a - \Omega_b)}{b^2 - a^2},$$

where Ω_a and Ω_b are the angular velocities of the inner and outer cylinders, respectively.

16.17 [Couette viscometer] Consider a torsion-wire Couette viscometer with inner and outer radii a and $b = a + d$. The gap is narrow ($d \ll a$), and the axial length long ($L \gg d$). Let the torque exerted by the torsion wire (from which the inner cylinder hangs) be $-\tau\phi$, where ϕ is the deflection angle and τ is a constant. Assume that the inner cylinder has total mass M and is made from a very thin shell of homogeneous material. Due to viscous dissipation in the fluid the cylinder performs damped torsion oscillations. **(a)** Show that the equation of motion for the torsion angle becomes

$$Ma^2 \frac{d^2\phi}{dt^2} = -\tau\phi - 2\pi\eta a^3 \frac{L}{d} \frac{d\phi}{dt},$$

under the assumption that the fluid flow may be considered quasi-steady at all times. **(b)** Find the solution to the equation of motion and discuss how the viscometer can be used to measure the viscosity.

16.18 Consider Couette flow with rotating inner and static outer cylinder of length L. Calculate **(a)** the azimuthal volume flux and **(b)** the kinetic energy.

16.19 The drive shaft in a truck has a diameter of 15 cm and rotates in a bearing of length 10 cm with 0.1 mm of lubricant of viscosity $\eta = 0.01$ Pa s. Calculate the rate of heat production when the shaft rotates 10 revolutions per second.

16.20 Show that the vorticity field of the spindle vortex (16.70) vanishes everywhere.

* **16.21** **(a)** Show that the local specific rate of dissipation is

$$\varepsilon(r) = \frac{G^2}{4\eta\rho_0} r^2$$

in laminar pipe flow. **(b)** Show the total dissipated power in a stretch of a circular tube is given by (16.38).

17 Creeping flow

Viscosity may be so large that a fluid only flows with difficulty. Heavy oils, honey, and even tight crowds of people show insignificant effects of inertia, and are instead dominated by internal friction. Such fluids do not make spinning vortices or become turbulent, but rather ooze or creep around obstacles. Fluid flow thus dominated by viscosity is quite appropriately called *creeping flow*. Since there is no absolute meaning to "large viscosity", creeping flow is more correctly characterized by the Reynolds number being small, $\mathsf{Re} \ll 1$.

Creeping flow may occur in any fluid, as long as the typical velocity and geometric extent of the flow combine to make a small Reynolds number. Blood flowing through a microscopic capillary can be as sluggish as heavy oil. Tiny organisms like bacteria live in air and water like ourselves, but theirs is a world of creeping and oozing rather than whirls and turbulence, and movement requires special devices [Vogel 1994]. Modern microtechnology has opened up a whole new field of *microfluidics* at very low Reynolds number [Bruus 2008].

For any creature operating in the creeping flow regime, the most important quantity is the fluid's resistance against motion, also called *drag*. The same contact forces may also produce *lift* orthogonally to the direction of motion, but far from containing boundaries lift is of the same order of magnitude as drag (in creeping flow). A tiny creature that must already overcome a drag many times its weight in order to move freely has no problem with swimming upward. That is why bacteria do not sprout wings but rely on oars or screws to move around.

In this chapter we shall first study creeping flow around moving bodies far from containing boundaries. The behavior of drag at higher Reynolds numbers is included for completeness. Afterward we turn to lubrication theory based on creeping flow in narrow gaps.

17.1 Stokes flow

At low Reynolds number, $\mathsf{Re} \ll 1$, the advective acceleration can be left out of the Navier–Stokes equations for incompressible flow (15.21), so that they for steady flow become

$$\boxed{\nabla p = \eta \nabla^2 \boldsymbol{v}, \qquad \nabla \cdot \boldsymbol{v} = 0,} \tag{17.1}$$

an approximation usually called *Stokes flow*. Gravity, $\boldsymbol{g} = -\nabla\Phi$, has for simplicity been left out but is easy to include by replacing p by $p + \rho_0\Phi$. From the divergence of the first equation it follows that the pressure must satisfy the Laplace equation, $\nabla^2 p = 0$.

Creeping flow is mathematically (and numerically) much easier to handle than general flow because of the absence of non-linear terms that tend spontaneously to break the natural symmetry of the solutions in time as well as space with turbulence as the extreme result. The linearity of the creeping flow equations may sometimes be used to express solutions to complicated flow problems as linear superpositions of simpler solutions.

Drag and lift on a moving body

Consider a solid body cruising with constant velocity U through a static fluid. The body creates a temporary disturbance in the fluid that disappears again some time after the body has passed a fixed observation point. Seen from the body, the fluid appears to move in a steady pattern that at sufficiently large distances becomes uniform, described by the vector $\boldsymbol{U}$ (see the margin figure). Newtonian relativity guarantees, as we have discussed before, that these situations are physically equivalent, so we may use the Stokes flow equations (17.1).

The only way the fluid can influence a solid body is through contact forces acting on its surface. The *reaction force* is defined as the total contact force that *the fluid exerts on the body* (see Chapter 21). It is convenient to resolve it into two components:

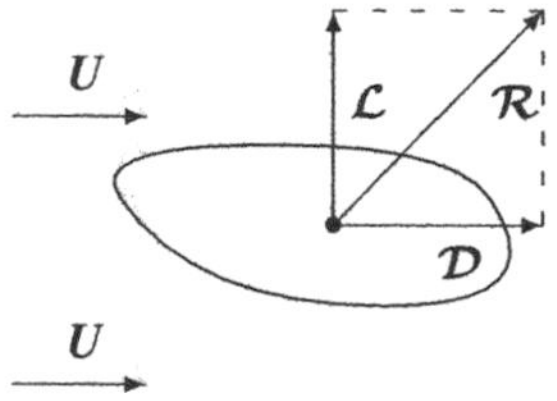

The total reaction force $\mathcal{R}$ from the fluid on the body is composed of lift $\mathcal{L}$ and drag $\mathcal{D}$.

$$\mathcal{R} \equiv \oint_S \boldsymbol{\sigma} \cdot d\boldsymbol{S} = \mathcal{D} + \mathcal{L}, \tag{17.2}$$

where S is the surface of the body. The first component $\mathcal{D}$ is called the *drag* and acts in the direction of the asymptotic flow, whereas the other is called the *lift* and acts orthogonally to this direction. We thus have

$$\mathcal{D} = \mathcal{D}\hat{\boldsymbol{e}}_U, \qquad \mathcal{L} \cdot \hat{\boldsymbol{e}}_U = 0, \tag{17.3}$$

where $\hat{\boldsymbol{e}}_U = \boldsymbol{U}/U$ is the direction of the asymptotic velocity. The drag $\mathcal{D}$ is always positive whereas the lift $\mathcal{L}$ can take any direction orthogonal to the asymptotic velocity.

Assume now that the body has a non-exceptional shape—think of a potato—with a typical size L. Since the pressure only appears in (17.1) in the combination p/η, and since the no-slip boundary condition on the surface of the body does not involve the pressure, *the velocity field cannot depend on the viscosity η, but only on the asymptotic velocity $\boldsymbol{U}$ and on the geometry and orientation of the body.* The linearity of creeping flow guarantees that the velocity field must everywhere be proportional to the magnitude of the asymptotic velocity, $|\boldsymbol{v}| \sim U$.

Skin drag is produced by viscous shear stresses acting on the surface of the body. The velocity gradients near the surface are of magnitude $|\nabla\boldsymbol{v}| \sim U/L$ so the magnitude of the shear stress becomes $|\boldsymbol{\sigma}| \sim \eta U/L$. Multiplying by the surface area of the body, which is of magnitude L^2, skin drag takes the form

$$\mathcal{D}_{\text{skin}} = f_{\text{skin}}\eta UL, \tag{17.4}$$

with an unknown dimensionless prefactor f_{skin} of order unity. Since it cannot depend on the velocity or the viscosity, it is only a function of the shape and orientation of the body.

Form drag is produced by pressure variations over the body. The pressure gradient may similarly be estimated from (17.1) to be $|\nabla p| \approx \eta U/L^2$. Multiplied by the size L of the body, the pressure variations over the body become of the same size as the shear stresses, $|\Delta p| \sim \eta U/L$, and multiplying by the surface area L^2, we get the form drag,

$$\mathcal{D}_{\text{form}} = f_{\text{form}}\eta UL. \tag{17.5}$$

In general the prefactor f_{form} differs from f_{skin}.

The *total drag* is the sum of skin and form drag, and becomes

$$\mathcal{D} = f_{\text{drag}}\, \eta U L, \tag{17.6}$$

with $f_{\text{drag}} = f_{\text{skin}} + f_{\text{form}}$.

Lift is produced by the same contact forces as drag and its magnitude is also of the form

$$|\mathcal{L}| = f_{\text{lift}}\, \eta U L, \tag{17.7}$$

with still another prefactor. Lift vanishes for a symmetric body like a sphere or a cylinder that is parallel or orthogonal to the asymptotic flow. Note that lift is not necessarily related to the direction of gravity (if present) but can point in any direction orthogonal to the direction of motion. Near a solid boundary lift can be much larger than drag and may in fact serve to keep the body away from the boundary. This is the secret behind lubrication, a subject taken up in Section 17.4. In nearly ideal fluids lift may also be much larger than drag, even far away from container boundaries, as will be discussed extensively in Chapter 29 on aerodynamics.

17.2 Creeping flow around a solid ball

A solid spherical ball of radius a moving at constant speed U through a viscous fluid is the centerpiece of creeping flow, going back to Stokes in 1851. Provided the Reynolds number

$$\mathsf{Re} = \frac{2aU}{\nu} \tag{17.8}$$

is sufficiently small, the pressure and velocity fields must satisfy the Stokes equations (17.1) in the rest frame of the ball. Being a fundamental problem, we shall present the road to the solution in full detail by means of the spherical formalism of Section D.2 on page 623.

Symmetry and linearity

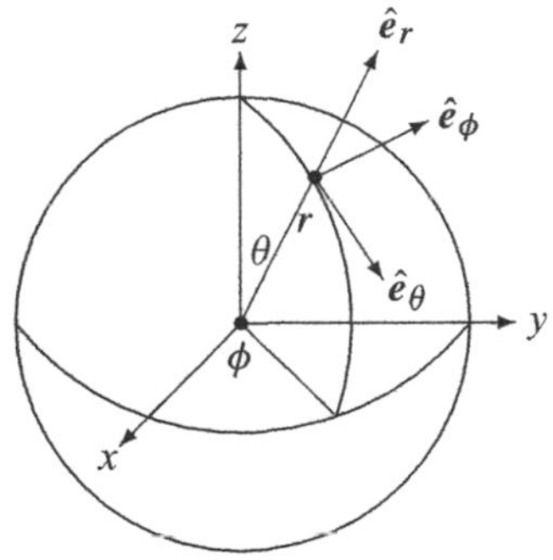

Spherical coordinates and their basis vectors.

The first step in obtaining the solution is to recognize that the perfect spherical symmetry of the sphere is only broken by the asymptotic velocity vector $\boldsymbol{U}$, and that the velocity field and the pressure must be linear in $\boldsymbol{U}$, as discussed above. Since the only other vector at play in the formulation of the problem is the radius vector $\boldsymbol{x}$ from the center of the sphere, the scalar pressure must be proportional to $\boldsymbol{U}\cdot\boldsymbol{x}$. In spherical coordinates with $\boldsymbol{U} = U\hat{\boldsymbol{e}}_z$, we have $\boldsymbol{U}\cdot\boldsymbol{x} = rU\cos\theta$, leading to a pressure field of the form

$$p = \eta U \cos\theta\, q(r), \tag{17.9}$$

where $q(r)$ is a function of only the radial distance $r = |\boldsymbol{x}|$. The explicit factor η is included because, as we have argued above, the pressure must be proportional to the viscosity.

The velocity field must similarly be a linear combination, $\boldsymbol{v} = A(r)\boldsymbol{U} + B(r)(\boldsymbol{U}\cdot\boldsymbol{x})\boldsymbol{x}$ of the vectors $\boldsymbol{U}$ and $(\boldsymbol{U}\cdot\boldsymbol{x})\boldsymbol{x}$ with coefficients that can only depend on r. Using that $\boldsymbol{x}\cdot\hat{\boldsymbol{e}}_r = r$ and $\boldsymbol{x}\cdot\hat{\boldsymbol{e}}_\theta = 0$, together with $\boldsymbol{U}\cdot\hat{\boldsymbol{e}}_r = U\cos\theta$ and $\boldsymbol{U}\cdot\hat{\boldsymbol{e}}_\theta = -U\sin\theta$, it follows that the spherical field components must be of the form

$$v_r = U\cos\theta\, f(r), \qquad v_\theta = -U\sin\theta\, g(r), \qquad v_\phi = 0, \tag{17.10}$$

where $f(r)$ and $g(r)$ are functions of r only.

Generic solution

Having determined the general form of the solution, we must now insert it into the Stokes equations and solve the resulting ordinary differential equations. There are several ways of doing this, but since the calculations can easily become messy we shall attempt to follow a fairly direct path of minimal effort.

Using that the Laplacian of the pressure has to vanish, we obtain from the spherical expression for the scalar Laplacian (D.40) on page 625,

$$\nabla^2 p = \eta U \cos\theta \left(\frac{d^2 q}{dr^2} + \frac{2}{r}\frac{dq}{dr} - \frac{2q}{r^2}\right) = 0. \tag{17.11}$$

Requiring the expression in parentheses to vanish,

$$\frac{d^2 q}{dr^2} + \frac{2}{r}\frac{dq}{dr} - \frac{2q}{r^2} = 0, \tag{17.12}$$

we have obtained an ordinary second-order differential equation that is homogeneous in r. One may verify that the two linearly independent solutions to this equation are $q \sim 1/r^2$ and $q \sim r$, so that we may write

$$\boxed{q = \frac{C}{r^2} + Dr,} \tag{17.13}$$

where C and D are integration constants.

The divergence of the velocity field must vanish. Using Equation (D.35) on page 624, we get

$$\nabla \cdot \boldsymbol{v} = \cos\theta \left(\frac{df}{dr} + \frac{2f}{r} - \frac{2g}{r}\right) U = 0, \tag{17.14}$$

leading to

$$g = f + \tfrac{1}{2} r \frac{df}{dr}. \tag{17.15}$$

This leaves only the function $f(r)$ to be determined.

From the Stokes equation (17.1) and the radial part of the vector Laplacian (D.41), we obtain

$$\frac{d^2 f}{dr^2} + \frac{2}{r}\frac{df}{dr} + \frac{4g}{r^2} - \frac{4f}{r^2} = \frac{dq}{dr}, \tag{17.16}$$

and inserting q and g from above, we arrive at

$$\frac{d^2 f}{dr^2} + \frac{4}{r}\frac{df}{dr} = -\frac{2C}{r^3} + D. \tag{17.17}$$

The general solution to this linear second-order equation for f is obtained by standard methods,

$$\boxed{f = A + \frac{B}{r^3} + \frac{C}{r} + \frac{1}{10} D r^2,} \tag{17.18}$$

where A and B are integration constants. Inserting this into (17.15), we find

$$\boxed{g = A - \frac{B}{2r^3} + \frac{C}{2r} + \frac{1}{5} D r^2.} \tag{17.19}$$

The framed formulas constitute the complete generic solution for this problem.

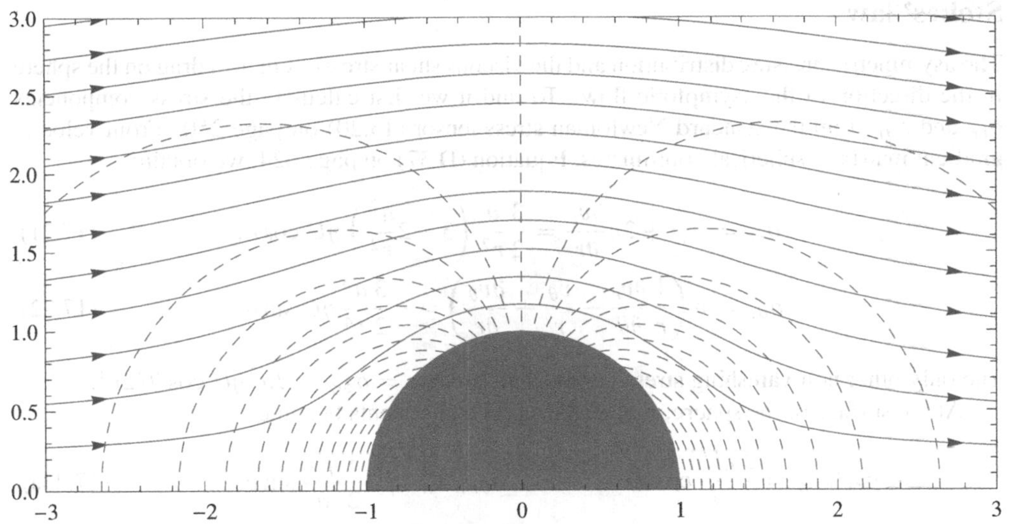

Figure 17.1. Stokes flow around a unit sphere with only the upper half shown. The figure must be rotated around the horizontal z-axis to obtain the full three-dimensional flow pattern. The fluid enters from the left and leaves to the right. The streamlines are started equidistantly far upstream and continued by means of the velocity field. The dashed curves are isobars with equidistant pressure values. This pattern is qualitatively different from the potential flow pattern shown in Figure 13.4R on page 224, because in creeping flow the influence of the sphere stretches much farther due to the $1/r$ terms in Equations (17.20).

The field

Finally we must apply boundary conditions to determine the four unknown integration constants, A, B, C, and D. For $r \to \infty$, the field must approach the homogeneous field $\boldsymbol{U}$, implying $f \to 1$ and $g \to 1$, or $A = 1$ and $D = 0$. At the surface of the sphere, $r = a$, the field must vanish because of the no-slip condition, leading to $B = \frac{1}{2}a^3$ and $C = -\frac{3}{2}a$.

Putting it all together, the spherical velocity components and the pressure become

$$v_r = \left(1 - \frac{3}{2}\frac{a}{r} + \frac{1}{2}\frac{a^3}{r^3}\right) U \cos\theta, \tag{17.20a}$$

$$v_\theta = -\left(1 - \frac{3}{4}\frac{a}{r} - \frac{1}{4}\frac{a^3}{r^3}\right) U \sin\theta, \tag{17.20b}$$

$$v_\phi = 0, \tag{17.20c}$$

$$p = -\frac{3}{2}\eta\frac{a}{r^2} U \cos\theta. \tag{17.20d}$$

The pressure is forward-backward asymmetric, highest on the part of the sphere that turns toward the incoming flow (at $\theta = \pi$). This asymmetry is in marked contrast to the potential flow solution (13.64), for which the pressure is symmetric. The flow pattern is independent of the viscosity of the fluid, whereas the pressure is proportional to η, as expected for creeping flow problems where the boundary conditions do not involve the pressure.

In Figure 17.1 the solution has been plotted in the xz-plane for a unit radius sphere ($a = 1$). The streamlines have been obtained by numeric integration of the streamline equation (12.2) on page 191 starting equidistantly far to the left (at $z = -100$) to avoid problems with the long-range terms. In Problem 17.7, the streamlines are obtained in a more efficient way. The isobars correspond to equidistant pressure values.

Stokes' law

The asymmetric pressure distribution and the viscous shear stresses create a drag on the sphere in the direction of the asymptotic flow. To find it we first calculate the stress components σ_{rr} and $\sigma_{\theta r}$ from the standard Newtonian stress tensor (15.20) on page 250. From velocity gradient matrix in spherical coordinates, Equation (D.37) on page 624, we obtain:

$$\sigma_{rr} = -p + 2\eta \frac{\partial v_r}{\partial r} = \frac{3}{2}\frac{a}{r^2}\left(3 - 2\frac{a^2}{r^2}\right)\eta U \cos\theta, \tag{17.21}$$

$$\sigma_{\theta r} = \eta\left(\frac{1}{r}\frac{\partial v_r}{\partial \theta} - \frac{v_\theta}{r} + \frac{\partial v_\theta}{\partial r}\right) = -\frac{3}{2}\frac{a^3}{r^4}\eta U \sin\theta. \tag{17.22}$$

The only other non-vanishing stress components are $\sigma_{\theta\theta} = \sigma_{\phi\phi} = 2a^3\eta U \cos\theta/2r^4$.

At the surface of the sphere, $r = a$, the stress components become

$$\sigma_{rr}|_{r=a} = \frac{3}{2}\frac{\eta U}{a}\cos\theta, \qquad \sigma_{\theta r}|_{r=a} = -\frac{3}{2}\frac{\eta U}{a}\sin\theta, \tag{17.23a}$$

from which we get the surface stress vector

$$\boldsymbol{\sigma}\cdot\hat{\boldsymbol{e}}_r|_{r=a} = \hat{\boldsymbol{e}}_r\sigma_{rr} + \hat{\boldsymbol{e}}_\theta\sigma_{\theta r} = \frac{3}{2}\frac{\eta U}{a}\hat{\boldsymbol{e}}_z = \frac{3}{2}\frac{\eta \boldsymbol{U}}{a}. \tag{17.24}$$

Surprisingly it is of constant magnitude and points everywhere in the direction of the asymptotic flow across the entire surface.

The total reaction force is simply obtained by multiplying the constant stress vector by the area $4\pi a^2$ of the sphere, but since the force is proportional to $\boldsymbol{U}$ it is a pure drag,

$$\boxed{\mathcal{D} = 6\pi\eta a U.} \tag{17.25}$$

This is the famous *Stokes' law* from 1851. The symmetry of the sphere could have told us in advance that there would be no lift because there is no global direction other than $\boldsymbol{U}$ defined for the problem. Taking $L = 2a$ to represent the size of the sphere, this form of the drag was already predicted in (17.6) with a geometric prefactor $f_{\text{drag}} = 3\pi$. In Problem 17.4 it is shown that 2/3 of the total is skin drag and 1/3 is form drag. Contrary to the inviscid (ideal) flow case where it was meaningful to calculate the lift on a half-buried sphere, this does not make much sense here because the presence of the solid bottom on which the half-sphere rests strongly modifies the viscous flow over the half-sphere.

Terminal velocity

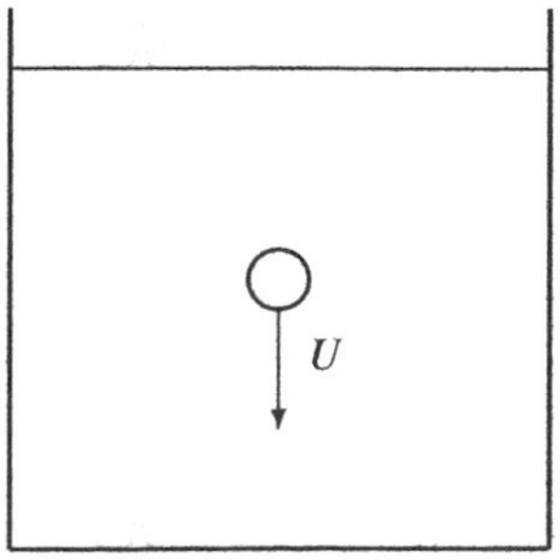

Solid sphere falling through viscous fluid at constant terminal speed U.

Although Stokes' law has been derived in the rest frame of the sphere, it is also valid in the rest frame of the asymptotic fluid. The terminal velocity of a falling solid sphere may be obtained by equating the force of gravity (minus buoyancy) with the Stokes' drag,

$$(\rho_1 - \rho_0)\frac{4}{3}\pi a^3 g_0 = 6\pi\eta a U, \tag{17.26}$$

where g_0 is the gravitational acceleration and ρ_1 is the average density of the sphere. Solving for U, we find

$$\boxed{U = \frac{2}{9}\left(\frac{\rho_1}{\rho_0} - 1\right)\frac{a^2 g_0}{\nu},} \tag{17.27}$$

where $\nu = \eta/\rho_0$ is the kinematic viscosity.

Example 17.1 [Sedimentation]: Sand grains of diameter 0.1 mm and density 2.5 times that of water float toward the bottom of the sea with a terminal velocity of 0.8 cm s^{-1}, as calculated from Stokes' law. The corresponding Reynolds number is about 0.8, which approaches the limit of the region of validity of Stokes' law.

Conversely, in the *falling sphere viscometer*, Equation (17.27) may be used to determine the viscosity ν from a measurement of the terminal velocity for a sphere of known radius and density. One may also, as Millikan did in his famous electron charge experiments (1909–1913), determine the radius of a freely falling sphere from a measurement of the terminal velocity, provided the viscosity of the fluid is known. Millikan used microscopic oil drops falling in air, and it was actually necessary to include corrections to Stokes law for the internal motion of the oil inside the tiny drops. The oil drops were electrically charged with a few electrons and could be made to hover, rise, or fall in the field of gravity as slowly as desired by means of an electric field of suitable strength. Knowing the radius, the electric force on an oil drop could be compared to gravity, allowing the electric charge of the electron to be determined[1].

Robert Andrews Millikan (1868–1953). American experimental physicist, prolific author, and excellent teacher. Awarded the Nobel prize in 1923 for the determination of the charge of the electron. Verified in 1916 Einstein's expressions for the photoelectric effect, and investigated the properties of cosmic radiation.

Limits to Stokes flow

The Reynolds number (17.8) is only a rough estimate for the ratio between advective and viscous terms in the Navier–Stokes equations for a particular geometry. Having obtained an explicit solution we may now calculate this ratio throughout the fluid to see if it is indeed small.

Close to the sphere, the velocity is very small because of the no-slip condition. Problems are only expected to arise at large distances, where the leading corrections to the uniform flow are provided by the a/r terms in the solution (17.1). Disregarding the angular dependence, the leading advective terms are at large distances,

$$|\rho_0 \boldsymbol{v} \cdot \nabla \boldsymbol{v}| \sim \rho_0 U^2 \frac{a}{r^2}, \tag{17.28}$$

whereas the viscous terms become

$$|\eta \nabla^2 \boldsymbol{v}| \sim \eta U \frac{a}{r^3}. \tag{17.29}$$

The actual ratio between advective and viscous terms is then

$$\frac{|\rho_0 \boldsymbol{v} \cdot \nabla \boldsymbol{v}|}{|\eta \nabla^2 \boldsymbol{v}|} \sim \frac{\rho_0 U r}{\eta} \sim \frac{r}{a}\mathsf{Re} \tag{17.30}$$

where Re is the global Reynolds number (17.8). As this expression grows with r, we conclude that however small the Reynolds number may be, there will always be a distance $r \gtrsim a/\mathsf{Re}$ where the advective force begins to dominate the viscous. This clearly illustrates that the global Reynolds number is just a guideline, not a guarantee that creeping flow will occur everywhere in a system. Since any finite potato-like body at sufficiently large distances looks like a sphere, this problem must in fact be present in all creeping flows.

The slowly decreasing a/r terms in the Stokes solution (17.20) also point to the influence of the containing vessel. The relative magnitude of these terms at the boundary of the vessel is estimated as $2a/D$, where D is the size (diameter) of the vessel. In the falling sphere viscometer, a 1% measurement of viscosity thus requires that the sphere diameter must be smaller than 1% of the vessel size. That can be hard to fulfill for highly viscous fluids, where a measurable terminal speed demands rather large and heavy spheres.

[1]The viscosity of air used in Millikan's experiment was determined by Couette viscometer measurements (p. 280). At that time, there was actually a systematic error of about 0.5% in the viscosity of air deriving mainly from the failure to take into account the end caps of the Couette viscometer, an error first corrected more than 20 years later [Bea39].

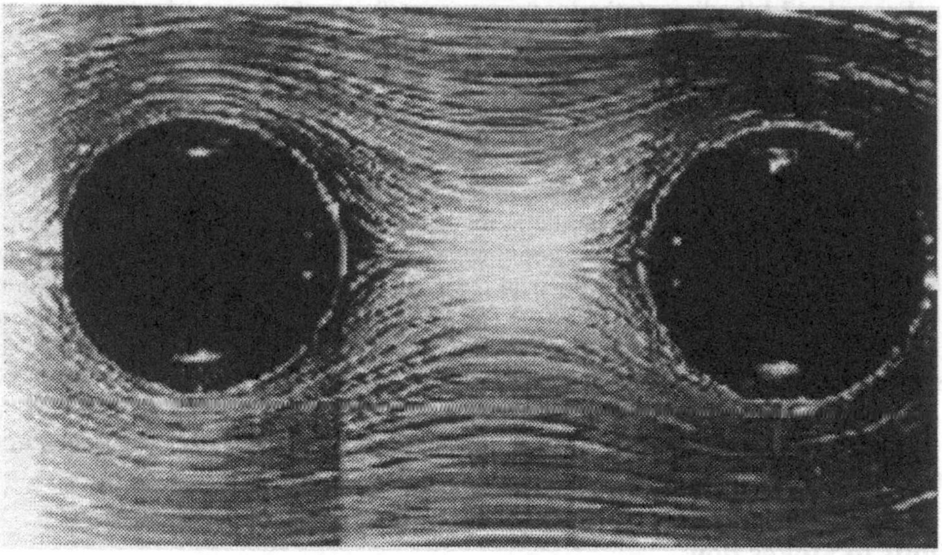

Figure 17.2. Creeping flow around two spheres at Re = 0.013, nearly indistinguishable from the flow in Figure 17.1. The gap between the spheres is one diameter and the fluid is glycerin. (Source: S. Taneda. Reproduced from [Tan79]. With permission.)

17.3 Beyond Stokes' law

Stokes' law has been derived in the limit of vanishing Reynolds number (17.8), and is empirically valid for Reynolds numbers less than unity, $\mathsf{Re} \lesssim 1$. For larger values of Re the simplicity of the problem nevertheless allows us to make a general analysis, as we did for turbulent pipe flow (Section 16.5 on page 274), even if we cannot solve the Navier–Stokes equations explicitly.

Drag factor

Since the only parameters defining the problem are the radius a, the velocity U, the viscosity η, and the density of the fluid ρ_0, we may for any Reynolds number write the drag on the sphere in the form of the Stokes law multiplied by a dimensionless *drag factor* $f(\mathsf{Re})$,

$$\mathcal{D} = 6\pi\eta a U f(\mathsf{Re}). \tag{17.31}$$

This factor accounts for the deviations from Stokes' law and is evidently anchored at unity for vanishing Reynolds number, that is, $f(0) = 1$. It is quite analogous to the drag factor in pipe flow.

Drag coefficient

In the opposite limit at large Reynolds number $\mathsf{Re} \gg 1$, far beyond the creeping flow region, the sphere literally plows its way through the fluid, leaving a wake of highly disturbed and turbulent fluid. The drag may now be estimated from the rate of loss of momentum from the incoming fluid that is disturbed by the sphere. The incoming fluid carries a momentum density $\rho_0 U$ and since the sphere presents a cross-sectional area πa^2 to the flow, the rate at which momentum impinges on the sphere is $\rho_0 U \cdot \pi a^2 U = \rho_0 \pi a^2 U^2$. Alternatively, the drag may be estimated from the dynamic pressure, $\Delta p = \frac{1}{2}\rho_0 U^2$, times the cross-sectional area πa^2.

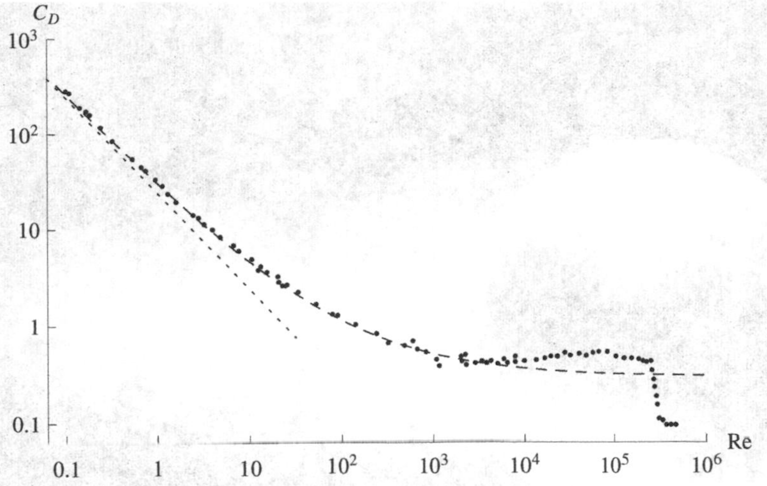

Figure 17.3. Drag coefficient for a smooth ball (data extracted from [Schlichting and Gersten 2000, Figure 1.19]). The dotted line corresponds to Stokes' law $C_D = 24/\text{Re}$ and the dashed curve is the interpolation (17.34). The sharp drop (the "drag crisis") at $\text{Re} = 2.5 \times 10^5$ signals the onset of turbulence in the boundary layer on the front half of the sphere and an accompanying shift in the shape of the trailing wake. After the drop, the drag coefficient slowly rises again (not shown here).

The drag at high Reynolds numbers is thus expected to grow with the square of the velocity (at subsonic speeds). But the fluid in the wake trailing the sphere is not completely at rest, such that only a certain fraction of the incoming momentum will be lost. It is for this reason customary to define the dimensionless *drag coefficient*,

$$C_D = \frac{\mathcal{D}}{\frac{1}{2}\rho_0 \pi a^2 U^2}, \tag{17.32}$$

with a conventional factor $1/2$ in the denominator.

Inserting the definition of the drag factor (17.31) we find

$$C_D = \frac{24}{\text{Re}} f(\text{Re}) \tag{17.33}$$

for all Reynolds numbers. It is a matter of taste whether one prefers to describe the drag on a sphere by means of the drag coefficient or the friction factor. At small Reynolds numbers, it seems a bit pointless to use the drag coefficient, because it introduces a strong artificial variation $C_D \approx 24/\text{Re}$ for $\text{Re} \to 0$ without a corresponding strong variation in the physics (which is simply described by Stokes' law). At large Reynolds numbers the drag coefficient becomes a constant, which is more convenient to use than the linearly rising friction factor, although its value ultimately depends on the roughness of the ball surface.

One may join the regions of low and high Reynolds numbers by the simple interpolating expression

$$C_D = \frac{24}{\text{Re}} + \frac{6}{\sqrt{\text{Re}}} + 0.3. \tag{17.34}$$

The first term is Stokes' result and the last is a constant terminal form drag. The middle term may be understood as due to friction in a thin laminar boundary layer (see Chapter 28) on the forward half of the sphere. There are a number of different interpolating expressions in the literature covering the same empirical data, for example [White 1991, p. 182]. As seen in Figure 17.3, the above formula agrees decently with data for all Reynolds numbers up to $\text{Re} \approx 10^4$, where the drag coefficient rises a bit to about 0.5.

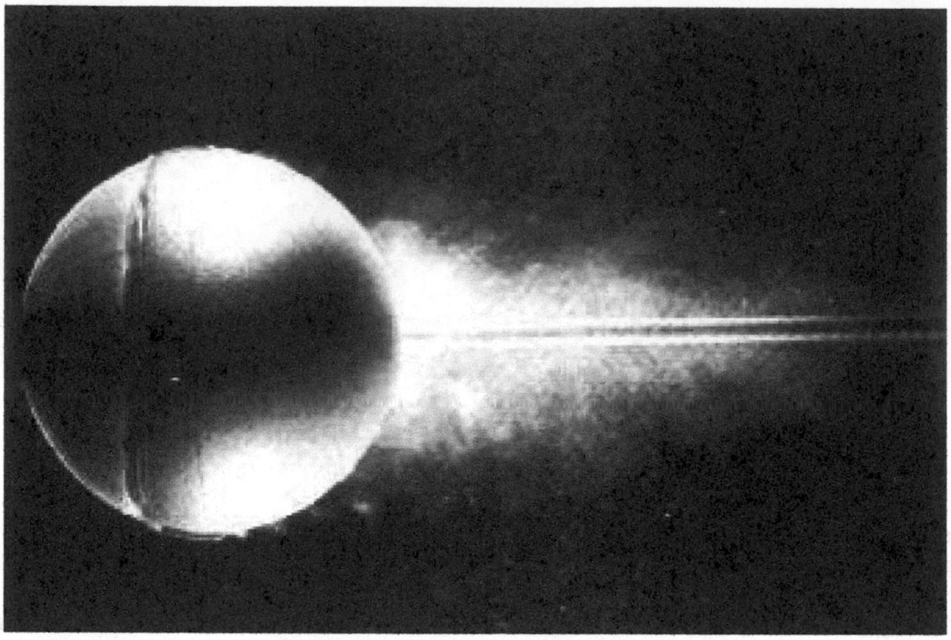

Figure 17.4. Turbulent boundary layer tripped by a wire ("seam") on a sphere at $\mathsf{Re} = 30{,}000$. The turbulent layer separates on the rear of the sphere whereas the laminar boundary layer separates slightly before the crest. See also Figure 28.1 on page 482. (Source: ONERA, the French Aerospace Lab. H. Werlé [Wer80]. With permission.)

Alexandre-Gustave Eiffel (1832 –1923). French structural engineer. Famous for building the Eiffel tower in Paris. Worked on aerodynamics for the last 21 years of his life. Built wind tunnels and performed numerous studies of drag and lift. (Source: Wikimedia Commons.)

The drag crisis

Above $\mathsf{Re} \approx 2.5 \times 10^5$ the drag coefficient drops sharply by more than a factor of 4, after which it begins to rise again, a phenomenon called the *drag crisis*. This dramatic behavior around $\mathsf{Re} \approx 2.5 \times 10^5$ was first observed by Gustave Eiffel in 1914 [Anderson 1997].

The "crisis" is caused by a transition from laminar to turbulent flow in the boundary layer of the forward-facing half of the sphere. This transition is accompanied by a front-to-back shift in the separation point for the turbulent wake that trails the sphere (see Figure 17.4 and the margin figure). At a Reynolds number just beyond the drag crisis, the wake is narrower than before, entailing a smaller loss of momentum, that is, smaller drag. At still higher Reynolds numbers the drag coefficient regains part of its former magnitude but for Mach numbers above 0.8, the drag crisis has essentially disappeared [Schlichting and Gersten 2000]. At supersonic speeds the terminal value of the drag coefficient increases with increasing Mach number (ratio of flow speed to speed of sound in the fluid).

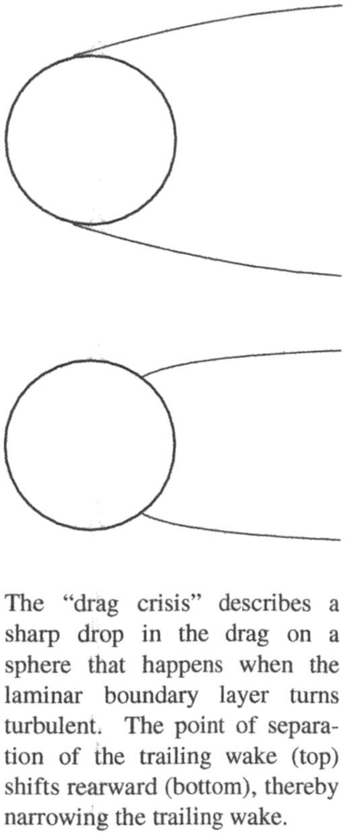

The "drag crisis" describes a sharp drop in the drag on a sphere that happens when the laminar boundary layer turns turbulent. The point of separation of the trailing wake (top) shifts rearward (bottom), thereby narrowing the trailing wake.

The Reynolds number at which the boundary layer actually becomes turbulent depends, as discussed before, on the surface properties of the sphere. Roughness tends to facilitate the generation of a turbulent boundary layer and moves the onset of turbulence and thus the drag crisis to sometimes much lower Reynolds numbers than for a smooth surface. This is the deeper reason for manufacturing golf balls with surface dimples [Meh85]. A golf ball flying at a typical speed of 30 m s^{-1} has $\mathsf{Re} \approx 1.6 \times 10^5$, which is below the drag crisis for a smooth ball but not for a dimpled one. The lower drag on a dimpled ball permits it to fly considerably longer for a given initial thrust with a club. The seams of a tennis ball serve the same function whereas a ping-pong ball is quite smooth. Even if it starts out fast, a ping-pong ball is so light that the drag quickly brings its velocity down. At 10 m s^{-1}, we have $\mathsf{Re} \approx 5 \times 10^4$, which is well below the drag crisis for a smooth ball. But even if turbulence could be triggered by providing it with seams or dimples they would probably interfere destructively with this swift game of skill and high precision.

Terminal speed

The variation in drag with Reynolds number makes the calculation of the terminal speed of a spherical ball somewhat more complicated than Equation (17.27). Equating the force of gravity with the drag we obtain instead of (17.26) for $\rho_1 > \rho_0$,

$$(\rho_1 - \rho_0)\tfrac{4}{3}\pi a^3 g_0 = \tfrac{1}{2}C_D \rho_0 \pi a^2 U^2,$$

and solving for the terminal speed, we get

$$U = \sqrt{\frac{8}{3}\left(\frac{\rho_1}{\rho_0} - 1\right)\frac{a g_0}{C_D}}. \tag{17.35}$$

Since C_D depends on the Reynolds number, this is an implicit equation for the terminal velocity except for very large Reynolds numbers where C_D is constant. The terminal velocity grows with the square root of the density, ρ_1, implying that the terminal kinetic energy density, $\frac{1}{2}\rho_1 U^2$, of a steadily falling sphere grows with the square of the density. This is why projectiles and bombs with heavy metal jackets (made, for example, from depleted uranium) are used to penetrate concrete structures and rock.

Example 17.2 [Falling skydiver]: A skydiver weighing 70 kg curls up roughly like a ball with radius $a = 50$ cm. The average density $\rho_1 \approx 133$ kg m^{-3} is smaller than that of water because the human body does not permit true morphing into a sphere. Taking $C_D \approx 0.3$ and $\rho_0 \approx 1$ kg m^{-3}, we obtain $U = 76$ m s^{-1} $= 274$ km h^{-1}. The corresponding Reynolds number is $\mathsf{Re} = 4.8 \times 10^6$, confirming the approximation used.

Example 17.3 [Falling soap bubble]: A freely falling spherical soap bubble drifts slowly toward the ground, because of the tiny mass excess due to the bubble fluid in its skin. Taking the skin thickness to be uniformly τ, the mass of the of the skin is $M_1 = 4\pi a^2 \tau \rho_1$. Equating the weight with the drag, $M_1 g_0 = \frac{1}{2}C_D \pi a^2 \rho_0 U^2$, the terminal speed becomes

$$U = \sqrt{\frac{8\tau g_0}{C_D}\frac{\rho_1}{\rho_0}}, \tag{17.36}$$

independent of the radius. Taking $\tau = 1$ μm and $C_D \approx 0.7$, we find $U = 33$ cm s^{-1}. For a bubble radius of $a = 1$ cm, the Reynolds number becomes $\mathsf{Re} = 418$, which roughly agrees with the value of C_D in Figure 17.3. The terminal speed is considerably lower than typical wind speeds or drafts, which explains why soap bubbles tend to move easily around.

Non-spherical shape

The general arguments given in Section 17.1 showed that in creeping flow, the drag on an arbitrary body is proportional to viscosity, velocity, size, and a shape-dependent dimensionless factor. Using the Stokes' drag (17.25) as a baseline, we may write the drag on an arbitrary body as

$$\mathcal{D} = 6\pi\eta a_S U, \tag{17.37}$$

where a_S is a characteristic length, sometimes called the *Stokes radius*. The Stokes radius may be determined from a measurement of the terminal speed of the falling object,

$$a_S = \frac{m' g_0}{6\pi\eta U}, \tag{17.38}$$

where m' is the mass of the body corrected for buoyancy.

The Stokes radius may be calculated analytically for some simple bodies, for example a circular disk of radius a. In the two major orientations of the disk [White 1991],

$$\frac{a_S}{a} = \begin{cases} \dfrac{8}{3\pi} \approx 0.85 & \text{disk orthogonal to flow,} \\ \dfrac{16}{9\pi} \approx 0.57 & \text{disk parallel to flow.} \end{cases} \tag{17.39}$$

In spite of the vast geometric difference between the two cases, the Stokes radii are of the same order of magnitude. For general bodies of non-exceptional geometry, one may estimate the Stokes radius from the surface area or the volume of the body.

The drag coefficient may be defined as

$$\boxed{C_D = \frac{\mathcal{D}}{\frac{1}{2}\rho_0 A U^2},} \tag{17.40}$$

where A is the area that the body presents to the flow. For blunt bodies the area may be taken to be the "shadow" of the body on a plane orthogonal to the direction of motion. For a given shape, the drag force,

$$\mathcal{D} = C_D \tfrac{1}{2}\rho_0 A U^2, \tag{17.41}$$

exposes the dependence on the fluid (ρ_0), the body size (A), and the state of motion (U). Except for the region of the drag crisis, typical values for C_D are of the order of unity for bodies of non-exceptional geometry. A flat circular disk orthogonal to the flow has $C_D \approx 1.17$, whereas a circular cup with its opening toward the flow has $C_D \approx 1.4$. If on the other hand the disk is infinitely thin and oriented parallel with the flow, there will be no asymptotic form drag, and the drag coefficient will vanish as $1/\sqrt{\mathsf{Re}}$ in the limit of infinite Reynolds number.

Streamlining

Drag reduction by *streamlining* is important in the construction of all kinds of moving vehicles, for example cars and airplanes. Car manufacturers have over the years reduced the drag coefficient to lower than 0.4, and there is still room for improvement. Modern drag-reducing helmets have also appeared on the heads of bicycle racers and speed skaters, giving the performers of these sports quite alien looks. Among animals, drag reduction yields evolutionary advantages that have led to the beautiful outlines of fast flyers and swimmers, like falcons and sharks. The density of water is about a thousand times that of air, implying a thousand times larger drag (17.41) in water than in air. This has forced swimming animals toward extremes of streamlined shapes. The mackerel has thus reduced the drag coefficient of its sleek form to the astonishingly low value of 0.0043 (at $\mathsf{Re} \approx 10^5$), which is a factor of eight lower than for a swimming human [Vogel 1994, p. 143]. Since the power output against drag grows as $\mathcal{D}U \sim U^3$, one expects that the density ratio of 1,000 should typically allow birds to move with speeds that are about $1{,}000^{1/3} = 10$ times higher than for fish of similar size and shape, a speed ratio that seems quite reasonable.

17.4 Lubrication

The most important technological invention of all time must be the wheel and its bearing. Early on it was realized that friction in the bearing was considerably lowered by lubricating it with viscous fluid. Wooden bearings might even catch fire if not lubricated. Fat from pigs, olive oil, and mineral oil turned out to work much better than water. The theoretical analysis of creeping flow in a narrow gap goes back to Reynolds (1886).

Estimates

In this chapter we have until now assumed that moving bodies stay well away from other bodies. Let us now consider a body moving with velocity U close to a static solid boundary. The width of the gap between body and boundary is assumed to be of size $d \ll L$, where L is the extent of the gap in the direction of motion. Since the flow across the gap is negligible, the advective acceleration $(\boldsymbol{v} \cdot \nabla)\boldsymbol{v}$ is dominated by the flow variation along the gap, whereas the Laplacian $\nabla^2 \boldsymbol{v}$ is dominated by the flow variation from 0 to U across the gap. The Reynolds number in the gap therefore becomes

$$\mathsf{Re}_{\text{gap}} \approx \frac{|\rho(\boldsymbol{v} \cdot \nabla)\boldsymbol{v}|}{|\eta \nabla^2 \boldsymbol{v}|} \approx \frac{\rho U^2/L}{\eta U/d^2} = \left(\frac{d}{L}\right)^2 \mathsf{Re}, \tag{17.42}$$

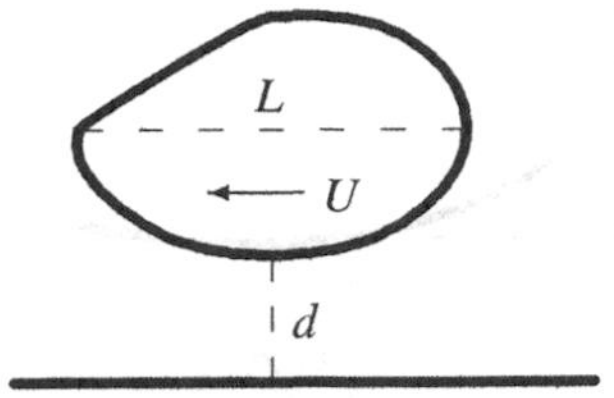

Object moving close to a nearly flat wall. The gap is assumed to be so narrow that creeping flow conditions prevail.

where $\mathsf{Re} = \rho U L/\eta$ is the free-flow Reynolds number far from the boundary. Even if Re is large, viscous forces will dominate the flow in the gap when the distance d becomes so small that $\mathsf{Re}_{\text{gap}} \ll 1$, or

$$d \ll \delta = \frac{L}{\sqrt{\mathsf{Re}}}. \tag{17.43}$$

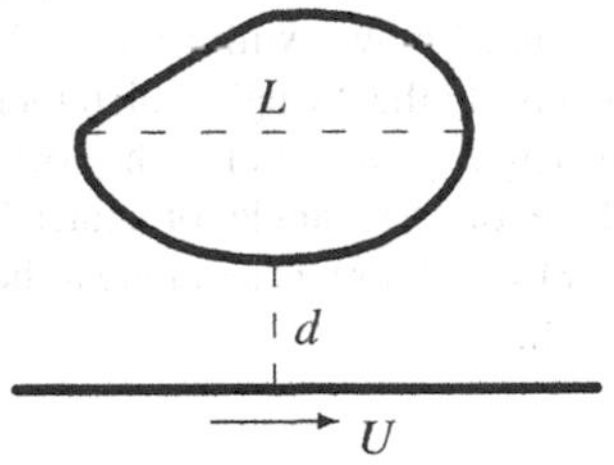

In the reference frame where the flat boundary moves with velocity U, the object is stationary and the flow is steady.

In Chapter 28 we shall see that δ is a measure of the thickness of the boundary layer surrounding a moving body. The conditions for creeping flow will in other words be fulfilled when the gap lies well inside the boundary layer. In the following we shall assume this to be the case.

The magnitude of the pressure gradient along the gap can be estimated from the equation for creeping flow (17.1), leading to $|\nabla p| = \eta \left|\nabla^2 \boldsymbol{v}\right| \approx \eta U/d^2$. Multiplying by the length of the gap L we obtain the pressure variation along the gap, and finally multiplying by the gap area A we get an estimate for the lift force,

$$\mathcal{L} \approx f_{\text{lift}} \frac{\eta U L A}{d^2}. \tag{17.44}$$

Here we have also put in a dimensionless prefactor f_{lift} of order unity that depends on the geometry and orientation of the body. We shall see (page 302) that for a flat wing we have $f_{\text{lift}} \approx \alpha L/2d$, where α is the angle of attack. The flow around the upper part of the body will also contribute to lift, but being independent of d it will in the limit of $d \to 0$ always be dominated by the lift in the gap, barring special configurations with vanishing gap lift.

Across the gap the velocity changes from 0 to U, so that the normal velocity gradient becomes U/d, implying that the shear stress is of magnitude $\eta U/d$. Multiplying with the area A of the gap we obtain an estimate for the magnitude of the *drag* from fluid friction in the gap (with an unknown dimensionless prefactor of order unity),

$$\mathcal{D} \approx f_{\text{drag}} \frac{\eta U A}{d}. \tag{17.45}$$

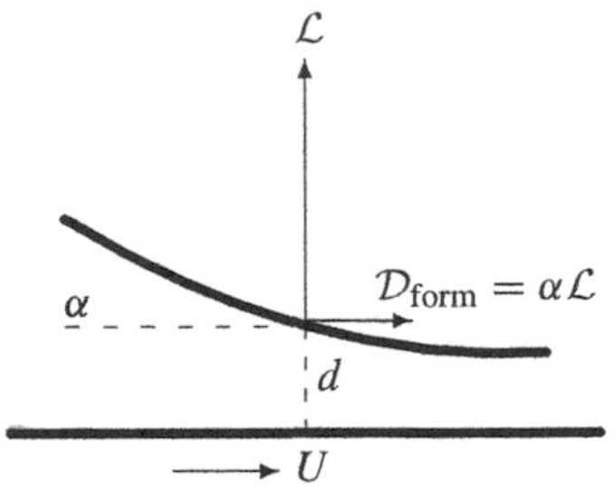

The pressure force acts orthogonally to the body surface and gives rise to lift as well as form drag. The drag is proportional to the angle of attack α, which is always assumed small, $|\alpha| \lesssim d/L$.

The ratio of lift to drag is normally large compared to unity, $\mathcal{L}/\mathcal{D} \sim \alpha L^2/d^2$. In addition to skin drag there will also be form drag of the same magnitude (see the margin figure). As for lift, the drag from the flow around the body outside the gap can be ignored in the limit of $d \to 0$.

Example 17.4 [Playing card]: When you deal a pack of cards on a table with a smooth surface, it is easy to overestimate the speed the cards must be thrown with. Suddenly, several cards skip easily over the surface and land on the floor. The reason is that a lubricating layer of air has formed between the card and the surface of the table, and has caused the friction you expected when throwing the card nearly to drop away. A typical playing card has size 7 cm × 10 cm, and since we do not know how it moves we shall take $L \approx 8$ cm. Skimming through the air above a horizontal table at $U \approx 1$ m s^{-1}, it has $\mathsf{Re} \approx 5 \times 10^3$ and thus $\delta \approx 1$ mm. Taking $d \approx \delta$ and the form factors equal to unity, the lift–to–drag ratio becomes about 80, explaining why the card skips so easily.

Creeping flow in a long narrow gap

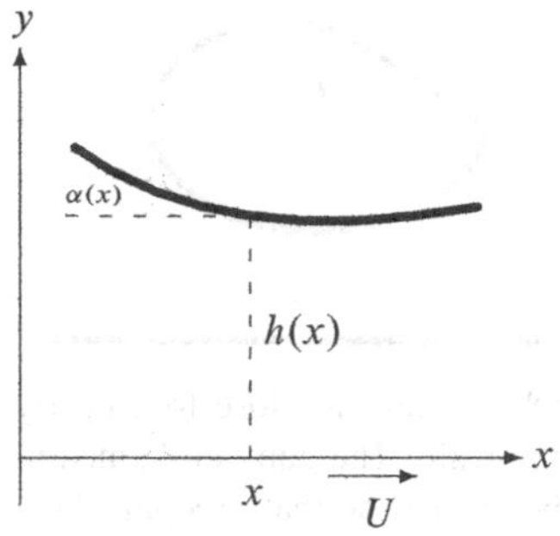

Local geometry of creeping "flight" in the xy-plane. The "ground" moves with velocity U relative to the "wing", which flies at height $y = h(x)$. The local slope of the "angle of attack", $\alpha(x) = -h'(x)$, is assumed to be small.

For simplicity we shall consider the essentially two-dimensional case of a stationary "wing" with "chord length" L along the x-axis. The wing shape is constant along z with "span" S such that the total wing area is $A = LS$. The "ground" is chosen to be perfectly flat at $y = 0$, and moves with constant velocity U along the x-axis. The height of the wing above the ground is for all z given by a positive function $y = h(x)$, which is assumed to be slowly varying $|h'(x)| \ll 1$. The free-flow Reynolds number is assumed to be so large and the height so small that the creeping approximation is valid, that is $h(x) \ll \delta \ll L$.

Under these conditions we need only to retain the dominant derivatives with respect to y in the Laplacian in (17.1). Dropping all second derivatives with respect to x, we get a simplified set of equations for the flow in the gap,

$$\frac{\partial p}{\partial x} = \eta \frac{\partial^2 v_x}{\partial y^2}, \qquad \frac{\partial p}{\partial y} = \eta \frac{\partial^2 v_y}{\partial y^2}, \qquad \frac{\partial v_y}{\partial y} = -\frac{\partial v_x}{\partial x}. \tag{17.46}$$

From the last equation we estimate $|v_y| \approx Ud/L$ where d is a measure of $h(x)$, for example an average along the gap. The two first equations now yield estimates of the pressure variations: $\Delta_x p = |\partial_x p| L \approx \eta U L/d^2$ and $\Delta_y p = |\partial_y p| d \approx \eta U/L$. Evidently, the variation along x completely dominates, and we may to this order of approximation take the pressure in the gap to be only a function of x.

Inserting $p = p(x)$ in the first gap equation it may immediately be integrated with the boundary conditions $v_x = U$ at $y = 0$ and $v_x = 0$ at $y = h(x)$, yielding

$$v_x = U\left(1 - \frac{y}{h(x)}\right) - \frac{p'(x)}{2\eta} y(h(x) - y), \tag{17.47}$$

where $p'(x)$ is the pressure gradient. The solution is a superposition of velocity-driven planar flow (15.6) and pressure-driven planar flow (16.5), with a variable plate distance $h(x)$ and pressure gradient $p'(x)$.

The moving ground at $y = 0$ drags fluid along in the direction of positive x. The discharge rate becomes (per unit of length in the z-direction)

$$Q = \int_0^{h(x)} v_x(x, y)\, dy = \frac{1}{2} U h(x) - \frac{p'(x) h(x)^3}{12\eta}. \tag{17.48}$$

Since the fluid is incompressible, Q is independent of x, so that we have

$$\boxed{p'(x) = 6\eta \left(\frac{U}{h(x)^2} - \frac{2Q}{h(x)^3}\right).} \tag{17.49}$$

Eliminating the pressure gradient in the velocity (17.47), we find

$$\boxed{v_x = U \frac{(h - y)(h - 3y)}{h^2} + Q \frac{6y(h - y)}{h^3},} \tag{17.50}$$

where we for clarity have suppressed the explicit dependence on x, now entirely due to the slowly varying gap height $h(x)$. Finally, we insert v_x into the continuity equation and integrate over y with the condition $v_y = 0$ for $y = 0$, and get

$$\boxed{v_y = -2h' \frac{Uh - 3Q}{h^4} y^2 (h - y).} \tag{17.51}$$

The appearance of the small height derivative $h'(x)$ confirms that $|v_y| \ll |v_x|$.

Effective gap width and flow reversal

The pressure change along the gap is calculated by integrating the pressure gradient,

$$\Delta p = \int_L p'(x)\,dx = 6\eta \left(U \int_L \frac{dx}{h(x)^2} - 2Q \int_L \frac{dx}{h(x)^3} \right). \tag{17.52}$$

The wing is supposed to move in a fluid that would have constant pressure, were it not for the disturbance created by the wing itself. From now on we assume that the pressure $p(x)$ is the same at both ends of the wing, $\Delta p = 0$. Thus we obtain the following relation between the discharge rate and the velocity,

$$Q = \frac{1}{2}Ud, \qquad\qquad d = \frac{\langle h^{-2} \rangle}{\langle h^{-3} \rangle}, \tag{17.53}$$

where $\langle F \rangle = \int_L F(x)\,dx/L$ denotes the average of a function $F(x)$ in the gap. The relation between Q and U depends on the shape of the gap only through the *effective gap width* d. For flat plates with constant gap height, $h(x) = h_0$, we get $d = h_0$, and this shows that d represents the width of a gap between parallel plates, carrying the same discharge rate Q as the actual gap.

The pressure gradient (17.49) may now be written

$$p' = 6\eta U \frac{h-d}{h^3}. \tag{17.54}$$

In regions where $h(x) > d$, the pressure will rise whereas it will fall if $h(x) < d$. The velocity field may also be expressed in terms of d,

$$v_x = U \left(1 - \frac{y}{h} \right)\left(1 - \frac{3y(h-d)}{h^2} \right), \qquad v_y = -Uh' \frac{2h-3d}{h^4} y^2(h-y). \tag{17.55}$$

Whereas the first parenthesis in v_x is always positive, the last may have either sign. It is positive for small y, but may for a given x vanish and become negative for $y > h^2/3(h-d)$. When this point lies inside the gap, $0 < y < h$, the fluid close to the wing will flow *against* the direction of ground motion. This is only possible for $3(h-d) > h$, or

$$h(x) > \frac{3}{2}d. \tag{17.56}$$

In regions where this condition is fulfilled, v_y will have opposite sign of h', showing that "rollers" of counter-rotating fluid will appear (see the margin figure).

Lift, drag, moment, and power

In steady flow, there is no loss of momentum from the fluid. Consequently, the total force exerted by the fluid on the body must be equal and opposite to the total force exerted by the fluid on the flat ground. Since the flow outside the gap is negligible for $d \to 0$, the lift and drag on the body become

$$\mathcal{L} = -\int_L \sigma_{yy}\big|_{y=0}\, S\,dx, \qquad \mathcal{D} = -\int_L \sigma_{xy}\big|_{y=0}\, S\,dx, \tag{17.57}$$

where S is the wing span (along z). Neither can there be loss of angular momentum from the fluid, implying that the total moment of force exerted on the body must be equal and opposite to the total moment of force exerted on the ground

$$\mathcal{M}_z = -\int_L \left[x\sigma_{yy} - y\sigma_{yx} \right]_{y=0} S\,dx = -\int_L x\,\sigma_{yy}\big|_{y=0}\, S\,dx. \tag{17.58}$$

Here the moment has been calculated with respect to the origin of the coordinate system.

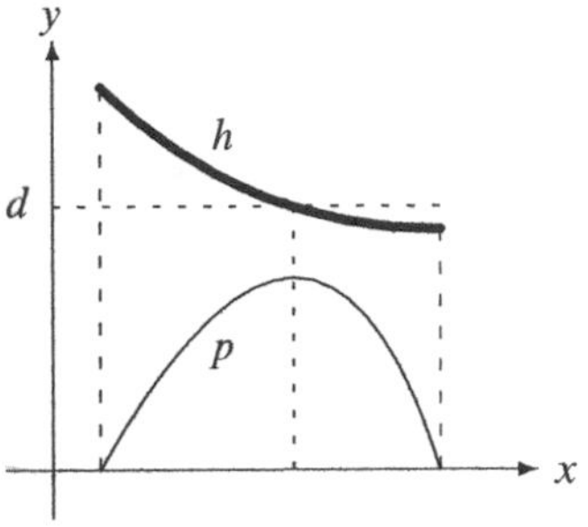

Sketch of the pressure $p(x)$ in relation to d for a wing with positive angle of attack. The ambient pressure outside the gap is set to zero.

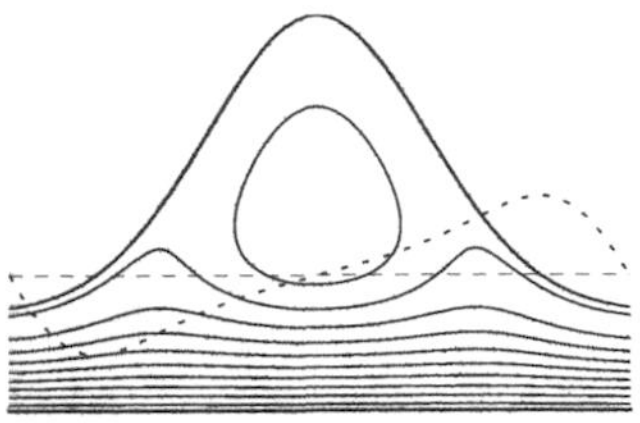

Flow reversal under a Gaussian bump $h = 1 + 3\exp(-x^2)$ in the interval $-2 < x < 2$. The effective height is $d \approx 1.38$ (dashed line). The streamlines are fully drawn, and the dotted curve indicates the pressure variation. The vertical scale is exaggerated.

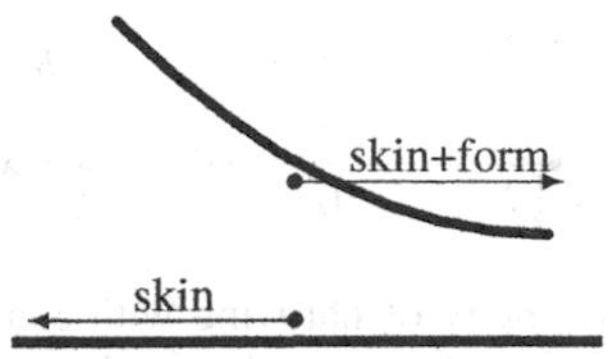

The drag on the wing is composed of skin and form drag, whereas the drag on the ground is only skin drag. In steady flow the total drag on the wing is equal and opposite to the total drag on the ground.

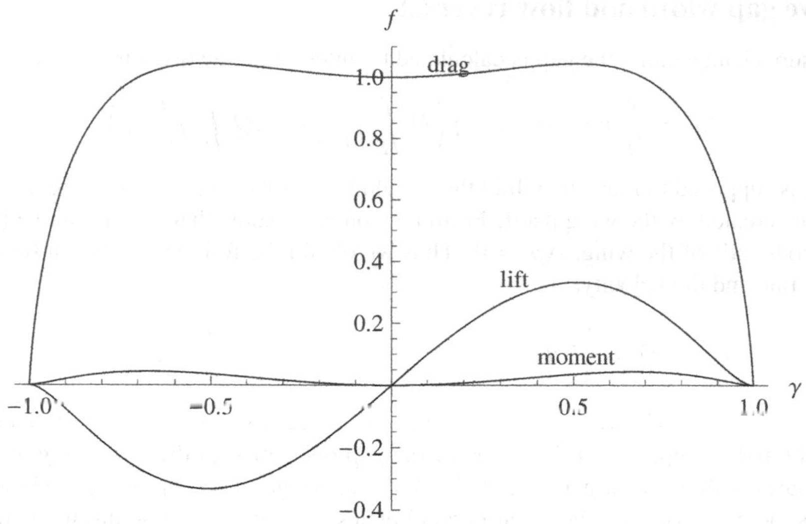

Figure 17.5. Flat-wing prefactors for lift, drag, and moment as functions of γ. All quantities vanish for $|\gamma| \to 1$. The lift vanishes linearly and the moment quadratically for $\gamma \to 0$.

The normal and shear stresses on the ground are obtained from the solution above,

$$\sigma_{yy}\big|_{y=0} = -p + \eta\, \nabla_y v_y\big|_{y=0} = -p, \tag{17.59a}$$

$$\sigma_{xy}\big|_{y=0} = \eta \left[\frac{\partial v_x}{\partial y} + \frac{\partial v_y}{\partial x}\right]_{y=0} = -\eta U \frac{4h - 3d}{h^2}. \tag{17.59b}$$

In accordance with the previous estimates their ratio is $\sigma_{xy}/\sigma_{yy} \approx \mathcal{O}(d/L)$.

Finally, the rate of work of those external forces $-\mathcal{D}$ that keep the ground moving at constant velocity U must be

$$P = \mathcal{D}U. \tag{17.60}$$

Since the wing is at rest, no other external forces perform work on the system, so this must equal the total rate of dissipation.

Case: Flat wing

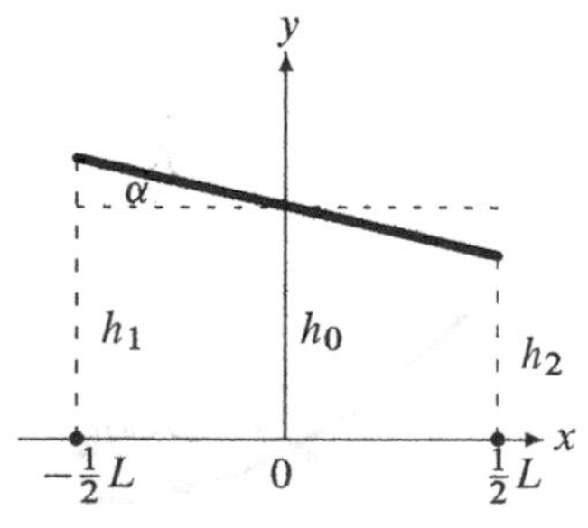

Geometry of flat wing with constant angle of attack α and average height h_0. Varying α rotates the wing around its midpoint.

A flat wing with constant angle of attack α and mean height h_0 has (see the margin figure),

$$h = h_0 - \alpha x \qquad \text{for } -\tfrac{1}{2}L \le x \le \tfrac{1}{2}L. \tag{17.61}$$

It is also convenient to introduce the front and back end heights, h_1 and h_2 with average $h_0 = \frac{1}{2}(h_1 + h_2)$ (see the margin figure). A convenient dimensionless parameter is the normalized angle of attack

$$\gamma = \frac{h_1 - h_2}{h_1 + h_2} = \frac{\alpha L}{2h_0}, \tag{17.62}$$

which ranges over the interval, $-1 < \gamma < 1$. Taking $dh = -\alpha dx$ as integration parameter we easily find the effective gap width (17.53),

$$d = \frac{h_1 h_2}{h_0} = (1 - \gamma^2) h_0. \tag{17.63}$$

It is never larger than h_0, and vanishes if either end of the wing touches the ground.

For positive angle of attack $\gamma > 0$, flow reversal takes place at the front of the wing for $\gamma > 1/3$. Likewise, for $\gamma < -1/3$, there will instead be flow reversal at the rear of the wing. Note, however, that a negative angle of attack is liable to create cavitation in the fluid.

The pressure gradient (17.54) may likewise be integrated to yield

$$p = \frac{3\eta U}{\alpha}\left(\frac{2}{h} - \frac{d}{h^2} - \frac{1}{h_0}\right), \tag{17.64}$$

where the last term secures that the pressure vanishes at both ends of the wing. The lift and drag and moment are obtained from (17.57)–(17.59):

$$\mathcal{L} = \int_{-L/2}^{L/2} p(x)\, S dx = \frac{3\left(1-\gamma^2\right)^2}{2\gamma^2}\left(\log\frac{1+\gamma}{1-\gamma} - 2\gamma\right)\cdot\frac{\eta U L A}{d^2}, \tag{17.65}$$

$$\mathcal{D} = -\int_{-L/2}^{L/2} \sigma_{xy}\big|_{y=0}\, S dx = \frac{1-\gamma^2}{\gamma}\left(2\log\frac{1+\gamma}{1-\gamma} - 3\gamma\right)\cdot\frac{\eta U A}{d}, \tag{17.66}$$

$$\mathcal{M}_z = \int_{-L/2}^{L/2} x p(x)\, S dx = \frac{3\left(1-\gamma^2\right)^2}{8\gamma^3}\left(\left(3-\gamma^2\right)\log\frac{1+\gamma}{1-\gamma} - 6\gamma\right)\cdot\frac{\eta U L^2 A}{d^2}. \tag{17.67}$$

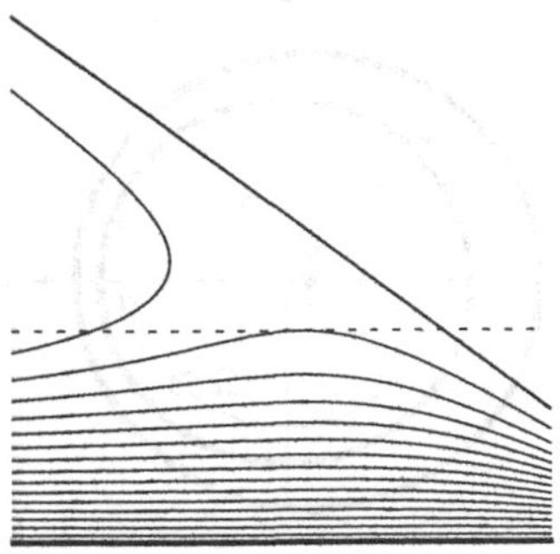

Flow reversal under flat wing for $L = 20h_0$ and $\gamma = 0.6$. The dashed line indicates the effective width d. The scale of the ordinate is exaggerated by a factor of 10.

The first two γ-dependent factors on the right-hand sides correspond to the prefactors f_{lift} and f_{drag} introduced in the estimates (17.44) and (17.45). All these prefactors are plotted in Figure 17.5 as functions of γ. For $|\gamma| \ll 1$ we have $f_{\text{lift}} \approx \gamma$ and $f_{\text{drag}} \approx 1$, which are good approximations in most cases of interest. The moment prefactor is always positive but mostly very small and vanishes like $f_{\text{moment}} \approx \frac{1}{5}\gamma^2$ for $\gamma \to 0$. If the angle of attack α is positive, the moment tends to rotate the wing toward the horizontal, whereas for negative α it tends to turn the front of the wing further into the ground, thereby destabilizing the flight.

Example 17.5 [Playing card]: A completely flat unbent playing card thrown with a positive angle of attack will slowly sink further and further toward the table surface while the tiny moment rotates it toward the horizontal and the lift becomes still smaller (here we ignore again any forces acting on the upper side of the card). Thrown with a negative angle of attack, the playing card will get sucked toward the table at an increasing rate because the positive moment makes the angle of attack even more negative. Eventually the front of the card may catch on surface irregularities and turn over, showing its value to the dismay of the players.

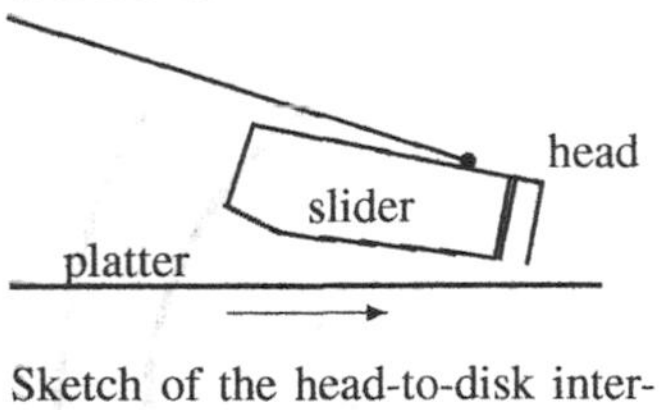

Sketch of the head-to-disk interface in a hard disk. The platter rotates toward the right and drags air into the gap between the slider and the surface, and thereby prevents the slider from touching the platter. The elastic actuator arm counteracts the lift force from the air in the gap. The head itself is here positioned at the rear end of the slider.

Example 17.6 [Hard-disk magnetic read/write heads]: The continued sophistication in the design of read/write heads and platter surfaces has been a major cause for the enormous improvement in hard disk performance over the past 30 years. A modern (2002) hard disk has a platter diameter of about 9 cm and runs at a speed of about 7,000 rpm, leading to average platter surface speeds of $U \approx 16\ \text{m s}^{-1} \approx 60\ \text{km h}^{-1}$. The read/write heads sit on the tip of an actuator arm that can roam over the rotating platters and exchange data with the magnetic surfaces. The read/write head is formed as a flat wing or "slider" of size $L \approx 0.5$ mm, for which $\mathsf{Re} \approx 500$, and the maximal gap size for creeping flow comes to $\delta \approx 22\ \mu\text{m}$. The need for increased data density demands smaller and smaller flying height for the actual read/write device sitting on the rear of the slider. Typically (2002) it is about $h_2 \approx 150$ nm, implying that the slider flies deeply inside the boundary layer. Taking the average slider gap height to be a conservative $h_0 = 1\ \mu\text{m}$, the Reynolds number in the gap becomes $\mathsf{Re}_{\text{gap}} = 2 \times 10^{-3}$, which is far within the creeping flow regime. The geometry makes the angle of attack $\alpha = 0.2$ degrees, $\gamma = 0.85$, and the effective height $d = 280$ nm. The geometric prefactors become $f_{\text{lift}} = 0.13$ and $f_{\text{drag}} = 0.81$, resulting in lift $\mathcal{L} = 0.6$ N and drag $\mathcal{D} = 2 \times 10^{-4}$ N, so that the lift-to-drag ratio becomes about 290. The average gap pressure excess is about 2.5 bar, and considerable adiabatic heating will take place in the gap. The lift force corresponds to a weight of 25 g, and this force must be countered by the elastic actuator arm to keep the flying height constant. If the total mass of the actuator arm and slider is of the order of a few grams, this explains why modern hard disks can tolerate rather large accelerations before the head crashes into the platter and destroys the disk.

* 17.5 Application: Loaded journal bearing

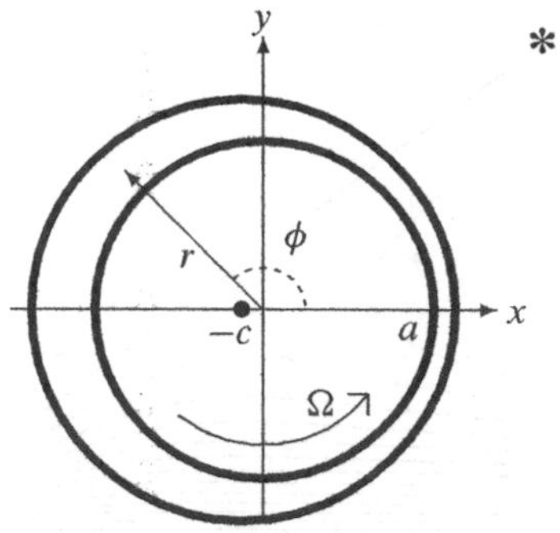

Geometry of off-center journal bearing. The inner cylinder has radius a and the outer b. The outer cylinder is shifted to the left by an amount c. The inner cylinder rotates with constant angular velocity Ω in the counter-clockwise direction.

In Section 16.6 we discussed the case of laminar flow between two concentric rotating cylinders, the prototypical *journal bearing*. If the inner shaft or the outer sleeve (bushing) carries a load orthogonal to the shaft, the cylinders will no more be concentric, although we shall assume that they are still parallel with respect to each other. In a non-rotating journal bearing, the lubricating fluid will be squeezed out and the shaft will come into direct contact with the sleeve. When the shaft (or the sleeve) is brought into rotation, fluid will be dragged along due to the no-slip condition and forced into the contact region of the gap, thereby creating a pressure that tends to lift the shaft away from the sleeve. In this section we shall only discuss laminar flow in the gap. A rotating journal bearing is also prone to the centrifugal instabilities discussed in Section 16.7 on page 281 with formation of Taylor vortices and more complicated structures.

Narrow gap approximation

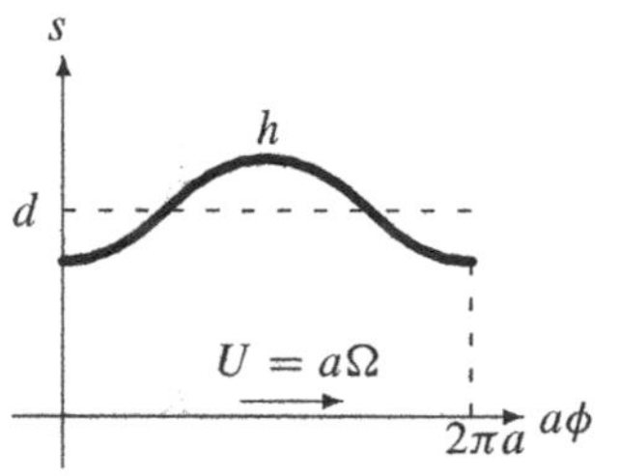

Flattened gap between the cylinders. The inner cylinder replaces the flat boundary along the x-axis with $x \to a\phi$, and the space between the cylinders is described by $y \to s = r - a$. The velocity of the inner cylinder is $U = a\Omega$

Let the inner cylinder have radius a and the outer radius $b > a$ with a difference $h_0 = b - a$ that is assumed to be tiny, $h_0 \ll a$. In a coordinate system with the z-axis coinciding with the axis of the inner cylinder, we may without loss of generality assume that the point of closest approach takes place on the positive x-axis. Denoting the center of the outer cylinder $x = -c$, the points of the outer cylinder are determined by the equation $(x + c)^2 + y^2 = b^2$, which in standard cylindrical coordinates becomes $r^2 + c^2 + 2rc\cos\phi = b^2$. Substituting $r = a + h$ we find to first order in the small quantities h and c,

$$h = h_0 - c\cos\phi. \tag{17.68}$$

It is convenient to introduce the dimensionless eccentricity parameter,

$$\gamma = \frac{c}{h_0}, \tag{17.69}$$

which must lie in the interval $-1 \le \gamma \le 1$.

The shaft has length S, circumference $L = 2\pi a$, and surface area $A = LS$. It rotates at constant angular velocity Ω with surface velocity $U = a\Omega$. Disregarding the possibility that lubricant may be squeezed out along the z-axis (a non-trivial technical detail), the problem is essentially two-dimensional. Under these assumptions, the narrow-gap approximation developed in the preceding section may be brought into play, replacing x by $a\phi$ and y by $s = r - a$. The narrow gap between the cylinders has essentially been converted to a flat-wall gap of length L with strictly periodic boundary conditions (at $\phi = 0$ and $\phi = 2\pi$).

Pressure and lift

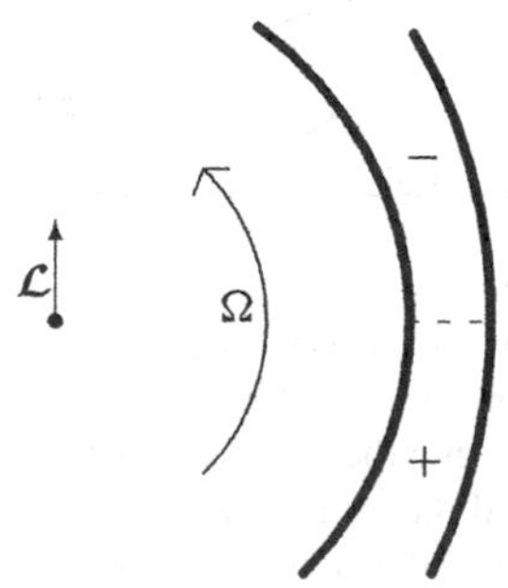

The pressure generated by the shaft's rotation is asymmetric with respect to the point of closest approach. The total lift $\mathcal{L}$ on the shaft is always parallel with the direction of motion at the point of closest approach.

The effective gap width defined in (17.53) may now be evaluated by averaging over ϕ, with the result (see Problem 17.16) that

$$d = h_0\frac{2(1-\gamma^2)}{2+\gamma^2}. \tag{17.70}$$

The pressure derivative is as before given by (17.54), and integrating over ϕ one obtains the expression (most easily checked by differentiation with respect to ϕ),

$$p = -\frac{\eta UL}{d^2}\cdot\frac{12\gamma(1-\gamma^2)^2(2-\gamma\cos\phi)\sin\phi}{\pi(2+\gamma^2)^3(1-\gamma\cos\phi)^2}. \tag{17.71}$$

Evidently the pressure vanishes at the point of closest approach $\phi = 0$ and at the opposite point $\phi = \pi$. If $\gamma > 0$, the pressure is positive in the lower half of the gap and negative in the

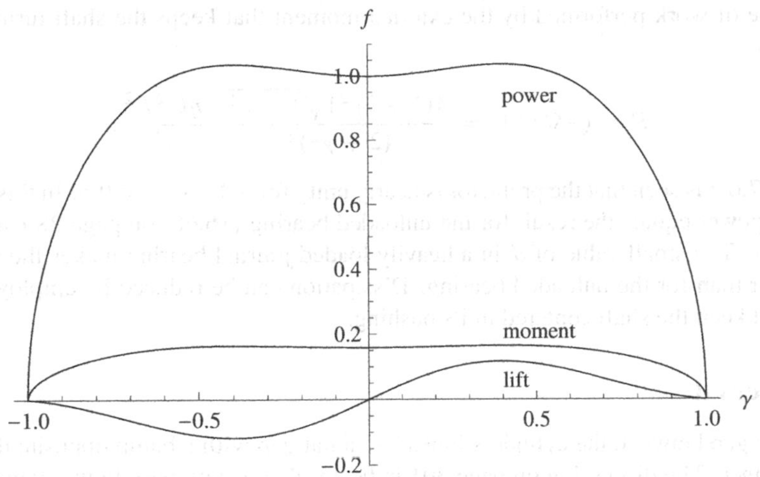

Figure 17.6. Prefactors for lift, moment, and dissipated power in a journal bearing as functions of γ.

upper. The up-down antisymmetry of the pressure under $\phi \to -\phi$ implies that the average pressure is zero, $\langle p \rangle = 0$.

In Cartesian coordinates the surface element of the inner cylinder expressed in polar variables becomes $d\boldsymbol{S} = (\cos\phi, \sin\phi, 0)\, a d\phi\, dz$. The up-down antisymmetry of p shows that the lift on the shaft in the x-direction vanishes,

$$\mathcal{L}_x = \oint_{r=a} (-p) dS_x = 0. \tag{17.72}$$

Replacing the cosine by a sine, the lift on the shaft along y becomes (Problem 17.16),

$$\mathcal{L}_y = \oint_{r=a} (-p) dS_y = \frac{12\gamma(1-\gamma^2)^{3/2}}{\pi(2+\gamma^2)^3} \cdot \frac{\eta U L A}{d^2}, \tag{17.73}$$

where $L = 2\pi a$, $A = LS$, and $U = \Omega a$. The γ-dependent expression represents the prefactor $f_{\rm lift}$ in our estimate (17.44). Its behavior is shown in Figure 17.6.

Why the strange behavior: It is perhaps a bit counterintuitive that the lift is always parallel with the direction of motion of the shaft at the point of closest approach, rather than orthogonal to it. This is, as we have seen, a consequence of the antisymmetry of the pressure with respect to the radial direction at closest approach. When the shaft just starts to rotate from rest, the direction of lift will thus be orthogonal to the direction of load, and with no other forces at play, this lift will tend to shift the point of closest approach sideways in the direction of motion of the shaft surface. As the point of closest approach moves away the lift changes direction, until it reaches a point where radial direction is orthogonal to the direction of the load. The actual distance at closest approach $h_0(1-\gamma)$ is determined by the balance between load and lift; and the larger the load, the closer the value of γ must be to unity.

Moment of drag and power

The shear stress $\sigma_{\phi r}$ at the surface of the shaft is given by the general expression (17.59). In this case it makes no sense to speak about a drag force, but rather the moment of the shear stress around the center of the shaft (Problem 17.16)

$$\mathcal{M}_z = \frac{A}{2\pi}\int_0^{2\pi} a\sigma_{\phi r} d\phi = -\frac{2(1+2\gamma^2)\sqrt{1-\gamma^2}}{\pi(2+\gamma^2)^2} \cdot \frac{\eta U L A}{d}. \tag{17.74}$$

It is negative, as one would expect.

The rate of work performed by the external moment that keeps the shaft turning against viscosity is

$$P = (-\Omega)\mathcal{M}_z = \frac{4(1+2\gamma^2)\sqrt{1-\gamma^2}}{(2+\gamma^2)^2} \cdot \frac{\eta U^2 L^2}{d}. \tag{17.75}$$

In Figure 17.6 it is seen that the prefactor is nearly unity for $-0.5 < \gamma < 0.5$. In this region the dissipated power equals the result for the unloaded bearing (16.68) on page 280, and outside it is smaller. The small value of d in a heavily loaded journal bearing makes the dissipation much larger than for the unloaded bearing. Dissipation can be reduced by employing rollers or balls that keep the shaft centered in its bushing.

Flow reversal

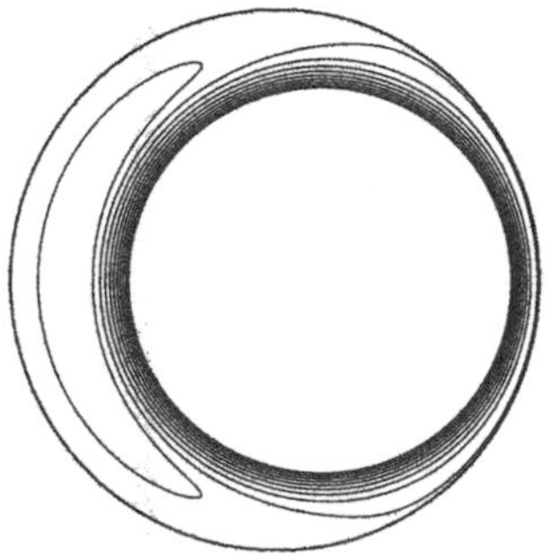

Creeping flow pattern in loaded journal bearing with $h_0/a = 0.4$ and $\gamma = 0.6$. The flow is reversed and forms a counter-rotating patch of fluid opposite the point of closest approach.

The narrow gap between the cylinders looks like a flat gap with a bump opposite the point of closest contact. The discussion on page 301 indicates that a stationary flow-reversed "roller" may arise at this point. The reversal condition (17.56) becomes

$$1 - \gamma \cos\phi > 3\frac{1-\gamma^2}{2+\gamma^2}. \tag{17.76}$$

Taking $\phi = \pi$ the roller will appear when $\gamma^2 + 3\gamma - 1 > 0$. Solving the quadratic inequality this is the case for $|\gamma| > (\sqrt{13} - 3)/2 \approx 0.303$ (see the margin figure).

Problems

17.1 A spherical particle begins to fall from rest in constant gravity. Determine how the particle reaches terminal speed (17.27) under the assumption that the Reynolds number is always small.

17.2 [Sedimentation in the sea] Tiny unicellular animals in the sea die and their carcasses settle slowly toward the bottom. Assume that an animal's carcass is spherical with diameter 10 μm and average density 1.2 times that of water. Calculate the Reynolds number and the time it takes for the carcass to settle to the bottom of the deep sea (depth 4 km).

17.3 [Atmospheric sedimentation] A microscopic spherical grain of vulcanic (or cosmic) dust of radius a enters the homentropic atmosphere (Section 2.6) at a height z. **(a)** Determine the time t_0 it takes for the grain to settle to the ground under the assumption that the density of the grain is much larger than the density of air, and that Stokes' law is valid at any height with the actual density and viscosity of air at that height. **(b)** Evaluate the settling time numerically for reasonable values of the parameters.

17.4 Calculate the pressure contribution to the drag from Stokes' solution for a sphere.

17.5 An object with constant drag coefficient C_D falls from rest in constant gravity. Determine how the particle reaches terminal speed under the assumption that buoyancy can be disregarded.

17.6 Consider Stokes' flow around a sphere of radius a. **(a)** Calculate the volume discharge of fluid passing a concentric annular disk of radius $b > a$ placed orthogonally to the flow. **(b)** Calculate the volume in relation to the volume that would pass through the same disk if the sphere were not present. **(c)** Justify qualitatively why the ratio vanishes for $b \to a$.

* **17.7** Consider spherical Stokes flow and **(a)** show that the streamlines are determined by the solutions to

$$\frac{dr}{dt} = v_r = A(r)U\cos\theta, \qquad \frac{d\theta}{dt} = \frac{v_\theta}{r} = -B(r)U\sin\theta,$$

with

$$A(r) = 1 - \frac{3}{2}\frac{a}{r} + \frac{1}{2}\frac{a^3}{r^3} = \left(1 - \frac{a}{r}\right)^2 \left(1 + \frac{1}{2}\frac{a}{r}\right),$$

$$B(r) = \frac{1}{r}\left(1 - \frac{3}{4}\frac{a}{r} - \frac{1}{4}\frac{a^3}{r^3}\right) = \frac{1}{r}\left(1 - \frac{a}{r}\right)\left(1 + \frac{1}{4}\frac{a}{r} + \frac{1}{4}\frac{a^2}{r^2}\right).$$

(b) Show that this leads to a solution of the form

$$\sin\theta = e^{-\int B/A\,dr} = \frac{d}{(r-a)\sqrt{1+a/2r}},$$

where d is an integration constant. This is the equation for the streamlines in polar coordinates.
(c) Show that d is the asymptotic distance of the flow line from the polar axis (also called the impact parameter).
(d) Find the relation between d and the point of closest approach of the flow line to the sphere.

* **17.8** Show that the rate of work of the contact forces exerted by a steadily moving body on the fluid through which it moves is $\mathcal{D}U$, where $\mathcal{D}$ is the total drag.

* **17.9** **(a)** Show that the creeping flow solution around a fixed body in an asymptotically uniform steady velocity field $\boldsymbol{U}$ must be of the form

$$\boldsymbol{v}(\boldsymbol{x}) = \mathsf{A}(\boldsymbol{x}) \cdot \boldsymbol{U}, \qquad p(\boldsymbol{x}) = \eta \boldsymbol{Q}(\boldsymbol{x}) \cdot \boldsymbol{U},$$

where $\mathsf{A}(\boldsymbol{x})$ is a tensor field and $\boldsymbol{Q}(\boldsymbol{x})$ is a vector field, both independent of $\boldsymbol{U}$.
(b) Determine the field equations and the boundary conditions for A and $\boldsymbol{Q}$.
(c) Calculate the total reaction force on the body.

17.10 A simple hydraulic clutch connecting two rotating shafts consists of two circular plates of radius a and variable distance d bathed in an incompressible oil of mass density ρ_0 and viscosity η. One shaft can exert a turning moment on the other through viscous forces, and in the limit of $d \to 0$ the moment becomes so large that two shafts are solidly connected. For simplicity we consider the only case where one shaft is rotating with constant angular velocity Ω while the other is fixed. Edge effects may be disregarded everywhere. For numeric calculations one may use $a = 10$ cm, $\rho_0 = 0.8$ g cm^{-3}, $\eta = 0.5$ Pa s, and $\Omega = 100$ rpm $= 100 \cdot 2\pi/60$ s^{-1}.

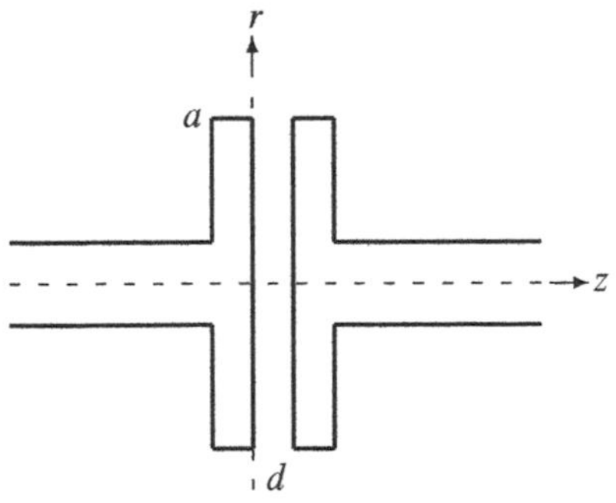

(a) Find the condition for creeping flow between the plates. What limitation does this condition set on d?
(b) Show that the fields (either in cylindrical or Cartesian coordinates)

$$\boldsymbol{v} = r\Omega\frac{z}{d}\boldsymbol{e}_\phi = (-y, x, 0)z\frac{\Omega}{d}, \qquad p = 0, \tag{17.77}$$

satisfy the Navier–Stokes equations for creeping flow in the gap (disregarding gravity) as well as the boundary conditions on the plates.
(c) Calculate the vorticity field and the stress tensor in the gap.
(d) Calculate the couple (turning moment of force) on the fixed shaft. How small must the distance between the plates be for the couple to be 100 Nm?
(e) Determine the dissipated power deposited in the oil.

17.11 Estimate the gap height for creeping horizontal flight in constant gravity when the angle of attack is α and the body has mass M.

17.12 Show that the fluid discharge rate Q (Equation (17.48)) is independent of x.

17.13 Find the conditions under which the velocity (17.50) has an extremum in the gap for a given x.

17.14 Define the gap average

$$\langle F \rangle = \frac{1}{L} \int_L F(x)\, dx,$$

and express all dynamic gap quantities in terms of averages.

17.15 Consider a nearly flat gap with $h(x) = d(1 + \chi(\xi))$, where $d = \langle h \rangle$ is the average height and $\xi = x/L$. **(a)** Using the results of Problem 17.14, calculate for $|\chi| \ll 1$ the leading non-trivial approximation to all the dynamic quantities. **(b)** Compare with the exact flat wing results.

* **17.16** Show that for $a > |b|$,

$$\frac{1}{2\pi} \int_0^{2\pi} \frac{d\phi}{a - b\cos\phi} = \frac{1}{\sqrt{a^2 - b^2}},$$

and use this to derive the integrals

$$\int_0^{2\pi} \frac{1}{1 - \gamma\cos\phi} \frac{d\phi}{2\pi} = \frac{1}{\sqrt{1 - \gamma^2}}, \qquad \int_0^{2\pi} \frac{1}{(1 - \gamma\cos\phi)^2} \frac{d\phi}{2\pi} = \frac{1}{(1 - \gamma^2)^{3/2}},$$

$$\int_0^{2\pi} \frac{1}{(1 - \gamma\cos\phi)^3} \frac{d\phi}{2\pi} = \frac{2 + \gamma^2}{2(1 - \gamma^2)^{5/2}}, \qquad \int_0^{2\pi} \frac{\cos\phi}{(1 - \gamma\cos\phi)^2} \frac{d\phi}{2\pi} = \frac{\gamma}{(1 - \gamma^2)^{3/2}},$$

$$\int_0^{2\pi} \frac{\cos\phi}{(1 - \gamma\cos\phi)^3} \frac{d\phi}{2\pi} = \frac{3\gamma}{2(1 - \gamma^2)^{5/2}}.$$

18 Rotating fluids

The conductor of a carousel knows about fictitious forces. Moving from horse to horse while collecting tickets, he not only has to fight the centrifugal force trying to kick him off, but also has to deal with the dizzying sideways Coriolis force. On a carousel of 4-m radius and turning once every 6 seconds, the centrifugal force is strongest at the rim where it amounts to 45% of gravity. Walking across the carousel at a normal speed of $1\ \mathrm{m\,s^{-1}}$, the conductor experiences a Coriolis force of about 21% of gravity. Provided the carousel turns anticlockwise seen from above, as most carousels seem to do, the Coriolis force always pulls the conductor off his course to the right. The conductor prefers to move from horse to horse against the rotation, and this is quite understandable, since the Coriolis force then counteracts the centrifugal force.

Gaspard Gustave Coriolis (1792 –1843). French mathematician. Worked on friction and hydraulics. Introduced the terms "work" and "kinetic energy" in the form used today. Defined what is now called the Coriolis force in 1835. (Source: Wikimedia Commons.)

Earth's rotation creates a centrifugal "antigravity" field, reducing gravity at the equator by 0.35%. This is hardly a worry, unless you have to adjust Olympic records for geographical latitude. The Coriolis force is even less noticeable; you have to move as fast as a jet aircraft for it to be comparable to the centrifugal force. Weather systems and sea currents are nevertheless so huge that even the weak Coriolis force has sufficient time to act and become a major player in their dynamics. The Coriolis force guarantees that low-pressure weather cyclones on the northern half of the globe always turn anticlockwise (seen from above), whereas high-pressure anticyclones always turn clockwise, and conversely south of the equator.

In this chapter we shall investigate the strange behavior of fluids in rotating reference frames, concentrating on the effects of the Coriolis force. At the end we shall debunk the persistent urban legend about the sense of rotation of ordinary bathtub vortices.

18.1 Fictitious forces

When dealing with a moving object, for example a rotating planet, it is often convenient to give up the inertial coordinate system and instead attach a fixed coordinate system to the moving object. In a non-inertial, accelerated coordinate system the laws of mechanics take a different form. The acceleration on the left-hand side of Newton's Second Law must be corrected by terms deriving from the motion of the coordinate system, and when these terms are shifted to the right-hand side they can be interpreted as forces. As these forces apparently have no objective cause, in contrast to forces caused by other bodies, they are called *fictitious*. A better name might be *inertial* forces, since there is nothing fictitious about the jerk you experience when the bus suddenly stops.

Velocity and acceleration in steady rotation

Consider a Cartesian coordinate system rotating with constant angular velocity Ω relative to an inertial system. Without loss of generality we align the z-axis with the axis of rotation in both systems. The coordinates $\boldsymbol{x} = (x, y, z)$ of a point particle in the rotating system are then related to its coordinates $\boldsymbol{x}' = (x', y', z')$ in the inertial system by a simple rotation through an angle Ωt (see the margin figure),

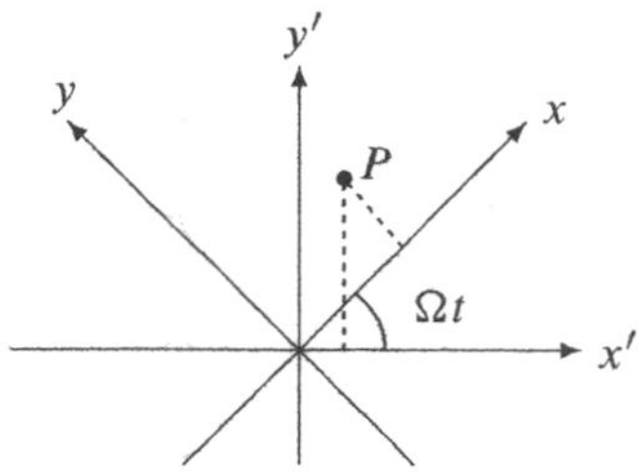

Transformation of coordinates of a point particle P from an inertial to a rotating coordinate system.

$$x = x' \cos \Omega t + y' \sin \Omega t, \tag{18.1a}$$
$$y = -x' \sin \Omega t + y' \cos \Omega t, \tag{18.1b}$$
$$z = z'. \tag{18.1c}$$

Differentiating once with respect to time t, we find the particle velocity $\boldsymbol{v} = d\boldsymbol{x}/dt$ in the rotating system expressed in terms of its velocity $\boldsymbol{v}' = d\boldsymbol{x}'/dt$ in the inertial system, plus extra terms arising from differentiation of the sines and cosines. Reexpressing these terms in the rotating coordinates $\boldsymbol{x}$, we get

$$v_x = v'_x \cos \Omega t + v'_y \sin \Omega t + \Omega y, \tag{18.2a}$$
$$v_y = -v'_x \sin \Omega t + v'_y \cos \Omega t - \Omega x, \tag{18.2b}$$
$$v_z = v'_z. \tag{18.2c}$$

Differentiating once more with respect to time, we obtain the particle acceleration $\boldsymbol{w} = d\boldsymbol{v}/dt = d^2\boldsymbol{x}/dt^2$ in the rotating system expressed in terms of its acceleration $\boldsymbol{w}' = d\boldsymbol{v}'/dt = d^2\boldsymbol{x}'/dt^2$ in the inertial system,

$$w_x = w'_x \cos \Omega t + w'_y \sin \Omega t + \Omega^2 x + 2\Omega v_y, \tag{18.3a}$$
$$w_y = -w'_x \sin \Omega t + w'_y \cos \Omega t + \Omega^2 y - 2\Omega v_x, \tag{18.3b}$$
$$w_z = w'_z. \tag{18.3c}$$

Finally, introducing the time-dependent rotation matrix (see Appendix B for more details)

$$\mathsf{A} = \begin{pmatrix} \cos \Omega t & \sin \Omega t & 0 \\ -\sin \Omega t & \cos \Omega t & 0 \\ 0 & 0 & 1 \end{pmatrix}, \tag{18.4}$$

these equations may be written in vector notation,

$$\boldsymbol{x} = \mathsf{A} \cdot \boldsymbol{x}', \tag{18.5a}$$
$$\boldsymbol{v} = \mathsf{A} \cdot \boldsymbol{v}' - \boldsymbol{\Omega} \times \boldsymbol{x}, \tag{18.5b}$$
$$\boldsymbol{w} = \mathsf{A} \cdot \boldsymbol{w}' - \boldsymbol{\Omega} \times (\boldsymbol{\Omega} \times \boldsymbol{x}) - 2\boldsymbol{\Omega} \times \boldsymbol{v}, \tag{18.5c}$$

where $\boldsymbol{\Omega} = \Omega\, \hat{\boldsymbol{e}}_z$ is the rotation vector.

Centrifugal and Coriolis forces

The force $\boldsymbol{f}'$ acting on a point particle in the inertial system depends in general on the position $\boldsymbol{x}'$ and velocity $\boldsymbol{v}' = d\boldsymbol{x}'/dt$ of the particle. Think, for example, of the force of gravity $\boldsymbol{f}' = -GmM\boldsymbol{x}'/|\boldsymbol{x}'|^3$ from a point particle of mass M. The force depends on $\boldsymbol{x}'$ (but not on $\boldsymbol{v}'$), and in the rotating system, the force becomes (like any other vector)

$$\boldsymbol{f} = \mathsf{A} \cdot \boldsymbol{f}'. \tag{18.6}$$

Since $|\boldsymbol{x}| = |\boldsymbol{x}'|$, the transformed gravitational force takes the same form in the rotating and inertial systems, namely $\boldsymbol{f} = -GmM\boldsymbol{x}/|\boldsymbol{x}|^3$. If the force, for example the Lorentz force in electrodynamics, depends on the velocity, it must be transformed according to (18.5b).

Multiplying (18.5c) by m, Newton's Second Law in the inertial system, $m\boldsymbol{w}' = \boldsymbol{f}'$, becomes in the rotating system,

$$m\boldsymbol{w} = \boldsymbol{f} - m\boldsymbol{\Omega} \times (\boldsymbol{\Omega} \times \boldsymbol{x}) - 2m\boldsymbol{\Omega} \times \boldsymbol{v}. \tag{18.7}$$

Although this equation has been derived for a special choice of coordinates, it must in this form be valid for any steadily rotating Cartesian coordinate systems with origin on the rotation axis, even if the rotation vector $\boldsymbol{\Omega}$ is not necessarily aligned with the z-axis.

The extra terms on the right-hand side are the usual fictitious forces: the *centrifugal force* $-m\boldsymbol{\Omega} \times (\boldsymbol{\Omega} \times \boldsymbol{x})$ and the *Coriolis force* $-2m\boldsymbol{\Omega} \times \boldsymbol{v}$. Both of these forces are proportional to the mass of the particle and resemble in this respect ordinary gravity. In a coordinate system that moves in a completely general way, there will appear two further fictitious forces (see Problem 18.5).

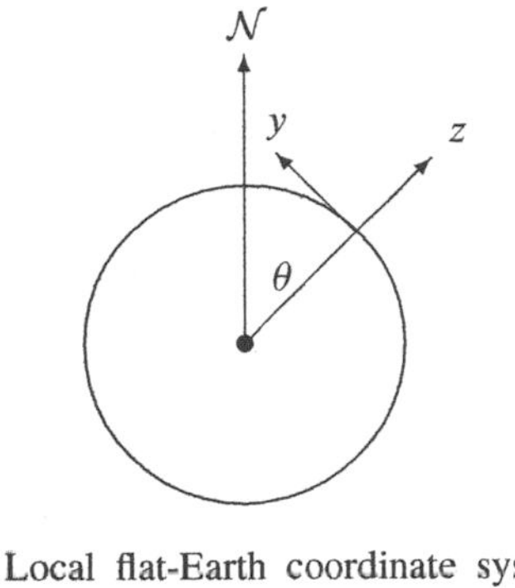

Local flat-Earth coordinate system and its relation to the polar angle.

Fictitious forces on Earth

The effect of the centrifugal force on Earth is primarily to flatten the spherical shape so that it conforms to an equipotential surface for the sum of the gravitational and centrifugal potentials (see Section 4.3 on page 60). This sum is usually called the *geopotential* and the equipotential surface the *geoid*. The gradient of the geopotential is everywhere orthogonal to the geoid and defines what is meant by the local *vertical*. Therefore, on the geoid there are no remaining horizontal components of the centrifugal force, only a reduction of the normal gravity by maximally 0.35%, an amount that is virtually negligible in all our daily doings.

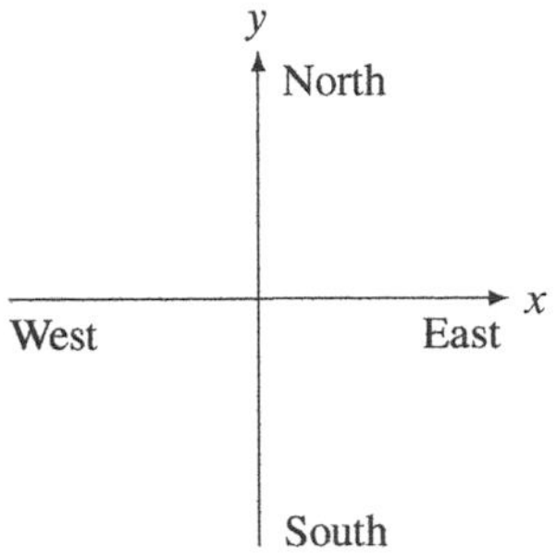

Flat-Earth local coordinate system at a point on the surface, seen from above. The z-axis is vertical and comes out of the paper.

The Coriolis force is different because it only acts on objects that move in the rotating coordinate system. Let us introduce a local flat-Earth coordinate system tangential to the surface at a given point with the x-axis toward the east and the y-axis toward the north. In this system, the Earth's rotation vector is $\boldsymbol{\Omega}_0 = \Omega_0(0, \sin\theta, \cos\theta)$, where θ is the polar angle and $\Omega_0 = 0.7292 \times 10^{-4}$ s $\approx 2\pi/24$ h the angular velocity in its rotation. The Coriolis acceleration, $\boldsymbol{g}^C = -2\boldsymbol{\Omega}_0 \times \boldsymbol{v}$, may then be written (disregarding the small corrections due to the slightly non-spherical shape of the geoid)

$$g_x^C = 2\Omega_0 \cos\theta\, v_y - 2\Omega_0 \sin\theta\, v_z, \tag{18.8a}$$

$$g_y^C = -2\Omega_0 \cos\theta\, v_x, \tag{18.8b}$$

$$g_z^C = 2\Omega_0 \sin\theta\, v_x. \tag{18.8c}$$

The vertical Coriolis acceleration g_z^C is always very small compared to local vertical gravity, g_0. Even for a jet aircraft flying on an east-west course at a speed approaching the velocity of sound, it is only of the same magnitude as the centrifugal force. So we may ignore the vertical Coriolis force along with the centrifugal force in normal ordinary earthly matters. The conclusion is that under normal steady-flow circumstances where the motion is nearly horizontal, $|v_z| \ll |v_y \cot\theta|$, the Coriolis acceleration is to a good approximation given by the leading terms in the two first equations above. In vector notation they may be written

$$\boldsymbol{g}^C \approx -2\boldsymbol{\Omega} \times \boldsymbol{v}, \tag{18.9}$$

where $\boldsymbol{\Omega} = \Omega_0 \cos\theta\, \hat{\boldsymbol{e}}_z$ is called the *local angular velocity*.

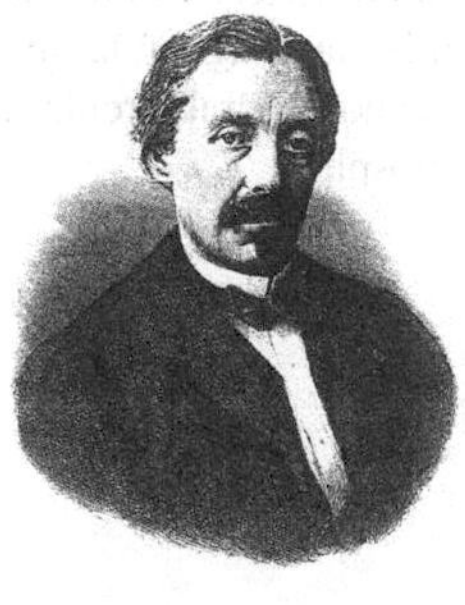

Jean-Bernard Léon Foucault (1819–1868). French physicist. Set up an enormous pendulum in the Paris Panthéon in 1851 to demonstrate Earth's rotation. Invented the gyroscope, improved the reflecting telescope, and measured the velocity of light in absolute units. (Source: Wikimedia Commons.)

Thus, for most practical purposes, the Coriolis force in a local region behaves as if the Earth were flat and rotated around the local vertical with angular velocity $\Omega = \Omega_0 \cos\theta$. The above formula shows that *on the northern hemisphere the direction of the Coriolis force is always to the right of the horizontal velocity* (and conversely at the southern where $\Omega < 0$). At middle latitudes $\theta \approx 45°$ the local angular velocity is $\Omega \approx 0.5 \times 10^{-4}\ \mathrm{s}^{-1}$, and a Foucault pendulum makes a full turn in $2\pi/\Omega \approx 34$ h, demonstrating to reasonable people and to the dismay of the Flat Earth Society that the ground indeed rotates under our feet.

18.2 Steady flow in a rotating system

In laboratory experiments with rotating vertical containers, for example Newton's bucket (page 59), the rotation vector is always vertical, $\mathbf{\Omega} = \Omega\, \hat{e}_z$, where Ω is the container's angular velocity. The total acceleration field is the sum of the gravitational and centrifugal fields, $g = (\Omega^2 x, \Omega^2 y, -g_0)$, and has both vertical and horizontal components that together are responsible for the curved equilibrium shape of the open surface in the rotating bucket. On the rotating Earth, however, the centrifugal force has permanently modified its figure and thereby absorbed the horizontal components of gravity, so that the total acceleration field close to the surface is always orthogonal to the geoid, that is, $g = -g_0 \hat{e}_z$ with g_0 depending slightly on the latitude. The analysis at the end of the preceding section shows that for steady and nearly horizontal flow, the local rotation vector may be also taken to be vertical, $\mathbf{\Omega} = \Omega\, \hat{e}_z$, where $\Omega = \Omega_0 \cos\theta$ is the local angular velocity. We shall not carry through a complete analysis of steady flow in spherical coordinates because it takes quite a bit of work to handle the approximations correctly [Pedlosky 1987].

In laboratory experiments as well as nature's large-scale experiments with air and water, the Reynolds number of the flows will typically be very high, so that we may disregard viscosity well away from solid boundaries. Adding the Coriolis acceleration to the right-hand side of the general Euler equation (14.1), we obtain the equations for nearly ideal steady flow in a rotating reference frame,

$$(\boldsymbol{v} \cdot \nabla)\boldsymbol{v} = \boldsymbol{g} - 2\mathbf{\Omega} \times \boldsymbol{v} - \frac{\nabla p}{\rho}, \qquad \nabla \cdot (\rho \boldsymbol{v}) = 0. \tag{18.10}$$

Here $\boldsymbol{g}$, as discussed above, includes the centrifugal force explicitly in laboratory experiments whereas it can be ignored for flow near the surface of the Earth. The modification due viscosity will be analyzed in the following section.

The Rossby number

Carl-Gustav Arvid Rossby (1898 –1957). Swedish born meteorologist who mostly worked in the United States. Contributed to the understanding of large-scale motion and general circulation of the atmosphere.

Let us as before characterize the flow by a length scale L and a velocity scale U. The ratio of the scale of advective acceleration and Coriolis acceleration is called the *Rossby number*,

$$\mathsf{Ro} = \frac{|(\boldsymbol{v} \cdot \nabla)\boldsymbol{v}|}{|2\mathbf{\Omega} \times \boldsymbol{v}|} \approx \frac{U^2/L}{2\Omega U} = \frac{U}{2\Omega L}. \tag{18.11}$$

The Coriolis force is insignificant if $\mathsf{Ro} \gg 1$, that is, when the flow velocity U is much greater than the variation ΩL in the rotation velocity across the system. When, on the other hand, the flow velocities are much smaller than the variation in rotation velocity, $\mathsf{Ro} \ll 1$, the Coriolis force exerts a dominant influence on the flow.

Ocean currents and weather cyclones are relatively steady phenomena, covering typical distances of 1,000 km. The characteristic speeds are meters per second for the ocean currents and tens of meters per second for the winds. With a local angular velocity $\Omega \approx 0.5 \times 10^{-4}\ \mathrm{s}^{-1}$, the Rossby number becomes $\mathsf{Ro} \approx 0.01$ for ocean currents and $\mathsf{Ro} \approx 0.1$ for weather cyclones. Both these phenomena are thus dominated by the Coriolis force, the winds the least because of their higher velocities.

When a human of size $L \approx 1$ m swims with a speed of $U \approx 1\ \mathrm{m\,s^{-1}}$, the Rossby number becomes $\mathsf{Ro} \approx 10^4$, and the Coriolis force can be completely neglected. The water draining out of your bathtub moves with similar speeds over similar distances, making the Rossby number just as large and the Coriolis force just as insignificant as for swimming (see Sections 18.4 and 18.5 for a discussion of this point). However, in space stations designed for long-term habitation, gravity must for health reasons be simulated by rotation, and the large Coriolis force will play havoc with ballistic games like Ping-Pong, football, and basketball (see Problem 18.1).

Geostrophic flow

In large-scale natural systems, the Rossby number is so small and the Reynolds number so large that one can ignore both the viscous and the advective terms. Such a flow, completely dominated by the Coriolis force, is said to be *geostrophic*. Dropping the advective acceleration on the left-hand side of (18.10), we arrive at the following remarkably simple equation of *geostrophic balance*,

$$\boxed{\frac{\nabla p}{\rho} = \boldsymbol{g} - 2\boldsymbol{\Omega} \times \boldsymbol{v}.} \tag{18.12}$$

For $\boldsymbol{v} = \boldsymbol{0}$ we recover the equation of local hydrostatic balance, $\nabla p = \rho \boldsymbol{g}$, as we should.

Since $\boldsymbol{\Omega} = \Omega\, \hat{\boldsymbol{e}}_z$ the above equation does not determine v_z at all. Taking $v_z = 0$ and using that $\boldsymbol{\Omega} \times (\boldsymbol{\Omega} \times \boldsymbol{v}) = -\Omega^2 \boldsymbol{v}$, we find the horizontal *geostrophic wind*,

$$\boldsymbol{v} = \frac{\boldsymbol{\Omega} \times (\nabla p - \rho \boldsymbol{g})}{2\Omega^2 \rho}. \tag{18.13}$$

At the Earth's surface where the combination of gravity and centrifugal forces results in vertical acceleration, $\boldsymbol{g} = -g_0 \hat{\boldsymbol{e}}_z$, the $\rho \boldsymbol{g}$-term yields no contribution. This shows that on the Earth the geostrophic wind is always orthogonal to the horizontal pressure gradient. In other words, *streamlines follow isobars in horizontal geostrophic flow*.

Were it not for the Coriolis force, air masses would simply stream along the negative pressure gradient from high pressure toward low pressure regions. On the northern hemisphere the Coriolis force generates cyclones by forcing the air toward the right when it begins to stream toward the low pressure, until in the end its velocity becomes aligned with an isobar. Funnily enough, exactly the same mechanism creates anticyclones when air streams away from a high pressure region. The simple weather-map rule that streamlines follow isobars can, however, not be exact because no air would be transported from high to low pressure regions by geostrophic flow. As we shall see in the following section, there is a viscous boundary layer, called the Ekman layer, near the ground in which the wind makes an acute angle with the isobars, and this mechanism provides the necessary inflow toward the depression.

Water level in an open canal

The Coriolis force may tilt the otherwise horizontal surface of a moving liquid. Suppose a constant water current flows with velocity U through a canal of width d. In a flat-Earth coordinate system with the x-axis along the middle of the canal and vertical z-axis, we get from the geostrophic equation (18.12) with $\boldsymbol{g} = -g_0 \hat{\boldsymbol{e}}_z$,

$$\frac{\partial p}{\partial x} = 0 \qquad \frac{\partial p}{\partial y} = -2\Omega \rho U, \qquad \frac{\partial p}{\partial z} = -\rho g_0. \tag{18.14}$$

Assuming constant density $\rho = \rho_0$, the solution is evidently

$$p = -\rho_0 g_0 z - 2\Omega \rho_0 U y. \tag{18.15}$$

At the open surface, $z = h(y)$, the pressure must be constant, and solving for $h(y)$ with the boundary condition $h(0) = 0$, the water level becomes

$$h = -\frac{2\Omega U}{g_0} y. \tag{18.16}$$

That the water level is highest on the right bank of the canal ($y = -d/2$) agrees with the general rule that the Coriolis force wants to turn the water toward the right on the northern hemisphere.

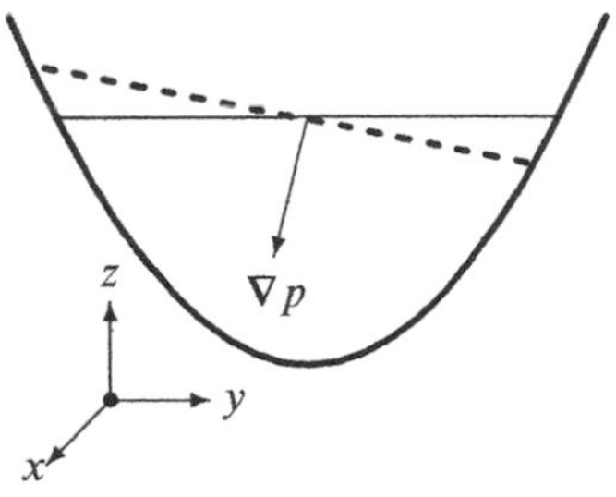

The water level in a horizontal canal is tilted by the Coriolis force, such that the surface pressure is constant. Here the angular velocity is positive around z and water flows out of the picture in the positive x-direction.

Example 18.1 [The Danish Great Belt]: The Danish Great Belt is a 20-km wide strait running roughly north-south at a latitude of 56° north. The local angular velocity is $\Omega = 0.6 \times 10^{-4}\ \mathrm{s}^{-1}$. A current flowing toward the north at $1\ \mathrm{m\,s}^{-1}$ raises the water level on the eastern bank by about $2\Omega U d/g_0 \approx 25$ cm relative to the western bank. Although $\mathsf{Ro} = U/2\Omega d \approx 0.4$, the use of the geostrophic equation (18.12) is nevertheless justified because the advective term vanishes for flow along the x-direction when the velocity is independent of x.

The Taylor–Proudman theorem

Geoffrey Ingram Taylor (1886–1975). British physicist, mathematician and engineer. Had great impact on all aspects of 20th century fluid mechanics from aircraft to explosions. Devised a method to determine the bulk viscosity of compressible fluids. Studied the movements of unicellular marine creatures.

Joseph Proudman (1888–1975). British mathematician and oceanographer. First proved the Taylor–Proudman theorem in 1915.

The Taylor–Proudman theorem concerns the *essential two-dimensionality of geostrophic flow*. Using that $\boldsymbol{g} = -\boldsymbol{\nabla}\Phi$ and that the curl of a gradient vanishes, it follows from the geostrophic equation that

$$2\boldsymbol{\nabla} \times (\boldsymbol{\Omega} \times \boldsymbol{v}) = -\boldsymbol{\nabla} \times \left(\frac{\boldsymbol{\nabla} p}{\rho}\right) = \frac{\boldsymbol{\nabla}\rho \times \boldsymbol{\nabla} p}{\rho^2}.$$

But we also have

$$\boldsymbol{\nabla} \times (\boldsymbol{\Omega} \times \boldsymbol{v}) = \boldsymbol{\Omega}\boldsymbol{\nabla} \cdot \boldsymbol{v} - \boldsymbol{\Omega} \cdot \boldsymbol{\nabla}\boldsymbol{v} = \Omega\left(\boldsymbol{\nabla} \cdot \boldsymbol{v}\hat{\boldsymbol{e}}_z - \frac{\partial \boldsymbol{v}}{\partial z}\right), \tag{18.17}$$

and solving the continuity equation, $\boldsymbol{\nabla} \cdot (\rho\boldsymbol{v}) = \rho\boldsymbol{\nabla} \cdot \boldsymbol{v} + \boldsymbol{v} \cdot \boldsymbol{\nabla}\rho = 0$ for the divergence, we arrive at

$$\boxed{\frac{\partial \boldsymbol{v}}{\partial z} = -\frac{\boldsymbol{\nabla}\rho \times \boldsymbol{\nabla} p}{2\Omega\rho^2} - \hat{\boldsymbol{e}}_z \frac{(\boldsymbol{v} \cdot \boldsymbol{\nabla})\rho}{\rho}.} \tag{18.18}$$

This clearly shows that for constant density $\rho = \rho_0$, that is, for an incompressible fluid, both terms on the right-hand side vanish, so that the velocity field must be independent of z and thus two-dimensional. For compressible fluids obeying a barotropic relationship, $\rho = \rho(p)$, we have $\boldsymbol{\nabla}\rho = (d\rho/dp)\boldsymbol{\nabla} p$ so that $\boldsymbol{\nabla}\rho \times \boldsymbol{\nabla} p = \boldsymbol{0}$, and since the second term only contributes to the vertical velocity, the horizontal velocity field will be independent of z and therefore also two-dimensional. The theorem fails if the surfaces of constant density and pressure do not coincide, a fluid state called *baroclinic flow* that may arise when heat conduction is neither absent (isentropic flow) nor infinitely fast (isothermal flow).

Taylor columns

The Taylor–Proudman theorem is a strange result, which predicts that if one disturbs the flow of a rotating fluid at, say, $z = 0$, then the pattern of the disturbance will, after all time-dependence has died away, have been copied to all other values of z. A so-called *Taylor column* of disturbed flow is thus created in the rotating fluid (Taylor columns are sometimes also called *Proudman pillars*). Even a body moving steadily along the axis of rotation will push a long column of fluid in front of itself, and trail another behind. The mechanics underlying the formation of Taylor columns derives from a complicated interplay between the advective and viscous terms left out in the geostrophic equation, and it is not easy to give an explanation in simple physical terms [Batchelor 67].

In spite of the strangeness, many laboratory experiments beginning with Taylor's own in 1923 have verified the existence of Taylor columns. Natural Taylor columns in the atmosphere have not been observed, but there are papers claiming that a Taylor column may have been observed in the Chukchi sea [MD97], and that Jupiter's Great Red Spot could be a Taylor column [TDH75].

18.3 The Ekman layer

Boundary layers arise around a body in nearly ideal flow, because viscous forces must necessarily come into play to secure the no-slip boundary condition. Steady boundary layers are normally asymmetric with respect to the direction of the "slip-flow" outside the layer, being thinnest at the leading edge of a body and thickening toward the rear. This happens even at an otherwise featureless body surface and may be understood as a cumulative effect of the slowing down of the fluid by the contact with the boundary. The further downstream from the leading edge of a body, the longer time will the fluid have been under the influence of shear forces from the boundary, and the thicker the boundary layer will be. In Chapter 28 we will analyze boundary layers in general, but in this section we shall solve the much simpler case of the Ekman boundary layer in a rotating system.

Vagn Walfrid Ekman (1874–1954). Swedish physical oceanographer. Contributed to the understanding of the dynamics of ocean currents. Described what is now called the Ekman layer in 1905. (Source: Wikimedia Commons.)

A fluid in geostrophic flow with small Rossby number is dominated by the Coriolis force, which also plays a major role in the formation of boundary layers. As the flow velocity rises from zero at the boundary to its asymptotic value in the geostrophic slip-flow outside the boundary, the Coriolis force becomes progressively stronger, making the flow veer more and more to the right (for anticlockwise rotation). This geostrophic crosswind effectively "blows away" accumulated fluid and prevents the downstream growth of the boundary layer. Such a boundary layer, confined to a finite thickness by the Coriolis force, is called an *Ekman layer*.

Thickness of the Ekman layer

Assuming for simplicity that the fluid is incompressible with constant density, $\rho = \rho_0$, the geostrophic equation (18.12) becomes, after addition of the usual viscous term,

$$\boxed{\frac{\nabla p}{\rho_0} = \boldsymbol{g} - 2\boldsymbol{\Omega} \times \boldsymbol{v} + \nu\nabla^2\boldsymbol{v}.} \tag{18.19}$$

Well away from solid boundaries the ratio of the viscous to Coriolis forces becomes

$$\mathsf{Ek} \equiv \frac{|\nu\nabla^2\boldsymbol{v}|}{|2\boldsymbol{\Omega} \times \boldsymbol{v}|} \approx \frac{\nu U/L^2}{2\Omega U} = \frac{\nu}{2\Omega L^2} = \frac{\mathsf{Ro}}{\mathsf{Re}}, \tag{18.20}$$

where $\mathsf{Ro} = U/2\Omega L$ is the Rossby number and $\mathsf{Re} = UL/\nu$ is the Reynolds number. Evidently, for geostrophic flow with $\mathsf{Ro} \ll 1$ and $\mathsf{Re} \gg 1$, this *Ekman number* is so small that viscosity is completely irrelevant.

Approaching a solid boundary the Laplacian will, however, grow inversely with the square of the distance δ from the boundary, $\nabla^2 \sim 1/\delta^2$, because it is the sum of double derivatives in all directions, and the largest double derivative is found toward the nearest boundary. The Ekman number near the boundary thus becomes $\mathsf{Ek} \approx \nu/2\Omega\delta^2$ and is of order unity for

$$\delta \approx \sqrt{\frac{\nu}{\Omega}}. \tag{18.21}$$

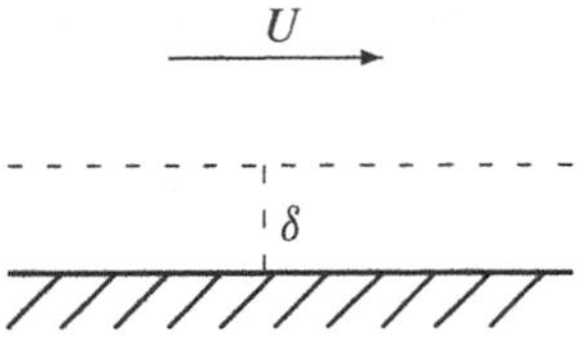

An Ekman layer of thickness δ interpolates between the geostrophic flow with velocity U and the zero velocity on the solid ground, demanded by the no-slip condition.

This is, as we shall see, a very good estimate of the thickness of the Ekman layer.

The presence of an Ekman layer of the right thickness has been amply confirmed by laboratory experiments. For the atmosphere at middle latitudes the thickness becomes $\delta =$ 55 cm when the diffusive viscosity $\nu = 1.54 \times 10^{-5}$ m^2 s^{-1} is used. This disagrees strongly with the measured thickness of the Ekman layer in the atmosphere, which is more like a kilometer. The reason is that atmospheric flow tends to be turbulent rather than laminar with an effective viscosity that can be up to a million times larger than the diffusive viscosity. We shall not go further into this question here; see however [Pedlosky 1987].

Structure of the Ekman layer

Above the thin Ekman layer we assume that there is a thicker horizontal layer of steady geostrophic flow in the x-direction with velocity $\boldsymbol{v} = U\hat{\boldsymbol{e}}_x$. The pressure in the geostrophic layer must be the same as we found for canal flow (page 313),

$$p = -\rho_0 g_0 z - 2\Omega\rho_0 U y. \qquad (18.22)$$

In looking for a solution to Equation (18.19), which interpolates between the static boundary and the geostrophic flow, we exploit the symmetry of the problem. As long as the centrifugal acceleration can be ignored (which it can at the surface of the Earth), the equations of motion as well as the boundary conditions are independent of the exact position in x and y. It is then natural to guess that there may be a maximally symmetric solution $\boldsymbol{v} = \boldsymbol{v}(z)$ that is also independent of x and y and only depends on the height z.

Mass conservation, $\nabla \cdot \boldsymbol{v} = \nabla_z v_z = 0$, tells us that the vertical velocity component is a constant. Since it has to vanish on the non-permeable boundary $z = 0$, it must vanish everywhere, $v_z = 0$. The flow in the transition layer is horizontal and independent of x and y, but its direction and magnitude varies with height z. From the z-component of (18.19) we find $\partial p/\partial z = -\rho_0 g_0$, which shows that $p = -\rho_0 g_0 z + p_0(x, y)$ for all z. Since this expression has to agree with the geostrophic pressure above the Ekman layer, it follows that the pressure is also given by (18.22) inside the Ekman layer. *The geostrophic pressure penetrates the Ekman layer all the way to the ground.*

Inserting the geostrophic pressure into (18.19), the horizontal velocity components must obey the following two coupled second-order linear differential equations:

$$\nu\nabla_z^2 v_x = -2\Omega v_y \qquad \nu\nabla_z^2 v_y = -2\Omega(U - v_x). \qquad (18.23)$$

Combining them we obtain a single fourth-order equation for $U - v_x$:

$$\nabla_z^4(U - v_x) = -\frac{4\Omega^2}{\nu^2}(U - v_x). \qquad (18.24)$$

The general solution to such a linear fourth-order differential equation is a linear combination of four terms of the form e^{kz}, where $k^4 = -4\Omega^2/\nu^2$. Setting $\delta = \sqrt{\nu/\Omega}$, the roots of $k^4 = -4/\delta^4$ are $k = \pm(1 \pm i)/\delta$. The roots with positive real part are of no use because the solution e^{kz} would grow exponentially for $z \to \infty$. The most general acceptable solution is then of the form

$$U - v_x = Ae^{-(1+i)z/\delta} + Be^{-(1-i)z/\delta}, \qquad (18.25)$$

$$v_y = i\left(Ae^{-(1+i)z/\delta} - Be^{-(1-i)z/\delta}\right), \qquad (18.26)$$

where A and B are integration constants and where the last expression is obtained from the first equation in (18.23). Applying finally the no-slip boundary condition, $v_x = v_y = 0$ for $z = 0$, we find $A = B = U/2$, and the solution becomes (see the margin figure)

$$\boxed{v_x = U\left(1 - e^{-z/\delta}\cos z/\delta\right), \qquad v_y = Ue^{-z/\delta}\sin z/\delta.} \qquad (18.27)$$

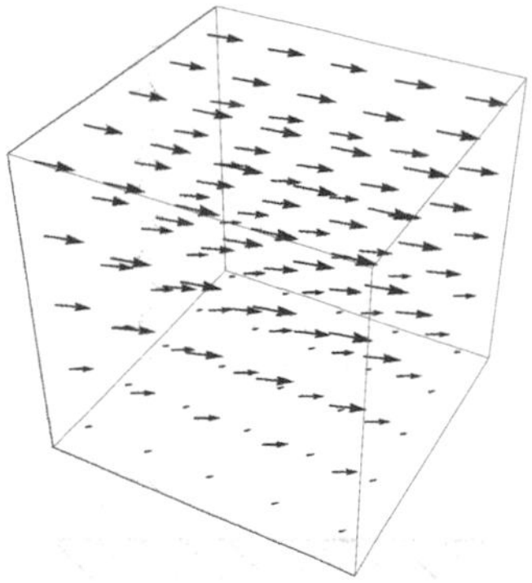

Ekman flow pattern in a vertical box on a planar surface with anticlockwise rotation. Note how the fluid close to the ground flows to the left of the geostrophic flow higher up by 45°. This is in accordance with the general rule that for anticlockwise rotation, the flow is forced toward the right by an amount that grows with the magnitude of the velocity.

Evidently $\delta = \sqrt{\nu/\Omega}$ is a natural measure of the thickness of the Ekman layer.

In Figure 18.1L the velocity components are plotted as a function of scaled height z/δ in units of the asymptotic flow U. One notes that v_x first overshoots its asymptotic value, and then quickly returns to it. The y-component also oscillates but is 90° out of phase with the x-component. The direction of the velocity close to $z = 0$ is 45° to the left of the asymptotic geostrophic flow. Plotted parametrically as a function of height, the velocity components create a characteristic spiral, called the *Ekman spiral*, shown in Figure 18.1R. The damping is however so strong that only the very first turn in this spiral is visible.

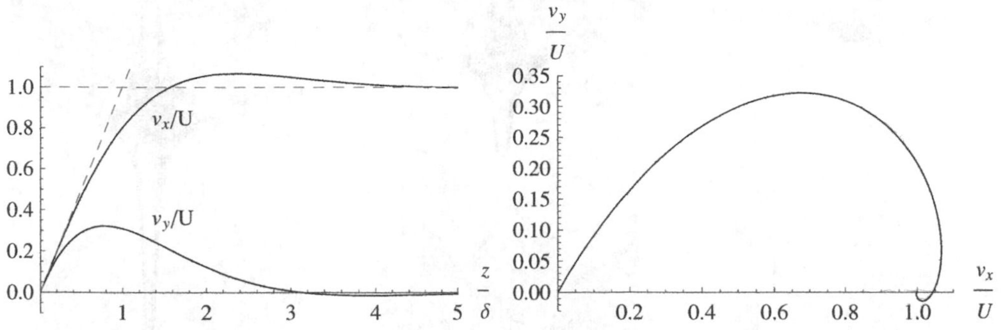

Figure 18.1. Plot of the Ekman layer field. **Left:** The velocity components as a function of height z. **Right:** Parametric plot of the velocities as a function of z leads to the characteristic Ekman spiral.

* Upwelling and suction

If the geostrophic flow does not run along the x-direction, but has the components $\boldsymbol{U} = (U_x, U_y, 0)$, the Ekman flow (18.27) becomes instead (as may easily be verified),

$$\boldsymbol{v} = \boldsymbol{U}\left(1 - e^{-z/\delta}\cos z/\delta\right) + \hat{\boldsymbol{e}}_z \times \boldsymbol{U}e^{-z/\delta}\sin z/\delta. \tag{18.28}$$

If furthermore the geostrophic velocity components change slowly with x and y on a large scale $L \gg \delta$, this expression should still be valid, because the thickness (18.21) is independent of the velocity of the geostrophic flow.

The presence of an Ekman layer underneath a slowly varying geostrophic flow generates in fact a non-vanishing asymptotic flow in the z-direction. To find it we calculate $\nabla_z v_z$ using mass conservation $\boldsymbol{\nabla}\cdot\boldsymbol{v} = \nabla_x v_x + \nabla_y v_y + \nabla_z v_z = 0$, and we find from (18.28)

$$\nabla_z v_z = -\nabla_x v_x - \nabla_y v_y = (\boldsymbol{\nabla}\times\boldsymbol{U})_z e^{-z/\delta}\sin z/\delta.$$

Here we have used that the geostrophic flow by itself has to satisfy the divergence condition, $\boldsymbol{\nabla}\cdot\boldsymbol{U} = \nabla_x U_x + \nabla_y U_y = 0$. The factor in parenthesis, $(\boldsymbol{\nabla}\times\boldsymbol{U})_z = \nabla_x U_y - \nabla_y U_x$, is the geostrophic vorticity in the z-direction. Integrating the above equation with respect to z and using the boundary condition $v_z = 0$ for $z = 0$, we obtain

$$v_z = \tfrac{1}{2}\delta\,(\boldsymbol{\nabla}\times\boldsymbol{U})_z\left(1 - e^{-z/\delta}(\cos z/\delta + \sin z/\delta)\right). \tag{18.29}$$

This equation is most easily verified by differentiating it with respect to z. Evidently, the vertical velocity is of size $U\delta/L$, which is always smaller than the geostrophic flow U by a factor $\delta/L \ll 1$. This confirms the soundness of the procedure used to obtain this result.

For $z \gg \delta$ the exponential falls away, and there remains a vertical component in the asymptotic geostrophic flow,

$$\boxed{U_z = \tfrac{1}{2}\delta\,(\boldsymbol{\nabla}\times\boldsymbol{U})_z.} \tag{18.30}$$

Since it is independent of z, it is neither at variance with the geostrophic nature of the exterior flow nor with the Taylor–Proudman "vertical copy" theorem.

If the geostrophic vorticity $(\boldsymbol{\nabla}\times\boldsymbol{U})_z$ is positive, that is, of the same sign as the global rotation, fluid wells up from the Ekman layer (without changing its thickness). This is, for example, the case for a low-pressure cyclone, where the cross-isobaric flow inside the Ekman layer toward the center of the cyclone is accompanied by upwelling of fluid. Conversely, if the geostrophic vorticity is negative, as in high-pressure anticyclones, fluid is sucked down into the Ekman layer from the geostrophic flow. Both of these effects tend to equalize the pressure between the center and the surroundings of these vast vortices.

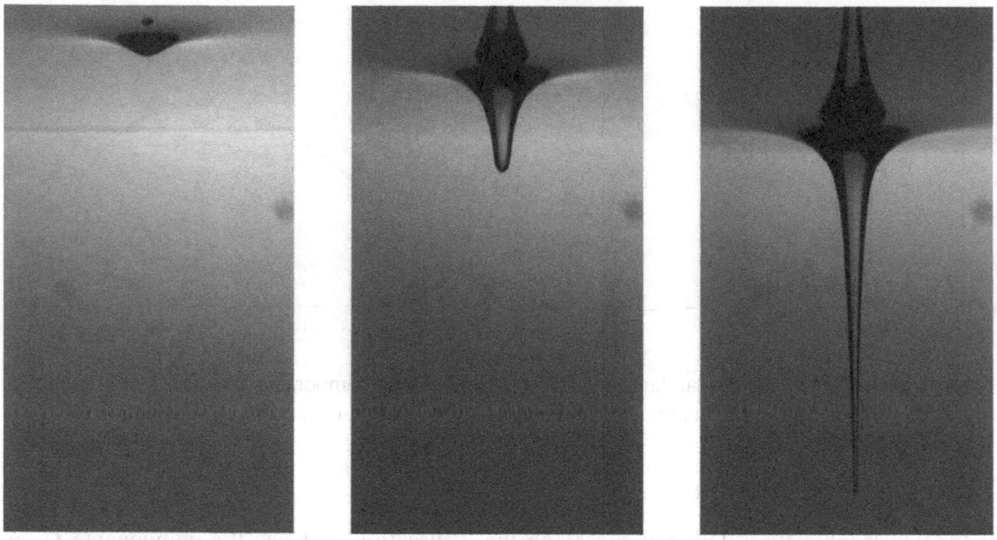

Figure 18.2. Water vortex in a container rotating at 6, 12, and 18 rpm. The upper part is the reflection of the vortex in the water surface. The radius of the drain-hole is $a = 1$ mm, the asymptotic water level is $L = 11$ cm, and the container radius $A = 20$ cm. At 18 rpm the central dip is 6 cm, and the volume discharge through the drain is $Q = 3.16\ \text{cm}^3\ \text{s}^{-1}$, corresponding to an average drain velocity of $101\ \text{cm s}^{-1}$. The circulation constant is measured to be $C = 16.0\ \text{cm}^2\ \text{s}^{-1}$ and the inner core rotates about 150 turns per second! (Source: Courtesy Anders Andersen, Technical University of Denmark.)

* 18.4 Application: Steady bathtub vortex

In the laboratory, gravity-sustained vortices may be created by letting a liquid, typically water, run freely out through a tiny drain-hole in the center of a slowly rotating cylindrical container. The liquid lost through the drain is constantly pumped back into the container. In the steady state the pump provides the kinetic energy of the liquid falling out through the drain, and its angular momentum is provided by the motor rotating the container. We shall for simplicity disregard the influence of the outer container wall.

In the following we shall repeatedly refer to the experiment [ABS&03, ABS&06] shown in Figure 18.2R with the parameters given in the figure caption. In this experiment, a steady flow pattern with a beautiful central vortex is established after about half an hour. The vortex is remarkably stable and its flow can be studied experimentally by modern imaging techniques. The needle-like central depression is accompanied by a very rapid central rotation of more than 150 turns per second, or nearly 10,000 rpm, which is as fast as a Formula-One car engine at full throttle!

Rossby radius

Empirically, the bulk of the vortex outside the surface depression is found to take the shape of a *line vortex* with azimuthal velocity, $v_\phi = C/r$ (in the rotating coordinate system of the container). The azimuthal Reynolds number of the line vortex is independent of r,

$$\mathsf{Re}_\phi = \frac{r v_\phi}{\nu} = \frac{C}{\nu}. \tag{18.31}$$

In the experiment it has the value $\mathsf{Re}_\phi \approx 1{,}600$, which is well below the onset of turbulence.

The local Rossby number of the line vortex at a distance r from its axis is

$$\text{Ro} = \frac{v_\phi}{2\Omega\, r} = \frac{C}{2\Omega\, r^2}. \tag{18.32}$$

It decreases rapidly with growing r and drops below unity for $r \gtrsim R$, where

$$R = \sqrt{\frac{C}{2\Omega}}. \tag{18.33}$$

We shall call this the *Rossby radius*, and in the experiment of Figure 18.2R we find $R = 2.1$ cm. Well inside the Rossby radius for $r \ll R$, the Coriolis acceleration is small compared to the advective acceleration, and the vortex resembles the bathtub-like vortices to be discussed in Chapter 27, except that there is marked upflow from the Ekman layer at the bottom outside the drain. At the other extreme, well beyond the Rossby radius for $r \gg R$, the flow will be purely geostrophic.

Cylindrical geostrophic flow

The Taylor–Proudman "copycat" theorem guarantees that in the geostrophic regime the flow cannot depend on the vertical height z well away from boundaries. Assuming further that the flow is rotationally invariant, it follows that the velocity components in cylindrical coordinates, v_r, v_ϕ, and v_z, are only functions of r. The geostrophic equation (18.12) now decomposes into the three equations,

$$\frac{\partial \Phi^*}{\partial r} = 2\Omega v_\phi, \qquad \frac{\partial \Phi^*}{\partial \phi} = -2\Omega v_r, \qquad \frac{\partial \Phi^*}{\partial z} = 0, \tag{18.34}$$

where

$$\Phi^* = g_0 z - \frac{1}{2}\Omega^2 r^2 + \frac{p}{\rho_0} \tag{18.35}$$

is the *effective potential* (see (2.36) on page 32), including the gravitational and centrifugal contributions. By the usual argument, the uniqueness of the pressure forbids any ϕ-dependence, so the second equation leads to $v_r = 0$. The effective potential is then be obtained by integrating the two other equations with the result

$$\Phi^* = 2\Omega \int v_\phi(r)\, dr. \tag{18.36}$$

Note that whereas the radial flow always has to vanish, we find apparently no restrictions on the azimuthal flow v_ϕ or the upflow v_z. In particular, it does not follow from this argument that the azimuthal flow must take the form of a line vortex, $v_\phi = C/r$.

The main conclusion is that *geostrophic flow can never carry any inflow* toward the drain. This is, in fact, equivalent to the previously derived result that flow lines and isobars coincide in geostrophic flow. The only way inflow can occur is through deviations from clean geostrophic flow, and that happens primarily in the Ekman layer close to the bottom of the container, and especially inside the Rossby radius.

The asymptotic upflow from the Ekman layer (18.30) is controlled by the vorticity of the geostrophic flow. Since the Taylor–Proudman theorem guarantees that the upflow is independent of z, and since there can be no geostrophic inflow, this upflow will, unabated, reach the nearly horizontal open surface at top. But there it has nowhere to go, so the only possibility that remains is for the upflow v_z to vanish. By (18.30) the vorticity of the geostrophic flow must also vanish, and this is only possible for $v_\phi = C/r$ (see Problem 16.20). To complete the argument, one must verify that the Ekman layer also forming at the upper open surface of the rotating water is incapable of diverting any large flow toward the drain (which is true). In conclusion, *it is essentially the finite vertical extent of the rotating fluid that forces the geostrophic flow to be that of a line vortex!*

The Ekman layer valve

The constant thickness of the Ekman layer $\delta = \sqrt{\nu/\Omega}$ becomes merely 0.7 mm for the third vortex in Figure 18.2. The Ekman fields may be directly taken over from (18.27) with the asymptotic azimuthal velocity $U \to C/r$,

$$v_\phi = \frac{C}{r}\left(1 - e^{-z/\delta}\cos z/\delta\right), \qquad v_r = -\frac{C}{r}e^{-z/\delta}\sin z/\delta. \tag{18.37}$$

The total radial inflow rate can now be calculated,

$$Q = \int_0^\infty (-v_r)2\pi r\, dz = \pi C\delta. \tag{18.38}$$

This is a fundamental result that connects the primary circulating flow with the secondary inflow toward the drain. The Ekman layer in effect acts as a valve, only allowing a certain amount of fluid to flow toward the drain per unit of time. In the third vortex of Figure 18.2 the circulation constant was measured to be $C \approx 16.0\ \mathrm{cm^2\,s^{-1}}$, leading to a total predicted inflow of $Q \approx 3.66\ \mathrm{cm^3\,s^{-1}}$. Since the whole inflow has to go down the drain, this is in fair agreement with the measured drain flow of $Q = 3.16\ \mathrm{cm^3\,s^{-1}}$, especially in view of the difficulties in actually determining the circulation constant. The discrepancy could in fact be caused by the finite radial size of the container, which has been ignored here.

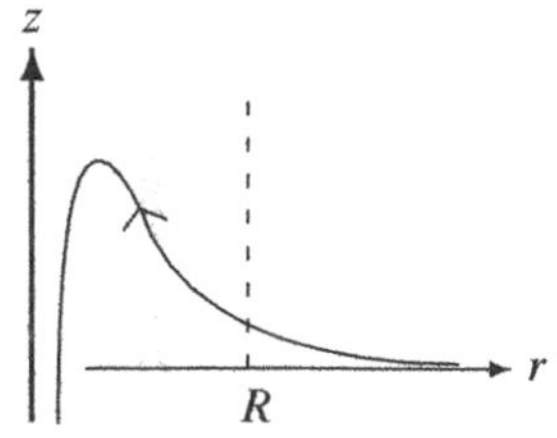

Typical flowline for fluid streaming in through the Ekman layer outside the Rossby radius R, welling up inside R, and finally falling through the drain.

Inner vortex

Inside the Rossby radius the non-linear advective forces take over, but there will still exist a thin—in fact thinner—Ekman-like layer close to the bottom (this layer has been analyzed in [ALB03]). But even if the circulating primary flow is that of a line vortex, the non-linearities will now cause an upwelling of fluid from the bottom layer. The bulk flow is no more geostrophic, so there is no injunction against the upflow turning into an inflow directed toward the center of the vortex. Thus, the general picture is that the secondary flow creeps inward along the bottom through the Ekman layer outside the Rossby radius, flares up from the bottom inside the Rossby radius and turns toward the center for finally to dive sharply down into the drain.

18.5 Debunking an urban legend

Let us now turn to the persistent "urban legend" concerning the direction of rotation of real bathtub vortices and their dependence on the Earth's rotation. The legend originates in the correct physical theory of the Coriolis force, amply confirmed by the everyday observation of weather cyclones. So the urban legend can only be debunked by quantitative arguments, usually not given much attention in urban circles.

Suppose to begin with that our bathtub is essentially infinitely large and that the water level is $L = 50$ cm. Bernoulli's theorem tells us that the drain velocity is at most $W = \sqrt{2g_0L} \approx 300\ \mathrm{cm\,s^{-1}}$. Taking the drain radius to be $a = 2.5$ cm, the maximal drain discharge rate becomes $Q \approx 6\ \mathrm{L\,s^{-1}}$. This seems not unreasonable for bathtubs that typically contain a couple of hundred liters of water. Assuming furthermore that the bulk of the flow is perfectly laminar, we find the Ekman thickness $\delta \approx 14$ cm due to Earth's rotation at middle latitudes. From (18.38) we get the vortex circulation constant, $C \approx 140\ \mathrm{cm^2\,s^{-1}}$, corresponding to a Rossby radius of $R = 12$ m. Most bathtubs are not that big and this shows that the Earth's rotation can only have little influence on a real bathtub vortex, in spite of the many claims to the contrary. A swimming pool of Olympic dimensions approaches, on the other hand, the right scale. What matters for the man-sized bathtub is much more the bather's accidental deposition of angular momentum in the water while getting out.

There is, however, the objection that the effect of the Earth's rotation could show up, if the water were left to settle down for some time before the plug is pulled. For that to happen, the Rossby number (18.11) would have to become comparable to unity. Taking the diameter of a real tub to be $d \approx 1$ m, this implies that the water velocity near the rim of the tub should not be larger than $2\Omega d \approx 0.1$ mm s^{-1}. This seems terribly small, not much larger than the thickness of a human hair per second!

The following argument indicates what patience is needed to carry out a successful experiment. After the initial turbulence from filling the tub has died out, the water settles into a laminar flow, which is further slowed down under the action of viscous forces (the Reynolds number for a flow with velocity 0.1 mm s^{-1} is about 100). Viscosity not only smooths out local velocity differences but also secures that the fluid eventually comes to rest with respect to the container. The typical viscous diffusion time over a distance L is $t \approx L^2/4\nu$, as we have seen for momentum diffusion on page 248. In a bathtub with a water level of $L = 50$ cm, the time it takes for the bottom of the container to influence the water at the top is about $t \approx 17$ h at middle latitudes! To make the experiment work, you must not only let the water settle for a few times 17 hours, but also secure that no heat is added to the water that may generate convection, and that no air drafts are present in and around the container.

Some experimenters *are* in fact patient and careful enough to observe the effect. See, for example, the very enjoyable paper by Trefethen et al. on an experiment carried out in the southern hemisphere where bathtub vortices tend to turn clockwise [TBF&65].

Problems

18.1 A space station is built in the form of a wheel with diameter 100 m. **(a)** Calculate the revolution time necessary to obtain standard gravity at the outer rim. **(b)** How and how much does the Coriolis force influence the game of Ping-Pong? **(c)** What about basketball?

18.2 The Danish Great Belt is a strait about $d = 20$ km wide with a typical surface current velocity of $U_0 = 1$ m s^{-1}. There are two layers of water, a lower and slower saline layer with a lighter and faster brackish layer on top. Assuming a density difference of about 4% and a velocity difference of about 25%, calculate the difference in water levels for the interface between saline and brackish water (disregarding viscosity).

18.3 An object is dropped from rest at a height $z = h$ and falls to the ground. **(a)** Determine the equations of motion for the fall, including the Coriolis acceleration (but disregarding air resistance). **(b)** Solve them with the correct initial conditions. **(c)** Determine the leading approximation when the fall takes much less than half a day ($2\Omega_0 t \ll 1$). **(d)** Calculate the deviation from the Galilean expression for the fall to the ground. **(e)** What does it become for $h = 20$ m.

* **18.4** Consider an arbitrary time-dependent orthogonal matrix $\mathbf{A}(t) = \{A_{ij}(t)\}$. Show that there exists a rotation vector $\mathbf{\Omega}(t) = \{\Omega_i(t)\}$ such that

$$\dot{\mathbf{A}} = -\mathbf{A} \times \mathbf{\Omega} \qquad \text{or} \qquad \dot{A}_{ij} = -\sum_{kl} \epsilon_{jkl} A_{ik} \Omega_l$$

and determine the form of $\mathbf{\Omega}$.

* **18.5** In an inertial Cartesian system, the coordinates of a point are denoted $\boldsymbol{x}'$, whereas in a generally non-inertial moving Cartesian system, the coordinates of the same point are denoted $\boldsymbol{x}$. In the inertial system the motion of the non-inertial system is described by the time-dependent coordinates of its origin $\boldsymbol{c}'(t)$ and basis vectors $\boldsymbol{a}_i(t)$. Hint: Use the preceding problem.

(a) Show that the transformation between the inertial and rotating coordinate systems is

$$\boldsymbol{x} = \mathbf{A} \cdot (\boldsymbol{x}' - \boldsymbol{c}').$$

(b) Show that the velocity $\dot{\boldsymbol{x}}$ in the rotating system is

$$\dot{\boldsymbol{x}} = \mathbf{A} \cdot (\dot{\boldsymbol{x}}' - \dot{\boldsymbol{c}}') - \boldsymbol{\Omega} \times \boldsymbol{x}.$$

(c) Show that the acceleration becomes

$$\ddot{\boldsymbol{x}} = \mathbf{A} \cdot (\ddot{\boldsymbol{x}}' - \ddot{\boldsymbol{c}}') - \dot{\boldsymbol{\Omega}} \times \boldsymbol{x} - 2\boldsymbol{\Omega} \times \dot{\boldsymbol{x}} - \boldsymbol{\Omega} \times (\boldsymbol{\Omega} \times \boldsymbol{x}).$$

The extra terms are the transformed acceleration of the origin $-\mathbf{A} \cdot \ddot{\boldsymbol{c}}'$ and the accelerating rotation $-\dot{\boldsymbol{\Omega}} \times \boldsymbol{x}$.

(d) Show that Newton's Second Law in the rotating system becomes

$$m\ddot{\boldsymbol{x}} = \boldsymbol{f} - m(\ddot{\boldsymbol{c}} + \dot{\boldsymbol{\Omega}} \times \boldsymbol{x} + 2\boldsymbol{\Omega} \times \dot{\boldsymbol{x}} + \boldsymbol{\Omega} \times (\boldsymbol{\Omega} \times \boldsymbol{x})),$$

where $\boldsymbol{f} = \mathbf{A} \cdot \boldsymbol{f}'$ is the force in the rotating sytem and $\ddot{\boldsymbol{c}} = \mathbf{A} \cdot \ddot{\boldsymbol{c}}'$ is the transformed acceleration of the origin of the rotating system ($\boldsymbol{c}$ is not a double time derivative!).

19 Computational fluid dynamics

Computational fluid dynamics (cfd) is a whole field in its own right. Swift modern computers have, to a large extent, replaced wind tunnels and wave tanks for the design of airplanes, ships, cars, bridges, and in fact any human construction that is meant to operate in a fluid. The same richness of phenomena that makes analytic solutions to the equations of fluid mechanics difficult to obtain also makes these equations hard to handle by direct numerical methods. Secondary flows, instabilities, vortices of all sizes, and turbulence complicate matters and may require numerical precision that can be hard to attain. The infinite speed of sound in incompressible fluids creates its own problems, and on top of that there are intrinsic errors and instabilities.

As in numeric elastostatics (Chapter 11), a number of steps must be carried out in any simulation. First, it is necessary to clarify which equations one wishes to solve and even there make simplifications to the problem or class of problems at hand. Second, continuous space must be discretized, and here there are a variety of methods based on finite differences, finite elements, or finite volumes [Versteeg and Malalasekera 1995]. Third, a discrete dynamic process must be set up that guides the initial field configuration toward the desired solution. Most often this process emulates the time evolution of fluid dynamics itself, as described by the Navier–Stokes equations. Finally, convergence criteria and error estimates are needed to monitor and gain confidence in the numerical solutions.

In this chapter we shall compute two-dimensional laminar flow in a channel of finite length and determine how it turns into the well-known parabolic (Poiseuille) profile downstream from the entry.

19.1 Unsteady, incompressible flow

In numeric elastostatics we were able to set up an artificial dissipative, dynamic gradient descent that guided the displacement field toward a static solution with minimal elastic energy. This technique cannot be transferred to computational fluid dynamics because a solution to the steady-state equations does not correspond to an extremum of any bounded quantity (Problem 19.1). Instead we shall attempt to copy Nature by simulating the complete set of time-dependent Navier–Stokes equations (page 251 or 256).

Appealing to the behavior of real fluids, the natural viscous dissipation built into these equations should hopefully guide the velocity field toward a steady-state solution. There is, however, no guarantee—neither from Nature nor from the equations—that the flow will always settle down and become steady, even when the boundary conditions are time independent. We are all too familiar with the unsteady and sometimes turbulent flow that may spontaneously arise under even the steadiest of circumstances, as for example a slow river narrowing down, or even worse coming to a waterfall. But this is actually not so bad, because a forced steady-state solution is completely uninteresting when the real fluid refuses to end up up in that state. We do not, for example, care much for the Poiseuille solution (page 269) to pipe flow at a Reynolds number beyond the transition to turbulence, nor for that matter the laminar flow around a sphere in an ideal fluid (page 225). If steady laminar flow is desired, one must keep the Reynolds number so low that there is a real chance for it to become established.

Field equations

Not to complicate matters we shall only consider incompressible fluids with constant density $\rho = \rho_0$, for which the divergence vanishes at all times. Taking for simplicity $\rho_0 = 1$, the Navier–Stokes equations (15.21) may be written as

$$\frac{\partial \boldsymbol{v}}{\partial t} = \boldsymbol{F} - \nabla p, \qquad \nabla \cdot \boldsymbol{v} = 0, \tag{19.1}$$

where we have defined the "force" as

$$\boldsymbol{F} = -(\boldsymbol{v} \cdot \nabla)\boldsymbol{v} + \nu \nabla^2 \boldsymbol{v}, \tag{19.2}$$

depending only on the velocity. For simplicity, volume forces are not included.

Poisson equation for pressure

In an incompressible fluid the pressure is determined indirectly through the vanishing of the divergence of the velocity field, as discussed in the comment on page 208. Calculating the divergence of both sides of (19.1), we find a Poisson equation for p,

$$\nabla^2 p = \nabla \cdot \boldsymbol{F}. \tag{19.3}$$

This constraint must be fulfilled for the velocity field to remain free of divergence at all times. Knowing the velocity field $\boldsymbol{v}$ at a given time we can calculate the right-hand side and solve this equation with suitable boundary conditions to determine the pressure everywhere in the fluid at that particular instant of time.

Solutions to the Poisson equation are, however, *non-local* functions of the source, basically of the same form as the electrostatic or gravitational potential,

$$p(\boldsymbol{x}, t) = p_0(\boldsymbol{x}, t) - \int \frac{\nabla' \cdot \boldsymbol{F}(\boldsymbol{x}', t)}{4\pi |\boldsymbol{x} - \boldsymbol{x}'|} dV',$$

where $p_0(\boldsymbol{x}, t)$ is a solution to the Laplace equation $\nabla^2 p_0 = 0$, determined by the boundary conditions. Any local change in the velocity field at a point $\boldsymbol{x}'$ is *instantaneously* communicated back to the pressure at any other point $\boldsymbol{x}$ in the fluid (although damped by the denominator $|\boldsymbol{x} - \boldsymbol{x}'|$). These changes in pressure are in turn communicated to the velocity field via the Navier–Stokes equation (19.1). The pressure thus links the velocity field at any instant of time non-locally to its immediately preceding values anywhere in the system, even for infinitesimally small time intervals. One might say that pressure instantly informs the world at large about the global state of the incompressible velocity field.

Physically, the unpleasant non-local behavior is caused by the assumption of absolute incompressibility, which is just as untenable in the real world of local interactions as absolute rigidity. As pointed out before, incompressibility should be viewed as a property of the flow rather than of the fluid itself. The unavoidable compressibility of real matter will in fact limit the rate at which pressure changes can propagate through a fluid to the speed of sound. Nevertheless, such a conclusion does not detract from the practical usefulness of the divergence condition, $\nabla \cdot \boldsymbol{v} = 0$, for flow speeds well below the speed of sound.

Boundary conditions

In many fluid dynamics problems, fixed impermeable walls guide the fluid between openings where it enters and leaves the system. At fixed walls the velocity field must vanish at all times, because of the impermeability and no-slip conditions, which respectively require the normal and tangential components to vanish. Setting $\boldsymbol{v} = \boldsymbol{0}$ on the left-hand side of the equation of motion (19.1), we obtain a boundary condition for pressure,

$$\nabla p = \boldsymbol{F}, \tag{19.4}$$

on any fixed boundary. The same is the case at a fluid inlet, where the velocity field is often fixed to an externally defined constant value, $\boldsymbol{v} = \boldsymbol{U}$. Outlet velocities are usually not controlled externally, and as a boundary condition on the velocity field one may choose the vanishing of the normal derivative, $(\hat{\boldsymbol{n}} \cdot \nabla)\boldsymbol{v} = \boldsymbol{0}$. Alternatively, the stress vector may be required to vanish at the exit, $\boldsymbol{\sigma} \cdot \hat{\boldsymbol{n}} = \boldsymbol{0}$, even if that is a bit harder to implement.

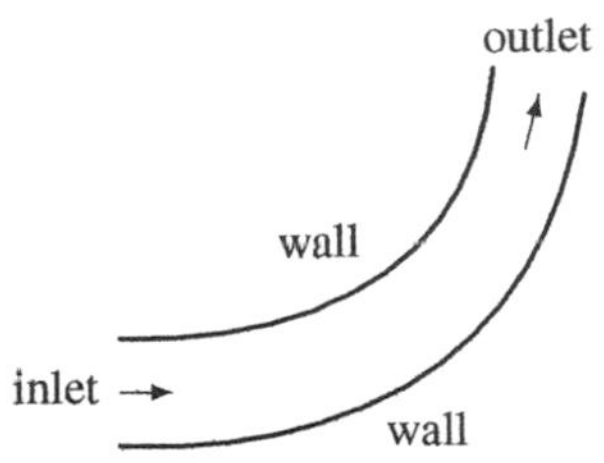

A flow guidance system with inlet, outlet, and walls.

19.2 Temporal discretization

Suppose the current velocity field is $\boldsymbol{v}(\boldsymbol{x},t)$ and the current pressure field $p(\boldsymbol{x},t)$. From these we can calculate the current acceleration field $\boldsymbol{F}(\boldsymbol{x},t)$ and then use the equation of motion (19.1) to move the velocity field forward in time through a small but finite time step Δt,

$$\boldsymbol{v}(\boldsymbol{x},t+\Delta t) = \boldsymbol{v}(\boldsymbol{x},t) + \big(\boldsymbol{F}(\boldsymbol{x},t) - \nabla p(\boldsymbol{x},t)\big)\Delta t. \tag{19.5}$$

Taylor expansion of the left-hand side shows that the error is $\mathcal{O}\left(\Delta t^2\right)$, but less error-prone higher-order schemes are also possible [Anderson 1995, Griebel et al. 1998]. Provided the velocity field is free of divergence and the pressure satisfies (19.3), it follows that the new velocity field obtained from this equation will also be free of divergence.

Divergence suppression

But approximation errors cannot be avoided in any finite step algorithm. Since for this reason the current velocity field $\boldsymbol{v}(\boldsymbol{x},t)$ may not be perfectly free of divergence, it is more appropriate to demand that the future velocity field be free of divergence, that is, $\nabla \cdot \boldsymbol{v}(\boldsymbol{x},t+\Delta t) = \boldsymbol{0}$. Calculating the divergence of both sides of (19.5) we obtain a modified Poisson equation for the pressure,

$$\nabla^2 p = \nabla \cdot \boldsymbol{F} + \frac{\nabla \cdot \boldsymbol{v}}{\Delta t}. \tag{19.6}$$

The factor $1/\Delta t$ amplifies the divergence errors, so that this Poisson equation will initially be concerned with correction of divergence errors, and only when they have been suppressed will the field $\boldsymbol{F}$ gain influence on the pressure. In practice, the stepping algorithm (19.5) can get into trouble if the divergence becomes too large. It is consequently important to secure that the initial velocity field is reasonably free of divergence. In a complicated flow geometry, that can in fact be quite hard to attain.

Stability conditions

There are essentially only two possibilities for what can happen to the approximation errors in the course of many time steps. Either the errors will become systematically larger, in which case the computation goes straight to the land of failed calculations, or the errors will diminish or at least stay constant and "small", thereby keeping the computation on track. Smaller time steps are usually better, but it takes careful mathematical analysis to determine a precise value for the upper limit to the size of the time step. The result depends strongly on both the spatial and temporal discretization methods (see for example [Anderson 1995, Press et al. 1992]), and may range from zero to infinity depending on the particular algorithm that is implemented.

Here we shall present an intuitive argument for the stability conditions that apply to the straightforward numerical simulation (19.5) on a spatial grid with typical coordinate spacings Δx, Δy, and Δz. We shall derive the conditions from the physical processes that compete in displacing fluid particles in, say, the x-direction. One is advection with velocity v_x, which displaces the particle a distance $|v_x|\,\Delta t$ in a time interval Δt. Another is momentum diffusion due to viscosity, which effectively displaces the particle by $\sqrt{\nu \Delta t}$ (see page 248). Intuitively, it seems reasonable to demand that these displacements be smaller than the grid spacing,

$$|v_x|\,\Delta t \lesssim \Delta x, \qquad \sqrt{\nu \Delta t} \lesssim \Delta x. \tag{19.7}$$

Taking into account that the maximal velocity in the flow provides the most stringent advective condition, the global condition may be taken to be

$$\Delta t \lesssim \min\left(\frac{\Delta x}{|v_x|_{\max}}, \frac{\Delta x^2}{\nu}\right), \tag{19.8}$$

and similarly for the other coordinate directions. This limit is of course only an estimate, and trial-and-error may be used to determine the largest possible value of Δt before instability sets in. For the first-order update rule (19.5), the time step typically has to be an order of magnitude smaller than the above limit. For higher-order methods it can be chosen much closer to the limit [Press et al. 1992].

The advective condition is most restrictive for high velocities, and the diffusive for large viscosity. The allowed time step is generally largest for the coarsest spatial grid, which on the other hand is blind to finer details of the flow. There is thus a payoff between the detail desired in the simulation and its rate of progress. *High detail entails slow progress and a high cost in computer time.*

19.3 Spatial discretization

In the discussion of numeric elastostatics we described a method based on finite differences with errors of only second order in the grid spacings (Section 11.2 on page 180). Although it is possible to solve simple flow problems using this method, most such problems will benefit from a somewhat more sophisticated treatment. The method of *staggered grids* to be presented here comes at essentially no cost in computer memory or time, but does complicate matters a bit. A number of applications of this method are given in [Griebel et al. 1998].

Restriction to two dimensions

For simplicity we shall limit the following discussion to two-dimensional flow in the xy-plane, exemplified by the flow in a channel between parallel plates (Section 16.2 on page 262). The restriction to two dimensions still leaves ample room for interesting applications. Generalization to three dimensions is straightforward.

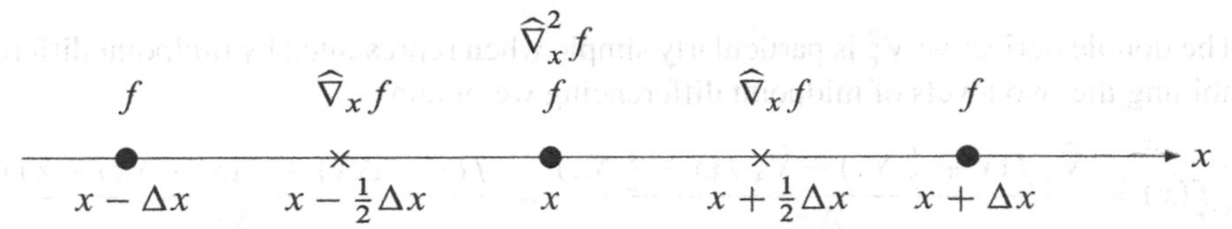

Figure 19.1. A double difference $\nabla_x^2 f(x)$ is represented by the midpoint difference of two neighboring single midpoint differences.

Two-dimensionality is taken to mean that the fields can only depend on x and y and that $v_z = 0$. The equations of motion (19.1) and (19.2) then simplify to

$$\frac{\partial v_x}{\partial t} = F_x - \nabla_x p, \qquad \frac{\partial v_y}{\partial t} = F_y - \nabla_y p, \tag{19.9}$$

with

$$F_x = -v_x \nabla_x v_x - v_y \nabla_y v_x + \nu \left(\nabla_x^2 + \nabla_y^2\right) v_x + g_x, \tag{19.10a}$$

$$F_y = -v_x \nabla_x v_y - v_y \nabla_y v_y + \nu \left(\nabla_x^2 + \nabla_y^2\right) v_y + g_y. \tag{19.10b}$$

The divergence condition becomes

$$\nabla_x v_x + \nabla_y v_y = 0, \tag{19.11}$$

and the stresses (using that $\eta = \nu$ for $\rho_0 = 1$)

$$\sigma_{xx} = -p + 2\nu \nabla_x v_x, \tag{19.12a}$$

$$\sigma_{yy} = -p + 2\nu \nabla_y v_y, \tag{19.12b}$$

$$\sigma_{xy} = \sigma_{yx} = \nu(\nabla_x v_y + \nabla_y v_x). \tag{19.12c}$$

They are not as important here as in computational elastostatics, because the boundary conditions in fluid mechanics are often formulated directly in terms of the velocities.

Midpoint differences

The main objection to the central difference, $\widehat{\nabla}_x f(x) = (f(x + \Delta x) - f(x - \Delta x))/2\Delta x$, defined in Equation (11.8) on page 180 is that it spans twice the interval Δx and does not use the value of the function in the central point x to which it "belongs". This opens for "leapfrog" or "flip-flop" instabilities in which neighboring grid points behave quite differently. The problem becomes particularly acute in the interplay between the equations of motion and the instantaneous Poisson equation for pressure.

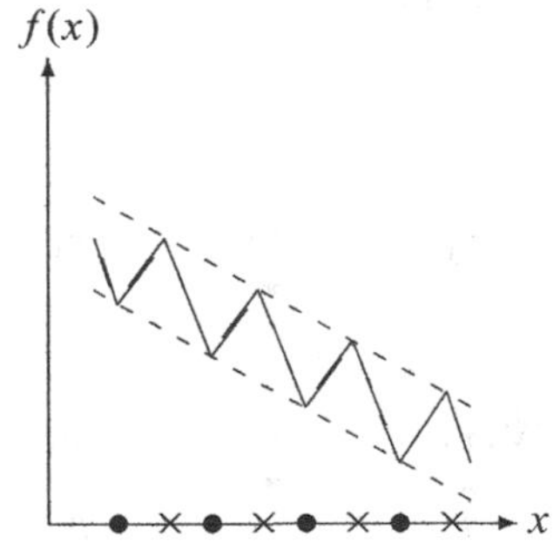

The solid zigzag curve $f(x)$ between the two straight lines has constant central difference on the grid! Such behavior is typical of the "leapfrog" errors that may (and will) arise when using naive central differencing.

One way out is to recognize that the difference in field values between two neighboring grid-points properly "belongs" to the midpoint of the line that connects them. If we denote the coordinates of the two points by x and $x + \Delta x$, the forward difference from x becomes the central difference around the midpoint $x + \frac{1}{2}\Delta x$ (see the margin figure),

$$\widehat{\nabla}_x^+ f(x) = \frac{f(x + \Delta x) - f(x)}{\Delta x} = \widehat{\nabla}_x f\left(x + \tfrac{1}{2}\Delta x\right), \tag{19.13}$$

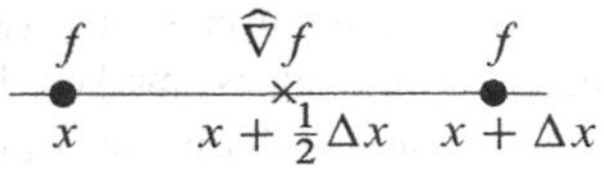

The finite difference between the grid points x and $x + \Delta x$ belongs to the point $x + \frac{1}{2}\Delta x$.

which has errors of only second order in $\Delta x/2$. But the midpoints between grid points are not themselves part of the grid. We could of course double the grid and use $\frac{1}{2}\Delta x$ as grid-spacing, thereby including the midpoints, but that would just bring us back to the situation we started out to correct. Instead we shall think of the midpoint values of first-order derivatives as being *virtual*, meaning that they may arise during a calculation but are not retained as part of the discrete information stored on the grid.

The double derivative ∇_x^2 is particularly simple when represented by midpoint differences. Combining the two levels of midpoint differencing we obtain

$$\widehat{\nabla}_x^2 f(x) = \frac{\widehat{\nabla}_x f(x+\frac{1}{2}\Delta x) - \widehat{\nabla}_x f(x-\frac{1}{2}\Delta x)}{\Delta x} = \frac{f(x+\Delta x) + f(x-\Delta x) - 2f(x)}{\Delta x^2}, \tag{19.14}$$

which has errors of second order in Δx only. Geometrically the two levels of midpoint differences bring the double derivative back to the point it "originated in", so that $\widehat{\nabla}_x^2 f(x)$ belongs to the same point as the original value $f(x)$ (see Figure 19.1).

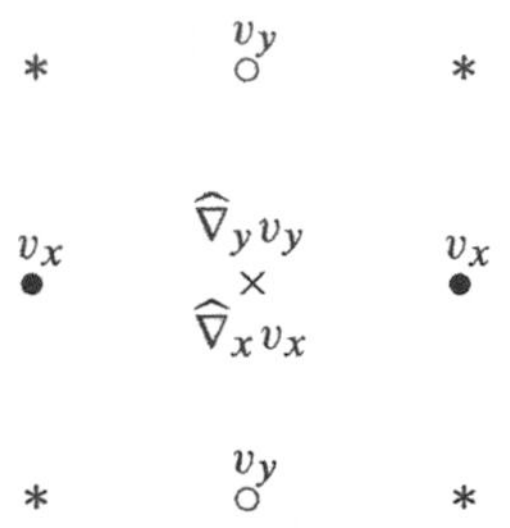

The grids for v_x (dots) and v_y (circles) do not overlap if their midpoint differences, $\widehat{\nabla}_x v_x$ and $\widehat{\nabla}_y v_y$, must belong to the same point (crosses). At the remaining points (asterisks) the other midpoint differences, $\widehat{\nabla}_x v_y$ and $\widehat{\nabla}_y v_x$, are found.

Staggered grids

To see how this works out, let us discretize the two-dimensional divergence condition (19.11) by expressing it in terms of midpoint differences,

$$\widehat{\nabla}_x v_x + \widehat{\nabla}_y v_y = 0. \tag{19.15}$$

To secure that errors are of second order in both Δx and Δy, the two midpoint differences must belong to the same point. For this to be possible, we must know the values of v_x and v_y in different points (as shown in the margin figure). This shows that the grids for v_x and v_y cannot overlap, but must be shifted by $(\pm\frac{1}{2}\Delta x, \pm\frac{1}{2}\Delta y)$ with respect to each other.

All together there are four different grids, each with coordinate spacings Δx and Δy, that are shifted or *staggered* by half lattice spacings with respect to each other. As shown in the margin figure, we use different symbols for these grids with dots and circles denoting the v_x and v_y grids, while crosses and asterisks denote respectively their diagonal and off-diagonal midpoint derivatives. A complete picture of the staggered grids is shown in Figure 19.2.

Discretized equations of motion

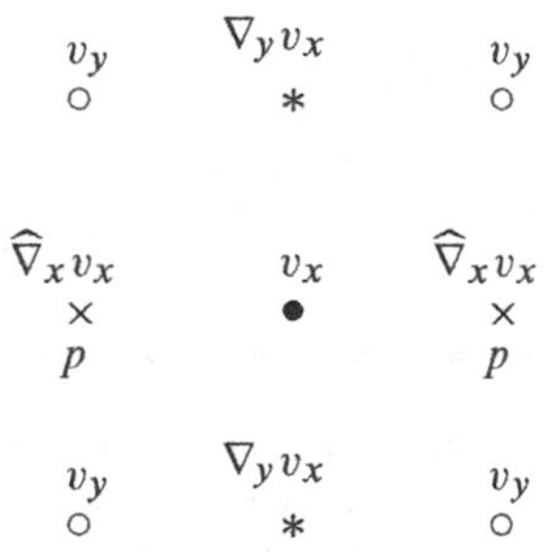

The pressure p belongs to the grid marked by crosses, such that $\nabla_x p$ belongs to the same grid as v_x. To calculate the difference $\widehat{\nabla}_x v_x$ (crosses) on the v_x-grid (dots) we must average the two neighboring values marked by crosses, and similarly for $\widehat{\nabla}_y v_x$ (marked by asterisks). The value of v_y on the v_x-grid (dots) is obtained by averaging over all four nearest neighbors on the v_y-grid (circles).

The left-hand sides of the equations of motion (19.9) belong to the same grids as v_x and v_y, that is, to the dot-grid and the circle-grid, respectively. To secure that errors are of second order in the grid spacings, the force fields F_x and F_y, as well as the pressure gradients $\nabla_x p$ and $\nabla_y p$, should respectively belong to the same grids as v_x and v_y. The pressure field p presents no problem, if it is assigned to the cross-grid, and neither do the double derivatives contained in F_x and F_y, since they belong naturally to the correct grids (as shown above).

The advective term, $-v_x \nabla_x v_x$, raises a problem because $\widehat{\nabla}_x v_x$ belongs to the cross-grid and not to the dot-grid as we would like it to. To keep errors to second order, one must form the average of $\widehat{\nabla}_x v_x$ over the two neighboring values on the cross-grid. Similarly, in the advective term, $-v_y \nabla_y v_x$, the derivative $\widehat{\nabla}_y v_x$ is naturally found on the asterisk-grid and must also be averaged over the two nearest neighbors to get its value on the dot-grid. The worst case is v_y, for which the value on the dot-grid is obtained as the average over the four nearest neighbors on the circle-grid.

Denoting averages over the relevant neighboring points by brackets $\langle\cdots\rangle$, we may write the discretized acceleration fields in the form (see Problem 19.4 for the explicit expressions),

$$F_x = -v_x \left\langle \widehat{\nabla}_x v_x \right\rangle - \langle v_y \rangle \left\langle \widehat{\nabla}_y v_x \right\rangle + \nu \left(\widehat{\nabla}_x^2 + \widehat{\nabla}_y^2 \right) v_x, \tag{19.16a}$$

$$F_y = - \langle v_x \rangle \left\langle \widehat{\nabla}_x v_y \right\rangle - v_y \left\langle \widehat{\nabla}_y v_y \right\rangle + \nu \left(\widehat{\nabla}_x^2 + \widehat{\nabla}_y^2 \right) v_y. \tag{19.16b}$$

The corresponding changes in the velocities during a time step Δt become

$$\Delta v_x = (F_x - \widehat{\nabla} p)\Delta t, \qquad \Delta v_y = (F_y - \widehat{\nabla}_y p)\Delta t, \tag{19.17}$$

also to be evaluated on the dot- and circle-grids, respectively.

Solving the discrete Poisson equation

During an iteration cycle, the accelerations (19.16) are calculated from the current values of the discrete fields at time t, and afterward the corresponding pressure at time t is calculated by solving the discretized version of the Poisson equation (19.6)

$$\left(\widehat{\nabla}_x^2 + \widehat{\nabla}_y^2\right) p = s \equiv \widehat{\nabla}_x F_x + \widehat{\nabla}_y F_y + \frac{\widehat{\nabla}_x v_x + \widehat{\nabla}_y v_y}{\Delta t}. \tag{19.18}$$

The solution to this equation may be obtained by relaxation methods, for example gradient descent (see Section 11.1 on page 177), in which the pressure $p(x, y, t)$ undergoes successive changes of the form

$$\Delta p = \epsilon \left(\left(\widehat{\nabla}_x^2 + \widehat{\nabla}_y^2\right) p - s\right), \tag{19.19}$$

where $\epsilon > 0$ is the step size. The relaxation algorithm converges toward a solution to $\nabla^2 p = s$ for sufficiently small ϵ because it descends along the steepest downward gradient toward the unique minimum of a quadratic "energy"-function (see Problem 19.2).

Denoting the n-th approximation to the solution by $p_n(x, y, t)$, the discretized relaxation process may be written explicitly as

$$p_{n+1}(x, y, t) = p_n(x, y, t) + \epsilon \left(\left(\widehat{\nabla}_x^2 + \widehat{\nabla}_y^2\right) p_n(x, y, t) - s(x, y, t)\right). \tag{19.20}$$

Starting with some field configuration $p_0(x, y, t)$ and imposing boundary conditions after each step, this process will eventually lead to the desired solution, $p(x, y, t)$. The problem is, however, that simple gradient descent is slow, too slow in fact to be applied in every time step of the solution to the equations of motion. *Conjugate gradient descent* [Press et al. 1992] offers considerable speed-up by calculating the optimal step-size directly, but as it turns out there are still faster methods.

From the double difference operator (19.14) we see that the coefficient of p_n in (19.20) is $1 - 2\epsilon(1/\Delta x^2 + 1/\Delta y^2)$, and this suggests the following reparametrization of the step-size in terms of a dimensionless *convergence parameter* ω,

$$\epsilon = \frac{\omega}{2}\left(\frac{1}{\Delta x^2} + \frac{1}{\Delta y^2}\right)^{-1} = \frac{\omega}{2}\frac{\Delta x^2 \Delta y^2}{\Delta x^2 + \Delta y^2}. \tag{19.21}$$

The coefficient of p_n now becomes $1 - \omega$, and this allows for a precise definition of what is meant by *underrelaxation* ($\omega < 1$) and *overrelaxation* ($\omega > 1$). Straightforward gradient descent (conjugate or not) in which the new field, $p_{n+1}(x, y, t)$, is calculated all over the grid before replacing the old, $p_n(x, y, t)$, only converges when underrelaxed.

In *successive overrelaxation* or SOR (see [Anderson 1995] and [Griebel et al. 1998]), the new value $p_{n+1}(x, y, t)$ at a grid point replaces the old value $p_n(x, y, t)$ as soon as it is calculated during a sweep of the grid. The method converges for $1 < \omega < 2$ and in practice the best value for ω may be located by trial-and-error, usually not far below the upper limit, say $\omega = 1.7$–1.9. Since the SOR algorithm sweeps sequentially through the grid, one should be aware that this procedure may create small asymmetry errors in an otherwise symmetric situation. But fast it is, on small grids often converging in just a few iterations after an initial phase has passed.

19.4 Application: Laminar channel entry flow

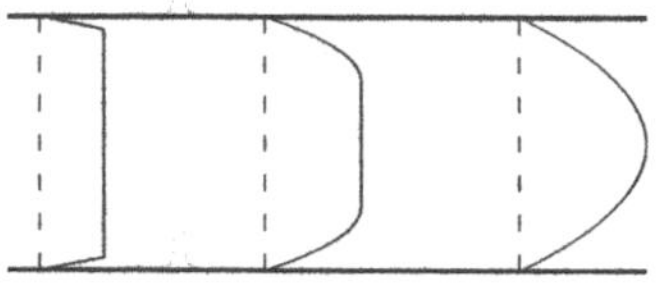

Sketch of the expected shape of the velocity profile at various distances downstream from the entry.

A simple and—from the look of it—well-behaved problem concerns the steady entry flow pattern in a channel between two parallel plates with gap width $d = 2a$. Directly at the entry, the flow is thought to be uniform with a flat velocity distribution that downstream smoothly turns into the characteristic parabolic (Poiseuille) shape. Due to the gradual approach to the parabolic shape, it is customary to define the *entry length* L_e as the downstream distance at which the central flow has reached 99% of its terminal value $\frac{3}{2}U$.

We have previously (page 264) estimated that the entry length grows proportionally with the Reynolds number, $L_e/d \approx k\mathsf{Re}$, with $k \approx 0.04$ and $\mathsf{Re} = Ud/\nu$. For small Reynolds number, the influence of the entry is always expected to reach to at a distance comparable to the channel width d, so that we expect that L_e/d becomes a constant h or order unit for $\mathsf{Re} \to 0$. A suitable expression that interpolates between the two behaviors is

$$\frac{L_e}{d} \approx \sqrt{h^2 + k^2\mathsf{Re}^2}, \tag{19.22}$$

which should be valid for laminar flow in the region $0 \lesssim \mathsf{Re} \lesssim 2{,}000$. After the onset of turbulence around $\mathsf{Re} \approx 2{,}000$, the entry length first drops dramatically and later grows slowly with the Reynolds number in the region of fully developed turbulence.

Boundary conditions

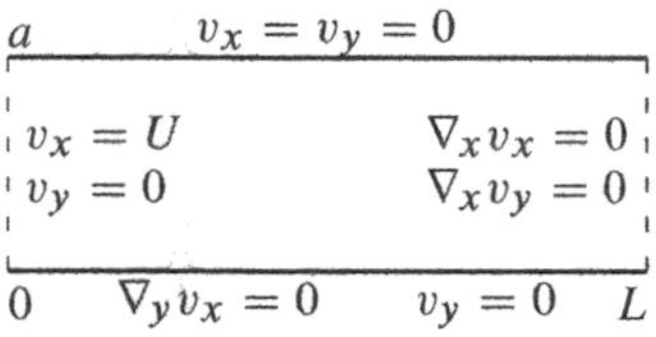

Boundary conditions for channel flow in the region $0 \le x \le L$ times $0 \le y \le a$. At the entry (west), the flow is uniform; at the exit (east), the flow is longitudinal; at the upper plate (north), the velocity must vanish, and at the midline (south), symmetry demands that v_y and $\nabla_y v_x$ must vanish.

In the laminar regime we expect the flow to be mirror symmetric around the midline, $y = 0$,

$$v_x(x, -y) = v_x(x, y), \qquad v_y(x, -y) = -v_y(x, y), \tag{19.23}$$

so that we only need to consider the rectangular region, $0 \le x \le L$ times $0 \le x \le a$, between the middle of the channel and one of the plates (see the margin figure). We choose the following boundary conditions:

$$v_x = U, \qquad v_y = 0, \qquad \text{for } x = 0, \tag{19.24a}$$

$$\nabla_x v_x = 0, \qquad \nabla_x v_y = 0, \qquad \text{for } x = L, \tag{19.24b}$$

$$\nabla_y v_x = 0, \qquad v_y = 0, \qquad \text{for } y = 0, \tag{19.24c}$$

$$v_x = 0, \qquad v_y = 0, \qquad \text{for } y = a. \tag{19.24d}$$

The first line expresses that the entry flow is uniform, and the second that the exit flow is independent of x. The third line follows from the symmetries at the midline, while the last expresses the no-slip condition on the solid wall. In a flow like this, controlled by the entry velocity, the boundary conditions on pressure follow from the velocity conditions, as will be discussed below.

Initial state

The equations of motion must be supplied with initial data that fulfill the spatial boundary conditions and the condition of vanishing divergence. This is not nearly as simple as it sounds, even if there is great freedom in the choice of initial data and even if the final steady state is supposed to be independent of this choice. The problem becomes acute for the boundaries of irregular shape which will nearly always be the case in any realistic flow problem.

Here we shall choose the initial velocity and pressure fields (at $t = 0$) to be

$$v_x = U, \qquad v_y = 0, \qquad p = 0 \tag{19.25}$$

everywhere inside the channel. This certainly fulfills the divergence condition, but has a discontinuous jump on the northern and southern boundaries. The initial fields also fulfill the Poisson equation for pressure (19.6).

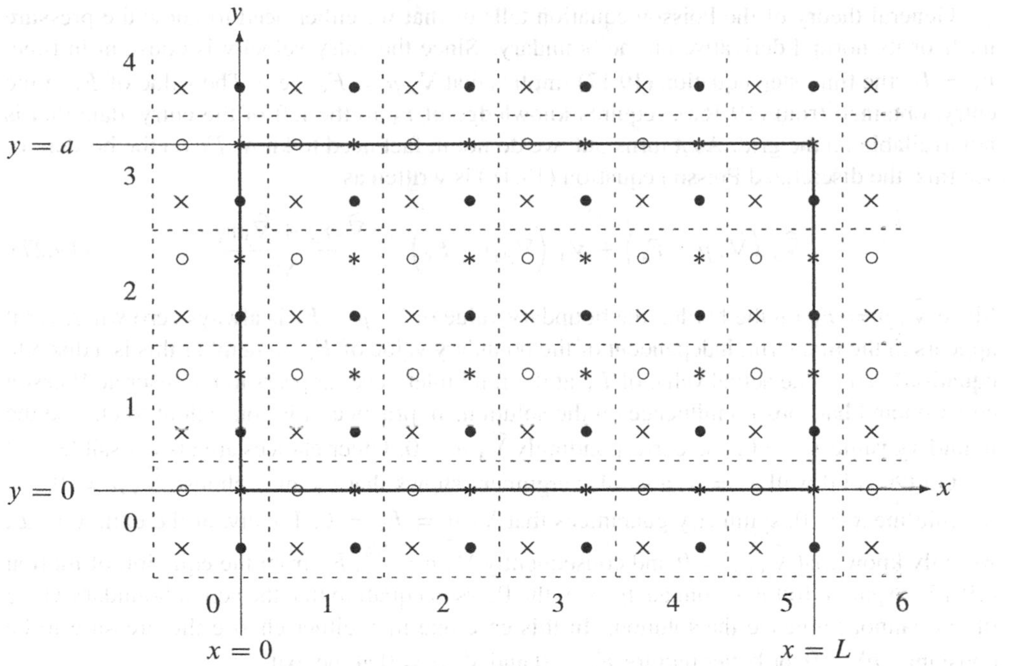

Figure 19.2. A rectangular region (solid line) of size $L \times a$ is discretized using staggered grids, here with $N_x = 5$ and $N_y = 3$, such that $L = 5\Delta x$ and $a = 3\Delta y$. The cells are numbered as shown. The boundary conditions necessitate one layer of nodes outside the region on all sides, thus requiring $7\times 5 = 35$ basic cells. Only the nodes that are actually used are shown here with the dashed lines to aid the eye. The velocity v_x is defined on the dots, the velocity v_y on the circles, and the pressure p on the crosses.

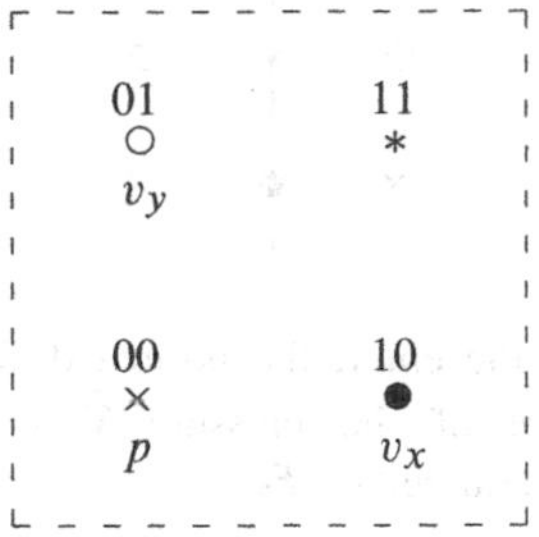

The basic cell of size $\Delta x \times \Delta y$ contains one point (node) from each of the four staggered grids, identified by the binary numbers. The three fundamental fields v_x, v_y, and p are located at separate grids.

Discretization by staggered grids

A rectangular region of size $L\times a$ is discretized using staggered grids with coordinate intervals $\Delta x = L/N_x$ and $\Delta y = a/N_y$, where N_x and N_y are integers. Implementation of the boundary conditions with midpoint differences necessitates an extra layer of nodes padding the boundary such that the we need an array of $(N_x+2)\times(N_y+2)$ basic cells of size $\Delta x\times\Delta y$ to tile the region of interest. We shall choose the basic cell type shown in the margin with cells arranged as in Figure 19.2. This choice allows for the simple parametrization of the discrete coordinates for the fields,

$$x = \left(i_x + \tfrac{1}{2}(j_x - 1)\right)\Delta x, \qquad y = \left(i_y + \tfrac{1}{2}(j_y - 1)\right)\Delta y, \tag{19.26}$$

with cells numbered $i_x = 0, 1, \cdots, N_x + 1$ and $i_y = 0, 1, \cdots, N_y + 1$, and nodes numbered $j_x = 0, 1$ and $j_y = 0, 1$ within each cell, as shown in the margin figure.

The discrete boundary conditions have to reflect that not all field values are known precisely on the border. At the entry $x = 0$, the field v_x belongs to the border (dots), so that the boundary condition remains $v_x = U$, while the v_y-condition must be implemented as an average, $\langle v_y\rangle = 0$, over the nearest v_y-nodes (circles) on both sides of the border. On the wall, $y = a$, the roles are reversed, and we have $\langle v_x\rangle = 0$ and $v_y = 0$, while at the symmetry border $y = 0$, both $v_y = 0$ and $\widehat{\nabla}_y v_x = 0$ belong to the border and are directly implemented. Finally, at the exit, $x = L$, the condition $\widehat{\nabla}_x v_y = 0$ can be directly implemented, but since there is no data on v_x beyond the exit, the condition $\widehat{\nabla}_x v_x = 0$ can only be approximatively implemented as a backward difference, $\widehat{\nabla}_x^- v_x = 0$, actually belonging to the crosses left of the eastern border.

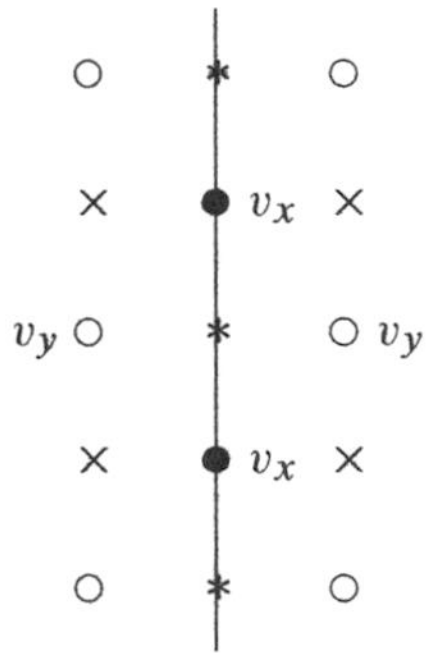

At the entry, $x = 0$, the value of the velocity v_x may be implemented directly, whereas the boundary value of the velocity v_y is calculated as an average over the two nearest neighbors on both sides.

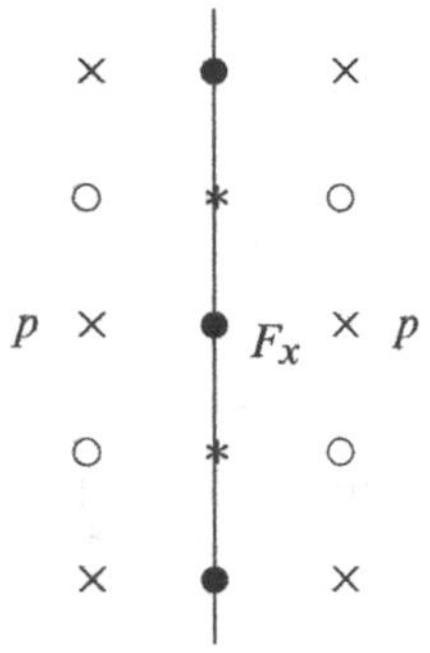

At the entry, the normal difference of the pressure $\widehat{\nabla}_x p$ is determined by F_x.

General theory of the Poisson equation tells us that we either need to know the pressure itself or its normal derivative on the boundary. Since the entry velocity is constant in time, $v_x = U$, the time-step equation (19.17) implies that $\widehat{\nabla}_x p = F_x$ there. The value of F_x at the entry, obtained from (19.16a), requires knowledge of v_x to the left of the entry, data that is not available on the grid. As it turns out, we do not in fact need to know F_x on the border. To see this, the discretized Poisson equation (19.18) is written as

$$\widehat{\nabla}_x\left(\widehat{\nabla}_x p - F_x\right) + \widehat{\nabla}_y\left(\widehat{\nabla}_y p - F_y\right) = \frac{\widehat{\nabla}_x v_x + \widehat{\nabla}_y v_y}{\Delta t}. \tag{19.27}$$

Since $\widehat{\nabla}_x p = F_x$ on the border, the boundary value of $\widehat{\nabla}_x p - F_x$ is always zero wherever it appears in the first term, independent of the boundary value of F_x (remember this is a discrete equation). Thus, the actual value of F_x at the fluid inlet never appears in the discrete Poisson equation and has thus no influence on the solution. In practice, it is convenient to choose the boundary value $F_x = 0$, and correspondingly $\widehat{\nabla}_x p = 0$. Other choices are also possible.

On the solid wall $y = a$, a similar argument shows that we may choose $\widehat{\nabla}_y p = 0$. At the midline $x = 0$, symmetry guarantees that $\widehat{\nabla}_y p = F_y = 0$. Finally, at the exit, $x = L$, we only know that $\widehat{\nabla}_x v_x = 0$ and consequently $\widehat{\nabla}_x^2 p = \widehat{\nabla}_x F_x$ from the equation of motion (19.17). Again it follows from analysis of the Poisson equation that the actual boundary value of F_x cannot influence the solution. In this case one may either choose the pressure to be constant, $\langle p \rangle = 0$, or better require $F_x = 0$ and $\widehat{\nabla}_x p = 0$ at the exit.

Summarizing, the discrete boundary conditions are taken to be

$$v_x = U, \quad \langle v_y \rangle = 0, \quad \widehat{\nabla}_x p = 0, \quad \text{for } x = 0, \tag{19.28a}$$

$$\widehat{\nabla}^-_x v_x = 0, \quad \widehat{\nabla}_x v_y = 0, \quad \widehat{\nabla}_x p = 0, \quad \text{for } x = L, \tag{19.28b}$$

$$\widehat{\nabla}_y v_x = 0, \quad v_y = 0, \quad \widehat{\nabla}_y p = 0, \quad \text{for } y = 0, \tag{19.28c}$$

$$\langle v_x \rangle = 0, \quad v_y = 0, \quad \widehat{\nabla}_y p = 0, \quad \text{for } y = a. \tag{19.28d}$$

In the second line, the difference $\widehat{\nabla}^-_x v_x$ actually belongs to the points half a grid spacing, $\frac{1}{2}\Delta x$, inside the channel. We shall ignore this small first-order error.

Monitoring the process

The convergence of the Poisson relaxation process (19.20) may be monitored by the dimensionless parameter

$$\chi = \sqrt{\frac{\sum\left(\widehat{\nabla}_x^2 p + \widehat{\nabla}_y^2 p - s\right)^2}{\sum s^2}}, \tag{19.29}$$

where the sum runs over all internal grid points. It is dimensionless, independent of the grid for $N_x, N_y \to \infty$. It measures how well the Poisson equation is fulfilled relative to the source field s.

One may also monitor the divergence, which ideally should vanish. A convenient convergence parameter is

$$\chi' = \sqrt{\frac{\sum\left(\widehat{\nabla}_x v_x + \widehat{\nabla}_y v_y\right)^2}{\sum\left(\widehat{\nabla}_x v_x\right)^2 + \left(\widehat{\nabla}_y v_y\right)^2}} \tag{19.30}$$

and measures how well the two differences cancel each other in the divergence.

Iteration cycle

The grid arrays for all the fields, $v_x[i_x, i_y]$, $v_y[i_x, i_y]$, $p[i_x, i_y]$, $F_x[i_x, i_y]$, and $F_y[i_x, i_y]$, are first cleared to zero, and then the velocity is initialized to $v_x[i_x, i_y] = U$ for $i_x = 0, \dots, N_x$ and $i_y = 1, \dots, N_y$.

Assuming that we have obtained a current set of discrete fields, v_x, v_y, and p at time t, the following iteration cycle produces a new set of fields at $t + \Delta t$.

1. Calculate the acceleration fields, F_x and F_y, at all internal points from (19.16). The boundary values of the accelerations remain zero (cf. the discussion above).

2. Calculate the source of the Poisson equation (19.18) from the new fields at all internal points.

3. Solve the Poisson equation iteratively by means of the following sub-algorithm.

   (a) Apply the boundary conditions to the pressure. Explicitly they are

$$\begin{aligned} &p[0, i_y] = p[1, i_y], \quad &&p[N_x + 1, i_y] = p[N_x, i_y], \quad &&i_y = 1, \dots, N_y, \\ &p[i_x, 0] = p[i_x, 1], \quad &&p[i_x, N_y + 1] = p[i_x, N_y], \quad &&i_x = 0, \dots, N_x + 1. \end{aligned}$$

   (b) Update the pressure in all internal points using successive overrelaxation (SOR) with $\omega < 2$.

   (c) Repeat from (a) until the desired precision (χ) or the iteration limit is reached.

4. Update the velocities v_x and v_y at time $t + \Delta t$ from (19.17) in all internal points (i.e., not on the boundary). The internal grid points are given by $i_x = 1, \dots, N_x - 1$ and $i_y = 1, \dots, N_y$ for v_x, and $i_x = 1, \dots, N_x$ and $i_y = 1, \dots, N_y - 1$ for v_y.

5. Apply the boundary conditions (19.28) to determine boundary values of the velocities. Explicitly they become

$$\begin{aligned} &v_x[0, i_y] = U, &&v_x[N_x, i_y] = v_x[N_x - 1, i_y], &&i_y = 1, \dots, N_y, \\ &v_y[0, i_y] = -v_y[1, i_y], &&v_y[N_x + 1, i_y] = v_y[N_x, i_y], &&i_y = 1, \dots, N_y - 1, \\ &v_x[i_x, 0] = v_x[i_x, 1], &&v_x[i_x, N_y + 1] = -v_x[i_x, N_y], &&i_x = 0, \dots, N_x, \\ &v_y[i_x, 0] = 0, &&v_y[i_x, N_y] = 0, &&i_x = 0, \dots, N_x + 1. \end{aligned}$$

   Note how much care is necessary in the specifications of index ranges.

6. Repeat from Step 1 until the iteration limits are reached.

At chosen intervals the grid arrays may be displayed graphically or written out to a file for later treatment and display.

It is still quite an art to make this simple computational fluid dynamics algorithm converge. In the first-order approximation to the time evolution used here, the time step cannot be chosen too close to the limit (19.8), but must typically be a factor $\tau \approx 0.2$ smaller for convergence to be ensured over a wide range of Reynolds numbers. The limit to the number of SOR iterations may typically be chosen to be at least N_y, and the optimal SOR convergence parameter can either be obtained by trial-and-error or from a theoretical value (see [Press et al. 1992]).

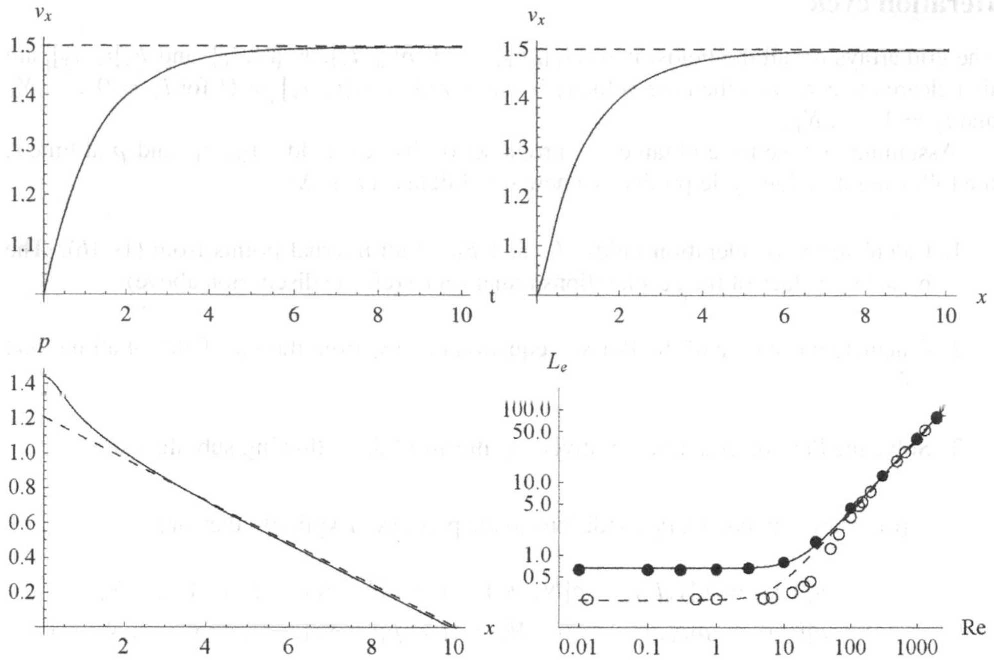

Figure 19.3. Channel entry flow at Reynolds number Re $= 100$. **Top left:** The unsteady rise of the central exit velocity $v_x(L, 0, t)$ as a function of time t. **Top right:** The rise of the steady-state central velocity $v_x(x, 0)$ as a function of the distance x from the entry. Since $U = 1$, the time and space scales are equal and the shapes of the curves quite similar. **Bottom left:** The steady-state pressure as a function of x. Its slope becomes constant (as in the Poiseuille case) at about $x = 4$. **Bottom right:** The 99% entry length plotted as a function of Reynolds number. The filled circles represent data from our simulations. The fully drawn line is the fit to (19.22) with $h = 0.66$ and $k = 0.041$. The open circles are taken from the careful theoretical calculations by Sadri and Floryan [SF02]. These authors drive their flow in a way that softens the flow near the sharp entry, and this modifies the entry length for Reynolds numbers less than 100. The dashed curve represents the fit of their data to (19.22), but one notices that this function cannot represent the data points in the interval $10 \lesssim$ Re $\lesssim 100$.

Results

Having already fixed the mass scale by choosing unit density $\rho_0 = 1$, the length scale is set by choosing unit plate distance $d = 1$, and the time scale by setting the entry velocity $U = 1$. With these units the only parameters left in the problem are the (now dimensionless) kinematic viscosity ν and the (also dimensionless) length L of the channel. The Reynolds number equals the reciprocal viscosity, Re $= 1/\nu$. The length of the channel should be chosen at least twice the expected entry length to secure that the parabolic profile becomes well established, $L \approx 2L_e$, where L_e is obtained from the estimate (16.13). In these simulations we have chosen $N_x = 40$ and $N_y = 20$ to cover the variation in both dimensions with the same density of grid points. The maximal number of SOR iterations is chosen to be 20. The algorithm runs fairly fast for $0 \lesssim$ Re $\lesssim 2{,}000$ and converges in a few thousand iterations, depending on the Reynolds number.

In Figure 19.3 (top left) the time evolution of the central exit velocity $v_x(L, 0, t)$ is plotted as a function of time for Re $= 100$. With a time interval $dt = 0.012$ it reaches 99% of the terminal velocity $v_x(L, 0, \infty)$ after about 350 time steps corresponding to $t = 4.37$. The downstream rise of the steady flow velocity $v_x(x, 0)$ is plotted in Figure 19.3 (top right), and reaches 99% of its maximum at $x = 4.49$.

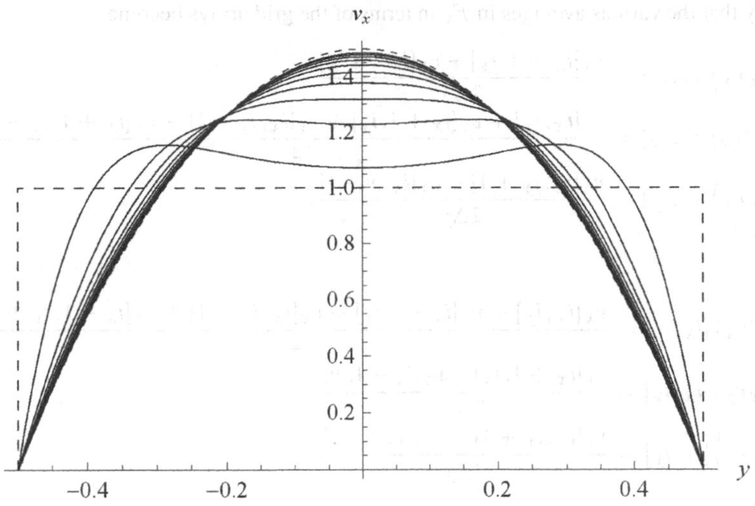

Figure 19.4. The downstream evolution of the velocity profile toward the parabolic shape is very similar to the simulation shown in Figure 16.1 on page 265.

In Figure 19.3 (bottom left) the pressure in the middle of the channel is plotted as a function of x, and it also reaches the Poiseuille form with constant gradient for $x \gtrsim 4$. Finally, in Figure 19.3 (bottom right) the entry length, defined as the point where v_x has reached 99% of maximum, is plotted as a function of Reynolds number. The data fits the interpolating formula (19.22) very well, with $h = 0.66$ and $k = 0.041$. A modern theoretical calculation by Sadri and Floryan [SF02] yields $h = 0.25$ and $k = 0.038$. The reason for the lower value of h lies in the softer entry flow used by these authors at $x = 0$, avoiding the otherwise infinitely high stresses reached at the channel walls at $x = 0$.

Problems

* **19.1** Show that it is impossible to find an integrated quantity F for which the equation for incompressible steady flow (16.1) corresponds to an extremum. Hint: Show that there is no integral F for which the variation is of the form

$$\delta F = \int [(\boldsymbol{v} \cdot \nabla)\boldsymbol{v} - \nu\nabla^2\boldsymbol{v} + \nabla p] \cdot \delta\boldsymbol{v}\, dV,$$

where $\delta\boldsymbol{v}$ vanishes at the boundaries.

19.2 **(a)** Show that the Poisson equation

$$\nabla^2 p = s$$

is the minimum of the quadratic "energy" function

$$\mathcal{E} = \int_V \left(\frac{1}{2}(\nabla p(\boldsymbol{x}))^2 + p(\boldsymbol{x})s(\boldsymbol{x})\right) dV,$$

under suitable boundary conditions. **(b)** Use this result to devise a gradient descent algorithm toward the minimum.

19.3 Indicate in Figure 19.2 which staggered grids naturally carry the various stress components, σ_{xx}, σ_{yy}, and σ_{xy}.

19.4 Verify that the various averages in F_x in terms of the grid arrays become

$$\left\langle \widehat{\nabla}_x v_x \right\rangle [i_x, i_y] = \frac{v_x[i_x+1, i_y] - v_x[i_x-1, y]}{2\Delta x},$$

$$\left\langle v_y \right\rangle [i_x, i_y] = \frac{v_y[i_x, i_y] + v_y[i_x+1, i_y] + v_y[i_x, i_y-1] + v_y[i_x+1, i_y-1]}{4},$$

$$\left\langle \widehat{\nabla}_y v_x \right\rangle [i_x, i_y] = \frac{v_x[i_x, i_y+1] - v_x[i_x, y-1]}{2\Delta y};$$

and in F_y

$$\left\langle v_x \right\rangle [i_x, i_y] = \frac{v_x[i_x, i_y] + v_x[i_x-1, i_y] + v_x[i_x, i_y+1] + v_x[i_x-1, i_y+1]}{4},$$

$$\left\langle \widehat{\nabla}_x v_y \right\rangle [i_x, i_y] = \frac{v_y[i_x+1, i_y] - v_y[i_x-1, y]}{2\Delta x},$$

$$\left\langle \widehat{\nabla}_y v_y \right\rangle [i_x, i_y] = \frac{v_y[i_x, i_y+1] - v_y[i_x, y-1]}{2\Delta y}.$$

Part IV

Balance and conservation

Balance and conservation

The overall mechanical behavior of any material body is constrained by eight mechanical laws of balance, two scalar and two vector. In many practical cases these laws of balance allow for quick estimates of how bodies of continuous matter move and interact.

None of the following chapters are absolutely necessary for a minimal curriculum, even if Chapters 20 and 21 give important insight into the practical applications of the global laws of balance.

List of chapters

20. **Mechanical balances:** A general theorem on global balance due to Reynolds is derived. Each of the laws of balance is analyzed and illustrative examples are presented.

21. **Action and reaction:** Reaction forces and moments are defined and analyzed. Momentum balance is applied to rocket propulsion and fire-hose nozzles. Euler's turbine equation is derived from angular momentum balance and applied to centrifugal and axial pumps, as well as to large-scale turbines.

22. **Energy:** Heat is kinetic energy residing in random molecular motions and must be included to establish full energy balance. The First Law of Thermodynamics is derived in both global and local forms. Fourier's heat equation is applied to a number of cases. Dissipation is calculated for generic flows, for example channel flow between parallel plates. The full set of coupled mechanical and heat equations is presented and basic thermodynamic quantities are defined.

23. **Entropy:** The concept of entropy is introduced. Local and global equations of entropy balance are derived. The Second Law of Thermodynamics, which expresses the inexorable growth of entropy in isolated systems, is discussed. Fluctuation analysis brings the estimates derived in Chapter 1 on a firm footing.

20 Mechanical balances

Momentum, angular momentum, and kinetic energy are purely mechanical quantities that, like mass, are carried along with the movement of material. This is also the case for certain thermodynamic quantities such as internal energy and entropy. These mechanical and thermodynamic quantities are all *extensive*, meaning that the amount in a composite body is the sum of the amounts in the parts. *Intensive* quantities, for example density, pressure, and temperature, are not additive.

The global equation of mass conservation expresses that mass is transported without actual loss or gain. Extensive quantities are, however, generally not conserved but have *sources* that create (and destroy) them. Thus, the source of momentum is force, the source of angular momentum is moment of force, and the source of kinetic energy is rate of work, also called power. The global laws express the *balance* between creation and accumulation of a quantity, and basically state the obvious: *In any volume of matter the net amount of a quantity produced by the source is either accumulated in the volume or leaves it through the surface.*

In this chapter we shall derive the global laws of mechanical balance in a form that clearly exposes their relation to the corresponding laws in Newtonian particle mechanics, and that is also suitable for practical applications. For each of the laws we shall discuss the circumstances under which the corresponding quantity is conserved. We shall see that momentum and angular momentum are in general conserved for isolated systems, whereas kinetic energy is not. Only when other forms of energy are included is it possible to define a total energy that *is* conserved for isolated systems. In Chapter 22 we shall turn to full-fledged thermal continuum physics, in which dissipation of mechanical energy into heat is also taken into account.

20.1 Quantities and sources

In Newtonian particle mechanics, a "body" is usually understood as a collection of a fixed number of point particles whereas in continuum mechanics the notion of a body is much more general. Any volume of material—usually called a *control volume*—may at any moment of time be viewed as a body. Intuitively we think of bodies as consisting of a number of parts with homogeneous material properties, but we shall not limit our volumes in this way although it may often be convenient to do so. In the course of time, the *control surface* $S(t)$ of the control volume $V(t)$ may be moved around and deformed *any way we desire*. This is presumably why the adjective *"control"* is used.

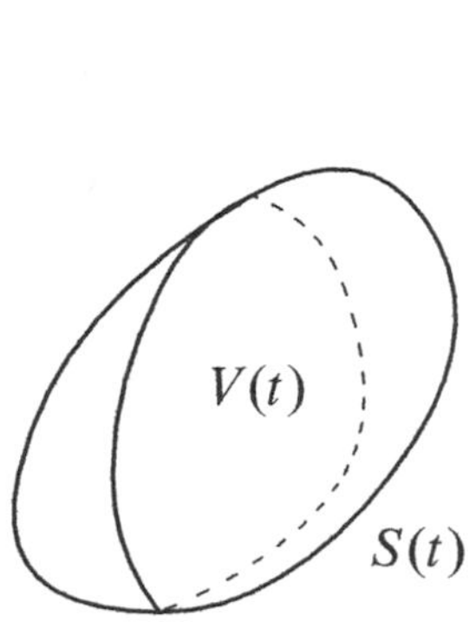

A general body consists of all the matter contained in an arbitrary time-dependent control volume $V(t)$ with control surface $S(t)$.

One may wonder whether it is really necessary to consider bodies this general. In the discussion of mass conservation (page 194), we only considered arbitrary fixed control volumes that did not change with time. Not permitting arbitrarily moving control volumes would however put unreasonable constraints on our freedom to analyze the physics of continuous matter.

Mechanical quantities

In any control volume V the global mechanical quantities are the total mass M, the total linear momentum $\mathcal{P}$, the total angular momentum $\mathcal{L}$, and the total kinetic energy $\mathcal{K}$. They are all defined as integrals over kinematic fields[1],

$$M = \int_V \rho \, dV, \qquad \mathcal{P} = \int_V \boldsymbol{v} \rho \, dV, \tag{20.1a}$$

$$\mathcal{L} = \int_V \boldsymbol{x} \times \boldsymbol{v} \rho \, dV, \qquad \mathcal{K} = \int_V \tfrac{1}{2}\boldsymbol{v}^2 \rho \, dV. \tag{20.1b}$$

The appearance of the mass density as a factor in all the integrands shows that each of these quantities may be understood as the sum of the amounts of the quantity carried by each material particle in the control volume of mass $dM = \rho \, dV$.

The remaining part of each integrand is called the *specific quantity* and represents the local amount of the quantity per unit of mass. Thus the *specific momentum* is $d\mathcal{P}/dM = \boldsymbol{v}$, the *specific angular momentum* $d\mathcal{L}/dM = \boldsymbol{x} \times \boldsymbol{v}$, and the *specific kinetic energy* $d\mathcal{K}/dM = \frac{1}{2}\boldsymbol{v}^2$. One may also add the trivial *specific mass* $dM/dM = 1$ and the *specific volume* $dV/dM = 1/\rho$ to the list.

Laws of balance

Each of the global mechanical quantities obeys a global law of balance analogous to the corresponding law of Newtonian particle mechanics (see Appendix A). Each law of balance is an equation that equals the material (comoving) time derivative of the global quantity with the corresponding source (to be defined later),

$$\frac{DM}{Dt} = 0, \qquad \frac{D\mathcal{P}}{Dt} = \mathcal{F}, \tag{20.2a}$$

$$\frac{D\mathcal{L}}{Dt} = \mathcal{M}, \qquad \frac{D\mathcal{K}}{Dt} = P. \tag{20.2b}$$

We shall soon see that $\mathcal{F}$ is the total force, $\mathcal{M}$ the total moment of force, and P the total rate of work (power). Mass has no source, and mass balance (or rather mass conservation) has already been established in Section 12.3. At this point we do not know how to define the material derivative, nor the form of the sources for linear momentum, angular momentum, and kinetic energy. In the remainder of this chapter we shall determine how these sources depend on the dynamic (force-carrying) fields.

The global laws all follow from Cauchy's equations of motion for continuous matter, and are thus automatically fulfilled for any solution to the general field equations (12.27) on page 199. *Having found such a solution, one should not worry about the balance of mass, momentum, angular momentum, or kinetic energy.* But when one is unable to solve the field equations, or when one has to guess an approximate form of the solution, the laws of balance impose two scalar and two vector—altogether eight—constraints on the system. In many cases this is sufficient for obtaining a decent understanding of its overall behavior.

[1]There is an unfortunate clash in the use of the same symbol $\mathcal{L}$ for angular momentum and lift (page 288). Total mass M and total moment of force $\mathcal{M}$, as well as total momentum $\mathcal{P}$ and total power P, are in this book distinguished by typography. There seems to be no universally accepted notation for the global quantities.

Reynolds' transport theorem

At this point it is necessary to make a small digression into a more abstract formalism that allows us once and for all to define the material (comoving) time derivative of a global quantity. There are eight individual components of the mechanical quantities (i.e., M, $\mathcal{P}_x$, $\mathcal{P}_y$, $\mathcal{P}_z$, $\mathcal{L}_x$, $\mathcal{L}_y$, $\mathcal{L}_z$, or $\mathcal{K}$), each given by a generic expression of the form,

$$Q(t) = \int_{V(t)} q(\boldsymbol{x},t)\, dM = \int_{V(t)} q(\boldsymbol{x},t)\, \rho(\boldsymbol{x},t)\, dV, \tag{20.3}$$

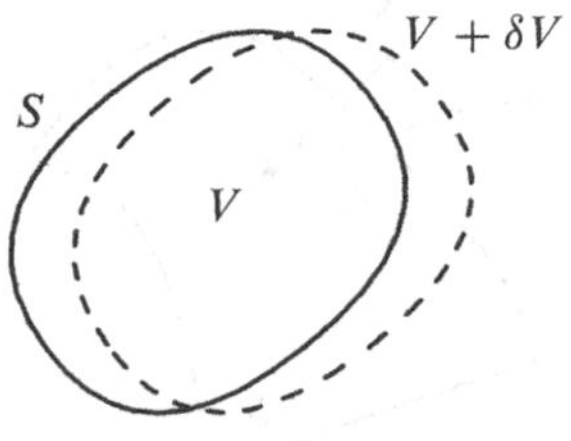

Movement of the control volume in a small time interval. The signed change in volume is given by the thin surface layer between the dashed and solid curves.

where the field $q = dQ/dM$ is the corresponding specific quantity. In the following we shall suppress the explicit space and time dependence.

Generally, we are not satisfied with only comoving control volumes, so we need to establish a relation between the material rate of accumulation, DQ/Dt, and the actual rate of accumulation, dQ/dt, in a general control volume. Let now $\boldsymbol{v}_S(\boldsymbol{x},t)$ denote the velocity of a surface element $d\boldsymbol{S}$ near the point $\boldsymbol{x}$ at time t (see the margin figure). The time dependence of the global quantity $Q(t)$ has two origins, namely the changing density of the quantity, $q(\boldsymbol{x},t)\rho(\boldsymbol{x},t)$, and the changing control volume, $V(t)$. In a small time interval δt, the volume changes to $V(t+\delta t) = V(t) + \delta V(t)$, and the global quantity changes by

$$\delta Q = \int_V \delta(q\rho)\, dV + \int_{\delta V} q\rho\, dV = \int_V \delta(q\rho)\, dV + \oint_S q\rho\, \boldsymbol{v}_S \delta t \cdot d\boldsymbol{S}.$$

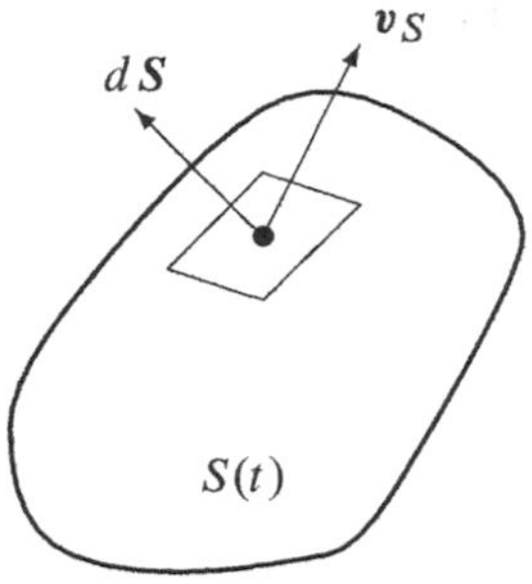

The surface element $d\boldsymbol{S}$ of the control volume moves with velocity $\boldsymbol{v}_S(\boldsymbol{x},t)$.

The second term is the (signed) amount of the quantity that the moving control surface scoops up per unit of time. Dividing by δt the actual rate of change of Q becomes

$$\frac{dQ}{dt} = \int_V \frac{\partial(q\rho)}{\partial t}\, dV + \oint_S q\rho\, \boldsymbol{v}_S \cdot d\boldsymbol{S}. \tag{20.4}$$

If the surface is fixed, $\boldsymbol{v}_S = \boldsymbol{0}$, the last term vanishes, as it should.

The surface of a *comoving* control volume always follows the material (in which case it is actually not under our control!). Setting $\boldsymbol{v}_S = \boldsymbol{v}$ in the equation above, we find

$$\frac{DQ}{Dt} = \int_V \frac{\partial(q\rho)}{\partial t}\, dV + \oint_S q\rho\, \boldsymbol{v} \cdot d\boldsymbol{S}, \tag{20.5}$$

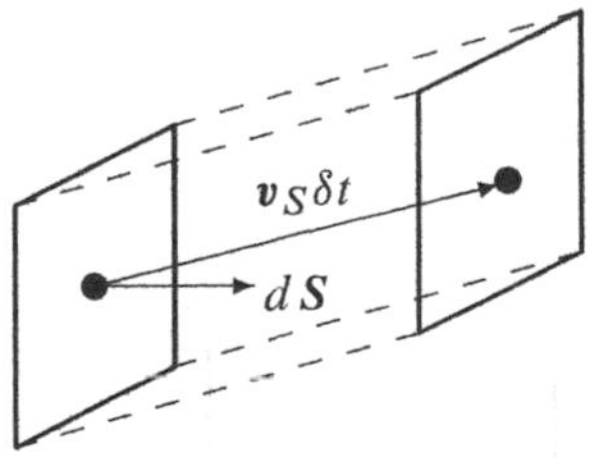

The volume $dV = \boldsymbol{v}_S \delta t \cdot d\boldsymbol{S}$ is scooped up by the control surface element $d\boldsymbol{S}$ in a small time interval δt.

and using Gauss' theorem the integrand of the resulting volume integral becomes

$$\frac{\partial(q\rho)}{\partial t} + \nabla\cdot(q\rho\boldsymbol{v}) = q\left(\frac{\partial\rho}{\partial t} + \nabla\cdot(\rho\boldsymbol{v})\right) + \rho\left(\frac{\partial q}{\partial t} + (\boldsymbol{v}\cdot\nabla)q\right).$$

The expression in the first parentheses vanishes because of the continuity equation (12.12) on page 195, and the expression in the second parentheses is recognized as the local material time derivative (12.19) of the specific quantity q. Putting it all together we arrive at the theorem

$$\boxed{\frac{DQ}{Dt} = \int_V \frac{Dq}{Dt}\rho\, dV = \int_V \frac{Dq}{Dt}\, dM.} \tag{20.6}$$

It provides an important link between the local and global material rates of change, and facilitates—as we shall see in the following sections—the derivation of the global laws of balance from the basic dynamical equations.

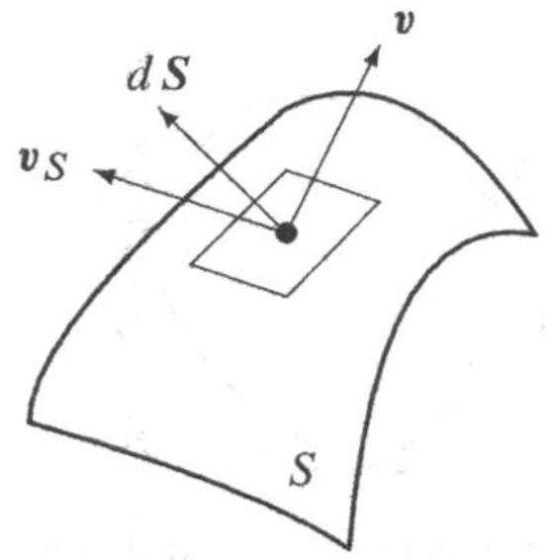

Matter moves with velocity v through the surface element dS, which itself moves with velocity v_S. The velocity of the matter relative to the moving surface is $v - v_S$.

It is often fairly easy to calculate the amount of a quantity Q in the chosen control volume and thereby its ordinary time derivative dQ/dt. Combining (20.4) and (20.5), the material derivative may be written as

$$\frac{DQ}{Dt} = \frac{dQ}{dt} + \dot{Q}, \tag{20.7}$$

where

$$\dot{Q} = \oint_V q\rho(\boldsymbol{v} - \boldsymbol{v}_S) \cdot d\boldsymbol{S}. \tag{20.8}$$

Since $\boldsymbol{v} - \boldsymbol{v}_S$ is the velocity of matter relative to the moving control surface (see the margin figure), $\dot{Q}$ is the (signed) rate at which the quantity leaves the control volume through its surface, also called the *outflow* of the quantity. It may conveniently be written

$$\dot{Q} = \oint_S q\, d\dot{M}, \qquad d\dot{M} = \rho(\boldsymbol{v} - \boldsymbol{v}_S) \cdot d\boldsymbol{S}, \tag{20.9}$$

where $d\dot{M}$ is the *mass flow rate* through a surface element.

The relation (20.7) goes under the name of *Reynolds' transport theorem* (1903) and constitutes together with (20.6) the general basis for global analysis in continuum physics. The theorem simply states that *the rate of accumulation of a quantity in a comoving volume equals the rate of accumulation in a general volume plus the rate at which the quantity leaves the general volume through its surface.* Combined with an equation of balance, this theorem proves the "obvious" statement made in the introduction to this chapter.

20.2 Mass balance

Specific mass corresponds to $q(\boldsymbol{x}, t) = 1$, implying $Dq/Dt = 0$. From (20.6) it follows that the mass source vanishes, leading to the *equation of global mass balance*,

$$\frac{DM}{Dt} = \frac{dM}{dt} + \dot{M} = 0, \qquad \dot{M} = \oint_S d\dot{M}. \tag{20.10}$$

The vanishing source expresses that mass is always conserved because its rate of accumulation, $dM/dt = -\dot{M}$, in any control volume equals the mass flow *into* the volume through its surface. Previously this relation was derived for a fixed volume (page 194), but now we see that it may actually be applied to an arbitrary moving control volume. For steady flow in a fixed volume, we have $dM/dt = 0$, so that the total mass flow must vanish, $\dot{M} = 0$.

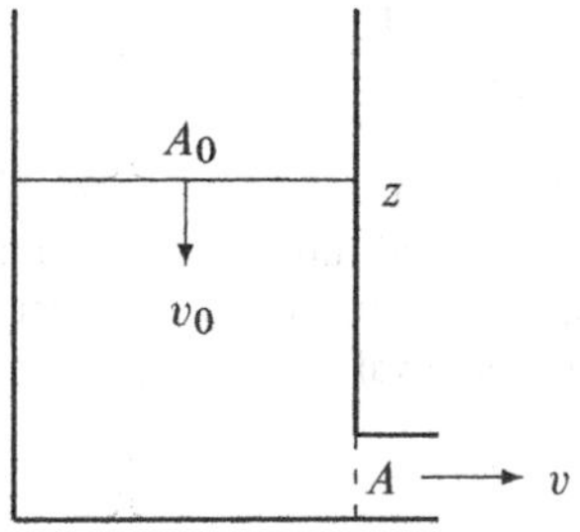

Container drained through a pipe with cross-sectional area A. The water streams into the drain with average velocity v. The control volume is bounded by the fixed cistern walls, the moving horizontal water surface, and the fixed entry to the pipe.

Application: Down the drain

Suppose a cistern filled with water is flushed through a drain pipe near the bottom of the container. In this case the flow is unsteady and we do not know the actual flow pattern, but mass conservation nevertheless allows us to draw an important conclusion that turns out to be rather trivial. Let the drain pipe have entry cross-section A, and the cistern be vertical with horizontal cross-section A_0. We choose a control volume encompassing all the water in the cistern but not the pipe (see the margin figure). As the cistern drains, the level of the horizontal water surface drops, whereas the other parts of the control surface are fixed. This is an example of a natural choice of control volume that is neither fixed nor comoving.

When the water level is z, the total mass of the water is $M = \rho_0 A_0 z$, and the rate of mass loss through the drain is $\dot{M} = \rho_0 A v$, where v is the average velocity of the drain flow. There is no mass flow through the moving open surface of the cistern. Global mass balance (20.10) and Reynolds' transport theorem (20.7) with $q = 1$ now leads to

$$\frac{DM}{Dt} = \frac{dM}{dt} + \dot{M} = \rho_0 A_0 \frac{dz}{dt} + \rho_0 A v = 0. \tag{20.11}$$

Since $dz/dt = -v_0$ where v_0 is the average downward velocity of the water in the cistern, we obtain $A_0 v_0 = Av$, which is nothing but Leonardo's law (12.9) on page 193. Although this looks like a bit of an overkill to get a well-known result, it illustrates the way the global law works for a non-trivial control volume.

20.3 Momentum balance

The total (linear) momentum in a control volume is

$$\mathcal{P} = \int_V \boldsymbol{v}\, \rho dV = \int_M \boldsymbol{v}\, dM, \tag{20.12}$$

representing the sum of the momenta of all the material particles in the control volume.

Total force

Using Cauchy's equation (12.26) on page 198, that is, $\rho D\boldsymbol{v}/Dt = \boldsymbol{f}^*$, it follows from (20.6) with $q \to \boldsymbol{v}$ that the *global momentum balance* takes the form

$$\boxed{\frac{D\mathcal{P}}{Dt} \equiv \frac{d\mathcal{P}}{dt} + \dot{\mathcal{P}} = \mathcal{F},} \tag{20.13}$$

with

$$\dot{\mathcal{P}} = \oint_S \boldsymbol{v}\, d\dot{M}, \qquad\qquad \mathcal{F} = \int_V \boldsymbol{f}^*\, dV. \tag{20.14}$$

It has been previously shown (page 102) that $\mathcal{F}$ is indeed the total force,

$$\boxed{\mathcal{F} = \int_V \boldsymbol{f}\, dV + \oint_S \boldsymbol{\sigma} \cdot d\boldsymbol{S},} \tag{20.15}$$

composed of body forces and external stresses acting on the control surface. Momentum balance thus tells us that *the total momentum of a comoving control volume is constant (i.e., conserved) if and only if the total force vanishes.*

* Internal and external forces

Whereas the contact forces acting on the surface of the control volume are always external, the volume forces acting in the volume V may be split into external and internal parts,

$$\boldsymbol{f} = \boldsymbol{f}_{\text{int}} + \boldsymbol{f}_{\text{ext}}, \tag{20.16}$$

where $\boldsymbol{f}_{\text{int}}$ is produced by the material inside the volume V, and $\boldsymbol{f}_{\text{ext}}$ by the material outside.

Static volume forces are normally *two-particle* forces, which means that a material particle of volume dV' situated at $\boldsymbol{x}'$ acts on a material particle of volume dV at $\boldsymbol{x}$ with a force $d^2\boldsymbol{\mathcal{F}} = \boldsymbol{f}(\boldsymbol{x}, \boldsymbol{x}')dVdV'$. The internal and external forces then become

$$\boldsymbol{f}_{\text{int}} = \int_V \boldsymbol{f}(\boldsymbol{x}, \boldsymbol{x}')\,dV', \qquad \boldsymbol{f}_{\text{ext}} = \int_{-V} \boldsymbol{f}(\boldsymbol{x}, \boldsymbol{x}')\,dV', \tag{20.17}$$

where $-V$ is shorthand for the volume of the material outside the volume V.

Static volume forces must also obey Newton's Third Law, $\boldsymbol{f}(\boldsymbol{x}, \boldsymbol{x}') = -\boldsymbol{f}(\boldsymbol{x}', \boldsymbol{x})$, and the antisymmetry of the integrand implies that the total internal force vanishes,

$$\boldsymbol{\mathcal{F}}_{\text{int}} = \int_V \boldsymbol{f}_{\text{int}}\,dV = \int_V \int_V \boldsymbol{f}(\boldsymbol{x}, \boldsymbol{x}')\,dV\,dV' = \boldsymbol{0}. \tag{20.18}$$

Only the external body forces thus contribute to the total force (20.15), so that *the total momentum of a comoving control volume is conserved if the total external force vanishes and all internal forces are two-particle forces obeying Newton's Third Law.*

Application: Launch of a small rocket

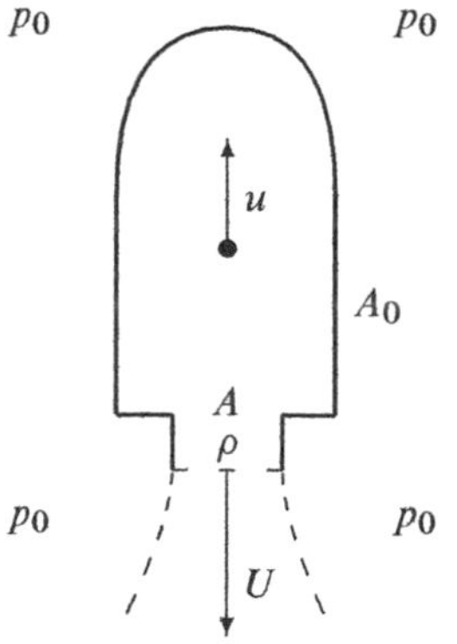

Sketch of a rocket moving upward with vertical velocity u while emitting material of density ρ and relative velocity U through an opening of cross-section A. The control volume moves with the rocket and follows the outside of the hull, cutting across the exhaust opening. The rocket moves through an atmosphere with ambient air pressure p_0 everywhere around the rocket.

As a demonstration of how to use global momentum balance, we consider the launch of a small fireworks rocket. The rocket accelerates vertically upward by burning chemical fuel and spewing the hot reaction gases downward. The gases are emitted from the rocket through an opening (nozzle) with cross-section A, density ρ, and velocity U relative to the rocket body. We assume that the gas velocity is much smaller than the velocity of sound (which it is in toy rockets), so that the pressure at the exit may be taken to be equal to the ambient atmospheric pressure p_0.

It is most convenient to choose a control volume that follows the *outside* of the rocket, cutting across the nozzle outlet (see the margin figure). Such a control volume always moves with the instantaneous speed u of the rocket relative to the ground. It contains at any moment all the material of the rocket, including the remaining fuel and the burning gases, but not the gases that have been exhausted through the nozzle. Assuming that the gas exits uniformly across the nozzle, the outflow of mass becomes $\dot{M} = \rho UA$ during the burn. After the burn the rocket moves ballistically with constant mass like a cannonball shot vertically upward, and eventually falls back to the ground.

To avoid fictitious forces we choose to analyze the problem in an inertial reference frame, the natural one being at rest with respect to the ground where the rocket's instantaneous velocity is u (see however Problem 20.2). In this frame, the vertical velocity of the exhaust gases is $v_z = u - U$. Assuming that the center of gravity remains fixed relative to the rocket during the burn (and that is by no means sure), the total momentum is $\mathcal{P}_z = Mu$, where M is the instantaneous mass of the rocket (i.e., the control volume). Its material derivative becomes, according to Reynolds' theorem (20.7),

$$\frac{D\mathcal{P}_z}{Dt} = \frac{d(Mu)}{dt} + v_z\dot{M} = M\frac{du}{dt} - U\dot{M}. \tag{20.19}$$

In the last step we have inserted $v_z = u - U$ and used mass balance $dM/dt + \dot{M} = 0$.

The total vertical force on the rocket is the sum of its weight $-Mg_0$ and the resistance or drag caused by the interaction of the rocket's hull with the air. Air drag has two components: *skin drag* from viscous friction between air and hull, and *form drag* from the changes in pressure at the hull caused by the rocket "punching" through the atmosphere. The magnitude of the form drag can be estimated from the stagnation pressure increase $\Delta p = \frac{1}{2}\rho_0 U^2$ at the tip of the rocket where ρ_0 is the air density (see page 215). Multiplying by the cross-section of the rocket A_0, the estimate of the form drag becomes $-\frac{1}{2}\rho_0 u^2 A_0$ (with opposite sign if u is negative). Form drag thus grows quadratically with rocket speed, and at high Reynolds

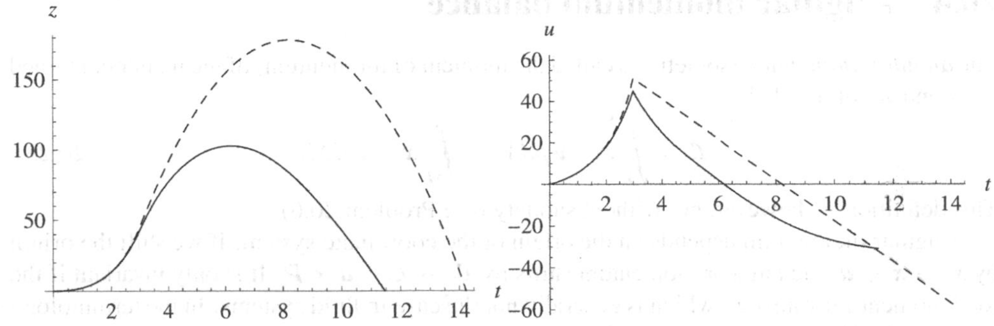

Figure 20.1. Vertical flight of a small fireworks rocket. Height (left) and velocity (right) as a function of time during with form drag (solid) and without (dashed). The cusp in the velocity happens at the end of the burn and signals the transition to ballistic flight. The rocket has diameter 6 cm, total mass 1 kg, payload 0.2 kg, and form drag coefficient $C_D = 1$. The rocket burns for 3 s, emitting gases at a speed of 50 m s^{-1}. The numeric solution of the rocket equation (20.21) shows that with form drag the rocket reaches a velocity of 45 m s^{-1} at the end of the burn and a height of 43 m. During the subsequent ballistic flight, it reaches a maximum height of 103 m about 6 s after start, and finally it falls back and hits the ground again with a speed of 30 m s^{-1} about 11.5 s after start. Without form drag the rocket would reach a maximum height of 178 m about 8 s after start, and it would hit the ground with a speed of 60 m s^{-1} after 14 s.

number it completely dominates skin drag, which only grows linearly. Leaving out skin drag, the total vertical force on the control volume may be written (with correct sign) as

$$\mathcal{F}_z = -Mg_0 - \mathcal{D}, \qquad \mathcal{D} = \tfrac{1}{2}\rho_0 \text{sign}(u)u^2 A_0 C_D. \tag{20.20}$$

Here we have also included a dimensionless factor C_D, called the *drag coefficient*, which takes the actual shape of the rocket into account. Typically, the drag coefficient is about 0.5 to 1.0 for plump bodies at high Reynolds numbers (page 298).

Momentum balance (20.13) equates the material rate of change of momentum with the total force, $D\mathcal{P}_z/Dt = \mathcal{F}_z$, and we arrive at the *rocket equations*

$$\boxed{M\frac{du}{dt} = U\dot{M} - Mg_0 - \mathcal{D}, \qquad \frac{dM}{dt} = -\dot{M}.} \tag{20.21}$$

Konstantin Eduardovich Tsiolkovsky (1857–1935). Russian rocket scientist. Considered the father of spaceflight, he early considered multistage rockets to reach Earth orbit and was the first to imagine a space elevator. Worked as a high-school mathematics teacher and was also an imaginative science fiction writer. (Source: Wikimedia Commons.)

If we ignore form drag by taking $\mathcal{D} = 0$, and assume that the gas density ρ and velocity U are constant in time, the mass outflow $\dot{M} = \rho U A$ will be constant. The equations are now easily solved during the burn, with the result that

$$u = U \log\frac{M_0}{M} - g_0 t, \qquad M = M_0 - \dot{M}t, \tag{20.22}$$

where M_0 is the initial mass at $t = 0$. Expanding for small t, it follows that the condition for take-off, $u > 0$, requires $U\dot{M} > M_0 g_0$. In the absence of gravity, $g_0 = 0$, the first equation is called *Tsiolkovsky's rocket equation (1903)*.

In general, for $C_D > 0$, the rocket equations can only be solved numerically. The results are shown for a typical fireworks rocket in Figure 20.1. One notes how the form drag reduces the maximum height for this rocket to about half of what it would be without drag (dashed curve). If the rocket shape were highly streamlined, the drag coefficient would be somewhat smaller, allowing it to attain greater heights for the same amount of fuel.

20.4 Angular momentum balance

The *angular momentum* (sometimes called the moment of momentum) of the matter contained in a control volume V is,

$$\mathcal{L} = \int_V \boldsymbol{x} \times \boldsymbol{v}\, \rho dV = \int_M \boldsymbol{x} \times \boldsymbol{v}\, dM. \tag{20.23}$$

This definition is, however, not without subtlety (see Problem 20.6).

Angular momentum depends on the origin of the coordinate system. If we shift the origin by $\boldsymbol{x} \to \boldsymbol{x} + \boldsymbol{a}$, the angular momentum shifts by $\mathcal{L} \to \mathcal{L} + \boldsymbol{a} \times \mathcal{P}$. It is only invariant if the total momentum vanishes, which is generally not the case in fluid systems. In the terminology of Appendix B, angular momentum is an improper vector.

Total moment of force

The material time derivative of the specific angular momentum $\boldsymbol{x} \times \boldsymbol{v}$ is

$$\rho\frac{D(\boldsymbol{x} \times \boldsymbol{v})}{Dt} = \rho\frac{D\boldsymbol{x}}{Dt} \times \boldsymbol{v} + \rho\boldsymbol{x} \times \frac{D\boldsymbol{v}}{Dt} = \boldsymbol{x} \times \boldsymbol{f}^*,$$

because $D\boldsymbol{x}/Dt = \boldsymbol{v}$ and $\rho D\boldsymbol{v}/Dt = \boldsymbol{f}^*$. Using (20.6) with $q \to \boldsymbol{x} \times \boldsymbol{v}$, we arrive at the equation of *global angular momentum balance,*

$$\boxed{\frac{D\mathcal{L}}{Dt} \equiv \frac{d\mathcal{L}}{dt} + \dot{\mathcal{L}} = \mathcal{M},} \tag{20.24}$$

with

$$\dot{\mathcal{L}} = \oint_S \boldsymbol{x} \times \boldsymbol{v}\, d\dot{M}, \qquad \mathcal{M} = \int_V \boldsymbol{x} \times \boldsymbol{f}^*\, dV. \tag{20.25}$$

The source $\mathcal{M}$ is the total moment of the effective forces acting on the material particles in the control volume. Under a shift $\boldsymbol{x} \to \boldsymbol{x} + \boldsymbol{a}$, the total moment transforms as $\mathcal{M} \to \mathcal{M} + \boldsymbol{a} \times \mathcal{F}$, where $\mathcal{F}$ is the total force. Combined with momentum balance (20.13), both sides of the angular momentum balance (20.24) thus shift in the same way so that the form of this relation is independent of the choice of origin.

It is, however, a bit more complicated to derive the global form of the total moment than it was for the total force. Inserting the effective force (6.15) on page 103,

$$f_i^* = f_i + \sum_j \nabla_j \sigma_{ij}, \tag{20.26}$$

we find the identity (see Appendix B)

$$(\boldsymbol{x} \times \boldsymbol{f}^*)_i - (\boldsymbol{x} \times \boldsymbol{f})_i = \sum_{jkl} \epsilon_{ijk} x_j \nabla_l \sigma_{kl} = \sum_{jkl} \epsilon_{ijk} \nabla_l (x_j \sigma_{kl}) - \sum_{jk} \epsilon_{ijk} \sigma_{kj}.$$

The first term in the previous expression is a divergence, which by Gauss' theorem leads to a surface integral, and the last term vanishes when the stress tensor is symmetric, which we assume it is. After integration, the total moment of force may be written

$$\boxed{\mathcal{M} = \int_V \boldsymbol{x} \times \boldsymbol{f}\, dV + \oint_S \boldsymbol{x} \times \boldsymbol{\sigma} \cdot d\boldsymbol{S}.} \tag{20.27}$$

The total moment thus consists of the moment of the body forces plus the moment of the external contact forces acting on the control surface. From angular momentum balance (20.24) it follows immediately that *the total angular momentum in a comoving control volume is constant (i.e., conserved) if and only if the total moment of force vanishes.*

* Internal moment of force

Splitting the body force density into internal and external contributions as in (20.16), and assuming again that we are dealing with two-particle body forces $\boldsymbol{f}(\boldsymbol{x}, \boldsymbol{x}')$ obeying Newton's Third Law, $\boldsymbol{f}(\boldsymbol{x}, \boldsymbol{x}') = -\boldsymbol{f}(\boldsymbol{x}', \boldsymbol{x})$, the moment of the internal body forces may be written as

$$\begin{aligned} \boldsymbol{\mathcal{M}}_{\text{int}} &= \int_V \boldsymbol{x} \times \boldsymbol{f}_{\text{int}}\, dV = \int_V \int_V \boldsymbol{x} \times \boldsymbol{f}(\boldsymbol{x}, \boldsymbol{x}')\, dV\, dV' \\ &= \int_V \int_V \boldsymbol{x}' \times \boldsymbol{f}(\boldsymbol{x}', \boldsymbol{x})\, dV\, dV' = \frac{1}{2} \int_V \int_V (\boldsymbol{x} - \boldsymbol{x}') \times \boldsymbol{f}(\boldsymbol{x}, \boldsymbol{x}')\, dV\, dV'. \end{aligned}$$

In the second line we interchanged $\boldsymbol{x}$ and $\boldsymbol{x}'$ and used the antisymmetry of $\boldsymbol{f}(\boldsymbol{x}, \boldsymbol{x}')$.

If the two-particle forces are furthermore *central*, $\boldsymbol{f}(\boldsymbol{x}, \boldsymbol{x}') \sim \boldsymbol{x} - \boldsymbol{x}'$, which is the case for gravity and electrostatics, the cross-product vanishes, and we conclude that the moment of the internal forces vanishes, $\boldsymbol{\mathcal{M}}_{\text{int}} = \boldsymbol{0}$. Under these conditions, internal forces can simply be ignored in the calculation of the moment of force, and we have thus shown that *the angular momentum of a comoving control volume is conserved for a system not subject to an external moment of force as long as the internal two-particle forces are central.*

Application: Spinning up a rotating lawn sprinkler

Commercial rotating lawn sprinkler with two arms. (Source: Image courtesy Dural Irrigation, Australia. With permission.)

As an illustration of the use of angular momentum balance we shall calculate how a rotating lawn sprinkler spins up after the water pressure is turned on. Some rotating lawn sprinklers are constructed with a small number n of horizontal arms of length R mounted on a common pivot. Each arm is a tube carrying water from the pivot toward a nozzle with outlet area A. The water leaves the nozzle with constant vector velocity $\boldsymbol{U}$ along the normal to the nozzle exit. When the water pressure is turned on at the pivot, the sprinkler begins to rotate and reaches after a while a steady state in which it rotates with constant angular velocity. The sprinkler is in fact a *radial outflow turbine*.

We choose a control volume that follows the outer surface of the sprinkler, cutting across the nozzle outlets and horizontally through the pivot. In cylindrical coordinates with the z-axis along the axis of rotation, the angular momentum of the sprinkler is $\boldsymbol{\mathcal{L}} = I\, \Omega\, \hat{\boldsymbol{e}}_z$, where I is the moment of inertia of the whole sprinkler plus water and Ω is the instantaneous angular velocity (see the margin figure). The moment of inertia should be calculated in the usual way from the mass distribution in the arms, the pivot, and the water contained in the system. For incompressible water, the moment of inertia is time-independent, and the amount of water entering the sprinkler equals the amount it discharges, so that the total mass outflow vanishes, $\dot{M} = 0$.

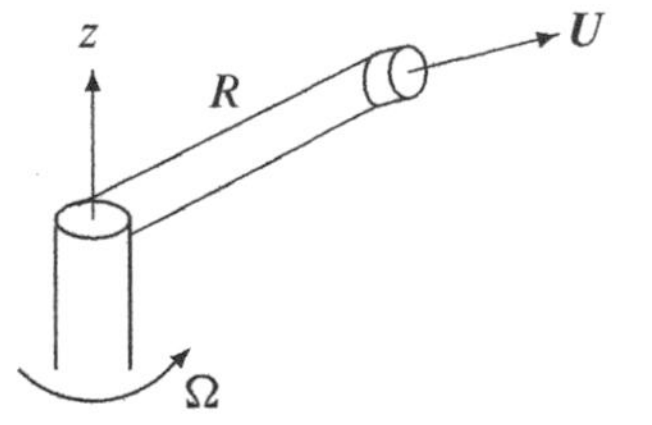

Arm of a lawn sprinkler. The short nozzle is both bent and elevated such that the water leaves with velocity $\boldsymbol{U}$ relative to the arm. The control volume envelops the rotating sprinkler and cuts across the nozzle outlet and the pivot inlet (not shown).

The water entering vertically through the pivot carries no angular momentum, so the pivot does not contribute to the surface integral in Reynolds' transport theorem. In cylindrical coordinates, the control surface moves with the velocity $\boldsymbol{v}_S = R\Omega\hat{\boldsymbol{e}}_\phi$ and the water leaves with velocity $\boldsymbol{v} = R\Omega\hat{\boldsymbol{e}}_\phi + \boldsymbol{U}$. Since $(\boldsymbol{x} \times \boldsymbol{v})_z = Rv_\phi = R(R\Omega + U_\phi)$ at the outlet, we get

$$\frac{D\mathcal{L}_z}{Dt} = \frac{d(I\Omega)}{dt} + R(R\Omega + U_\phi)Q, \tag{20.28}$$

where $Q = n\rho_0 UA$ is the total mass discharge from the n identical arms of the sprinkler. Disregarding friction in the pivot (and air resistance on the arms), the moment of force along z vanishes, $\mathcal{M}_z = 0$, and we get

$$I\frac{d\Omega}{dt} + R(R\Omega + U_\phi)Q = 0, \tag{20.29}$$

because of angular momentum conservation.

With initial condition $\Omega = 0$ at $t = 0$, the solution to this differential equation is

$$\Omega = -\frac{U_\phi}{R}\left(1 - e^{-t/\tau}\right), \qquad \tau = \frac{I}{R^2 Q}. \tag{20.30}$$

The sign shows, not unsurprisingly, that the sprinkler rotates in the opposite direction of the tangential component of the water jet. For $t \gg \tau$, the rotation becomes steady with angular velocity $\Omega_0 = -U_\phi/R$, and the tangential velocity component vanishes, $v_\phi = 0$, relative to the ground. The water is thrown away from the sprinkler nozzle with radial velocity $v_r = U_r$ and vertical velocity $v_z = U_z$, depending on the construction of the sprinkler (see Problem 20.5). This is of course an ideal case; in reality there will be viscous losses and friction in the pivot, making v_ϕ non-zero in steady rotation, as well as air resistance, which influences the shape of the ballistic orbit of the emitted water.

20.5 Kinetic energy balance

The total kinetic energy of the material in a control volume is defined to be

$$\mathcal{K} = \int_V \tfrac{1}{2}\boldsymbol{v}^2 \rho\, dV = \int_M \tfrac{1}{2}\boldsymbol{v}^2\, dM. \tag{20.31}$$

It must be emphasized that this is only the kinetic energy of the mean flow of matter represented by the velocity field $\boldsymbol{v}$, and that there is a "hidden" kinetic energy associated with the thermal motion of the molecules relative to the mean flow. We shall return to this question in Chapter 22 (see also Problem 20.6).

Total power

The material derivative of the specific kinetic energy $\frac{1}{2}\boldsymbol{v}^2$ is calculated by means of the fundamental dynamical equation $\rho D\boldsymbol{v}/Dt = \boldsymbol{f}^*$,

$$\rho \frac{D\left(\frac{1}{2}\boldsymbol{v}^2\right)}{Dt} = \rho\boldsymbol{v}\cdot\frac{D\boldsymbol{v}}{Dt} = \boldsymbol{v}\cdot\boldsymbol{f}^*.$$

Using Reynolds relation (20.6) with $q \to \frac{1}{2}\boldsymbol{v}^2$, we arrive at the *global equation of kinetic energy balance*,

$$\boxed{\frac{D\mathcal{K}}{Dt} \equiv \frac{d\mathcal{K}}{dt} + \dot{\mathcal{K}} = P,} \tag{20.32}$$

where

$$\dot{\mathcal{K}} = \oint_S \frac{1}{2}\boldsymbol{v}^2\, d\dot{M}, \qquad P = \int_V \boldsymbol{v}\cdot\boldsymbol{f}^*\, dV. \tag{20.33}$$

Thus we conclude that the source of kinetic energy P is the total *rate of work*, or *power*, of the effective forces $d\boldsymbol{\mathcal{F}} = \boldsymbol{f}^*\, dV$ acting on the moving material particles in the control volume. Formally, we may say that the kinetic energy is conserved when the total power vanishes, but that is—as we shall see—rarely the case.

To derive a global expression for the power we again use Equation (6.16) to get

$$\boldsymbol{v} \cdot \boldsymbol{f}^* - \boldsymbol{v} \cdot \boldsymbol{f} = \sum_{ij} v_i \nabla_j \sigma_{ij} = \sum_{ij} \nabla_j (v_i \sigma_{ij}) - \sum_{ij} \sigma_{ij} \nabla_j v_i .$$

Using Gauss' theorem on the first contribution, we find, with $\boldsymbol{\sigma} : \nabla \boldsymbol{v} \equiv \sum_{ij} \sigma_{ij} \nabla_j v_i$,

$$P = \int_V \boldsymbol{v} \cdot \boldsymbol{f} \, dV + \oint_S \boldsymbol{v} \cdot \boldsymbol{\sigma} \cdot d\boldsymbol{S} - \int_V \boldsymbol{\sigma} : \nabla \boldsymbol{v} \, dV. \tag{20.34}$$

The total power thus consists of the power of the volume forces (external as well as internal), the power of the contact forces (always external), plus a third contribution (including the sign) that we interpret as the *power of the internal stresses.* Although Newton's Third Law guarantees that the contact forces between neighboring material particles cancel, the rate of work of these forces does not cancel because neighboring particles have slightly different velocities, as witnessed by the appearance of the velocity gradients $\nabla_j v_i$ in the last term.

There is *in general no conservation of kinetic energy*, even in the absence of external body forces and stresses. As long as there are velocity differences, the contribution from the last term will in general not vanish. For incompressible viscous fluid we have seen (page 252) that it is always negative and therefore represents a continued drain of kinetic energy from the control volume, called *dissipation.* In Chapter 22 we shall see that the dissipated kinetic energy is not lost but reappears as internal energy (heat).

Application: Pulling the plug

As an illustration of the use of kinetic energy balance, we return to the familiar example of a cistern being drained through a narrow pipe. Initially the pipe is full, and all the water is assumed to be at rest. When the plug is pulled at the bottom of the cistern, the fluid goes through a short initial acceleration before steady flow is established.

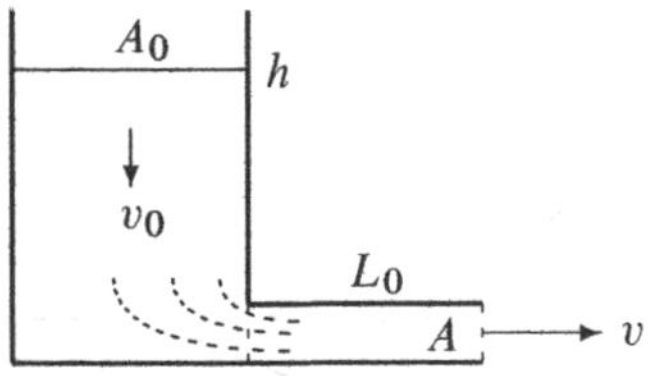

The water first accelerates in the pipe while pouring out through the exit. The kinetic energy of the water in the transition region between cistern and pipe (dashed flowlines) is assumed to be negligible.

The drain is a horizontal pipe of length L_0 and cross-section A, and the cistern is vertical with large cross-section $A_0 \gg A$. For simplicity the cistern is assumed to be continuously refilled to maintain a constant water level h, and the water is assumed to be non-viscous and incompressible, so that $\sigma_{ij} = -p\delta_{ij}$ and $\nabla \cdot \boldsymbol{v} = 0$. The control volume is fixed and contains all the water in the system between the open surface, the pipe outlet, and the walls of the cistern and pipe.

To estimate the various terms in the kinetic energy balance we shall again make the *plug flow* approximation in which the flow is assumed to have the same vertical velocity throughout the cistern, and the same horizontal velocity everywhere in the pipe. Ignoring the transition region from the cistern to the pipe (see Problem 20.7), the total kinetic energy of the water in the system is approximatively given by the sum of the kinetic energies of the water moving vertically down through the cistern with average velocity v_0 and the water moving horizontally through the pipe with average velocity v. Using mass conservation (Leonardo's law), $v_0 A_0 = vA$, to eliminate v_0, we find

$$\mathcal{K} \approx \frac{1}{2}\rho_0 A_0 h \, v_0^2 + \frac{1}{2}\rho_0 A L_0 v^2 = \frac{1}{2}\rho_0 A v^2 \left(L_0 + h \frac{A}{A_0} \right) = \frac{1}{2}\rho_0 A L v^2, \tag{20.35}$$

where

$$L = L_0 + h \frac{A}{A_0}. \tag{20.36}$$

Even if $A/A_0 \ll 1$, the water level may be so high and the pipe so short that the second term cannot be ignored.

Kinetic energy is only lost through the outlet A and gained through the inlet A_0, and in the same approximation as above, the net rate of loss of kinetic energy from the system is

$$\dot{\mathcal{K}} = \oint_S \tfrac{1}{2} \boldsymbol{v}^2 \, d\dot{M} \approx \frac{1}{2}\rho_0 A v^3 - \frac{1}{2}\rho_0 A_0 v_0^3 = \frac{1}{2}\rho_0 A v^3 \left(1 - \frac{A^2}{A_0^2}\right) \approx \frac{1}{2}\rho_0 A v^3,$$

where we in the last step have used $A \ll A_0$. Consequently, the material derivative of the kinetic energy becomes

$$\frac{D\mathcal{K}}{Dt} = \rho_0 A L v \frac{dv}{dt} + \frac{1}{2}\rho_0 A v^3, \tag{20.37}$$

to be used as the left-hand side of kinetic energy balance (20.32).

For non-viscous incompressible fluids, the rate of work of the internal contact forces vanishes because $\sum_{ij} \sigma_{ij} \nabla_j v_i = -p \nabla \cdot \boldsymbol{v} = 0$. At both the open surface and at the pipe outlet, there is atmospheric pressure p_0, such that rate of work of the external pressure becomes

$$-\oint p \, \boldsymbol{v} \cdot d\boldsymbol{S} = -p_0(-v_0)A_0 - p_0 v A = 0. \tag{20.38}$$

Gravity only performs work on the descending water in the cistern at the rate

$$P = \int_V \rho_0 \boldsymbol{v} \cdot \boldsymbol{g} \, dV \approx \rho_0(-v_0)(-g_0)A_0 h = \rho_0 g_0 v A h. \tag{20.39}$$

If the pipe is not horizontal, the vertical drop in the pipe should also be included.

Kinetic energy balance, $D\mathcal{K}/Dt = P$, therefore results in the following differential equation for the drain pipe velocity,

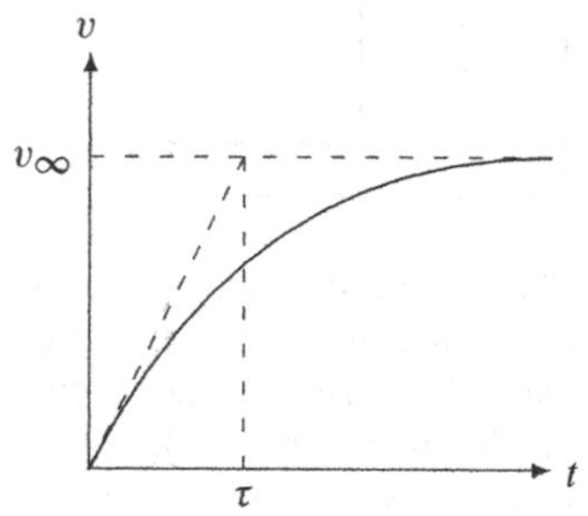

Rise of the velocity in the drain pipe toward Torricelli's terminal velocity. As shown in the drawing, the characteristic risetime can also be obtained by a geometric construction.

$$\frac{dv}{dt} = \frac{v_\infty^2 - v^2}{2L}, \qquad v_\infty = \sqrt{2g_0 h}. \tag{20.40}$$

The right-hand side of the differential equation vanishes for $v = v_\infty$, which is the well-known terminal velocity obtained from Torricelli's law (page 214). Solving the differential equation with the initial condition $v = 0$ at $t = 0$, one finds

$$v(t) = v_\infty \tanh \frac{t}{\tau}, \qquad \tau = \frac{2L}{v_\infty}. \tag{20.41}$$

Not unreasonably, the characteristic rise time τ equals the time it takes the water to pass through the pipe at half the terminal speed. At $t = 2\tau$, the water has reached 96% of the terminal speed.

Example 20.1 [Barrel of wine]: For a cylindrical barrel of wine (Example 13.3 on page 215) with diameter 1 m and height 2 m, emptied through a spout with diameter 5 cm and length $L_0 =$ 20 cm, the effective length becomes $L = 20.5$ cm. The terminal speed is about 6 m s^{-1}, and the water acquires 96% of this speed after just 0.13 s! If you try to put back the plug in the spout after merely half a second, there will nevertheless be about 6 liters of wine on the floor.

20.6 Mechanical energy balance

Even if kinetic energy is not conserved in general, kinetic energy balance can—as shown in the preceding section—sometimes be useful in obtaining an estimate of the solution to a simple fluid mechanics problem by calculating the rate of work of the external forces. In this section we shall see that under quite general circumstances it is possible to avoid this calculation.

Mechanical energy

Part of the body force is often due to a *static* field of gravity, $\boldsymbol{g} = -\nabla\Phi(\boldsymbol{x})$. Splitting off this part from the total body force, we write

$$\boldsymbol{f} = \rho\boldsymbol{g} + \tilde{\boldsymbol{f}}, \tag{20.42}$$

where $\tilde{\boldsymbol{f}}$ is the residual body force, due to time-dependent gravitational fields and to electromagnetic fields. Since the potential energy of a material particle in the static gravitational field is $\Phi(\boldsymbol{x})\,dM$, the *potential energy* of the body becomes

$$\mathcal{V} = \int_V \Phi\, dM = \int_V \Phi\,\rho\, dV. \tag{20.43}$$

As in ordinary mechanics it is useful to extend the concept of energy by defining the *total mechanical energy,*

$$\mathcal{E} = \mathcal{K} + \mathcal{V} = \int_V E\, dM, \qquad E = \tfrac{1}{2}\boldsymbol{v}^2 + \Phi, \tag{20.44}$$

where the integrand $E = d\mathcal{E}/dM$ is the *specific mechanical energy*. In Chapter 22 we shall further extend the concept of energy by including the *internal energy*.

Residual power

From Reynolds' transport theorem (20.6) with $q \to \Phi$, we find the material rate of change,

$$\frac{D\mathcal{V}}{Dt} = \int_V \frac{D\Phi}{Dt}\, dM = \int_V (\boldsymbol{v}\cdot\nabla)\Phi\, dM = -\int_V \boldsymbol{v}\cdot\boldsymbol{g}\, dM,$$

where we have used that the potential is static, $\partial\Phi/\partial t = 0$. Finally, applying kinetic energy balance (20.32), the *global equation of mechanical energy balance* becomes

$$\boxed{\frac{D\mathcal{E}}{Dt} \equiv \frac{d\mathcal{E}}{dt} + \dot{\mathcal{E}} = \tilde{P},} \tag{20.45}$$

where

$$\dot{\mathcal{E}} = \oint_S E\, d\dot{M} \tag{20.46}$$

is the flux of mechanical energy and

$$\tilde{P} = \int_V \boldsymbol{v}\cdot\tilde{\boldsymbol{f}}\, dV + \oint_S \boldsymbol{v}\cdot\boldsymbol{\sigma}\cdot d\boldsymbol{S} - \int_V \boldsymbol{\sigma}\colon\nabla\boldsymbol{v}\, dV \tag{20.47}$$

is the *residual power*. It must be emphasized that mechanical energy balance is an exact rewriting of kinetic energy balance. Provided there are no residual forces, and provided we can disregard the last term, the mechanical energy is conserved when the total contact force acting on the surface of the control volume vanishes.

Application: Connected tubes

Mechanical energy balance may be used to estimate how water moves in one of the most basic experiments in fluid dynamics, analyzed by Newton himself in *Principia* [Newton 1999, proposition 44]. Consider a long straight tube with cross-section area A, bent through 180° somewhere in the middle and placed with the open ends vertically upward (see the margin sketch). Water is poured into the system, and as everybody knows, gravity will eventually make the levels of water equal in the two vertical tubes. Before reaching equilibrium, the water sloshes back and forth with diminishing amplitude. Basic physics tells us that the energy originally given to the water oscillates between being kinetic and potential, while slowly draining away because of internal friction in the water.

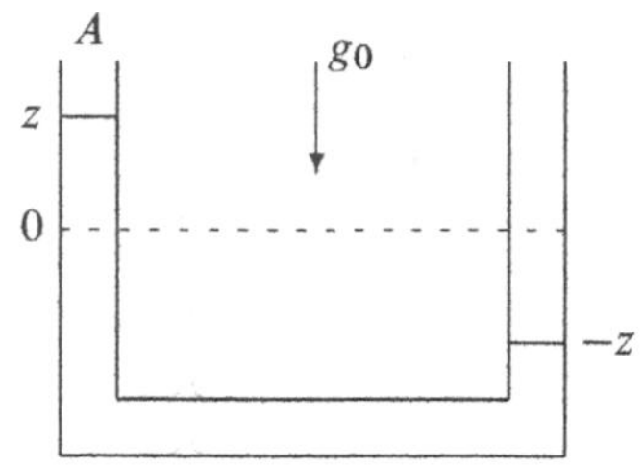

When the water is at rest, the water level will be the same in connected vertical tubes. The tube may have any shape between the two vertical sections.

Even if we do not know the exact solution to the fluid flow problem, we are nevertheless able to make a reasonably quantitative estimate of the behavior of the incompressible water. The Reynolds number will be fairly high for water in normal-size tubes, say $\mathsf{Re} \approx 1{,}000$, but for simplicity we shall assume that it is still below the onset of turbulence. Provided that the entry length is much greater than the length of the tube, it seems justified to assume that the flow is nearly uniform across the tube, that is, plug flow[2].

The control volume is in this case chosen to encompass all the water in the tube, so that it is explicitly comoving. Denoting the instantaneous plug flow velocity of the water in the tube by v and the full length of the water column by L, the total mass of the water column will be $M = \rho_0 AL$, where ρ_0 is the constant density of water, and the kinetic energy becomes $\mathcal{K} = \frac{1}{2}Mv^2 = \frac{1}{2}\rho_0 ALv^2$. When the water level in one vertical tube is raised by z relative to the equilibrium level, mass conservation tells us that it is lowered by the same amount in the other vertical tube. Since a water column of height z and weight $g_0\rho_0 Az$ has effectively been moved from one vertical tube to the other (see the margin figure) while raised by z relative to equilibrium, the potential energy becomes $\mathcal{V} = g_0\rho_0 Az^2$. Adding these contributions we get an estimate for the total mechanical energy,

$$\mathcal{E} = \mathcal{K} + \mathcal{V} = \tfrac{1}{2}\rho_0 ALv^2 + g_0\rho_0 Az^2. \tag{20.48}$$

The plug flow velocity is of course related to the height z by $v = dz/dt$.

The ambient atmospheric pressure on the water is the same in both vertical tubes and does not perform any work, and for simplicity we also assume that there is no viscous dissipation, so that $\boldsymbol{\sigma}\colon \nabla \boldsymbol{v} = -p\nabla \cdot \boldsymbol{v} = 0$. The residual power thus vanishes, $\tilde{P} = 0$; and since the control volume is comoving, there is no flow through its surface, $\dot{\mathcal{E}} = 0$. The total energy must therefore be constant,

$$\frac{d\mathcal{E}}{dt} = \rho_0 ALv\frac{dv}{dt} + 2g_0\rho_0 Az\frac{dz}{dt} = 0. \tag{20.49}$$

Using that $v = dz/dt$, this becomes a simple harmonic equation

$$\frac{d^2z}{dt^2} = -\frac{2g_0}{L}z. \tag{20.50}$$

The solution is of the form

$$z = a\cos\omega t, \tag{20.51}$$

with amplitude a and circular frequency $\omega = \sqrt{2g_0/L}$. As noted by Newton, this is the frequency of the small-amplitude oscillations of a pendulum with length $L/2$, but in contrast to the pendulum, the motion of the water is purely harmonic, also for large amplitudes.

[2] If the entry length were much shorter than the tube length, one would have to use a Poiseuille-type flow.

Bernoulli's theorem and mechanical energy balance

In Chapter 13 we derived Bernoulli's theorem for an incompressible, inviscid fluid in steady flow. Let us now consider a control volume in the form of a tiny *stream tube* consisting of all the streamlines that go into a tiny surface S_1 and leave through the tiny surface S_2. The volume flow rate Q through these areas must be the same because no fluid passes through the walls of the stream tube,

$$Q = \int_{S_1} \boldsymbol{v} \cdot d\boldsymbol{S} = \int_{S_2} \boldsymbol{v} \cdot d\boldsymbol{S}. \tag{20.52}$$

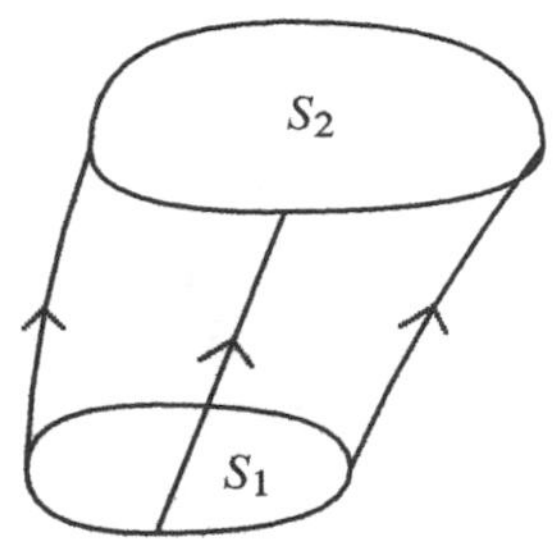

A stream tube consisting of all the streamlines that enter the area S_1 and exit through S_2.

Since the mass flow is $\rho_0 Q$, the mechanical energy flow rate becomes

$$\dot{\mathcal{E}} = \oint_S E \, d\dot{M} \approx E_2 \rho_0 Q - E_1 \rho_0 Q. \tag{20.53}$$

The residual power (20.47) only gets contributions from the pressure at exit and entry,

$$\tilde{P} = -\oint_S p\boldsymbol{v} \cdot d\boldsymbol{S} \approx -p_2 Q + p_1 Q, \tag{20.54}$$

because the flow is parallel with the walls of the stream tube (so that $\boldsymbol{v} \cdot d\boldsymbol{S} = 0$).

In steady flow, the mechanical energy of the fluid in the stream tube must be constant, that is, $d\mathcal{E}/dt = 0$, which simplifies mechanical energy balance to $\dot{\mathcal{E}} = \tilde{P}$. Using the above results, this equation may be conveniently rewritten as

$$H_1 = H_2, \tag{20.55}$$

where

$$H = E + \frac{p}{\rho_0} = \frac{1}{2}v^2 + \Phi + \frac{p}{\rho_0} \tag{20.56}$$

is identical to the Bernoulli function (13.13) on page 212.

As the two tiny areas cutting the stream tube are quite arbitrary, it follows that H must be constant along the narrow stream tube. This shows that *mechanical energy balance in steady, inviscid, and incompressible flow is equivalent to Bernoulli's theorem.* In Chapter 22 we shall derive an even more general form of Bernoulli's theorem.

Problems

20.1 Prove the relation (20.4) by carefully calculating the difference $Q(t + \delta t) - Q(t)$ to first order in δt.

20.2 Derive the rocket equations (20.21) in the accelerated frame of the rocket.

20.3 Consider a rocket in vertical flight without air resistance under the same conditions as on page 344. The burn starts at $t = 0$, and when the burn ends the rocket mass is reduced to $M_1 < M_0$. **(a)** Calculate the time t_1, the velocity u_1 and the height z_1 of the rocket at the end of the burn. **(b)** Calculate the time t_2 and the maximal height z_2 attained by the rocket in free flight after the burn. **(c)** Determine the time t_3 and the velocity u_3 when rocket arrives back on the ground.

20.4 Show that Tsiolkovsky's rocket solution (20.22) only requires the exhaust velocity U to be time independent.

20.5 A two-armed lawn sprinkler with horizontal arm length $R = 20$ cm and nozzle outlet diameter $d = 4$ mm must be designed to water the grass within a diameter of $D = 10$ m, rotating steadily at $\Omega_0 = 10$ rpm. Friction and air resistance can be disregarded. It can be assumed that a spoiler mechanism spreads the water uniformly over the area. **(a)** Determine the velocity components U_r, U_ϕ, and U_z of the water relative to the nozzle in the optimal case. **(b)** What is the total mass flow of water? **(c)** How many millimeters of artificial rain are deposited per hour in the area?

20.6 A material particle of volume V contains N point-like particles of mass m_n with instantaneous positions $\boldsymbol{x}_n$ and velocities $\boldsymbol{v}_n$. The total mass is $M = \sum_n m_n$ and the center of mass is $\boldsymbol{x} = \sum_n m_n \boldsymbol{x}_n / M$. Define the relative positions $\boldsymbol{x}'_n = \boldsymbol{x}_n - \boldsymbol{x}$ and the relative velocities $\boldsymbol{v}'_n = \boldsymbol{v}_n - \boldsymbol{v}$, where $\boldsymbol{v} = \sum_n m_n \boldsymbol{v}_n / M$ is the center-of-mass velocity.

(a) Show that the total angular momentum $\boldsymbol{\mathcal{L}} = \sum_n m_n \boldsymbol{x}_n \times \boldsymbol{v}_n$ may be written as

$$\boldsymbol{\mathcal{L}} = M\boldsymbol{x} \times \boldsymbol{v} + \sum_n m_n \boldsymbol{x}'_n \times \boldsymbol{v}'_n.$$

If a particle's relative position is uncorrelated with its relative velocity, $\langle \boldsymbol{x}'_n \boldsymbol{v}'_n \rangle = \boldsymbol{0}$, which is normally the case, the last term averages out. This justifies the expression (20.23).

(b) Show that the total kinetic energy $\mathcal{K} = \sum_n \frac{1}{2} m_n \boldsymbol{v}_n^2$ may be written as

$$\mathcal{K} = \frac{1}{2} M \boldsymbol{v}^2 + \frac{1}{2} \sum_n m_n (\boldsymbol{v}'_n)^2.$$

The last term represents the kinetic energy residing in the relative motion, that is, the heat, which can never average to zero.

20.7 Consider a circular drain of radius a in the bottom of a large circular cistern of radius $b \gg a$. Assume that the average velocity of the water through a half-sphere of radius r centered at the drain is $v(r) = (a/r)^2 v$, where v is the average velocity at the drain, so that the same amount of water passes through the half-sphere for all $r > a$. Calculate the kinetic energy associated with this velocity distribution and compare it with the kinetic energies of the water in **(a)** the cistern and in **(b)** the pipe.

20.8 Use mechanical energy balance (20.45) rather than kinetic energy balance to analyze the acceleration of the water in a cistern, when the plug is pulled (under the same conditions as on page 349).

20.9 A container with (constantly refilled) water level h and cross-section A_0 is drained through a very long thin pipe of cross-section $A \ll A_0$ forming an angle α below the horizon. A valve is placed a good distance L_0 down the pipe, but the pipe continues downward with the same slope far beyond the valve. Initially there is water in the pipe up to the valve. Find the equation of motion for the water front after the valve is opened and the front has progressed a distance x past the valve. One may assume plug-flow everywhere.

20.10 A body rotates as a solid with instantaneous rotation vector $\boldsymbol{\Omega}$ so that its instantaneous velocity field is $\boldsymbol{v} = \boldsymbol{\Omega} \times \boldsymbol{x}$. Show that the instantaneous power is given by $P = \boldsymbol{\Omega} \cdot \boldsymbol{\mathcal{M}}$.

20.11 Before the advent of modern high-precision positioning systems, standard height levels could be transmitted across a strait by means of a long tube filled with water. **(a)** Calculate the oscillation time of the water in the tube for a strait about 20 km wide when viscosity is disregarded. **(b)** Use energy balance to calculate the influence of viscosity on the oscillations for a pipe of radius 1 cm. (Hint: Use Poiseuille dissipation.) **(c)** How long does it take for the water to come to rest?

21

Action and reaction

A rocket accelerating in empty space while spewing hot gases in the opposite direction is at first sight rather mysterious because there are no external forces to account for its acceleration. On reflection we understand that the high-speed gas streaming out of the rocket nozzle carries momentum away from the rocket and thereby adds momentum to the rocket itself. But why are rockets and jet aircraft said to be "reaction-driven" while cars, ships, and propeller-driven aircraft are not? What is actually reacting to what?

Pipes, ducts, and other conduits carrying fluids are important components of many machines and installations. Everywhere in such a system the conduit walls *act* on the fluid with contact forces that confine and guide the fluid, and by Newton's Third Law the fluid must *react* back on the walls with equal and opposite contact forces. It is the fluid *reaction force* that lifts a rocket, keeps the jet aircraft flying, pushes the pistons of a car engine, and accelerates the projectile in a gun barrel. This is, however, purely a matter of convention, and there is no deeper distinction between the agents in a Newtonian action/reaction pair.

Lawn sprinklers, fans, windmills, hairdryers, propellers, pumps, water turbines, and compressors are examples of turbomachinery (or rotodynamic machinery) used everywhere today. In such machines the fluid exerts a *reaction moment of force* on the main rotating part, called the rotor. Depending on the sign of the reaction moment, energy is either taken out of the fluid as in a hydraulic turbine or put into it as in a pump or air blower.

In this chapter we shall develop the systematic formalism for reaction forces and moments, based on the global laws derived in the preceding chapter.

21.1 Reaction force

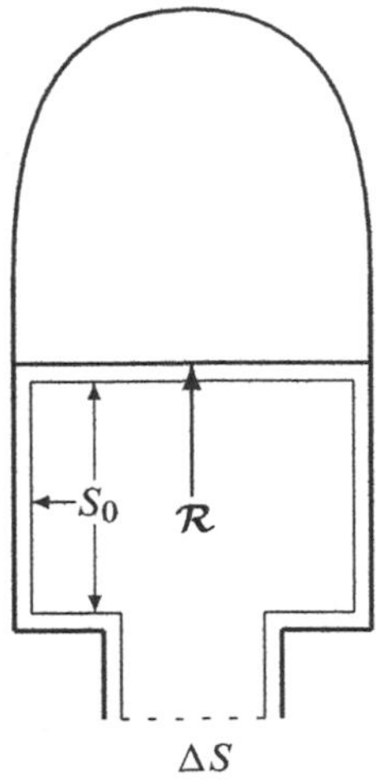

The burning gas in the combustion chamber of a rocket engine exerts a total contact force $\mathcal{R}$ on the open surface S_0 (fully drawn) of the walls of the rocket engine that confine the gas. The gas escapes through the opening ΔS (dashed) at the bottom.

The total surface of a fluid control volume may always be split into two parts, $S = S_0 + \Delta S$, where S_0 is the interface between the fluid and impenetrable walls while ΔS consists of all the openings where fluid may enter or leave the control volume (see the margin figure). The *absolute reaction force* is defined as the *total contact force that the fluid exerts on the walls*,

$$\mathcal{R} = -\int_{S_0} \sigma \cdot dS, \tag{21.1}$$

where the sign reflects *Newton's Third Law*. Note that the sign is different from that in Equation (17.2) on page 288 because we now refer to the surface of the fluid control volume rather than the surface of the body.

If we know the wall stresses everywhere, the reaction force can of course be calculated directly from this definition. But when that is not possible, momentum balance often allows us to determine the reaction force indirectly from the flow conditions that reign at the openings ΔS. To see that we rewrite the definition of the total force (20.15) and find

$$\mathcal{F} = \int_V \boldsymbol{f}\, dV + \int_{S_0+\Delta S} \boldsymbol{\sigma} \cdot d\boldsymbol{S} = \int_V \boldsymbol{f}\, dV + \int_{\Delta S} \boldsymbol{\sigma} \cdot d\boldsymbol{S} - \mathcal{R}.$$

Momentum balance (20.13), that is, $D\mathcal{P}/Dt = \mathcal{F}$ where $\mathcal{P}$ is the *momentum of the fluid in the control volume*, allows us to rewrite the reaction force as,

$$\mathcal{R} = \int_V \boldsymbol{f}\, dV + \int_{\Delta S} \boldsymbol{\sigma} \cdot d\boldsymbol{S} - \int_{\Delta S} \boldsymbol{v}\, d\dot{M} - \frac{d\mathcal{P}}{dt}, \tag{21.2}$$

where $d\dot{M} = \rho(\boldsymbol{v} - \boldsymbol{v}_S) \cdot d\boldsymbol{S}$ is the rate of mass flow through a surface element. The next to the last term (including the sign) represents the momentum *inflow* to the control volume, and needs only to run over the openings, ΔS, because of the no-slip condition: $\boldsymbol{v} = \boldsymbol{v}_S$ on S_0.

On the other side of the impenetrable walls there is of course an external body confining and guiding the fluid. Although momentum inflow in itself is not a true force, momentum balance makes it appear as a contribution to the above indirect expression for the reaction force. If momentum inflow is the dominant contribution, one may with some justification say that the external body is driven by the reaction to the net outflow of momentum.

Nearly ideal, steady flow in constant gravity

If the only body force is constant gravity, $\boldsymbol{f} = \rho \boldsymbol{g}_0$, the first term in (21.2) is simply the weight of the fluid in the control volume, $M\boldsymbol{g}_0$. In nearly ideal flow the stress tensor is dominated by the pressure, $\sigma_{ij} \approx -p\delta_{ij}$, and in steady flow we naturally choose a fixed (static) control volume, so that its total momentum is constant, $d\mathcal{P}/dt = \boldsymbol{0}$. Under these conditions the absolute reaction force becomes

$$\mathcal{R} = M\boldsymbol{g}_0 - \int_{\Delta S} p\, d\boldsymbol{S} - \int_{\Delta S} \boldsymbol{v}\, d\dot{M}, \tag{21.3}$$

where $d\dot{M} = \rho \boldsymbol{v} \cdot d\boldsymbol{S}$ because $\boldsymbol{v}_S = \boldsymbol{0}$. For incompressible flow, the mass M of the fluid is independent of the velocity whereas for compressible flow it could in principle depend on the velocity. Thus, for example, if you increase the speed of the gas in a natural gas pipeline by increasing the inlet pressure, the gas in the pipeline becomes denser and thus weighs more.

Most machines are immersed in an ambient "atmosphere" of air or water that we shall assume to be at rest with hydrostatic pressure p_0 (which in principle could vary across the machine). If the fixed control volume were filled with ambient fluid at rest, the absolute (and static) reaction force would be, $\mathcal{R}_0 = M_0\boldsymbol{g}_0 - \int_{\Delta S} p_0\, d\boldsymbol{S}$, where M_0 is the mass of the ambient fluid in the control volume. Usually, we are interested in the change, $\Delta R = \mathcal{R} - \mathcal{R}_0$, in the reaction force that occurs when the ambient fluid is replaced by the intended moving fluid, as for example when water enters the nozzle of a firehose. This *relative reaction force* becomes

$$\Delta\mathcal{R} = (M - M_0)\boldsymbol{g}_0 - \int_{\Delta S} (p - p_0)\, d\boldsymbol{S} - \int_{\Delta S} \boldsymbol{v}\, d\dot{M}. \tag{21.4}$$

Evidently, the relative reaction force is obtained by reducing the mass of the fluid by the mass of the displaced ambient fluid, and calculating the pressure relative to the ambient pressure, also called the *gauge* (or *gage*) pressure.

Case: Thrust of a rocket engine

When the external body is kept at rest, as it will normally be when the flow is steady, the fluid reaction force—including the weight of the fluid—has to be countered by other forces, usually supplied by supports and fixations. Thus, a rocket engine with its fuel container must be placed on a horizontal test bed that is firmly anchored to the ground to prevent it from flying off during tests (see the margin figure).

Assuming that the exhaust velocity, $v_x = -U$, is nearly constant across the outlet area A (i.e., plug flow), the rate of mass loss through the exhaust becomes $\dot{M} = \rho U A$, where ρ is the gas density. Consequently the relative reaction force or *thrust* is

$$\Delta\mathcal{R}_x = (p - p_0)A + U\dot{M} = (p - p_0 + \rho U^2)A. \tag{21.5}$$

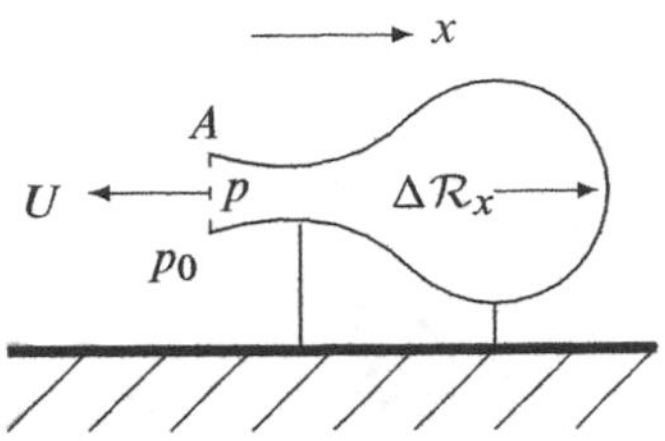

Sketch of a supersonic rocket engine mounted on a test bed in ambient atmospheric pressure p_0. The thrust $\Delta\mathcal{R}_x$ could be determined from the strain it produces in the supports.

The exhaust pressure will only be significantly different from the ambient pressure in supersonic flow, as discussed before in the analysis of rocket engines (page 240).

The flow can, however, never be truly steady because there is a continuous mass drain on the stored fuel. Since most of the stored fuel is at rest in a high density form (liquid or solid), the total momentum of the combustion gases can be kept nearly constant for a long time. The condition for the above calculation to be valid is that the rate of change of momentum be much smaller than the rate of momentum outflow, $|d\mathcal{P}/dt| \ll \dot{M}U$.

Case: Nozzle puzzle

A *nozzle*—for example sitting at the end of a firehose—is formed by narrowing down the cross-section of a pipe from A_1 to $A_0 < A_1$ over a fairly short distance (see the margin figure). Mass conservation guarantees that for incompressible water with density ρ_0, the volume inflow and outflow are the same,

$$Q = v_1 A_1 = v_0 A_0, \tag{21.6}$$

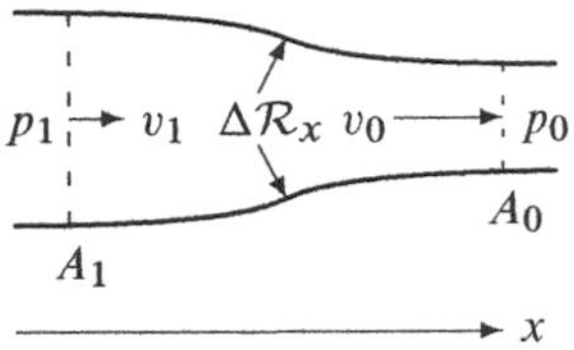

The relative reaction force on a contracting nozzle points along the direction of flow because there is a higher than atmospheric pressure in the constriction.

where v_1 and v_0 are the average inflow and outflow velocities. Assuming plug flow and disregarding gravity, the relative reaction force along the nozzle becomes

$$\Delta\mathcal{R}_x = (p_1 - p_0)A_1 + (v_1 - v_0)\rho_0 Q, \tag{21.7}$$

where p_1 is the inlet pressure and p_0 the outlet pressure (equal to the ambient pressure).

Bernoulli's theorem allows us to eliminate the pressure difference between inlet and outlet,

$$\frac{1}{2}v_1^2 + \frac{p_1}{\rho_0} = \frac{1}{2}v_0^2 + \frac{p_0}{\rho_0}, \tag{21.8}$$

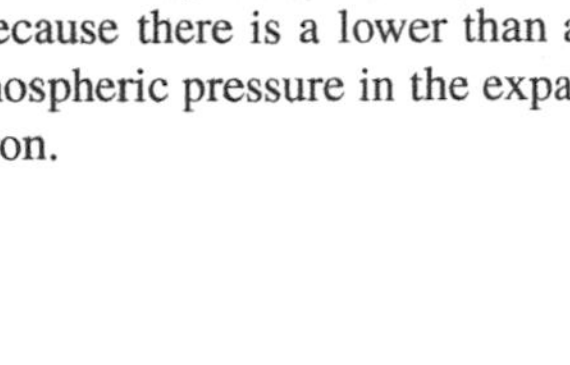

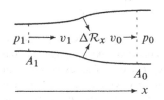

The relative reaction force on an expanding nozzle points also points along the direction of flow because there is a lower than atmospheric pressure in the expansion.

and using mass conservation, both velocities can be eliminated with the result that

$$\Delta\mathcal{R}_x = \frac{(A_1 - A_0)^2}{2A_1 A_0}\frac{\rho_0 Q^2}{A_0}. \tag{21.9}$$

It should not be surprising that the reaction force vanishes for a straight pipe with $A_1 = A_0$, because we have ignored viscosity. It is, however, a bit surprising that the fluid appears to exert a positive reaction force on the nozzle along the direction of flow whether the nozzle is converging ($A_1 > A_0$) or diverging ($A_1 < A_0$).

On reflection this is in fact in agreement with our intuition because Bernoulli's theorem tells us that the pressure on the inside of the converging nozzle is larger than the ambient pressure surrounding the nozzle and therefore pushes the nozzle forward. This is the force that slides the nozzle off your garden hose if you have not attached it firmly enough. Conversely, the pressure on the inside of an diverging nozzle is lower than atmospheric, leading to a forward pull on the nozzle itself. Although not part of common garden experience, we conclude that a badly attached diverging nozzle should also slide off.

Figure 21.1. Three firefighters handling nozzle and hose in a training exercise at Shaw Airforce Base (South Carolina, U.S.). The dramatic stance of the middle "hose man" testifies to the strong forces at play. (Source: Wikimedia Commons.)

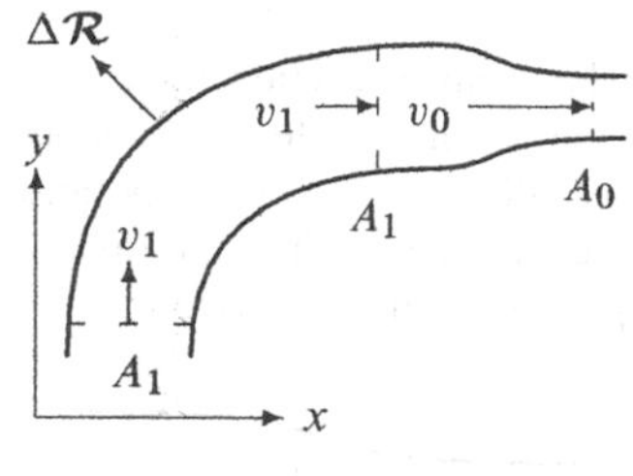

A nozzled firehose with an unsupported 90° bend in the xy-plane. The extra force in the x-direction required by the bend must be countered by the hands that hold the static nozzle through the bendable but unstretchable material of the hose.

But why does it take one or more firemen to press forward on the nozzle of a firehose while combating a fire (see Figure 21.1)? Shouldn't they instead pull back to counter the reaction force on the nozzle? If they loosen their grip, everybody knows that the firehose will, like an ordinary garden hose, flail back and drench everybody around. The puzzle is first resolved when one realizes that a soft but unstretchable firehose always meanders through bends and turns upstream of the nozzle. The water passing through a bend will react to the changing direction with a reaction force pointing away from the center of the bend. For the case of a 90° bend (see the margin figure), the component of the relative reaction force (21.4) along the direction of the nozzle simply equals minus the momentum outflow at the exit,

$$\Delta\mathcal{R}_x = -v_0\rho_0 Q = -\frac{\rho_0 Q^2}{A_0}. \tag{21.10}$$

Due to the unstretchable material, this force (a pull!) is transmitted along the hose from the bend to the nozzle, and that is why you have to grasp the nozzle handle firmly and push forward. Left to itself, a straight firehose is completely unstable to any amount of bending.

Example 21.1 [Firehose]: Older firehoses with 2.5-inch diameter equipped with a 1.5-inch nozzle can typically deliver 20 to 40 liters of water per second. At a volume discharge of 20 liters per second, the reaction force is −350 N, which is just about manageable for a single firefighter. If the hose breaks at the nozzle, the required handle force will change instantaneously from +350 N to −200 N. Since these forces are rather large, there is a real risk that the fireman will fly off together with the nozzle. A modern 5-inch firehose can deliver up to 100 liters of water per second through a 3-inch nozzle, leading to a reaction force of about −2,200 N. This nozzle cannot even be handled by three firefighters, but should be firmly anchored in a boat or truck.

Figure 21.2. Runner from a Francis turbine at the Grand Coulee Dam (Washington, USA). The runner rotates anticlockwise seen from above, and is here shown in the workshop with markings indicating how the blades continue under the "skirt". (Source: Wikimedia Commons.)

21.2 Reaction moment

The absolute reaction moment may be defined in analogy with the reaction force (21.1). As before S_0 denotes the part of the control surface that covers impenetrable walls, and the *reaction moment* is defined as total moment of force that the fluid imposes on the walls,

$$\mathcal{N} = -\int_{S_0} \boldsymbol{x} \times \boldsymbol{\sigma} \cdot d\boldsymbol{S}. \tag{21.11}$$

Using angular momentum balance (20.24), this may be rewritten as

$$\mathcal{N} = \int_V \boldsymbol{x} \times \boldsymbol{f}\, dV + \int_{\Delta S} \boldsymbol{x} \times \boldsymbol{\sigma} \cdot d\boldsymbol{S} - \int_{\Delta S} \boldsymbol{x} \times \boldsymbol{v}\, d\dot{M} - \frac{d\mathcal{L}}{dt}, \tag{21.12}$$

where ΔS denotes the openings in the control surface where fluid may enter or leave.

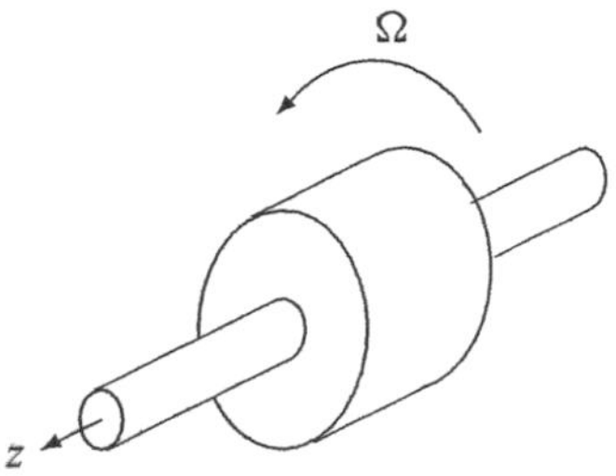

Rotor on its shaft, rotating with angular velocity Ω around the z-axis.

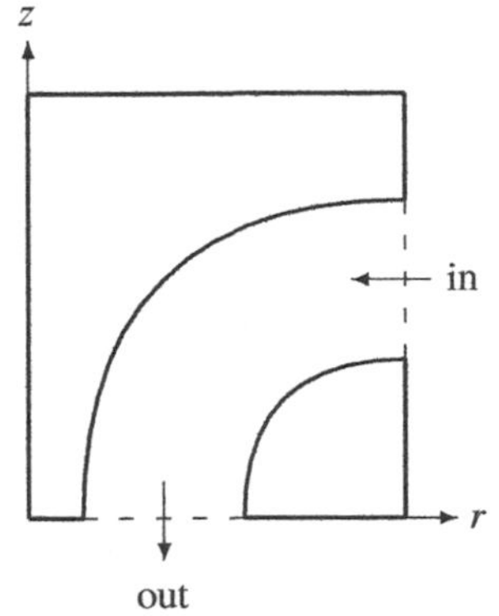

Basic rotor geometry in cylindrical coordinates, and a channel with radial inflow and axial outflow, as in a Francis turbine. The azimuthal flow that actually drives the turbine, is not shown here.

Turbomachinery

In turbomachinery there is always a solid rotating part, called the *rotor*, *runner*, or *impeller*, which diverts the flow and either puts kinetic energy into the fluid (as in a pump) or takes it away (as in a turbine). The rotor shaft is mounted in bearings on a non-rotating *stator* that only give it freedom to rotate around a fixed axis. It is carefully designed with a number of channels that guide the fluid, often just thin *blades* that block the flow through the rotor as little as possible. The purpose of the moving blades is to interact with the circulating fluid so that it can perform (signed) work on the rotor. Rotors essentially only differ in the way the channels are designed. At one extreme we find the *radial flow rotor* where the channels are orthogonal to the cylinder axis, at another the *axial flow rotor* where the channels are orthogonal to the radial direction. Rotor design is a highly evolved engineering discipline and most rotors in real turbomachinery employ a mixture of axial and radial flow. In, for example, the Francis turbine, fluid enters the rotor radially from the outside and exits along the axis (see Figure 21.2). Extensive engineering discussions of rotodynamic machines are found in [Fox and McDonald 1985, Massey 1998, Douglas et al. 2001].

Rotor power

The control volume is naturally chosen to be fixed with respect to the rotor and encompasses all the fluid inside the cylindrical outline of the rotor. We shall always choose cylindrical coordinates with the z-axis along the rotor axis. The reaction moment along the rotor axis then becomes

$$\mathcal{N}_z = -\int_{S_0} (\boldsymbol{x} \times \boldsymbol{\sigma} \cdot d\boldsymbol{S})_z = -\int_{S_0} r(\boldsymbol{\sigma} \cdot d\boldsymbol{S})_\phi. \tag{21.13}$$

As the last expression shows, there must be non-vanishing contact forces in the (azimuthal) ϕ-direction, since otherwise there will be no rotor moment and thus no rotor power. At low Reynolds number, one could conceivably drive a rotor alone by the moment of the shear stresses, $\sigma_{\phi r}$ and $\sigma_{\phi z}$, but at high Reynolds number, pressure is by far the most important stress. This is in fact the deeper reason for always designing the channels to block the azimuthal flow so that the diagonal stress, $\sigma_{\phi\phi} \approx -p$, can act on their inner surfaces and perform positive (or negative) work on the rotor.

If the rotor spins with angular velocity $\boldsymbol{\Omega} = \Omega\,\hat{\boldsymbol{e}}_z$ in the inertial "laboratory" frame of reference, the velocity of the control surface as well as of the solid rotor body is

$$\boldsymbol{v}_S = \boldsymbol{\Omega} \times \boldsymbol{x} = r\Omega\hat{\boldsymbol{e}}_\phi. \tag{21.14}$$

Using that $\boldsymbol{v}_S \cdot \boldsymbol{\sigma} = \boldsymbol{\Omega} \times \boldsymbol{x} \cdot \boldsymbol{\sigma} = \boldsymbol{\Omega} \cdot \boldsymbol{x} \times \boldsymbol{\sigma}$, the rate of work of the fluid reaction on the rotor—the *rotor power*—becomes

$$P = -\int_{S_0} \boldsymbol{v}_S \cdot \boldsymbol{\sigma} \cdot d\boldsymbol{S} = -\boldsymbol{\Omega} \cdot \int_{S_0} \boldsymbol{x} \times \boldsymbol{\sigma} \cdot d\boldsymbol{S} = \boldsymbol{\Omega} \cdot \boldsymbol{\mathcal{N}} = \Omega\mathcal{N}_z. \tag{21.15}$$

If the rotor power is positive, the fluid performs work on the rotor (as in a turbine), whereas for for negative rotor power, the rotor performs work on the fluid (as in a pump).

Euler's turbine equation

Suppose now that the cylindrical rotor has a number of similarly shaped blades placed regularly around the axis. In the laboratory frame, the whirling blades make the velocity field in the rotor explicitly time dependent. The total angular momentum of the fluid inside the rotor will nevertheless for symmetry reasons be time independent when the rotor spins steadily with constant angular velocity Ω, so that we may take $d\mathcal{L}_z/dt = 0$. The center of mass of the fluid must—also for symmetry reasons—lie on the cylinder axis where it cannot produce a moment of force along the axis (in constant gravity).

With these simplifications, the z-component of the reaction moment (21.12) becomes

$$\mathcal{N}_z = \int_{\Delta S} r(\boldsymbol{\sigma} \cdot d\boldsymbol{S})_\phi - \int_{\Delta S} r v_\phi \, d\dot{M}, \tag{21.16}$$

where the mass flow element is

$$d\dot{M} = \rho\boldsymbol{u} \cdot d\boldsymbol{S}, \qquad \boldsymbol{u} = \boldsymbol{v} - \boldsymbol{v}_S = \boldsymbol{v} - \Omega r\,\hat{\boldsymbol{e}}_\phi, \tag{21.17}$$

where $\boldsymbol{u}$ is the flow velocity relative to the spinning rotor. The first term in $\mathcal{N}_z$ will only contribute if there are azimuthal contact forces acting in the channel openings. For cylindrical rotors with channel openings in the radial and axial directions, $d\boldsymbol{S} = \hat{\boldsymbol{e}}_r dS_r + \hat{\boldsymbol{e}}_z dS_z$, only the viscous shear stresses, $\sigma_{\phi r}$ and $\sigma_{\phi z}$, can contribute to the first term. We shall in the following assume that the flow is so nearly ideal that we may ignore the first term.

The openings ΔS in the rotor control volume are now divided between inlets S_{in} and outlets S_{out}, all oriented along the direction of flow so that we may write symbolically $\Delta S = S_{\text{out}} - S_{\text{in}}$ (see the margin figure). Since the rotor contains the same amount of fluid at all times, mass conservation tells us that equal amounts of fluid pass through the inlets and outlets per unit of time. Assuming that the fluid is effectively incompressible with constant density $\rho = \rho_0$, and introducing the volumetric discharge Q, the positive *mass throughput* becomes

$$\rho_0 Q = \int_{S_{\text{out}}} d\dot{M} = \int_{S_{\text{in}}} d\dot{M}. \tag{21.18}$$

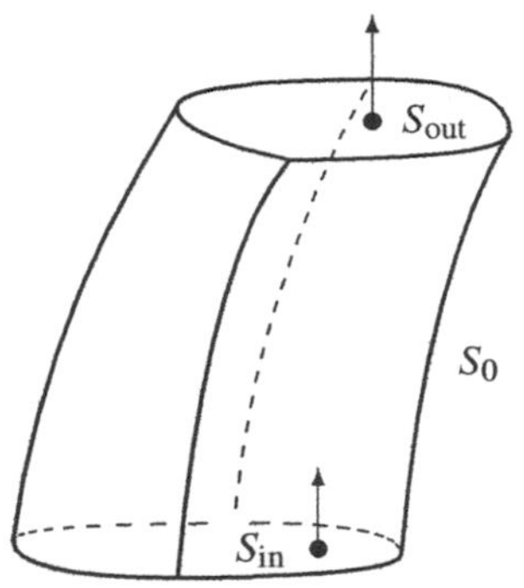

A rotor channel with inlet S_{in} and outlet S_{out}, both oriented along the flow.

The *mass flow average* of the specific angular momentum over inlets or outlets is defined by

$$\langle r v_\phi \rangle_{\text{in,out}} = \frac{1}{\rho_0 Q} \int_{S_{\text{in,out}}} r v_\phi d\dot{M}, = \frac{1}{Q} \int_{S_{\text{in,out}}} r v_\phi \boldsymbol{u} \cdot d\boldsymbol{S}, \tag{21.19}$$

and similarly for other quantities. The rotor power, $P = \Omega \mathcal{N}_z$, may then be written in the convenient form, called *Euler's turbine equation* (1754),

$$W \equiv \frac{P}{\rho_0 Q} = \Omega \langle r v_\phi \rangle_{\text{in}} - \Omega \langle r v_\phi \rangle_{\text{out}}. \tag{21.20}$$

The quantity W is the (signed) work that the fluid performs on the rotor per unit of fluid mass passing through it. We shall call it the *specific rotor work*.

Euler's equation is a very clean result, subject only to the condition that the moment of shear stresses on inlets and outlets be negligible. The specific work and the rotor power are positive if the average specific angular momentum $\langle r v_\phi \rangle$ is larger at the inlets than at the outlets, and conversely. Thus, in a turbine ($W > 0$), the flow through the rotor takes angular momentum out of the fluid whereas in a pump ($W < 0$), angular momentum is added to it.

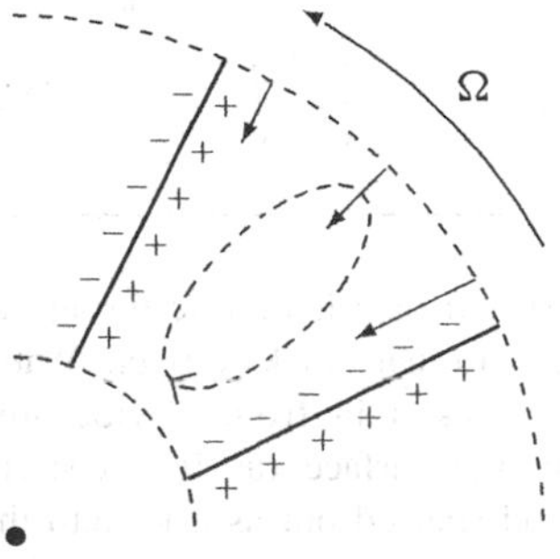

Radial inflow to a turbine rotor. There must be suction on one side of a blade and pressure on the other for the fluid to perform positive work on the rotor blades. The arrows indicate the secondary variation in radial inflow velocities caused by the pressure differences (using Bernoulli's principle). Effectively, it superposes a counter-rotating whirl on top of the otherwise uniform inflow.

Secondary flow: There is actually a correction to the clean Euler result. Since pressure is the main azimuthal stress component at high Reynolds number, there must always be a pressure difference between the two sides of a rotor blade (see the margin figure). It will by Bernoulli's theorem drive a secondary circulation opposite to the rotor's spin, thereby diminishing the specific angular momentum of the inflow. This so-called *slip* correction is inversely proportional to the number of blades, because the pressure jump across each blade is smaller the more blades there are. We shall not discuss this issue further; see however [Douglas et al. 2001, p. 728].

Bernoulli's theorem

The flow through the rotor is as mentioned unsteady seen from the laboratory frame because of the whirling blades, but in the corotating rest frame of the rotor, the flow through the rotor channels may be taken to be nearly steady. Assuming that the rotor axis is vertical, the Bernoulli function becomes (including the pseudo-gravitational centrifugal potential),

$$H_0 = \tfrac{1}{2} u^2 + g_0 z - \tfrac{1}{2} \Omega^2 r^2 + \frac{p}{\rho_0}. \tag{21.21}$$

The constancy of H_0 along any streamline in ideal steady flow implies that it will be constant for any material particle transported through the system. As a consequence, the mass flow average of H_0 must be the same at inlet and outlet,

$$\langle H_0 \rangle_{\text{in}} = \langle H_0 \rangle_{\text{out}}. \tag{21.22}$$

Given the turbine geometry and the inlet and outlet velocities, this equation may be used to calculate the average pressure difference between inlets and outlets.

Substituting $\boldsymbol{u} = \boldsymbol{v} - \Omega r \hat{\boldsymbol{e}}_\phi$ in H_0, we find $H_0 = H - \Omega r v_\phi$, where

$$H = \tfrac{1}{2}v^2 + g_0 z + \frac{p}{\rho_0} \tag{21.23}$$

is the Bernoulli function in the laboratory frame. Consequently, Bernoulli's theorem in the rotating frame (21.22) becomes, when transposed to the laboratory,

$$\boxed{\langle H \rangle_{\rm in} = \langle H \rangle_{\rm out} + W,} \tag{21.24}$$

where the specific rotor work W is given by Euler's turbine equation (21.20). This equation expresses energy balance and will, of course, automatically be fulfilled if we know an exact solution to Euler's field equations. If we do not, it provides an important relation among the flow parameters.

In all rotodynamic machines the fluid is guided toward the rotor through suitable piping that takes the fluid from the actual machine entry to the rotor inlet. Similarly, there is always installed piping to connect the rotor outlet with the actual machine exit. In a hydraulic turbine these pipes are respectively called the *penstock* and the *draft tube* (see the margin figure). Under the same conditions as above, we may apply Bernoulli's theorem in the laboratory frame to the steady flow from entry to inlet and from outlet to exit,

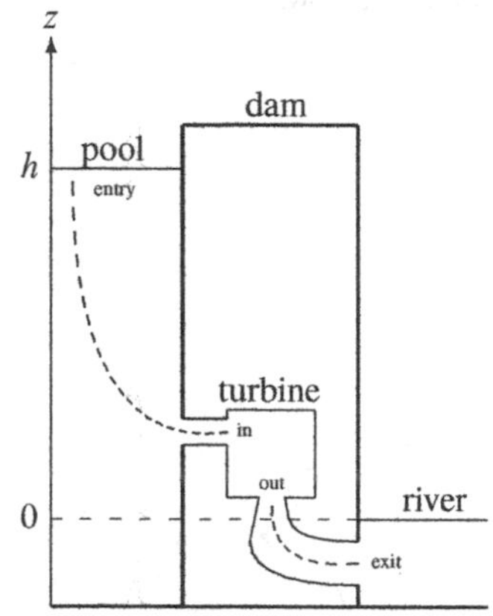

Sketch of a hydraulic turbine in a dam that blocks a river. Water is sucked in from a wide area of the surface (as in a cistern) and spewed out as a jet into the river below the dam. In ideal flow the Bernoulli function in the laboratory frame is constant along the dashed streamline connecting entry with inlet (through the penstock) and the outlet with exit (through the draft tube).

$$\langle H \rangle_{\rm entry} = \langle H \rangle_{\rm in}, \qquad \langle H \rangle_{\rm out} = \langle H \rangle_{\rm exit}, \tag{21.25}$$

so that the Bernoulli relation (21.24) may be written as

$$\boxed{\langle H \rangle_{\rm entry} = \langle H \rangle_{\rm exit} + W.} \tag{21.26}$$

This is a very useful relation because it allows the specific rotor work to be related to the external parameters characterizing the machine performance, such as the water head that drives a hydraulic turbine.

Losses

In the laboratory frame, the Bernoulli equation (21.26) basically says that in nearly ideal, incompressible fluid, the average specific enthalpy entering the machine either leaves it again or is converted to specific rotor work (positive or negative). If for other reasons energy is lost on the way, for example through viscous dissipation and turbulence, these losses must be included on the right-hand side, so that the equation becomes

$$\langle H \rangle_{\rm entry} = \langle H \rangle_{\rm exit} + W + W_{\rm loss}, \tag{21.27}$$

where $W_{\rm loss}$ is the sum of the specific losses suffered along the way from the entry to the rotor inlet, from the rotor inlet to outlet, and from the rotor outlet to the exit. These losses occur along the flow direction and are thus all positive.

Writing the above equation as

$$W = \langle H \rangle_{\rm entry} - \langle H \rangle_{\rm exit} - W_{\rm loss}, \tag{21.28}$$

it follows—for unchanged entry and exit flow conditions—that the losses always diminish the specific rotor work W. Thus, in agreement with intuition the losses in a turbine reduce the specific work, $W > 0$, that can be delivered by the rotor, whereas in a pump the losses augment the specific external work, $-W > 0$, that is needed to drive it. In the following applications we shall mostly ignore the losses.

Case: Radial flow rotor

Radial flow rotors are mostly used in high-pressure turbomachinery, such as pumps, compressors, and hydraulic turbines. Although we shall only discuss a pump, the formalism is easily transcribed to a turbine by changing the sign of the flow and the direction of rotation.

The radial flow rotor has inner radius a, outer radius $b > a$, and axial length L. The blades are mounted radially on a hub and are assumed to be closely spaced with negligible thickness and the same shape for all z. In general the blades are curved with fixed geometric slopes, α at $r = a$ and β at $r = b$, with respect to the radial direction (see the margin figures). We do not need to know precisely how the blades are designed inside the rotor, except that they should interpolate smoothly between inlets and outlets to diminish internal losses.

We shall assume that the fluid enters tangentially along the blades at $r = a$ and leaves tangentially at $r = b$ such that the flow in the rotating frame fulfills the *smoothness conditions*

$$u_\phi^{\text{in}} = \alpha u_r^{\text{in}}, \qquad u_\phi^{\text{out}} = \beta u_r^{\text{out}}. \tag{21.29}$$

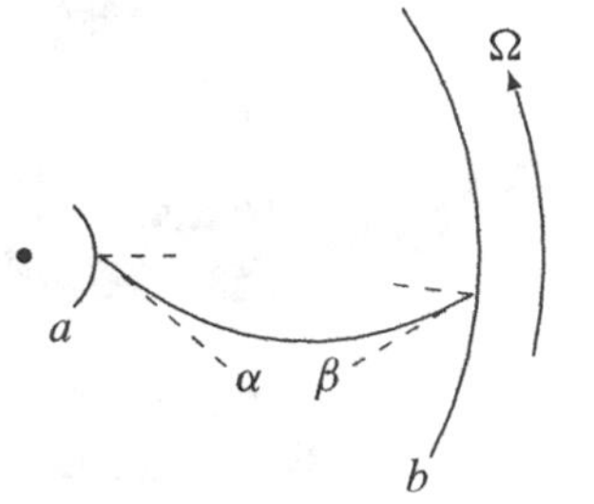

Sketch of a curved blade with slope α at $r = a$ and β at $r = b$ relative to the radial direction. Here, α is negative and β positive.

Whereas the outlet condition is always fulfilled because the fluid leaves as a jet (see page 212), the pump will still function if the inlet condition is violated, so that the incoming fluid hits the inlet edge of the blades at a non-vanishing *angle of attack*. This might happen if the flow through the pump were throttled while the motor kept the rotor churning at the same speed. Provided the angle of attack is not too large, the flow will in fact quickly adapt itself to the blade direction a bit downstream from the inlet. Such a rapid change of the inlet flow direction is called an *inlet shock* although it has nothing to do with the supersonic shocks discussed in Chapter 26. An inlet shock always increases the dissipative loss, and at large angles of attack it may also generate turbulence and even cavitation. We shall from now on assume that the pump is operated so that an inlet shock does not arise.

A centrifugal ventilator driven by an electrical motor. Air enters axially through the circular opening and exits through the square opening. Notice the widening outlet scroll, which permits homogeneous rotor outflow. (Source: Image courtesy Juvenco A/S, Denmark. With permission.)

Mass conservation guarantees (for incompressible fluid) that the volumetric discharge rate Q is the same at inlet and outlet. The radial inflow and outflow are assumed to be homogeneous across the inlet and outlet areas, $2\pi a L$ and $2\pi b L$, and orthogonal to the z-axis. In the laboratory frame the velocity field $\boldsymbol{v} = \Omega r \hat{\boldsymbol{e}}_\phi + \boldsymbol{u}$ becomes, at the inlet and outlet,

$$v_r^{\text{in}} = \frac{Q}{2\pi a L}, \qquad v_\phi^{\text{in}} = \Omega a + \alpha \frac{Q}{2\pi a L}, \qquad v_z^{\text{in}} = 0, \tag{21.30a}$$

$$v_r^{\text{out}} = \frac{Q}{2\pi b L}, \qquad v_\phi^{\text{out}} = \Omega b + \beta \frac{Q}{2\pi b L}, \qquad v_z^{\text{out}} = 0. \tag{21.30b}$$

Inserting these flows into Euler's turbine equation (21.20) we obtain the specific work,

$$\boxed{W = -\Omega^2 (b^2 - a^2) - (\beta - \alpha) \frac{\Omega Q}{2\pi L}.} \tag{21.31}$$

The first term can be traced back to the Coriolis force (see Problem 21.4).

In many centrifugal pumps, for example circulation pumps for central heating systems, the fluid enters and exits through pipes having the same cross-section A, so that the average entry and exit velocities are the same, $v_{\text{entry}} = v_{\text{exit}} = U = Q/A$. Assuming plug flow, the Bernoulli averages become

$$\langle H \rangle_{\text{entry}} = \frac{1}{2} U^2 + \frac{p_{\text{entry}}}{\rho_0}, \qquad \langle H \rangle_{\text{exit}} = \frac{1}{2} U^2 + \frac{p_{\text{exit}}}{\rho_0}.$$

Bernoulli's theorem (21.26) in the laboratory frame then leads to a pressure difference,

$$\boxed{\Delta p = p_{\text{exit}} - p_{\text{entry}} = -\rho_0 W.} \tag{21.32}$$

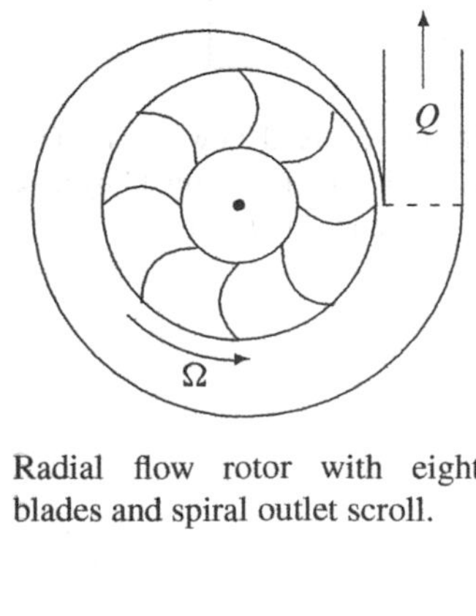

Radial flow rotor with eight blades and spiral outlet scroll.

Since W is negative for a pump, the outflow has a pressure excess that allows the pump to lift fluid through a height (or head) $h = \Delta p / \rho_0 g_0 = -W/g_0$. If losses $W_{\text{loss}} > 0$ are added to W, the available head is diminished, as one would expect.

Components of the Duraheart centrifugal pump used for assisting the left heart ventricle in providing adequate blood flow to the body. (Source: Image courtesy Terumo Heart, Inc. With permission.)

Parameter	Value	Unit
$2a$	15	mm
$2b$	40	mm
L	2.4	mm
α	−2.0	
β	−0.36	

A set of geometric parameters comparable to those of the Duraheart Pump depicted above. The rotor is assumed to have straight blades (see Problem 21.5).

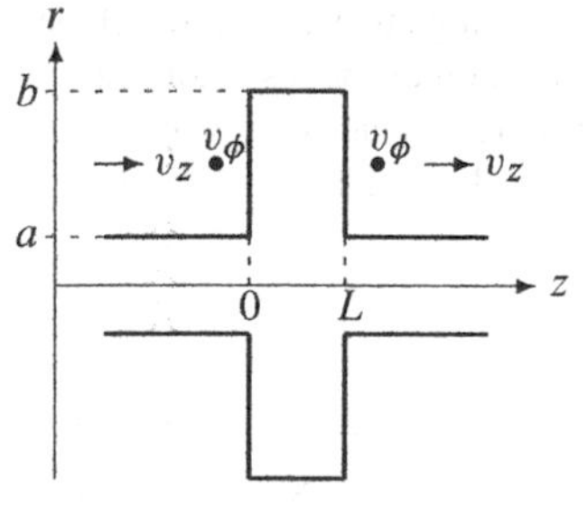

Sketch of axial flow rotor. The control volume is defined by $0 < z < L$ and $a < r < b$.

Most often, the fluid enters the rotor without any swirl,

$$v_\phi^{\text{in}} = \Omega a + \alpha \frac{Q}{2\pi a L} = 0. \tag{21.33}$$

To be able to fulfill this condition, the rotor must be constructed with negative inlet slope, $\alpha < 0$, provided Q and Ω have the same sign. At the inlet, the blades thus have to lean backward with respect to the direction of rotation so that they literally "shovel in" the fluid. To secure uniform outflow in all directions, the rotor is normally surrounded by a spiral-shaped *outlet scroll* that accumulates the rotor outflow and "straightens it out" (see the margin figure on the preceding page). Evidently, the two terms in the outflow collaborate for $\beta > 0$, so that the fluid is also "shoveled out" of the rotor, but the pump will also work for a wide range of negative values of β. Under the condition of no inlet swirl, the specific work (21.31) may be written as

$$W = -\Omega^2 b^2 \left(1 - \frac{a^2 \beta}{b^2 \alpha}\right). \tag{21.34}$$

Provided a is considerably smaller than b and $|\beta/\alpha|$ is not too large, we have $W \approx -\Omega^2 b^2$. The specific work is thus dominated by the outer perimeter speed Ωb of the rotor.

Example 21.2 [Centrifugal heart assist pump]: The human heart is a pulsed pump that delivers blood to the body with a typical average volume rate of 5 liters per minute, or $Q = 83\ \text{cm}^3\,\text{s}^{-1}$. During a heartbeat cycle the arterial pressure head varies between 80 and 140 mmHg (above atmospheric pressure) with an average around 110 mmHg, corresponding to $\Delta p = 14.7$ kPa. A centrifugal pump designed to deliver a steady performance at this level must according to (21.32) produce the specific work $-W = \Delta p/\rho_0 = 14.7\ \text{J}\,\text{kg}^{-1}$, and thus the power $-P = Q\Delta p = 1.2$ W. The perimeter speed, $\Omega b \approx \sqrt{-W} = 3.8\ \text{m}\,\text{s}^{-1}$, is surprisingly high and requires the rotor to spin quite fast in a tiny ventricular assistance pump where b can only be a few centimeters. In the margin table is shown a set of geometric parameters consistent with the above values, leading to $\Omega \approx 1{,}850$ rpm for smooth inflow. This is similar to the angular velocities actually used in such pumps (a review of heart pumps may be found in [HST06]).

Case: Axial flow rotor

Axial flow rotors are typically used in high-volume turbomachinery, such as fans, ventilators, and low-pressure turbines. Here we shall only analyze a pump, but the formalism may easily be transcribed to turbines.

The rotor is assumed to be a cylinder with inner radius a, outer radius b, and length L (see the margin figure). The blades are assumed to have negligible thickness and extend through the length of the rotor. A major simplifying assumption is that there is no radial flow anywhere in the rotor, $v_r = 0$, so that the flow through the cylinder shell between r and $r + dr$ does not mix with the flow in neighboring shells. In particular it follows from mass conservation and cylinder symmetry that the axial velocity $v_z(r)$ can only depend on r, whereas $v_\phi(r, z)$ may depend on both r and z. These assumptions allow us to first analyze the single shell at r and only later integrate over all shells in the interval $a \le r \le b$.

As for the radial rotor, we assume that the pump is operated so that the fluid flows smoothly in and out of the rotor with $u_\phi^{\text{in}} = \alpha u_z^{\text{in}}$ and $u_\phi^{\text{out}} = \beta u_z^{\text{out}}$, where α and β are the slopes of the inlet and outlet blades with respect to the z-axis. The inlet and outlet velocity fields become, for the shell at r,

$$v_r^{\text{in}} = 0, \qquad v_\phi^{\text{in}} = \Omega r + \alpha U, \qquad v_z^{\text{in}} = U, \tag{21.35}$$

$$v_r^{\text{out}} = 0, \qquad v_\phi^{\text{out}} = \Omega r + \beta U, \qquad v_z^{\text{out}} = U. \tag{21.36}$$

Note that the velocity U as well as the slopes α and β may depend on r.

From Euler's turbine equation (21.20) we obtain the specific work of the fluid,

$$W = (\alpha - \beta)\Omega r U. \tag{21.37}$$

For a pump it must be negative, that is, $\alpha < \beta$ for $\Omega U > 0$. The inlet and outlet Bernoulli functions in the laboratory frame are (assuming effectively incompressible flow)

$$H_{\text{in}} = \frac{1}{2}(\Omega r + \alpha U)^2 + \frac{1}{2}U^2 + \frac{p_{\text{in}}}{\rho_0}, \qquad H_{\text{out}} = \frac{1}{2}(\Omega r + \beta U)^2 + \frac{1}{2}U^2 + \frac{p_{\text{out}}}{\rho_0}.$$

From (21.24) the pressure difference is obtained,

$$\Delta p = p_{\text{out}} - p_{\text{in}} = \frac{1}{2}\left(\alpha^2 - \beta^2\right)\rho_0 U^2. \tag{21.38}$$

For a pump the pressure difference must be positive, $|\alpha| > |\beta|$, and since $\alpha < \beta$, it follows that α must be negative, whereas β can be both negative and positive in the range $\alpha < \beta < -\alpha$. Both slopes are negative for the air blower shown in the margin figures.

If the inlet and outlet slopes are the same, $\alpha = \beta$, the specific rotor work and the pressure difference appear to vanish. This seems a bit puzzling because propellers with flat blades also require engine power to turn steadily. The explanation is that for such devices the smoothness condition is generally not fulfilled, and the fluid hits the propeller blades at a non-vanishing *angle of attack*, resulting in aerodynamic forces (see Chapter 29) that also require engine power to overcome.

For the case of an axial pump, say an air blower, the inlet flow is normally without swirl,

$$v_\phi^{\text{in}} = \Omega r + \alpha U = 0. \tag{21.39}$$

Using this to eliminate $U = -\Omega r/\alpha$, we obtain

$$W = -\left(1 - \frac{\beta}{\alpha}\right)\Omega^2 r^2, \qquad \Delta p = \frac{1}{2}\left(1 - \frac{\beta^2}{\alpha^2}\right)\rho_0 \Omega^2 r^2. \tag{21.40}$$

The non-vanishing swirl in the outlet flow may be "straightened out" by suitable fixed *guide vanes* downstream from the rotor.

The actual geometric design of the blades provides us with the slopes. For uniform axial flow U, the inlet slope becomes linear in r, that is, $\alpha = -\Omega r/U$. Assuming furthermore that the outlet slope is also linear (which is not required), the ratio β/α becomes a constant. The volumetric discharge rate is simply $Q = UA$, where $A = \pi(b^2 - a^2)$ is the inlet area, and the average specific work and pressure head can now be calculated by integrating over this area,

$$\langle W\rangle = -\frac{1}{2}\left(1 - \frac{\beta}{\alpha}\right)(a^2 + b^2)\Omega^2, \qquad \langle \Delta p\rangle = -\frac{1}{2}\left(1 + \frac{\beta}{\alpha}\right)\rho_0 \langle W\rangle. \tag{21.41}$$

As shown in the following example, these expressions are quite useful.

Example 21.3 [Air blower]: The axial air blower in the margin figure has $a = 60$ mm and $b = 125$ mm and rotates at $\Omega = 2{,}500$ rpm $= 262\ \text{s}^{-1}$. Choosing $\alpha = -2r/b$ and $\beta = 0.4\alpha$, we find $U = 16\ \text{m}\,\text{s}^{-1}$ and $Q = 0.62\ \text{m}^3\,\text{s}^{-1}$. The specific work becomes $W = -395\ \text{m}^2\,\text{s}^{-2}$ and the power $P = -0.29$ kW. These values agree decently with the product specifications [11]. The pressure drop across the blower becomes merely $\Delta p = 332$ Pa $= 3.32$ millibar.

Axial air blower with seven blades rotating clockwise in this image. The z-axis goes into the paper so that the velocity U and the angular velocity Ω are both positive. (Source: Image courtesy Juvenco A/S, Denmark. With permission.)

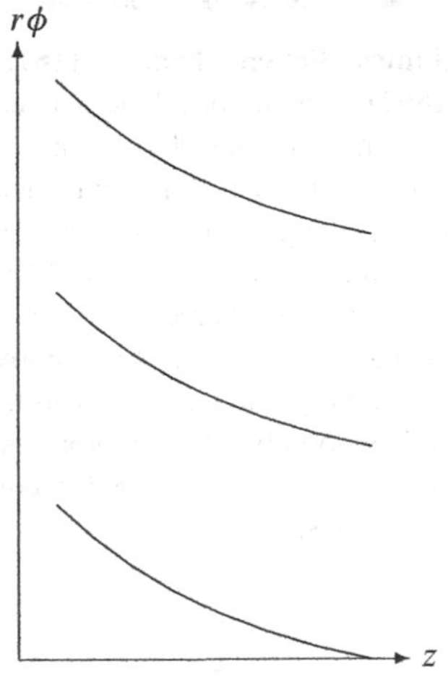

Sketch of three blades in the air blower above folded out along the azimuthal arc length $r\phi$ for fixed radial distance r. The blades move upward in this picture with negative inlet and outlet slopes.

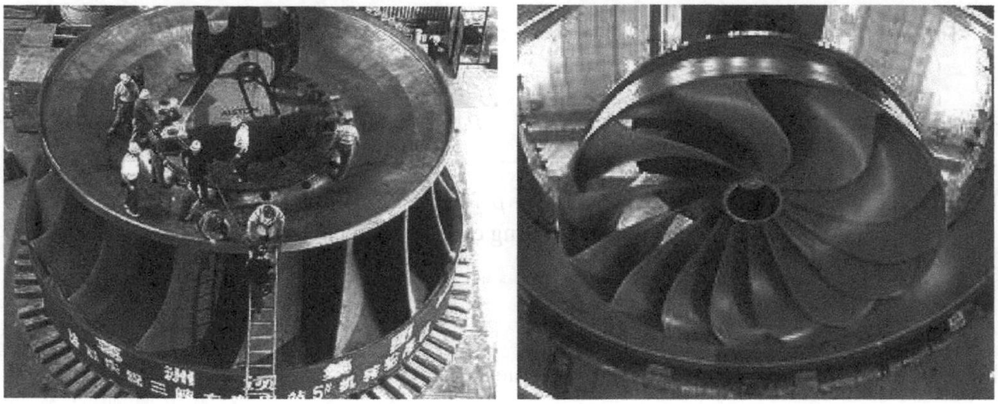

Figure 21.3. Runner from one of the 700-MW main hydraulic turbines in the Three Gorges Dam (China). In this Francis-type runner, water enters radially from the outside and exits axially downward. The runner turns clockwise, seen from above, so that the water flows "backward" into the runner. **Left:** Top view of the runner before installation showing only the radial inlets with a "skirt" covering the outlets. Notice the peculiar backward leaning, so-called X-blades; a fairly recent invention. (Source: Trelleborg AB. With permission.) **Right:** View up under the skirt of the runner during installation. (Source: Chinese Government publication.)

James Bicheno Francis (1815–1892). British born American engineer. He based his analysis and testing of turbines on scientific principles to obtain higher hydraulic efficiency. In 1849 he patented improvements on earlier designs, in particular by employing static guide vanes to lead the fluid smoothly into the rotor from the outside. (Source: Wikimedia Commons.)

21.3 Application: The Francis turbine

The Francis turbine is a hybrid construction in which fluid enters the rotor radially from the outside and exits along the axis. Francis turbines are highly efficient and mostly used in high-pressure turbomachinery, such as the hydraulic turbines shown in Figures 21.2 and 21.3. Here we shall only consider an idealized model, which nevertheless captures the essentials of this design. The classical treatment of reaction turbines is given in [Bodmer 2003] published in 1895. A short but general discussion of hydroelectric power may be found in [BH06a].

Inlet and outlet fields

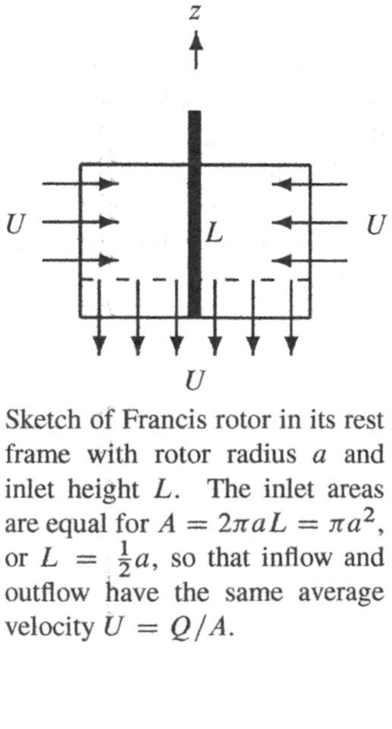

Sketch of Francis rotor in its rest frame with rotor radius a and inlet height L. The inlet areas are equal for $A = 2\pi a L = \pi a^2$, or $L = \frac{1}{2}a$, so that inflow and outflow have the same average velocity $U = Q/A$.

The rotor is cylindrical with radius a and inlet height L. We shall again ignore the thickness of the turbine blades and assume that the inflow takes place uniformly over the whole inlet area $A = 2\pi a L$ with velocity $U = Q/A$, where $Q > 0$ is the total volumetric flow rate through the turbine. The outlet is for simplicity required to have the same radius a and the same area $A = \pi a^2$ (ignoring the rotor shaft), such that the height is $L = \frac{1}{2}a$. We assume that the rotor blades are fixed to the rotor and carefully constructed such that the axial outflow velocity is also uniformly U (see Figure 21.3). At the inlet the blades have a constant slope α with respect to the radial direction and a variable outlet slope $\beta = \beta(r)$ with respect to the axial direction. Assuming smooth inflow, the inlet and outlet velocity fields become

$$v_r^{\text{in}} = -U, \qquad v_\phi^{\text{in}} = \Omega a - \alpha U, \qquad v_z^{\text{in}} = 0, \tag{21.42}$$

$$v_r^{\text{out}} = 0, \qquad v_\phi^{\text{out}} = \Omega r - \beta U, \qquad v_z^{\text{out}} = -U. \tag{21.43}$$

If we require that there be no swirl in the outflow, that is, $v_\phi^{\text{out}} = 0$, the outlet slope must be designed with a linear radial growth, $\beta = r\Omega/U$. This is basically what lies behind the elegantly curved shape of the rotor blades seen in Figure 21.3 (right). Evidently, since β is fixed by the rotor design, this condition can only be fulfilled for pairs of values of Ω and U having a fixed ratio. For electrical power production, the angular velocity Ω is determined by the AC frequency, so that there is only one so-called *design value* of U producing no outlet swirl. For any other velocity, say U', the azimuthal outflow becomes $v_\phi^{\text{out}} = (1 - U'/U)\Omega r$, describing a solid rotation with angular velocity $\Omega' = (1 - U'/U)\Omega$.

Specific work and static head

The specific rotor work is calculated from Euler's turbine equation (21.20) and becomes for the design velocity, where $v_\phi^{\text{out}} = 0$,

$$W = \Omega a(\Omega a - \alpha U). \tag{21.44}$$

This must be positive for a turbine, that is, $\alpha U < \Omega a$. Even if the specific work is largest for $\alpha < 0$, hydraulic turbines like the ones shown in Figures 21.2 and 21.3 are in practice designed with $0 < \alpha < \Omega a/U$, so that the water flows "backward" into the rotor (see the margin figure).

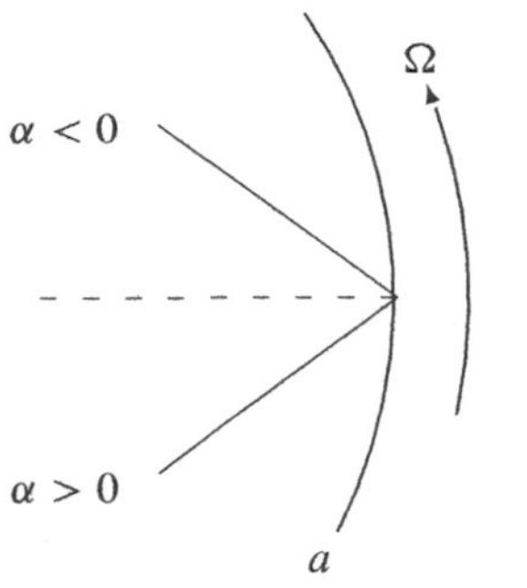

Blades with positive and negative inlet slopes with respect to the radial direction and the sense of rotation.

A hydraulic turbine is normally placed in a dam blocking the outflow from a large reservoir (or pool) of river water, as shown in the margin figure. The *static head* h of the reservoir is the vertical distance between the pool surface and the surface of the river downstream from the dam, also called the tailwater. In steady, nearly ideal flow, the laboratory Bernoulli field (21.23) is constant along the streamlines connecting the pool surface at $z = h$ and the turbine rotor inlet. Since the inflowing water is collected from a wide area (as in the emptying of a cistern), the entry velocity is negligible, $\boldsymbol{v}_{\text{entry}} \approx \mathbf{0}$. Averaging over the streamlines, we find

$$\langle H \rangle_{\text{entry}} \approx g_0 h + \frac{p_0}{\rho_0}, \tag{21.45}$$

where p_0 is atmospheric pressure. After passing the turbine, the water passes through a widening *draft tube* until it exits as a jet of radius $c > a$ at depth d under the tailwater surface. At the exit the pressure is higher, $p = p_0 + \rho_0 g_0 d$, and the flow is uniform without swirl. Since the discharge rate Q is the same at outlet and exit, we have $v_{\text{exit}} = (a^2/c^2)U$, so that exit average becomes

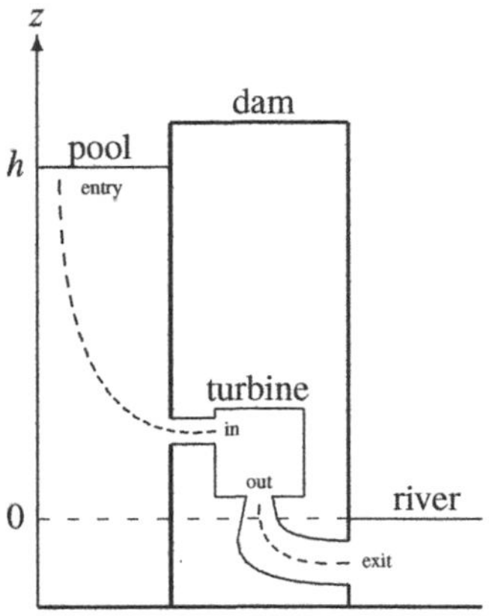

The Bernoulli function in the laboratory frame is constant along the dashed streamlines, but cannot be continued through the turbine because the flow is unsteady in the laboratory frame.

$$\langle H \rangle_{\text{exit}} = \frac{1}{2}v_{\text{exit}}^2 + g_0(-d) + \frac{p}{\rho_0} = \frac{1}{2}\frac{a^4}{c^4}U^2 + \frac{p_0}{\rho_0}. \tag{21.46}$$

Note that the exit depth cancels in this expression.

Finally, we obtain from the Bernoulli relation (21.26),

$$g_0 h = W + \frac{1}{2}\frac{a^4}{c^4}U^2. \tag{21.47}$$

The left-hand side is the specific gravitational energy at the pool surface (i.e., the gravitational energy per unit of mass transported through the installation). The first term on the right is the specific rotor work, and the last represents the specific loss of kinetic energy to the tailwater flow. If the turbine is operated with a discharge rate that differs from the design value, a swirl will be left in the draft tube and its kinetic energy must be added to the right-hand side, together with the viscous and turbulent losses we have systematically left out. What actually happens in the draft tube is, however, very complex and still an active subject of research.

Turbine efficiency

The maximal work one can possibly extract from the dam must equal the potential energy of the raised pool water relative to the tailwater. It can in theory be obtained by hoisting water gently down from rest at the surface of the pool and releasing it at rest to the tailwater surface. The work performed by gravity while lowering an amount M of water through the height h is of course Mg_0h.

The turbine *efficiency* is defined as the ratio of the actual turbine rate of work (power) $P = \rho_0 Q W$ and the theoretical maximum rate of work of gravity $\rho_0 Q g_0 h$,

$$\eta = \frac{P}{\rho_0 Q g_0 h} = \frac{W}{g_0 h} = 1 - \frac{a^4}{c^4}\frac{U^2}{2g_0 h}. \tag{21.48}$$

In the last step we divided (21.47) by $g_0 h$. It appears that the efficiency in principle may be driven as close to 100% as we wish by increasing the draft tube radius c until the water exits with near-zero velocity. But that creates other problems because the outlet pressure by Bernoulli's theorem will then always be lower than atmospheric. Such suction at the turbine outlet increases the risk of cavitation, setting a practical limit to how much the draft tube can be allowed to expand.

Optimal design parameters

The power $P = \rho_0 Q W$, with $Q = UA$ and $W = \Omega a(\Omega a - \alpha U)$, is for $\alpha > 0$ a quadratic function of the rotor flow velocity with maximum for $U = \Omega a/2\alpha$. Given the design parameters (α, a, c) and the angular velocity (Ω) of the rotor, this defines the optimal work point for the turbine,

$$\begin{aligned} U &= \frac{1}{2\alpha}\Omega a, & Q &= \frac{\pi}{2\alpha}\Omega a^3, & W &= \frac{1}{2}\Omega^2 a^2, \\ h &= \frac{1}{2g_0}\Omega^2 a^2\left(1 + \frac{a^4}{4c^4}\right), & P &= \frac{\pi}{4\alpha}\rho_0 \Omega^3 a^5, & \eta &= \left(1 + \frac{a^4}{4c^4}\right)^{-1}. \end{aligned} \tag{21.49}$$

Conversely, if the goal is to obtain a particular turbine performance for a particular installation, these expressions can be used to calculate the optimal design parameters.

Case: The Three Gorges Dam Project

The Yangtze River in China carries an average (yearly) discharge of 14,400 $m^3\,s^{-1}$ with large excursions during a year. During the rainy season the discharge may reach several times the average, and there may be large variations between years that have historically resulted in extensive flooding accompanied by massive loss of life and property. The main purpose of the Three Gorges Dam Project is in fact to prevent flooding at the cost of permanently inundating an area of about a 1,000 km^2. The secondary purpose is, of course, to produce hydraulic power.

During the flood season, which lasts from June to September, the river discharge is on average about 28,000 $m^3\,s^{-1}$. The static head is deliberately kept low, on average 78 m, making room in the pool for those sudden increases in the discharge that formerly caused flooding. The theoretically maximum power that can be produced at this head and discharge rate is 21.4 GW. At the end of the flood season the pool level is allowed to rise quickly through the month of October to reach a maximal head of about 110 m, where it is kept for about 2 months. During the remainder of the dry season from January to May, the head is slowly lowered until at the beginning of the rainy season in June it is again below 80 m. This strategy allows the hydraulic turbines to draw a higher discharge rate than the average available supply of 7,500 $m^3\,s^{-1}$ during the dry season, thereby producing power at a higher rate than the average theoretical maximum of 7.8 GW.

After becoming fully operational in late 2009, the installation comprises 26 main turbines (later to be increased to 32), each with a rated capacity of 710 MW at a rated head of 81 m. The huge runner from one of these turbines is shown in Figure 21.3. The total rated capacity of the dam is thus 18.5 GW, covering about 6% of China's average rate of electrical energy consumption in 2006 [12]. The rotor of the electric generator has 40 pole pairs and rotates at a fixed speed of 75 rpm, thus producing power at $40 * 75/60 = 50$ Hz frequency. A video animation of the turbine is found in [13].

Some of the design parameters may be found in [YTW06] and [14]. As input values we shall use the fixed angular velocity $\Omega = 75$ rpm, and the quoted values for the design head $h = 81$ m, the design power $P = 710$ MW, and the design efficiency $\eta = 93\%$. From (21.49) we obtain the values shown in the margin table. First, $\Omega a = \sqrt{2g_0 h\eta}$ is calculated, yielding a, and c then follows from η. Next, W and $Q = P/\rho_0 W$ are calculated. Finally, $U = Q/A$ and $\alpha = \Omega a/2U$ are obtained.

The derived parameters agree decently with the actual values. It should, however, be borne in mind that the turbines actually used in the project are much more complicated than the primitive cylindrical model used here. As seen in Figure 21.3 the blades have non-constant inlet slope, and the outlet radius is larger than the inlet radius. The installation is also required to provide high efficiency in both the dry and wet seasons with widely different values of the average head and river discharge rate. To permit control of the discharge rate of the turbines, a "wicket gate" of adjustable vanes is built into the inlet scroll (see the margin figure). Simulations are necessary to obtain reliable performance and design values that take into account the various sources of power losses due to viscosity, turbulence, inlet shock, and water leaks. Some of these losses take the same form as the draft tube loss, and are in our model lumped together in the effective draft tube radius c, determined above. Others take a different form and must be properly included as extra terms in Bernoulli's equation, in particular when operating the turbines off the optimal design values.

Parameter	Value	Unit
Ω	75	rpm
h	81	m
P	710	MW
η	93	%
a	4.9	m
A	75.2	m^2
L	2.4	m
c	5.7	m
W	740	m^2s^{-2}
Q	960	m^3s^{-1}
U	12.8	ms^{-1}
α	1.5	
$\tan^{-1}\alpha$	56	°

Design parameters for the Three Gorges turbines. Above the horizontal line, the input parameters are obtained from [YTW06]. Below the line the parameters are calculated as described in the text.

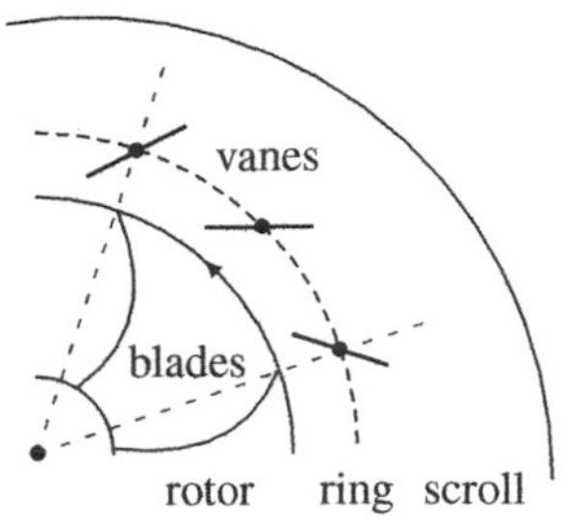

Inlet scroll and "wicket gate" ring of adjustable guide vanes (normally more closely spaced than shown here). The rotor turns anticlockwise in this drawing and has fixed (unadjustable) blades with positive inlet slope.

Problems

21.1 Calculate the reaction from a steadily draining cistern in **(a)** the horizontal and **(b)** the vertical direction. The cistern area is A_0 and the pipe exit area A, and one may assume plug flow. The cistern is being refilled to keep its level h constant.

21.2 Derive the rocket equations (20.21) for vertical flight by means of the reaction force (21.2). The mass of the gases in the combustion chamber is ΔM (assumed to be constant) and the total mass of the rocket plus burning gases is M, so that the mass of the rocket body itself is $M - \Delta M$. The rocket loses fuel mass at the constant rate $\dot{M}$, and the gases are exhausted with a constant velocity U relative to the rocket. There is no pressure jump at the exit.

21.3 A horizontal 1-inch water pipe bends horizontally 180°. Estimate the magnitude of the reaction force from the water on the pipe, when the steady water speed is 1 m s^{-1}.

* **21.4** Show that in the rotating frame of a rotor, the $\Omega^2 r^2$ terms in Euler's turbine equation (21.20) can be traced back to the Coriolis force.

21.5 The blades in a radial rotor have slopes α at $r = a$ and β at $r = b$. Calculate β when the blade is straight without curvature.

22 Energy

In continuum physics the macroscopic realm is separated from the microscopic by length scales set by the desired precision of the continuum description, as was discussed in Chapter 1. The smallest macroscopic body is called a material particle, and its mass and momentum can be calculated by adding the contributions from the molecules it contains. But while the molecular masses are all positive and simply add up, the nearly random molecular momenta almost cancel in the sum, leaving only the product of the mass of the particle and its center-of-mass velocity. The velocity field of continuum physics is based on the center-of-mass velocity because that yields the correct momentum of the material particles and brings the continuum description into contact with Newtonian mechanics.

The kinetic energy of a material particle, calculated from the velocity field, clearly misses the kinetic energy residing in the random molecular velocities relative to the center of mass. Loosely speaking, this "hidden" kinetic energy, plus contributions from other molecular fluctuations, constitutes the *heat content* of the material particle, while the average kinetic energy per molecule in the particle is a measure of its *absolute temperature*.

Although the idea that heat is a mechanical phenomenon had been around for centuries, other models of heat held the day until well into the nineteenth century. After James Joule in 1843 determined a precise value for the mechanical equivalent of heat, the intimate relation between mechanics and heat became firmly established. A couple of decades later, thermodynamics had reached the form that we use today. Avoiding all reference to the microscopic structure of matter, classical thermodynamics focuses entirely on the macroscopic interplay of heat, work, energy, and entropy.

In this and the following chapter we shall apply thermodynamics to continuous systems, thereby completing the analysis of the laws of balance begun in Chapter 20. The elements of thermodynamics presented here can in no way replace a proper course.

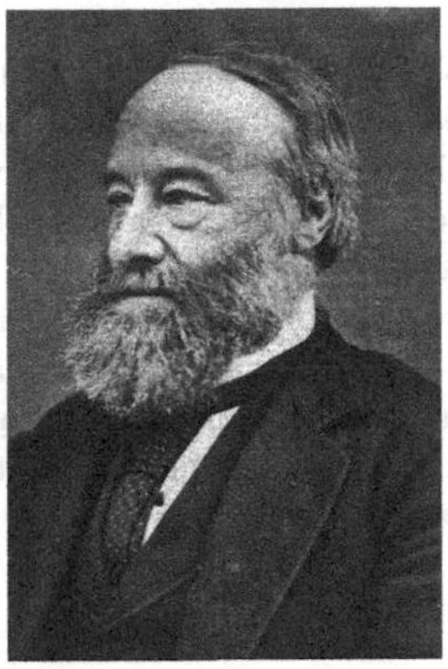

James Prescott Joule (1818–1889). English physicist. Gifted experimenter who in 1843 was the first to demonstrate convincingly the equivalence of mechanical work and heat, a necessary step on the road to thermodynamics. (Source: Wikimedia Commons.)

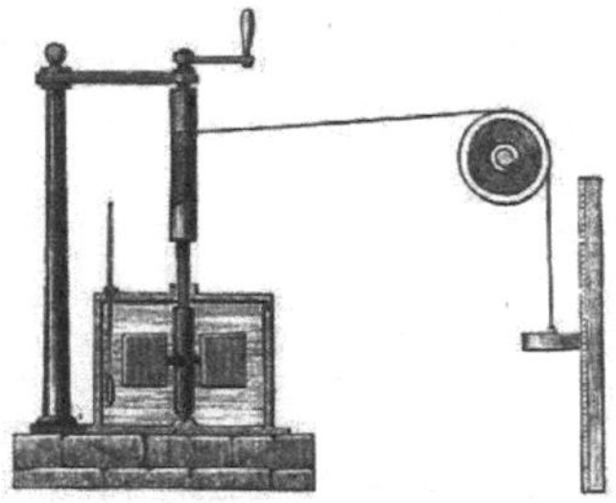

Joule's experiment on the mechanical equivalent of heat. A slowly falling weight drives a paddle wheel that agitates the water. The work is measured by the ruler on the right and the heat is measured by a thermometer inserted into the water. (Source: Wikimedia Commons.)

22.1 First Law of Thermodynamics

The conservation of the mechanical energy of a system not subject to external forces had been known since Newton's time. Joule's demonstration of the equivalence of heat and mechanical work opened the road for a wider principle of energy conservation that included the energy contained in the microscopic thermal fluctuations. The *First Law of Thermodynamics* postulates that for any state of a system there exists a quantity, called the *energy*, which is conserved (i.e., stays constant) when the system is completely isolated from its environment.

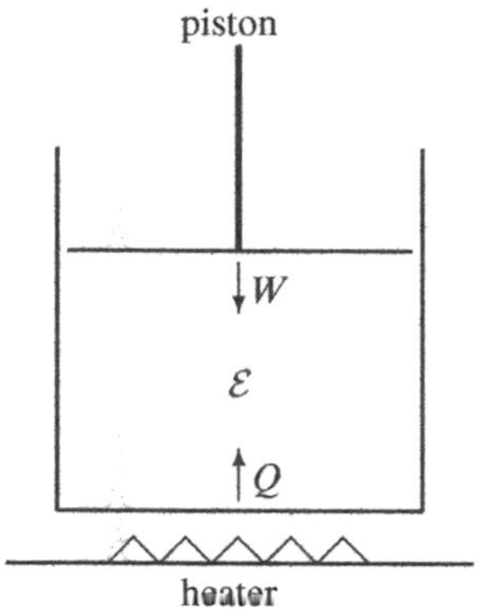

The archetypal thermodynamics thought experiment: A gas-filled cylindrical chamber with a moveable piston and a controllable heater. Mechanical work W is performed on the gas by pushing the piston into the chamber, and heat Q is transferred to the gas by raising the temperature of the heater.

Energy balance

As a consequence of the First Law, the only way to change the energy $\mathcal{E}$ of a system is by breaking its isolation and letting it interact with the environment for some time. Assuming that the system remains *closed* during this process, so that no matter flows in or out, macroscopic amounts of energy can essentially only enter or leave in two ways. The environment can either perform (signed) mechanical work W *on* the system, or it can transfer (signed) heat Q *to* the system (as illustrated in the margin figure). From the equivalence of heat and work, it follows that the change in energy becomes (ignoring chemical transformations)

$$\boxed{\Delta\mathcal{E} = Q + W.} \tag{22.1}$$

The First Law furthermore postulates, in analogy with Newtonian mechanics, that the energy only depends on the current state of the system (however that may be described). This is quite reasonable, considering that heat is identified with the mechanical energy of microscopic fluctuations. So *even if heat and work may each depend on the detailed history of the interaction, their sum depends only on where it starts and ends.*

In continuum physics, any (control) volume of matter may be viewed as a body or system. Since a comoving volume does not exchange matter with its environment, we shall recast the above relation as an equation of balance for the material rate of change of energy,

$$\boxed{\frac{D\mathcal{E}}{Dt} = \dot{Q} + \dot{W}.} \tag{22.2}$$

Here $\dot{W}$ is the (signed) rate at which the environment performs work *on* the control volume, and $\dot{Q}$ is the (signed) rate at which the environment transfers heat *to* it.

Rate of work

The environment can perform work on a system through short-range contact forces, $\boldsymbol{\sigma} = \{\sigma_{ij}\}$, acting on the surface of the control volume, as well as through long-range body forces $\boldsymbol{f}$ acting throughout the control volume. In Section 20.6 on page 351 we saw how a static gravitational field could be "moved over" to become potential energy, leaving only the residual body forces $\tilde{\boldsymbol{f}} = \boldsymbol{f} - \rho\boldsymbol{g}$ to contribute to the rate of work of the environment,

$$\dot{W} = \oint_S \boldsymbol{v}\cdot\boldsymbol{\sigma}\cdot d\boldsymbol{S} + \int_V \boldsymbol{v}\cdot\tilde{\boldsymbol{f}}\,dV. \tag{22.3}$$

The residual body forces, if present, stem from possibly time-dependent gravitational or electromagnetic fields. As long as their energy is not included in the total energy, they must be treated as external.

Rate of heat transfer

The everyday experience of handling a hot potato tells us that heat can be carried along with the movement of matter but that possibility is already taken care of in the equation of energy balance (22.2) through the use of the material derivative, effectively turning the arbitrary control volume into a comoving volume. When we heat water, it becomes equally clear that heat can also be conducted through the solid bottom of the kettle. Heat transfer by conduction is not an advective transport phenomenon tied to the motion of material but rather a diffusion process by which the warmer material of the kettle bottom "infects" the colder water with heat. In the molecular picture, the faster molecules of the kettle bottom collide with the slower molecules of the water and share their kinetic energy.

Macroscopically we describe the conduction of heat by the current density of heat, $\boldsymbol{q}(\boldsymbol{x}, t)$, defined so that the amount of heat passing through a surface element $d\boldsymbol{S}$ in a tiny time interval δt is $\boldsymbol{q}\delta t \cdot d\boldsymbol{S}$. If heat is also produced locally inside the body by chemical, nuclear, or other processes at the rate h per unit of volume, the total rate of heat conducted *into* the system becomes

$$\dot{Q} = -\oint_S \boldsymbol{q} \cdot d\boldsymbol{S} + \int_V h\, dV. \tag{22.4}$$

As long as the energy produced in local processes is not included in the total energy, they must be viewed as external to the body even if the processes take place everywhere inside.

Example 22.1 [Body heat]: The basal metabolic rate for men is about 2,000 kcal day^{-1}, corresponding about 100 W, a typical number also used for dimensioning cooling systems for concert halls. Only a small part of this energy is used for work. With a body surface area of about 1.7 m^2, the average heat current density through the skin becomes $q \approx 56$ W m^{-2}. Since no heat is accumulating in the body ($\dot{Q} \approx 0$), the average specific rate of heat production (h/ρ) for a man weighing 70 kg becomes about 1.4 W kg^{-1}.

Example 22.2 [Geothermal heat]: The average geothermal heat current density at the surface is $q_c \approx 0.06$ W m^{-2} for the continents and $q_s \approx 0.10$ W m^{-2} for the seas. Given that the continents cover a fraction $\gamma \approx 29\%$ of the surface (with mean radius $a = 6{,}371$ km), the total rate of heat flux out of the Earth becomes $\oint_S \boldsymbol{q} \cdot d\boldsymbol{S} = 4\pi a^2(\gamma q_c + (1-\gamma)q_s) \approx 45 \times 10^{12}$ W. This is three times larger the average world power consumption, which in 2006 was about 16×10^{12} W [15]. Assuming that the Earth is in a steady state (not cooling down or heating up), we infer that the average rate of (radioactive) heat production is only 7.6×10^{-12} W kg^{-1}, which is the same as 0.24 joule per metric ton per year.

Local energy balance

The total energy of a body is, like the mechanical quantities introduced in Chapter 20, assumed to be additive over the material particles making up the body,

$$\mathcal{E} = \int_V E\, dM = \int_V E\, \rho\, dV. \tag{22.5}$$

As for all other quantities, mass conservation makes the *specific energy* $E = d\mathcal{E}/dM$ a somewhat better behaved quantity than the energy density $d\mathcal{E}/dV = \rho E$.

We now apply Gauss' theorem to the surface integrals in $\dot{Q}$ and $\dot{W}$, and using Reynolds theorem (20.6) on page 341, we may cast energy balance (22.2) into the local form,

$$\rho \frac{DE}{Dt} = -\nabla \cdot \boldsymbol{q} + h + \nabla \cdot (\boldsymbol{v} \cdot \boldsymbol{\sigma}) + \boldsymbol{v} \cdot \tilde{\boldsymbol{f}}. \tag{22.6}$$

The stress term may now be rewritten as

$$\nabla \cdot (\boldsymbol{v} \cdot \boldsymbol{\sigma}) = \sum_{ij} \nabla_j (v_i \sigma_{ij}) = \sum_{ij} \sigma_{ij} \nabla_j v_i + v_i \nabla_j \sigma_{ij} = \boldsymbol{\sigma} : \nabla \boldsymbol{v} + \boldsymbol{v} \cdot (\nabla \cdot \boldsymbol{\sigma}^\mathsf{T}),$$

so that *local energy balance* finally takes the form

$$\boxed{\rho \frac{DE}{Dt} = -\nabla \cdot \boldsymbol{q} + h + \boldsymbol{\sigma} : \nabla \boldsymbol{v} + \boldsymbol{v} \cdot (\tilde{\boldsymbol{f}} + \nabla \cdot \boldsymbol{\sigma}^\mathsf{T}).} \tag{22.7}$$

The right-hand side is also a sum of heat and work contributions, although the stresses now give rise two distinct contributions.

Internal energy balance

Part of the energy of a body resides in its mechanical (i.e., kinetic plus potential) energy. Separating out this contribution, we write the specific energy as

$$E = \tfrac{1}{2}v^2 + \Phi + E_{\text{int}}, \tag{22.8}$$

where E_{int} is called the *specific internal energy.*

Using Cauchy's equation, $\rho D\boldsymbol{v}/Dt = \boldsymbol{f} + \nabla\cdot\boldsymbol{\sigma}^\top$, and $\boldsymbol{f} = \rho\boldsymbol{g} + \tilde{\boldsymbol{f}}$ we find

$$\rho\frac{DE}{Dt} = \boldsymbol{v}\cdot(\boldsymbol{f} + \nabla\cdot\boldsymbol{\sigma}^\top) + \rho(\boldsymbol{v}\cdot\nabla)\Phi + \rho\frac{DE_{\text{int}}}{Dt} = \boldsymbol{v}\cdot(\tilde{\boldsymbol{f}} + \nabla\cdot\boldsymbol{\sigma}^\top) + \rho\frac{DE_{\text{int}}}{Dt},$$

and finally, inserting local energy balance (22.7) on the left-hand side, we get

$$\rho\frac{DE_{\text{int}}}{Dt} = -\nabla\cdot\boldsymbol{q} + h + \boldsymbol{\sigma}:\nabla\boldsymbol{v}. \tag{22.9}$$

Notice that this equation of *local internal energy balance* neither depends on the static gravitational potential Φ, nor on the residual body forces $\tilde{\boldsymbol{f}}$. The terms of the right-hand side represent respectively the local convergence of heat flow $-\nabla\cdot\boldsymbol{q}$, the local heat production h, and the local rate of work of the stresses $\boldsymbol{\sigma}:\nabla\boldsymbol{v}$, including dissipation. However, having been derived by purely formal manipulation of energy balance (22.2), it contains neither more nor less. To go further we need one more assumption.

Local thermodynamic equilibrium

The total energy of a material particle,

$$d\mathcal{E} = E\,dM = \tfrac{1}{2}v^2\,dM + \Phi\,dM + E_{\text{int}}\,dM, \tag{22.10}$$

consists of the kinetic energy of its center-of-mass motion, the potential energy of its center-of-mass position, plus its internal energy. Very general principles of physics (i.e., translational, rotational, and Galilean invariance) tell us that the internal properties of a tiny body cannot depend on its center-of-mass position, its spatial orientation, or its center-of-mass velocity. The internal energy of a material particle can in other words only depend on the state of the matter contained in the particle. Notice that this is confirmed by the right-hand side of internal energy balance (22.9), which does not depend explicitly on the position or orientation. It does depend on the velocity field through the velocity gradient tensor $\nabla\boldsymbol{v}$, but that is evidently invariant under a Galilean boost, $\boldsymbol{v} \to \boldsymbol{v} + \boldsymbol{c}$, to another reference system moving with a constant velocity $\boldsymbol{c}$ relative to the first.

In continuum physics it is, partly for this reason, always assumed that *the specific internal energy can be locally identified with the specific internal energy of classical thermodynamics.* For an isotropic fluid of homogeneous composition, the thermodynamic specific internal energy may be viewed as a function $U(T, \rho)$ of temperature and density, and the specific internal energy field is identified with the local function of the corresponding fields,

$$E_{\text{int}}(\boldsymbol{x}, t) = U\big(T(\boldsymbol{x}, t), \rho(\boldsymbol{x}, t)\big). \tag{22.11}$$

For elastic matter the internal energy will also depend on the state of deformation.

We shall return to the question of the limits to the validity of this assumption and the broader assumption of *local thermodynamic equilibrium* when we discuss entropy in Chapter 23. For now it allows us to apply the local equation of internal energy balance (22.9) to a number of illustrative cases.

* Bernoulli's theorem revisited

Bernoulli's theorem has always been extremely useful for analyzing nearly inviscid flow and was on page 353 shown to be related to mechanical energy balance. With the more general energy concept now at hand, Bernoulli's theorem is under certain conditions valid even in the presence of viscous dissipation. The simple reason is that all dissipated kinetic energy must reappear as internal energy, as long as it cannot leave the system through its surface.

In steady flow the energy of a *fixed control volume* is always constant, $d\mathcal{E}/dt = 0$. Assuming that no heat is transferred to the control volume, $\dot{Q} = 0$, and that there are no residual body forces, $\tilde{\boldsymbol{f}} = \boldsymbol{0}$, energy balance (22.2) takes the form

$$\oint_S E\, d\dot{M} = \oint_S \boldsymbol{v}\cdot\boldsymbol{\sigma}\cdot d\boldsymbol{S} = -\oint_S p\boldsymbol{v}\cdot d\boldsymbol{S} + \oint_S \boldsymbol{v}\cdot\boldsymbol{\tau}\cdot d\boldsymbol{S}. \tag{22.12}$$

In the last step we have written $\boldsymbol{\sigma} = -p\mathbf{1} + \boldsymbol{\tau}$, where $\boldsymbol{\tau}$ represents the viscous stresses. Since $\boldsymbol{v}\cdot d\boldsymbol{S} = d\dot{M}/\rho$, the pressure term may be pulled over to the left-hand side so that

$$\oint_S H\, d\dot{M} = \oint_S \boldsymbol{v}\cdot\boldsymbol{\tau}\cdot d\boldsymbol{S}, \tag{22.13}$$

where

$$\boxed{H = E + \frac{p}{\rho} = \tfrac{1}{2}v^2 + \Phi + U + \frac{p}{\rho}} \tag{22.14}$$

is the *specific enthalpy*. Since $1/\rho$ is the specific volume, the sum of the two last terms, $H_{\text{int}} = U + p/\rho$, is the *specific internal enthalpy* of classical thermodynamics.

Consider now a stream tube connecting the inlet area A_{in} with the outlet area A_{out} (see the margin figure). Along such a tube the mass flow rate, $Q = \int_A d\dot{M}$, is the same for any cross-section A, so that it is meaningful to define the mass flow average, $\langle H\rangle = Q^{-1}\int_A H\, d\dot{M}$. When the right-hand side of (22.13) vanishes, we arrive at the most general version of Bernoulli's theorem,

$$\boxed{\langle H\rangle_{\text{in}} = \langle H\rangle_{\text{out}}\,.} \tag{22.15}$$

Provided the viscous stresses $\boldsymbol{\tau}$ can be ignored at inlet and outlet, this relation would be fulfilled for a duct with solid insulating walls where the no-slip condition requires the velocity to vanish on the sides. In Section 26.3 we shall encounter another case where it is valid.

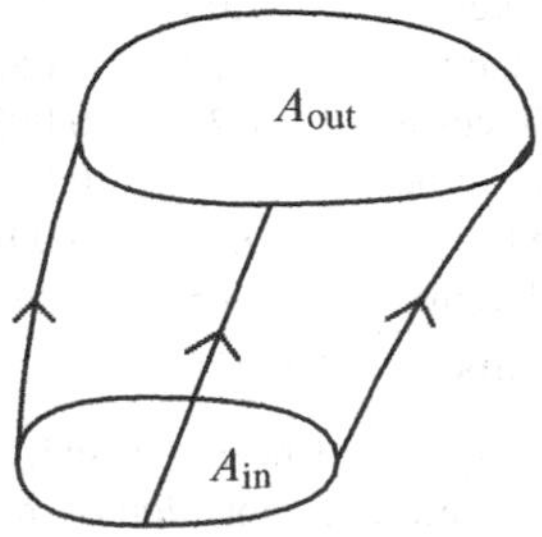

A stream tube consisting of all the streamlines that enter the area A_{in} and exit at A_{out}.

22.2 Incompressible fluid at rest

The simplest of all materials is an incompressible homogeneous isotropic fluid with constant mass density, $\rho(\boldsymbol{x}, t) = \rho_0$. At rest the pressure obeys the local equation of hydrostatic equilibrium, $\boldsymbol{\nabla} p = \rho_0 \boldsymbol{g}$, and is completely decoupled from thermodynamics. The only free thermodynamic variable is the temperature T, and all other local thermodynamic quantities, such as the specific internal energy $U(T)$, are functions of the local temperature $T(\boldsymbol{x}, t)$.

The temperature derivative of the specific energy,

$$c = \frac{dU}{dT}, \tag{22.16}$$

is called the *specific heat capacity*—or just the *specific heat*—of the material. It represents the amount of heat necessary to raise one unit of mass by one unit of temperature.

Being at rest with strictly constant density is, however, impossible to maintain in the face of the uneven expansion of the fluid caused by temperature variations. In Section 22.4 we shall see that for fluid in motion with flow velocities well below the speed of sound, it is most appropriate to identify c with the heat capacity at constant pressure, c_p.

Liquid	$10^{-3}c_p$	k
Water	4.18	0.61
Ethanol	2.44	0.17
Glycerol	2.43	0.29
Gasoline	2.22	0.15
Oil	1.67	0.15
Mercury	0.14	8.25
Units	$\frac{\text{J}}{\text{K kg}}$	$\frac{\text{W}}{\text{K m}}$

Specific heat capacity c_p and heat conductivity k for a few common liquids near normal temperature and pressure.

Fourier's law of heat conduction

Jean Baptiste Joseph Fourier (1768–1830) French scientist who made fundamental contributions to mathematics (Fourier series) and to the theory of heat. (Source: Wikimedia Commons.)

It is a common and completely general experience that heat is always conducted "downhill" from higher to lower temperatures. In isotropic matter at rest, there are no internal directions defined, so the only way to relate the vector current density of heat $\boldsymbol{q}$ to the scalar temperature field T is via *Fourier's law of heat conduction*,

$$\boxed{\boldsymbol{q} = -k\nabla T.} \tag{22.17}$$

The positive coefficient k is called the *thermal conductivity* of the material. At room temperature the thermal conductivity of water is 0.61 $\text{W K}^{-1}\text{m}^{-1}$.

Example 22.3 [Geothermal temperature gradient]: The average geothermal heat current density at the surface is $q_c \approx 0.06\ \text{W m}^{-2}$ for the continents [Kaye and Laby 1995]. Taking the thermal conductivity of bedrock to be $k \approx 2\ \text{W K}^{-1}\ \text{m}^{-1}$ [Lide 1996], the average geothermal temperature gradient in the upper crust becomes $|\nabla T| \approx q_c/k \approx 0.03\ \text{K m}^{-1}$, or 30° per kilometer. It must be emphasized that the geothermal current varies from place to place and with depth due to variations in the thermal conductivity of the upper crust [WHN09].

Fourier's heat equation

Liquid	$10^{-3}\rho$	$10^{6}\kappa$
Water	1.00	0.146
Ethanol	0.79	0.088
Glycerol	1.26	0.095
Gasoline	0.69	0.098
Oil	0.93	0.097
Mercury	13.5	4.37
Units	kg m^{-3}	$\text{m}^2\ \text{s}^{-1}$

Density ρ and heat diffusivity κ for a few common liquids near normal temperature and pressure.

Although the specific heat c and the heat conductivity k may both depend on the thermodynamic variables (i.e., the temperature), we shall for simplicity assume that they are constants. Setting $\boldsymbol{v} = \boldsymbol{0}$, $U = cT$, and $h = 0$ in the equation of internal energy balance (22.9), and using Fourier's law of heat conduction, we arrive at *Fourier's heat equation*,

$$\boxed{\frac{\partial T}{\partial t} = \kappa \nabla^2 T,} \tag{22.18}$$

where the constant

$$\kappa = \frac{k}{\rho_0 c} \tag{22.19}$$

is called the *heat diffusivity*. For water we get $\kappa \approx 0.146 \times 10^{-6}\ \text{m}^2\ \text{s}^{-1}$, which is about 7 times smaller than the kinematic viscosity $\nu \approx 1.00 \times 10^{-6}\ \text{m}^2\ \text{s}^{-1}$. The heat diffusivities for common liquids are shown in the margin table.

Case: Steady planar heat flow

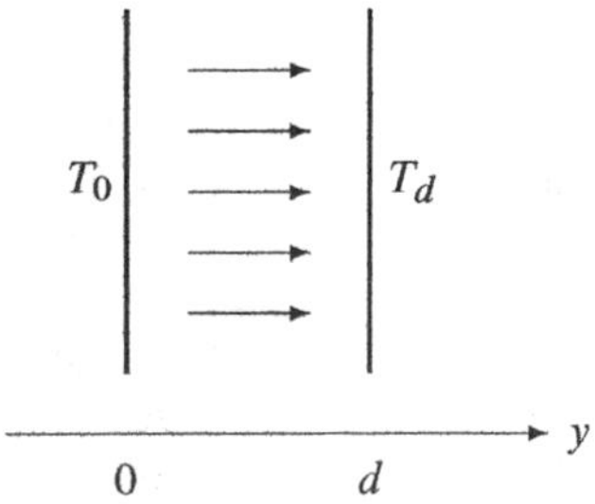

Steady heat flow between parallel plates with different temperatures, $T_0 > T_d$. The plates continue far above and below the section shown here. The fluid is at rest and there is no gravity.

A slab of isotropic material at rest is enclosed between two infinitely extended flat plates held at different fixed temperatures, T_0 at $y = 0$ and $T_d = T_0 - \Theta$ at $y = d$. Eventually, a steady heat flow will arise with a time-independent temperature field that only depends on the transverse coordinate, $T = T(y)$. Fourier's equation becomes $\nabla_y^2 T = 0$, which with the given boundary conditions has the solution

$$T = T_0 - \Theta \frac{y}{d}. \tag{22.20}$$

Notice that the steady temperature distribution does not depend on the value of the thermal diffusivity. We shall later investigate how fast the steady state is established.

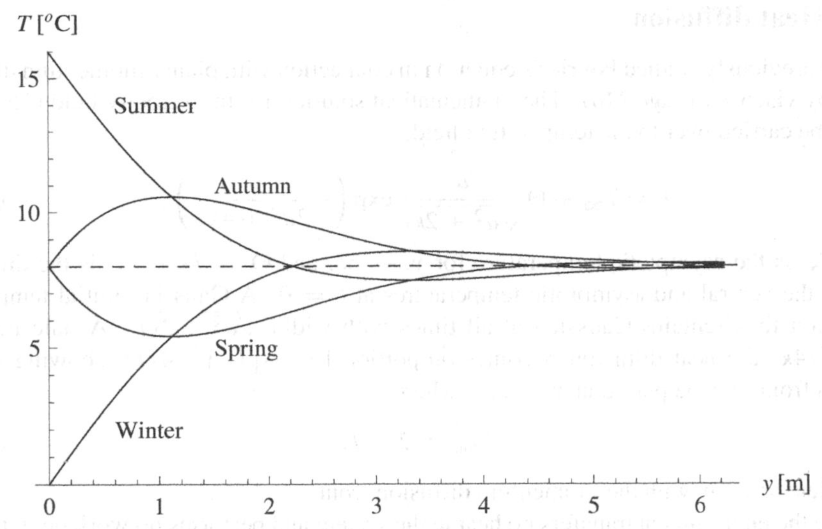

Figure 22.1. Estimated mean ground temperatures in Denmark as a function of depth in the middle of each of the four seasons (Example 22.4). The dashed line indicates the yearly mean temperature. It takes more than a month for the damped surface temperature to penetrate 1 m of depth. Beyond $y = 6$ m there is essentially no yearly variation. Notice how the temperature gradient—which determines the direction of the heat flow—changes sign roughly at the penetration depth at mid spring and autumn.

Case: Planar heat wave

Suppose the temperature is forced to oscillate sinusoidally in the plane $y = 0$ with period τ and amplitude Θ around the mean temperature T_0, such that the boundary condition becomes $T = T_0 + \Theta\cos(2\pi t/\tau)$ at $y = 0$. Direct insertion into Fourier's equation shows that the solution is a damped heat wave in which the temperature at a depth $y > 0$ is

$$T = T_0 + \Theta\, e^{-y/d} \cos\left(2\pi\frac{t}{\tau} - \frac{y}{d}\right), \qquad d = \sqrt{\frac{\kappa\tau}{\pi}}, \tag{22.21}$$

where d is the so-called *penetration depth*.

The wavelength of the damped oscillation is $\lambda = 2\pi d$, and at a depth of half a wavelength the damping factor has reduced the amplitude to $e^{-\pi} = 4.3\%$ of its surface value. So it is not really much of a wave (see Figure 22.1). The argument of the cosine shows that the surface temperatures reach the depth y with a time delay $\Delta t = \tau y/2\pi d$.

Example 22.4 [Annual soil temperature variation]: The mean surface temperature of soil follows the annual variations in atmospheric temperature with period $\tau = 1$ year. The thermal diffusivity of soil, consisting of moist sand and clay, is probably not unlike that of water, $\kappa \approx 0.2 \times 10^{-6}$ m^2 s^{-1}, making the soil penetration depth $d = 1.4$ m, such that the amplitude of the temperature variations at this depth has decreased to $e^{-1} = 37\%$ of the surface amplitude (see Figure 22.1). The wavelength is $\lambda = 8.9$ m. At $z = d$, variations in the surface temperatures will be delayed by $\Delta t = \tau/2\pi$, that is, nearly 2 months.

Denmark, for example, has a temperate Northern climate with mean temperature $T_0 \approx 8$°C and a yearly variation of $\Theta \approx 8$°C. Freezing temperatures occur often in winter, even if the average barely gets below 0°C. A freezing spell would have to be severe and last more than a month to provoke subzero temperatures at a depth of 1 m, and that is quite improbable. Frost-free depth is accordingly defined to be 90 cm for water mains and house foundations.

On top of the yearly variation there is also a daily temperature variation with somewhat smaller amplitude around the daily mean temperature. Using the same diffusivity as above, we find $d = 7.4$ cm with a time delay of about 4 h.

Case: Heat diffusion

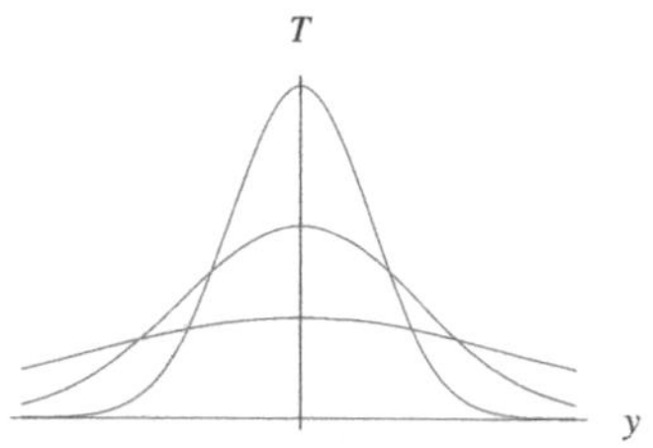

Heat diffusion makes a Gaussian temperature distribution widen and diminish in height as time goes by.

We have previously studied Fourier's equation in connection with planar momentum diffusion caused by viscosity (page 246). The mathematical solution for the velocity field (15.13) can directly be carried over to the temperature field,

$$T = T_\infty + \Theta \frac{a}{\sqrt{a^2 + 2\kappa t}} \exp\left(-\frac{y^2}{2a^2 + 4\kappa t}\right), \tag{22.22}$$

where T_∞ is the asymptotic temperature for $y \to \infty$, and $\Theta = T_0 - T_\infty$ is the difference between the central and asymptotic temperatures at $t = 0$. A Gaussian initial temperature distribution thus remains Gaussian at all times with width $\sqrt{a^2 + 2\kappa t}$. At late times for $t \gg a^2/4\kappa$, the heat diffusion becomes proportional to $\exp(-y^2/4\kappa t)$, showing that the diffusion front may be placed at $y = \delta_{\text{front}}$ where

$$\delta_{\text{front}} \approx 2\sqrt{\kappa t}, \tag{22.23}$$

in complete analogy with the momentum diffusion front.

Since the environment transfers no heat to the system and performs no work on it, the total internal energy in the material must remain constant while it spreads away from the central region. The entropy must as always increase in an isolated system (see Chapter 23).

Case: Growth of a thermal boundary layer

Consider an infinite heat-conducting plate at $y = 0$ in thermal contact with a fluid filling the half space $y > 0$. Initially the temperature of both plate and fluid is T_∞, but at time $t = 0$ the plate temperature is suddenly raised to $T_0 = T_\infty + \Theta$. Heat now flows out of the plate into the fluid (or conversely for $\Theta < 0$) and in the end the fluid assumes the plate temperature T_0 everywhere.

This problem was solved for the case of the boundary layer that arises when the fluid is suddenly set in motion (page 248). The solution can immediately be taken over,

$$T(y,t) = T_0 - \Theta f\left(\frac{y}{\sqrt{\kappa t}}\right), \tag{22.24}$$

where $f(s)$ is defined in (15.17) on page 249. For $s \to \infty$, the dominant behavior is Gaussian, $1 - f(s) \sim \exp(-y^2/4\kappa t)$, confirming that the heat front is given by (22.23).

In this case the internal energy is not conserved but increases with time for $\Theta > 0$ because heat is conducted into the fluid from the plate. From Fourier's law we find the (signed) heat flux out of the plate (using that $f'(0) = 1/\sqrt{\pi}$),

$$q_x(0,t) = -k \left.\frac{\partial T}{\partial y}\right|_{y=0} = \frac{k\Theta}{\sqrt{\pi\kappa t}}. \tag{22.25}$$

The total amount of heat transferred to the fluid from an area A of the wall during the time interval $0 \le t' \le t$ becomes

$$Q = \int_0^t q_x(0,t')\, A\, dt' = 2\rho_0 c \Theta A \sqrt{\frac{\kappa t}{\pi}}. \tag{22.26}$$

It diverges for $t \to \infty$ because all the fluid in the half-space in the end must acquire the plate temperature.

Example 22.5: A kettle of water of depth 10 cm and temperature 20°C is placed on a heater at 80°C. Taking $\delta = 2\sqrt{\kappa t} = 10$ cm the heat transferred per unit of bottom area becomes a reasonable 1,400 J cm^{-2}, whereas the diffusion time from the bottom to the surface becomes a whopping $t \approx 4.75$ h. Nobody waits that long to get hot water, so convection must play an important role (Chapter 30).

Case: Steady heat production

If heat is produced at a rate h per unit of volume, local internal heat balance (22.9) for a fluid at rest takes the form

$$\rho_0 c \frac{\partial T}{\partial t} = k\nabla^2 T + h. \tag{22.27}$$

Provided h is time independent and the environment has constant properties, an equilibrium will eventually be attained in which the heat produced in a volume equals the amount of heat that leaves the volume through its surface. The temperature field will then become time independent and obey the steady heat equation,

$$k\nabla^2 T + h = 0. \tag{22.28}$$

Often h will itself vary with temperature because the heat producing processes are temperature dependent.

Suppose nevertheless that heat is produced at a fixed rate $h(\boldsymbol{x}, t) = h_0$ inside a homogeneous sphere of radius a, making the total heat production rate $\dot{Q} = \frac{4}{3}\pi a^3 h_0$. In spherical coordinates the steady heat equation now takes the form

$$\frac{1}{r^2}\frac{d}{dr}\left(r^2 \frac{dT}{dr}\right) = -\frac{h_0}{k}. \tag{22.29}$$

Requiring the surface temperature to be constant $T(a) = T_0$, the solution is

$$T = T_0 + \frac{h_0}{6k}\left(a^2 - r^2\right). \tag{22.30}$$

As expected, the central temperature $T_c = T_0 + h_0 a^2/6k$ is higher than the surface temperature for $h_0 > 0$. Averaging over the sphere we find $\langle r^2 \rangle = \frac{3}{5}a^2$, and the average temperature becomes $\langle T \rangle = T_0 + a^2 h_0/15k$.

Example 22.6 [Geothermal heat production]: The total geothermal heat flow from the Earth is $\dot{Q} \approx 42 \times 10^{12}$ W, implying an average heat production rate $h_0 \approx 4 \times 10^{-8}$ W m^{-3}. If the Earth were made from uniform material with $k \approx 2$ W K^{-1} m^{-1}, the central temperature excess would be $T_c - T_0 \approx 130{,}000$ K.

This estimate fails miserably in comparison with geophysical models, which place the central temperature around 6,000 K. The reason is that the Earth is a highly non-uniform object with respect to heat transfer from the inside to the surface. In the liquid mantle, convection rather than conduction is the dominant mechanism, even though the thermal conductivity also increases with depth. Both of these effects lead to a redistribution of heat so that the temperature gradient in the mantle becomes much smaller.

Example 22.7 [Human skin temperature]: The average heat output from a human being is $\dot{Q} \approx 100$ W, a number typically used for dimensioning cooling systems for concert halls. With a typical mass of $M \approx 70$ kg and density equal to that of water $\rho_0 \approx 1{,}000$ kg m^{-3}, the volume becomes $V = 0.07$ m^3, implying an average heat production density rate $h_0 \approx 1.4$ kW m^{-3}. In the "spherical approximation", a human body of this mass would have radius $a \approx 26$ cm. Taking the thermal conductivity equal to that of water, $k = 0.6$ W K^{-1} m^{-1}, we obtain the difference between the average temperature and the skin temperature $\langle T \rangle - T_0 \approx 10$ K. Taking $\langle T \rangle = 37$°C, the skin temperature is predicted to be $T_0 = 27$°C. The conclusion is that a naked human being should be able to survive "indefinitely" in water of this temperature, a result that agrees decently with common experience.

Unfortunately the central temperature becomes 53°C in this model, which is far from reality. The reason is in this case that the active circulation of blood redistributes heat by advection so that the temperature is kept at 37°C in most of the body, except near the surface. In Problem 22.3 a simple model of this kind is explored.

22.3 Incompressible fluid in motion

Advective transport of internal energy by a moving fluid is included in the material derivative on the left-hand side of the internal energy equation (22.9). The last term on the right-hand side also depends on the velocity and represents for incompressible fluid with constant density the contribution of viscous friction, that is, *dissipation*, to the internal energy. In the everyday world of water and air it is, however, not so easy to come up with good examples of that. You can wave your arms in the air or stir a pot of water as much as you wish without generating a noticeable rise in temperature. Only if you insulate a system against heat loss and go at it for a long time, is it possible to accumulate the dissipated heat until the temperature change becomes measurable. This is exactly what Joule did in his stirring experiments in the 1840s.

Coupled equations of motion

We assume as before that the density ρ_0, the viscosity η, the specific heat c, and the heat conductivity k are all constant. Substituting the stress tensor, $\boldsymbol{\sigma} = -p\mathbf{1} + \eta(\boldsymbol{\nabla}\boldsymbol{v} + \boldsymbol{\nabla}\boldsymbol{v}^\mathsf{T})$, we arrive at the set of coupled partial differential equations consisting of the Navier–Stokes equations and the heat equation, including advection and viscosity,

$$\rho_0\left(\frac{\partial \boldsymbol{v}}{\partial t} + (\boldsymbol{v}\cdot\boldsymbol{\nabla})\boldsymbol{v}\right) = \rho_0\boldsymbol{g} - \boldsymbol{\nabla}p + \eta\nabla^2\boldsymbol{v}, \qquad \boldsymbol{\nabla}\cdot\boldsymbol{v} = 0, \tag{22.31a}$$

$$\rho_0 c\left(\frac{\partial T}{\partial t} + (\boldsymbol{v}\cdot\boldsymbol{\nabla})T\right) = k\nabla^2 T + h + \eta\,(\boldsymbol{\nabla}\boldsymbol{v} + \boldsymbol{\nabla}\boldsymbol{v}^\mathsf{T}) : \boldsymbol{\nabla}\boldsymbol{v}, \tag{22.31b}$$

where $(\boldsymbol{\nabla}\boldsymbol{v} + \boldsymbol{\nabla}\boldsymbol{v}^\mathsf{T}) : \boldsymbol{\nabla}\boldsymbol{v} = \sum_{ij}(\nabla_i v_j + \nabla_j v_i)\nabla_j v_i$.

With the assumption of constant values for ρ_0, c, η, and k, the temperature has no influence on the Navier–Stokes equations, and the flow of heat takes place against the background of a mass flow completely controlled by the external forces that drive the fluid. This limit is often called *forced convection* to distinguish it from the opposite limit, *free convection*, where the motion of the fluid is entirely caused by temperature differences (see Chapter 30).

Dimensionless numbers

Several dimensionless numbers are used to characterize the coupled equations of motion.

The Reynolds number: The Reynolds number was already introduced in Chapter 15 as the ratio of advective to viscous terms in the Navier–Stokes equation,

$$\mathsf{Re} = \frac{|\rho(\boldsymbol{v}\cdot\boldsymbol{\nabla})\boldsymbol{v}|}{|\eta\nabla^2\boldsymbol{v}|} \approx \frac{UL}{\nu}, \tag{22.32}$$

where U is the velocity[1] scale, L the length scale for major variations in the flow, and $\nu = \eta/\rho$ the momentum diffusivity (kinematic viscosity).

The Péclet number: In analogy with the Reynolds number, the ratio of the advective to conduction terms in the heat equation may be used to characterize heat transfer. This ratio is called the *Péclet number*,

Jean-Claude Eugene Péclet (1793–1857). French physicist. One of the first scholars of Ecole Normale (Paris); known for his clarity of style, sharp-minded views, and well-performed experiments.

$$\mathsf{Pe} = \frac{|\rho c(\boldsymbol{v}\cdot\boldsymbol{\nabla})T|}{|k\nabla^2 T|} \approx \frac{UL}{\kappa}, \tag{22.33}$$

where L is the length scale for major temperature variations, and $\kappa = k/\rho c$ is the heat diffusivity. For small Péclet number, advection of heat can be disregarded, whereas for large Péclet number, heat conduction is negligible.

[1] Apologies for the clash in the use of the symbol U for both the velocity scale and the specific internal energy. This should not give rise to problems in this section, where the specific energy is simply cT.

Prandtl number: The *Prandtl number* is the ratio of momentum to heat diffusivities,

$$\mathsf{Pr} = \frac{\nu}{\kappa} = \frac{\eta c}{k}. \tag{22.34}$$

The length scales for major velocity and temperature variations are usually the same, so that the Prandtl number becomes the ratio of the Péclet to Reynolds numbers, $\mathsf{Pr} \approx \mathsf{Pe}/\mathsf{Re}$.

In contrast to other dimensionless numbers, the Prandtl number is a property of the fluid rather than the flow. In gases it is slightly less than unity, for example $\mathsf{Pr} = 0.73$ for diatomic gases such as air. In liquids it may take a wide range of values: in water it is about 7, whereas in liquid metals it is quite small, for example 0.025 for mercury. For insulating liquids like oil, the Prandtl number may be 1,000 or larger (see also Table 22.1 on page 386).

Brinkman number: The *Brinkman number* is ratio of dissipation to heat conduction,

$$\mathsf{Br} = \frac{|\eta(\nabla \boldsymbol{v} + \nabla \boldsymbol{v}^{\top}) : \nabla \boldsymbol{v}|}{|k\nabla^2 T|} \approx \frac{\eta U^2}{k\Theta}, \tag{22.35}$$

where Θ is the scale of temperature variations. It is most important when the Reynolds and Péclet numbers are both small, so that advection of momentum and heat can be disregarded.

Eckert number: The *Eckert number* is a measure for the ratio of kinetic to internal energy,

$$\mathsf{Ec} \approx \frac{U^2}{c\Theta} \approx \frac{\mathsf{Br}}{\mathsf{Pr}}. \tag{22.36}$$

It is useful for characterizing the relative importance of mass and heat flow.

Case: Dissipation in planar Poiseuille flow

Steady laminar pressure-driven (Poiseuille) flow between parallel plates was analyzed in Section 16.2 on page 262. The velocity field was found to be

$$v_x = \tfrac{3}{2}U\left(1 - \frac{y^2}{a^2}\right) \qquad \text{with } -a \le y \le a, \tag{22.37}$$

where $d = 2a$ is the distance between the plates and U is the average velocity.

Conducting plates: When the plates are perfect heat conductors held at a fixed temperature T_0, all the dissipated heat will be led away through the walls. We may assume that the temperature only depends on the transversal coordinate, $T = T(y)$, so that the advective term in the heat equation (22.31b) vanishes. Since $\nabla_y v_x = -3Uy/a^2$, the only contribution to the dissipative term, is $\eta(\nabla_y v_x)^2$, and the heat equation reduces to

$$k\frac{d^2T}{dy^2} + 9\eta\frac{U^2y^2}{a^4} = 0. \tag{22.38}$$

The general solution is

$$T = -\frac{3\eta}{4k}U^2\frac{y^4}{a^4} + A + By,$$

where A and B are integration constants.

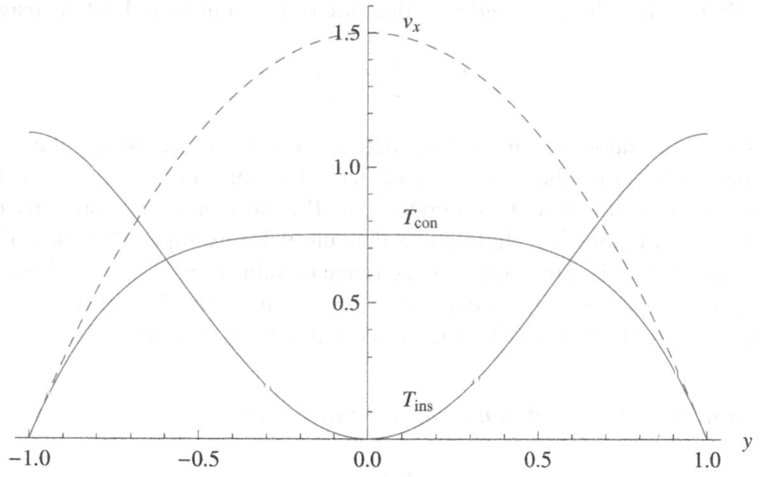

Figure 22.2. Dissipation in planar Poiseuille flow in units $a = 1$, $U = 1$, and $\Theta = 1$. The curves labeled T_{con} and T_{ins} apply, respectively, to perfectly conducting and insulating plates. The velocity profile labeled v_x is dashed.

If the conducting plates are maintained at the same temperature T_0, the boundary conditions become $T(\pm a) = T_0$, and we find immediately (see Figure 22.2) that

$$T = T_0 + \frac{3}{4}\Theta\left(1 - \frac{y^4}{a^4}\right) \qquad \text{with } \Theta = \frac{\eta U^2}{k}. \tag{22.39}$$

The maximal temperature rise is found midway between the plates at $y = 0$. The Brinkman number $\mathsf{Br} = \eta U^2/k\Theta$ is order unity because all the heat is conducted away. If the plates keep different temperatures, $T(\pm a) = T_0 \pm \Delta T$, one simply adds $\Delta T y/a$ to the solution. For water flowing with $U = 1\ \text{m s}^{-1}$ we get $\Theta \approx 1.5$ mK, independent of the plate distance.

Insulating plates: In this case heat cannot leave the fluid through the walls, but will instead be advected with the flow and accumulated downstream until it is discharged with the fluid. The temperature will now depend on both x and y, and the heat equation takes the form

$$\rho_0 c v_x \frac{\partial T}{\partial x} = k\left(\frac{\partial^2 T}{\partial x^2} + \frac{\partial^2 T}{\partial y^2}\right) + 9\eta\frac{U^2 y^2}{a^4}, \tag{22.40}$$

with v_x given by (22.37). Since there should be no heat conduction through the walls, the boundary conditions are $q_y = -k\partial T/\partial y = 0$ for $y = \pm a$.

The above partial differential equation looks rather formidable. We know, however, from Equation (16.12) on page 264 that the rate of work of the pressure on the fluid is proportional to the length L of the plates. Since this work is completely converted into heat, we shall assume that there is a solution linear in x,

$$T(x, y) = F(y) + xG(y). \tag{22.41}$$

Both F and G must satisfy the insulation conditions, $dF/dy = dG/dy = 0$ for $y = \pm a$.

Inserting this trial form into the heat equation, and separating the x dependence, it reduces to two coupled ordinary differential equations

$$\rho_0 c v_x G = k\frac{d^2 F}{dy^2} + 9\eta\frac{U^2 y^2}{a^4}, \qquad \frac{d^2 G}{dy^2} = 0. \tag{22.42}$$

The last equation has a trivial linear solution, and applying the insulating boundary conditions it follows that G must be constant, and the first equation may now be solved for F.

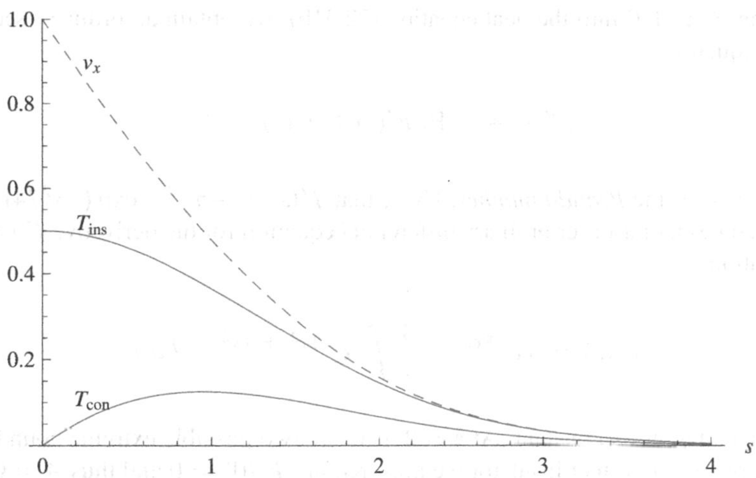

Figure 22.3. Dissipation in the Stokes layer for $\mathsf{Pr} = 1$ in units $U = 1$ and $\Theta = 1$. The abscissa is $s = y/2\sqrt{\nu t}$. The curves labeled T_{con} and T_{ins} apply, respectively, to perfectly conducting and insulating plates. The velocity profile labeled v_x is dashed.

Applying again the insulating boundary conditions, the final result becomes

$$T = T_0 + \frac{9}{8}\Theta\left(2 - \frac{y^2}{a^2}\right)\frac{y^2}{a^2} + \Theta' x, \qquad \Theta = \frac{\eta U^2}{k}, \qquad \Theta' = \frac{3\eta U}{\rho_0 c a^2}. \tag{22.43}$$

The transverse shape of the temperature field (see Figure 22.2) is the same for all x, so the linear x-dependence only serves to increase the overall temperature level as heat accumulates downstream. Similar results are obtained for Poiseuille flow in pipes with circular cross-section.

Example 22.8: Water flowing at $U = 1\ \mathrm{m\,s^{-1}}$ has $\Theta \approx 1.5$ mK. Taking $2a = 1$ mm we obtain $\Theta' \approx 2.5\ \mathrm{mK\,m^{-1}}$ and a Reynolds number $\mathsf{Re} \approx 1{,}100$, which is inside the laminar region. It takes a plate length of nearly 400 m just to raise the overall temperature by one degree, so don't propose to produce hot water this way!

* Case: Dissipation in the Stokes layer

Momentum diffuses away from an infinitely extended plate that is suddenly set into motion with velocity U in a fluid at rest. We found earlier that the velocity field takes the form

$$v_x(y,t) = U f(s) \qquad \text{with } s = \frac{y}{\sqrt{\nu t}} \tag{22.44}$$

and $f(s)$ given in Equation (15.17) on page 249.

Let us now assume that before the plate was set into motion the temperature of the fluid was everywhere T_∞, and let us look for a temperature field that is independent of x and depends on y and t in the same way as the velocity field,

$$T(y,t) = T_\infty + \Theta\, F(s) \qquad \text{with } \Theta = \frac{\eta U^2}{k}. \tag{22.45}$$

As above the dissipation temperature scale Θ is the only possible combination of viscosity, heat conductivity, and velocity. The asymptotic boundary condition is $F(\infty) = 0$.

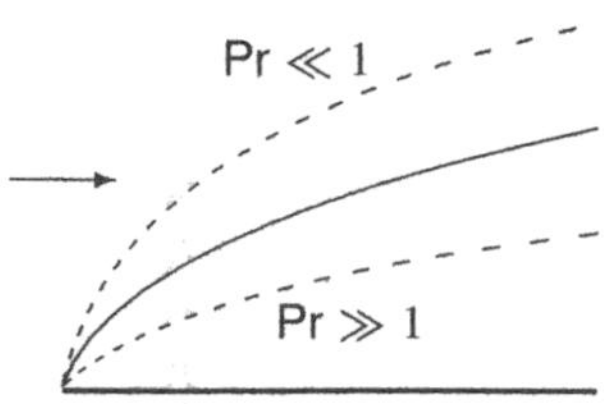

Advective cooling of a plate with wind coming in from the left. The boundary for the mass flow (solid line) and the heat fronts (dashed) for large and small Prandtl numbers.

Inserting v_x and T into the heat equation (22.31b), we obtain an ordinary second-order differential equation,

$$F''(s) + \tfrac{1}{2}s\,\mathsf{Pr}\,F'(s) + f'(s)^2 = 0, \tag{22.46}$$

where $\mathsf{Pr} = \nu/\kappa$, is the *Prandtl number*. Using that $f'(s) = -\pi^{-1/2}\exp\left(-s^2/4\right)$, the above equation becomes a first-order ordinary differential equation for the derivative $F'(s)$, with the general solution

$$F'(s) = A\,e^{-\mathsf{Pr}\,s^2} - \frac{4}{\pi}\int_0^s e^{-2u^2-\mathsf{Pr}\,(s^2-u^2)}\,du, \tag{22.47}$$

where A is an integration constant. At $s = 0$ there are two possible extreme boundary conditions. If the plate is a perfect insulator we must require $F'(0) = 0$ and thus $A = 0$. Alternatively, if the plate is a perfect conductor with temperature T_∞, we must require $F(0) = 0$, and that determines A as a function of Pr. In Figure 22.3, the numeric solutions to these extreme cases are plotted for $\mathsf{Pr} = 1$.

22.4 General homogeneous isotropic fluids

In the following we shall limit ourselves to fluids with material properties that are independent of any particular position (homogeneous) and of any particular direction (isotropic). For such fluids the thermodynamic state is described by only two variables, for example the temperature T and the density ρ. All other thermodynamic quantities are functions of these, such as the pressure $p(T,\rho)$ and the specific internal energy $U(T,\rho)$. It is often convenient to reexpress thermodynamic quantities as functions of other pairs of thermodynamic variables, and most of the mathematical intricacies in thermodynamics arise from such reformulations.

Coupled equations of motion

It is convenient to split the stress tensor into a part due to the thermodynamic pressure and a remainder due to viscosity, $\boldsymbol{\sigma} = -p\mathbf{1} + \boldsymbol{\tau}$. The coupled equations of mass and heat flow may then be written compactly,

$$\frac{D\rho}{Dt} = -\rho\nabla\cdot\boldsymbol{v} \qquad \text{(continuity)}, \tag{22.48a}$$

$$\rho\frac{D\boldsymbol{v}}{Dt} = \rho\boldsymbol{g} - \nabla p + \nabla\cdot\boldsymbol{\tau}^\mathsf{T} \qquad \text{(Cauchy)}, \tag{22.48b}$$

$$\rho\frac{DU}{Dt} = -\nabla\cdot\boldsymbol{q} + h - p\nabla\cdot\boldsymbol{v} + \boldsymbol{\tau}:\nabla\boldsymbol{v} \qquad \text{(heat)}. \tag{22.48c}$$

To complete these equations we need to specify the external gravitational field $\boldsymbol{g}(\boldsymbol{x},t)$, the state equations for the thermodynamic pressure $p = p(T,\rho)$, and the specific internal energy $U = U(T,\rho)$, together with constitutive equations for the viscous stresses $\boldsymbol{\tau}$, the heat current $\boldsymbol{q}$, and the heat production h. When all these are provided, we have obtained a closed set of coupled equations of motion for the velocity field $\boldsymbol{v}(\boldsymbol{x},t)$, the density field $\rho(\boldsymbol{x},t)$, and the temperature field $T(\boldsymbol{x},t)$. The pressure and specific energy may afterward be calculated from these fields.

Constitutive equations

Although the following analysis does not depend on the explicit form of the constitutive relations we shall mostly think of Fourier's law of heat conduction (22.17) and the Stokes tensor (15.29) on page 256,

$$\boldsymbol{q} = -k\nabla T \qquad \text{(Fourier)}, \qquad (22.49a)$$

$$\boldsymbol{\tau} = \eta\left(\nabla\boldsymbol{v} + \nabla\boldsymbol{v}^\top - \tfrac{2}{3}\nabla\cdot\boldsymbol{v}\,\mathbf{1}\right) + \zeta\nabla\cdot\boldsymbol{v}\,\mathbf{1} \qquad \text{(Stokes)}. \qquad (22.49b)$$

The heat conductivity k, the shear viscosity η, and the bulk viscosity ζ may in general depend on the temperature and pressure. The bulk viscosity is frequency dependent but may usually be taken to be of the same order of magnitude as the shear viscosity. For water at 25°C and 100 MHz, the ratio of bulk to shear viscosity has been determined by acoustic measurement to be $\zeta/\eta \approx 2.7$ [DG09].

We shall mainly leave undefined gravity $\boldsymbol{g}$, external heat production h, and the equation of states for the pressure, $p(\rho, T)$, and internal energy, $U(T, \rho)$. An important subclass of fluids, the ideal gases, form as always a theoretical benchmark that we shall repeatedly return to. They obey the simple state equations (see Appendix E),

$$p = R\rho T, \qquad U = c_V T, \qquad (22.50)$$

where $R = R_{\text{mol}}/M_{\text{mol}}$ is the specific gas constant and $c_V = R/(\gamma - 1)$ is the specific heat at constant volume (γ being the adiabatic index).

The thermodynamic properties are shown for some liquids in Table 22.1 and for some gases in Table 22.2. Somewhat surprisingly, the heat diffusivity of air at rest ($\kappa_{\text{air}} = 31 \times 10^{-6}\ \text{m}^2\,\text{s}^{-1}$) turns out to be about 200 times larger than that of water ($\kappa_{\text{water}} = 0.14 \times 10^{-6}\ \text{m}^2\,\text{s}^{-1}$). Why is it then that we use air for insulation in thermoglass windows, bed covers, and fur coats—rather than sleeping and walking in wet suits? The explanation is that for a given temperature gradient, the actual heat current (22.49a) is not determined by the diffusivity but by the thermal conductivity, which is about 23 times larger in water than in air. The role of fur coats or wet suits is mainly to prevent advection of heat by air or water currents, which will otherwise rapidly remove the warm fluid adjacent to your skin, thereby increasing the temperature gradient at the skin and thus the heat flow from your body.

Expansivity and compressibility

Viewing the density as a function of temperature and pressure, $\rho = \rho(T, p)$, its first-order differential may be written as

$$d\rho = \left(\frac{\partial\rho}{\partial T}\right)_p dT + \left(\frac{\partial\rho}{\partial p}\right)_T dp, \qquad (22.51)$$

where the index on the derivative as usual indicates which quantity must be held fixed during differentiation.

The relative changes in density per unit of temperature or pressure are called[2] the *(isobaric) thermal expansivity* α and the *(isothermal) compressibility* β,

$$\alpha = -\frac{1}{\rho}\left(\frac{\partial\rho}{\partial T}\right)_p, \qquad \beta = \frac{1}{\rho}\left(\frac{\partial\rho}{\partial p}\right)_T. \qquad (22.52)$$

The signs have been chosen to reflect that the density normally decreases with rising temperature and increases with rising pressure.

[2]In other texts they are called the *coefficients of thermal expansion and compression.* In this book we shall mostly use the shorter and more systematic names *expansivity* and *compressibility*.

Table 22.1. Thermodynamic properties of a few common liquids at normal temperatures and pressures.

Liquid	$10^{-3}\rho$	$10^3\eta$	k	$10^{-3}c_p$	$10^3\alpha$	$10^9\beta$	Pr	γ
Water	1.00	1.00	0.61	4.18	0.21	0.49	6.85	1.01
Ethanol	0.79	1.08	0.17	2.44	0.75	0.76	15.5	1.13
Glycerol	1.26	1410	0.29	2.43	0.49	0.25	11800	1.10
Gasoline	0.70	0.50	0.12	2.22	0.95	1.30	9.25	1.16
Oil	0.92	67.	0.15	1.67	0.64	0.61	746.	1.15
Mercury	13.5	1.53	8.25	0.14	0.18	0.037	0.026	1.16
Bromine	3.1	0.94	0.122	0.474	1.12	1.45	3.67	1.22
Units	$\mathrm{kg\,m^{-3}}$	Pa s	$\mathrm{J\,K^{-1}m^{-1}}$	$\mathrm{J\,K^{-1}kg^{-1}}$	$\mathrm{K^{-1}}$	$\mathrm{Pa^{-1}}$		

The Prandtl number is calculated as $\mathrm{Pr} = c_p\eta/k$ and the adiabatic index from (22.61). The heat capacity at constant volume is $c_v = c_p/\gamma$ and the heat diffusivity $\kappa = k/\rho c_p$. Notice the explicit scale factors on the top and the (SI) units at the bottom. Some of liquids (gasoline and oil) have variable composition and the values should be seen as typical, but not exact.

The density differential may now be written as

$$\frac{d\rho}{\rho} = -\alpha\, dT + \beta\, dp. \tag{22.53}$$

The isothermal compressibility is the inverse of the isothermal bulk modulus, $\beta = 1/K_T$, defined in Equation (2.43) on page 33.

Typical values: Generally, the expansivity of liquids grows with the temperature while the compressibility decreases (slowly). For ideal gases the equations of state (22.50) immediately yields $\alpha = 1/T$ and $\beta = 1/p$. At normal temperature, and pressure the ideal gas values become $\alpha \approx 3.3 \times 10^{-3}\ \mathrm{K^{-1}}$ and $\beta \approx 10^{-5}\ \mathrm{Pa^{-1}}$. Whereas the expansivity of gases is comparable to that of liquids (within an order of magnitude), the compressibility of gases is about 10,000 times larger!

Specific heat capacity at constant volume

Viewing the specific energy as a function of temperature and specific volume, $U = U(T, \rho)$, the first-order differential becomes

$$dU = \left(\frac{\partial U}{\partial T}\right)_\rho dT + \left(\frac{\partial U}{\partial \rho}\right)_T d\rho. \tag{22.54}$$

The first term is the amount of heat transferred to a unit-mass material particle during a small (reversible) change of temperature dT while its mass density is held constant. The second is the work performed on the particle under a small (reversible) change of density $d\rho$ while its temperature is held constant.

The first coefficient above is the *specific heat capacity at constant density*. Since the volume of a unit-mass particle is $1/\rho$, it is normally called the *specific heat capacity at constant volume* and denoted c_v. In the following chapter we shall use the mere existence of entropy to derive a relation (due to Helmholtz) that expresses the second coefficient in terms of the thermal expansivity and isothermal compressibility,

$$c_v = \left(\frac{\partial U}{\partial T}\right)_\rho, \qquad \rho^2\left(\frac{\partial U}{\partial \rho}\right)_T = p - \frac{\alpha}{\beta}T. \tag{22.55}$$

Table 22.2. Thermodynamic properties of a few common gases at 300 K and 1 bar.

Gas	$10^3 M_{\rm mol}$	$10^{-3} R$	γ	$10^{-3} c_p$	$10^6 \eta$	$10^3 k$	Pr
H_2	2.0	4.16	1.41	14.30	9.0	187	0.69
He	4.0	2.08	1.63	5.38	20.0	157	0.69
Ne	20.2	0.41	1.64	1.05	32.1	50	0.68
N_2	28.0	0.30	1.40	1.04	17.9	26	0.72
O_2	32.0	0.26	1.40	0.91	20.8	26	0.73
Ar	39.9	0.21	1.67	0.52	22.9	18	0.66
CO_2	44.0	0.19	1.30	0.82	15.0	17	0.72
Air	29.0	0.29	1.40	1.00	18.6	26	0.72
Units	$\rm kg\,mol^{-1}$	$\rm J\,K^{-1}kg^{-1}$		$\rm J\,K^{-1}kg^{-1}$	Pa s	$\rm W\,K^{-1}m^{-1}$	

The Prandtl number is calculated as $\mathsf{Pr} = c_p\eta/k$. The heat capacity at constant volume is $c_v = c_p/\gamma$. Notice the explicit scale factors on the top and the SI-units at the bottom.

The energy differential now takes the form

$$dU = c_v dT + \left(p - \frac{\alpha}{\beta}T\right)\frac{d\rho}{\rho^2}. \tag{22.56}$$

For an ideal gas we have $\alpha = 1/T$ and $\beta = 1/p$, and the expression in parentheses vanishes. The specific energy of an ideal gas is thus independent of the density and can only depend on the temperature. We shall, however, always assume that the specific heat is a constant $c_v = R/(\gamma - 1)$.

The above relation also holds for the material time derivatives. Making use of the continuity equation (22.48a) to eliminate the divergence on the right-hand side of the heat equation (22.48c), we obtain

$$\rho c_v \frac{DT}{Dt} - \frac{\alpha T}{\beta \rho}\frac{D\rho}{Dt} = -\nabla\cdot\boldsymbol{q} + h + \boldsymbol{\tau}\colon\nabla\boldsymbol{v}. \tag{22.57}$$

Given the terms on the right-hand side, this is an explicit form of the general heat equation in terms of temperature, density, and velocity. The coefficients α, β, and c_v are obtained as derivatives of the state functions and depend in general also on temperature and density. Typically, they are slowly varying at ordinary temperatures and pressures.

Thermal pressure coefficient

Keeping the volume fixed while heating a gas is usually not a big problem because gases are easily compressed. But for liquids at normal temperatures and pressures, the expansivity is comparable to that of gases while the compressibility is 10,000 times smaller. When one attempts to raise the temperature while keeping the volume fixed, heat expansion will for this reason generate a much higher pressure in a liquid than in a gas. We can actually calculate how much by setting $d\rho = 0$ in (22.53) to obtain

$$\left(\frac{\partial p}{\partial T}\right)_\rho = \frac{\alpha}{\beta}. \tag{22.58}$$

This quantity is called the *thermal pressure coefficient*.

Typical values: For an ideal gas we have $\alpha/\beta = p/T$, so the pressure coefficient becomes about 330 Pa K^{-1} at normal temperature and pressure. For a liquid it is huge, typically $\alpha/\beta \approx 10^6$ Pa K^{-1}, corresponding to about 10 bar K^{-1}. Thus, heating a confined liquid by 100 K will generate a pressure of 1,000 bar—and break most vessels.

Specific heat capacity at constant pressure

In most practical situations, for example in central heating systems, the fluid will not be rigidly confined but instead allowed to expand or contract freely under nearly constant pressure. It is for this reason better to express the left-hand side of the heat equation (22.57) in terms of the temperature and the pressure. Substituting the density differential (22.53) into the differential corresponding to the left-hand side of the heat equation, it becomes

$$\rho c_v dT - \frac{\alpha T}{\beta\rho} d\rho = \rho\left(c_v + \frac{\alpha^2 T}{\beta\rho}\right) dT - \alpha T dp. \tag{22.59}$$

The last term vanishes for constant pressure, showing that the expression in parentheses,

$$c_p = c_v + \frac{\alpha^2 T}{\beta\rho}, \tag{22.60}$$

must be the *specific heat at constant pressure*. For an ideal gas where $\alpha = 1/T$ and $\beta = 1/p$, we find the well-known result $c_p = c_v + R$ where $R = R_{\text{mol}}/M_{\text{mol}}$ is the specific gas constant.

The ratio of specific heats is called the *adiabatic index*,

$$\boxed{\gamma \equiv \frac{c_p}{c_v} = 1 + \frac{\alpha^2 T}{\beta\rho c_v} = \left(1 - \frac{\alpha^2 T}{\beta\rho c_p}\right)^{-1}.} \tag{22.61}$$

For the ideal gases in Table 22.2 it is a constant ranging between 1.3 and 1.7, whereas for the liquids (except water) in Table 22.1 it is about 1.15 at normal temperature and pressure. Water has an exceptionally low expansivity at room temperature, leading to an adiabatic index very close to unity. This reflects that the expansivity of water vanishes at 4°C, where the density has a maximum (at normal pressure).

The heat equation now takes the form

$$\boxed{\rho c_p \frac{DT}{Dt} - \alpha T \frac{Dp}{Dt} = -\nabla \cdot \boldsymbol{q} + h + \boldsymbol{\tau} : \nabla \boldsymbol{v}.} \tag{22.62}$$

This equation has of course exactly the same content as (22.57). The advantage is that under quite general and often occurring conditions, the second term on the left-hand side may be disregarded.

Nearly incompressible flow

Using the definition (22.61) of the adiabatic index, the ratio of the coefficients of the two terms on the left-hand side of (22.62) may be rewritten as

$$\frac{\alpha T}{\rho c_p} = \left(1 - \frac{1}{\gamma}\right)\frac{\beta}{\alpha}. \tag{22.63}$$

Since $1 - 1/\gamma$ is always smaller than unity, the condition

$$\left|\frac{Dp}{Dt}\right| \ll \frac{\alpha}{\beta}\left|\frac{DT}{Dt}\right|, \tag{22.64}$$

is sufficient for disregarding the second term on the left-hand side of (22.62).

When this condition is fulfilled, the heat equation simplifies to

$$\rho c_p \frac{DT}{Dt} \simeq -\nabla \cdot \boldsymbol{q} + h + \boldsymbol{\tau} : \nabla \boldsymbol{v}. \tag{22.65}$$

Under the same condition the density differential (22.53) implies that

$$\frac{1}{\rho}\frac{D\rho}{Dt} \simeq -\alpha \frac{DT}{Dt}. \tag{22.66}$$

This shows that the comoving changes in density due to heat expansion can in general not be ignored. But if the scale of global temperature variations in the flow is $|\Delta T| \simeq \Theta$, the relative global density variations become $|\Delta\rho|/\rho \simeq \alpha\Theta$, which may be ignored provided $\alpha\Theta \ll 1$. In that case the flow may be called *nearly incompressible*, and the density may normally be replaced by a constant, $\rho = \rho_0$.

The exception is *free convection* where buoyancy variations caused by small temperature variations can generate important vertical flows in a gravitational field. The so-called Boussinesq approximation for convective flows is derived below, and used in Chapter 30.

Role of the Mach number

In most practical cases, fluid flows freely through ducts that are designed to accommodate heat expansion, the typical example being a central heating system equipped with an expansion tank. Since the pressure is communicated through the fluid at the speed of sound, while heat is merely advected and conducted, we would expect the condition (22.64) be fulfilled, as long as the flow velocity is much smaller than the speed of sound.

Assuming that the flow is nearly steady with Reynolds number so large that the pressure obeys the Bernoulli principle

$$\frac{Dp}{Dt} \simeq -\rho v \frac{Dv}{Dt} \tag{22.67}$$

we find

$$\frac{\beta}{\alpha}\frac{Dp}{Dt} = \frac{\gamma}{\rho\alpha c_S^2}\frac{Dp}{Dt} \approx -\frac{\gamma}{\alpha}\frac{v}{c_S^2}\frac{Dv}{Dt} = -\frac{\gamma}{\alpha}\mathsf{Ma}\frac{D\mathsf{Ma}}{Dt}, \tag{22.68}$$

where $c_S = \sqrt{\gamma/\rho\beta}$ is the isentropic sound speed and $\mathsf{Ma} = v/c_S$ the Mach number.

Finally, since γ is of order unity, the heat condition (22.64) may now be written as

$$\mathsf{Ma}\left|\frac{D\mathsf{Ma}}{Dt}\right| \ll \alpha \left|\frac{DT}{Dt}\right|. \tag{22.69}$$

For a typical flow through a duct or around obstacles at rest, the global changes in the Mach number are comparable to the Mach number itself, $\Delta\mathsf{Ma} \approx \mathsf{Ma}$, so that we may write the global version of the above condition as

$$\mathsf{Ma}^2 \ll \alpha\Theta \tag{22.70}$$

where $|\Delta T| \approx \Theta$ as before is the magnitude of the global changes in temperature.

Consequently, in nearly ideal flow with small temperature variations, $\alpha\Theta \ll 1$, and smallish Mach number, $\mathsf{Ma} \lesssim \sqrt{\alpha\Theta}$, we are permitted to use the heat equation in the form (22.65). This argument finally justifies[3] the earlier calculations on incompressible fluids at rest and in motion (Sections 22.2 and 22.3).

[3] The admissibility of the incompressibility condition in thermal hydrodynamics is a long-standing problem which has only recently been formally justified; see [AKO05].

* The Boussinesq approximation

Valentin Joseph Boussinesq (1842–1929). French physicist and mathematician. Contributed to many aspects of hydrodynamics: whirls, solitary waves, drag, advective cooling, and turbulence.

In a gravitational field, small temperature variations may give rise to vertical buoyancy forces that dominate an otherwise nearly incompressible flow. Including this possibility leads to a hybrid formalism for incompressible flow, called the Boussinesq approximation.

Consider a nearly incompressible fluid at rest with uniform density ρ_0 and temperature T_0 in a time-independent field of gravity $\boldsymbol{g}_0 = -\nabla\Phi_0$. In mechanical equilibrium the pressure is purely hydrostatic $p_0 + \rho_0\Phi_0$. Suppose now that the equilibrium is disturbed by changing the temperature by $\Delta T(\boldsymbol{x},t)$, the density by $\Delta\rho(\boldsymbol{x},t)$, and the pressure by $\Delta p(\boldsymbol{x},t)$. The Navier–Stokes equation (22.48b) then becomes

$$\rho\frac{D\boldsymbol{v}}{Dt} = \rho\boldsymbol{g}_0 - \nabla p + \nabla\cdot\boldsymbol{\tau}^{\mathsf{T}} = \Delta\rho\boldsymbol{g}_0 - \nabla\Delta p + \nabla\cdot\boldsymbol{\tau}^{\mathsf{T}}. \tag{22.71}$$

Provided the disturbance satisfies $\beta\,|\Delta p| \ll \alpha\,|\Delta T| \ll 1$, the change in the density field follows from the density differential (22.53),

$$\frac{\Delta\rho}{\rho_0} \simeq -\alpha\Delta T. \tag{22.72}$$

In the leading approximation, the Navier–Stokes equation becomes

$$\rho_0\frac{D\boldsymbol{v}}{Dt} \simeq -\alpha\Delta T\rho_0\boldsymbol{g}_0 - \nabla\Delta p + \nabla\cdot\boldsymbol{\tau}^{\mathsf{T}}, \tag{22.73}$$

and in the same approximation the heat equation (22.65) simplifies to

$$\rho_0 c_p\frac{D\Delta T}{Dt} \simeq k\nabla^2\Delta T + \boldsymbol{\tau}\colon\nabla\boldsymbol{v}, \tag{22.74}$$

where we have used Fourier's law and dropped heat production h. What remains is now to show that the last (dissipative) term in this equation can be ignored, and then use this result to determine the form of the continuity equation.

In steady flow with $D\boldsymbol{v}/Dt = (\boldsymbol{v}\cdot\nabla)\boldsymbol{v}$, it follows from the Navier–Stokes equation (22.73) that the velocity scale of buoyancy is $U \sim \sqrt{\alpha\Theta g_0 L}$, where Θ is the scale of the temperature variations and L is the vertical length scale of the flow geometry. The ratio of dissipation to heat conduction (the Brinkman number) becomes

$$\mathsf{Br} = \frac{|\boldsymbol{\tau}\colon\nabla\boldsymbol{v}|}{|k\nabla^2\Delta T|} \sim \frac{\eta U^2/L^2}{k\Theta/L^2} \sim \mathsf{Pr}\frac{\alpha}{c_p}g_0 L, \tag{22.75}$$

where $\mathsf{Pr} = \nu/\kappa$ is the Prandtl number. Assuming macroscopic geometry with $g_0 L \sim \mathrm{m}^2/\mathrm{s}^2$, we find $\mathsf{Br} \sim 10^{-6}$ for gases and $\mathsf{Br} \sim 10^{-4}$–10^{-6} for typical liquids. This shows that it is in general a good approximation to disregard dissipation in the convective heat equation.

From the continuity equation (12.16) on page 196 we find

$$\nabla\cdot\boldsymbol{v} = -\frac{1}{\rho}\frac{D\rho}{Dt} \approx -\frac{1}{\rho_0}\frac{D\Delta\rho}{Dt} = \alpha\frac{D\Delta T}{Dt} = \alpha\kappa\nabla^2\Delta T,$$

where we in the last step have used the heat equation without dissipation. Relative to the velocity gradient, the divergence is estimated to be

$$\frac{|\nabla\cdot\boldsymbol{v}|}{|\nabla\boldsymbol{v}|} \sim \frac{\alpha\kappa\Theta/L^2}{U/L} \sim \sqrt{\alpha\Theta}\frac{\kappa}{\sqrt{g_0 L^3}}. \tag{22.76}$$

Assuming again a macroscopic geometry with $\sqrt{g_0 L^3} \sim m^2/s$, the right-hand side is of order 10^{-6} or smaller for ordinary liquids and gases. The continuity equation may consequently be replaced by the incompressibility condition $\nabla \cdot \boldsymbol{v} \approx 0$. This result also simplifies the diffusive term in the Navier–Stokes equation to the usual $\nabla \cdot \boldsymbol{\tau}^\top \approx \eta \nabla^2 \boldsymbol{v}$.

Collecting everything, we arrive at the *Boussinesq approximation* for convective mass and heat flow,

$$\boxed{\begin{aligned} \frac{D\Delta T}{Dt} &= \kappa \nabla^2 \Delta T, \\ \frac{D\boldsymbol{v}}{Dt} &= -\alpha \Delta T\, \boldsymbol{g}_0 - \frac{\nabla \Delta p}{\rho_0} + \nu \nabla^2 \boldsymbol{v}, \\ \nabla \cdot \boldsymbol{v} &= 0. \end{aligned}} \tag{22.77}$$

We shall return to and use this approximation in Chapter 30.

Problems

22.1 Consider two plates at $y = y_1$ and $y = y_2$ and fixed temperatures T_1 and T_2. **(a)** Show that if there is incompressible fluid at rest between the plates, the temperature in the fluid is

$$T = T_1 + (T_2 - T_1)\frac{y - y_1}{y_2 - y_1}.$$

(b) Show that this is also a solution if the fluid is inviscid and moves steadily along x.

22.2 Consider two coaxial cylinders with radii a_1 and a_2 and incompressible fluid at rest between. **(a)** Show that the temperature distribution between the cylinders is

$$T = T_1 + (T_2 - T_1)\frac{\log(r/a_1)}{\log(a_2/a_1)}.$$

(b) Show that this is also true if the fluid is inviscid and moves steadily along z.

22.3 Consider a spherical body of radius a with constant heat production everywhere. The body consists of a core of radius $c < a$ with heat conductivity k_c and a skin in the form of a spherical shell $c < r < a$ with heat conductivity k_s. **(a)** Calculate the steady temperature field in the whole body, given that the central temperature is T_c.

In the human body the core temperature is maintained at $T_c = 37°\text{C}$ by active blood circulation. For a "spherical human" this may be simulated by choosing essentially infinite central heat conductivity, $k_c \gg k_s$. **(b)** Use the parameters of Example 22.7 to determine the temperature drop in the skin when its thickness is $s = a - c \approx 2$ cm and $k_c = 10k_S$.

22.4 **(a)** Show that the equation of *local enthalpy balance* may be written as

$$\rho\frac{DH}{Dt} = -\nabla \cdot \boldsymbol{q} + h + \nabla \cdot (\boldsymbol{v} \cdot \boldsymbol{\tau}) + \boldsymbol{v} \cdot \tilde{\boldsymbol{f}} + \frac{\partial p}{\partial t},$$

with $\boldsymbol{\tau}$ defined by $\boldsymbol{\sigma} = -p\mathbf{1} + \boldsymbol{\tau}$. **(b)** What is the corresponding local equation of internal enthalpy balance?

22.5 Show that the advective term on the left-hand side of the incompressible heat equation (22.31b) can be moved over and included in a generalized heat flow,

$$\boldsymbol{q}' = -k\nabla T + \rho_0 c T \boldsymbol{v}.$$

23 Entropy

The identification of heat with the mechanical energy residing in random molecular fluctuations was fairly straightforward once Joule had determined the mechanical equivalent of heat. In 1850, Rudolf Clausius unearthed another macroscopic quantity connected with heat, and later (in 1865) named it *entropy*. It is without doubt one of the most subtle concepts in physics. The entropy of a material body may be thought of as a quantitative measure of the macroscopic lack of detailed knowledge about the microscopic state of matter. Entropy may also be thought of as a measure of the effective number of independent fluctuating molecular variables. Since the absolute temperature is universally proportional to the average fluctuation energy in each independent molecular variable, entropy may be calculated from the ratio of heat to absolute temperature. Entropy is often just viewed as a measure of internal *disorder*.

Rudolf Julius Emmanuel Clausius (1822–1888). German physicist who became one of the fathers of classical thermodynamics (the other being William Thomson, aka Lord Kelvin). In his famous paper from 1850, he not only rejects the earlier caloric theory of heat and replaces it by Joule's concept of energy, but also identifies the new concept of entropy, and formulates the two laws of thermodynamics. (Source: Wikimedia Commons.)

Entropy is not conserved for an isolated system. Sadi Carnot's theoretical analysis in 1824 of simple heat engines indicated, however, that independent of what goes on inside an isolated system, its entropy can never decrease. Generalized to become a fundamental postulate, this insight is known as the *Second Law of Thermodynamics*. Surprisingly, the Second Law defines a direction of time where Newtonian mechanics offers none. It is paradoxical that we, on the one hand, view heat as the macroscopic expression of microscopic mechanical energy but on the other have to give up the Newtonian indifference to the direction of time. This is, however, the price to pay for working with systems about which we have only partial knowledge of the actual state. The Second Law claims that for a system in isolation, our knowledge of the state can never become better than it was at the start. How the inexorable growth of entropy should be understood in the context of the whole universe is still being debated [Teg08].

23.1 Entropy in classical thermodynamics

A non-rotating macroscopic body at rest in an inertial system has vanishing total linear and angular momenta. When such a body is completely isolated from the environment, the linear and angular momenta will remain zero, and its energy will stay constant whatever goes on inside. Common experience tells us that all internal macroscopic mass and heat flows must eventually die out because of the unavoidable viscosity and thermal conductivity of all materials. Microscopically there will still be a lot of action but macroscopically all internal forces will in the end be balanced and all internal temperature differences will have gone away. The body has reached *global thermodynamic equilibrium*, in which the temperature and pressure are uniformly constant across the whole body.

Reversible processes

Nicolas Léonard Sadi Carnot (1796–1832). French physicist who in 1824 published a book in which he analyzed theoretical models of heat engines, now known as Carnot cycles. He died young, and his work was nearly forgotten until William Thomson (aka Lord Kelvin) in 1849 pointed out its importance. (Source: Wikipedia.)

Classical thermodynamics only deals with relations between global equilibrium states. Physically, to relate different global equilibrium states, the body must be allowed to interact with the environment. Generally, such a process will create all kinds internal mass and heat flows that may bring the body far away from global equilibrium. Think, for example, of a hammer blow to a nail or an explosion in a rigid container. After the interaction, global equilibrium will of course eventually be reestablished.

If, however, the interaction is "sufficiently slow and weak", the body can remain near global equilibrium all the way with almost constant and uniform thermodynamic parameters that nevertheless slowly change during the interaction. Although there will unavoidably arise weak internal heat and mass flows, the velocity and temperature gradients are assumed to be so small that dissipation is negligible. Such a process is said to be *reversible* because in the absence of dissipation one can always—step by step—backtrack the original interaction by reversing all external driving forces and temperature gradients. As is evident from this description, a truly reversible interaction is an idealization that cannot be reached in practice. Real-world interactions can only be approximatively (or nearly) reversible.

The fundamental thermodynamic relation

During a tiny step of a reversible interaction, the body receives a tiny amount of heat dQ and work dW (but no material) from the environment. According to the First Law (22.1) on page 372, the change in energy becomes $d\mathcal{E} = dQ + dW$. Flow velocities are, however, so small in nearly reversible processes that the kinetic energy can be ignored relative to the internal energy, making $d\mathcal{E} = d\mathcal{U}$. Furthermore, since the pressure p is nearly constant across the body, and since viscous stresses can be disregarded, the work must be of the form $dW = -p\,dV$, where dV is the change in volume. As mentioned earlier, entropy may loosely be thought of as the ratio of heat and absolute temperature, which suggests that it should change by[1] (ignoring chemical transformations)

$$d\mathcal{S} \equiv \frac{dQ}{T} = \frac{1}{T}d\mathcal{U} + \frac{p}{T}dV. \tag{23.1}$$

This is the fundamental entropy relation for infinitesimal reversible interactions without mass exchange. Keeping a tally of the quantities entering on the right-hand side, it allows us to track the changes in entropy during any reversible interaction without mass exchange. It is, however, not evident that the entropy relation is a *perfect differential* that can be integrated to make the entropy a proper function of the state, say $\mathcal{S} = \mathcal{S}(\mathcal{U}, V)$ or $\mathcal{S} = \mathcal{S}(T, V)$.

Case: Entropy of an ideal gas

Consider a volume V of an ideal gas with equation of state, $pV = nR_{\text{mol}}T$. The internal energy is $\mathcal{U} = C_v T$, where $C_v = Mc_v$ is the (total) heat capacity at constant volume. Eliminating the pressure by means of the equation of state, the entropy differential takes the form

$$d\mathcal{S} = C_v\frac{dT}{T} + nR_{\text{mol}}\frac{dV}{V}. \tag{23.2}$$

This expression is perfectly integrable with the result

$$\mathcal{S} = C_v \log T + nR_{\text{mol}} \log V + \text{const.} \tag{23.3}$$

The integrability condition for arbitrary homogeneous isotropic matter will be derived below.

[1]The use of script capitals to denote entropy ($\mathcal{S}$) and internal energy ($\mathcal{U}$) is a bit atypical for thermodynamics, but is consistent with the continuum physics notation used in this book.

Since $T \sim pV$ this becomes $\mathcal{S} = C_v \log p + C_p \log V + \text{const}$, where $C_p = C_v + nR_{\text{mol}}$ is the heat capacity at constant pressure. Finally, introducing $\gamma = C_p/C_v$ we may write

$$\boxed{\mathcal{S} = C_v \log (pV^\gamma) + \text{const.}} \tag{23.4}$$

Constant entropy thus implies $pV^\gamma = $ const, which is the well-known relation (2.48) on page 35 valid for adiabatic reversible (i.e., isentropic) processes.

Integrability condition

It is not the intention here to repeat the intricate physical arguments (going back to Carnot) that justify the universal integrability of entropy. They may be found in all textbooks on classical thermodynamics, for example [Kondepudi and Prigogine 1998]. Here we shall only derive the necessary condition for integrability.

Let us assume that the energy is a true function of the state, $\mathcal{U}(T, V)$, with differential

$$d\mathcal{U} = \frac{\partial \mathcal{U}}{\partial T} dT + \frac{\partial \mathcal{U}}{\partial V} dV. \tag{23.5}$$

Inserting this into the entropy differential (23.1), we obtain

$$d\mathcal{S} = \frac{1}{T}\frac{\partial \mathcal{U}}{\partial T} dT + \frac{1}{T}\left(\frac{\partial \mathcal{U}}{\partial V} + p\right) dV. \tag{23.6}$$

A necessary condition for this to be a perfect differential is that the cross-derivatives must be equal:

$$\frac{\partial}{\partial V}\left(\frac{1}{T}\frac{\partial \mathcal{U}}{\partial T}\right) = \frac{\partial}{\partial T}\left(\frac{1}{T}\frac{\partial \mathcal{U}}{\partial V} + \frac{p}{T}\right). \tag{23.7}$$

This simply expresses the symmetry of the double derivative, $\partial^2\mathcal{S}/\partial T\partial V = \partial^2\mathcal{S}/\partial V\partial T$.

Carrying through the outer differentiations, the double derivatives of the energy cancel, and the condition takes the form (in proper thermodynamics notation)

$$\boxed{\left(\frac{\partial \mathcal{U}}{\partial V}\right)_T = T^2\left(\frac{\partial (p/T)}{\partial T}\right)_V = T\left(\frac{\partial p}{\partial T}\right)_V - p.} \tag{23.8}$$

This is the *Helmholtz equation* which allows us to calculate the energy derivative with respect to volume given the equation of state, $p = p(\rho, T)$. Setting $\mathcal{U} = MU$ and $V = M/\rho$ and using Equation (22.58) on page 387, this leads indeed to the expression (22.55).

Hermann Ludwig Ferdinand von Helmholtz (1821–1894). German physician and physicist. Invented in 1851 the ophthalmoscope (for studying the eye) and published an influential *Handbook of Physiological Optics*. Was deeply engaged in the physiology of perception and the physiological basis for the theory of music. In physics his main contribution was in vortex theory. Worked as professor of physics in Berlin from 1871 to 1888. (Source: Wikimedia Commons.)

23.2 Entropy balance

Consider an isolated body in global thermodynamic equilibrium. Any (smaller) part of this body is definitely not isolated, even if the rest of the body performs no work on it and transfers no mass or heat to it, because all the body's matter is uniformly at rest with uniform temperature and pressure. The part is said to be in *thermodynamic equilibrium with its environment*. In continuum physics where body parts may be as small as material particles, it is always assumed that each material particle relaxes so quickly that it is effectively all the time in thermodynamic equilibrium with its local environment, even when the whole body is not in global equilibrium. This assumption of *local thermodynamic equilibrium* allows us to view intensive thermodynamic parameters, such as the temperature and pressure, as defined everywhere at all times, that is, as fields that vary smoothly in space and time.

Smoothness conditions: There are, however, certain conditions that necessarily must be fulfilled for this picture to hold. In the ideal gas model we have used so often, the temperature of a material particle is proportional to the average molecular kinetic energy relative to the particle's center of mass. Since this average is a sum over (positive) contributions from each molecule, the temperature will—like the mass of the particle—fluctuate as molecules randomly enter and leave the volume. Thus, to secure the relative temperature fluctuations be smaller than a desired precision of the continuum description, a material particle must have a linear size larger than the micro scale L_{micro}, defined in Equation (1.6) on page 6. Furthermore, to secure that the temperature of the particle is indistinguishable from the temperatures of its immediate neighbors within the desired precision, the temperature gradient must obey $|\nabla T| \lesssim T/L_{\text{macro}}$ with L_{macro} defined in (1.8). The concept of local thermodynamic equilibrium thus depends on the desired precision of the continuum description. But that should not surprise us. Continuum physics requires always less than perfect eyesight! We shall return to this point at the end of the chapter.

Local entropy balance

Appealing to the principle of local thermodynamic equilibrium we can recast the fundamental equation (23.1) as a local relation valid for *comoving* material particles (with constant mass). Expressed in terms of the specific entropy $S = \mathcal{S}/M$, the specific internal energy $U = \mathcal{U}/M$, and the specific volume $V/M = 1/\rho$, it becomes

$$dS = \frac{1}{T}dU + \frac{p}{T}d\left(\rho^{-1}\right) = \frac{1}{T}dU - \frac{p}{T\rho^2}d\rho. \tag{23.9}$$

This relation must also hold for the corresponding material time derivatives. Using the heat equation (22.48c) for isotropic fluid to eliminate DU/Dt and the continuity equation (22.48b) to eliminate $D\rho/Dt$, we finally arrive at the *equation of local entropy balance*,

$$\boxed{\rho T\frac{DS}{Dt} = -\boldsymbol{\nabla}\cdot\boldsymbol{q} + h + \boldsymbol{\tau}:\boldsymbol{\nabla}\boldsymbol{v}.} \tag{23.10}$$

The right-hand side is the sum of the three local sources of heat: convergence of heat flow, local heat production, and dissipation. If the sum of the heat sources vanishes (or nearly does), so that $DS/Dt \simeq 0$, the specific entropy will be constant along the path of any material particle. In that case we speak about *isentropic flow*. The entropy will in general be different for different material paths unless all paths start out with the same value for the entropy (in which case the flow is called *homentropic*).

Global entropy balance

The total entropy of a body is obtained by integrating the specific entropy,

$$\mathcal{S} = \int_V S\,dM = \int_V \rho S\,dV. \tag{23.11}$$

Making use of Reynolds' relation (20.6) on page 341 between the local and global rates, we find from local entropy balance the following expression for the material rate of entropy change of an arbitrary body in local thermodynamic equilibrium,

$$\frac{D\mathcal{S}}{Dt} = \int_V \frac{-\boldsymbol{\nabla}\cdot\boldsymbol{q} + h + \boldsymbol{\tau}:\boldsymbol{\nabla}\boldsymbol{v}}{T}\,dV. \tag{23.12}$$

Using the identity

$$\frac{\boldsymbol{\nabla}\cdot\boldsymbol{q}}{T} = \boldsymbol{\nabla}\cdot\left(\frac{\boldsymbol{q}}{T}\right) - \boldsymbol{q}\cdot\boldsymbol{\nabla}\frac{1}{T} = \boldsymbol{\nabla}\cdot\left(\frac{\boldsymbol{q}}{T}\right) + \frac{\boldsymbol{q}\cdot\boldsymbol{\nabla}T}{T^2},$$

together with Gauss' theorem, the *equation of global entropy balance* takes the form

$$\frac{DS}{Dt} = -\oint_S \frac{\boldsymbol{q}}{T} \cdot d\boldsymbol{S} + \int_V \frac{h}{T} dV - \int_V \frac{\boldsymbol{q} \cdot \nabla T}{T^2} dV + \int_V \frac{\boldsymbol{\tau} : \nabla \boldsymbol{v}}{T} dV. \tag{23.13}$$

Recapitulating, this equation has been derived from the equation of local entropy balance (23.10), which in turn has been derived from the classical thermodynamic entropy differential (23.1) by means of the principle of local thermodynamic equilibrium and the equations of motion (22.48) for isotropic fluid.

It is in fact a bit surprising that we have arrived at an equation for the rate of change of entropy rather than an inequality. At this point, entropy seems to be nothing but an extra field added on top of the already complete set of equations of motion (22.48). What happened to the Second Law?

The Second Law of Thermodynamics

The Second Law claims that the entropy cannot decrease for a body that is isolated from all external influences. In isolation, the current density of heat vanishes at the surface of the body, $\boldsymbol{q} \cdot d\boldsymbol{S} = 0$, and the local rate of heat production vanishes everywhere inside the body, $h = 0$. Demanding that the right-hand side of the global entropy balance (23.13) be non-negative for an arbitrary isolated body, the two last contributions to the rate of change must both be everywhere non-negative:

$$-\boldsymbol{q} \cdot \nabla T \geq 0, \qquad \boldsymbol{\tau} : \nabla \boldsymbol{v} \geq 0. \tag{23.14}$$

From the constitutive equations (22.49) of isotropic homogeneous fluids, we get

$$-\boldsymbol{q} \cdot \nabla T = k(\nabla T)^2, \qquad \boldsymbol{\tau} : \nabla \boldsymbol{v} = \tfrac{1}{2}\eta \sum\nolimits_{ij} v_{ij}^2 + \zeta(\nabla \cdot \boldsymbol{v})^2, \tag{23.15}$$

where $v_{ij} = \nabla_i v_j + \nabla_j v_i - \frac{2}{3}\nabla \cdot \boldsymbol{v}\delta_{ij}$. For both of these expressions to be always non-negative we must require that the heat conductivity k, the shear viscosity η, and the bulk viscosity ζ are all non-negative.

In continuum physics, where the complete set of equations of motion is usually known, *the Second Law mainly imposes constraints on the form of the constitutive equations and their parameters*. Had we, for example, included a term of the form $-k'\nabla p$ in the heat flow $\boldsymbol{q}$, it would be virtually impossible to secure that $k'\nabla p \cdot \nabla T \geq 0$ everywhere. The Second Law thus tells us that Fourier's law (22.49a) is the only real possibility for isotropic fluids.

The entropy conditions (23.14) in turn imply that for an arbitrary unrestricted interaction with the environment the Second Law implies the *Clausius-Duhem inequality*,

$$\frac{DS}{Dt} \geq -\oint_S \frac{\boldsymbol{q}}{T} \cdot d\boldsymbol{S} + \int_V \frac{h}{T} dV. \tag{23.16}$$

Since both integrals can be viewed as being a sum of local contributions of the form dQ/T on the surface and in the volume, the inequality expresses that in any local interaction the contribution to entropy must fulfill the inequality

$$dS \geq \frac{dQ}{T}. \tag{23.17}$$

This is the *Clausius inequality*, which is equivalent to the Clausius-Duhem inequality and thus to the Second Law.

* 23.3 Fluctuations

Continuum physics is a highly useful approximation because the molecular granularity of matter is never a concern in our daily doings. As discussed already in Chapter 1, the desired precision of the continuum description sets a limit to the length scale of the smallest bodies that can be considered part of the continuum description. Below that scale, random molecular fluctuations become observable as demonstrated, for example, by Brownian motion of small particles in water.

Although the subject of fluctuations strictly speaking falls outside continuum physics, it is as we shall see nevertheless possible to calculate their magnitude from classical thermodynamics augmented with a single input from statistical mechanics.

Thermodynamic model of fluctuations

According to the Second Law, any process taking place in an isolated body can never decrease its entropy. Consequently the entropy must be maximal in global equilibrium where all internal processes have stopped. Any local fluctuation that brings a small amount of matter out of equilibrium with the rest of the body can only diminish the total entropy. The fluctuation itself must from the point of view of classical thermodynamics be caused by "agents" external to the otherwise isolated body. These agents, or "demons" as Maxwell called them, are of course the random molecular motions that indeed do not belong to the macroscopic thermodynamic description. After the fluctuation has taken place, "natural" irreversible processes will on their own relax the body back to global equilibrium.

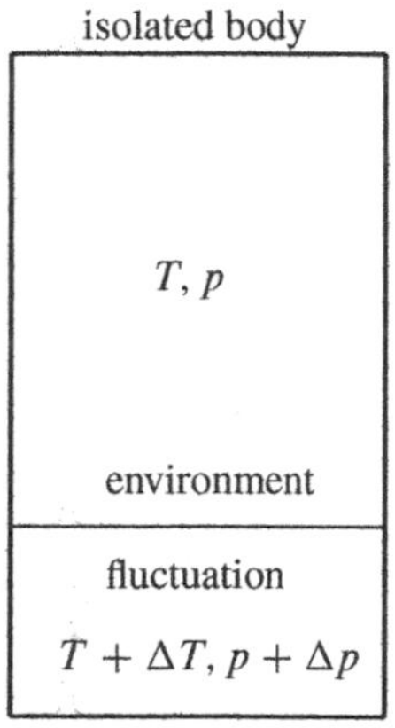

Thermodynamic model of a fluctuation. The environment is so large that its temperature and pressure do not change significantly compared to the changes due to the fluctuations.

Let the isolated body have uniform temperature T and pressure p in global equilibrium, and suppose the fluctuation takes place in a material particle, such that its temperature changes from T to $T + \Delta T$ and its pressure from p to $p + \Delta p$, while its mass M remains unchanged. The fluctuation changes simultaneously the particle's energy from $\mathcal{U}$ to $\mathcal{U} + \Delta\mathcal{U}$, its volume from V to $V + \Delta V$, and its entropy from $\mathcal{S}$ to $\mathcal{S} + \Delta\mathcal{S}$. Since the energy and volume of the total isolated body is constant, the environment of the particle must at the same time suffer a change in energy $\Delta\mathcal{U}_{\text{env}} = -\Delta\mathcal{U}$ and volume $\Delta V_{\text{env}} = -\Delta V$. Assuming that these small changes cannot influence the temperature T and pressure p of the huge environment, the change in the total entropy of particle plus environment becomes

$$
\begin{aligned}
\Delta\mathcal{S}_{\text{tot}} &= \Delta\mathcal{S} + \Delta\mathcal{S}_{\text{env}} \\
&= \mathcal{S}(\mathcal{U} + \Delta\mathcal{U}, V + \Delta V) - \mathcal{S}(\mathcal{U}, V) + \frac{1}{T}\Delta\mathcal{U}_{\text{env}} + \frac{p}{T}\Delta V_{\text{env}} \\
&= \frac{1}{T}\Delta\mathcal{U} + \frac{p}{T}\Delta V + \frac{1}{2}\frac{\partial^2\mathcal{S}}{\partial\mathcal{U}^2}\Delta\mathcal{U}^2 + \frac{1}{2}\frac{\partial^2\mathcal{S}}{\partial V^2}\Delta V^2 + \frac{\partial^2\mathcal{S}}{\partial\mathcal{U}\partial V}\Delta\mathcal{U}\Delta V \\
&\quad + \frac{1}{T}\Delta\mathcal{U}_{\text{env}} + \frac{p}{T}\Delta V_{\text{env}}.
\end{aligned}
$$

In the first line we have used that the entropy is additive. In the second we have expressed the change in entropy of the particle in terms of the changes in its energy and volume, while the change in entropy of the environment due to the its unchangeable temperature and pressure is given by the entropy differential (23.1). Finally, in the third line we have expanded the particle entropy to second order, and used that $\partial\mathcal{S}/\partial\mathcal{U} = 1/T$ and $\partial\mathcal{S}/\partial V = p/T$.

The linear terms in the expansion cancel between the particle and the environment, so that

$$
\Delta\mathcal{S}_{\text{tot}} = \frac{1}{2}\frac{\partial^2\mathcal{S}}{\partial\mathcal{U}^2}\Delta\mathcal{U}^2 + \frac{1}{2}\frac{\partial^2\mathcal{S}}{\partial V^2}\Delta V^2 + \frac{\partial^2\mathcal{S}}{\partial\mathcal{U}\partial V}\Delta\mathcal{U}\Delta V. \tag{23.18}
$$

The change in the total entropy is a second-order effect.

The change in inverse temperature may be written as

$$\Delta\left(\frac{1}{T}\right) = \frac{\partial(1/T)}{\partial \mathcal{U}}\Delta\mathcal{U} + \frac{\partial(1/T)}{\partial V}\Delta V = \frac{\partial^2 \mathcal{S}}{\partial \mathcal{U}^2}\Delta\mathcal{U} + \frac{\partial^2 \mathcal{S}}{\partial \mathcal{U}\partial V}\Delta V. \tag{23.19}$$

Using a similar expression for p/T, the change in total entropy (23.18) takes the simple form

$$\Delta\mathcal{S}_{\text{tot}} = \frac{1}{2}\Delta\left(\frac{1}{T}\right)\Delta\mathcal{U} + \frac{1}{2}\Delta\left(\frac{p}{T}\right)\Delta V. \tag{23.20}$$

Expressing the leading factors in terms of ΔT and Δp, and replacing the energy change by $\Delta\mathcal{U} = T\Delta\mathcal{S} - p\Delta V$, we finally arrive at the important expression

$$\boxed{\Delta\mathcal{S}_{\text{tot}} = \frac{-\Delta T\Delta\mathcal{S} + \Delta p\Delta V}{2T}} \tag{23.21}$$

The Second Law tells us that this change in entropy can never be positive but must in general be negative for any local fluctuation that brings the system out of global equilibrium.

The fluctuations in entropy $\Delta\mathcal{S}$ and pressure Δp may be expressed in terms of the fluctuations in temperature ΔT and volume ΔV by means of the fundamental relation (23.1) and the equation of state, $p = p(T, V)$,

$$\Delta\mathcal{S} = \frac{1}{T}\left(\frac{\partial\mathcal{U}}{\partial T}\right)_V \Delta T + \frac{1}{T}\left(\left(\frac{\partial\mathcal{U}}{\partial V}\right)_T + p\right)\Delta V, \tag{23.22}$$

$$\Delta p = \left(\frac{\partial p}{\partial T}\right)_V \Delta T + \left(\frac{\partial p}{\partial V}\right)_T \Delta V. \tag{23.23}$$

Inserting this into (23.21) and using Helmholtz equation (23.8), we find

$$\Delta\mathcal{S}_{\text{tot}} = -\frac{1}{2T^2}\left(\frac{\partial\mathcal{U}}{\partial T}\right)_V \Delta T^2 + \frac{1}{2T}\left(\frac{\partial p}{\partial V}\right)_T \Delta V^2. \tag{23.24}$$

Conveniently, the $\Delta T\Delta V$ cross-terms canceled, leaving a perfect quadratic form in ΔT and ΔV. The coefficients,

$$C_v = \left(\frac{\partial\mathcal{U}}{\partial T}\right)_V, \qquad K_T = -V\left(\frac{\partial p}{\partial V}\right)_T, \tag{23.25}$$

are respectively the heat capacity at constant volume of the material particle and the isothermal bulk modulus. Since $\Delta\mathcal{S}_{\text{tot}}$ cannot be positive, both of these coefficients must be non-negative.

The probability distribution of fluctuations

In 1904, Einstein proposed[2] that the probability of observing a fluctuation $\Delta\mathcal{S}_{\text{tot}}$ in the total entropy of a macroscopic body is

$$\boxed{\Pr(\Delta\mathcal{S}_{\text{tot}}) = Z\exp\left(\frac{\Delta\mathcal{S}_{\text{tot}}}{k_B}\right),} \tag{23.26}$$

where $k_B = 1.38065 \times 10^{-23}$ J K^{-1} is Boltzmann's constant and Z is a normalization factor. Since the entropy is maximal for an isolated body in global thermodynamic equilibrium, the fluctuation entropy must be negative, $\Delta\mathcal{S}_{\text{tot}} < 0$; and since k_B is a molecular scale parameter, the probability of finding any macroscopic deviation from thermodynamic equilibrium will always be strongly damped.

[2]Einstein obtained his formula by inverting Boltzmann's famous expression for the entropy, $\mathcal{S} = k_B \log \mathcal{W}$, where $\mathcal{W}$ is the number of microstates that are consistent with a given macrostate.

For the temperature and volume fluctuations we found above[3],

$$\Delta S_{\text{tot}} = -\frac{C_v}{2T^2}\Delta T^2 - \frac{K_T}{2TV}\Delta V^2. \tag{23.27}$$

Evidently, the probability distribution (23.26) is independently Gaussian in both temperature and volume fluctuations with variances

$$\langle \Delta T^2 \rangle = k_B \frac{T^2}{C_v}, \qquad \langle \Delta V^2 \rangle = k_B \frac{TV}{K_T}. \tag{23.28}$$

Using that $C_v = Mc_v$ and $V = M/\rho$ combined with the trivial relation $N_A k_B = R_{\text{mol}} = RM_{\text{mol}}$, we obtain the relative fluctuation variances in temperature and density $\rho = M/V$,

$$\frac{\langle \Delta T^2 \rangle}{T^2} = \frac{R}{Nc_v}, \qquad \frac{\langle \Delta \rho^2 \rangle}{\rho^2} = \frac{R\rho T}{NK_T}, \tag{23.29}$$

where $N = nN_A$ is the number of molecules and $n = M/M_{\text{mol}}$ the number of moles.

For the root-mean-square fluctuations $\overline{\Delta T} = \sqrt{\langle \Delta T^2 \rangle}$ etc., we thus have

$$\frac{\overline{\Delta T}}{T} = \frac{A_T}{\sqrt{N}}, \qquad A_T = \sqrt{\frac{R}{c_v}},$$
$$\frac{\overline{\Delta \rho}}{\rho} = \frac{A_\rho}{\sqrt{N}}, \qquad A_\rho = \sqrt{\frac{R\rho T}{K_T}}. \tag{23.30}$$

	A_T	A_ρ
Air	0.63	1.00
Water	0.33	0.26
Ethanol	0.29	0.18
Glycerol	0.20	0.09
Mercury	0.59	0.08
Bromine	0.36	0.26

Fluctuation constants for air and some liquids at normal temperature and pressure.

For ideal gases, $c_v = R/(\gamma - 1)$, $p = R\rho T$, and $K_T = p$ so that $A_T = \sqrt{\gamma - 1}$ and $A_\rho = 1$. For liquids, the fluctuation constants are of order unity, but generally smaller than for gases (see the margin table). These results finally confirm the simple estimates of density and temperature fluctuations made on page 6 and 396, respectively.

Problems

23.1 **(a)** Show that

$$dS = C_v \frac{dT}{T} + \alpha K_T \, dV,$$

where $K_T = 1/\beta$ is the isothermal bulk modulus.

(b) Show that the isentropic bulk modulus is

$$K_S = \gamma K_T,$$

where γ is the adiabatic index (22.61).

[3] See [Landau and Lifshitz 1980] for calculation of fluctuations in other variables.

Part V

Selected topics

Selected topics

In this final part a few selected topics are presented in some detail. The level of difficulty is fairly uniform but somewhat higher than in the first parts of the book.

None of the following chapters are necessary for a minimal curriculum, and all are fairly demanding. Some, for example Chapters 25, 27, and 29 may be included in part, depending on individual preferences.

List of chapters

24. **Elastic vibrations:** Vibrations or waves in elastic solids are known from the ringing of a church bell to the propagation of earthquakes. They have a fairly complex structure with both longitudinal and transverse components. Special types arise at open surfaces or material interfaces.

25. **Gravity waves:** Surface waves in the sea are primarily driven by gravity and inertia, but for small wavelengths, surface tension also plays a role. Viscous attenuation can be disregarded except for the smallest wavelengths.

26. **Jumps and shocks:** Hydraulic jumps in the open surfaces of incompressible liquids under the influence of gravity are known from the kitchen sink as well as from river bores. Superficially they bear resemblance to the sharp compressional shocks found in supersonic booms, cracks of a whip, and violent explosions.

27. **Whirls and vortices:** The vortex in the bathtub drain is easy to see because of its deformation of the water surface, and the same is true of the vortices that arise when paddling a canoe. Less conspicuous are the trail of alternating vortices cast off from a taut wire humming in the wind, the turbulent whirls that trail your car, or the bound vortices that encircle aircraft wings and trail from the wingtips.

28. **Boundary layers:** Near any solid surface, boundary layers arise that interpolate between the velocity of the surface and the flow at large. Boundary layer separation from the solid surface often takes place, but is still hard to predict.

29. **Subsonic flight:** Airplanes have been fascinating to all, since they first lifted off about 100 years ago. The need for reliable air transport has driven the development of modern aerodynamics, and in this chapter the first steps for understanding the basic theory are taken.

30. **Convection:** Buoyant bubbles of warm fluid rise like hot-air balloons in a field of gravity. Convection stirs the pots on the kitchen stove, and drives vertical mixing in the atmosphere as well in the interior of planets and stars.

31. **Turbulence:** Everywhere around us we meet turbulence, usually without noticing it. A comprehensive theory of turbulence is still the most important outstanding problem in fluid mechanics, even if the basic equations of motion have been known for centuries. Only the very first steps in understanding the theory and phenomenology of turbulence are taken in this last chapter.

24 Elastic vibrations

The magnificent ringing of a church bell originates in small-amplitude elastic oscillations of its shape, transmitted to the air as ordinary sound waves that travel far and wide to announce important events. The present-day shape of bells goes back a thousand years and has been developed through trial and error by bell founders. The scientific study of bells and their relation to the musical quality of their sound was initiated by Lord Rayleigh [Ray90].

All solid bodies are elastic to some extent and can vibrate in different ways around their relaxed shapes. Vibrations in ideal elastic materials do not lose energy but in reality energy will always be lost to internal friction (heat). A vibrating body in contact with a much larger material environment, say the atmosphere, will unavoidably lose energy through radiation of sound. A church bell or tuning fork may thus ring for a long time but eventually stops because of dissipative and radiative losses. Disregarding such losses, any free, small-amplitude vibration can be resolved into a superposition of independent *standing waves* or *modes* with precisely defined harmonic vibration frequencies. For finite bodies the frequency spectrum is always discrete but at sufficiently high frequencies the spectrum becomes approximatively continuous. Alternatively, any elastic vibration may be viewed as a superposition of elementary monochromatic *elastic waves*, coupled to each other by the boundary conditions.

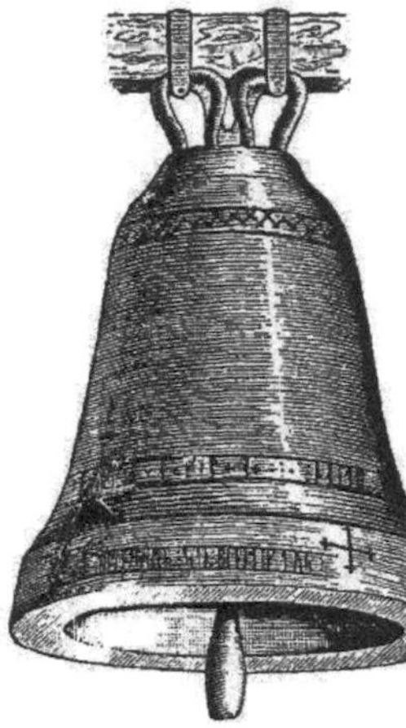

Church bell with runic inscription from 1228 (Saleby, Västergötland, Sweden). The inscription reads: *When I was completed, then there were thousand two hundred twenty winters and eight from the birth of God. AGLA. Ave Maria gracia plena. Dionysius sit benedictus.* In spite of the evident Christian message, partly in Latin, Roman lettering had apparently not yet eliminated the old Norse way of writing. (Source: Wikimedia Commons.)

In this chapter we shall set up the basic equations for small-amplitude elastic vibrations and waves in homogeneous isotropic elastic materials. Vibrations constitute a large subfield of continuum physics (see, for example, [Hagedorn and DasGupta 2007]) and cannot be given just treatment in a single chapter. We shall for this reason focus on the propagation of elastic bulk waves in solids, and their reflection and refraction at planar interfaces. Special surface and interface waves known from earthquakes are analyzed toward the end of the chapter.

24.1 Elastodynamics

As in elastostatics (Chapter 9), the displacement gradients are also here assumed to be so small that all differences between the Eulerian and Lagrangian descriptions can be disregarded. The state of a deformable material is best described by a—now time-dependent—Lagrangian displacement field $\boldsymbol{u}(\boldsymbol{x}, t)$ that indicates how much a material particle at time t is displaced from its *original* position $\boldsymbol{x}$ in a chosen static reference state (in Chapter 7 denoted by $\boldsymbol{X}$). This state may itself already be highly stressed and deformed. There are, for example, huge static stresses in balance with gravity in the pylons and girders of a bridge, but when the wind acts on the bridge, small-amplitude vibrations may arise around the static state.

Equation of motion and boundary conditions

The actual position of a displaced particle is $\boldsymbol{x} + \boldsymbol{u}(\boldsymbol{x}, t)$, and since its original position $\boldsymbol{x}$ is time independent, its actual velocity becomes $\boldsymbol{v}(\boldsymbol{x}, t) = \partial \boldsymbol{u}(\boldsymbol{x}, t)/\partial t$ and its acceleration $\boldsymbol{w}(\boldsymbol{x}, t) = \partial^2 \boldsymbol{u}(\boldsymbol{x}, t)/\partial t^2$. Newton's Second Law—mass times acceleration equals force—applied to every material particle in the body takes the usual form, $dM\,\boldsymbol{w} = \boldsymbol{f}^* dV$. Dividing by dV and reusing the effective force density obtained on the left-hand side of the equilibrium equation (9.2) on page 140, we arrive at *Navier's equation of motion* (1821),

$$\rho \frac{\partial^2 \boldsymbol{u}}{\partial t^2} = \boldsymbol{f} + \mu \nabla^2 \boldsymbol{u} + (\lambda + \mu) \nabla \nabla \cdot \boldsymbol{u}, \tag{24.1}$$

where $\boldsymbol{f}$ is the true body force density, and λ, μ, and ρ are assumed to be material parameters that do not depend on space and time. In case they do depend on the spatial position $\boldsymbol{x}$, as they for example do in Earth's solid mantle, Navier's equation of motion takes a slightly different form (see Problem 24.1). For time-independent displacements, the above equation of motion reduces of course to the equilibrium equation.

The boundary conditions are

$$[\boldsymbol{u}] = \boldsymbol{0}, \qquad [\boldsymbol{\sigma} \cdot \hat{\boldsymbol{n}}] = \boldsymbol{0}, \tag{24.2}$$

and express that the time-dependent displacement field and the stress vector must be continuous across material interfaces. A surface to vacuum imposes no condition on the displacement field, but requires that the stress vector vanish. We shall always disregard surface tension in interfaces between elastic solids.

Driving forces and dissipation

Time-dependent displacements are often caused by contact forces that—like the wind on a bridge—impose time-dependent stresses on the surface of a body. If you hit a nail with a hammer or stroke the strings of a violin, time-varying displacement fields are likewise set up in the material. Body forces may also drive time-dependent displacements. The moving Moon and the Earth's rotation adds time-dependent gravitational forces on top of the static gravity and centrifugal forces from Earth itself. Magnetostrictive, electrostrictive, and piezoelectric materials deform under the influence of electromagnetic fields, and are, for example, used in loudspeakers to set up vibrations that can be transmitted to air as sound.

The omnipresent dissipation of kinetic energy will in the end make all vibrations die out and turn their energy into heat. Sustained vibrations in any body can strictly speaking only be maintained by time-dependent external forces continually performing work by interacting with the body. Dissipation is nevertheless so small in most elastic materials that it, to a very good approximation, can be omitted, as it is in Navier's equation of motion.

Violin paradox: How can stroking a violin string with a horsehair bow moving at constant speed make the string vibrate at a nearly constant frequency when we claim that sustained vibration demands time-dependent driving forces?

Although the external force delivered by your arm to the bow is nearly constant for the length of the stroke, the interaction between the bow and the string develops time dependence because of the finite difference between static and dynamic friction forces (see Section 6.1). The string sticks to the bow when it starts to move until the restoring elastic force in the string surpasses the static friction force. At that moment the string slips and begins to move with much smaller dynamic friction, or even no friction if it lifts off the bow. Swinging once back and forth, the string eventually again nearly matches the speed of the bow and sticks because of static friction. Since it only sticks for a very short time, the frequency generated in this way is nearly equal to the natural oscillation frequency of a taut but otherwise free string. This *stick-slip* mechanism underlies many oscillatory phenomena apparently generated by steady driving agents (see Problem 6.9).

Longitudinal and transverse displacement fields

Navier's equation of motion (24.1) without the volume force ($\boldsymbol{f} = \boldsymbol{0}$) is essentially of the same form as Equation (14.5) on page 230 for small-amplitude sound waves in compressible fluids. The main difference is that in fluids the sound field is a scalar, represented by the pressure or density, whereas in solids the displacement field is a vector, allowing for the extra term on the right-hand side of Navier's equation. Not unsurprisingly, elastic waves in elastic solids have two independent components corresponding to, respectively, compressional and shear deformation. In finite bodies the (static) boundary conditions on the surface usually couple these components and create standing waves that contain both.

A vector field $\boldsymbol{u}$ may always be resolved into *longitudinal* component $\boldsymbol{u}_L$ with no curl and a *transverse* component $\boldsymbol{u}_T$ with no divergence (see Problem 24.4),

$$\boldsymbol{u} = \boldsymbol{u}_L + \boldsymbol{u}_T, \qquad \nabla \times \boldsymbol{u}_L = \boldsymbol{0} \qquad \nabla \cdot \boldsymbol{u}_T = 0. \tag{24.3}$$

Using that the longitudinal component satisfies $\boldsymbol{0} = \nabla \times (\nabla \times \boldsymbol{u}_L) = \nabla(\nabla \cdot \boldsymbol{u}_L) - \nabla^2 \boldsymbol{u}_L$, or $\nabla\nabla \cdot \boldsymbol{u}_L = \nabla^2 \boldsymbol{u}_L$, it follows that Navier's equation of motion (24.1) for purely longitudinal and transverse components take the form (for $\boldsymbol{f} = \boldsymbol{0}$),

$$\rho \frac{\partial^2 \boldsymbol{u}_L}{\partial t^2} = (\lambda + 2\mu)\nabla^2 \boldsymbol{u}_L, \qquad \rho \frac{\partial^2 \boldsymbol{u}_T}{\partial t^2} = \mu \nabla^2 \boldsymbol{u}_T. \tag{24.4}$$

Both of the above equations are examples of the standard wave equation for non-dispersive waves with longitudinal and transverse phase velocities

$$\boxed{c_L = \sqrt{\frac{\lambda + 2\mu}{\rho}}, \qquad c_T = \sqrt{\frac{\mu}{\rho}}.} \tag{24.5}$$

In typical elastic materials the phase velocities are a few kilometers per second, which is an order of magnitude greater than the velocity of sound in air but roughly of the same magnitude as the sound velocity in liquids, such as water (see the margin table).

Material	c_L	q
Aluminum	6.4	0.48
Titanium	6.1	0.51
Iron	5.9	0.54
Nickel	5.8	0.52
Magnesium	5.8	0.54
Quartz	5.5	0.63
Tungsten	5.2	0.55
Copper	4.7	0.49
Silver	3.7	0.45
Gold	3.6	0.33
Lead	2.1	0.33
Units	km s^{-1}	

Longitudinal sound speed and the ratio of transverse to longitudinal speed, $q = c_T/c_L$, for various isotropic materials. The lightest and hardest materials generally have the largest longitudinal sound speed.

The ratio of the transversal and longitudinal velocities is a useful dimensionless parameter,

$$q = \frac{c_T}{c_L} = \sqrt{\frac{\mu}{\lambda + 2\mu}} = \sqrt{\frac{1 - 2\nu}{2(1 - \nu)}}. \tag{24.6}$$

It depends only on Poisson's ratio ν, and is a monotonically decreasing function of ν. Its maximal value $\frac{1}{2}\sqrt{3} \approx 0.87$ is obtained for $\nu = -1$, implying that the transverse velocity is always smaller than the longitudinal one. In practice there are no materials with $\nu < 0$, so the realizable upper limit to the ratio is instead $\frac{1}{2}\sqrt{2} \approx 0.71$. For the typical value $\nu = \frac{1}{3}$ we get $q = \frac{1}{2}$ so transverse waves typically propagate with half the speed of longitudinal waves.

The tiny pressure change generated by the displacement field is $\Delta p = -K\nabla \cdot \boldsymbol{u}$, where $K = \lambda + \frac{2}{3}\mu$ is the bulk modulus (see page 130). Since $\nabla \cdot \boldsymbol{u} = 0$ for transverse waves, only the longitudinal waves are accompanied by an oscillating pressure. They are for this reason also called *pressure* waves or *compressional* waves. Transverse waves generate no pressure changes in the material, only shear, and are accordingly called *shear* waves.

Earthquake wave types: In earthquakes (see Figure 24.1), longitudinal pressure waves are denoted P (for *primary*), because they arrive first due to the higher longitudinal phase velocity in any material. Typically they move at speeds of 4 to 7 km s^{-1} in the Earth's crust. Transverse shear waves move at roughly half the speed and thus arrive later at the seismometer. They are for this reason denoted by S (for *secondary*). In fluid material, such as the Earth's liquid core, shear waves cannot propagate far (see page 248). Besides these *body waves*, earthquakes are also accompanied by *surface waves* to be discussed in Section 24.4.

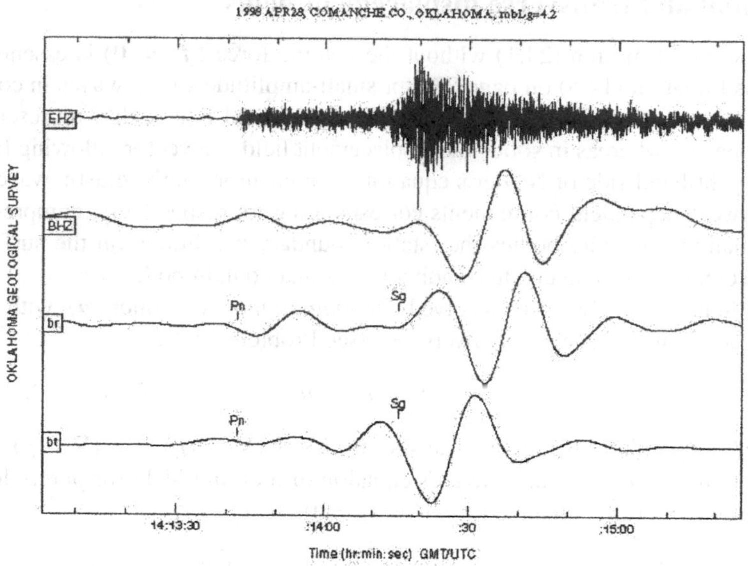

Figure 24.1. Seismogram of an earthquake of strength 4.2 that took place in Comanche county, Oklahoma, on April 28, 1998. The four traces are **EHZ** vertical earth velocity at all frequencies, **BHZ** the low-frequency vertical component, **br** the low-frequency horizontqal compressional component (Rayleigh waves), and **bt** the low-frequency horizontal shear component (Love waves). The times labeled Pn and Sg represent the onset of the primary and secondary disturbances. (Source: Oklahoma Geological Survey, University of Oklahoma, USA. With permission.)

24.2 Harmonic vibrations

The linearity of Navier's equation of motion (24.1) with time-independent material constants and boundary conditions allows us to exploit a general mathematical theorem due to Fourier, which tells us that any time-dependent field may be expanded in a linear superposition of *harmonic* (or *monochromatic*) vibrations, each oscillating with a single frequency. In the following analysis we assume for simplicity that there is no volume force, $\boldsymbol{f} = \boldsymbol{0}$.

Oscillation modes

A harmonic displacement field with *circular frequency* ω must satisfy the harmonic equation $\partial_t^2 \boldsymbol{u} = -\omega^2 \boldsymbol{u}$. The general solution is a linear superposition of two real time-independent vector fields,

$$\boldsymbol{u}(\boldsymbol{x}, t) = \boldsymbol{u}_1(\boldsymbol{x}) \cos \omega t + \boldsymbol{u}_2(\boldsymbol{x}) \sin \omega t. \tag{24.7}$$

The component fields generally also depend on ω but we suppress it here for ease of notation.

Instead of working with two real fields, it is convenient to collect them in a single *complex* time-independent field,

$$\boldsymbol{u}(\boldsymbol{x}) = \boldsymbol{u}_1(\boldsymbol{x}) + i\boldsymbol{u}_2(\boldsymbol{x}). \tag{24.8}$$

The harmonic time-dependent displacement field and its time derivative then become respectively the real and imaginary parts of the complex field $\boldsymbol{u}(\boldsymbol{x}) \exp(-i\omega t)$,

$$\boldsymbol{u}(\boldsymbol{x}, t) = \mathcal{R}e \left[\boldsymbol{u}(\boldsymbol{x})\, e^{-i\omega t}\right], \qquad \frac{\partial \boldsymbol{u}(\boldsymbol{x}, t)}{\partial t} = \omega\, \mathcal{I}m \left[\boldsymbol{u}(\boldsymbol{x})\, e^{-i\omega t}\right]. \tag{24.9}$$

Since both the real and imaginary parts satisfy Navier's equation of motion, it follows that the field $\boldsymbol{u}(\boldsymbol{x}) \exp(-i\omega t)$ and any superposition of such fields must also satisfy it.

Inserting $\boldsymbol{u}(\boldsymbol{x})\exp(-i\omega t)$ in Navier's equation (24.1) with $\boldsymbol{f} = \boldsymbol{0}$, we obtain a single time-independent equation for the complex amplitude field $\boldsymbol{u}(\boldsymbol{x})$,

$$\mu\nabla^2\boldsymbol{u} + (\lambda+\mu)\nabla\nabla\cdot\boldsymbol{u} = -\rho\omega^2\boldsymbol{u}. \tag{24.10}$$

Here we have suppressed the explicit dependence on both $\boldsymbol{x}$ and ω in $\boldsymbol{u} = \boldsymbol{u}(\boldsymbol{x},\omega)$. Provided we can solve this equation with the actual boundary conditions and for all relevant values of ω, we may by Fourier's theorem construct any solution to Navier's equation of motion with these boundary conditions by suitable superposition of the harmonic solutions.

The above equation is an *eigenvalue equation* for the tensor differential operator

$$\mu\mathbf{1}\nabla^2 + (\lambda+\mu)\nabla\nabla = \{\mu\delta_{ij}\nabla^2 + (\lambda+\mu)\nabla_i\nabla_j\} \tag{24.11}$$

with eigenvalue $-\rho\omega^2$. Quite generally it may be shown for a finite body with time-independent boundary conditions that there are only a discrete spectrum of real eigen-frequencies, which define the allowed *modes of oscillation*, whereas in an infinite elastic medium the eigen-frequencies form a continuous spectrum.

Example 24.1 [Thickness oscillations of a plate]: Consider a plate of thickness d in the x-direction ($0 \le x \le d$) and infinite extent in the y- and z-directions. Suppose the slab is deformed only along the x-direction with displacement field $\boldsymbol{u} = u_x(x)\hat{\boldsymbol{e}}_x$. In the absence of body forces, the eigenvalue equation (24.10) takes the form

$$(\lambda+2\mu)\nabla_x^2 u_x = -\rho\omega^2 u_x. \tag{24.12}$$

The general solution to this harmonic equation is $u_x = A\cos(kx) + B\sin(kx)$ with $k = \omega/c_L$ and arbitrary coefficients A and B. If no external stresses act on the sides of the slab, we must have $\sigma_{xx} = (\lambda+2\mu)u_{xx} = (\lambda+2\mu)\nabla_x u_x = 0$ for $x = 0$ and d. It then follows that $B = 0$ and $kd = n\pi$, where n is a positive integer, so that the discrete eigenfrequencies are

$$\omega_n = n\frac{\pi c_L}{d} \qquad \text{for } n = 1, 2, \ldots. \tag{24.13}$$

Notice that the shear stresses, $\sigma_{yx} = \mu(\nabla_y u_x + \nabla_x u_y)$ and $\sigma_{zx} = \mu(\nabla_z u_x + \nabla_x u_z)$, vanish on the two faces as they should.

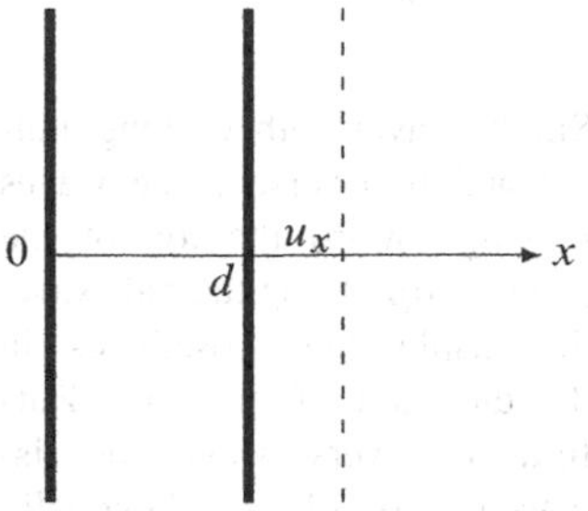

Sketch of uniform plate deformation along x. The displacement u_x is exaggerated.

Plane waves

Infinitely extended material bodies do not exist. Nevertheless, deeply inside a finite body, far from the boundaries, conditions are almost as if the body were infinite. Provided the typical wavelengths contained in the displacement field are much smaller than the distance to the boundaries, the displacement field may meaningfully be resolved into a superposition of independent plane waves.

A plane harmonic wave is described by (the real part of) the complex harmonic field,

$$\boldsymbol{u}(\boldsymbol{x},t) = \boldsymbol{a}\,e^{i(\boldsymbol{k}\cdot\boldsymbol{x}-\omega t)}. \tag{24.14}$$

Here $\boldsymbol{a}$ is the complex *amplitude* or *polarization vector*, and $\boldsymbol{k}$ the *wave vector*. The wave's *direction of propagation* is $\hat{\boldsymbol{k}} = \boldsymbol{k}/|\boldsymbol{k}|$, its *wavelength* $\lambda = 2\pi/|\boldsymbol{k}|$, and its *period* $\tau = 2\pi/\omega$. The *phase* of the wave is $\phi = \boldsymbol{k}\cdot\boldsymbol{x} - \omega t$ and its *phase velocity* $c = \omega/|\boldsymbol{k}|$. Inserting $\boldsymbol{u}(\boldsymbol{x}) = \boldsymbol{a}e^{i\boldsymbol{k}\cdot\boldsymbol{x}}$ into (24.10), we obtain an algebraic eigenvalue equation,

$$\mu\boldsymbol{k}^2\boldsymbol{a} + (\lambda+\mu)\boldsymbol{k}\boldsymbol{k}\cdot\boldsymbol{a} = \rho\omega^2\boldsymbol{a}, \tag{24.15}$$

for the 3×3 matrix $\mu\boldsymbol{k}^2\mathbf{1} + (\lambda+\mu)\boldsymbol{k}\boldsymbol{k} = \{\mu\boldsymbol{k}^2\delta_{ij} + (\lambda+\mu)k_ik_j\}$.

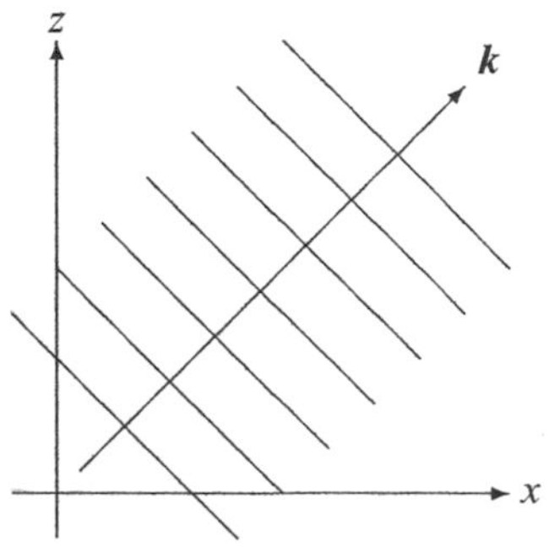

A plane wave has constant phase on planes orthogonal to the wave vector $\boldsymbol{k}$ (here with $k_y = 0$). The phase is spatially periodic with wavelength $\lambda = 2\pi/|\boldsymbol{k}|$.

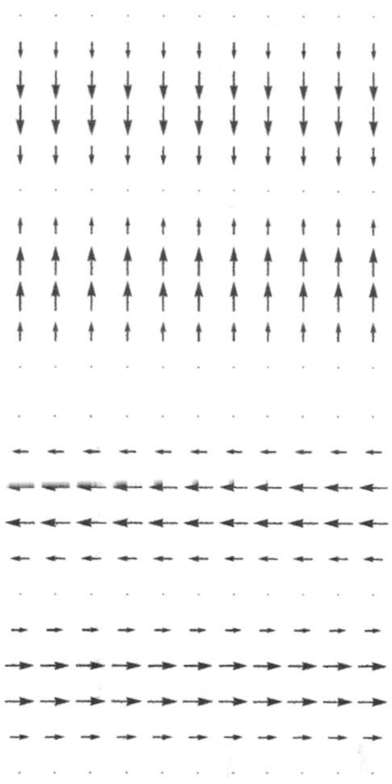

Single wavelength of longitudinal and transverse plane waves moving toward the top of the page. **Top:** Longitudinal wave; the displacement oscillates in the direction of motion. **Bottom:** Transverse wave; the displacement oscillates orthogonally to the direction of motion.

The eigenvectors are easily found. One is longitudinal with amplitude proportional to the direction of propagation, $\boldsymbol{a} \sim \hat{\boldsymbol{k}}$, and eigenvalue $\rho\omega^2 = (\lambda + 2\mu)\boldsymbol{k}^2$, so that the longitudinal plane wave mode takes the form

$$\boldsymbol{u}_L(\boldsymbol{x}, t) = a\,\hat{\boldsymbol{k}}\,e^{i(\boldsymbol{k}\cdot\boldsymbol{x}-\omega t)}, \qquad \boldsymbol{k} = \frac{\omega}{c_L}\hat{\boldsymbol{k}}, \tag{24.16}$$

where the amplitude a is a constant. The two other eigenvectors are transverse with amplitudes orthogonal to the direction of propagation, $\boldsymbol{a} \perp \hat{\boldsymbol{k}}$, and identical eigenvalues, $\rho\omega^2 = \mu\boldsymbol{k}^2$. The most general form of the transverse plane wave mode is

$$\boldsymbol{u}_T(\boldsymbol{x}, t) = a\hat{\boldsymbol{t}}\,e^{i(\boldsymbol{k}\cdot\boldsymbol{x}-\omega t)}, \qquad \boldsymbol{k} = \frac{\omega}{c_T}\hat{\boldsymbol{k}}, \tag{24.17}$$

where $\hat{\boldsymbol{t}}$ is any unit vector orthogonal to $\hat{\boldsymbol{k}}$. Notice that for given frequency ω and propagation direction $\hat{\boldsymbol{k}}$, the *longitudinal and transverse waves have different wave numbers* $|\boldsymbol{k}|$, and thus different wavelengths.

Wave intensity

Elastic waves carry energy along with the material flow. The rate of work by elastic contact forces on a control volume is $\oint_S \boldsymbol{v}\cdot\boldsymbol{\sigma}\cdot d\boldsymbol{S}$. Conversely, the rate of work performed by the control volume on its environment is $\oint_S \boldsymbol{I}\cdot d\boldsymbol{S}$, where

$$\boldsymbol{I} = -\boldsymbol{v}\cdot\boldsymbol{\sigma} \tag{24.18}$$

is called the *intensity vector* and its magnitude the *intensity*. Notice that both $\boldsymbol{v}$ and $\boldsymbol{\sigma}$ must be real fields for the expression to make physical sense.

For the generic harmonic plane wave (24.14) with real polarization vector $\boldsymbol{a}$ and phase $\phi = \boldsymbol{k}\cdot\boldsymbol{x} - \omega t$, we find for real $\boldsymbol{k}$ and $\boldsymbol{a}$ (see Problem 24.7 for the complex case),

$$\begin{aligned}\boldsymbol{v}\cdot\boldsymbol{\sigma} &= \mathcal{R}e\left[-i\omega\boldsymbol{a}e^{i\phi}\right]\cdot\mathcal{R}e\left[\left(\mu(i\boldsymbol{k}\boldsymbol{a}+\boldsymbol{a}i\boldsymbol{k})+\lambda(i\boldsymbol{k}\cdot\boldsymbol{a})\mathbf{1}\right)e^{i\phi}\right]\\ &= -\omega\boldsymbol{a}\sin\phi\cdot\left(\mu(\boldsymbol{k}\boldsymbol{a}+\boldsymbol{a}\boldsymbol{k})+\lambda(\boldsymbol{k}\cdot\boldsymbol{a})\mathbf{1}\right)\sin\phi\\ &= -\omega\left((\lambda+\mu)(\boldsymbol{a}\cdot\boldsymbol{k})\boldsymbol{a}+\mu\boldsymbol{a}^2\boldsymbol{k}\right)\sin^2\phi.\end{aligned}$$

The average of the oscillating factor over a period is $\langle\sin^2\phi\rangle = \frac{1}{2}$. Specializing to purely longitudinal and transverse waves, the average intensities become

$$\boxed{\langle\boldsymbol{I}_L\rangle = \tfrac{1}{2}\rho\omega^2 a^2 c_L\hat{\boldsymbol{k}}, \qquad \langle\boldsymbol{I}_T\rangle = \tfrac{1}{2}\rho\omega^2 a^2 c_T\hat{\boldsymbol{k}},} \tag{24.19}$$

where we have replaced the elasticity parameters by the phase velocities (24.5).

Waves carry both kinetic and potential (deformation) energy. The average total energy densities are identical for longitudinal and transverse waves (see Problem 24.6),

$$\boxed{\langle\varepsilon\rangle = \tfrac{1}{2}\rho\omega^2 a^2.} \tag{24.20}$$

Evidently, the intensities may be written $\langle\boldsymbol{I}_L\rangle = \langle\varepsilon\rangle\, c_L\hat{\boldsymbol{k}}$ and $\langle\boldsymbol{I}_T\rangle = \langle\varepsilon\rangle\, c_T\hat{\boldsymbol{k}}$, demonstrating that the energy in longitudinal or transverse waves on average is transported with the respective phase velocities along the propagation direction.

24.3 Refraction and reflection

The simplest system, which differs from an infinitely extended elastic medium, consists of two semi-infinite elastic media interfacing along a plane. The materials on both sides of the interface are homogeneous and isotropic, but have different longitudinal and transverse phase velocities, c_L, c_T and c'_L, c'_T. A plane wave incident on one side of the interface will give rise to both a *refracted* wave on the other side and a *reflected* wave on the same side. Even if the incident wave is purely longitudinal or purely transverse, the refracted and reflected waves will in general be superpositions of longitudinal and transverse waves propagating in different directions.

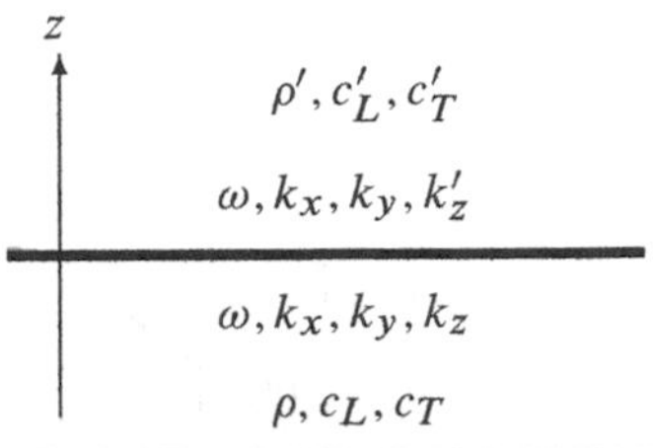

A plane interface between two media. The material properties are different on the two sides of the interface, but the the boundary conditions demand that the frequency and the wavenumbers components along the interface are the same.

Boundary conditions

Taking the interface to be the xy-plane of the coordinate system, the boundary conditions demand continuity of the displacement fields and the stress vectors on the two sides of the interface at $z = 0$,

$$u'_x = u_x, \qquad u'_y = u_y, \qquad u'_z = u_z, \tag{24.21a}$$
$$\sigma'_{xz} = \sigma_{xz}, \qquad \sigma'_{yz} = \sigma_{yz}, \qquad \sigma'_{zz} = \sigma_{zz}. \tag{24.21b}$$

The other three stress components are not required to be continuous.

The planar geometry is invariant under translations in time and in the xy-plane, but not along the z-direction because of the presence of the interface. The invariance guarantees that the displacement field can be written as a superposition of plane waves that are all proportional to $\exp[i(k_x x + k_y y - \omega t)]$ for $z < 0$ and to $\exp[i(k'_x x + k'_y y - \omega' t)]$ for $z > 0$. The boundary conditions demand that for $z = 0$ these fields must coincide for all values of x, y, and t, implying that

$$k'_x = k_x, \qquad k'_y = k_y, \qquad \omega' = \omega. \tag{24.22}$$

This shows that refracted and reflected waves must all propagate in the plane containing the incident wave vector and the normal to the interface. In the following we shall without loss of generality choose the waves to propagate in the xz-plane with $k_y = 0$ and $k_x \geq 0$.

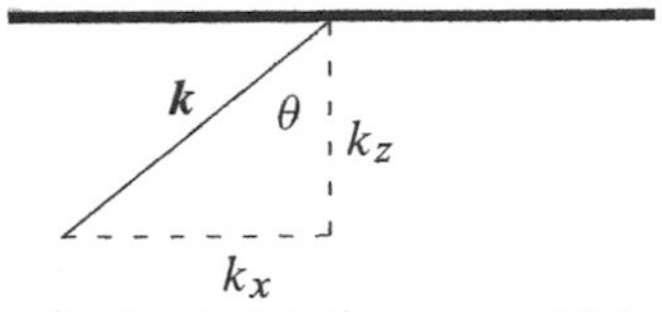

Geometry for determining the angle of incidence between the direction of propagation and the normal to the interface.

Willebrord Snellius (1580–1626). Dutch mathematician, born Willebrord Snel van Royen. Contributed to geodesy (triangulation), and is credited with the discovery of the law of refraction, although it had been known since the early middle ages.

Snell's law

A simple geometric construction (see the margin figure) shows that the angle between the direction of propagation of a plane wave and the normal to the interface is given by

$$\sin\theta = \frac{k_x}{|\boldsymbol{k}|} = \frac{k_x c}{\omega}, \tag{24.23}$$

where $c = \omega/|\boldsymbol{k}|$ is the phase velocity. Since k_x and ω are the same for any plane wave component, the propagation angles of two wave components with phase velocities c_1 and c_2 must be related by *Snell's law*,

$$\boxed{\frac{\sin\theta_2}{\sin\theta_1} = \frac{c_2}{c_1}.} \tag{24.24}$$

This relation applies to any combination of plane wave components whether they are longitudinal or transverse, on the same side (ipsilateral) as for reflection, or on opposite sides (contralateral) as for refraction. Since $c_L > c_T$, we always have $\theta_L > \theta_T$ for the longitudinal and transverse components of a refracted or reflected wave. A reflected wave of the same type as the incident wave, that is, both longitudinal or both transverse, will always have equal angles of incidence and reflection. The angles of the refraction are determined by the different material properties of the interfacing media and cannot be generally characterized, except that the angle of refraction is always larger for longitudinal than for transverse waves.

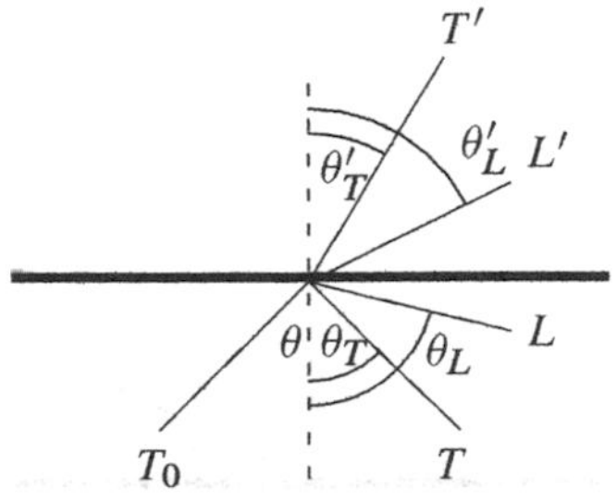

An incident transverse wave T_0 produces a reflected transverse wave having the same angle with the normal, $\Theta_T = \theta$, but may also produce a longitudinal wave with larger angle.

Snell's law takes the same form for all kinds of waves, be they elastic, acoustic, or electromagnetic. The fact that light moves with higher phase velocity in air than in water immediately tells us that a light ray passing from water to air is refracted into a larger angle with the normal. This explains the familiar observation that a partially immersed pencil apparently breaks at the water surface.

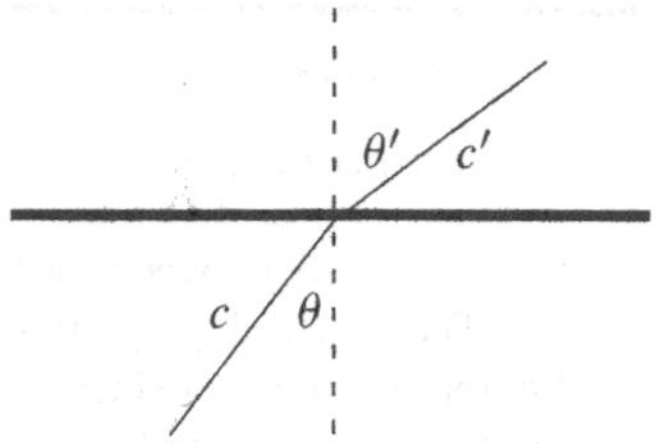

Refraction into a medium with higher phase velocity $c' > c$.

Total reflection: A peculiar thing happens when a wave passes from a material with phase velocity c into another material with larger phase velocity $c' > c$ and thus $\theta' > \theta$. The angle of refraction will reach the maximal value, $\theta'_{\text{max}} = 90°$, for an angle of incidence given by $\sin\theta_{\text{max}} = c/c'$. The refracted wave appears to crawl along the interface. For $\theta > \theta_{\text{max}}$ the incident wave is completely unable to penetrate the interface, a phenomenon called *total reflection* or sometimes *total internal reflection*.

Total reflection of light is well known to divers looking at the water surface from below, or to fish looking at you from inside their aquarium. Since water has higher refractive index than air, the speed of light in water will be lower than in the air outside. Seen from the water, all of the outside world will be imaged within a circular region of the interface (called Snell's window) with angular radius θ_{max}. Total reflection is also of great importance for the functioning of optical fibers where it guarantees that light sent down the fiber stays inside even if the fiber bends and winds slightly.

Classification of elastic waves

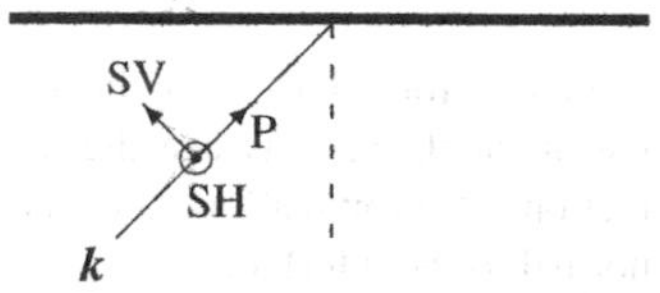

Incident elastic wave. The polarization of a P-wave is parallel to the wave vector. The polarization of an SV-wave lies in the propagation plane. The polarization of an SH-wave points out of the paper.

So far we have only used the general concepts of longitudinal (L) and transverse (T) to classify the elastic waves. Another elastic wave terminology derives from the study of earthquakes. A point-like seismic disturbance within the Earth will emit waves in all directions with some distribution of frequencies. Because of their larger phase velocity, the longitudinal wave components will arrive at a seismic observation station on the horizontal surface before the transverse ones. Longitudinal waves are accordingly said to be *primary* (P) while the transverse are *secondary* (S). Relative to the horizontal surface of the Earth, the polarization vector of a transverse wave may naturally be resolved into a *horizontal* component (SH) or *vertical* one (SV). The polarization of the SH component is orthogonal to the propagation plane, and always points in the horizontal direction. The polarization of the SV component lies in the propagation plane and is never truly vertical. In the special case of vertical propagation it will in fact be horizontal.

An incident P-wave or SV-wave can generate two similar waves on both sides of the interface, but no SH-waves. Consequently, there are in the worst case four unknown amplitudes to be determined by the boundary conditions. To avoid having to solve four linear equations with four unknowns, we shall limit the following analysis to a couple of cases resulting in only two equations with two unknowns. But even these simplest cases lead to rather complicated expressions.

Case: Reflection and refraction of incident SH-wave

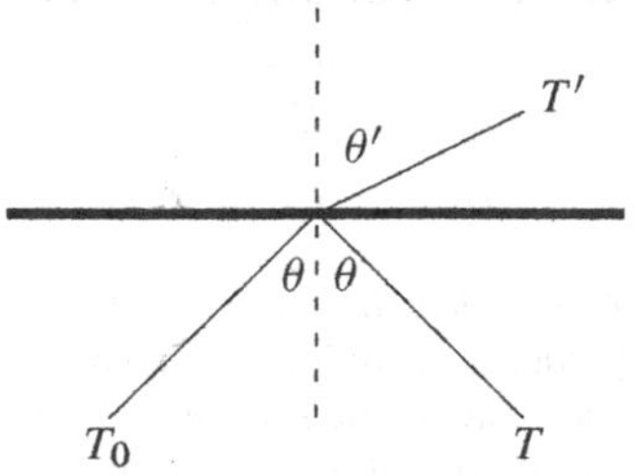

An incident SH-wave T_0 is reflected and refracted into similar wave types, T and T'.

It is not hard to see that an incident SH-wave can only give rise to one reflected and one refracted SH-wave. SH-waves propagating in the xz-plane have polarization vector along the y-axis. Defining $k = k_x$, $k_T = k_z = \sqrt{(\omega/c_T)^2 - k^2}$ and $k'_T = k'_z = \sqrt{(\omega/c'_T)^2 - k^2}$, and leaving out a common factor $e^{i(kx-\omega t)}$, we find

$$u_y = A_0 e^{ik_T z} + A e^{-ik_T z}, \qquad u'_y = A' e^{ik'_T z}, \tag{24.25}$$

where A_0 is the amplitude of the incident wave, A the amplitude of the reflected wave, and A' the amplitude of the refracted wave.

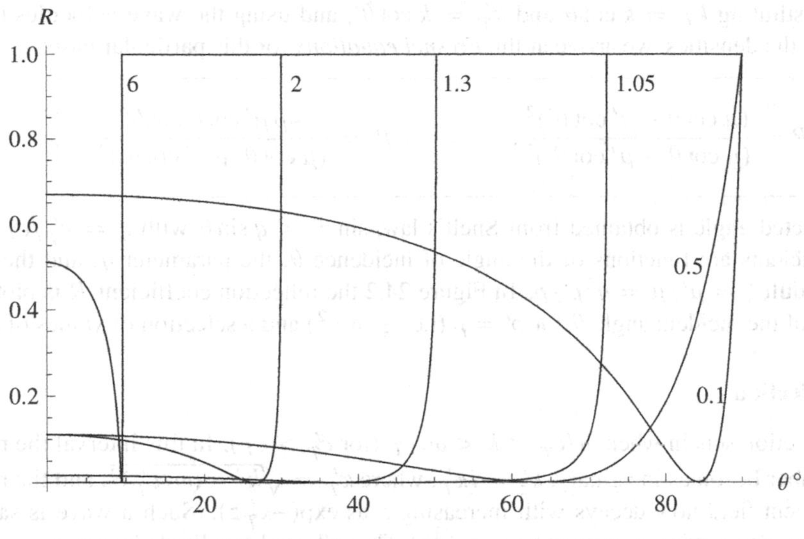

Figure 24.2. Coefficient of reflection for SH-wave incident at an interface with $\rho' = \rho$ as a function of the incidence angle θ (in degrees) for a selection of values of $q = c'_T/c_T = \sqrt{\mu'/\mu}$. One notices the appearance of an upper cut-off to the angle for $q > 1$, where total reflection sets in.

The boundary conditions are $u'_y = u_y$ and $\sigma'_{yz} = \sigma_{yz}$ at $z = 0$. Using that $\sigma_{yz} = \mu\nabla_z u_y$, and $\sigma'_{yz} = \mu'\nabla_z u'_y$ we are led to the equations

$$A' = A_0 + A, \qquad \mu' k'_T A' = \mu k_T (A_0 - A). \tag{24.26}$$

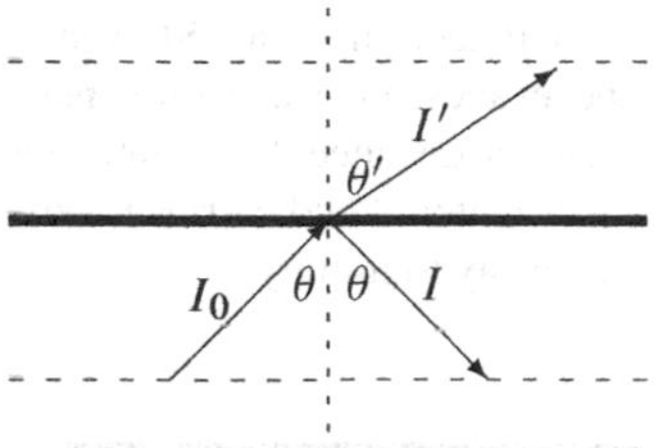

The total energy flux through the box bounded by the dashed horizontal lines must vanish.

Solving for A and A', we find

$$\frac{A}{A_0} = \frac{\mu k_T - \mu' k'_T}{\mu k_T + \mu' k'_T}, \qquad \frac{A'}{A_0} = \frac{2\mu k_T}{\mu k_T + \mu' k'_T}. \tag{24.27}$$

What is interesting is not the amplitudes themselves but the partioning of the energy of the incident wave into the energy of the reflected and refracted waves. Energy conservation implies that on average the energy flux through any closed surface must vanish, since energy cannot be accumulated anywhere. We choose a surface consisting of infinitely extended planes orthogonal to the z-axis above and below the interface (see the margin figure). Taking into account the opposite normals on the upper and lower planes of the bounding box, and dropping the brackets $\langle\cdots\rangle$, we have

$$-I_0 \cos\theta + I \cos\theta + I' \cos\theta' = 0,$$

where I_0, I, and I' are the magnitudes of the wave intensities. Dividing this equation by $I_0 \cos\theta$, we find

$$R + R' = 1, \tag{24.28}$$

where

$$R = \frac{I}{I_0} = \frac{A^2}{A_0^2}, \qquad R' = \frac{I' \cos\theta'}{I_0 \cos\theta} = \frac{\rho' A'^2 c'_T \cos\theta'}{\rho A_0^2 c_T \cos\theta}, \tag{24.29}$$

are respectively called the *coefficients of reflection and refraction.*

Augustin-Jean Fresnel (1788–1827). French physicist. Derived the equations for the amplitudes of reflected and transmitted light. Rejected Newton's corpuscular theory of light in favor of an ether theory. (Source: Wikimedia Commons.)

Substituting $k_T = k\cot\theta$ and $k'_T = k\cot\theta'$, and using the wave velocities (24.5) to eliminate the densities, we arrive at the *Fresnel equations* for this particular case:

$$R = \frac{(\mu\cot\theta - \mu'\cot\theta')^2}{(\mu\cot\theta + \mu'\cot\theta')^2}, \qquad R' = \frac{4\mu\mu'\cot\theta\cot\theta'}{(\mu\cot\theta + \mu'\cot\theta')^2}. \tag{24.30}$$

The refracted angle is obtained from Snell's law, $\sin\theta' = q\sin\theta$ with $q = c'_T/c_T$, so that the coefficients are functions of the angle of incidence θ, the parameter q, and the ratio of shear moduli $\xi = \mu'/\mu = q^2\rho'/\rho$. In Figure 24.2 the reflection coefficient R is plotted as a function of the incident angle θ for $\rho' = \rho$ (i.e., $\xi = q^2$) and a selection of values of q.

Total reflection

Total reflection sets in when $\omega/c'_T < k < \omega/c_T$ (for $c'_T > c_T$). In this interval the refracted wavenumber becomes imaginary, $k'_z = i\kappa'_T$, where $\kappa'_T = \sqrt{k^2 - (\omega/c'_T)^2}$, and the refracted displacement field now decays with increasing z as $\exp(-\kappa'_T z)$. Such a wave is said to be *evanescent*, and carries no energy to $z = +\infty$. The reflected amplitude becomes

$$\frac{A}{A_0} = \frac{\mu k_T - i\mu'\kappa'_T}{\mu k_T + i\mu'\kappa'_T} = e^{-i\psi}, \tag{24.31}$$

where $\tan\frac{1}{2}\psi = \mu'\kappa'_T/\mu k_T$. Since the complex modulus is unity, $|A/A_0| = 1$, the reflected wave has the same intensity as the incident wave, although *phase shifted* by ψ relative to it. The intensity of the evanescent wave points along the surface (see Problem 24.7).

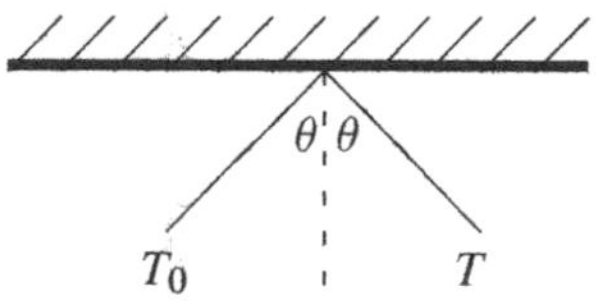

Total reflection of an SH-wave. The evanescent wave only penetrates a distance $1/\kappa'_T$ into the upper material and carries no energy away to infinity.

Case: Reflection of incident SV-wave from a free surface

Refraction is of course impossible at a free surface against vacuum. In the preceding case the interface becomes formally free for $\mu' \to 0$. In this limit, the upper material becomes effectively vacuum and there is no refracted wave, only a reflected one that carries all the incident energy. It resembles total reflection although there is no evanescent wave.

An SV-wave incident on a free surface can only be reflected into another SV-wave or a P-wave. Defining as before $k = k_x$, $k_T = \sqrt{(\omega/c_T)^2 - k^2}$, and $k_L = \sqrt{(\omega/c_L)^2 - k^2}$, and leaving out the common factor $e^{i(kx-\omega t)}$, the field becomes

$$\boldsymbol{u} = A_0(-k_T, 0, k)e^{ik_T z} + A_T(k_T, 0, k)e^{-ik_T z} + A_L(k, 0, -k_L)e^{-ik_L z}, \tag{24.32}$$

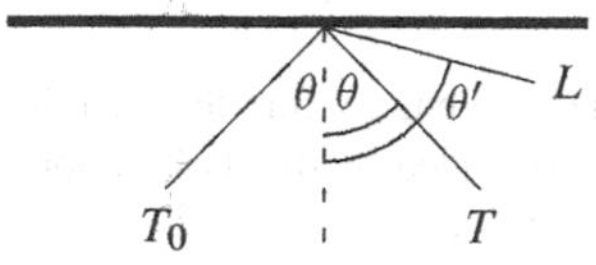

Reflection of a transverse SV-wave at a free surface. The transverse (SV) reflection angle is always the same as the angle of incidence, whereas the longitudinal (P) reflection angle is larger. There is a maximal incident angle $\theta_{\max} = \arcsin q$ for which a longitudinal reflection is possible. Above that value the longitudinal wave becomes evanescent.

where A_0, A_T, and A_L are constants. It may readily be verified that the two first terms have vanishing divergence and that the last term is a gradient.

Since there are no continuity requirements on the displacement field, the boundary conditions reduce in this case to the vanishing of the stress vector,

$$\sigma_{xz} = \sigma_{yz} = \sigma_{zz} = 0, \qquad \text{for } z = 0. \tag{24.33}$$

Inserting the displacement field (24.32), we find the non-vanishing strains at $z = 0$:

$$\begin{aligned} u_{xx} &= \nabla_x u_x = -ikk_T(A_0 - A_T) + ik^2 A_L, \\ u_{zz} &= \nabla_z u_z = ikk_T(A_0 - A_T) + ik_L^2 A_L, \\ 2u_{xz} &= \nabla_x u_z + \nabla_z u_x = i(k^2 - k_T^2)(A_0 + A_T) - 2ikk_L A_L. \end{aligned}$$

From these we obtain the non-vanishing stress components:

$$\sigma_{xz} = 2\mu u_{xz} = i\mu\left[(k^2 - k_T^2)(A_T + A_0) - 2kk_L A_L\right], \tag{24.34a}$$

$$\sigma_{zz} = (2\mu + \lambda)u_{zz} + \lambda u_{xx} = -i\mu\left[2kk_T(A_T - A_0) + (k^2 - k_T^2)A_L\right]. \tag{24.34b}$$

In the last step we have used the relation $\rho\omega^2 = (\lambda + 2\mu)(k_L^2 + k^2)$ to eliminate λ.

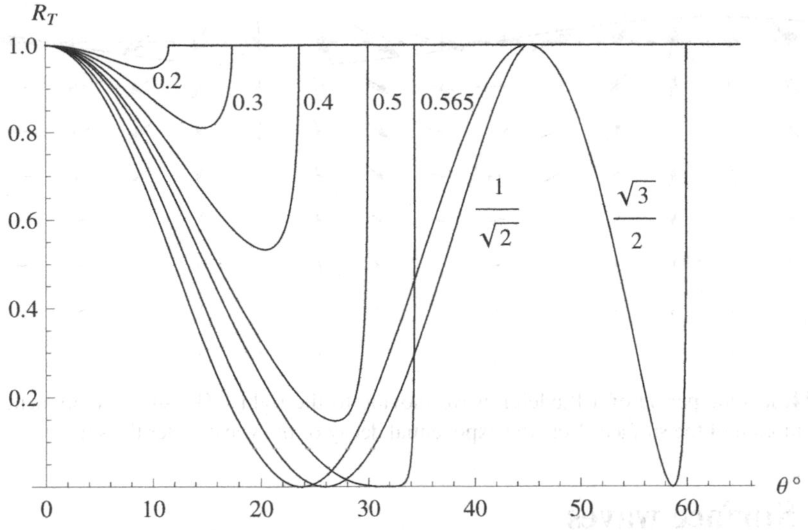

Figure 24.3. Reflection coefficient for SV-incidence as a function of the incidence angle θ for a selection of values for the ratio $q = c_T/c_L$ (shown on the plot). One notices the cut-off in θ where the reflected P-wave becomes evanescent, and all the energy of the incident SV-wave is found in the reflected SV-wave. There is always a minimum of the transverse reflection coefficient corresponding to maximal longitudinal reflection. From $q = 0.565$ this minimum reaches $R_T = 0$ at $\theta = 31°$ and the incoming SV-wave is completely converted to a P-wave. The theoretical maximal possible ratio is $q = \frac{1}{2}\sqrt{3} = 0.866$ and has actually two conversion points and one maximum at $\theta = 45°$. Normal materials can only reach $q = \frac{1}{2}\sqrt{2} = 0.707$, below which there is only one minimum and no maximum of the transverse reflection coefficient.

Requiring these stresses to vanish, we obtain two linear equations for A_T and A_L that are easily solved, and we finally obtain the Fresnel equations,

$$\frac{A_T}{A_0} = \frac{4k^2 k_L k_T - (k^2 - k_T^2)^2}{4k^2 k_L k_T + (k^2 - k_T^2)^2}, \qquad \frac{A_L}{A_0} = \frac{4k k_T (k^2 - k_T^2)}{4k^2 k_L k_T + (k^2 - k_T^2)^2}. \tag{24.35}$$

Now put $k_T = k \cot\theta$ and $k_L = k \cot\theta'$, and the solution may be cast into a form depending on only the two reflection angles,

$$\frac{A_T}{A_0} = \frac{4\cot\theta\cot\theta' - (1-\cot^2\theta)^2}{4\cot\theta\cot\theta' + (1-\cot^2\theta)^2}, \qquad \frac{A_L}{A_0} = \frac{4\cot\theta(1-\cot^2\theta)}{4\cot\theta\cot\theta' + (1-\cot^2\theta)^2}. \tag{24.36}$$

Snell's law, $\sin\theta/\sin\theta' = c_T/c_L = q$, connects as before the two angles. This clearly shows that a transverse incident wave produces a transverse and longitudinal reflected wave for $\sin\theta < q$, whereas for $\sin\theta > q$ the longitudinal wave is evanescent.

In calculating the reflection coefficients, it must be remembered to include the norms of the polarization vectors in the amplitudes of the displacement field (24.32). The reflection coefficients become

$$R_T = \frac{A_T^2}{A_0^2}, \qquad R_L = \frac{A_L^2 \cot\theta'}{A_0^2 \cot\theta}. \tag{24.37}$$

One may verify that these indeed satisfy $R_T + R_L = 1$. In Figure 24.3 the transverse reflection coefficient is plotted as a function of the angle of incidence for some values of $q = c_T/c_L$.

The case of a longitudinal incident wave is very similar and is analyzed in Problem 24.2.

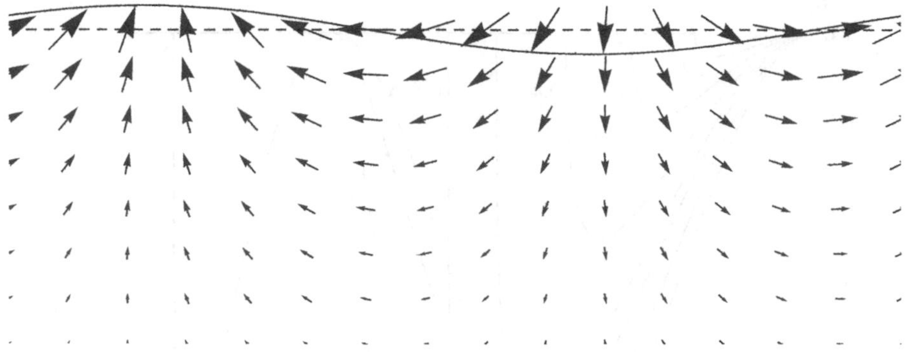

Figure 24.4. One period of a Rayleigh wave moving to the right. The shear component creates a wave-like motion of the surface. Note the exponential decay of the wave under the surface.

∗ 24.4 Surface waves

At material interfaces there are special types of waves that do not penetrate into the bulk of the materials but decay exponentially with the distance from the interface. We have already seen in the preceding section how such wave components can arise in the refraction of an incident wave into a material with larger phase velocity when the angle of incidence becomes large enough. In this section we shall consider two kinds of free surface waves, called *Rayleigh waves* and *Love waves*. Both have geophysical significance in relation to earthquakes where they arise when seismic bulk waves encounter the surface of the Earth.

Rayleigh waves

Elastic surface waves are similar to the well-known deep-water surface waves that will be analyzed in Chapter 25. Instead of gravity, elasticity is the restoring force that wants to move the wave's material back to the undisturbed state. Like gravity waves, elastic surface waves propagate without the need for an incident wave, and we therefore expect that an elastic surface wave should be a superposition of an SV- and P-wave of the form (24.32) with $A_0 = 0$. Leaving out the common factor $e^{i(kx-\omega t)}$, we have

$$\boldsymbol{u} = A_T(k_T, 0, k)e^{-ik_T z} + A_L(k, 0, -k_L)e^{-ik_L z}. \tag{24.38}$$

The condition for this to be a surface wave that decays exponentially with depth $z < 0$ is that $k_T = i\kappa_T$ and $k_L = i\kappa_L$, where $\kappa_T = \sqrt{k^2 - (\omega/c_T)^2}$ and $\kappa_L = \sqrt{k^2 - (\omega/c_L)^2}$. As $c_T < c_L$ in all materials, both κ_T and κ_L will be real for $\omega < kc_T$.

The free surface boundary conditions express again that the stresses must vanish for $z = 0$. Setting $A_0 = 0$ in (24.34) we obtain the two coupled linear equations for A_T and A_L,

$$\left(k^2 - k_T^2\right) A_T - 2kk_L A_L = 0, \qquad 2kk_T A_T + \left(k^2 - k_T^2\right) A_L = 0.$$

The condition for these equations to have a non-trivial solution is that the determinant vanishes. In terms of κ_T and κ_L, this condition becomes

$$\boxed{(k^2 + \kappa_T^2)^2 = 4k^2\kappa_T\kappa_L,} \tag{24.39}$$

which is a second-order algebraic equation for k^2.

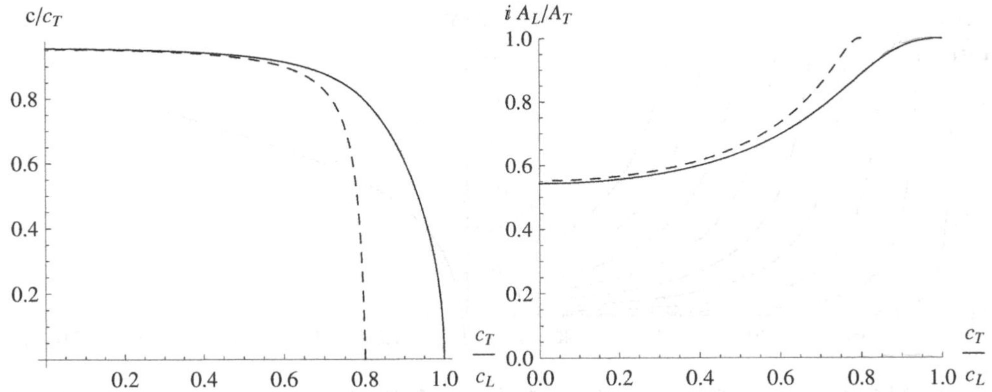

Figure 24.5. Rayleigh waves. **Left:** Phase velocity $\xi = c/c_T$ in units of c_T as a function of $q = c_T/c_L$. For all possible physical values, $q < \frac{1}{2}\sqrt{2} \approx 0.7$, the phase velocity lies less than 10% below the phase velocity of transverse waves. The approximation (24.42) is quite precise for physical values of q (dashed curve). **Right:** Ratio of longitudinal and transverse amplitudes as a function of q. Again, the approximation (24.42) is quite good (dashed curve).

Defining the phase velocity along the surface $c = \omega/k$, this condition can be expressed entirely in terms of phase velocities,

$$\left(2 - \frac{c^2}{c_T^2}\right)^2 = 4\sqrt{\left(1 - \frac{c^2}{c_T^2}\right)\left(1 - \frac{c^2}{c_L^2}\right)}. \tag{24.40}$$

The simplest way to solve this equation is to square it and solve for the ratio $q = c_T/c_L$ in terms of the ratio $\xi = c/c_T$,

$$q = \sqrt{\frac{16 - 24\xi^2 + 8\xi^4 - \xi^6}{16(1 - \xi^2)}}. \tag{24.41}$$

In Figure 24.5L the phase velocity ξ of Rayleigh waves has been plotted as a function of q. The maximal value $\xi_0 = 0.955313\ldots$ at $q = 0$ is the only real positive root of the polynomial in the numerator under the square root.

Typical physical values of q are around 0.5, showing that the value of ξ is close to unity in all practical cases. Normally, the phase velocity of Rayleigh waves thus lies just a little below the phase velocity of free transverse waves. Expanding (24.41) to lowest order near $\xi = 1$, we find the approximation

$$\xi = 1 - \frac{2}{43 - 64q^2}, \tag{24.42}$$

which for $q = 0.5$ is better than 1%.

Seismic surface waves: Seismic waves created deep inside the Earth's crust are reflected from the surface. If the angle of incidence is large enough, the transverse components will excite Rayleigh waves running along the surface. Since their speed is slightly lower than the transverse waves, they arrive even later than S-waves at a seismometer (if they originate in the same point). During the passing of a Rayleigh wave, the surface suffers a combination of horizontal compressional and vertical shear displacements, much like a wave moving across the sea (see Figures 24.4 and 24.1). Horizontal shear displacements are absent in a Rayleigh wave.

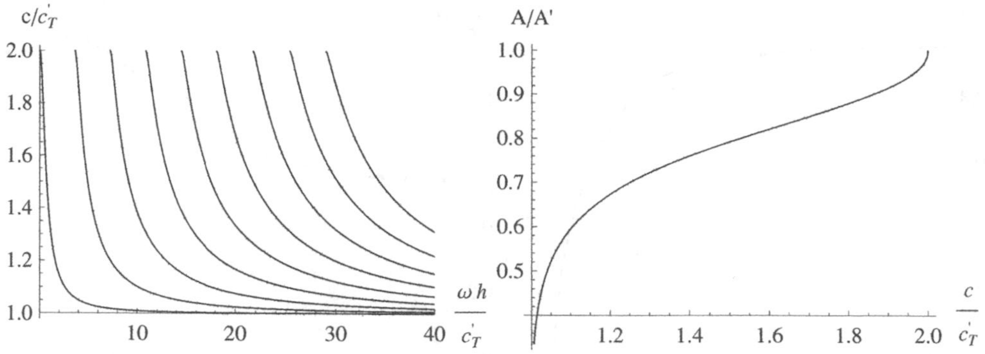

Figure 24.6. Love waves for $c'_T = 0.5c_T$ and $\mu' = \mu$. **Left:** Phase velocity c/c'_T as a function of $\omega h/c'_T$. For a given frequency there is a finite number of possible phase velocities. **Right:** Amplitude ratio A/A' as a function of phase velocity c/c'_T.

Love waves

Augustus Edward Hough Love (1863–1940). British scholarly physicist. Contributed to the mathematical theory of elasticity, and to the understanding and analysis of the waves created by earthquakes.

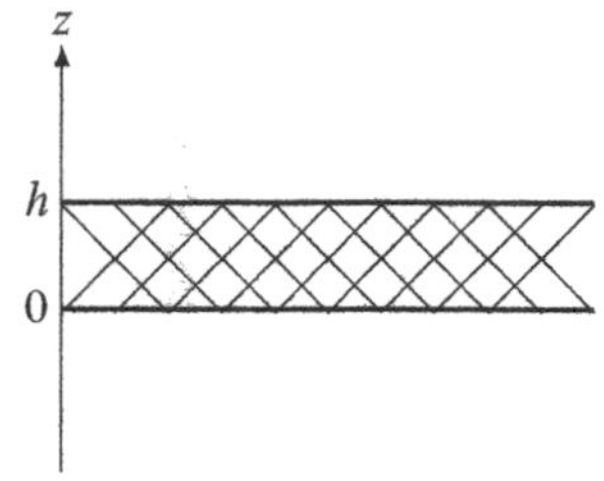

Love waves may arise in a layer of thickness h on top of a half-space $z < 0$ of other material.

One could think that there might be SH polarized surface waves, either at a free surface or at a material interface, but neither of these types are in fact possible (see Problem 24.3). Love found, however, in 1911 that free SH-waves may be created if the surface material is heterogeneous with elastic properties that change with depth z.

The simplest geometry is obtained by placing a layer of material of thickness h situated on top of a material filling the half-space $z < 0$. Provided the phase velocity in the lower half-space is greater than in the upper layer, $c_T > c'_T$, and that the horizontal wave vector falls in the interval $\omega/c_T < k < \omega/c'_T$, there will be a solution that is exponentially damped in the lower half-space and has running waves in the upper layer. Since the stress in the upper layer, $\sigma'_{yz} = \mu'\nabla_z u'_y$, must vanish at the top of the layer, $z = h$, the SH-fields take the following form in the lower and upper layers:

$$u_y = Ae^{\kappa_T z}, \qquad u'_y = A'\cos k'_T(h - z), \tag{24.43}$$

where $\kappa_T = \sqrt{k^2 - (\omega/c_T)^2}$ and $k'_T = \sqrt{(\omega/c'_T)^2 - k^2}$. The remaining boundary conditions are as before $u'_y = u_y$ and $\sigma'_{yz} = \sigma_{yz}$ at $z = 0$, leading to

$$A = A'\cos(k'_T h), \qquad \mu\kappa_T A = \mu' k'_T A' \sin(k'_T h). \tag{24.44}$$

A non-trivial solution can only exist for

$$\mu\kappa_T = \mu' k'_T \tan(k'_T h). \tag{24.45}$$

Introducing $c = \omega/k$, we have $\kappa_T = k\sqrt{1 - (c/c_T)^2}$ and $k'_T = k\sqrt{(c/c'_T)^2 - 1}$, and solving the above equation for kh, we obtain

$$kh = \frac{1}{\sqrt{\frac{c^2}{c'^2_T} - 1}} \arctan \frac{\mu\sqrt{1 - \frac{c^2}{c_T^2}}}{\mu'\sqrt{\frac{c^2}{c'^2_T} - 1}}. \tag{24.46}$$

For any value of the phase velocity in the interval $c'_T < c < c_T$, this permits us to calculate the value of kh and—since h is assumed known—of $\omega = ck$. Conversely, for a given frequency ω, one may solve this equation for the allowed values of c. As shown in Figure 24.6L, the solution is multi-valued.

The infinity of branches of the arctan function yields an infinite number of possible frequencies for a given phase velocity. Conversely, there are only a finite number of allowed phase velocities for a given frequency. The solutions c/c'_T are plotted in Figure 24.6L as functions of the dimensionless frequency parameter $\omega h/c'_T$ for a convenient choice of material parameters. In Figure 24.6R, the amplitude ratio A/A' is plotted as a function of the phase velocity parameter c/c'_T.

A surface layer acts like a waveguide for Love waves. Contrary to Rayleigh waves, Love waves are *dispersive*, with phase velocity that depends on the wavelength (or frequency). Since $c'_T < c < c_T$, Love waves move faster than Rayleigh waves in the surface layer, but slower than shear waves in the bulk. Love waves thus arrive before the Rayleigh waves, if they originate in the same point (see Figure 24.1). In earthquakes, Love waves are the most destructive because of the shearing horizontal motion of the surface layer, which is not well tolerated by buildings.

Problems

24.1 Derive Navier's equation of motion when the Lamé coefficients may depend on position and time.

24.2 Show that the reflection amplitudes for a longitudinal incident wave on a free surface are

$$\frac{A_L}{A_0} = \frac{4\cot\theta\cot\theta' - (1-\cot^2\theta')^2}{4\cot\theta\cot\theta' + (1-\cot^2\theta')^2}, \qquad \frac{A_T}{A_0} = -\frac{4\cot\theta(1-\cot^2\theta')}{4\cot\theta\cot\theta' + (1-\cot^2\theta')^2},$$

where θ is the angle of incidence and $\cot\theta' = \sqrt{-1+(q\sin\theta)^{-2}}$.

24.3 [Free Love waves] **(a)** Show that SH interface waves or **(b)** free SH surface waves cannot exist.

* **24.4** **(a)** Show that an arbitrary vector field may be resolved into (not necessarily unique) longitudinal and transverse components, $\boldsymbol{u} = \boldsymbol{u}_L + \boldsymbol{u}_T$, and that the longitudinal component may be chosen to be a gradient. Hint: First solve the equation $\nabla^2\psi = \nabla\cdot\boldsymbol{u}$, and then use ψ to construct $\boldsymbol{u}_L$ and $\boldsymbol{u}_T$. **(b)** Let now $\boldsymbol{u} = \boldsymbol{u}_L + \boldsymbol{u}_T$ be a solution to Navier's equation of motion (24.1) for $\boldsymbol{f} = \boldsymbol{0}$. Show that the longitudinal and transverse components may each be chosen to satisfy Navier's equation of motion.

24.5 Show that the time average of a product of the real parts of two complex oscillating quantities is

$$\left\langle \mathcal{R}e\left[Ae^{-i\omega t}\right]\mathcal{R}e\left[Be^{-i\omega t}\right]\right\rangle = \frac{1}{2}\mathcal{R}e\left[A^\times B\right].$$

24.6 Calculate the **(a)** kinetic and **(b)** elastic energy densities for the generic wave (24.14) with complex amplitude and complex wave vector.

24.7 **(a)** Show that for complex amplitude $\boldsymbol{a}$ and wave vector $\boldsymbol{k}$, the time average of the intensity is

$$\langle \boldsymbol{I}\rangle = \tfrac{1}{2}\omega\mathcal{R}e\left[\boldsymbol{a}^\times\cdot(\mu(\boldsymbol{k}\boldsymbol{a}+\boldsymbol{a}\boldsymbol{k})+\lambda\boldsymbol{a}\cdot\boldsymbol{k}\boldsymbol{1})\right]e^{-2\mathcal{I}m[\boldsymbol{k}]\cdot\boldsymbol{x}}.$$

(b) Find the average intensity of an evanescent SH-wave with $k_z = i\kappa_T$ and $\kappa_T = \sqrt{k^2-\omega^2/c_T^2}$.

25 Gravity waves

Surface waves in the sea are typically created by the interaction of wind and water, which somehow transforms the steady motion of the streaming air into the nearly periodic swelling and subsiding of the water. The waves appear to roll toward the coast in fairly orderly sequence of crests and troughs that is translated into the quick run-up and run-off at the beach. On top of that there is, of course, the slow ebb and flow of the tides.

In constant flat-Earth gravity, the interface between immiscible fluids at rest is always flat and horizontal, whereas for moving fluids it can take a very complex instantaneous shape under the combined influence of inertia, pressure, gravity, container shape, surface tension, and viscosity. Interface waves dominated by gravity and pressure are generally called *gravity waves*. If the interfacing fluids have very different densities, as is the case for the sea and the atmosphere, one may simply disregard the lighter fluid and consider open surface waves of the heavier fluid toward vacuum. For fluids of nearly equal density, such as a saline bottom layer in the sea overlayed with a brackish layer, one speaks about *internal waves*.

Waves can be created in many ways. In this chapter we shall not be concerned with the mechanisms by which waves are created, but rather with their internal dynamics after they have somehow been brought into existence. The chapter is restricted to the linear theory of small-amplitude surface waves in incompressible fluids under constant gravity. When the amplitude grows so large that the nonlinear aspects of fluid mechanics come into play, new phenomena arise, such as the choppy shape of ocean waves, the overturning and breaking of waves at the beach, and the sonic boom of an aircraft overhead (see Chapter 26).

25.1 Basic wave concepts

A typical non-breaking wave consists of mounds and hollows of roughly similar shape and size in the otherwise smooth equilibrium surface. Although a general wave can be very complex, it is convenient to describe its local features in terms of scale parameters that are normally reserved for harmonic waves:

- a—amplitude. The length scale for vertical variations in the surface, for example the height of a mound above the equilibrium level, or equivalently the depth of a hollow below.
- λ—wavelength. The horizontal length scale of the wave, typically the distance between neighboring mounds or hollows.

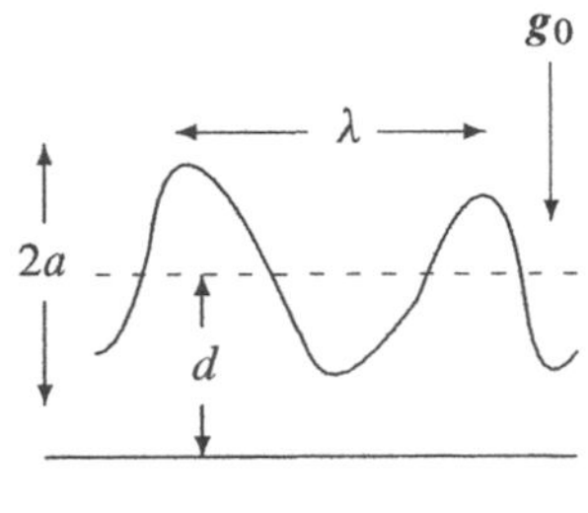

Mounds and hollows in a sea of depth d. A mound is described by amplitude a, wavelength λ, and period τ.

- τ—period. The time scale for major changes in the wave pattern, for example the time it takes a mound to become a hollow and turn back into a mound.

The ratio $c = \lambda/\tau$ is called the *celerity* or *phase velocity* of the wave, and characterizes the speed with which the waveform moves or changes shape.

We shall mostly assume that the sea has a flat, horizontal bottom at depth d below the equilibrium surface. If the wavelength is much greater than the depth, $\lambda \gg d$, we shall speak about *long waves* or more graphically *shallow-water waves*. Similarly, for wavelength smaller than the depth, $\lambda \ll d$, the waves are called *short waves* or *deep-water waves*. Waves with amplitude much smaller than both wavelength and depth, $a \ll \lambda, d$, are called *small-amplitude* waves, and they are the ones we shall focus on in the following.

Dispersion relation for shallow-water gravity waves

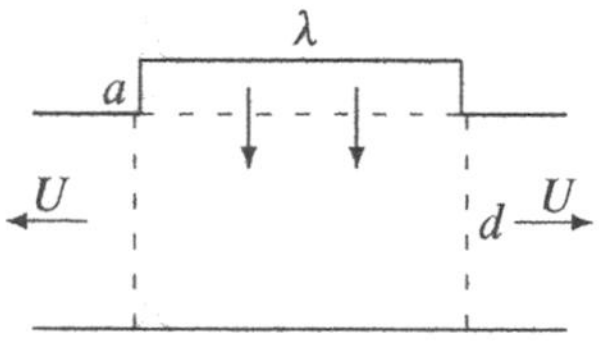

A small rectangular mound of height a, width λ, and length L (into the paper) rising out of a shallow sea of depth d. When the mound collapses vertically, all the water underneath has to disperse in the horizontal direction with a typical velocity U.

Suppose a rectangular "waterberg" of height a, width λ and length L is somehow created in the surface of a the otherwise flat sea[1]. We shall for simplicity assume that these quantities satisfy the inequalities $a \ll d \ll \lambda \ll L$, which means that the amplitude is small, the wavelength is long, and the mound is essentially two-dimensional (see the margin figure). Such a waterberg cannot be stable, but will quickly collapse under gravity, creating instead a smaller hollow that may later rise again to make an even smaller mound that in turn collapses, and so on. Eventually all traces of the initial waterberg will have disappeared while secondary waves run away over the surface in both directions orthogonal to the mound.

The mass of the waterberg is $M_0 \sim \rho_0 a\lambda L$ while the mass of the sea underneath is much greater, $M_1 \sim \rho_0 d\lambda L$. During the collapse, all the water in the waterberg will move vertically downward and merge with the sea in a certain time τ. Due to the proximity of the bottom of the shallow sea, all the water underneath will be set into motion out of this region with a certain average speed U. Disregarding all purely numeric factors, the mass that leaves the region in the time τ must equal the mass that comes down, $\rho_0 U\tau L d \sim \rho_0 a\lambda L$, so that

$$U \sim \frac{a\lambda}{\tau d}. \tag{25.1}$$

Since the wavelength is much greater than the depth, $\lambda \gg d$, the horizontal flow velocity U is much greater than the downward vertical velocity, which is of magnitude a/τ.

During the collapse, gravitational potential energy is converted into kinetic energy (assuming dissipation can be ignored). Since the potential energy, $M_0 g_0 a \sim \rho_0 g_0 a^2 \lambda L$, is quadratic in a, a hollow ($a < 0$) will have the same potential energy as a mound ($a > 0$) and thus fill out in the same time τ. The simple reason is that while gravity presses a mound downward, buoyancy correspondingly presses a hollow upward. Disregarding again numeric factors, the equality of potential and kinetic energy, $M_0 g_0 a \sim M_1 U^2$, yields an estimate of the collapse time,

$$\boxed{\tau \sim \frac{\lambda}{\sqrt{g_0 d}}.} \tag{25.2}$$

A relation of this kind that expresses the conservation of mechanical (i.e., kinetic plus potential) energy of a wave is called the *dispersion law* for the wave. Its precise form will in general contain a factor of order unity that can depend on the dimensionless ratios a/λ and d/λ, and possibly on other dimensionless parameters characterizing the actual shape of the wave. We shall later see that for shallow-water gravity waves, the above relation is in fact an equality.

[1]Such a mound can arise when an earthquake suddenly lifts the flat sea-bottom by an amount a across a width λ along a straight fault line of length L. Provided the earthquake is nearly instantaneous, all the water above is also lifted by the same amount a over the same footprint area $\lambda \times L$. This is in fact one of the ways a *tsunami* may start out.

From the dispersion relation we immediately get the celerity,

$$c = \frac{\lambda}{\tau} \sim \sqrt{g_0 d}, \tag{25.3}$$

which is also exact for shallow-water waves. Like an echo of Toricelli's law (page 214) it is of the same order of magnitude as the free-fall velocity $\sqrt{2g_0 d}$ from the height d.

In shallow water, $\lambda \gg d$, the celerity is independent of both amplitude and wavelength. In deep water with $\lambda \ll d$, there is no bottom to divert the water so that the horizontal velocity tends to be of the same order of magnitude as the vertical, that is, $U \sim a/\tau$ near the surface. From the expression (25.1) for U, it follows that $d \sim \lambda$, and we may thus conclude that the deep sea must have an effective depth of the same magnitude as the wavelength. We shall soon see that the effective depth of the deep sea is in fact $\lambda/2\pi$.

Example 25.1 [Tsunami]: In a sea of depth $d \approx 1$ km, a long fault slip deforms the sea bottom vertically across a region of width $\lambda \approx 100$ km. The celerity of the resulting tsunami becomes $c \approx 100\ \text{m s}^{-1}$, or 360 km per hour, with a period $\tau \approx 1{,}000$ s or 20 min. Since the amplitude is typically $a \sim 1$ m in the open sea, you can neither see a tsumani nor outrun it!

Small-amplitude waves are nearly linear

The shape of a surface wave is only a manifestation of the (literally) underlying hydrodynamics, governed by the Navier–Stokes equations. The nonlinearity of these equations makes general surface waves much more complex than, for example, electromagnetic waves governed by the linear Maxwell equations. But if the advective acceleration $(\boldsymbol{v} \cdot \nabla)\boldsymbol{v}$ can be disregarded in comparison with the local acceleration $\partial\boldsymbol{v}/\partial t$, the Navier–Stokes equations also become linear. For shallow-water waves we obtain the ratio of advective to local acceleration,

$$\frac{|(\boldsymbol{v} \cdot \nabla)\boldsymbol{v}|}{|\partial\boldsymbol{v}/\partial t|} \approx \frac{U^2/\lambda}{U/\tau} \approx \frac{U}{c} \approx \frac{a}{d}. \tag{25.4}$$

It thus appears that the advective term plays no role for small-amplitude waves with $a/d \ll 1$.

Nonlinear gravity waves are however quite subtle and have been intensely discussed since Stokes published his famous analysis of oscillatory waves in 1847. Stokes showed that the linear approximation requires $a/d \ll (kd)^2$, where $k = 2\pi/\lambda$. This is stricter than the naive condition $a/d \ll 1$ for shallow-water waves (see also [Urs53]). In the tsunami example above, Stokes' condition leads to $a \ll 4$ m, which is normally fulfilled.

The role of viscosity

When the advective term can be disregarded relative to the local acceleration, the relevant dimensionless quantity for assessing the importance of viscosity is the ratio of local to viscous acceleration (similar to the Reynolds number),

$$\frac{|\partial\boldsymbol{v}/\partial t|}{\nu\,|\nabla^2\boldsymbol{v}|} \sim \frac{U/\tau}{\nu U/d^2} = \frac{d^2}{\nu\tau}. \tag{25.5}$$

Here we have assumed that the viscous acceleration is determined by the vertical variation in the horizontal flow over the depth d. For deep-water waves, d may be replaced by $\lambda/2\pi$.

The typical sea swells we encounter when swimming close to the shore at a depth of a meter or two have periods from a few seconds to a fraction of a minute. With $\nu \approx 10^{-6}\ \text{m}^2\ \text{s}^{-1}$ for water, the above ratio will be in the hundred thousands, and viscosity plays essentially no role for such waves. In daily life we are otherwise quite familiar with viscous flow, for example while pouring syrup or stirring porridge, but waves in syrup or porridge are not so interesting because they quickly die out. In water, gravity waves simply keep rolling along, although eventually viscosity will also make these waves die away if left on their own. That problem will be dealt with separately in Section 25.6.

25.2 Harmonic surface waves

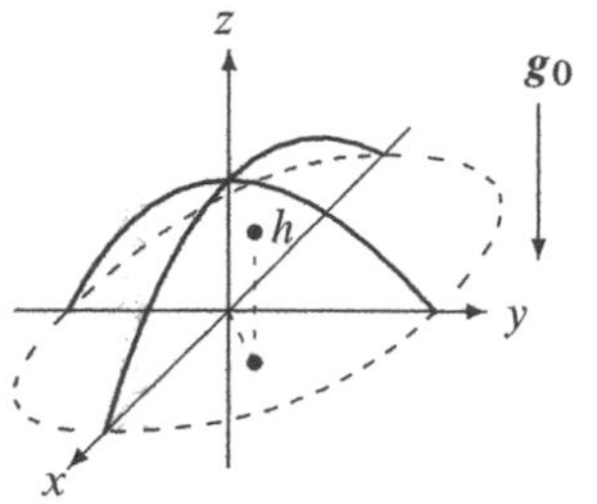

Part of a surface wave in the flat-Earth coordinate system. The dashed curve is the intersection of the wave with $z = 0$.

In flat-Earth coordinates a wave will distort the equilibrium surface $z = 0$ of the infinitely extended sea, so that its instantaneous height becomes a function of the horizontal coordinates and time,

$$z = h(x, y, t). \tag{25.6}$$

Here we shall focus on trains of waves that progress periodically and in a regular pattern across the horizontal equilibrium surface.

Line waves

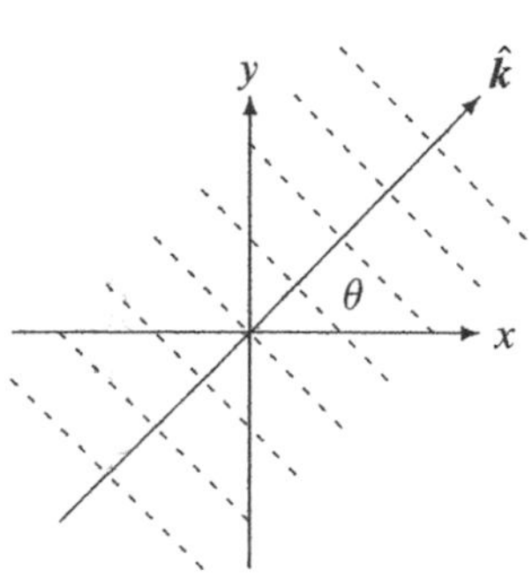

A periodic line wave on the surface. The crests are parallel lines orthogonal to the direction of the wave vector $\hat{\boldsymbol{k}} = (\cos\theta, \sin\theta, 0)$, forming an angle θ with the x-axis.

We shall nearly always assume that the sea has a flat bottom at constant depth, $z = -d$. The translational invariance in the horizontal directions and the linearity of the Navier–Stokes equations in the small-amplitude limit allows us to resolve the velocity field $\boldsymbol{v}$ and the associated surface deformation h into independent Fourier components (running waves) along the horizontal directions. An elementary harmonic or monochromatic *line wave* on the surface takes the form

$$h = a\cos(k_x x + k_y y - \omega t + \chi), \tag{25.7}$$

where $\boldsymbol{k} = (k_x, k_y, 0)$ is the *wave vector*, $\omega = 2\pi/\tau$ the *circular frequency*, and χ the *phase shift*. The magnitude of the wave vector, $k = |\boldsymbol{k}| = 2\pi/\lambda$, is called the *wavenumber*. The argument of the cosine, $\phi = k_x x + k_y y - \omega t + \chi$, is called the *phase* of the wave. The phase shift χ can for a single wave always be absorbed in the choice of origin of the coordinate system or of time, but differences between phase shifts may become of physical importance when waves are superposed.

The maxima or minima, or *crests* and *troughs* as they are called for surface waves, move steadily along in the direction $\hat{\boldsymbol{k}} = \boldsymbol{k}/k = (\cos\theta, \sin\theta, 0)$ with the *phase velocity*,

$$c = \frac{\lambda}{\tau} = \frac{\omega}{k}. \tag{25.8}$$

In general the dispersion law will be nonlinear, $\tau = \tau(\lambda)$, or equivalently $\omega = \omega(k)$, and the phase velocity will depend on the wavenumber.

Group velocity

Consider now two harmonic line waves, both for simplicity chosen to run along the x-axis with the same amplitudes. Their phases are $\phi_1 = k_1 x - \omega_1 t + \chi_1$ and $\phi_2 = k_2 x - \omega_2 t + \chi_2$. Using the trigonometric relation,

$$\cos\phi_1 + \cos\phi_2 = 2\cos\frac{\phi_1 + \phi_2}{2}\cos\frac{\phi_1 - \phi_2}{2}, \tag{25.9}$$

the superposition $h = h_1 + h_2$ may be written as

$$h = 2a\cos(kx - \omega t + \chi)\cos(\Delta k\, x - \Delta\omega\, t + \Delta\chi), \tag{25.10}$$

where $k = \frac{1}{2}(k_1 + k_2)$, $\omega = \frac{1}{2}(\omega_1 + \omega_2)$, and $\chi = \frac{1}{2}(\chi_1 + \chi_2)$ are the average quantities for the two waves, and $\Delta k = \frac{1}{2}(k_1 - k_2)$, $\Delta\omega = \frac{1}{2}(\omega_1 - \omega_2)$, and $\Delta\chi = \frac{1}{2}(\chi_1 - \chi_2)$ are the half differences. The first oscillating factor evidently describes a harmonic line wave moving along the x-axis with the average values of the wavenumbers, frequencies, and phase shifts, but the amplitude of this wave is now *modulated* by the second factor.

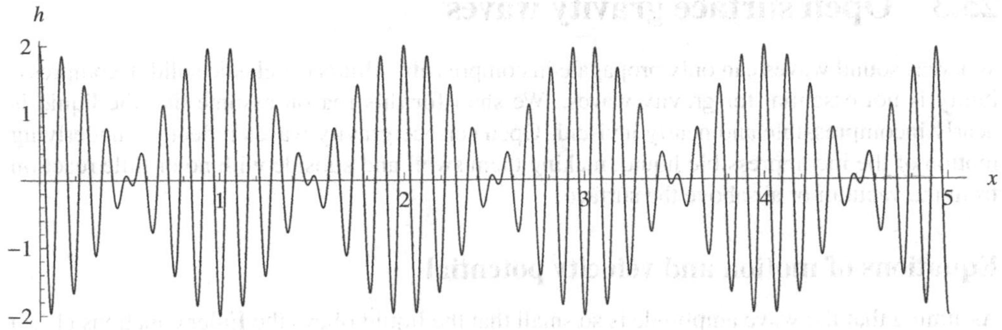

Figure 25.1. Superposition of two harmonic line waves with nearly equal wavenumbers. Here with $k = 16\pi$, $\Delta k = \pi$, $a = 1$, and $\chi = 0$. The rapid oscillation of the "carrier" wave is modulated and broken into a "beat pattern" of identical wave packets of length $\Delta x = \pi/\Delta k = 1$ centered at $x = n$ for all integer n.

If the differences are much smaller than the averages, $|\Delta k| \ll |k|, |\Delta\omega| \ll |\omega|$, and $|\Delta\chi| \ll |\chi|$, the second cosine factor will only slowly modulate the rapid oscillations of the first. Since the second cosine vanishes when its arguments pass through $\frac{\pi}{2} + n\pi$, where n is an arbitrary integer, it will chop up the wave into a string of *wave packets* of typical length $\Delta x = \pi/|\Delta k|$, as pictured in Figure 25.1. Inside each wave packet, the crests will move with the phase velocity $c = \omega/k$, whereas the center of each wave packet will move with the speed $\Delta\omega/\Delta k \approx d\omega(k)/dk$. Thus, the propagation speed of a wave packet is determined by the derivative of the dispersion law,

$$\boxed{c_g = \frac{d\omega}{dk},} \tag{25.11}$$

also called the *group velocity*.

Any superposition of waves with nearly equal wave vectors will in fact form one or more wave packets moving with the group velocity (see Problem 25.6). If the dispersion law is linear, as it is for shallow-water waves, the group and phase velocities will be equal; but if the dispersion law is non-linear, as for deep-water waves, they will be different. Deep-water waves are said to be *dispersive* because the shorter wavelengths move slower than the longer and make the harmonic components of a wave packet split up or disperse according to wavelength. If the group velocity is smaller than the phase velocity, $c_g < c$, the wave crests will move forward inside a wave packet as it proceeds across the surface, and conversely if it is larger.

Energy transport and group velocity

In a single wave packet, the velocity field is only non-zero in the region covered by the wave packet, so that the energy of the wave must be concentrated here and transported along the surface with the group velocity rather than with the phase velocity. The same must be true for any superposition of single wave packets with wavenumbers taken from a narrow band of width Δk around k, such as the one shown in Figure 25.1. In the limit where the bandwidth Δk narrows down to nothing, the energy must still be transported with the group velocity, so in the end we reach the slightly strange conclusion that even in a purely monofrequent harmonic line wave, the energy in the wave must also be transported with the group velocity. We shall later confirm this by an explicit calculation (page 436).

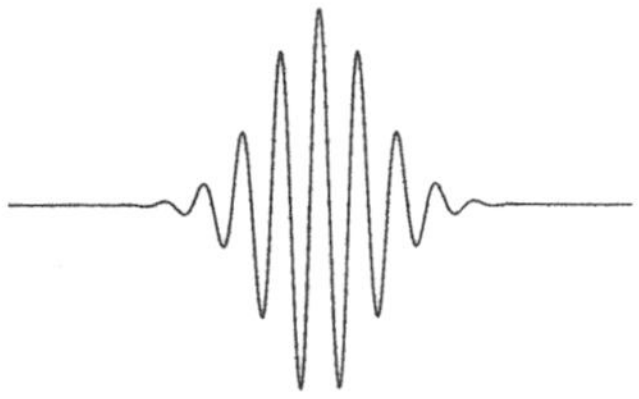

A single Gaussian wave packet.

25.3 Open surface gravity waves

Whereas sound waves can only propagate in compressible fluids (or elastic solids), compressibility is not essential for gravity waves. We shall for this reason assume that the liquid is nearly incompressible and nearly inviscid. Open surface gravity waves reflect the underlying motion of the incompressible liquid, making them swell and subside with no or little reaction from the vacuum or air above the surface.

Equations of motion and velocity potential

Assuming that the wave amplitude is so small that the liquid obeys the Euler equations (13.1) on page 207 without the advective acceleration term, we have

$$\frac{\partial \boldsymbol{v}}{\partial t} = -\frac{1}{\rho_0}\nabla p + \boldsymbol{g}, \qquad \nabla \cdot \boldsymbol{v} = 0, \tag{25.12}$$

where $\boldsymbol{g} = -g_0 \hat{\boldsymbol{e}}_z$. Both terms on the right-hand side of the Euler equation are gradients of scalar fields, so if the velocity field is initially a gradient field, it will continue to be so. Introducing the *velocity potential* Ψ (Section 13.6 on page 220), we have

$$\boldsymbol{v} = \nabla\Psi, \qquad \nabla^2\Psi = 0; \tag{25.13}$$

and inserting the gradient field into the Euler equation and solving for the pressure, we get

$$p = p_0 - \rho_0\left(g_0 z + \frac{\partial \Psi}{\partial t}\right). \tag{25.14}$$

One might in principle add an arbitrary function of time to the right-hand side, but such a function could without loss of generality be absorbed into Ψ.

Boundary conditions

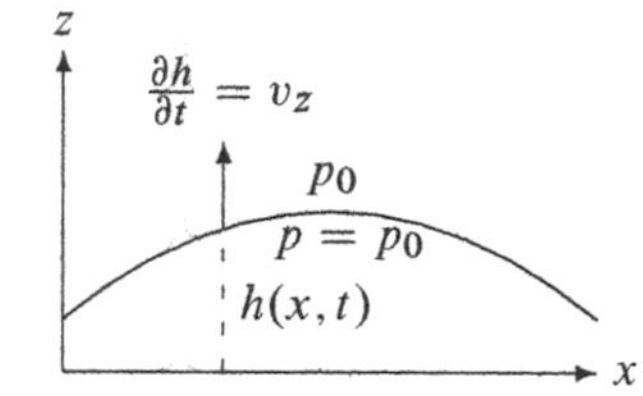

Boundary conditions for a small-amplitude wave along x.

At the open liquid surface, two boundary conditions must be applied. The first is purely kinematic and expresses that a fluid particle sitting on the surface should follow the surface motion. For small amplitudes the wave surface may be viewed as nearly horizontal everywhere, implying that the vertical velocity of the surface should equal the vertical velocity of the fluid just below the surface, or

$$\frac{\partial h}{\partial t} = \nabla_z\Psi \qquad \text{for } z = h. \tag{25.15}$$

The second boundary condition is dynamic and expresses the continuity of the pressure across the surface in the absence of surface tension. Assuming that there is vacuum (or air) with constant pressure p_0 above the surface, this boundary condition becomes $p = p_0$ or

$$g_0 h = -\frac{\partial \Psi}{\partial t} \qquad \text{for } z = h. \tag{25.16}$$

We shall see that in the linear approximation, we must, for consistency, take $z = 0$ in these equations rather than $z = h$.

Besides the open surface conditions, there will be further boundary conditions set by the shape of the container. We shall assume that the liquid is infinitely extended in the horizontal directions and that the bottom of the sea is flat, so that the vertical velocity must vanish,

$$\nabla_z\Psi = 0 \qquad \text{for } z = -d. \tag{25.17}$$

If the liquid is confined in the horizontal directions, boundary conditions implementing the existence of walls must be applied (see Problem 25.1).

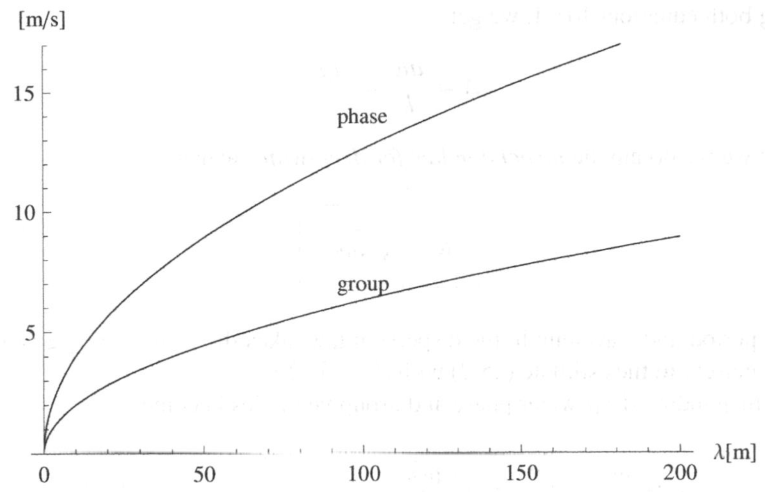

Figure 25.2. Phase and group velocities of deep-water waves as a function of wavelength. This figure corresponds to the small-wavelength part of Figure 25.3. Typical ocean swells have wavelengths of $\lambda \approx 150$ m, period $\tau \approx 10$ s, and phase velocity $c \approx 15\ \mathrm{m\,s^{-1}} \approx 55\ \mathrm{km\,h^{-1}}$. The group velocity equals half of this.

General solution

The linearity of the Laplace equation and translational invariance in the horizontal directions as well as in time allows us to consider only a single Fourier component in the form of a running wave,

$$\Psi = f(z)\sin(k_x x + k_y y - \omega t + \chi), \tag{25.18}$$

where k_x, k_y, ω, and χ are arbitrary constants. The most general solution is a superposition of such waves. The Laplace equation $\nabla^2\Psi = 0$ demands $f''(z) - (k_x^2 + k_y^2)f(z) = 0$, which has the solution

$$f(z) = Ae^{kz} + Be^{-kz}, \tag{25.19}$$

where $k = \sqrt{k_x^2 + k_y^2}$, and A and B are constants.

Deep-water waves

In dealing with a single running wave, we may without loss of generality take $k_x = k, k_y = 0$, and $\chi = 0$. In deep water, $d \to \infty$, the velocity field must be finite for $z \to -\infty$, implying that $B = 0$. Taking $h = a\cos(kx - \omega t)$, the open surface boundary conditions (25.15) and (25.16) become

$$a\omega\sin(kx - \omega t) = k\,Ae^{kh}\sin(kx - \omega t),$$
$$g_0 a\cos(kx - \omega t) = \omega Ae^{kh}\cos(kx - \omega t).$$

As they stand, these equations cannot be satisfied because of the exponential factor containing the oscillating height $h(x,t)$. In a small-amplitude wave, the amplitude is, however, assumed much smaller than the wavelength, $a \ll \lambda$, which implies that $|kh| \ll 1$. Consequently we have $\exp(kh) \approx 1$, so that the above conditions become $a\omega = Ak$ and $g_0 a = A\omega$. This demonstrates that in the linear approximation we must for consistency put $z = 0$ on the right-hand side of the boundary conditions instead of $z = h$, because only then do we obtain a surface height with the expected harmonic form.

Solving both equations for A, we get

$$A = \frac{a\omega}{k} = \frac{ag_0}{\omega}, \tag{25.20}$$

from which we we obtain the *dispersion law for deep-water waves*,

$$\boxed{\omega = \sqrt{g_0 k}.} \tag{25.21}$$

In terms of period and wavelength, the dispersion law takes the form $\tau = \sqrt{2\pi\lambda/g_0}$, corresponding precisely to the estimate (25.2) with $d \to \lambda/2\pi$.

The corresponding deep-water phase and group velocities become

$$c \equiv \frac{\omega}{k} = \sqrt{\frac{g_0}{k}} = \sqrt{\frac{g_0\lambda}{2\pi}}, \qquad c_g = \tfrac{1}{2}c, \tag{25.22}$$

and are plotted in Figure 25.2. Since the phase velocity is double the group velocity, the wave crests will always move forward inside a wave packet with double the speed of the wave packet itself.

Dispersive separation of wavelengths: The dispersive nature of deep-water waves has important consequences. A local surface disturbance in deep water—for example created by a storm far out at sea—always contains many different wavelengths. The long-wave components are faster and will run ahead to arrive at the beach, maybe a day or so, before the slower short-wave components. The long-distance separation of wavelengths causes all the waves that arrive on the beach to be nearly monofrequent, rolling in at regular time intervals that slowly become shorter as the smaller wavelengths take over.

For a deep-water wave, the complete solution is

$$h = a\cos(kx - \omega t), \tag{25.23a}$$

$$\Psi = ace^{kz}\sin(kx - \omega t), \tag{25.23b}$$

$$v_x = a\omega e^{kz}\cos(kx - \omega t), \tag{25.23c}$$

$$v_z = a\omega e^{kz}\sin(kx - \omega t), \tag{25.23d}$$

$$p = p_0 - \rho_0 g_0\left(z - ae^{kz}\cos(kx - \omega t)\right). \tag{25.23e}$$

Integrating the velocities with respect to time, we find the corresponding (Lagrangian) displacement fields,

$$u_x = -ae^{kz}\sin(kx - \omega t), \tag{25.24a}$$

$$u_z = ae^{kz}\cos(kx - \omega t). \tag{25.24b}$$

The horizontal and vertical displacements have the same scale, ae^{kz}, but are 90 ° out of phase. The fluid particles thus move through orbits that are approximatively circles of radius ae^{kz} at depth z. The circle radii thus diminish exponentially with depth. For $z = 0$, we get $u_z = h$, as we should.

Due to the exponential, a deep-water surface wave only influences the flow to a depth $|z| \approx 1/k = \lambda/2\pi$. The shape of the bottom has no influence on the surface waves, as long as the wavelength satisfies $\lambda \ll 2\pi d$, that is, as long as the waves are short.

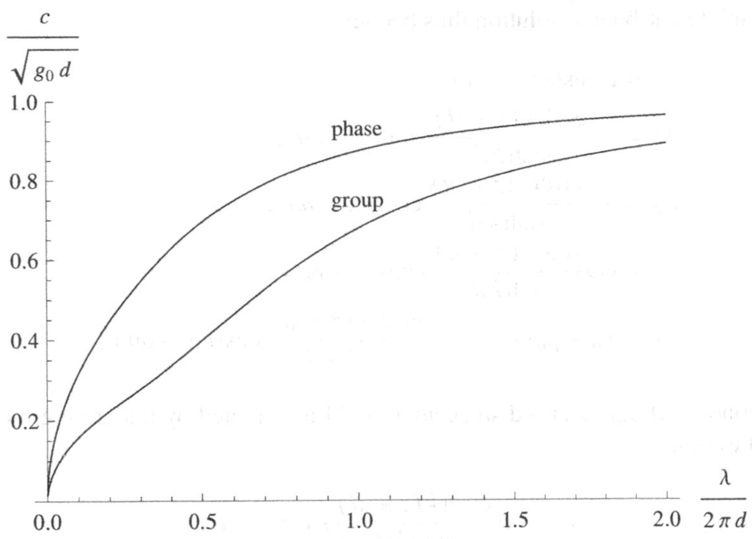

Figure 25.3. Phase and group velocities of flat-bottom gravity waves as functions of $\lambda/2\pi d$. The phase and group velocities level out and become equal for large wavelengths, approaching the common shallow-water value, $c = c_g = \sqrt{g_0 d}$ for $\lambda \gg 2\pi d$. The influence of the finite depth is clearly notable in the group velocity for $\lambda/2\pi d \gtrsim 0.5$.

Waves at finite depth

When wavelengths become comparable to the depth of the ocean, we must take into account the shape of the bottom. For a horizontally infinite ocean with a perfectly flat impermeable bottom at constant depth $z = -d$, the condition the vertical velocity must vanish at the bottom, as expressed in (25.17). In the absence of viscosity, we are not at liberty to impose a no-slip condition on the horizontal velocities.

The flat-bottom boundary condition implies that the function $f(z)$ in (25.19) must satisfy $f'(-d) = 0$, or $Ae^{-kd} = Be^{kd}$, so that

$$f(z) = C \cosh k(z+d), \tag{25.25}$$

where C is another constant. As for deep-water waves, it is determined by the open surface boundary conditions (25.15) and (25.16), and we find for $k\,|h| \ll 1$,

$$C = \frac{a\omega}{k \sinh kd} = \frac{a g_0}{\omega \cosh kd}. \tag{25.26}$$

The last equality yields the dispersion law,

$$\boxed{\omega = \sqrt{g_0 k \tanh kd},} \tag{25.27}$$

with the corresponding phase and group velocities,

$$c = \sqrt{\frac{g_0}{k} \tanh kd}, \qquad c_g = \tfrac{1}{2} c \left(1 + \frac{2kd}{\sinh 2kd}\right). \tag{25.28}$$

They are plotted in Figure 25.3 in dimensionless form. The phase velocity appears to curve downward for all wavelengths, whereas the group velocity changes curvature twice. Generally, the finite depth becomes important from $\lambda \gtrsim \pi d$. For large wavelengths, $\lambda \gg 2\pi d$, that is, $k \to 0$, both velocities approach the common shallow-water value $c = c_g = \sqrt{g_0 d}$.

The complete flat-bottom solution thus becomes

$$h = a\cos(kx - \omega t), \tag{25.29a}$$

$$\Psi = ac\frac{\cosh k(z+d)}{\sinh kd}\sin(kx - \omega t), \tag{25.29b}$$

$$v_x = a\omega\frac{\cosh k(z+d)}{\sinh kd}\cos(kx - \omega t), \tag{25.29c}$$

$$v_z = a\omega\frac{\sinh k(z+d)}{\sinh kd}\sin(kx - \omega t), \tag{25.29d}$$

$$p = p_0 - \rho_0 g_0\left(z - a\frac{\cosh k(z+d)}{\cosh kd}\cos(kx - \omega t)\right). \tag{25.29e}$$

The corresponding (Lagrangian) displacement field is obtained by integrating the velocities with respect to time,

$$u_x = -a\frac{\cosh k(z+d)}{\sinh kd}\sin(kx - \omega t), \tag{25.30a}$$

$$u_z = a\frac{\sinh k(z+d)}{\sinh kd}\cos(kx - \omega t). \tag{25.30b}$$

This shows that the fluid particle orbits are ellipses of general magnitude a, which become progressively flatter as the bottom is approached for $z \to -d$. Evidently, there is no net mass transport in the direction of motion of the wave. Note that for $z = 0$, we get $u_z = h = a\cos(kx - \omega t)$, as we should.

Shallow-water limit

Long waves have wavelength much greater than the depth, that is, $\lambda \gg 2\pi d$ or equivalently $kd \ll 1$. Since $\tanh kd \approx kd$ we obtain the *shallow-water dispersion law* from (25.27),

$$\boxed{\omega = \sqrt{g_0 d}\, k,} \tag{25.31}$$

which agrees with the estimate (25.2). Shallow-water waves are ideally *non-dispersive* with common phase and group velocities,

$$c = c_g = \sqrt{g_0 d}. \tag{25.32}$$

In the shallow-wave limit, $kd \ll 1$, the leading terms in the flat-bottom solution become

$$h = a\cos(kx - \omega t), \tag{25.33a}$$

$$\Psi \approx \frac{ag_0}{\omega}\sin(kx - \omega t), \tag{25.33b}$$

$$v_x \approx \frac{ac}{d}\cos(kx - \omega t), \tag{25.33c}$$

$$v_z \approx 0, \tag{25.33d}$$

$$p \approx p_0 - \rho_0 g_0 (z - a\cos(kx - \omega t)). \tag{25.33e}$$

The horizontal velocity is the same for all z, so that all the water underneath sloshes back and forth in unison as the wave proceeds. The vertical velocity vanishes in this approximation because $|v_z/v_x| \sim kd$. The pressure at any depth z is just the hydrostatic pressure from the water column above, including the height of the wave.

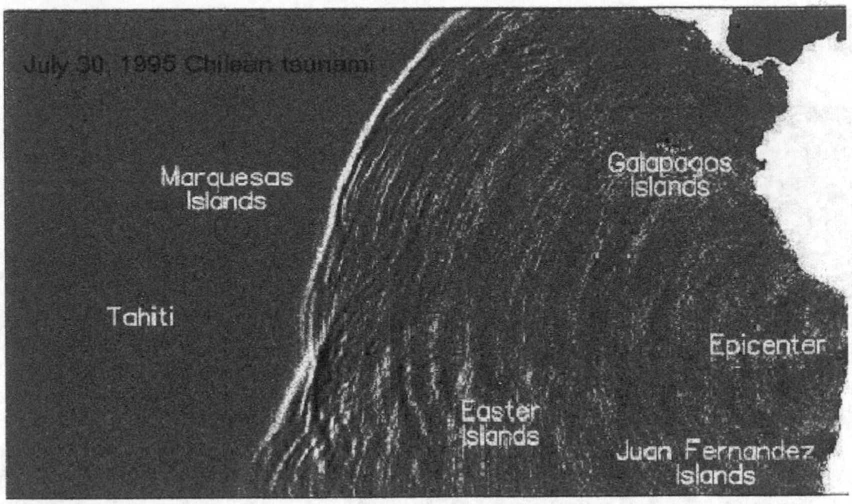

Figure 25.4. Chilean tsunami on July 30, 1995. The tsunami spread from the epicenter of an earthquake on the Chilean pacific coast. (Source: Satellite image. Provenance unknown.)

Example 25.2 [Tsunami]: Tsunamis (meaning harbor waves) are huge shallow-water wave trains with wavelengths up to 500 km, generated by underwater earthquakes, landslides, volcanic eruptions, or large meteorite impacts. The average depth of the oceans is about 4000 m so typical tsunamis move with speeds of a passenger jet plane, $c \approx 200 \text{ m s}^{-1} \approx 710 \text{ km h}^{-1}$. For $\lambda = 500$ km, the period becomes $\tau = \lambda/c \approx 2500 \text{ s} \approx 42$ min, but since the amplitude is small, say $a \lesssim 1$ m, a tsunami will be nearly imperceptible to a ship at sea. But when the tsunami approaches a coastline, the water depth decreases and the wave slows down while increasing its amplitude with sometimes devastating effect on the shore. One of the deadliest tsunamis in history hit the coasts around the Indian Ocean on December 26, 2004.

25.4 Capillary waves

Surface tension was introduced in Chapter 5. Its strength is characterized by a material constant α, representing the attractive force per unit of length of the surface, or equivalently an extra positive energy per unit of surface area. Surface tension attempts to diminish the total surface area, and thus collaborates with gravity in restoring the flat, horizontal equilibrium shape of the surface of the sea (see Figure 25.5).

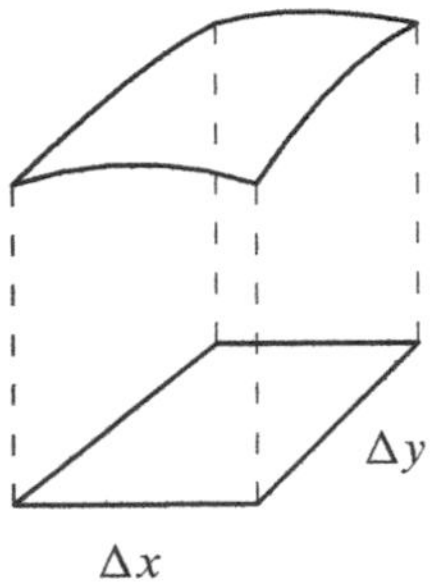

The small rectangle in the xy-plane defines a piece of the wave surface of area $A = \Delta x \Delta y$.

Pressure jump across a nearly flat surface

Surface tension gives rise to a pressure jump across any interface between two fluids. In the case of an open horizontal surface toward air or vacuum with constant pressure p_0, the dynamic boundary condition expresses the pressure just below the surface as

$$p = p_0 + \Delta p \qquad \text{for } z = h. \tag{25.34}$$

For a nearly flat surface, the inverse radius of curvature in any horizontal direction is given by the double derivative (5.8) on page 76. Combined with the Young–Laplace law (5.11), we obtain

$$\boxed{\Delta p = -\alpha(\nabla_x^2 + \nabla_y^2)h.} \tag{25.35}$$

The sign reflects that a mound must yield a positive pressure below (see the margin figure).

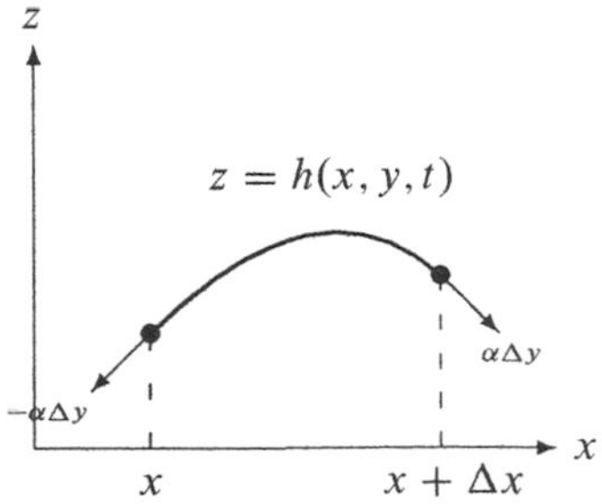

The total vertical force on a small patch is determined by projecting the nearly horizontal pull of surface tension on the vertical (here downward). By Newton's Third Law, the pressure in the water just below must be larger than above.

Figure 25.5. Capillary waves in front of ordinary gravity waves at small depth. (Source: Fabrice Neyret. With permission.)

Deep-water capillary waves

We saw in Section 5.1 that the relative influence of surface tension and gravity in a liquid/air interface is characterized by a length scale, called the *capillary length*, which is $L_c = \sqrt{\alpha/\rho_0 g_0} \approx 2.7$ mm for the water/air interface. Surface tension only plays a role for waves of this length scale and below.

Due to the smallness of the capillary length, the water may normally be assumed to be infinitely deep. The velocity potential is in that case $\Psi = Ae^{kz}\sin(kx - \omega t)$, and from the kinematic boundary condition (25.15) and the modified dynamic boundary condition,

$$g_0 h + \frac{\Delta p}{\rho_0} = -\frac{\partial \Psi}{\partial t} \qquad \text{for } z = h, \tag{25.36}$$

we obtain, in the usual way,

$$A = a\frac{\omega}{k} = a\frac{\rho_0 g_0 + \alpha k^2}{\rho_0 \omega}. \tag{25.37}$$

The second equality determines as always the frequency,

$$\boxed{\omega = \sqrt{g_0 k + \frac{\alpha}{\rho_0} k^3} = \sqrt{g_0 k\left(1 + k^2 L_c^2\right)}.} \tag{25.38}$$

This dispersion law agrees very well with experiments [CAL95]. As foreseen, surface tension collaborates with gravity and becomes more important than gravity for $kL_c \gtrsim 1$ or $\lambda \lesssim \lambda_c = 2\pi L_c$, which has the value $\lambda_c = 1.7$ cm in water.

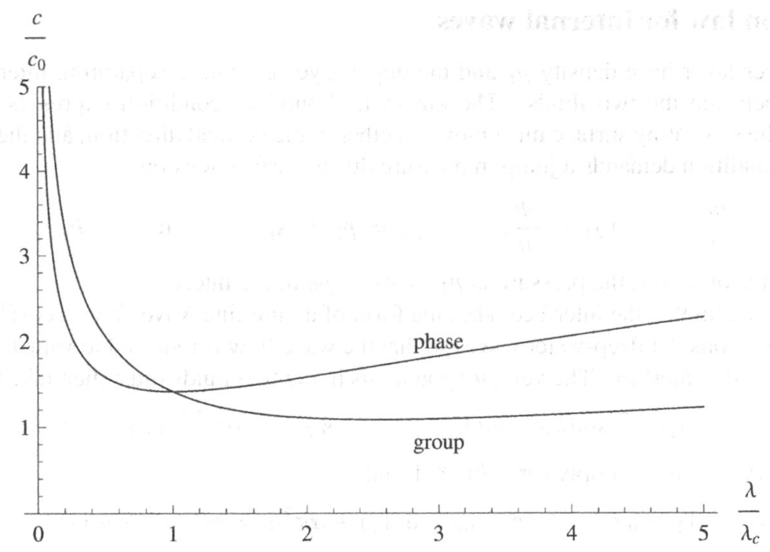

Figure 25.6. Phase and group velocities for deep-water capillary gravity waves as functions of the wavelength. The velocities are normalized by $c_0 = \sqrt{g_0 L_c}$ and the wavelength by $\lambda_c = 2\pi L_c$. In water we have $\sqrt{g_0 L_c} \approx 16$ cm s^{-1} and $\lambda_c \approx 1.7$ cm. Phase and group velocities cross each other at $\lambda = \lambda_c$, where the phase velocity is minimal. The group velocity has a minimum at $\lambda \approx 2.54\,\lambda_c$.

The phase and group velocities become, respectively,

$$c = \sqrt{\frac{g_0}{k}\left(1 + k^2 L_c^2\right)}, \qquad c_g = \frac{1}{2} c \frac{1 + 3k^2 L_c^2}{1 + k^2 L_c^2}, \tag{25.39}$$

and are plotted for water in Figure 25.6. Note that the phase velocity has a minimum for $kL_c = 1$, or $\lambda = \lambda_c$, where the group velocity also equals the phase velocity (see Problem 25.2). Capillary wave crests normally move ahead of normal gravity waves (see Figure 25.5). The minimum of the group velocity is found at a somewhat larger wavelength, $\lambda/\lambda_c = 2.54\ldots$.

For very small wavelengths, $\lambda \ll \lambda_c$ or $kL_c \gg 1$, surface tension dominates completely, and we find the dispersion law for purely capillary waves,

$$\omega = \sqrt{\frac{\alpha k^3}{\rho_0}}, \qquad c = \sqrt{\frac{\alpha k}{\rho_0}}, \qquad c_g = \frac{3}{2} c. \tag{25.40}$$

In purely capillary waves the phase velocity is only 2/3 of the group velocity, so the wave crests appear to move backward inside a capillary wave packet!

25.5 Internal waves

In the ocean a heavier saline layer of water may often be found below a lighter more brackish layer, allowing so-called *internal waves* to arise at the interface. Even if the difference in density between the fluids is small, the equilibrium interface will always be horizontal with the lighter liquid situated above the heavier (as discussed previously in section 4.1). Were it somehow possible to invert the ocean so that the lighter fluid came to lie below the heavier, instability would surely arise, and the liquids would after some time find their way back to their "natural" order. As we shall see, surface tension can in fact stabilize the inverted situation in sufficiently small containers.

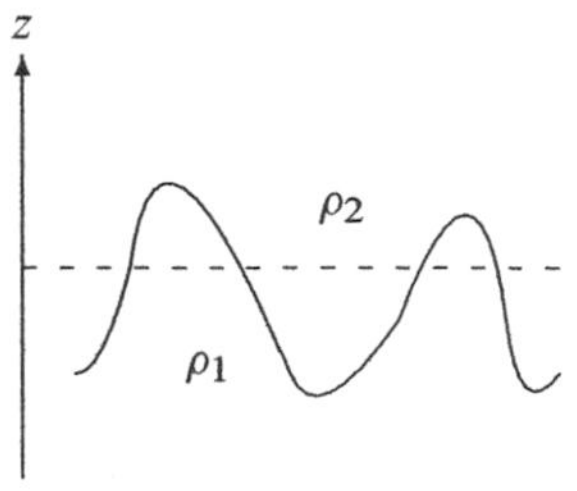

Internal waves at an interface with a heavier liquid below and a lighter above ($\rho_1 > \rho_2$).

Dispersion law for internal waves

Let the lower layer have density ρ_1 and the upper layer ρ_2 with a separating interface $z = h(x, y, t)$ between the two fluids. The kinematic boundary condition expresses that both fluids and the separating surface must move together in the vertical direction, and the dynamic boundary condition demands a jump in pressure due to surface tension,

$$v_{1z} = \frac{\partial h}{\partial t}, \qquad v_{2z} = \frac{\partial h}{\partial t}, \qquad p_1 = p_2 + \Delta p, \qquad \text{for } z = h. \tag{25.41}$$

In the absence of waves, the pressure is $p_1 = p_2 = p_0$ in the interface.

Suppose again that the interface takes the form of a pure line wave, $h = a\cos(kx - \omega t)$. We shall only consider deep-water waves so that the wave flow is required to vanish far below and far above the interface. The velocity potentials in the two fluids must then take the form

$$\Psi_1 = A_1 e^{kz}\sin(kx - \omega t), \qquad \Psi_2 = A_2 e^{-kz}\sin(kx - \omega t).$$

The boundary conditions imply for $|kh| \ll 1$ that

$$kA_1 = -kA_2 = a\,\omega, \qquad \rho_1(g_0 a - \omega A_1) + \alpha k^2 a = \rho_2(g_0 a - \omega A_2). \tag{25.42}$$

Solving these we find that

$$A_1 = -A_2 = a\frac{\omega}{k} = \frac{a}{\omega}\frac{g_0(\rho_1 - \rho_2) + \alpha k^2}{\rho_1 + \rho_2}. \tag{25.43}$$

From the last equality we obtain the *dispersion law for deep-water internal waves*,

$$\boxed{\omega = \sqrt{\frac{g_0 k(\rho_1 - \rho_2) + \alpha k^3}{\rho_1 + \rho_2}}.} \tag{25.44}$$

When the densities are nearly equal, $\rho_2 \lesssim \rho_1$, the internal waves have generally much lower frequencies (and velocities) than waves of the same wavelength at the surface.

The capillary length is as before defined as the length scale where the gravity contribution to the frequency is of the same magnitude as the surface tension contribution,

$$L_c = \sqrt{\frac{\alpha}{|\rho_1 - \rho_2|\, g_0}}. \tag{25.45}$$

It diverges when the densities become (nearly) equal. In this limit, gravity plays very little role, and the internal waves have become almost purely capillary waves, described by (25.40) with $\rho_0 = 2\rho_1 = 2\rho_2$.

> **Example 25.3 [Brackish/saline interface in the sea]:** Let a brackish surface layer lie above a saline layer with 4% higher density. The capillary wavelength for internal waves becomes $\lambda_c = 2\pi L_c = 8.5$ cm. A wave with this wavelength has period $\tau = 1.2$ s and the crests move with phase velocity $c = 7.2 \text{ cm s}^{-1}$.

The Rayleigh–Taylor instability

If the heavier fluid lies below the lighter, $\rho_1 > \rho_2$, the frequency ω of an internal wave is always real, but if the heavier fluid is on top, the dispersion law may be written as

$$\omega = \sqrt{g_0 \frac{\rho_2 - \rho_1}{\rho_1 + \rho_2} k(k^2 L_c^2 - 1)}. \tag{25.46}$$

The argument of the square root will be negative for $kL_c < 1$, or $\lambda > \lambda_c = 2\pi L_c$. In that case ω becomes imaginary, and the otherwise sinusoidal form is replaced by an exponential growth $e^{|\omega| t}$ in time, signaling an instability called the *Rayleigh–Taylor instability*.

In an infinitely extended ocean, there is room for waves with wavelengths of any size, and the inverted situation will always be unstable. In a finite container there is an upper limit to the allowed wavelengths because the boundary conditions require the horizontal velocities to vanish at the vertical walls surrounding the fluids. Any flow in a finite box-shaped container of horizontal length L must obey the boundary conditions $v_x = 0$ for both $x = 0$ and $x = L$. Linear Euler flow in a box can, like the other flows we are studying here, always be resolved into a superposition of standing waves with horizontal velocity $v_x \sim \sin kx \cos \omega t$. The boundary conditions now select the allowed wavenumbers to be $k = n\pi/L$, where $n = 1, 2, \ldots$ is an arbitrary positive integer. For $n = 1$ we obtain the largest wavelength, $\lambda = 2\pi/k = 2L$, and this shows that as long as

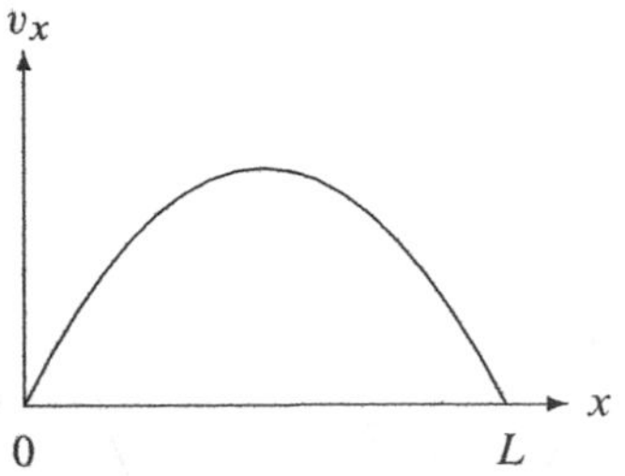

Any flow in a box of finite length must always stop at the container walls.

$$L < \frac{1}{2}\lambda_c = \pi L_c = \pi\sqrt{\frac{\alpha}{(\rho_2 - \rho_1)g_0}}, \qquad (25.47)$$

unstable wave modes with $\lambda > \lambda_c$ cannot occur. Thus, if you invert a container with horizontal size smaller than half the capillary wavelength, the heavier liquid will remain stably on top of the lighter.

Example 25.4 [Home experiment]: Air against water has as we have seen before a capillary wavelength of $\lambda_c = 1.7$ cm, so the inverted situation is stable for $L < 0.85$ cm. Try it yourself with any glass or metal tube of, for example, 5 mm diameter. Block one end of the tube with your finger and fill it with water. When the tube is inverted, the water does not fall out.

* The Kelvin–Helmholtz instability

Layers of inviscid fluids may slide past each other with a finite slip-velocity. When a small disturbance arises in the interface between the fluids, it will so to speak "get in the way" of the smooth flow, leading us to expect instability at a sufficiently high slip-velocity. In this case we assume that the heavier fluid is below the lighter, that is, $\rho_1 > \rho_2$.

Suppose the upper layer is moving with velocity U in the rest frame of the lower layer. Taking into account the advection with velocity U, the vertical velocity and the pressure in the fluid just above the interface become (in the rest frame of the lower layer),

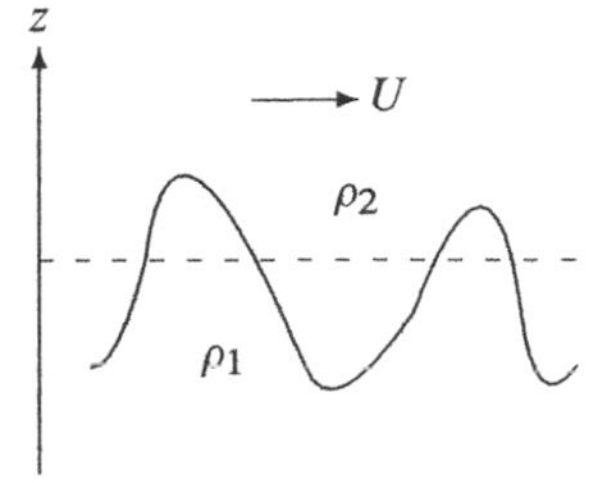

The lighter layer above the interface moves horizontally with velocity U relative to the lower.

$$v_{2z} = \frac{\partial h}{\partial t} + U\nabla_x h, \qquad p_2 = p_0 - \rho_0\left(g_0 h + \frac{\partial \Psi_2}{\partial t} + U\nabla_x\Psi_2\right). \qquad (25.48)$$

Putting it all together, the boundary conditions (25.41) become

$$kA_1 = a\omega, \qquad -kA_2 = a(\omega - kU),$$

and

$$\rho_1(g_0 a - \omega A_1) + \alpha k^2 a = \rho_2(g_0 a - (\omega - kU)A_2).$$

Combined, they lead to a quadratic equation for the frequency

$$(\rho_1 + \rho_2)\omega^2 - 2\rho_2 kU\omega + \rho_2 k^2 U^2 = k((\rho_1 - \rho_2)g_0 + \alpha k^2). \qquad (25.49)$$

Given the wavenumber k, the roots are real (and the system stable) for

$$U^2 \le \left(\frac{1}{\rho_1} + \frac{1}{\rho_2}\right)\frac{(\rho_1 - \rho_2)g_0 + \alpha k^2}{k}. \qquad (25.50)$$

For $\rho_1 > \rho_2$, the right-hand side has an absolute minimum when $kL_c = 1$, where L_c is the capillary length for internal waves (25.45). Selecting the minimum of the right-hand side by setting $k = 1/L_c$, the condition for absolute stability becomes

$$U \le U_c = \sqrt{2g_0 L_c \left(\frac{\rho_1}{\rho_2} - \frac{\rho_2}{\rho_1}\right)}. \tag{25.51}$$

For air flowing over water, the critical velocity is $U_c = 6.7 \text{ m s}^{-1}$.

The general stability condition (25.50) may now be written as

$$\frac{U}{U_c} \le \sqrt{\frac{1}{2}\left(\frac{\lambda}{\lambda_c} + \frac{\lambda_c}{\lambda}\right)}. \tag{25.52}$$

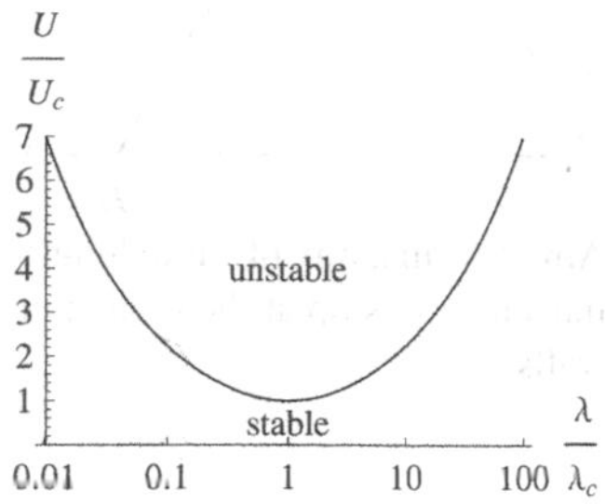

Plot of U/U_c as a function of λ/λ_c. For the water/air interface, the capillary wavelength is $\lambda_c = 1.7$ cm and the critical velocity is $U_c = 6.7 \text{ m s}^{-1}$. For a given velocity U, one can read off the range of unstable wavelengths from this figure.

In the margin figure the separatrix corresponding to equality is plotted as a function of λ/λ_c. For $U > U_c$ there will be a range of wavelengths around the capillary wavelength $\lambda = \lambda_c$ for which small disturbances will diverge exponentially with time. This is the *Kelvin–Helmholtz* instability, which permits us at least in principle to understand how a steadily streaming, sufficiently strong wind is able to generate waves from tiny disturbances.

What actually happens to the unstable waves with their exponentially growing amplitudes, for example how they grow into the larger waves created by a storm, cannot be predicted from linear theory. It is, however, possible to say something about the statistics of wind-generated ocean waves without going into the nonlinear theory (see Section 25.7).

25.6 Global wave properties

Surface waves contain mass, momentum, and energy, and the movement of fluid also transports these quantities around. To calculate the flux of these quantities for gravity waves at finite depth, we shall use the flat-bottom fields (25.29), remembering that they represent only the linear approximation to wave motion. It is for this reason imperative systematically to keep only the leading order in the amplitude.

Total energy

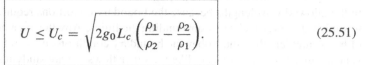

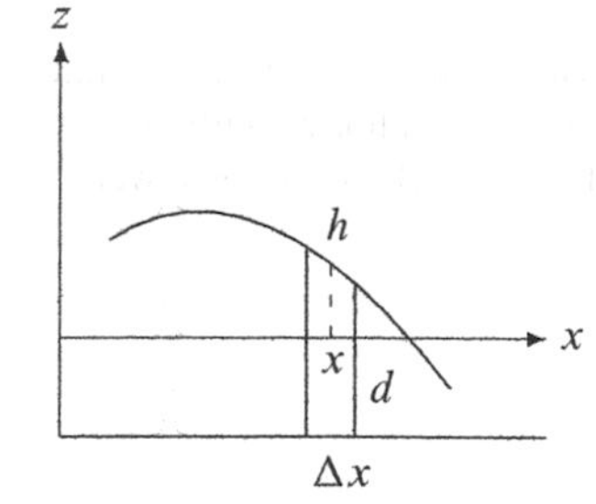

A thin water column of "footprint" area $A = L\Delta x$ and height h in a sea of depth d.

The first goal is to calculate the total energy in a wave. Consider a thin column of water width Δx along x and length Δy along y, so that its "footprint" area is $A = \Delta x \Delta y$. Relative to the static water level $z = 0$, its potential energy is

$$\mathcal{V} = \int_V \rho_0 g_0 z \, dV = \int_0^h \rho_0 g_0 z \, A dz = \tfrac{1}{2}\rho_0 g_0 h^2 A. \tag{25.53}$$

The potential energy is always positive, and rises and falls in tune with the square of the wave height. Note that this is an exact result, also valid when the amplitude is large. In the linearized approximation where $h = a\cos(kx - \omega t)$, we find the average of the potential energy over a full period,

$$\langle \mathcal{V} \rangle = \frac{1}{\tau}\int_0^\tau \mathcal{V} \, dt = \tfrac{1}{4}\rho_0 g_0 a^2 A, \tag{25.54}$$

because the average over a squared cosine is $1/2$.

The kinetic energy of the water in the column becomes

$$\mathcal{K} = \int_V \tfrac{1}{2}\rho_0 \boldsymbol{v}^2 \, dV = \int_{-d}^h \tfrac{1}{2}\rho_0 \left(v_x^2 + v_z^2\right) A dz \approx \int_{-d}^0 \tfrac{1}{2}\rho_0 \left(v_x^2 + v_z^2\right) A dz. \tag{25.55}$$

In the last step we have again assumed that the amplitude is small, $|h| \ll d, \lambda$, and replaced h by 0 in the upper limit of the integral.

Inserting the explicit gravity wave solution (25.29), we obtain the time average of the integrand,

$$\langle v_x^2 + v_z^2 \rangle = \frac{1}{2}\left(\frac{a\omega}{\sinh kd}\right)^2 \left(\cosh^2 k(z+d) + \sinh^2 k(z+d)\right).$$

Finally, using the relation $\cosh^2\phi + \sinh^2\phi = \cosh 2\phi$, the integral over z can be carried out, and we find after using the dispersion relation (25.27),

$$\langle \mathcal{K} \rangle = \frac{1}{4}\rho_0 A \left(\frac{a\omega}{\sinh kd}\right)^2 \frac{\sinh 2kd}{2k} = \frac{1}{4}\rho_0 g_0 a^2 A. \tag{25.56}$$

As expected in the general estimates of Section 25.1, we have $\langle \mathcal{K} \rangle = \langle \mathcal{V} \rangle$.

The average of the total mechanical energy over a full period thus becomes

$$\boxed{\langle \mathcal{E} \rangle = \langle \mathcal{K} \rangle + \langle \mathcal{V} \rangle = \tfrac{1}{2}\rho_0 g_0 a^2 A.} \tag{25.57}$$

Surprisingly, the average energy per unit of surface area, $\langle \mathcal{E} \rangle / A$, only depends on the amplitude and not on the depth or wavelength. A small-amplitude harmonic surface wave in water with amplitude $a = 1$ m contains a wave energy per unit of surface area of about $\langle \mathcal{E} \rangle / A \approx 5000\ \mathrm{J\ m^{-2}}$.

Mass flux

The mass flux through a vertical cut S through the wave of length L in the y-direction is,

$$\dot{M} = \int_S \rho_0 \boldsymbol{v} \cdot d\boldsymbol{S} = \int_{-d}^{h} \rho_0 v_x\, L dz \approx \int_{-d}^{0} \rho_0 v_x\, L dz. \tag{25.58}$$

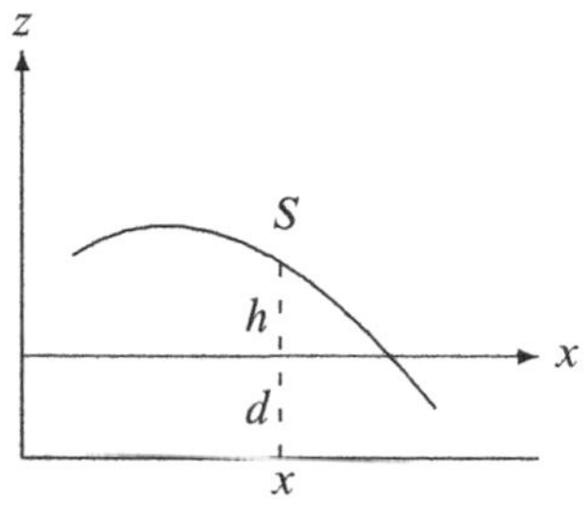

A vertical cut S through the wave (dashed). Its length is L in the y-direction.

The velocity field is of first order in a and for consistency we have in the last step replaced h by 0. Carrying out the integral by means of (25.29a), we find

$$\dot{M} = \rho_0 a c L \cos(kx - \omega t). \tag{25.59}$$

The time average of the flux over a full period evidently vanishes, $\langle \dot{M} \rangle = 0$, showing that the water merely sloshes back and forth. The actual volume of water that is displaced during half a period of forward motion is $Q = aL\lambda/\pi$. The amount of sloshing water in a deep-water wave with amplitude 1 m and wavelength 100 m is nearly 32 m^3 per meter of transverse crest length.

Momentum flux

The total horizontal momentum transport through the vertical cut is, similarly,

$$\dot{\mathcal{P}}_x = \int_S \rho_0 v_x \boldsymbol{v} \cdot d\boldsymbol{S} = \int_{-d}^{h} \rho_0 v_x^2\, L dz \approx \int_{-d}^{0} \rho_0 v_x^2\, L dz. \tag{25.60}$$

Averaging over a full period, we get

$$\langle v_x^2 \rangle = \frac{1}{2}\left(\frac{a\omega}{\sinh kd}\right)^2 \cosh^2 k(z+d), \tag{25.61}$$

and find upon integration,

$$\langle \dot{\mathcal{P}}_x \rangle = \tfrac{1}{2}\rho_0 g_0 a^2 L \frac{c_g}{c}, \tag{25.62}$$

where c and c_g are the phase and group velocities (25.28), respectively.

The average momentum flux is a measure of the average force of a wave hitting an obstacle, although the true force is strongly complicated by the shape of the obstacle and by reflected waves [Mei 1989].

Example 25.5: Wading near the shore in water to your waist, you are hit by harmonic waves of amplitude $a = 10$ cm. If you are $L = 50$ cm wide, a deep-water wave with $c_g = \frac{1}{2}c$ will act on you with an average landward force of $\frac{1}{4}\rho_0 g_0 L a^2 = 12$ N. That will not topple you, but a wave with five times higher amplitude easily could.

Wave power

Waves are able to do work and many ingenious schemes have been thought up for the exploitation of wave power. The power of a wave may be calculated from the rate of work performed by the pressure on a vertical cut S through the wave,

$$\dot{W} = \int_S p\,\boldsymbol{v} \cdot d\boldsymbol{S} = \int_{-d}^{h} p\, v_x\, L dz \approx \int_{-d}^{0} p\, v_x\, L dz. \tag{25.63}$$

Writing the pressure (25.29d) as $p = p_0 - \rho_0 g_0 z + \rho_0 c v_x$ and using $\langle v_x \rangle = 0$, we find the average $\langle p\, v_x \rangle = \rho_0 c \langle v_x^2 \rangle$, and thus the average power,

$$\boxed{\langle \dot{W} \rangle = \langle \dot{\mathcal{P}}_x \rangle c = \tfrac{1}{2}\rho_0 g_0 a^2 L c_g,} \tag{25.64}$$

where the group velocity is given by (25.28). Comparing with (25.57) we see that that the factor in front of c_g is the average wave energy $d\langle \mathcal{E} \rangle / dx$ per unit of length in the propagation direction. Interpreting $\langle \dot{W} \rangle$ as the average energy flux along x, this result confirms that the energy in a harmonic wave indeed propagates with the group velocity.

Example 25.6 [Tsunami]: The tsunami of Example 25.2 with $\lambda = 500$ km, $c = c_g = 200$ m s^{-1}, and $a = 1$ m carries an average power per unit of transverse length of $\langle \dot{W} \rangle / L \approx 1$ MW m^{-1}. Since it takes only 10 joule to lift one kilogram through one meter, this energy corresponds to lifting 100 metric ton per second. A Tsunami can really wreak havoc at a coast!

Rise of a shallow-water swell

When a small-amplitude wave train (see Figure 25.7) approaches a gently sloping beach, the phase and group velocity of the waves will decrease with diminishing water depth according to the shallow-water expression (24.37). It is well-known that the amplitude grows at the same time, but how fast does it grow? And what about the wavelength?

Suppose the waves roll steadily in from afar with constant period τ. In the steady situation, wave crests cannot accumulate anywhere, so that the same number of waves must hit the coast in a given time interval as roll in from far away, implying that the period τ between successive wave crests must be the same everywhere, independently of the bottom depth. The constancy of τ in turn implies that the wavelength must scale with depth like the phase velocity, $\lambda = c\tau \sim \sqrt{d}$. Similarly, energy cannot accumulate anywhere in the steady situation, so that the average wave power $\langle \dot{W} \rangle \sim a^2 c_g \sim a^2 \sqrt{d}$ must be independent of d, implying that $a \sim d^{-1/4}$ (Greens Law, see [Lamb 1993, p. 273]). Altogether, these considerations show that a shallow-water wave starting out at depth d_0 with amplitude a_0 and wavelength λ_0, will develop according to

$$a = a_0 \left(\frac{d_0}{d} \right)^{1/4}, \qquad \lambda = \lambda_0 \sqrt{\frac{d}{d_0}}, \tag{25.65}$$

when the depth is reduced to d (see Figure 25.7).

Figure 25.7. Sketch of gravity waves coming in from the right toward a coast with a gently sloping flat bottom. The waves increase in amplitude as they decrease in wavelength. When the amplitude becomes comparable to the depth (on the left), non-linear effects take over and cause the waves to break.

These expressions are only valid for very gently sloping beaches that may be viewed as locally flat so that the waves propagate according to the shallow-water expressions everywhere. When this is not fulfilled, the bottom boundary condition must take into account the shape of the actual beach and a much more complicated formalism ensues [Stoker 1992].

Example 25.7 [Ocean swell]: Typical wind-generated oceanic swells have wavelengths of $\lambda_0 = 150$ m, velocity $c_0 = 15\ \mathrm{m\,s^{-1}}$, period $\tau = 10$ s in deep water, and perhaps an amplitude of $a_0 = 1$ m. When the depth decreases to about $d_0 \approx \lambda_0/2\pi \approx 25$ m, the wave becomes a shallow-water wave. At a depth of $d \approx 2$ m, the wave characteristics are $a \approx 2$ m, $\lambda \approx 43$ m, $c = 4.5\ \mathrm{m\,s^{-1}}$. At this point, the amplitude has become equal to the depth and strong non-linear effects will take over the wave, so that it breaks and produces a surf.

Example 25.8 [Tsunami]: The tsunami discussed in Example 25.2 with $\lambda_0 = 500$ km, $\tau = 2500$ s, $a_0 = 1$ m, $c_0 = 200\ \mathrm{m\,s^{-1}}$ at a water depth $d_0 = 4000$ m rises to $a = 4.5$ m, $\lambda = 25$ km, and $c = 10\ \mathrm{m\,s^{-1}}$ at a depth of $d = 10$ m. Beyond this point strong non-linear effects set in and the tsunami will break. At the coast this particular tsunami appears as a sequence of powerful "tidal surges" arriving every 42 minutes. Their detailed behavior depends on the local topography of the coast.

Rate of viscous dissipation

Surface waves contain both potential and kinetic energy, and this energy is attenuated by several effects. First, there is viscous attenuation due to internal friction in the bulk of the fluid. Second, there is attenuation from viscous and turbulent losses in the boundary layer that necessarily forms near the bottom. Finally there is dissipation due to deviations of the value of surface tension from its equilibrium value, an effect that for example plays a role when oil is poured on troubled waters. Here we shall focus only on viscous attenuation in the bulk.

The rate of mechanical energy loss due to viscous dissipation in an incompressible liquid is given by (15.24), which for a thin vertical column of footprint area $A = \Delta x \Delta y$, simplifies to

$$\dot{W}_{\text{int}} = 2\eta \int_{-d}^{h} \sum_{ij} v_{ij}^2\, A dz \approx 2\eta \int_{-d}^{0} \sum_{ij} v_{ij}^2\, A dz, \tag{25.66}$$

where $v_{ij} = \frac{1}{2}(\nabla_i v_j + \nabla_j v_i)$ is the strain rate and η is the viscosity. Evaluating it for a standard harmonic wave propagating along x, the integrand may be recast as

$$\sum_{ij} v_{ij}^2 = (\nabla_x v_x)^2 + (\nabla_z v_z)^2 + \frac{1}{2}(\nabla_x v_z + \nabla_z v_x)^2 = 2(\nabla_x v_x)^2 + 2(\nabla_x v_z)^2.$$

In the last step er have used mass conservation $\nabla_z v_z = -\nabla_x v_x$ and irrotationality $\nabla_z v_x = \nabla_x v_z$. For a harmonic wave (25.29), we may replace ∇_x by k in the time average, so that it

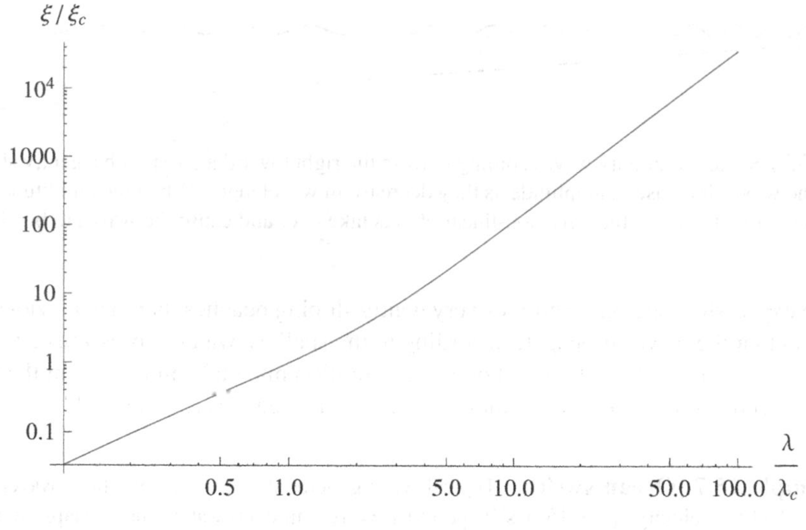

Figure 25.8. The viscous amplitude attenuation length ξ/ξ_c plotted as a function of wavelength λ/λ_c, where $\lambda_c = 2\pi L_c$ and $\xi_c = (L_c^2/\nu)\sqrt{g_0 L_c/2}$ (see Problem 25.7). In water the viscosity is $\nu = 10^{-6}\ \mathrm{m^2\,s^{-1}}$ and the surface tension $\alpha = 0.073\ \mathrm{N\,m^{-1}}$, leading to $\lambda_c = 1.7$ cm and $\xi_c = 86$ cm. Evidently, viscous attenuation only plays a role for small wavelengths.

becomes

$$\left\langle \sum_{ij} v_{ij}^2 \right\rangle = 2k^2 \left\langle v_x^2 + v_z^2 \right\rangle. \tag{25.67}$$

This is proportional to the integrand in the kinetic energy (25.55) and taking over the result (25.56) this leads to

$$\left\langle \dot{W}_{\text{int}} \right\rangle = 8\nu k^2 \left\langle \mathcal{K} \right\rangle, \tag{25.68}$$

where $\nu = \eta/\rho_0$ is the kinematic viscosity.

Relative to the average of the total energy $\langle \mathcal{E} \rangle = 2\langle \mathcal{K} \rangle$, the rate of dissipation finally becomes

$$\boxed{\frac{\left\langle \dot{W}_{\text{int}} \right\rangle}{\langle \mathcal{E} \rangle} = 4\nu k^2,} \tag{25.69}$$

where ν is the kinematic viscosity. The dissipative energy loss grows quadratically with the wavenumber and is most important for small wavelengths, that is, capillary waves.

The loss of energy over a wave period $\tau = 2\pi/\omega$ is $\dot{W}_{\text{int}}\,\tau$ and becomes, relative to the average energy of the wave (25.57),

$$\frac{\left\langle \dot{W}_{\text{int}} \right\rangle \tau}{\langle \mathcal{E} \rangle} = 4\nu k^2 \tau = \frac{32\pi^3 \nu}{\omega \lambda^2}. \tag{25.70}$$

The condition for the calculation to be valid is that the relative attenuation be small, or $\omega\lambda^2 \gg 32\pi^3\nu$. In deep water with $\nu \approx 10^{-6}\ \mathrm{m^2\,s^{-1}}$, $\alpha = 0.073\ \mathrm{N\,m^{-1}}$, and ω given by (25.38), this is fulfilled for $\lambda \gg \lambda_{\min} \approx 54\ \mu$m, so that the condition should be satisfied under any ordinary circumstances.

Energy and amplitude attenuation coefficients

The energy propagates, as we have argued on page 423, with the group velocity c_g rather than the phase velocity c. Dividing the energy dissipation rate (25.69) by c_g from (25.39), we obtain the spatial *energy attenuation coefficient* 2κ, defined as the fractional loss of energy per unit of length,

$$\boxed{2\kappa \equiv \frac{\langle \dot{W}_{\text{int}} \rangle}{\langle \mathcal{E} \rangle \, c_g} = \frac{4\nu k^2}{c_g}.} \tag{25.71}$$

Since the energy is quadratic in the amplitude, the energy attenuation coefficient is twice the spatial *amplitude attenuation coefficient* κ. The inverse amplitude attenuation coefficient

$$\xi = \frac{1}{\kappa} = \frac{c_g}{2\nu k^2} \tag{25.72}$$

is called the *amplitude attenuation length* and indicates the distance over which the surface wave amplitude falls to about $e^{-1} \approx 37\%$ of its initial value. The amplitude attenuation length is plotted in Figure 25.8 as a function of the wavelength (see Problem 25.7).

Example 25.9: In water for $\lambda = 1$ m, one finds $\xi \approx 8$ km whereas for $\lambda = \lambda_c \approx 1.7$ cm, one gets $\xi = \xi_c \approx 86$ cm. A raindrop hitting a lake surface thus creates a disturbance that dies out after propagating through a meter or less whereas a big object, like the human body, will make waves with much longer wavelength that continues essentially unattenuated right across the lake. In both these cases, however, the amplitude of the ring-shaped surface waves will also diminish for purely geometric reasons.

* 25.7 Statistics of wind-generated ocean waves

Waves may arise spontaneously from tiny perturbations at the wind/water interface when the wind speed surpasses the Kelvin–Helmholtz instability threshold (page 433). The continued action of the wind and non-linear wave interactions raise the waves further, until a kind of dynamic equilibrium is reached in which the surface may be viewed as a statistical ensemble of harmonic waves with a wide spectrum of periods, wavelengths, and amplitudes. Even if we do not understand the mechanism at play, it is nevertheless possible to draw some quite general conclusions about the wave statistics and compare with observations.

The power spectrum

A ship or buoy bobbing at a fixed position (x, y) of the ocean surface reflects the local surface height, $h(t) = h(x, y, t)$, as a function of time. The variations in surface height may be determined by many different techniques, for example based on accelerometers, radar, or satellites. While the wind blows steadily, a long record of measurements can be collected. The underlying wave structure of the surface creates strong correlations between successive measurements of the local height. Short waves are carried on top of larger waves and so on.

Whereas the average surface height vanishes, $\langle h \rangle = 0$, the average of its square $\langle h^2 \rangle$ is non-vanishing. Using discrete Fourier analysis, the value of $\langle h^2 \rangle$ can be resolved according to frequency, such that $S(\omega)\, d\omega$ represents the contribution from the interval $\omega|\omega + d\omega$ and

$$\langle h^2 \rangle = \int_0^\infty S(\omega)\, d\omega. \tag{25.73}$$

Since the energy is proportional to the square of the amplitude, $S(\omega)$ is called the *power spectrum* of the waves. From the empirical statistics of wave heights through a range of frequencies, the power spectrum can be obtained.

The canonical form of the spectrum

The empirical spectra have a single peak at a certain frequency ω_p with a long tail toward higher frequencies and a sharp drop-off below. The position of the peak depends strongly on the wind velocity U whereas the high-frequency tail appears to be the same for all U (see Figure 25.9). We shall now see that it is possible to understand the general form of the spectrum using statistical mechanics.

The wind speed U sets the level of excitation of the ocean surface at large, but cannot control what happens locally so that the local wave energy E in a small neighborhood of a fixed point in principle can take any value. But because the energy has to come from the huge reservoir of wave energy in the surrounding ocean, the probability that the local energy actually has the value E is suppressed by a canonical Boltzmann factor $e^{-\beta E}$, where the "inverse temperature" β is a measure of the level of excitation of the ocean. The power spectrum may accordingly be written as

$$S \sim e^{-\beta E} \frac{dE}{d\omega}, \tag{25.74}$$

where $dE/d\omega$ is the energy per unit of frequency.

Provided the nonlinearity is not excessive, the local energy is proportional to the square of the amplitude $E \sim a^2$. Since the local energy should depend on U, the amplitude must take the form $a \sim g_0/\omega^2$, because that is the only length scale that may be constructed from g_0 and ω. Taking $E \sim g_0^2/\omega^4$ and normalizing the frequency in the exponent by g_0/U, we get the following model for the spectrum (after redefining β),

$$\boxed{S(\omega) = \alpha \frac{g_0^2}{\omega^5} \exp\left[-\beta \left(\frac{g_0}{U\omega}\right)^4\right],} \tag{25.75}$$

where α and β are dimensionless parameters. This spectrum has indeed a sharp low-frequency cut-off, a single peak, and a high-frequency tail proportional to ω^{-5} that is independent of U (assuming that α is independent of U).

The root-mean-square amplitude and the peak frequency are easily evaluated,

$$\sqrt{\langle h^2 \rangle} = \sqrt{\frac{\alpha}{\beta}} \frac{U^2}{2g_0}, \qquad \omega_p = \left(\frac{4\beta}{5}\right)^{1/4} \frac{g_0}{U}. \tag{25.76}$$

Note that these quantities are scaled by the only possible combinations of U and g_0 that have the right dimensions.

The Pierson–Moskowitz empirical spectrum

Assuming that the statistical equilibrium is the same everywhere on the ocean surface, the dimensionless parameters α and β can only depend on U and g_0, but since there is no dimensionless combination of U and g_0, both α and β must be constants, independent of U. Pierson and Moskowitz [PM64, Pie64, Mos64] fitted empirical spectra for a range of wind velocities and found the values

$$\alpha = 8.1 \times 10^{-3}, \qquad \beta = 0.74. \tag{25.77}$$

The actual spectrum is quite sensitive to the height at which the wind speed is determined because of air turbulence close to the surface. In the data used in the fit, the wind speed was measured about 20 m above the average surface level.

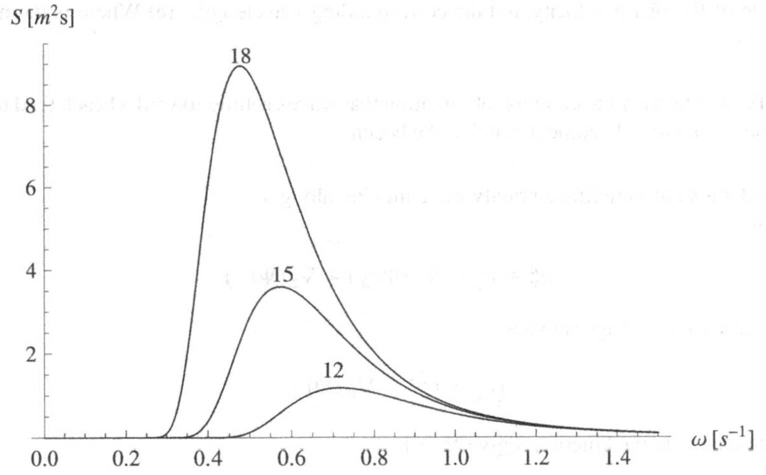

Figure 25.9. Pierson–Moskowitz wave spectrum S as a function of circular frequency ω for three different wind speeds $U = 12, 15, 18\ \mathrm{m\,s^{-1}}$. Note that the high-frequency tail is independent of U.

The root-mean-square amplitude and spectrum peak position are then

$$\sqrt{\langle h^2 \rangle} = 0.052\,\frac{U^2}{g_0}, \qquad \omega_p = 0.88\,\frac{g_0}{U}. \tag{25.78}$$

In Figure 25.9 the Pierson–Moskowitz spectrum is shown for three different wind speeds. One notes how the high-frequency tails coincide and how the low-frequency cut-off becomes sharper as the wind speed increases.

Example 25.10: At a wind speed of $U = 15\ \mathrm{m\ s^{-1}}$, the period at peak is about $\tau_p = 11$ s, corresponding to a deep-water wavelength of $\lambda_p = 186$ m and a phase velocity of $c_p = 17\ \mathrm{m\ s^{-1}}$. The root-mean-square amplitude of the waves raised by this wind is $a = 1.2$ m. As a measure of the nonlinearity, one may take $k_p a = 0.04$, where $k_p = \omega_p^2/g_0 = 0.03\ \mathrm{m^{-1}}$ is the peak wavenumber, determined from the deep-water dispersion relation (25.21). According to this estimate, the average nonlinearity at play at this wind speed is indeed quite small.

Fetch

The assumed dynamic equilibrium of the ocean surface takes a long time to develop, and more so the higher the wind speed. A wind with velocity U lasting time t must have traveled over an upwind distance $L = Ut$, called the *fetch*. Even for a moderate wind at $15\ \mathrm{m\ s^{-1}}$, the statistical equilibrium of the waves takes about 8 h to develop, so the fetch is about 500 km. Empirically the fetch grows roughly like U^3 (see the margin figure), so that a strong wind at $30\ \mathrm{m\ s^{-1}}$ has a fetch of 4,000 km. Such winds rarely manage to fully develop the equilibrium power spectrum of the sea because of the finite distance to the lee shore and the finite size of weather systems.

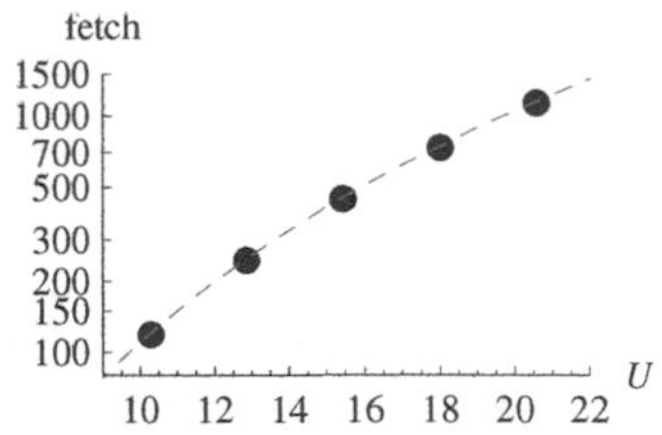

Logarithmic plot of Pierson-Moskowitz data on fetch, fitted to a curve of the form AU^α with $A = 0.064$ and $\alpha = 3.24$. The velocity is the wind speed in units of meter per second and the fetch is in units of kilometers.

Problems

25.1 Consider a liquid in a rectangular container with horizontal side length's a and b and vertical depth d. Determine the most general harmonic standing gravity wave in the container.

25.2 **(a)** Determine where the phase and group velocities (25.39) for deep-water capillary waves cross each other, and determine the common value at the crossing (use $\alpha = 0.073\ \mathrm{N\ m^{-1}}$). **(b)** Determine the minimal value of the phase velocity and the corresponding wavelength. **(c)** Where is the minimum of the group velocity?

25.3 Justify qualitatively the common observation that waves rolling toward a beach tend to straighten out so that the wave crests become parallel to the beach.

25.4 Consider a small-amplitude gravity wave moving along x.
(a) Show that

$$v_x^2 + v_z^2 = \nabla_x(\Psi v_x) + \nabla_z(\Psi v_z).$$

(b) Show that the time average satisfies

$$\left\langle v_x^2 + v_z^2 \right\rangle = \nabla_z \left\langle \Psi v_z \right\rangle.$$

(c) Use this to calculate the kinetic energy (25.56).

25.5 A square jar is half filled with water of density 1 g cm^{-3} lying below oil of density 0.8 g cm^{-3}. The interface has surface tension 0.3 N m^{-1}. Determine the largest horizontal size of the jar that permits the jar to be turned around with the oil remaining stably below the water.

* **25.6 [Gaussian wave packet]** Calculate explicitly the form of a superposition of harmonic waves

$$h = \int_{-\infty}^{\infty} a(k)\cos[kx - \omega(k)t + \chi(k)]\,dk,$$

where

$$a(k) = \frac{1}{\Delta k\sqrt{\pi}}\exp\left(-\frac{(k-k_0)^2}{\Delta k^2}\right), \tag{25.79a}$$

$$\omega(k) = \omega_0 + c_g(k - k_0), \tag{25.79b}$$

$$\chi(k) = \chi_0 - x_0(k - k_0). \tag{25.79c}$$

Describe its form and determine what x_0 represents. Hint: Write the wave as the real part of a complex wave and use the known Gaussian integrals.

25.7 **(a)** Show that the wave attenuation length (25.72) may be written in dimensionless form as

$$\frac{\xi}{\xi_c} = \frac{2(1 + 3k^2L_c^2)}{(2kL_c)^{5/2}\sqrt{1 + k^2L_c^2}}, \qquad \xi_c = \frac{L_c^2}{\nu}\sqrt{\frac{g_0 L_c}{2}}. \tag{25.80}$$

(b) Calculate ξ_c for water.

26

Jumps and shocks

Wading in the water near a beach and fighting to stay upright in the surf, you are evidently under the influence of non-linear dynamics, simply because the breaking waves look so different from the smooth swells in the open sea that gave rise to them. Equally non-linear are the dynamics behind the hydraulic jumps observed every day in the kitchen sink, and the closely related dramatic river bores created by the rising tide at the mouth of a river and rolling far upstream. Non-linear dynamics also lie behind the familiar sonic boom caused by a high-speed airplane passing overhead, and the hopefully less familiar short-range shock wave created by an exploding grenade. At much larger scales one encounters the huge atmospheric shock waves released by nuclear explosions, or the enormous shock waves from supernova explosions or galaxy collisions that may trigger star formation in the tenuous clouds of interstellar matter, a phenomenon to which we may ultimately owe our own existence.

The beauty of fluid mechanics lies in the knowledge that all these effects stem from the same non-linear aspects of the Navier–Stokes equations. In this chapter we shall mainly use the global laws of balance derived in Chapter 20 to analyze the physics of fluids in the extreme limit where the non-linearity may create discontinuities, or near-discontinuities, in the properties of the fluid. There are, as mentioned, two major classes of such phenomena: hydraulic jumps and shock fronts. In an incompressible liquid under the influence of gravity, a hydraulic jump is signaled by a fairly abrupt rise in the height of the open surface of a fast stream. In any fluid, including the ones that we under normal circumstances would call incompressible, a sufficiently violent event may create a supersonic shock front across which all fluid properties change sharply.

26.1 Hydraulic jumps

A stationary *hydraulic jump* or *step* is most easily observed in a kitchen sink (Figure 26.4). The column of water coming down from the tap splays out from the impact region in a roughly circular flow pattern, and at a certain radius the thinning sheet of water abruptly thickens and stays thick beyond. The flow in the transition region may be quite complicated, even turbulent, as can be seen in the Qiantang river bore (Figure 26.1), which may be viewed as a moving hydraulic jump. Stationary hydraulic jumps may also arise in spillways that channel surplus water from a dam into the river downstream. In this section we analyze the global features of hydraulic jumps, following Rayleigh [Ray14] who was the first to recognize their resemblance to shocks in compressible fluids (to be analyzed in the following section).

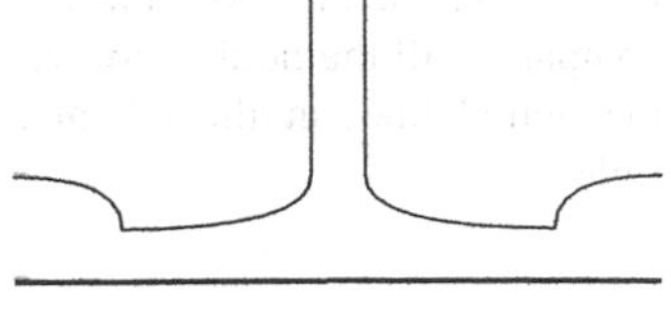

Sketch of the stationary hydraulic jump in a kitchen sink. The water coming down from the tap splays out in a horizontal sheet that abruptly thickens.

Figure 26.1. The Qiantang tidal river bore is the largest in the world and may be understood as a moving straight-line hydraulic jump. Its height can be 4 m, its width 3 km, and its speed more than 30 km h^{-1}. (Source: Eric Jones, Proudman Oceanographic Laboratory. With permission.)

Mass conservation

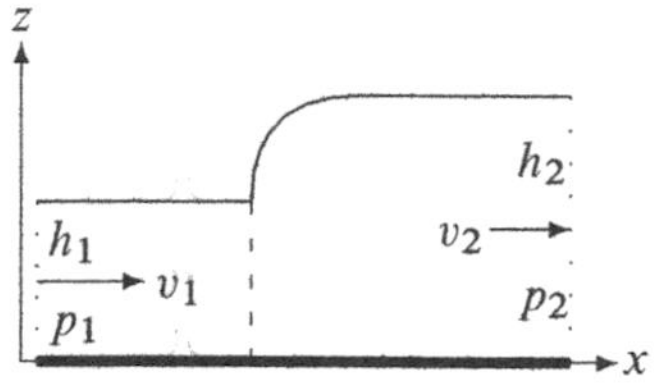

Stationary straight-line hydraulic jump. Incompressible liquid enters from the left with uniform velocity v_1 and height h_1 and exits on the right at a lower uniform velocity v_2 and greater height h_2. The entry and exit pressures, p_1 and p_2, are hydrostatic. Right at the front (dashed line), the flow pattern is complicated, often turbulent. The control volume encompasses all the liquid between the dotted lines at the left and right.

Consider a stationary jump along a straight line orthogonal to a uniform planar horizontal flow, a kind of static non-linear gravity wave. The incompressible flow is assumed to be steady before and after the jump, whereas in the transition region there may be intermittency and turbulence. Far upstream of the jump, the incompressible liquid is assumed to flow uniformly along the x-direction with constant velocity $v_x = v_1$ and constant water level $z = h_1$ above the bottom at $z = 0$. Far downstream the flow is likewise assumed to be uniform with constant velocity $v_x = v_2$ and constant water level $z = h_2$. The dimensionless *strength* of the jump is defined as

$$\sigma = \frac{h_2 - h_1}{h_1} = \frac{h_2}{h_1} - 1. \tag{26.1}$$

In principle the strength could be negative but energy balance will later show that the strength is always positive, $\sigma > 0$, so that $h_2 > h_1$. A jump is said to be *weak* when $\sigma \ll 1$ and *strong* when $\sigma \gg 1$.

The control volume is chosen with vertical sides and horizontal width L in the y-direction (see the margin figure). The upstream and downstream sides of the control volume are orthogonal to the direction of flow and placed so far from the transition region that both inflow and outflow are steady and uniform with velocities v_1 and v_2. Mass conservation then guarantees that the total volume discharge rate at the outlet is the same as at the inlet,

$$Q = Lh_1v_1 = Lh_2v_2. \tag{26.2}$$

From this we get the ratio of heights and velocities,

$$\frac{h_2}{h_1} = \frac{v_1}{v_2} = 1 + \sigma. \tag{26.3}$$

Since $\sigma > 0$, the upstream asymptotic velocity is always larger than the downstream one.

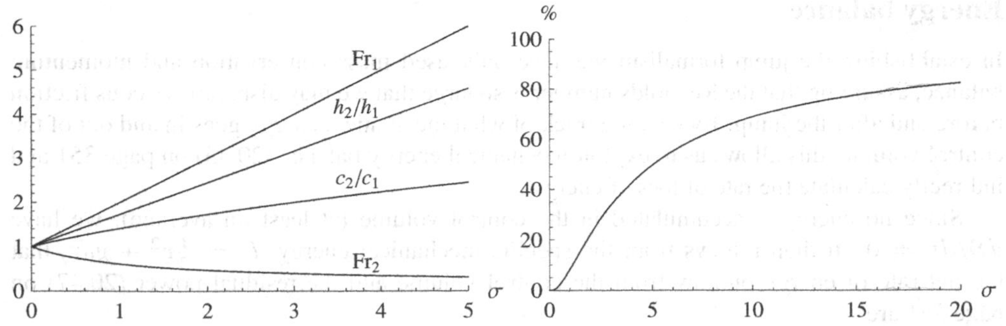

Figure 26.2. **Left:** Various dimensionless quantities as functions of the strength of the jump. **Right:** The fractional kinetic energy loss in the jump as a function of the strength.

Momentum balance

In steady flow the total momentum of the control volume is constant, $d\mathcal{P}_x/dt = 0$, and momentum balance (page 343) implies that the net momentum out of the control volume must equal the total external force acting on the control volume, $\dot{\mathcal{P}}_x = \mathcal{F}_x$. Disregarding viscosity, the horizontal force is entirely due to the pressure acting on the two vertical sides of the control volume where the liquid enters or leaves. The pressure in the uniform flow that reigns in these regions is hydrostatic, given by $p_1 = p_0 + \rho_0 g_0 (h_1 - z)$ at the inlet and $p_2 = p_0 + \rho_0 g_0 (h_2 - z)$ at the outlet, where p_0 is the constant (atmospheric) pressure on the open liquid surface. Carrying out the integrals, we have

$$\dot{\mathcal{P}}_x = \rho_0 Q v_2 - \rho_0 Q v_1, \qquad \mathcal{F}_x = -\tfrac{1}{2}\rho_0 g_0 L h_2^2 + \tfrac{1}{2}\rho_0 g_0 L h_1^2. \tag{26.4}$$

Eliminating the outlet parameters by means of (26.2), momentum balance, $\dot{\mathcal{P}}_x = \mathcal{F}_x$, may for $\sigma > 0$ be solved for the inlet velocity (from which the outlet velocity follows),

$$v_1 = \sqrt{g_0 h_1}\sqrt{(1+\sigma)\left(1+\tfrac{1}{2}\sigma\right)}, \qquad v_2 = \frac{v_1}{1+\sigma}. \tag{26.5}$$

The inlet velocity scale is set by $c_1 = \sqrt{g_0 h_1}$, which we recognize as the small-amplitude shallow-water wave speed (25.32) on page 428 in the inflow region. Similarly, the shallow-water wave speed in the outflow region is $c_2 = \sqrt{g_0 h_2} = c_1\sqrt{1+\sigma}$. Since $\sigma > 0$, it follows immediately that the velocities obey the inequalities $v_1 > c_2 > c_1 > v_2$.

The dimensionless ratios between a flow velocity and the corresponding shallow-water wave speed is called the *Froude number*. For inlet and outlet, they are

$$\mathsf{Fr}_1 \equiv \frac{v_1}{c_1} = \sqrt{(1+\sigma)\left(1+\tfrac{1}{2}\sigma\right)}, \qquad \mathsf{Fr}_2 \equiv \frac{v_2}{c_2} = \frac{\mathsf{Fr}_1}{(1+\sigma)^{3/2}}. \tag{26.6}$$

Since $\sigma > 0$, we have $\mathsf{Fr}_1 > 1 > \mathsf{Fr}_2$, and the flow is said to be *supercritical* before the jump, and *subcritical* after. The various quantities are plotted in Figure 26.2L.

Conversely, we may determine the jump strength from the inlet or outlet Froude numbers,

$$\sigma = \frac{\sqrt{1+8\mathsf{Fr}_1^2}-3}{2} = \frac{1-4\mathsf{Fr}_2^2+\sqrt{1+8\mathsf{Fr}_2^2}}{4\mathsf{Fr}_2^2}. \tag{26.7}$$

These relations are useful when the height and velocity are known for only one side of the jump, as in Example 26.2.

William Froude (1810–79). English engineer and naval architect. Discovered what are now called scaling laws, allowing predictions of ship performance to be made from studies of much smaller model ships. (Source: Wikimedia Commons.)

Energy balance

In establishing the jump formalism we have only used mass conservation and momentum balance, assuming that the Reynolds number is so large that we may disregard viscous friction before and after the jump. By keeping track of what mechanical energy goes in and out of the control volume, this allows us to exploit mechanical energy balance (20.45) on page 351 and indirectly calculate the rate of loss of energy.

Since no energy is accumulated in the control volume (at least on average), we have $d\mathcal{E}/dt = 0$. It then follows from the specific mechanical energy, $E = \frac{1}{2}\boldsymbol{v}^2 + g_0 z$, that the net rate of energy outflow from the control volume and the residual power (20.47) on page 351 are

$$\dot{\mathcal{E}} = \oint_S E\, d\dot{M} = \rho_0 Q\left(\tfrac{1}{2}v_2^2 + \tfrac{1}{2}g_0 h_2\right) - \rho_0 Q\left(\tfrac{1}{2}v_1^2 + \tfrac{1}{2}g_0 h_1\right),$$

$$\tilde{P} = -\oint_S p\boldsymbol{v}\cdot d\boldsymbol{S} - \int_V \boldsymbol{\sigma}:\nabla\boldsymbol{v}\, dV = -\tfrac{1}{2}\rho_0 g_0 L h_2^2 v_2 + \tfrac{1}{2}\rho_0 g_0 L h_1^2 v_1 - \dot{W}_{\text{int}}.$$

In both expressions the factors 1/2 arise from integrating over inlet and outlet.

The rate of dissipation $\dot{W}_{\text{int}}$ is given by (15.24) on page 252 but cannot be directly evaluated without a model for the flow in the transition region. Applying mechanical energy balance, $\dot{\mathcal{E}} = \tilde{P}$, we nevertheless obtain an indirect expression for the dissipation rate,

$$\dot{W}_{\text{int}} = \rho_0 Q\left(\frac{1}{2}v_1^2 + g_0 h_1\right) - \rho_0 Q\left(\frac{1}{2}v_2^2 + g_0 h_2\right) = \rho_0 Q(H_1 - H_2). \tag{26.8}$$

In the last step we introduced the values H_1 and H_2 of the Bernoulli function (13.13) on page 212, calculated at the surface of the water at inlet and outlet. This clearly demonstrates that the jump in the Bernoulli function is directly related to dissipation.

Dividing $\dot{W}_{\text{int}}$ by rate of kinetic energy inflow, $\dot{\mathcal{K}}_1 = \frac{1}{2}\rho_0 Q v_1^2$, and eliminating the downstream quantities h_2 and v_2 by means of (26.3), the *fractional dissipative loss* of kinetic energy becomes (see Figure 26.2R)

$$\boxed{\frac{\dot{W}_{\text{int}}}{\dot{\mathcal{K}}_1} = \frac{\sigma^3}{(1+\sigma)^2(2+\sigma)}.} \tag{26.9}$$

Since the dissipation rate necessarily must be positive (page 252), we conclude that a hydraulic jump as promised will always have positive strength, $\sigma > 0$, implying that $h_2 > h_1$ and $v_2 < v_1$.

For a weak jump with $\sigma = 0.1$, the energy loss is merely 0.04%, while for $\sigma = 1$ it is 8.3%, and for $\sigma = 10$ it becomes 69%. Hydraulic jumps are efficient dissipators of kinetic energy, and this is in fact their function in dam spillways where high-speed surplus water must be slowed down before it is released into the river downstream of the dam.

Moving jumps

Moving hydraulic jumps are seen on the beach when waves roll in, sometimes in several layers on top of each other. More dramatic *river bores* may be formed by the rising tide near the mouth of a river where the water becomes shallower. When the circumstances are right, such waves can roll far up the river with a foaming turbulent front (see Figure 26.1). In the laboratory, a river bore can be created in a long canal with water initially at rest. When the wall in one end of the canal is set into motion with constant velocity, a bore will form and move along the canal with constant speed and constant water level.

In Rayleigh's global model viscosity is ignored, and the difference between a river bore and a stationary hydraulic jump lies entirely in the frame of reference (see the margin figure). The river bore is obtained in the moving frame where the fluid in front of the jump is at rest. Subtracting v_1 from all velocities, the front itself will move with negative velocity $-v_1$, and the fluid behind the jump will move with negative velocity $v_2 - v_1$. In the margin figure the vector arrows of the negative velocities have been reversed.

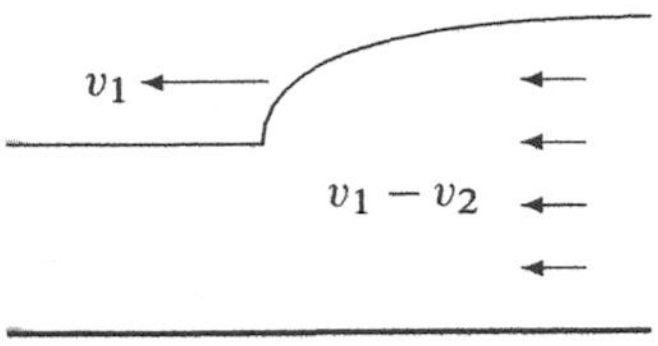

A river bore moving in from the right in water initially at rest. The water behind the front moves slower than the front itself. The velocities are obtained by subtracting v_1 from all the velocities of the stationary jump, and reversing the arrows of the negative velocities.

It is also possible to choose a reference frame in which the fluid behind the jump is at rest (see the margin figure). Subtracting v_2 from the velocities of the stationary jump, this describes a stationary flow being reflected in a closed canal. Such a *reflection bore* moves with negative velocity $-v_2$ while the inflow has positive velocity $v_1 - v_2$.

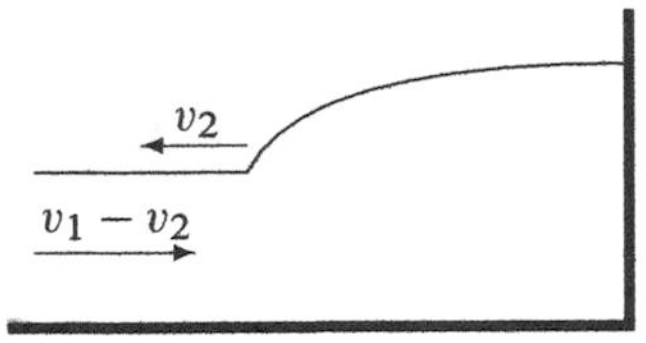

A reflection bore in a closed canal is obtained by subtracting v_2 from all velocities of the stationary jump.

Example 26.1 [The Qiantang river bore]: The Qiantang river bore is the largest in the world (see Figure 26.1). "Guesstimating" $h_1 = 4$ m and $h_2 - h_1 = 4$ m, the jump becomes of medium strength, $\sigma = 1$. The Froude numbers are $\mathsf{Fr}_1 = 1.73$ and $\mathsf{Fr}_2 = 0.61$. The front velocity $v_1 = 11\ \mathrm{m\,s^{-1}} = 39\ \mathrm{km\,h^{-1}}$ agrees decently with the reported speed of this bore (considering the simplicity of the calculation). The Reynolds number $\mathsf{Re} = 4.3 \times 10^7$ is so large that bottom friction can presumably be ignored whereas the front itself is strongly turbulent.

What causes the jump?

A thin layer of *inviscid* liquid in uniform horizontal planar flow has absolutely no physical reason to suddenly become thicker a certain distance downstream from the inlet. But real liquids are viscous and must fulfill the no-slip condition at the bottom. This slows down the layer of inflowing liquid, which by mass conservation must slowly increase its thickness[1].

In steady flow let the height of the inflow layer be $h(x)$ and the average velocity $v(x)$, so that the total flux $q = h(x)v(x)$ (per unit of transverse distance) is constant. To estimate the thickening we recall (page 248) that momentum diffusion orthogonal to the flow reaches the open surface height $z = h(x)$ in the time $t = h(x)^2/4\nu$. A small distance dx further downstream, it takes a little extra time $dt = h(x)h'(x)dx/2\nu$. The time it takes for the fluid to stream through the distance dx is on average $dt = dx/v(x)$, and in the stationary situation where momentum diffusion determines the shape of the open surface, the two ways of calculating the time interval must yield the same result, so that the slope $h'(x) = 2\nu/v(x)h(x) = 2\nu/q$ is constant. The fluid level thus rises linearly with x,

$$\boxed{h(x) = h_0 + \frac{2\nu}{q}(x - x_0),} \tag{26.10}$$

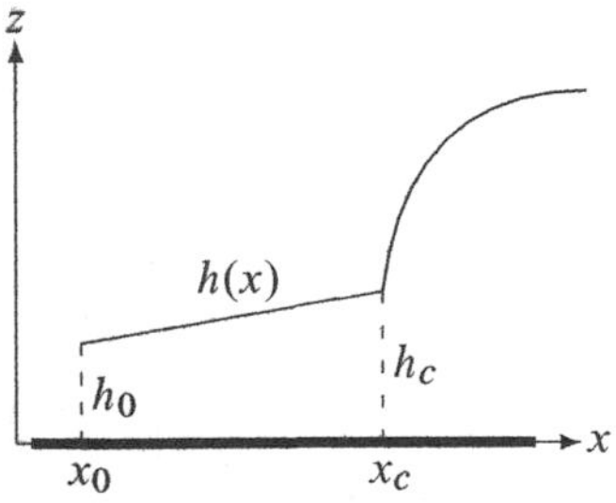

The height $h(x)$ rises y linearly until the jump takes place at the position $x = x_c$.

where x_0 is the point at which momentum diffusion first reaches the surface. In a recent laboratory experiment [BAB09], it was indeed observed that the inflow level rises with nearly constant positive slope up to the point where the jump is triggered. A reasonable numerical agreement with the above expression could, however, only be obtained by considering the inflow to be turbulent with effective viscosity proportional to q and about an order of magnitude larger than the laminar viscosity.

During the nearly 100 years since Rayleigh's seminal paper, a large body of literature has dealt with the question of why and where the jump takes place. The consensus today is that for shallow-water flows, the stationary jump is closely related to boundary layer separation (see Chapter 28). But neither for ordinary boundary layers nor for the hydraulic jump does there exist a simple and efficient analytic method for prediction of the actual point of separation [BPW97, BAB09].

[1]For river bores this argument is not valid because the inflow is at rest relative to the bottom.

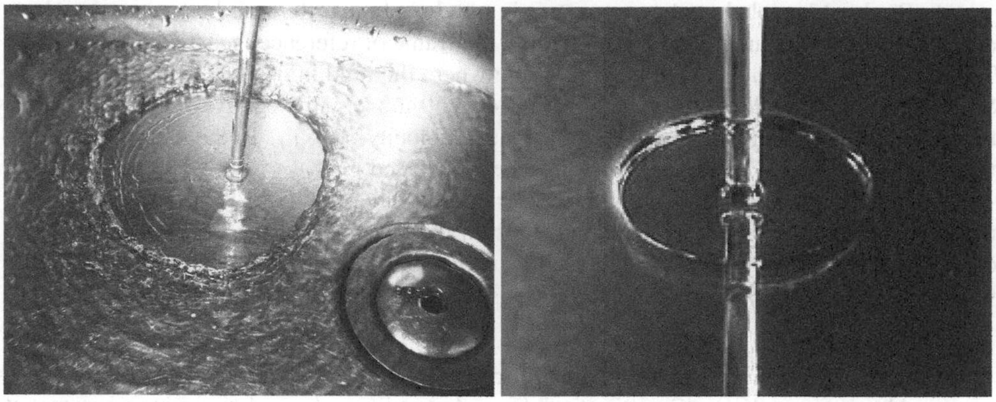

Figure 26.3. **Left:** Hydraulic jump in the kitchen sink with irregular turbulent flow in the jump and downstream from it. (Source: Philippe Belleudy, Université Joseph Fourier. With permission.) **Right:** Hydraulic jump in the laboratory with perfectly circular shape and nearly laminar flow. (Source: Reproduced from [BPW97]. With permission.)

26.2 Circular jump

In the kitchen sink it is possible to observe nearly circular stationary hydraulic jumps, as for example in Figure 26.3L. The actual shape deviates from perfect circularity because of obstacles and the slightly curved shape of the bottom of a real-world kitchen sink. In laboratory experiments it is, under controlled circumstances, possible to create nearly perfectly circular jumps like the one shown in Figure 26.3R.

Inviscid flow before and after the jump

The position of the circular jump is as difficult to predict as for a straight-line jump, so here we shall simply place the jump at a certain radius, $r = R$, from the center and calculate the flow before and after the jump. We shall as before assume that the steady, inviscid flow over the kitchen sink bottom is independent of z, that is, of the form $v_r = v(r)$. The general constancy of the mass flux Q, together with the constancy of the Bernoulli function H outside the jump region, then leads to the equations

$$2\pi r h(r) v(r) = Q, \qquad \tfrac{1}{2} v(r)^2 + g_0 h(r) = \begin{cases} H_1 & \text{for } r < R \\ H_2 & \text{for } r > R. \end{cases} \tag{26.11}$$

Combined, they yield a third-degree equation for either $v(r)$ and $h(r)$ before and after the jump. The solution is however so messy that we shall avoid it.

The jump at $r = R$ is assumed to be locally straight, so that the formalism of the preceding section can be applied. For a strong jump with $\sigma \gg 1$, the solution is easy to obtain. Since the local Froude number $\mathsf{Fr}(r) = v(r)/\sqrt{g_0 h(r)}$ will be large just before the jump and small behind it, it follows that $v(r) \approx \sqrt{2H_1}$ will be nearly constant before the jump, whereas $h(r) = H_2/g_0$ will be nearly constant behind it. Thus, we have approximately

$$\begin{aligned} v(r) &= v_1, & h(r) &= \frac{Q}{2\pi r v_1}, & &\text{for } r < R; \\ v(r) &= \frac{Q}{2\pi r h_2}, & h(r) &= h_2, & &\text{for } r > R, \end{aligned} \tag{26.12}$$

where v_1 and h_2 are constants. The Reynolds number $\mathsf{Re} = v(r)h(r)/\nu = Q/2\pi r\nu$ is continuous across the jump and decreases inversely with the radius.

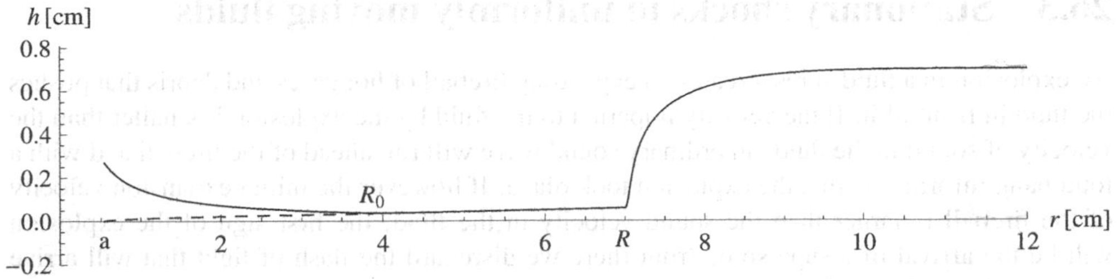

Figure 26.4. Outline of the hydraulic jump in the kitchen sink discussed in the text and in Example 26.2. Note that the height is on another scale than the radius. The solid line is the water level given by the model (26.12), whereas the dashed line is the estimated thickness of the boundary layer. The thickness of the jump is *ad hoc* set comparable to the jump height.

Influence of viscosity

As for the straight-line jump, viscous friction will decelerate the flow and tend to thicken the layer. Assuming that the liquid enters as plug flow and splays out at radius $r = a$, Bernoulli's theorem tells us that the velocity $v_1 = Q/\pi a^2$ is nearly constant until viscosity sets in. Viscosity causes momentum diffusion from the bottom, creating a boundary layer of vertical thickness $\delta = 2\sqrt{\nu t}$ where $t = (r - a)/v_1$ is the time it takes the liquid to stream from the inlet to radius r. One may estimate the radius $r = R_0$ where the boundary layer reaches the surface by solving the equation $\delta(R_0) = h(R_0)$. For $a \ll R_0$ the solution is easily found,

$$R_0 = \left(\frac{Qa^2}{16\pi\nu}\right)^{1/3}. \tag{26.13}$$

For $r < R_0$ the flow is essentially inviscid and follows the solution in (26.12). For $r > R_0$ viscosity determines the shape of inflow. The slope of the surface height is estimated in the same way as on page 447 to be $h'(r) = 2\nu/h(r)v(r) = 4\pi r\nu/Q$. Since it grows linearly with the radius (rather than being constant), we obtain a quadratic expression,

$$h(r) = h_0 + 2\pi\frac{\nu}{Q}(r^2 - R_0^2), \tag{26.14}$$

where $h_0 = a^2/2R_0$ is the height at $r = R_0$. The average velocity in this region is obtained from mass conservation $v(r) = Q/2\pi r h(r)$.

Example 26.2 [A kitchen sink experiment]: In a casual home-made kitchen sink experiment the discharge rate was measured to $Q = 100\ \mathrm{cm^3\,s^{-1}}$ by collecting water from the tap in a standard kitchen measure for a while. The radius of the jet coming down from the tap was $a = 0.5$ cm. The jump radius $R = 7$ cm and the constant outlet height $h_2 = 0.7$ cm were also measured. The rest now follows from theory.

First we calculate everything for inviscid flow. The outlet velocity just after the jump becomes, $v_2 = Q/2\pi R h_2 = 3.2\ \mathrm{cm\,s^{-1}}$ and the corresponding Froude number $\mathsf{Fr}_2 = v_2/\sqrt{g_0 h_2} = 0.12$. The strength of the jump, $\sigma = 33$, is obtained from (26.7), and from it we obtain the inlet Froude number $\mathsf{Fr}_1 = 24$. Next we calculate the constant velocity before the jump $v_1 = \mathsf{Fr}_1^{2/3}(g_0 Q/2\pi R)^{1/3} = 109\ \mathrm{cm\,s^{-1}}$ and the height just before the jump $h_1 = Q/2\pi R v_1 = 0.02$ cm. Its smallness explains that it cannot be determined in a kitchen experiment.

Next, taking viscosity into account we find from (26.13) the viscous transition radius $R_0 = 3.9$ cm, which does satisfy $R_0 \gg a$. The water level at $r = R_0$ is $h_0 = 0.036$ cm and rises according to (26.14) to 0.059 cm just before the jump which is still too small to measure, and the velocity drops to $38\ \mathrm{cm\,s^{-1}}$.

The solution is plotted in Figure 26.4. The shape of the water surface before the jump region is obtained from the above approximative theory whereas the shape in the jump region is a guess using the height just before the jump and the constant asymptotic height after the jump.

26.3 Stationary shocks in uniformly moving fluids

An explosion in a fluid at rest creates an expanding fireball of hot gases and debris that pushes the fluid in front of it. If the velocity imparted to the fluid by the explosion is smaller than the velocity of sound in the fluid, an ordinary sound wave will run ahead of the fireball and with a loud bang inform you that the explosion took place. If however the initial expansion velocity of the fireball is larger than the sound velocity in the fluid, the first sign of the explosion will be the arrival of a supersonic front (here we disregard the flash of light that will arrive much earlier). The sudden jump in the properties of a fluid at the passage of a supersonic front is called a *shock*. The understanding of shocks is of great importance for the design of supersonic aircraft, and of jet and rocket engines.

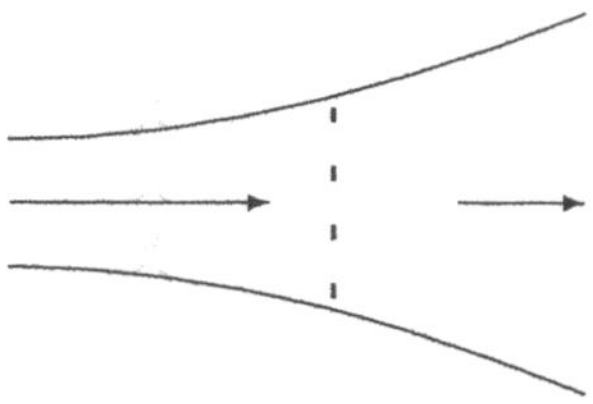

A stationary planar shock (dashed) in an expanding nozzle. The inflow is supersonic and the outflow subsonic. See Section 14.2 on page 233.

We shall begin by analyzing stationary planar (normal) shocks with planar isobars orthogonal to the direction of the flow, analogous to planar sound waves. Normal shocks in nearly ideal fluids are, as will be shown below, extremely thin, which allows us to view any shock as singular and therefore locally planar. We shall only analyze normal shocks that have somehow been brought into existence, but refrain from going into the question of what triggers the formation of a shock. Since a uniformly moving fluid filling all of space is equivalent to a fluid at rest, it must be completely stable and will not go into shock by itself. This is quite different from a uniformly flowing horizontal river of liquid, which as we have seen may be triggered into a hydraulic jump by viscous friction between the liquid and the static river bottom. Since viscosity alone is incapable of generating a shock in the uniformly flowing fluid, other external causes must be invoked to understand how a shock is triggered, for example the gently widening exit cone in a Laval nozzle (page 240), or the supersonically expanding blast sphere from an explosion analyzed in the following section.

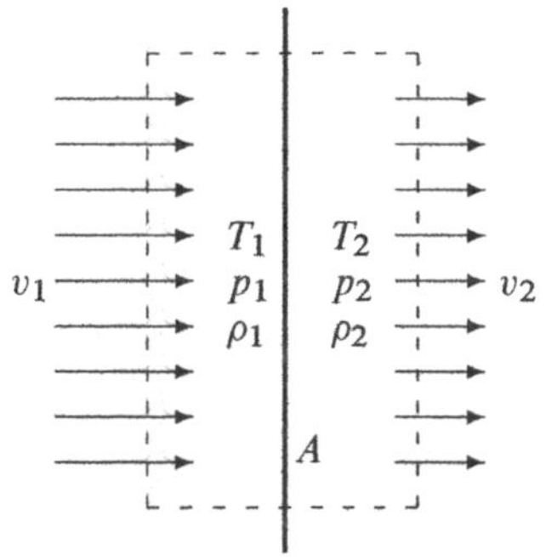

Control volume covering an area A of a stationary normal planar shock.

The Rankine–Hugoniot equations

In the rest system of the planar shock we choose as control volume a rectangular box containing an area A of the shock front. We shall for simplicity disregard gravity and limit the analysis to an ideal gas with adiabatic index γ, but we shall not assume that the flow is inviscid. The planar symmetry of the flow guarantees that the velocity field $\boldsymbol{v} = v(x)\,\hat{\boldsymbol{e}}_x$ only depends on x. Consequently, there can be no shear stresses acting on the sides of the control volume parallel with the flow, even where they pass through the shock itself.

Far upstream from the shock, the gas has uniform constant velocity v_1, temperature T_1, pressure p_1, and density ρ_1, and far downstream it has uniform velocity v_2, temperature T_2, pressure p_2, and density ρ_2 (see the margin figure). Since the total mass and momentum of the control volume must be constant in steady flow, $d\mathcal{M}/dt = 0$ and $d\mathcal{P}_x/dt = 0$, mass conservation takes the form $\dot{M} = \rho_2 A v_2 - \rho_1 A v_1 = 0$, and momentum balance becomes $\dot{\mathcal{P}}_x = \mathcal{F}_x$ where $\dot{\mathcal{P}}_x = \rho_2 A v_2^2 - \rho_1 A v_1^2$ and $\mathcal{F}_x = -p_2 A + p_1 A$. The planar symmetry also secures that the conditions for the generalized Bernoulli theorem (22.15) on page 375 are fulfilled, even if dissipation in the shock front converts some of the kinetic energy of the gas to heat. Using the specific internal enthalpy (or pressure potential) of an ideal gas, $w = U + p/\rho = c_p T = RT\gamma/(\gamma - 1)$, we finally arrive at the so-called *Rankine–Hugoniot* equations,

William John Macquorn Rankine (1820–1872). Scottish civil engineer and physicist. Worked on thermodynamics (with Kelvin), in particular steam engine theory. Created a now defunct absolute temperature based on Fahrenheit degrees. (Source: Wikimedia Commons.)

$$\rho_1 v_1 = \rho_2 v_2 \qquad \text{(mass)}, \qquad (26.15a)$$

$$\rho_1 v_1^2 + p_1 = \rho_2 v_2^2 + p_2 \qquad \text{(momentum)}, \qquad (26.15b)$$

$$\frac{1}{2}v_1^2 + \frac{\gamma}{\gamma - 1}\frac{p_1}{\rho_1} = \frac{1}{2}v_2^2 + \frac{\gamma}{\gamma - 1}\frac{p_2}{\rho_2} \qquad \text{(Bernoulli)}. \qquad (26.15c)$$

If the gas molecules dissociate in the shock, the adiabatic indices would have to be different on the two sides (see Problem 26.6).

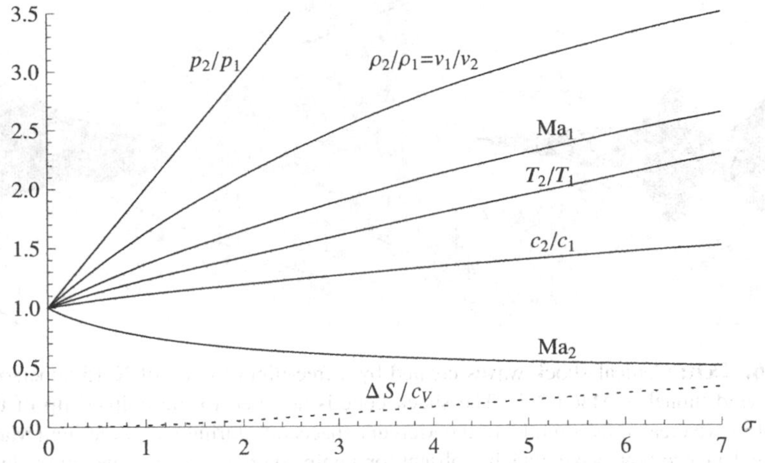

Figure 26.5. Dimensionless shock parameters as functions of the strength σ for $\gamma = 7/5$. The dotted curve is the specific entropy jump across the shock, $\Delta S/c_v = \Delta S/C_v$, from (26.23).

Parameterizing the shock by its strength

Given the inlet parameters, the Rankine–Hugoniot equations may be solved explicitly for the outlet parameters (see Problem 26.3). As for hydraulic jumps, it is however much easier to parameterize the solutions in terms of a suitably defined shock strength.

Using the first Rankine–Hugoniot equation to eliminate v_2 in the second, we obtain

$$v_1^2 = \frac{\rho_2}{\rho_1} \cdot \frac{p_2 - p_1}{\rho_2 - \rho_1}, \qquad v_2^2 = \frac{\rho_1}{\rho_2} \cdot \frac{p_2 - p_1}{\rho_2 - \rho_1}. \tag{26.16}$$

Inserted into the third, the ratio of densities may be written as

$$\frac{\rho_2}{\rho_1} = \frac{\gamma(p_1 + p_2) + p_2 - p_1}{\gamma(p_1 + p_2) + p_1 - p_2}. \tag{26.17}$$

Since this depends only on the ratio of pressures, we shall put

$$\frac{p_2}{p_1} = 1 + \sigma, \tag{26.18}$$

where σ is the *strength* of the shock. The density and velocity ratios now follow,

$$\frac{\rho_2}{\rho_1} = \frac{v_1}{v_2} = \frac{2\gamma + (\gamma + 1)\sigma}{2\gamma + (\gamma - 1)\sigma}, \tag{26.19}$$

and the temperature ratio may be obtained from the ideal gas law. A shock is said to be strong when $\sigma \gg 1$ and weak when $\sigma \ll 1$. We shall see below that the Second Law of Thermodynamics requires $\sigma > 0$.

The isentropic sound velocities before and after the shock are

$$c_1 = \sqrt{\gamma \frac{p_1}{\rho_1}}, \qquad c_2 = \sqrt{\gamma \frac{p_2}{\rho_2}}, \tag{26.20}$$

and using (26.16) this allows us to obtain the Mach numbers,

$$\mathsf{Ma}_1 \equiv \frac{v_1}{c_1} = \sqrt{1 + \frac{\gamma + 1}{2\gamma}\sigma}, \qquad \mathsf{Ma}_2 \equiv \frac{v_2}{c_2} = \sqrt{1 - \frac{\gamma + 1}{2\gamma}\frac{\sigma}{1 + \sigma}}. \tag{26.21}$$

For $\sigma > 0$ we have $\mathsf{Ma}_1 > 1 > \mathsf{Ma}_2$. In Figure 26.5 the various quantities are plotted as functions of the strength for $\gamma = 7/5$.

Pierre Henri Hugoniot (1851–1887). French engineer; worked mainly in the marine artillery service. Wrote only one (long) article, finally published in the year of his untimely death, on "the propagation of movement in bodies, particularly in perfect gases".

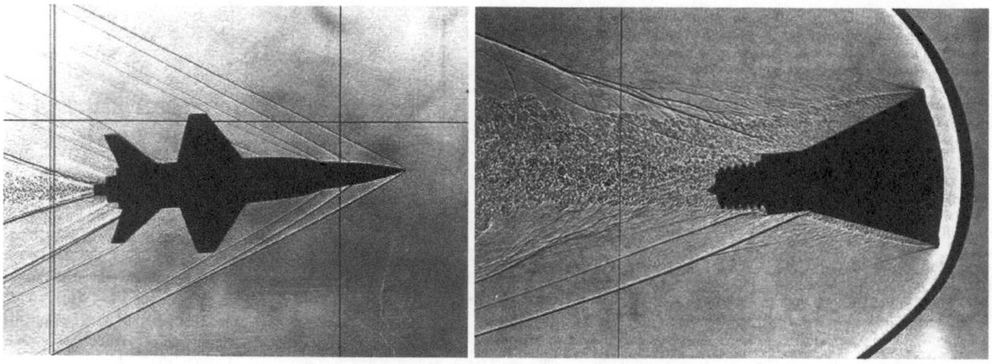

Figure 26.6. **Left:** Conical shock-waves created by a free-flight model of X-15 supersonic aircraft fired into a wind tunnel at Mach 3.5. The shock cone is attached to the pointed tip of the aircraft. **Right:** Shock-wave created by a model of the Mercury spacecraft during reentry at high Mach number. One notes the detached bow wave, which is absent for a pointed object like the one on the left. The rear of the objects generate secondary weaker shock fronts. (Sources: Courtesy NASA.)

Strong shock limit: In the limit of a strong shock, $\sigma \to \infty$, some quantities approach constant values:

$$\frac{\rho_2}{\rho_1} = \frac{v_1}{v_2} \to \frac{\gamma+1}{\gamma-1}, \qquad \frac{p_2}{\rho_1 v_1^2} \to \frac{2}{\gamma+1}, \qquad \mathsf{Ma}_2 \to \sqrt{\frac{\gamma-1}{2\gamma}}. \tag{26.22}$$

For a diatomic gas with $\gamma = 7/5$, we find $\rho_2/\rho_1 \to 6$ and $\mathsf{Ma}_2 \to 1/\sqrt{7} \approx 0.378$.

Entropy change in a shock

Since gas pressure is always positive, the strength of the shock could in principle range over both negative and positive values in the interval $-1 < \sigma < \infty$. The physical asymmetry between positive and negative strength becomes apparent when we calculate the entropy change, $\Delta\mathcal{S} = \mathcal{S}_2 - \mathcal{S}_1$. From the entropy of an ideal gas (23.4) on page 395 with $V = M/\rho$, we find

$$\frac{\Delta\mathcal{S}}{C_v} = \log\left[\frac{p_2\rho_2^{-\gamma}}{p_1\rho_1^{-\gamma}}\right] = \log(1+\sigma) - \gamma\log\frac{2\gamma+(\gamma+1)\sigma}{2\gamma+(\gamma-1)\sigma}. \tag{26.23}$$

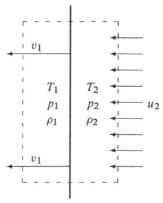

A planar shock front moving toward the left with supersonic velocity v_1 into a gas at rest. The fluid behind the front moves to the left with velocity $u_2 = v_1 - v_2 > 0$, which may or may not be supersonic.

The right-hand side is a monotonically increasing function of σ that vanishes for $\sigma = 0$ and is thus negative for $-1 < \sigma < 0$ (see Problem 26.5). By the Clausius–Duhem inequality (23.16), which embodies the Second Law of Thermodynamics, the entropy change cannot be negative in the absence of external heat sources, and consequently we must have $\sigma > 0$ for a non-trivial shock. We have thus shown that *in a stationary normal shock, the flow must go from supersonic to subsonic in the downstream direction.*

Normal shock wave

A violent explosion in the atmosphere (at rest) creates a moving spherical shock front, which we shall study in the following section. In view of the extreme thinness of the shock front under normal atmospheric pressure and temperature (see below), we may view any shock front as locally flat and use the results obtained for the normal shock.

It is as always possible freely to choose another steadily moving reference frame without changing the physics (Newtonian relativity). For the reference frame moving with velocity v_1 along the x-direction, the previously incoming gas will now be at rest whereas the shock front itself moves in the opposite direction with supersonic velocity v_1. The gas behind the front moves in the same direction with velocity $u_2 = v_1 - v_2$, which may or may not be supersonic, depending on the strength of the shock. For a diatomic gas with $\gamma = 7/5$, the flow will be supersonic, that is, $u_2 > c_2$ when $\sigma > 3.82$, corresponding to $\mathsf{Ma}_1 > 2.07$.

Oblique shock

An *oblique* planar shock front may be obtained in a reference frame moving with constant velocity V tangentially to the plane of the stationary normal shock. In the moving frame the flow velocities are denoted u_1 and u_2 with incidence angles ϕ_1 and ϕ_2. Using that the tangential velocity is the same on both sides of the shock, $V = v_1 \tan\phi_1 = v_2 \tan\phi_2$, we obtain from (26.19) the following relation,

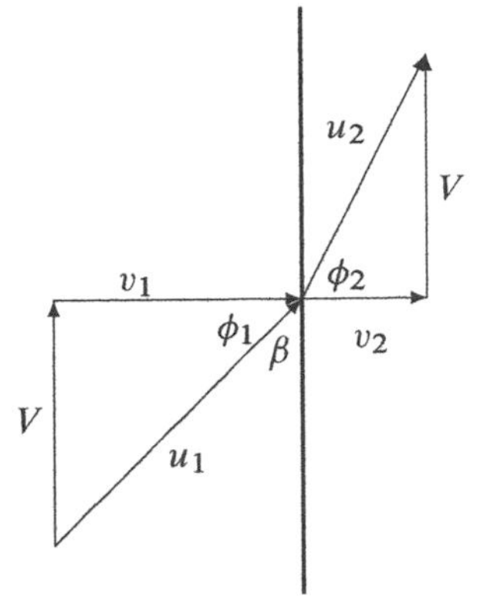

Geometry of an oblique stationary shock in a reference frame moving tangentially (downward) with velocity V.

$$\frac{\tan\phi_2}{\tan\phi_1} = \frac{v_1}{v_2} = \frac{\rho_2}{\rho_1} = \frac{2\gamma + (\gamma+1)\sigma}{2\gamma + (\gamma-1)\sigma}. \tag{26.24}$$

Since $\sigma > 0$, the right-hand side is always larger than unity, implying that $\phi_2 > \phi_1$.

Consider now an object, for example the model aircraft in Figure 26.6L, at rest in an atmosphere moving with constant supersonic velocity. The picture shows how the pointed tip of the aircraft creates a perfectly conical shock wave that trails the aircraft. A similar shock wave is created by the speeding blunt object in Figure 26.6R, but the shock front is now detached from the blunt object. It is, however, quite complicated to calculate the half opening angle of the shock cone from the Mach number and the shape of the object, even as simple as a pointed cone with a given half opening angle.

One result can, however, be obtained without much calculation. In the notation of the oblique shock, the Mach number of the incoming supersonic flow is $\mathsf{Ma} = u_1/c_1$, forming an angle $\beta = 90° - \phi_1$ with the shock front. From the margin figure it is seen that $v_1 = u_1 \sin\beta$ and dividing by c_1 we get $\mathsf{Ma}_1 = \mathsf{Ma}\sin\beta$. In the limit of a weak shock, $\mathsf{Ma}_1 \to 1$, we thus have

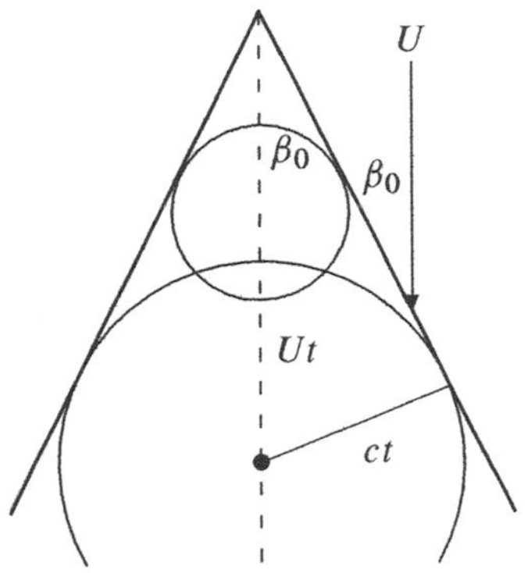

$$\boxed{\beta \to \beta_0 = \arcsin\frac{1}{\mathsf{Ma}}.} \tag{26.25}$$

The Mach construction of the shock cone for a supersonic object. In a time t, the object moves forward a distance Ut while the sound emitted at $t = 0$ forms a sphere of radius ct. The opening angle of the envelope of all spheres is $\sin\beta_0 = ct/Ut = c/U = 1/\mathsf{Ma}$.

The angle β_0 is called the *Mach angle*. A simple geometric construction due to Mach shows that the weak shock cone may be understood as the envelope of all the spherical sound waves emitted earlier from the nose of the object. If the shock is strong, the relation between the shape of the object and the half opening angle of the shock cone is more complicated (see for example [Anderson 2004]).

Example 26.3: An aircraft moving at $\mathsf{Ma} = 2$ has Mach angle $\beta_0 = 30°$, so when you hear the sonic boom of the aircraft and think it is right overhead at $h = 20$ km altitude, it is already a distance $d = h\cot\beta_0 \approx 35$ km beyond your position.

* Thickness of a shock

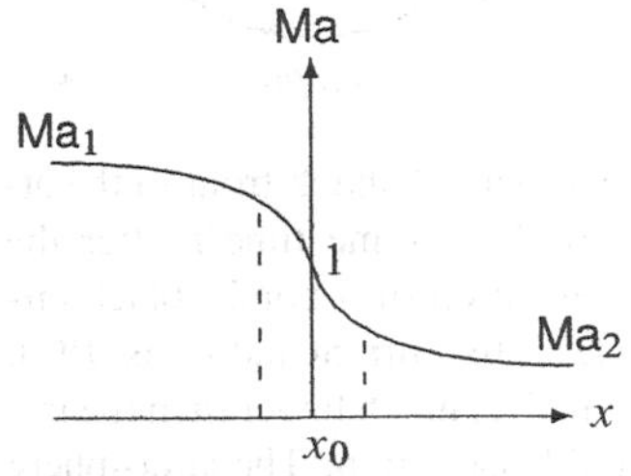

The Mach number must descend through the sonic point at $x = x_0$. The thickness δ is the distance between the dashed lines.

A shock cannot truly be a mathematical singularity. In the shock front itself the large velocity and temperature gradients will cause momentum and heat diffusion along the direction of motion, and that will set a limit to the sharpness of the shock. We shall now estimate the thickness of the shock from dimensional considerations using the longitudinal viscosity $\xi = \zeta + \frac{4}{3}\eta$ (see page 258) and the heat conductivity k. A complication is that both of these quantities may depend on the thermodynamic parameters and thus vary along the flow.

As the flow proceeds smoothly through the shock, the Mach number Ma descends from $\mathsf{Ma}_1 > 1$ to $\mathsf{Ma}_2 < 1$, and passes through $\mathsf{Ma} = 1$ at the sonic point, $x = x_0$, where velocity and temperature gradients are largest. At this point the only quantities with the dimension of length that may naturally be constructed are

$$\delta_{\rm mom} = \frac{\xi_0}{q}, \qquad \delta_{\rm heat} = \frac{k_0}{c_{p0} q}, \tag{26.26}$$

where $q = \rho v$ is the mass flux, which is constant along the flow. Apart from numerical prefactors that we cannot obtain from dimensional considerations alone, these length scales determine the momentum and heat diffusion thicknesses.

In ideal gases the longitudinal viscosity ξ and the thermal conductivity k both vary with the square root of the temperature, while the isobaric specific heat capacity c_p is constant. This means that $\xi_0 = \xi_1\sqrt{T_0/T_1} = \xi_2\sqrt{T_0/T_2}$, and similarly for k_0. If the sonic temperature is estimated as the geometric mean of the inlet and outlet temperatures, $T_0 \approx \sqrt{T_1 T_2}$, we have $\xi_0 = \xi_1(T_2/T_1)^{1/4} = \xi_2(T_1/T_2)^{1/4}$ and similarly for k_0. The ratio of the two thickness scales thus becomes independent of the temperature,

$$\frac{\delta_{\rm mom}}{\delta_{\rm heat}} = \frac{\xi_1 c_p}{k_1} = \frac{\xi_2 c_p}{k_2}. \tag{26.27}$$

It is closely related to the Prandtl number (22.34) on page 381, which for ideal gases is of order unity.

Example 26.4 [Thickness of atmospheric shocks]: Under normal conditions the atmosphere has $\rho_1 = 1.2$ kg m^{-3}, $c_1 = 343$ m s^{-1}, $\xi_1 \approx \frac{7}{3}\eta_1 = 4.3 \times 10^{-5}$ Pa s, and $k_1 = 26 \times 10^{-3}$ W K^{-1}m^{-1}. For a weak shock with $\sigma \approx 0$, we find $v_1 \approx c_1$, $q \approx 410$ kg m^{-2}s^{-1}, $\xi_0 \approx \xi_1$, and $k_0 \approx k_1$, so that $\delta_{\rm mom} \approx 103$ nm and $\delta_{\rm heat} \approx 63$ nm. A strong shock in the atmosphere with $\sigma = 40$ has $\mathsf{Ma}_1 = 5.9$, $v_1 = 2.0$ km s^{-1}, and $\xi_0 \approx \xi_1(\sigma(\gamma-1)/(\gamma+1))^{1/4} \approx 6.9 \times 10^{-5}$ Pa s, leading to $q \approx 2400$ kg m^{-2}s^{-1}, and $\delta_{\rm mom} \approx 28$ nm and $\delta_{\rm heat} \approx 17$ nm. The thickness scales are all comparable to the mean free path in the gas, thereby raising legitimate worries about the validity of the calculation.

26.4 Application: Atmospheric blast wave

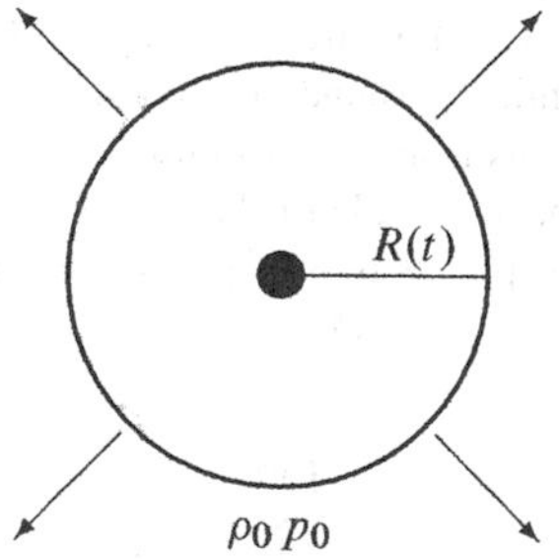

A spherical shock front in the atmosphere some time t after the detonation of a bomb (black circle). Its current radius is $R(t)$, which is much larger than the initial blast region. The atmosphere is initially at rest with density ρ_0 and pressure p_0. The volume of the sphere consists almost entirely of air.

A large explosion in the atmosphere generates a *blast wave* bounded by a spherical supersonic shock front. Such blast waves are mostly invisible, but films of nuclear bomb explosions show that the physical conditions can become so extreme that the blast wave can be seen as a rapidly expanding, almost perfectly spherical fireball, appearing right after the initial flash but before the mushroom cloud erupts. In this section we shall investigate the time evolution of such blast waves in the atmosphere, following the road laid out by L.I. Sedov [Sed46], G.I. Taylor [Tay50], and others.

Radius of the strong shock front

Let the atmosphere initially be at rest with constant density ρ_0, pressure p_0, and temperature T_0. The blast adds almost instantly a huge amount of energy $\Delta\mathcal{E}$ to a tiny region of radius a, which contains the possibly ionized gases produced in the blast, as well as the remains of the bomb and other debris. The huge pressure in the initial fireball creates a shock front expanding at supersonic speed. After some time t, the shock front has become nearly spherical with a radius $R(t)$ that is very large compared to the initial size a. The volume inside the front contains essentially all of the initial energy $\Delta\mathcal{E}$ in the form of "shocked" air with only little contamination from the debris. The shock has now become a purely atmospheric phenomenon and all details about its origin in any particular explosion have been "forgotten".

Figure 26.7. Trinity fireball 16 ms after detonation where the radius is about 110 m. The sharp outline of the fireball indicates that it coincides with the shock front. Later the shock front will detach from the fireball. (Source: Los Alamos National Laboratory, New Mexico, USA.)

Under these conditions the expanding radius of the shock front $R(t)$ will be determined by the equations of gas dynamics, and can only depend on time t, the energy $\Delta\mathcal{E}$, and the atmospheric density ρ_0 and pressure p_0. In a strong shock the ambient pressure p_0 is negligible compared to the pressures inside, implying that $R(t)$ should be finite in the limit of $p_0 \to 0$, and thus only depend on t, $\Delta\mathcal{E}$ and ρ_0. Since $\Delta\mathcal{E}/\rho_0$ is measured in units of of $\mathrm{J\,kg^{-1}\,m^3 = m^5\,s^{-2}}$, the only possible relationship between the radius and time is

$$\boxed{R(t) = A\left(\frac{\Delta\mathcal{E}t^2}{\rho_0}\right)^{1/5},} \tag{26.28}$$

where A is a numerical constant of order unity (which we shall calculate below).

The first nuclear bomb test, codenamed *Trinity*, was carried out at Alamogordo in New Mexico on July 16, 1945. Photographs of the event (one shown in Figure 26.7) and some data were released by the US Atomic Energy Commission in 1947, although the bomb's yield $\Delta\mathcal{E}$ remained classified. Taylor used the above formula to estimate it to be about 10^{14} J from the data accompanying the photographs. This feat created some embarrassment with the security authorities and earned him a mild admonishment, even though he had done absolutely nothing wrong [Batchelor 1996].

Example 26.5 [Trinity blast]: Repeating Taylor's calculation using the fireball in Figure 26.7 with $t = 16$ ms and radius $R \approx 110$ m, it follows indeed from (26.28) with $A = 1$ that $\Delta\mathcal{E} \approx 7.6 \times 10^{13}$ J, corresponding to 18 metric kilotons of the high explosive TNT. The front moves at speed $v_1 = dR/dt = \frac{2}{5}R/t \approx 2750\ \mathrm{m\,s^{-1}}$, implying $\mathsf{Ma}_1 \approx 8.0$ and $\sigma \approx 74$ (for $\gamma = 7/5$). Just inside the front (in the strong shock limit) we find from (26.22) the pressure $p_2 \approx 76$ bar, density $\rho_2 \approx 7.2\ \mathrm{kg\,m^{-3}}$, flow velocity $v_1 - v_2 = 2{,}300\ \mathrm{m\,s^{-1}}$, and temperature $T_2 \approx 3{,}700$ K.

Isentropic radial gas dynamics

A time-dependent spherically invariant flow in an ideal gas is described by a purely radial velocity field $\boldsymbol{v} = u(r,t)\hat{\boldsymbol{e}}_r$, a density field $\rho(r,t)$, a pressure field $p(r,t)$, and the field of specific entropy $S = c_v \log(p\rho^{-\gamma})$. In the absence of gravity, viscosity, and heat conduction, the fields obey the radial equations of motion (see Appendix D),

$$\frac{\partial u}{\partial t} + v\frac{\partial u}{\partial r} = -\frac{1}{\rho}\frac{\partial p}{\partial r}, \tag{26.29a}$$

$$\frac{\partial \rho}{\partial t} + v\frac{\partial \rho}{\partial r} = -\rho\frac{\partial u}{\partial r} - \frac{2\rho u}{r}, \tag{26.29b}$$

$$\frac{\partial S}{\partial t} + u\frac{\partial S}{\partial r} = 0. \tag{26.29c}$$

The first is the Euler equation, the second the radial continuity equation, and the last expresses that the flow is isentropic, so that the specific entropy is constant along a particle orbit. The last equation also shows that if the initial state were *homentropic* with spatially constant specific entropy $\partial S/\partial r = 0$, it would remain so forever. A strong blast can, however, not be homentropic because entropy is produced right at the front and afterward incorporated in the region behind the front as it runs ahead of the mass flow.

Boundary conditions

The boundary conditions are fixed by the shock relations derived in the preceding section. Everywhere outside the shock front, that is for $r > R(t)$, the density and pressure are always equal to the ambient values whereas the front velocity decreases because of the expansion. Pretending that we do not know the time dependence of the shock radius $R(t)$, we have in the notation of the preceding section,

$$\rho_1 = \rho_0, \qquad p_1 = p_0, \qquad v_1 = \dot{R}(t), \tag{26.30}$$

where a dot indicates differentiation with respect to time.

Just inside the shock front for $r \lesssim R(t)$, the pressure p_2, density ρ_2, and flow velocity $u_2 = v_1 - v_2$ are given in terms of the time-dependent strength $\sigma(t)$ by (26.18) and (26.19). It is, however, more convenient to use the expressions derived in Problem 26.3, to write the boundary conditions in the form,

$$u(R(t),t) = \frac{2}{\gamma+1}\dot{R}(t)\left(1 - \frac{1}{\mathsf{Ma}_1(t)^2}\right), \tag{26.31a}$$

$$p(R(t),t) = \frac{2}{\gamma+1}\rho_0\dot{R}(t)^2\left(1 - \frac{\gamma-1}{2\gamma}\frac{1}{\mathsf{Ma}_1(t)^2}\right), \tag{26.31b}$$

$$\rho(R(t),t) = \frac{\gamma+1}{\gamma-1}\rho_0\left(1 + \frac{2}{\gamma-1}\frac{1}{\mathsf{Ma}_1(t)^2}\right)^{-1}, \tag{26.31c}$$

where

$$\mathsf{Ma}_1(t) = \frac{\dot{R}(t)}{c_0}, \qquad c_0 = \sqrt{\frac{\gamma p_0}{\rho_0}}, \tag{26.32}$$

is the Mach number of the shock front.

Strong self-similar shock

In principle what remains is to solve the radial gas dynamics equations (26.29) with the boundary conditions (26.31). The solution must also provide an expression for the expansion radius $R(t)$, which for a strong shock has to agree with Taylor's estimate (26.28). It is, however, as always non-trivial to solve the coupled partial differential equations.

Things simplify decisively in the strong shock limit, $\mathsf{Ma}_1 \to \infty$. Since p_0 becomes negligible, the only parameter with dimension of length is the radius of the shock front $R(t)$, and the properly scaled non-dimensional radial variable is therefore

$$\xi = \frac{r}{R(t)}. \tag{26.33}$$

We shall assume that we may write each field as a dimensional expression times a spatial form factor, depending only on this variable,

$$u = \dot{R}(t)w(\xi), \qquad \rho = \rho_0 f(\xi), \qquad p = \rho_0 \dot{R}(t)^2 q(\xi). \tag{26.34}$$

The boundary conditions at $\xi = 1$ are determined from the strong shock limit (26.22):

$$f(1) = \frac{\gamma+1}{\gamma-1}, \qquad w(1) = q(1) = \frac{2}{\gamma+1}. \tag{26.35}$$

It may be shown that the self-similar fields cannot satisfy the boundary conditions (26.31) for a weaker shock with a Mach number approaching unity.

Inserting these fields into (26.29) we obtain three coupled ordinary first-order differential equations, called the *Taylor–Sedov equations*, for the form factors (with a prime for differentiation with respect to ξ),

$$\alpha w + (w - \xi)\, w' = -\frac{q'}{f}, \tag{26.36a}$$

$$(w - \xi)\, f' = -\left(w' + \frac{2w}{\xi}\right) f, \tag{26.36b}$$

$$2\alpha + (w - \xi)\left(\frac{q'}{q} - \gamma\frac{f'}{f}\right) = 0, \tag{26.36c}$$

$$\alpha = \frac{R\ddot{R}}{\dot{R}^2}. \tag{26.36d}$$

Leonid Ivanovich Sedov (1907–1999). Russian physicist with broad interests in mechanics. Contributed in particular to the theory of high-speed motion of bodies in water, and to the theory of intense explosions. He was the first to identify the field of continuum mechanics and recognize its general importance in physics research and teaching.

Since α only depends on t and the other functions only on ξ, it follows from the above equations that α must be a constant, independent of time. The solution to (26.36d) is then a power law $R(t) \sim t^{1/(1-\alpha)}$. The value of α cannot be determined from the self-similar equations of motion and the boundary conditions. We shall determine it below by requiring that the excess energy inside the shock front be constant.

Numerical solution

Although it is possible to find an analytic solution [Sedov 1959], it turns out to be quite complicated, and it is much easier to integrate the differential equations numerically. The numeric solution is plotted in Figure 26.8L for $\gamma = 7/5$ with the functions normalized by their boundary values at $\xi = 1$. Evidently there are two distinct regions in a strong shock. For $\xi \lesssim 0.7$ a *core* is formed with linearly rising velocity, nearly vanishing density and spatially constant pressure with $q(0)/q(1) \approx 0.37$ or $q(0) = 0.30$. For $\xi \gtrsim 0.7$ the density and pressure rise rapidly to meet the required values at the front.

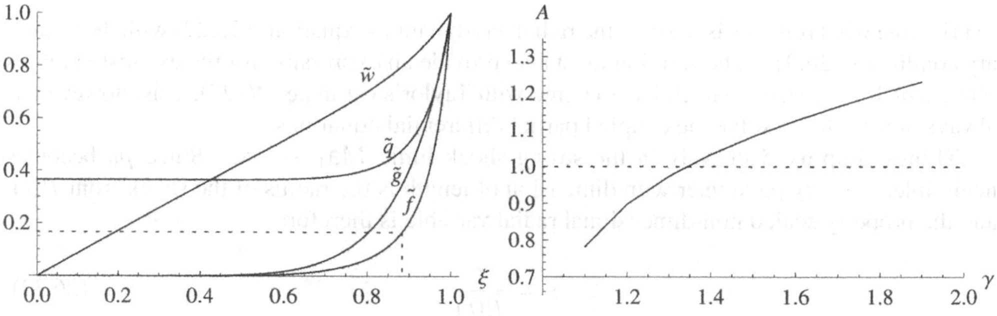

Figure 26.8. Left: The numerical solution to the dynamic equations (26.36) for $\gamma = 7/5$ as a function of $\xi = r/R(t)$. All form factors, $\tilde{q} = q(\xi)/q(1)$ etc., are normalized to their values at $\xi = 1$. The curve $\tilde{g} = \tilde{f}(\xi)/\tilde{q}(\xi)$ is the form factor of the inverse temperature. Notice the distinct core with linearly rising velocity, constant pressure, and nearly vanishing density. The dashed line indicates where atmospheric density is passed (at $\xi = 0.88$). **Right:** The dimensionless coefficient A in the blast radius (26.28) as a function of γ.

Excess energy

The excess energy inside the shock front consists of the kinetic energy of the moving gas plus its internal energy minus the internal energy of the initial undisturbed atmosphere. Using the specific internal energy of an ideal gas, $U = p/\rho(\gamma - 1)$, the excess energy in the shock sphere becomes

$$\Delta\mathcal{E} = \int_0^{R(t)} \left(\frac{1}{2}\rho(r,t)u(r,t)^2 + \frac{p(r,t) - p_0}{\gamma - 1}\right) 4\pi r^2\, dr. \tag{26.37}$$

Barring radiative losses, this energy of the comoving control volume must be conserved and equal to the energy added to the atmosphere in the initial point-like explosion. It takes some algebraic work to demonstrate explicitly from the dynamics (26.29) that the time derivative of this expression indeed vanishes, but for general reasons it is so.

Inserting the self-similar fields (26.34), the excess energy becomes (for $p_0 \to 0$)

$$\Delta\mathcal{E} = \rho_0 R^3 \dot{R}^2 K(\gamma), \tag{26.38}$$

where $K(\gamma)$ is the dimensionless integral,

$$K(\gamma) = 4\pi \int_0^1 \left(\frac{1}{2} f(\xi) w(\xi)^2 + \frac{q(\xi)}{\gamma - 1}\right) \xi^2\, d\xi. \tag{26.39}$$

Since $R(t) \sim t^{1/1-\alpha}$, it follows that $\mathcal{E} \sim R^3 \dot{R}^2 \sim t^{(3+2\alpha)/(1-\alpha)}$, so for the excess energy to be constant the exponent must be $\alpha = -3/2$, and thus $R \sim t^{2/5}$. This confirms the dimensional analysis (26.28), and inserting $\dot{R} = 2R/5t$ in (26.38) we also obtain an expression for the dimensionless constant,

$$A(\gamma) = \left(\frac{25}{4K(\gamma)}\right)^{1/5}. \tag{26.40}$$

In Figure 26.8R the numeric solution is plotted for a range of γ values. For $\gamma = 7/5$ we have $K = 5.3$, and thus $A = 1.03$, which justifies taking $A \approx 1$ in our earlier estimates.

The weakening shock

As the shock front expands, it decreases in strength until it no longer satisfies the conditions for the strong shock approximation used above. The characteristic Mach number for the breakdown of the strong shock approximation may be estimated as $\mathsf{Ma}_1 = \sqrt{2\gamma/(\gamma-1)}$, where the two terms in the density function (26.31c) become equal. For $\gamma = 7/5$ we find at this point that $\mathsf{Ma}_1 = \sqrt{7} = 2.65$, and thus $v_1 = \mathsf{Ma}_1 c_1 = 900\ \mathrm{m\,s^{-1}}$. The strength becomes $\sigma = 7$ and thus the pressure $p_2 = 8$ atm, density $\rho_2 = 4.2\ \mathrm{kg\,m^{-3}}$, flow velocity $u_2 = 650\ \mathrm{m\,s^{-1}}$, and temperature $T_2 = 670\,\mathrm{K}$. Since the front velocity scales as $v_1 = \dot{R} \sim t^{-3/5}$, the breakdown time ratio may be estimated from the front velocity ratio. In the *Trinity* case (Example 26.5), the breakdown happens $t \approx 100$ ms when the front radius is $R \approx 230$ m. The Mach number of the flow just behind the front is now $\mathsf{Ma} = 1.25$.

Continuing the expansion beyond this point, the core pressure will decrease until it reaches the ambient pressure p_0. At this point the core is in force balance with the surrounding atmosphere and will stop expanding. The core is very hot with a correspondingly low density, and its buoyancy will make it rise like a thermal bubble, creating thereby the well-known stalk of the mushroom cloud. We may estimate the time when the core expansion ceases by equating the core pressure $p(0) = q(0)\rho_0 v_1^2$ with atmospheric pressure p_0. This is probably a rather bad approximation, but in lieu of a better one we find in the *Trinity* case (Example 26.5) that the front velocity at this point is $v_1 \approx 530\ \mathrm{m\,s^{-1}}$. Using the above scaling relations, this happens at $t \approx 250$ ms when the front radius is $R \approx 330$ m. After this point the shock continues as a spherical wave in the form of a thin shell while the core rises.

Problems

26.1 Express the strength of a hydraulic jump in terms of the Froude number. Show that it is possible to make a linear approximation that is better than 3.3% for $\mathsf{Fr} > 2$.

26.2 Discuss mass conservation for **(a)** the river bore and **(b)** the reflection bore, expressed in terms of the actual velocities in the relevant frames of reference (see the margin figures on page 447).

26.3 **(a)** Verify that the solution to the Rankine–Hugoniot equations (26.15) may be written as

$$\frac{v_2}{v_1} = \frac{\rho_1}{\rho_2} = \frac{\gamma-1}{\gamma+1}\left(1 + \frac{2}{\gamma-1}\frac{1}{\mathsf{Ma}_1^2}\right), \qquad \frac{p_2}{\rho_1 v_1^2} = \frac{2}{\gamma+1}\left(1 - \frac{\gamma-1}{2\gamma}\frac{1}{\mathsf{Ma}_1^2}\right).$$

(b) Show that for a normal wave (with $u_2 = v_1 - v_2$),

$$\frac{u_2}{v_1} = \frac{2}{\gamma+1}\left(1 - \frac{1}{\mathsf{Ma}_1^2}\right).$$

26.4 In a general material, the Bernoulli relation takes the form

$$\frac{1}{2}v_1^2 + U_1 + \frac{p_1}{\rho_1} = \frac{1}{2}v_2^2 + U_2 + \frac{p_2}{\rho_2},$$

where U_1 and U_2 are the specific internal energies before and after the shock front. Show that

$$U_2 - U_1 = \frac{(p_1 + p_2)(\rho_2 - \rho_1)}{2\rho_1\rho_2}.$$

This identity goes under the name of Hugoniot's equation.

26.5 Show that the entropy change (26.23) is a growing function of σ for $\sigma > -1$.

$*$ **26.6** Solve the Rankine–Hugoniot equations with different adiabatic indices γ_1 and γ_2.

26.7 The largest hydrogen bomb ever detonated in the atmosphere had a yield of 50 megatons of TNT (about 2.5×10^{17} J). **(a)** Estimate the radius and velocity of the shock front 1 s after the explosion. **(b)** Estimate the pressure, density, and temperature just behind the shock front at this time. **(c)** Show that the strong shock approximation is still valid at this time. **(d)** Calculate the parameters when the strong shock approximation breaks down. **(e)** Estimate the parameters when the core stops expanding.

27 Whirls and vortices

Whirls and vortices are common features of real fluids. Stirring the coffee, you create a circulating motion, a whirl that dies out after some time when you stop stirring. A fast spinning vortex may form at the drain of a bathtub when it is emptied, and can remain quasi-stable as long as there is enough water in the tub. Normally, whirls and vortices are invisible far away from open surfaces, but in the bathtub the drain vortex is made visible by soap remains clouding the water and the depression it creates at the surface.

In the atmosphere above heated ground, whirls and vortices arise all the time. Mostly they are small and invisible, but sometimes they pick up dust and debris and appear as dust devils. Larger dust devils, in some countries called sky-pumps, are known to pick up haystacks and scatter them, or to throw tables around in sidewalk cafes. When the heat-driven air vortices grow really big and nasty, they become tornadoes. The force powering heat-driven vortices is fundamentally the same as the force that maintains a bathtub vortex, namely gravity. Whereas a bathtub vortex is driven by gravity acting on the water going down the drain, a tornado is maintained by the buoyancy of hot air draining skyward.

Vortices are also found in the wakes of moving objects. From the tips of the wings of aircraft there will always trail long mostly invisible vortices (see Chapter 29). Large passenger aircraft taking off or landing create strong and fairly stable vortices, capable of overturning smaller planes following after. "Beware of vortex" is a common warning issued by flight controllers to small aircraft taking off or landing behind heavy passenger planes.

The chapter begins with a discussion of the structure of free cylindrical vortices in an incompressible fluid, and continues with an analysis of singular vortex dynamics in non-viscous fluids. The remainder of the chapter concerns non-singular laminar vortices in viscous fluids driven by secondary flow. Extensive treatment of vortex structure and dynamics may be found in [Saffman 1992] and [Maurel and Petitjeans 2000].

27.1 Free cylindrical vortices

The spindle-driven vortex discussed on page 281 was powered by a solid cylindrical spindle of radius a, which delivered the work necessary to overcome viscous friction in the fluid by rotating with angular velocity Ω. Although the flow pattern, $v_\phi = \Omega a^2/r$, did not depend explicitly on the viscosity of the fluid, a finite viscosity was nevertheless necessary for the spindle to be able to "crank up" the vortex starting from fluid at rest.

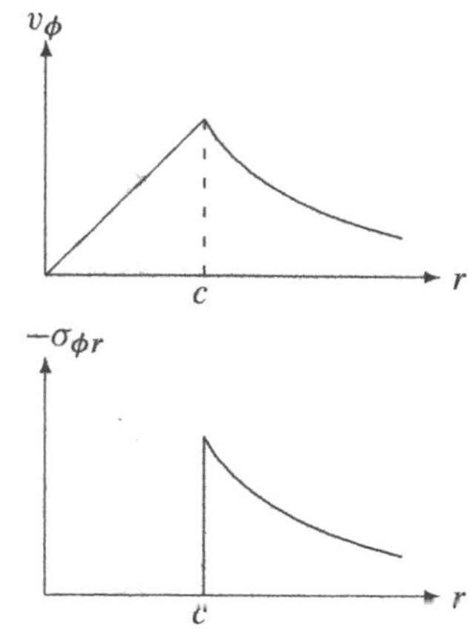

Sketch of the velocity field and shear stress for the Rankine vortex.

Rankine vortex

In a truly inviscid fluid, we would not be able to crank up the vortex, but once it had somehow been established, the vortex would keep spinning forever. The solid spindle could then with impunity be replaced by a *core* made from the same fluid as the vortex and rotating as a solid body with the same angular velocity Ω as the spindle.

The *Rankine vortex* with core radius c is constructed in precisely this way,

$$v_\phi = \begin{cases} \Omega r & 0 \leq r \leq c \\ \dfrac{\Omega c^2}{r} & c \leq r < \infty. \end{cases} \tag{27.1}$$

The vorticity, $\omega_z = dv_\phi/dr + v_\phi/r$, vanishes for $r > c$ and is constant for $r < c$.

In the absence of gravity, the pressure $p = p(r)$ can only depend on r, and the radial pressure gradient $-dp/dr$ must provide the centripetal force $-\rho_0 v_\phi^2/r$ necessary to maintain the circular fluid motion (sometimes called *cyclostrophic balance*),

$$\frac{dp}{dr} = \rho_0 \frac{v_\phi^2}{r}. \tag{27.2}$$

Integrating this equation and requiring the pressure to vanish at infinity, we obtain

$$p = -\rho_0 \int_r^\infty \frac{v_\phi^2}{r}\, dr = -\frac{1}{2}\rho_0 \Omega^2 c^2 \begin{cases} 2 - \dfrac{r^2}{c^2} & r < c \\ \dfrac{c^2}{r^2} & r > c. \end{cases} \tag{27.3}$$

The cusp in the Rankine vortex velocity at the core boundary generates a finite jump in the shear stress, $\sigma_{\phi r} = \eta(dv_\phi/dr - v_\phi/r)$, which is physically unacceptable because it violates Newton's Third Law. The Rankine vortex is only meaningful for truly inviscid flow.

Horace Lamb (1849–1934). English physicist, who contributed mainly to acoustics and fluid mechanics. His textbooks remained for many years the standard in these fields. (Source: Wikimedia Commons.)

Lamb vortex

There is an infinity of smooth interpolating functions that coincide with the Rankine vortex for both $r \ll c$ and $r \gg c$, for example the *Lamb vortex* shown in Figure 27.1L for $t = 0$,

$$v_\phi = \frac{\Omega c^2}{r}\left(1 - e^{-r^2/c^2}\right). \tag{27.4}$$

The vorticity is nearly constant for $r \lesssim 0.5c$ and vanishes rapidly for $r \gtrsim 2c$. The shear stress is continuous everywhere and does not violate Newton's Third Law.

Integrating (27.2) with the interpolating vortex and requiring the pressure to vanish at infinity, one gets

$$p = -\frac{1}{2}\rho_0 \Omega^2 c^2 F\left(\frac{r^2}{c^2}\right), \tag{27.5}$$

where $F(\xi)$ is a purely mathematical function,

$$F(\xi) = \int_\xi^\infty (1 - e^{-x})^2 \frac{dx}{x^2}. \tag{27.6}$$

The pressure is shown in Figure 27.1R for $t = 0$.

The resolution of the stress problem comes, however, at a price, because the interpolating vortex cannot be stable on its own in a viscous fluid. The shear stress will cause dissipation of kinetic energy everywhere in the fluid and make the vortex unavoidably spin down. The numerically largest shear stress is found just outside the core and will slow down the core's rotation while expanding its radius to conserve angular momentum. This suggests that the Lamb vortex with a time-dependent core radius might be an exact solution to the Navier–Stokes equations.

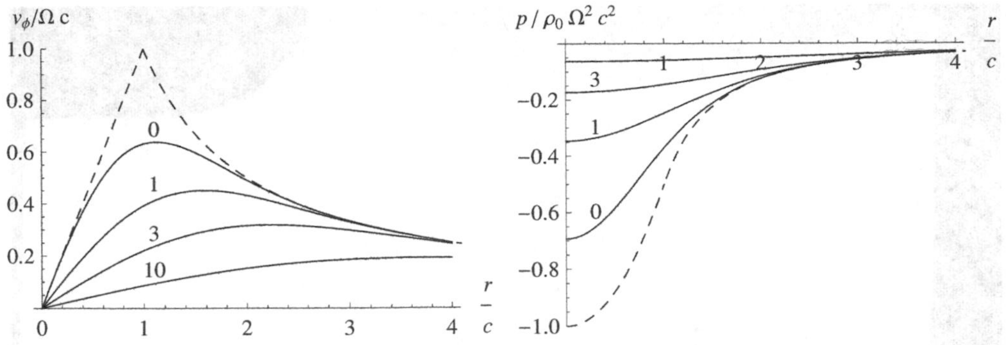

Figure 27.1. Time evolution of the velocity (left) and pressure (right) for the Oseen–Lamb vortex at selected times $t/\tau = 0, 1, 3, 10$. The core grows larger with time and the central angular velocity (i.e., the slope of v_ϕ at $r = 0$) becomes smaller. The dashed curves are for the Rankine vortex, and the curves marked 0 are for the Lamb vortex.

Oseen–Lamb vortex

Under the influence of viscosity, a free vortex will always decay and its swirling fluid will in the end come to rest. Assuming that the time-dependent velocity field retains its cylindrical form $\boldsymbol{v} = v_\phi(r,t)\,\hat{\boldsymbol{e}}_\phi$, the Navier–Stokes equations are (see Appendix D)

$$\frac{\partial v_\phi}{\partial t} = \nu \frac{\partial}{\partial r}\left(\frac{1}{r}\frac{\partial (r v_\phi)}{\partial r}\right), \qquad \frac{\partial p}{\partial r} = \rho_0 \frac{v_\phi^2}{r}. \tag{27.7}$$

The first equation bears a strong resemblance to the momentum diffusion equation (15.5) on page 246, although in this case it is angular momentum that diffuses. The second expresses cyclostrophic balance as in the Rankine and Lamb vortices.

Given any initial field, $v_\phi(r, 0)$, the above dynamic equation will always produce a solution. By direct insertion one may verify that the following expression, which starts out at $t = 0$ as the Lamb vortex, is a solution:

Carl Wilhelm Oseen (1879–1944). Swedish mathematical physicist. Worked on relativity. Influential member of the Nobel committee, he was instrumental in awarding Einstein the prize in 1921 for the photoelectric effect, rather than for relativity.

$$v_\phi(r,t) = \frac{\Omega c^2}{r}\left(1 - \exp\left(-\frac{r^2}{c^2 + 4\nu t}\right)\right). \tag{27.8}$$

The pressure is obtained as before,

$$p = -\rho_0 \int_r^\infty \frac{v_\phi^2}{r}\,dr = -\frac{1}{2}\rho_0\Omega^2 \frac{c^4}{c^2 + 4\nu t} F\left(\frac{r^2}{c^2 + 4\nu t}\right), \tag{27.9}$$

where $F(\xi)$ as before is given in (27.6). This solution to a vortex at peace with viscosity is attributed to Oseen and Lamb. The shape and time evolution of its velocity and pressure is shown in Figure 27.1.

The time-dependent core radius and angular velocity of the core become

$$c(t) = \sqrt{c^2 + 4\nu t}, \qquad \Omega(t) = \left.\frac{\partial v_\phi}{\partial r}\right|_{r=0} = \Omega \frac{c^2}{c^2 + 4\nu t}. \tag{27.10}$$

Core expansion sets in at time $\tau = c^2/4\nu$, and for $t \gg \tau$ the expanding core grows like $c(t) = 2\sqrt{\nu t}$, which is recognized as the typical diffusion scale (see page 248).

Finally, it should be mentioned that there are many other analytic solutions to (27.7), some of which are explored in Problems 27.13 and 27.14.

Figure 27.2. Tornado over water, also called a water spout. (Source: US National Oceanic and Atmospheric Administration's National Weather Service.)

27.2 Basic vortex theory

In 1858, Helmholtz presented a set of theoretical insights into the dynamics of vortices in inviscid fluids[1]. These insights showed that vortices in ideal flow may be thought of as a kind of indestructible objects. To see that, we need to develop the concept of vorticity and circulation a bit further than in Section 13.5 on page 219.

Vortex tubes

The set of vortex lines that pass through the points of a closed curve at a given time form together a *vortex tube* that may curve and bend in space. As it consists entirely of vortex lines, a vortex tube cannot terminate anywhere in the fluid but must connect to the boundaries (possibly at infinity) or perhaps close on itself in particular simple geometries.

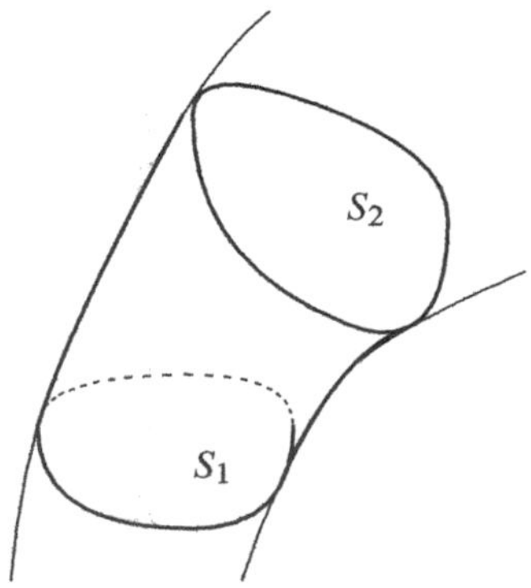

The flux of vorticity is the same in every cross-section of a vortex tube.

We shall now prove that whatever shape it takes, *the total flux of vorticity through every cross-section of a vortex tube is the same.* Consider a closed surface formed by a stretch of a vortex tube between two cross-sections, S_1 and S_2. Since the divergence of the vorticity field $\boldsymbol{\omega} = \nabla \times \boldsymbol{v}$ always vanishes, $\nabla \cdot \boldsymbol{\omega} = 0$, Gauss' theorem tells us that the surface integral over this stretch of tube must always vanish,

$$\oint_S \boldsymbol{\omega} \cdot d\boldsymbol{S} = \int_{S_2} \boldsymbol{\omega} \cdot d\boldsymbol{S} - \int_{S_1} \boldsymbol{\omega} \cdot d\boldsymbol{S} = 0. \tag{27.11}$$

The sides of the tube do not contribute to the integral because the surface elements are orthogonal to the vortex lines, that is, to the vorticity field, and that proves the claim.

[1] Vortex theory is still a very active subject; see for example [Aref 2010].

Kelvin's circulation theorem

To see that vortices in ideal fluids almost behave as physical objects that move along with the motion of the fluid around them, we first need to prove a theorem due to Kelvin. Kelvin's circulation theorem states that *in an inviscid fluid, the circulation around any closed material (comoving) curve is independent of time*[2], or

$$\boxed{\frac{D\Gamma}{Dt} \equiv \frac{d\Gamma(C(t),t)}{dt} = 0.} \tag{27.12}$$

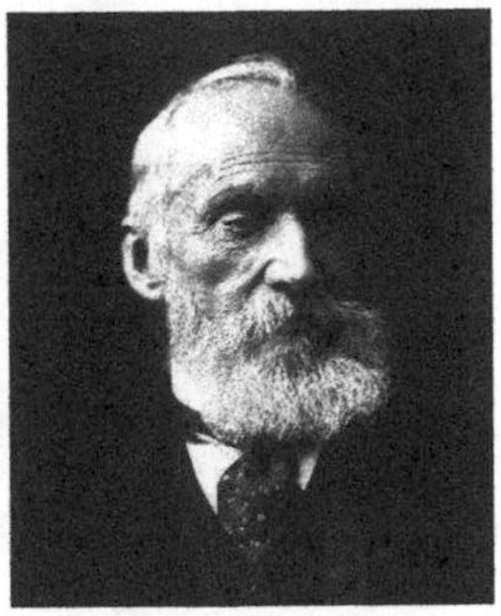

William Thomson, alias Lord Kelvin (1824–1907). Scottish mathematician and physicist. Instrumental in the development of thermodynamics, in particular the relation between the Second Law and irreversibility. Viewed electromagnetic forces as elastic strains in the ether. (Source: Wikimedia Commons.)

The closed material curve $C(t)$ is thus washed along with the fluid and may change shape dramatically without any change in the circulation of the velocity field. Kelvin's theorem applies only to ideal flow. In a viscous fluid, the circulation will change at a rate proportional to the viscosity, which both dissipates and generates vorticity. It is as mentioned before impossible to generate vorticity—or to get rid of it—without the aid of viscosity.

Proof: The proof of the theorem is fairly straightforward. Let $C(t)$ be the comoving closed curve. In a small time interval δt, the circulation along this curve changes by

$$\begin{aligned}\delta\Gamma(C(t),t) &= \Gamma(C(t+\delta t), t+\delta t) - \Gamma(C(t),t)\\ &= \oint_{C(t+\delta t)} \boldsymbol{v}(\boldsymbol{x}', t+\delta t)\cdot d\boldsymbol{\ell}' - \oint_{C(t)} \boldsymbol{v}(\boldsymbol{x},t)\cdot d\boldsymbol{\ell}\\ &= \oint_{C(t)} \boldsymbol{v}(\boldsymbol{x}+\boldsymbol{v}(\boldsymbol{x},t)\delta t, t+\delta t)\cdot(d\boldsymbol{\ell} + (d\boldsymbol{\ell}\cdot\nabla)\boldsymbol{v}\delta t) - \oint_{C(t)} \boldsymbol{v}(\boldsymbol{x},t))\cdot d\boldsymbol{\ell}.\end{aligned}$$

In the last step we have used that the transformation $\boldsymbol{x}' = \boldsymbol{x} + \boldsymbol{v}(\boldsymbol{x},t)\delta t$ maps the curve $C(t)$ on $C(t+\delta t)$, thereby transforming the line element to $d\boldsymbol{\ell}' = d\boldsymbol{\ell} + (d\boldsymbol{\ell}\cdot\nabla)\boldsymbol{v}\delta t$.

Expanding everything to first order in δt, we have

$$\begin{aligned}\delta\Gamma(C(t),t) &= \delta t \oint_{C(t)} \left(\frac{\partial \boldsymbol{v}}{\partial t} + (\boldsymbol{v}\cdot\nabla)\boldsymbol{v}\right)\cdot d\boldsymbol{\ell} + \delta t \oint_{C(t)} \boldsymbol{v}\cdot(d\boldsymbol{\ell}\cdot\nabla)\boldsymbol{v}\\ &= \delta t \oint_{C(t)} \left(\boldsymbol{g} - \frac{\nabla p}{\rho_0}\right)\cdot d\boldsymbol{\ell} + \delta t \oint_{C(t)} \frac{1}{2}\nabla\left(\boldsymbol{v}^2\right)\cdot d\boldsymbol{\ell}\\ &= 0.\end{aligned}$$

In the second step the Euler equation was used in the first term and the second term was rewritten conveniently. Since all the terms are gradient fields, their integrals around the closed curve vanish, and Kelvin's theorem now follows.

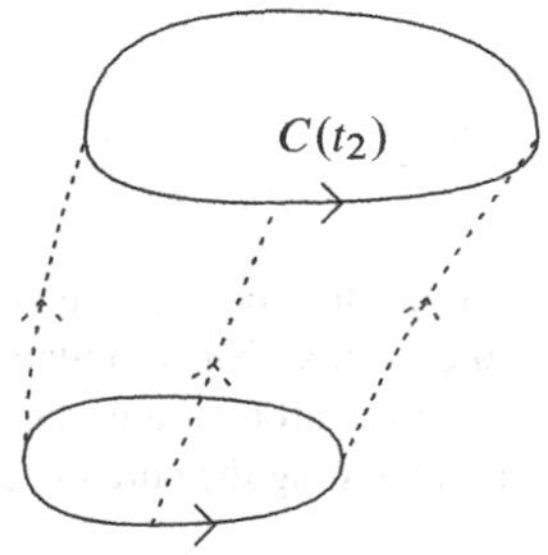

In ideal flow, the circulation around the comoving curve $C(t)$ is the same at t_1 as at t_2. The dashed lines indicate material particle trajectories.

Although we have carried it through for incompressible fluid, the proof may be extended to barotropic compressible fluid (Problem 27.5).

Vortex core

Vortices typically have a central vortex tube (see Figure 27.2) carrying almost all of the vorticity with nearly vanishing vorticity outside, as we saw for the Oseen–Lamb vortex. This tube defines the *core* of the vortex, and the vorticity flux carried by the core is called the *strength* of the vortex. By Stokes' theorem, the strength may also be calculated from the circulation of any closed curve that encircles the core once without going through it. By Kelvin's theorem, the strength remains unchanged in an inviscid fluid, independent of how the core of vortex may twist and turn as it is blown along with the ambient flow.

[2] Kelvin's theorem is not valid for viscous fluids or for inviscid relativistic fluids. The latter may be of importance for understanding the otherwise mysterious origin of galactic magnetic fields [MY10].

The core and with it the whole vortex thus acquires an identity of its own. Nevertheless, the unavoidable viscosity—how small it may be—will slowly sap the strength of a free vortex and unrelentingly spin it down while the core expands, until the vortex in the end is completely gone.

27.3 Line vortices

In the extreme limit where the vortex core becomes infinitely thin and its vorticity infinitely large while the vortex strength remains constant, the thread-like core is called a *vortex filament*, and the singular vortex itself is called a *line vortex* (not to be confused with a vortex line). A line vortex is entirely characterized by its strength and the curve that describes the whereabouts of the filament[3].

Viscosity will of course make the core of any free vortex expand and eventually cause the whole vortex to disappear, but if the core radius is small compared to the radius of curvature of the filament and other length scales in the flow, a real vortex may behave as a line vortex over longer periods of time. In dealing with line vortices, one should never forget that the core radius does not truly vanish. Although the formalism may be prettier, singular quantities are anathema to physics, and there will unavoidably arise paradoxes and meaninglessness. Some of problems with the singular vortex formalism are discussed in [Saffman 1992].

In this section we shall only consider straight parallel line vortices that give rise to an efficient and well-behaved two-dimensional formalism.

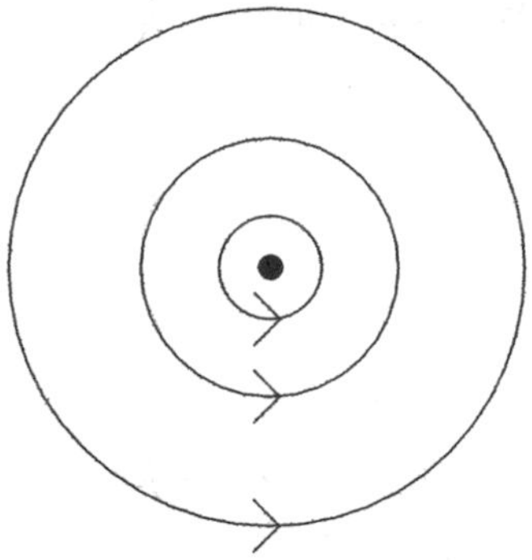

Irrotational flow outside the core of a line vortex. The circulation is the same around any curve circumnavigating the core once.

Straight line vortex

The velocity field of a straight line vortex along the z-axis is, in both cylindrical or Cartesian coordinates,

$$\boldsymbol{v} = \frac{C}{r}\,\hat{\boldsymbol{e}}_\phi = C\frac{(-y, x, 0)}{x^2 + y^2}. \qquad (27.13)$$

Comparing with the Rankine vortex (27.1) we see that the straight-line vortex is obtained in the limit $\Omega \to \infty$ and $c \to 0$ such that $C = \Omega c^2$ remains finite. Integrating around the vortex along a circle of radius r, we find the circulation $\Gamma = 2\pi r v_\phi = 2\pi C$, which by Stokes' theorem is also the strength of the line vortex. The circulation per unit of angle, $C = \Gamma/2\pi$, shall in the following be called the *reduced strength*.

Outside the singular core the velocity field has vanishing vorticity and must therefore be a gradient field, $\boldsymbol{v} = \boldsymbol{\nabla}\Psi$, where Ψ is the velocity potential. Using the gradient operator expressed in cylindrical coordinates (see Appendix D), it follows immediately that the gradient of the azimuthal angle ϕ is $\boldsymbol{\nabla}\phi = \hat{\boldsymbol{e}}_\phi/r$. Consequently, the velocity potential of a line vortex may be written as

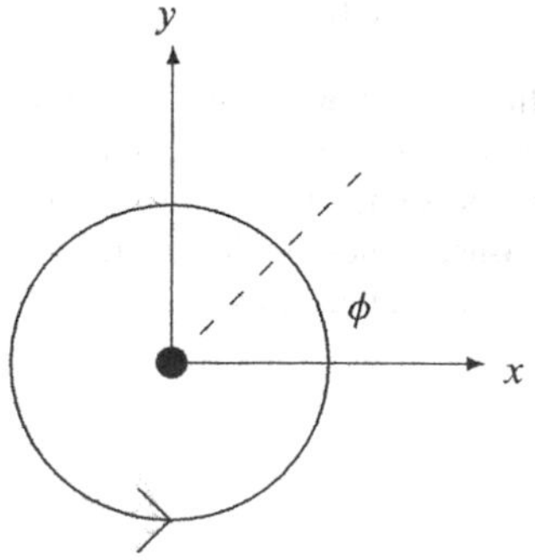

The multivalued angular dependence of the velocity potential of an elementary vortex filament reflects the non-vanishing vorticity carried in its core.

$$\Psi = C\phi = C\arctan\frac{y}{x}. \qquad (27.14)$$

The azimuthal angle ϕ is a multivalued function that increases by 2π each time the curve circles around the singular vortex core, reflecting of course the non-vanishing value of the circulation $\Gamma = 2\pi C$ arising from the infinite vorticity in the core.

The singular core carrying the vorticity makes the space around a straight-line vortex multiple-connected. The circulation should always be calculated for a curve that encircles the core only once.

[3]The indestructibility of vortex filaments in ideal flow inspired Lord Kelvin in 1867 to suggest that atoms could be microscopic ring vortices in the ideal ether, and that vortex theory could be the foundation for what today would be called a "theory of everything". This early "string theory" of matter was quickly abandoned but inspired instead the development of mathematical *knot theory*.

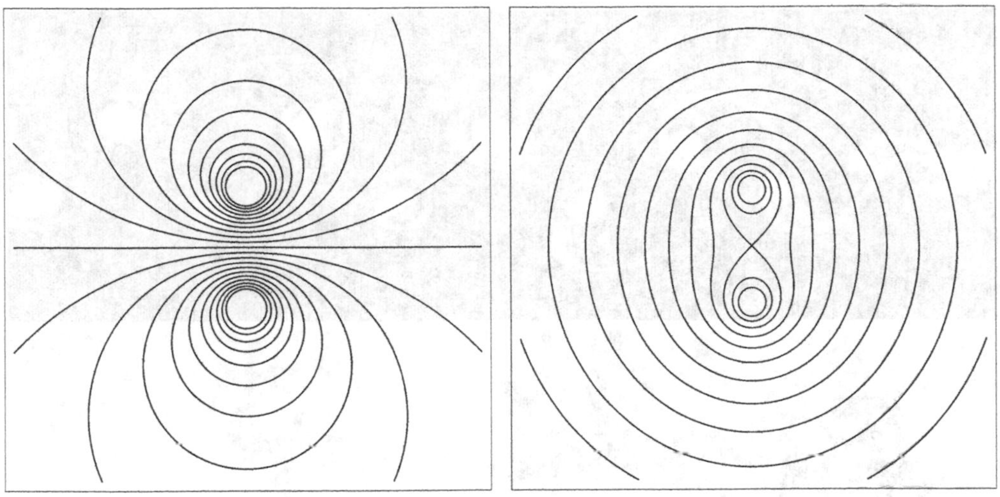

Figure 27.3. Streamlines around a pair of parallel straight vortex filaments in a fluid asymptotically at rest. **Left:** Counter-rotating vortices moving horizontally at the same speed. **Right:** Corotating vortice moving in a circle around the center of the figure.

Complex notation

A straight vortex filament is essentially two-dimensional, and two-dimensional problems are often facilitated by complex notation. Defining the complex position and velocity fields,

$$z = x + iy, \qquad w = v_x - iv_y, \tag{27.15}$$

it follows from the Cartesian form of the line vortex (27.13) that the complex velocity field of a vortex centered at the origin is a simple pole,

$$w = C\frac{-y - ix}{x^2 + y^2} = -i\frac{C}{x + iy} = -i\frac{C}{z}. \tag{27.16}$$

The field of a vortex centered in another point in the complex plane is obtained by shifting the pole position to that point.

Two interacting line vortices

Vortices are not always loners like the bathtub vortex or the tornado. The advection of one vortex filament in the flow created by the others gives rise to a number of interesting, beautiful, and often quite counter-intuitive phenomena. Although the linearity of potential flow allows us to add the velocity fields of the filaments, the end result can become quite complex. For parallel straight filaments, the analysis is much simpler when using complex notation.

Consider now two counter-rotating parallel filaments of opposite reduced strengths, C and $-C$, a distance $2b$ apart. Each vortex will blow the other along with the same velocity $C/2b$, and thus keep a constant distance $2b$ between their centers. If they are positioned at $y = \pm b$, the instantaneous complex velocity field becomes

$$w = -i\frac{C}{z - ib} + i\frac{C}{z + ib} = \frac{2Cb}{z^2 + b^2}. \tag{27.17}$$

The velocity fields v_x and v_y may be calculated from the real and imaginary parts of this expression (Problem 27.6), and the streamlines are shown in Figure 27.3L. At the origin the velocity is $2C/b$, which is four times the drift velocity, $C/2b$, whereas at long distance from the vortices, $r \gg b$, the velocity field vanishes as $1/r^2$.

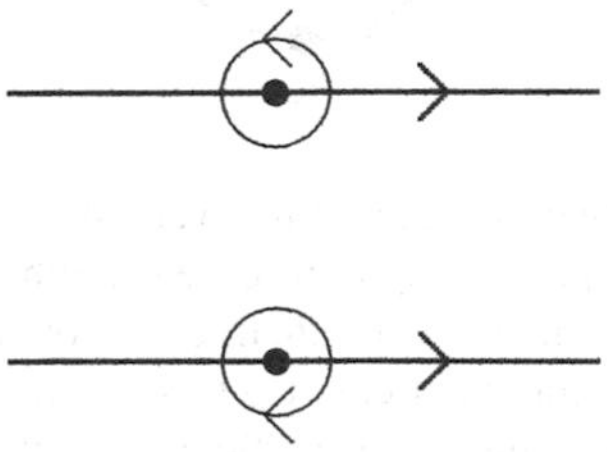

Two straight vortex filaments of equal but opposite strength blow each other in the same direction with the same speed, here toward the right. The same picture could also describe a cross-section through the center of a smoke ring.

Theodore von Kármán (1881–1963). Influential Hungarian-American engineer. Lived from 1930 in the United States, and became in 1944 co-founder of the Jet Propulsion Laboratory at the California Institute of Technology. Made major contributions to the understanding of fluid mechanics, aircraft structures, rocket propulsion, and soil erosion. A crater on the Moon bears his name today. (Source: Wikimedia Commons.)

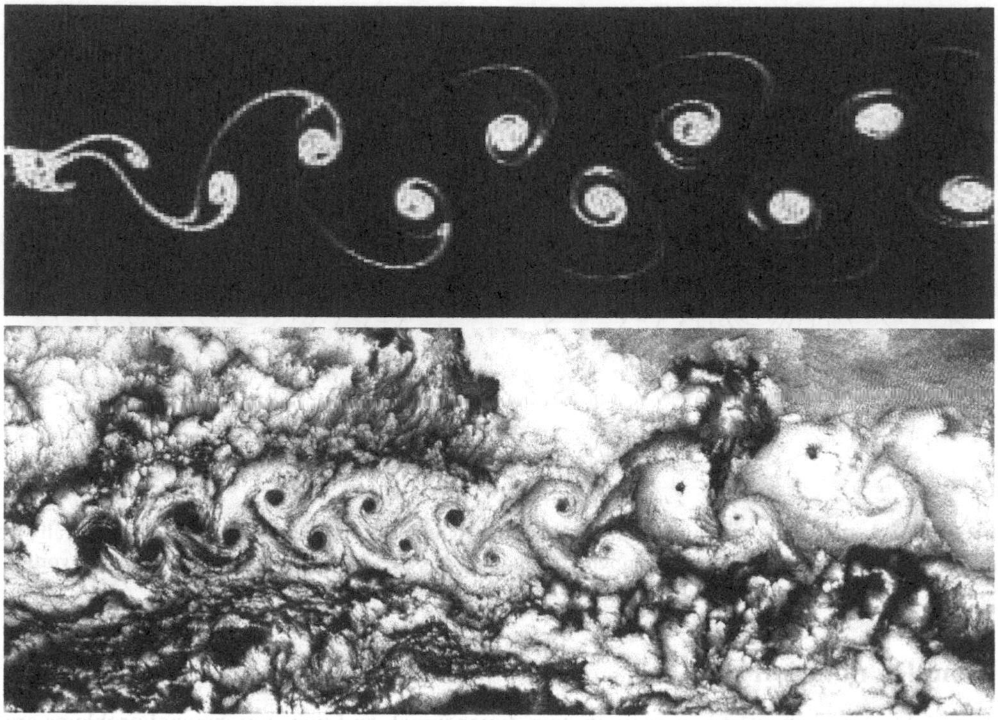

Figure 27.4. Karman vortex streets. **Top:** Photograph of the vortex street behind circular cylinder at Re = 105. (Source: Sadatoshi Taneda. Reproduced from [Dyke 1982].) **Bottom:** Arctic vortex street captured on June 6, 2001. The 300 km long north–south vortex street is formed downwind from the island Jan Mayen situated 650 km northeast of Iceland. The vortex street is created by the 2.2 km high Beerenberg volcano on the island. (Source: NASA/GSFC/LaRC/JPL, MISR Team.)

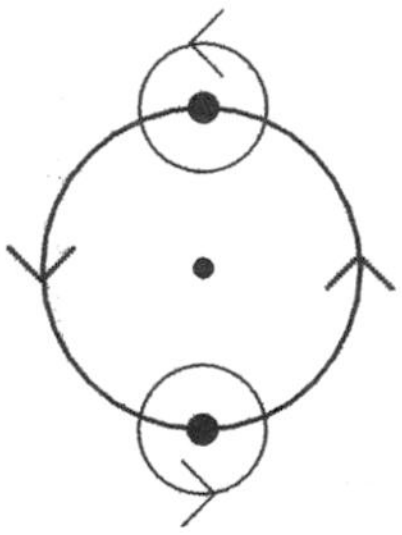

Two parallel line vortices of equal strength force each other to move with their cores on a common circle. Up to six vortex filaments may participate in the dance.

The instantaneous field of a pair of corotating straight vortex filaments of equal reduced strength C a distance $2b$ apart will instead blow each other in opposite directions, making the two vortex cores dance around on a circle of radius b with angular velocity satisfying $\Omega b = C/2b$, or $\Omega = C/2b^2$. Positioned at $y = \pm b$ the instantaneous field is

$$w = -i\frac{C}{z - ib} - i\frac{C}{z + ib} = -i\frac{2Cz}{z^2 + b^2}. \tag{27.18}$$

The streamlines of this field are shown in Figure 27.3R. At the origin the velocity field vanishes, whereas at long distances from the vortices it becomes the field of a single vortex with strength $2C$.

The equations of motion for an arbitrary collection of line vortices may easily be written down (Problem 27.10) and solved for many symmetric initial configurations, but even if the vortices move in a regular fashion, their orbits are not necessarily stable toward small perturbations. For example, if n corotating vortices are positioned regularly around a circle of radius b, the configuration is only stable for $2 \leq n \leq 6$.

* The Karman vortex street

Consider a long object, for example a cylinder, moving through a fluid at rest with constant velocity and its axis orthogonal to the direction of motion. At fairly low Reynolds numbers (well above unity), the boundary layers will detach and form two counter-rotating eddies behind the object. At higher Reynolds numbers around 100, the flow behind the object becomes

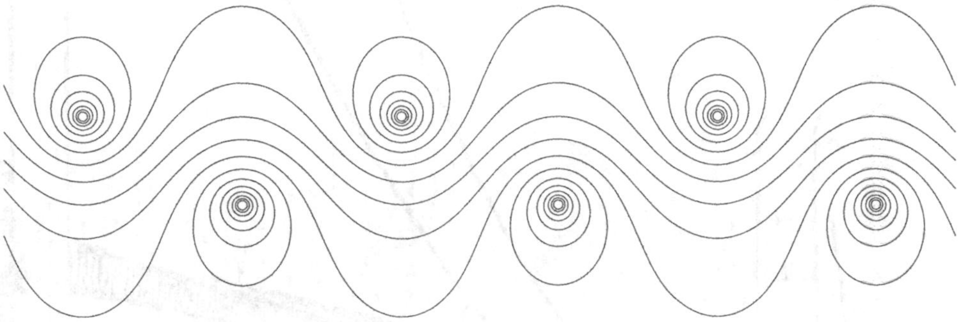

Figure 27.5. Streamlines for the von Karman vortex street in a fluid asymptotically at rest with $b/a = 0.28$. The whole pattern blows itself horizontally at constant speed.

unstable, causing the eddies alternately to detach at regular time intervals. Thus, in the wake of the moving object one may in a certain region of Reynolds numbers observe a beautiful alternating pattern of free counter-rotating vortices shed by the object. This pattern is called the *Karman vortex street*. In Figure 27.4, two examples of real-world vortex streets are shown. At still higher Reynolds numbers the instabilities become chaotic and lead to turbulence.

To analyze the vortex street we first consider an infinite collection of corotating vortices of unit reduced strength spaced regularly with interval $\Delta x = \pi$ along the x-axis. The opposite flows from the infinity of neighbors to the left and right of any given vortex guarantees that they are not mutually advected by each other, and the complex velocity field becomes

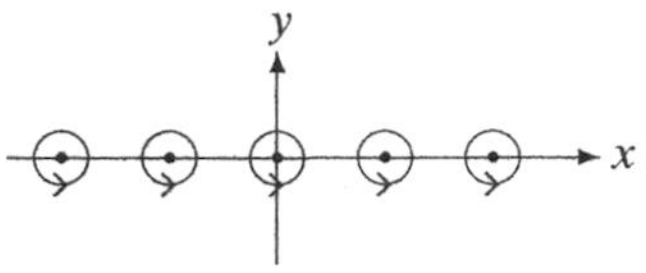

A chain of corotating vortices of equal strength placed at regular intervals along the x-axis.

$$w = -i \sum_{n=-\infty}^{\infty} \frac{1}{z - n\pi} = -i \cot z. \tag{27.19}$$

The proof that the sum equals the cotangent only requires complex function theory at a fairly elementary level (Problem 27.9).

The vortex street is now modeled by two such vortex chains, one at $y = b$ with reduced strength C and the other at $y = -b$ with reduced strength $-C$. In each chain the vortices are spaced regularly with interval $2a$, but one of the chains is shifted by a with respect to the other. The total field becomes

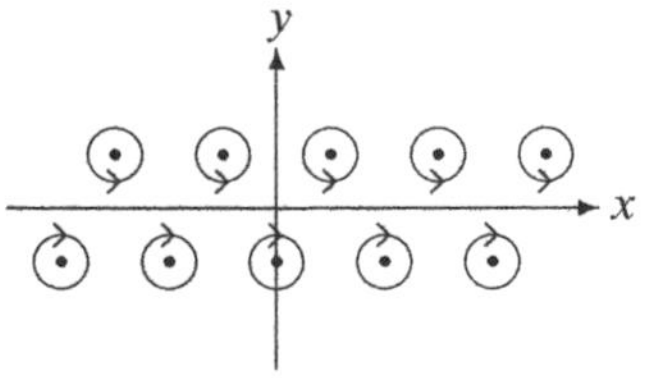

Modeling the von Karman vortex street as two vortex chains of opposite strength.

$$w = -iC \sum_n \frac{1}{z - (2n+1)a - ib} + iC \sum_n \frac{1}{z - 2na + ib}. \tag{27.20}$$

Using $\cot(z + \pi/2) = -\tan z$ we find the total field,

$$\boxed{w = i\frac{C\pi}{2a}\left(\cot\frac{\pi(z+ib)}{2a} + \tan\frac{\pi(z-ib)}{2a}\right).} \tag{27.21}$$

The streamlines of this field are plotted in Figure 27.5 for aspect ratio $b/a = 0.28$.

Evaluating the velocity field of the upper chain on the positions $z = 2na - ib$ of the vortices of the lower chain, we obtain the velocity of the lower chain,

$$W = v_x(2na, -b) = \frac{C\pi}{2a}\tanh\frac{\pi b}{a}. \tag{27.22}$$

This is also the velocity of the upper chain, so that in a fluid at rest the whole vortex system propagates along x with this speed. A closer analysis of the equations of motion for the

Figure 27.6. Tacoma Narrows bridge in torsional oscillation. (Source: University of Washington Libraries, Special Collections, PH Coll 290-31. With permission.)

vortex street [Milne-Thomson 1968, Saffman 1992] reveals that it is unstable to infinitesimal perturbations of the vortex positions, except for

$$\sinh \frac{\pi b}{a} = 1, \tag{27.23}$$

corresponding to the aspect ratio $b/a = 0.2805\ldots$, which is close to the observed values. This is in fact expected to hold for a large class of similar vortex streets [Jim87].

Vibrations driven by periodic vortex shedding

The vortices periodically shed from a moving object, or a stationary object in a uniform flow, act back on the object with a force of alternating sign. Such a force is capable of driving sustained vibrations in the object as well as in the fluid surrounding it. When the driving frequency is close to a natural vibration frequency of the object, strong vibrations may be resonantly excited, a phenomenon known from the "singing" of a taut wire in cross wind.

If the object creating the vortex street is moving with velocity U along x through the fluid, the velocity of the vortex street becomes $U - W$ relative to the object. The (non-circular) frequency of vortex shedding from the moving object becomes

$$f = \frac{U - W}{2a} = \frac{U}{2a}\left(1 - \frac{\pi C}{2Ua} \tanh \frac{\pi b}{a}\right). \tag{27.24}$$

The expression in parenthesis is in fact independent of the velocity U of the object because the eddies giving rise to the detached vortices have a core size comparable to b and thus a reduced strength $C \approx Ub$.

The *Strouhal number* [Str78] is a dimensionless measure of the vortex shedding frequency, defined from the velocity U of the object and its effective diameter d,

$$\text{Sr} = \frac{fd}{U} \qquad (27.25)$$

For the Karman vortex street with aspect ratio $b/a = 0.28$, diameter $d \approx 2b$, strength $C \approx Ub$, and frequency f from (27.24), we find the value $\text{Sr} \approx 0.2$, which is typical.

Vincenc Strouhal (1850–1922). Czech mathematical physicist. Carried out experiments on the aeolian harp in 1878. Founded the physics department of the Karls-Universität in Prague and served as rector.

Example 27.1 [Piano wire]: A piano wire of diameter $d = 1$ mm in a cross wind of $U = 3\text{ m s}^{-1}$ has $\text{Re} \approx 200$. Taking $\text{Sr} \approx 0.2$, the vortex shredding frequency becomes $f = 600$ Hz, which will become audible if close to the natural frequency of the piano wire.

Example 27.2 [Tacoma Narrows bridge]: A spectacular case of large-amplitude wind-driven vibrations caused the collapse of the Tacoma Narrows bridge (Puget Sound, Washington, USA) on November 7, 1940 (see Figure 27.6). Although the vibrations were originally attributed to resonant vortex shedding, it was later realized that this could not be the case. The collapse happened when the amplitude of a torsional mode of the bridge grew beyond the limit of the structural tolerance. On the day of the collapse, the wind speed was $U \approx 20\text{ m s}^{-1}$. With a span thickness of $d \approx 4$ m and a Strouhal number 0.2, this leads to a vortex shedding period of about $1/f = 1$ s, which is too far from the observed 5 s period of the torsional mode.

The mechanism actually driving the amplitude of the torsional oscillation toward collapse was later shown to be a kind of aerodynamic flutter. A fairly recent account of the underlying physics may be found in [BS91]. Photographs and a dramatic film clip of the collapse are readily available at many internet sites.

27.4 Advective vortex spin-up

It is a common observation that a vortex can be "spun up" by letting fluid out through a drain hole in the bottom of a bathtub. It is, of course, unavoidable that the process of getting out of a real bathtub leaves a certain amount of swirl in the water. A circulating fluid particle of constant mass dM moving radially inward will (in nearly ideal flow) conserve its angular momentum $d\mathcal{L}_z = rv_\phi\, dM$, so that its azimuthal velocity increases as $v_\phi \sim 1/r$ while r diminishes. In a bathtub of ordinary size, this incidental initial angular momentum distribution created by the bather is normally much greater than the angular momentum continually supplied by the slowly rotating Earth (see Sections 18.4 and 18.5).

Inflow of angular momentum

To analyze this process we shall assume a steady uniform radial inflow toward the center of the vortex and no outflow outside the drain hole at $r = 0$. Mass conservation secures that the fluid discharge $Q = -2\pi r v_r L$ over a stretch of length L of the vortex core must be constant. Consequently, the radial and axial fields take the form

$$v_r = -\frac{q}{r}, \qquad v_z = 0, \qquad (27.26)$$

where $q = Q/2\pi L$ is a constant.

Assuming that the radial Reynolds number $\text{Re}_r = r\,|v_r|/\nu = q/\nu$ is large, we may disregard viscosity, and use the time-dependent Euler equation for the azimuthal velocity,

$$\frac{\partial v_\phi}{\partial t} + \frac{v_r}{r}\frac{\partial (r v_\phi)}{\partial r} = 0. \qquad (27.27)$$

Notice that there is never a pressure term on the right-hand side because of the cylindrical symmetry. With the given radial inflow (27.26), the above equation may be cast in the form

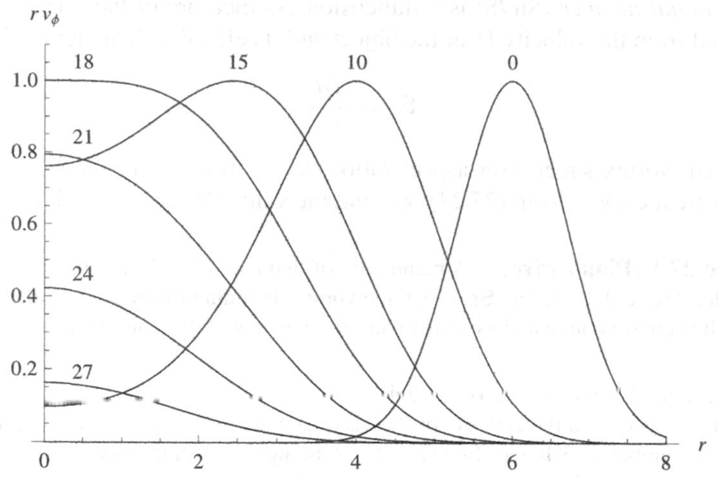

Figure 27.7. "Movie" of advective spin-up of a vortex from an initial Gaussian specific angular momentum distribution. Its evolution is plotted as a function of radial distance for selected times (shown above the curves). As it approaches the drain, the shape is progressively flattened until the peak passes $r = 0$ for $t = 18$, and the vortex quickly disappears down the drain.

$\partial(rv_\phi)/\partial\tau = \partial(rv_\phi)/\partial\xi$, where $\tau = 2qt$ and $\xi = r^2$. This shows that rv_ϕ must be a function of $r^2 + 2qt$, which we shall write in the form

$$v_\phi(r,t) = \frac{C\left(\sqrt{r^2+2qt}\right)}{r}, \tag{27.28}$$

where $C(\cdot)$ is the initial distribution of specific angular momentum $C(r) = rv_\phi(r,0)$ well away from the drain ($r \gg a$).

At any time $t > 0$, the specific angular momentum $rv_\phi(r,t)$ is nearly independent of r in the region $0 < r \ll \sqrt{2qt}$, and the velocity field is nearly that of a line vortex $v_\phi \approx C\left(\sqrt{2qt}\right)/r$. As time advances, the initial circulation profile is probed to farther and farther distances (see Figure 27.7). The size of the region of line vortex shape grows as $\sqrt{2qt}$, reminiscent of a diffusive process although it has nothing to do with that. This shows that a real bathtub vortex may change speed and even stop-up and reverse its sense of rotation, depending on the details of how you got out of the water.

A steady vortex is never reached unless the initial specific angular momentum distribution $C(r)$ approaches a constant for $r \to \infty$. These arguments show that it is in fact impossible to spin-up a truly steady vortex by means of steady radial inflow, for the simple reason that it requires infinite angular momentum to be present in the flow to begin with. If the angular momentum is initially finite, all of it will sooner or later go down the drain in a non-rotating container. In a rotating container, angular momentum may be continually supplied from the moment of force exerted on the fluid by the container walls, resulting eventually in a steady flow. This case was discussed in some detail in Section 18.4.

27.5 Steady vortex sustained by secondary flow

Intuitively, it seems as if the natural tendency for the free vortex core to expand under the influence of viscosity could be counteracted by a sufficiently strong steady radial inflow v_r. Since the fluid cannot accumulate at the center of the vortex, it must be accompanied by a steady axial outflow v_z. We shall again assume that the azimuthal velocity, $v_\phi = v_\phi(r)$, only depends on r, but that $v_r = v_r(r,z)$ and $v_z = v_z(r,z)$ in principle may also depend on z.

Vortex equations

In the presence of secondary flow, the azimuthal steady flow equation contains non-vanishing advective terms on the left-hand side. Using the by-now-familiar methods of Appendix D, it is seen that the advective part of the azimuthal equation only has two non-vanishing terms,

$$\hat{e}_\phi \cdot (\boldsymbol{v} \cdot \nabla)\boldsymbol{v} = v_r \nabla_r v_\phi + \frac{v_\phi}{r} v_r = \frac{v_r}{r} \frac{d(r v_\phi)}{dr},$$

such that the azimuthal equation may now be written as

$$\boxed{\frac{v_r}{r} \frac{d(r v_\phi)}{dr} = \nu \frac{d}{dr}\left(\frac{1}{r}\frac{d(r v_\phi)}{dr}\right).} \tag{27.29}$$

This equation allows us to determine the radial inflow $v_r(r)$ necessary to maintain *any* azimuthal flow $v_\phi(r)$. Conversely, given $v_r(r)$, the azimuthal field $v_\phi(r)$ may be calculated from (27.29), in the form of a double integral:

$$v_\phi = \frac{1}{r} \int \exp\left(\frac{1}{\nu} \int v_r \, dr\right) r \, dr, \tag{27.30}$$

which is well suited for numerical evaluation.

The relation between radial and axial flow is given by the equation of continuity, which in cylindrical coordinates takes the form

$$\frac{1}{r}\frac{\partial (r v_r)}{\partial r} + \frac{\partial v_z}{\partial z} = 0. \tag{27.31}$$

Solving for v_z, we obtain

$$v_z(r, z) = w(r) - \frac{z}{r}\frac{d(r v_r(r))}{dr}, \tag{27.32}$$

where $w(r)$ is an arbitrary function that specifies the axial flow at $z = 0$. The second term represents the accumulated radial inflow.

Finally, the pressure is determined from the radial Navier-Stokes equation. The axial Navier-Stokes equation cannot always be fulfilled, so this procedure does not guarantee an exact solution to the Navier-Stokes equations, although the next two cases are.

Case: The Burgers vortex

The complete flow field that maintains the shape of the Lamb vortex (27.4) is

$$v_\phi = \frac{\Omega c^2}{r}\left(1 - e^{-r^2/c^2}\right), \tag{27.33a}$$

$$v_r = -\frac{2\nu}{c^2} r, \tag{27.33b}$$

$$v_z = \frac{4\nu}{c^2} z. \tag{27.33c}$$

Johannes Martinus Burgers (1895–1981). Dutch physicist. Worked on turbulence, vortex theory, sedimentation, gas dynamics, shock waves, and plasma physics. The Burgers equation, the Burgers vortex, and the Burgers vector are today standard terms in the vocabulary of physics.

It is called the *Burgers vortex* [Nieuwstadt and Steketee 1995] and is in fact an exact solution to the full set of steady-flow Navier–Stokes equations (see Problem D.1). The pressure is

$$\frac{p}{\rho_0} = -\frac{1}{2}\Omega^2 c^2 F\left(\frac{r^2}{c^2}\right) - \frac{2\nu^2}{c^4}\left(r^2 + 4z^2\right), \tag{27.34}$$

with $F(\xi)$ given in (27.6).

The scale of the secondary flow is set by the viscous core decay time, $\tau = c^2/4\nu$, for the Oseen-Lamb vortex, but that is not particularly surprising, since the purpose of the secondary flow is precisely to counteract viscous core expansion.

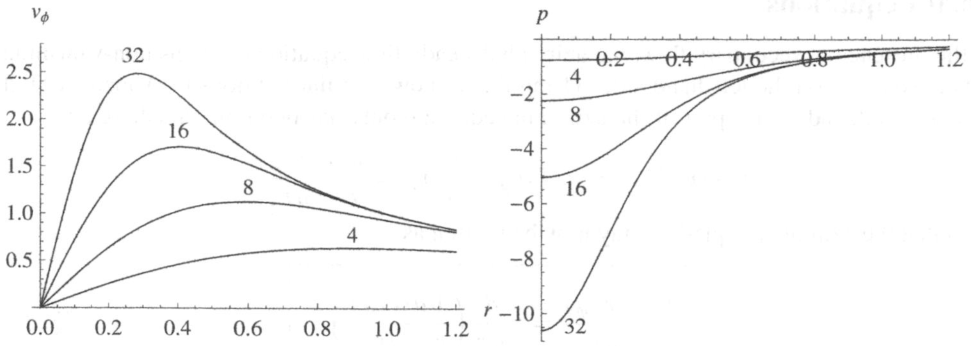

Figure 27.8. Vortex with extended axial jet for $\alpha = 4, 8, 16, 32$. The parameters are $a = 1$, $C = 1$. **Left:** Azimuthal flow v_ϕ. **Right:** Cyclostrophic pressure p, normalized to vanish at infinity.

Case: Localized axial jet

In the Burgers vortex, the axial flow v_z is independent of r, and that makes it very different from natural-born vortices, such as the bathtub vortex, where the axial downflow must converge upon a narrow drain hole, or the tornado, where the upflow primarily takes place inside a narrow funnel. Here we shall first consider the extreme case of a vortex with a *localized axial jet* at $r = 0$. This vortex may be viewed as a generalization of the line vortex (27.13) to include a steady inflow. It is also an exact solution to the Navier–Stokes equations for $r > 0$.

Demanding that $v_z = 0$ outside the vortex core $r > 0$, it follows from (27.32) that $w = 0$ and $d(rv_r)/dr = 0$, such that the radial inflow must be of the form[4]

$$v_r = -\frac{\alpha \nu}{r}, \tag{27.35}$$

where α is a positive dimensionless number (in fact equal to the radial Reynolds number $\mathsf{Re}_r = r\,|v_r|\,/\nu$). The total volume flux of fluid flowing inward through a stretch of the vortex of length L is $Q = 2\pi r L(-v_r) = 2\pi L\alpha\nu$.

Solving the vortex equation (27.29) with this radial inflow, we obtain

$$v_\phi = \frac{C}{r} + A\,r^{1-\alpha}, \tag{27.36}$$

where C and A are constants. For $\alpha > 2$, the second term decays faster than the first, and this confirms that a certain radial inflow is needed to maintain a line vortex at large distances.

The pressure can only depend on r and takes the form:

$$\frac{p}{\rho_0} = -\frac{\alpha^2\nu^2 + C^2}{2r^2} - \frac{2AC}{\alpha} r^{-\alpha} + \frac{A^2}{2(1-\alpha)} r^{2(1-\alpha)}, \tag{27.37}$$

apart from a constant.

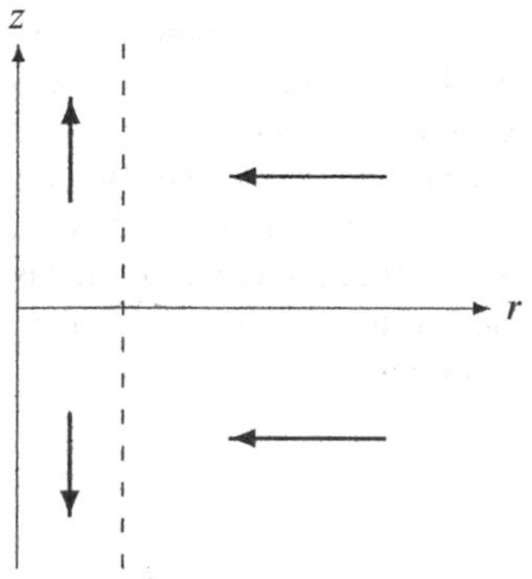

Sketch of the general structure of the secondary flow pattern in a vortex with an extended axial jet.

Case: Extended axial jet

The inflow for a vortex with an extended axial jet of radius a may, for example, be chosen as

$$v_r = -\frac{\alpha\nu}{r}\left(1 - e^{-r^2/a^2}\right), \tag{27.38}$$

which for $a \to 0$ converges toward that of the localized jet. Conversely, for $a \to \infty$ the inflow becomes $-\alpha\nu r/a^2$ which for $a = c\sqrt{\alpha/2}$ is identical to that of the Burgers' vortex. The extended jet thus interpolates smoothly between the two cases.

[4] A number of exact and approximative solutions of this general family are analyzed in [SBH97].

The axial flow now follows from (27.32) with $w = 0$,

$$v_z = z\,\frac{2\alpha\nu}{a^2}\,e^{-r^2/a^2}, \tag{27.39}$$

which vanishes rapidly for $r > a$, as it should.

The azimuthal field and the pressure can only be calculated numerically and are pictured in Figure 27.8 for a selection of values of α. Although the solution interpolates perfectly between two exact steady-flow solutions to the Navier–Stokes equations, it is itself not an exact solution in the transition region near $r = a$.

27.6 Application: The bathtub vortex

A bathtub vortex is an isolated open surface liquid vortex powered by gravity through the loss of liquid through its drain[5]. The most conspicuous feature of such a vortex is the central depression, which may even penetrate the drain and make audible sounds. In the laboratory steady vortices can be created in rotating containers (see Section 18.4), but here we shall ignore this complication and simply assume that the asymptotic flow has the shape of a line vortex $rv_\phi \to C$ with uniform radial inflow $rv_r \to -q$ and constant height $h \to L$.

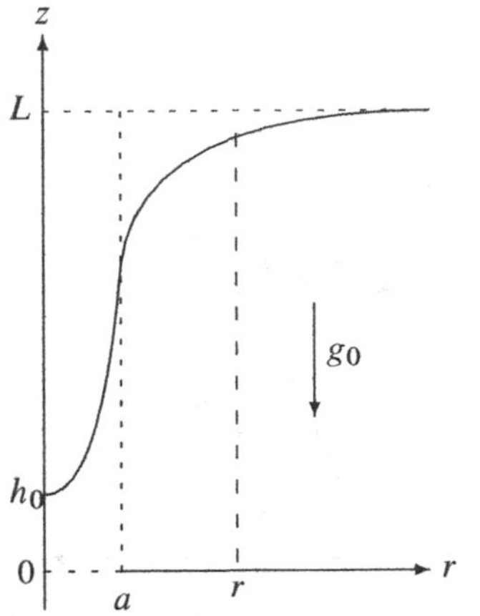

Bathtub-like liquid vortex with open surface. At $z = 0$ there is a drain-hole of radius a. There is a central depression of height h_0 and the asymptotic liquid level is $z = L$.

Bathtub equations

In a flat-Earth coordinate system with gravity directed toward negative z, the vortex is drained through a circular region, a "drain hole" of radius $r = a$ situated at $z = 0$. The open liquid surface of the vortex is assumed to be rotationally invariant, $z = h(r)$, and we shall again assume that the primary flow is cylindrical, $v_\phi = v_\phi(r)$, at least well above the bottom of the container. It follows as before that the azimuthal flow must obey (27.29) with a cylindrical radial flow, $v_r = v_r(r)$, and the axial flow given by (27.32) with $w(r)$ being the drain flow at $z = 0$. Besides these equations there are two boundary conditions on the open surface, one kinematic and one dynamic.

The central depression of the open liquid surface forces the radial inflow into a region of smaller height and thereby speeds it up in comparison with flow in the axial jet vortex. Consequently, there must exist a relation between the height of the surface, $h(r)$, and the radial inflow, $v_r(r)$. The quantitative form of this relation is derived from the fact that the streamlines must follow the surface,

$$\frac{dh(r)}{dr} = \frac{v_z(r, h(r))}{v_r(r)}.$$

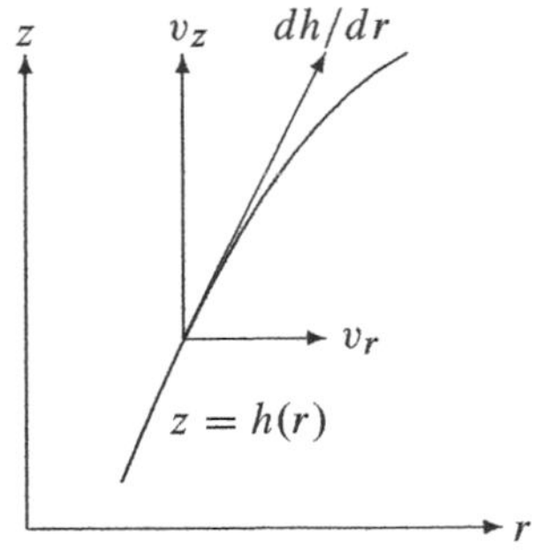

The streamlines must have the same slope as the free surface, $v_z/v_r = dh/dr$.

Using (27.32) with $z = h$, this *kinematic condition* may be written as

$$\boxed{\frac{1}{r}\frac{d(rv_rh)}{dr} = w,} \tag{27.40}$$

which is exact in the cylindrical approximation used here.

In nearly ideal flow, the pressure gradients are the only term in the Navier–Stokes equation that can balance gravity and deliver the centripetal force to maintain the circulating flow,

$$\frac{\partial p}{\partial z} = -\rho_0 g_0, \qquad \frac{\partial p}{\partial r} = \rho_0 \frac{v_\phi^2}{r}. \tag{27.41}$$

Secondary flow is expected to be proportional to the kinematic viscosity ν.

[5]This section is partly inspired by Lundgren's analysis [Lun85]. An extensive discussion of the bathtub vortex, including surface tension, which has been neglected here, may be found in [SY08].

Given that the pressure is constant $p = p_0$ at the open surface $z = h(r)$, it follows from the first of these equations that the hydrostatic pressure takes the usual form,

$$p(r,z) = p_0 + \rho_0 g_0 (h(r) - z), \tag{27.42}$$

well above the drain hole where the secondary velocities are small. Inserting this into cyclostrophich balance, we obtain

$$\boxed{\frac{dh}{dr} = \frac{v_\phi^2}{g_0 r}.} \tag{27.43}$$

A simple geometrical construction reveals the meaning of this *dynamic condition*: the tangential component of vertical gravity must cancel the tangential component of the horizontal centripetal force acting on a liquid particle sitting on the rotating surface.

Choice of drain flow

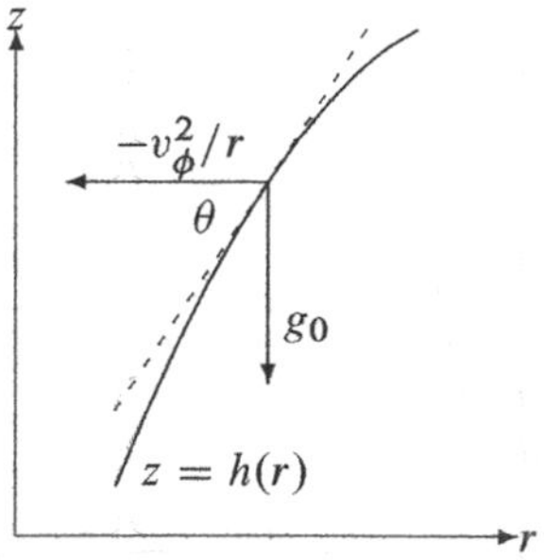

The projections of gravity and the centripetal acceleration on the surface tangent (dashed) cancel, $g_0 \sin\theta + (-v_\phi^2/r)\cos\theta = 0$, where $\tan\theta = dh/dr$.

The three bathtub equations (27.29), (27.40), and (27.43) connect the four fields, v_ϕ, v_r, w, and h. Although the Navier–Stokes equation in principle could also provide a fourth equation (see Problem D.1), it is in view of the many approximations most convenient to impose a reasonable choice for the drain flow $w(r)$, and then solve the bathtub equations for the remaining three fields. Given any such choice of $w(r)$, the kinematic condition (27.40) can be directly integrated,

$$r v_r(r) h(r) = \int_0^r w(s)\, s\, ds, \tag{27.44}$$

where we have used that v_r and h both must be finite for $r \to 0$. Multiplied by 2π this equation simply expresses that the amount of fluid $2\pi r h v_r$ flowing in through the cylinder of height $h(r)$, must equal the amount $\int_0^r w(s) 2\pi s\, ds$ draining out inside r.

To simulate a drain hole of radius a we shall choose a "soft plug flow"

$$w = -W e^{-r^2/a^2}, \tag{27.45}$$

where W is the outflow velocity at the center of the drain. Carrying out the integral we find

$$r v_r h = -\tfrac{1}{2} W a^2 \left(1 - e^{-r^2/a^2}\right). \tag{27.46}$$

For $r \to \infty$ we obtain a relation between the asymptotic inflow $q = \alpha\nu$ and W,

$$2L\alpha\nu = W a^2, \tag{27.47}$$

expressing that the total asymptotic influx $2\pi\alpha\nu L$ must equal the total discharge $\pi a^2 W$.

Non-dimensional formulation

It is convenient to introduce the dimensionless radial variable,

$$\xi = \frac{r^2}{a^2}, \tag{27.48}$$

and replace the field variables by functions of ξ,

$$r v_r = -\alpha\nu u(\xi), \qquad r v_\phi = C v(\xi), \qquad h = L f(\xi), \tag{27.49}$$

normalized by the asymptotic values.

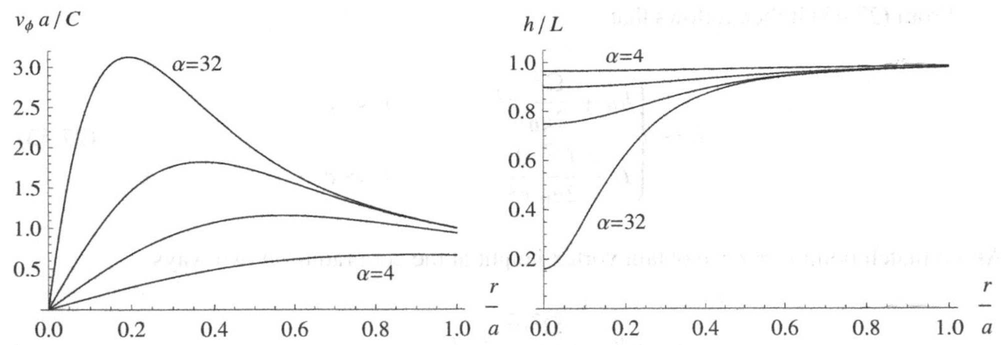

Figure 27.9. Bathtub vortex at $\beta = 0.208$ and $\alpha = 4, 8, 16, 32$, in dimensionless units. The core radius corresponds roughly to the top in v_ϕ. **Left:** Azimuthal velocity. **Right:** Surface height.

The choice of drain flow (27.46) now yields a relation u and f,

$$f(\xi)u(\xi) = 1 - e^{-\xi}, \tag{27.50}$$

and from the azimuthal vortex equation (27.29) and the cyclostrophic condition (27.43), we finally arrive at the closed set of differential equations,

$$v''(\xi) = -\alpha \frac{v'(\xi)u(\xi)}{2\xi}, \qquad f'(\xi) = \beta^2 \frac{v(\xi)^2}{2\xi^2}, \tag{27.51}$$

where

$$\beta = \frac{C}{a\sqrt{g_0 L}} \tag{27.52}$$

is another dimensionless constant. Whereas α equals the asymptotic radial Reynolds number, $\mathsf{Re}_r = [r v_r/\nu]_{r\to\infty} = \alpha$, the constant β is proportional to the azimuthal Froude number $\mathsf{Fr}_\phi = v_\phi/\sqrt{g_0 h}$ at the edge of the drain ($r = a$).

The coupled differential equations must now be solved with the boundary conditions

$$u(0) = v(0) = 0, \qquad u(\infty) = v(\infty) = f(\infty) = 1. \tag{27.53}$$

The results are shown in Figure 27.9 for a particular choice of β and a selection of radial Reynolds numbers. In its gross features the shape is virtually indistinguishable from the vortex with extended axial jet in Figure 27.8.

Estimating the depth of the depression

The core of the bathtub vortex is assumed to rotate as a solid body with angular velocity Ω, whereas far from the core the flow is assumed to be that of a line vortex with circulation constant C, approximately a Rankine vortex,

$$v_\phi \approx \begin{cases} \Omega r & r \ll c \\ \dfrac{C}{r} & r \gg c. \end{cases} \tag{27.54}$$

The core radius c is estimated by matching these expressions,

$$c \approx \sqrt{\frac{C}{\Omega}}, \tag{27.55}$$

and is assumed to lie inside the drain radius, $c < a$.

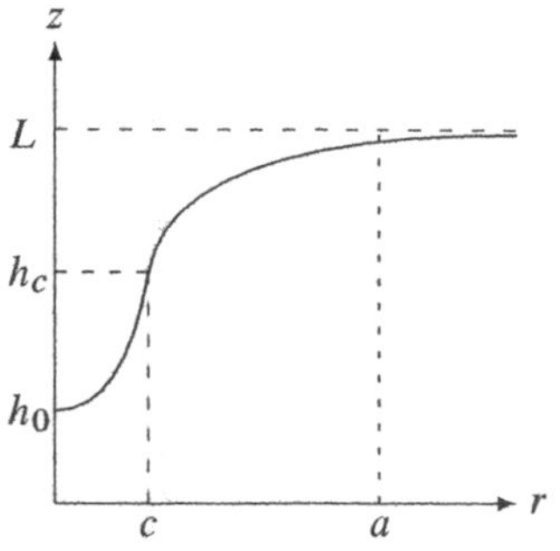

Estimated shape of the bathtub vortex.

From (27.43) it then follows that

$$h \approx \begin{cases} h_0 + \dfrac{\Omega^2}{2g_0} r^2 & r \ll c \\ L - \dfrac{C^2}{2g_0} \dfrac{1}{r^2} & r \gg c. \end{cases} \tag{27.56}$$

At the match point $r = c$ we obtain vortex height at the core radius in two ways

$$h_c \approx h_0 + \frac{\Omega^2 c^2}{2g_0} \approx L - \frac{C^2}{2g_0 c^2},$$

and using the core radius estimate (27.55) we find the depth of the depression and the height at the core radius,

$$L - h_0 \approx \frac{C^2}{g_0 c^2}, \qquad h_c \approx \frac{L + h_0}{2}. \tag{27.57}$$

This makes the estimate of the vortex surface shape (27.56) as well as its derivative continuous at the match point.

Knowing h we may obtain v_r from (27.46) and insert it into (27.29). Quite generally the transition between the core and the outer vortex happens when $\alpha = q/\nu = 2$ (as was the case for the Burgers vortex), or in other words for $q \approx -rv_r \approx 2\nu$. Taking $r = c$ in (27.46) and expanding the exponential to lowest order, we find

$$c^2 \approx \frac{4\nu h_c}{W} \approx \frac{2\nu(L + h_0)}{W}, \tag{27.58}$$

Inserting this result in the expression for the depth of the depression, we get

$$L - h_0 \approx \frac{C^2}{g_0} \frac{W}{2\nu(L + h_0)}, \tag{27.59}$$

and solving for h_0, we finally arrive at

$$\boxed{h_0 \approx \sqrt{L^2 - \frac{C^2 W}{2g_0 \nu}} = L\sqrt{1 - \alpha\beta^2}.} \tag{27.60}$$

Evidently we must require $\alpha\beta^2 < 1$. For if the argument of the square root becomes negative, the central dip in the surface has plunged right through the drain, thereby violating the basic assumptions behind the model (see also Problem 27.18).

Example 27.3 [Bathtub vortex]: A bathtub is filled with water to a height $L = 50$ cm. On exiting, the bather accidentally imparts a small rotation to the water, corresponding to an average specific angular momentum of $C \approx 15\ \text{cm}^2\ \text{s}^{-1}$. When the plug of radius $a = 2$ cm is pulled, the water is found to drain at a rate of $Q \approx 1\ \text{L}\,\text{s}^{-1}$. A needle-like vortex forms, which according to the above expression is $L - h_0 \approx 24$ cm deep. The core diameter becomes $c \approx 1$ mm and the core is estimated to rotate about 250 turns per second!

Problems

27.1 Calculate the **(a)** angular momentum and **(b)** kinetic energy of the core of the Rankine vortex per unit of axial length. **(c)** Calculate the same quantities outside the core for $c < r < R$.

27.2 A vortex in a liquid deforms its open surface that would otherwise be at $z = 0$. Assume that the Rankine vortex may be used as an approximative model for such a vortex. **(a)** Find the surface shape of a liquid in constant gravity. **(b)** What is the depth of the depression? **(c)** Calculate the depth for a Rankine vortex with core radius $c = 1$ cm, rotating 600 times a minute.

27.3 **(a)** Calculate the vorticity in the Oseen–Lamb vortex. **(b)** What is its total flux (i.e., its strength)?

27.4 The naive expectation that the total angular momentum does not change with time for the Oseen–Lamb vortex (27.8) is problematic because the total angular momentum of the infinitely extended vortex is infinite (see Problem 27.13 for an example of a vortex with finite angular momentum). Consider instead angular momentum balance (20.24) on page 346 for a static cylindrical control volume of radius R and length L. **(a)** Verify angular momentum balance $d\mathcal{L}_z/dt + \dot{\mathcal{L}}_z = \mathcal{M}_z$. **(b)** Discuss what happens for $R \to \infty$.

27.5 Show that Kelvin's theorem (27.12) is valid for an ideal barotropic compressible fluid.

27.6 Calculate the real velocity fields of the **(a)** counter- and **(b)** corotating vortex pairs (see (27.17) and (27.18)).

27.7 Determine the stream function for a singular line vortex (27.13) .

* **27.8** The streamline plots in Figure 27.3 are made as contour plots of the stream function ψ, defined such that $v_x = \nabla_y \psi$ and $v_y = -\nabla_x \psi$. Determine the stream function for **(a)** counter- and **(b)** corotating vortex pairs of equal strength (Hint: Use the answers to Problems 27.6 and 27.7).

* **27.9** Define the function

$$f_N(z) = \cot z - \sum_{n=-N}^{N} \frac{1}{z - \pi n}$$

and show that $f(z) = \lim_{N\to\infty} f_N(z)$ vanishes everywhere in the complex plane. Hints: **(a)** Show that f is antisymmetric, $f(-z) = -f(z)$. **(b)** Show that $f(z)$ is periodic with period $\Delta x = \pi$. **(c)** Show that $f(z)$ is holomorphic in the strip $-\pi/2 < x < \pi/2$. **(d)** Show that $f(z)$ vanishes at the boundaries of the strip. Finally, a general theorem of complex analysis states that a function that is holomorphic in a region and vanishes on the boundaries vanishes identically. By periodicity it also vanishes in all other strips.

27.10 Consider a collection of parallel vortices enumerated by n with constant strengths C_n and the cores moving along individual orbits $z = z_n(t)$. **(a)** Show that the equations of motion of the centers are

$$\frac{dz_n^\times(t)}{dt} = -i \sum_{m\neq n} \frac{C_m}{z_n(t) - z_m(t)},$$

where $^\times$ stands for complex conjugation. **(b)** Verify that the orbits of a pair of vortices of opposite strength satisfy these equations of motion. **(c)** The same for a pair of the same strength.

* **27.11** Verify that the Burgers vortex (27.33) and the vortex with a localized axial jet (page 474) are exact solutions. Hint: Use cylindrical coordinates and the Navier–Stokes equations given in Problem D.1.

27.12 Calculate the streamlines for the Burgers vortex.

27.13 (**a**) Show that the so-called Taylor vortex,

$$v_\phi(r,t) = \tau \frac{r}{t^2} e^{-r^2/4\nu t},$$

is a solution to (27.7), where τ is a constant with dimension of time. (**b**) Calculate the total angular momentum per unit of axial length of the Taylor vortex and show that it is constant in time.

27.14 Assume that a free cylindrical vortex takes the form

$$v_\phi(r,t) = \frac{F(t)}{r} f(\xi), \qquad \xi = \frac{r^2}{4\nu t}.$$

(**a**) Find a differential equation for $f(\xi)$.
(**b**) Show that $F(t) \sim t^{-\alpha}$, where α is a dimensionless parameter.
(**c**) Show that the solution is

$$f_\alpha(\xi) = \sum_{n=0}^{\infty} \frac{(n+\alpha)!}{n!\,\alpha!} (-1)^n \frac{\xi^{n+1}}{(n+1)!}.$$

(**d**) Find a closed form for $\alpha = 0, 1, 2$ and draw the vortex shapes.

27.15 Assume that a vortex at $t = 0$ has the shape of an interpolating Lamb vortex (27.4) with core radius c, and that there is a radial inflow identical to that of the Burgers vortex (27.33) corresponding to a core radius a. Show that the azimuthal flow of the original vortex converges upon a Lamb vortex of radius a and determine the instantaneous radius as a function of time.

27.16 Assume that the water in a bathtub originally rotates like a solid body, $v_\phi = \Omega r$, far from the drain. (**a**) Determine the far field at a later time and (**b**) the rate at which angular momentum flows in toward the drain.

27.17 Show that the vortex equation (27.29) may be solved explicitly by quadrature when $v_r(r)$ is given.

27.18 Show that the depth of the depression is

$$L - h_0 = \int_0^\infty \frac{v_\phi^2}{g_0 r}\, dr = k \frac{C^2}{g_0 c^2}$$

and determine a value for k from the interpolating vortex. How does this change the estimate (27.60) for the central height of the vortex?

28 Boundary layers

At large Reynolds number, a fluid will behave as ideal nearly everywhere, except close to solid boundaries where the no-slip condition requires the speed of the fluid to match the speed of the boundary wall. Here, transition layers will arise in which the velocity of the flow changes quickly from the velocity of the wall to the velocity in the fluid at large. Boundary layers are typically thin compared to the radii of curvature of the solid walls.

Ludwig Prandtl (1875–1953). German physicist, often called the father of aerodynamics. Developed boundary layer theory, and contributed to wing theory, streamlining, compressible subsonic airflow, and turbulence [And05, O'M10]. (Source: Wikimedia Commons.)

In a boundary layer the character of the flow thus changes from creeping near the boundary to ideal well outside. It is a general trait that the most interesting physics takes place in such transition regions. Humans, living out their lives in nearly ideal flows of air and water at Reynolds numbers in the millions with boundary layers only millimeters thick, are normally not conscious of them. Smaller animals eking out an existence at the surface of a stone in a river may be much more aware of the vagaries of boundary layer physics, which may even influence their body shapes and internal layout of organs.

Boundary layers serve to "insulate" bodies from the ideal flow that surrounds them. They have a "life of their own" and may separate from the solid walls and wander into regions containing only fluid. Detached layers may again split up, creating complicated unsteady turbulent patterns of whirls and eddies. Systematic boundary layer theory was initiated by Prandtl in 1904 and has in the twentieth century become a major subtopic of fluid mechanics. Advanced understanding of fluid mechanics begins with an understanding of boundary layers, although there is still no simple theory capable of predicting where separation takes place.

In this chapter we shall focus on Prandtl's theory of steady incompressible boundary layers in the absence of gravity. Under the assumption that boundary layers are always thin, the bounding walls shall always be viewed as effectively flat, which is a major theoretical simplification. Comprehensive treatments of boundary layers may be found in [White 1991], [Schlichting and Gersten 2000], and [Sobey 2000].

28.1 Basic physics of boundary layers

Before turning to Prandtl's boundary layer theory it is useful to make some simple estimates. Boundary layers are, as mentioned, only meaningful in nearly ideal flow at high Reynolds numbers, and serve to soften the singular velocity jumps that otherwise would be allowed in truly ideal flow. For simplicity we shall not consider mid-stream interface layers but limit the discussion to those boundary layers that envelop the surfaces of solid undeformable bodies at rest in steady inviscid flow (see Figure 28.1).

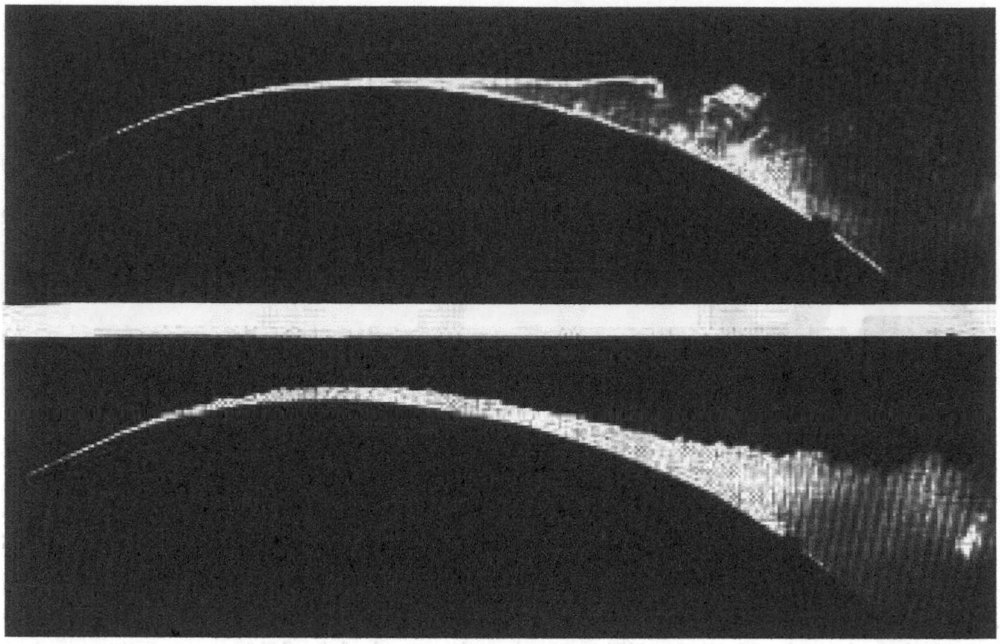

Figure 28.1. Laminar and turbulent boundary layers on a convex wall. Both layers show downstream thickening (toward the right). The laminar layer (top) separates on the crest of the wall while the turbulent boundary layer separates further downstream. This is the phenomenon behind the drag crisis discussed on page 296. (Source: Reproduced from [Hea82].)

Initial growth

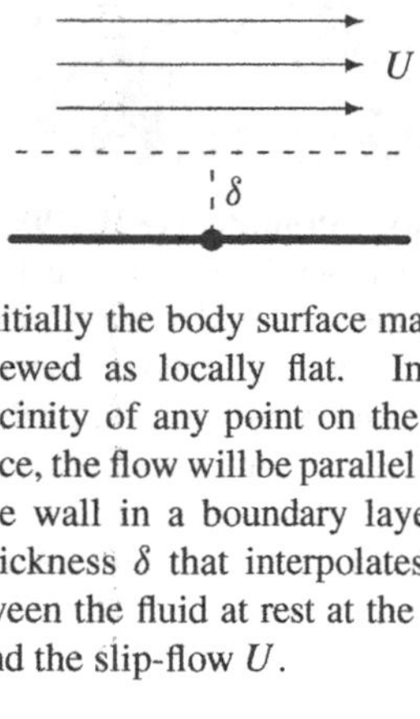

Initially the body surface may be viewed as locally flat. In the vicinity of any point on the surface, the flow will be parallel with the wall in a boundary layer of thickness δ that interpolates between the fluid at rest at the wall and the slip-flow U.

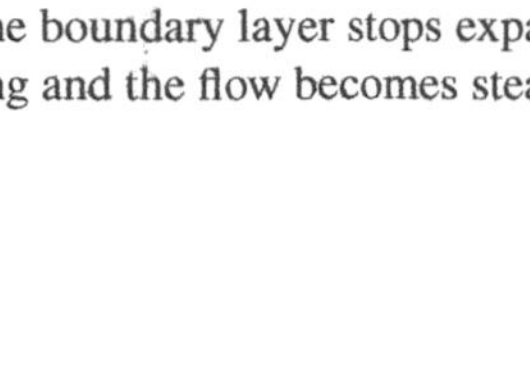

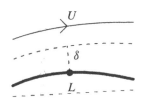

The body geometry scale L sets the time scale $t \sim L/U$ when the boundary layer stops expanding and the flow becomes steady.

In a truly ideal fluid the boundary conditions only forbid the fluid to flow along the normal to the impenetrable surface of the body, but allow the fluid to *slip* tangentially along the surface with some local velocity U near any given point, determined by the Euler equations and the geometry of the body. If viscosity is suddenly turned on, the tangential velocity will also have to vanish (the no-slip condition), implying that an extremely thin boundary layer of thickness δ must have sprung into existence between the surface of the body and the steady inviscid slip-flow.

A short time t after viscosity has been turned on, the local acceleration (actually deceleration) of the fluid near the boundary is of magnitude $|\partial \boldsymbol{v}/\partial t| \sim U/t$. Normal bodies have finite geometries with built-in length scales. If the smallest length scale for significant changes in slip-velocity (due to the body geometry) is L, the advective acceleration becomes of magnitude $|(\boldsymbol{v}\cdot\nabla)\boldsymbol{v}| \sim U^2/L$. At the earliest times, $t \ll L/U$, the local acceleration dominates the advective acceleration, and the geometry is irrelevant.

The viscous acceleration is of magnitude $\left|\nu\nabla^2\boldsymbol{v}\right| \sim \nu U/\delta^2$ because the velocity field varies from 0 to U across the boundary layer of thickness δ. In the absence of advection, the Navier–Stokes equation requires the local and viscous accelerations to be of the same magnitude, $U/t \sim \nu U/\delta^2$, or

$$\delta \sim \sqrt{\nu t}. \tag{28.1}$$

As for the Stokes layer (page 249), the constant of proportionality will depend on the precise definition of thickness, but the thickness scale will always be $\sqrt{\nu t}$. Any suddenly created boundary layer starts out like this, everywhere growing thicker with the square root of time *independent of the local slip-flow velocity*. For the geometrically featureless Stokes layer, the thickening continues forever.

Going steady

As the boundary layer grows thicker everywhere along the body surface, the local acceleration diminishes relative to the advective. At late times, $t \gg L/U$, the advective acceleration completely dominates and the local acceleration can be ignored. If the local acceleration is dropped, the Navier–Stokes equation formally loses its time dependence and the flow in the boundary layer becomes steady with no further local thickening. It takes, however, a lot of careful analysis to see whether the flow actually "goes steady", or whether instabilities occur that may lead to a radical change in the character of the flow, such as boundary layer separation or turbulence.

Downstream thickening

A steady boundary layer has a natural tendency to *downstream thickening*, which may be understood as a cumulative effect of the slow-down of the fluid from the contact with the boundary wall (see the margin figures and Figure 28.1). The further downstream from the leading edge of a body, the longer time the fluid will have been under the influence of the shear forces from the wall, and the thicker the boundary layer becomes.

The generic example is a semi-infinite thin plate subject to a constant slip-velocity $v_x = U$ orthogonal to the leading edge. The time it takes for the slip-flow to pass from the leading edge to a point as distance x downstream is $t = x/U$, so that

$$\delta \sim \sqrt{\nu t} = \sqrt{\frac{\nu x}{U}}. \tag{28.2}$$

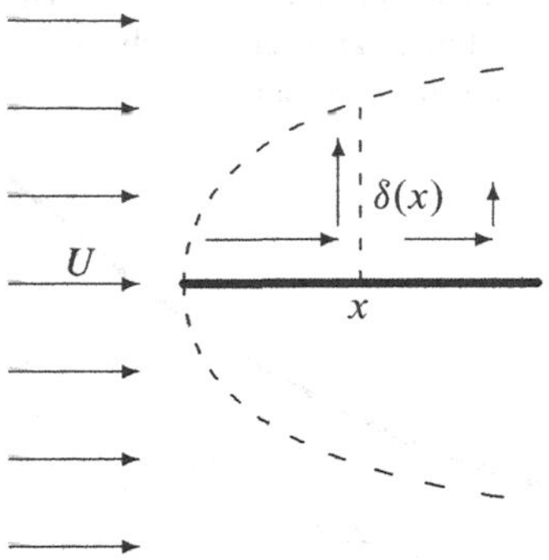

A semi-infinite plate in an otherwise uniform flow. The dashed curve is the estimated parabolic boundary layer shape. The arrows inside the boundary layer indicate the decelerating flow and the compensating upwelling.

The downstream thickening takes the shape of a parabola (see the margin figure).

Inside this boundary layer the flow decelerates downstream due to the viscous drag from the plate, which spreads into the fluid by momentum diffusion. Mass conservation requires a compensating *upwelling* of fluid into the mainstream, and this upwelling may be seen as the cause of the thickening. The amount of fluid $v_y dx$ that wells up through a small interval dx must equal the amount $U d\delta$ that passes through the extra thickness, so that

$$v_y \sim U\frac{d\delta}{dx} \sim \sqrt{\frac{U\nu}{x}}. \tag{28.3}$$

Boundary layers always modify the mainstream flow by adding such an intrinsic or "natural" upflow orthogonal to the boundary.

It is almost as if the boundary itself were permeable and fluid were pumped up through it into the mainstream, although this fluid actually stems from the slip-flow that has been *displaced* by the boundary layer. If in reality the boundary is permeable and fluid is sucked down through it at a constant rate, the upwelling can be avoided, and a steady boundary layer of constant thickness can be created (Problem 28.1).

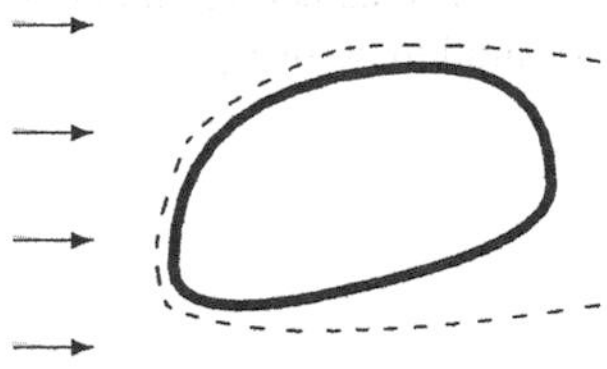
Sketch of the boundary layer around a bluff body in steady uniform flow. On the windward side the boundary layer is thin, whereas it widens and tends to separate on the lee side. In the channel formed by the separated boundary layer, unsteady flow patterns may arise.

Slip-flow variations

The mainstream flow is determined by the bodies and container walls that guide the fluid, and the slip-velocity will generally not be constant but vary along the boundaries. If $U(x)$ denotes the varying slip-velocity along a streamline parameterized by its path-length x, we may estimate the thickness as

$$\delta \sim \sqrt{\nu t}, \qquad t = \int \frac{dx}{U(x)}, \tag{28.4}$$

where t (up to an additive constant) is the time it takes for a material particle in the slip-flow to arrive at the downstream point x.

If the slip-flow *accelerates* downstream, $dU/dx > 0$, it will counteract the natural deceleration in the boundary layer and may even overwhelm the upwelling, leading to a net *downwash* toward the boundary. Slip-flow acceleration thus tends to stabilize a boundary layer by reducing its tendency to thicken, and may lead to constant or even diminishing thickness. Conversely, if the slip-flow flow *decelerates* downstream, $dU/dx < 0$, this will add to the deceleration in the boundary layer and increase its thickness as well as the upwelling.

The estimate (28.4) does in fact hold for the class of self-similar boundary layers analyzed in Section 28.6 (see Problem 28.8). But boundary layer theory is not always that simple. Most importantly, the integral may diverge at a point x_0 where the slip-velocity vanishes, $U(x_0) = 0$. Such a *stagnation point* is always found at the front end of a bluff body (see Section 28.5), and possibly further downstream if the boundary layer separates.

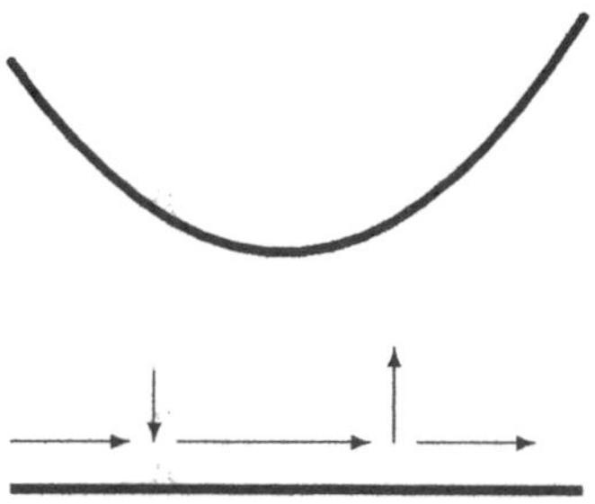

Sufficiently strong mainstream acceleration in a converging channel produces a downwash of fluid, and conversely in a diverging channel.

Separation

Even at very moderate mainstream deceleration, the upwelling can become so strong that at some point the fluid flowing in the mainstream direction cannot feed it. Some of the fluid in the boundary layer will then have to flow against the mainstream flow to supply the upwelling. Between the forward and reversed flows there will be a separation line ending in a *separation point* on the wall. Such *flow reversal* was also noted in lubrication (page 301), although there are no boundary layers in creeping flow.

The velocity has of course to vanish everywhere at the boundary wall. Moving a bit up into the reversed flow, the velocity has to become negative with respect to the mainstream flow. Letting y denote a local coordinate orthogonal to the wall, the orthogonal velocity gradient $\partial v_x/\partial y$ must therefore be negative near the boundary wall in the reversal region, and the same goes for the *wall shear stress*,

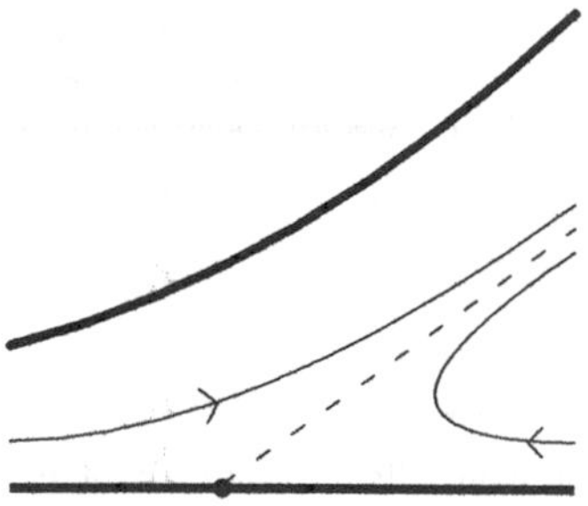

Flow reversal and boundary layer separation in a diverging channel with decelerating flow.

$$\sigma_{\text{wall}} = \eta \left.\frac{\partial v_x}{\partial y}\right|_{y=0}. \tag{28.5}$$

Right at the separation point the wall stress must accordingly vanish, $\sigma_{\text{wall}} = 0$, because it is positive before and negative after. This is the true criterion for separation.

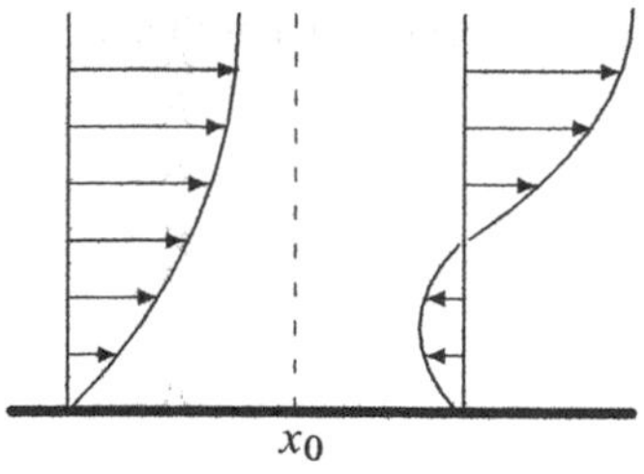

Velocity profiles before and after the separation point x_0.

Turbulence

Boundary layers may also become turbulent. Turbulence typically sets in some distance downstream from the front where the laminar boundary layer has grown so thick that the *local Reynolds number*,

$$\mathsf{Re}_\delta = \frac{U\delta}{\nu}, \tag{28.6}$$

becomes large enough. The flat plate estimate (28.2) implies that

$$\mathsf{Re}_\delta \sim \sqrt{\mathsf{Re}_x}, \tag{28.7}$$

where $\mathsf{Re}_x = Ux/\nu$ is the *downstream Reynolds number*. The actual value of Re_δ depends on how the thickness is defined (see page 488) and is not so easy to measure. It is easier to determine the downstream point where turbulence sets in, typically for $\mathsf{Re}_x \approx 5 \times 10^5$. Incidentally this is just about the range in which humans and their machines operate.

Turbulence efficiently mixes fluid in all directions while the orderly layers of fluid that otherwise isolate the wall from the mainstream flow all but disappear. There will, however, always remain a thin viscous *laminar sublayer* closest to the wall, in which the average velocity rises linearly with distance. Since the turbulent velocity fluctuations allow the slip-flow to press closer to the wall, the wall stress will be larger than in a completely laminar boundary layer. The skin drag on a body is consequently expected to increase when the boundary layer becomes turbulent.

28.2 Boundary layer theory

When Prandtl introduced the concept of boundary layers he pointed out that there were simplifying features, allowing for a less complicated formalism than the full Navier–Stokes equations. The greater simplicity comes from the assumption of inviscid mainstream flow. If L is the length scale for significant variations in the mainstream flow around a body, the largest downstream Reynolds number is $\mathsf{Re} = UL/\nu$, so that $\delta \lesssim L/\sqrt{\mathsf{Re}}$. For $\mathsf{Re} \gg 1$, the boundary layer thickness is always much smaller than the size of the body. Assuming it has an unexceptional geometry with radii of curvature not much smaller than L, the body surface may for boundary layer analysis be taken to be essentially flat. *At high Reynolds number the actual geometry of the body and other nearby bodies only shows up in the variations of the slip-flow along the unfolded (flattened) body surface.*

Justified by this analysis, and by the potential applications to wing theory, we shall limit the analysis to the idealized two-dimensional case of an incompressible fluid flowing along an infinitely extended planar boundary wall at $y = 0$.

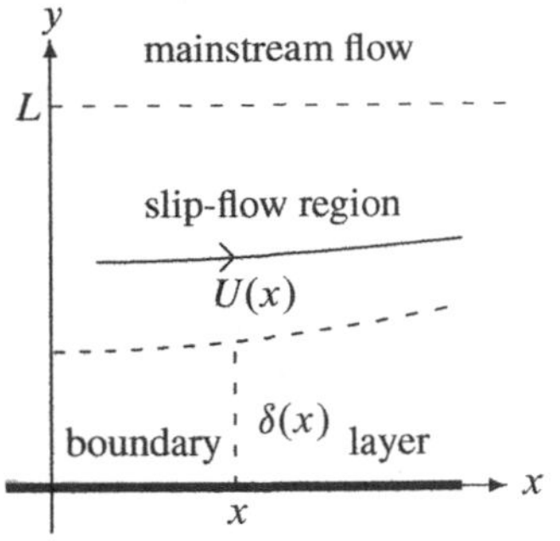

Geometry of two-dimensional planar boundary flow. In the absence of viscosity there would be a slowly varying slip-flow $U(x)$ along the boundary. Viscosity interposes a thin boundary layer of thickness $\delta(x)$ between the slip-flow and the boundary.

The slip-flow

The slip-flow velocity $U(x)$ should be understood as being valid in the region $\delta \lesssim y \ll L$, well outside the boundary layer but still so close to the boundary that the fluid streams primarily along the x-axis. The continuity equation, $\partial v_x/\partial x + \partial v_y/\partial y = 0$, requires that the slip-flow field takes the general form

$$v_x = U(x), \qquad v_y = V(x) - y\frac{dU(x)}{dx}, \tag{28.8}$$

where $V(x)$ represents the natural upflow from the boundary layer. The boundary layer prevents us from imposing the boundary condition $v_y = 0$ at $y = 0$ on the slip-flow.

In the absence of gravity, the x-component of the Euler equation (13.1) is solved by the pressure,

$$p = p_0 - \frac{1}{2}\rho_0 U(x)^2, \tag{28.9}$$

which may also derived from Bernoulli's principle.

The Prandtl equations

In two dimensions the steady-flow dynamics is governed by the continuity equation and the incompressible Navier–Stokes equations,

$$v_x\frac{\partial v_x}{\partial x} + v_y\frac{\partial v_x}{\partial y} = -\frac{1}{\rho_0}\frac{\partial p}{\partial x} + \nu\left(\frac{\partial^2 v_x}{\partial x^2} + \frac{\partial^2 v_x}{\partial y^2}\right), \tag{28.10a}$$

$$v_x\frac{\partial v_y}{\partial x} + v_y\frac{\partial v_y}{\partial y} = -\frac{1}{\rho_0}\frac{\partial p}{\partial y} + \nu\left(\frac{\partial^2 v_y}{\partial x^2} + \frac{\partial^2 v_y}{\partial y^2}\right). \tag{28.10b}$$

The Prandtl equations are obtained as an approximation to these equations in the limit where all terms of relative magnitude $1/\mathsf{Re}$ are dropped.

In both the Navier–Stokes equations the double derivative with respect to y is proportional to $1/\delta^2$, whereas the double derivative with respect to x is proportional to $1/L^2$ and may immediately be dropped. From the continuity equation, $\partial v_x/\partial x + \partial v_y/\partial y = 0$, we get $v_x/L \sim v_y/\delta$, or $v_y \sim U\delta/L$, and this makes both terms on the left-hand side of the second Navier–Stokes equation of magnitude $U^2\delta/L^2$. Using the estimate $\delta \sim \sqrt{\nu L/U}$, the viscous term on the right-hand side, $\nu\partial^2 v_y/\partial y^2 \sim \nu U/\delta L \sim U^2\delta/L^2$, is of the same magnitude as the left-hand side.

Consequently, the normal pressure gradient is also of magnitude, $\partial p/\partial y \sim \rho_0 U^2\delta/L^2$, and multiplying by δ we obtain the pressure variation across the boundary layer $\Delta_y p \sim \rho_0 U^2\delta^2/L^2 \sim \rho_0 U^2/\mathsf{Re}$. This is negligible compared to the typical variation in slip-flow pressure, which by Bernoulli's theorem is $\Delta_x p \sim \rho_0 U^2$. In this approximation, the true pressure in the boundary layer simply equals the slip-flow pressure (28.9) everywhere. The slip-flow pressure is said to be "stiff" because it penetrates right through the boundary layer and acts directly on the boundary.

Finally inserting the slip-flow pressure into the first equation above, and dropping the second derivative with respect to x, we arrive at *Prandtl's momentum equation*,

$$v_x\frac{\partial v_x}{\partial x} + v_y\frac{\partial v_x}{\partial y} = U\frac{dU}{dx} + \nu\frac{\partial^2 v_x}{\partial y^2}. \tag{28.11}$$

Together with the continuity equation, expressing local mass conservation,

$$\frac{\partial v_x}{\partial x} + \frac{\partial v_y}{\partial y} = 0, \tag{28.12}$$

we have obtained two coupled partial differential equations, which for any given $U(x)$ in principle determine $v_x(x, y)$ and $v_y(x, y)$ in the boundary layer, subject to the no-slip boundary conditions $v_x = v_y = 0$ for $y = 0$ and the asymptotic boundary condition $v_x \approx U(x)$ in the slip-flow region $\delta \lesssim y \ll L$.

Upflow

Integrating the continuity equation over the interval $0 \leq y' \leq y$, we obtain the exact relation

$$v_y = -\frac{\partial}{\partial x}\int_0^y v_x(x, y')\, dy'. \tag{28.13}$$

The two coupled differential equations have thus been converted to a single integro-differential equation. It is useful to rewrite this equation by substituting $v_x = U - (U - v_x)$ to get

$$v_y = -y\frac{dU(x)}{dx} + \frac{\partial q(x, y)}{\partial x}, \qquad q = \int_0^y (U(x) - v_x(x, y'))\, dy'. \tag{28.14}$$

The definition of the field $q(x, y)$ shows that it is a measure of the slip-flow flux that has been *displaced* by the presence of the boundary in the interval $0 \leq y' \leq y$.

In the slip-flow region $\delta(x) \lesssim y \ll L$, we have $v_x(x, y) \approx U(x)$, so that the displaced flux becomes independent of y, that is, $q(x, y) \approx Q(x)$. In this region the upflow becomes

$$v_y \approx \frac{dQ(x)}{dx} - y\frac{dU(x)}{dx}. \tag{28.15}$$

As foreseen in (28.8), the upflow in the slip-flow region has two contributions, one $V(x) = dQ(x)/dx$ from the boundary layer itself and one $-y\,dU(x)/dx$ due to variations in the slip-flow. Both terms are of magnitude $U\delta/L \sim U/\sqrt{\mathsf{Re}}$ and thus small in comparison with $U(x)$.

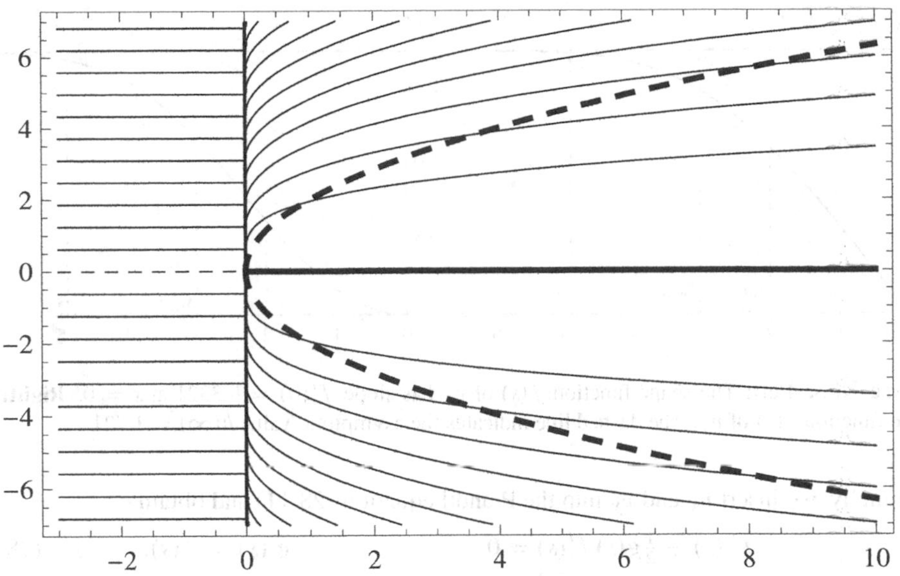

Figure 28.2. Streamlines around a semi-infinite thin plate with fluid flowing uniformly in from the left. The heavy dashed curve indicates the thickness scale, $|y| = \delta = 5\sqrt{x}$. The kink in the streamlines at $x = 0$ signals breakdown of the Prandtl approximation in this region. The plot has been created from the contours of the stream function $\psi(x, y) = \sqrt{U\nu x}\, g(y\sqrt{U/\nu x})$.

28.3 The Blasius layer

The most important example of a two-dimensional steady laminar boundary layer is furnished by a semi-infinite plate at $y = 0$ with $0 < x < \infty$. Subject to a uniform slip-flow $\boldsymbol{v} = U\hat{\boldsymbol{e}}_x$, we previously estimated the boundary layer thickness to be $\delta \sim \sqrt{\nu x/U}$ and we shall now confirm this estimate. The solution was first found by Blasius in 1908.

Paul Richard Heinrich Blasius (1883–1970). German engineer, a student of Prandtl's. Contributed to boundary-layer drag and smooth pipe resistance. Spent only 6 years in research before moving to education. (Source: Reproduced from [Hag03].)

Self-similar solution

The main variation in v_x happens across the boundary layer, suggesting that it will be convenient to write v_x in dimensionless form (in analogy with the Stokes layer (15.15) on page 249),

$$v_x = Uf\left(y\sqrt{\frac{U}{\nu x}}\right), \tag{28.16}$$

where $f(s)$ satisfies the boundary conditions $f(0) = 0$ and $f(\infty) = 1$. In principle this function could also depend on the downstream Reynolds number $\mathsf{Re}_x = Ux/\nu$, but the correctness of the assumption will be justified by a solution satisfying the boundary conditions.

From the equation of continuity (28.13) we obtain

$$v_y = -\frac{\partial}{\partial x}\left(\sqrt{U\nu x}\, g(s)\right), \qquad g(s) = \int_0^s f(s')ds'.$$

Carrying out the differentiation, we obtain

$$v_y = \sqrt{\frac{U\nu}{4x}}\, h(s), \qquad h(s) = sf(s) - g(s). \tag{28.17}$$

From the general asymptotic expression (28.15) with $dU/dx = 0$, it follows that $h(s)$ must be finite for $s \to \infty$, so that $g(s) \approx s - h(\infty)$ for $s \gg 1$.

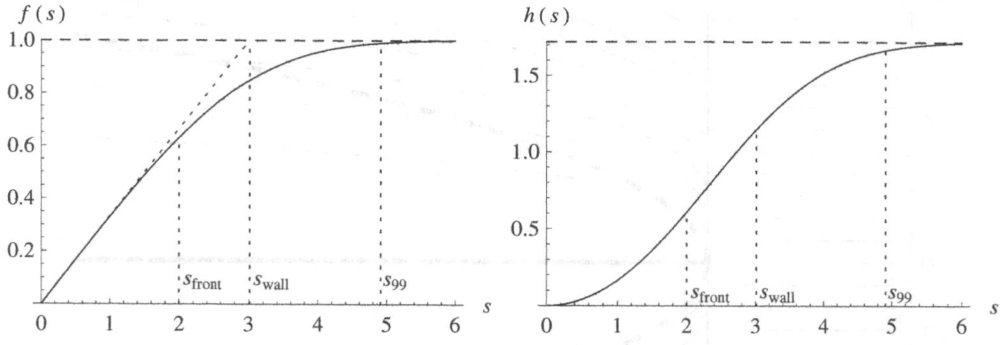

Figure 28.3. **Left:** The shape function $f(s)$ of v_x has slope $f'(0) = 0.3321$ at $s = 0$. **Right:** The shape function $h(s)$ of v_y. The dashed line indicates the asymptotic value $h(\infty) = 1.721$.

Finally, we insert v_x and v_y into the Prandtl equation (28.11) and obtain

$$f''(s) + \tfrac{1}{2}g(s)f'(s) = 0, \qquad g'(s) = f(s). \tag{28.18}$$

These coupled equations can trivially be converted into a single differential equation[1],

$$\boxed{g'''(s) + \tfrac{1}{2}g(s)g''(s) = 0,} \tag{28.19}$$

called the *Blasius equation.* It must be solved with the boundary conditions $g(0) = 0$, $f(0) = g'(0) = 0$, and $f(\infty) = g'(\infty) = 1$.

Numeric integration is straightforward by "ballistics", starting with $f'(0) = \mu$ and afterward searching for the value of μ that yields $f(\infty) = 1$. The result is $\mu = 0.3321\ldots$ and the functions $f(s)$ and $h(s)$ are shown in Figures 28.2 and 28.3.

Thickness definitions

The leading approximation fulfilling the boundary conditions for $s \gg 1$ is $g(s) \approx s$, and it follows from the Blasius equation that $g''(s) \sim \exp(-s^2/4) = \exp(-y^2U/4\nu x)$. This shows that the front thickness takes the universal value, valid for all laminar momentum diffusion processes (see page 248),

$$\delta_{\text{front}} = 2\sqrt{\frac{\nu x}{U}}. \tag{28.20}$$

At $y = \delta_{\text{front}}$, the velocity is roughly 63% of the slip-flow velocity.

The velocity slope at the wall, $\partial v_x/\partial y|_{y=0}$, defines another thickness,

$$\delta_{\text{wall}} = \left.\frac{U}{\partial v_x/\partial y}\right|_{y=0} = \frac{1}{f'(0)}\sqrt{\frac{\nu x}{U}} = 3.01\sqrt{\frac{\nu x}{U}}, \tag{28.21}$$

because $f'(0) = 0.3321$. At $y = \delta_{\text{wall}}$, the velocity is 85% of the slip-flow velocity.

In technical contexts one often prefers to define the laminar boundary layer thickness *ad hoc* as the distance from the wall where the flow has reached 99% of the slip-flow velocity. Since $f(s) = 0.99$ for $s = 4.91$, this thickness becomes

$$\delta_{99} = 4.91\sqrt{\frac{\nu x}{U}}. \tag{28.22}$$

Typically one simply uses $\delta \approx 5\sqrt{\nu x/U}$ as the "true" boundary layer thickness.

[1] This is one version of the Blasius equation. The precise choice of the dimensionless parameter s influences the coefficient of the second term (here 1/2), although the physics does not depend on it.

In dimensionless form, the thickness scale may be expressed in terms of the cross-flow and downstream Reynolds numbers,

$$\mathsf{Re}_\delta \equiv \frac{U\delta_{99}}{\nu} \approx 5\sqrt{\mathsf{Re}_x}, \tag{28.23}$$

where as before $\mathsf{Re}_x = Ux/\nu$. Turbulence typically sets in around $\mathsf{Re}_x \approx 5 \times 10^5$, corresponding to $\mathsf{Re}_\delta \approx 3{,}500$.

Wall shear stress and laminar friction coefficient

The shear stress acting on the wall becomes

$$\sigma_{\text{wall}} = \eta \left.\frac{\partial v_x}{\partial y}\right|_{y=0} = \frac{\eta U}{\delta_{\text{wall}}(x)} = f'(0)\,\eta U \sqrt{\frac{U}{\nu x}}. \tag{28.24}$$

Dividing by $\frac{1}{2}\rho_0 U^2$ we arrive at the dimensionless *local laminar friction coefficient*,

$$\boxed{C_f \equiv \frac{\sigma_{\text{wall}}}{\frac{1}{2}\rho_0 U^2} = \frac{2f'(0)}{\sqrt{\mathsf{Re}_x}} = \frac{0.664}{\sqrt{\mathsf{Re}_x}}.} \tag{28.25}$$

The divergence for $x \to 0$ signals breakdown of the boundary layer approximation at the leading edge of the plate.

The average of the friction coefficient over a downstream length L becomes

$$\langle C_f \rangle = \frac{1}{L}\int_0^L C_f(x)\,dx = \frac{4f'(0)}{\sqrt{\mathsf{Re}_L}} = 1.328\,\mathsf{Re}_L, \tag{28.26}$$

where as before $\mathsf{Re}_L = UL/\nu$. It may be used to calculate the viscous skin drag by multiplying with $\frac{1}{2}\rho_0 U^2$ times the total plate area. Since skin drag decreases with the square root of the Reynolds number, it is of little importance in most everyday situations with Reynolds numbers in the millions. For most objects, skin drag is overwhelmed by form drag, which becomes constant at large Reynolds numbers (see Chapter 29).

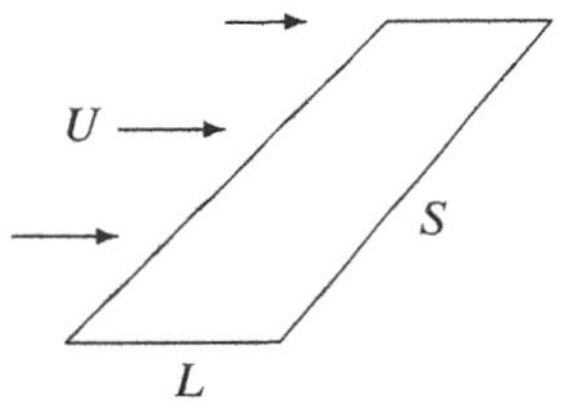

A flat, thin wing aligned with the flow only experiences skin drag.

Example 28.1 [Weather vane]: A little rectangular metal weather vane with chord length $L = 25$ cm and span $S = 20$ cm has the total area $A = LS = 500\ \text{cm}^2$. In a $U = 10\ \text{m s}^{-1}$ wind, the downstream Reynolds number is $\mathsf{Re} = UL/\nu \approx 1.6 \times 10^5$, the average friction coefficient $\langle C_f \rangle = 0.003$, and the wall shear stress $\langle \sigma_{\text{wall}} \rangle = 0.2$ Pa. The total skin drag on both sides becomes a measly $\mathcal{D}_{\text{skin}} = 2A\,\langle \sigma_{\text{wall}} \rangle = 0.02$ N if the vane is aligned with the wind.

Advective heat flow

Heat is advected as well as dissipated in boundary layers[2]. Here we shall only consider the advection of heat along a flat plate held at a constant temperature $T_0 = T_\infty + \Theta$, which is different than the ambient slip-flow temperature T_∞. Dropping dissipation and the double derivative with respect to x in the Laplacian, the steady flow heat equation (22.31b) on page 380 takes the form

$$v_x \frac{\partial T}{\partial x} + v_y \frac{\partial T}{\partial y} = \kappa \frac{\partial^2 T}{\partial y^2}. \tag{28.27}$$

As the mass flow is not influenced by the heat flow in an incompressible fluid, we may insert the exact Blasius solution (28.16) and (28.17) for v_x and v_y.

[2] A general discussion of heat flow in boundary layers is found in [Schlichting and Gersten 2000].

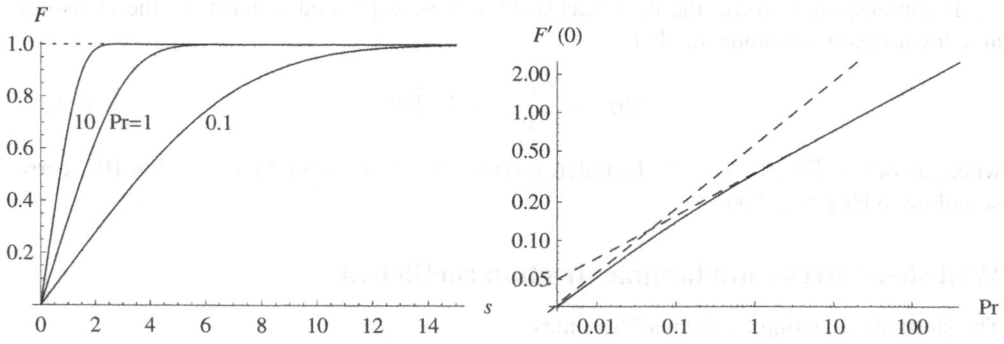

Figure 28.4. **Left:** The function $F(s)$ for three different values of the Prandtl number. **Right:** The slope $F'(0)$ as a function of Pr. The dashed lines are the asymptotic approximations (Problem 28.4).

Assuming that the temperature field is also a function of the Blasius variable,

$$T(x, y) = T_0 - \Theta F\left(y\sqrt{\frac{U}{\nu x}}\right), \tag{28.28}$$

we obtain the equation

$$\boxed{F''(s) + \tfrac{1}{2}\mathsf{Pr}\, g(s) F'(s) = 0,} \tag{28.29}$$

where $g(s)$ is the Blasius function satisfying (28.19) and $\mathsf{Pr} = \nu/\kappa$ is the Prandtl number (page 381). The boundary conditions are $F(0) = 0$ and $F(\infty) = 1$.

One may verify by insertion that the solution is

$$F(s) = \frac{H(s)}{H(\infty)}, \tag{28.30}$$

with

$$H(s) = \int_0^s e^{-\frac{1}{2}\mathsf{Pr}\, G(u)} du, \qquad G(s) = \int_0^s g(u)\, du. \tag{28.31}$$

In Problem 28.3 it is shown that one may write $H(s)$ as a single integral that is, however, less friendly to numeric integration. In Figure 28.4L the result of the numeric integration is plotted for three different values of the Prandtl number.

Total rate of heat transfer

The gradient of the temperature field on the plate determines the current density of heat flow orthogonal to the plate,

$$q_y = -k\, \nabla_y T\big|_{y=0} = F'(0) k \Theta \sqrt{\frac{U}{\nu x}}. \tag{28.32}$$

The average of q_y over a downstream length L becomes

$$\langle q_y \rangle = \frac{1}{L}\int_0^L q_y(x) dx = 2F'(0) k \Theta \sqrt{\frac{U}{\nu L}}. \tag{28.33}$$

The total heat transfer from a flat plate is obtained by multiplying $\langle q_y \rangle$ with the area of the plate. In Figure 28.4R the slope $F'(0) = 1/H(\infty)$ is plotted as a function of the Prandtl number.

One may wonder why the viscosity ν appears at all in the above expression for heat transfer. In Problem 28.4 it is shown that

$$F'(0) \approx \begin{cases} 0.56\,\mathsf{Pr}^{1/2} & \text{for } \mathsf{Pr} \ll 1 \\ 0.34\,\mathsf{Pr}^{1/3} & \text{for } \mathsf{Pr} \gg 1. \end{cases} \tag{28.34}$$

For $\mathsf{Pr} \ll 1$ the factor $\sqrt{\mathsf{Pr}}$ cancels the dependence on ν in (28.33)—as one would expect because the viscous boundary layer lies deep inside the thermal boundary layer (see the margin figure). This is not the case for $\mathsf{Pr} \gg 1$ because the heat flow now takes place deep inside the viscous boundary layer.

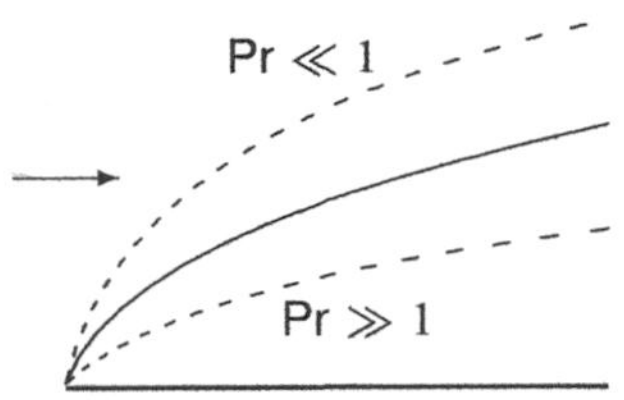

Advective cooling of a plate with wind coming in from the left. The boundaries for mass flow (solid line) and heat flow (dashed) at large and small Prandtl numbers.

Case: Wind chill

When you take a walk on a cold day, your heat loss is amplified by a cold wind that removes the warm air near your body and creates a steeper temperature gradient at the surface of the skin, resulting in a larger conductive heat transfer from your body to the air. In meteorology this is known as *wind chill*, and the local "wind chill temperature" is often announced by weather forecasters during winter. A similar phenomenon must occur in a hot desert wind, although the local "wind burn temperature" is not a regular part of a summer weather forecast. In a sauna with air at 120°C, it is well known that one should not move around too fast.

Water chill is even more important. A cold water current moving with the same speed and temperature as a cold wind results in much stronger cooling because of the 25 times higher thermal conductivity of water. This is why you are only able to survive naked for minutes in streaming water at 0°C, whereas you may survive for hours in a wind of that temperature. Similarly, hot water scalds you much faster than hot air of the same temperature.

Example 28.2 [Human heat loss]: The grown-up human body has a total skin surface area of about $A \approx 2\ \mathrm{m}^2$. A naked human standing with shoulders aligned with the wind will roughly present a (two-sided) plate area with $L \approx 0.5$ m. For air with $\mathsf{Pr} = 0.73$ we have $F'(0) = 0.30$. In a wind with $U = 1\ \mathrm{m\,s^{-1}}$ and temperature $\Theta = 10\,\mathrm{K}$ below the skin temperature, the heat loss rate becomes $\dot{Q} = \langle q \rangle A \approx 100$ W. Since the grown-up human body produces heat at this rate, a skin temperature around 33°C can thus be maintained essentially indefinitely in a gentle breeze with temperature 23°C, not unlike what you find on a Scandinavian beach on a summer day. In this calculation we have ignored natural convection and evaporative heat losses.

Under the same conditions in water with $\mathsf{Pr} = 6.85$ and $F'(0) = 0.64$, the rate of heat loss becomes enormous, $\dot{Q} \approx 23$ kW, and will almost instantly cool the skin to the temperature of the water. Everybody is familiar with the (relatively mild) skin shock that is experienced even when one jumps into water as warm as 23°C.

Instead of expressing wind chill by heat loss, weather forecasters prefer to announce an equivalent fictitious *wind chill temperature*. Let the actual wind be speed U and its temperature T_∞, so that the skin temperature excess is $\Theta = T_0 - T_\infty$. Since the average heat current density (28.33) only depends on the combination $\Theta\sqrt{U}$, the heat loss would be the same for a (possibly fictitious) wind speed U^* and skin temperature excess $\Theta^* = T_0 - T_\infty^*$, satisfying $\Theta^*\sqrt{U^*} = \Theta\sqrt{U}$. Solving for the fictitious wind chill temperature T_∞^*, we find an expression of the form

Paul Allman Siple (1908–1968). American Antarctic explorer. Accompanied the first Byrd expedition to Antarctica in 1928–1930 (as an Eagle Scout). Participated in Byrd's second expedition 1933–1935 as a chief biologist. Coined the term "wind chill" in 1939.

$$T_\infty^* = T_0 - (T_0 - T_\infty)\sqrt{\frac{U}{U^*}}. \tag{28.35}$$

This is the essential part of the first *wind chill formula* by Siple and Passel from 1945 [SP45], although they expressed their calculation in terms of the actual heat loss.

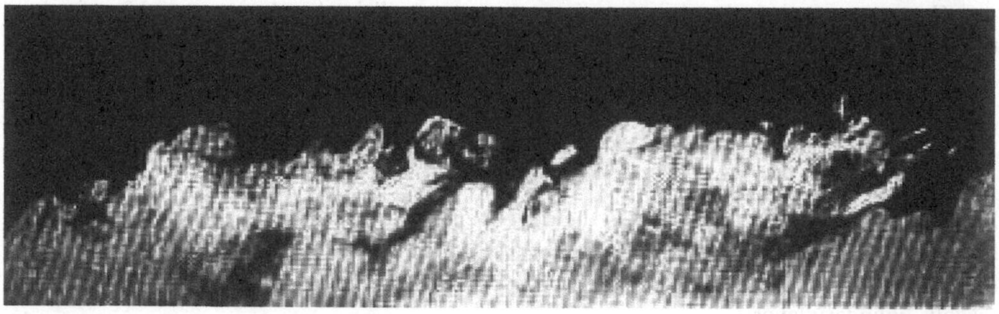

Figure 28.5. Turbulent boundary layer in a wind tunnel at $U\delta_{99}/\nu = 4{,}000$, showing the strong irregularity of the instantaneous interface with the mainstream flow. (Source: Reproduced from [Fal77]. With permission.)

After measuring cooling rates of water in small plastic containers, Siple and Passel included a couple of correction terms to get

$$T_\infty^* = T_0 - (T_0 - T_\infty)\left(\sqrt{\frac{U}{U^*}} - 0.22\frac{U}{U^*} + 0.47\right). \tag{28.36}$$

The parameters were chosen to be $T_0 = 33°\text{C}$ and $U^* = 5\ \text{m s}^{-1}$. For $T_\infty = 0°\text{C}$ and a wind speed $U = 10\ \text{m s}^{-1}$, this yields the wind chill temperature $T_\infty^* \approx -14°\text{C}$.

Over the years the Siple–Passel formula has been criticized [Osc95], in particular because it has an unphysical minimum at $U = 4.91U^* \approx 25\ \text{m s}^{-1}$, and because it does not take into account that the skin will not remain at $T_0 = 33°\text{C}$ when the wind blows. It was nevertheless used by the US National Weather Service from 1973 until 2001, when it was replaced by a somewhat more conservative—though less transparent—formula based on physiology and experiments. For the above example it yields the smaller value $T_\infty^* = -7°\text{C}$ [16].

28.4 Turbulence in the Blasius layer

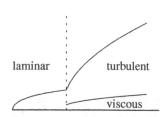

Sketch of the thickness of the boundary layer from the leading edge through the transition region. At the transition (dashed line), the turbulent layer grows rapidly whereas the viscous sublayer only grows slowly.

Sufficiently far downstream from the leading edge, the Reynolds number, $\mathsf{Re}_x = Ux/\nu$, will eventually grow so large that the boundary layer becomes turbulent. Empirically, the transition happens for $5 \times 10^5 \lesssim \mathsf{Re}_x \lesssim 3 \times 10^6$, depending on the circumstances, for example the uniformity of the mainstream flow and the roughness of the plate surface. We shall in the following discussion take $\mathsf{Re}_x = 5 \times 10^5$ as the nominal transition point fro the Blasius layer.

The transition line across the plate is not a straight line but rather an irregular, time-dependent, jagged, even fractal interface between the laminar and turbulent regions. This is also the case for the extended, nearly "horizontal" interface between the turbulent boundary layer and the fluid at large (see Figure 28.5). Such intermittent and fractal behavior is common to the onset of turbulence in all systems.

Friction coefficient

In a turbulent boundary layer, the true velocity field fluctuates in all directions and in time around some mean velocity field $\boldsymbol{v}$. Even very close to the wall, there will be notable fluctuations. The no-slip condition nevertheless has to be fulfilled and a thin sublayer dominated by viscous stresses must exist close to the wall. In this viscous sublayer the average velocity v_x rises linearly from the surface with a slope, $\partial v_x/\partial y|_{y=0} = \sigma_{\text{wall}}/\eta$, which can be determined from drag measurements.

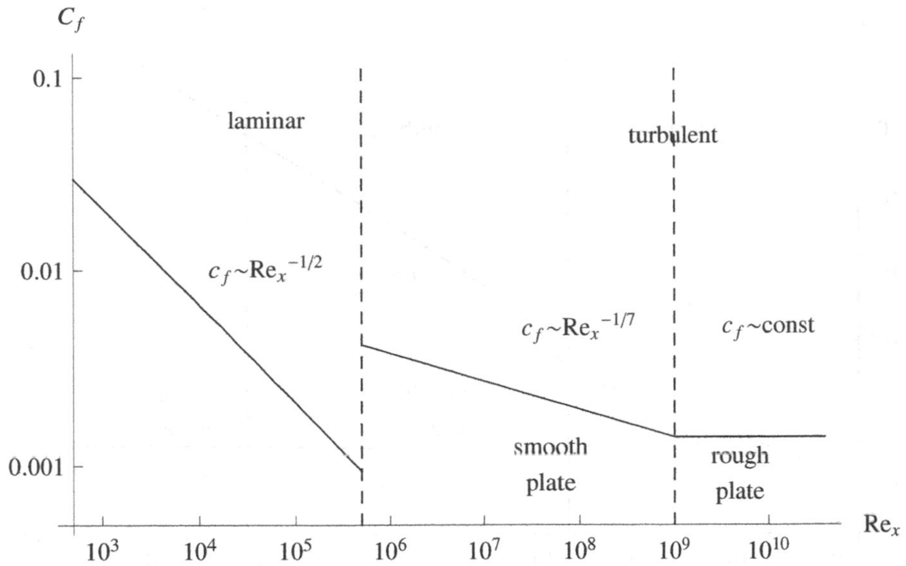

Figure 28.6. Schematic plot of the local friction coefficient C_f across both laminar and turbulent regions as a function of the downstream Reynolds number $\mathsf{Re}_x = Ux/\nu$. The transitions at the nominal points $\mathsf{Re}_x = 5 \times 10^5$ and at $\mathsf{Re}_x = 10^9$ are in reality softer than shown here. The terminal value of C_f depends on the roughness of the plate surface (see [White 1999] for details).

A decent semi-empirical expression for the *turbulent friction coefficient* was already given by Prandtl,

$$C_f \equiv \frac{\sigma_{\text{wall}}}{\frac{1}{2}\rho_0 U^2} = 0.027\,\mathsf{Re}_x^{-1/7}, \tag{28.37}$$

where, as before, $\mathsf{Re}_x = Ux/\nu$. The turbulent friction coefficient thus decreases much slower than the corresponding laminar friction coefficient (28.25). The two expressions cross each other at $\mathsf{Re}_x \approx 7{,}800$, which is far below the transition to turbulence, implying a jump from $C_f \approx 0.94 \times 10^{-3}$ to $C_f \approx 4.1 \times 10^{-3}$ at $\mathsf{Re}_x \approx 5 \times 10^5$. A turbulent boundary layer thus causes much more skin drag than a laminar boundary one (by a factor of more than 4 at the nominal transition point).

In Figure 28.6 the friction coefficient is plotted across the laminar and turbulent regimes. The transition from laminar to turbulent is in reality not nearly as sharp as shown here, partly because of the average over the jagged transition line. Eventually, for sufficiently large Re_x, the roughness of the plate surface makes the friction coefficient nearly independent of viscosity and thus of Re_x (see page 276).

As for the Blasius layer, it is convenient to calculate the average turbulent friction coefficient over a downstream length L:

$$\langle C_f \rangle = \frac{1}{L}\int_0^L C_f(x)\,dx = 0.0315\,\mathsf{Re}_L^{-1/7}. \tag{28.38}$$

If we take into account that the flow may be laminar up to the nominal transition point $x = x_0$ where $\mathsf{Re}_{x_0} = 5 \times 10^5$, we obtain instead

$$\langle C_f \rangle = 0.0315\,\mathsf{Re}_L^{-1/7} + \frac{x_0}{L}\left(1.328\,\mathsf{Re}_{x_0}^{-1/2} - 0.0315\,\mathsf{Re}_{x_0}^{-1/7}\right). \tag{28.39}$$

Note that this expression coincides with the fully turbulent expression for $x_0 = 0$ and the fully laminar for $x_0 = L$.

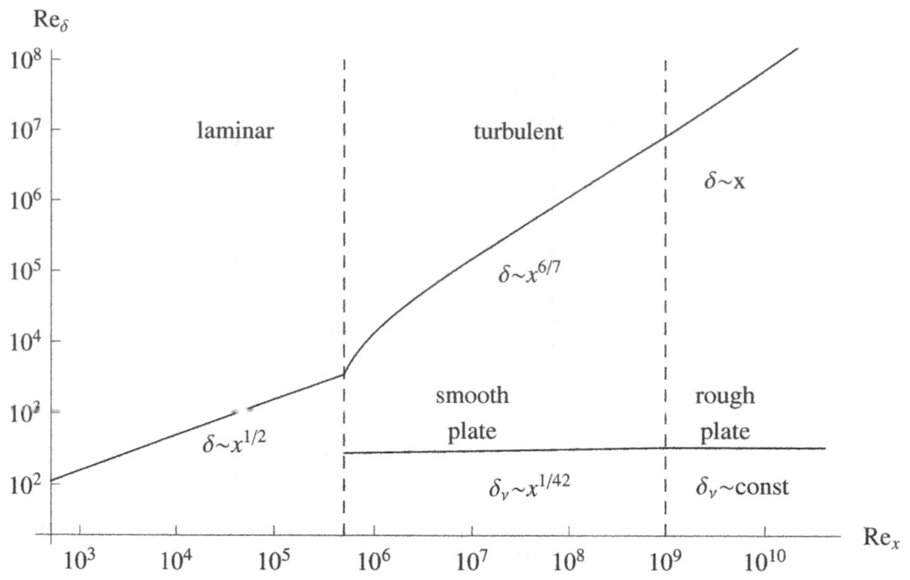

Figure 28.7. Schematic plot of the true thickness, represented by $\mathsf{Re}_\delta = U\delta/\nu$, as a function of downstream distance x, represented by $\mathsf{Re}_x = Ux/\nu$. Also shown is the thickness of the viscous sublayer δ_ν. The real transition at the nominal value, $\mathsf{Re}_x = 5 \times 10^5$, is even softer than shown here. The transition from smooth to rough plate at $\mathsf{Re} = 10^9$ is barely visible.

Turbulent mean flow

Empirically, the flat-plate turbulent boundary layer profile outside the thin viscous sublayer is decently described by the simple model for the mean flow (also due to Prandtl),

$$\frac{v_x}{U} = \left(\frac{y}{\delta}\right)^{1/7} \qquad \text{for } 0 \lesssim y \leq \delta, \tag{28.40}$$

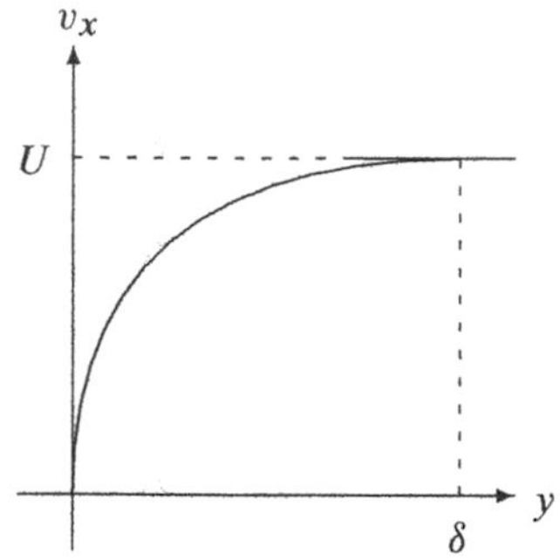

The turbulent mean velocity profile is approximately a power law that diverges at the wall. A finite wall stress is provided by a thin viscous sublayer.

where δ is the true boundary layer thickness. Note that $\partial v_x/\partial y$ diverges for $y \to 0$ but we shall now see that it is nevertheless possible to determine the shear wall stress and thus the friction coefficient indirectly from Prandtl's expression by a method going back to Karman (1921). We shall use this relation in reverse to determine the boundary layer thickness from the empirical friction coefficient (28.37).

Karman's relation: For constant slip-flow U, it follows from the Prandtl equations (28.12) and (28.11) that

$$-\nu \frac{\partial^2 v_x}{\partial y^2} = \frac{\partial((U - v_x)v_x)}{\partial x} + \frac{\partial(v_y(U - v_x))}{\partial y}. \tag{28.41}$$

Integrating over all y and using the boundary conditions $v_y = 0$ for $y = 0$ and $v_x = U$ for $y = \infty$, we see that the second term on the right-hand side does not contribute at all. On the left-hand side we use $\partial v_x/\partial y = 0$ for $y = \infty$ to obtain

$$\boxed{\nu \left.\frac{\partial v_x}{\partial y}\right|_{y=0} = \frac{d}{dx}\int_0^\infty (U - v_x)v_x \, dy.} \tag{28.42}$$

The quantity on the left-hand side is simply $\sigma_{\text{wall}}/\rho_0$, and the integral on the right-hand side may be understood as the *flux of lost momentum* of the displaced slip-flow (Karman 1921).

Using Prandtl's expression (28.40) for the velocity field, the lost momentum flux becomes

$$\int_0^\delta (U - v_x) v_x \, dy = \frac{7}{72} U^2 \delta; \tag{28.43}$$

and using Karman's relation (28.42) together with the definition $C_f = \sigma_{\text{wall}}/\frac{1}{2}\rho_0 U^2$ with $\sigma_{\text{wall}} = \rho_0 \nu \partial v_x/\partial y|_{y=0}$, we obtain

$$\frac{d\delta}{dx} = \frac{36}{7} C_f. \tag{28.44}$$

In a fully turbulent boundary layer, this equation is integrated with the empirical expression (28.37) and the initial value $\delta = 0$ at $x = 0$, which yields

$$\delta = \frac{36}{7} \int_0^x C_f \, dx \approx 0.16 \frac{\nu}{U} \mathsf{Re}_x^{6/7}. \tag{28.45}$$

The thickness of the turbulent boundary layer thus grows much faster than the thickness of the laminar (Blasius) layer (28.22).

If we take into account that the flow is laminar up to the nominal transition point $x = x_0$ where $\mathsf{Re}_{x_0} = 5 \times 10^5$, we obtain instead from (28.22),

$$\frac{U\delta}{\nu} = \begin{cases} 4.91\sqrt{\mathsf{Re}_x} & x < x_0 \\ 4.91\sqrt{\mathsf{Re}_{x_0}} + 0.16\left(\mathsf{Re}_x^{6/7} - \mathsf{Re}_{x_0}^{6/7}\right) & x > x_0, \end{cases} \tag{28.46}$$

which is continuous across the nominal transition point, as seen in Figure 28.7.

The viscous sublayer

It is also possible to get an estimate of the thickness δ_ν of the viscous sublayer from the intercept between the linearly rising field, $v_x = y\,\sigma_{\text{wall}}/\eta$, in the sublayer and Prandt's expression (28.40). This will of course leave an unphysical kink in the velocity field at $y = \delta_\nu$, which gives rise to a small jump in shear stress that violates Newton's Third Law (slightly). The real transition from the viscous sublayer to turbulent main layer is softer than shown here.

Demanding continuity at $y = \delta_\nu$, we get

$$\frac{\sigma_{\text{wall}}}{\eta} \delta_\nu = U \left(\frac{\delta_\nu}{\delta}\right)^{1/7}.$$

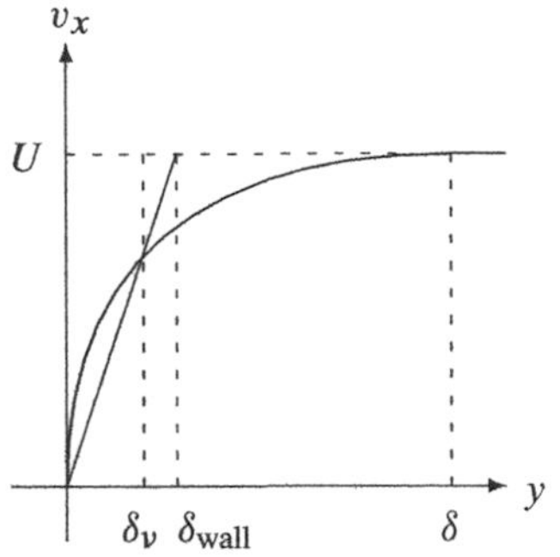

The thickness δ_ν of the viscous sublayer is obtained from the intercept between Prandtl's expression (28.40) and the linearly rising velocity in the sublayer.

Solving this equation for δ_ν and inserting σ_{wall} from (28.37) and δ from (28.45), we obtain the remarkable expression

$$\frac{U\delta_\nu}{\nu} = 206\, \mathsf{Re}_x^{1/42}. \tag{28.47}$$

The small power makes δ_ν nearly constant. At the nominal transition point $\mathsf{Re}_x = 5 \times 10^5$, its value is 282, growing to 337 at $\mathsf{Re}_x = 10^9$. The sublayer thickness is also plotted in Figure 28.7 and its variation with Re_x is in fact barely perceptible.

28.5 Planar stagnation flow

Boundary layers similar to the Blasius layer arise around any body with a sharp edge that simply parts the incoming flow without giving rise to stagnation. Physically, no body is perfect, and an edge always has some curvature; but as long as its radius of curvature is much smaller than the boundary layer thickness, the edge may be considered to be sharp.

In this section we shall analyze the other extreme case of a bluff body where the local radius of curvature at the stagnation point is much larger than the boundary layer thickness. Such a body may be considered flat in the vicinity of the stagnation point, but since the slip-flow velocity has to vanish at the stagnation poin,t we cannot use Prandtl's boundary layer theory, which is only valid for large Reynolds number.

Slip-flow near stagnation

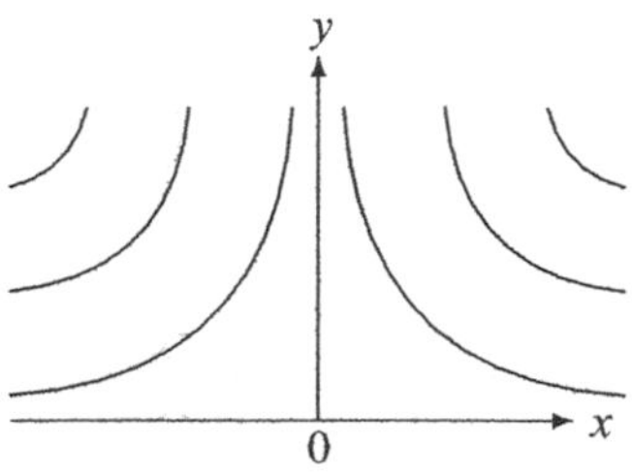

Inviscid streamlines near the stagnation point at $x = y = 0$.

In the absence of viscosity, v_x will in general be non-vanishing on the plate, except at the stagnation point $x = y = 0$, where it must vanish. To first non-trivial order, the simplest field obeying the continuity equation and Bernoulli's theorem is of the form

$$v_x = \frac{x}{\tau}, \qquad v_y = -\frac{y - y_0}{\tau}, \qquad \frac{p}{\rho_0} = H_\infty - \frac{x^2 + (y - y_0)^2}{2\tau^2}, \tag{28.48}$$

where τ has the dimension of time. One may readily verify that this field satisfies the exact Navier–Stokes equations (28.10). The constant y_0 represents the possible upflow from the boundary layer, and the constant H_∞ represents the value of the upstream Bernoulli field.

Viscous stagnation flow

Viscosity requires the boundary flow along x to vanish everywhere. Since the only quantity with the dimension of length that can be formed for this problem is $\sqrt{\nu\tau}$, we shall assume

$$v_x = \frac{x}{\tau} f\left(\frac{y}{\sqrt{\nu\tau}}\right), \tag{28.49}$$

with some function $f(s)$ satisfying $f(0) = 0$ and $f(\infty) = 1$. Using the continuity equation in the form (28.13), we find the simple result

$$v_y = -\sqrt{\frac{\nu}{\tau}}\, g\left(\frac{y}{\sqrt{\nu\tau}}\right), \qquad g(s) = \int_0^s f(s')ds'. \tag{28.50}$$

Inserting the velocities into the exact Navier–Stokes equations (28.10), the pressure gradients become

$$\frac{1}{\rho_0}\frac{\partial p}{\partial x} = \frac{x}{\tau^2}\left(f''(s) + g(s)f'(s) - f(s)^2\right), \tag{28.51a}$$

$$\frac{1}{\rho_0}\frac{\partial p}{\partial y} = -\sqrt{\frac{\nu}{\tau^3}}\left(f'(s) + g(s)f(s)\right). \tag{28.51b}$$

The symmetry of the double derivative $\partial^2 p/\partial x\partial y = \partial^2 p/\partial y\partial x$ demands that the expression in the first parentheses, $f''(s) + g(s)f'(s) - f(s)^2$, must be a constant independent of y. Since $f(s) \to 1$ for $s \to \infty$ with a Gaussian tail, this constant must be -1, and we finally arrive at the coupled equations

$$f''(s) + g(s)f'(s) - f(s)^2 = -1, \qquad g'(s) = f(s). \tag{28.52}$$

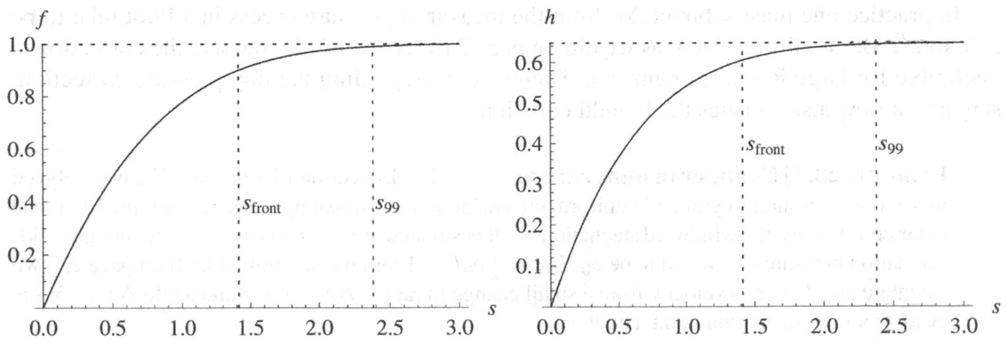

Figure 28.8. Planar stagnation flow. **Left:** The shape function $f(s)$ of v_x has slope $f'(0) = 1.2326$ at $s = 0$. **Right:** The shape function $h(s) = s - g(s)$.

Combining them, we arrive at the single equation (sometimes called the *Hiemenz equation*)

$$\boxed{g'''(s) + g(s)g''(s) + 1 - g'(s)^2 = 0.} \tag{28.53}$$

The structure of this equation resembles the Blasius equation (28.19) and may be solved numerically in the same way. The solution is shown in Figure 28.8. In the following section we shall see that both equations are members of a one-parameter family of similar equations.

Upflow and stagnation pressure

In the limit of $s \to \infty$, we have $f(s) \to 1$ and $g(s) \to s$. Inserted into (28.53), we find $g''(s) \sim \exp(-s^2/2)$, which tells us that the momentum diffusion front is at $s = \sqrt{2}$. Accordingly we have

$$\delta_{\text{front}} = 1.414\sqrt{\nu\tau}, \qquad \delta_{99} = 2.379\sqrt{\nu\tau}. \tag{28.54}$$

The difference $h(s) = s - g(s)$ converges at infinity to $h(\infty) \approx 0.6479$. Asymptotically for $s \gg 1$ or $y \gg \sqrt{\nu\tau}$, the velocity field thus becomes

$$v_x \approx \frac{x}{\tau}, \qquad v_y \approx -\frac{y - y_0}{\tau}, \qquad y_0 = h(\infty)\sqrt{\nu\tau}. \tag{28.55}$$

The upflow from the boundary layer modifies—as expected—the slip-flow but the correction is small compared to the downflow for $y \gg \sqrt{\nu\tau}$.

The pressure gradient (28.51) may immediately be integrated to yield

$$p = p_0 - \rho_0 \frac{x^2 + \nu\tau\left(g(s)^2 + 2f(s)\right)}{2\tau^2} = p_0 - \tfrac{1}{2}\rho_0\left(\frac{x^2}{\tau^2} + v_y^2\right) - \rho_0\frac{\nu f(s)}{\tau},$$

where p_0 is the true stagnation pressure. Asymptotically, for $y \gg \sqrt{\nu\tau}$, the pressure must coincide with the slip-flow pressure (28.48), implying that the true stagnation pressure is

$$p_0 = \rho_0\left(H_\infty + \frac{\nu}{\tau}\right). \tag{28.56}$$

Now $\rho_0 H_\infty$ would be the stagnation pressure calculated for inviscid flow, so the presence of the boundary layer increases the true stagnation pressure by

$$\boxed{\Delta p = \frac{\eta}{\tau},} \tag{28.57}$$

where $\eta = \rho_0\nu$ is the shear viscosity.

In practice one must subtract Δp from the measured pressure excess in a Pitot tube to be able to use Bernoulli's theorem as we did on page 215. As we shall now see, the correction is negligible for large Reynolds numbers. Therefore, disregarding the tiny pressure correction, stagnation flow also satisfies the Prandtl equations.

Example 28.3 [Cylinder in uniform crosswind]: In Section 13.7 on page 222 we analyzed inviscid flow around a cylinder in uniform crosswind U and vanishing pressure at infinity. In plane polar coordinates, the windward stagnation point is situated at $r = a$ and $\phi = \pi$, and the (inviscid) stagnation pressure was found to be $\rho_0 H_\infty = \frac{1}{2}\rho_0 U^2$. From the solution (13.52) on page 223 we calculate the change in velocity from a small change in radius $\Delta r = r - a$ and angle $\Delta\phi = \phi - \pi$ away from the stagnation point, and find

$$v_r = -2\frac{U}{a}\Delta r, \qquad v_\phi = 2U\Delta\phi. \tag{28.58}$$

In the boundary layer notation we substitute $\Delta r \to y$ and $\Delta\phi \to x/a$ and the time constant may be read off, $\tau = a/2U$. The relative correction to the stagnation pressure and the thickness scale then become

$$\frac{\Delta p}{\frac{1}{2}\rho_0 U^2} = \frac{4\nu}{Ua} = \frac{8}{\mathsf{Re}}, \qquad \frac{\delta_{99}}{a} = \frac{2.4}{\sqrt{\mathsf{Re}}}, \tag{28.59}$$

where $\mathsf{Re} = 2aU/\nu$. For Reynolds number 1,600, the pressure correction is 0.5% of the stagnation pressure and the thickness is 6% of the radius.

* 28.6 Self-similar boundary layers

In this section we shall analyze self-similar boundary layers created by slip-flows $U(x)$ that vary along the surface of the body. The assumption of self-similarity will as before allow us to convert the partial differential equations of fluid mechanics into ordinary differential equations. We shall learn that self-similarity essentially only allows for power law slip-flows of the form $U \sim x^m$. The class of such slip-flows is, on the other hand, sufficiently general to illustrate what happens when the slip-flow accelerates ($m > 0$) or decelerates ($m < 0$), although it cannot handle boundary layer separation. As before, we assume that the mainstream Reynolds number is so large that we may use the Prandtl approximation.

Self-similarity

A self-similar boundary layer with slip-flow $U(x)$ and thickness scale $\lambda(x)$ is of the form

$$v_x = U(x) f\left(\frac{y}{\lambda(x)}\right), \tag{28.60}$$

where $f(s)$ is a dimensionless function of the dimensionless variable satisfying the boundary conditions $f(0) = 0$ and $f(\infty) = 1$. It follows immediately from the equation of continuity (28.13) that

$$v_y = -\frac{\partial}{\partial x}\big(U(x)\lambda(x)g(s)\big), \qquad g(s) = \int_0^s f(s')\,d's, \tag{28.61}$$

where the differentiation with respect to x must be done for fixed y.

The Prandl equation (28.11) only contains first-order derivatives with respect to x. We may without loss of generality represent the derivatives of U and λ by

$$\frac{d\lambda}{dx} = (\alpha - \beta)\frac{\nu}{\lambda U}, \qquad \frac{dU}{dx} = \beta\frac{\nu}{\lambda^2}, \tag{28.62}$$

where α and β are dimensionless functions of x.

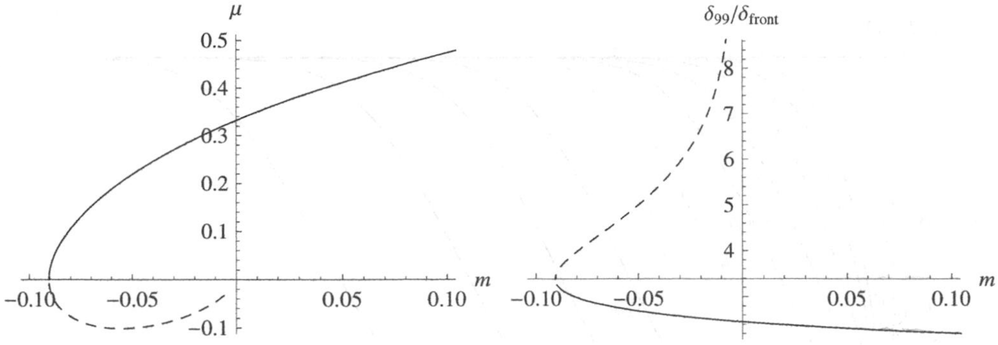

Figure 28.9. Left: The wall slope $\mu = f'(0)$ as a function of m in the interval $-0.1 \leq m \leq 0.1$. It is double-valued for negative m, and the solutions join at the critical point $m_c = -0.0904$ where $\mu = 0$. The critical curvature is $\kappa_c = \left[\partial^2 m/\partial \mu^2\right]_{\mu=0} = 2.320$. **Right:** The ratio $\delta_{99}/\delta_{\text{front}}$ plotted in the same interval. It is also double-valued for negative m and takes the value $\delta_{99}/\delta_{\text{front}} = 3.386$ for $m = m_c$.

Substituting v_x and v_y into the Prandl equation (28.11), we obtain the following coupled ordinary differential equations,

$$f''(s) + \alpha g(s) f'(s) + \beta\left(1 - f(s)^2\right) = 0, \qquad g'(s) = f(s). \tag{28.63}$$

There is clearly not much room for an x-dependence in α and β (see Problem 28.6), and we shall in the remainder of this section take both to be constants.

Combining the coupled equations into a single third-order equation, we arrive at the most general form of the famous *Falkner–Skan equation* from 1931 [FS31],

$$\boxed{g'''(s) + \alpha g(s) g''(s) + \beta\left(1 - g'(s)^2\right) = 0.} \tag{28.64}$$

Evidently, it contains the Blasius equation (28.19) for $\alpha = \frac{1}{2}$ and $\beta = 0$, and the Hiemenz equation (28.53) for $\alpha = \beta = 1$. Since we are free to choose the overall scale of $\lambda(x)$, the Falkner–Skan equation has in fact only one free parameter. We shall, however, avoid making a particular choice in order to demonstrate that it cannot influence the physics.

The solutions

Combining the definitions (28.62), we obtain $d(\lambda^2 U)/dx = \nu(2\alpha - \beta)$. Apart from the special case $\beta = 2\alpha$, the solution is $\lambda^2 U = \nu(2\alpha - \beta)x$ when the origin of x is chosen conveniently. Using this to eliminate λ^2 in the second definition, we get the simple equation

$$\frac{dU}{dx} = \frac{\beta}{2\alpha - \beta}\frac{U}{x}, \tag{28.65}$$

which integrates to

$$\boxed{U = U_0\left|\frac{x}{x_0}\right|^m, \qquad m = \frac{\beta}{2\alpha - \beta},} \tag{28.66}$$

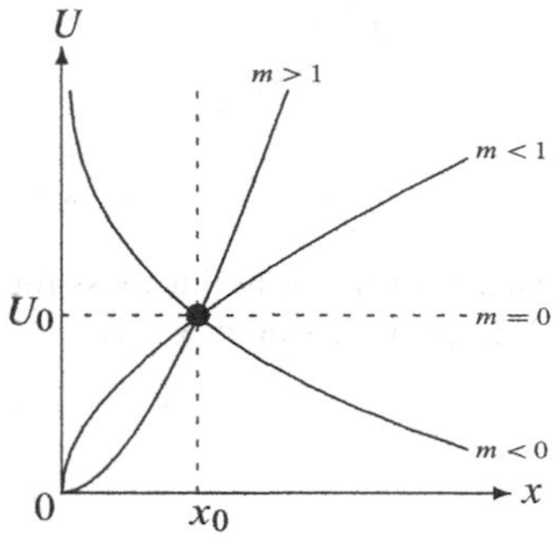

Shapes of the slip-velocity for different values of $m > -1$.

where U_0 is a constant velocity and x_0 is a constant length[3]. This demonstrates that the solutions physically are distinguished by a single parameter m. The Blasius solution corresponds to $m = 0$ and the Hiemenz solution to $m = 1$.

[3]There is in fact only one constant $A = U_0 |x_0|^{-m}$, but we use this notation for dimensional clarity.

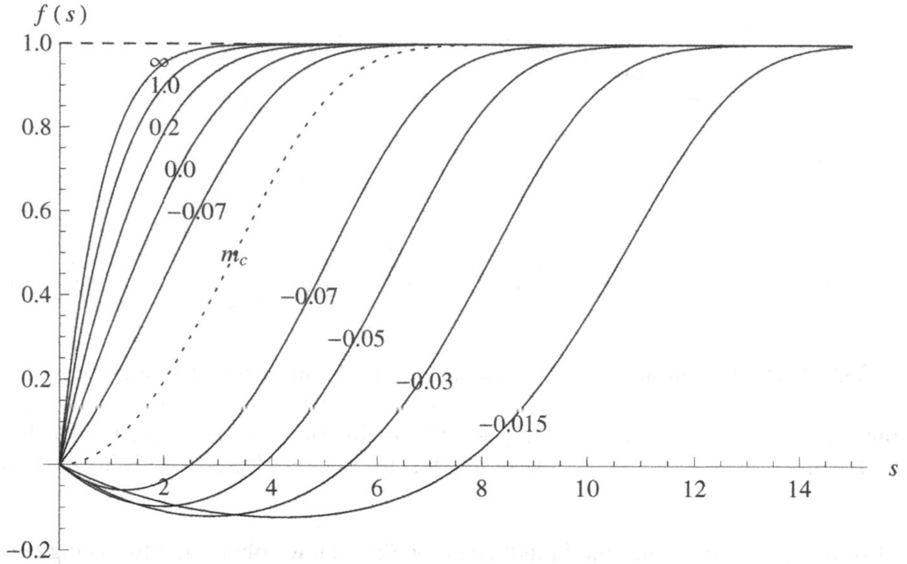

Figure 28.10. Self-similar Falkner–Skan velocity profiles $f(s)$ for $\alpha = 1/2$ and $\beta = m/(1+m)$. The values are selected from the interval $m_c < m < \infty$, where $m_c = -0.0904286$. The Blasius profile is obtained for $m = 0$ and the Hiemenz profile (Section 28.5) for $m = 1$. Note that there are two solutions for each value of m in the interval $m_c < m < 0$, one of which has reversed flow close to the wall. The critical profile with $m = m_c$ is shown dashed.

The thickness scale may now be calculated from $\lambda^2 U = \nu(2\alpha - \beta)x$. To relate it to a physical thickness, we need to determine the asymptotic behavior of v_x for $y \to \infty$. Asymptotically for $s \to \infty$, the boundary condition $f(s) \to 1$ implies that $g(s) \to s$ so that (28.63) asymptotically takes the form $f''(s) + \alpha s f'(s) \to 0$. Solving for $f'(s)$, we get the asymptotic behavior $f'(s) \sim \exp\left(-\frac{1}{2}\alpha s^2\right)$, which shows that we must require $\alpha > 0$ to get a proper Gaussian tail. The physically meaningful momentum diffusion front then becomes

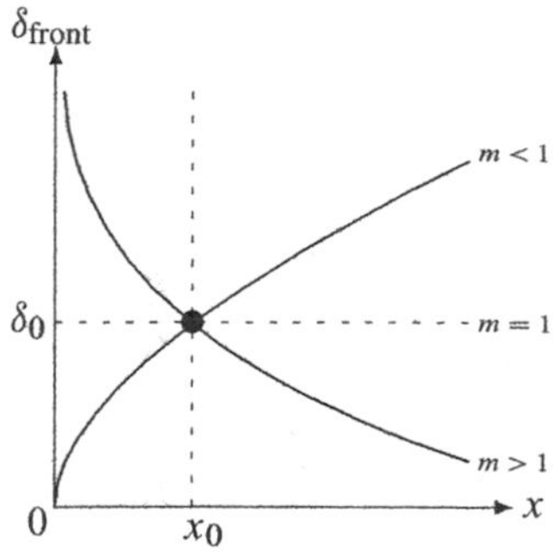

Shapes of the front thickness for different values of $m > -1$.

$$\delta_{\text{front}} = \lambda\sqrt{\frac{2}{\alpha}} = 2\sqrt{\frac{\nu x}{(1+m)U(x)}}. \tag{28.67}$$

As expected, it depends only on m but not on α and β individually. Up to this point we have restricted the parameter space to $\alpha > 0$, but it now appears that for the normal case $U > 0$ and $x > 0$, we must also have $m > -1$, that is, $\beta < 2\alpha$.

To plot the velocity profiles we need to make a choice, for example $\alpha = 1/2$ and $\beta = m/(1+m)$, which makes $\delta_{\text{front}} = 2\lambda$. A selection is shown in Figure 28.10 for some values of m. For $m > 1$, the growth of the boundary layer is suppressed by the accelerating flow and the thickness $\delta_{\text{front}} \sim x^{(1-m)/2}$ decreases downstream. In the interval $0 < m < 1$, the slip-velocity still accelerates downstream, the solutions are unique and all resemble the Blasius solutions, except that the thickness grows slower than $\sqrt{x}$. In the small interval $m_c < m < 0$ where $m_c = -0.0904\ldots$, the slip-flow decelerates and the thickness grows slightly faster than $\sqrt{x}$. There are precisely two solutions in the interval $m_c < m < 0$, one with positive and one with negative wall slope $\mu = f'(0)$. In Figure 28.9 the wall slope and the physical ratio $\delta_{99}/\delta_{\text{front}}$ are plotted as functions of m. There are also boundary layer solutions for $-1 < m < m_c$, but they oscillate in the asymptotic region and are presumably of less interest to physics [Sobey 2000].

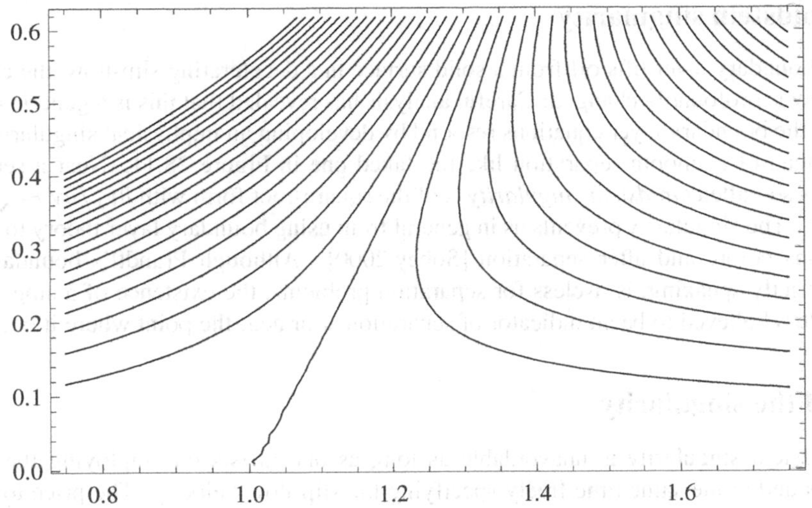

Figure 28.11. Boundary layer separation faked by varying the Falkner–Skan power $m = m_c + \frac{1}{2}\kappa_c \mu^2$ with $\mu = (x_c - x)/L_c$. The streamlines are obtained by choosing $x_c = 1$ and $L_c = 20$. The power goes from $m = -0.089$ at $x = 0.75$ through $m = m_c = -0.09$ at $x = 0$ and back to $m = -0.080$ at $x = 1.75$. The axes have equal scales.

Faked separation

The family of self-similar Falkner–Skan solutions *cannot* be used to model boundary layer separation, because the velocity profile for given m by construction has the same shape for all x. We may, however, fake separation by letting the wall slope μ slide very slowly with x along the positive branch toward the critical point $x = x_c$, where $\mu = 0$ and continue along the negative branch. Near the critical point, $m = m_c$, where the positive branch joins with the negative, the exponent m is quadratic in μ,

$$m = m_c + \frac{1}{2}\kappa_c \mu^2, \tag{28.68}$$

with $\kappa_c = 2.320$. If we choose

$$\mu = \frac{x_c - x}{L_c}, \tag{28.69}$$

where L_c is a "large" length, we obtain the impressive streamline plot in Figure 28.11. But even if it looks right, this picture is a fraud, because the Prandtl equations, as we shall see below, do not allow a smooth passage through the separation point. Instead they force the slope parameter to develop an impassable square-root singularity at the separation point of the form, $\mu = \sqrt{(x_c - x)/L_c}$.

28.7 Laminar boundary layer separation

During the twentieth century the problem of predicting the point of boundary layer separation around variously shaped objects has been of great importance to fluid mechanics for fundamental as well as technological reasons. It has proven to be a challenging problem, to say the least [Schlichting and Gersten 2000, Sobey 2000]. A number of approximative schemes have been proposed and tried out, and with suitable empirical input they compare reasonably well with analytic or numeric calculations [White 1991]. We shall analyze some of these in the following sections.

The Goldstein singularity

When a boundary layer lifts off from a solid surface in a decelerating slip-flow, the character of the flow is profoundly changed. Careful analysis has revealed that this is a generic problem to which the boundary layer equations respond by developing an unphysical singularity of the form instead of a smooth separation like the faked one in Figure 28.11. Near a separation point, this so-called *Goldstein singularity* is of the square-root form with $\partial v_x/\partial y \sim \sqrt{x_c - x}$ for $y = 0$. The singularity prevents us in general from using boundary layer theory to connect the regions before and after separation [Sobey 2000]. Although Prandtl's boundary layer theory, strictly speaking, is useless for separation problems, the existence of a singularity is nevertheless believed to be an indicator of separation at or near the point where it occurs.

Sydney Goldstein (1903–1989). British mathematical physicist. Worked on numerical solutions to the steady-flow laminar boundary layer equations.

Beyond the singularity

The Goldstein singularity is unavoidable as long as one insists on employing the Prandtl equations and at the same time freely specifying the slip-flow velocity. The price to pay for avoiding the singularity is that the Prandtl equations must be replaced by the Navier–Stokes equations, and that the mainstream flow cannot be fully specified in advance but has to be allowed to be influenced by what happens deep inside the boundary layer. Since separation originates in the innermost viscous "deck" of the boundary layer, viscosity thus takes a decisive part in selecting the nearly inviscid flow at large. Again this emphasizes that inviscid flow solutions are not unique, and that truly inviscid flow is a flawed ideal.

In the last half of the twentieth century, it has been conclusively demonstrated through theoretical analysis and numerical simulation that the Navier–Stokes equations do not lead to any boundary layer singularities and do in fact smoothly connect the regions before and after separation. The most successful method is a natural extension of Prandtl's method of dividing the flow into two "decks", a viscous transition layer and an inviscid slip-flow layer that in the end are "stitched together" to form the complete boundary layer. In the "triple deck" approach, the transition layer is further subdivided into a near-wall viscous sublayer and a second viscous layer interpolating between the sublayer and the slip-flow. Unfortunately, there does not seem to be any simple intuitive way of presenting this modern "interactive" approach [Sychev et al. 1998, Sobey 2000].

28.8 Wall-anchored model

In this section we shall first justify that the Goldstein singularity exists by an expansion in exact field derivatives at the wall, and then determine its position in a collection of (rather academic) flows where the exact position is known. The result of this approach is not spectacular because it is too dependent on the behavior of the velocity field right at the wall. In the following sections we shall see how this is improved by giving up the wall conditions and instead using momentum and energy balance.

Exact wall derivatives

Let us write the Prandtl equations in the form

$$\nu \frac{\partial^2 v_x}{\partial y^2} = v_x \frac{\partial v_x}{\partial x} + v_y \frac{\partial v_x}{\partial y} - U\frac{dU}{dx}, \qquad \frac{\partial v_y}{\partial y} = -\frac{\partial v_x}{\partial x}, \tag{28.70a}$$

with boundary conditions $v_x = v_y = 0$ at $y = 0$ and $v_x \to U$ for $y \to \infty$. These equations appear well suited for a Taylor expansion in powers of y by successively calculating the derivatives $\partial^n v_x/\partial y^n$ at $y = 0$, for $n = 2, 3, \ldots$.

Provided we know the shear strain rate at the wall,

$$\omega = \left.\frac{\partial v_x}{\partial y}\right|_{y=0}, \tag{28.71}$$

the next three coefficients are quickly calculated from (28.70a),

$$\left.\frac{\partial^2 v_x}{\partial y^2}\right|_{y=0} = -\frac{U\dot{U}}{\nu}, \qquad \left.\frac{\partial^3 v_x}{\partial y^3}\right|_{y=0} = 0, \qquad \left.\frac{\partial^4 v_x}{\partial y^4}\right|_{y=0} = \frac{\omega\dot{\omega}}{\nu}, \tag{28.72}$$

where a dot indicates differentiation with respect to x. The wall derivatives can be calculated to any order, and depend only on the functions $U(x)$ and $\omega(x)$ and their derivatives.

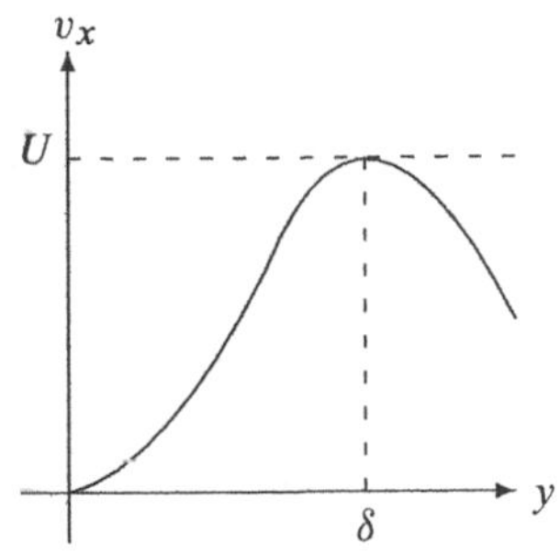

The fourth-order wall expansion joins continuously with $v_x = U$ at its maximum $y = \delta$. The continuation of the polynomial $y > \delta$ is unphysical.

Wall expansion to fourth order

Any exact solution to the Prandtl equations must to fourth order in the distance from the wall be of the form

$$\boxed{v_x = \omega\, y - \frac{U\dot{U}}{2\nu} y^2 + \frac{\omega\dot{\omega}}{24\nu} y^4 + \mathcal{O}\left(y^5\right).} \tag{28.73}$$

To make it resemble the field of a boundary layer, we shall require that v_x has a maximum at $y = \delta(x)$, which is interpreted as the limit to the boundary layer. Setting $v_x = U(x)$ and $\partial v_x/\partial y = 0$ for $y = \delta$, we derive the equations

$$\omega\,\delta - \frac{U\dot{U}}{2\nu}\delta^2 + \frac{\omega\dot{\omega}}{24\nu}\delta^4 = U, \qquad \omega - \frac{U\dot{U}}{\nu}\delta + \frac{\omega\dot{\omega}}{6\nu}\delta^3 = 0. \tag{28.74}$$

Multiplying the second equation by $\delta/4$ and subtracting it from the first, we find

$$\omega = \frac{4\nu + \dot{U}\delta^2}{3\delta}. \tag{28.75}$$

This shows that there is in fact only one free function in the problem, satisfying a first-order differential equation that we shall derive below.

Behavior near the separation point

Since $\omega = 0$ at the separation point $x = x_c$, the separation thickness δ_c will be finite,

$$\delta_c = 2\sqrt{-\frac{\nu}{\dot{U}_c}}. \tag{28.76}$$

Clearly the velocity must be decelerating at separation, $\dot{U}_c < 0$, for this to make sense. From the second condition (28.74), it follows that $\omega\dot{\omega} \approx -3U_c\dot{U}_c^2/2\nu$ in the vicinity of the separation point. Integrating we find $\omega^2 \approx -3U_c\dot{U}_c^2(x - x_c)/\nu$, which only makes sense for $x < x_c$ when $U_c > 0$, as is normally the case for slip-flows that are specified in advance (like the ones in Table 28.1). The solution is thus

$$\omega \approx \sqrt{\frac{3U_c\dot{U}_c^2}{\nu}(x_c - x)} \tag{28.77}$$

just before the separation point, and no solution after. This is the Goldstein singularity.

For $y \ll \delta$, the velocity field components become $v_x \sim y\sqrt{x_c - x}$ and $v_y \sim y^2/\sqrt{x_c - x}$. The upflow diverges for $x \to x_c$, clearly violating one the assumptions, $|v_y| \ll |v_x|$, underlying the Prandtl equations. This makes the whole procedure suspect.

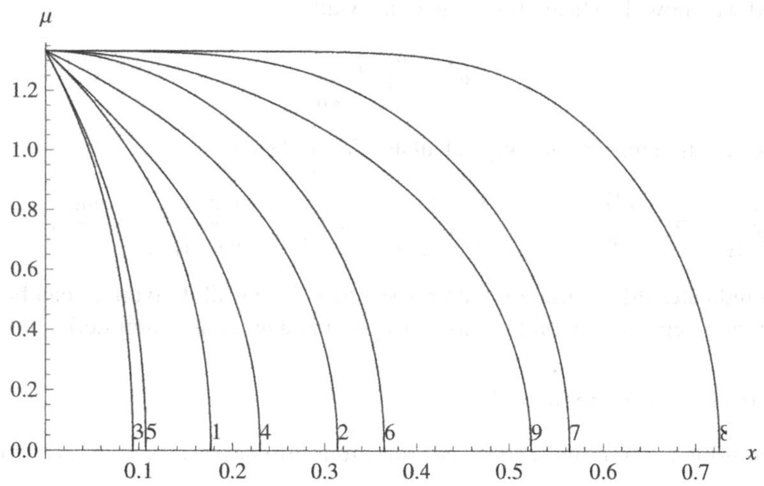

Figure 28.12. Plot of the solutions to Equation (28.80) for the decelerating slip-flows enumerated in Table 28.1. Notice the vertical tangents at $\mu = 0$, signaling the Goldstein singularity.

Complete solution

The fourth-order wall expansion (28.73) not only allows us to determine the behavior just before the critical point, but also the complete dependence on x as well as the position of the singularity. It is convenient to replace ω by a dimensionless parameter μ, so that (28.75) becomes a parameterization in this variable,

$$\omega = \mu \frac{U}{\delta}, \qquad \delta = \sqrt{-(4-3\mu)\frac{\nu}{\dot{U}}}. \tag{28.78}$$

If the flow is accelerating ($\dot{U} > 0$), we must have $\mu > 4/3$ to make δ real, and $\mu < 4/3$ if decelerating. A smooth transition from acceleration to deceleration or conversely only takes place where $\dot{U} = 0$, and requires that μ at the same time passes smoothly through $4/3$.

With this parameterization the wall expansion takes the much simpler form

$$\boxed{\frac{v_x}{U} \equiv f(s) = \mu\, s + \tfrac{1}{2}(4-3\mu)\, s^2 + \tfrac{1}{2}(\mu - 2)\, s^4 + \mathcal{O}\left(s^5\right),} \tag{28.79}$$

where $s = y/\delta$. Now inserting ω and δ into either of the equations (28.74), we find the following first-order differential equation for the dimensionless slope,

$$\dot{\mu} = \frac{24}{\mu}\frac{2-\mu}{8-3\mu}\frac{\dot{U}}{U} - \mu\frac{4-3\mu}{8-3\mu}\left(\frac{\ddot{U}}{\dot{U}} + 2\frac{\dot{U}}{U}\right). \tag{28.80}$$

Given $U(x)$, the dimensionless slope $\mu(x)$ may now be calculated for all x. Notice that $\mu\dot{\mu} \to 6\dot{U}_c/U_c$ for $\mu \to 0$, confirming the presence of the Goldstein singularity.

The numeric solutions to the differential equation are shown in Figure 28.12 for the nine decelerating slip-flows in Table 28.1 which all start out with $\delta = 0$, that is, $\mu = 4/3$. The separation points are listed in the third column (marked "Wall"). They agree coarsely with the exact results (second column) but overshoot by up to 50%. The poor performance of the wall-model must be ascribed to the much too solid anchoring of the polynomial to the wall.

Table 28.1. Table of decelerating slip-flows.

	$U(x)$	x_c	Wall	Wall+mom	Mom+ener-3	Mom+ener-4	Approx
1.	$1-x$	0.120	0.176	0.132	0.117	0.121	0.123
2.	$\sqrt{1-x}$	0.218	0.314	0.237	0.213	0.219	0.221
3.	$(1-x)^2$	0.064	0.094	0.070	0.062	0.064	0.065
4.	$(1+x)^{-1}$	0.151	0.230	0.172	0.146	0.154	0.158
5.	$(1+x)^{-2}$	0.071	0.107	0.080	0.069	0.072	0.074
6.	$1-x^2$	0.271	0.365	0.281	0.269	0.270	0.268
7.	$1-x^4$	0.462	0.565	0.461	0.465	0.460	0.449
8.	$1-x^8$	0.640	0.729	0.629	0.651	0.643	0.621
9.	$\cos x$	0.389	0.523	0.402	0.384	0.386	0.383
	max error	0	52%	14%	3.7%	2.1%	5%
	average	0	40%	9%	2.5%	1.0%	3%

The positions of the separation points are calculated in the various models discussed in the text. The slip-velocities and the exact positions x_c in the second column are taken from [White 1991]. The separation points determined by the wall-anchored fourth-degree polynomial (28.73) are listed in the column marked "Wall". The column marked "Wall+mom" lists the results of the third-degree polynomial (28.81) with mixed wall plus momentum balance conditions. In the column marked "Mom+ener-3" and "Mom+ener-4", the separation points are determined alone from momentum and energy balance with third- and fourth-degree polynomials. Finally, in the last column, the separation points are obtained from the simple approximation (28.105).

28.9 Wall derivative plus momentum balance

If we give up all but the first wall derivative in (28.72), we need only a third-degree polynomial to fulfill the two slip-flow conditions $v_x = U$ and $\partial v_x/\partial y = 0$ for $y = \delta$,

$$\boxed{\frac{v_x}{U} \equiv f(s) = \mu s + (3-2\mu)s^2 + (\mu-2)s^3 + \mathcal{O}\left(s^4\right),} \tag{28.81}$$

where, as before, $s = y/\delta$. One may easily verify that $f(1) = 1$ and $f'(1) = 0$.

Momentum balance

We have already derived a relation (28.42) from momentum balance in uniform flow. To generalize it to varying slip-flow, we first rewrite the Prandtl equation (28.11) as

$$-\nu\frac{\partial^2 v_x}{\partial y^2} = (U - v_x)\frac{dU}{dx} + \frac{\partial[v_x(U-v_x)]}{\partial x} + \frac{\partial[v_y(U-v_x)]}{\partial y}. \tag{28.82}$$

Integrating over y from 0 to ∞, and using the boundary values $v_y \to 0$ for $y \to 0$ and $v_x \to U$ for $y \to \infty$, we obtain the general Karman relation

$$\nu\left.\frac{\partial v_x}{\partial y}\right|_{y=0} = \frac{dU}{dx}\int_0^\infty (U - v_x)\,dy + \frac{d}{dx}\int_0^\infty (U-v_x)v_x\,dy. \tag{28.83}$$

The expression $\rho_0\nu\partial v_x/\partial y|_{y=0}$ is the shear stress on the plate, $\rho_0(U - v_x)$ is the flux of displaced mass, and $\rho_0(U - v_x)v_x$ is the flux of displaced momentum. Multiplied by dx, the above relation states that the force on a slice dx of the plate equals the change in velocity across the slice times the rate of mass displacement plus the change in the rate of momentum displacement across the slice.

To put this relation into a simpler form we define the intrinsic thickness measures,

$$\delta_{\text{mass}} = \frac{1}{U}\int_0^\infty (U - v_x)\,dy, \qquad \delta_{\text{mom}} = \frac{1}{U^2}\int_0^\infty (U - v_x)v_x\,dy, \tag{28.84}$$

so that momentum balance becomes

$$\boxed{\nu\,\omega = U\frac{dU}{dx}\delta_{\text{mass}} + \frac{d\left(U^2\delta_{\text{mom}}\right)}{dx},} \tag{28.85}$$

where we, as before, have defined $\omega = \partial v_x/\partial y|_{y=0}$.

Solution

Demanding $\partial^2 v_x/\partial y^2 = -U\dot{U}/\nu$ for $y = 0$, the slope and thickness may in analogy with the preceding case be parameterized as

$$\omega = \mu\frac{U}{\delta}, \qquad \delta = \sqrt{-2(3-2\mu)\frac{\nu}{\dot{U}}}. \tag{28.86}$$

This shows that for decelerating flow, $\dot{U} < 0$, we must have $\mu < 3/2$.

The mass and momentum thicknesses can now be calculated as integrals over $f(s)$,

$$\frac{\delta_{\text{mass}}}{\delta} = \frac{1}{2}\left(1 - \frac{\mu}{6}\right), \qquad \frac{\delta_{\text{mom}}}{\delta} = \frac{9}{70}\left(1 + \frac{\mu}{6} - \frac{2\mu^2}{27}\right). \tag{28.87}$$

Inserted into momentum balance we obtain (after a bit of algebra) the differential equation

$$\dot{\mu} = \frac{35\left(18 - 9\mu + 2\mu^2\right)}{(3-\mu)(9+20\mu)}\frac{\dot{U}}{U} - \frac{(3-2\mu)\left(54 + 9\mu - 4\mu^2\right)}{2(3-\mu)(9+20\mu)}\left(\frac{\ddot{U}}{\dot{U}} - 4\frac{\dot{U}}{U}\right). \tag{28.88}$$

Solving this equation numerically with $\mu(0) = 3/2$ for the decelerating slip-flows in Table 28.1, we obtain the separation points in the column marked "Wall+mom". They are clearly much better than for the pure wall expansion. The slope $\dot{\mu}$ is large and negative but finite for $\mu = 0$. It diverges, as can be seen, shortly after for $\mu = -9/20 = -0.45$. Giving up all but one wall derivative shifts the singularity a bit beyond the separation point.

28.10 Momentum plus energy balance

Abandoning all conditions on the wall derivatives, we now apply only momentum and energy balance to determine the unknown functions ω and δ.

Energy balance

A relation expressing kinetic energy balance is obtained by multiplying the Prandtl equation with v_x, and rewriting it in the form

$$\nu\left(\frac{\partial v_x}{\partial y}\right)^2 = \frac{1}{2}\nu\frac{\partial^2(v_x^2)}{\partial y^2} + \frac{1}{2}\frac{\partial((U^2 - v_x^2)v_x)}{\partial x} + \frac{1}{2}\frac{\partial(v_y(U^2 - v_x^2))}{\partial y}. \tag{28.89}$$

Integrating this over all y and using the boundary conditions, we get

$$\nu\int_0^\infty \left(\frac{\partial v_x}{\partial y}\right)^2 dy = \frac{1}{2}\frac{d}{dx}\int_0^\infty (U^2 - v_x^2)v_x\,dy. \tag{28.90}$$

Here, $\rho_0\nu(\partial v_x/\partial y)^2$ is the density of heat dissipation (in the Prandtl approximation where

$\partial v_y/\partial x$ can be ignored), and $\frac{1}{2}\rho_0(U^2 - v_x^2)v_x$ is the flux of displaced kinetic energy. Multiplied by dx this relation states that the rate of heat dissipation in the slice dx equals the change in the rate of kinetic energy displacement across the slice.

As for momentum balance, we introduce intrinsic thickness measures,

$$\frac{1}{\delta_{\text{heat}}} = \frac{1}{U^2}\int_0^\infty \left(\frac{\partial v_x}{\partial y}\right)^2 dy, \qquad \delta_{\text{ener}} = \frac{1}{U^3}\int_0^\infty (U^2 - v_x^2)v_x\, dy, \tag{28.91}$$

and obtain energy balance in the form

$$\boxed{\frac{\nu U^2}{\delta_{\text{heat}}} = \frac{1}{2}\frac{d\left(U^3\delta_{\text{ener}}\right)}{dx}.} \tag{28.92}$$

It is also possible to derive further relations for angular momentum balance and thermal energy balance [White 1991, Schlichting and Gersten 2000].

Polynomial of third degree

Reusing the third-degree polynomial (28.81), the mass and momentum thicknesses are as before given by (28.87), and the heat and energy thicknesses become

$$\frac{\delta}{\delta_{\text{heat}}} = \frac{6}{5}\left(1 - \frac{\mu}{6} + \frac{\mu^2}{9}\right), \qquad \frac{\delta_{\text{ener}}}{\delta} = \frac{27}{140}\left(1 + \frac{\mu}{6} - \frac{4\mu^2}{81} - \frac{\mu^3}{162}\right). \tag{28.93}$$

Since we do not know the relation between δ and μ, we shall parameterize by

$$\omega = \mu\frac{U}{\delta}, \qquad \delta = \sqrt{\lambda\frac{\nu}{U}}, \tag{28.94}$$

where $\lambda(x)$ is a new parameter with dimension of length. A convenient dimensionless scale parameter $\Lambda = \lambda\dot{U}/U$ was introduced by Pohlhausen in 1921 [Poh21].

Inserting the thicknesses into momentum and energy balance, these equations may be recast into two coupled first-degree differential equations in λ and μ of the form

$$\lambda\mu\dot{\mu} = A(\mu) + B(\mu)\frac{\lambda\dot{U}}{U}, \qquad \mu\dot{\lambda} = C(\mu) + D(\mu)\frac{\lambda\dot{U}}{U}, \tag{28.95}$$

where the coefficients are rational functions of μ given in Problem 28.10.

For the decelerating flows the initial values are $\lambda = 0$ and $\mu = \mu_0$, which therefore must be a solution to $A(\mu_0) = 0$. The numerator of A has one negative and three positive roots, the smallest positive being $\mu_0 = 1.56463$. At the critical point we have $\mu = 0$ and $\lambda = \lambda_c$. From the first equation it then follows that $\mu \sim \sqrt{x_c - x}$, and from the second equation that $\lambda - \lambda_c \sim \sqrt{x_c - x}$. Both μ and $\lambda - \lambda_c$ are singular with vertical tangents at the critical point.

The coupled equations are fairly easy to solve numerically with the results shown in the column marked "Mom+ener-3" of Table 28.1. The separation points agree well with the exact separation points with typical errors of about 3%.

Polynomial of fourth degree

In the third-degree polynomial approach, there is an abrupt change in curvature $\partial^2 v_x/\partial y^2$ at the point $y = \delta$ where the velocity field v_x attaches to the slip-flow. This can be remedied by using a fourth-degree polynomial that attaches at an inflection point.

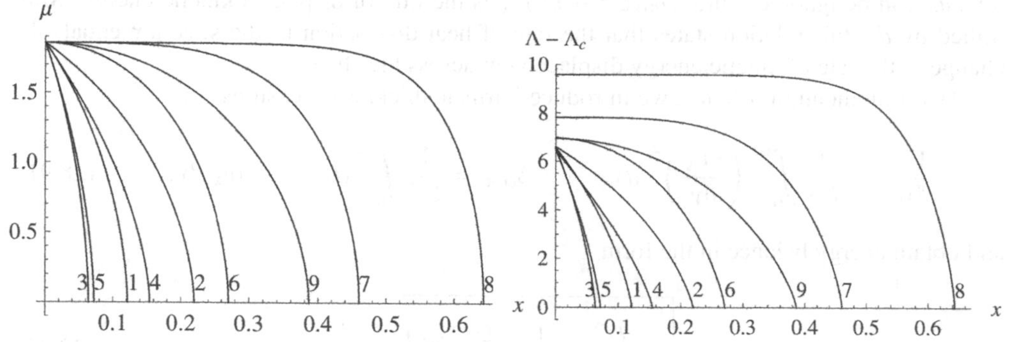

Figure 28.13. Momentum plus energy balance solutions to the fourth-degree polynomial (28.96) for the decelerating slip-flows enumerated in Table 28.1. **Left:** The dimensionless slope μ. Notice the vertical tangents at $\mu = 0$, signaling the Goldstein singularity. **Right:** The deviation of the dimensionless Pohlhausen parameter $\Lambda = \lambda \dot{U}/U$ from its critical value.

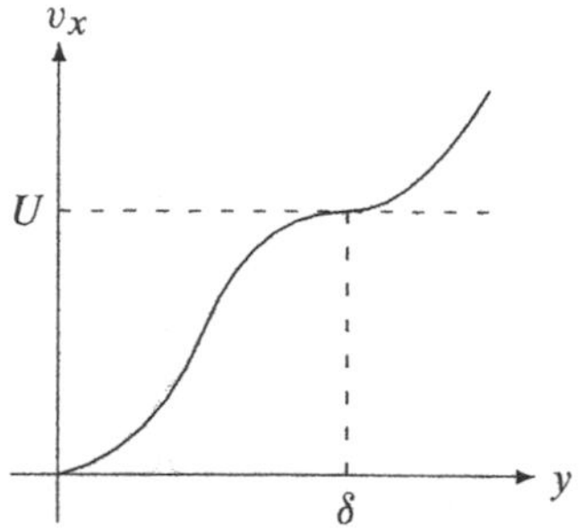

The fourth-order polynomial joins smoothly with $v_x = U$ with a horizontal inflection point at $y = \delta$. The continuation beyond $y = \delta$ is not used.

One may verify that the following polynomial of fourth degree,

$$\frac{v_x}{U} \equiv f(s) = \mu\, s + (6 - 3\mu)\, s^2 - (8 - 3\mu)\, s^3 + (3 - \mu)\, s^4, \tag{28.96}$$

is the only one that satisfies the conditions $f(0) = 0$, $f(1) = 1$ and $f'(1) = f''(1) = 0$. The various thicknesses become

$$\frac{\delta_{\text{mass}}}{\delta} = \frac{2}{5}\left(1 - \frac{\mu}{8}\right), \qquad \frac{\delta_{\text{mom}}}{\delta} = \frac{4}{35}\left(1 + \frac{\mu}{12} - \frac{5\mu^2}{144}\right), \tag{28.97a}$$

$$\frac{\delta}{\delta_{\text{heat}}} = \frac{48}{35}\left(1 - \frac{\mu}{12} + \frac{\mu^2}{16}\right), \qquad \frac{\delta_{\text{ener}}}{\delta} = \frac{876}{5005}\left(1 + \frac{\mu}{12} - \frac{253\mu^2}{10512} - \frac{7\mu^3}{3504}\right). \tag{28.97b}$$

With the same parameterization as (28.94) we arrive again at coupled differential equations of the form (28.95), but with different expressions for the coefficients functions A, B, C, and D and the initial value $\mu_0 = 1.8569$ (see Problem 28.11).

The results of the numeric integration are plotted in figure 28.13. The separation points are singular and shown in column marked "Mom+ener-4" of Table 28.1. The agreement with the exact separation points is almost perfect (better than 1% in nearly all cases). This demonstrates that—at least in these contrived examples—momentum and energy balance for the Prandtl approximation are together capable of predicting the separation points with acceptable precision and in addition confirm that they are singular.

∗ 28.11 Integral approximation to separation

Is it possible to simplify the determination of the separation points and still retain a reasonable predictive ability? The following approximation is based on the near constancy of the ratios $\delta_{\text{mom}}/\delta_{\text{ener}}$ and $\delta_{\text{ener}}/\delta_{\text{heat}}$ plotted in Figure 28.14. The reason is that the linear terms in μ cancel in these quantities, as may be seen from (28.97).

Approximative energy balance

Multiplied by $4U^3\delta_{\text{ener}}$, energy balance (28.92) may be rewritten as

$$4\nu U^5 \frac{\delta_{\text{ener}}}{\delta_{\text{heat}}} = \frac{d(U^6\delta_{\text{ener}}^2)}{dx}. \tag{28.98}$$

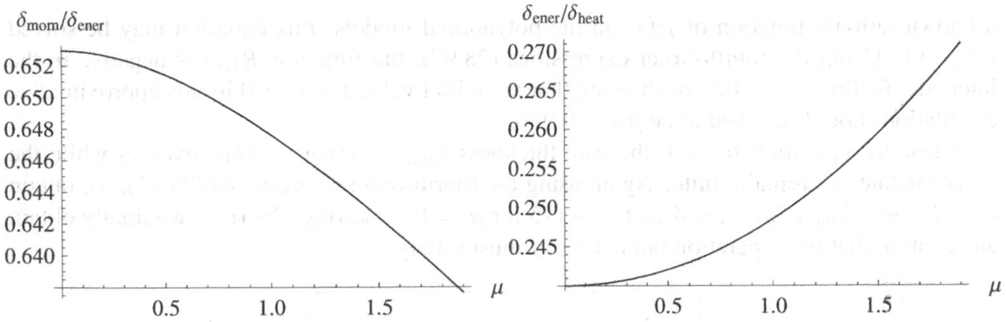

Figure 28.14. Variation of the thickness ratios $\delta_{mom}/\delta_{ener}$ and $\delta_{ener}/\delta_{heat}$ with the slope parameter μ. Note that both parameters have a quite narrow range of variation.

Integrating from $x = 0$ and using $\delta_{ener} = 0$ for $x = 0$, we obtain

$$U^6\delta_{ener}^2 = 4\nu \int_0^x U^5 \frac{\delta_{ener}}{\delta_{heat}}\, dx. \tag{28.99}$$

The ratio $\delta_{ener}/\delta_{heat}$ is fairly constant in the interval $\mu = 0$ to $\mu = \mu_0 \approx 1.86$, as shown in Figure 28.14. Taking it outside the integral, we arrive at

$$\boxed{\delta_{heat}\delta_{ener} \approx 4\nu \frac{L(x)}{U(x)},} \tag{28.100}$$

where

$$L(x) = U(x)^{-5} \int_0^x U(x')^5\, dx' \tag{28.101}$$

is the velocity-weighted distance along x.

Approximative momentum balance

Multiplying momentum balance in (28.85) by $2U^4\delta_{mom}$, it may be rewritten to become

$$2\nu U^5 \frac{\delta_{mom}}{\delta_{wall}} = 2U^5\dot{U}(\delta_{mass} - \delta_{mom})\delta_{mom} + \frac{d(U^6\delta_{mom}^2)}{dx}, \tag{28.102}$$

where we for convenience have introduced the wall thickness $\delta_{wall} = U/\omega$. As may be seen from Figure 28.14, the ratio $\delta_{mom}/\delta_{ener}$ is very nearly constant. Taking the factor $\delta_{mom}^2/\delta_{ener}^2$ outside the differentiation in the last term, we may use energy balance (28.98) to get the following algebraic relation between the various thicknesses:

$$2\nu \frac{\delta_{mom}}{\delta_{wall}} = 2\dot{U}(\delta_{mass} - \delta_{mom})\delta_{mom} + 4\nu \frac{\delta_{mom}^2}{\delta_{ener}\delta_{heat}}. \tag{28.103}$$

Rewriting this expression by means of (28.100), it becomes a dimensionless relation,

$$\boxed{\frac{\dot{U}L}{U} = \frac{\delta_{ener}\delta_{heat} - 2\delta_{mom}\delta_{wall}}{4\delta_{wall}(\delta_{mass} - \delta_{mom})} \equiv R(\mu).} \tag{28.104}$$

Since the left-hand side depends only on the given velocity $U(x)$, and the right-hand side is a known rational function of $\mu(x)$ in the polynomial models, this equation may be solved for $\mu(x)$. Using the fourth-order expressions (28.97), the function $R(\mu)$ is negative in the interval $-6.206 < \mu < 0.966$, showing that the initial value for $x = 0$ in this approximative calculation should be taken to be $\mu_0 = 0.966$.

Close to separation $\mu \to 0$ the wall thickness $\delta_{\text{wall}} = U/\omega = \delta/\mu$ diverges while the other thicknesses remain finite. Again using the fourth-order expressions (28.97), we obtain $\dot{U}L/U \to -\delta_{\text{mom}}/2(\delta_{\text{mass}} - \delta_{\text{mom}}) = -1/5$ for $\mu = 0$. Inserting (28.101), we finally obtain an equation that the separation point $x = x_c$ must satisfy

$$\boxed{\frac{\dot{U}(x_c)}{U(x_c)^6} \int_0^{x_c} U(x)^5\, dx = -\frac{1}{5}.} \tag{28.105}$$

Solving this equation for the usual test cases we find the results shown in the last column of Table 28.1. The errors are, on average, 3%, which is as good as the semi-empirical methods [White 1991]. The model also predicts $\mu(x)$ by solving (28.104) and thus all the thickness ratios (28.97) in units of $\delta(x)$, which in the end is determined from (28.100).

Separation from cylinder

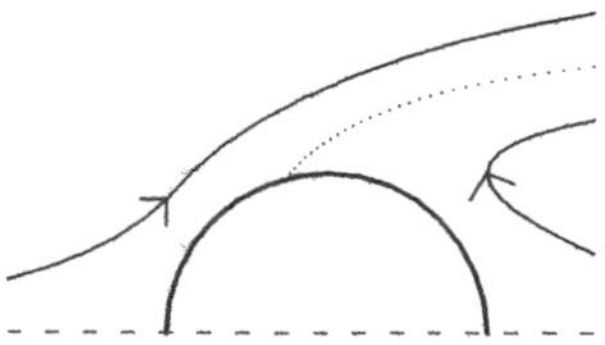

Sketch of the much-studied separation flow around a cylinder (of which only half is shown here). The separating streamline is dotted.

Some of the more interesting slip-flows first accelerate and then decelerate. Among them the notorious cylinder in a uniform crosswind, which has been the favorite target for boundary layer research for nearly a hundred years. The cylinder presents two difficulties. The first is the question of what is the correct slip-flow. Since separation interacts with the mainstream flow, this is not a simple question. The second is the technical calculation of a precise value for the separation point in a slip-flow that may not be purely decelerating.

If the external flow around the cylinder of unit radius is taken to be potential flow, the slip-flow is given by (13.53b), that is, $U(x) = 2U_0 \sin x$, where U_0 is the strength of the uniform crosswind and x is the angle of observation (previously called ϕ). The approximative equation(28.105) yields $x_c = 1.800 = 103°$. The exact separation point for this slip-flow (obtained by numerical calculation) is known to be $x_c = 1.823 \approx 104°$, so in this case the approximation also works to a precision about 1%.

The real flow around the cylinder is, however, not potential flow because of the vorticity induced into the main flow by the boundary layer, especially in the separation region and beyond. Hiemenz (1911) determined experimentally that the slip-flow at a Reynolds number $\mathsf{Re} = U_0 a/\nu = 9{,}500$ was well described by the odd polynomial $U = U_0(1.814x - 0.271x^3 - 0.0471x^5)$. Now the approximative equation (28.105) predicts separation at $x_c = 1.372 = 78.6°$, which is only about 2% away from the value $x_c = 80.5°$ measured by Himenz. Even if a decent agreement is obtained, this is not so impressive because the slip-flow itself has not been calculated from first principles. Most of the information about the separation from the cylinder lies in fact in the measured slip-flow.

Problems

28.1 A fluid flows with constant slip-flow velocity along a half-infinite plate, as in the Blasius setup. **(a)** Show that it is possible to obtain a boundary layer of constant thickness, if fluid is sucked through the plate at a constant rate. **(b)** Discuss what happens if fluid is pushed through instead.

28.2 Show that the Blasius solution satisfies

$$\int_0^\infty (1 - f(s))\, ds = h(\infty), \qquad \int_0^\infty f(s)(1 - f(s))\, ds = 2f'(0).$$

28.3 Show that one may write the function in (28.31) as a single integral (apart from an unimportant constant factor)

$$H(s) = \int_0^s f'(u)^{\mathsf{Pr}}\, du,$$

where Pr is the Prandtl number.

28.4 Show (for heat flow in the Blasius layer) that the asymptotic behavior of $F'(0) = 1/H(\infty)$ is

$$F'(0) \approx \begin{cases} \sqrt{\dfrac{\mathsf{Pr}}{\pi}} = 0.56\sqrt{\mathsf{Pr}} & (\mathsf{Pr} \to 0), \\ \dfrac{\left(\frac{9}{4} f'(0)\mathsf{Pr}\right)^{1/3}}{\Gamma\left(\frac{1}{3}\right)} = 0.34\mathsf{Pr}^{1/3} & (\mathsf{Pr} \to \infty). \end{cases}$$

* **28.5** Assume that a two-dimensional flow is of the form

$$v_x = U(x), \qquad v_y = V(x) - y\frac{dU(x)}{dx}$$

all over space (and not just near a boundary). **(a)** Determine the differential equations that must be satisfied by U and V. **(b)** Determine V for constant U.

28.6 In the self-similarity equations (28.63), assume that α and β actually depend on x. Show that there are no solutions satisfying the boundary conditions $g(0) = g'(0) = 0$.

28.7 In the Falkner–Skan equation (28.64), put $g(s) = s - A + u(s - A)$ with arbitrary A and assume that $u(s)$ decays exponentially for $s \to \infty$. **(a)** Show that in the leading exponential approximation $u(s)$ must satisfy the linear differential equation

$$u''' + \alpha s u'' - 2\beta u' = 0.$$

(b) Assume that the leading approximation is $u[s] \sim s^{-k} \exp(-\alpha s^2/2)$ for $s \to \infty$ and determine k.

28.8 Show that for constant α and β we may write the solution to the first equation (28.62) as

$$\lambda(x) = \sqrt{2(\alpha - \beta)\nu t}, \qquad t = \int \frac{dx}{U(x)}.$$

This confirms the estimate (28.4) .

* **28.9** **(a)** Justify that the wall-anchored dimensionless slope (28.80) in the interval $0 < \mu < 4/3$ may be approximated by $\dot{\mu} = 6\dot{U}/\mu U$ and **(b)** solve it. **(c)** Obtain an equation for the separation point, and **(d)** compare the resulting values for the nine cases in Table 28.1.

28.10 For the third-degree polynomial (28.81), the rational functions in (28.95) become

$$A = -\frac{14(\mu + 3)\left(17\mu^3 - 291\mu^2 + 1242\mu - 1296\right)}{(\mu - 3)\left(2\mu^2 - 3\mu - 72\right)},$$

$$B = -\frac{2(\mu + 9)\left(\mu^2 - 11\mu + 39\right)\left(\mu^2 - \mu - 18\right)}{(\mu - 3)\left(2\mu^2 - 3\mu - 72\right)},$$

$$C = \frac{28\left(19\mu^3 - 408\mu^2 + 1089\mu - 648\right)}{(\mu - 3)\left(2\mu^2 - 3\mu - 72\right)},$$

$$D = \frac{2\mu^4 - 23\mu^3 - 29\mu^2 + 2604\mu - 4212}{(\mu - 3)\left(2\mu^2 - 3\mu - 72\right)}.$$

(a) Determine all the roots of A.

28.11 For the fourth-degree polynomial (28.96), the rational functions in (28.95) become

$$A = -\frac{4\left(2085\mu^4 - 42581\mu^3 + 50796\mu^2 + 1103760\mu - 1976832\right)}{7(\mu-4)\left(5\mu^2 - 4\mu - 384\right)},$$

$$B = -\frac{5\left(\mu^2 - 15\mu + 72\right)\left(21\mu^3 + 253\mu^2 - 876\mu - 10512\right)}{21(\mu-4)\left(5\mu^2 - 4\mu - 384\right)},$$

$$C = \frac{8\left(1965\mu^3 - 74866\mu^2 + 242988\mu - 164736\right)}{7(\mu-4)\left(5\mu^2 - 4\mu - 384\right)},$$

$$D = \frac{5\left(21\mu^4 - 374\mu^3 - 132\mu^2 + 66888\mu - 126144\right)}{21(\mu-4)\left(5\mu^2 - 4\mu - 384\right)}.$$

(a) Determine all the roots of A.

* **28.12** **(a)** Show that the right-hand side of (28.104) is the rational function

$$R = \frac{-1976832 + 1103760\,\mu + 50796\,\mu^2 - 42581\,\mu^3 + 2085\,\mu^4}{2860\,\left(72 - 15\,\mu + \mu^2\right)\left(48 - 4\,\mu + 3\,\mu^2\right)}.$$

(b) Find the lowest positive root in the numerator.

29 Subsonic flight

The take-off of a large airplane never fails to impress passengers and bystanders alike. After building up speed during a brief half-minute run, gravity lets go and the plane marvelously lifts off. For this to happen, it is obvious that the engines and the airflow around the wings must together generate a vertical force that is larger than the weight of the aircraft. After becoming airborne, the airplane accelerates further for a while, and then goes into a fairly steep steady climb until it levels off at its cruising altitude. Aloft in horizontal flight at constant speed, the aerodynamic lift must balance the weight, whereas the engine thrust goes to oppose the drag. What is not obvious to most passengers is how the lift depends on the forward speed, the angle of attack, and the shape and attitude of the airframe, especially the wings.

Wilbur and Orville Wright (1867–1912, 1871–1948). American flight pioneers. From their bicycle shop in Dayton, Ohio, the inseparable brothers carried out systematic empirical investigations of the conditions for flight, beginning in 1896. Built gliders and airfoil models, wind tunnels, engines, and propellers. They finally succeeded in performing the first heavier-than-air, manned, powered flight on December 17, 1903, at Kitty Hawk, North Carolina. (Source: Wikimedia Commons.)

The explanation of aerodynamic lift is in fact quite simple, even if it was only in the beginning of the twentieth century—about the same time as the first generation of airplanes were built—that the details became understood. In nearly ideal flow, pressure is by far the dominant stress acting on any surface, so to get lift the pressure must on average be higher underneath the wing than above. Bernoulli's theorem then implies that the airspeed must be higher above the wing than below, effectively creating a circulation around the wing, a kind of bound vortex superimposed on the general airflow. Without this circulation, caused by the flying attitude and shape of the wing, there can be no lift.

In this chapter we shall only study the most basic theory for subsonic flight with an emphasis on concepts and estimates. Aerodynamics is a huge subject (see for example [Anderson 2001] and [17]) of importance for all objects moving through the air, such as rifle bullets, rockets, airplanes, cars, birds, and sailing ships, and even for submarines moving through water. The age-old human dream of flying like birds, and the potential for triumphant rise and tragic fall unavoidably associated with flying machines, makes aerodynamics different from most other branches of science.

29.1 Aircraft controls

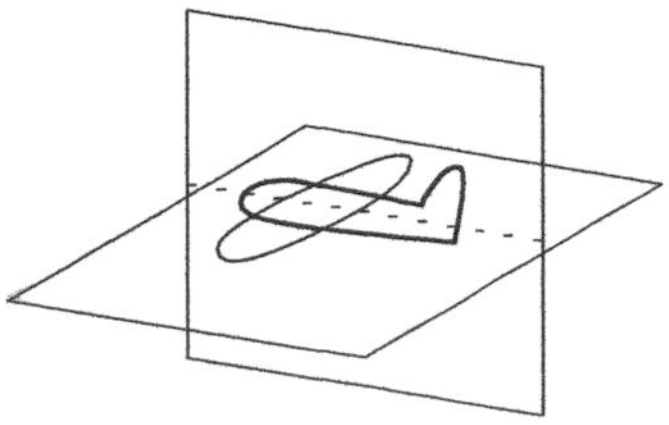

The symmetry plane and wing plane of normal winged aircraft.

Historically, aircraft design went through many phases with sometimes weird shapes emerging, especially during the end of the nineteenth century. In the twentieth century where sustained powered flight was finally attained, most of the design problems were solved through systematic application of theory and experiment. The history of the evolution of aerodynamics, the courageous men and their wonderful flying machines, is dramatic to say the least (see for example [Anderson 1997]).

Control surfaces

The majority of all winged aircraft ever built are symmetric under reflection in a midplane. The wings are typically placed in a plane orthogonal to the midplane, but often swept somewhat backward and a bit upward. On the wings, and also on the horizontal and vertical stabilizing wing-like surfaces found at the tail-end of most aircraft, there are smaller movable *control surfaces*, connected physically or electronically to the "stick" and the "pedals" in the cockpit. Still smaller moveable sections of the control surfaces allow the pilot to *trim* the aircraft. When an aircraft is trimmed for steady flight, the cockpit controls are relaxed and do not require constant application of force to keep the airplane steady. At cruising speed, the aircraft is typically handled with quite small movements of the controls, often carried out by the autopilot, whereas at low speeds, for example during take-off and landing, much larger moves are necessary.

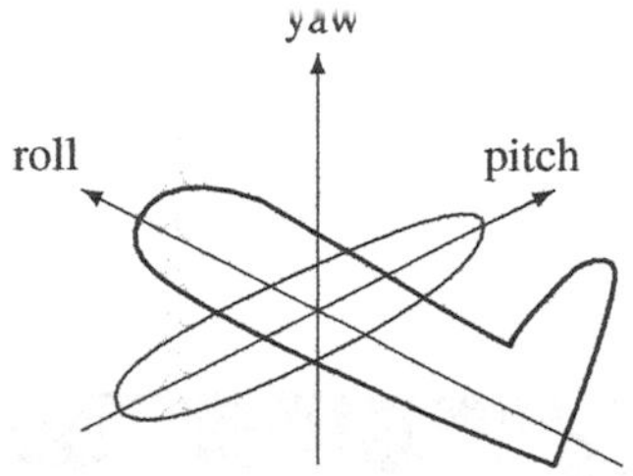

The three axes of a normal winged airplane. The aircraft rolls around the longitudinal axis, pitches around the lateral axis, and yaws around the directional axis. The nose and tail are found at the ends of the longitudinal axis, and the wings lie in the plane of the longitudinal and lateral axes.

Main aircraft axes

The symmetry plane and the wing plane of normal aircraft define three orthogonal axes. The first is the *longitudinal* axis, running along the body of the aircraft in the intersection of the midplane and the wing plane. Rotation around this axis is called *roll*, and is controlled by the *ailerons* usually found at the trailing edge of the wings near the wing tips. When the pilot moves the stick from side to side, the ailerons move opposite to each other and create a rolling moment around the longitudinal axis. The second is the *lateral* axis, which lies in the wing plane and is orthogonal to the midplane. Rotation around this axis is called *pitch* and is normally controlled by the *elevator*, usually found in the tail of the aircraft. The pilot moves the stick forward and backward to create a pitching moment around the lateral axis. The third is the *directional* axis, which is orthogonal to both the wings and the body, and thus vertical in straight level flight. Rotation around this axis is called *yaw* and normally controlled by the *rudder*, usually also placed in the tail end. In conventional aircraft, the pilot presses foot pedals to move the rudder and create a yawing moment around the directional axis. The Wright brothers were the first to introduce controls for all three axes of their aircraft.

Take-off, cruise, and landing

The take-off of a normal passenger aircraft begins with a *run* typically lasting half a minute. In modern aircraft with a nose wheel, the body stays horizontal during the whole run, whereas in older aircraft with a tail wheel or slider, the tail would first lift up well into the run and then make the body nearly horizontal. Having reached sufficient speed for flight, about 250–300 km h^{-1} for large passenger jet planes, the pilot gently pulls the stick back and thereby raises the elevator, creating a pitching moment that lifts the nose wheel off the runway while the main undercarriage stays in contact with the ground. After a bit of acceleration in this attitude, the undercarriage also leaves the runway, and the aircraft is airborne. For safety reasons the aircraft should not lift off until the speed is somewhat above the minimal speed for flight. In older airplanes the actual lift-off was almost imperceptible, whereas the powerful engines of modern aircraft make the lift-off much more notable through the rather steep climb angle that the aircraft is capable of assuming immediately afterward. The climb normally lasts until the aircraft has reached cruising altitude, typically 10,000 m, at which point it levels off and accelerates further until it reaches cruising speed, around 800–900 km h^{-1} for a modern jet. At normal temperature and pressure, the sound speed is about 1,200 km h^{-1} but at cruising altitude the fall in temperature has reduced it by about 10%—as can be seen from Equation (14.8) on page 231. The airspeed is thus about 85% of sound speed, also called Mach 0.85.

Landing is by far the hardest part of flying. The aircraft has to be brought down to the ground and the speed must be reduced. At low speed, the aircraft controls need to be worked harder than at high speed, and random winds and turbulence influence the aircraft much more. Keeping the air speed above stall speed is uppermost in the pilot's mind, because a stall at low altitude makes the airplane crash into the ground. Landing speeds are comparable to take-off speeds, but the aircraft has to be maneuvered into the narrow space that the runway presents, and in all kinds of weather. Landing lengths can be made shorter than take-off lengths by diverting jet exhaust into the forward direction or reversing propeller blades, in addition to the application of wheel brakes.

Example 29.1 [Boeing 747-200]: Jet engines develop nearly constant thrust (force) at a given altitude such that their power (energy output per unit of time) increases proportionally with airspeed, all other factors being equal. Propeller engines yield instead roughly constant power so that the thrust decreases inversely with airspeed. Ignoring air resistance (drag), the constant thrust from jet engines translates into nearly constant acceleration during the take off run. A typical large passenger jet airliner (Boeing 747-200) has a maximal weight of 374,000 kg and four engines that together yield a maximum thrust of 973,000 N, corresponding to a runway acceleration of 2.6 m s^{-2} when fully loaded. At this acceleration, the plane reaches take off speed of 290 km h^{-1} in 31 s after a run of about 1,250 m. Actual take-off length is somewhat larger because of drag and rolling friction. For safety reasons, runways are required to be at least twice that length, typically between 3 and 4 km.

Extreme flying

An aircraft can in principle move through the air in any attitude—and some pilots enjoy making their planes do exactly that—but there is an intended normal flying attitude with the wings nearly horizontal and orthogonal to the airflow. In this attitude, the aircraft is designed such that the flow of air over the wings and body of the aircraft is as laminar as possible, because laminar flow yields the largest lift force and smallest drag.

In other attitudes, steep climb, dive, roll, loop, tight turn, spin, tail-glide, sideways crabbing, and whatnot, the airflow over the wings may become turbulent, resulting in almost complete loss of lift. When that happens, the aircraft is said to have *stalled.* Stalling an aircraft in level flight at sufficient altitude is a common—and fun—training exercise. First, the engine power is cut to make the aircraft slow down. While the airspeed is dropping, the pilot slowly pulls back the stick to pitch the nose upward so that the aircraft keeps constant altitude. This can of course not continue, and at a critical point the laminar flow over the wings is lost and replaced by turbulence. The aircraft suddenly and seemingly by its own volition pitches its nose downward and begins to pick up speed in a dive. A modern aircraft normally recovers all by itself and goes into a steady glide at a somewhat lower altitude. A stall close to the ground can be catastrophic, as the many hang-glider accidents can confirm. The first fatal one happened in 1896 and cost the life of flight pioneer Otto Lilienthal.

Otto Lilienthal (1848–1896). German engineer. One of the great pioneers of manned flight. Over more than two decades he carried out systematic studies of lift and drag for many types of wing surfaces and demonstrated among other things the superiority of cambered airfoils. Constructed (and exported!) manned gliders, and also took out patents on such flyers in 1893. Stalled and crashed from a height of about 17 m outside Berlin on August 9, 1896. Whether he would actually have invented powered flight before the Wright brothers did in 1903 is not clear [Anderson 1997]. (Source: Wikimedia Commons.)

Stall warning: Most aircraft are today equipped with mechanical stall detection devices near the leading edge of the wings, and audible stall warnings are frequently heard in aircraft cockpits during landing, just before touchdown. The warnings indicate that a stall in the wing flow is imminent, although the aircraft will normally not go into a proper stall before touching down.

Other situations may arise in which only a part of the lifting surfaces stalls. In a tight turn at low speed, the inner wing may stall whereas the outer wing keeps flying, and the aircraft goes into a vertical *spin.* In the early days of flight it was nearly impossible to recover from such a situation, which could easily arise if the aircraft was damaged, for example in air combat. In those days, pilots were not equipped with parachutes, and they often saw no other way out than jumping from the airplane, rather than burn with it. Today's passenger aircraft are not cleared for spin, but it can be fun to take a modern small aircraft that *is* cleared for aerobatics into a spin at sufficient altitude, for example by pulling hard back and sideways at the stick just before it otherwise would go into a normal stall. Most people find the experience quite unpleasant and disorienting, especially due to the weightlessness that is felt while the aircraft slowly tumbles over before it goes into a proper spin. Again, modern aircraft are so stable that they tend to slip out of a spin by themselves if the controls are left free.

29.2 Aerodynamic forces and moments

There are several stages in the process of getting to understand flight. The first of these concerns the global forces and moments that act on a moving body completely immersed in a nearly ideal fluid such as air. Initially we put no constraints on the shape of the object or on the motion of the air relative to the object and discuss only the forces acting on it, although mostly we shall think of an aircraft under normal flight conditions. Apart from scattered comments we leave the discussion of moments to more specialized textbooks [Anderson 2001].

Total force

The total force $\mathcal{F}$ acting on a body determines the acceleration of its center of mass. The only way a fluid can act on an immersed body is through contact forces, described by the stress tensor $\boldsymbol{\sigma} = \{\sigma_{ij}\}$. Including the weight $M\boldsymbol{g}_0$ and engine thrust $\boldsymbol{T}$, the total force becomes

$$\boxed{\mathcal{F} = \boldsymbol{T} + M\boldsymbol{g}_0 + \mathcal{R},} \tag{29.1}$$

where

$$\mathcal{R} = \oint_{\text{body}} \boldsymbol{\sigma} \cdot d\boldsymbol{S} \tag{29.2}$$

is the resultant of all contact forces acting on the body, also called the *reaction force* (see Chapter 21). In principle this includes hydrostatic buoyancy forces (Chapter 3), which serve to diminish the effective gravitational mass of a body. For heavier-than-air flying, buoyancy can normally be disregarded.

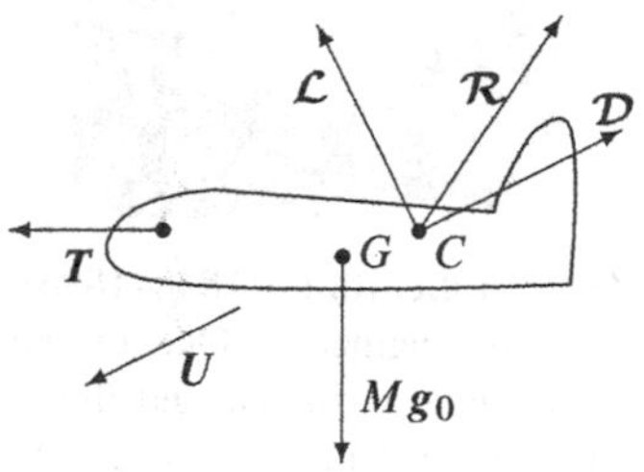

Sketch of the forces acting on a body moving with instantaneous center-of-mass velocity $\boldsymbol{U}$. The thrust $\boldsymbol{T}$ propels the object forward, gravity $M\boldsymbol{g}_0$ pulls it down, and the aerodynamic reaction force $\mathcal{R}$ may be resolved into lift $\mathcal{L}$ and drag $\mathcal{D}$. Note that the reaction $\mathcal{R}$ and its components are plotted as acting on a single point C, called the aerodynamic center, although this concept may not always be meaningful. For stability, the center of thrust should lie forward of the aerodynamic center.

Lift and drag

It is convenient to resolve the reaction force into two components: the *lift*, which is orthogonal to the instantaneous center-of-mass velocity $\boldsymbol{U}$ of the aircraft, and the *drag*, which is parallel with it,

$$\mathcal{R} = \mathcal{L} + \mathcal{D}. \tag{29.3}$$

Lift and drag thus satisfy

$$\mathcal{L} \cdot \boldsymbol{U} = 0, \qquad \mathcal{D} \times \boldsymbol{U} = \boldsymbol{0}. \tag{29.4}$$

The drag always acts in the opposite direction of the center-of-mass velocity whereas the lift may take any direction orthogonal to it.

Downward lift?: Lift may even point directly downward. It is for this reason dangerous to fly an aircraft inverted close to the ground, because the gut reaction of pulling the stick toward you to get away from the ground will generate an extra lift that sends you directly into the ground. During banked turns, an airplane also generates lift away from the vertical, creating in this way the centripetal force necessary to change its direction.

* Total moment of force

The total moment of all forces acting on the body is

$$\mathcal{M} = \mathcal{M}_T + x_G \times M g_0 + \mathcal{M}_R, \tag{29.5}$$

where $\mathcal{M}_T$ is the moment of thrust, x_G the center of gravity of the body, and the moment of the contact forces is

$$\mathcal{M}_R = \oint_{\text{body}} x \times \sigma \cdot dS. \tag{29.6}$$

The total moment depends on the choice of origin for the coordinate system; but if the total force $\mathcal{F}$ on the body vanishes, the total moment becomes independent of the origin. In that case one may calculate the total moment around any convenient point, for example the center of gravity. The individual contributions to the total moment will depend on the choice of origin, even if their sum is independent.

29.3 Steady flight

In steady flight the aircraft moves with constant center-of-mass velocity in a non-accelerated frame of reference, so that the sum of all forces must vanish,

$$\mathcal{F} = T + M g_0 + \mathcal{L} + \mathcal{D} = 0. \tag{29.7}$$

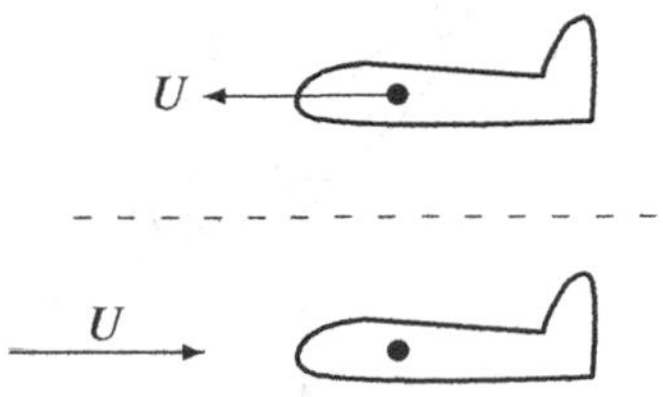

The two situations above are physically equivalent. In the upper drawing the center-of-mass moves with velocity U to the left. In the lower, the center-of-mass does not move, but the surrounding air moves with velocity U to the right at great distances from the object.

Even if passenger comfort demands that the pilot tries to achieve nearly vanishing total force on an airplane, irregular motion of the air may buffet the plane around. In extreme cases, unannounced clear air turbulence may suddenly cause unfettered passengers to fly around inside the cabin. We shall disregard such phenomena and assume that the aircraft is capable of flying with a steady velocity through an atmosphere that would have been at rest were it not for the moving aircraft. Since forces in Newtonian mechanics are the same in all inertial reference frames, we are free to work in the rest-frame of the aircraft where the air asymptotically moves at constant speed.

The orientation of an aircraft with respect to the direction of flight (and the vertical) is called its *flying attitude*. The main attitude parameter responsible for lift is the *angle of attack* α, also called the *angle of incidence*, between the airflow and the base plane of the aircraft, formed by the longitudinal and lateral axes. In normal flight at high speed, the angle of attack is usually quite small, typically a couple of degrees.

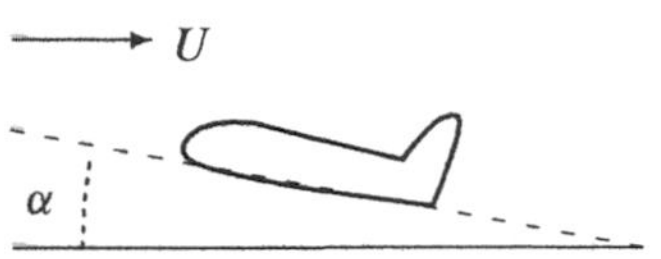

The angle of attack α is the angle between airflow U and the plane of the aircraft.

In the same way as floating bodies, ships, and icebergs should be in stable hydrostatic equilibrium, aircraft should preferably also be dynamically stable in steady flight, meaning that a small perturbation of the aircraft's steady flying attitude should generate a moment counteracting the perturbation. In general this requires the center of thrust to lie forward of the aerodynamic center.

Most modern aircraft are dynamically stable when properly trimmed, and that is very good for amateur pilots; but in military fighter planes, dynamic stability is sometimes traded for maneuverability. Certain modern fighter planes can in fact only maintain a stable attitude through corrections continually applied to the control surfaces by a fast computer.

Steady climb

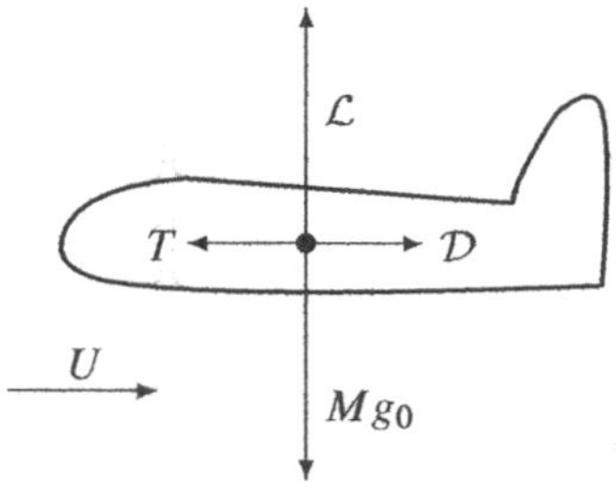

Sketch of forces in steady horizontal powered flight with a small angle of attack. Lift balances weight, and drag balances thrust.

After acceleration and take-off, a powered aircraft normally goes into a steady climb forming a constant positive *climb angle* (or *angle of ascent*) θ with the horizon. Cockpit instruments usually indicate the vertical velocity, called the *rate of ascent* or *climb rate* $v_z = U \sin\theta$, rather than the climb angle. Having reached cruising altitude, the pilot reduces power to a fuel-economic setting and the aircraft levels off with $\theta = 0$ and a tiny angle of attack consistent with steady flight. In this phase of flight, lift must nearly balance weight and thrust must nearly balance drag. Approaching its destination, the power is reduced, though usually not cut off completely, and the aircraft goes into a powered descent with negative θ. Just before landing, power is lowered to near zero, the aircraft flares out before the final touchdown with nearly vanishing angle of ascent, $\theta \approx 0$, and nose up under a fairly large angle of attack.

Assuming that the thrust is directed along the longitudinal axis, as is normally the case for fixed-wing aircraft[1], we obtain the following expressions for lift and drag by projecting the equation of force balance (29.7) on the direction of motion and the direction orthogonal to it (see the margin figure),

$$\mathcal{D} = T\cos\alpha - Mg_0\sin\theta, \qquad \mathcal{L} = -T\sin\alpha + Mg_0\cos\theta. \tag{29.8}$$

Given the values of all the parameters on the right-hand sides, we may calculate the values of lift and drag that are required to keep the aircraft in steady flight.

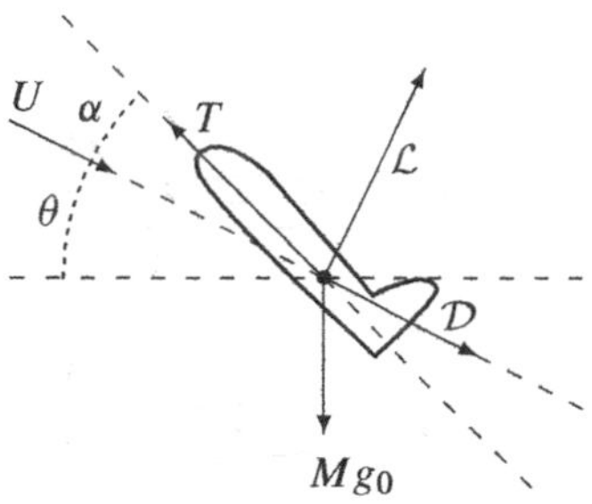

Forces acting on an aircraft in powered steady climb at an angle θ with angle of attack α. All the forces are assumed to lie in the symmetry plane of the aircraft. For convenience we have moved all forces to the center of gravity.

This is, however, not really the way a pilot operates an aircraft. Typically, the pilot selects a power setting for the engine(s) and a certain rate of climb, and then waits until the aircraft steadies on a certain airspeed, angle of attack, and angle of climb. This procedure, of course, presupposes that there are such solutions within the aircraft's flight envelope for the specified values of power and climb rate (if not, the aircraft will stall). We shall see later that aerodynamic theory allows us to calculate lift and drag for a given aircraft in terms of the airspeed, the angle of attack, and the air density. The steady flight equations (29.8) may then be solved for the airspeed U and the angle of attack α, given the air density, the weight, the engine power, and the climb rate. There is the further complication that for a given engine setting, the thrust T tends to fall inversely with airspeed for propeller engines, whereas it stays more or less constant for jets. On top of that, the air temperature, pressure, and density all vary with altitude.

During steady climb with a small angle of attack, $\alpha \ll 1$, the steady flight equations (29.8) may be written as

$$\sin\theta \approx \frac{T - \mathcal{D}_0}{Mg_0}, \qquad \mathcal{L} \approx Mg_0\cos\theta, \tag{29.9}$$

where $\mathcal{D}_0$ is the residual drag at zero angle of attack. Thus, the ratio of the *excess of power* $T - \mathcal{D}_0$ to the weight of the aircraft Mg_0 determines the climb angle, as one might expect. To get a finite positive angle of climb, the thrust must not only overcome the drag but also part of the weight of the airplane. From the climb angle one can afterward calculate the lift that the airflow over the wings and body of the aircraft necessarily must generate to obtain a steady climb.

Example 29.2 [Boeing 747]: During initial climb, speed is fairly low, and if the residual drag can be ignored relative to thrust it follows that $\sin\theta \lesssim T/Mg_0$. For the fully loaded Boeing 747-200 of Example 29.1 we find $T/Mg_0 \approx 0.27$ and thus $\theta \lesssim 15°$.

[1] In other types of aircraft, the engine thrust can also have a component orthogonal to the longitudinal axis that also contributes to lift. In the extreme case of a VTOL (vertical take-off and landing) aircraft, for example a helicopter, there is almost no other lift, and the engine thrust balances by itself both drag and weight.

Unpowered steady descent

Most freely falling objects quickly reach a constant terminal velocity. Stones fall vertically, whereas aircraft, paper gliders, paragliders, and parachutists in free fall will attain sometimes large horizontal speeds. An aircraft in unpowered flight is able to glide toward the ground with constant velocity and constant rate of descent. It is part of early training for pilots to learn how to handle their craft in unpowered steady descent, and usually the aircraft is so dynamically stable that it, by itself, ends up in a steady glide if the engine power is cut and the controls are left free. Paper gliders, on the other hand, often go through a series of swooping dives broken by stalls, or spiral toward the ground in a spin.

During steady unpowered descent the air hits the aircraft from below at an angle γ with the horizontal, called the *glide angle*, corresponding to a *glide slope* $\tan\gamma$. The ratio of horizontal to vertical air speed is called the *glide ratio*, and equals the inverse of the glide slope, that is, $\cot\gamma$. An aircraft can glide steadily with different airspeeds for a large range of angles of attack. The angle of attack that yields maximal glide ratio determines how far an aircraft at best can reach by gliding down from a given altitude h, also called its *glide range*, $x = h\cot\gamma$.

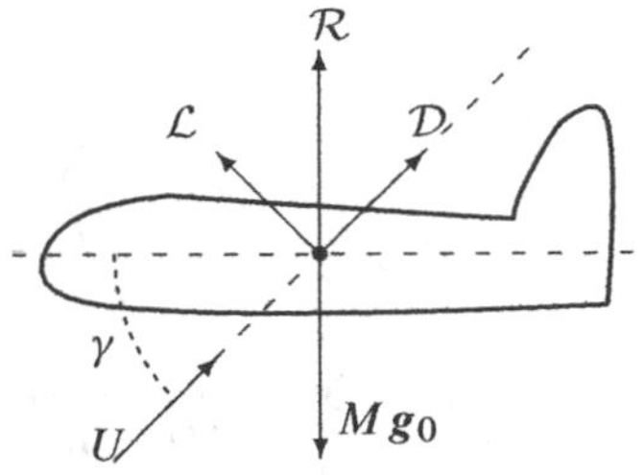

Sketch of forces in unpowered steady descent at a glide angle γ. Lift and drag collaborate to balance the weight. The airplane can glide with many different angles of attack (here shown with $\alpha = \gamma$) but there is a best angle of attack that yields the smallest glide angle, or equivalently the highest glide ratio.

Glide ratios: Typical commercial aircraft have best glide ratios of 10 to 20, with 17 for the Boeing 747 and 8 for the Concorde. So, if the engines cut out at an altitude of 10 km, the pilot has to look for a place to land inside 100 to 200 km, depending on the aircraft. These glide ratios are comparable to those of gliding birds like the swift with glide ratio 10 and soaring birds like the albatross with glide ratio 20. Modern sailplanes may reach glide ratios around 30 to 55 and in extreme cases even higher. The space shuttle with its stubbed wings and large weight is a rather bad glider, and approaches the runway at a glide angle $\gamma = 19°$, corresponding to a glide ratio of around 3, comparable to that of a sparrow. The human body is even worse, with a best glide ratio of about unity.

In steady unpowered descent, the aerodynamic reaction force must be equal and opposite to the weight of the aircraft, or in size $\mathcal{R} = Mg_0$. Resolving the reaction force into lift and drag, we find

$$\mathcal{L} = Mg_0\cos\gamma, \qquad \mathcal{D} = Mg_0\sin\gamma. \tag{29.10}$$

These equations could also have been obtained from the steady flight equations (29.8) with $T = 0$ and $\theta = -\gamma$. From the glide angle and the weight of an aircraft, we may thus determine both the lift and drag that acts on it in this flight condition. The glide ratio evidently equals the ratio of lift to drag in unpowered descent,

$$\cot\gamma = \frac{\mathcal{L}}{\mathcal{D}}. \tag{29.11}$$

Aerodynamics tells us (see the following section) that the ratio of lift to drag essentially only depends on the angle of attack, so the best glide ratio is obtained by choosing that angle of attack which maximizes $\mathcal{L}/\mathcal{D}$.

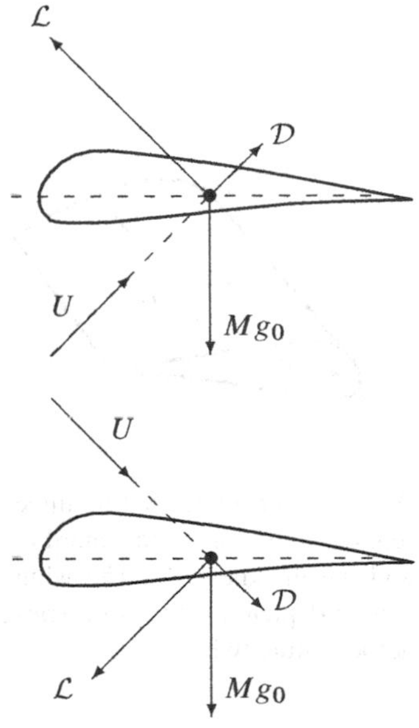

Sketch of forces during bird wing downstroke (top) and upstroke (bottom) in forward flight. The total force does not vanish but accelerates the bird. The lift provides forward thrust in both cases when the drag is not too large.

How birds get thrust: Note that neither lift nor drag are horizontal in unpowered descent. The lift is tilted forward and the drag backward. The forward tilt of the lift also allows us to understand broadly how birds and insects generate thrust in level flight by flapping their wings straight up and down. In this unsteady flight mode there is no instantaneous balance of aerodynamic forces and gravity, but during the downstroke the air will hit the wing from below and generate a tilted lift, propelling the bird forward (and upward), provided the drag is not too large. During upstroke the picture is inverted, and air hits the wing from above, but the lift is still tilted forward and thus again propels the bird forward, provided drag does not overwhelm it. Insects and birds that hover instead of flying horizontally get lift by interacting with vortices created at the leading edge of the wings during the downstroke and by other mechanisms. See for example [ST02] and references therein, as well as [Dudley 2000] and [Tennekes 1997].

Horizontal banked turn

Consider an aircraft flying steadily under power with velocity U in a horizontal circle of radius R. From the (rotating) rest frame of the aircraft, the air again flows steadily past with velocity U, but now there is also a centrifugal force MU^2/R directed away from the center of the circle. The engine thrust is assumed to balance the drag, and the lift must therefore balance the vector sum of the weight and the centrifugal force.

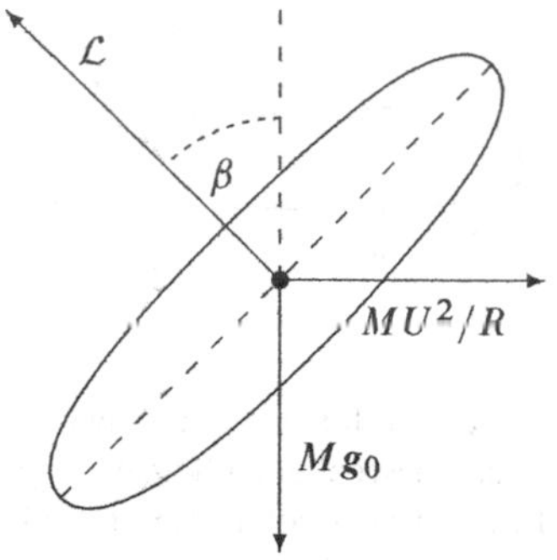

Sketch of forces in the transverse plane during a steady horizontal banked turn, here with tilt angle $\beta \approx 45°$.

Denotingthe tilt angle between the vertical and the lift vector by β, and projecting the lift on the horizontal as well as the vertical directions, we obtain the equations

$$\mathcal{L}\cos\beta = Mg_0, \qquad \mathcal{L}\sin\beta = \frac{MU^2}{R}. \tag{29.12}$$

The lift divides out in the ratio of these equations, and we get

$$\tan\beta = \frac{U^2}{g_0 R}. \tag{29.13}$$

Airplanes are normally tilted (banked) through precisely this angle β during turns, such that the floor of the aircraft remains orthogonal to the lift. In such a *clean turn*, the effective gravity experienced inside the airplane is (in units of standard gravity)

$$\frac{g_{\rm eff}}{g_0} = \frac{\mathcal{L}}{Mg_0} = \frac{1}{\cos\beta}, \tag{29.14}$$

also called the *load factor* or the *g-factor*.

Example 29.3 [g-factors]: In a clean 60° banked turn, one pulls a g-factor of 2. Fighter jets may generate g-factors up toward 10, corresponding to bank angles $\beta \lesssim 84°$. To avoid passenger discomfort, most commercial aircraft rarely bank beyond 15° with a nearly imperceptible increase in load factor of about 4%. At a speed of about 900 km h^{-1} and $\beta \approx 15°$, the clean turn diameter is $2R = 2U^2 \cot\beta/g_0 \approx 48$ km, and a full turn at this speed takes $2\pi R/U \approx 10$ min.

29.4 Estimating lift

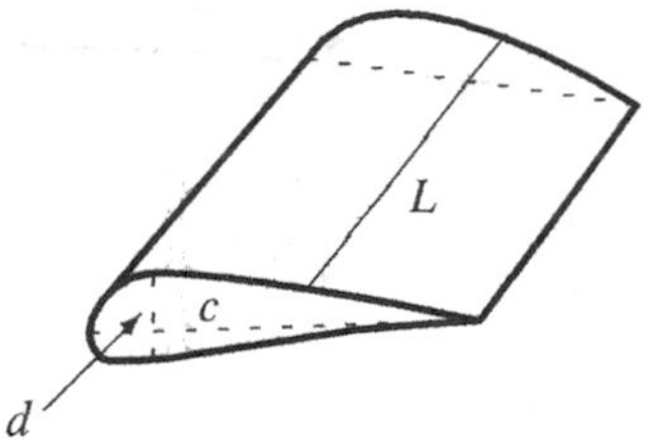

A wing is characterized by three lengths: the span L, the chord c, and the thickness d. The wing profile depicted here also carries aerodynamic twist.

Aerodynamic lift in nearly ideal flow is almost entirely caused by pressure differences between the upper and lower wing surfaces. In this section we shall describe the basic physics of lift and estimate its properties from relatively simple physical arguments, and in the following section we shall make similar estimates of the various contributions to drag. It should, however, be borne in mind that we would rather calculate lift and drag from fluid mechanics, in terms of the angle of attack, velocity, air density, and the shape of the wing. Such theoretical knowledge makes it possible to predict which parameter intervals allow an aircraft to become airborne and sustain steady flight. In Section 29.7 we explicitly calculate the lift for thin airfoils.

Wing and airfoil geometry

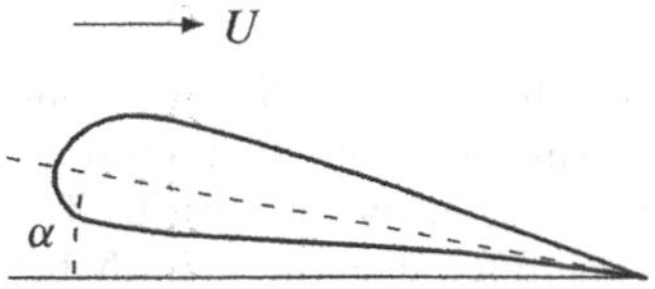

The angle of attack for an airfoil is conventionally defined to be the angle between the airflow and chord line.

An airplane wing may be characterized by three different length scales: the tip-to-tip length or *span* L, the transverse width or *chord* c, and the thickness d. A wing can only in the coarsest of approximations be viewed as a rectangular box. Typical wings are both thin and long, $d \ll c \ll L$. Many wings *taper* toward the tip and are swept back toward the rear. Other wing shapes are also found, for example the delta-wing of supersonic aircraft such as Concorde. For a rectangular wing, the dimensionless number L/c is called the *aspect ratio*. For tapering and unusually shaped wings where the true chord length may be ill-defined, one

Figure 29.1. NASA's Helios prototype electrically powered flying wing. (Source: Photo by Carla Thomas. NASA Dryden Flight Research Center.)

may instead use the average chord length $c = A/L$, where A is the planiform area of the wing (the area of the wing's "shadow" on the wing plane), so that the aspect ratio becomes $L/c = L^2/A$. The wing may furthermore twist slightly along the span leading to a varying angle of attack. This is, in particular, true for propellers that basically are wings mounted on a rotating shaft.

The transverse wing profile, also called the *airfoil*, is normally slightly curved, or *cambered*, along the chord, with a soft leading edge and a sharp trailing edge. The angle of attack of an airfoil is defined to be the angle between the asymptotic airflow and the *chord line*, which is a straight line of length c connecting the leading and trailing edges. Depending on how the wings are attached to the aircraft, there may be a small difference between the angles of attack of the wing and the aircraft.

> **Example 29.4 [Aspect ratios]:** The Boeing 747-400 has a wing span of $L \approx 64$ m and a wing area $A \approx 520$ m^2, leading to an average chord length of $c \approx 8$ m and an aspect ratio of $L/c \approx 8$. For comparison, the albatross with its narrow long wings has an aspect ratio of about 20, at par with modern sailplanes. At the extreme end, one finds NASA's solar-cell powered flyer Helios, which has an aspect ratio of nearly 31. Incidentally, a man with his arms stretched out as wings has an aspect ratio of about 20 but he cannot fly, so aspect ratio is not everything.

Wing loading

Suppose an aircraft with thin and almost planar wings flies horizontally under a small angle of attack. The chordwise Reynolds number,

$$\mathsf{Re}_c = \frac{Uc}{\nu}, \tag{29.15}$$

is a dimensionless combination of the asymptotic flow speed U, the chord c, and the kinematic viscosity of air, $\nu = \eta/\rho$, which characterizes the airflow around the wing. It will always be assumed to be very large, of the order of many millions, so that the slip-flow pattern around the wings just outside the omnipresent boundary layers may be taken to be very nearly ideal. In Section 28.2 it was shown that the slip-flow pressure then penetrates the boundary layer and acts directly on the upper and lower wing surfaces.

Figure 29.2. The famous Cessna 150 from the First Production in 1959. (Source: The Cessna 150–152 Club. With permission.)

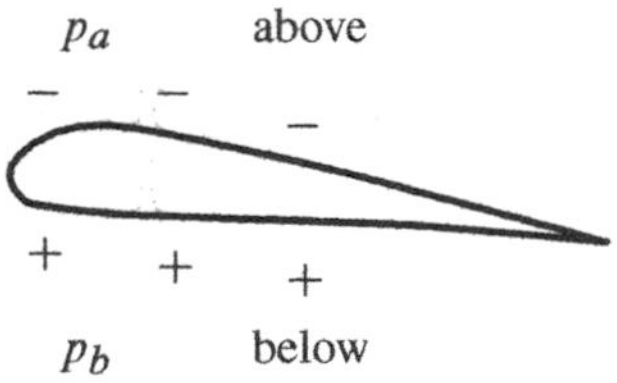

The pressure is on average lower above the wing than below. This is what carries the aircraft.

The only way to obtain lift is by the slip-flow pressure p_a immediately above the wing being generally lower than the slip-flow pressure p_b immediately below the wing (see Figure 29.3). The difference in average pressures above and below the wing may be estimated from the total lift per unit of wing area $\mathcal{L}/A$, also called the *wing loading*,

$$\Delta p_{ab} \equiv \langle p_a \rangle - \langle p_b \rangle \approx -\frac{\mathcal{L}}{A}, \tag{29.16}$$

where the brackets indicate averaging over the upper or lower wing surfaces. In steady level flight, lift equals weight $\mathcal{L} = Mg_0$, and wing loading is easy to calculate from the aircraft parameters.

Example 29.5 [Cessna 150]: The popular Cessna 150 two-seater has a wing span of $L = 10.2$ m, wing area $A = 14.8$ m^2, and thus an average chord of $c \approx 1.45$ m, and an aspect ratio of $L/c \approx 7.0$. With a maximum take-off mass of $M = 681$ kg, the wing loading becomes $\mathcal{L}/A \approx 451$ Pa, which is merely 0.45% of atmospheric pressure at sea level. Cruising with $U = 180$ km h^{-1} at sea level, the Mach number is $\mathsf{Ma} \approx 0.15$ and the Reynolds number $\mathsf{Re}_c \approx 5\times10^6$. The approximation of nearly ideal flow is very well fulfilled with boundary layers only millimeters thick.

Example 29.6 [Boeing 747-400]: The cruising speed for the common long-distance aircraft, Boeing 747-400, is $U \approx 900$ km h^{-1} ≈ 250 m s^{-1}, which corresponds to $\mathsf{Ma} = 0.89$ at the normal cruising altitude of $z = 10$ km. With an average chord length of $c \approx 8$ m, the chordwise Reynolds number becomes $\mathsf{Re}_c \approx 6 \times 10^7$ when we take $\nu \approx 3.4 \times 10^{-5}$ m^2 s^{-1} at the cruising altitude. The maximal take-off mass is $M \approx 400{,}000$ kg distributed over a wing area of about $A \approx 520$ m^2, leading to a wing loading of about $\mathcal{L}/A \approx 7{,}500$ Pa, which is only 7.5% of atmospheric pressure at sea level, but about 31% of the actual pressure at the normal cruising altitude (see Section 2.6).

Velocity differences

In nearly ideal flow, Bernoulli's theorem may be used to relate the pressure and flow velocity well outside the boundary layers. As previously discussed (Section 14.2 on page 233) , air is effectively incompressible at speeds much lower than the speed of sound c_S, that is, for small Mach number $\mathsf{Ma} = U/c_S$, say $\mathsf{Ma} \lesssim 0.3$. The following analysis of lift in incompressible flow with constant density ρ_0 thus excludes passenger jets cruising at $\mathsf{Ma} \approx 0.8$–0.9 but we shall return to this question toward the end of the section.

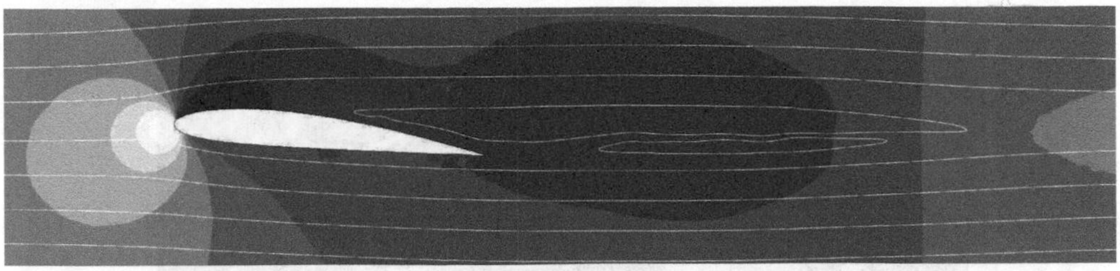

Figure 29.3. Pressure distribution around a NACA2412 airfoil at Reynolds number $\mathsf{Re}_c = 10{,}000$ and angle of attack $\alpha \approx 6°$, obtained by numeric simulation. The stagnation pressure is high (light colored) at the leading edge and the lifting pressure low (dark colored) above the the wing. The lifting pressure typically acts on top of the wing about one quarter of the chord length downstream from the leading edge, as can be seen by the darker coloring in this region. The rest of the airfoil is basically there to secure an orderly departure of the air from the wing. The streamlines (light) show that the high angle of attack has caused boundary layer separation with reversed flow above the trailing edge and quite far behind.

Any streamline coming in from afar will start out with the same velocity U and pressure P, and Bernoulli's theorem (13.13) on page 212 relates these values to the velocity and pressure along the streamline. For a streamline passing immediately above the wing well outside the boundary layer, we find

$$\frac{1}{2}U^2 + \frac{P}{\rho_0} = \frac{1}{2}v_a^2 + \frac{p_a}{\rho_0}, \tag{29.17}$$

where v_a is the flow velocity. A thin wing will not change the flow velocity much, and using $v_a \approx U$ we find to leading order in $v_a - U$,

$$p_a - P = \tfrac{1}{2}\rho_0(U^2 - v_a^2) = \tfrac{1}{2}\rho_0(U + v_a)(U - v_a) \approx \rho_0 U(U - v_a).$$

The assumption $|v_a - U| \ll U$ must necessarily break down near the leading edge of the wing where there is a stagnation point with vanishing flow speed, but for a sufficiently thin planar wing, these end effects can be disregarded. Averaging this relation over the upper wing surface, and subtracting a similar expression for a streamline running immediately below the wing, we get

$$\langle p_a \rangle - \langle p_b \rangle \approx -\rho_0 U(\langle v_a \rangle - \langle v_b \rangle). \tag{29.18}$$

Combining this result with (29.16) we obtain an estimate of the difference in average flow velocities above and below the wing in units of the mainstream flow,

$$\Delta v_{ab} \equiv \langle v_a \rangle - \langle v_b \rangle \approx -\frac{\Delta p_{ab}}{\rho_0 U} \approx \frac{\mathcal{L}}{\rho_0 U A}. \tag{29.19}$$

In the preceding examples, the Cessna 150 cruising at $U \approx 50\ \mathrm{m\ s^{-1}}$ has $\Delta v_{ab} \approx 7.5\ \mathrm{m\ s^{-1}}$, whereas formally the Boeing 747-400 cruising at 10 km with $250\ \mathrm{m\ s^{-1}}$ has $\Delta v_{ab} = 70\ \mathrm{m\ s^{-1}}$.

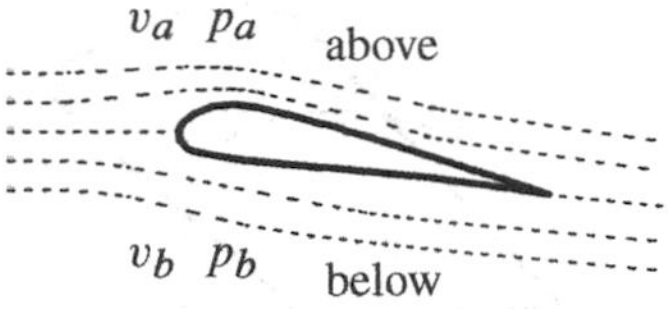

Sketch of streamlines around a wing with positive lift. The wing profile and angle of attack accelerates the flow across the top of the wing and retards it below, and thus creates a lower pressure above than below. Note the presence of front and rear stagnation points with vanishing airspeed. Note also the downwash behind the wing.

Circulation and lift

The difference in average flow velocities above and below the wing may also be estimated from the circulation around the wing along a contour C that encircles the wing in the direction of the asymptotic airflow on top and against it below,

$$\Gamma = \oint_C \boldsymbol{v} \cdot d\boldsymbol{\ell} \approx c\ \langle v_a \rangle - c\ \langle v_b \rangle = c\ \Delta v_{ab}. \tag{29.20}$$

For this to make sense the contour must run in the slip-flow just outside the boundary layers, except where it crosses the thin trailing wake.

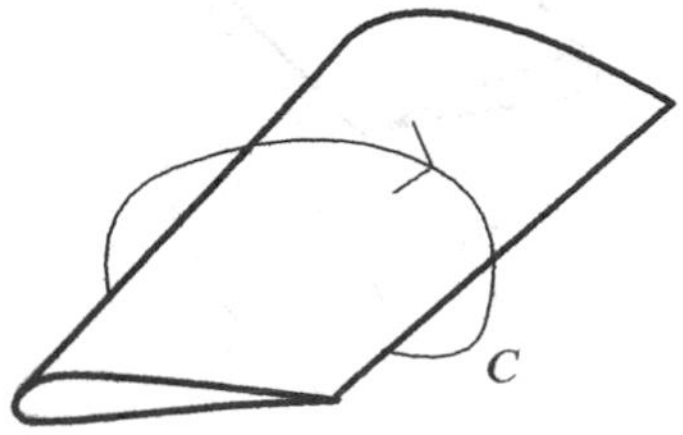

The integration contour for calculating the wing circulation.

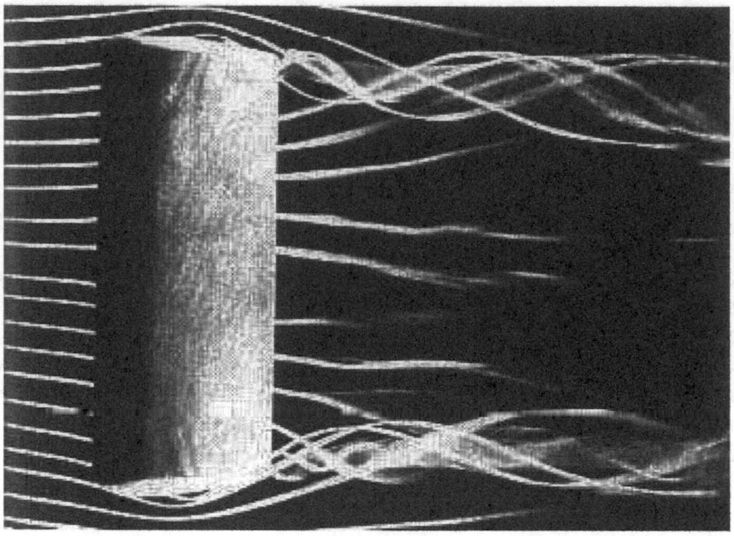

Figure 29.4. Trailing wing tip vortices at Re $= 10^5$. (Source: Reproduced from [Hea82].)

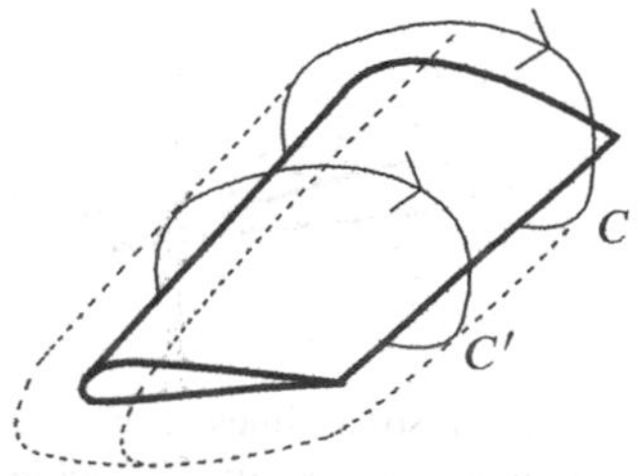

Sketch of vortex bound to the (left) wing of an aircraft. The circulation is the same for the two curves C and C', provided there is no flux of vorticity through the surface bounded by these curves. Note how C' can be slid off the tip of the wing and shrunk to a point if the flow is truly irrotational (which it therefore cannot be).

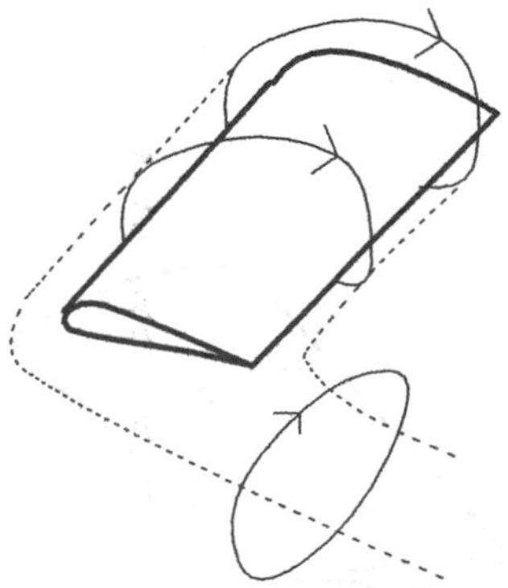

The bound vortex turning into a trailing vortex at the wing tip.

Using (29.19) with $A = cL$, we finally obtain the relation

$$\boxed{\mathcal{L} \approx \rho_0 U L \Gamma.} \tag{29.21}$$

This is the famous *Kutta–Joukowsky* theorem from the beginning of the twentieth century. Here derived from estimates, we shall see in Section 29.6 that in nearly ideal and nearly irrotational flow, this relation is in fact exact when one averages circulation along the wing.

The realization that lift and circulation are two sides of the same coin was probably *the single most important insight into the mechanics of flight.* With this in hand, the road was opened for calculating the lift produced by any specific airfoil for which the circulation could be evaluated in asymptotically uniform, irrotational ideal flow.

The horseshoe vortex system

In nearly ideal irrotational flow, the circulation is the same around any curve encircling the wing because Stokes' theorem relates the difference in circulation between two such curves to the flux of vorticity (which is assumed to vanish) through the surface bounded by the two curves. The lift-generating circulation thus forms a *bound vortex* that ideally cannot leave the wing. For an infinitely long wing, this creates no problem; but for a wing of finite span, the assumption of vanishing vorticity has to break down, because one of the curves may be "slid over" the tip of the wing and shrunk to a point with no circulation. The inescapable conclusion is that since lift requires non-zero circulation, vorticity must come off somewhere along the finite span of a real wing.

The shedding of vorticity from a wing of finite span depends strongly on its shape. A wing that tapers toward the tip will shed vorticity everywhere along its trailing edge, though most near the tip. If the wing is rectangular with constant chord, the vorticity will tend to appear very close to the tip. In any case, the vorticity coming off the wing is blown backward with respect to the direction of flight, forming a *trailing vortex* in continuation of the bound vortex. Alternatively one may see the trailing vortex as created by the flow around the tip seeking to equalize the higher pressure underneath the wing with the lower pressure above. Together with the bound vortex, the two trailing vortices coming off the wing tips form a horseshoe-shaped vortex system accompanying all winged aircraft in flight.

Lift coefficient

A wing's ability to produce lift is characterized by the dimensionless *lift coefficient*,

$$C_L = \frac{\mathcal{L}}{\frac{1}{2}\rho_0 U^2 A}. \tag{29.22}$$

The denominator is proportional to the stagnation pressure $\frac{1}{2}\rho_0 U^2$ at the leading edge of the wing. Knowing the lift coefficient, we can immediately determine the average velocity difference from (29.19): $\Delta v_{ab} = \frac{1}{2} C_L U$.

Being dimensionless, the lift coefficient can only depend on dimensionless quantities, such as the angle of attack α, the Reynolds number $\mathsf{Re}_c = Uc/\nu$, the aspect ratio L/c, and other dimensionless quantities characterizing the shape of the wing. The mainly empirical studies of wing behavior in the last half of the nineteenth century, up to and including the Wright brothers, led to the understanding that the angle of attack was the most important parameter in the lift coefficient. The dependence on the other dimensionless parameters was found to be weaker, in fact so weak that it was mostly ignored before 1900.

The weak dependence of the lift coefficient on the Reynolds number and wing shape parameters allows us to conclude that the lift itself,

$$\mathcal{L} = \frac{1}{2}\rho_0 U^2 A C_L, \tag{29.23}$$

is directly proportional to the air density, the wing area, and the square of the velocity. At take-off and especially during approach to landing, where speeds are low, the pilot can increase the wing area by means of *flaps*. Since lift always nearly equals the constant weight of the aircraft, it follows that the increase in area can be compensated by a decrease in airspeed (for a fixed angle of attack). Fully extended flaps also have a considerably larger angle of attack than the wing itself, thereby increasing the lift coefficient and leading to a further reduction in the landing speed.

Dependence on angle of attack

Empirically, the lift coefficient is surprisingly linear in the angle of attack,

$$C_L \approx \lambda(\alpha - \alpha_0), \tag{29.24}$$

where $\lambda = dC_L/d\alpha$ is called the *lift slope*, and α_0 is the angle of attack at which the lift vanishes. In Section 29.7 we shall see theoretically that for thin airfoils the slope is universally $\lambda = 2\pi$ (with the angle of attack measured in radians). The zero-lift angle α_0 depends mainly on the shape of the airfoil, and is usually small and negative, for example $\alpha_0 \approx -2°$ for the Cessna 150 of Example 29.5.

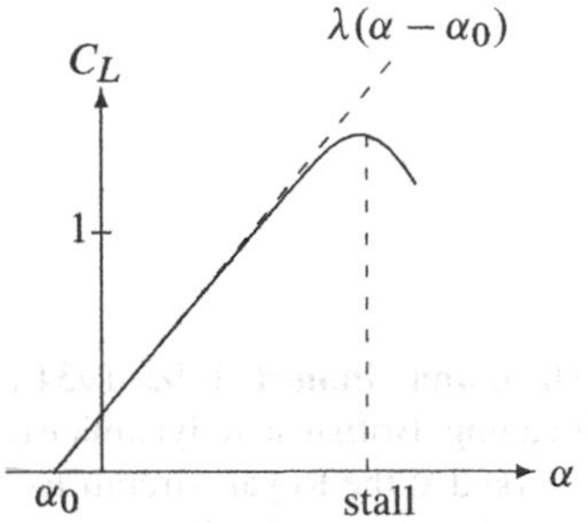

Sketch of a lift curve rising linearly until it veers off rather sharply at an angle of typically 15–20°, signaling stall. Beyond this angle, lift drops precipitously, and so does the airplane.

It follows from the above equations and the constancy of required lift (equal to the weight of the aircraft) that the relative angle of attack $\alpha - \alpha_0$ must vary inversely with the square of the velocity, $\alpha - \alpha_0 \sim 1/U^2$. With decreasing airspeed the required angle of attack rises rapidly until at some critical value, boundary layer separation no longer takes place at the trailing edge of the wing but instead suddenly moves forward toward the leading edge, accompanied by turbulence over nearly all of the upper wing surface. The end result is a dramatic loss of lift and a large increase in drag, a phenomenon called *stall*, which was described on page 515.

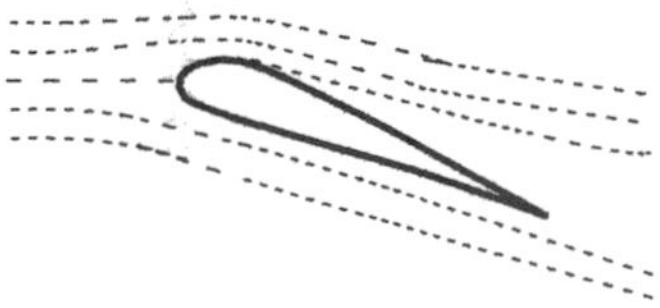

Sketch of the flow around an airfoil at large angle of attack. The boundary layer separates near the leading edge and replaces the previously laminar flow above the wing with turbulent flow, yielding small lift and large drag.

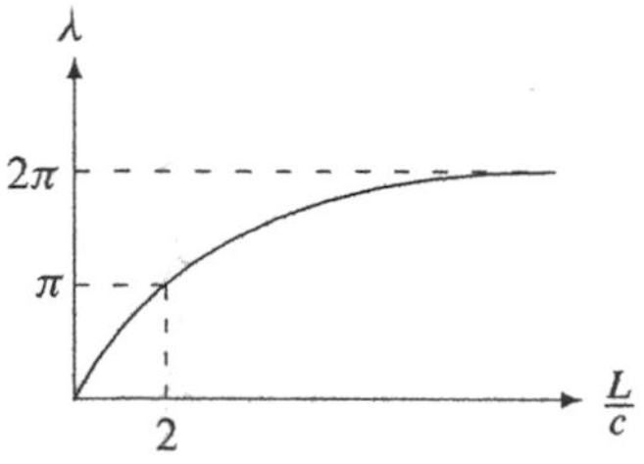

Sketch of the variation of the lift slope with aspect ratio.

Stall typically happens at a critical angle of attack, called the *stall angle* , of the order of $\alpha_{\rm stall} \approx 15°\text{–}20°$ for normal aircraft. Whereas the lift slope and zero-lift angle are essentially independent of the Reynolds number in the linear regime, the stall angle increases a bit with the increasing Reynolds number. For special aircraft the stall angle can be fairly high, for example 35 ° for delta-winged aircraft such as the Space Shuttle or Concorde. In such aircraft the higher stall angle is offset by a smaller lift slope, say $\lambda \approx 3$ (per radian) rather than 2π. In order to get sufficient lift, these aircraft are forced to take off and land under remarkably high angles of attack.

Dependence on aspect ratio

The shedding of vorticity from finite wings makes the lift slope depend on the aspect ratio L/c. An expression useful for estimating this effect for thin airfoils is [Anderson 2001]

$$\lambda \approx \frac{2\pi}{1 + 2\dfrac{c}{L}}, \tag{29.25}$$

which in the limit of the infinite aspect ratio, $L/c \to \infty$, converges upon 2π.

Example 29.7 [Cessna 150]: Consider again the Cessna 150 in Example 29.5. At cruising speed $U = 180\ \text{km h}^{-1}$, we have $\Delta v_{ab}/U = 0.15$, implying a lift coefficient $C_L = 0.30$. With aspect ratio $L/c \approx 7$ the lift slope is $\lambda = 4.9$, and the relative angle of attack becomes $\alpha - \alpha_0 = 3.5°$. For this airfoil $\alpha_0 = -2°$, and the true angle of attack becomes $\alpha = 1.5°$. This airfoil has a stall angle of $\alpha_{\rm stall} = 16°$, corresponding to a stall velocity $U_{\rm stall} \approx 80\ \text{km h}^{-1}$, although stall can be delayed somewhat by enlarging the wing area by means of flaps.

* Dependence on Mach number

Even if modern passenger jets fly below the speed of sound, their Mach number $\mathsf{Ma} = U/c_s$ is so close to unity at cruising speed that there will be major corrections to the lift. To find these corrections we go back to the expression for the divergence of the velocity field (14.19) on page 233 in compressible Euler flow. Inserting $\boldsymbol{v} = \boldsymbol{U} + \Delta\boldsymbol{v}$ and expanding to first order in the presumed small velocity $\Delta\boldsymbol{v}$, we find

$$\nabla \cdot \Delta\boldsymbol{v} \approx \frac{\boldsymbol{U} \cdot (\boldsymbol{U} \cdot \nabla)\Delta\boldsymbol{v}}{c_s^2} = \frac{U^2}{c_s^2}\nabla_x \Delta v_x, \tag{29.26}$$

where in the last step we used $\boldsymbol{U} = U\hat{\boldsymbol{e}}_x$. In ideal irrotational flow the velocity field is the gradient of the velocity potential, $\Delta\boldsymbol{v} = \nabla\Psi$. Inserting this in the above equation, we find

$$(1 - \mathsf{Ma}^2)\nabla_x^2\Psi + \nabla_y^2\Psi + \nabla_z^2\Psi = 0. \tag{29.27}$$

This shows that Ψ is a solution to the incompressible Laplace equation (13.33), provided the x-coordinate is replaced by $x \to x' = x/\sqrt{1 - \mathsf{Ma}^2}$.

Since lift stems from an integral of the pressure with respect to x along the chord, this rule immediately yields the lift in subsonic compressible flow,

$$\boxed{\mathcal{L} = \frac{\mathcal{L}_0}{\sqrt{1 - \mathsf{Ma}^2}},} \tag{29.28}$$

where $\mathcal{L}_0$ is the lift in incompressible flow for $\mathsf{Ma} \to 0$. This *Prandtl–Glauert rule* is valid up to a critical Mach number of about 0.7 where the flow over part of the airfoil may become sonic [Anderson 2001, Anderson 1997]. Applying it nevertheless to the Boeing 747 of Example 29.6 we find $\mathcal{L} \approx 2.2\mathcal{L}_0$ at cruising speed, a quite sizable increase in lift, which at a fixed velocity can be compensated by diminishing the angle of attack and thereby the drag.

Hermann Glauert (1892–1934). Leading British aerodynamicist. Worked at the Royal Aircraft Establishment on airfoil and propeller theory, and on the autogyro. Derived the Mach number correction in 1927 in the way done here, independent of Prandtl, who had discussed such a rule in the early 1920s.

Figure 29.5. Prandtl–Glauert condensation around a B-1B bomber "breaking the sound barrier". Crossing the Prandtl–Glauert "singularity" at Mach 1, the pressure drops sharply and creates this condensation cloud in humid air. (Source: Gregg Stansbery via Wikipedia.)

29.5 Estimating drag

Whereas lift has only one cause, namely the pressure difference between the upper and lower wing surfaces, drag has several. First, there is skin friction from the air flowing over the wing. Second, there is form drag due to the wing obstructing the free airflow and leaving a trail of turbulent air behind, and third there is induced drag from the vortices that always trail the wing tips. For real aircraft, the body shape and various protrusions (radio antenna, Pitot tube, rivet heads, etc.) also add to drag.

It is convenient to discuss drag in terms of the dimensionless drag coefficient,

$$C_D = \frac{\mathcal{D}}{\frac{1}{2}\rho_0 U^2 A}, \tag{29.29}$$

which has the same denominator as the lift coefficient (29.22).

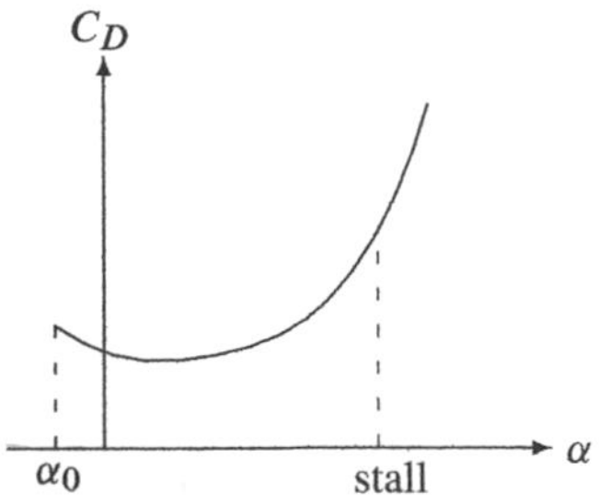

Sketch of a typical drag curve for a cambered airfoil as a function of the angle of attack. Note that the drag of such an airfoil actually decreases for small angles of attack. Beyond the stall angle, the drag coefficient rises steeply.

Skin friction

Close to the wing surfaces there are thin boundary layers in which the flow velocity changes rapidly from zero right at the skin of the wing to the mainstream airspeed just outside. The boundary layer thickness may be estimated from the flat plate laminar Blasius solution (28.22) on page 488 and from semi-empirical turbulent expression (28.45),

$$\frac{U\delta}{\nu} \approx \begin{cases} 4.91\,\mathsf{Re}_x^{1/2} & \text{fully laminar} \\ 0.16\,\mathsf{Re}_x^{6/7} & \text{fully turbulent,} \end{cases} \tag{29.30}$$

where $\mathsf{Re}_x = Ux/\nu$ is the chord-wise Reynolds number at the downstream position x.

Boundary layers do not have the same thickness across the wing surface, but are generally thinnest at the leading edge of the wing and become thicker toward the rear. Usually, the Reynolds number is so high that the boundary layers pass from laminar to turbulent somewhere downstream from the leading edge. For aircraft with chord-wise Reynolds numbers in the millions and chords of the order of meters, a fully laminar boundary layer is only millimeters thick whereas a fully turbulent layer is an order of magnitude thicker.

To estimate the skin friction we again use Blasius' result (28.26) on page 489 for the average friction coefficient in the laminar regime and the semi-empirical expression (28.38) in the turbulent. Since the total wing area is $2A$, we find

$$C_D^{\rm skin} = \frac{\mathcal{D}_{\rm skin}}{\frac{1}{2}\rho_0 U^2 A} = \frac{2\langle\sigma_{\rm wall}\rangle}{\frac{1}{2}\rho_0 U^2} = 2\langle C_f\rangle \approx \begin{cases} 2.65\ \mathsf{Re}_c^{-1/2} & \text{fully laminar} \\ 0.063\ \mathsf{Re}_c^{-1/7} & \text{fully turbulent,} \end{cases} \tag{29.31}$$

where $\langle C_f\rangle = \frac{1}{c}\int_0^c C_f(x)\,dx$ is the average local friction coefficient on one side of the wing.

The skin drag coefficient always decreases with increasing Reynolds number but varies much more slowly in the turbulent region than in the laminar. Turbulent drag is considerably larger than laminar drag, although precise theoretical calculation of skin drag (using Equation (28.39)) is quite hard because it is difficult to predict the line $x = x_0$ along the span where the boundary layer becomes turbulent. This is one of the reasons why wind tunnel experiments, and in more recent times numeric simulations, are still so important for aerodynamics engineering.

Example 29.8 [Cessna 150]: For the Cessna 150 of Example 29.5 at cruising speed with $\mathsf{Re}_c \approx 5 \times 10^6$, the estimate of the maximal laminar boundary layer thickness becomes $\delta \approx$ 3.4 mm, whereas the estimate of the maximal turbulent thickness becomes $\delta \approx$ 26 mm. The corresponding laminar skin drag coefficient is $C_D^{\rm skin} \approx 0.0012$, whereas the turbulent one is about six times larger, $C_D^{\rm skin} \approx 0.0070$. The true skin drag is probably closer to this value.

Form drag

The flow around a highly streamlined body, such as a thin wing narrowing down into a sharp trailing edge, will be nearly ideal everywhere, except in the boundary layers. It has been pointed out before (and we shall prove it in the following section) that a body in a truly ideal, irrotational flow does not experience any drag at all, independent of its shape. Skin friction and form drag must therefore owe their existence to viscosity, but where skin friction is due to shear stresses in the boundary layer, form drag arises from changes in the pressure distribution over the body caused by the presence of boundary layers.

Airfoil boundary layers tend to become turbulent at some point downstream from the leading edge of the wing. At or near the sharp trailing edge, the boundary layers separate from the wing and continue as a *trailing wake* (see Figures 29.6 and 29.7). The unsteady turbulent wake found immediately behind the trailing edge of a wing expands slowly and eventually calms down and becomes steady and laminar at some downstream distance from the wing. Further downstream, the laminar wake continues to expand by viscous diffusion at a considerably faster rate than the turbulent wake. In Section 29.8 we shall determine the general form of the field in the distant laminar wake.

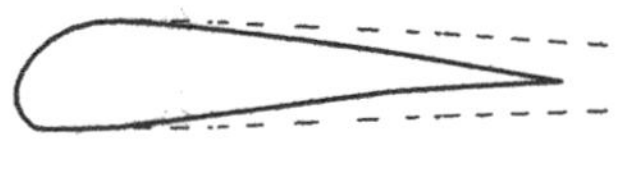

Boundary layers thicken and become turbulent toward the rear of the wing and leave a trailing turbulent wake behind. The initial thickness of the trailing wake is comparable to the boundary layer thickness (here strongly exaggerated).

Inside the trailing wake immediately behind the body, the pressure will be considerably lower than the stagnation pressure $\Delta p = \frac{1}{2}\rho_0 U^2$ at the leading edge (see figure 29.3), and this pressure drop is the cause of form drag. The thickness of the turbulent wake immediately after the trailing edge of the wing may be estimated from the boundary layer thickness δ_c at $x = c$, leading to the form drag estimate $\mathcal{D}_{\rm form} \sim \Delta p L \delta_c$. In terms of the drag coefficient, we thus find

$$C_D^{\rm form} = \frac{\mathcal{D}_{\rm form}}{\frac{1}{2}\rho_0 U^2 A} \sim \frac{\delta_c}{c} \approx \frac{U\delta_c}{\nu}\frac{1}{\mathsf{Re}_c} \approx 0.16\,\mathsf{Re}_c^{-1/7}, \tag{29.32}$$

which is comparable to skin drag.

Figure 29.6. Velocity distribution ($|v|$) around an airfoil at Reynolds number $\mathsf{Re}_c = 10{,}000$ and $\alpha \approx 1°$, obtained by numeric simulation. Note the faster flow above the wing (lighter), the stagnating flow at the leading edge (darker), and the strong slowdown of the flow in the boundary layers and the trailing wake (very dark). A few isobars are also shown (dark). The boundary layers are thin and laminar and thicken toward the rear, especially on the upper surface where the initial acceleration of the air is followed by deceleration. There is essentially no turbulence in the boundary layers at a Reynolds number as low as this. At more realistic Reynolds numbers in the millions, the boundary layers are mostly turbulent and about an order of magnitude thinner than here. Well behind the airfoil the wake has a thickness comparable to the boundary layers. The slow viscous expansion of the laminar wake is not visible at the scale of this figure.

With growing angle of attack, flow separation may occur on the upper side of the wing at some point before the trailing edge of the airfoil, thereby increasing form drag and diminishing lift. At a certain angle of attack, the point of separation for the turbulent boundary layer on the top side of the wing may suddenly shift forward from the trailing edge, creating a highly turbulent region over most of the wing. This leads to loss of almost all of the lift and at the same time an increased drag. The wing and the aircraft are then said to have *stalled*.

The efforts of aircraft designers between the World Wars in the twentieth century were mainly directed toward form drag reduction by streamlining. A smaller drag generally implies higher top speed, greater payload capacity, and better fuel economy. Besides streamlining of lift surfaces, drag reduction was also accomplished by internalizing the wing support structure and the undercarriage, and providing the engines with carefully designed cowlings.

Induced drag

The two vortices trailing from the wing tips of an aircraft rotate in opposite directions and carry roughly the same circulation Γ as the vortex bound to the wing. They are created at a rate determined by the speed U of the airplane and persist indefinitely in a truly ideal fluid. In a viscous fluid they spin down and dissolve after a certain time.

The process of "spinning up" and "feeding out" the trailing vortices from the wing tips of the aircraft is accompanied by a continuous loss of energy that causes an extra drag on the aircraft. We can estimate the order of magnitude of this drag from the kinetic energy contained in the core of a vortex with circulation Γ and core radius a. Since the maximal flow speed is of order $v_\phi \sim \Gamma/2\pi a$, the kinetic energy of two vortex segments of length Δx becomes of magnitude $\Delta\mathcal{K} \sim \rho_0 v_\phi^2 \pi a^2 \Delta x \sim \rho_0 \Gamma^2 \Delta x$ (dropping all purely numeric factors). This loss of energy corresponds to a drag on the aircraft of magnitude

$$\mathcal{D}_{\text{induced}} = \frac{\Delta\mathcal{K}}{\Delta x} \sim \rho_0 \Gamma^2 \sim \rho_0 U^2 c^2 C_L^2,$$

where we in the last estimate used $\Gamma = \frac{1}{2} U c\, C_L$, obtained from the relation between lift and circulation (29.21) and the definition of the lift coefficient (29.22). The estimate of the induced drag coefficient thus becomes

$$C_D^{\text{induced}} = \frac{\mathcal{D}_{\text{induced}}}{\frac{1}{2}\rho_0 U^2 A} \sim \frac{c}{L} C_L^2. \tag{29.33}$$

Classical wing theory yields an expression of precisely this form but roughly a factor π smaller. Since induced drag is a by-product of the lift-generating flow around a finite wing, it is also called *drag due to lift*. It is the unavoidable price we must pay for a finite wing span.

Induced drag is normally smaller than skin drag, but grows rapidly with increased angle of attack and may win over skin drag at low speeds. This happens, for example, at take-off and landing where the angle of attack is large and the skin friction small. Most importantly, induced drag decreases with increasing aspect ratio L/c, explaining why large aspect ratios are preferable, up to the point where the sheer length of the wing begins to compromise the strength of the wing structure.

Example 29.9 [Cessna 150]: We return again to the Cessna 150 of Example 29.8 and earlier, cruising in level flight with $C_L \approx 0.3$ and aspect ratio 7 we find $C_{\text{induced}} = 0.004$ (including the factor $1/\pi$). The induced drag is thus about half of the turbulent skin drag. At half this speed the relative angle of attack is 4 times bigger, so that the induced drag coefficient becomes 16 times bigger whereas the turbulent skin drag coefficient stays roughly constant.

Lift-to-drag ratio

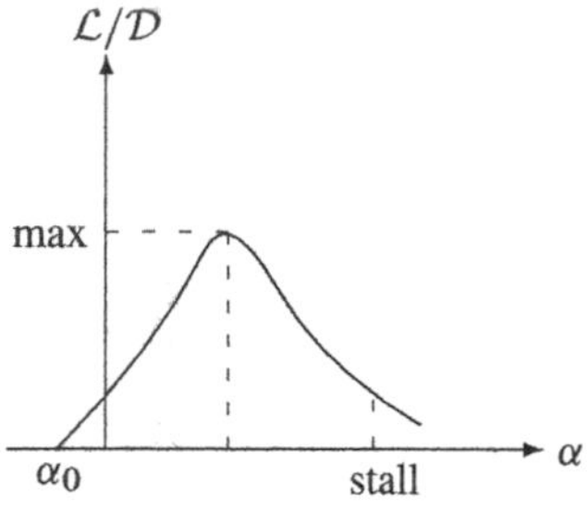

Sketch of a typical lift-to-drag curve for an airplane. The maximal value of lift to drag determines the best glide ratio and the best glide speed.

The total drag coefficient C_D is the sum of all the contributions from various sources: skin drag, form drag, induced drag, etc. The lift-to-drag ratio, in French called *la finesse*,

$$\frac{\mathcal{L}}{\mathcal{D}} = \frac{C_L}{C_D}, \tag{29.34}$$

is a measure of the *aerodynamic efficiency* of an airplane. Like lift and drag, the lift-to-drag ratio is strongly dependent on the angle of attack and less on the Reynolds number and aspect ratio. The quadratic growth of induced drag as a function of angle of attack normally overcomes the linear rise in lift and creates a maximum in $\mathcal{L}/D$ for a certain angle of attack, typically at about half the stall angle. In view of the difficulty in making theoretical estimates of drag on the airframe, empirical lift-to-drag curves are usually provided for a particular aircraft to document its performance.

Example 29.10 [Cessna 150]: The quoted maximal lift-to-drag ratio for the Cessna 150 is 8.4 at a speed $U_{\text{glide}} \approx 110$ km h^{-1}. According to (29.11), the corresponding best glide angle is about $\gamma \approx 6.8°$. From the known lift $\mathcal{L} \approx Mg_0 \cos\gamma$ it then follows that the lift coefficient is $C_L \approx 0.8$, and using the known lift slope $\lambda = 4.9$, the angle of attack becomes $\alpha \approx 7.2°$, making the aircraft nearly horizontal in the glide. Cruising instead in level flight with $U = 180$ km h^{-1}, the lift coefficient is instead $C_L \approx 0.3$, and the quoted lift-to-drag ratio is reduced to about 4.5.

Louis-Charles Breguet (1880–1955). French engineer. First to recognize the importance of the lift-to-drag ratio for airplane performance. He built an airplane assembly factory that manufactured the famous Breguet 14 bomber for the French forces during World War I. In 1919 he founded the airline company that later became Air France.

* Breguet's range equation

Given an amount of fuel, the endurance and range of an aircraft depend on the aerodynamic efficiency of the airframe and the engine efficiency. The latter is usually measured by the *specific fuel consumption*, defined as the rate of fuel consumption (by weight) per unit of produced thrust T. Since the aircraft carries its fuel along, we have

$$f = -\frac{1}{T}\frac{d(Mg_0)}{dt}, \tag{29.35}$$

where M is the mass of the aircraft.

We shall assume that f, which has dimension of inverse time, is a constant characterizing the engine performance, independent of the thrust it delivers. Since $\mathcal{L} \approx Mg_0$ and $\mathcal{D} \approx T$ in level flight, the thrust can be determined from the lift-to-drag ratio $T = Mg_0/(\mathcal{L}/\mathcal{D})$, and inserting this expression, we find

$$f \approx -\frac{\mathcal{L}}{\mathcal{D}} \frac{1}{M} \frac{dM}{dt}. \tag{29.36}$$

As the fuel is spent, the mass of the aircraft decreases and with it the required lift. If the angle of attack is kept constant, the velocity has to decrease, but the lift-to-drag ratio will be constant. Integrating the above equation from $t = 0$, where the fuel tank is full and the mass of the aircraft is M_0, to time t, where the tank is empty and the mass is $M_0 - \Delta M$, we obtain the aircraft's *endurance*, that is, the length of time it can fly on a given amount of fuel,

$$t = \frac{1}{f} \frac{\mathcal{L}}{\mathcal{D}} \log \frac{M_0}{M_0 - \Delta M}. \tag{29.37}$$

The gentle growth of the logarithm shows that the endurance is not increased significantly even if the fuel is a sizable fraction of the aircraft's total mass.

Similarly, writing $dM/dt = U dM/dx$ in (29.36) and integrating, we obtain an equation for the distance an aircraft can fly on a given amount of fuel, called *Breguet's range equation*,

$$\boxed{x = \frac{U}{f} \frac{\mathcal{L}}{\mathcal{D}} \log \frac{M_0}{M_0 - \Delta M}.} \tag{29.38}$$

In deriving this equation we have assumed constancy of $U\mathcal{L}/\mathcal{D}$ as the fuel is being spent, such that U and $\mathcal{L}/\mathcal{D}$ may be taken to be the initial values for the fully loaded aircraft.

Example 29.11 [Cessna 150]: The Cessna 150 has a usable fuel capacity of $\Delta M = 61$ kg and maximal mass of $M = 681$ kg. Cruising at $U = 180$ km h^{-1}, the fully loaded aircraft uses fuel at a rate of about 16 kg h^{-1}. Taking $\mathcal{L}/\mathcal{D} \approx 5$, the total thrust is $T \approx Mg_0/(\mathcal{L}/\mathcal{D}) \approx 1300$ N, and the specific fuel consumption becomes $f \approx 3.3 \times 10^{-5}$ s^{-1}. The predicted endurance becomes $t \approx 4$ h and the range $x \approx 720$ km, in decent agreement with quoted values.

The "sound barrier"

We have previously derived the Prandtl–Glauert rule (29.28) for the dependence of the lift on the Mach number. That result was based on the assumption of ideal flow and does not apply to skin and form drag that, as we have discussed, is always caused by viscosity.

Empirically, the drag coefficient is constant up to the critical Mach number $\mathsf{Ma}_c \approx 0.7$ where part of the accelerated flow above the wing becomes sonic. After this point the drag begins to increase rapidly with increasing speed. At the time of the Second World War and just after, aircraft came close to the speed of sound and became exposed to the violent stresses that reign here, stresses that could lead to breakup in the air. Although common sense told the engineers that the drag could not actually diverge when the aircraft reached sound speed, it was not clear whether it could mount to such high values that sound velocity would become a barrier, in practice impenetrable with the engines and airframes available at that time. History of course tells us that the "sound barrier" was passed on October 14, 1947, with a rocket-propelled experimental aircraft (see [Anderson 2001] for more details).

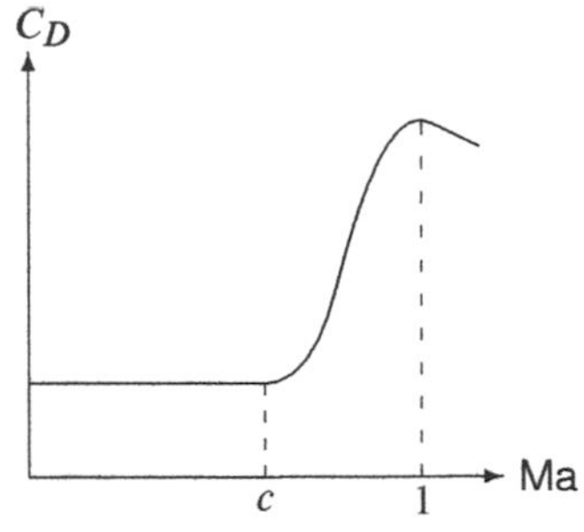

Sketch of the drag coefficient as a function of Mach number. The drag is essentially constant up to the critical point Ma_c, after which it rises dramatically toward the "sound barrier" at $\mathsf{Ma} = 1$.

* 29.6 Lift, drag, and the trailing wake

An often recurring question that can lead to heated discussions is whether an airplane stays aloft in steady flight because of the pressure difference between the lower and upper wing surfaces, or whether it gets lift from diverting momentum downward. Either position is in fact tenable in a discussion, but as we shall see the correct answer is more subtle than might be guessed at first glance.

Momentum balance in a box

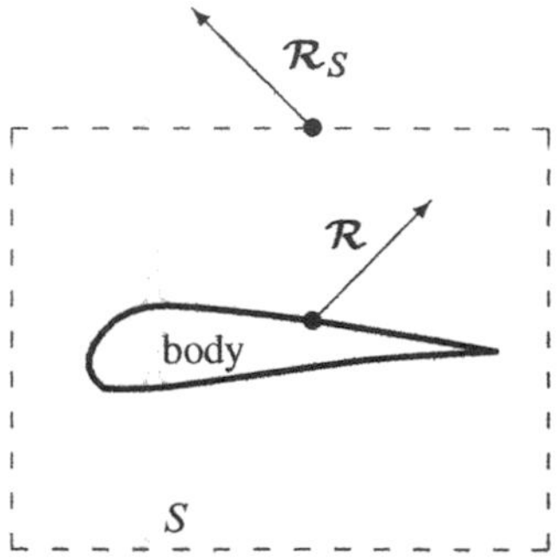

Streamlined body and enclosing box S. The box does not need to be rectangular as it is here but can be a volume of any shape. The reaction forces on body and box are for illustration purposes set to act on arbitrarily chosen points on the surfaces.

The general treatment of momentum balance in Section 20.3 on page 343 indicates that the total contact force on the airplane should be balanced by an opposite momentum flux at great distances, where all stresses have died away.

Let the steadily moving body—an aircraft or wing—be surrounded with a huge imagined box of any shape S, and let the volume of air between the body surface and the box be our control volume. As can be seen from Figure 29.7, this box will cut through the trailing wake somewhere behind the body. Disregarding gravity, the total force acting on the control volume of air consists of the contact forces on the two bounding surfaces,

$$\mathcal{F} = -\mathcal{R} + \oint_S \boldsymbol{\sigma} \cdot d\boldsymbol{S}, \tag{29.39}$$

where $\mathcal{R}$ is the air's reaction force (29.2) on the body. For the moment we do not split up the reaction force into lift and drag.

In steady flow the total momentum of the air contained in the control volume remains constant, $d\mathcal{P}/dt = \boldsymbol{0}$, apart from small time-dependent contributions from the fluctuating velocity field in the turbulent wake (which we shall ignore here). Since there can be no momentum flux through the impermeable surface of the body, it follows from momentum balance (20.13) on page 343 that the total force on the air in the control volume must equal the flux of momentum out of the box, $\mathcal{F} = \oint_S \rho \boldsymbol{v}\boldsymbol{v} \cdot d\boldsymbol{S}$. Assuming for simplicity that the air is effectively incompressible[2], $\rho = \rho_0$, we find the reaction force,

$$\boxed{\mathcal{R} = -\rho_0 \oint_S \boldsymbol{v}\boldsymbol{v} \cdot d\boldsymbol{S} + \oint_S \boldsymbol{\sigma} \cdot d\boldsymbol{S},} \tag{29.40}$$

where $\boldsymbol{\sigma} = \{\sigma_{ij}\}$ is the incompressible stress tensor $\sigma_{ij} = -p\delta_{ij} + \eta(\nabla_i v_j + \nabla_j v_i)$.

The total reaction force on the body can thus be calculated from the pressure and velocity fields, all evaluated at the surface of any box surrounding the body. It must be emphasized that this result is exact, valid for any shape and size of body and box (as long as the box engulfs the body).

Box at spatial infinity

Now let the box expand to huge distances in all directions such that the velocity field on its surface approaches the asymptotic value, $\boldsymbol{v} \to \boldsymbol{U}$. Setting $\boldsymbol{v} = \boldsymbol{U} + \Delta\boldsymbol{v}$ in the first (momentum) term of (29.40), it becomes to first order in $\Delta\boldsymbol{v}$,

$$-\rho_0 \oint_S (\boldsymbol{U}\boldsymbol{U} + \boldsymbol{U}\Delta\boldsymbol{v} + \Delta\boldsymbol{v}\boldsymbol{U}) \cdot d\boldsymbol{S}.$$

Given that the total vector area of a closed surface always vanishes, $\oint d\boldsymbol{S} = \boldsymbol{0}$, together with global mass conservation, $\oint \Delta\boldsymbol{v} \cdot d\boldsymbol{S} = 0$, the first two terms in the integrand do not contribute to the integral.

[2]The analysis can also be carried through for barotropic compressible air (Problem 29.5).

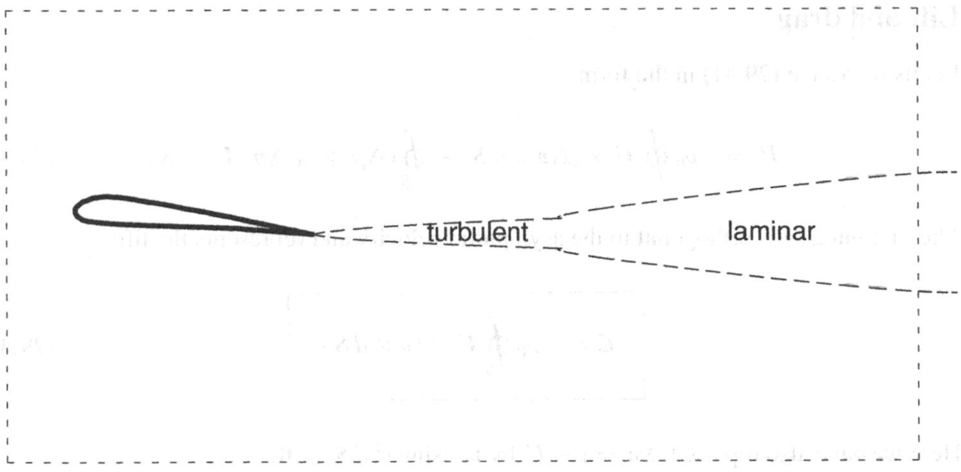

Figure 29.7. Sketch of a body (airfoil) and its trailing wake. Initially the trailing wake is turbulent, but expands slowly and becomes laminar some distance downstream. The thickness of the wake is greatly exaggerated compared to the distance from the body. The dashed box surrounding the system and crossing through the trailing wake is used in the text to define a control volume of air between the surfaces of the body and the box.

If we think of the box as a huge sphere with radius r and surface area $4\pi r^2$, the velocity field correction must decay like $|\Delta \boldsymbol{v}| \sim 1/r^2$ at great distances outside the trailing wake for any contribution to survive in the limit. Inside the wake the velocity field behaves differently (see Section 29.8) but the general conclusions remain valid. The velocity derivatives in the stress tensor must consequently vanish like $|\nabla \Delta \boldsymbol{v}| \sim 1/r^3$, and cannot contribute to the second term in (29.40). Pressure is thus the only stress component that has a possibility of surviving in this limit, and the reaction force on the body may be written as

$$\mathcal{R} = -\rho_0 \oint_S \Delta \boldsymbol{v}\, \boldsymbol{U} \cdot d\boldsymbol{S} - \oint_S \Delta p\, d\boldsymbol{S}. \tag{29.41}$$

Since a constant pressure yields no contribution, we have replaced the pressure by the residual pressure $\Delta p = p - p_0$, where p_0 is the constant asymptotic pressure. The residual pressure can only contribute to the integral if it decays as $\Delta p \sim 1/r^2$ outside the trailing wake, which we shall see that it does.

We are now in a position to answer the question of whether there remains a pressure contribution to lift far from the moving body. Although the derivation of the above equation shows that the *sum* of the momentum flow and pressure contributions is independent of the choice of the box shape, each term by itself may (and will) depend on it. *The limiting value of the pressure contribution may depend on how the box is taken to infinity.* One may fear that this answer will not resolve the heated discussion unless the participants appreciate the following argument!

If, for example, we choose a cube or sphere and let it expand uniformly in all directions, there will usually be a residual pressure contribution to lift, even in the limit of an infinite box (see Section 29.8 for an explicit calculation). Alternatively, one may choose a box in the form of a huge cylinder with radius R and length L, oriented with its axis parallel to the asymptotic flow $\boldsymbol{U}$. The pressure integral over the end caps cannot contribute to lift, because their normals are parallel to the velocity. If we now let the radius R become infinite, *before* the end caps are moved off to infinity, the pressure integral over the cylinder surface will behave like the area $2\pi L R$ times the pressure $p \sim 1/R^2$. It thus vanishes like L/R for $R \to \infty$, leaving no pressure contribution to lift in the limit [Landau and Lifshitz 1987].

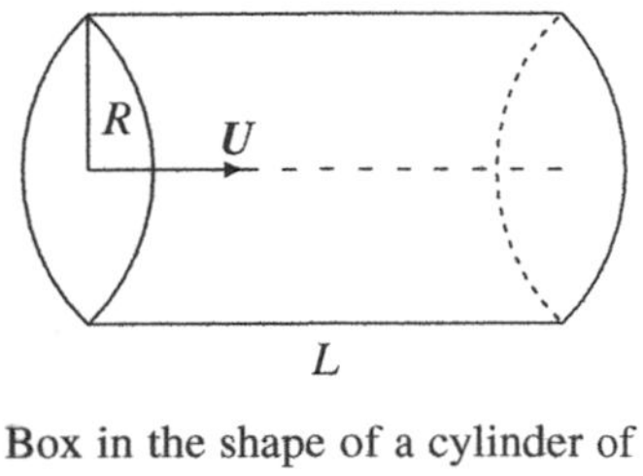

Box in the shape of a cylinder of radius R and length L with axis parallel to the asymptotic velocity $\boldsymbol{U}$.

Lift and drag

Let us rearrange (29.41) in the form

$$\mathcal{R} = -\rho_0 \oint_S \boldsymbol{U} \times (\Delta\boldsymbol{v} \times d\boldsymbol{S}) - \oint_S (\Delta p + \rho_0 \Delta\boldsymbol{v} \cdot \boldsymbol{U})\, d\boldsymbol{S}. \tag{29.42}$$

The first integral is orthogonal to the asymptotic velocity and represents the lift:

$$\boxed{\mathcal{L} = -\rho_0 \oint_S \boldsymbol{U} \times (\boldsymbol{v} \times d\boldsymbol{S}).} \tag{29.43}$$

Here we have also replaced $\Delta\boldsymbol{v} = \boldsymbol{v} - \boldsymbol{U}$ by $\boldsymbol{v}$, using $\oint d\boldsymbol{S} = \boldsymbol{0}$.

In regions where the flow is inviscid and all streamlines connect to spatial infinity, the pressure excess is determined by Bernoulli's theorem,

$$\Delta p = \tfrac{1}{2}\rho_0(U^2 - \boldsymbol{v}^2) \approx -\rho_0 \boldsymbol{U} \cdot \Delta\boldsymbol{v}. \tag{29.44}$$

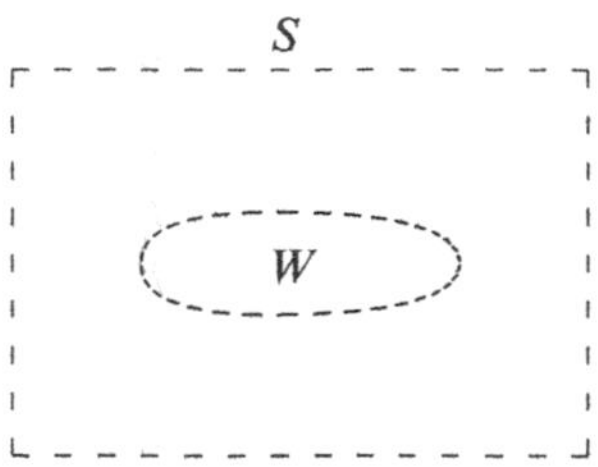

The downstream face of the box cuts the wake in a region W chosen to be orthogonal to the asymptotic flow $\boldsymbol{U}$.

The integrand of the second term in (29.42) vanishes everywhere outside the wake and can only come from the region W where the box cuts through the wake. Choosing W to be planar and orthogonal to $\boldsymbol{U}$, its surface element $d\boldsymbol{S}$ will be parallel with $\boldsymbol{U}$, and the second term becomes a pure drag,

$$\boxed{\mathcal{D} = -\int_W (\Delta p + \rho_0 \Delta\boldsymbol{v} \cdot \boldsymbol{U})\, dS,} \tag{29.45}$$

where dS is the area element of W. If we lift the restriction that W is a planar part of S orthogonal to $\boldsymbol{U}$, the formulae for lift and drag become slightly different (Problem 29.7). Otherwise they are valid for all kinds of bodies moving steadily through an incompressible fluid at subsonic speed.

d'Alembert's paradox: A gift to powered flight

We have previously on page 222 and 225 shown that a cylinder or sphere in potential flow experiences no drag (d'Alembert's paradox). We can now see that d'Alembert's paradox is valid for any object in asymptotically uniform, truly ideal flow because Bernoulli's theorem (29.44) is then fulfilled everywhere, implying that the drag (29.45) must vanish. Drag arises because a body moving through nearly ideal fluid cannot help creating viscous boundary layers that continue into the narrow trailing wake. Equation (29.45) thus resolves the paradox: *The drag on a body can be calculated from the viscous flow in the distant trailing wake.* This will be analyzed in detail in Section 29.8.

In general, the narrower the wake, the smaller the drag will be. As we have seen in the estimates of the preceding section, drag is typically an order of magnitude smaller than lift for properly streamlined bodies, such as airfoils. This indicates that one should rather view d'Alembert's "paradox" as a statement about the near vanishing of the drag for streamlined bodies at high Reynolds number. This is in fact what makes flying technically possible with engines only capable of producing a thrust much smaller than the weight of the aircraft. Without d'Alembert's "paradox", the Wright brothers would never have had a chance of flying at Kitty Hawk in December 1903, given the puny engine power then available to them.

Lift and vorticity

The box S used in the lift integral (29.43) is assumed to be of essentially infinite size. Now let S' be another closed surface surrounding the body somewhere inside the box S. From Gauss' theorem we obtain (Problem 29.6)

$$\left(\oint_S - \oint_{S'}\right) \boldsymbol{U} \times (\boldsymbol{v} \times d\boldsymbol{S}) = -\int_V \boldsymbol{U} \times (\nabla \times \boldsymbol{v})\, dV = -\int_V \boldsymbol{U} \times \boldsymbol{\omega}\, dV, \tag{29.46}$$

where V is the region between the two surfaces, and $\boldsymbol{\omega} = \nabla \times \boldsymbol{v}$ is the vorticity field. In the extreme case we may take S' to be the body surface itself, where the velocity and thus the integral over S' vanishes because of the no-slip condition. It then follows from the above equation that the lift (29.43) is also given by the integral of the vorticity field over the air,

$$\boxed{\mathcal{L} = \rho_0 \boldsymbol{U} \times \int_{\text{air}} \boldsymbol{\omega}\, dV.} \tag{29.47}$$

This integral can only receive contributions from the regions of non-vanishing vorticity, in other words from the boundary layers and the trailing wake. Again we conclude that *without vorticity created by friction, there can be no lift*!

More generally, if there is no vorticity found in V, the integral over the box S equals the integral over S'. The original box may in other words be deformed into any other closed surface as long as it crosses no region containing vorticity. The box at infinity has now served its purpose and may be forgotten. In the following, the surface S in (29.43) may be taken to be simply any surface surrounding the body, as long as there is no vorticity *outside* S.

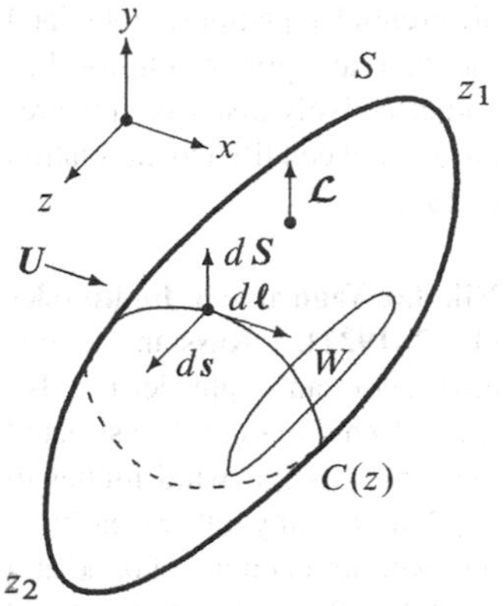

The surface S may always be sliced into a sequence of oriented planar curves $C(z)$ parallel with the xy-plane for $z_1 \leq z \leq z_2$.

Lift and circulation

It is useful to introduce a "natural" coordinate system with the x-axis along the direction of the asymptotic velocity $\boldsymbol{U} = U\hat{\boldsymbol{e}}_x$ and the y-axis along the lift $\mathcal{L} = \mathcal{L}\hat{\boldsymbol{e}}_y$. Working out the cross-products, the lift (29.43) becomes

$$\mathcal{L} = \rho_0 U \oint_S (\boldsymbol{v} \times d\boldsymbol{S})_z = \rho_0 U \oint_S v_x dS_y - v_y dS_x, \tag{29.48}$$

together with the condition

$$\oint_S v_x dS_z - v_z dS_x = 0, \tag{29.49}$$

expressing that there should be no lift along the z-direction. For a symmetric aircraft in normal horizontal flight, this condition is automatically fulfilled.

The closed surface S may always be sliced into a set of planar closed contours $C(z)$ parallel with the xy-plane, and parametrized by the z-coordinate in some interval $z_1 \leq z \leq z_2$. The contours are given negative orientation in the xy-plane, that is, clockwise as seen from positive z-values. Let now $d\boldsymbol{\ell} = (dx, dy, 0)$ be a line element on a point of the curve $C(z)$; it is evidently a tangent vector to the surface S. Let $d\boldsymbol{s} = (0, dy, dz)$ be another tangent vector to the surface in the yz-plane with the same y-coordinate dy. Then the outward-pointing surface element becomes $d\boldsymbol{S} = d\boldsymbol{s} \times d\boldsymbol{\ell} = (-dydz, dzdx, -dxdy)$ and thus

$$v_x dS_y - v_y dS_x = (v_x dx + v_y dy)\, dz = \boldsymbol{v} \cdot d\boldsymbol{\ell}\, dz.$$

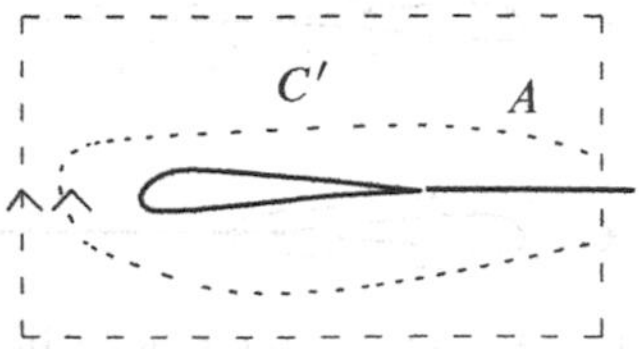

The contours C and C' cross the trailing wake along the same curve (at infinity) and have the same circulation as long as the area A between the curves carries no flux or vorticity.

This shows that the lift (29.47) may be written as an integral over z,

$$\mathcal{L} = \rho_0 U \int_{z_1}^{z_2} \Gamma(z)\, dz, \tag{29.50}$$

with an integrand given by the circulation around $C(z)$,

$$\Gamma(z) = \oint_{C(z)} \boldsymbol{v} \cdot d\boldsymbol{\ell}. \tag{29.51}$$

If the contour is deformed, Stokes' theorem (13.31) tells us that the circulation is unchanged provided the contour is swept through an area A devoid of vorticity. Since we have assumed that vorticity is only found in the boundary layers and the trailing wake, the contour may be freely deformed, as long as the piece of the contour that crosses the wake is kept fixed, and the new contour does not pass through the wake or into the boundary layers. This is of course the same conclusion as was reached in the preceding subsection.

Again it should be emphasized that no approximations have been made, and that this result is exactly valid, as long as the wake-crossing takes place at great distance from the body along a line parallel with the y-axis.

The general Kutta–Joukowsky theorem

The lift integral (29.50) may be written in the form of a generalized version of the Kutta–Joukowsky theorem (29.21),

$$\boxed{\mathcal{L} = \rho_0 U L \langle \Gamma \rangle,} \tag{29.52}$$

where

$$\langle \Gamma \rangle = \frac{1}{L} \int_{z_1}^{z_2} \Gamma(z) \, dz \tag{29.53}$$

is the circulation along z averaged over the wing span $L = z_2 - z_1$. The only difference is that the integration contour $C(z)$ used to calculate $\Gamma(z)$ in (29.51) has to cross the wake far away from the airfoil, whereas in the original Kutta–Joukowsky theorem it is supposed to hug the airfoil tightly all the way around. We shall now see how to get rid of the "tail" of the contour when the chord-wise Reynolds number $\mathsf{Re}_c = Uc/\nu$ and the aspect ratio L/c are very large.

In this limit the flow around the wing becomes nearly ideal and irrotational, and the boundary layers turn into a "skin" of vorticity covering the airfoil with nearly vanishing thickness $\delta \sim c/\sqrt{\mathsf{Re}_c}$. Downstream from the airfoil, the skin continues into the trailing wake, which forms a horizontal sheet, also of nearly vanishing thickness δ. Physically, the flow velocity in the wake cannot become infinite, so that the downstream volume flux in the wake, which per unit of span is of order $v_x \delta$, must itself vanish in the limit of infinite Reynolds number. It then follows from mass conservation that the orthogonal velocity v_y must be the same above and below the sheet, for otherwise fluid would accumulate in the wake.

The pressure must also be the same above and below the trailing wake sheet because of Newton's Third Law. Combining these two results with Bernoulli's theorem, which states that $p + \frac{1}{2}\rho_0(v_x^2 + v_y^2 + v_z^2)$ takes the same value everywhere outside the wake, we conclude that $v_x^2 + v_z^2$ must be the same just above and below the wake. The span-wise induced flow v_z is connected to the shedding of vorticity along the span, especially the wing-tip vortices. When the aspect ratio is large, this flow will be tiny compared to the downstream flow, that is, $|v_z| \sim v_x c/L \ll v_x$, so that it may be ignored in the Bernoulli function. Consequently, v_x itself takes the same value just above and below the trailing sheet. The two oppositely directed contributions to the integral (29.51) running along the tail of C thus tend to cancel each other in the limit of infinite Reynolds number and aspect ratio.

Since the orthogonal velocity v_y may not be infinite inside the wake, the part of the integral from the contour passing through the sheet will be of order of magnitude $v_y \delta$ and thus vanish in the limit. The contribution from the tail of the integration contour can thus be ignored in the leading approximation and we may let it encircle the wing while hugging the airfoil profile. Finally, we have arrived at the (generalized) Kutta–Joukowsky theorem.

Martin Wilhelm Kutta (1867–1944). German mathematician. Probably best known for his extension of a method developed by Runge for numeric solutions to differential equations. Obtained the first analytic result for lift, and effectively discovered the relation between lift and circulation in 1902.

Nikolai Yegorovich Joukowsky (1847–1921). Russian mathematician and physicist (also spelled Zhukovskii). Constructed the first Russian wind tunnel in 1902 and many others early in the twentieth century. Found and used the relation between lift and circulation in 1906.

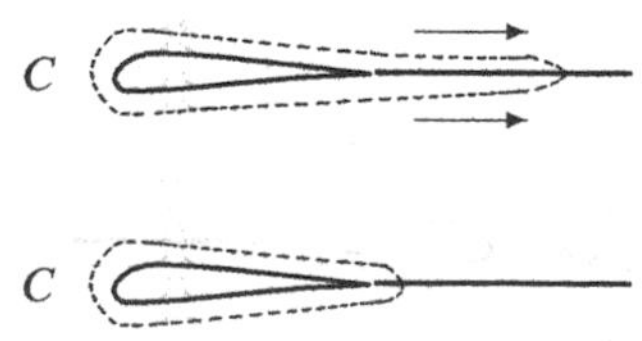

The dashed contour C hugs the wing profile closely, but is still attached to the distant part where it crosses the wake (top). For infinite Reynolds number, the velocity is the same above and below the wake, allowing us to cut off the tail (bottom).

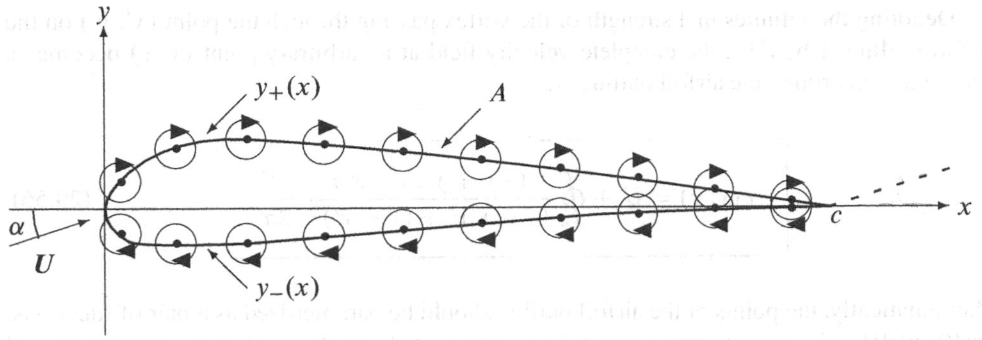

Figure 29.8. The irrotational velocity field of the wing is viewed as a superposition of elementary line vortices with singular cores arranged around the outline A of the airfoil. On the top of the airfoil, the vortices tend to increase the local velocity above the asymptotic flow $\boldsymbol{U}$, and conversely at the bottom. The airfoil is positioned with the chord-line on the x-axis and the y-axis at the leading edge. Its geometry is described by two functions $y_\pm(x)$ with $0 \le x \le c$, where c is the chord length. The asymptotic flow is $\boldsymbol{U} = U(\cos\alpha, \sin\alpha)$, where α is the angle of attack. The trailing wake, indicated by the dashed line, also forms an angle α with the x-axis. The z-axis comes out of the plane.

* 29.7 Two-dimensional airfoil theory

Most wings have fairly large aspect ratios in the vicinity of $L/c \approx$7–20 with airfoil cross-sections that taper gently toward the wing tips. For nearly infinite aspect ratio and nearly constant cross-section, there is very little induced flow along z toward the wing tips, so that the flow becomes essentially two-dimensional,

$$\boldsymbol{v} = (v_x(x,y), v_y(x,y), 0). \tag{29.54}$$

For such an airfoil, the circulation will be independent of z. Although it would be possible to simplify the following calculations using complex notation as in Section 27.3, we shall use the physically more transparent real notation.

The field of the vortex sheet

In the limit of nearly infinite Reynolds number, vorticity only exists in infinitesimally thin boundary skins. Outside these skins and outside the infinitesimal sheet of the trailing wake, we assume that the flow is irrotational, described by a velocity potential Ψ that satisfies Laplace's equation $\nabla^2\Psi = 0$. Being a linear equation, its solutions may be superposed. All the nonlinearity of the original Euler equation has been collected in the Bernoulli pressure $p = \frac{1}{2}\rho_0(U^2 - \boldsymbol{v}^2)$. The additivity of potential flows makes it possible to view the irrotational flow outside the boundary layers as arising from a superposition of the asymptotic flow $\boldsymbol{U}$ and the field generated by the sheet of vorticity covering the wing surface.

Due to the two-dimensional nature of the flow, the vortex sheet making up the skin may be understood as a collection of elementary line vortex cores running parallel with the z-axis (see Figure 29.8). The contribution from the velocity field of a line vortex passing through the origin of the coordinate system with the core parallel with the z-axis is of the well-known form (27.13),

$$\boldsymbol{v} = \frac{\Gamma}{2\pi}\frac{(-y, x)}{x^2 + y^2}, \tag{29.55}$$

where Γ is its circulation.

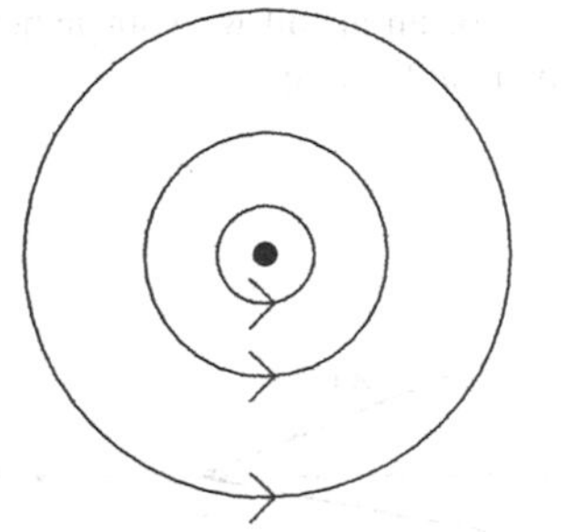

Outside the singular core of a line vortex the flow is irrotational with circular streamlines. The field around any collection of line vortices is obtained by adding their individual fields together.

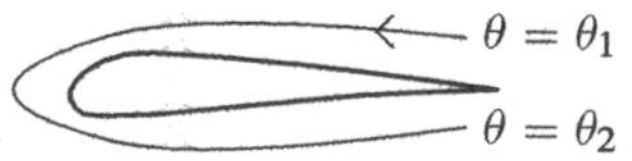

The airfoil outline may be parametrized with a parameter θ running over the interval $\theta_1 \leq \theta \leq \theta_2$.

Denoting the infinitesimal strength of the vortex passing through the point (x', y') on the airfoil outline A by $d\Gamma'$, the complete velocity field at an arbitrary point (x, y) becomes a curve integral around the airfoil outline A,

$$\boxed{\boldsymbol{v}(x, y) = \boldsymbol{U} + \oint_A \frac{(-y + y', x - x')}{(x - x')^2 + (y - y')^2} \frac{d\Gamma'}{2\pi}.} \qquad (29.56)$$

Mathematically, the points of the airfoil outline should be parametrized as a pair of functions, $(x(\theta), y(\theta))$, of a running parameter θ in some interval, say $\theta_1 \leq \theta \leq \theta_2$, beginning and ending at the cusp. The circulation element then becomes $d\Gamma = \gamma(\theta)\, d\theta$, where $\gamma(\theta)$ is the circulation density. Wherever possible we shall suppress this elaborate, but mathematically more concise notation.

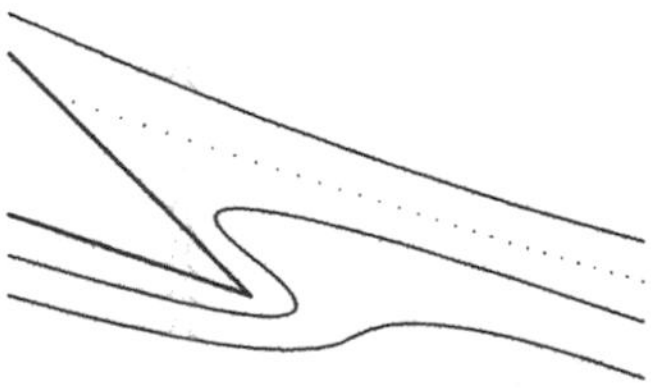

Separating flow pattern near the trailing end of an airfoil with stagnation point before the cusp. Such a flow may have vanishing lift.

The Kutta condition

Near the front and rear ends of the airfoil there are stagnation points where the velocity field vanishes. In ideal flow there may be more than one velocity field solution satisfying the Euler equation (13.1) and the boundary conditions. Such solutions may have different stagnation points and thus different circulation. It is even possible to find a solution with vanishing circulation (and lift). We shall see below that if the rear stagnation point is not situated right at the cusp of the trailing edge, unacceptable infinite velocity field values will arise at the cusp. The *Kutta condition* (1902) enforces that the trailing edge cusp (c in Figure 29.8) is actually a stagnation point. The condition thus repairs a mathematical problem in truly ideal flow by selecting a particular solution. In the real world, a streamlined airfoil under a small angle of attack in nearly ideal flow will, in fact, fulfill the Kutta condition because of viscous friction in the boundary layers that selects a unique laminar solution.

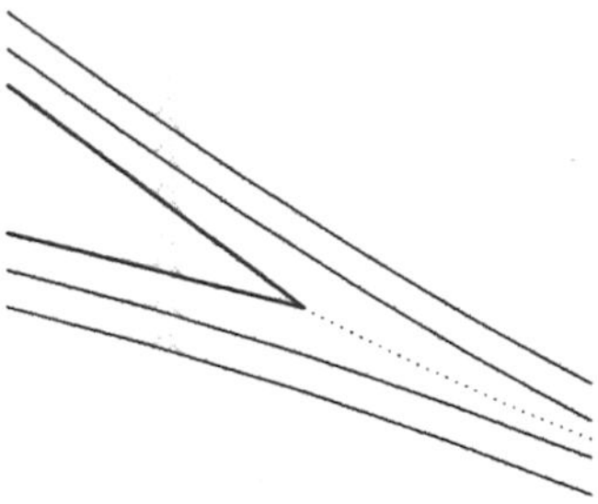

Laminar flow near the trailing end of an airfoil with stagnation point at the cusp.

The problem arises when we attempt to calculate the velocity field at a point (x, y) on the airfoil outline itself, because the integrand of (29.56) is formally infinite for $(x', y') = (x, y)$. At any point where the airfoil is smoothly varying, there is in fact no problem and the integral is finite. To see this we cut out a small parameter interval $\theta - \epsilon < \theta' < \theta + \epsilon$ around the singularity in the integrand. Assuming smoothness in this interval we may expand $x - x' \approx (\theta - \theta')\dot{x}$ and $y - y' = (\theta - \theta')\dot{y}$, where $\dot{x} = dx(\theta)/d\theta$ and $\dot{y} = dy(\theta)/d\theta$. The leading contribution to the integral from this interval then becomes

$$\Delta\boldsymbol{v} \approx \gamma(\theta)\frac{(-\dot{y}, \dot{x})}{\dot{x}^2 + \dot{y}^2} \int_{\theta-\epsilon}^{\theta+\epsilon} \frac{1}{\theta - \theta'} \frac{d\theta'}{2\pi}. \qquad (29.57)$$

Here the integral vanishes because of symmetry around the singularity (mathematically it is a principal-value integral).

The argument fails, however, at the trailing edge for $\theta = \theta_{1,2}$ where the airfoil has a sharp cusp. Here we find instead the contributions,

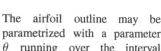

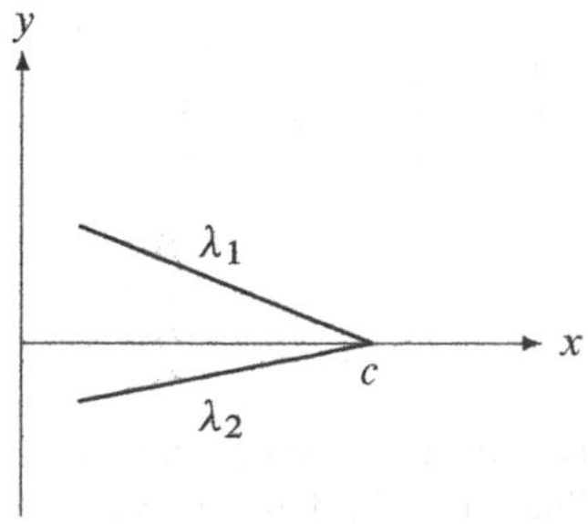

The cusp consists of two straight lines with different slopes $dy/dx = \lambda_1 = \dot{y}_1/\dot{x}_1$ and $dy/dx = \lambda_2 = \dot{y}_2/\dot{x}_2$.

$$\Delta\boldsymbol{v} \approx \gamma_1 \frac{(-\dot{y}_1, \dot{x}_1)}{\dot{x}_1^2 + \dot{y}_1^2} \int_{\theta_1}^{\theta_1+\epsilon} \frac{1}{\theta_1 - \theta'} \frac{d\theta'}{2\pi} + \gamma_2 \frac{(-\dot{y}_2, \dot{x}_2)}{\dot{x}_2^2 + \dot{y}_2^2} \int_{\theta_2-\epsilon}^{\theta_2} \frac{1}{\theta_2 - \theta'} \frac{d\theta'}{2\pi},$$

where the coefficients are evaluated at $\theta = \theta_{1,2}$. Since the slopes $(\dot{x}_1, \dot{y}_1)$ and $(\dot{x}_2, \dot{y}_2)$ are different, the divergent integrals will in general not cancel. Only if the vortex density vanishes from both sides of the cusp, that is, $\gamma_1 = \gamma_2 = 0$, will the singular contribution disappear.

Given that in ideal flow the velocity must be tangential to the airfoil outline A, the circulation becomes

$$\Gamma = \oint_A \boldsymbol{v}(x, y) \cdot d\boldsymbol{\ell} = \oint_A |\boldsymbol{v}|\, d\ell. \tag{29.58}$$

Locally, each little line element $d\ell$ of A contributes an infinitesimal amount,

$$d\Gamma = \gamma\, d\theta = |\boldsymbol{v}|\, d\ell, \tag{29.59}$$

to the circulation. Since $d\ell/d\theta$ is regular on each side of the cusp, the vanishing of the vortex density γ is equivalent to the vanishing of the velocity field $\boldsymbol{v}$ at the cusp, which is the Kutta condition.

The fundamental airfoil equation

The vortex density $\gamma(\theta)$ must be chosen such that the streamlines follow the airfoil outline. This is equivalent to requiring the normal component of the velocity field to vanish on the impermeable airfoil surface,

$$\boldsymbol{v} \cdot \boldsymbol{n} = 0, \tag{29.60}$$

where $\boldsymbol{v} = \boldsymbol{v}(x, y)$ is the slip-velocity and $\boldsymbol{n} = \boldsymbol{n}(x, y)$ is the normal at the point (x, y). For every point on the airfoil outline, we thus get one scalar condition, and together with the Kutta condition this is sufficient to determine the vortex density $\gamma(\theta)$.

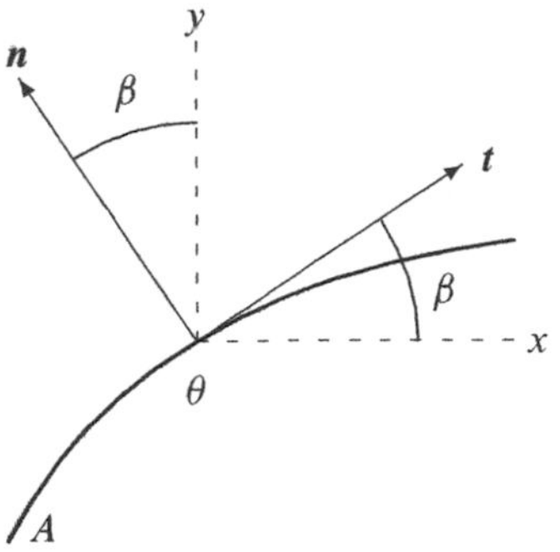

At any point $(x(\theta), y(\theta))$ of the airfoil outline A, the tangent angle is denoted β and the slope $\lambda = \tan\beta = \dot{y}/\dot{x}$. The tangent vector may be chosen to be $\boldsymbol{t} = (\dot{x}, \dot{y})$ and the normal $\boldsymbol{n} = (-\dot{y}, \dot{x})$.

No no-slip: We are not at liberty to impose a no-slip condition on the field, because the Euler equation (13.1) is only of first order in spatial derivatives. Although the field (29.56) exists both inside and outside the airfoil outline, the outside solution now fulfills the Euler equation and obeys the correct boundary conditions for inviscid flow around a solid body. Consequently, we may with impunity replace the region inside the airfoil outline with a solid body.

For convenience, the airfoil is positioned with its chord-line on the x-axis such that the asymptotic velocity becomes $\boldsymbol{U} = U(\cos\alpha, \sin\alpha)$ (see Figure 29.8). In the θ-parametrization, the tangent vector to the airfoil outline at the point θ is $\boldsymbol{t} = (\dot{x}, \dot{y}) = d(x, y)/d\theta$, and the normal may be taken to be $\boldsymbol{n} = (-\dot{y}, \dot{x})$. The boundary condition (29.60) then takes the explicit form

$$(-\dot{x}\sin\alpha + \dot{y}\cos\alpha)U = \oint \frac{\dot{x}(x - x') + \dot{y}(y - y')}{(x - x')^2 + (y - y')^2} \frac{d\Gamma'}{2\pi}, \tag{29.61}$$

where now both $(x, y) = (x(\theta), y(\theta))$ and $(x', y') = (x(\theta'), y(\theta'))$ are points on the airfoil outline and $d\Gamma' = \gamma(\theta')d\theta'$.

Marvellously this equation can be integrated with respect to θ. Using that for $\theta = \theta_1$ we must have $x = c$ and $y = 0$, it may be verified by differentiation with respect to θ that the following expression is the correct integral:

$$\boxed{\big((c - x)\sin\alpha + y\cos\alpha\big)U = \frac{1}{2}\oint \log\frac{(x - x')^2 + (y - y')^2}{(c - x')^2 + y'^2} \frac{d\Gamma'}{2\pi}.} \tag{29.62}$$

This is the *fundamental equation of two-dimensional airfoil theory*. Given the parametrized airfoil geometry through the functions $(x(\theta), y(\theta))$, this integral equation should be solved for the vortex density $\gamma(\theta) = d\Gamma/d\theta$.

Having done that, the total circulation may afterward be obtained by integrating the result, $\Gamma = \int_{\theta_1}^{\theta_2} \gamma(\theta)\, d\theta$. Finally, inserting this into the Kutta–Joukowsky theorem (29.52), we obtain the lift. We have thus established a precise analytic or numeric procedure that for ideal flow will yield the lift as a function of the geometry of the airfoil.

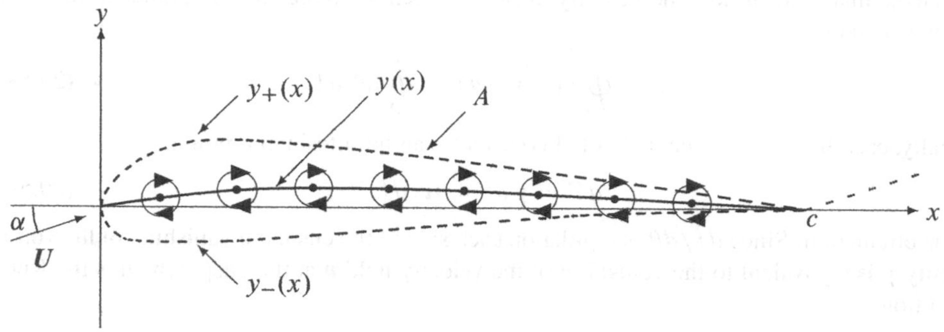

Figure 29.9. The irrotational velocity field of the wing is viewed as a superposition of elementary line vortices with singular cores arranged around the outline A of the airfoil. On the top of the airfoil, the vortices tend to increase the local velocity above the asymptotic flow $\boldsymbol{U}$, and conversely at the bottom. The airfoil is positioned with the chord-line on the x-axis and the y-axis at the leading edge. Its geometry is described by two functions $y_\pm(x)$ with $0 \le x \le c$, where c is the chord length. The asymptotic flow is $\boldsymbol{U} = U(\cos\alpha, \sin\alpha)$, where α is the angle of attack. The trailing wake, indicated by the dashed line, also has the angle α with the x-axis. The z-axis comes out of the plane.

Thin airfoil approximation

Even if modern airfoils are much thicker than the airfoils of the early airplanes, the ratio d/c of thickness to chord is rarely more than 10%–15%. Using x as the parameter, a decent approximation for thin airfoils is obtained by replacing the double layer of circulation density $\gamma_\pm(x)$ on the two halves of the airfoil outline by a single layer with circulation density $\gamma(x)$ distributed along a *camber line* $y(x)$, where

$$\gamma(x) = \gamma_+(x) + \gamma_-(x), \qquad y(x) = \tfrac{1}{2}(y_+(x) + y_-(x)). \tag{29.63}$$

For a thin airfoil with $d/c \ll 1$, we always have $|y(x) - y(x')| \ll |x - x'|$, so that the fundamental airfoil equation (29.62) reduces to the much simpler equation,

$$\big((c - x)\sin\alpha + y(x)\cos\alpha\big)U = \frac{1}{2\pi}\int_0^c \log\frac{|x - x'|}{c - x'}\,\gamma(x')\,dx'. \tag{29.64}$$

Given the camber line $y(x)$, this linear integral equation must be solved for $\gamma(x)$.

The circulation can be obtained directly by using the relation

$$\int_0^c \frac{c}{(c - x)\sqrt{x(c - x)}}\log\frac{|x - x'|}{c - x'}\,dx = -2\pi, \tag{29.65}$$

which is true for all $c > 0$ and all x' between 0 and c. It takes a fair bit of complex analysis to prove this (see Problem 29.9), but it may easily be checked numerically. Using this result in (29.64), we get the circulation

$$\Gamma = \int_0^c \gamma(x')dx' = -Uc\int_0^c \frac{(c - x)\sin\alpha + y(x)\cos\alpha}{(c - x)\sqrt{x(c - x)}}\,dx. \tag{29.66}$$

Since $\int_0^c dx/\sqrt{x(c - x)} = \pi$, we finally obtain

$$\Gamma = -Uc\left(\pi\sin\alpha + \cos\alpha\int_0^c \frac{y(x)}{(c - x)\sqrt{x(c - x)}}\,dx\right). \tag{29.67}$$

The integral converges because $y(x) \sim c - x$ for $x \to c$.

Taking into account that the contour of integration in the generalized Kutta–Joukowsky theorem (29.52) is clockwise and not counter-clockwise as assumed in the above calculation, the lift is $\mathcal{L} = -\rho_0 U L \Gamma$. The lift coefficient may now be written as[3]

$$\boxed{C_L = \frac{\mathcal{L}}{\frac{1}{2}\rho_0 U^2 c L} = 2\pi \frac{\sin(\alpha - \alpha_0)}{\cos\alpha_0},} \tag{29.68}$$

where the zero-lift angle α_0 is defined from the integral

$$\tan\alpha_0 = -\frac{1}{\pi}\int_0^c \frac{y(x)}{(c-x)\sqrt{x(c-x)}}\,dx. \tag{29.69}$$

Normally, airfoils have positive camber, $y(x) > 0$, so that $\alpha_0 < 0$. For a flat plate we evidently have $\alpha_0 = 0$ because $y(x) = 0$. For small angles of attack, $|\alpha|, |\alpha_0| \ll 1$, the lift coefficient indeed takes the form (29.24) with lift slope $\lambda = 2\pi$. In Problem 29.10 the integral is worked out for a simple non-trivial case.

* 29.8 The distant laminar wake

At short distances the velocity field is strongly dependent on the shape and attitude of a moving body, but far from the body such details are lost. It is, as we shall now see, possible to determine the general form of the laminar velocity field at large distances from the body in terms of the lift and drag that the body produces (see also [Landau and Lifshitz 1987]). The analysis in this section should be viewed as the natural continuation of d'Alembert's theorem to fluids that are not perfectly inviscid. Such fluids will not "close up" behind the moving body, but instead—as we have discussed above—leave a trailing wake, a disturbance that never dies out completely even at huge distance behind the body. In the real, unruly, and turbulent atmosphere, the trailing wake from a passing airplane will of course only be notable for a finite distance.

Oseen's approximation

Sufficiently far from the body, the velocity field is laminar and approximatively equal to the asymptotic value $\boldsymbol{U}$ both inside and outside the trailing wake. Inserting $\boldsymbol{v} = \boldsymbol{U} + \Delta\boldsymbol{v}$ and $p = p_0 + \Delta p$ into the steady flow Navier–Stokes equation without gravity,

$$(\boldsymbol{v}\cdot\nabla)\boldsymbol{v} = -\frac{1}{\rho_0}\nabla p + \nu\nabla^2\boldsymbol{v}, \tag{29.70}$$

we obtain to first order in $\Delta\boldsymbol{v}$,

$$(\boldsymbol{U}\cdot\nabla)\Delta\boldsymbol{v} = -\frac{1}{\rho_0}\nabla\Delta p + \nu\nabla^2\Delta\boldsymbol{v}. \tag{29.71}$$

The linearity of this equation allows us to superpose its solutions.

[3]This formula appears to be valid for all values of the angle of attack, but that is of course nonsense. It can only be used as long as the flow obeys the Kutta condition. At a not very large angle of attack, the boundary layers will detach and invalidate the whole calculation; and at the stall angle, the flow completely changes character.

Let us write the velocity difference as a sum,

$$\Delta \boldsymbol{v} = \boldsymbol{u} + \nabla\Psi, \tag{29.72}$$

where Ψ is a generalized velocity potential and $\boldsymbol{u}$ is the contribution from the vorticity in the trailing wake. Choosing Ψ as a solution to

$$(\boldsymbol{U}\cdot\nabla)\Psi = -\frac{\Delta p}{\rho_0} + \nu\nabla^2\Psi, \tag{29.73}$$

and using this equation to eliminate p in (29.71), it follows that the field $\boldsymbol{u}$ must satisfy

$$\boxed{(\boldsymbol{U}\cdot\nabla)\boldsymbol{u} = \nu\nabla^2\boldsymbol{u}.} \tag{29.74}$$

The incompressibility condition, $\nabla\cdot\Delta\boldsymbol{v} = 0$, yields a further relation,

$$\boxed{\nabla^2\Psi = -\nabla\cdot\boldsymbol{u}.} \tag{29.75}$$

This equation determines Ψ, given a solution $\boldsymbol{u}$ to (29.74), and then the pressure is obtained from (29.73).

Flow inside the wake

The trailing wake is assumed to be narrow compared to the distance to the body. In a coordinate system with the x-axis along the asymptotic velocity, $\boldsymbol{U} = U\hat{\boldsymbol{e}}_x$, we may assume that $x \gg |y|, |z|$ inside the wake. In the now-familiar way, it follows that the double x-derivative in the Laplacian of (29.74) is small compared with the y- and z-derivatives, and the equation for $\boldsymbol{u}$ becomes

$$U\frac{\partial\boldsymbol{u}}{\partial x} = \nu\left(\frac{\partial^2\boldsymbol{u}}{\partial y^2} + \frac{\partial^2\boldsymbol{u}}{\partial z^2}\right). \tag{29.76}$$

This is a standard diffusion equation of the same form as the momentum diffusion equation (15.5) with two transverse dimensions and "time" $t = x/U$. Note that t is also the time it takes for the asymptotic flow to reach the position x downstream from the body.

At distances much larger than the body size, $x \gg L$, the body appears as a point particle with no discernible shape, situated at the origin of the coordinate system. By insertion into the above equation, one may verify that the following expression is an exact "shapeless" solution,

$$\boldsymbol{u} = \frac{\boldsymbol{A}}{x}\exp\left(-U\frac{y^2+z^2}{4\nu x}\right), \tag{29.77}$$

where $\boldsymbol{A} = (A_x, A_y, A_z)$ is a constant vector. It is in fact also the most general solution at large downstream distance x (see Problem 29.11). Evidently, the distant wake has a Gaussian shape in the transverse directions with a narrow front width $\delta = \sqrt{4\nu x/U}$. The width of the laminar wake is, however, the same in both transverse directions, confirming that there is no imprint of the original shape of the object on the Gaussian form of the distant wake. For $|y|, |z| \lesssim \delta$, the solution decays as x^{-1} along the wake, rather than the expected $r^{-2} \approx x^{-2}$. This is consistent with the area of the wake being of order $\delta^2 \sim x$ such that the volume flux, that is, the integral of u_x over the cross-section of the wake, remains finite for $x \to \infty$. The terms that have been left out in the above solution by dropping the x-derivatives in the Laplacian are a further factor x^{-1} smaller than the above solution and cannot contribute in the limit.

The generalized potential is determined by solving (29.75). Consistently leaving out the double derivatives with respect to the Laplacian, it becomes

$$\frac{\partial^2 \Psi}{\partial y^2} + \frac{\partial^2 \Psi}{\partial z^2} = -\nabla \cdot \boldsymbol{u}. \tag{29.78}$$

On the right-hand side one cannot leave out the $\nabla_x u_x$ contribution to the divergence because it is only a factor $x^{-1/2}$ smaller than the others. It may—with some effort—be verified by insertion that the following potential is an exact solution to this equation inside the wake:

$$\Psi = \frac{2\nu}{U}\left(-\frac{A_y y + A_z z}{y^2 + z^2} + \frac{A_x}{2x}\right)\left(1 - \exp\left(-U\frac{y^2 + z^2}{4\nu x}\right)\right). \tag{29.79}$$

Here the first term in the first parentheses is of order $x^{-1/2}$ and the second of order x^{-1}. The leading corrections from leaving out the double x-derivatives in the Laplacian are of order $x^{-3/2}$. Only the exponential in the second parenthesis represents a true solution to the inhomogeneous equation (29.78), to which one may add an arbitrary solution to Laplace's equation $\nabla^2\Psi = 0$. Here we have added the solution that makes the potential non-singular for $y, z \to 0$.

Drag and lift

The pressure is obtained in the same approximation from (29.73),

$$\Delta p = -\rho_0 U \frac{\partial \Psi}{\partial x} + \rho_0 \nu \left(\frac{\partial^2 \Psi}{\partial y^2} + \frac{\partial^2 \Psi}{\partial z^2}\right) = \rho_0 \nu \frac{A_x}{x^2}. \tag{29.80}$$

Since it decays like x^{-2} and the area of the wake is $\delta^2 \sim x$, it cannot contribute to drag for $x \to \infty$, so that the leading contribution to the integrand of (29.45) becomes

$$\Delta p + \rho_0 U \Delta v_x \approx \rho_0 U u_x. \tag{29.81}$$

In the last step we have dropped the pressure and the potential derivative $\nabla_x \Psi \sim x^{-3/2}$, which are both negligible compared to $u_x \sim x^{-1}$. Integrating over all y, z, we find from (29.45)

$$\mathcal{D} = -\rho_0 U \iint u_x \, dy dz = -4\pi \rho_0 \nu A_x, \tag{29.82}$$

which fixes the coefficient A_x. The errors committed in extending the integral over the wake to all values of y and z are exponentially small.

The lift is obtained from the complete circulation integral (29.51),

$$\Gamma(z) = \oint_{C(z)} (\boldsymbol{u} + \nabla\Psi) \cdot d\boldsymbol{\ell} = \oint_{C(z)} \boldsymbol{u} \cdot d\boldsymbol{\ell} = -\int u_y dy.$$

In the second step we have used that Ψ is single-valued so that $\oint \nabla\Psi \cdot d\boldsymbol{\ell} = 0$, and in the third that $\boldsymbol{u}$ vanishes outside the wake. The minus sign stems from the contour running through the wake against the direction of the y-axis. Inserting this result into (29.50), we find

$$\mathcal{L} = -\rho_0 U \iint u_y \, dy dz = -4\pi \nu \rho_0 A_y, \tag{29.83}$$

which fixes A_y. Similarly, since there is no lift in the z-direction, we must have $A_z = 0$.

The complete three-dimensional field configuration inside the wake has now been obtained in terms of the lift and drag that the body generates:

$$\boxed{\begin{aligned} \boldsymbol{u} &= -\frac{(\mathcal{D},\mathcal{L},0)}{4\pi\rho_0\nu x}\exp\left(-U\frac{y^2+z^2}{4\nu x}\right), \\ \Psi &= \frac{1}{4\pi\rho_0 U x}\left(\frac{2xy}{y^2+z^2}\mathcal{L}-\mathcal{D}\right)\left(1-\exp\left(-U\frac{y^2+z^2}{4\nu x}\right)\right). \end{aligned}} \tag{29.84}$$

The correction terms are all of order x^{-1} relative to the leading terms.

Flow outside the wake

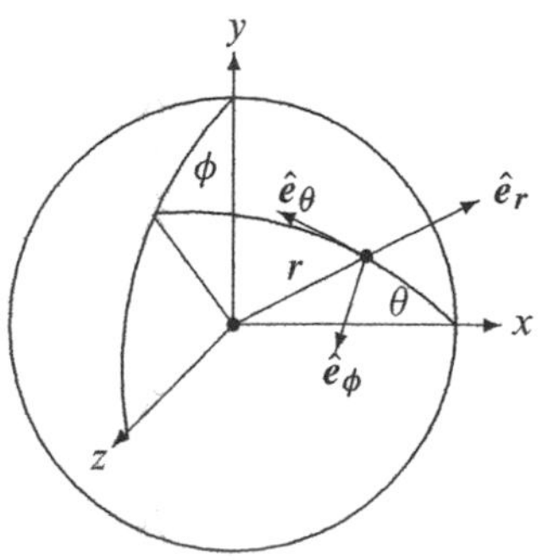

Spherical coordinates and their tangent vectors for the far field outside the wake.

Outside the wake, the flow is assumed to be irrotational with $\Delta \boldsymbol{v} = \boldsymbol{\nabla}\Psi$ and $\boldsymbol{\nabla}^2\Psi = 0$. We have before argued that $\Delta\boldsymbol{v} \sim 1/r^2$ at large distances, and consequently we must have $\Psi \sim 1/r$. In spherical coordinates with the polar axis in the x-direction and the null-meridian in the xy-plane, we may thus write

$$\Psi = \frac{F(\theta,\phi)}{r}. \tag{29.85}$$

The spherical Laplacian (D.40) implies that F has to satisfy

$$\left(\sin^2\theta\frac{\partial^2}{\partial\theta^2} + \cos\theta\sin\theta\frac{\partial}{\partial\theta} + \frac{\partial^2}{\partial\phi^2}\right)F = 0. \tag{29.86}$$

In view of the periodicity in ϕ, the complete solution may be written as a Fourier series,

$$F = A_0(\theta) + \sum_{n=1}^{\infty} A_n(\theta)\cos n\phi + B_n(\theta)\sin n\phi, \tag{29.87}$$

where the coefficients A_n and B_n satisfy the equation

$$\left(\sin^2\theta\frac{d^2}{d\theta^2} + \cos\theta\sin\theta\frac{d}{d\theta} - n^2\right)A_n = 0. \tag{29.88}$$

Surprisingly, this equation has a complete set of exact solutions $\left(\tan\frac{1}{2}\theta\right)^{\pm n}$ (see Problem 29.12). Since F has to be regular at $\theta = \pi$, where the tangent diverges, the exponents must be non-positive, $n \leq 0$. Furthermore, for $\theta \to 0$, where $A_n \sim \theta^{-n}$, this solution has to join continuously with the inside solution (29.84), which for $\delta \ll y, z \ll x$ behaves like

$$\Psi \approx \frac{1}{4\pi\rho_0 U x}\left(\frac{2xy}{y^2+z^2}\mathcal{L}-\mathcal{D}\right) \approx \frac{\mathcal{L}}{2\pi\rho_0 U}\frac{\cos\phi}{r\,\theta} - \frac{\mathcal{D}}{4\pi\rho_0 U r}. \tag{29.89}$$

This shows that the only possible exponents are $n = 0, 1$ with $A_0 = -\mathcal{D}/4\pi\rho_0 U$, $A_1 = \mathcal{L}/4\pi\rho_0 U$, and $B_1 = 0$. Thus, the potential far from the body outside the wake becomes

$$\boxed{\Psi = \frac{\mathcal{L}\cos\phi\cot\frac{1}{2}\theta - \mathcal{D}}{4\pi\rho_0 U r}.} \tag{29.90}$$

It joins continuously with the field inside the wake.

Pressure and lift

Since the leading contribution to pressure vanishes far downstream inside the wake, it plays no role in the drag, which as shown by (29.82) is entirely due to a loss of fluid momentum in the trailing wake. The lift contribution from pressure stems, on the other hand, entirely from the outside solution. Using Bernoulli's theorem and the spherical derivatives (D.32), the outside pressure becomes, $\Delta p = -\rho_0 U \nabla_x \Psi = -\rho_0 U(\cos\theta \nabla_r - \sin\theta \nabla_\theta)\Psi$. After a bit of algebra this reduces to

$$\boxed{\Delta p = -\frac{\mathcal{D} + \mathcal{L}\sin\theta\cos\phi}{4\pi r^2}.} \tag{29.91}$$

It is immediately clear that the spherically symmetric first term cannot contribute to the total pressure force. Also, since $dS_x = \cos\theta \cdot r^2 \sin\theta d\theta d\phi$, the second term, which is linear in $\cos\phi$, cannot produce a force in the x-direction, that is, a drag.

Due to its ϕ-dependence, the second term is, however, negative for $y > 0$ and positive for $y < 0$, and must therefore produce a lift. Using $dS_y = \sin\theta\cos\phi \cdot r^2 \sin\theta d\theta d\phi$, we find

$$-\oint \Delta p \, dS_y = \frac{\mathcal{L}}{4\pi}\iint \sin^3\theta \cos^2\phi \, d\theta d\phi = \frac{1}{3}\mathcal{L}. \tag{29.92}$$

Pressure thus produces one third of the lift, even for an infinite sphere. The remaining two thirds of lift stems from momentum flux in the trailing wake. As discussed before (page 532), the partition of lift between pressure and momentum flux depends crucially on the choice of integration surface at infinity.

Problems

29.1 The Concorde airliner has a powerplant of four engines that together develop 677 kN with afterburner. Its maximal take-off mass is 185,000 kg and its take-off speed 360 km h^{-1}. Ignore drag and estimate the **(a)** runway acceleration, **(b)** take-off time, and **(c)** runway length.

29.2 Show that the tangent to the bank angle in a horizontal banked turn is 2π times the ratio between the time it takes to fall freely from rest to velocity U (with no air resistance) divided by the time T it takes to make a complete turn.

29.3 The quoted take-off speed for the Cessna 150 with mass 681 kg is about 120 km h^{-1}, and the take-off length is 225 m in about 20 s. The engine power is maximally 75 kW. Assuming constant power and disregarding drag, **(a)** estimate the fraction of this power that is converted into thrust. **(b)** Estimate the take-off time and compare with the quoted value.

29.4 Show that the induced drag is smaller if the single trailing vortex is divided into a number of smaller vortices coming off the wing.

29.5 Determine how the calculations in Section 29.6 are modified when air is assumed to be a barotropic compressible fluid that asymptotically has constant density ρ_0.

29.6 Show that Equation (29.46) follows from Gauss' theorem. Hint: Dot with a constant vector.

29.7 Calculate lift and drag when it is not assumed that the wake is cut orthogonally to the asymptotic velocity (page 534).

*** 29.8** Show explicitly that the sheet vortex field (29.56) has circulation

$$\Gamma = \oint_C \boldsymbol{v} \cdot d\boldsymbol{\ell} = \oint_A d\Gamma,$$

where C is an arbitrary curve completely surrounding the airfoil outline A.

*** 29.9** Show that

$$\int_0^1 \frac{1}{(1-t)\sqrt{t(1-t)}} \log \frac{|t-x|}{1-x}\, dt = -2\pi$$

for $0 < x < 1$.

*** 29.10** Assume that the chord line beginning in $x = 0$ has vertical tangent. **(a)** Show that if the airfoil is smooth near $x = 0$, then

$$y_\pm(x) = \pm a\sqrt{x} + 2\lambda x + \mathcal{O}\left(x^{3/2}\right),$$

where a and λ are constants.

The simplest example of a thin wing camber function fulfilling the various conditions is therefore

$$y(x) = \tfrac{1}{2}(y_+(x) + y_-(x)) = 2\frac{\lambda}{c}x(c-x),$$

where λ sets the scale of the camber.

(b) Calculate the zero-lift angle α_0 for this airfoil as a function of λ.

*** 29.11** Consider the N-dimensional diffusion equation in the N real variables, $x_1, \dots, x_N$,

$$\frac{\partial F}{\partial t} = \sum_{n=1}^{N} \frac{\partial^2 F}{\partial x_n^2}.$$

(a) Show that with initial data $F(x,0) = F_0(x)$, where $x = (x_1, \dots, x_N)$, the solution at time t is

$$F(x,t) = (4\pi t)^{-N/2} \int F_0(y) \exp\left(-\frac{(x-y)^2}{4t}\right) d^N y,$$

where $(x-y)^2 = \sum_n (x_n - y_n)^2$.

(b) Show that if $F_0(y)$ is bounded (or decreases at least as rapidly as a Gaussian for $|y| \to \infty$), the solution for $t \to \infty$ is

$$F(x,y) = (4\pi t)^{-N/2} \exp\left(-\frac{x^2}{4t}\right) \int F_0(y)\, d^N y.$$

*** 29.12** Find all solutions to

$$\left(\sin^2\theta \frac{d^2}{d\theta^2} + \cos\theta \sin\theta \frac{d}{d\theta} - n^2\right) f = 0.$$

Hint: Use the variable $\xi = \log\tan(\theta/2)$.

30 Convection

Convection is a major driving agent behind most of the weather phenomena in the atmosphere, from ordinary cyclones to hurricanes, thunderstorms, and tornadoes. Continental drift on Earth as well as transport of heat from the inside of a star to the surface are also driven by convection. On a smaller scale we use convection in the home to create a natural circulation that transports heat around the rooms from the radiators of the central heating system. The circulation of water in the central heating system also used to be driven by convection but is today mostly driven by pumps.

The mechanism behind convection rests on a combination of material properties and gravity. Most fluids, also those we call incompressible, tend to expand when the temperature is raised, leading to a slight decrease in density. Were it not for gravity, the minuscule changes in density caused by local temperature variations would be of very little consequence, but gravity makes the warmer and lighter fluid buoyant relative to the colder and heavier, and the buoyancy forces will attempt to set the fluid into motion. Concentration gradients in mixed fluids can also drive convective flow. We shall in this book reserve the word "convection" to denote a flow driven by temperature differences in conjunction with buoyancy (in many texts called "free" convection), whereas "advection" is used to denote transport of any quantity—also heat—by a flow driven by other forces (also called"forced" convection; see Chapter 22). In practical situations, both mechanisms are of course often at play.

In this chapter we shall first discuss some examples of steady laminar convection flows driven by time-independent temperature differences on the container boundaries. Afterward we shall address the thermal instabilities characterizing the onset of convection. Of particular interest are the so-called Rayleigh–Bénard instabilities in a horizontal layer of fluid heated from below.

30.1 Heat-driven convection

Convection is caused by buoyancy forces in combination with the tendency for most materials to expand when heated. The coupled partial differential equations controlling the interplay of heat and motion are generally so complex that exact solutions are completely out of the question. Numeric simulations are, however, possible and used wherever practical problems have to be solved. Analytic insight into convection is mainly obtained from an approximation developed by Boussinesq in 1902.

The Boussinesq approximation

The proper derivation of the Boussinesq approximation was carried out in Chapter 22 (see page 390). Here we shall quickly rederive and to some extent justify it. The main ingredient is the (nearly) universal expansion of fluids subject to a temperature increase. Quantitatively, it is described by the *thermal expansivity* parameter α, which stands for the relative volume expansion (or density reduction) per unit of temperature, and typically is of order $10^{-3}\ \mathrm{K}^{-1}$ (see page 386).

Initially the nearly incompressible fluid is in hydrostatic equilibrium with uniform temperature T_0 and uniform density ρ_0 in a constant gravitational field $\boldsymbol{g}_0 = -g_0\hat{\boldsymbol{e}}_z$. The hydrostatic pressure is as usual, $p_0 - \rho_0 g_0 z$. If the fluid is unconfined and able to expand freely, the density change caused by raising the temperature locally by $\Delta T = T - T_0$ becomes $\Delta\rho = -\alpha\Delta T\rho_0$ when $|\alpha\Delta T| \ll 1$. This density change gives, in turn, rise to a local buoyancy contribution $\Delta\rho\boldsymbol{g}_0 = -\alpha\Delta T\rho_0\boldsymbol{g}_0$, which is added to the right hand side of the incompressible Navier–Stokes equation (22.31a) on page 380. To get rid of the hydrostatic pressure, we set $p = p_0 - \rho_0 g_0 z + \Delta p$ in (22.31a) and drop heat production and dissipation in the heat equation (22.31b) to arrive at the *Boussinesq approximation* for convective mass and heat flow:

$$\frac{\partial\Delta T}{\partial t} + (\boldsymbol{v}\cdot\nabla)\Delta T = \kappa\nabla^2\Delta T, \tag{30.1a}$$

$$\frac{\partial\boldsymbol{v}}{\partial t} + (\boldsymbol{v}\cdot\nabla)\boldsymbol{v} = -\alpha\Delta T\boldsymbol{g}_0 - \frac{\nabla\Delta p}{\rho_0} + \nu\nabla^2\boldsymbol{v}, \tag{30.1b}$$

$$\nabla\cdot\boldsymbol{v} = 0. \tag{30.1c}$$

If you feel uneasy about this quick derivation, in particular the dismissal of the dissipative term in the heat equation, consult the proper treatment in Chapter 22.

Case: Steady convection in open vertical slot heated on one side

We have previously (page 376) discussed steady heat flow in a fluid at rest between two plates, one of which was situated at $x = 0$ with temperature T_0 and the other at $x = d$ with higher temperature $T_1 = T_0 + \Theta$. We found there that the temperature rises linearly across the slot.

If the plates are vertical, buoyancy forces will act on the heated fluid and unavoidably set it into motion. For definiteness, the plates are assumed to be large but finite with the openings at the top and bottom connected to a reservoir of the same fluid at the same temperature T_0 as the cold plate. This provides the correct hydrostatic pressure at the top and bottom of the slot, a pressure that is necessary to prevent the fluid in the slot from "falling out" under its own weight, even before it is heated. Heating will cause the fluid in the slot to rise and merge with the fluid in the reservoir, and it seems reasonable to expect that a steady flow of heat and fluid may come about, in which the fluid rises fastest near the warm plate.

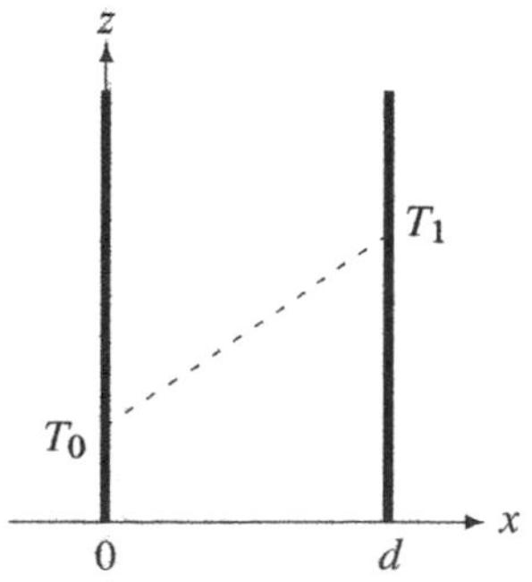

Steady heat flow between vertical parallel plates held at different temperatures ($T_1 > T_0$). The temperature rises linearly between the plates, $T = T_0 + x(T_1 - T_0)$, as shown by the dashed line.

From the planar symmetry of the configuration, we expect that the steady velocity field is vertical everywhere, and that it and the temperature field depend only on x,

$$\boldsymbol{v} = (0, 0, v_z(x)), \qquad \Delta T = \Delta T(x). \tag{30.2}$$

Under these assumptions, there will be no advective contribution to (30.1a), which reduces $\nabla_x^2\Delta T = 0$. The temperature thus varies linearly with x across the slot, and using the boundary conditions, we get

$$\Delta T = \Theta\frac{x}{d}, \tag{30.3}$$

the same result as in the static case.

It also follows that the advective term in (30.1b) vanishes, and from the x- and y-components of this equation we conclude that $\nabla_x \Delta p = \nabla_y \Delta p = 0$, such that the pressure can only depend on z. From the z-component we then get

$$\frac{1}{\rho_0}\nabla_z \Delta p(z) = \nu \nabla_x^2 v_z(x) + \alpha\Theta\frac{x}{d}g_0.$$

Since the left-hand side depends only on z and the right-hand side only on x, both sides of this equation are constant; and since the pressure excess Δp must vanish at the top and bottom of the slot, it must vanish everywhere, $\Delta p = 0$. Using the no-slip boundary conditions on the plates, the solution becomes

$$\boxed{v_z = \frac{\alpha\Theta g_0 d^2}{6\nu}\frac{x}{d}\left(1 - \frac{x^2}{d^2}\right).} \tag{30.4}$$

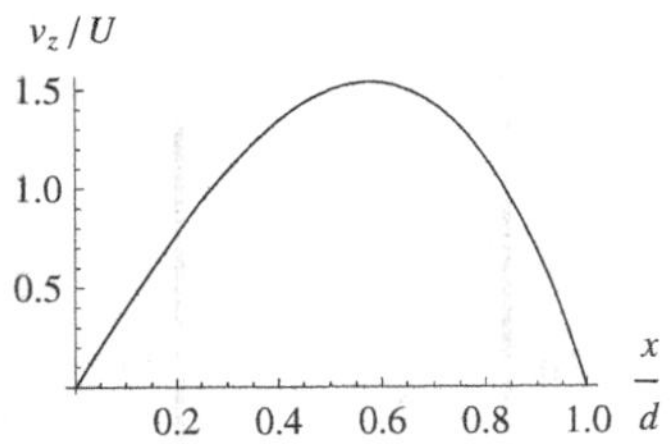

Convective velocity profile in units of the width of the slot and the average velocity. Its shape is reminiscent of the pressure-driven planar flow (16.5) although skewed a bit toward the right, that is, the hot plate.

Note that the steady flow pattern is independent of the heat diffusivity κ because ΔT is independent of it. The velocity field scale is set by viscosity rather than by heat conductivity.

The maximal velocity in the slot is found a bit to the right of the middle, at $x = d/\sqrt{3}$, and the average velocity in the slot becomes

$$U = \frac{1}{d}\int_0^d v_z(x)\,dx = \frac{\alpha\Theta g_0 d^2}{24\nu}. \tag{30.5}$$

The Reynolds number corresponding to this velocity is

$$\mathsf{Re} = \frac{Ud}{\nu} = \frac{1}{24}\cdot\frac{\alpha\Theta g_0 d^3}{\nu^2}. \tag{30.6}$$

The second factor on the right is called the *Grashof number* and denoted Gr.

Franz Grashof (1826–1893). German engineer who sought to transform machine building into a proper science.

Example 30.1 [Slot with water or air]: A temperature difference $\Theta \approx 10\,\text{K}$ is imposed on a water-filled slot of width $d \approx 1$ cm. The mean velocity becomes $U \approx 12\ \text{cm s}^{-1}$, and the corresponding Reynolds number is $\mathsf{Re} \approx 1{,}400$, indicating that the flow should be laminar, as was assumed implicitly in the above calculation. If the slot instead contains air, the velocity is $8\ \text{cm s}^{-1}$ and the Reynolds number 50.

Entry length for convection

The rate at which heat is transported by convection from the slot into the reservoir may be calculated from the internal enthalpy carried by the fluid as it exits the slot,

$$\dot{Q} = \int_0^d \rho_0 c_p \Delta T\, v_z L dx = \frac{\rho_0 c_p \alpha\Theta^2 g_0 d^3 L}{45\nu}, \tag{30.7}$$

where L is the size of the slot in the y-direction. This raises a puzzle because the temperature gradient is constant, $\nabla_x \Delta T = \Theta/d$, across the slot, and Fourier's law (22.17) on page 376 then implies that the same amount of heat is added to the fluid at the cold plate as is removed from the hot. Consequently, no net heat is added to the fluid from the plates, in blatant contradiction with the above calculation and with common experience.

What is wrong is the assumption that the solutions (30.3) and (30.4) are valid at the bottom of the slot where the fluid enters. Here the temperature gradient cannot be constant, because mass conservation in the steady state forces the cold fluid to enter with the same average velocity U across the slot. Since it takes a certain amount of time, $t \approx d^2/4\kappa$, for the heat

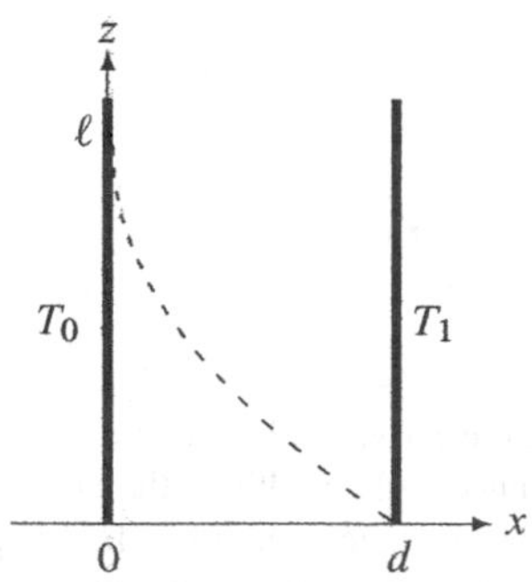

The heat front (dashed) at the entry to the vertical slot. The heat transferred to the fluid at the warm plate must diffuse across the slot but is at the same time advected upward with average velocity U. It makes contact with the cold plate after having moved a vertical distance ℓ, called the entry length for heat.

supplied by the warm plate to diffuse across the slot (see Equation (22.23) on page 378), the fluid will have moved through a vertical distance

$$\ell \approx Ut = \frac{\alpha\Theta g_0 d^4}{96\kappa\nu} \tag{30.8}$$

before the heat comes into contact with the cold plate. For consistency we should compare the heat loss at the exit (30.7) with the total rate of heat transferred into the fluid in the entry region. Applying Fourier's law to the entry area $L\ell$, the heat transfer is estimated as $\dot{Q} = \ell L k \nabla_x T \approx \ell L \cdot k\,\Theta/d$, and using $k = \rho_0 c_p \kappa$ this becomes indeed of the same magnitude as the heat loss (30.7) at the exit. The conclusion is that heat is only added to the fluid in the entry region.

The entry length may also be written as

$$\ell \approx \frac{\mathsf{Ra}}{96} d, \tag{30.9}$$

where the dimensionless quantity

$$\mathsf{Ra} = \frac{\alpha\Theta g_0 d^3}{\nu\kappa} \tag{30.10}$$

is the famous *Rayleigh number*. In units of the slot width, the relative entry length for heat, ℓ/d, is about 1% of the Rayleigh number, whereas the viscous entry length (16.13) on page 265 in units of the slot width is estimated to be about 6% of the Reynolds number.

Example 30.2 [Slot with water or air, continued]: In Example 30.1 the Rayleigh number for water becomes $\mathsf{Ra} = 2 \times 10^5$, and we get a huge entry length $\ell \approx 21$ m, whereas for air we find $\mathsf{Ra} \approx 900$ and a much more manageable $\ell \approx 10$ cm. For comparison, the viscous entry length is 82 cm in water and 3 cm in air.

* Case: Thermal boundary layer

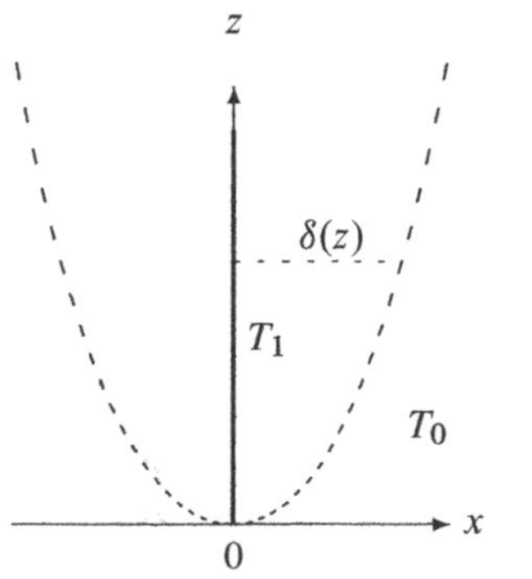

Outline of thermal boundary layers forming on both sides of a thin plate with constant temperature $T_1 = T_0 + \Theta$, placed vertically in an infinite sea of fluid originally at rest with temperature T_0. The layer has a z-dependent thickness $\delta(z)$.

If the slot height is much smaller than the entry length , $h \ll \ell$, the heated fluid will never reach the cold plate before it exits from the slot. In this limit, the cold plate can be ignored, and the appropriate model is instead that of a warm vertical plate with constant temperature placed in a sea of cold fluid. The heated fluid rising along the plate then forms a *thermal boundary layer*, and we shall now determine the steady laminar flow pattern in such a boundary layer by combining the Boussinesq approximation with Prandtl's boundary layer approximation (Section 28.2 on page 485).

The coordinate system is chosen with vertical z-axis along the plate. Replacing ℓ by z and d by $\delta(z)$ in the estimate of the heat entry length (30.8), we obtain an estimate of the z-dependent thickness of the boundary layer (apart from a dimensionless numerical factor),

$$\delta(z) \sim \left(\frac{\kappa\nu z}{\alpha\Theta g_0}\right)^{1/4}. \tag{30.11}$$

Since for $z \to \infty$ we have $\delta/z \propto z^{-3/4}$, the boundary layer may indeed be viewed as thin, except for a region near the lower edge of the plate.

Under these circumstances we may apply the Prandtl formalism to the Boussinesq equations (30.1) and discard the double derivative with respect to z in the Laplace operators, as well as the pressure excess Δp. These simplifications lead to the following coupled (Boussinesq–Prandtl) equations for the velocity field, $\boldsymbol{v} = (v_x, 0, v_z)$, and the temperature

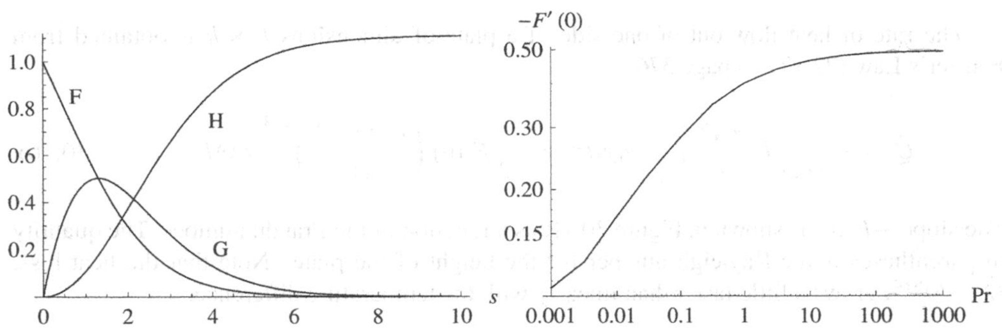

Figure 30.1. Left: Structure of the self-similar thermal boundary layer for $\mathsf{Pr} = 1$. Note that asymptotically for $s \to \infty$ there is a horizontal flow toward the plate that feeds the convective upflow. **Right:** The heat slope at the plate, $-F'(0)$, as a function of the Prandtl number. Notice the doubly logarithmic plot that reveals a power law (straight line) for small Prandtl numbers (the wiggles reflect numeric precision errors).

excess, $\Delta T = T - T_0$:

$$(v_x \nabla_x + v_z \nabla_z)\Delta T = \kappa \nabla_x^2 \Delta T, \tag{30.12a}$$

$$(v_x \nabla_x + v_z \nabla_z) v_z = \nu \nabla_x^2 v_z + \alpha \Delta T g_0, \tag{30.12b}$$

$$\nabla_x v_x + \nabla_z v_z = 0. \tag{30.12c}$$

These equations must be solved with the boundary conditions that $\Delta T = \Theta$ and $v_x = v_z = 0$ for $x = 0$, and $\Delta T, v_z \to 0$ for $x \to \infty$.

Since there is no other possible length scale for x than $\delta(z)$ we shall assume that the fields only depend on x through the variable $x/\delta(z)$, or

$$s = \left(\frac{\alpha \Theta g_0}{\kappa \nu z}\right)^{1/4} x. \tag{30.13}$$

The fields are parametrized with dimensionless functions of this dimensionless variable times suitable dimensional prefactors,

$$\Delta T = \Theta F(s), \qquad v_z = \sqrt{\alpha \Theta g_0 z}\, G(s), \qquad v_x = -\left(\frac{\alpha \Theta \kappa \nu g_0}{z}\right)^{1/4} H(s). \tag{30.14}$$

The field equations become coupled differential equations in s alone,

$$4F'' + \sqrt{\mathsf{Pr}}\,(sG + 4H)F' = 0, \tag{30.15a}$$

$$4\sqrt{\mathsf{Pr}}\,G'' + (sG + 4H)G' - 2G^2 + 4F = 0, \tag{30.15b}$$

$$4H' + sG' - 2G = 0, \tag{30.15c}$$

where $\mathsf{Pr} = \nu/\kappa$ is the Prandtl number. These equations can be solved numerically with the boundary conditions $F(0) = 1$, $G(0) = H(0) = 0$, and $F(\infty) = G(\infty) = 0$. The result is shown in Figure 30.1L for $\mathsf{Pr} = 1$. Interestingly, the solution has an asymptotic horizontal flow toward the plate (represented by $H(\infty) = 1.10941\ldots$ for $\mathsf{Pr} = 1$), rather than a vertical upflow from below, as might have been expected. The divergence of the horizontal velocity $v_x \sim -H(\infty) z^{-1/4}$ for $z \to 0$ is a spurious consequence of the Prandtl approximation.

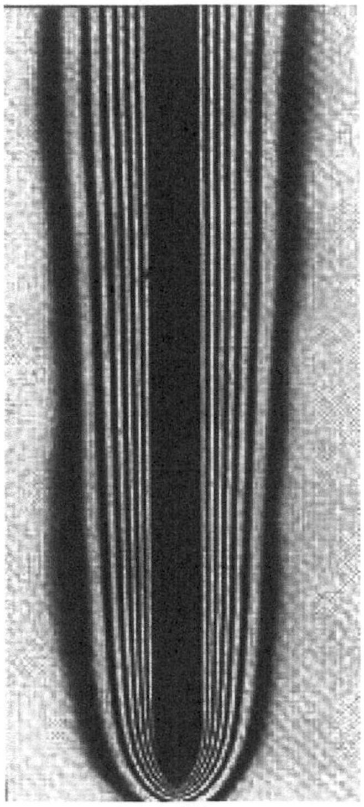

Photograph of isotherms around a thermal boundary layer in air. The Grashof number is about $\mathsf{Gr} \approx 5 \times 10^6$. (Source: Reproduced from [ES48].)

The rate of heat flow out of one side of a plate of dimensions $L \times h$ is obtained from Fourier's Law (22.17) on page 376

$$\dot{Q} = -\int_{L\times h} k \left.\frac{\partial \Delta T}{\partial x}\right|_{x=0} dy dz = -\frac{4}{3} F'(0) \left(\frac{\alpha \Theta g_0 h^3}{\kappa \nu}\right)^{1/4} k\, \Theta L. \tag{30.16}$$

The slope $-F'(0)$ is shown in Figure 30.1R as a function of the Prandtl number. The quantity in parentheses is the Rayleigh number for the height of the plate. Note that the heat loss, $\dot{Q} \sim \Theta^{5/4}$, grows a little faster than linearly with the temperature difference.

Example 30.3 [Heat radiator]: A heat radiator consisting of a single plate of height $h = 70$ cm and width $L = 1$ m is kept at $T = 65°$C and placed in a room at $T_0 = 20°$C. For air we have Pr $= 0.73$ and $F'(0) = -0.388$ leading to a total heat flow $\dot{Q} = 235$ W from the radiator (including both sides). The maximal vertical velocity $U = \max v_z$ is quite naturally found at the top of the radiator, $z = h$, and inspection of Figure 30.1L yields $U = \max_x v_z(x, h) \approx 0.5\sqrt{\alpha \Theta g_0 h} \approx 50$ cm s^{-1} at $s = 1.5$, corresponding to a boundary layer width $\delta \approx 1$ cm. The horizontal Reynolds number becomes $\mathsf{Re}_\delta = U\delta/\nu \approx 340$, so there should be no turbulence in this boundary layer. The asymptotic horizontal inflow is merely $v_x|_{x\to\infty} \approx -1$ cm s^{-1} at $z = 5$ cm above the bottom of the radiator.

Example 30.4 [Human heat loss]: The skin surface area of a naked grown-up human is $A \approx 2$ m^2, and standing up the height is $h \approx 2$ m and the width $L \approx 1$ m. Taking $\Theta = 10$ K we find the heat loss $\dot{Q} \approx 40$ W. Since this is less than half the heat production of 100 W, a human being should be able to maintain a skin temperature of 27°C in calm air at 17°C. As soon as there is wind, the temperature gradient at the skin becomes much larger, and the increased heat loss rapidly begins to chill the body (see Example 28.2 on page 491).

30.2 Convective instability

Fluids with horizontal temperature variations, such as the vertical slot discussed above, cannot remain in hydrostatic equilibrium but must immediately start to convect (see Problem 30.1). The situation is, however, quite different if the fluid is subject to vertical temperature variations, $T = T(z)$. In the following argument we assume that the fluid is nearly incompressible, so that we do not run into problems with the natural (homentropic) temperature lapse of the "atmospheric" kind, discussed in Section 2.6.

If the ambient temperature increases with height ($dT/dz > 0$), hydrostatic equilibrium is stable because a blob of fluid that is quickly displaced upward will arrive with lower temperature and higher density than its new surroundings, and thus experience a downward residual weight, tending to bring it down again. Hydrostatic equilibrium may, however, not be stable if the temperature falls with height ($dT/dz < 0$) because a blob that is suddenly displaced upward into a region of lower temperature will arrive with lower density than its new surroundings and consequently experience an upward buoyancy force that tends to drive it further up. Were it not for drag and heat loss, a buoyant blob would rise with ever-increasing velocity.

Drag from the surrounding fluid grows with the upward blob velocity, linearly to begin with, quadratically later. Conductive heat loss lowers the excess temperature of the blob and thereby its buoyancy. Both of these effects are proportional to the surface area of the blob whereas buoyancy is proportional to the volume. Consequently, we expect that large blobs of fluid tend to be more unstable and rise faster than small. This also indicates that there is a critical blob size below which blobs are not capable of rising at all.

Here we shall estimate the critical blob size, and see that it may be expressed as a critical Rayleigh number. In the following sections we shall calculate the critical value of the Rayleigh number for the onset of instability in a horizontal slot under various conditions.

Stability estimate for spherical blob of fluid

For simplicity we begin with an incompressible fluid at rest in constant gravity g_0 with a constant negative vertical temperature gradient, $G = -dT/dz$, so that the temperature field is of the form $T(z) = T_0 - Gz$. A constant gradient can, as we have seen, be created in a horizontal slot with a fixed temperature difference between the lower and upper plates.

Imagine now that a spherical blob of fluid of radius a is set into upward motion with a small *steady* velocity $U > 0$. This is, of course, a thought experiment, and we do not speculate on the technological difficulties in creating and maintaining such a blob. While it slowly rises toward lower and lower temperatures, the warmer blob will transfer its excess heat to the colder environment over a typical diffusion time $t \sim a^2/\kappa$. In this time the blob rises through the height $\Delta z \approx Ut \sim Ua^2/\kappa$, and the environment cools by $\Delta T \sim -G\Delta z \sim -GUa^2/\kappa$. In the steady state, the competition between the falling temperature of the environment and the loss of heat from the blob should lead to a time-independent temperature excess of size ΔT, which in turn determines the buoyancy force.

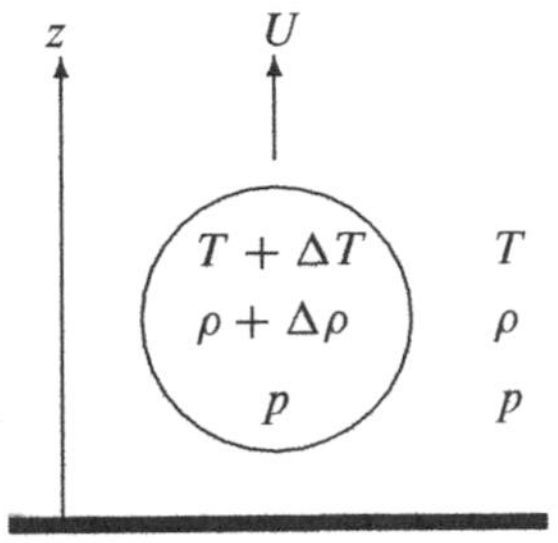

A spherical blob of fluid moving upward with constant velocity U. If the temperature gradient is negative ($dT/dz < 0$), the temperature of the moving blob will be larger than its surroundings ($\Delta T > 0$). The pressure is assumed to be the same inside and outside the blob.

Unfortunately the above estimate of ΔT is a bit weak because advection of heat for sufficiently small velocity U will be negligible compared to diffusion. The heat escaping from the blob will quickly spread far beyond the blob radius and thereby raise the temperature of the environment. The rising sphere thus finds itself surrounded by a "cocoon" of fluid (of its own making) that is warmer than the environment and therefore provides smaller buoyancy than would be the case if the temperature of the environment reigned all the way to the surface of the sphere. In order to calculate the buoyancy force we must know the temperature distribution inside the blob relative to the temperature at its surface.

It is most convenient to go to the rest frame of the blob where the flow outside the blob is steady, although the temperature of the environment is dropping at a constant rate. The true temperature field inside the blob at rest must then be of the form

$$T' = T_0 - G(z + Ut) + \Delta T, \tag{30.17}$$

where by assumption the temperature excess field ΔT is time independent. Inside the blob, T' must obey Fourier's heat equation (22.18) on page 376 for fluids at rest, which under the given assumptions becomes

$$-GU = \kappa\nabla^2\Delta T. \tag{30.18}$$

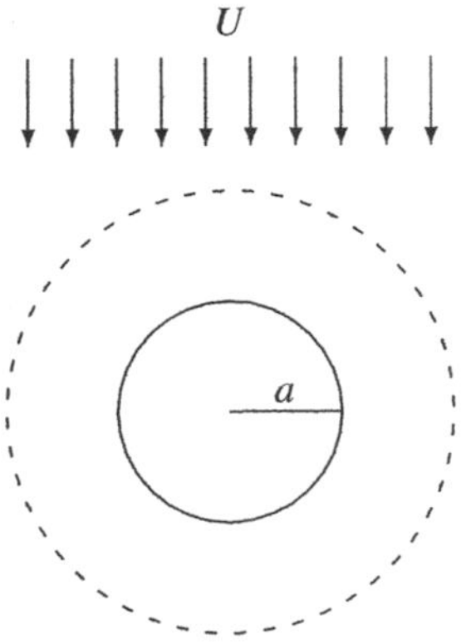

The heat front reaches far beyond the blob radius when the blob velocity is tiny. Here the blob is viewed in its rest frame, where asymptotically there is a uniform downward wind U carrying a decreasing temperature.

Seeking a spherical solution we find $\Delta T = -(GU/6\kappa)r^2 + \text{const}$, and the difference between the temperature inside the blob and on its surface becomes

$$\delta T = \Delta T - \Delta T|_{r=a} = \frac{GU}{6\kappa}(a^2 - r^2). \tag{30.19}$$

This is indeed of the same order of magnitude as the previous estimate $\Delta T \sim GUa^2/\kappa$. The total upward buoyancy force is obtained from the integrated density change $\delta\rho = -\alpha\delta T\rho_0$ inside the blob,

$$\mathcal{F}_B = \int_V \delta\rho(-g_0)dV = \rho_0 g_0 \alpha \int_0^a \delta T(r) 4\pi r^2\, dr = \frac{4\pi\rho_0 g_0 \alpha G a^5 U}{45\kappa}. \tag{30.20}$$

Evidently, the buoyancy grows like the fifth power of the radius because its volume grows like the third power and the diffusion time like the second.

The viscous drag on a solid sphere in slow steady motion is given by Stokes' law (17.25) on page 292. Disregarding any internal flow in the blob, we estimate that

$$\mathcal{F}_D = 6\pi\eta a U. \tag{30.21}$$

This is valid for small Reynolds number $\mathsf{Re} = 2aU/\nu \ll 1$, a condition that is always fulfilled when the blob starts to rise from rest.

Notice that both buoyancy and drag are proportional to the velocity so that the competition between them is "fair". If the buoyancy is smaller than the drag, $\mathcal{F}_B < \mathcal{F}_D$, the sphere cannot continue to rise on its own. In dimensionless form, the *stability condition* becomes

$$\boxed{\frac{\mathcal{F}_B}{\mathcal{F}_D} = \frac{2}{135}\frac{g_0\alpha G a^4}{\kappa\nu} < 1,} \tag{30.22}$$

where $\nu = \eta/\rho_0$ is the kinematic viscosity (momentum diffusivity) of the fluid. Evidently, this inequality puts an upper limit on the size of stable blobs.

The critical Rayleigh number

In terms of the blob diameter $d = 2a$, the stability condition (30.22) may be written as a condition on the Rayleigh number,

$$\mathsf{Ra} \equiv \frac{g_0\alpha G d^4}{\kappa\nu} < 1{,}080. \tag{30.23}$$

Tracing back over the preceding calculation, we see that the large critical value $\mathsf{Ra}_c = 1{,}080$ on the right-hand side is mainly due to the "cocoon" of warm fluid carried along with the blob and diminishing its buoyancy. The critical Rayleigh number for blobs of general globular shape may always be taken to be around 1,000; whereas blobs with radically different shapes, for example long cylinders, will have quite different critical Rayleigh numbers, although typically they will be large.

If the geometry of a fluid container cannot accommodate blobs larger than a certain diameter d, and if the Rayleigh number for this diameter is below the critical value, the fluid in the container will be stable with the given negative temperature gradient. The critical Rayleigh number depends, however, strongly on the geometry of the container, and cannot in general be calculated analytically. In the following sections we shall determine it for the simplest of all geometries, the horizontal slot.

Estimate of terminal blob speed

If the Rayleigh number for a blob is larger than the critical value, $\mathsf{Ra} > \mathsf{Ra}_c$, the blob will accelerate upward with ever larger speed. For large Reynolds numbers, form drag on a sphere, $\mathcal{F}_D \approx \frac{1}{4}\rho_0\pi a^2 U^2$ (see page 294), grows quadratically with velocity, and will eventually overtake the buoyancy force $\mathcal{F}_B$. The terminal speed, determined by solving $\mathcal{F}_D = \mathcal{F}_B$, becomes, for $\mathsf{Ra} \gg \mathsf{Ra}_c$,

$$U \approx \sqrt{\frac{2}{45}\frac{g_0\alpha G d^3}{\kappa}}, \tag{30.24}$$

where d is the blob diameter. The shape of a rising blob is, however, like a raindrop strongly influenced by the motion, so this is only a coarse estimate.

Example 30.5 [Pot of water]: Consider a pot of water of depth $h = 10$ cm on a warm plate held at 50°C in a room with temperature 20°C. The temperature difference is $\Theta \approx 30$ K and the gradient $G = \Theta/h = 300\,\text{K m}^{-1}$. The stability limit for spherical blobs is then obtained from (30.23) and becomes $d \lesssim 3.7$ mm. Convective currents are thus expected to arise spontaneously everywhere in the pot. A water blob with diameter $d = 1$ cm will reach a terminal speed of $U \approx 23\text{ cm s}^{-1}$, in reasonable agreement with daily experience.

If instead there is heavy porridge in the pot with heat properties like water but kinematic viscosity, say $\nu \approx 1\text{ m}^2\text{ s}^{-1}$, the critical diameter becomes $d \approx 12$ cm. Such blobs cannot find room in the container, and only little convection is expected.

* 30.3 Linear stability analysis of convection

The onset of instability in a dynamical system is usually determined by linearizing the dynamical equations around a particular "baseline" state that may or may not be stable. The solutions to the linearized dynamics represent the possible fluctuations around the baseline state; and if no fluctuation can grow indefinitely with time, the baseline state is said to be stable. Conversely, the existence of a single run-away fluctuation indicates that the baseline state is unstable. In the space of parameters that control the system, the condition that all fluctuations be damped leads to an inequality like (30.23), which in the limit of equality defines a *critical surface*, separating the stable region in parameter space from the unstable.

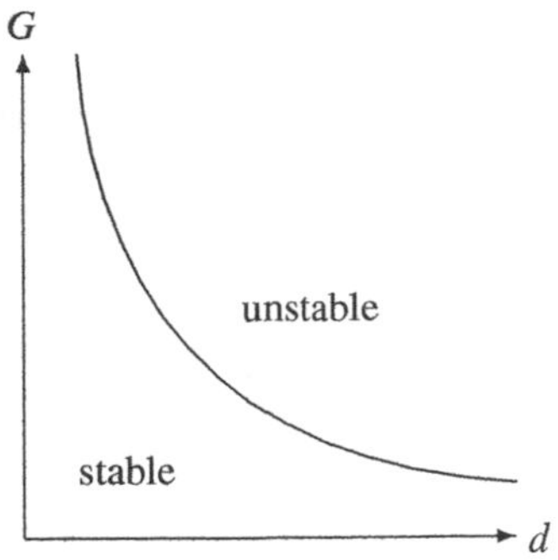

Sketch of the stability plot for heat convection in a system with negative temperature gradient G and size d. The critical surface is $G \sim d^{-4}$.

Linearized dynamics of flow and heat

In the case of an infinitely extended horizontal slot, the baseline state is simply a nearly incompressible fluid at rest in hydrostatic equilibrium with a vertical temperature distribution of constant negative gradient, $T = T_0 - Gz$. Assuming constant expansivity α, the pressure must obey the equations of hydrostatic equilibrium (2.23) on page 28 with density $\rho = \rho_0(1 - \alpha(T - T_0)) = \rho_0(1 + \alpha G z)$ of the heated fluid. Solving these equations we find $p = p_0 - \rho_0 g_0 z - \frac{1}{2}\rho_0 g_0 \alpha G z^2$, where p_0 is the pressure at $z = 0$.

A tiny velocity perturbation $\boldsymbol{v}$ will generate tiny corrections, ΔT and Δp, so that the true temperature and pressure fields become

$$T = T_0 - Gz + \Delta T, \qquad p = p_0 - \rho_0 g_0 z - \tfrac{1}{2}\rho_0 g_0 \alpha G z^2 + \Delta p.$$

To first order in the small quantities ΔT, Δp, and $\boldsymbol{v}$, the linearized dynamic equations become (under the same considerations as for the Boussinesq equations)

$$\frac{\partial \Delta T}{\partial t} - G v_z = \kappa \nabla^2 \Delta T,$$
$$\frac{\partial \boldsymbol{v}}{\partial t} = -\frac{\nabla \Delta p}{\rho_0} + \nu \nabla^2 \boldsymbol{v} + \alpha \Delta T g_0 \hat{\boldsymbol{e}}_z, \tag{30.25}$$
$$\nabla \cdot \boldsymbol{v} = 0.$$

These five coupled partial linear differential equations should now be solved for the five fluctuation fields, ΔT, Δp, and $\boldsymbol{v}$, with the appropriate boundary conditions for the particular geometry under study.

Fourier transformation

Fourier transformation is the method of choice for solving homogeneous linear partial differential equations with constant coefficients. All fields are assumed to be superpositions of elementary harmonic waves of the form $\exp(\lambda t + i\boldsymbol{k} \cdot \boldsymbol{x})$, where $\boldsymbol{k}$ is a real wave vector and λ may be a complex number. For a single harmonic wave with amplitudes $\Delta\tilde{T}, \Delta\tilde{p}$, and $\tilde{\boldsymbol{v}}$, we obtain from the linearized dynamics,

$$\lambda \Delta\tilde{T} - G\tilde{v}_z = -\kappa k^2 \Delta\tilde{T}, \tag{30.26a}$$
$$\lambda \tilde{\boldsymbol{v}} = -\frac{i\boldsymbol{k}}{\rho_0}\Delta\tilde{p} - \nu k^2 \tilde{\boldsymbol{v}} + \alpha \Delta\tilde{T} g_0 \hat{\boldsymbol{e}}_z, \tag{30.26b}$$
$$\boldsymbol{k} \cdot \tilde{\boldsymbol{v}} = 0. \tag{30.26c}$$

These algebraic equations should now be solved for the Fourier amplitudes of the fields.

Solving the first equation for $\tilde{v}_z$, and dotting the second equation with $\boldsymbol{k}$ (using the third), we obtain

$$\tilde{v}_z = \frac{\lambda + \kappa \boldsymbol{k}^2}{G} \Delta \tilde{T}, \qquad \frac{\Delta \tilde{p}}{\rho_0} = -\alpha \Delta \tilde{T} g_0 \frac{i k_z}{\boldsymbol{k}^2}.$$

Inserting this into the z-component of (30.26b), we find a linear equation for $\Delta \tilde{T}$, which only has a non-trivial solution for

$$(\lambda + \nu \boldsymbol{k}^2)(\lambda + \kappa \boldsymbol{k}^2) = \alpha G g_0 \left(1 - \frac{k_z^2}{\boldsymbol{k}^2}\right). \tag{30.27}$$

This equation expresses simply that the determinant of the system of five homogeneous linear algebraic equations (30.26) must vanish.

Being a quadratic equation in λ, it always has two roots:

$$\lambda = -\frac{1}{2}(\nu + \kappa)\boldsymbol{k}^2 \pm \frac{1}{2}\sqrt{(\nu - \kappa)^2 \left(\boldsymbol{k}^2\right)^2 + 4\alpha G g_0 \left(1 - \frac{k_z^2}{\boldsymbol{k}^2}\right)}. \tag{30.28}$$

Both roots are real and one of the roots is evidently negative, whereas the other may be positive. For stability this root should also be negative, which means that

$$(\nu + \kappa)\boldsymbol{k}^2 > \sqrt{(\nu - \kappa)^2 \left(\boldsymbol{k}^2\right)^2 + 4\alpha G g_0 \left(1 - \frac{k_z^2}{\boldsymbol{k}^2}\right)}.$$

Squaring this inequality and rearranging, it may be written

$$\frac{\alpha G g_0}{\kappa \nu} < \frac{\left(\boldsymbol{k}^2\right)^3}{\boldsymbol{k}^2 - k_z^2}. \tag{30.29}$$

The right-hand side depends on the geometry of the fluid container. If d is the typical length scale for the geometry, the right-hand side scales as $|\boldsymbol{k}|^4 \sim d^{-4}$. If there is no intrinsic length scale ($d = \infty$), the minimum of the right-hand side is zero, and there can be no stability, just as we concluded for the rising blobs.

For finite d we multiply both sides of (30.29) by d^4. The inequality now becomes a condition on the dimensionless Rayleigh number of the same form as (30.23),

$$\mathsf{Ra} \equiv \frac{g_0 \alpha G d^4}{\kappa \nu} < \frac{(k_x^2 + k_y^2 + k_z^2)^3}{k_x^2 + k_y^2} d^4. \tag{30.30}$$

The finite geometry imposes constraints on the possible wave vectors $\boldsymbol{k}$ on the right-hand side. Among all the allowed fluctuations, the one that leads to the smallest value of the right-hand side is called the *critical fluctuation*, and the corresponding value of the right-hand side is called the *critical Rayleigh number* Ra_c. It defines the upper limit to the convective stability of the baseline state.

Critical fluctuations

When the stability condition (30.30) is fulfilled with a non-vanishing right-hand side, all fluctuations are exponentially damped in time, and the fluid will essentially stay at rest if perturbed slightly. If a fluctuation violates the stability condition, its amplitude will grow exponentially with time, resulting in more complicated flow patterns, even turbulence, about which linear stability analysis has nothing to say.

Right at the critical point where the inequality becomes an equality, the largest of the two stability exponents (30.28) must vanish, $\lambda = 0$. This indicates that the critical fluctuations are time independent (for a rigorous proof of this assertion, see [Landau and Lifshitz 1986]). In this case it is better to revert to ordinary space where the critical fluctuations must obey the steady-flow versions of the linearized dynamic equations (30.25),

$$-Gv_z = \kappa \nabla^2 \Delta T, \tag{30.31a}$$

$$\frac{\nabla \Delta p}{\rho_0} = \nu \nabla^2 \boldsymbol{v} + \alpha \Delta T g_0 \hat{\boldsymbol{e}}_z, \tag{30.31b}$$

$$\nabla \cdot \boldsymbol{v} = 0. \tag{30.31c}$$

These equations may in fact be combined into a single equation for the temperature excess ΔT. Calculating the divergence of the second equation, we first get

$$\frac{1}{\rho_0} \nabla^2 \Delta p = \alpha g_0 \nabla_z \Delta T, \tag{30.32}$$

and using this equation together with (30.31a), both v_z and Δp can be eliminated from the z-component of the second equation, and we obtain

$$\boxed{(\nabla^2)^3 \Delta T = \frac{\alpha G g_0}{\kappa \nu} (\nabla^2 - \nabla_z^2) \Delta T.} \tag{30.33}$$

This equation is equivalent to the condition of the vanishing determinant (30.27) for $\lambda = 0$, and should be solved with the boundary conditions for the geometry of the system. Being a kind of eigenvalue equation, each solution determines a value of the coefficient $\alpha G g_0 / \kappa \nu$. The smallest of these values determines the onset of convection, that is, the critical Rayleigh number.

Non-linear terms: A critical fluctuation is apparently another steady solution to the combined heat and mass flow problem. It must, however, be kept in mind that an essential assumption behind linear stability analysis is that the fluctuation amplitudes are infinitesimal so that non-linear terms can be disregarded. Non-linear terms tend in fact to be beneficial and exert a stabilizing influence on the critical fluctuation such that it is able to persist somewhat above the critical value without diverging exponentially. This is in fact why critical fluctuations can be observed at all.

Henri Bénard **(1874–1939).** French physicist. Discovered hexagonal convection patterns in thin layers of whale oil in 1900. Such cellular convective structures were later named Bénard cells.

* 30.4 Application: Rayleigh–Bénard convection

Warming a horizontal layer of fluid from below is a common task in the kitchen as well as in industry. It was first investigated experimentally by Bénard in 1900 and later analyzed theoretically by Rayleigh in 1916. The most conspicuous feature of the heated fluid is that convection breaks the original planar symmetry, thereby creating characteristic convection patterns. That the symmetry must break is fairly clear, because it is impossible for all the fluid in the layer to start to rise simultaneously. A localized fluctuation that begins to rise will have to veer off into the horizontal direction because of the horizontal boundaries. We shall see below that at the onset of convection, the flow breaks up into an infinite set of "rollers" with alternating sense of rotation.

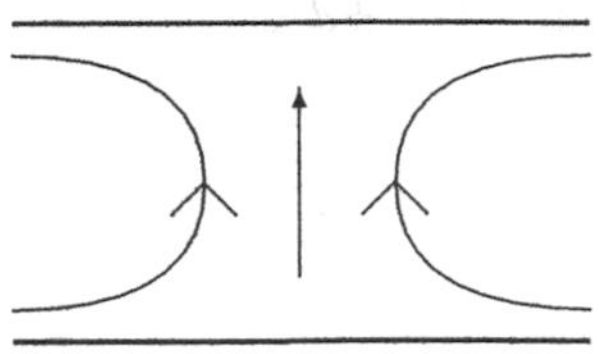

A rising flow in a horizontal layer of fluid has to veer off horizontally both at the top and the bottom.

Much later, in 1956, it was understood that the beautiful hexagonal surface tessellation (Figure 30.2) observed by Bénard in a thin layer of heated whale oil was not caused by buoyancy alone but was driven by the interplay between buoyancy and temperature-dependent surface tension, a phenomenon now called Bénard–Marangoni convection. Here we shall only discuss clean Rayleigh–Bénard convection in layers of fluid so thick that the Marangoni effect can be disregarded.

Figure 30.2. Bénard–Marangoni convection pattern. (Source: Manuel G. Velarde, Universidade Complutense, Madrid. With permission.)

General solution

Let the horizontal layer of incompressible fluid have thickness d and be subject to a constant negative temperature gradient G. The boundaries are chosen symmetrically at $z = \pm d/2$, for reasons that will become clear in the following. Since the flow has to veer off at the boundaries, the fields must depend on z, implying that $k_z \neq 0$ in the stability condition (30.30). The wavenumbers k_x and k_y can in principle take any real values because of the infinitely extended planar symmetry; but since the right-hand side of the stability condition diverges for both $k_x^2 + k_y^2 \to 0$ and $k_x^2 + k_y^2 \to \infty$, there must be a minimum at a finite value of $k_x^2 + k_y^2$. This argument demonstrates that the critical solution must have a periodic horizontal structure.

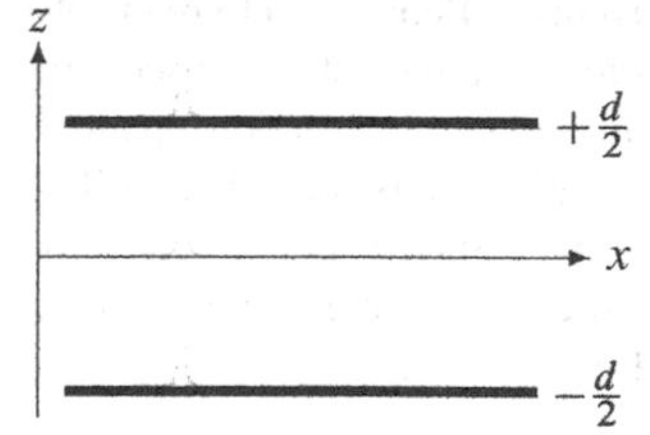

The plates are placed symmetrically at $z = \pm d/2$.

For given k_x and k_y, we may without loss of generality rotate the coordinate system to obtain $k_y = 0$ and $k_x > 0$, implying that the fields only depend on x and z but not on y. Since there is no geometric constraint in the x-direction, we are still free to Fourier transform along x, such that the most general form of the temperature excess takes the form

$$\Delta T = \Theta \cos k_x x f(z), \tag{30.34}$$

where Θ is a constant and $f(z)$ is a (so far unknown) dimensionless function of z. From (30.31a) we find the vertical velocity,

$$v_z = -\frac{\kappa\Theta}{G} \cos k_x x (\nabla_z^2 - k_x^2) f(z), \tag{30.35}$$

and from the divergence condition (30.31c) we get the horizontal velocity,

$$v_x = \frac{\kappa\Theta}{G k_x} \sin k_x x (\nabla_z^2 - k_x^2) f'(z), \tag{30.36}$$

where $f'(z) = df(z)/dz$. The third velocity component v_y does not participate, and may be chosen to vanish, $v_y = 0$.

Inserting ΔT into the determinant equation (30.33), we get a sixth-order ordinary differential equation for this function,

$$\left(\nabla_z^2 - k_x^2\right)^3 f(z) = -\frac{\alpha G g_0}{\kappa \nu} k_x^2 f(z). \tag{30.37}$$

Using this relation one may verify that the pressure excess,

$$\frac{\Delta p}{\rho_0} = -\frac{\kappa \nu \Theta}{G k_x^2} \cos k_x x (\nabla_z^2 - k_x^2)^2 f'(z), \tag{30.38}$$

satisfies (30.32).

In a given physical context the actual solution depends on the boundary conditions imposed on the fields. We shall always assume that the boundaries are perfect conductors of heat such that $\Delta T = 0$ for $z = \pm d/2$. For the velocity fields the boundary conditions depend on whether the boundaries are solid plates or free open surfaces. We shall (as Rayleigh did in 1916) first analyze the latter case, which is by far the simplest.

Case: Two free boundaries

The simplest choice that satisfies the boundary condition $\Delta T = 0$ for $z = \pm d/2$ is

$$f(z) = \cos k_z z, \qquad k_z = \frac{(1+2n)\pi}{d}, \tag{30.39}$$

where $n = 0, 1, 2, \ldots$ is an integer. Inserting this into (30.37) and solving for the Rayleigh number, we obtain

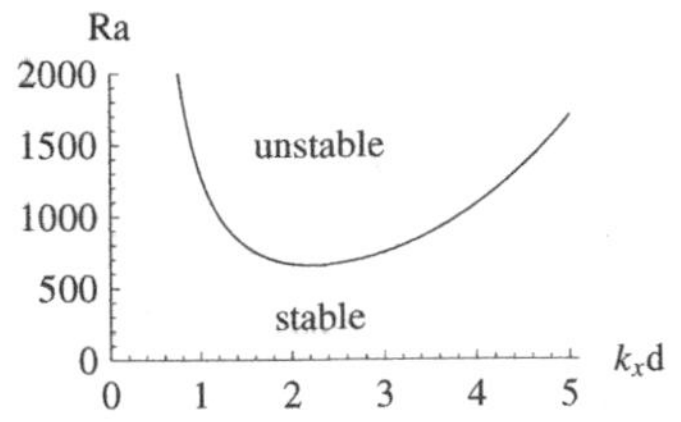

Plot of Ra versus $k_x d$ for $n = 0$. The minimum $\mathsf{Ra} = 27\pi^4/4 \approx 658$ is found at $k_x d = \pi/\sqrt{2} \approx 2.22$.

$$\mathsf{Ra} \equiv \frac{\alpha G g_0 d^4}{\kappa \nu} = \frac{(k_z^2 + k_x^2)^3}{k_x^2} d^4. \tag{30.40}$$

The minimum of the right-hand side is found for $k_x = k_z/\sqrt{2}$ and $n = 0$, such that the critical Rayleigh number is

$$\boxed{\mathsf{Ra}_c = \frac{27}{4}\pi^4 \approx 657.511\ldots.} \tag{30.41}$$

The complete critical solution becomes (with $k_x = \pi/d\sqrt{2}$ and $k_z = \pi/d$)

$$\Delta T = \Theta \cos k_x x \cos k_z z, \tag{30.42a}$$

$$v_x = \sqrt{2}\, U \sin k_x x \sin k_z z, \tag{30.42b}$$

$$v_y = 0, \tag{30.42c}$$

$$v_z = U \cos k_x x \cos k_z z, \tag{30.42d}$$

where $U = 3\pi^2 \kappa \Theta / 2 G d^2$.

The solution is pictured in the three panels of Figure 30.3. The flow pattern (middle panel) consists of an infinite sequence of nearly elliptic "rollers" with aspect ratio $\sqrt{2}$. The temperature pattern (top panel) is 90 ° out of phase with the flow pattern. This confirms the intuition that the central temperature should be higher when fluid transports heat from the warm lower boundary toward the cold upper boundary, and conversely.

Figure 30.3. Critical fields in the horizontal layer of fluid with free boundaries (Equation (30.42)). The steady flow pattern consists of an infinity of approximatively elliptical rolls of rotating fluid with aspect ratio $\sqrt{2}$ and alternating sense of rotation and temperature excess. **Top:** Contour plot of the temperature field ΔT with high temperature indicated by white. **Middle:** Streamlines for the steady flow (v_x, v_z) with white indicating clockwise rotation. **Bottom:** Deformation of the originally parallel boundaries (strongly exaggerated).

From the above solution we immediately obtain the shear stress,

$$\sigma_{xz} = \eta(\nabla_x v_z + \nabla_z v_x) = \frac{\pi U \eta}{\sqrt{2}d} \sin k_x x \cos k_z z. \tag{30.43}$$

It evidently vanishes for $z = \pm d/2$, and since we trivially have $\sigma_{yz} = 0$, both boundaries are completely free of shear. There is no practical problem in arranging the upper boundary to be shear-free; that is in fact what we do when we cook. A shear-free lower boundary is, on the contrary, rather unphysical, so the main virtue of the shear-free model is that it is easy to solve.

The pressure excess in the critical solution may be calculated from (30.38),

$$\Delta p = \frac{2}{3\pi} \alpha \Theta \rho_0 g_0 d \cos k_x x \sin k_z z, \tag{30.44}$$

so the excess in the normal stress becomes

$$\Delta \sigma_{zz} = -\Delta p + 2\eta \nabla_z v_z = -\frac{10}{9\pi} \alpha \Theta \rho_0 g_0 d \cos k_x x \sin k_z z. \tag{30.45}$$

It does not vanish at the boundaries, showing that the solution is not perfect. The non-vanishing normal stress can, however, be compensated by hydrostatic pressure if the layer thickness is allowed to vary a bit. Dividing by $\rho_0 g_0$ we find the required shift at the two boundaries,

$$\Delta z = -\left.\frac{\Delta \sigma_{zz}}{\rho_0 g_0}\right|_{z=\pm d/2} = \pm \frac{10}{9\pi} \alpha \Theta d \cos k_x x. \tag{30.46}$$

The shape of the deformed layer is shown in the bottom panel of Figure 30.3.

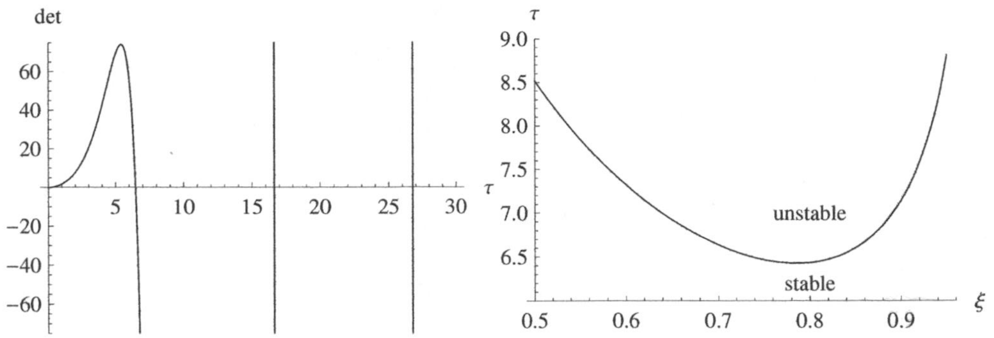

Figure 30.4. Solution for horizontal slot with solid boundaries. **Left:** The value of the determinant as a function of $\tau = \mathsf{Ra}^{1/4}$ for $\xi = 0.75$. One notes the regularly spaced solutions where the determinant crosses zero. **Right:** Stability plot for the lowest branch as a function of τ. The minimum $\mu_c = 6.42846\ldots$ for $\xi_c = 0.785559\ldots$ determines the critical Rayleigh number.

Case: Two solid boundaries

A horizontal slot bounded by two solid plates is easy to set up experimentally. Numerous experiments were carried out in the twentieth century and agree very well with the theoretical results [Chandrasekhar 1981].

For simplicity we choose the plate distance $d = 1$ in the following analysis. The fundamental equation (30.37) is an ordinary sixth-order differential equation with constant coefficients, implying that the solution is a superposition of exponentials $e^{\mu z}$, where μ is a root of the sixth-order algebraic equation

$$(\mu^2 - k_x^2)^3 = -\mathsf{Ra}\, k_x^2. \tag{30.47}$$

The six roots are

$$\mu = \pm\sqrt{k_x^2 + \mathsf{Ra}^{1/3} k_x^{2/3} \sqrt[3]{-1}}, \tag{30.48}$$

where $\sqrt[3]{-1} = -1$, and $(1 \pm i\sqrt{3})/2$ is any one of the three third roots of -1. Parameterizing $k_x = \tau\xi^3$ with $\tau = \mathsf{Ra}^{1/4}$, the roots may be written $\mu = \pm\tau\xi\sqrt{\xi^4 + \sqrt[3]{-1}}$. Writing the six roots in the form $\mu = \pm i\mu_0$ and $\mu = \pm\mu_1 \pm i\mu_2$, we find

$$\mu_0 = \tau\xi\sqrt{1 - \xi^4}, \tag{30.49a}$$

$$\mu_1 = \tfrac{1}{2}\tau\xi\sqrt{1 + 2\xi^4 + 2\sqrt{1 + \xi^4 + \xi^8}}, \tag{30.49b}$$

$$\mu_2 = \tfrac{1}{2}\tau\xi\sqrt{-1 - 2\xi^4 + 2\sqrt{1 + \xi^4 + \xi^8}}. \tag{30.49c}$$

These quantities are all real for $0 < \xi < 1$.

The boundary conditions are as before $\Delta T = 0$, together with $v_x = v_z = 0$ at $z = \pm 1/2$. Since the boundary conditions as well as the fundamental equation (30.37) are invariant under a change of sign of z, it follows that the solutions are either symmetric (even) or antisymmetric (odd) in z. In the even case, we have

$$f(z) = A\cos\mu_0 z + B\cosh\mu_1 z\cos\mu_2 z + C\sinh\mu_1 z\sin\mu_2 z, \tag{30.50}$$

where A, B, and C are constants. From the general solution (30.34), (30.35), and (30.36), we see that $f(z)$, $f''(z)$, and $f'''(z) - k_x^2 f'(z)$ must vanish for $z = 1/2$.

Figure 30.5. Critical fields in a horizontal layer with solid boundaries. The steady flow pattern consists of an infinity of approximately circular cylindrical rolls of fluid with alternating sense of rotation and alternating temperature excess. **Top:** Contour plot of the temperature field ΔT with high temperature indicated by white. **Bottom:** Streamlines for the steady flow (v_x, v_z) with white indicating clockwise rotation. Note how the streamlines "shy away" from the solid walls because of the no-slip condition.

These conditions provide three homogeneous equations for the determination of A, B, and C. Such equations only have a non-trivial solution if their (3×3) determinant vanishes. It takes a bit of algebra to show that the determinant is proportional to

$$\det(\tau, \xi) \propto \mu_0(\cosh \mu_1 + \cos \mu_2) \sin \frac{\mu_0}{2} + \left((\mu_1 + \sqrt{3}\,\mu_2) \sinh \mu_1 + (\mu_2 - \sqrt{3}\,\mu_1) \sin \mu_2\right) \cos \frac{\mu_0}{2}. \tag{30.51}$$

Solving the transcendental equation, $\det(\tau, \xi) = 0$, yields a family of solutions $\tau = \tau(\xi)$, as shown in Figure 30.4L. The minimum of the lowest branch determines the critical values $\tau_c = \mathsf{Ra}_c^{1/4} = 6.42846\ldots$ and $\xi_c = 0.785559\ldots$ (see Figure 30.4R). The critical Rayleigh number becomes (see Problem 30.3 for an approximative calculation)

$$\boxed{\mathsf{Ra}_c = 1707.76\ldots} \tag{30.52}$$

and the corresponding wavenumbers

$$\mu_0 = 3.9737\ldots, \qquad \mu_1 = 5.19439\ldots, \qquad \mu_2 = 2.12587\ldots. \tag{30.53}$$

Finally, solving the boundary conditions at the critical point, the coefficients become (apart from an overall scale factor)

$$A = 1, \qquad B = 0.120754\ldots, \qquad C = 0.00132946\ldots. \tag{30.54}$$

From these values the actual fields ΔT, v_z, and v_x may be determined. As can be seen from Figure 30.5, the critical flow pattern consists of an infinity of roughly circular cylindrical rolls, which due to the no-slip conditions appear to "shy away" from the boundaries.

It should be mentioned that in the antisymmetric (odd) case, we have

$$f(z) = D \sin \mu_0 z + E \cosh \mu_1 z \sin \mu_2 z + F \sinh \mu_1 z \cos \mu_2 z. \tag{30.55}$$

It is treated in the same way and leads to a critical Rayleigh number $\mathsf{Ra}_c \approx 17610.4\ldots$, which is uninteresting because it is (much) larger than for the even solution.

Figure 30.6. Critical fields in the horizontal layer of fluid with solid bottom and free top. **Top:** Contour plot of the temperature field ΔT with high temperature indicated by white. **Middle:** Streamlines for the steady flow with white indicating clockwise rotation. Note how the no-slip condition makes the streamlines "shy away" from the bottom while it "hugs" the free surface at the top. **Bottom:** Deformation of the originally flat upper boundary (strongly exaggerated).

Case: Solid bottom and free top

This is the situation most often found in the household and industry. Since the boundary conditions are asymmetric, the solution is a superposition of all six possibilities,

$$\begin{aligned} f(z) = {} & A\cos\mu_0 z + B\cosh\mu_1 z\cos\mu_2 z + C\sinh\mu_1 z\sin\mu_2 z \\ & + D\sin\mu_0 z + E\cosh\mu_1 z\sin\mu_2 z + F\sinh\mu_1 z\cos\mu_2 z. \end{aligned} \tag{30.56}$$

Although more complicated, the solution is found in the same way as before, and the critical values are $\tau_c = 5.75986\ldots$ and $\xi_c = 0.775115\ldots$. Consequently, the critical Rayleigh number is

$$\boxed{\mathsf{Ra}_c = 1100.65\ldots} \tag{30.57}$$

This value is probably by accident nearly the same value as the estimate for a rising bubble (30.23). The wavenumbers for this solution are

$$\mu_0 = 3.56895\ldots, \qquad \mu_1 = 4.55531\ldots, \qquad \mu_2 = 1.8947\ldots, \tag{30.58}$$

and the coefficients are

$$\begin{aligned} &A = 1, && B = 0.086726\ldots, && C = -0.00956513\ldots, \\ &D = 0.216993\ldots, && E = 0.00778275\ldots, && F = -0.08632\ldots, \end{aligned} \tag{30.59}$$

again apart from an overall factor. The flow pattern is shown in Figure 30.6.

Energy balance?

Where does the energy to drive the rolls come from? The steadily rotating fluid could in principle be set to do useful work, and according to the First Law of Thermodynamics this work must be taken from the heat flowing between the plates. In effect, the plates act as heat reservoirs and the convection as a heat engine converting heat to work by means of the buoyancy of warm fluid. In the present setup, all the work done by the rotating fluid is actually dissipated back into heat by internal viscous forces, so in the steady state the energy of the fluid is constant, and no steady inflow of heat into the system is required. The local heat flow through the boundaries will, however, be uneven because of the local variations in the temperature gradient.

Convective pattern formation

The spontaneous formation of convection patterns in otherwise featureless geometries is a common occurrence. The non-linear terms that have been left out in the linear approximation will exert a stabilizing influence on the patterns such that they are able to persist at Rayleigh numbers somewhat larger than the critical one. At still larger Rayleigh numbers, the rolls of the critical pattern will develop further instabilities and eventually turbulent convection may result. Photographs of convection patterns may be found in [Tritton 1988, Dyke 1982, Ball 1997]).

Problems

30.1 **(a)** Show that there cannot be hydrostatic equilibrium in vertical gravity with horizontal temperature differences. **(b)** Estimate the speed with which the fluid rises.

30.2 Consider the convective flow in a vertical slot with constant temperature difference Θ (page 549). Show that the ratio between the exit heat flow $\dot{Q}$ and the heat flow $\dot{Q}_0$ out of the warm plate in a vertical slot is

$$\mathsf{Nu} = \frac{\dot{Q}}{\dot{Q}_0} = \frac{1}{45} \cdot \frac{d}{h} \cdot \mathsf{Ra},$$

where Ra is the Rayleigh number. This number is called the *Nusselt number*.

* **30.3** **(a)** Show that an approximate solution to the vanishing of the determinant (30.51) for Rayleigh–Bénard flow in a horizontal slot is given by the equation

$$\tau = 2\frac{\tau}{\mu_0}\left(\pi - \arctan\frac{\mu_1 + \sqrt{3}\mu_2}{\mu_0}\right),$$

where the right-hand side is only a function of ξ. Hint: Use that $\cosh \mu_1 \gg 1$ and $\tanh \mu_1 \approx 1$, and drop the smaller terms in the determinant. **(b)** Show that numerical minimization of the right-hand side leads to $\tau = 6.44397\ldots$ for $\xi = 0.787942\ldots$, corresponding to a critical Rayleigh number $\mathsf{Ra}_c = 1{,}724$.

31 Turbulence

Turbulence is so commonplace that we hardly notice it in our daily lives. The size of humans and their natural speed of locomotion in air or water bring the typical Reynolds numbers into the millions, but most of us never think of the trail of turbulent air we leave in our wake when we jog. Our vehicles move at much higher speeds and generate more turbulence, although great efforts have been made to limit the extent of the turbulent wake by streamlining cars, airplanes, ships, and submarines. In strong winds the lower part of the atmosphere may become strongly turbulent, but patches of turbulent air can in fact be found at any altitude, normally in connection with cloud formations, although also in clear air.

Even under the steadiest of conditions, for example the draining of a large water cistern through a pipe, turbulence will completely fill the pipe when the Reynolds number becomes sufficiently large. The road from laminar flow to turbulence as a function of the slowly increasing Reynolds number is complex and goes through many stages of sometimes intermittent flow patterns, not yet fully understood. At sufficiently high Reynolds number, the turbulent flow seems to settle down into a nearly homogeneous and isotropic state where the small-scale fluctuations in the flow appear to be of the same nature throughout the pipe, except for a stretch near the entrance and a thin layer very close to the wall. Such featureless, *fully developed turbulence* is much more amenable to analysis than the flow stages leading up to it.

In this closing chapter of the book we shall only touch on the most basic and most robust aspects of turbulence, a subject that for a long time has been and still is an important topic of basic research. The many modern textbooks on the subject may be consulted to obtain a deeper insight than can be provided here ([Batchelor 1953], [Townsend 1956], [Libby 1996], [Frisch 1995], [Lesieur 1997], [Pope 2003], and [Davidson 2004]).

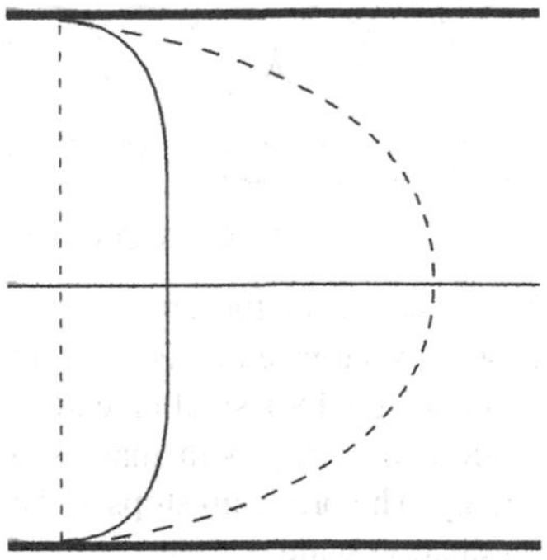

Sketch of the velocity profile for fully developed turbulent pipe flow compared with the parabolic Poiseuille profile (dashed). Apart from thin boundary layers (and a region downstream from the entrance), the mean velocity field is approximately constant across the pipe.

31.1 Scaling in fully developed turbulence

Turbulence typically appears as a fluctuating velocity component on top of a mainstream flow with high Reynolds number. In the discussion of the phenomenology of stationary turbulent pipe flow (Section 16.5 on page 274), it was pointed out that after turbulence has become fully developed some distance downstream from the pipe entrance, it proceeds plug-like down the pipe with a fairly uniform mainstream velocity distribution across the pipe, apart from a thin layer near the wall where the velocity drops to zero. In the reference frame of the mainstream flow and at length scales much smaller than the diameter of the pipe, the bulk of the fluid appears to be in a state of uniform agitation.

Specific rate of dissipation

A stationary flow can only be maintained if the environment continually performs work on the fluid to balance the unavoidable viscous dissipation. In pipe flow the external work is performed by the pressure drop Δp maintained between its ends, and if the pipe carries a volume flux Q, the rate of work (power) of the pressure drop is $P = \Delta p\, Q$ (see page 271). Dividing by the total mass M of the fluid in the pipe, the *specific rate of dissipation* averaged over the whole pipe becomes

$$\varepsilon = \frac{P}{M} = \frac{\Delta p\, Q}{M} = \frac{U^3}{a} C_f(\mathsf{Re}). \tag{31.1}$$

In the last step we have used that the total mass is $M = \pi a^2 L \rho_0$; that the pressure drop is balanced by the wall stress, $\pi a^2 \Delta p = 2\pi a L \sigma_{\text{wall}}$; and that the volume flux $Q = \pi a^2 U$, where a is the pipe radius and U the average velocity of the flow. The Reynolds number is $\mathsf{Re} = 2aU/\nu$ and from the definition of the dimensionless friction coefficient (16.47) on page 275, the last expression follows.

In the region of smooth-pipe turbulent flow, $4\times 10^3 \lesssim \mathsf{Re} \lesssim 10^5$, we have seen (page 276) that the friction coefficient decreases slowly as $\mathsf{Re}^{-1/4}$ with increasing Reynolds number. Beyond this region, the friction coefficient becomes constant, although its actual value depends on the unavoidable roughness of the pipe surface. The astonishing conclusion following from (31.1) is that *although ultimately caused by viscous friction, the specific rate of dissipation is finite in the limit of vanishing viscosity (i.e., infinite Reynolds number)!*

Example 31.1 [Concrete waterpipe]: A large concrete pipe with diameter $2a = 1$ m carries water at a discharge rate of $Q \approx 10\ \text{m}^3\ \text{s}^{-1}$. The average velocity is $U \approx 13\ \text{m s}^{-1}$, corresponding to a Reynolds number of $\mathsf{Re} = 1.3 \times 10^7$ well into the rough pipe region. With a roughness scale of 1 mm, the friction coefficient calculated from Equation (16.50) on page 276 is $C_f(\infty) \approx 5 \times 10^{-3}$, leading to a specific dissipation $\varepsilon \approx 21\ \text{W kg}^{-1}$. Perhaps it does not sound like much, but the dissipation rate (per unit of pipe length) is $\rho_0 \pi a^2 \varepsilon \approx 16\ \text{kW m}^{-1}$, which is about 4,000 times larger than it would have been had the pipe flow remained laminar with a friction coefficient $C_f \approx 16/\mathsf{Re} \approx 1.3 \times 10^{-6}$ (see page 275).

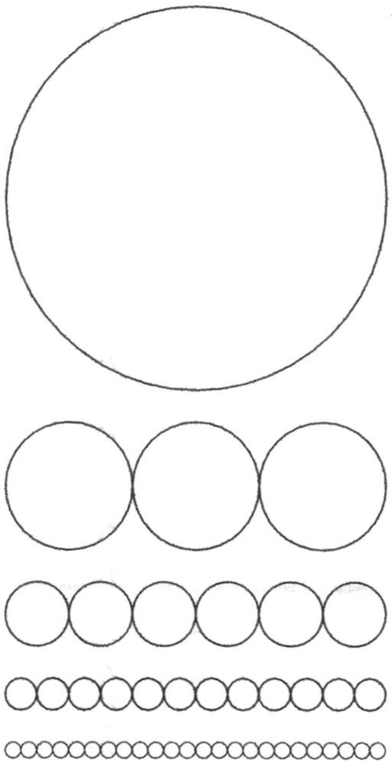

Naive sketch of the energy cascade. A large eddy is thought to break up into smaller eddies while conserving both mass and energy. The break-up stops at the dissipation scale.

Richardson's energy cascade

In the model we shall study here, the fluctuating velocity component is viewed as composed of fairly localized flow structures, generically called "eddies", covering a wide range of sizes. The non-linearity of fluid mechanics is assumed to transform ("break up") the eddies into smaller eddies while conserving mass and kinetic energy, until they become so small that they are wiped out by viscous friction (a picture owed to Richardson, 1922).

In an isolated patch of turbulence, the kinetic energy will in this way be continually sapped "from below" until the turbulence dies away and the flow becomes laminar for in the end to stop completely. We have all seen this happen after rapidly filling a bucket with water. Alternatively, as we saw for the pipe, the kinetic energy that is lost to heat under steady external conditions will instead be continually re-supplied by the work of the external forces.

Lewis Fry Richardson (1881–1953). British physicist. The first to apply the method of finite differences to predict weather.

Kolmogorov's scaling law

The energy cascade plays out in an interval of eddy sizes $\lambda_d \lesssim \lambda \lesssim L$, where the upper cut-off L is related to the flow geometry (for example, the pipe diameter), and the lower cut-off λ_d represents the dissipation length scale where viscosity "kicks in". Far above the dissipation cut-off, $\lambda \gg \lambda_d$, the eddy dynamics should be independent of the viscosity, whereas far below the geometry cut-off, $\lambda \ll L$, the actual flow geometry should not matter for the eddy dynamics.

In the so-called *inertial range* satisfying both conditions, $\lambda_d \ll \lambda \ll L$, the only parameters that can be at play in the eddy dynamics are the size λ of an eddy and the time τ it takes for an eddy of this size to transform into smaller eddies. In the inertial range the magnitude of any other quantity must be related to these parameters, essentially by dimensional arguments. The break-up of an eddy is, for example, bound to move some fluid a distance of magnitude λ in time τ, leading to an estimate of the velocity variations inside an eddy, $u \sim \lambda/\tau$.

The energy-cascade model yields a relation between the time scale τ and the length scale λ by way of the specific dissipation rate ε. To see this we focus on a collection of eddies of size λ with total mass dM and total kinetic energy $d\mathcal{T} \sim dMu^2$. In a time interval τ, all these eddies break up into smaller eddies, making the rate of energy turnover $d\dot{\mathcal{T}} \sim d\mathcal{T}/\tau \sim dMu^2/\tau$. Since both mass and energy are assumed to be conserved in the cascade that follows, the energy turnover rate must be the same all the way down to the smallest scales, and thus in the end nearly equal to the viscous dissipation rate in the original amount of fluid, $dP = \varepsilon dM$. Using $dP \approx d\dot{\mathcal{T}}$ and $u \sim \lambda/\tau$, we find

$$\varepsilon = \frac{dP}{dM} \approx \frac{d\dot{\mathcal{T}}}{dM} \sim \frac{u^2}{\tau} \sim \frac{\lambda^2}{\tau^3}. \tag{31.2}$$

Since ε has dimension $[\varepsilon] = \text{W kg}^{-1} = \text{m}^2\,\text{s}^{-3}$, one might think that this estimate could have been derived from a dimensional argument alone, but that would be wrong. The crucial input from the energy-cascade model is that the terminal rate of dissipation dP may be set equal to the kinetic energy turnover rate $d\dot{\mathcal{T}}$ for any collection of eddies, independent of their size.

Solving (31.2) for the time scale, we find the λ-dependent result,

$$\boxed{\tau \sim \varepsilon^{-1/3}\lambda^{2/3}, \qquad u \sim \frac{\lambda}{\tau} \sim (\varepsilon\lambda)^{1/3},} \tag{31.3}$$

also known as *Kolmogorov's scaling law*.

Andrei Nikolaevich Kolmogorov (1903–1987). Russian mathematician. Made major contributions to a wide range of subjects: Markov processes, Lebesgue measure theory, axiomatic foundation of probability theory, turbulence, and dynamical systems. (Source: Wikimedia Commons.)

A basic assumption in the energy-cascade model is that the turbulence is homogeneous and isotropic in the rest frame of the mainstream flow. Even if ultimately maintained by the work of steady external forces, the physics in the inertial range only depends on the background velocity U through the specific dissipation rate ε, which for pipe flow is given by (31.1). At the scale L of the geometry, it follows from the scaling law that the velocity of the largest eddies is $u_L \sim (\varepsilon L)^{1/3}$. This is, however, not necessarily the same as the mainstream velocity U. Inserting the pipe flow expression (31.1) with $L \sim 2a$, we find $u_L/U \sim C_f^{1/3}$. In Example 31.1 with $C_f = 5 \times 10^{-3}$, this makes $u_L/U \approx 0.2$.

Statistics

The fundamental concept necessary for establishing a statistical model of turbulence is the fraction $dF = dM/M$ of the total fluid mass (or volume) residing in eddies with sizes in a small interval $d\lambda$ around λ. The eddy distribution over sizes, $dF/d\lambda$, will in general depend on λ as well as on the system scale L and the viscosity ν; but in the inertial range, $\lambda_d \ll \lambda \ll L$, the eddy distribution should only depend on the basic scale parameters, λ and τ. Since $dF/d\lambda$ has dimension of inverse length, it must be of the form

$$\boxed{\frac{dF}{d\lambda} \sim \frac{1}{\lambda} \qquad (\lambda_d \ll \lambda \ll L).} \tag{31.4}$$

The integral of this distribution diverges, however, for both $\lambda \to 0$ and $\lambda \to \infty$. So although this eddy distribution looks universal, it cannot be properly normalized without introducing a slowly varying non-universal normalization factor depending logarithmically on the limits of the inertial range. Such factors have been investigated by Gagne and Castaing [GC91]; see also Problem 31.2.

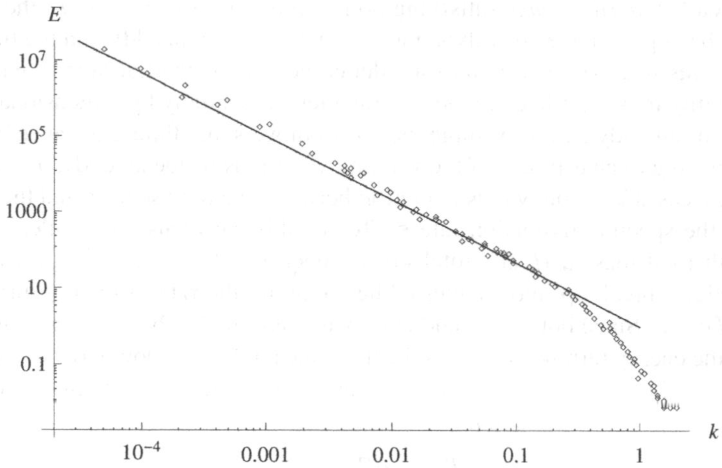

Figure 31.1. Doubly logarithmic plot of the turbulent energy spectrum E as a function of the wavenumber k (suitably normalized). The largest Reynolds number is about 3×10^8, yielding more than three decades in the inertial range. The data agrees very well with Kolmogorov's $-5/3$ law (solid line) up to the beginning of the dissipation range at abscissa 0.1. Data extracted from the classical experiment carried out by towing a probe after a ship through the Discovery Passage of Western Canada [GSM62]. This paper contains a delightful account of the difficulties encountered while attempting experimentally to verify a simple theoretical prediction in the real world of ships at sea.

Energy spectrum

The kinetic energy residing in the eddies with size λ may now be estimated as $d\mathcal{T} \sim u^2 dM$. Using $dM = M dF$, and dividing $d\mathcal{T}$ by $M d\lambda$ we obtain the distribution of the specific kinetic energy according to eddy size,

$$\frac{1}{M}\frac{d\mathcal{T}}{d\lambda} \sim \frac{u^2}{\lambda} \sim \frac{\lambda}{\tau^2} \sim \varepsilon^{2/3}\lambda^{-1/3}. \tag{31.5}$$

To find the total kinetic energy, we must integrate this expression over all λ. Since the integral of this expression is convergent for $\lambda \to 0$ and diverges for $\lambda \to \infty$, we cut it off at the system scale L and find the total kinetic energy per unit of mass,

$$\frac{\mathcal{T}}{M} \sim (\varepsilon L)^{2/3} \sim u_L^2, \tag{31.6}$$

where u_L is the eddy velocity for $\lambda = L$. The dependence on the system scale shows that in spite of the fact that the distribution function (31.5) decreases slowly with wavelength, most of the turbulent kinetic energy resides in the largest eddies near the system scale cut-off.

This is seen even better by expressing the kinetic energy distribution in terms of the inverse length scale, the wavenumber $k = 2\pi/\lambda$. Using $d\lambda \sim dk/k^2$ we obtain the specific energy spectrum of turbulence as a function of wavenumber,

$$\boxed{E(k) \equiv \frac{1}{M}\frac{d\mathcal{T}}{dk} \sim \varepsilon^{2/3}k^{-5/3}.} \tag{31.7}$$

This is Kolmogorov's famous $-5/3$ power law from 1941. The negative exponent shows that the spectrum is indeed dominated by the smallest wavenumbers, $k \sim 1/L$, corresponding to the largest length scales, $\lambda \approx L$. The left-out dimensionless constant in front of the spectrum is called the *Kolmogorov constant* and was empirically determined to be about 1.5. Many experiments have confirmed the validity of the Kolmogorov spectrum in the inertial range, for example the one plotted in Figure 31.1.

The dissipation scale

The rate of viscous dissipation is calculated from the velocity gradients in the flow in Equation (15.24) on page 252. Estimating the dissipation rate in a collection of eddies of mass dM and size λ, we find

$$dP \sim \eta\,|\nabla v|^2\,dV = \nu\,|\nabla v|^2\,dM \sim \nu\left(\frac{u}{\lambda}\right)^2 M\frac{d\lambda}{\lambda} \sim M\frac{\nu}{\lambda\tau^2}\,d\lambda.$$

Dividing by $M d\lambda$ we find the specific dissipation spectrum (per unit of size),

$$\frac{1}{M}\frac{dP}{d\lambda} \sim \frac{\nu}{\lambda\tau^2} \sim \nu\,\varepsilon^{2/3}\lambda^{-7/3}. \tag{31.8}$$

The largest dissipation evidently takes place at the smallest scales. To get the total specific dissipation ε, we must integrate over λ; and since the integral diverges at the smallest scales, we cut it off at λ_d and get

$$\varepsilon = \frac{P}{M} \approx \int_{\lambda_d}^{\infty}\frac{1}{M}\frac{dP}{d\lambda}\,d\lambda \sim \nu\varepsilon^{2/3}\lambda_d^{-4/3}. \tag{31.9}$$

Solving for λ_d we arrive at the *Kolmogorov dissipation scale*,

$$\boxed{\lambda_d \sim \varepsilon^{-1/4}\nu^{3/4}.} \tag{31.10}$$

In practice the inertial scaling laws may be assumed to be valid about an order of magnitude inside inertial range, that is, for $10\lambda_d \lesssim \lambda \lesssim 0.1L$. The time and velocity scales at the dissipation scale become

$$\tau_d \sim \varepsilon^{-1/2}\nu^{1/2}, \qquad u_d \sim \varepsilon^{1/4}\nu^{1/4}. \tag{31.11}$$

At this scale, everything can be expressed in terms of the specific rate of dissipation ε and the viscosity ν.

Example 31.2 [Concrete water pipe, continued]: For the pipe flow in Example 31.1 we find $\lambda_d \sim 15\ \mu$m, $\tau_d \sim 220\ \mu$s, and $u_d \sim 7$ cm s^{-1}. The geometric scale is $L \approx 1$ m, implying that the inertial range should be 15 μm $\ll \lambda \ll$ 1 m. With a safety factor of about 10 in both ends of the range, the inertial range covers less than three full decades. This illustrates that it may be necessary to go to very large Reynolds numbers to get three or more decades of length scales in the inertial range. The broadest inertial ranges have in fact been reached in large natural systems, such as the ocean and the atmosphere (see Figure 31.1).

Eddy viscosity

The random turbulent motion also transfers momentum between adjacent regions of fluid in a way that resembles molecular momentum diffusion, estimated in Equation (15.2) on page 245. This naturally leads to the idea that turbulent diffusion at a length scale λ likewise might be described by a kind of effective *eddy viscosity* ν_λ, which in the inertial range is determined by the eddy length and time scales,

$$\nu_\lambda \sim \frac{\lambda^2}{\tau} \sim u\lambda \sim \varepsilon^{1/3}\lambda^{4/3}. \tag{31.12}$$

The eddy viscosity grows monotonically with the size of the eddies and is maximal at the system scale, $\nu_L \sim \varepsilon^{1/3}L^{4/3}$, a value that must also characterize the average eddy viscosity and equals about 3 m^{2}s^{-1} in Example 31.1. The smallest eddy viscosity $\nu_\lambda \sim \nu$ is obtained for $\lambda \sim \lambda_d$, showing that at the dissipation scale, momentum diffusion due to eddies is of the same magnitude as molecular momentum diffusion, which is not so surprising.

Non-dimensional formulation

It is sometimes useful to use a dimensionless formalism by introducing the Reynolds number for eddies of size λ,

$$\mathsf{Re}_\lambda = \frac{u\lambda}{\nu} \sim \frac{\nu_\lambda}{\nu} \sim \frac{\epsilon^{1/3}\lambda^{4/3}}{\nu}. \tag{31.13}$$

At the dissipation scale $\lambda \sim \lambda_d$, we find as expected $\mathsf{Re}_\lambda \sim 1$, such that the ratios between the scale parameters and the dissipation scales may be written as

$$\frac{\lambda}{\lambda_d} \sim \mathsf{Re}_\lambda^{3/4}, \qquad \frac{\tau}{\tau_d} \sim \mathsf{Re}_\lambda^{1/2}, \qquad \frac{u}{u_d} \sim \mathsf{Re}_\lambda^{1/4}. \tag{31.14}$$

These results are valid in the inertial range $1 \ll \mathsf{Re}_\lambda \ll \mathsf{Re}_L = u_L L/\nu$.

Numerical simulation of turbulence

In a numerical simulation of a turbulent flow, the grid should be so fine-grained that the dissipation scale can be resolved in all directions. Consequently, the number N of grid points must obey

$$N \gtrsim \left(\frac{L}{\lambda_d}\right)^3 \sim \mathsf{Re}_L^{9/4}. \tag{31.15}$$

Direct numeric simulation of realistic turbulent flows belongs among the hardest computational problems. Fortunately the need to find practical solutions to the technical problems presented by cars, ships, and airplanes have led to a number of more or less *ad hoc* approximations that yield quite decent results.

Example 31.3 [Concrete water pipe, continued]: In the pipe flow Example 31.1 we have $\mathsf{Re}_L \approx 2.7 \times 10^6$, so that one would have to use at least 3×10^{14} grid points to simulate turbulent pipe flow. Even with a computer executing 10^{12} floating point operations per second (a teraflop), it would still take hours to do a single update of the whole grid, because each grid point requires quite a few operations. A complete simulation probing all time scales would require at least $\tau_L/\tau_d = \sqrt{\mathsf{Re}_L} \approx 1{,}600$ updates, and might easily take a full year of computation.

Critique of Kolmogorov's theory

The derivation of the energy spectrum from scaling considerations depends only on the existence of a process by which the energy of turbulent motion at a given length scale is continually redistributed over smaller length scales until it is, in the end, dissipated by viscous forces. The detailed mechanism underlying this process is not known for sure, although more than 50 years of intense research has produced a wealth of ideas and results. Numerical simulations have revealed that fully developed turbulence contains vortex filaments, and it is believed that dynamic stretching of these vortices plays a central role in transferring the vortex energy to smaller length scales.

Many objections, both mathematical and physical, have been raised against the energy-cascade model (see for example [Frisch 1995]). One physical objection is that fully developed turbulent flow is not as featureless as assumed here, but is observed experimentally to contain large-scale coherent structures, not included in the Kolmogorov model (see for example [Hunt and Vassilicos 2000]). Another is that the Kolmogorov theory utterly fails to take into account the conspicuous intermittency of turbulence, especially in the unstable regime near the transition from laminar to turbulent flow.

In view of the numerous experimental verifications of the Kolmogorov energy spectrum, one is, however, driven to conclude that the law must at least contain an element of truth, whatever its origin.

31.2 Mean flow and fluctuations

In this section we shall formalize the analysis of turbulence, and for simplicity we continue to assume that the turbulent flow is driven by steady boundary conditions, necessary for obtaining a statistically stationary state. In the preceding scaling analysis it was tacitly assumed that the mainstream flow did not influence the fluctuating component of the velocity field. We shall now see that the fluctuations do in fact exert a decisive influence on the mainstream field, although mathematically it is fraught with difficulties.

Mean values

The mainstream field is defined as a suitable local average of the true velocity field, for example over a time interval T starting at any time t,

$$\langle \boldsymbol{v}(\boldsymbol{x},t)\rangle = \lim_{T\to\infty}\frac{1}{T}\int_0^T \boldsymbol{v}(\boldsymbol{x},t+s)\,ds. \tag{31.16}$$

All other fields, for example the pressure, may similarly be averaged over time,

$$\langle p(\boldsymbol{x},t)\rangle = \lim_{T\to\infty}\frac{1}{T}\int_0^T p(\boldsymbol{x},t+s)\,ds. \tag{31.17}$$

These averages are assumed to exist and be time independent in the limit, although that is by no means a foregone conclusion. Defined in this way, the mean value of any field can in principle be determined experimentally with any desired precision from the average of a sufficiently large number of measurements of the instantaneous field value near the point $\boldsymbol{x}$ over a sufficiently long time interval T. The meaning of "near" and "sufficiently" depends on the dynamics and the desired measurement precision, as was discussed in Chapter 1.

The fluctuating part of a field is defined as the difference between the true field and its mean value. For the velocity and pressure fields we introduce special symbols for the fluctuations:

$$\boldsymbol{u}(\boldsymbol{x},t) = \boldsymbol{v}(\boldsymbol{x},t) - \langle \boldsymbol{v}(\boldsymbol{x},t)\rangle\,, \tag{31.18a}$$

$$q(\boldsymbol{x},t) = p(\boldsymbol{x},t) - \langle p(\boldsymbol{x},t)\rangle\,. \tag{31.18b}$$

By definition these fluctuations have vanishing means, $\langle \boldsymbol{u}\rangle = \boldsymbol{0}$ and $\langle q\rangle = 0$.

As in statistical mechanics, it is often convenient for more formal analysis to replace the time average of a field by the average of the field over a suitable statistical ensemble of fields. That this leads to the same mean values as time averages is a deep result that we shall not discuss here. In the following we only assume that there is a well-defined statistical procedure for calculating mean values, denoted by angular brackets $\langle\cdots\rangle$. The system is said to be in *statistical equilibrium* when all mean values are time independent.

Mean field equations

Even if quite different from laminar flow, turbulent flow is also assumed to be governed by the usual Navier–Stokes equations. To avoid cumbersome notation we shall in the following shift the fluctuations out of the basic fields by substituting $\boldsymbol{v} \to \boldsymbol{v}+\boldsymbol{u}$ and $p \to p+q$, such that the incompressible Navier–Stokes equations (in the absence of gravity) become

$$\frac{\partial(\boldsymbol{v}+\boldsymbol{u})}{\partial t} + (\boldsymbol{v}+\boldsymbol{u})\cdot\nabla(\boldsymbol{v}+\boldsymbol{u}) = -\frac{1}{\rho_0}\nabla(p+q) + \nu\nabla^2(\boldsymbol{v}+\boldsymbol{u}), \tag{31.19a}$$

$$\nabla\cdot(\boldsymbol{v}+\boldsymbol{u}) = 0. \tag{31.19b}$$

It must be emphasized that in these equations $\boldsymbol{v} = \langle \boldsymbol{v}\rangle$ and $p = \langle p\rangle$ now denote the mean fields, whereas $\boldsymbol{u}$ and q denote time-dependent fluctuations with vanishing means. For generality we here also allow the mean fields to be time dependent.

Taking the mean of the divergence condition (31.19b), we get $\nabla \cdot \boldsymbol{v} = 0$. This shows that the mean of an incompressible flow is also incompressible, a consequence of the linearity of the divergence condition. From the divergence condtion it follows in turn that the fluctuation field is incompressible, $\nabla \cdot \boldsymbol{u} = 0$. Taking the mean of the Navier–Stokes equation (31.19a), we get, similarly,

$$\frac{\partial \boldsymbol{v}}{\partial t} + (\boldsymbol{v} \cdot \nabla)\boldsymbol{v} + \langle (\boldsymbol{u} \cdot \nabla)\boldsymbol{u} \rangle = -\frac{1}{\rho_0}\nabla p + \nu \nabla^2 \boldsymbol{v}. \tag{31.20}$$

The last term on the left-hand side stems from the non-linear advective acceleration, and creates a coupling between the fluctuations and the mean fields. Using that $\nabla \cdot \boldsymbol{u} = 0$, it may be rewritten as a tensor divergence,

$$(\boldsymbol{u} \cdot \nabla)u_i = \sum_j u_j \nabla_j u_i = \sum_j \nabla_j (u_i u_j) = \nabla \cdot (\boldsymbol{uu}), \tag{31.21}$$

and moving the mean of this term to the right-hand side of (31.20), we arrive at the *mean field equations* (or *Reynolds equations*) for turbulent flow,

$$\boxed{\begin{aligned} \frac{\partial \boldsymbol{v}}{\partial t} + (\boldsymbol{v} \cdot \nabla)\boldsymbol{v} &= -\frac{1}{\rho_0}\nabla p + \nu \nabla^2 \boldsymbol{v} - \nabla \cdot \langle \boldsymbol{uu} \rangle . \\ \nabla \cdot \boldsymbol{v} &= 0. \end{aligned}} \tag{31.22}$$

The fluctuation field satisfies an equation that may be derived from the Navier–Stokes equation (31.19a) by subtracting the first mean field equation. Rewriting the advective term using (31.21), we obtain the fluctuation equations,

$$\frac{\partial \boldsymbol{u}}{\partial t} + (\boldsymbol{v} \cdot \nabla)\boldsymbol{u} + (\boldsymbol{u} \cdot \nabla)\boldsymbol{v} + \nabla \cdot (\boldsymbol{uu} - \langle \boldsymbol{uu} \rangle) = -\frac{1}{\rho_0}\nabla q + \nu \nabla^2 \boldsymbol{u}, \tag{31.23a}$$

$$\nabla \cdot \boldsymbol{u} = 0. \tag{31.23b}$$

The split into mean fields and fluctuations has thus led to two sets of coupled differential equations rather than one. They are of course together equivalent to the original Navier–Stokes equations so nothing has really been gained, except increased complexity!

The closure problem

The mean field equations (31.22) would by themselves form a closed system of dynamic equations, if only we could somehow express the second-order fluctuation moment $\langle u_i u_j \rangle$ in terms of the mean field and its derivatives. Unfortunately, as we shall now see, that is in general not possible.

One might, for example, attempt to determine the second-order moment $\langle u_i u_j \rangle$ from first principles by using the above fluctuation field equations (31.23) to calculate $\partial(u_i u_j)/\partial t$, and afterward take the mean. But such a procedure would introduce an unknown third-order moment of the form $\langle u_i u_j u_k \rangle$ through the non-linear fluctuation term $\nabla \cdot (\boldsymbol{uu})$ (see Problem 31.3 for details). Again one could use (31.23) to get an equation for the third-order moment, only to find that it involves a fourth-order moment, and so on. Even if in this way an infinite set of equations can in principle be written down for the moments, each equation will refer to a moment of higher order than the one it is meant to determine. *The set of moment equations does not close.*

The closure problem is inherent in the statistical treatment of non-linear systems of dynamic equations, and a large part of the theoretical studies of turbulence have focused on modeling the moments of the fluctuations to obtain closure. Sometimes symmetries and dimensional considerations permit general conclusions to be drawn in simple flow geometries with high symmetry and small number of parameters, but a fundamental theory of turbulence allowing for effective closure in any geometry is yet to be discovered.

Reynolds stresses

The first Reynolds equation takes the same form as the Navier–Stokes equation with an effective stress tensor,

$$\sigma^*_{ij} = -p\,\delta_{ij} + \eta(\nabla_i v_j + \nabla_j v_i) - \rho_0 \left\langle u_i u_j \right\rangle. \tag{31.24}$$

The last term was introduced by Reynolds in 1894 and is appropriately called the *Reynolds stress tensor*.

Apart from the fact that we cannot in general calculate the Reynolds stresses, they appear in the Reynolds equation for the mean velocity field on equal footing with the mean pressure and the mean viscous stresses. Since the Reynolds equation in analogy with the Navier–Stokes equation may be understood as expressing local momentum balance, the Reynolds stresses also act as forces by transferring momentum across internal surfaces cutting through the fluid. The only difference is that the Reynolds stresses must vanish at solid boundaries due to the no-slip condition for the fluctuations $\boldsymbol{u} = \boldsymbol{0}$, and that prevents us from measuring them directly. There are, however, many ways of measuring the velocity fluctuations over a long time interval, and from such data the mean values $\left\langle u_i u_j \right\rangle$, as well as the mean value of any other product, can be obtained.

Velocity defect

It has been mentioned before that in fully developed turbulent flow, the mean velocity does not vary much across the geometry, except very close to the walls where the no-slip condition has to be fulfilled. If $\boldsymbol{U}$ is a typical mean velocity in the bulk of the turbulent flow, the quantity

$$\Delta\boldsymbol{v} = \boldsymbol{v} - \boldsymbol{U} \tag{31.25}$$

is called the *velocity defect*. In the bulk of the turbulent flow, it is expected to be small, $|\Delta\boldsymbol{v}| \ll U$, while it will be large near a solid wall where the no-slip condition requires $\boldsymbol{v} = \boldsymbol{0}$. The magnitude of the velocity defect therefore determines the magnitude of the velocity gradients, $|\boldsymbol{\nabla v}| \sim \Delta v/L$, in the bulk of the turbulent flow.

From the Kolmogorov scaling analysis in the preceding section we know that the average eddy velocity must be of the same order of magnitude as the maximal eddy velocity u_L such that $|\langle \boldsymbol{uu} \rangle| \sim u_L^2$. The effective turbulent viscosity ν_{turb} can now be estimated from the ratio of the Reynolds stresses to the viscous stresses,

$$\nu_{\text{turb}} \sim \frac{|\langle \boldsymbol{uu} \rangle|}{|\boldsymbol{\nabla v}|} \sim \frac{u_L^2}{\Delta v/L} = \frac{u_L^2 L}{\Delta v}. \tag{31.26}$$

Since the eddy viscosity (31.12) grows rapidly with the eddy scale λ, the largest eddies will dominate the turbulent viscosity. Setting $\nu_{\text{turb}} \sim \nu_L = L u_L$ in the above estimate, we conclude that $\Delta v \sim u_L$.

These arguments support that *the velocity defect in the bulk of the turbulent flow is of roughly the same magnitude as the turbulent velocity fluctuations*. We shall verify this claim in the discussion of simple turbulent flows in Sections 31.5 and 31.6.

Effective Reynolds number for turbulence

The condition for turbulent viscosity to dominate over the molecular viscosity is that the Reynolds number of the system-scale eddies is large, $\mathsf{Re}_L = u_L L/\nu \sim \nu_{\text{turb}}/\nu \gg 1$. There is in fact no turbulence unless this condition is fulfilled.

We are now also in a position to estimate the effective turbulent Reynolds number, defined as the ratio between the advective mean field contribution and the dominant Reynolds stress contribution,

$$\mathsf{Re}_{\text{turb}} = \frac{|(\boldsymbol{v}\cdot\nabla)\boldsymbol{v}|}{|\nabla\cdot\langle\boldsymbol{u}\boldsymbol{u}\rangle|} \sim \frac{U\Delta v/L}{u_L^2/L} \sim \frac{U}{u_L} = \frac{\mathsf{Re}}{\mathsf{Re}_L}, \tag{31.27}$$

where as before $\mathsf{Re} = UL/\nu$. In the following we shall see that the right-hand side is not usually very large, in realistic systems somewhere between 10 and 40. This is comforting, since it would be completely meaningless if $\mathsf{Re}_{\text{turb}}$ were to become so large that the mean field equations became unstable and themselves developed turbulence!

31.3 Universal inner layer near a smooth wall

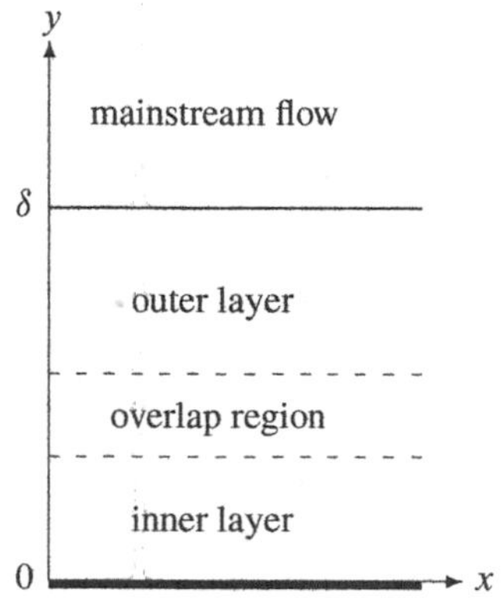

Sketch of the turbulent boundary layer structure near a solid wall. The inner layer interfaces to the wall, whereas the outer layer interfaces to the mainstream flow. The inner and outer layers overlap in a finite region between the dashed lines. There is no clear break between the inner and outer layers.

Turbulent flow along a solid wall appears at first sight to be rather uniform and reach all the way to the wall, but that cannot be completely true. Viscosity demands that the mean velocity field as well as the fluctuation field must vanish exactly on the wall. The conflict between viscosity's insistence on a no-slip condition and turbulence's tendency to even out mean velocity differences leads to the formation of a two-decked boundary layer structure. Thus, in any turbulent boundary flow there will be an *inner layer* that secures the fulfillment of the no-slip condition, and an *outer layer* that implements the interface to the mainstream flow. The inner and outer layers must *overlap* in a finite region, because in a viscous flow there cannot be a sharp break anywhere in the true velocity field or in any of its derivatives. It is the requirement of a finite overlap region that connects the inner and outer layers, and thereby the main flow with the stresses that the fluid exerts on the wall.

In this section we shall carry out a semi-empirical analysis of the inner layer and show that it has a universal structure valid for all wall-bounded flows. The analysis of the mainflow-dependent outer layer is taken up in the following section. We shall make the same basic assumptions as for laminar boundary flow (Chapter 28). The total thickness δ of the boundary layer will thus be assumed to be much smaller than the mainstream geometry scale L. The wall may then be viewed as locally flat, coinciding with the xz-plane, $y = 0$, in a Cartesian coordinate system. The x-axis is placed along the mainstream mean flow direction such that the mean flow along z vanishes throughout the boundary layer, $v_z = 0$. The non-vanishing velocity components v_x and v_y depend mainly on y, but may change slowly with x over the mainstream geometry scale L.

Linear law of the wall

Very close to the wall, the fundamental local quantity is the *mean shear wall stress*

$$\boxed{\sigma_{\text{wall}} = \left\langle\sigma_{xy}\right\rangle_{y=0} = \eta\, \nabla_y v_x\big|_{y=0}\,.} \tag{31.28}$$

The mean turbulent drag on the wall is obtained by integrating the wall stress over the wall.

For physical reasons σ_{wall} must be finite, so that the near-wall velocity field becomes

$$v_x = \frac{\sigma_{\text{wall}}}{\eta}\, y, \tag{31.29}$$

independent of what happens in the bulk of the flow, be it laminar or turbulent. Turbulent fluctuations are in fact, as we shall see later, negligible in this *viscous sublayer*.

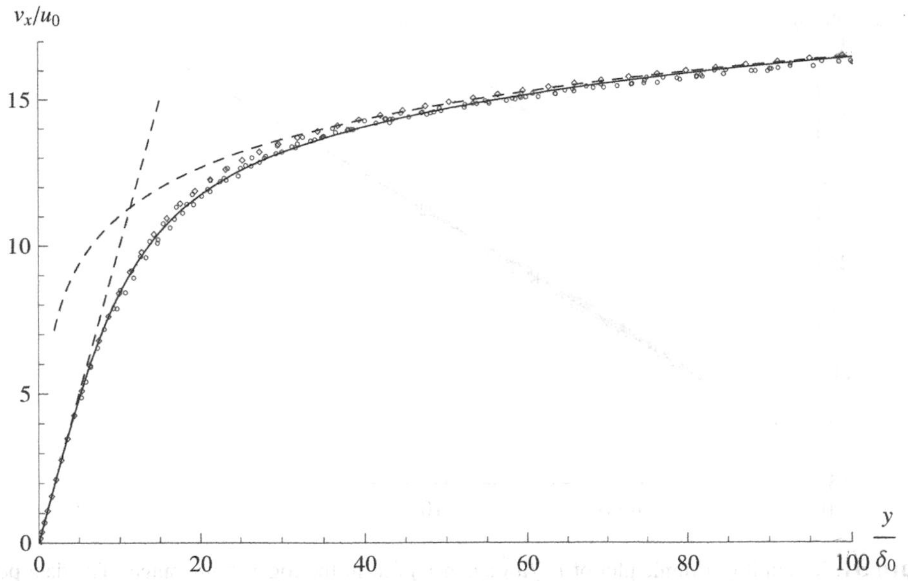

Figure 31.2. Plot of v_x/u_0 against y/δ_0 in the near-wall region $0 \leq y/\delta_0 \lesssim 100$. The data points are obtained from direct numerical simulation of channel flow [MKM99] (diamonds) and from experiments on the Princeton SuperPipe facility [MLJ&04] (circles). Both datasets are publicly available at respectively [18] and [19]. There is a slight disagreement between the datasets in the transition region $10 < y/\delta_0 < 50$. The dashed straight line is the velocity in the linear sublayer (31.32), and the dashed curve is the velocity in the logarithmic sublayer (31.35) with $A = 2.37$ and $B = 5.55$ obtained from a fit to the SuperPipe data in the logarithmic region $y/\delta_0 > 300$ (see Figure 31.3). The solid line is the universal interpolation (31.47) between the linear and logarithmic laws of the wall with $\alpha = 1.070 \times 10^{-3}$ and $\beta = 1.277 \times 10^{-4}$, obtained by fitting the SuperPipe data.

Since stress and energy density are measured in the same units, Pa = kg m^{-1}s^{-2} = J m^{-3}, we may translate the wall stress into a velocity u_0, called the *friction velocity*, by setting

$$\sigma_{\text{wall}} = \rho_0 u_0^2. \tag{31.30}$$

This only makes sense for positive wall-stress and therefore excludes the discussion of turbulent boundary layer separation. Introducing the *viscous length scale*

$$\delta_0 = \frac{\nu}{u_0}, \tag{31.31}$$

the near-wall velocity field may now be written conveniently as[1]

$$\boxed{\frac{v_x}{u_0} = \frac{y}{\delta_0}.} \tag{31.32}$$

Empirically, this *linear law of the wall* is found to be quite accurate for $0 \leq y/\delta_0 \lesssim 5$, as may be seen in Figure 31.2. The limit to the boundary layer in units of the friction scale defines the dimensionless *Karman number*

$$\mathsf{Ka} = \frac{\delta}{\delta_0} = \frac{u_0 \delta}{\nu}, \tag{31.33}$$

often also denoted Re_τ. In Figure 31.2 the channel flow data have $\mathsf{Ka} = 590$ whereas the SuperPipe data range from $\mathsf{Ka} = 1{,}800$ to 529,000.

[1] We shall not use the customary but rather unnatural notation $y^+ = y/\delta_0$ and $u^+ = v_x/u_0$, etc.

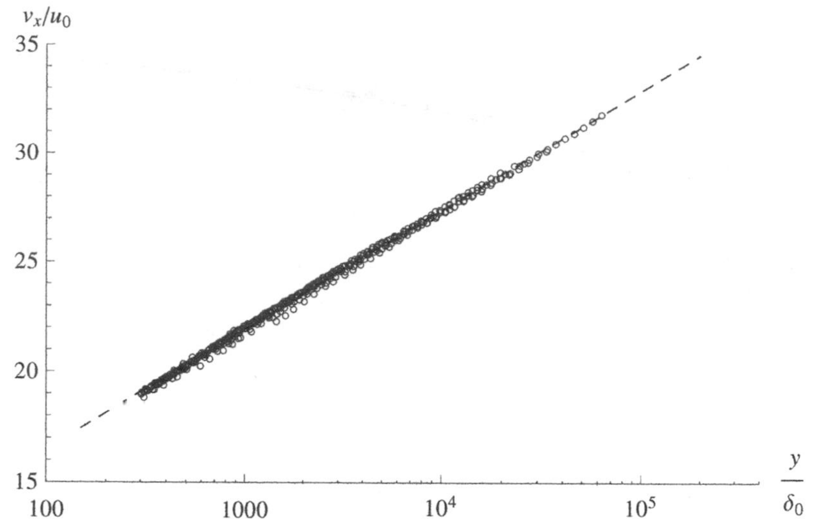

Figure 31.3. Semilogarithmic plot of v_x/u_0 against y/δ_0 in the logarithmic range. The data points are obtained at the Princeton SuperPipe facility [MLJ&04] (publicly available at [19]), and consist of 19 different sets from $\mathsf{Ka} \approx 1{,}800$ to 529,000. The dashed line is a fit to the logarithmic law (31.35) in the interval $300 < y/\delta_0 < 0.12\,\mathsf{Ka}$, yielding $A = 2.37$ and $B = 5.55$.

Logarithmic law of the wall

Deep inside the boundary layer but well beyond the linear region, that is, for $\delta_0 \ll y \ll \delta$, the flow will be turbulent; but because of the proximity of the wall, the friction velocity u_0 still sets the velocity scale. In keeping with the discussion in the preceding sections we expect that the mean velocity slope $\nabla_y v_x$ in this region is essentially independent of the viscosity. It can then only depend on u_0 and y, and must for dimensional reasons be of the form

$$\nabla_y v_x = A \frac{u_0}{y}, \tag{31.34}$$

where A is a dimensionless constant. Integrating this expression, we obtain the *logarithmic law of the wall*,

$$\boxed{\frac{v_x}{u_0} = A \log \frac{y}{\delta_0} + B \qquad (\delta_0 \ll y \ll \delta),} \tag{31.35}$$

where B is another dimensionless constant, and where we have normalized the logarithm by the viscous length scale δ_0 to make its argument dimensionless.

The logarithmic law goes back to von Karman and Prandtl in the early 1930s, and $\kappa = 1/A$ is called the *Karman constant.* In Figure 31.3 it is compared with the Princeton SuperPipe experiments, yielding $A = 2.37$ and $B = 5.55$.

Universal inner law of the wall

In the inner layer $0 \leq y \ll \delta$, the flow should not depend on δ, so that we may express the streamwise mean velocity v_x in terms of the friction velocity u_0 and the dimensionless distance from the wall,

$$\boxed{\frac{v_x}{u_0} = f\left(\frac{y}{\delta_0}\right).} \tag{31.36}$$

Here $f(s)$ is a purely numeric function of its variable $s = y/\delta_0$ in the interval $0 \leq s \ll \delta/\delta_0$.

From the preceding analysis we already know the asymptotic behavior of $f(s)$ for small and not too large s,

$$f(s) = \begin{cases} s & (0 \le s \lesssim 5) \\ A \log s + B & (50 \lesssim s \ll \delta/\delta_0). \end{cases} \tag{31.37}$$

These expressions cross each other at $s \approx 11$. Although we do not know the exact shape of the function in the crossover region, $5 \lesssim s \lesssim 50$, this anchoring in the extremes does not leave much liberty for variation in between. Below we shall present the simplest possible interpolation that connects the extremes.

The mean field in the y-direction is determined by the divergence condition $\nabla_x v_x + \nabla_y v_y = 0$, and may in the complete inner wall layer be verified to be (Problem 31.4)

$$v_y = -y\, f\left(\frac{y}{\delta_0}\right) \nabla_x u_0. \tag{31.38}$$

The ratio of the two velocities is consequently

$$\frac{v_y}{v_x} = -\frac{y \nabla_x u_0}{u_0}. \tag{31.39}$$

Since $\nabla_x u_0 \sim u_0/L$, the magnitude of the ratio is $v_y/v_x \sim y/L$, making the upflow negligible in the whole inner wall region $0 \le y \ll \delta$.

Reynolds shear stress in the inner layer

In the preceding analysis we have only used dimensional arguments and mass conservation, and completely ignored the Navier–Stokes equation, which after all controls the dynamics. Since all the x-dependence in the inner wall layer stems from the wall friction velocity u_0, any x-derivative must be of magnitude $\nabla_x \sim 1/L$. Similarly, all y-dependence is governed by δ_0 such that $\nabla_y \sim 1/\delta_0$. In the limit of $L \to \infty$, we may drop all x-derivatives (and therefore also v_y) in the first Reynolds equation (31.22), and get

$$0 = \nu \nabla_y^2 v_x - \nabla_y \left\langle u_x u_y \right\rangle, \tag{31.40a}$$

$$\nabla_y p = -\rho_0 \nabla_y \left\langle u_y^2 \right\rangle. \tag{31.40b}$$

Integrating these equations with respect to y and using the boundary condition given by the wall shear stress (31.28), they become

$$u_0^2 = \nu \nabla_y v_x - \left\langle u_x u_y \right\rangle, \tag{31.41a}$$

$$p = p_0 - \rho_0 \left\langle u_y^2 \right\rangle, \tag{31.41b}$$

where p_0 is the pressure on the wall. The last equation shows that in contrast to the laminar case, turbulence causes a (small) drop in the pressure as we move away from the wall.

Finally, inserting the mean field (31.36) into (31.41a), we get

$$\boxed{\left\langle u_x u_y \right\rangle = -u_0^2 \left(1 - f'\left(\frac{y}{\delta_0}\right)\right).} \tag{31.42}$$

The slope $f'(s)$ decreases from $f' = 1$ at $s = 0$ to $f' = 0$ for $s = \infty$; so for $\delta_0 \ll y \ll \delta$, the average is constant $\left\langle u_x u_y \right\rangle \approx -u_0^2$. Due to the negative sign, the fluctuations u_x and u_y are always statistically anti-correlated, meaning that if $u_x > 0$, then $u_y < 0$, and conversely.

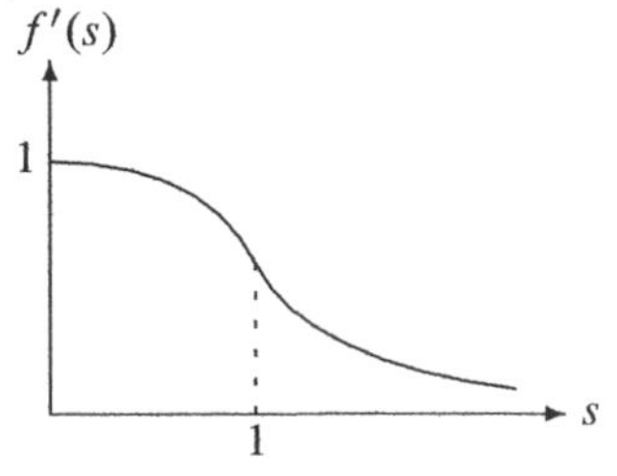

Sketch of the slope $f'(s)$.

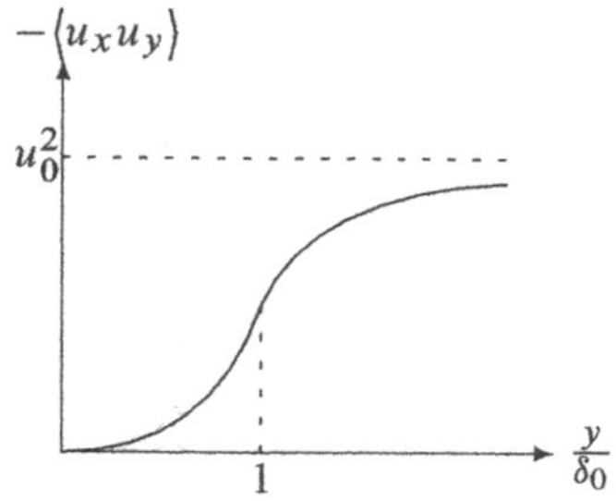

Sketch of $-\langle u_x u_y\rangle$ as a function of y/δ_0.

How fast does $\langle u_x u_y\rangle$ vanish in the viscous sublayer? For physical reasons all velocity derivatives must be finite at the wall, so that all components of the fluctuation field $\boldsymbol{u} = (u_x, u_y, u_z)$ must vanish at least as fast as y. Combined with the divergence condition $\nabla\cdot\boldsymbol{u} = \nabla_x u_x + \nabla_y u_y + \nabla_z u_z = 0$, it follows that $u_x \sim y$ and $u_z \sim y$, whereas $\nabla_y u_y \sim y$ and therefore $u_y \sim y^2$. From this we get $\langle u_x u_y\rangle \sim y^3$, and putting in the dimensional factors this may be written as

$$\langle u_x u_y\rangle \approx -\alpha\, u_0^2 \left(\frac{y}{\delta_0}\right)^3 \qquad (0 \leq y \lesssim \delta_0), \tag{31.43}$$

where α is a numerical constant. In terms of the dimensionless shape function $f(s)$, this translates into $f'(s) \approx 1 - \alpha\, s^3$ and thus $f(s) = s - \frac{1}{4}\alpha s^4$.

The local turbulent shear viscosity ν_{turb} is defined by writing

$$-\langle u_x u_y\rangle = \nu_{\text{turb}} \nabla_y v_x. \tag{31.44}$$

Inserting this into (31.41a) and making use of (31.42), we find

$$\frac{\nu_{\text{turb}}}{\nu} = \frac{1}{f'(y/\delta_0)} - 1 \approx \begin{cases} \alpha\left(\dfrac{y}{\delta_0}\right)^3 & (y \ll \delta_0) \\[2ex] \dfrac{1}{A}\dfrac{y}{\delta_0} & (\delta_0 \ll y \ll \delta). \end{cases} \tag{31.45}$$

In accordance with intuition, the turbulent viscosity in the inner layer thus vanishes rapidly at the wall but increases linearly at greater distance.

Interpolation between linear and logarithmic sublayers

There are many ways of generating an interpolating function $f(s)$ connecting the extremes $s \ll 1$ and $s \gg 1$. A minimal interpolation for the slope that takes into account both $f' \approx 1 - \alpha s^3$ for $s \ll 1$ and $f' \approx A/s$ for $s \gg 1$, for example,

$$f'(s) = \frac{1}{1 + (\alpha + \beta)\, s^3} + \frac{A\beta\, s^3}{A + \beta\, s^4}, \tag{31.46}$$

where α and β are numeric constants. One may verify that it integrates to

$$\begin{aligned} f(s) &= \frac{1}{4} A \log\left(1 + \frac{\beta}{A} s^4\right) \\ &+ \frac{1}{18\gamma}\left(\pi\sqrt{3} - 6\sqrt{3}\arctan\frac{1 - 2\gamma s}{\sqrt{3}} + 3\log\frac{(1+\gamma s)^2}{1 - \gamma s + \gamma^2 s^2}\right), \end{aligned} \tag{31.47}$$

where $\gamma = (\alpha + \beta)^{1/3}$.

For $s \to \infty$, this approaches the logarithmic law of the wall (31.35) with

$$B = \frac{2\pi}{3\sqrt{3}}\frac{1}{\gamma} - \frac{1}{4} A \log\frac{A}{\beta}. \tag{31.48}$$

As shown in Figure 31.3 the Princeton SuperPipe data [MLJ&04] yield a good fit for $A = 2.37$, $B = 5.55$, $\gamma = 0.1062$, $\beta = 1.277 \times 10^{-4}$, and thus $\alpha = 1.070 \times 10^{-3}$.

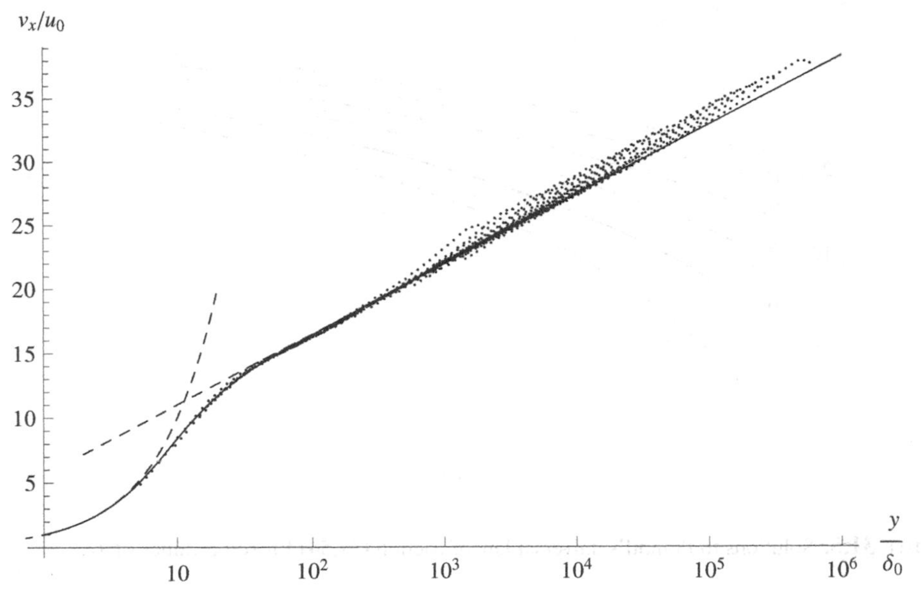

Figure 31.4. Semilogarithmic plot of v_x/u_0 against y/δ_0 over a wide range. The data points are obtained at the Princeton SuperPipe facility [MLJ&04] and are publicly available [19]. The deviations from the logarithmic law represent the outer layer wake function for pipe flow (see Section 31.6). The data points are obtained at 19 different Karman numbers ranging from Ka $\approx$1,800 to 529,000. The solid curve is the interpolation function (31.47) with $A = 2.37$ and $B = 5.55$, and the fitted values are $\alpha = 1.070 \times 10^{-3}$ and $\beta = 1.277 \times 10^{-4}$. The dashed curve is the velocity in the linear sublayer (31.32), and the dashed straight line the velocity in the logarithmic sublayer (31.35).

31.4 The outer layer

Eventually, at a scale set by the boundary layer thickness δ, the logarithmic growth of the mean velocity field v_x will come under the influence of the mainstream flow, causing deviations from the universal inner law of the wall (31.36). It is clearly not possible to give a general description of all mainstream flows, but as long as the boundary layer is thin compared to the mainstream geometry, $\delta \ll L$, one may nevertheless say something quite general about the flow in the outer layer.

The law of the wake

It was previously (on page 573) pointed out that in a fully developed turbulent flow, the mean velocity only varies slowly across the geometry, and that the velocity defect is of the same order of magnitude as the turbulent fluctuations. We shall interpret this to mean that the difference between the velocity and the logarithmic law (31.35) is itself of magnitude u_0, so that the velocity field in the complete boundary layer, $0 < y < \delta$, must be of the form

$$\boxed{\frac{v_x}{u_0} = f\left(\frac{y}{\delta_0}\right) + w\left(\frac{y}{\delta}\right),} \tag{31.49}$$

where $f(s)$ is the inner wall law (31.36), and $w(s)$ is the *wake function* introduced by Coles in 1956 [Col56]. The wake function must vanish for $s \to 0$ in order not to interfere with the inner law. In Figure 31.4 the wake function only shows up as the little bump toward the end of each of the 19 velocity profiles.

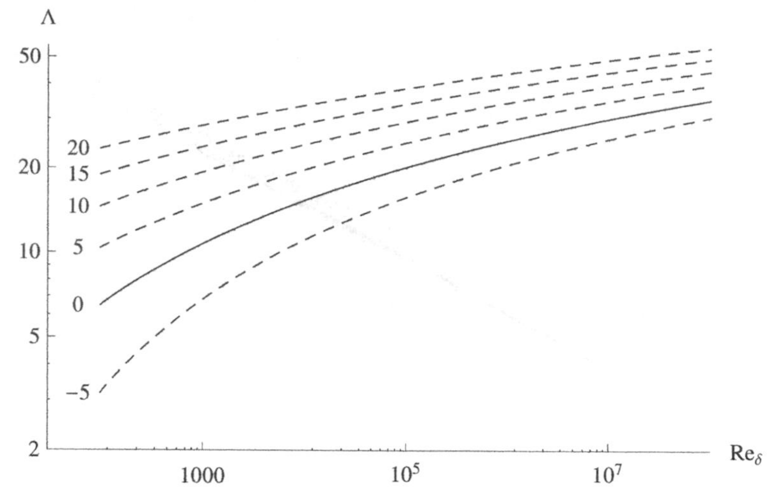

Figure 31.5. Solutions to Prandtl's universal law of friction (31.54) for some values of D.

Suppose that we know the velocity $v_x = U_\delta$ for $y = \delta \gg \delta_0$, so that

$$\frac{U_\delta}{u_0} = A \log \frac{\delta}{\delta_0} + B + C, \tag{31.50}$$

where $C = w(1)$. Subtracting, we obtain the *velocity defect function* in the outer layer $\delta_0 \ll y \lesssim \delta$,

$$\boxed{\frac{v_x - U_\delta}{u_0} = A \log \frac{y}{\delta} + w\left(\frac{y}{\delta}\right) - C.} \tag{31.51}$$

This expression depends only on y/δ and vanishes by construction for $y = \delta$. The constant B has fallen out here and is of no importance for the velocity defect in the outer layer.

Prandtl's universal law of friction

In the outer layer, the local Reynolds number and Fanning friction coefficient take the form

$$\mathsf{Re}_\delta \equiv \frac{U_\delta \delta}{\nu} = \Lambda \mathsf{Ka}, \qquad C_f \equiv \frac{\sigma_{\text{wall}}}{\frac{1}{2}\rho_0 U_\delta^2} = \frac{2}{\Lambda^2}, \tag{31.52}$$

where $\mathsf{Ka} = \delta/\delta_0 = u_0\delta/\nu$ is the Karman number (31.33) and

$$\Lambda \equiv \frac{U_\delta}{u_0} = A \log \mathsf{Ka} + B + C. \tag{31.53}$$

Together these equations define a relation between C_f and Re_δ parameterized by Ka.

Setting $\mathsf{Ka} = \mathsf{Re}_\delta/\Lambda$, we arrive at *Prandtl's universal law of friction,*

$$\boxed{\Lambda = A \log\left(\frac{\mathsf{Re}_\delta}{\Lambda}\right) + D,} \tag{31.54}$$

where $D = B + C$. It determines Λ and thereby C_f implicitly as function of Re_δ. Whereas A and B are universal, the constant C, and thus D, depend on the problem. The solution is plotted in Figure 31.5 for some values of D.

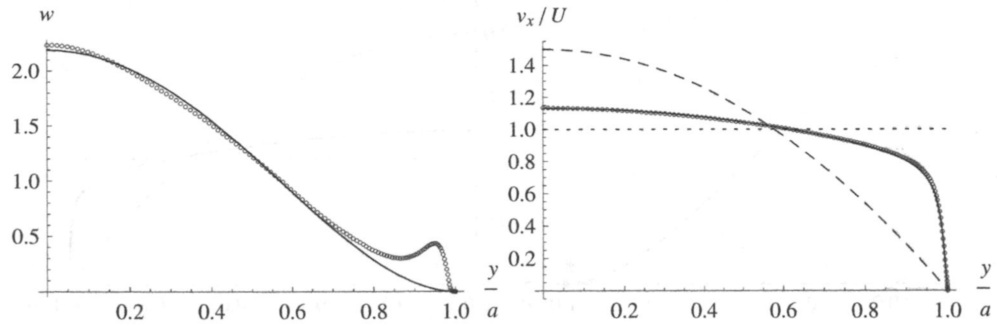

Figure 31.6. Pressure-driven channel flow in the interval $0 < y < a$ with the wall at $y = a$. The data points are from the simulation [MKM99] at $\mathsf{Ka} = 587.19$, corresponding to $\mathsf{Re} \approx 22{,}000$. The parameters are those determined in Figure 31.4. **Left:** The wake function. The solid curve is $w(s) = C\left(1 - s^2\right)^2$ with $C = 2.19$. The wiggle in the data near the wall is caused by the tiny deviations from the universal fit in Figure 31.2. **Right:** The velocity profile in units of the average velocity. The solid curve is obtained from (31.56) and the procedure described in the text. The dashed curve is the laminar solution $v_x/U = \frac{3}{2}\left(1 - (y/a)^2\right)$.

31.5 Application: Turbulent channel flow

Symmetry and dimensional arguments play an important role in the analysis of stationary turbulence in simple flow geometries where the number of parameters is small. In this section we shall apply the semi-empirical theory of turbulent wall-bounded flows to model the flow profile in planar pressure-driven channel flow. The laminar solution for this geometry has previously been obtained in Section 16.2. Since the geometry is infinitely extended along the flow direction, we shall assume that the outer boundary layers have merged to form a *turbulent core* that smoothly connects the outer layer of one wall with that of the opposite wall.

Mean velocity profile

In channel flow, both plates are fixed at distance $d = 2a$, and the fluid is driven between them by a constant pressure gradient. As before, we shall position the plates at $y = \pm a$ and modify the preceding formalism accordingly. The stream-wise mean velocity field must obey

$$v_x(-y) = v_x(y) \tag{31.55}$$

because of the geometric symmetry around the mid-plane. A representation of the same nature as (31.49) that respects this symmetry and has the correct near-wall behavior (31.37) for $y \to \pm a$ is, for example,

$$\boxed{\frac{v_x}{u_0} = f\left(\frac{a^2 - y^2}{2a\delta_0}\right) + w\left(\frac{y}{a}\right).} \tag{31.56}$$

The wake function must be symmetric, $w(-s) = w(s)$ with $w(0) = 1$ and $w(\pm 1) = 0$.

In Figure 31.6L, the data points for the wake function are obtained from the high-precision simulation data [MKM99] at $\mathsf{Ka} \equiv a/\delta_0 = 587.19$ with the parameters determined from the fit in Figure 31.2, that is, $A = 2.37$, $B = 5.55$, $\alpha = 1.070 \times 10^{-3}$, and $\beta = 1.277 \times 10^{-4}$. The wake function $w(s) = C(1 - s^2)^2$ with $C = 2.19$ yields a decent fit in the interval $y/a < 0.8$.

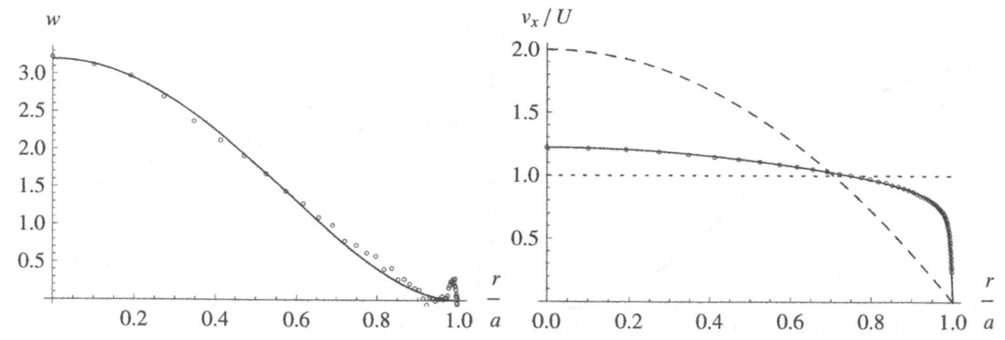

Figure 31.7. Pressure-driven pipe flow for Ka $= 1{,}825$ corresponding to Re $= 77{,}300$. The data are taken from the Princeton Superpipe results [MLJ&04] and [19]. **Left:** The wake function. Data on the wake function are calculated from (31.61) with the parameters determined in Figure 31.3. The solid profile is $w = C\left(1 - r^2/a^2\right)^2$ with $C = 3.20$. **Right:** The velocity profile in units of the average velocity. The solid curve is the fitted velocity profile (31.61). The dashed curve is the laminar profile $v_x/U = 2(1 - r^2/a^2)$ carrying the same volume discharge.

Average velocity

The average velocity becomes

$$U = \frac{1}{a}\int_0^a v_x(y)\,dy = u_0 \int_0^1 \left[f\left(\tfrac{1}{2}\mathsf{Ka}\left(1 - s^2\right)\right) + w(s)\right] ds. \tag{31.57}$$

Disregarding the inner linear region and inserting the logarithmic law $f(s) = A\log s + B$, the integral becomes elementary, with the result that

$$\boxed{\Lambda \equiv \frac{U}{u_0} \approx A\log(2\mathsf{Ka}) - 2A + B + \tfrac{8}{15}C.} \tag{31.58}$$

For the simulation data [MKM99] we have Ka $= 587$ and $C = 2.19$ and find $\Lambda = 18.73$. Precise numerical integration of (31.57) yields essentially the same result. The conventional Reynolds number for channel flow can now be calculated,

$$\mathsf{Re} = \frac{2aU}{\nu} = 2\mathsf{Ka}\,\Lambda, \tag{31.59}$$

which evaluates to Re $\approx 22{,}000$ for the simulation. In Figure 31.6R, the resulting velocity field is plotted in units of the average velocity (see below).

In terms of the conventional Reynolds number, Prandtl's universal friction law takes the same general form as before (31.54),

$$\Lambda = A\log\frac{\mathsf{Re}}{\Lambda} + D, \tag{31.60}$$

where $D = -2A + B + \frac{8}{15}C = 1.98$ in this case.

31.6 Application: Turbulent pipe flow

The basic phenomenology of turbulent pipe flow was discussed in Section 16.5 on page 274. In a pipe of radius a, the longitudinal extent of the flow is essentially infinite whereas the orthogonal extent $2\pi a$ is finite; but as long as $\delta_0 \ll a$, the flat-wall universal law of the wall should also apply to this case. The following analysis proceeds essentially along the same lines as in channel flow with the results shown in Figure 31.7.

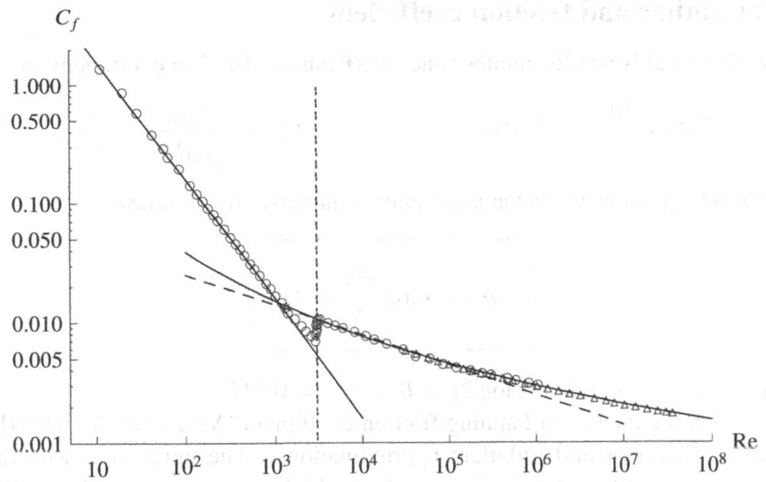

Figure 31.8. The friction coefficient for smooth-pipe flow as a function of the Reynolds number. The solid straight line is the laminar result $C_f = 16/\mathsf{Re}$, and the solid curve obtained from the solution to (31.65). The dashed straight line is the Blasius smooth-pipe expression $C_f = 0.079\mathsf{Re}^{-1/4}$ (page 276). The data points are extracted from [MSZ&04] and stem from two independent experiments (circles and triangles). The transition to turbulence (the vertical dashed line) takes place at $\mathsf{Re} = 3{,}000$.

Mean velocity profile

Symmetry demands that the mean velocity field v_z depends only on r and has vanishing derivative $\partial v_z/\partial r = 0$ for $r = 0$. A representation that satisfies these conditions and agrees with the universal inner layer shape near the wall $r = a$ is of the same form as the channel flow expression (31.56),

$$\boxed{\frac{v_x}{u_0} = f\left(\frac{a^2 - r^2}{2a\delta_0}\right) + w\left(\frac{r}{a}\right).} \tag{31.61}$$

The wake function $w(s)$ must vanish at the wall, $w(1) = 0$, and have vanishing derivative at the center, $w'(0) = 0$.

In Figure 31.7L, the wake function data is plotted for one of the profiles obtained from the Princeton SuperPipe [MLJ&04] at $\mathsf{Ka} = 1{,}825$ corresponding to $\mathsf{Re} = 77{,}300$. The parameters are as in Figure 31.4, that is, $A = 2.27$, $B = 5.55$, $\alpha = 1.070 \times 10^{-3}$,and $\beta = 1.277\times10^{-4}$. The wake function function is fitted to $w = C(1-r^2/a^2)^2$ with $C = 3.20$.

Average velocity

The average velocity becomes

$$U = \frac{1}{\pi a^2}\int_0^a v_z(r)2\pi r\,dr = u_0\int_0^1 \left(f\left(\tfrac{1}{2}\mathsf{Ka}(1-s^2)\right) + w(s)\right)2s\,ds. \tag{31.62}$$

The value obtained by numerical integration is $U/u_0 = 18.505$. A good approximation is obtained by $f(s) \approx A\log s + B$,

$$\boxed{\Lambda \equiv \frac{U}{u_0} = A\log\mathsf{Ka} - A(1+\log 2) + B + \tfrac{1}{3}C,} \tag{31.63}$$

and yields $\Lambda \approx 31.04$. The velocity profile v_x/U is plotted in Figure 31.7R.

Reynolds number and friction coefficient

In pipe flow, the usual Reynolds number and the (Fanning) friction coefficient are

$$\mathsf{Re} = \frac{2aU}{\nu} = 2\Lambda\mathsf{Ka}, \qquad C_f = \frac{\sigma_{\text{wall}}}{\frac{1}{2}\rho_0 U^2} = \frac{2}{\Lambda^2}. \tag{31.64}$$

Combining these equations we get again Prandtl's universal friction law,

$$\boxed{\Lambda = A \log \frac{\mathsf{Re}}{\Lambda} + D,} \tag{31.65}$$

where in this case $D = -A(1 + 2\log 2) + B + \frac{1}{3}C = 0.915$.

In Figure 31.8 the measured Fanning friction coefficient [MSZ&04, MLJ&04] is plotted together with the laminar and turbulent approximations. The agreement with data is fine in both the laminar and turbulent regions up to the highest experimental Reynolds number, $\mathsf{Re} = 3 \times 10^7$, whereas the Blasius power law (16.49) on page 276 only fits well up to $\mathsf{Re} = 10^5$.

31.7 Turbulence modeling

In the preceding sections we have analyzed a couple of systems so simple that their behavior in the turbulent regime is severely constrained by symmetry and dimensional arguments. It must be emphasized that this approach is not fundamental but rather semi-empirical, as witnessed by the appearance of numerical constants that have to be determined from comparison with data. Some of these constants are justifiably believed to be universal, whereas others depend on the general flow configuration.

Ideally, we would like to calculate these constants from first principles. Since we have at hand the fundamental Navier–Stokes equations governing all Newtonian fluids, it is "just" a question of solving them for any particular flow geometry. Direct numerical simulation of turbulent flow is indeed possible and yields believable results with high precision as in the case of channel flow (Section 31.5). Combined with symmetry and dimensional analysis, the numerical constants may in this sense be determined from first principles without any recourse to real experiments. Unfortunately, the hoped-for universality is not perfect, and different numerical and real experiments disagree somewhat on the precise values of the constants.

Theoretically, the closure problem is the real obstacle to solving the Navier–Stokes equations with turbulence. Barring direct numerical simulation and "cheaper" variants thereof such as large eddy simulation, the only way out of this conundrum seems to be to break off the otherwise infinite set of fluctuation equations, and thereby enforce closure. There are, however, many ways of doing so, and in the last 100 years numerous closure modeling schemes have appeared, sometimes introducing many new adjustable parameters. It is not the intention here to go further into this huge field, which is well presented in the large number of modern textbooks on turbulence.

Problems

31.1 Show that the mean value of the time derivative vanishes, $\langle \partial \boldsymbol{v}/\partial t \rangle = \boldsymbol{0}$, when the velocity field is bounded, $|\boldsymbol{v}(\boldsymbol{x}, t)| \leq C$ for all $\boldsymbol{x}$ and t. This condition is normally fulfilled for physical systems of finite size.

31.2 **(a)** Normalize the eddy probability distribution (31.4) in the inertial range $\lambda_d < \lambda < L$. **(b)** How does it change the Kolmogorov law?

31.3 **(a)** Show that the following equation is satisfied by the second-order fluctuation moment tensor $\langle u_i u_j \rangle$:

$$\frac{\partial \langle u_i u_j \rangle}{\partial t} + (\boldsymbol{v} \cdot \nabla) \langle u_i u_j \rangle + \langle u_i \boldsymbol{u} \rangle \cdot \nabla v_j + \langle u_j \boldsymbol{u} \rangle \cdot \nabla v_i + \nabla \cdot \langle \boldsymbol{u} u_i u_j \rangle$$
$$= -\frac{1}{\rho_0} \left(\langle u_i \nabla_j q \rangle + \langle u_j \nabla_i q \rangle \right) + \nu \left(\nabla^2 \langle u_i u_j \rangle - 2 \langle \nabla u_i \cdot \nabla u_j \rangle \right).$$

(b) Discuss the closure problem.

31.4 Prove that $v_y = -y f(y/\delta_0) \nabla_x u_0$ is the correct expression for the inner layer upflow (see page 577).

31.5 **(a)** Investigate the convergence of the following iteration scheme for the universal friction law (31.54):

$$\Lambda_{n+1} = A \log \frac{\mathsf{Re}_\delta}{\Lambda_n} + D, \qquad \Lambda_0 = 1,$$

where Λ_n is the nth approximant to $\Lambda = U_\delta / u_0$. **(b)** Find the approximant sequence for $A = 2.37$, $D = 0$, and $\mathsf{Re}_\delta = 10^6$.

Part VI

Appendices

Appendices

The following appendices support the main text and may be sampled as the need arises.

List of appendices

A. **Newtonian mechanics:** This subject is the foundation of continuum mechanics, and a certain familiarity is assumed throughout. In this chapter the basic principles are outlined with an eye on the global laws of balance.

B. **Cartesian coordinates:** Vector and tensor algebra in Cartesian coordinates is presented in a systematic fashion with emphasis on coordinate transformations.

C. **Field calculus:** Fields are scalar-, vector-, and tensor-functions of space and time. The vector differential operator *nabla* is the basis for the efficient formulation of the equations of continuum physics. The fundamental integral theorems due to Gauss and Stokes are presented and proved.

D. **Curvilinear coordinates:** Non-Cartesian coordinates are useful for systems with geometric symmetries. Only cylindrical and spherical coordinates are discussed.

E. **Ideal gases:** Ideal gases form a convenient "laboratory" for understanding thermodynamics in a non-trivial case. The equation of state, energy and entropy of ideal gases are presented.

A

Newtonian mechanics

The equations of continuum mechanics are derived by a systematic application of Newton's laws to systems that nearly behave as if they consisted of idealized point particles. It is for this reason useful here to recapitulate the basic mechanics of point particles, and to derive the global laws that so often are found to be useful.

The global laws state for any collection of point particles that

- *The rate of change of momentum equals force*
- *The rate of change of angular momentum equals the moment of force*
- *The rate of change of kinetic energy equals power*

Even if these laws are not sufficient to determine the dynamics of a physical system, they represent seven individual constraints on the motion of any system of point particles, independent of how complex it is. They are equally valid for continuous systems when it is properly taken into account that the number of particles in a body may change with time.

A.1 Dynamic equations

In Newtonian mechanics, a physical system or *body* is understood as a collection of a certain number N of point particles numbered $n = 1, 2, \ldots, N$. Each particle obeys Newton's Second Law,

$$m_n \frac{d^2 \boldsymbol{x}_n}{dt^2} = \boldsymbol{f}_n, \tag{A.1}$$

where m_n denotes the (constant) mass of the nth particle, $\boldsymbol{x}_n$ its instantaneous position, and $\boldsymbol{f}_n$ the instantaneous force acting on the particle. Due to the mutual interactions between the particles, the forces may depend on the instantaneous positions and velocities of all the particles, including themselves,

$$\boldsymbol{f}_n = \boldsymbol{f}_n\left(\boldsymbol{x}_1, \ldots, \boldsymbol{x}_N, \frac{d\boldsymbol{x}_1}{dt}, \ldots, \frac{d\boldsymbol{x}_N}{dt}, t\right). \tag{A.2}$$

The forces will in general also depend on parameters describing the external influences from the system's environment, for example the Earth's gravity. The explicit dependence on t in the last argument of the force usually derives from such time-dependent external influences. It is, however, often possible to view the environment as just another collection of particles and include it in a larger *isolated* body not subject to any environmental influences.

The dynamics of a collection of particles thus becomes a web of coupled second-order differential equations in time. In principle, these equations may be solved numerically for all values of t, given initial positions and velocities for all particles at a definite instant of time, say $t = t_0$. Unfortunately, the large number of molecules in any macroscopic body usually presents an insurmountable obstacle to such an endeavor. Even for smaller numbers of particles, *deterministic chaos* may effectively prevent any long-term numeric integration of the equations of motion.

A.2 Force and momentum

A number of quantities describe the system as a whole. The total mass of the system is

$$M = \sum_n m_n, \tag{A.3}$$

and the total force is

$$\mathcal{F} = \sum_n f_n. \tag{A.4}$$

Note that these are true definitions. Nothing in Newton's laws tells us that it is physically meaningful to add masses of different particles, or worse, forces acting on different particles. As shown in Problem A.1, there is no obstacle to making a different definition of total force.

Alternative mechanics: The choice made here is particularly convenient for particles moving in a constant field of gravity, such as we find on the surface of the Earth, because the gravitational force on a particle is directly proportional to the mass of the particle. With the above definitions, the total gravitational force, the weight, becomes proportional to the total mass. This additivity of weights, the observation that a volume of flour balances an equal volume of flour, independent of how it is subdivided into smaller volumes, goes back to the dawn of history.

Having made these definitions, the form of the equations of motion (A.1) tells us that we should also define the average of the particle positions weighted by the corresponding masses, called the *center of mass*,

$$\boldsymbol{x}_M = \frac{1}{M} \sum_n m_n \boldsymbol{x}_n, \tag{A.5}$$

which obeys the equation

$$M \frac{d^2 \boldsymbol{x}_M}{dt^2} = \mathcal{F}. \tag{A.6}$$

Formally, this equation is of the same form as Newton's Second Law for a single particle, so *the center of mass moves like a point particle under the influence of the total force.* But before we get completely carried away, it should be remembered that the total force depends on the positions and velocities of all the particles, not just on the center of mass position $\boldsymbol{x}_M$ and its velocity $d\boldsymbol{x}_M/dt$. The above equation is in general *not* a solvable equation of motion for the center of mass.

Stiff bodies: There are, however, important exceptions. The state of a stiff body is characterized by the position and velocity of its center of mass, together with the body's orientation and its rate of change. If the total force on the body does not depend on the orientation, the above equation truly becomes an equation of motion for the center of mass. It is fairly easy to show that for a collection of spherically symmetric stiff bodies, the gravitational forces can only depend on the positions of the centers of mass, even if the bodies rotate. It was Newton's good fortune that planets and stars to a good approximation behave like point particles.

The above equation may be reformulated by defining the *total momentum* of the body,

$$\mathcal{P} = \sum_n m_n \frac{d\boldsymbol{x}_n}{dt} = M \frac{d\boldsymbol{x}_M}{dt}. \tag{A.7}$$

Like the total force it is a purely *kinematic* quantity, depending only on the particle velocities, calculated as the sum over the individual momenta $m_n d\boldsymbol{x}_n/dt$ of each particle. The equation of motion (A.6) implies that the total momentum obeys the equation

$$\boxed{\frac{d\mathcal{P}}{dt} = \mathcal{F}.} \tag{A.8}$$

Again it should be noted that this equation cannot be taken as an equation of motion, except in very special circumstances. It should rather be viewed as a *constraint* (or rather three since it is a vector equation) that follows from the true equations of motion, independent of what form the forces take. This constraint is particularly useful if the total momentum is known to be constant, or equivalently the center of mass has constant velocity, because then the total force must vanish.

A.3 Moment of force and angular momentum

Similarly, the *total moment of force* acting on the system is defined as

$$\mathcal{M} = \sum_n \boldsymbol{x}_n \times \boldsymbol{f}_n. \tag{A.9}$$

Like the total force, it is a *dynamic* quantity calculated from the sum of the individual moments of force acting on the particles.

The corresponding kinematic quantity is the *total angular momentum*,

$$\mathcal{L} = \sum_n \boldsymbol{x}_n \times m_n \frac{d\boldsymbol{x}_n}{dt}. \tag{A.10}$$

Differentiating with respect to time we find

$$\frac{d\mathcal{L}}{dt} = \sum_n m_n \left(\frac{d\boldsymbol{x}_n}{dt} \times \frac{d\boldsymbol{x}_n}{dt} + \boldsymbol{x}_n \times \frac{d^2\boldsymbol{x}_n}{dt^2} \right).$$

The first term in parenthesis vanishes because the cross-product of a vector with itself always vanishes, and using the equations of motion in the second term, we obtain

$$\boxed{\frac{d\mathcal{L}}{dt} = \mathcal{M}.} \tag{A.11}$$

Like the equation for total momentum and total force, (A.8), this equation is also a constraint that must be fulfilled, independent of the nature of the forces acting on the particles.

Angular momentum has to do with the state of rotation of the body as a whole. If the total angular momentum is known to be constant, as for a non-rotating body, the total moment of force must vanish. This is also what lies behind the lever principle.

Levers: From the earliest times levers have been used to lift and move heavy weights, such as the stones found in stone-age monuments. A primitive lever is simply a long stick with one end wedged under a heavy stone. Applying a small "man-sized" force orthogonal to the other end of the stick, the product of the long arm and the small force translates into a much larger force at the end of the small arm wedged under the stone.

The moment of force and the angular momentum both depend explicitly on the choice of origin of the coordinate system. These quantities might as well have been calculated around any other fixed point $\boldsymbol{c}$, leading to

$$\mathcal{M}(\boldsymbol{c}) = \sum_n (\boldsymbol{x}_n - \boldsymbol{c}) \times \boldsymbol{f}_n = \mathcal{M} - \boldsymbol{c} \times \mathcal{F}, \tag{A.12a}$$

$$\mathcal{L}(\boldsymbol{c}) = \sum_n (\boldsymbol{x}_n - \boldsymbol{c}) \times m_n \frac{d(\boldsymbol{x}_n - \boldsymbol{c})}{dt} = \mathcal{L} - \boldsymbol{c} \times \mathcal{P}. \tag{A.12b}$$

This shows that if the total force vanishes, the total moment of force becomes independent of the choice of origin, and similarly if the total momentum vanishes, the total angular momentum will be independent of the choice of origin.

If a point $\boldsymbol{c}$ exists such that $\mathcal{M}(\boldsymbol{c}) = \boldsymbol{0}$, we get $\mathcal{M} = \boldsymbol{c} \times \mathcal{F}$. In this case, the point $\boldsymbol{c}$ is called the *center of action* or *point of attack* for the total force $\mathcal{F}$. In general, there is no guarantee that a center of action exists, since it requires the total force $\mathcal{F}$ to be orthogonal to the total moment $\mathcal{M}$. Even if the center of action exists, it is not unique because any other point $\boldsymbol{c} + k\mathcal{F}$ with arbitrary k is as good a center of action as $\boldsymbol{c}$.

Center of gravity: In constant gravity where $\boldsymbol{f}_n = m_n \boldsymbol{g}_0$, and therefore $\mathcal{F} = M\boldsymbol{g}_0$, it follows immediately that the $\mathcal{M} = \sum_n \boldsymbol{x}_n \times \boldsymbol{f}_n = \sum_n m_n \boldsymbol{x}_n \times \boldsymbol{g}_0 = \boldsymbol{x}_M \times M\boldsymbol{g}_0 = \boldsymbol{x}_M \times \mathcal{F}$. The center of mass is also the point of attack for gravity, also called the *center of gravity*.

A.4 Power and kinetic energy

Forces generally perform work on the particles they act on. The total rate of work or *total power* performed by the forces acting on all the particles making up a body is

$$P = \sum_n \boldsymbol{f}_n \cdot \frac{d\boldsymbol{x}_n}{dt}. \tag{A.13}$$

Note that there is a dot-product between the force and the velocity. In non-Anglo-Saxon countries this is called *effect* rather than power.

The corresponding kinematic quantity is the *total kinetic energy*,

$$\mathcal{K} = \frac{1}{2} \sum_n m_n \left(\frac{d\boldsymbol{x}_n}{dt}\right)^2, \tag{A.14}$$

which is the sum of individual kinetic energies of each particle. Differentiating with respect to time and using the equations of motion (A.1), we find

$$\boxed{\frac{d\mathcal{K}}{dt} = P.} \tag{A.15}$$

The rate of change of the total kinetic energy equals the total power.

It is convenient to split off the center-of-mass position in the kinetic energy by writing $\boldsymbol{x}_n = \boldsymbol{x}_M + \boldsymbol{x}'_n$. Since $\sum_n m_n \boldsymbol{x}'_n = \boldsymbol{0}$ there will be no cross-terms, and we find

$$\mathcal{K} = \frac{1}{2} M \left(\frac{d\boldsymbol{x}_M}{dt}\right)^2 + \frac{1}{2}\sum_n m_n \left(\frac{d\boldsymbol{x}'_n}{dt}\right)^2. \tag{A.16}$$

The *total kinetic energy is the sum of the kinetic energy of the center-of-mass motion and the kinetic energies of the particle motions relative to the center of mass.* In statistical physics this theorem allows us to make a clear distinction between the macroscopic kinetic energy of a material particle and the microscopic kinetic energy contained in random molecular motion, also called heat.

A.5 Internal and external forces

The force acting on a particle may often be split into an internal part due to the other particles in the same system and an external part due to the system's environment,

$$\boldsymbol{f}_n = \boldsymbol{f}_n^{\text{int}} + \boldsymbol{f}_n^{\text{ext}}. \tag{A.17}$$

The internal forces, in particular gravitational forces, are often two-particle forces with $\boldsymbol{f}_{n,n'}$ denoting the force that particle n' exerts on particle n. The total internal force on particle n thus becomes

$$\boldsymbol{f}_n^{\text{int}} = \sum_{n'} \boldsymbol{f}_{n,n'}. \tag{A.18}$$

Most two-particle forces also obey Newton's Third Law, which states that the force from n' on n is equal and opposite to the force from n on n',

$$\boldsymbol{f}_{n,n'} = -\boldsymbol{f}_{n',n}. \tag{A.19}$$

Strong theorems follow from the assumption of two-particle internal forces. The first is that the total internal force vanishes,

$$\mathcal{F}^{\text{int}} = \sum_n \boldsymbol{f}_n^{\text{int}} = \sum_{n,n'} \boldsymbol{f}_{n,n'} = \boldsymbol{0}, \tag{A.20}$$

because of the antisymmetry (A.19). This expresses the simple fact that you cannot lift yourself up by your bootstraps. Although the external forces may themselves be due to two-particle forces, there is no corresponding theorem about the total external force, at least as long as the environment is not included in the description to make the total system isolated.

The momentum rate equation (A.8) now takes the form

$$\boxed{\frac{d\mathcal{P}}{dt} = \mathcal{F}^{\text{ext}},} \tag{A.21}$$

showing that it is sufficient to know the total external force acting on a system in order to calculate its rate of change of total momentum. The details of the internal forces can be ignored as long as they are of the two-particle kind and obey Newton's Third Law.

Under the same assumptions, the internal moment of force becomes

$$\mathcal{M} = \sum_{n,n'} x_n \times f_{n,n'} = \frac{1}{2} \sum_{n,n'} (x_n - x_{n'}) \times f_{n,n'}. \tag{A.22}$$

This does not in general vanish, except for the case of *central forces* where

$$f_{n,n'} \sim x_n - x_{n'}. \tag{A.23}$$

Gravitational forces and electrostatic forces are of this type. Provided the internal forces stem from central two-particle forces, the total moment of force equals the external moment,

$$\boxed{\frac{d\mathcal{L}}{dt} = \mathcal{M}^{\text{ext}}.} \tag{A.24}$$

This rule is, however, not on nearly the same sure footing as the corresponding equation for the momentum rate (A.21).

Finally, there is not much to be said about the total power (A.13), which in general has non-vanishing internal and external contributions.

A.6 Hierarchies of particle interactions

Under what circumstances can a collection of point particles be viewed as a point particle? The dynamics of the solar system may to a good approximation be described by a system of interacting point particles, although the planets and the Sun are in no way point-like at our own scale. At the scale of the whole universe, even galaxies are sometimes treated as point particles.

A point particle approximation may be justified as long as the internal cohesive forces that keep the interacting bodies together are much stronger than the external forces. In addition to mass and momentum, such a point particle can also be endowed with an intrinsic angular momentum (spin), and an intrinsic energy. The material world appears in this way as a hierarchy of approximately point-like interacting particles, from atoms to galaxies, at each level behaving as if they had no detailed internal structure. Corrections to the ideal point-likeness can later be applied to add more detail to this overall picture. Over the centuries this extremely reductionist method has shown itself to be very fruitful, but it is an open (scientific) question whether it can continue indefinitely.

Problems

A.1 Try to define the total force as $\mathcal{F}' = \sum_n m_n f_n$ rather than (A.4). **(a)** Investigate what this entails for the global properties of a system. **(b)** Can you build consistent global mechanics on this definition?

B

Cartesian coordinates

The space in which we live is (nearly) *flat* everywhere. Its geometry is Euclidean, meaning that Euclid's axioms and the theorems deduced from them are valid everywhere. One of the consequences of the Euclidean geometry is Pythagoras' theorem, which in the well-known way relates the lengths of the sides of any right-angled triangle. The simplicity of Pythagoras' theorem favors the use of right-angled *Cartesian* coordinate systems in which the distance between two points in space is the square root of the sum of the squares of their coordinate differences. In Cartesian coordinates, vector algebra also finds its simplest form.

One should be aware that the emphasis on strictly algebraic treatment of geometric concepts differs from what is usually seen in texts at this level. As described in Chapter 1, all of the usually ill-defined concepts of Euclidean geometry are relegated to the practical operational procedures, the "reference frame", relative to which actual measurements are carried out.

This appendix serves in most respects to define the mathematical notation and present the efficient modern methods of Cartesian vector and tensor algebra. In Appendix C vector notation is extended to include differentiation and integration of fields, and in Appendix D cylindrical and spherical (non-Cartesian) curvilinear coordinate systems are introduced and related to Cartesian coordinates.

B.1 Cartesian vectors

In a Cartesian coordinate system the distance between any two points, $\boldsymbol{x} = (x, y, z)$ and $\boldsymbol{x}' = (x', y', z')$, is given by the expression[1]

$$d(\boldsymbol{x}', \boldsymbol{x}) = \sqrt{(x'-x)^2 + (y'-y)^2 + (z'-z)^2}. \tag{B.1}$$

This distance function implies that space is Euclidean, and therefore has all the properties one learns about in elementary geometry. Although we could prove this claim right here, it becomes nearly trivial after vector algebra has been established.

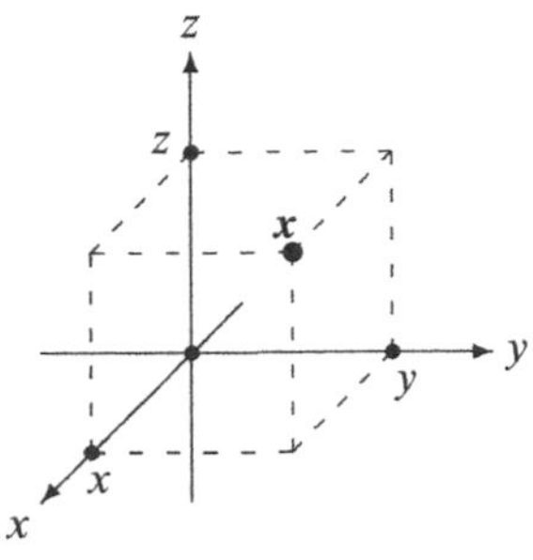

A Cartesian coordinate system with axes labeled x, y, and z, looks just like the one you got to know (and love) in high school.

[1] In this book we have chosen to follow the physics tradition in which the Cartesian coordinates are labeled x, y, and z. Mathematicians would prefer instead to label the coordinates x_1, x_2, and x_3, which is definitely a more systematic notation. Boldface symbols, $\boldsymbol{x} = (x, y, z)$, are used to denote Cartesian coordinate triplets and vectors (see below). For calculations with pencil on paper several different notations can be used to distinguish a triplet from other symbols, for example a bar ($\overline{x}$), an arrow ($\vec{x}$), or underlining ($\underline{x}$).

Definition of a vector

The Cartesian distance function (B.1) depends only on the coordinate differences between two points, not on their individual values. Since all geometry is contained in the distance function, coordinate differences will be of such importance in the analytic description of space in Cartesian coordinate systems that a special notation and a special type of algebra is necessary to deal efficiently with them.

A *vector* is a triplet of Cartesian coordinate differences[2], for example between the points $\boldsymbol{x}'$ and $\boldsymbol{x}$,

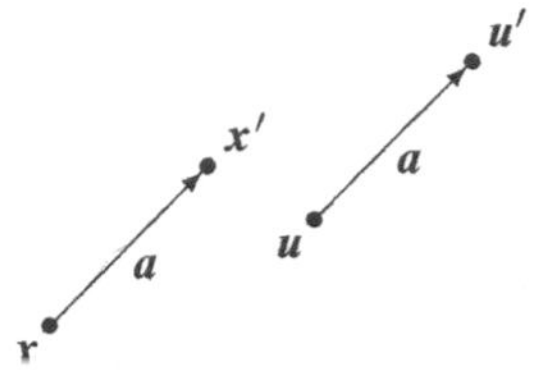

The vector $\boldsymbol{a}$ connects the point $\boldsymbol{x}$ with the point $\boldsymbol{x}'$. The same vector also connects the points $\boldsymbol{u}$ and $\boldsymbol{u}'$.

$$\boldsymbol{a} \equiv (a_x, a_y, a_z) = (x' - x, y' - y, z' - z). \tag{B.2}$$

This particular vector is naturally visualized by drawing a straight arrow from $\boldsymbol{x}$ to $\boldsymbol{x}'$. Since a vector only depends on the coordinate differences between its endpoints, the same symbol $\boldsymbol{a}$ can be used for the vector connecting other pairs of points, say $\boldsymbol{u} = (u, v, w)$ and $\boldsymbol{u}' = (u', v', w')$, as long as they have the same coordinate differences. A vector has, like a real arrow, both direction and length, but *no fixed origin*. The same vector $\boldsymbol{a}$ will *carry* you from $\boldsymbol{x}$ to $\boldsymbol{x}'$, and from $\boldsymbol{u}$ to $\boldsymbol{u}'$.

Notice how the three components of the vector, a_x, a_y, and a_z, are labeled by the coordinate symbols x, y, and z. This is a general rule that will also be used for curvilinear coordinates.

Position vectors

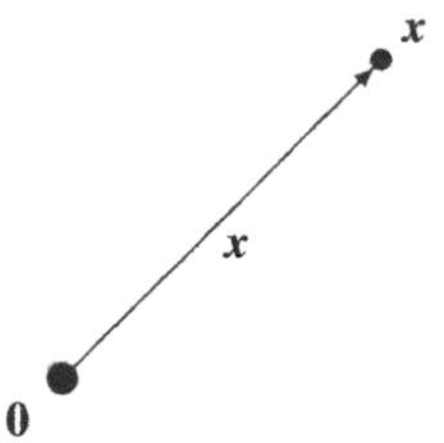

The position vector $\boldsymbol{x}$ connects the origin $\boldsymbol{0}$ with the point $\boldsymbol{x}$. This notational ambiguity does not cause any problems.

Conceptually there is a great difference between the triplet of real numbers making up the coordinates of a point, also called its *position*, and the triplet of coordinate differences making up a vector. Whereas it a priori makes no geometric sense to add the coordinates of two points, the sum of the components of two vectors is just another vector. The distinction is, however, not so clear in Cartesian coordinate systems, because the three real numbers in the triplet $\boldsymbol{x}$ can be formally viewed as the difference between the true coordinates of a point and the coordinates, $\boldsymbol{0} = (0, 0, 0)$, of the origin of the coordinate system. We shall for this reason permit ourselves the ambiguity of also calling $\boldsymbol{x}$ the *position vector*, thereby allowing the rules of vector algebra defined below also to be applied to coordinate triplets.

It must be stressed that the identification of positions and vectors is *absolutely not* possible in curvilinear coordinates, for example cylindrical or spherical, and even less so in non-Euclidean spaces. In those cases, true vectors can only be defined from infinitesimal coordinate differences in the local Cartesian coordinate systems that always exist in the neighborhood of any point.

B.2 Vector algebra

In keeping with the algebraization of geometry initiated four hundred years ago by Descartes, we shall focus on the algebraic rather than the geometric properties of vectors. As discussed in Section 1.4 on page 11 the coordinate system is viewed as the physical reference frame for the determination of the coordinates of any point in space according to well-defined operational procedures. In this way we avoid all undefined geometrical primitives, such as the points, lines, and circles of Euclidean geometry.

The following definitions endow vectors with the usual properties of the familiar geometric vectors. Geometric visualization is of course as useful as ever, and we shall whenever possible use simple sketches to illustrate what is meant.

[2] Apologies for the slight shift in notation between Section 1.5 on page 14 and this appendix.

Linear operations

Linear operations lie at the core of vector algebra,

$$k\,\boldsymbol{a} = (ka_x, ka_y, ka_z) \qquad \text{(scaling by a factor)}, \qquad (B.3)$$

$$\boldsymbol{a}+\boldsymbol{b} = (a_x+b_x, a_y+b_y, a_z+b_z) \qquad \text{(addition)}, \qquad (B.4)$$

$$\boldsymbol{a}-\boldsymbol{b} = (a_x-b_x, a_y-b_y, a_z-b_z) \qquad \text{(subtraction)}. \qquad (B.5)$$

Mathematically, these rules show that the set of all vectors constitute a three-dimensional vector space.

A set of vectors $\boldsymbol{a}_1, \boldsymbol{a}_2, \ldots, \boldsymbol{a}_N$ are said to be *linearly dependent* if there exists a vanishing linear combination with non-zero coefficients, $k_1\boldsymbol{a}_1 + k_2\boldsymbol{a}_2 + \cdots + k_N\boldsymbol{a}_N = \boldsymbol{0}$. More than three vectors are always linearly dependent because space is three-dimensional.

A *straight line* with origin $\boldsymbol{a}$ and direction vector $\boldsymbol{b} \neq \boldsymbol{0}$ is described by the linear vector function $\boldsymbol{x}(s) = \boldsymbol{a}+\boldsymbol{b}\,s$ with $-\infty < s < \infty$. Note that the origin $\boldsymbol{a} = \boldsymbol{x}(0)$ is a position vector whereas the direction vector $\boldsymbol{b} = \boldsymbol{x}(1)-\boldsymbol{x}(0)$ is the difference between the coordinates of two points, and thus a true vector. The straight line is in fact the shortest path between any two points, and its length is equal to the distance between them, as it must in Euclidean geometry (see Problem B.2).

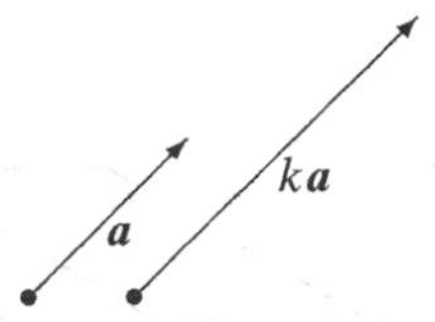

Geometric scaling of a vector by a factor k.

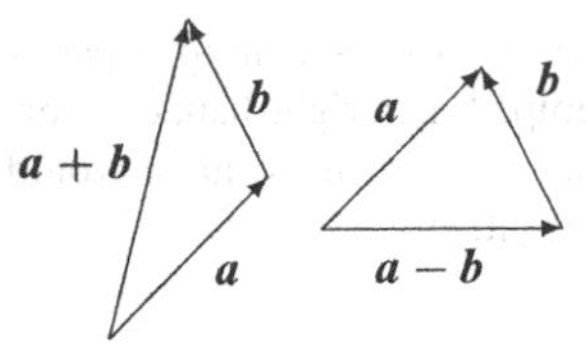

Geometric addition and subtraction of vectors.

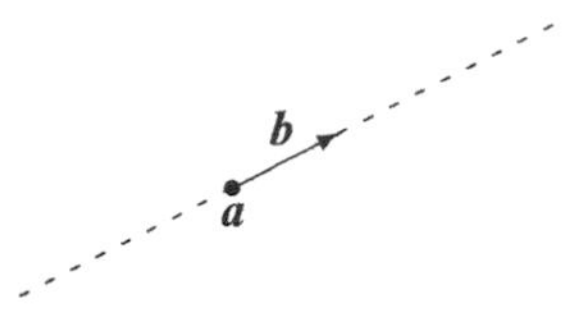

A straight line with origin $\boldsymbol{a}$ and direction vector $\boldsymbol{b}$.

Dot-product

The *dot-product* or *scalar product* of two vectors is familiar from geometry,

$$\boldsymbol{a}\cdot\boldsymbol{b} = a_x b_x + a_y b_y + a_z b_z \qquad \text{(dot-product)}. \qquad (B.6)$$

Two vectors are said to be *orthogonal* when their dot-product vanishes, $\boldsymbol{a}\cdot\boldsymbol{b} = 0$. The *square* of a vector equals its dot-product with itself, $\boldsymbol{a}^2 = \boldsymbol{a}\cdot\boldsymbol{a} = a_x^2 + a_y^2 + a_z^2$. The length of a vector is $|\boldsymbol{a}| = \sqrt{\boldsymbol{a}^2}$.

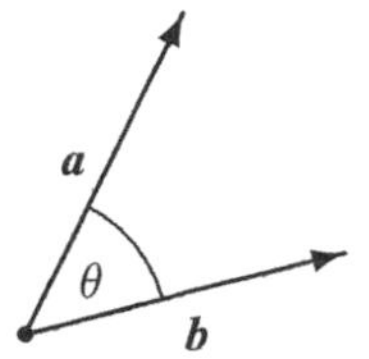

Geometrically, the dot-product of two vectors is $\boldsymbol{a}\cdot\boldsymbol{b} = |\boldsymbol{a}|\,|\boldsymbol{b}|\cos\theta$, where $|\boldsymbol{a}|$ and $|\boldsymbol{b}|$ are the lengths of $\boldsymbol{a}$ and $\boldsymbol{b}$, and θ is the angle between them.

Cross-product

The *cross-product* or *vector product* is also familiar from geometry,

$$\boldsymbol{a}\times\boldsymbol{b} = (a_y b_z - a_z b_y, a_z b_x - a_x b_z, a_x b_y - a_y b_x) \qquad \text{(cross-product)}. \qquad (B.7)$$

It is defined entirely in terms of the coordinates and we do not in the rule itself distinguish between left-handed and right-handed coordinate systems. Whether you use your right or left hand when you draw a cross-product on paper does not matter for the vector-product rule, as long as you consistently use the same hand for all such drawings. The cross-product $\boldsymbol{a}\times\boldsymbol{b}$ is also called the *area vector* of the parallelogram spanned by the vectors $\boldsymbol{a}$ and $\boldsymbol{b}$, because the length $|\boldsymbol{a}\times\boldsymbol{b}|$ is equal to its area.

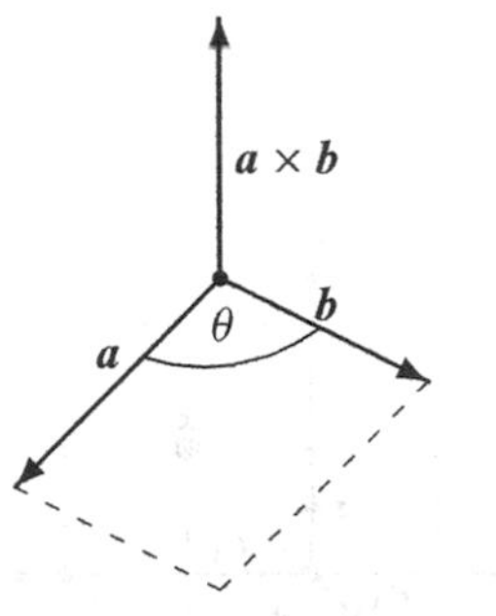

Geometrically, the cross-product of two vectors is a vector orthogonal to both, here drawn using a right-hand rule. Its length $|\boldsymbol{a}\times\boldsymbol{b}| = |\boldsymbol{a}|\,|\boldsymbol{b}|\sin\theta$ equals the area of the parallelogram (dashed) spanned by $\boldsymbol{a}$ and $\boldsymbol{b}$.

Tensor product

The *tensor product* is not familiar from geometry,

$$\boldsymbol{a}\boldsymbol{b} = \begin{pmatrix} a_x \\ a_y \\ a_z \end{pmatrix} (b_x, b_y, b_z) = \begin{pmatrix} a_x b_x & a_x b_y & a_x b_z \\ a_y b_x & a_y b_y & a_y b_z \\ a_z b_x & a_z b_y & a_z b_z \end{pmatrix} \qquad \text{(tensor product)}. \qquad (B.8)$$

It is unusual in that it produces a 3×3 matrix from two vectors, but otherwise it is perfectly well defined and quite useful to have around. It is nothing but an ordinary matrix product of a column-matrix and a row-matrix, also called the *direct product* and sometimes in the older literature the *dyadic product*. Later we shall introduce more general geometric objects, called tensors, of which the simplest are represented by matrices of this kind. Tensor products—and tensors in general—cannot be given a simple visualization.

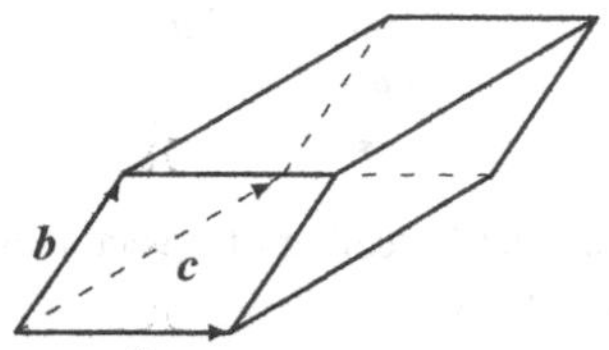

Three vectors spanning a parallelepiped. In a right-handed coordinate system, its volume would be negative.

Volume product

The trilinear product of three vectors obtained by combining the cross-product and the dot-product is called the *volume product*,

$$\boldsymbol{a} \times \boldsymbol{b} \cdot \boldsymbol{c} = a_x b_y c_z + a_y b_z c_x + a_z b_x c_y - a_x b_z c_y - a_y b_x c_z - a_z b_y c_x. \tag{B.9}$$

The right-hand side shows that the volume product equals the determinant of the matrix constructed from the three vectors,

$$\boldsymbol{a} \times \boldsymbol{b} \cdot \boldsymbol{c} = \begin{vmatrix} a_x & b_x & c_x \\ a_y & b_y & c_y \\ a_z & b_z & c_z \end{vmatrix} \qquad \text{(volume product)}. \tag{B.10}$$

The volume product is like the determinant antisymmetric under exchange of any pair of vectors (here columns). The volume product equals the signed volume of the parallelepiped spanned by the vectors $\boldsymbol{a}$, $\boldsymbol{b}$, and $\boldsymbol{c}$.

Norms

The *norm* or *length* of a vector is defined by

$$|\boldsymbol{a}| = \sqrt{\boldsymbol{a}^2} = \sqrt{a_x^2 + a_y^2 + a_z^2} \qquad \text{(norm or length)}. \tag{B.11}$$

The Cartesian distance (1.14) can now be written $d(\boldsymbol{a}, \boldsymbol{b}) = |\boldsymbol{a} - \boldsymbol{b}|$, and this is of course the real reason why the norm is defined as it is.

The norm of an arbitrary 3×3 matrix $\mathbf{A} = \{a_{ij}\}$ is defined as[3]

$$|\mathbf{A}| = \sqrt{\sum_{ij} a_{ij}^2} \qquad \text{(matrix norm)}, \tag{B.12}$$

where both sums run over the coordinate labels x, y, and z. It follows that the norm of a tensor product equals the product of the norms of the vector factors, $|\boldsymbol{ab}| = |\boldsymbol{a}|\,|\boldsymbol{b}|$, so this definition makes good sense.

B.3 Basis vectors

The *coordinate axes* of a Cartesian coordinate system are straight lines with a common origin $\mathbf{0} = (0, 0, 0)$ and direction vectors of unit length, called *basis vectors*[4],

$$\hat{\boldsymbol{e}}_x = (1, 0, 0), \qquad \hat{\boldsymbol{e}}_y = (0, 1, 0), \qquad \hat{\boldsymbol{e}}_z = (0, 0, 1). \tag{B.13}$$

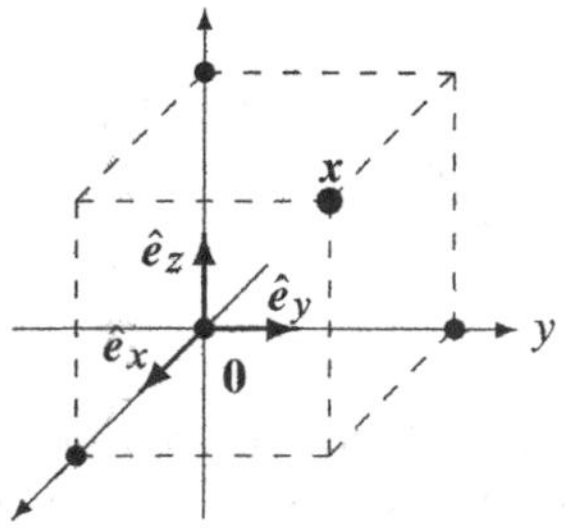

A Cartesian coordinate system and its basis vectors.

Every position $\boldsymbol{x}$ may trivially be written as a linear combination of the basis vectors with the coordinates as coefficients,

$$\boldsymbol{x} = x\hat{\boldsymbol{e}}_x + y\hat{\boldsymbol{e}}_y + z\hat{\boldsymbol{e}}_z, \tag{B.14}$$

and similarly for any vector: $\boldsymbol{a} = a_x\hat{\boldsymbol{e}}_x + a_y\hat{\boldsymbol{e}}_y + a_z\hat{\boldsymbol{e}}_z$.

[3] To distinguish a matrix from a vector, the matrix symbol will be written in heavy boldface font. The distinction is unfortunately not particularly visible in print. With pencil on paper, 3×3 matrices are sometimes marked with a double bar, $\overline{\overline{1}}$, or a double arrow, $\overleftrightarrow{1}$.

[4] We shall generally use a "hat" to indicate that a vector has unit length. In some textbooks basis vectors are symbolized by "hatted" coordinate labels: $\hat{\boldsymbol{x}}$, $\hat{\boldsymbol{y}}$, and $\hat{\boldsymbol{z}}$. Although without doubt an efficient notation, it can become quite unreadable and will not be used here.

The basis vectors are *normalized* and mutually *orthogonal*,

$$|\hat{e}_x| = |\hat{e}_y| = |\hat{e}_z| = 1, \qquad \hat{e}_x \cdot \hat{e}_y = \hat{e}_y \cdot \hat{e}_z = \hat{e}_z \cdot \hat{e}_x = 0. \tag{B.15}$$

Using these relations and (B.14), we find

$$x = \hat{e}_x \cdot \boldsymbol{x}, \qquad y = \hat{e}_y \cdot \boldsymbol{x}, \qquad z = \hat{e}_z \cdot \boldsymbol{x}, \tag{B.16}$$

showing that the coordinates of a point may be understood as the normal projections of the point on the axes of the coordinate system.

Completeness

Combining (B.14) with (B.16) we obtain the identity

$$\hat{e}_x(\hat{e}_x \cdot \boldsymbol{x}) + \hat{e}_y(\hat{e}_y \cdot \boldsymbol{x}) + \hat{e}_z(\hat{e}_z \cdot \boldsymbol{x}) = \boldsymbol{x},$$

valid for all $\boldsymbol{x}$. Since this is a linear identity, we may remove $\boldsymbol{x}$ and express this *completeness* relation in a compact form by means of the tensor product (B.8),

$$\hat{e}_x\hat{e}_x + \hat{e}_2\hat{e}_2 + \hat{e}_3\hat{e}_3 = \mathbf{1}, \tag{B.17}$$

where on the right-hand side the symbol $\mathbf{1}$ stands for the 3×3 unit matrix.

Handedness

It must be emphasized that the handedness of the coordinate system has not entered the formalism. Correspondingly, the volume of the unit cube,

$$\hat{e}_x \times \hat{e}_y \cdot \hat{e}_z = 1, \tag{B.18}$$

is always unity, independent of whether you call your coordinate system right-handed or left-handed. Handedness first shows up when you try to understand the world through a looking glass (see Section B.6).

B.4 Index notation

Vector notation is sufficient for most areas of physics because physical quantities are mostly scalars like mass and charge, or vectors such as velocity and force. Sometimes it is, however, necessary to use a more powerful and transparent notation that generalizes better to complex expressions. It is called *index notation* or *tensor notation*, and consists in all simplicity of writing out the coordinate indices explicitly wherever they occur. Instead of thinking of a Cartesian position as a triplet $\boldsymbol{x}$ endowed with algebraic rules, we think of it as the set of coordinates x_i with the index i implicitly running over the coordinate labels, for example $i = x, y, z$ or $i = 1, 2, 3$ or whatever, without having to state it explicitly every time.

Algebraic operations

Vector and index notations coexist quite peacefully as witnessed by the linear operations

$$(k\boldsymbol{a})_i = ka_i, \tag{B.19}$$

$$(\boldsymbol{a} + \boldsymbol{b})_i = a_i + b_i, \tag{B.20}$$

$$(\boldsymbol{a} - \boldsymbol{b})_i = a_i - b_i. \tag{B.21}$$

In each of these equations it is tacitly understood that the index runs over the coordinate labels, for example $i = x, y, z$.

For the scalar product we also let the sum range implicitly over the coordinate labels,

$$\boldsymbol{a} \cdot \boldsymbol{b} = \sum_i a_i b_i . \tag{B.22}$$

In full-fledged tensor calculus even the summation symbol is left out and understood as implicitly present for all indices that occur precisely twice in a term. Although this efficient notation was introduced by Einstein himself in 1916, we shall nevertheless refrain from using it here.

Kronecker delta

Leopold Kronecker (1823–1891). German mathematician, contributed to the theory of elliptic functions, algebraic equations, and algebraic numbers.

The nine dot-products of the three basis vectors with themselves forms a 3×3 matrix with two indices that each run implicitly over the three coordinate labels,

$$\delta_{ij} \equiv \hat{\boldsymbol{e}}_i \cdot \hat{\boldsymbol{e}}_j = \begin{cases} 1 & \text{for } i = j, \\ 0 & \text{otherwise.} \end{cases} \tag{B.23}$$

It is called the *Kronecker delta*, and the corresponding matrix $\mathbf{1} = \{\delta_{ij}\}$ is the well-known unit matrix.

The Kronecker delta is the first example of a true *tensor* of rank 2. Another is the tensor product (B.8) of two vectors, which in index notation takes the form

$$(\boldsymbol{ab})_{ij} = a_i b_j . \tag{B.24}$$

Having two (or more) indices is, however, not enough for a collection of values to earn the right to be called a tensor. In the following sections we shall see that what really characterizes such collections is the way they behave under coordinate transformations.

Levi-Civita epsilon

Tullio Levi-Civita (1873–1941). Italian mathematician, contributed to differential calculus, relativity, and founded (with Ricci) tensor analysis in curved spaces.

The volume products of the three basis vectors with themselves constitute a collection of 27 values, called the *Levi-Civita symbol*,

$$\epsilon_{ijk} \equiv \hat{\boldsymbol{e}}_i \times \hat{\boldsymbol{e}}_j \cdot \hat{\boldsymbol{e}}_k = \begin{cases} +1 & \text{for } ijk = xyz \;\; yzx \;\; zxy, \\ -1 & \text{for } ijk = xzy \;\; yxz \;\; zyx, \\ 0 & \text{otherwise.} \end{cases} \tag{B.25}$$

It is $+1$ for even permutations of xyz and -1 for odd permutations. It vanishes if any two indices coincide.

The bilinear cross-product (B.7) can now be written as a double sum over two indices,

$$(\boldsymbol{a} \times \boldsymbol{b})_i = \sum_{jk} \epsilon_{ijk} a_j b_k , \tag{B.26}$$

while the trilinear volume product (B.9) becomes a triple sum,

$$\boldsymbol{a} \times \boldsymbol{b} \cdot \boldsymbol{c} = \sum_{ijk} \epsilon_{ijk}\, a_i b_j c_k . \tag{B.27}$$

The Levi-Civita symbol ϵ_{ijk} is in fact a tensor of third rank, completely antisymmetric in the three indices. We shall rarely use this notation, although it does come in handy when everything else fails.

B.5 Cartesian coordinate transformations

The same Euclidean world may be described geometrically by different observers with different Cartesian reference frames. Each observer constructs his own preferred Cartesian coordinate system and determines all positions relative to that. Every observer thinks that his basis vectors have the simple form (B.13) and satisfy the same orthogonality and completeness relations. Every observer believes he is right-handed. How can they ever agree on anything with such a self-centered view of the world?

The answer is that the two descriptions are related by a *coordinate transformation*. Since the length of a curve connecting any two points can be determined by laying out agreed-upon unit rulers along its path, it follows that both observers will agree on what is the shortest path, and thus on what is the distance between the points. Seen from one Cartesian coordinate system, which we shall call the "old", the axes of another Cartesian coordinate system, called the "new", will therefore also appear to be straight lines with a common origin. Furthermore, since the scalar product of two vectors can be expressed in terms of the norm (Problem B.5), it must—like distance—be independent of the specific coordinate system, such that the new axes will also appear to be orthogonal in the geometry of the old coordinate system. Different observers will thus agree that their respective coordinate systems are indeed Cartesian.

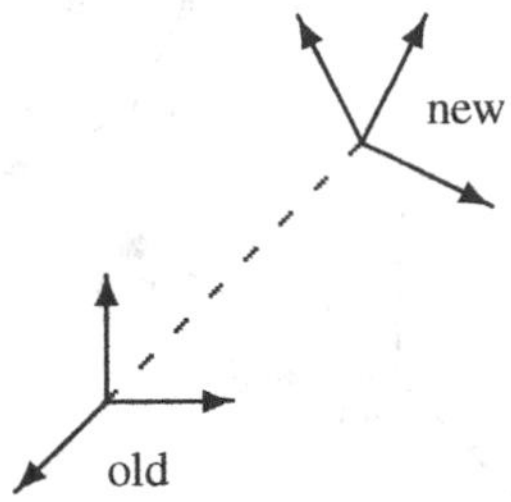

The old and the new Cartesian systems.

Simple transformations

We begin the analysis of coordinate transformations with the familiar elementary ones: translation, rotation, and reflection. These transformations express the coordinates of a point in the new system as a function of the coordinates of the same point in the old. The simple transformations defined below refer each to a special situation of the two coordinate systems. The most general transformation can be constructed from a combination of simple transformations (Problem B.17).

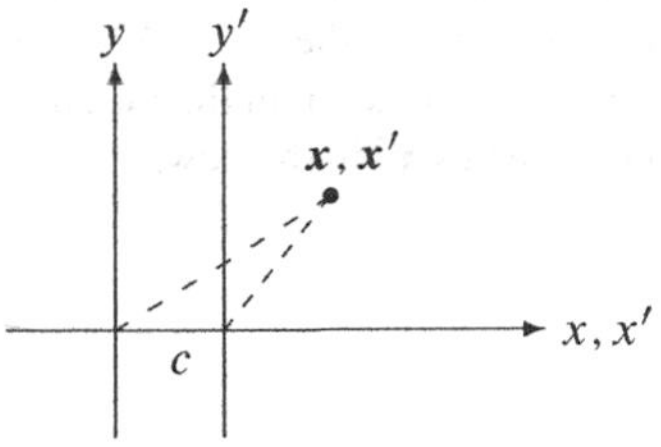

Simple translation of the coordinate system by c along the x-axis.

Simple translation: A simple *translation* of the origin of coordinates along the x-axis by a constant amount c is given by

$$x' = x - c, \tag{B.28a}$$
$$y' = y, \tag{B.28b}$$
$$z' = z. \tag{B.28c}$$

The axes of the new coordinate system are in this case parallel with the axes of the old.

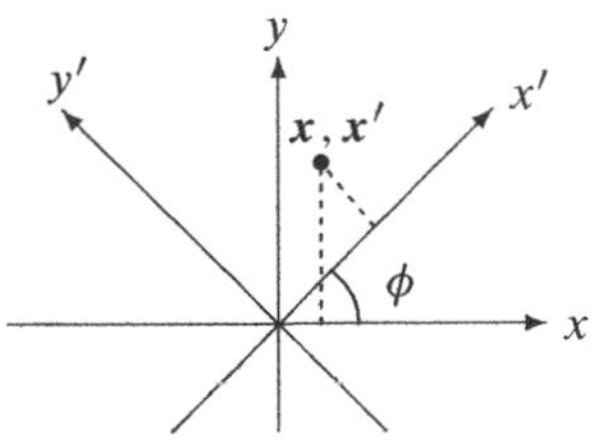

Simple rotation of the coordinate system around the z-axis (pointing out of the paper) through an angle ϕ.

Simple rotation: A simple *rotation* of the coordinate system through an angle ϕ around the z-axis is described by the transformation

$$x' = x\cos\phi + y\sin\phi, \tag{B.29a}$$
$$y' = -x\sin\phi + y\cos\phi, \tag{B.29b}$$
$$z' = z. \tag{B.29c}$$

In this case the z-axes are parallel in the old and the new systems.

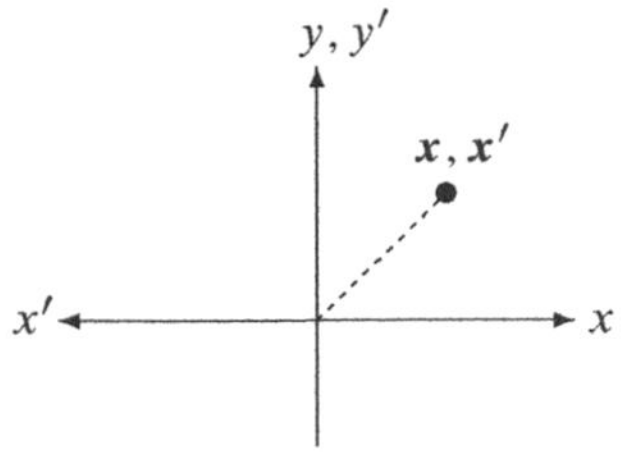

A simple reflection in the yz-plane.

Simple reflection: A simple *reflection* of the x-axis in the yz-plane is described by

$$x' = -x, \quad y' = y, \quad z' = z. \tag{B.30}$$

A simple reflection always transforms a right-handed coordinate system into a left-handed one, and vice versa, independent of which hand you may claim to be the right one.

General transformations

In the *coordinates of the old system*, the new Cartesian coordinate system can be characterized by its origin $\boldsymbol{c}$ and its three orthogonal and normalized basis vectors $\hat{\boldsymbol{a}}_x$, $\hat{\boldsymbol{a}}_y$, and $\hat{\boldsymbol{a}}_z$, satisfying the usual relations,

$$|\hat{\boldsymbol{a}}_x| = |\hat{\boldsymbol{a}}_y| = |\hat{\boldsymbol{a}}_z| = 1, \qquad \hat{\boldsymbol{a}}_x \cdot \hat{\boldsymbol{a}}_x = \hat{\boldsymbol{a}}_y \cdot \hat{\boldsymbol{a}}_z = \hat{\boldsymbol{a}}_z \cdot \hat{\boldsymbol{a}}_x = 0. \tag{B.31}$$

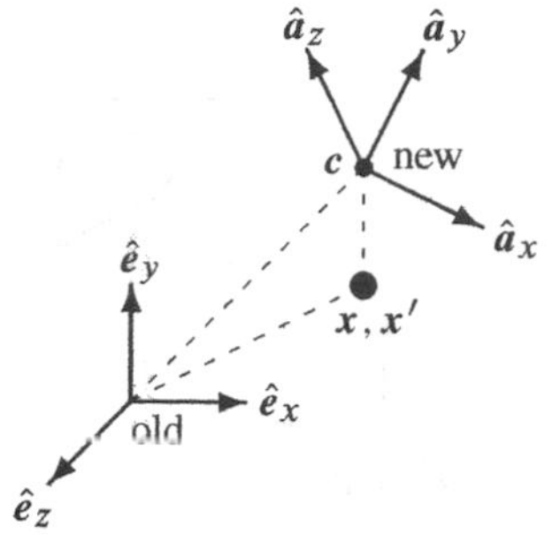

Arrangement of the old and new coordinate systems. The origin of the new system is $\boldsymbol{c}$ in the old, and the new basis vectors are $\hat{\boldsymbol{a}}_x$, $\hat{\boldsymbol{a}}_y$, and $\hat{\boldsymbol{a}}_z$ in the old. A point with coordinates $\boldsymbol{x}$ in the old, has coordinates $\boldsymbol{x}'$ in the new.

The position $\boldsymbol{x}' = (x', y', z')$ of a point in the new coordinate system must then correspond to following position in the old,

$$\boldsymbol{x} = \boldsymbol{c} + x'\hat{\boldsymbol{a}}_x + y'\hat{\boldsymbol{a}}_y + z'\hat{\boldsymbol{a}}_z. \tag{B.32}$$

The new coordinates are determined by multiplying with the new basis vectors and using orthonormality (B.31)

$$\begin{aligned} x' &= \hat{\boldsymbol{a}}_x \cdot (\boldsymbol{x} - \boldsymbol{c}) = a_{xx}(x - c_x) + a_{xy}(y - c_y) + a_{xz}(z - c_z), \\ y' &= \hat{\boldsymbol{a}}_y \cdot (\boldsymbol{x} - \boldsymbol{c}) = a_{yx}(x - c_x) + a_{yy}(y - c_y) + a_{yz}(z - c_z), \\ z' &= \hat{\boldsymbol{a}}_z \cdot (\boldsymbol{x} - \boldsymbol{c}) = a_{zx}(x - c_x) + a_{zy}(y - c_y) + a_{zz}(z - c_z), \end{aligned}$$

where $a_{ij} = \hat{\boldsymbol{a}}_i \cdot \hat{\boldsymbol{e}}_j = (\hat{\boldsymbol{a}}_i)_j$ are the coordinates of the new basis vectors in the old system. This is the most general coordinate transformation that can connect any two Cartesian coordinate systems.

Using index notation, the general coordinate transformation may also be written (it is perhaps better here to think of integer indices, $i = 1, 2, 3$),

$$\boxed{x'_i = \sum_j a_{ij}(x_j - c_j).} \tag{B.33}$$

In matrix notation the transformation becomes even more compact[5],

$$\boldsymbol{x}' = \mathsf{A} \cdot (\boldsymbol{x} - \boldsymbol{c}), \tag{B.34}$$

where $\mathsf{A} = \{a_{ij}\}$ is called the *transformation matrix*.

For a true vector $\boldsymbol{u} = \boldsymbol{x}_1 - \boldsymbol{x}_2$, defined as the difference between the coordinates of two positions, the transformation does not depend on the origin $\boldsymbol{c}$ of the new system, so that the true vector transformation rule becomes

$$\boxed{\boldsymbol{u}' = \mathsf{A} \cdot \boldsymbol{u}.} \tag{B.35}$$

Since the basis vectors of the old system are true vectors, they become $\hat{\boldsymbol{e}}'_i = \mathsf{A} \cdot \hat{\boldsymbol{e}}_i$ in the new, with coordinates $(\hat{\boldsymbol{e}}'_i)_j = \hat{\boldsymbol{e}}'_i \cdot \hat{\boldsymbol{e}}_j = a_{ji}$. This is quite different from the coordinates of the new basis in the old, $\boldsymbol{a}_i \cdot \hat{\boldsymbol{e}}_j = a_{ij}$. Keeping track of the various coordinates in the two systems can in fact be quite subtle!

Example B.1: The transformation matrix for a simple translation along the x-axis (B.28) is just the unit matrix, $\mathsf{A} = \mathbf{1}$, whereas for a simple rotation around the z-axis (B.29) we obtain

$$\mathsf{A} = \begin{pmatrix} \cos\phi & \sin\phi & 0 \\ -\sin\phi & \cos\phi & 0 \\ 0 & 0 & 1 \end{pmatrix}. \tag{B.36}$$

A simple reflection in the yz-plane (B.30) is characterized by a diagonal transformation matrix with $(-1, 1, 1)$ along the diagonal.

[5] In ordinary mathematical vector and matrix calculus one would not use the dot to indicate matrix multiplication (nor to indicate a scalar product), but this notation is quite natural for the three-dimensional vectors and matrices that we encounter so often in physics.

Orthogonality and completeness of the new basis

The orthogonality and completeness of the new basis vectors imply that

$$\hat{\boldsymbol{a}}_i \cdot \hat{\boldsymbol{a}}_j = \delta_{ij}, \qquad \sum_i \hat{\boldsymbol{a}}_i \hat{\boldsymbol{a}}_i = \mathbf{1}. \tag{B.37}$$

In index notation these two relations take the form

$$\sum_k a_{ik} a_{jk} = \sum_k a_{ki} a_{kj} = \delta_{ij}, \tag{B.38}$$

which in matrix notation becomes the usual conditions for the matrix to be orthogonal,

$$\boxed{\mathbf{A} \cdot \mathbf{A}^\top = \mathbf{A}^\top \cdot \mathbf{A} = \mathbf{1}.} \tag{B.39}$$

Here $(\mathbf{A}^\top)_{ij} = a_{ji}$ is the transposed matrix having the new basis vectors as columns.

The transposed matrix has the same determinant as the original matrix. Using that the determinant of a product of matrices is the product of the determinants and that the transposed matrix $\mathbf{A}^\top$ has the same determinant as $\mathbf{A}$, it follows from (B.39) that $(\det \mathbf{A})^2 = 1$, or

$$\det \mathbf{A} = \pm 1. \tag{B.40}$$

The transformation matrices are thus divided into two completely separate classes, those with determinant $+1$, generically called *rotations*, and those with determinant -1, generically called *reflections*. Since the simple reflection (B.30) has determinant -1, all reflections may be composed from a simple reflection followed by a rotation.

* Infinitesimal rotations

Suppose the new basis lies very close to the old, such that we may write

$$a_{ij} = \delta_{ij} + b_{ij} \qquad \text{with } |b_{ij}| \ll 1. \tag{B.41}$$

Inserting this into the orthogonality relation (B.38) we find to first order,

$$\delta_{ij} = \sum_k a_{ik} a_{jk} = \delta_{ij} + b_{ij} + b_{ji}, \tag{B.42}$$

which shows that *the infinitesimal transformation matrix is always antisymmetric*,

$$b_{ij} = -b_{ji}. \tag{B.43}$$

Using the Levi-Civita symbol to write it as $b_{ij} = \sum_k \epsilon_{ijk} \phi_k$, it follows that

$$\boxed{\hat{\boldsymbol{a}}_i = \hat{\boldsymbol{e}}_i + \boldsymbol{\phi} \times \boldsymbol{e}_i,} \tag{B.44}$$

showing that the new basis vectors are simply obtained by a common (solid) rotation of the old basis vectors through an infinitesimal angle $|\boldsymbol{\phi}|$ around the axis $\hat{\boldsymbol{\phi}} = \boldsymbol{\phi}/|\boldsymbol{\phi}|$.

B.6 Scalars, vectors, and tensors

When you change the coordinate system the world stays the same; it is only the way you describe it that changes. Some geometrical quantities, for example the distance between two points, are unaffected by any coordinate transformation; others, like your current position or velocity, will change in a characteristic way. In physics we shall only use quantities that transform in a regular way under coordinate transformations and therefore may be viewed as representations of geometrical objects. This guarantees that physical laws take the same form in any coordinate system and therefore places all observers on an equal footing. The geometric objects are generically called tensors, although the most common and important ones have been given separate names.

Classification by pure rotations

Geometric quantities are primarily classified according to their behavior under pure rotations of the form

$$\boldsymbol{x}' = \mathbf{A} \cdot \boldsymbol{x} \tag{B.45}$$

with the condition that the determinant is unity, $\det \mathbf{A} = +1$. The classes are

Scalars: A single quantity S is called a *scalar* if it is invariant under pure rotations

$$\boxed{S' = S.} \tag{B.46}$$

The distance, the norm, and the dot-product are examples of scalars. In physics the natural constants, material constants, as well as mass and charge are scalars.

Vectors: Any triplet of quantities $\boldsymbol{U}$ is called a *vector* if it transforms in the same way as the coordinates (B.33) under a pure rotation,

$$\boxed{U_i' = \sum_j a_{ij} U_j,} \tag{B.47}$$

or equivalently in matrix form,

$$\boldsymbol{U}' = \mathbf{A} \cdot \boldsymbol{U}. \tag{B.48}$$

In physics, velocity, acceleration, momentum, angular momentum, force, and many other quantities are vectors. This definition of a vector requires triplets to have special transformation properties to qualify as vectors. A triplet containing your weight, your height, and your age is not a vector but a collection of three scalars.

Tensors: Using the vector transformation (B.47) the tensor product $\boldsymbol{V}\boldsymbol{W}$ of two vectors is seen to transform according to the rule

$$(\boldsymbol{V}'\boldsymbol{W}')_{ij} \equiv V_i' W_j' = \left(\sum_k a_{ik} V_k\right)\left(\sum_l a_{jl} W_l\right) = \sum_{kl} a_{ik} a_{jl} (\boldsymbol{V}\boldsymbol{W})_{kl},$$

where in the last step we have reordered the two sums into a convenient form.

More generally, any set of nine quantities (arranged as a matrix)

$$\mathbf{T} = \{T_{ij}\} = \begin{pmatrix} T_{xx} & T_{xy} & T_{xz} \\ T_{yx} & T_{yy} & T_{yz} \\ T_{zx} & T_{zy} & T_{zz} \end{pmatrix} \tag{B.49}$$

is called a *tensor of rank 2*, provided it obeys the transformation law,

$$\boxed{T'_{ij} = \sum_{kl} a_{ik} a_{jl} T_{kl}.} \tag{B.50}$$

In matrix form this may be written

$$\mathbf{T}' = \mathbf{A} \cdot \mathbf{T} \cdot \mathbf{A}^{\mathsf{T}}. \tag{B.51}$$

In physics, the moment of inertia of an extended body and the quadrupole moment of an electric charge distribution are well-known tensors of second rank.

Tensors of higher rank may be constructed in a similar way. A tensor of rank r has r indices and is a collection of 3^r quantities that transform as the direct product of r vectors. A scalar is a tensor of rank 0, and a vector a tensor of rank 1. We have so far only met one third rank tensor, the Levi-Civita symbol (B.25) (see Problem B.18). When nothing else is said, a tensor is always assumed to be of rank 2.

∗ Subclassification by pure reflections

Geometric quantities may be further subclassified according to their behavior under a pure reflection, defined as the transformation

$$\boldsymbol{x}' = -\boldsymbol{x}. \tag{B.52}$$

Instead of a simple reflection in the yz-plane, we have chosen a complete reflection of the coordinates through the origin. Geometrically, the reflection in the origin may be viewed as a composite of three simple reflections along each of the three coordinate axes, or as a simple reflection of a coordinate axis followed by a simple rotation through 180° around the same axis.

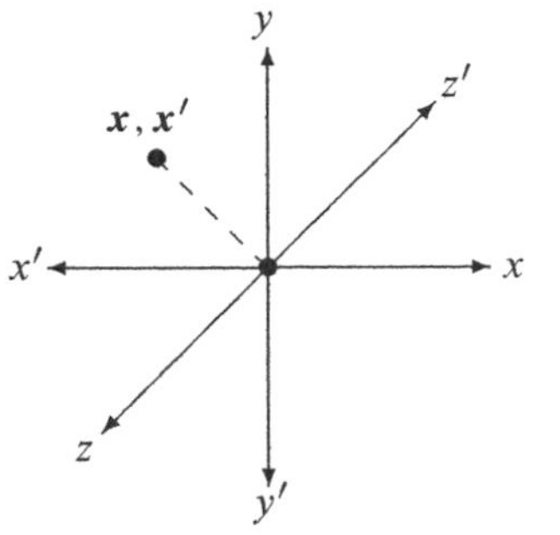

A complete reflection of the coordinate system in the origin. A rotation through 180° around the x-axis converts this to a simple reflection of the x-axis (in the yz-plane).

Polar vectors: A vector that obeys the transformation equation (B.47) under pure rotations as well as under pure reflections is called a *polar* vector. Under a pure reflection in the origin, the coordinates of a polar vector change sign just like the coordinates of a point,

$$\boldsymbol{U}' = -\boldsymbol{U}. \tag{B.53}$$

Since the coordinate axes all reverse direction, the geometrical position in space of a polar vector is unchanged by a reflection of the coordinate system and the vector may faithfully be represented by an arrow, also under reflection. In physics, acceleration, force, velocity, momentum, and electric dipole moments are all polar vectors.

Evidently, the basis vectors of the old coordinate system have the coordinates, $\hat{\boldsymbol{e}}'_i = -\hat{\boldsymbol{e}}_i$, in the new (reflected) coordinate system. Basis vectors always transform as polar vectors under reflection. One should not get confused by the fact that the new basis vectors in the new system have (by definition) the same coordinates $\hat{\boldsymbol{e}}_i$ as the old basis vectors in the old system, because they refer to different geometric objects in the two coordinate systems.

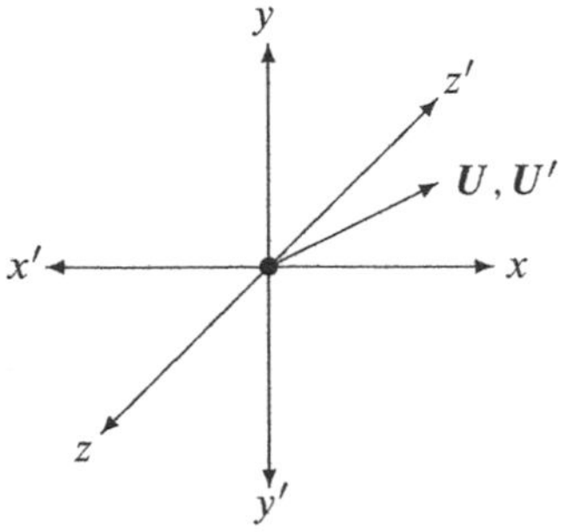

A polar vector retains its geometric placement under a reflection of the coordinate system in the origin.

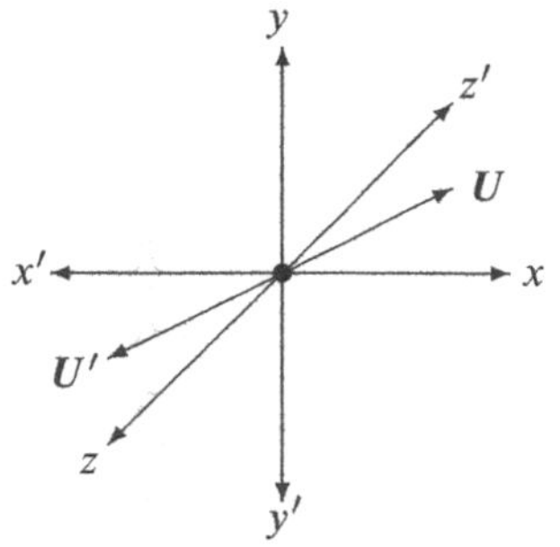

Geometrically, an axial vector has its direction reversed under a reflection of the coordinate system in the origin because it has the same coordinates in the reflected system as in the original.

Axial vectors: The cross-product of two polar vectors, $\boldsymbol{U} = \boldsymbol{V} \times \boldsymbol{W}$, behaves differently than a polar vector under a reflection. According to our rules for calculating the cross-product, which are the same in all coordinate systems, we find

$$\boldsymbol{U}' \equiv \boldsymbol{V}' \times \boldsymbol{W}' = (-\boldsymbol{V}) \times (-\boldsymbol{W}) = \boldsymbol{V} \times \boldsymbol{W} = \boldsymbol{U}, \tag{B.54}$$

without the expected change of sign. Since $\boldsymbol{U}$ behaves as a polar vector under a proper rotation with determinant $+1$, we conclude that the missing minus sign is associated with any transformation with determinant -1, in other words with any reflection. Generalizing, we define an *axial* vector $\boldsymbol{U}$ as a set of three quantities that transform according to the rule

$$\boxed{U_i' = \det \mathbf{A} \sum_j a_{ij} U_j,} \tag{B.55}$$

under an arbitrary Cartesian coordinate transformation without translation. The extra determinant factor counteracts the overall sign change otherwise associated with reflections of polar vectors. In physics, angular momentum, moment of force, and magnetic dipole moments are all axial vectors. Axial vectors are also called *pseudovectors*

The geometric direction of an axial vector depends on what we choose to be right and left. It is for this reason wrong to think of an axial vector as an arrow in space. It has magnitude and direction like a cylinder axis, but it is not oriented like a polar vector. The positive direction of an axial vector is not a geometric property, but a property fixed by convention. For consistency all humans (even the British) have agreed that one particular coordinate system and all coordinate systems that are obtained from it by proper rotation are right-handed, whereas coordinate systems that are related to this class by reflection are left-handed. We do not know whether non-human aliens would have adopted the same convention, but should we ever meet such beings we would be able to find the correct transformation between our reference frames and theirs.

Pseudo-scalars: The volume product of three polar vectors $P = \boldsymbol{U} \cdot \boldsymbol{V} \times \boldsymbol{W}$ is a scalar quantity that changes sign under a pure reflection of the coordinate system. More generally a *pseudo-scalar* transforms like

$$\boxed{P' = \det \mathbf{A} P,} \tag{B.56}$$

under a Cartesian coordinate transformation without translation.

The sign of a pseudo-scalar is not absolute, but depends on the handedness of the coordinate system, and thus on convention. One might think that physics had no use for such quantities, because after all physics itself does not depend on coordinate systems, only its mathematical description does. Nevertheless, magnetic charge, if it is ever found, would be pseudo-scalar, and more importantly some of the familiar elementary particles, for example the pi-mesons, are described by pseudo-scalar fields.

Pseudo-tensors: The transformation properties of *pseudo-tensors* of higher rank also include an extra factor of the determinant of the transformation matrix. The Levi-Civita symbol is a pseudo-tensor of third rank (Problem B.18).

* Subclassification by pure translations

A pure translation takes the form

$$x' = x - c. \tag{B.57}$$

Translationally invariant vectors (and more generally, tensors) are called *proper*, whereas tensors that change under translations are called *improper*. In physics the center of mass and the electric dipole moment are improper polar vectors, whereas angular momentum, moment of force, and magnetic dipole moment are improper axial vectors. The moment of inertia and all kinds of higher multipole moments are improper tensors.

Problems

B.1 Calculate the distance between two points on the spherical Earth of radius a in terms of longitude α, latitude δ, and height h over the surface.

B.2 Show using Cartesian coordinates that the shortest path (geodesic) between two arbitrary points $\boldsymbol{a}$ and $\boldsymbol{b}$ is a described by the linear function $(1-\xi)\boldsymbol{a} + \xi\boldsymbol{b}$ with $0 \leq \xi \leq 1$.

B.3 Let $\boldsymbol{a} = (2, 3, -6)$ and $\boldsymbol{b} = (3, -4, 0)$. Calculate **(a)** the lengths of the vectors, **(b)** the dot-product, **(c)** the cross-product, **(d)** and the tensor product.

B.4 Are the vectors $\boldsymbol{a} = (3, 1, -2)$, $\boldsymbol{b} = (4, -1, -1)$, and $\boldsymbol{c} = (1, -2, 1)$ linearly dependent (meaning that there exists a non-trivial set of coefficients such that $\alpha\boldsymbol{a} + \beta\boldsymbol{b} + \gamma\boldsymbol{c} = \boldsymbol{0}$)?

B.5 Prove that

$$|\boldsymbol{a}\cdot\boldsymbol{b}| \leq |\boldsymbol{a}|\,|\boldsymbol{b}|, \qquad |\boldsymbol{a}+\boldsymbol{b}| \leq |\boldsymbol{a}| + |\boldsymbol{b}|.$$

B.6 Prove that

$$(\boldsymbol{a}\times\boldsymbol{b}\cdot\boldsymbol{c})\,(\boldsymbol{d}\times\boldsymbol{e}\cdot\boldsymbol{f}) = \begin{vmatrix} \boldsymbol{a}\cdot\boldsymbol{d} & \boldsymbol{a}\cdot\boldsymbol{e} & \boldsymbol{a}\cdot\boldsymbol{f} \\ \boldsymbol{b}\cdot\boldsymbol{d} & \boldsymbol{b}\cdot\boldsymbol{e} & \boldsymbol{b}\cdot\boldsymbol{f} \\ \boldsymbol{c}\cdot\boldsymbol{d} & \boldsymbol{c}\cdot\boldsymbol{e} & \boldsymbol{c}\cdot\boldsymbol{f} \end{vmatrix}.$$

B.7 Show that

$$\sum_i \delta_{ii} = 3, \qquad \sum_j \delta_{ij}\delta_{jk} = \delta_{ik}$$

B.8 Show that

$$\epsilon_{ijk}\,\epsilon_{lmn} = \delta_{il}\delta_{jm}\delta_{kn} + \delta_{im}\delta_{jn}\delta_{kl} + \delta_{in}\delta_{jl}\delta_{km} - \delta_{in}\delta_{jm}\delta_{kl} - \delta_{il}\delta_{jn}\delta_{km} - \delta_{im}\delta_{jl}\delta_{kn}$$

$$\sum_k \epsilon_{ijk}\,\epsilon_{lmk} = \delta_{il}\delta_{jm} - \delta_{im}\delta_{jl},$$

$$\sum_{jk} \epsilon_{ijk}\,\epsilon_{ljk} = 2\delta_{il},$$

$$\sum_{ijk} \epsilon_{ijk}\,\epsilon_{ijk} = 6.$$

Hint: Use Problem B.6.

B.9 Prove the relations

$$a \times b \cdot c = a \cdot b \times c = b \cdot c \times a,$$
$$(a \times b) \times c = (a \cdot c)\, b - (b \cdot c)\, a,$$
$$(a \times b) \cdot (c \times d) = (a \cdot c)(b \cdot d) - (a \cdot d)(b \cdot c),$$
$$|a \times b|^2 = |a|^2\, |b|^2 - (a \cdot b)^2.$$

Hint: Use Problem B.8.

B.10 Show that

$$(a \times b \cdot c) d = a \times b (c \cdot d) + b \times c (a \cdot d) + c \times a (b \cdot d) \tag{B.58}$$

for arbitrary vectors a, b, c, and d.

B.11 Show that the following association rules hold for the tensor product

$$(ab) \cdot c = a(b \cdot c), \qquad a \cdot (bc) = (a \cdot b)c.$$

B.12 Show that the trace $\sum_i T_{ii}$ of a tensor is invariant under a rotation.

B.13 Show that the Kronecker delta transforms as a tensor.

B.14 Show that the distance $|x - y|$ is invariant under any transformation between Cartesian coordinate systems.

* **B.15** Consider two Cartesian coordinate systems and make no assumptions about the transformation $x' = f(x)$ between them. Show that the invariance of the distance,

$$\left|x' - y'\right| = |x - y|,$$

implies that the transformation is of the form $x' = \mathbf{A} \cdot x + b$ where $\mathbf{A}$ is an orthogonal matrix.

* **B.16** Show that if $W_i = \sum_j T_{ij} V_j$ and if it is known that W is a vector for any vectors V, then T_{ij} must be a tensor. This is called the *quotient rule.*

* **B.17** Show that the general Cartesian coordinate transformation may be built up from a combination of simple translations, rotations, and reflections.

* **B.18** Show that

(a) The Levi-Civita symbol satisfies

$$\sum_{lmn} a_{il} a_{jm} a_{kn} \epsilon_{lmn} = \det \mathbf{A}\, \epsilon_{ijk},$$

where $\mathbf{A} = \{a_{ij}\}$ is an arbitrary matrix.

(b) The Levi-Civita symbol (which by the definition of the cross-product must be invariant, $\epsilon'_{ijk} = \epsilon_{ijk}$) in fact transforms as a pseudo-tensor

$$\epsilon'_{ijk} = \det \mathbf{A} \sum_{lmn} a_{il} a_{jm} a_{kn} \epsilon_{lmn} = \epsilon_{ijk}$$

for an arbitrary transformation matrix. This confirms the invariance.

(c) The cross-product of two vectors $W = U \times V$ must transform as a pseudo-vector

$$W'_i = \det \mathbf{A} \sum_j a_{ij} W_j.$$

∗ B.19 Verify that under a simple rotation, a tensor T_{ij} transforms into

$$
\begin{aligned}
T'_{xx} &= \cos\phi(T_{xx}\cos\phi + T_{xy}\sin\phi) + \sin\phi(T_{yx}\cos\phi + T_{yy}\sin\phi),\\
T'_{xy} &= \cos\phi(-T_{xx}\sin\phi + T_{xy}\cos\phi) + \sin\phi(-T_{yx}\sin\phi + T_{yy}\cos\phi),\\
T'_{xz} &= \cos\phi T_{xz} + \sin\phi T_{yz},\\
T'_{yx} &= -\sin\phi(T_{xx}\cos\phi + T_{xy}\sin\phi) + \cos\phi(T_{yx}\cos\phi + T_{yy}\sin\phi),\\
T'_{yy} &= -\sin\phi(-T_{xx}\sin\phi + T_{xy}\cos\phi) + \cos\phi(-T_{yx}\sin\phi + T_{yy}\cos\phi),\\
T'_{yz} &= -\sin\phi T_{xz} + \cos\phi T_{yz},\\
T'_{zx} &= T_{zx}\cos\phi + T_{zy}\sin\phi,\\
T'_{zy} &= -T_{zx}\sin\phi + T_{zy}\cos\phi,\\
T'_{zz} &= T_{zz}.
\end{aligned}
$$

C Field calculus

In continuum physics the basic mathematical objects are *fields*. Like the geometric objects discussed in Appendix B, fields are classified as scalar, vector, and tensor fields, each type having a different number of components. A field component is simply a real-valued function of the three spatial coordinates and time. A scalar field has only one component, a vector field has three, and a general tensor field of rank r has 3^r components. In quantum physics there are also important complex fields of half-integer rank, called spinor fields, but we do not need them in classical physics.

The invention of field calculus (or vector calculus as it is mostly called) goes back to Gibbs at the end of the nineteenth century [Crowe 1994]. Field calculus revolutionized the mathematical treatment of the coupled partial differential equations that seemed to turn up in all branches of physics. When the fundamental tools of calculus, differentiation and integration, are combined with vector algebra, new types of differential operators and new types of integrals naturally arise. In combination, the formalism can still become quite complicated, and one may eventually have to resort to index notation to avoid ambiguity.

In this appendix we develop only the most basic tools of field calculus, some of which have already been given in the main text in a simpler form. The presentation is precise but not mathematically rigorous.

Josiah Willard Gibbs (1839–1903). American engineer and theoretical physicist. In spite of his unassuming manner, he was a towering scientist contributing to thermodynamics, electromagnetism, mathematics, and astronomy. In his last publication he put statistical mechanics on a firm foundation.

C.1 Spatial derivatives

The triplet of spatial derivatives is given a special symbol, called *nabla*[1],

$$\nabla \equiv \frac{\partial}{\partial \boldsymbol{x}} = \left(\frac{\partial}{\partial x}, \frac{\partial}{\partial y}, \frac{\partial}{\partial z}\right). \tag{C.1}$$

It is a *differential operator* that in most respects also behaves like a vector (Problem C.5). Some of the common rules for differentiation, for example linearity and the chain rule, can be directly used in combination with the nabla operator, whereas others such as the product rule become more complicated (see below).

[1] Although it is sometimes natural to use the shorthand $\partial_u = \partial/\partial u$ for the partial derivative with respect to any variable u, the corresponding vector symbol $\boldsymbol{\partial} = (\partial_x, \partial_y, \partial_z)$ is rarely used instead of ∇. In the older literature the gradient, divergence, and curl are often denoted by $\mathsf{grad}\, S = \nabla S$, $\mathsf{div}\, \boldsymbol{U} = \nabla \cdot \boldsymbol{U}$, and $\mathsf{curl}\, \boldsymbol{U} = \nabla \times \boldsymbol{U}$ or in continental Europe $\mathsf{rot}\, \boldsymbol{U} = \nabla \times \boldsymbol{U}$. This notation is now all but obsolete.

Gradient

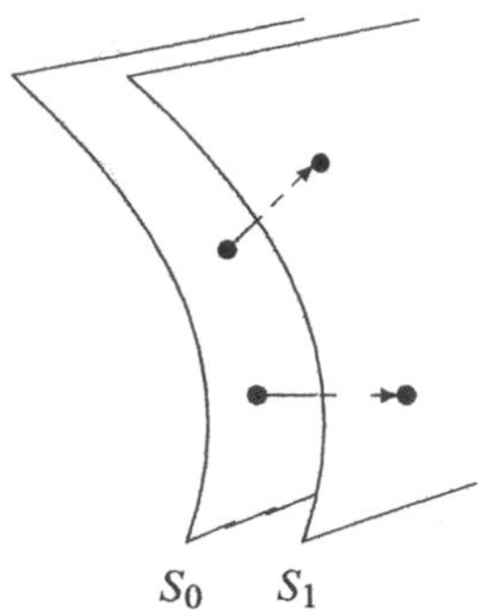

Two contour surfaces defined by $S(\boldsymbol{x}) = S_0$ and $S(\boldsymbol{x}) = S_1$ with $S_1 > S_0$. The gradient at S_0 is always orthogonal to the contour surface and is oriented from S_0 to S_1.

Operating on a scalar field $S(\boldsymbol{x})$, nabla creates a vector field called the *gradient* of S,

$$\nabla S = (\nabla_x S, \nabla_y S, \nabla_z S) = \left(\frac{\partial S}{\partial x}, \frac{\partial S}{\partial y}, \frac{\partial S}{\partial z}\right). \tag{C.2}$$

Here we have for clarity suppressed the explicit dependence on the spatial coordinates.

The gradient is useful for calculating the difference in field values at two nearby points. Expanding to first order in each of the coordinate differentials, $d\boldsymbol{x} = (dx, dy, dz)$, we get

$$S(\boldsymbol{x} + d\boldsymbol{x}) - S(\boldsymbol{x}) = dx\frac{\partial S(\boldsymbol{x})}{\partial x} + dy\frac{\partial S(\boldsymbol{x})}{\partial y} + dz\frac{\partial S(\boldsymbol{x})}{\partial z}.$$

The left-hand side is the differential $dS(\boldsymbol{x})$ of the field, so that this important rule can be written

$$\boxed{dS(\boldsymbol{x}) = d\boldsymbol{x} \cdot \nabla S(\boldsymbol{x}).} \tag{C.3}$$

A scalar field is often pictured by means of surfaces of constant value, $S(\boldsymbol{x}) =$ const, also called *contour surfaces*. It is now easy to see that *the gradient in a given point is always orthogonal to the contour surface containing this point.* For if both $\boldsymbol{x}$ and $\boldsymbol{x} + d\boldsymbol{x}$ lie on the same contour surface, the differential dS must vanish, or $d\boldsymbol{x} \cdot \nabla S(\boldsymbol{x}) = \boldsymbol{0}$; and since all differentials $d\boldsymbol{x}$ are tangential to the contour surface, it follows from the vanishing of the dot product that the gradient ∇S must itself be orthogonal to the surface.

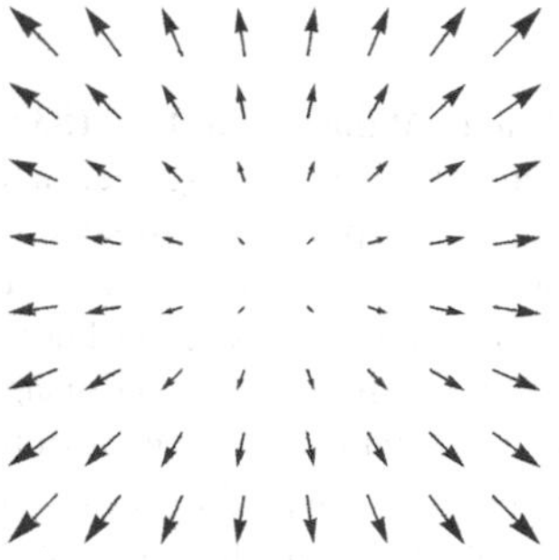

Diverging vector field $\boldsymbol{U} = \boldsymbol{x}$ plotted in the xy-plane. Its divergence is $\nabla \cdot \boldsymbol{U} = 3$.

Divergence

Dotting ∇ with a vector field $\boldsymbol{U}(\boldsymbol{x})$ we obtain a scalar field, called the *divergence* of $\boldsymbol{U}$,

$$\nabla \cdot \boldsymbol{U} = \nabla_x U_x + \nabla_y U_y + \nabla_z U_z = \frac{\partial U_x}{\partial x} + \frac{\partial U_y}{\partial y} + \frac{\partial U_z}{\partial z}. \tag{C.4}$$

If you plot a vector field with a positive divergence by means of small arrows, the arrows will have a tendency to diverge from each other, or converge if the divergence is negative.

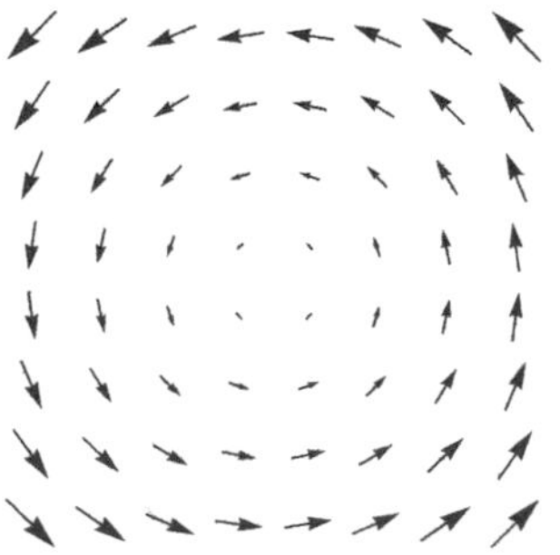

Curling vector field $\boldsymbol{U} = \hat{\boldsymbol{e}}_z \times \boldsymbol{x}$ plotted in the xy-plane. Its curl is $\nabla \times \boldsymbol{U} = 2\hat{\boldsymbol{e}}_z$.

Curl

Using the gradient operator as the left-hand component in the cross-product (B.7) with a vector field $\boldsymbol{U}(\boldsymbol{x})$, we obtain another vector field, called the *curl* of $\boldsymbol{U}$,

$$\begin{aligned}\nabla \times \boldsymbol{U} &= (\nabla_y U_z - \nabla_z U_y, \nabla_z U_x - \nabla_x U_z, \nabla_x U_y - \nabla_y U_x) \\ &= \left(\frac{\partial U_z}{\partial y} - \frac{\partial U_y}{\partial z}, \frac{\partial U_x}{\partial z} - \frac{\partial U_z}{\partial x}, \frac{\partial U_y}{\partial x} - \frac{\partial U_x}{\partial y}\right).\end{aligned} \tag{C.5}$$

If you plot a vector field with a non-vanishing curl by means of small arrows, the arrows will have a tendency to circulate.

Product rules

Combining these operators with products of various fields leads to a number of useful rules, which we list here. Parentheses are put in liberally to facilitate understanding, even if in practice one would use fewer.

- Rules with two nablas:

$$\nabla \cdot (\nabla \times \boldsymbol{U}) = 0, \tag{C.6a}$$
$$\nabla \times (\nabla S) = \boldsymbol{0}, \tag{C.6b}$$
$$\nabla \times (\nabla \times \boldsymbol{U}) = \nabla(\nabla \cdot \boldsymbol{U}) - \nabla^2 \boldsymbol{U}. \tag{C.6c}$$

- Simple product rules:

$$\nabla(S_1 S_2) = (\nabla S_1) S_2 + S_1(\nabla S_2), \tag{C.7a}$$
$$\nabla \cdot (S\boldsymbol{U}) = (\nabla S) \cdot \boldsymbol{U} + S(\nabla \cdot \boldsymbol{U}), \tag{C.7b}$$
$$\nabla \times (S\boldsymbol{U}) = (\nabla S) \times \boldsymbol{U} + S(\nabla \times \boldsymbol{U}). \tag{C.7c}$$

- Complex product rules:

$$\nabla \times (\boldsymbol{U}_1 \times \boldsymbol{U}_2) = (\boldsymbol{U}_2 \cdot \nabla)\boldsymbol{U}_1 - (\boldsymbol{U}_1 \cdot \nabla)\boldsymbol{U}_2 + \boldsymbol{U}_1(\nabla \cdot \boldsymbol{U}_2) - \boldsymbol{U}_2(\nabla \cdot \boldsymbol{U}_1), \tag{C.8a}$$
$$\nabla(\boldsymbol{U}_1 \cdot \boldsymbol{U}_2) = \boldsymbol{U}_1 \times (\nabla \times \boldsymbol{U}_2) + \boldsymbol{U}_2 \times (\nabla \times \boldsymbol{U}_1) + (\boldsymbol{U}_1 \cdot \nabla)\boldsymbol{U}_2 + (\boldsymbol{U}_2 \cdot \nabla)\boldsymbol{U}_1. \tag{C.8b}$$

- The most complex rule:

$$\boldsymbol{a} \times \boldsymbol{b} \cdot (\boldsymbol{c} \cdot \nabla)\boldsymbol{U} + \boldsymbol{b} \times \boldsymbol{c} \cdot (\boldsymbol{a} \cdot \nabla)\boldsymbol{U} + \boldsymbol{c} \times \boldsymbol{a} \cdot (\boldsymbol{b} \cdot \nabla)\boldsymbol{U} = (\boldsymbol{a} \times \boldsymbol{b} \cdot \boldsymbol{c})\, \nabla \cdot \boldsymbol{U} \tag{C.9}$$

is valid for arbitrary vector fields $\boldsymbol{a}$, $\boldsymbol{b}$, and $\boldsymbol{c}$. This rule, which follows from (??) with $\boldsymbol{d}$ replaced by ∇, expresses that in three dimensions any set of four vectors is always linearly dependent.

C.2 Spatial integrals

In physics—where nothing is truly infinite or truly infinitesimal—spatial integrals are best understood in the Riemannian sense where the integration domain is subdivided into a huge number of tiny subdomains, and the integral is approximated by a sum over these subdomains. This procedure can be carried to any precision for integrands that vary smoothly over the integration domain.

Curve integral

The curve integral of a smoothly varying vector field $\boldsymbol{U}(\boldsymbol{x})$ along an oriented curve C running from $\boldsymbol{a}$ to $\boldsymbol{b}$ is defined by

$$\int_C \boldsymbol{U}(\boldsymbol{x}) \cdot d\boldsymbol{\ell} = \lim_{N\to\infty} \sum_{n=1}^{N} \boldsymbol{U}(\boldsymbol{x}_n) \cdot d\boldsymbol{\ell}_n. \tag{C.10}$$

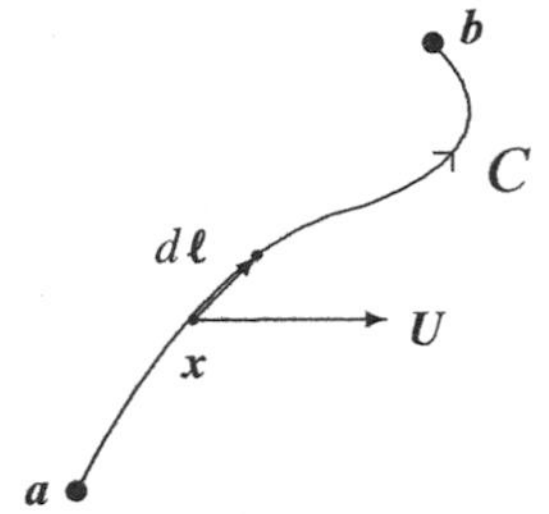

An oriented curve C running from the point $\boldsymbol{a}$ to $\boldsymbol{b}$ with a straight-line curve element $d\boldsymbol{\ell}$ near $\boldsymbol{x}$.

As shown on the right, the integral should be understood as the limit of a huge sum over tiny straight-line pieces, each represented by its vector *line (or curve) element* $d\boldsymbol{\ell}$ near $\boldsymbol{x}$, which is parallel with the curve and has length $d\ell = |d\boldsymbol{\ell}|$. In Cartesian coordinates the vector line element is equal to the position differential $d\boldsymbol{\ell} = d\boldsymbol{x} = (dx, dy, dz)$. In curvilinear coordinates it is more complicated.

Surface integral

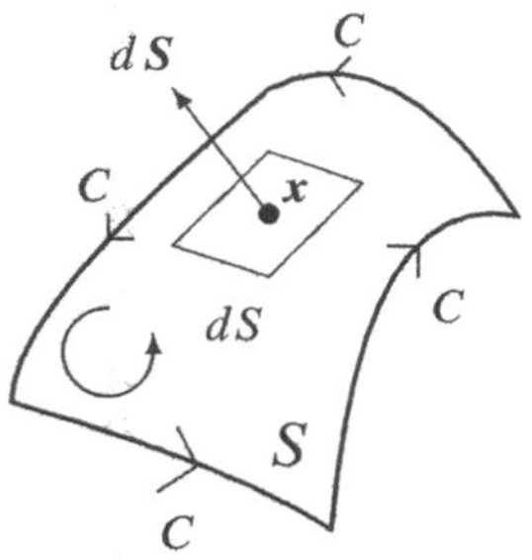

An oriented surface S with a tiny flat surface patch dS near x. All normals point to the same side of the surface. The perimeter curve C is oriented consistently with the surface.

The surface integral of a smoothly varying vector field $U(x)$ over an oriented surface S is defined by

$$\int_S U(x) \cdot dS = \lim_{N \to \infty} \sum_{n=1}^{N} U(x_n) \cdot dS_n. \tag{C.11}$$

The integral is the limit of a huge sum over tiny flat surface patches, each represented by its area vector or *surface element* dS near x, orthogonal to the surface with area $dS = |dS|$. On an oriented surface, neighboring surface elements must consistently point to the same side of the surface (thereby excluding non-orientable surfaces such as the Möbius band and the Klein bottle). By universal convention, *the normals of a closed surface are always chosen to be directed out of the enclosed volume.* For reasons of "symbol economy" we have here used the same letter S for both the domain of integration and the infinitesimal element. In Cartesian coordinates a surface element in the xy-plane becomes $dS = \hat{e}_z\, dxdy$.

Volume integral

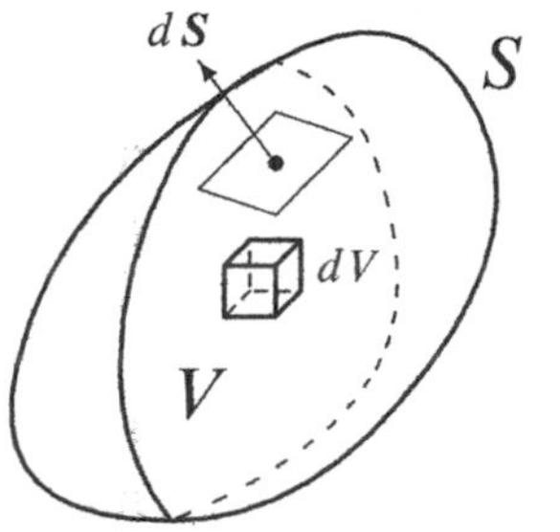

A volume V with its enclosing surface S, a volume element dV near x and a surface element dS.

The volume integral of a smoothly varying scalar field $S(x)$ over a volume V is defined by

$$\int_V S(x)\, dV = \lim_{N \to \infty} \sum_{n=1}^{N} S(x_n) dV_n. \tag{C.12}$$

The integral is a huge sum over tiny *volume elements*, each represented by its volume dV near x. In Cartesian coordinates the volume element takes the form $dV = dxdydz$.

C.3 Fundamental integral theorems

For each of the above types of integral there is a fundamental mathematical theorem relating it to an integral of lower dimension. We have collected them here because they will be useful for nearly all of the topics covered in this book. Proofs are found in the following section.

Gradient theorem

Let C be an arbitrary oriented curve with endpoints a and b, and let $S(x)$ be a scalar field. Then the integral of the gradient of S along C equals endpoints difference of S,

$$\boxed{\int_C \nabla S \cdot d\ell = S(b) - S(a).} \tag{C.13}$$

Note that the right-hand side does not depend on the curve connecting the endpoints a and b.

Stokes' theorem

Let S be an arbitrary oriented surface with the closed curve C as perimeter, oriented in the same way as the surface. Then the surface integral of the curl of U over S equals the curve integral of the vector field U around the perimeter C,

$$\boxed{\int_S \nabla \times U \cdot dS = \oint_C U \cdot d\ell.} \tag{C.14}$$

The circle in the curve integral on the right is just there to remind us that the curve is closed. Note that the right-hand side takes the same value for any surface that has C as perimeter.

Gauss' theorem

Although this theorem was discovered several times more than 200 years ago, it is usually attributed to Gauss in 1813. Let V be an arbitrary volume bounded by the closed surface S, oriented with all normals pointing out of the volume. Then the volume integral over V of the divergence of $\boldsymbol{U}$ equals the surface integral of the vector field $\boldsymbol{U}$ over S,

$$\boxed{\int_V \nabla \cdot \boldsymbol{U}\, dV = \oint_S \boldsymbol{U} \cdot d\boldsymbol{S},} \tag{C.15}$$

where the circle indicates that the surface is closed.

* C.4 Proofs of the fundamental integral theorems

In each case we first prove that the theorem is *additive*, meaning that if it holds for all the parts of an integration domain, it holds for the whole domain. Since the integrals are defined as huge sums over tiny curve, surface, or volume elements, we only need to prove the theorem for such elements.

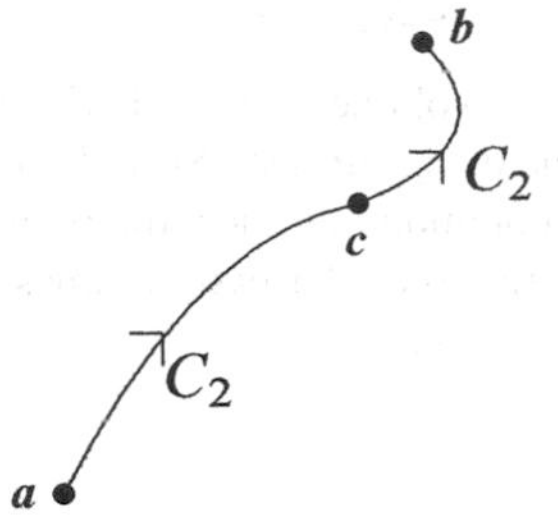

A an oriented curve C divided into two curves, C_1 and C_2, joined at the point $\boldsymbol{c}$.

Gradient theorem

To prove additivity, we first divide a curve C into two curves, C_1 and C_2, joined at point $\boldsymbol{c}$. The integral is the sum of the two contributions; and if it holds for each part, it holds for the complete curve because the two endpoint contributions at $\boldsymbol{c}$ cancel. Thus, the theorem holds in general if it holds for the tiny straight-line element $d\boldsymbol{\ell} = d\boldsymbol{x}$. But that has already been shown in (C.3).

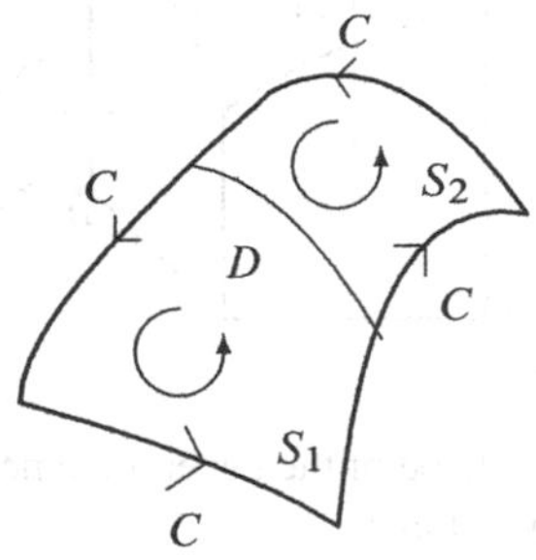

An oriented surface S divided into two surfaces S_1 and S_2 joined at the curve D.

Stokes' theorem

Let the oriented surface S be divided into two parts S_1 and S_2 with the same orientation by means of a curve D (see the margin figure). If the theorem is valid for each part, it will also be valid for the whole because the two boundary curves pass through the common piece D with opposite orientation, such that the two curve integrals along D will cancel. Consequently, the theorem holds in general if it holds for infinitesimal planar surface elements.

It is well-known that every surface can be triangulated, meaning that it can be approximated by tiny adjacent planar triangles that may be chosen to be right-angled. Without loss of generality we may place any particular right-angled triangle in the xy-plane with sides a and b along the axes. Such a triangle is described by the conditions (see the margin figure),

$$0 \le x \le a, \qquad 0 \le y \le b, \qquad \frac{x}{a} + \frac{y}{b} = 1. \tag{C.16}$$

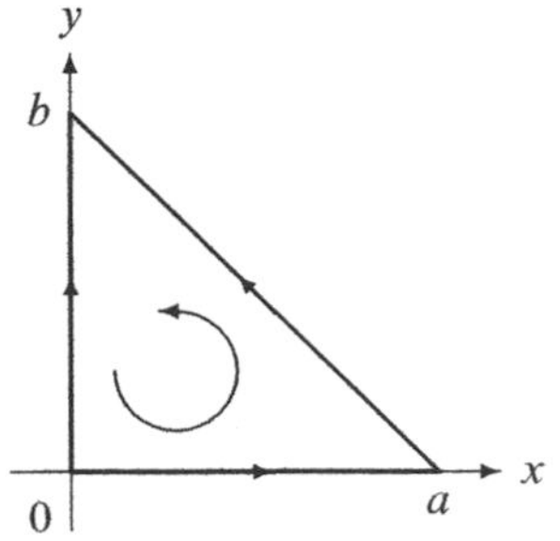

An oriented right-angled triangle with sides a and b along the coordinate axes.

Using that $d\boldsymbol{S} = \hat{\boldsymbol{e}}_z dxdy$, and defining $y(x) = b(1 - x/a)$ and $x(y) = a(1 - y/b)$, we get

$$\begin{aligned}
\int_S \nabla \times \boldsymbol{U} \cdot d\boldsymbol{S} &= \int_S \left[\nabla_x U_y(x, y) - \nabla_y U_x(x, y)\right] dxdy \\
&= \int_0^b \left[U_y(x(y), y) - U_y(0, y)\right] dy - \int_0^a \left[U_x(x, y(x)) - U_x(x, 0)\right] dx \\
&= \int_0^a U_x(x, 0)\, dx + \int_0^b U_y(x(y), y)\, dy + \int_a^0 U_x(x, y(x))\, dx + \int_b^0 U_y(0, y)\, dy \\
&= \oint_C \boldsymbol{U} \cdot d\boldsymbol{\ell}.
\end{aligned}$$

In the second step we have integrated over x in the first term and over y in the second. In the last step we have collected the four terms into the curve integral around the perimeter according to its orientation.

Gauss' theorem

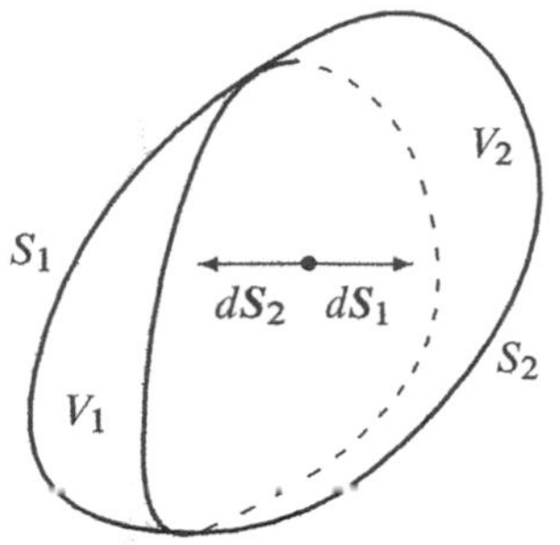

Two volumes V_1 and V_2 with enclosing surfaces S_1 and S_2. At a common interface, the outward normals of the two volumes are opposite.

If the volume V is divided into two volumes V_1 and V_2 that both satisfy this theorem, then the volume V will also satisfy the theorem because the outward normals from V_1 and V_2 have opposite directions at the common interface, and the contributions to the surface integrals from the common interface therefore cancel each other. Consequently, by the definition of the volume integral (C.12), the theorem holds in general if it holds for infinitesimal volume elements.

Carving the volume into smaller pieces along the coordinate planes, one can convince oneself that the general volume element can be chosen to be a rectangular box, except near the surface where the box must have a curved "lid" (see the margin figure). Choosing a suitable coordinate system, it is described by the conditions,

$$0 \le x \le a, \qquad 0 \le y \le b, \qquad 0 \le z \le c(x,y), \tag{C.17}$$

where $c(x,y)$ describes the lid.

For such a volume we prove the theorem for a special vector field only having a z-component, $\boldsymbol{U} = U_z \hat{\boldsymbol{e}}_z$. Integrating over z we obtain

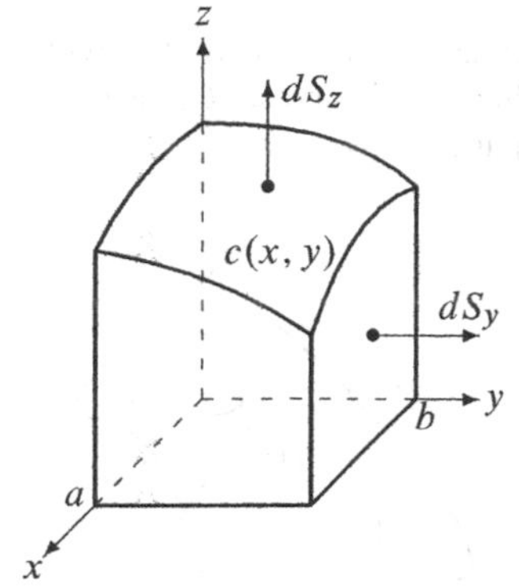

The shape of the general elementary volume.

$$\begin{aligned}\int_V \nabla \cdot \boldsymbol{U}\, dV = \int_V \nabla_z U_z\, dV &= \int_0^a \int_0^b \int_0^{c(x,y)} \nabla_z U_z\, dx dy dz \\ &= \int_0^a \int_0^b \left[U_z(x,y,c(x,y)) - U_z(x,y,0)\right] dx dy \\ &= \oint_S U_z\, dS_z = \oint_S \boldsymbol{U} \cdot d\boldsymbol{S}.\end{aligned}$$

Here we used that all surface elements are normal to the box and directed outward, such that $dS_z = 0$ on all faces of the box, except for the one in the xy-plane at $z = 0$ where $dS_z = -dxdy$ and the one at $z = c(x,y)$ where $dS_z = dxdy$. Adding all the little volumes together, we arrive at the theorem for the special vector field. The two other integrals over $\boldsymbol{U} = U_x \hat{\boldsymbol{e}}_x$ and $\boldsymbol{U} = U_y \hat{\boldsymbol{e}}_y$ yield analogous results, and adding all three concludes the proof.

C.5 Field transformations

A field $S(\boldsymbol{x})$, which like the mass density takes a single value in each point of space, is called a *scalar field*. A field $\boldsymbol{U}(\boldsymbol{x})$ that takes a vector value in each point is called a *vector field*, and a field $\mathbf{T}(\boldsymbol{x})$ that takes tensor values, a *tensor field*.

The transformation laws for fields are quite similar to the ones in Section B.6, the only difference being that the coordinates of the spatial position must also transform. For scalar, vector, and tensor fields, the transformation rules under pure rotations are

$$S'(\boldsymbol{x}') = S(\boldsymbol{x}), \tag{C.18a}$$

$$\boldsymbol{U}'(\boldsymbol{x}') = \mathbf{A} \cdot \boldsymbol{U}(\boldsymbol{x}), \tag{C.18b}$$

$$\mathbf{T}'(\boldsymbol{x}') = \mathbf{A} \cdot \mathbf{T}(\boldsymbol{x}) \cdot \mathbf{A}^\top, \tag{C.18c}$$

where $\boldsymbol{x}' = \mathbf{A} \cdot \boldsymbol{x}$. These definitions express that the new fields in the new position are obtained from the old fields in the old position by transforming them according to their tensor type. As mentioned before, such relations express the unique reality of physical quantities.

Problems

C.1 Calculate **(a)** the divergence and **(b)** the curl of the vector field $\boldsymbol{a} \times \boldsymbol{x}$.

C.2 Show that

$$\sum_i \nabla_i x_i = 3, \qquad \nabla_i x_j = \delta_{ij}, \qquad \nabla_i \nabla_j (x_k x_l) = \delta_{ik}\delta_{jl} + \delta_{il}\delta_{jk}.$$

C.3 Show that for arbitrary vector fields $\boldsymbol{U}$ and $\boldsymbol{V}$, we have

$$\nabla \times (\boldsymbol{U} \times \boldsymbol{V}) = \boldsymbol{V} \cdot \nabla \boldsymbol{U} - \boldsymbol{U} \cdot \nabla \boldsymbol{V} + \boldsymbol{U} \nabla \cdot \boldsymbol{V} - \boldsymbol{V} \nabla \cdot \boldsymbol{U}.$$

C.4 **(a)** Prove the following relations involving the nabla operator twice (here S is a scalar field and $\boldsymbol{V}$ a vector field),

$$\nabla \cdot (\nabla \times \boldsymbol{V}) = \boldsymbol{0}, \qquad \nabla \times (\nabla S) = 0, \qquad \nabla \times (\nabla \times \boldsymbol{V}) = \nabla(\nabla \cdot \boldsymbol{V}) - (\nabla \cdot \nabla)\boldsymbol{V}.$$

(b) Where in these relations does it make sense to remove the parentheses?

∗ **C.5** Show that the nabla operator transforms as a vector, $\nabla'_i = \sum_j a_{ij} \nabla_j$ under an arbitrary rotation.

D

Curvilinear coordinates

The distance between two points in Euclidean space takes the simplest form (B.1) in Cartesian coordinates. The geometry of concrete physical problems may, however, make non-Cartesian coordinates more suitable as a basis for analysis, even if the distance function becomes more complicated. The new coordinates are defined as functions of the Cartesian coordinates (and conversely). Together they define in each point three intersecting curves that correspond to the local curvilinear coordinate axes passing through the point.

At a deeper level it is often the *symmetry* of a physical problem that points to the most convenient choice of coordinates. Cartesian coordinates are well suited to problems with translational invariance, cylindrical coordinates for problems that are invariant under rotations around a fixed axis, and spherical coordinates for problems that are invariant or partially invariant under arbitrary rotations. In this section we shall only discuss cylindrical and spherical coordinates.

D.1 Cylindrical coordinates

In a flat Euclidean space it is always most convenient to view curvilinear coordinate systems through the "eyes" of a particular global Cartesian coordinate systems. The transformation from cylindrical coordinates r, ϕ, z to Cartesian coordinates x, y, z is given by given by[1]

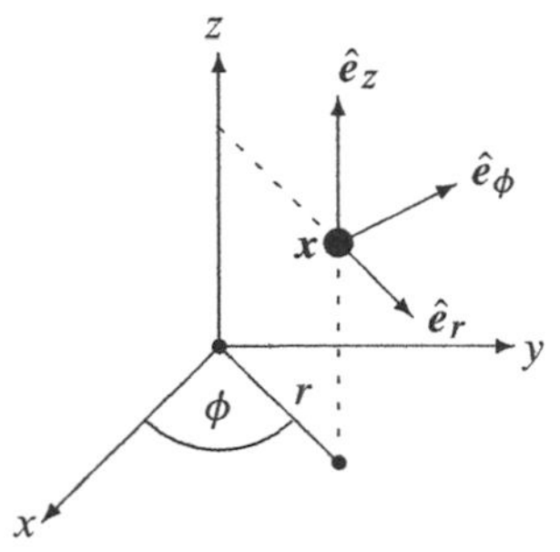

Cylindrical coordinates and their basis vectors.

$$x = r\cos\phi, \qquad y = r\sin\phi, \qquad z = z, \tag{D.1}$$

with domain of variation $0 \leq r < \infty$, $0 \leq \phi < 2\pi$, and $-\infty < z < \infty$. The inverse transformation takes the form

$$r = \sqrt{x^2 + y^2}, \qquad \phi = \arctan\frac{y}{x}, \qquad z = z, \tag{D.2}$$

with the domain of variation $-\infty < x, y, z < \infty$. Note that the reciprocal transformations $(x, y) \longleftrightarrow (r, \phi)$ simply define polar coordinates in the xy-plane. The last, $z \longleftrightarrow z$, is redundant but included to emphasize that these are to be understood as transformations in three-dimensional space.

[1]Some texts use θ instead of ϕ as the conventional name for the polar angle in the plane. Similarly, the radial variable r is sometimes denoted by s to distinguish it from the spherical radial distance. Rational arguments can be given for all choices.

Local basis vectors

The *cylindrical basis vectors* are defined from the tangent vectors, obtained by differentiating the Cartesian position with respect to the cylindrical coordinates,

$$\hat{\boldsymbol{e}}_r = \frac{\partial \boldsymbol{x}}{\partial r} = (\cos\phi, \sin\phi, 0), \tag{D.3a}$$

$$\hat{\boldsymbol{e}}_\phi = \frac{1}{r}\frac{\partial \boldsymbol{x}}{\partial \phi} = (-\sin\phi, \cos\phi, 0), \tag{D.3b}$$

$$\hat{\boldsymbol{e}}_z = \frac{\partial \boldsymbol{x}}{\partial z} = (0, 0, 1). \tag{D.3c}$$

They are orthogonal, normalized, and satisfy $\hat{\boldsymbol{e}}_r \times \hat{\boldsymbol{e}}_\phi \cdot \hat{\boldsymbol{e}}_z = 1$. They thus define a local basis with an orientation that changes from place to place.

An arbitrary vector field may be resolved in this basis

$$\boldsymbol{U} = \hat{\boldsymbol{e}}_r U_r + \hat{\boldsymbol{e}}_\phi U_\phi + \hat{\boldsymbol{e}}_z U_z, \tag{D.4}$$

where

$$U_r = \boldsymbol{U} \cdot \hat{\boldsymbol{e}}_r, \qquad U_\phi = \boldsymbol{U} \cdot \hat{\boldsymbol{e}}_\phi, \qquad U_z = \boldsymbol{U} \cdot \hat{\boldsymbol{e}}_z \tag{D.5}$$

are the projections of $\boldsymbol{U}$ on the local basis vectors. A tensor field $\mathbf{T}$ may similarly be resolved in dyadic products of the local basis vectors.

Line, surface, and volume elements

The differentials along the local coordinate axes,

$$d_r\boldsymbol{x} = \hat{\boldsymbol{e}}_r dr, \qquad d_\phi\boldsymbol{x} = \hat{\boldsymbol{e}}_\phi r d\phi, \qquad d_z\boldsymbol{x} = \hat{\boldsymbol{e}}_z dz, \tag{D.6}$$

allow us to resolve the Cartesian line, surface, and volume elements in the local basis,

$$d\boldsymbol{\ell} \equiv d_r\boldsymbol{x} + d_\phi\boldsymbol{x} + d_z\boldsymbol{x} = \hat{\boldsymbol{e}}_r\, dr + \hat{\boldsymbol{e}}_\phi\, r d\phi + \hat{\boldsymbol{e}}_z\, dz, \tag{D.7}$$

$$\begin{aligned} d\boldsymbol{S} &\equiv d_\phi\boldsymbol{x} \times d_z\boldsymbol{x} + d_z\boldsymbol{x} \times d_r\boldsymbol{x} + d_r\boldsymbol{x} \times d_\phi\boldsymbol{x} \\ &= \hat{\boldsymbol{e}}_r\, r\, d\phi dz + \hat{\boldsymbol{e}}_\phi\, dr dz + \hat{\boldsymbol{e}}_z\, r\, d\phi dr, \end{aligned} \tag{D.8}$$

$$dV \equiv d_r\boldsymbol{x} \times d_\phi\boldsymbol{x} \cdot d_z\boldsymbol{x} = r\, d\phi dr dz. \tag{D.9}$$

Using these infinitesimals, all integrals can be converted to cylindrical coordinates.

Resolution of the gradient

The derivatives with respect to the cylindrical coordinates are obtained by differentiation through the Cartesian coordinates,

$$\frac{\partial}{\partial r} = \frac{\partial x}{\partial r}\frac{\partial}{\partial x} + \frac{\partial y}{\partial r}\frac{\partial}{\partial y} = \cos\phi\frac{\partial}{\partial x} + \sin\phi\frac{\partial}{\partial y} = \hat{\boldsymbol{e}}_r \cdot \nabla,$$

$$\frac{\partial}{\partial \phi} = \frac{\partial x}{\partial \phi}\frac{\partial}{\partial x} + \frac{\partial y}{\partial \phi}\frac{\partial}{\partial y} = -r\sin\phi\frac{\partial}{\partial x} + r\cos\phi\frac{\partial}{\partial y} = r\hat{\boldsymbol{e}}_\phi \cdot \nabla.$$

Defining the projections of the nabla operator on the cylindrical basis,

$$\nabla_r = \hat{\boldsymbol{e}}_r \cdot \nabla = \frac{\partial}{\partial r}, \qquad \nabla_\phi = \hat{\boldsymbol{e}}_\phi \cdot \nabla = \frac{1}{r}\frac{\partial}{\partial \phi}, \qquad \nabla_z = \hat{\boldsymbol{e}}_z \cdot \nabla = \frac{\partial}{\partial z}, \tag{D.10}$$

it may be resolved on the basis

$$\nabla = \hat{e}_r\nabla_r + \hat{e}_\phi\nabla_\phi + \hat{e}_z\nabla_z = \hat{e}_r\frac{\partial}{\partial r} + \hat{e}_\phi\frac{1}{r}\frac{\partial}{\partial\phi} + \hat{e}_z\frac{\partial}{\partial z}. \tag{D.11}$$

The only non-vanishing derivatives of the basis vectors are

$$\frac{\partial\hat{e}_r}{\partial\phi} = \hat{e}_\phi, \qquad \frac{\partial\hat{e}_\phi}{\partial\phi} = -\hat{e}_r. \tag{D.12}$$

These are all the tools necessary to convert from Cartesian to cylindrical coordinates and back.

First-order expressions

Expressions involving only a single nabla factor, such as gradient of a scalar ∇S, divergence of a vector $\nabla\cdot\boldsymbol{U}$, and the curl $\nabla\times\boldsymbol{U}$, are fairly simple to evaluate by resolving the vector quantities into the curvilinear basis and making use of (D.12). In many practical cases with intrinsic symmetries, the task is quite easy. In laminar pipe flow (Section 16.4 on page 268) the initial assumption is that the velocity field is of the form $\boldsymbol{v} = v_z(r)\hat{e}_z$, and the divergence vanishes trivially, $\nabla\cdot\boldsymbol{v} = \hat{e}_r\nabla_r\cdot\hat{e}_z v_z = 0$. The advective acceleration also vanishes, $(\boldsymbol{v}\cdot\nabla)\boldsymbol{v} = v_z\nabla_z\hat{e}_z v_z = 0$. Notice that we in such calculations treat the nablas as operators acting on everything to the right of their position. This notation avoids many parentheses.

Now follows a list of various combinations of a single nabla and various fields, derived by such methods. The three basic first-order expressions are the gradient, divergence, and curl,

$$\nabla S = \hat{e}_r\frac{\partial S}{\partial r} + \hat{e}_\phi\frac{1}{r}\frac{\partial S}{\partial\phi} + \hat{e}_z\frac{\partial S}{\partial z}, \tag{D.13}$$

$$\nabla\cdot\boldsymbol{U} = \frac{\partial U_r}{\partial r} + \frac{1}{r}\frac{\partial U_\phi}{\partial\phi} + \frac{\partial U_z}{\partial z} + \frac{U_r}{r}, \tag{D.14}$$

$$\nabla\times\boldsymbol{U} = \hat{e}_r\left(\frac{1}{r}\frac{\partial U_z}{\partial\phi} - \frac{\partial U_\phi}{\partial z}\right) + \hat{e}_\phi\left(\frac{\partial U_r}{\partial z} - \frac{\partial U_z}{\partial r}\right) + \hat{e}_z\left(\frac{\partial U_\phi}{\partial r} - \frac{1}{r}\frac{\partial U_r}{\partial\phi} + \frac{U_\phi}{r}\right). \tag{D.15}$$

The gradient of a vector field is also very useful. In dyadic notation (see Appendix B), we have

$$\begin{aligned}\nabla\boldsymbol{U} = {} & \hat{e}_r\hat{e}_r\frac{\partial U_r}{\partial r} + \hat{e}_r\hat{e}_\phi\frac{\partial U_\phi}{\partial r} + \hat{e}_r\hat{e}_z\frac{\partial U_z}{\partial r} \\ & + \hat{e}_\phi\hat{e}_r\left(\frac{1}{r}\frac{\partial U_r}{\partial\phi} - \frac{U_\phi}{r}\right) + \hat{e}_\phi\hat{e}_\phi\left(\frac{1}{r}\frac{\partial U_\phi}{\partial\phi} + \frac{U_r}{r}\right) + \hat{e}_\phi\hat{e}_z\frac{1}{r}\frac{\partial U_z}{\partial\phi} \\ & + \hat{e}_z\hat{e}_r\frac{\partial U_r}{\partial z} + \hat{e}_z\hat{e}_\phi\frac{\partial U_\phi}{\partial z} + \hat{e}_z\hat{e}_z\frac{\partial U_z}{\partial z}.\end{aligned} \tag{D.16}$$

In solid and fluid mechanics it is used for calculating the stress tensor.

The dot-product with the vector $\boldsymbol{V}$ from the left becomes

$$\begin{aligned}(\boldsymbol{V}\cdot\nabla)\,\boldsymbol{U} = {} & \hat{e}_r\left(V_r\frac{\partial U_r}{\partial r} + \frac{V_\phi}{r}\frac{\partial U_r}{\partial\phi} + V_z\frac{\partial U_r}{\partial z} - \frac{V_\phi U_\phi}{r}\right) \\ & + \hat{e}_\phi\left(V_r\frac{\partial U_\phi}{\partial r} + \frac{V_\phi}{r}\frac{\partial U_\phi}{\partial\phi} + V_z\frac{\partial U_\phi}{\partial z} + \frac{V_\phi U_r}{r}\right) \\ & + \hat{e}_z\left(V_r\frac{\partial U_z}{\partial r} + \frac{V_\phi}{r}\frac{\partial U_z}{\partial\phi} + V_z\frac{\partial U_z}{\partial z}\right).\end{aligned} \tag{D.17}$$

In the Navier–Stokes equation it is used for calculating the advective acceleration.

Finally, the divergence (from the left) of a tensor field becomes (also in dyadic notation)

$$\begin{aligned}\nabla\cdot\mathbf{T} = \hat{e}_r &\left(\frac{\partial T_{rr}}{\partial r} + \frac{1}{r}\frac{\partial T_{\phi r}}{\partial \phi} + \frac{\partial T_{zr}}{\partial z} + \frac{T_{rr}}{r} - \frac{T_{\phi\phi}}{r}\right)\\ + \hat{e}_\phi &\left(\frac{\partial T_{r\phi}}{\partial r} + \frac{1}{r}\frac{\partial T_{\phi\phi}}{\partial \phi} + \frac{\partial T_{z\phi}}{\partial z} + \frac{T_{r\phi}}{r} + \frac{T_{\phi r}}{r}\right)\\ + \hat{e}_z &\left(\frac{\partial T_{rz}}{\partial r} + \frac{1}{r}\frac{\partial T_{\phi z}}{\partial \phi} + \frac{\partial T_{zz}}{\partial z} + \frac{T_{rz}}{r}\right). \end{aligned} \tag{D.18}$$

This may be used to formulate the equations of motion for continuum physics, although it is normally not necessary.

Second-order expressions

Expressions involving two nabla factors also turn up everywhere in continuum physics. They can of course be derived by combinations of first-order expressions, but it is useful to list them in their full glory.

The Laplacian of a scalar field is calculated from the divergence of the gradient, $\nabla^2 S = \nabla\cdot(\nabla S)$, and becomes, when the dust has settled,

$$\nabla^2 S = \frac{\partial^2 S}{\partial r^2} + \frac{1}{r^2}\frac{\partial^2 S}{\partial \phi^2} + \frac{\partial^2 S}{\partial z^2} + \frac{1}{r}\frac{\partial S}{\partial r}. \tag{D.19}$$

The Laplacian can also be applied to a vector field, and may be calculated as the divergence of the gradient of the vector field, $\nabla^2 S = \nabla\cdot(\nabla \boldsymbol{U})$. It is somewhat more complicated:

$$\begin{aligned}\nabla^2 \boldsymbol{U} = \hat{e}_r &\left(\frac{\partial^2 U_r}{\partial r^2} + \frac{1}{r^2}\frac{\partial^2 U_r}{\partial \phi^2} + \frac{\partial^2 U_r}{\partial z^2} + \frac{1}{r}\frac{\partial U_r}{\partial r} - \frac{2}{r^2}\frac{\partial U_\phi}{\partial \phi} - \frac{U_r}{r^2}\right)\\ + \hat{e}_\phi &\left(\frac{\partial^2 U_\phi}{\partial r^2} + \frac{1}{r^2}\frac{\partial^2 U_\phi}{\partial \phi^2} + \frac{\partial^2 U_\phi}{\partial z^2} + \frac{1}{r}\frac{\partial U_\phi}{\partial r} + \frac{2}{r^2}\frac{\partial U_r}{\partial \phi} - \frac{U_\phi}{r^2}\right)\\ + \hat{e}_z &\left(\frac{\partial^2 U_z}{\partial r^2} + \frac{1}{r^2}\frac{\partial^2 U_z}{\partial \phi^2} + \frac{\partial^2 U_z}{\partial z^2} + \frac{1}{r}\frac{\partial U_z}{\partial r}\right). \end{aligned} \tag{D.20}$$

Another second-order expression is the gradient of a divergence,

$$\begin{aligned}\nabla(\nabla\cdot\boldsymbol{U}) = \hat{e}_r &\left(\frac{\partial^2 U_r}{\partial r^2} + \frac{1}{r}\frac{\partial^2 U_\phi}{\partial r\partial \phi} + \frac{\partial^2 U_z}{\partial r\partial z} + \frac{1}{r}\frac{\partial U_r}{\partial r} - \frac{1}{r^2}\frac{\partial U_\phi}{\partial \phi} - \frac{U_r}{r^2}\right)\\ + \hat{e}_\phi &\left(\frac{1}{r}\frac{\partial^2 U_r}{\partial \phi\partial r} + \frac{1}{r^2}\frac{\partial^2 U_\phi}{\partial \phi^2} + \frac{1}{r}\frac{\partial^2 U_z}{\partial \phi\partial z} + \frac{1}{r^2}\frac{\partial U_r}{\partial \phi}\right)\\ + \hat{e}_z &\left(\frac{\partial^2 U_r}{\partial z\partial r} + \frac{1}{r}\frac{\partial^2 U_\phi}{\partial z\partial \phi} + \frac{\partial^2 U_z}{\partial z^2} + \frac{1}{r}\frac{\partial U_r}{\partial z}\right). \end{aligned} \tag{D.21}$$

The curl-of-curl is obtained from the "double-cross" relation (Problem B.9),

$$\nabla\times(\nabla\times\boldsymbol{U}) = \nabla(\nabla\cdot\boldsymbol{U}) - \nabla^2\boldsymbol{U}, \tag{D.22}$$

and may be evaluated using the preceding results.

D.2 Spherical coordinates

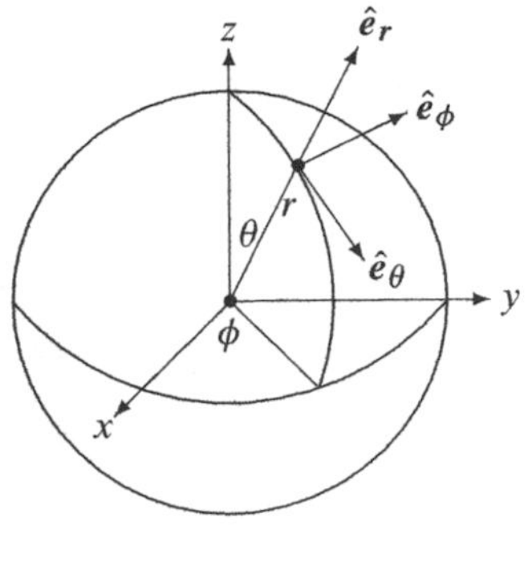

Spherical coordinates and their basis vectors.

The analysis of spherical coordinates follows much the same pattern as cylindrical coordinates, but we repeat the arguments here to make this section self-contained. Spherical or polar coordinates consist of the radial distance r, the polar angle θ, and the azimuthal angle ϕ. If the z-axis is chosen as the polar axis and the x-axis as the origin for the azimuthal angle, the transformation from spherical to Cartesian coordinates becomes

$$x = r\sin\theta\cos\phi, \qquad y = r\sin\theta\sin\phi, \qquad z = r\cos\theta, \tag{D.23}$$

with domain of variation $0 \le r < \infty$, $0 \le \theta \le \pi$ and $0 \le \phi < 2\pi$. Conversely, we have

$$r = \sqrt{x^2+y^2+z^2}, \qquad \theta = \arccos\frac{z}{\sqrt{x^2+y^2+z^2}}, \qquad \phi = \arctan\frac{y}{x}, \tag{D.24}$$

with the domain of variation $-\infty < x, y, z < \infty$.

Local basis vectors

The normalized tangent vectors along the directions of the spherical coordinate are

$$\hat{e}_r = \frac{\partial \boldsymbol{x}}{\partial r} = (\sin\theta\cos\phi, \sin\theta\sin\phi, \cos\theta), \tag{D.25a}$$

$$\hat{e}_\theta = \frac{1}{r}\frac{\partial \boldsymbol{x}}{\partial \theta} = (\cos\theta\cos\phi, \cos\theta\sin\phi, -\sin\theta), \tag{D.25b}$$

$$\hat{e}_\phi = \frac{1}{r\sin\theta}\frac{\partial \boldsymbol{x}}{\partial \phi} = (-\sin\phi, \cos\phi, 0). \tag{D.25c}$$

They are orthogonal, such that an arbitrary vector field may be resolved in these directions,

$$\boldsymbol{U} = \hat{e}_r U_r + \hat{e}_\theta U_\theta + \hat{e}_\phi U_\phi, \tag{D.26}$$

where

$$U_r = \hat{e}_r \cdot \boldsymbol{U}, \qquad U_\theta = \hat{e}_\theta \cdot \boldsymbol{U}, \qquad U_\phi = \hat{e}_\phi \cdot \boldsymbol{U} \tag{D.27}$$

are the projections of $\boldsymbol{U}$ on the spherical basis. A tensor field $\mathbf{T}$ may similarly be resolved in dyadic products of the local basis vectors.

Line, surface, and volume elements

The differentials along the local coordinate axes,

$$d_r\boldsymbol{x} = \hat{e}_r dr, \qquad d_\theta\boldsymbol{x} = \hat{e}_\theta r d\theta, \qquad d_\phi\boldsymbol{x} = \hat{e}_\phi r r\sin\theta d\phi, \tag{D.28}$$

allow us to resolve the Cartesian line, surface, and volume elements in the local basis,

$$d\boldsymbol{\ell} \equiv d_r\boldsymbol{x} + d_\theta\boldsymbol{x} + d_\phi\boldsymbol{x} = \hat{e}_r\, dr + \hat{e}_\theta\, r d\theta + \hat{e}_\phi\, r\sin\theta d\phi, \tag{D.29}$$

$$\begin{aligned} d\boldsymbol{S} &\equiv d_\theta\boldsymbol{x} \times d_\phi\boldsymbol{x} + d_\phi\boldsymbol{x} \times d_r\boldsymbol{x} + d_r\boldsymbol{x} \times d_\theta\boldsymbol{x} \\ &= \hat{e}_r\, r^2\sin\theta\, d\theta\, d\phi + \hat{e}_\theta r\sin\theta d\phi dr + \hat{e}_\phi\, r dr d\theta, \end{aligned} \tag{D.30}$$

$$dV \equiv d_r\boldsymbol{x} \times d_\theta\boldsymbol{x} \cdot d_\phi\boldsymbol{x} = r^2\sin\theta\, dr d\theta d\phi. \tag{D.31}$$

Using these infinitesimals, all integrals can be converted to spherical coordinates.

Resolution of the gradient

To find the relation between the Cartesian and curvilinear gradient, we differentiate through the Cartesian coordinates

$$\frac{\partial}{\partial r} = \frac{\partial \boldsymbol{x}}{\partial r} \cdot \nabla = \hat{\boldsymbol{e}}_r \cdot \nabla = \nabla_r,$$

$$\frac{\partial}{\partial \theta} = \frac{\partial \boldsymbol{x}}{\partial \theta} \cdot \nabla = r\hat{\boldsymbol{e}}_\theta \cdot \nabla = r\nabla_\theta,$$

$$\frac{\partial}{\partial \phi} = \frac{\partial \boldsymbol{x}}{\partial \phi} \cdot \nabla = r \sin\theta \hat{\boldsymbol{e}}_\phi \cdot \nabla = r \sin\theta \nabla_\phi.$$

This allows us to resolve the nabla operator in the curvilinear basis

$$\boxed{\nabla = \hat{\boldsymbol{e}}_r \nabla_r + \hat{\boldsymbol{e}}_\theta \nabla_\theta + \hat{\boldsymbol{e}}_\phi \nabla_\phi = \hat{\boldsymbol{e}}_r \frac{\partial}{\partial r} + \hat{\boldsymbol{e}}_\theta \frac{1}{r}\frac{\partial}{\partial \theta} + \hat{\boldsymbol{e}}_\phi \frac{1}{r\sin\theta}\frac{\partial}{\partial \phi}.} \tag{D.32}$$

Finally, the non-vanishing derivatives of the basis vectors are

$$\frac{\partial \hat{\boldsymbol{e}}_r}{\partial \theta} = \hat{\boldsymbol{e}}_\theta, \qquad \frac{\partial \hat{\boldsymbol{e}}_\theta}{\partial \theta} = -\hat{\boldsymbol{e}}_r, \tag{D.33a}$$

$$\frac{\partial \hat{\boldsymbol{e}}_\theta}{\partial \phi} = \cos\theta \hat{\boldsymbol{e}}_\phi, \qquad \frac{\partial \hat{\boldsymbol{e}}_r}{\partial \phi} = \sin\theta\, \hat{\boldsymbol{e}}_\phi, \qquad \frac{\partial \hat{\boldsymbol{e}}_\phi}{\partial \phi} = -\sin\theta\, \hat{\boldsymbol{e}}_r - \cos\theta\, \hat{\boldsymbol{e}}_\theta. \tag{D.33b}$$

These are all the relations necessary to convert differential equations from Cartesian to spherical coordinates.

First-order expressions

The three basic first-order expressions are the gradient, divergence, and curl,

$$\nabla S = \hat{\boldsymbol{e}}_r \frac{\partial S}{\partial r} + \hat{\boldsymbol{e}}_\theta \frac{1}{r}\frac{\partial S}{\partial \theta} + \hat{\boldsymbol{e}}_z \frac{1}{r\sin\theta}\frac{\partial S}{\partial \phi}, \tag{D.34}$$

$$\nabla \cdot \boldsymbol{U} = \frac{\partial U_r}{\partial r} + \frac{1}{r}\frac{\partial U_\theta}{\partial \theta} + \frac{1}{r\sin\theta}\frac{\partial U_\phi}{\partial \phi} + \frac{2U_r}{r} + \frac{U_\theta}{r\tan\theta}, \tag{D.35}$$

$$\begin{aligned} \nabla \times \boldsymbol{U} = {} & \hat{\boldsymbol{e}}_r \left(\frac{1}{r}\frac{\partial U_\phi}{\partial \theta} - \frac{1}{r\sin\theta}\frac{\partial U_\theta}{\partial \phi} + \frac{U_\phi}{r\tan\theta}\right) \\ & + \hat{\boldsymbol{e}}_\theta \left(\frac{1}{r\sin\theta}\frac{\partial U_r}{\partial \phi} - \frac{\partial U_\phi}{\partial r} - \frac{U_\phi}{r}\right) \\ & + \hat{\boldsymbol{e}}_\phi \left(\frac{\partial U_\theta}{\partial r} - \frac{1}{r}\frac{\partial U_r}{\partial \theta} + \frac{U_\theta}{r}\right). \end{aligned} \tag{D.36}$$

The gradient of a vector field becomes

$$\begin{aligned} \nabla \boldsymbol{U} = {} & \hat{\boldsymbol{e}}_r\hat{\boldsymbol{e}}_r \frac{\partial U_r}{\partial r} + \hat{\boldsymbol{e}}_r\hat{\boldsymbol{e}}_\theta \frac{\partial U_\theta}{\partial r} + \hat{\boldsymbol{e}}_r\hat{\boldsymbol{e}}_\phi \frac{\partial U_\phi}{\partial r} \\ & + \hat{\boldsymbol{e}}_\theta\hat{\boldsymbol{e}}_r \left(\frac{1}{r}\frac{\partial U_r}{\partial \theta} - \frac{U_\theta}{r}\right) + \hat{\boldsymbol{e}}_\theta\hat{\boldsymbol{e}}_\theta \left(\frac{1}{r}\frac{\partial U_\theta}{\partial \theta} + \frac{U_r}{r}\right) + \hat{\boldsymbol{e}}_\theta\hat{\boldsymbol{e}}_\phi \frac{1}{r}\frac{\partial U_\phi}{\partial \theta} \\ & + \hat{\boldsymbol{e}}_\phi\hat{\boldsymbol{e}}_r \left(\frac{1}{r\sin\theta}\frac{\partial U_r}{\partial \phi} - \frac{U_\phi}{r}\right) + \hat{\boldsymbol{e}}_\phi\hat{\boldsymbol{e}}_\theta \left(\frac{1}{r\sin\theta}\frac{\partial U_\theta}{\partial \phi} - \frac{U_\phi}{r\tan\theta}\right) \\ & + \hat{\boldsymbol{e}}_\phi\hat{\boldsymbol{e}}_\phi \left(\frac{1}{r\sin\theta}\frac{\partial U_\phi}{\partial \phi} + \frac{U_\theta}{r\tan\theta} + \frac{U_r}{r}\right), \end{aligned} \tag{D.37}$$

and is used for calculating the stress tensor.

Dotting from the left with $\boldsymbol{U}$, we get

$$(\boldsymbol{V}\cdot\nabla)\boldsymbol{U} = \hat{\boldsymbol{e}}_r\left(V_r\frac{\partial U_r}{\partial r} + \frac{V_\theta}{r}\frac{\partial U_r}{\partial\theta} + \frac{V_\phi}{r\sin\theta}\frac{\partial U_r}{\partial\phi} - \frac{V_\theta U_\theta}{r} - \frac{V_\phi U_\phi}{r}\right)$$
$$+\hat{\boldsymbol{e}}_\theta\left(V_r\frac{\partial U_\theta}{\partial r} + \frac{V_\theta}{r}\frac{\partial U_\theta}{\partial\theta} + \frac{V_\phi}{r\sin\theta}\frac{\partial U_\theta}{\partial\phi} + \frac{V_\theta U_r}{r} - \frac{V_\phi U_\phi}{r\tan\theta}\right)$$
$$+\hat{\boldsymbol{e}}_\phi\left(V_r\frac{\partial U_\phi}{\partial r} + \frac{V_\theta}{r}\frac{\partial U_\phi}{\partial\theta} + \frac{V_\phi}{r\sin\theta}\frac{\partial U_\phi}{\partial\phi} + \frac{V_\phi U_r}{r} + \frac{V_\phi U_\theta}{r\tan\theta}\right). \tag{D.38}$$

Finally, the left-divergence of a tensor field becomes

$$\nabla\cdot\mathbf{T} = \hat{\boldsymbol{e}}_r\left(\frac{\partial T_{rr}}{\partial r} + \frac{1}{r}\frac{\partial T_{\theta r}}{\partial\theta} + \frac{1}{r\sin\theta}\frac{\partial T_{\phi r}}{\partial\phi} + \frac{2T_{rr}}{r} + \frac{T_{\theta r}}{r\tan\theta} - \frac{T_{\theta\theta}}{r} - \frac{T_{\phi\phi}}{r}\right)$$
$$+\hat{\boldsymbol{e}}_\theta\left(\frac{\partial T_{r\theta}}{\partial r} + \frac{1}{r}\frac{\partial T_{\theta\theta}}{\partial\theta} + \frac{1}{r\sin\theta}\frac{\partial T_{\phi\theta}}{\partial\phi} + \frac{2T_{r\theta}}{r} + \frac{T_{\theta r}}{r} + \frac{T_{\theta\theta}}{r\tan\theta} - \frac{T_{\phi\phi}}{r\tan\theta}\right)$$
$$+\hat{\boldsymbol{e}}_\phi\left(\frac{\partial T_{r\phi}}{\partial r} + \frac{1}{r}\frac{\partial T_{\theta\phi}}{\partial\theta} + \frac{1}{r\sin\theta}\frac{\partial T_{\phi\phi}}{\partial\phi} + \frac{2T_{r\phi}}{r} + \frac{T_{\phi r}}{r} + \frac{T_{\theta\phi}}{r\tan\theta} + \frac{T_{\phi\theta}}{r\tan\theta}\right). \tag{D.39}$$

This may be used in formulating the equations of motion for continuum physics, although it is normally not necessary.

Second-order expressions

The Laplacian of a scalar field becomes

$$\nabla^2 S = \frac{\partial^2 S}{\partial r^2} + \frac{1}{r^2}\frac{\partial^2 S}{\partial\theta^2} + \frac{1}{r^2\sin^2\theta}\frac{\partial^2 S}{\partial\phi^2} + \frac{2}{r}\frac{\partial S}{\partial r} + \frac{1}{r^2\tan\theta}\frac{\partial S}{\partial\theta}. \tag{D.40}$$

The Laplacian of a vector field becomes

$$\nabla^2\boldsymbol{U} = \hat{\boldsymbol{e}}_r\left(\frac{\partial^2 U_r}{\partial r^2} + \frac{1}{r^2}\frac{\partial^2 U_r}{\partial\theta^2} + \frac{1}{r^2\sin^2\theta}\frac{\partial^2 U_r}{\partial\phi^2}\right.$$
$$\left.+\frac{2}{r}\frac{\partial U_r}{\partial r} + \frac{\cot\theta}{r^2}\frac{\partial U_r}{\partial\theta} - \frac{2}{r^2}\frac{\partial U_\theta}{\partial\theta} - \frac{2}{r^2\sin\theta}\frac{\partial U_\phi}{\partial\phi} - 2\frac{U_r}{r^2} - \frac{U_\theta}{r^2\tan\theta}\right)$$
$$+\hat{\boldsymbol{e}}_\theta\left(\frac{\partial^2 U_\theta}{\partial r^2} + \frac{1}{r^2}\frac{\partial^2 U_\theta}{\partial\theta^2} + \frac{1}{r^2\sin^2\theta}\frac{\partial^2 U_\theta}{\partial\phi^2}\right.$$
$$\left.+\frac{2}{r}\frac{\partial U_\theta}{\partial r} + \frac{\cot\theta}{r^2}\frac{\partial U_\theta}{\partial\theta} + \frac{1}{r^2}\frac{\partial U_r}{\partial\theta} - \frac{2\cos\theta}{r^2\sin^2\theta}\frac{\partial U_\phi}{\partial\phi} - \frac{U_\theta}{r^2\sin^2\theta}\right)$$
$$+\hat{\boldsymbol{e}}_\phi\left(\frac{\partial^2 U_\phi}{\partial r^2} + \frac{1}{r^2}\frac{\partial^2 U_\phi}{\partial\theta^2} + \frac{1}{r^2\sin^2\theta}\frac{\partial^2 U_\phi}{\partial\phi^2}\right.$$
$$\left.+\frac{2}{r}\frac{\partial U_\phi}{\partial r} + \frac{\cot\theta}{r^2}\frac{\partial U_\phi}{\partial\theta} + \frac{2\cos\theta}{r^2\sin^2\theta}\frac{\partial U_\theta}{\partial\phi} + \frac{2}{r^2\sin\theta}\frac{\partial U_r}{\partial\phi} - \frac{U_\phi}{r^2\sin^2\theta}\right). \tag{D.41}$$

Finally, the gradient of the divergence is

$$\begin{aligned}\nabla(\nabla\cdot\boldsymbol{U}) = \hat{\boldsymbol{e}}_r\bigg(&\frac{\partial^2 U_r}{\partial r^2} + \frac{1}{r}\frac{\partial^2 U_\theta}{\partial r\partial\theta} + \frac{1}{r\sin\theta}\frac{\partial^2 U_\phi}{\partial r\partial\phi}\\ &+\frac{2}{r}\frac{\partial U_r}{\partial r} + \frac{\cot\theta}{r}\frac{\partial U_\theta}{\partial r} - \frac{1}{r^2}\frac{\partial U_\theta}{\partial\theta} - \frac{1}{r^2\sin\theta}\frac{\partial U_\phi}{\partial\phi} - \frac{2U_r}{r^2} - \frac{U_\theta}{r^2\tan\theta}\bigg)\\ + \hat{\boldsymbol{e}}_\theta\bigg(&\frac{1}{r}\frac{\partial^2 U_r}{\partial\theta\partial r} + \frac{1}{r^2}\frac{\partial^2 U_\theta}{\partial\theta^2} + \frac{1}{r^2\sin\theta}\frac{\partial^2 U_\phi}{\partial\theta\partial\phi}\\ &+\frac{2}{r^2}\frac{\partial U_r}{\partial\theta} + \frac{\cot\theta}{r^2}\frac{\partial U_\theta}{\partial\theta} - \frac{\cos\theta}{r^2\sin^2\theta}\frac{\partial U_\phi}{\partial\phi} - \frac{U_\theta}{r^2\sin^2\theta}\bigg)\\ + \hat{\boldsymbol{e}}_\phi\bigg(&\frac{1}{r\sin\theta}\frac{\partial^2 U_r}{\partial\phi\partial r} + \frac{1}{r^2\sin\theta}\frac{\partial^2 U_\theta}{\partial\phi^2} + \frac{1}{r^2\sin^2\theta}\frac{\partial^2 U_\phi}{\partial\phi\partial\theta}\\ &+\frac{2}{r^2\sin\theta}\frac{\partial U_r}{\partial\phi} + \frac{\cos\theta}{r^2\sin^2\theta}\frac{\partial U_\theta}{\partial\theta}\bigg).\end{aligned}\tag{D.42}$$

The curl-of-curl follows from the "double-cross" relation (Problem B.9),

$$\nabla\times(\nabla\times\boldsymbol{U}) = \nabla(\nabla\cdot\boldsymbol{U}) - \nabla^2\boldsymbol{U},\tag{D.43}$$

and may be evaluated using the preceding results.

Problems

D.1 Assume that the velocity fields $v_{r,\phi,z}$ in cylindrical coordinates only depend on r, z and t. Verify that the Navier–Stokes equations for incompressible flow with constant density ρ_0 and constant gravity g_0 along the negative cylinder axis become

$$\begin{aligned}&\frac{\partial v_r}{\partial t} + v_r\frac{\partial v_r}{\partial r} - \frac{v_\phi^2}{r} + v_z\frac{\partial v_r}{\partial z} = \nu\left(\frac{\partial^2 v_r}{\partial r^2} + \frac{1}{r}\frac{\partial v_r}{\partial r} - \frac{v_r}{r^2} + \frac{\partial^2 v_r}{\partial z^2}\right) - \frac{1}{\rho_0}\frac{\partial p}{\partial r},\\ &\frac{\partial v_\phi}{\partial t} + v_r\frac{\partial v_\phi}{\partial r} + \frac{v_r v_\phi}{r} + v_z\frac{\partial v_\phi}{\partial z} = \nu\left(\frac{\partial^2 v_\phi}{\partial r^2} + \frac{1}{r}\frac{\partial v_\phi}{\partial r} - \frac{v_\phi}{r^2} + \frac{\partial^2 v_\phi}{\partial z^2}\right),\\ &\frac{\partial v_z}{\partial t} + v_r\frac{\partial v_z}{\partial r} + v_z\frac{\partial v_z}{\partial z} = \nu\left(\frac{\partial^2 v_z}{\partial r^2} + \frac{1}{r}\frac{\partial v_z}{\partial r} + \frac{\partial^2 v_z}{\partial z^2}\right) - \frac{1}{\rho_0}\frac{\partial p}{\partial z} - g_0,\\ &\frac{\partial v_r}{\partial r} + \frac{v_r}{r} + \frac{\partial v_z}{\partial z} = 0.\end{aligned}$$

E

Ideal gases

An ideal gas is a convenient "laboratory" for understanding the thermodynamics of a fluid with a non-trivial but simple equation of state. In this section we shall recapitulate the conventional thermodynamics of an ideal gas with constant heat capacity. For more extensive treatments, see for example [Kondepudi and Prigogine 1998, Cengel and Boles 2002].

E.1 Internal energy

In Section 2.1 on page 21 we analyzed Bernoulli's model of a gas consisting of essentially non-interacting point-like molecules, and found the pressure $p = \frac{1}{3}\rho\, v^2$, where v is the root-mean-square average molecular speed. Using the ideal gas law (2.26), which we repeat here,

$$pV = nR_{\text{mol}}T, \tag{E.1}$$

the average molecular kinetic energy contained in an amount $M = \rho V$ of the gas becomes

$$\tfrac{1}{2}M v^2 = \tfrac{3}{2}pV = \tfrac{3}{2}nR_{\text{mol}}T, \tag{E.2}$$

where $n = M/M_{\text{mol}} = N/N_A$ is the number of moles in the gas and R_{mol} is the universal molar gas constant. The derivation in Section 2.1 shows that the factor 3 stems from the three independent translational degrees of freedom available to point-like molecules. The above formula thus expresses that in a mole of a gas there is an internal kinetic energy $\frac{1}{2}R_{\text{mol}}T$ per mole associated with each translational degree of freedom of the point-like molecules.

Whereas monatomic gases like helium, neon, and argon have spherical molecules and thus only three translational degrees of freedom, diatomic gases like hydrogen, nitrogen, and oxygen have stick-like molecules with two extra rotational degrees of freedom orthogonal to the covalent bond connecting the atoms. Multiatomic gases like carbon dioxide and methane possess the complete set of three extra rotational degrees of freedom. According to the *equipartition theorem* of statistical mechanics, these degrees of freedom will each carry a kinetic energy $\frac{1}{2}R_{\text{mol}}T$ per mole. Molecules also possess vibrational degrees of freedom in the bond structure that tie them together, and they may become excited at high temperatures. We shall generally ignore the vibrational modes.

The *internal energy* of n moles of an ideal gas is defined to be

$$\boxed{\mathcal{U} = \tfrac{1}{2}k\, nR_{\rm mol}T,} \tag{E.3}$$

where k is the number of molecular degrees of freedom. A general result of thermodynamics (Helmholtz' theorem; see page 395) guarantees that the internal energy of an ideal gas cannot depend on the volume, but only on the temperature. Physically a gas may dissociate or even ionize when heated, and thereby change its value of k, but we shall for simplicity assume that k is in fact constant with $k = 3$ for monatomic, $k = 5$ for diatomic, and $k = 6$ for multiatomic gases. For mixtures of gases, the number of degrees of freedom is the molar average of the degrees of freedom of the pure components (see Problem 2.6).

Heat capacity

Suppose that we raise the temperature of the gas by δT without changing its volume. Since no work is performed, and since energy is conserved, it takes the amount of heat $\delta Q = \delta\mathcal{U} = C_V\delta T$, where the constant

$$C_V = \tfrac{1}{2}k\, nR_{\rm mol} \tag{E.4}$$

is called the *heat capacity at constant volume.*

If instead the pressure of the gas is kept constant while the temperature is raised by δT, we must also take into account that the volume expands by a certain amount δV and thereby performs work on its surroundings. The necessary amount of heat is now larger by this work, $\delta Q = \delta U + p\delta V$. Using the ideal gas law (E.1), we have for constant pressure $p\delta V = \delta(pV) = nR_{\rm mol}\delta T$. Consequently, the amount of heat that must be added per unit of temperature increase at constant pressure is

$$C_p = C_V + nR_{\rm mol}, \tag{E.5}$$

called the *heat capacity at constant pressure.* It is always larger than C_V because it includes the work of expansion.

The adiabatic index

The dimensionless ratio of the heat capacities,

$$\gamma = \frac{C_p}{C_V} = 1 + \frac{2}{k}, \tag{E.6}$$

is for reasons that will become clear in the following called the *adiabatic index.* It is customary to express the heat capacities in terms of γ rather than k,

$$C_V = \frac{1}{\gamma - 1}nR_{\rm mol}, \qquad C_p = \frac{\gamma}{\gamma - 1}nR_{\rm mol}. \tag{E.7}$$

Given the adiabatic index, all thermodynamic quantities for n moles of an ideal gas are completely determined. The value of the adiabatic index is $\gamma = 5/3$ for monatomic gases, $\gamma = 7/5$ for diatomic gases, and $\gamma = 4/3$ for multiatomic gases.

E.2 Entropy

When neither the volume nor the pressure are kept constant, the heat that must be added to the system in an infinitesimal reversible process is

$$\delta Q = \delta U + p\delta V = C_V \delta T + nR_{\text{mol}} T \frac{\delta V}{V}. \tag{E.8}$$

It is a mathematical fact that there exists no function, $Q(T, V)$, for which this expression is its differential (see Problem E.2). It may on the other hand be directly verified that

$$\delta S \equiv \frac{\delta Q}{T} = C_V \frac{\delta T}{T} + nR_{\text{mol}} \frac{\delta V}{V}, \tag{E.9}$$

can be integrated to yield a function

$$\boxed{S = C_V \log T + nR_{\text{mol}} \log V + \text{const}} \tag{E.10}$$

called the *entropy* of the amount of ideal gas. Being an integral the entropy is only defined up to an arbitrary constant. The entropy of the gas is, like its energy, an abstract quantity that cannot be directly measured. But since both quantities depend on the measurable thermodynamic quantities, ρ, p, and T, that characterize the state of the gas, we can calculate the value of energy as well as entropy in any state. But why bother to do so?

The laws of thermodynamics

The reason is that the two fundamental laws of thermodynamics are formulated in terms of the energy and the entropy. Both laws concern processes that may take place in an *isolated* system that is not allowed to exchange heat with the environment or perform work on it.

The *First Law* states that the energy is unchanged under any process in an isolated system (see Chapter 22). This implies that the energy of an open system can only change by exchange of heat and work with the environment. We actually used this law implicitly in deriving the heat capacities and entropy.

The *Second Law* states that the entropy cannot decrease (see Chapter 23). In the real world the entropy of an isolated system must in fact grow. Only if all the processes taking place in the system are completely *reversible* at all times, will the entropy stay constant. Reversibility is an ideal that can only be approached by very slow *quasi-static* processes, consisting of infinitely many infinitesimal reversible steps. Essentially all real-world processes are *irreversible* to some degree and must for this reason increase the entropy of an isolated system.

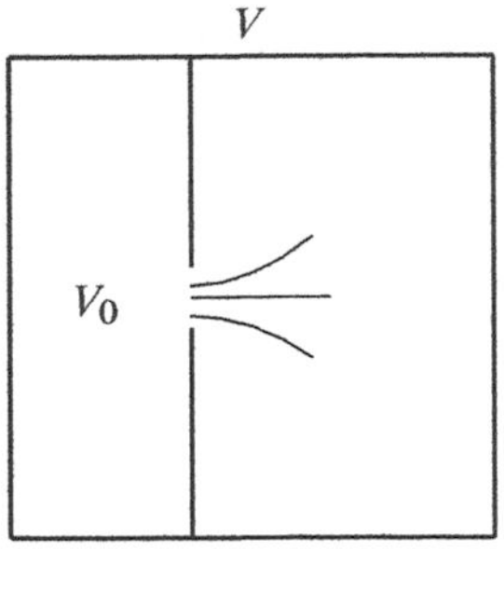

A compartment of volume V_0 inside an isolated box of volume V. Initially, the compartment contains an ideal gas with vacuum in the remainder of the box. When the wall breaks, the gas expands by itself to fill the whole box. The reverse process would entail a decrease in entropy and never happens by itself.

Example E.1 [Joule's expansion experiment]: An isolated box of volume V contains an ideal gas in a walled-off compartment of volume V_0. When the wall is opened, the gas expands into vacuum and fills the full volume V. The box is completely isolated from the environment, and since the internal energy only depends on the temperature (Helmholtz theorem), it follows from the First Law that the temperature must be the same before and after the event. The change in entropy then becomes

$$\Delta S = (C_V \log T + nR_{\text{mol}} \log V) - (C_V \log T + nR_{\text{mol}} \log V_0) = nR_{\text{mol}} \log(V/V_0),$$

which is evidently positive (because $V > V_0$). This result agrees with the Second Law, which thus appears to be unnecessary.

The strength of the Second Law becomes apparent when we ask the question of whether the air in the box could ever—perhaps by an unknown process to be discovered in the far future—by itself enter the compartment of volume V_0, leaving vacuum in the box around it. Since such an event would entail a negative change in entropy—which is forbidden by the Second Law—it will never happen for an isolated system like this.

Isentropic processes

Any process in an open system that does not exchange heat with the environment is said to be *adiabatic*. If the process is furthermore reversible, it follows that $\delta Q = 0$ in each infinitesimal step, so that $\delta S = \delta Q/T = 0$. The entropy (E.10) must in other words stay constant in any reversible, adiabatic process. Such a process is for this reason called *isentropic*.

The relation (E.5) and the definition of the adiabatic index (E.6) allow us to write the entropy (E.10) as

$$S = C_V \log\left(T V^{\gamma-1}\right) + \text{const.} \tag{E.11}$$

From this it follows that

$$T\, V^{\gamma-1} = \text{const} \tag{E.12}$$

for any isentropic process in an ideal gas. Using the ideal gas law to eliminate $V \sim T/p$, this may be written equivalently as

$$T^{\gamma} p^{1-\gamma} = \text{const.} \tag{E.13}$$

Eliminating instead $T \sim pV$, the isentropic condition takes its most common form,

$$\boxed{p\, V^{\gamma} = \text{const.}} \tag{E.14}$$

Note that the constants are different in these three equations.

Example E.2 [Bicycle pump]: When the air in a bicycle pump is compressed from V_0 to V_1 (while you block the valve with your finger), the adiabatic law implies that $p_1 V_1^{\gamma} = p_0 V_0^{\gamma}$. For $p_0 = 1$ atm and $V_1 = V_0/2$, we find $p_1 = 2.6$ atm. The temperature simultaneously rises about 110°C, but the hot air quickly becomes cold again during the backstroke. One may wonder why the fairly rapid compression stroke may be assumed to be reversible, but as long as the speed of the piston is much smaller than the velocity of sound, this is in fact a reasonable assumption. Conversely, we may conclude that the air expands with a velocity close to the speed of sound when the wall is opened in Example E.1, and this is definitely irreversible.

Problems

E.1 Calculate **(a)** specific gas constant and **(b)** the adiabatic index for a mixture of pure gases.

E.2 **(a)** Show that for a function $Q = Q(T, V)$, the differential takes the form $dQ = A\,dT + B\,dV$ with $\partial A/\partial V = \partial B/\partial T$. **(b)** Prove that this is not fulfilled for (E.8).

Answers to problems

Answers are provided to odd-numbered problems. An answer may not represent an explicit or complete solution to a problem, but only some useful indications on how to get there.

1 Continuous matter

1.1 **(a)** Use that $e^x = \sum_{n=0}^{\infty} x^n/n!$.
(b) The mean value is

$$\langle n \rangle = \sum_{n=0}^{\infty} n \Pr(n|N) = \sum_{n=1}^{\infty} \frac{N^n}{(n-1)!} e^{-N} = \sum_{n=0}^{\infty} \frac{N^{n+1}}{n!} e^{-N} = N.$$

(c) Similarly,

$$\langle n(n-1) \rangle = \sum_{n=0}^{\infty} n(n-1) \Pr(n|N) = \sum_{n=2}^{\infty} \frac{N^n}{(n-2)!} e^{-N} = N^2.$$

Then $\langle n^2 \rangle = N(N+1)$ and $\langle (n-N)^2 \rangle = N(N+1) - 2N^2 + N^2 = N$.

1.3 **(a)** The CMS-velocity is $\boldsymbol{v} = N^{-1} \sum_n \boldsymbol{v}_n$, and the average of its square becomes

$$\left\langle \boldsymbol{v}^2 \right\rangle = \frac{1}{N^2} \sum_{n,m} \langle \boldsymbol{v}_n \cdot \boldsymbol{v}_m \rangle = \frac{1}{N^2} \sum_n \left\langle \boldsymbol{v}_n^2 \right\rangle = \frac{1}{N^2} \sum_n v_0^2 = \frac{1}{N} v_0^2.$$

2 Pressure

2.1 **(a)** The range of the manometer is from 0 to 240 mmHg. The distance between the pressurized and open mercury surfaces must therefore at least be 240 mm = 24 cm. This is indeed the typical size of clinical mercury manometers.

2.3 The horizontal pressure force on the hemisphere must be equal to the pressure force on the vertical plane through the center of the sphere. The linear rise of pressure with depth makes the hydrostatic pressure act with its average value $\Delta p = \rho_0 g_0 h$ at the center. So the horizontal force becomes $\Delta p \pi a^2 \approx 2100$ N, corresponding to the weight of 210 kg.

If you do not like this argument, it is also possible with some effort to integrate the pressure force directly in spherical coordinates with the x-axis orthogonal to the wall,

$$\begin{aligned}\mathcal{F} &= -\int_{\text{half-sphere}} (p - p_0)\, d\boldsymbol{S} = -\int_{\text{half-sphere}} (p - p_0) \hat{\boldsymbol{e}}_r \, dS \\ &= -a^2 \int_0^{\pi} d\theta \int_{-\pi/2}^{\pi/2} d\phi \, \sin\theta \, (p - p_0) \, \hat{\boldsymbol{e}}_r.\end{aligned}$$

Using $p = p_0 + \rho_0 g_0(h - a\cos\theta)$, the x-component of the force becomes

$$\mathcal{F}_x = -a^2 \int_0^\pi d\theta \int_{-\pi/2}^{\pi/2} d\phi \, \sin\theta \, \rho_0 g_0 (h - a\cos\theta) \sin\theta \cos\phi$$
$$= -2a^2 \rho_0 g_0 \int_0^\pi d\theta \sin^2\theta (h - a\cos\theta) = -\rho_0 g_0 h \pi a^2.$$

2.5 (**a**) The surface inside the tube will be at the same level $h_1 + h_2$ as in the jar. (**b**) The heavy liquid in the tube must initially rise to the same level h_1 as in the jar. When the light liquid is poured in, the surface of the heavy liquid in the tube must rise further to balance the weight of the light and rise to a height $h_1 + h_2\rho_2/\rho_1 < h_1 + h_2$.

2.7 Solving for the pressure, we find

$$P = \frac{nRT}{V - nb} - \frac{n^2 a}{V^2}.$$

(**a**) Differentiating, we get

$$K_T = -V\left(\frac{\partial p}{\partial V}\right)_T = \frac{nRTV}{(V - nb)^2} - \frac{2an^2}{V^2}.$$

(**b**) It can become negative for sufficiently low temperature, satisfying

$$RT < \frac{2an(V - nb)^2}{V^3},$$

which means that the gas must have condensed.

* **2.9** Typical data for Mars are $g_0 = 3.73 \text{ m s}^{-1}$, $T_0 = 220\,\text{K}$, $M_{\text{mol}} = 44 \text{ g mol}^{-1}$, and $\gamma = 4/3$ (carbon dioxide). Thus $h_2 = \frac{\gamma}{\gamma - 1}\frac{RT_0}{g_0} \approx 45$ km.

3 Buoyancy and stability

3.1 (**a**) The weight of the displaced water is $\mathcal{F}_1 - \mathcal{F}_0 = \rho_0 V g_0$, so that $V = (\mathcal{F}_1 - \mathcal{F}_0)/\rho_0 g_0 = 0.04 \text{ m}^3$. (**b**) The weight of the stone is $\mathcal{F}_0 = \rho_1 V g_0$ so that $\rho_1 = \mathcal{F}_0/V g_0 = g_0 \rho_0 \mathcal{F}_1/(\mathcal{F}_1 - \mathcal{F}_0) = 2500 \text{ kg m}^{-3}$.

3.3 (**a**) Displacement $M_1 + M_2 = \rho_0((1 - f)V_1 + V_2)$ with $M_1 = \rho_1 V_1$ and $M_2 = \rho_2 V_2$. Solving for the volume ratio

$$\frac{V_1}{V_2} = \frac{\rho_2 - \rho_0}{(1 - f)\rho_0 - \rho_1},$$

the mass ratio becomes $M_1/M_2 = 2.36$.
(**b**) The denominator must be positive: $f < 1 - \rho_1/\rho_0 = 0.35$.

3.5 The total force is $\boldsymbol{\mathcal{F}} = (M_{\text{body}} - M_{\text{fluid}})\boldsymbol{g}_0$. The total moment is $\boldsymbol{\mathcal{M}} = \boldsymbol{x}_G \times M_{\text{body}}\boldsymbol{g}_0 - \boldsymbol{x}_B \times M_{\text{fluid}}\boldsymbol{g}_0 = \boldsymbol{x}_0 \times \boldsymbol{\mathcal{F}}$, where

$$\boldsymbol{x}_0 = \frac{M_{\text{body}}\boldsymbol{x}_G - M_{\text{fluid}}\boldsymbol{x}_B}{M_{\text{body}} - M_{\text{fluid}}}.$$

*** 3.7** The half upper side is $b = h \tan\alpha$. We define $a = L/2$ and let d be the draft.
(a) The origin of the coordinates is chosen at the peak with area function

$$A(z) = 2Lz\tan\alpha \qquad (0 < z < h).$$

The total volume of the ship is $V = \int_0^h A(z)\,dz = Lhb = Lh^2\tan\alpha$, and its center of gravity $z_G = \frac{1}{V}\int_0^h zA(z)\,dz = \frac{2}{3}h$. Putting $h = d$, the submerged volume becomes $V_0 = Ld^2\tan\alpha$ and the center of buoyancy becomes $z_B = \frac{2}{3}d$. The second-order moment is $I_0 = \frac{2}{3}L(d\tan\alpha)^3$ according to (3.24), and the metacentric height $z_M = \frac{2}{3}d + \frac{2}{3}d\tan^2\alpha = \frac{2}{3}d/\cos^2\alpha$. The stability condition (3.28) takes the form $d/h > \cos^2\alpha$. Archimedes' law yields $\rho_1 V = \rho_0 V_0$ or $\rho_1/\rho_0 = (d/h)^2$. The condition on the density ratio is $\rho_1/\rho_0 > \cos^4\alpha$.
(b) The origin of coordinates is again chosen at the peak of the cone. The area function is now $-A(z)$ in the interval $-h < z < 0$. The parameter d is for convenience chosen to be the "antidraught" such that the true draught is $h - d$. The total volume is again $V = Lh^2\tan\alpha$, and the center of gravity $z_G = -\frac{2}{3}h$. The submerged volume is $V_0 = L(h^2 - d^2)\tan\alpha$ and the center of buoyancy $z_B = -\frac{2}{3}(h^3 - d^3)/(h^2 - d^2)$. The waterline integral is the same as before, and the metacentric height $z_M = -\frac{2}{3}(h^3 - d^3/\cos^2\alpha)/(h^2 - d^2)$. The stability condition becomes again $d/h > \cos^2\alpha$, but since $\rho_1/\rho_0 = V_0/V = 1 - (d/h)^2$, the density condition becomes $\rho_1/\rho_0 < 1 - \cos^4\alpha$.
(c) Combining the two conditions we get $\cos^4\alpha < \rho_1/\rho_0 < 1 - \cos^4\alpha$. This is only possible for $\cos\alpha < 2^{-1/4}$ or $\alpha > 33\,°$.

*** 3.9** **(a)** The effective potential $H = g_0 z + w(p)$ is constant before the body enters. Afterward $H + \Delta H = g_0 z + \Delta\Phi + w(p + \Delta p)$ is constant. Expanding w to lowest order, it follows that $\Delta H = \Delta\Phi + \Delta p/\rho$ is constant. At large distances $\Delta\Phi \to 0$ and $\Delta p \to 0$ so that $\Delta H = 0$. This proves the claim. **(b)** On the surface of a spherical body we have $\Delta\Phi = -GM/a = -g_1 a$. **(c)** Since the initial fluid density $\rho(z)$ always decreases linearly with increasing z, Δp will be smaller on the top of the sphere than on the bottom, and thus increase the buoyancy force.

3.11 Let the small angle of tilt around the principal axis be α. The angular momentum along the tilt axis is $J\dot{\alpha}$, and using (3.27) the equation of motion becomes a harmonic equation,

$$J\ddot{\alpha} = -\alpha(z_M - z_G)Mg_0,$$

from which we read off the oscillation frequency.

3.13 **(a)** Under a rotation of the ship through α, the center-of-mass coordinates change as

$$y'_G = y_G\cos\alpha - z_G\sin\alpha, \qquad z'_G = y_G\sin\alpha + z_G\cos\alpha,$$

from which we get

$$\delta z_G \equiv z'_G - z_G = y_G\sin\alpha - z_G(1 - \cos\alpha) = \alpha y_G - \frac{1}{2}\alpha^2 z_G + \mathcal{O}\left(\alpha^3\right).$$

This expression would also be valid for the center-of-buoyancy, if it were fixed in the ship. Taking into account that the forces of buoyancy act opposite to gravity, the effective vertical shift becomes

$$\delta z_G - \delta z_B \approx -\frac{1}{2}\alpha^2(z_G - z_B),$$

where we have also used that $y_B = y_G$ before the rotation. Multiplying the vertical shift with the common weight of the ship and its displacement, $M_0 g_0$, we arrive at the first term in W.

The second term in W is obtained by moving the tilted displacement into place. Each little column dA of displaced water at y in the right wedge has negative mass $dM = -\rho_0 g_0 z\,dA$, where $z = \alpha y$. When the column is moved over to the left and placed in the correct position, its center of mass is changed by $-z$, which requires the work $dW = dMg_0 \times (-z) = \rho_0 g_0 z^2 dA$. Integrating over both wedges and dividing by 2 to compensate for double counting, we arrive at the second term.

(b) Finally, calculating the moment by differentiating with respect to the angle, we obtain (3.27),

$$M_x = -\frac{\partial W}{\partial\alpha} = \alpha M_0 g_0(z_G - z_B) - \alpha\rho_0 g_0 I_0, = \alpha M_0 g_0(z_G - z_M)$$

where we have used that $M_0 = \rho_0 V_0$.

4 Hydrostatic shapes

4.1 Let Δh be the change in sea level due to Δp. Use the constancy of the effective potential (2.36) in the water close to the surface to get $\Delta\Phi^* = \Delta p/\rho_0 + g_0\Delta h = 0$. Then $\Delta h = -\Delta p/\rho_0 g_0 \approx -20$ cm.

4.3 **(a)** Averaging over a period we get from (4.21) that the average is $\langle h\rangle = h_0 A_0$.
(b) Writing $\psi = \Omega t + \phi$, we find

$$\frac{\partial h}{\partial \psi} = -h_0(A_1 \sin\psi + 2A_2 \sin 2\psi) = -h_0(A_1 + 4A_2\cos\psi)\sin\psi.$$

It vanishes for $\psi = 0, \pi$, and for $\psi = \psi_0$, where $\cos\psi_0 = -A_1/4A_2 = -\frac{1}{4}\tan\delta\tan\delta_0$. This corresponds to the maximal and minimal tidal heights (when $|\delta - \delta_0| < 90°$). The difference becomes

$$\Delta h = h_{\text{max}} - h_{\text{min}} = h_0(A_1(1-\cos\psi_0) + A_2(1-\cos 2\psi_0)),$$

from which the range can be calculated, given the latitudes of Moon and observer.

5 Surface tension

5.1 The ratio of weight to equatorial surface tension becomes

$$\frac{4\pi R^2 \tau\rho_1 g_0}{2\pi R 2\alpha} = \frac{R\tau\rho_1 g_0}{\alpha} = \frac{R\tau}{L_c^2} = \text{Bo}.$$

5.3 **(a)** The volume of a wavelength becomes

$$V = \int_0^\lambda \pi r^2\,dz = \pi\lambda\left(a^2 + \frac{1}{2}b^2\right).$$

(b) The area of a wavelength becomes, to lowest significant order in b/λ,

$$A = \int_0^\lambda 2\pi r\sqrt{dz^2 + dr^2} = \int_0^\lambda 2\pi r\sqrt{1 + \left(\frac{dr}{dz}\right)^2}\,dz \approx 2\pi a\lambda\left(1 + \frac{\pi^2 b^2}{\lambda^2}\right).$$

(c) To keep the volume constant, we must vary a together with b so that

$$\frac{dV}{db} = \pi\lambda\left(2a\frac{da}{db} + b\right), \qquad \frac{dA}{db} = 2\pi\lambda\frac{da}{db}\left(1 + \frac{\pi^2 b^2}{\lambda^2}\right) + 2\pi a\lambda\frac{2\pi^2 b}{\lambda^2}.$$

Requiring $dV/db = 0$, we have $da/db = -b/2a$, and to leading order in b,

$$\frac{dA}{db} \approx \pi\lambda\frac{b}{a}\left(\frac{4\pi^2 a^2}{\lambda^2} - 1\right).$$

For $\lambda < 2\pi a$, the right-hand side is positive, and the area grows with the perturbation b. For $\lambda > 2\pi a$, the area diminishes and surface tension will diminish it further.

5.5 The radius R must satisfy $\frac{4}{3}\pi R^3 = \pi a^2\lambda_{\text{min}}$, or $R = \left(3\pi/\sqrt{2}\right)^{1/3} a \approx 1.88a$.

5.7 **(a)** For $x = 0$, the pressure is $p = p_0 - \rho_0 g_0 z$ with $0 \le z \le d$, where d is given by (5.36). Integrating from $z = 0$ we get the extra force in the x-direction per unit of length

$$\mathcal{F}_x = -\int_0^d (p - p_0)\,dz = \frac{1}{2}\rho_0 g_0 d^2.$$

The result is independent of the sign of d, that is, whether the angle is acute or obtuse.
(b) Surface tension makes the pressure on the wall in the raised meniscus lower than atmospheric, so the force is a pull that makes floating objects attract.

5.9 Use the result of Problem 5.8 with $\frac{1}{R_1} + \frac{1}{R_2} = 0$, and verify that the given function is indeed a solution.

5.11 Tate's Law becomes

$$\frac{4}{3}\pi R^3 \rho_0 g_0 = 2\pi a\alpha,$$

from which it follows that

$$R = \left(\frac{3}{2} a L_c^2\right)^{\frac{1}{3}} = \left(\frac{3a}{2L_c}\right)^{\frac{1}{3}} L_c.$$

6 Stress

6.1 The normal reaction is the weight N and the tangential reaction is $T = \mu N$. The angle is given by $\tan\alpha = T/N = \mu$.

6.3 **(a)** $\sigma = F/NA = 391$ Pa. **(b)** $\sigma = 80{,}000$ Pa $= 0.8$ bar.

6.5 The characteristic equation is $-\lambda^3 + 3\tau\lambda^2 = 0$. Eigenvalues $\lambda = 3\tau$ and $\lambda = 0$ (doubly degenerate). Eigenvectors $\hat{e}_1 = (1,1,1)/\sqrt{3}$, $\hat{e}_2 = (-2,1,1)/\sqrt{6}$, and $\hat{e}_3 = (0,-1,1)/\sqrt{2}$, or any linear combination of the last two.

6.7

$$\boldsymbol{\sigma}' = \mathbf{A}\cdot\boldsymbol{\sigma}\cdot\mathbf{A}^\top = \begin{pmatrix} 91/5 & -6 & -12/5 \\ -6 & 5 & -8 \\ -12/5 & -8 & 84/5 \end{pmatrix}.$$

* **6.9** **(a)** The body starts to move when the elastic force equals the maximal static friction, that is, for $ks = \mu_0 m g_0$ or $s = \mu_0 m g_0/k$.
(b) When the body is at the point x at time t, the actual stretch is $s + vt - x$. The equation of motion becomes

$$m\ddot{x} = k(s + vt - x) - \mu m g_0.$$

(c) Define $y = x - vt - s + \mu m g_0/k = x - vt - (1-r)s$ and $r = \mu/\mu_0$. Then $\ddot{y} = -\omega^2 y$, which has the solution $y = A\cos\omega t + B\sin\omega t + Ct + D$. The particular solution follows from the initial conditions $x = \dot{x} = 0$ for $t = 0$.
(d) The velocity is

$$\dot{x} = v(1 - \cos\omega t) + (1-r)s\omega\sin\omega t = 2v\sin^2\frac{\omega t}{2} + 2(1-r)s\omega\sin\frac{\omega t}{2}\cos\frac{\omega t}{2},$$

which vanishes for the first time after start when

$$\tan\frac{\omega t}{2} = -\frac{(1-r)s\omega}{v},$$

so that $\omega t_0 = 2\pi - 2\alpha$, where

$$\alpha = \arctan\frac{(1-r)s\omega}{v}.$$

The other possibility, $\sin(\omega t/2) = 0$, happens later, for $\omega t = 2\pi$.
(e) The stretch is

$$\begin{aligned} s + vt - x &= rs + \frac{v}{\omega}\sin\omega t + (1-r)s\cos\omega t \\ &= rs + \frac{v}{\omega\cos\alpha}\sin(\omega t + \alpha). \end{aligned}$$

The minimum stretch is $s_1 = rs - v/\omega \cos\alpha = s(r - (1-r)/\sin\alpha)$ at $t = t_1$, where $\omega t_1 + \alpha = 3\pi/2$. For the minimum stretch to be positive, one must require that $r > (1-r)/\sin\alpha$, and since $\sin\alpha < 1$, this inequality can only be fulfilled for $r > 1 - r$ or $r > 1/2$. Notice that $\omega(t_1 - t_0) = \alpha - \pi/2$ and $\alpha < \pi/2$, so that the minimum always happens before the block stops.
(f) When the block stops at $t = t_0$ the stretch is $s_0 = s(2r-1)$. The stretch grows to $s_0 + v\Delta t$ a time Δt after the body stops. The body starts to move again when the stretch becomes s, or $v\Delta t = s - s_0 = 2(1-r)s$, which is positive as expected. The circular frequency of the jumping motion is

$$\Omega = \frac{2\pi}{t_0 + \Delta t} = \frac{\omega}{1 + (\tan\alpha - \alpha)/\pi}.$$

When $v \to 0$, we have $\alpha \to \pi/2$ and $\Omega \to 0$, in accordance with intuition.

∗ 6.11 **(a)** At the height z above the ground, the force on a small piece dz of the line is

$$d\mathcal{F} = \left(-g_0 \frac{a^2}{(a+z)^2} + (a+z)\Omega^2\right) \rho A\, dz,$$

where $\Omega = 2\pi/24$ h is the angular velocity in the geostationary orbit and the second term represents the centrifugal force. The height of the geostationary orbit is obtained by setting $d\mathcal{F}(h) = 0$ and becomes $h/a = \left(g_0/a\Omega^2\right)^{1/3} - 1$. Since $d\mathcal{F}(z)$ only vanishes for $z = h$, the total force is, not surprisingly, maximal at the satellite. Integrating the force from 0 to h, we find the maximal force

$$\mathcal{F} = \int_0^h d\mathcal{F}(z) = \rho A h \left(-g_0 \frac{a}{a+h} + \Omega^2 \left(a + \frac{1}{2}h\right)\right).$$

The absolute value of the tension-to-density ratio becomes

$$\frac{\sigma}{\rho} = h\left(g_0 \frac{a}{a+h} - \Omega^2 \left(a + \frac{1}{2}h\right)\right) \approx 4.8 \times 10^7 \text{ m}^2\text{s}^{-2}.$$

(b) The tensile strength of a beryllium-copper alloy of density $\rho = 8230$ kg m^{-3} can go as high as $\sigma \approx 1.4$ GPa, leading to $\sigma/\rho \approx 1.7 \times 10^5$ m^{2}s^{-2}, a factor nearly 300 below the required value. Ropes based on carbon fibers are expected to approach the required value.

7 Strain

7.1 Put $(X, Y) = a(\cos\theta, \sin\theta)$ and $(x, y) = a(\cos(\theta + \phi), \sin(\theta + \phi))$, and use the well-known trigonometric relations for sums.

7.3 **(a)**

$$\nabla \cdot \boldsymbol{u} = 3\alpha + 2\beta(x + y + z), \qquad \nabla \times \boldsymbol{u} = (-2\alpha, -2\alpha, -2\alpha).$$

(b)

$$\{u_{ij}\} = \begin{pmatrix} \alpha + 2\beta x & \alpha & \alpha \\ \alpha & \alpha + 2\beta y & \alpha \\ \alpha & \alpha & \alpha + 2\beta z \end{pmatrix} = \alpha \begin{pmatrix} 1 & 1 & 1 \\ 1 & 1 & 1 \\ 1 & 1 & 1 \end{pmatrix} + 2\beta \begin{pmatrix} x & 0 & 0 \\ 0 & y & 0 \\ 0 & 0 & z \end{pmatrix}.$$

7.5 The strain tensor becomes

$$u_{ij} = \alpha \begin{pmatrix} 0 & 1 & 0 \\ 1 & 0 & 0 \\ 0 & 0 & 0 \end{pmatrix}.$$

The characteristic polynomial is $\det(\mathbf{u} - \lambda\mathbf{1}) = -\lambda(\lambda^2 - \alpha^2)$. The eigenvalues are $\lambda = \pm\alpha$ and $\lambda = 0$. The (unnormalized) eigenvectors are $(1, 1, 0)$, $(1, -1, 0)$, and $(0, 0, 1)$. The eigenvalues equal the relative changes in length along these directions.

7.7 Let $\boldsymbol{c} = \boldsymbol{a} + \boldsymbol{b}$. Then $2\boldsymbol{a} \cdot \boldsymbol{b} = \boldsymbol{c}^2 - \boldsymbol{a}^2 - \boldsymbol{b}^2$ and thus

$$\delta(ab) = |\boldsymbol{c}|\,\delta\,|\boldsymbol{c}| - |\boldsymbol{a}|\,\delta\,|\boldsymbol{a}| - |\boldsymbol{b}|\,\delta\,|\boldsymbol{b}| = \boldsymbol{c}^2 u_{cc} - \boldsymbol{a}^2 u_{aa} - \boldsymbol{b}^2 u_{bb}.$$

7.9 We must solve

$$\begin{aligned} &\nabla_x u_x = \nabla_y u_y = \nabla_z u_z = 0, \\ &\nabla_y u_z + \nabla_z u_y = \nabla_z u_x + \nabla_z u_z = \nabla_x u_y + \nabla_y u_x = 0. \end{aligned}$$

From the first we get that u_x can only depend on y and z, and for the second derivatives we get

$$\begin{aligned} \nabla_y^2 u_x &= -\nabla_y \nabla_x u_y = -\nabla_x \nabla_y u_y = 0, \\ \nabla_z^2 u_x &= -\nabla_z \nabla_x u_z = -\nabla_x \nabla_z u_z = 0, \\ \nabla_y \nabla_z u_x &= -\nabla_y \nabla_x u_z = -\nabla_x \nabla_y u_z = \nabla_x \nabla_z u_y = \nabla_z \nabla_x u_y = -\nabla_z \nabla_y u_x. \end{aligned}$$

From the last equation we get $\nabla_y \nabla_z u_x = 0$. Consequently, we must have $u_x = A + Dy + Ez$ and similar results for u_y and u_z. The vanishing of the shear strains relates some of the constants.

7.11 Trivial.

8 Hooke's Law

8.1 Expanding to second order around $r = a$, we find for r close to a,

$$V(r) = V(a) + (r-a)V'(a) + \tfrac{1}{2}(r-a)^2 V''(a).$$

The force between the particles is $-V'(r) = -V'(a) - (r-a)V''(a)$. In equilibrium, $V'(a) = 0$ and the force is linear in the distance from equilibrium.

8.3 The boundary conditions are fulfilled with the constant values $\sigma_{xx} = P$, $u_{yy} = 0$, and $\sigma_{zz} = 0$. Inserting this into (8.17b), we get $\sigma_{yy} = \nu P$; and using (8.17a,c), we get $u_{xx} = (1-\nu^2)P/E$ and $u_{zz} = -\nu(1+\nu)P/E$. The displacement field becomes $u_x = x u_{xx}$ and $u_z = z u_{zz}$. Since it satisfies the boundary conditions and the equilibrium equations, it is the right solution.

* **8.5** **(a)** Use that

$$u_{xx} = \nabla_x u_x, \qquad u_{yy} = \nabla_y u_y, \qquad u_{xy} = \tfrac{1}{2}(\nabla_x u_y + \nabla_y u_x).$$

(b) The equilibrium equations become

$$\nabla_x \sigma_{xx} + \nabla_y \sigma_{xy} = 0, \qquad \nabla_x \sigma_{yx} + \nabla_y \sigma_{yy} = 0.$$

(c) The solutions to these equations can always be found of the form

$$\begin{aligned} \sigma_{xx} &= \nabla_y \psi_x, & \sigma_{xy} &= -\nabla_x \psi_x, \\ \sigma_{yx} &= -\nabla_y \psi_y, & \sigma_{yy} &= \nabla_x \psi_y. \end{aligned}$$

The symmetry of the stress tensor requires that $\nabla_x \psi_x = \nabla_y \psi_y$, so that $\psi_x = \nabla_y \phi$ and $\psi_y = \nabla_x \phi$.
(d) The strain tensor becomes

$$\begin{aligned} E u_{xx} &= \sigma_{xx} - \nu\sigma_{yy} = \nabla_y^2 \phi - \nu \nabla_x^2 \phi, \\ E u_{yy} &= \sigma_{yy} - \nu\sigma_{xx} = \nabla_x^2 \phi - \nu \nabla_y^2 \phi, \\ E u_{zz} &= -\nu(\sigma_{xx} + \sigma_{yy}) = -\nu(\nabla_x^2 \phi + \nabla_y^2 \phi), \\ E u_{xy} &= (1+\nu)\sigma_{xy} = -(1+\nu)\nabla_x \nabla_y \phi. \end{aligned}$$

The equation now follows by insertion into (8.39).

(e) The equations to be solved are

$$u_{xx} = 2x, \qquad u_{yy} = -2\nu x, \qquad u_{zz} = -2\nu x,$$
$$u_{xy} = -2(1+\nu)y, \qquad u_{xz} = 0, \qquad u_{yz} = 0.$$

First the diagonal elements are solved on their own:

$$u_x = x^2, \qquad u_y = -2\nu xy, \qquad u_z = -2\nu xz.$$

Here, $u_{yz} = 0$ automatically. To fulfill $u_{xz} = 0$, an extra term is necessary in u_x, so that

$$u_x = x^2 + \nu z^2, \qquad u_y = -2\nu xy, \qquad u_z = -2\nu xz.$$

Finally, add another term to u_x to obtain the correct expression for u_{xy}:

$$u_x = x^2 + \nu z^2 - (2+\nu)y^2, \qquad u_y = -2\nu xy, \qquad u_z = -2\nu xz.$$

To this may be added an infinitesimal solid translation or rotation.

8.7 The magnitude of the second-order contribution to the gravitational energy density is

$$\left| \frac{1}{2}\rho \sum_{ij} u_i u_j \nabla_i g_j \right| \sim \rho\, |\boldsymbol{u}|^2\, |\nabla \boldsymbol{g}| .$$

The elastic energy density is similarly

$$\left| \frac{1}{2} \sum_{ijkl} E_{ijkl} u_{ij} u_{kl} \right| \sim E\, |\nabla \boldsymbol{u}|^2 ,$$

where E is a typical elasticity coefficient. The ratio of the two becomes

$$\frac{\rho\, |\boldsymbol{u}|^2\, |\nabla \boldsymbol{g}|}{E\, |\nabla \boldsymbol{u}|^2} \sim \frac{\rho\, |\boldsymbol{u}|^2\, |\boldsymbol{g}|/L}{E\, |\boldsymbol{u}|^2/L^2} \sim \frac{\rho g L}{E} \sim \frac{\Delta p}{E},$$

where $\Delta p \sim \rho g L$ is an estimate of the hydrostatic pressure change across the object. For normal materials we have $E \approx 100$ GPa $= 1$ MBar, so the ratio will only be approach unity for hydrostatic pressures of this magnitude, which are only found in planet-size bodies.

8.9 **(a)** Assume that $u_x = Ay$, $u_y = By$ and $u_z = 0$. Then we get from (7.46) that $u_{xx} = 0$, $u_{yy} = B - \frac{1}{2}(A^2 + B^2)$, and $u_{xy} = \frac{1}{2}A$. From this we get

$$A = \frac{P}{\mu}, \qquad B = 1 - \sqrt{1 - \frac{P^2}{\mu^2}}.$$

(b) The solution is only valid for $-\mu < P < \mu$.

9 Basic elastostatics

9.1 Use (8.11).

∗ 9.3 **(a)** Let the rear end of the bullet be flat and moving according to $x = x(t)$ with $x = 0$ for $t = 0$ reckoned from the back end $x = 0$ of the barrel. The volume is $V = Ax$, where $A = \pi a^2$ is the cross-section of the barrel. The equation of motion for the bullet is $m\ddot{x} = pA$. The isentropic expansion obeys $pV^\gamma = \text{const}$ or $px^\gamma = p_0 x_0^\gamma$, where p_0 is the initial pressure. The equation of motion takes the form

$$m\ddot{x} = p_0 A \left(\frac{x_0}{x}\right)^\gamma .$$

Multiplying by $\dot{x}$ and integrating, we get

$$\dot{x}^2 = \frac{2p_0 A x_0}{(\gamma - 1)m}\left(1 - \left(\frac{x_0}{x}\right)^{\gamma-1}\right),$$

where the constant has been determined such that $\dot{x} = 0$ for $x = x_0$.
(b) For $x = L$ we have $\dot{x} = U$, leading to

$$p_0 = \frac{(\gamma - 1)mU^2}{2Ax_0}\left(1 - \left(\frac{x_0}{L}\right)^{\gamma-1}\right)^{-1} .$$

Numerically this becomes $p_0 \approx 3{,}600$ bar.
(c) The exit pressure is $p_1 = p_0(x_0/L)^\gamma \approx 27$ bar.
(d) The initial temperature is $T_0 = p_0 M_{\text{mol}}/R\rho_0 \approx 1{,}300$ K and the final temperature $T_1 \approx 320$ K.
(e) The unclamped strains (9.84) become for $r = a$ (the inside), $b = a + d$, $P = p_0$, $E = 205$ GPa, and $\nu = 29\%$, $u_{rr} \approx -0.0026$, $u_{\phi\phi} \approx 0.0034$, and $u_{zz} \approx -0.00034$.
(f) The corresponding stresses are $\sigma_{rr} = -3{,}600$ bar and $\sigma_{\phi\phi} = 6{,}000$ bar, while $\sigma_{zz} = 0$ because the barrel is unclamped. The tensile strength of steel is 5,500 bar, which is comparable to $\sigma_{\phi\phi}$, so the barrel should be close to blowing up.

9.5 Letting $y \to y - \alpha$ in (9.19), we get

$$\begin{aligned}
u_x &= & -\frac{\nu}{R}\alpha x & \quad + \frac{\nu}{R}xy, \\
u_y &= \frac{\nu}{2R}\alpha^2 & -\frac{\nu}{R}\alpha y & \quad + \frac{1}{2R}(z^2 + \nu(y^2 - x^2)), \\
u_z &= & \frac{1}{R}\alpha z & \quad - \frac{1}{R}yz.
\end{aligned}$$

The first column represents a simple translation and the second a uniform stretching along z of the form given in (8.23).

9.7 **(a)** The gravitational force density points radially inward (see Equation (12.38)),

$$f_r = -\frac{GM(r)\rho_0}{r^2}, \qquad M(r) = \frac{4\pi}{3}r^3\rho_0 .$$

Integrate (9.47) to get the general solution to the radial displacement field $\boldsymbol{u} = u_r\hat{\boldsymbol{e}}_r$,

$$u_r = \frac{2\pi}{15}\frac{G\rho_0^2}{\lambda + 2\mu}r^3 + Ar + \frac{B}{r^2},$$

where A and B are constants. Clearly $B = 0$ so that the displacement is finite at $r = 0$. The strain tensor is obtained from (D.37) and has the non-vanishing components $u_{rr} = du_r/dr$ and $u_{\theta\theta} = u_{\phi\phi} = u_r/r$. The stresses become

$$\sigma_{rr} = (2\mu + \lambda)\frac{du_r}{dr} + 2\lambda\frac{u_r}{r}, \qquad \sigma_{\theta\theta} = \sigma_{\phi\phi} = \lambda\frac{du_r}{dr} + 2(\mu + \lambda)\frac{u_r}{r}.$$

Requiring $\sigma_{rr} = 0$ for $r = a$, we get

$$A = -\frac{6\mu + 5\lambda}{(2\mu + 3\lambda)(\lambda + 2\mu)}\frac{2\pi}{15}G\rho_0^2 a^2 .$$

The displacement is always negative, as expected.
(b) $u_r|_{r=a} = -350$ km. **(c)** $u_{rr}|_{r=0} = -0.1$. **(d)** $\sigma_{rr}|_{r=0} = -1.3$ MBar.

9.9 **(a)** The centrifugal force density is radial and given by $f_r = \rho_0 \Omega^2 r$.
(b) The general solution to (9.66) is

$$u_r = Ar + \frac{B}{r} - \frac{1}{8}\frac{\rho_0 \Omega^2}{\lambda + 2\mu} r^3,$$

where A and B are integration constants. Use that u_r must be finite for $r = 0$ and $\sigma_{rr} = 0$ for $r = a$. The final solution becomes

$$u_r = \frac{1}{8}\frac{\rho_0 \Omega^2 a^2}{\lambda + 2\mu} r \left(3 - 2\nu - \frac{r^2}{a^2}\right).$$

The displacement is always positive.
(c) The strains are

$$u_{rr} = \frac{1}{8}\frac{\rho_0 \Omega^2 a^2}{\lambda + 2\mu}\left(3 - 2\nu - 3\frac{r^2}{a^2}\right), \qquad u_{\phi\phi} = \frac{1}{8}\frac{\rho_0 \Omega^2 a^2}{\lambda + 2\mu}\left(3 - 2\nu - \frac{r^2}{a^2}\right).$$

The radial strain vanishes for $r = a\sqrt{1 - 2\nu/3}$, which lies inside the cylinder for $\nu > 0$. It is positive for $r = 0$ and negative for $r = a$.
(d) Breakdown happens for $r = 0$, where the strains and stresses are maximal and positive.

10 Slender rods

10.1 Integrating (10.3) with $\mathcal{F}_y = 0$ and $\mathcal{F}_z = -\mathcal{F}$, and using that $\mathcal{M}_x = 0$ at $y = L$, one gets

$$-EI\frac{d^2 y}{dz^2} = \mathcal{F}(y - y(L)),$$

which has the solution $y = y(L) + A\sin kz + B\cos kz$ with $k = \sqrt{\mathcal{F}/EI}$. Requiring $y(0) = y'(0) = 0$, the solution becomes $y(z) = y(L)(1 - \cos kz)$. Taking $z = L$, we get $\cos kL = 0$ or $kL = \frac{1}{2}\pi + n\pi$ with $n = 0, 1, \ldots$. The threshold is $\mathcal{F} = k^2 EI = \pi^2 EI/4L^2$, which is 4 times smaller than the Euler threshold.

* **10.3** **(a)** Each cross-section of the stringed bow is subject to a compression force $\mathcal{F}\cos\theta$ and thus a stress $\sigma_{ss} = -\mathcal{F}\cos\theta/A$. The corresponding strain is $u_{ss} = \sigma_{ss}/E = -\mathcal{F}\cos\theta/EA$. Since $u_{ss} = du_s/ds$, the total displacement becomes

$$\Delta L = \int_0^L u_{ss} ds = -\frac{\mathcal{F}}{EA}\int_{-\alpha}^{\alpha}\cos\theta\frac{ds}{d\theta}d\theta = -\frac{\mathcal{F}L_z}{EA},$$

where $L_0 = z(L)$ is the length of the string given by (10.25).

(b) Since the force on the stringed bow is comparable to the Euler threshold (see Figure 10.4), $\mathcal{F} \approx \pi^2 EI/L^2$ and $L_z \approx L$, we get

$$\frac{\Delta L}{L_0} \approx \pi^2\frac{I}{AL^2} \approx \frac{\pi}{4}\frac{A}{L^2} \approx \frac{A}{L^2}.$$

In the last step we have used that for a circular bow cross-section $I = \frac{\pi}{4}a^4 = A^2/4\pi$. Clearly, the right-hand expression is small for a long slender bow.

* **10.5** **(a)** Varying $y(x)$ to $y(x) + \delta y(x)$, one finds

$$\delta W = EI\int_0^L y''\delta y''\,dz - \mathcal{F}\int_0^L y'\delta y'\,dz = \int_0^L (EIy'''' + \mathcal{F}y'')\delta y\,dz.$$

Since this is true for all variations, the integrand has to vanish.

(b) Integrating twice, we get

$$EI\frac{d^2 y}{dz^2} = -\mathcal{F}y + Az + B,$$

where A and B are constants. Now replace y by $y + (Az + B)/\mathcal{F}$. Since the left-hand side is unchanged, we arrive at (10.13).

* **10.7** To see this, use (10.29) to prove that $\boldsymbol{\Omega} = \kappa \boldsymbol{b} + \tau \boldsymbol{t}$ is independent of s, that is, a constant vector $d\boldsymbol{\Omega}/dt = (\boldsymbol{0})$. It now follows that

$$\frac{d^2\boldsymbol{n}}{ds^2} = -\Omega^2\boldsymbol{n},$$

where $\Omega = \sqrt{\kappa^2 + \tau^2}$. The general solution to this equation is $\boldsymbol{n} = \boldsymbol{A}\cos\Omega s + \boldsymbol{B}\sin\Omega s$, where $\boldsymbol{A}$ and $\boldsymbol{B}$ are constant vectors, orthogonal to $\boldsymbol{\Omega}$. Squaring, we obtain $1 = |\boldsymbol{A}|^2\cos^2\Omega s + |\boldsymbol{B}|^2\sin^2\Omega s + 2\boldsymbol{A}\cdot\boldsymbol{B}\sin\Omega s\cos\Omega s$. Since it has to be true for all s, we must have $\boldsymbol{A}\cdot\boldsymbol{B} = 0$ and $|\boldsymbol{A}| = |\boldsymbol{B}| = 1$. Let us choose a coordinate system where $\hat{\boldsymbol{e}}_x = -\boldsymbol{A}$, $\hat{\boldsymbol{e}}_y = -\boldsymbol{B}$ and $\hat{\boldsymbol{e}}_z = \boldsymbol{\Omega}/\Omega$. In cylindrical coordinates with $\phi = \Omega s$ and $r = \sqrt{x^2 + y^2}$, we then have $\boldsymbol{n} = -\hat{\boldsymbol{e}}_r$ with $\hat{\boldsymbol{e}}_r = (\cos\phi, \sin\phi)$, and that proves the point.

11 Computational elastostatics

11.1 We assume a linear combination

$$\nabla_x^+ f(x) = af(x) + bf(x + \Delta x) + cf(x + 2\Delta x).$$

Expand to second order and require the coefficient of $f(x)$ and $\nabla_x^2 f(x)$ to vanish and the coefficient of $\nabla_x f(x)$ to be 1, to get

$$a + b + c = 0, \qquad \frac{1}{2}b + 2c = 0, \qquad b + 2c = 1.$$

The solution is $a = -3/2$, $b = 2$, and $c = -1/2$.

12 Continuum dynamics

12.1 **(a)** Streamlines through $(x_0, y_0, 0)$ at $t = t_0$ are straight lines,

$$x = x_0 + a(t - t_0), \qquad y = y_0 + bt_0(t - t_0).$$

(b) Particle orbits starting at $(x_0, y_0, 0)$ are parabolas turning upward in the direction of the y-axis,

$$x = x_0 + a(t - t_0), \qquad y = y_0 + \tfrac{1}{2}b(t^2 - t_0^2).$$

(c) Streaklines are obtained from the particle orbits by varying t_0 in the interval $-\infty < t_0 < t$. They are seen to be parabolas curving the opposite way of the particle orbits.

12.3 Before the event, streamlines and particle orbits point straight north, while streaklines point south. After the event, streamlines and particle orbits point east while streaklines point west for a part and then turn south.

12.5 Differentiating through all the time dependence, one gets

$$\frac{d\rho(\boldsymbol{x}(t), t)}{dt} = \frac{d\boldsymbol{x}(t)}{dt}\cdot\frac{\partial\rho(\boldsymbol{x}, t)}{\partial\boldsymbol{x}} + \frac{\partial\rho(\boldsymbol{x}, t)}{\partial t} = \boldsymbol{v}(\boldsymbol{x}, t)\cdot\nabla\rho(\boldsymbol{x}, t) + \frac{\partial\rho(\boldsymbol{x}, t)}{\partial t} = \frac{D\rho}{Dt}.$$

12.7 **(a)** Let Q be the total volume of flow in the stream. Then the average velocity in the x-direction is $v_x(x) = Q/\pi a(x)^2$. **(b)** The advective acceleration is estimated as $w \approx v_x dv_x/dx$. **(c)** Constant acceleration w implies $v_x \approx \sqrt{2wx}$ for a suitable choice of origin and orientation of the x-axis. Hence, $a(x) \sim 1/x^{1/4}$.

* **12.9** Trivial.

*** 12.11** Defining the primed time derivative $\partial'_t = (\partial_t)_{x'}$ as the derivative with respect to t with x' held fixed, we get

$$\partial'_t \rho'(x', t) = \partial'_t \rho(x' + Ut, t) = \partial_t \rho(x, t) + (U \cdot \nabla)\rho(x, t).$$

The advective term is simpler:

$$(v' \cdot \nabla')\rho'(x', t) = (v - U) \cdot \nabla\rho(x, t).$$

Adding these equation, the U terms cancel and we have $D'_t \rho' = D_t \rho$, and similarly $D'_t v' = D_t v$; whereas for the displacement we get $D'_t u' = D_t u - U$. The form invariance of the equations of motion now follows.

13 Nearly ideal flow

*** 13.1** **(a)** For $R \ll a$, we find from (13.9) that

$$t_0 - t = \frac{a}{c}\sqrt{\frac{3}{2}}\int_0^{R/a} \frac{dx}{\sqrt{x^{-3} - 1}} \approx \frac{a}{c}\sqrt{\frac{3}{2}}\int_0^{R/a} x^{3/2}\, dx = \frac{a}{c}\sqrt{\frac{3}{2}}\frac{2}{5}\left(\frac{R}{a}\right)^{5/2},$$

which exposes the 2/5 power. Dividing by t_0, the constant becomes

$$C = \left(\sqrt{\pi}\frac{5\Gamma\left(\frac{5}{6}\right)}{2\Gamma\left(\frac{1}{3}\right)}\right)^{2/5} = 1.28371\ldots.$$

(b) The surface pressure in the liquid becomes $P = -2\alpha/R$. The solution to (13.6) becomes

$$\dot{R}^2 = \frac{2p_0}{3\rho_0}\left(\frac{a^3}{R^3} - 1\right) + \frac{2\alpha}{\rho_0 R}\left(\frac{a^2}{R^2} - 1\right).$$

which can be integrated numerically.

(c) The first term in $\dot{R}^2$ dominates the second for $p_0 \gg 3\alpha/a$.

13.3 Leonardo's law tells us that the average velocity at the top of the barrel is $v_0 = (A/A_0)v$, where v is the average velocity in the spout. Bernoulli's theorem now says

$$\frac{1}{2}v_0^2 + \frac{p_0}{\rho_0} + g_0 h = \frac{1}{2}v^2 + \frac{p_0}{\rho_0},$$

so that with $\kappa = A/A_0$,

$$v = \sqrt{\frac{2g_0 h}{1 - \kappa^2}}.$$

The speed is slightly higher than the speed of free fall.

13.5 **(a)** Leonardo's law says $\pi a^2 U_0 = \pi b^2 U_1$, so that

$$U_1 = \frac{a^2}{b^2}U_0 = 5.74\ \mathrm{m\,s^{-1}}.$$

(b) Bernoulli's theorem applied to a streamline through the center leads to

$$\frac{1}{2}U_0^2 + \frac{p_0}{\rho_0} = \frac{1}{2}U_1^2 + \frac{p_1}{\rho_0}.$$

Consequently, the pressure drop in the center becomes

$$\Delta p = p_0 - p_1 = \frac{1}{2}\rho_0\left(\frac{a^4}{b^4} - 1\right)U_0^2 = 3{,}973\ \mathrm{Pa}.$$

(c) The pressure difference must balance the weight of the mercury minus its buoyancy, $\Delta p = (\rho_1 - \rho_0)g_0 h$, so that

$$h = \frac{\Delta p}{(\rho_1 - \rho_0)g_0} = 3.2 \text{ cm}.$$

(d) The mercury level is $h/2$ over the initial level. The condition becomes $h < 2(c + d)$, or

$$U_0 < 2\sqrt{\frac{b^4(c+d)g_0}{b^4 - a^2}\frac{\rho_1 - \rho_0}{\rho_0}} = 9.7 \text{ m s}^{-1}.$$

* **13.7** Consider two surfaces S_1 and S_2 with the same curve C as perimeter and oriented consistently with each other. Then we have

$$\int_{S_1} \nabla \times \boldsymbol{v} \cdot d\boldsymbol{S} - \int_{S_2} \nabla \times \boldsymbol{v} \cdot d\boldsymbol{S} = \oint_S \nabla \times \boldsymbol{v} \cdot d\boldsymbol{S} = \int_V \nabla \cdot (\nabla \times \boldsymbol{v})\, dV = 0.$$

Here, $S = S_1 - S_2$ is the closed surface formed by the two open surfaces with an extra minus sign because of the requirement that the closed surface should have an outwardly oriented normal. In the last step, Gauss' theorem has been used to convert the integral over S to an integral over the volume V contained in S.

13.9 From (13.57) we get the critical velocity

$$U = \sqrt{\frac{3\pi}{5}\left(\frac{\rho_1}{\rho_0} - 1\right) a g_0},$$

which becomes merely $U \approx 7 \text{ cm s}^{-1}$ for the worm.

* **13.11** **(a)** In spherical coordinates the integral becomes

$$\begin{aligned}
\int_{r \geq a} (\boldsymbol{v} - \boldsymbol{U})^2\, dV &= \int_{r \geq a} \left((v_r - U_r)^2 + (v_\theta - U_\theta)^2\right) dV \\
&= \int_a^\infty dr \int_0^\pi r d\theta \int_0^{2\pi} r \sin\theta d\phi\, U^2 \frac{a^6}{4r^6}\left(1 + 3\cos^2\theta\right) \\
&= \frac{2}{3}\pi a^3 U^2.
\end{aligned}$$

(b) The momentum of the fluid is

$$\int_{r \geq a} \rho_0 (\boldsymbol{v} - \boldsymbol{U})\, dV \sim \int_{r \geq a} \frac{dV}{r^3} \sim \int_a^\infty \frac{dr}{r}$$

which diverges logarithmically at infinity.

14 Compressible flow

14.1 **(a)** Carrying out the time derivatives, we get

$$\frac{\partial}{\partial t}\left(\frac{1}{2}\rho_0 \boldsymbol{v}^2\right) = -\boldsymbol{v} \cdot \nabla \Delta p = -\nabla \cdot (\Delta p \boldsymbol{v}) + \Delta p \nabla \cdot \boldsymbol{v},$$

$$\frac{\partial}{\partial t}\left(\frac{\Delta p \Delta \rho}{2\rho_0}\right) = \frac{\partial}{\partial t}\left(\frac{c_0^2 \Delta \rho^2}{2\rho_0}\right) = \frac{c_0^2 \Delta \rho}{\rho_0}\frac{\partial \Delta \rho}{\partial t} = -\Delta p \nabla \cdot \boldsymbol{v}.$$

Adding the equations, we get the desired result.

(b) Integrating over a constant volume V, we get

$$\frac{d}{dt}\int_V \varepsilon\, dV = -\oint \Delta p \boldsymbol{v} \cdot d\boldsymbol{S}.$$

The equation makes sense if the left-hand side is interpreted as the rate of change of the total energy, and the right-hand side as the rate at which thermodynamic work $-\Delta p \delta(dV)/\delta t$ with $\delta(dV) = -\boldsymbol{v}\delta t \cdot d\boldsymbol{S}$ is performed on the volume.

(c) Averaging over a period, one gets a factor 1/2 in each term:

$$\langle \varepsilon \rangle = \frac{1}{4}\rho_0 v_1^2 + \frac{c_0^2 \rho_1^2}{4\rho_0} = \frac{1}{2}\rho_0 v_1^2,$$

where $v_1 = c_0 \rho_1/\rho_0$ is the velocity amplitude.

14.3 The proof is straightforward. First we show that (13.29) is also valid for compressible fluid with H is given by (14.21). For potential flow where $\boldsymbol{\omega} = \boldsymbol{0}$, we get $\nabla(\partial\Psi/\partial t + H) = \boldsymbol{0}$ so that $\partial\Psi/\partial t + H$ is a constant that may only depend on time, and which may be absorbed into Ψ.

* **14.5** **(a)** Define $\rho = \rho_0 + \Delta\rho$ and $p = p_0 + \Delta p$. Using that $\nabla p_0 = \rho_0 \boldsymbol{g}$, the Euler and continuity equations become, to first order in the small quantities,

$$\rho_0 \frac{\partial \boldsymbol{v}}{\partial t} = \boldsymbol{g}\Delta\rho - \nabla\Delta p, \qquad \frac{\partial \Delta\rho}{\partial t} = -\nabla \cdot (\rho_0 \boldsymbol{v}).$$

Using $\Delta p = c_0^2 \Delta\rho$, we get

$$\frac{1}{c_0^2}\frac{\partial^2 \Delta p}{\partial t^2}\frac{\partial^2 \Delta\rho}{\partial t^2} = -\nabla \cdot \left(\rho_0 \frac{\partial \boldsymbol{v}}{\partial t}\right) = \nabla^2 \Delta p - \nabla \cdot (\boldsymbol{g}\Delta\rho) = \nabla^2 \Delta p - \nabla \cdot \left(\boldsymbol{g}\frac{\Delta p}{c_0^2}\right);$$

and using $\nabla \cdot \boldsymbol{g} = 0$, the wave equation follows.

(b) The ratio of the two terms on the right-hand side of the wave equation

$$\frac{|c_0^2 \nabla^2 \Delta p|}{|c_0^2(\boldsymbol{g}\cdot\nabla)(\Delta p/c_0^2)|} \approx \frac{c_0^2 |\Delta p|/\lambda^2}{g_0 |\Delta p|/\lambda} \approx \frac{c_0^2}{g_0 \lambda}.$$

The condition for ignoring gravity is that $g_0\lambda \ll c_0^2$ or $\lambda \ll c_0^2/g_0$. In the atmosphere this becomes $\lambda \ll 12$ km, which is quite reasonable in view of the height of the atmosphere.

* **14.7** **(a)** We have $Y_{\text{eth}} = 33\%$, $Y_{\text{h2o}} = 11\%$ and $Y_{\text{o2}} = 56\%$. Using (1.3), one finds the propellant molar mass $M_{\text{mol}}^{\text{entry}} = 32.6\ \text{g mol}^{-1}$ and $X_{\text{eth}}^{\text{entry}} = 23\%$, $X_{\text{o2}}^{\text{entry}} = 57\%$, and $X_{\text{h2o}}^{\text{entry}} = 20\%$.

(b) Each mole of oxygen consumes 1/3 of a mole of ethanol and produces 2/3 mole of carbon dioxide and 1 mole of water. The exhaust molar fractions are thus proportional to $X_{\text{eth}}^{\text{exit}} \sim X_{\text{eth}}^{\text{entry}} - \frac{1}{3}X_{\text{o2}}^{\text{entry}}$, $X_{\text{h2o}}^{\text{exit}} \sim X_{\text{h2o}}^{\text{entry}} + X_{\text{o2}}^{\text{entry}}$, and $X_{\text{co2}}^{\text{exit}} \sim \frac{2}{3}X_{\text{o2}}^{\text{entry}}$. Normalizing by the sum, we arrive at $X_{\text{eth}}^{\text{exit}} = 4\%$, $X_{\text{h2o}}^{\text{exit}} = 64\%$, and $X_{\text{co2}}^{\text{exit}} = 32\%$.

(c) Using (1.2), we find $M_{\text{mol}}^{\text{exit}} = 27.3\ \text{g mol}^{-1}$, and thereby $Y_{\text{eth}}^{\text{exit}} = 6\%$, $Y_{\text{h2o}}^{\text{exit}} = 42\%$, and $Y_{\text{co2}}^{\text{exit}} = 52\%$.

15 Viscosity

15.1 **(a)** For such processes we have $p \sim \rho^\gamma$ and $p \sim \rho T$, so that $\rho \sim T^{1/(\gamma-1)}$ and $\nu = \eta/\rho \sim T^{1/2-1/(\gamma-1)}$. **(b)** For monatomic gases $\gamma = 5/3$ and $\nu \sim T^{-1}$, for diatomic $\gamma = 7/5$ and $\nu \sim T^{-2}$, and for multiatomic $\gamma = 4/3$ and $\nu \sim T^{-5/2}$.

15.3 (a) The total volume flux along x per unit of length along z is

$$Q(t) = \int_{-\infty}^{\infty} v_x(y,t)\,dy.$$

From (15.5) we get, by integrating over y,

$$\frac{dQ}{dt} = \int_{-\infty}^{\infty} \frac{\partial v_x(y,t)}{\partial t}\,dy = \nu \int_{-\infty}^{\infty} \frac{\partial^2 v_x}{\partial y^2}\,dy = 0,$$

because $\partial v_x/\partial y$ vanishes at infinity.
(b) The total momentum per unit of length in the x- and z-directions is

$$\mathcal{P} = \int_{-\infty}^{\infty} \rho_0 v_x(y,t)\,dy = \rho_0 Q$$

and is constant because Q is.
(c) The kinetic energy per unit of length in both x and z is

$$\mathcal{K} = \tfrac{1}{2}\rho_0 \int_{-\infty}^{\infty} v_x(y,t)^2\,dy.$$

Its time derivative becomes

$$\frac{d\mathcal{K}}{dt} = \rho_0 \int_{-\infty}^{\infty} v_y \frac{\partial v_y}{\partial t}\,dy = \rho_0\eta \int_{-\infty}^{\infty} v_y \frac{\partial^2 v_y}{\partial y^2}\,dy = -\rho_0\eta \int_{-\infty}^{\infty} \left(\frac{\partial v_y}{\partial y}\right)^2 dy,$$

where in the last step we have performed a partial integration with respect to y, using that the velocity and its derivatives vanish at infinity. Since the last expression is negative, the kinetic energy always decreases.
(d) In the Gaussian case this becomes

$$\mathcal{K} = \tfrac{1}{2}\rho_0 U^2 \frac{a^2}{a^2+4\nu t}\int_{-\infty}^{\infty} \exp\left(-2\frac{y^2}{a^2+4\nu t}\right) dy = \tfrac{1}{2}\rho_0 U^2 \sqrt{\frac{\pi}{2}}\frac{a^2}{\sqrt{a^2+4\nu t}}.$$

It vanishes like $t^{-1/2}$ for $t \to \infty$.

*** 15.5** (a) One verifies explicitly that field satisfies diffusion equation, and that for all t,

$$\int_{-\infty}^{\infty} \frac{1}{2\sqrt{\pi\nu t}} \exp\left(-\frac{(y-y')^2}{4\nu t}\right) dy = 1.$$

For $t \to 0$, the Gaussian factor therefore becomes infinitely narrow (i.e., it becomes a δ-function) and thus $v_x(y,t) \to v_x(y,0)$.
(b) If $v_x(y',0) = 0$ for $|y'| > a$, one gets for $|y| \to \infty$ and $4\nu t \gg a^2$

$$v_x(y,t) \approx \frac{1}{2\sqrt{\pi\nu t}} \exp\left(-\frac{y^2}{4\nu t}\right) \int_{-a}^{a} v_x(y',0)\,dy'.$$

*** 15.7** (a) Since the interface is quadratic in x and y, and the velocity field $\boldsymbol{v}$ must be continuous across the interface, we have near the origin $\boldsymbol{v} = \boldsymbol{v}_0 + \nabla_x \boldsymbol{v}_0 + \nabla_y \boldsymbol{v}_0 + \mathcal{O}(x^2, y^2)$, so that the first-order derivatives must also be continuous.

Use the continuity of:
(b) $\boldsymbol{\sigma}\cdot\hat{\boldsymbol{e}}_z = \sigma_{xz}\hat{\boldsymbol{e}}_x + \sigma_{yz}\hat{\boldsymbol{e}}_y + \sigma_{zz}\hat{\boldsymbol{e}}_z$.
(c) $\nabla_x v_x$ and $\nabla_y v_y$ in the divergence condition $\nabla_z v_z = -\nabla_x v_x - \nabla_y v_y$.
(d) $\sigma_{zz} = -p + 2\eta\nabla_z v_z$ and of $\nabla_z v_z$.
(e) σ_{xz} and of $\nabla_x v_z$.
(f) $\nabla_x v_y$ and $\nabla_y v_x$.
(g) $\nabla_x v_x$ and of $\nabla_z v_z$.

16 Channels and pipes

16.1 Use the general solution (16.4) and the no-slip conditions $v_x = 0$ for $x = 0$ and $v_x = U$ for $y = d$ to get

$$v_x = \frac{G}{2\eta} y(d-y) + U\frac{y}{d}.$$

This is also a consequence of the linearity of the Navier–Stokes equation for planar flow. The maximum is found at $y = y_{\max} = \dfrac{d}{2} + \dfrac{U\eta}{Gd}$, which lies between the plates for $0 < y_{\max} < d$.

16.3 If pressure were used to drive the planar sheet, there would have to be a decreasing pressure along the open surface. But that is impossible because the boundary conditions on the open surface require constant pressure (zero in fact) toward the vacuum.

16.5 **(a)** Drag becomes $\mathcal{D} = \Delta p W d = GLWd = 12\eta ULW/d$ and **(b)** issipation $P = U\mathcal{D} = 12\eta U^2 LW/d$.

16.7 **(a)** The effective pressure gradient is $G = \rho_0 g_0$. Using (16.32) and the Reynolds number (16.33), we get $\mathsf{Re} = g_0 a^3/4\nu^2$. Solving for a, we find

$$a = \left(\frac{4\nu^2}{g_0}\mathsf{Re}\right)^{1/3} = a_0 \mathsf{Re}^{1/3},$$

where $a_0 = (4\nu^2/g_0)^{1/3}$. For water it becomes $a_0 \approx 74\ \mu\text{m}$.
(b) For $\mathsf{Re} = 2{,}300$ one finds $a = 1$ mm. **(c)** The entry length becomes $L_{\text{entry}} = 25$ cm.

16.9 **(a)** Since pressure differences $\Delta p = \rho_0 QR$ are additive when put in series whereas Q is the same, R is additive in series. **(b)** Since the mass flux $Q = \Delta p/\rho_0 R$ is additive over the branches of a parallel connection, whereas Δp is the same, reciprocal resistance is additive in parallel connections.

16.11 Straightforward.

16.13 **(a)** Insert the fields into the steady-flow equations and verify that they are fulfilled and that the boundary conditions are fulfilled. **(b)** For $a = b$ we get the circular pipe field, and for $b \to \infty$ we get the planar field. **(c)** Using the area element in elliptic coordinates the flux becomes

$$Q = \int v_z dA = \int_{r=0}^{1}\int_{\theta=0}^{2\pi} v_z(r) abr dr d\theta = \frac{\pi G a^3 b^3}{4\eta(a^2+b^2)}.$$

(d) $U = Q/\pi ab$. **(e)** The drag is $\mathcal{D} = GL\pi ab$.

16.15 In cylindrical coordinates assume that the flow field is radial, $\boldsymbol{v} = v_r(r)\,\hat{\boldsymbol{e}}_r$ outside the pipe. Volume conservation implies that $v_r 2\pi r L$ is the same for all r. Hence $v_r(r) = Q/2\pi r L$, where Q/L is the volume discharge through the pipe wall per unit of pipe length.

16.17 **(a)** The torque on the cylinder is the sum of the torque $-\tau\phi$ exerted by the wire and by the fluid. For $d \ll a$ we find from (16.63) (with the opposite sign) that

$$M_z = -\tau\phi - 2\pi\eta\Omega a^3 \frac{L}{d},$$

where $\Omega = \dot\phi = d\phi/dt$. The moment of inertia of the thin shell of the inner cylinder is $I = Ma^2$ and its angular momentum $\mathcal{L}_z = I\Omega = Ma^2\dot\phi$. The equation of motion then becomes $d\mathcal{L}_z/dt = \mathcal{M}_z$, which is indeed the desired expression.

(b) Insert $\phi \sim e^{\lambda t}$ to get

$$Ma^2\lambda^2 = -\tau - 2\pi\eta a^3 \frac{L}{d}\lambda,$$

which has the solutions

$$\lambda = -\gamma \pm \sqrt{\gamma^2 - \frac{\tau}{Ma^2}}, \qquad \gamma = \frac{\pi \eta a L}{Md}.$$

For $\gamma < \tau/Ma^2$, the solutions are oscillating but decay with the rate γ. For $\gamma > \tau/Ma^2$, the solutions are exponentially damped with the leading exponent $\lambda = -\gamma + \sqrt{\gamma^2 - \tau/Ma^2}$. Critical damping occurs for $\gamma = \tau/Ma^2$. A measurement of the damping rate allows the determination of γ and thereby η from the parameters of the apparatus.

16.19 $P = 2\pi\eta\Omega^2 a^3 L/d \approx 100$ W.

*** 16.21** **(a)** Using (16.29) and (15.24) we obtain the density of dissipation,

$$2\eta \sum_{ij} v_{ij}^2 = 2\eta(v_{rz}^2 + v_{zr}^2) = \eta(\nabla_r v_z)^2 = \frac{G^2}{4\eta} r^2;$$

and dividing by ρ_0 we obtain the local specific rate of dissipation.

(b) Integrating, we get the total rate of dissipation

$$P = 2\eta \int_V \sum_{ij} v_{ij}^2 2\pi r L \, dr = \frac{\pi G^2 a^4 L}{8\eta} = QGL = Q\Delta p.$$

17 Creeping flow

17.1 The equation of motion for a particle of mass m_1 and radius a is

$$m_1 \ddot{z} = -(m_1 - m_0) g_0 - 6\pi\eta a \dot{z},$$

where z is a vertical coordinate, $m_1 = 4/3\pi a^3 \rho_1$, and $m_0 = 4/3\pi a^3 \rho_0$. The solution is

$$\dot{z} = -U\left(1 - e^{-t/\tau}\right),$$

where U is the terminal velocity (17.27) and the time constant is,

$$\tau = \frac{2}{9}\frac{a^2}{\nu}\frac{\rho_1}{\rho_0} = \frac{U}{g_0}\frac{\rho_1}{\rho_1 - \rho_0}.$$

17.3 **(a)** From (17.27) we get for $\rho_1 \gg \rho_0$,

$$-\frac{dz}{dt} = \frac{2}{9}\rho_1 \frac{a^2 g_0}{\eta(z)},$$

where the height-dependent viscosity $\eta(z)$ is found from (15.3) and (2.52),

$$\eta(z) = \eta_0 \sqrt{1 - \frac{z}{h_2}}.$$

From the differential equation, we get

$$t = \int_0^z \frac{9}{2} \frac{\eta_0}{\rho_1 a^2 g_0} \sqrt{1 - \frac{z'}{h_2}} \, dz' = \frac{3\eta_0 h_2}{\rho_1 a^2 g_0}\left(1 - \left(1 - \frac{z}{h_2}\right)^{3/2}\right).$$

(b) Taking $2a = 1\ \mu$m, $z = 10$ km, and $\rho_1 = 2.5$ g cm^{-3}, we get $t \approx 4$ years.

17.5 The equation of motion for a particle of mass m_1 and radius a is

$$m_1\ddot{z} = -m_1 g_0 + \tfrac{1}{2} C_D \rho_0 A \dot{z}^2,$$

where z is a vertical coordinate, $m_1 = 4/3\pi a^3 \rho_1$, and $A = \pi a^2$. The downward velocity $v = -\dot{z}$ may be written

$$\dot{v} = g_0 \left(1 - \frac{v^2}{U^2}\right),$$

where U is given by (17.35). The solution is

$$v = U \tanh \frac{g_0 t}{U},$$

with an exponential time constant of $\tau = U/2g_0$.

* **17.7** **(a)** Write $\boldsymbol{x} = r\hat{\boldsymbol{e}}_r$ and use (D.33) to obtain

$$\boldsymbol{v} = \frac{d\boldsymbol{x}}{dt} = \frac{dr}{dt}\hat{\boldsymbol{e}}_r + r\frac{d\theta}{dt}\hat{\boldsymbol{e}}_\theta,$$

so that $v_r = dr/dt$ and $v_\theta = r d\theta/dt$.
(b) Combine the differential equations for the streamlines to obtain

$$\frac{d\theta}{dr} = -\frac{B}{A}\tan\theta,$$

which is a solvable first-order equation. The integral over r is carried out using partial fractions,

$$\frac{B}{A} = \frac{r^2 + \frac{1}{4}ar + \frac{1}{4}a^2}{r(r-a)(r+\frac{1}{2}a)} = -\frac{1}{2r} + \frac{1}{r-a} + \frac{1}{2r+a}.$$

(c) For $r \to \infty$, we get $d \to r\sin\theta = \sqrt{x^2+y^2}$.
(d) Put $\theta = \pi/2$ to get $d = (r-a)\sqrt{1+a/2r}$ where r is the point of closest approach. This is a third-order algebraic equation for r.

* **17.9** **(a)** Follows from linearity of the field equations and the pressure independence of the boundary conditions.
(b) The field equations are

$$\nabla^2 A_{ij} = \nabla_i Q_j, \qquad \sum_i \nabla_i A_{ij} = 0.$$

The boundary conditions are, for $|\boldsymbol{x}| \to \infty$,

$$A_{ij}(\boldsymbol{x}) \to \delta_{ij}, \qquad Q_i(\boldsymbol{x}) \to 0.$$

At the surface of the body the velocity field must vanish, $\boldsymbol{n} \cdot \mathbf{A}(\boldsymbol{x}) = 0$.
(c) The stress tensor is

$$\sigma_{ij} = -p\delta_{ij} + \eta(\nabla_i v_j + \nabla_j v_i) = \eta \sum_k \tau_{ijk} U_k,$$

where

$$\tau_{ijk} = -\delta_{ij} Q_k + \nabla_i A_{jk} + \nabla_j A_{ik}.$$

The total reaction force on the body with surface S is

$$\mathcal{R}_i = \oint_S \sum_j \sigma_{ij} dS_j = \eta \sum_k U_k \oint_S \tau_{ijk} dS_j.$$

The tensor,

$$\mathcal{R}_i = \eta \sum_k S_{ik} U_k, \qquad S_{ik} = \oint_S \tau_{ijk} dS_j.$$

17.11 Equate lift and weight:

$$\alpha\frac{L}{2d}\frac{\eta ULA}{d^2} = Mg_0.$$

Solve for d:

$$d = \left(\frac{\alpha\eta UL^2A}{2Mg_0}\right)^{1/3}.$$

17.13 The extremum of v_x as a function of y is found at

$$y = \frac{h(4h-3d)}{6(h-d)},$$

which must lie in the interval $0 < y < h$. This is only the case when $h < 3d/4$ or $h > 3d/2$.

17.15 **(a)** They become

$$d_0 \approx d\left(1 - 3\left\langle\chi^2\right\rangle\right),$$
$$\mathcal{L} \approx -6\frac{\eta UAL}{d^2}\langle\xi\chi\rangle,$$
$$\mathcal{D} \approx \frac{\eta UA}{d}\left(1 + 4\left\langle\chi^2\right\rangle\right),$$
$$\mathcal{M}_z \approx -3\frac{\eta UAL^2}{d^2}\left\langle\xi^2\left(\chi - 3\left(\chi^2 - \left\langle\chi^2\right\rangle\right)\right)\right\rangle.$$

(b) For the flat wing we have $\chi = -2\gamma\xi$ with $-1/2 < \xi < 1/2$,

$$\left\langle\xi^2\right\rangle = \frac{1}{12}, \qquad \left\langle\xi^3\right\rangle = 0, \qquad \left\langle\xi^4\right\rangle = \frac{1}{80}.$$

Then it follows that

$$\left\langle\chi^2\right\rangle = \frac{1}{3}\gamma^2, \qquad \langle\xi\chi\rangle = -\frac{1}{6}\gamma, \qquad \left\langle\xi^2\chi\right\rangle = 0, \qquad \left\langle\xi^2\chi^2\right\rangle = \frac{1}{20}\gamma^2,$$

and from these the leading approximations are obtained.

18 Rotating fluids

18.1 **(a)** The centrifugal acceleration is $g_0 = a\Omega^2$ so $\Omega = \sqrt{g_0/a} \approx 0.44\ \mathrm{s}^{-1}$ and $\tau = 2\pi/\Omega \approx 14$ s.
(b) For a horizontal table, the Coriolis force is generally vertical. At typical velocity of $v \approx 4\ \mathrm{m\,s}^{-1}$, the Coriolis acceleration is $g_c = 2\Omega v \approx 3.6\ \mathrm{m\,s}^{-2}$, which is 1/3 of standard gravity. It does not make (a horizontal) ball deviate to the side, but changes the vertical centrifugal acceleration considerably.
(c) For horizontal motion the game is influenced in much the same way as for Ping-Pong, but for basketball vertical motion is important and the Coriolis force makes the ball deviate horizontally.

18.3 **(a)** Including the gravitational acceleration and the Coriolis acceleration (18.8) leads to the equations of motion

$$\ddot{x} = 2\Omega_0\cos\theta\dot{y} - 2\Omega_0\sin\theta\dot{z}, \qquad \ddot{y} = -2\Omega_0\cos\theta\dot{x}, \qquad \ddot{z} = 2\Omega_0\sin\theta\dot{x} - g_0.$$

(b) Using the initial conditions $x = y = 0$ and $\dot{x} = \dot{y} = \dot{z} = 0$ for $t = 0$, the solutions to the last two equations are $\dot{y} = -2\Omega_0\cos\theta x$ and $\dot{z} = 2\Omega_0\sin\theta x - g_0t$. Inserted into the first it becomes

$\ddot{x} = -4\Omega_0^2 x + 2\Omega_0 \sin\theta g_0 t$, which is easily solved with the result

$$x = \sin\theta \frac{g_0}{4\Omega_0^2}(2\Omega_0 t - \sin(2\Omega_0 t)),$$
$$y = \sin\theta\cos\theta \frac{g_0}{4\Omega_0^2}(1 - 2\Omega_0^2 t^2 - \cos(2\Omega_0 t)),$$
$$z = h - \frac{1}{2} g_0 t^2 - \sin^2\theta \frac{g_0}{4\Omega_0^2}(1 - 2\Omega_0^2 t^2 - \cos(2\Omega_0 t)).$$

(c) For $2\Omega_0 t \ll 1$, this becomes

$$x \approx \tfrac{1}{3}\sin\theta g_0 \Omega_0 t^3, \qquad y \approx -\tfrac{1}{6}\sin\theta\cos\theta g_0 \Omega_0^2 t^4, \qquad z \approx h - \tfrac{1}{2} g_0 t^2 + \tfrac{1}{6}\sin^2\theta g_0 \Omega_0^2 t^4.$$

(d) For $t = \sqrt{2h/g_0}$ we get $2\Omega_0 t \approx 3 \times 10^{-4}$, so the approximation is valid. The deviation in x becomes 1.4 mm whereas the deviations in y and z become merely 72 nm (at middle latitudes).

∗ **18.5** **(a)** As on page 602 with primed and unprimed variables interchanged.
(b) Differentiating with respect to time, the velocity becomes

$$\dot{\boldsymbol{x}} = \mathbf{A}\cdot(\dot{\boldsymbol{x}}' - \dot{\boldsymbol{c}}') + \dot{\mathbf{A}}\cdot(\boldsymbol{x}' - \boldsymbol{c}').$$

The second term is now rewritten using Problem 18.4 with the rotation vector $\boldsymbol{\Omega}'$ in the inertial system,

$$\dot{\mathbf{A}}\cdot(\boldsymbol{x}' - \boldsymbol{c}') = -\mathbf{A}\times\boldsymbol{\Omega}'\cdot(\boldsymbol{x}' - \boldsymbol{c}') = -\mathbf{A}\times\boldsymbol{\Omega}'\cdot\mathbf{A}^{-1}\cdot\boldsymbol{x} = -\boldsymbol{\Omega}\times\boldsymbol{x}.$$

In the next to last step we assumed that $\det\mathbf{A} = 1$ such that the transformed cross-product becomes the cross-product of the transformed vectors.
(c) Differentiate once more after time and repeat the above steps to get the acceleration in the moving system

$$\begin{aligned}
\ddot{\boldsymbol{x}} &= \mathbf{A}\cdot(\ddot{\boldsymbol{x}}' - \ddot{\boldsymbol{c}}') + \dot{\mathbf{A}}\cdot(\dot{\boldsymbol{x}}' - \dot{\boldsymbol{c}}') - \dot{\boldsymbol{\Omega}}\times\boldsymbol{x} - \boldsymbol{\Omega}\times\dot{\boldsymbol{x}} \\
&= \mathbf{A}\cdot(\ddot{\boldsymbol{x}}' - \ddot{\boldsymbol{c}}') - \mathbf{A}\times\boldsymbol{\Omega}'\cdot(\dot{\boldsymbol{x}}' - \dot{\boldsymbol{c}}') - \dot{\boldsymbol{\Omega}}\times\boldsymbol{x} - \boldsymbol{\Omega}\times\dot{\boldsymbol{x}} \\
&= \mathbf{A}\cdot(\ddot{\boldsymbol{x}}' - \ddot{\boldsymbol{c}}') - \boldsymbol{\Omega}\times(\dot{\boldsymbol{x}} + \boldsymbol{\Omega}\times\boldsymbol{x}) - \dot{\boldsymbol{\Omega}}\times\boldsymbol{x} - \boldsymbol{\Omega}\times\dot{\boldsymbol{x}} \\
&= \mathbf{A}\cdot(\ddot{\boldsymbol{x}}' - \ddot{\boldsymbol{c}}') - \dot{\boldsymbol{\Omega}}\times\boldsymbol{x} - 2\boldsymbol{\Omega}\times\dot{\boldsymbol{x}} - \boldsymbol{\Omega}\times(\boldsymbol{\Omega}\times\boldsymbol{x}).
\end{aligned}$$

(d) Newton's Second Law in the inertial system is $m\ddot{\boldsymbol{x}}' = \boldsymbol{f}'$.

19 Computational fluid dynamics

∗ **19.1** The last term is easily integrated, because

$$\delta\int \boldsymbol{v}\cdot\nabla p\, dV = \int dV[\delta\boldsymbol{v}\cdot\nabla p + \boldsymbol{v}\cdot\nabla\delta p] = \int dV\delta\boldsymbol{v}\cdot\nabla p,$$

where we have used Gauss' theorem in the last step, and dropped the surface terms.

The middle term is also easily integrated because

$$\delta\int dV \tfrac{1}{2}\sum_{ij}(\nabla_i v_j)^2 = \int dV \sum_{ij}\nabla_i v_j \nabla_i \delta v_j = \int dV[-\nabla^2\boldsymbol{v}]\cdot\delta\boldsymbol{v},$$

where we again have used Gauss' theorem and discarded boundary terms.

The problem arises from the inertia term $\delta\boldsymbol{v}\cdot(\boldsymbol{v}\cdot\nabla)\boldsymbol{v} = \sum_{ij}\delta v_i v_j \nabla_j v_i$. Assume that the integral is an expression of the form $\sum_{ijkl} a_{ijkl} v_i v_j \nabla_k v_l$ with suitable coefficients a_{ijkl} satisfying $a_{ijkl} = a_{jikl}$. Varying the velocity and again dropping boundary terms, we get (suppressing the integral as well as the sums over repeated indices)

$$\begin{aligned}
\delta(a_{ijkl} v_i v_j \nabla_k v_l) &= 2a_{ijkl}\delta v_i v_j \nabla_k v_l + a_{ijkl} v_i v_j \nabla_k \delta v_l \\
&= 2a_{ijkl}\delta v_i v_j \nabla_k v_l - 2a_{ijkl}\delta v_l v_j \nabla_k v_i \\
&= 2(a_{ijkl} - a_{ljki})\delta v_i v_j \nabla_k v_l.
\end{aligned}$$

In order for this to reproduce the desired result $\delta v_i v_j \nabla_j v_i$, we must have

$$a_{ijkl} - a_{ljki} = \tfrac{1}{2}\delta_{il}\delta_{kj}$$

but that is impossible because the left-hand side is antisymmetric under interchange of i and l whereas the right-hand side is symmetric.

19.3 In the pressure stress $\sigma_{xx} = -p + 2\eta\nabla_x v_x$, both terms belong to the 00-grid, and similarly for σ_{yy}. The shear stress $\sigma_{xy} = \eta(\nabla_x v_y + \nabla_y v_x)$ belongs to the 11-grid.

20 Mechanical balances

20.1 The time dependence of a global quantity $Q(t)$ has two origins: the changing density $\rho q(\boldsymbol{x}, t)$ and the changing control volume $V(t)$. In a small time interval δt, the control volume changes from $V(t)$ to $V(t + \delta t)$, and expanding to first order in δt, we find

$$\begin{aligned} Q(t+\delta t) - Q(t) &= \int_{V(t+\delta t)} \rho q(\boldsymbol{x}, t+\delta t)\, dV - \int_{V(t)} \rho q(\boldsymbol{x}, t)\, dV \\ &= \int_{V(t+\delta t)} \big(\rho q(\boldsymbol{x}, t+\delta t) - \rho q(\boldsymbol{x}, t)\big)\, dV + \int_{V(t+\delta t)-V(t)} \rho q(\boldsymbol{x}, t)\, dV \\ &\approx \delta t \int_{V(t)} \frac{\partial(\rho q(\boldsymbol{x}, t))}{\partial t}\, dV + \delta t \oint_{S(t)} \rho q(\boldsymbol{x}, t)\, \boldsymbol{v}_S(\boldsymbol{x}, t) \cdot d\boldsymbol{S}, \end{aligned}$$

where $\boldsymbol{v}_S(\boldsymbol{x}, t)$ is the velocity of a surface element $d\boldsymbol{S}$ near the point $\boldsymbol{x}$ at time t.

20.3 **(a)** Let the mass loss rate be $Q = \dot{M}$. With initial conditions $v = 0$ for $t = 0$, the solution to (20.21) becomes

$$u = -g_0 t - U \log\left(1 - \frac{Qt}{M_0}\right), \qquad M = M_0 - Qt.$$

Integrating $u = dz/dt$ with initial condition $z = 0$ at $t = 0$, the attained height becomes

$$z = -\frac{1}{2} g_0 t^2 + Ut + \frac{UM_0}{Q}\left(1 - \frac{Qt}{M_0}\right) \log\left(1 - \frac{Qt}{M_0}\right).$$

At the end of the burn at $t = t_1 = (M_0 - M_1)/Q$, the velocity and height are

$$u_1 = -g_0 t_1 - U \log \frac{M_1}{M_0}, \qquad z_1 = -\frac{1}{2} g_0 t_1^2 + U t_1 + \frac{UM_1}{Q} \log \frac{M_1}{M_0}.$$

(b) After this moment, the ballistic orbit becomes

$$z = z_1 + u_1(t - t_1) - \tfrac{1}{2} g_0 (t - t_1)^2.$$

The height is maximal at

$$t_2 = t_1 + \frac{u_1}{g_0}, \qquad z_2 = z_1 + \frac{u_1^2}{2g_0}.$$

(c) After t_2 the orbit is

$$z = z_2 - \tfrac{1}{2} g_0 (t - t_2)^2$$

so that the rocket reaches the ground again at

$$t_3 = t_2 + \sqrt{\frac{2z_2}{g_0}}, \qquad u_3 = -\sqrt{2 g_0 z_2}.$$

20.5 **(a)** For optimal ballistic flight over a distance $D/2$ with $v_\phi = 0$, we must have $v_r = v_z = U_r = U_z = \frac{1}{2}\sqrt{g_0 D}$. The steady flow azimuthal velocity becomes $U_\phi = -R\Omega_0 = -0.2\ \mathrm{m\,s^{-1}}$.

(b) The total velocity relative to the nozzle is $U = \sqrt{U_r^2 + U_\phi^2 + U_z^2} = 7\ \mathrm{m\,s^{-1}}$. The nozzle area is $A = \frac{1}{4}\pi d^2 = 12.5\ \mathrm{mm}^2$ and the total mass flux becomes $Q = n\rho_0 AU = 634\ \mathrm{kg\,h^{-1}}$.
(c) The total watered area is $A_0 = \frac{1}{4}\pi D^2$ so that the rainfall becomes a gentle $Q/\rho_0 A = 8\ \mathrm{mm\,h^{-1}}$.

20.7 **(a)** Integrating over the half-space, the kinetic energy of the transition region is

$$\mathcal{K}' = \int_a^\infty \tfrac{1}{2}\rho_0 v(r)^2\, 2\pi r^2\, dr = \pi\rho_0 a^3 v^2.$$

The ratio of this kinetic energy to the kinetic energy in the cistern, $\mathcal{K}_0 = \frac{1}{2}\rho_0 A_0 h v_0^2 = \rho_0 A^2 h v^2/2A_0$, is

$$\frac{\mathcal{K}'}{\mathcal{K}_0} = 2\frac{b^2}{ah},$$

which can be large or small, depending on h.
(b) The ratio to the kinetic energy in the pipe $\mathcal{K}_1 = \frac{1}{2}\pi a^2 L\rho_0 v^2$ is

$$\frac{\mathcal{K}'}{\mathcal{K}_1} = \frac{2a}{L},$$

which is generally small.

20.9 Choose a control volume consisting of all the water in the system between the two open surfaces. The open surface in the cistern is fixed, whereas the open surface in the pipe is moving. The total kinetic energy in the system is (under the usual assumptions)

$$\mathcal{K} = \tfrac{1}{2}\rho_0 A_0 h v_0^2 + \tfrac{1}{2}\rho_0 A(L_0 + x)v^2.$$

Using Reynolds' transport theorem (20.7), the material derivative of the kinetic energy is

$$\frac{D\mathcal{K}}{Dt} = \frac{d\mathcal{K}}{dt} + \dot{\mathcal{K}} = \frac{d\mathcal{K}}{dt} - \frac{1}{2}\rho_0 A_0 v_0^3$$

because the system only gains kinetic energy through the water that is refilled to keep constant the open water surface of the cistern. The total power of gravity is

$$P = \rho_0 g_0 A_0 h v_0 + \rho_0 g_1 A(L_0 + x)v,$$

where $g_1 = g_0 \sin\alpha$ is the projection of gravity on the pipe direction. The total power of the pressure on the open surfaces $p_0 A_0 v_0 - p_0 A v$ vanishes as always because of Leonardo's law, $A_0 v_0 = Av$.

Kinetic energy balance, $D\mathcal{K}/Dt = P$, now leads to the differential equation

$$\frac{d^2x}{dt^2} = \frac{2g_0(h + (L_0 + x)\sin\alpha) - (dx/dt)^2}{2(L + x)},$$

where $L = L_0 + hA/A_0$. This equation can only be solved numerically.

20.11 **(a)** Use the result from the application of connected tubes (page 352) to get $\omega = \sqrt{2g_0/L} = 0.03\ \mathrm{s}^{-1}$ and thus $T = 2\pi/\omega \approx 200$ s.
(b) Taking the dissipation rate from Poiseuille flow (16.38), mechanical energy balance becomes (since there is no outflow)

$$\frac{d\mathcal{E}}{dt} = -8\pi\eta v^2 L.$$

Using the mechanical energy (20.48),

$$\frac{d^2z}{dt^2} = -\omega^2 z - \frac{8\nu}{a^2}\frac{dz}{dt}.$$

Inserting $z \sim e^{-\lambda t}$, the solution to the characteristic equation is

$$\lambda = \gamma \pm \sqrt{\gamma^2 - \omega^2},$$

where $\gamma = 4\nu/a^2 \approx 0.04\ \mathrm{s}^{-1}$. Since $\gamma > \omega$, the water will not oscillate at all but get to rest exponentially fast.
(c) The slowest decay rate is obtained for the negative sign and becomes $\lambda = 0.015\ \mathrm{s}^{-1}$, corresponding to an exponential decay time of $1/\lambda \approx 66$ s.

21 Action and reaction

21.1 We choose a control volume encompassing all water in the cistern.

(a) By Torricelli's law, the velocity of the horizontal outflow is $v = \sqrt{2g_0 h}$, and the horizontal reaction force becomes $\mathcal{R}_x = -v\dot{M} = -\rho_0 v^2 A = -2\rho_0 g_0 h A$.

(b) In the vertical direction there is a contribution from the weight $M g_0$ of the water and the inflow of water with velocity $v_0 = vA/A_0$ from the refill process,

$$\mathcal{R}_z = -Mg_0 + v_0\dot{M} = -Mg_0 - v_0^2\rho_0 A_0 = -Mg_0 - 2\rho_0 g_0 h A^2/A_0.$$

The movement of the replenished water increases the weight of the water by a tiny amount equal to a fraction A/A_0 of the horizontal reaction.

21.3 The force is found from (21.2). Gravity has no influence because the pipe is horizontal. The pressures at the entry and exit are equal, so there is no stress contribution. Due to steady flow, the rate of momentum change vanishes, so that the only contribution to the reaction force stems from the momentum inflow and outflow,

$$\mathcal{R} = 2v\dot{M} = 2\rho_0 v^2 A.$$

The cross-section of the pipe is $A \approx 5\ \mathrm{cm}^2$, so we get $\mathcal{F} \approx 1$ N.

21.5 Consider the triangle formed by the center, the inner radius blade attachment point and the outer radius blade attachment point. Use the sine relations for triangles $\sin\phi_\beta/a = \sin\phi_\alpha/b$ to derive

$$\beta = \alpha\frac{a}{\sqrt{b^2 + \alpha^2(b^2 - a^2)}},$$

where $\alpha = \tan\phi_\alpha$ and $\beta = \tan\phi_\beta$.

22 Energy

22.1 **(a)** The temperature does not depend on x (or z) because of symmetry, so that Fourier's equation becomes

$$\frac{d^2T}{dy^2} = 0,$$

which has the solution $T = A + By$. The constants A and B are determined by the boundary conditions.
(b) If the fluid moves steadily with velocity $v_x(y)$, the advective term vanishes, $(\boldsymbol{v}\cdot\nabla)T = 0$, and Fourier's equation takes the same form.

22.3 (**a**) The general solution to the steady spherical heat equation (22.28) is

$$T = -\frac{h_0}{6k_s}r^2 + \frac{A}{r} + B,$$

where A and B are integration constants. In the core, the temperature has to be T_c so that $A = 0$ and $B = T_c + h_0 c^2/6k_c$. In the shell, A could in principle be non-zero, but requiring continuity of the heat flow at the interface, it follows that $A = 0$ also in the shell. The complete solution is thus in the core and skin, respectively,

$$T = T_1 + \frac{h_0}{6k_c}(c^2 - r^2), \qquad T = T_1 - \frac{h_0}{6k_s}(r^2 - c^2),$$

where

$$T_1 = T_c - \frac{h_0 c^2}{6k_c} = T_0 + \frac{h_0(a^2 - c^2)}{6k_s}$$

is the interface temperature expressed in the central temperature T_c and the surface temperature T_0.

(**b**) For $k_c \to \infty$ we have $T_1 = T_c$ and the drop becomes $\Delta T = T_c - T_0 = 6\,\text{K}$.

22.5 The divergence of the general heat flow vector field becomes

$$\nabla \cdot \boldsymbol{q}' = -k\nabla^2 T + \rho_0 c(\boldsymbol{v} \cdot \nabla)T,$$

and the second term is the advective term that is normally found on the left-hand side of the heat equation.

23 Entropy

23.1 (**a**) In proper thermodynamic notation we get from (23.6) and (23.8),

$$dS = \frac{1}{T}\left(\frac{\partial \mathcal{U}}{\partial T}\right)_V dT + \frac{1}{T}\left(\left(\frac{\partial \mathcal{U}}{\partial V}\right)_T + p\right) dV = C_v \frac{dT}{T} + \left(\frac{\partial p}{\partial T}\right)_V dV.$$

Finally, using (22.58) and $\alpha/\beta = \alpha K_T$, we get the result.

(**b**) For $dS = 0$, we have

$$dT = -\frac{\alpha K_T T}{C_V} dV = -\frac{\alpha K_T T}{\rho V c_v} dV = \frac{\alpha T}{\rho^2 c_v \beta} d\rho.$$

Using (22.53), we get

$$\frac{d\rho}{\rho} = -\alpha dT + \beta dp = -\frac{\alpha^2 T}{\rho^2 c_v \beta} d\rho + \frac{dp}{K_T}.$$

Dividing by $d\rho$ and using that $dp/d\rho = K_S/\rho$, we obtain the result with γ given by (22.61).

24 Elastic vibrations

24.1 The stress tensor is unchanged, but the space-time dependence of the Lamé coefficient gives rise to gradients in the Cauchy equation. In index notation, Navier's equation becomes

$$\rho\frac{\partial^2 u_i}{\partial t^2} = f_i + \mu\nabla^2 u_i + (\lambda + \mu)\nabla_i \nabla \cdot \boldsymbol{u} + (\nabla \cdot \boldsymbol{u})\nabla_i \lambda + (\nabla_i \boldsymbol{u}) \cdot \nabla\mu + (\nabla\mu \cdot \nabla)u_i.$$

24.3 (a) The fields have the form

$$u_y = Ae^{\kappa_T z}, \qquad u'_y = A'e^{-\kappa'_T z}.$$

The boundary conditions $u'_y = u_y$ and $\sigma'_{yz} = \sigma_{yz}$ at $z = 0$ lead to the equations

$$A' = A, \qquad \mu'\kappa'_T A' = -\mu\kappa_T A.$$

In view of κ_T and κ'_T being positive, these equations do not have a solution. Thus, there are no interface waves of this kind.

(b) At a free surface, we would instead have $A' = 0$, but can only require $\sigma_{yz} = 0$, which leads to $A = 0$.

24.5 Use that $e^{-i\omega t} = \cos\omega t - i\sin\omega t$ and that $\langle\cos^2\omega t\rangle = \langle\sin^2\omega t\rangle = \frac{1}{2}$ and $\langle\cos\omega t\sin\omega t\rangle = 0$.

24.7 (a) Use $\boldsymbol{I} = -\boldsymbol{v}\cdot\boldsymbol{\sigma} = i\omega\boldsymbol{u}\cdot\boldsymbol{\sigma}$ and Problem 24.5 with $A \to i\omega\boldsymbol{u} = i\omega\boldsymbol{a}e^{i\boldsymbol{k}\cdot\boldsymbol{x}}$ and $B \to \boldsymbol{\sigma} = \mu(i\boldsymbol{k}\boldsymbol{a} + i\boldsymbol{a}\boldsymbol{k}) + \lambda i\boldsymbol{k}\cdot\boldsymbol{a}\boldsymbol{1}$.

(b) An evanescent SH wave has $\boldsymbol{a} = A\hat{\boldsymbol{e}}_y$ and $\boldsymbol{k} = (k, 0, i\kappa'_T)$ so that

$$\langle\boldsymbol{I}\rangle = \frac{1}{2}\rho\omega c_T^2 a^2 k e^{-2k_T z}\hat{\boldsymbol{e}}_x.$$

25 Gravity waves

25.1 In addition to the usual boundary conditions at the surface of the infinite sea, we must also require $v_x = 0$ for $x = 0, a$ and $v_y = 0$ for $y = 0, b$. The velocity potential for the standing wave takes the form (absorbing a phase shift in the time parameter)

$$\Psi = f(z)\cos k_x x\cos k_y y\cos\omega t.$$

The boundary conditions on the velocity require the wave vector to be of the form $k_x = m\pi/a$ and $k_y = n\pi/b$, where m and n are integers. The velocity potential must satisfy Laplace's equation, leading again to the expression (25.19), that is, $f(z) \sim \cosh k(z+d)$ with $k = \sqrt{k_x^2 + k_y^2}$. The complete solution becomes to first order in the amplitude a,

$$\begin{aligned}
h &= a\cos k_x x\cos k_y y\sin\omega t,\\
\Psi &= ac\frac{\cosh k(z+d)}{\sinh kd}\cos k_x x\cos k_y y\cos\omega t,\\
v_x &= -ack_x\frac{\cosh k(z+d)}{\sinh kd}\sin k_x x\cos k_y y\cos\omega t,\\
v_y &= -ack_y\frac{\cosh k(z+d)}{\sinh kd}\cos k_x x\sin k_y y\cos\omega t,\\
v_z &= a\omega\frac{\sinh k(z+d)}{\sinh kd}\cos k_x x\cos k_y y\cos\omega t,\\
p &= p_0 - \rho_0 g_0\left(z - a\frac{\cosh k(z+d)}{\cosh kd}\cos k_x x\cos k_y y\sin\omega t\right).
\end{aligned}$$

with c given by (25.28) and $\omega = ck$.

25.3 Consider a wave rolling in at an angle toward the beach. Since for shallow-water waves we have $c \sim \sqrt{d}$, the phase velocity of the part of a wave farther from the beach is greatest, causing the part of the crest that is farther out to approach the coastline faster than the crest closer to the beach.

25.5 Using (25.47) we get $L < 3.9$ cm.

25.7 (a) Straightforward. (b) $\xi_c = 86$ cm.

26 Jumps and shocks

26.1 Expand the equation to first order in $1/\mathsf{Fr}^2$:

$$\sigma = \frac{\sqrt{1+8\mathsf{Fr}^2}-3}{2} \approx \sqrt{2}\mathsf{Fr} - \frac{3}{2} + \frac{\sqrt{2}}{16\mathsf{Fr}} + \cdots .$$

The last term is 3.3% of the leading terms for $\mathsf{Fr} = 2$.

26.3 **(a)** Use (26.19) and eliminate σ by means of (26.21). **(b)** Trivial.

26.5 The derivative

$$\frac{d(\Delta s/c_V)}{d\sigma} = \frac{(\gamma^2-1)\sigma^2}{(1+\sigma)(\gamma^2(\sigma+2)^2-\sigma^2)}$$

has singularities for $\sigma = -1, -2\gamma/(\gamma+1), -2\gamma/(\gamma-1)$. The last two are always smaller than -1, so the entropy change is a growing function for $\sigma > -1$.

26.7 **(a)** $R = 2900$ m, $U = 1160$ m s^{-1}.
(b) $p_1 = 13$ atm, $\rho_1 = 7.2$ kg m^{-3}, $T_1 = 650$ K.
(c) Since $p_1 > 8p_0$, the strong shock approximation should still be valid.
(d) $U \approx 900$ m s^{-1}, $t \approx 1.5$ s, $R \approx 3.5$ km.
(e) $U \approx 500$ m s^{-1}, $t \approx 4$ s, $R \approx 5$ km

27 Whirls and vortices

27.1 **(a)** Inside the core we have $v_\phi = \Omega r$, and the angular momentum becomes, per unit of axial length,

$$\frac{d\mathcal{L}_z}{dz} = \int_0^c \rho_0 r v_\phi \, 2\pi r dr = \frac{1}{2}\pi\rho_0\Omega c^4.$$

(b) The kinetic energy becomes similarly

$$\frac{d\mathcal{T}}{dz} = \int_0^c \frac{1}{2}\rho_0 v_\phi^2 \, 2\pi r dr = \frac{1}{4}\pi\rho_0\Omega^2 c^4.$$

(c) Outside the core the field is $v_\phi = \Omega c^2/r$ and the angular momentum becomes

$$\frac{d\mathcal{L}_z}{dz} = \int_c^R \rho_0 r v_\phi \, 2\pi r dr = \pi\rho_0\Omega c^2(R^2 - c^2).$$

The kinetic energy becomes

$$\frac{d\mathcal{T}}{dz} = \int_c^R \frac{1}{2}\rho_0 v_\phi^2 \, 2\pi r dr = \pi\rho_0\Omega^2 c^4 \log\frac{R}{c}.$$

27.3 **(a)** The non-vanishing vorticity component is

$$\omega_z = \frac{dv_\phi}{dr} + \frac{v_\phi}{r} = 2\Omega\frac{c^2}{c^2+4\nu t}\exp\left(-\frac{r^2}{c^2+4\nu t}\right).$$

(b) The total flux of vorticity is $\int_0^\infty \omega_z 2\pi r \, dr = 2\pi\Omega c^2$.

27.5 The proof runs in the same way as on page 465 and using that for a barotropic fluid $\nabla p/\rho = \nabla w$, where w is the pressure potential.

27.7 $\psi = -\frac{1}{2}C\log(x^2+y^2)$.

∗ 27.9 **(a)** Changing $n \to -n$ in the sum, we get

$$f_N(-z) + f_N(z) = \sum_{n=-N}^{N} \frac{1}{-z - \pi n} - \sum_{n=-N}^{N} \frac{1}{-z + \pi n} = 0\,.$$

(b) Shifting the sum, one gets

$$f_N(z+\pi) - f_N(z) = \sum_{n=-N}^{N} \left(\frac{1}{z + \pi - \pi n} - \frac{1}{z - \pi n} \right) = \frac{1}{z + \pi(N+1)} - \frac{1}{z - \pi N}.$$

For fixed z, the right-hand side vanishes for $N \to \infty$.
(c) In the strip the only singularities are the pole $1/z$ of the cotangent and the pole $-1/z$ from the sum, and they cancel.
(d) We know from periodicity and antisymmetry that $f(\pi/2) = f(-\pi/2) = -f(\pi/2)$ and thus $f(\pi/2) = 0$. For arbitrary fixed x in the strip, the leading behavior for fixed $|y| \gg 1$ is

$$f_N(x+iy) \approx \cot(iy) + \sum_{n=1}^{N} \frac{2iy}{y^2 + \pi^2 n^2} \approx \mp i + 2iy \int_0^N \frac{dn}{y^2 + \pi^2 n^2}$$
$$\approx \mp i + \frac{2iy}{\pi\,|y|} \arctan\frac{N\pi}{|y|} \approx \mp i \pm i \frac{2}{\pi} \arctan\frac{N\pi}{|y|},$$

which converges to 0 for $N \to \infty$.

∗ 27.11 Straightforward, although a bit cumbersome.

27.13 **(a)** Insert and verify.
(b) The angular momentum per unit of axial length is

$$\frac{d\mathcal{L}_z}{dz} = \int_0^\infty r\rho_0 v_\phi(r,t) 2\pi r\, dr = 16\pi\tau\nu^2\rho_0.$$

27.15 The equation of motion is

$$\frac{\partial v_\phi}{\partial t} + \frac{v_r}{r}\frac{\partial(rv_\phi)}{\partial r} = \nu \frac{\partial}{\partial r}\left(\frac{1}{r}\frac{\partial(rv_\phi)}{\partial r}\right).$$

Assume that the azimuthal flow is of the form

$$rv_\phi = C(1 - e^{-r^2/\lambda(t)}).$$

Then λ must satisfy

$$\dot\lambda + \frac{4\nu}{a^2}\lambda = 4\nu.$$

Solving this with the boundary condition $\lambda(0) = c^2$, we find

$$\lambda(t) = a^2 + (c^2 - a^2)e^{-4\nu t/a^2}.$$

The radius at time t is $b(t) = \sqrt{\lambda(t)}$.

27.17 Integrating the vortex equation (27.29) once, one gets

$$\frac{1}{r}\frac{d(rv_\phi)}{dr} = 2\Omega \exp\left[\frac{1}{\nu}\int_0^r v_r(r')\,dr'\right],$$

where the normalization has been fixed by $v_\phi/r \to \Omega$ for $r \to 0$. Integrating once more, using $v_\phi = 0$ for $r = 0$, one gets

$$v_\phi = \frac{\Omega}{r}\int_0^r \exp\left[\frac{1}{\nu}\int_0^{r_2} v_r(r_1)\,dr_1\right] 2r_2 dr_2.$$

28 Boundary layers

28.1 (a) Assume that $v_x(y)$ and $v_y(y)$ only depend on y, not on x. The boundary conditions are $v_x(0) = 0$, $v_x(\infty) = U$ and $v_y(0) = -V$. The continuity equation shows that $dv_y(y)/dy = 0$ so $v_y(y) = -V$ for all y. The Prandtl equation for v_x becomes

$$-V\frac{dv_x}{dy} = \nu\frac{d^2v_x}{dy^2}$$

with the solution

$$v_x(y) = U(1 - e^{-Vy/\nu}).$$

The boundary layer thickness is $\delta = \nu/V$.

(b) For $V < 0$, the v_x grows exponentially for $y \to \infty$. It is therefore impossible to maintain the boundary condition of asymptotically uniform flow (that $v_x \to U$ for $y \to \infty$).

28.3 The normalization of $H(s)$ is irrelevant because $F(s)$ is renormalized by $H(\infty)$. Inserting $g(s) = -2f''(s)/f'(s)$, we get from (28.29)

$$\frac{F''}{F'} = \mathsf{Pr}\frac{f''}{f'},$$

which integrates to $\log F' = \mathsf{Pr}\log f' + \text{const}$. Thus $F' \sim f'^{\mathsf{Pr}}$, and integrating we arrive at the desired result.

∗ **28.5** (a) Insert the field in the Navier–Stokes equations (28.10) to get (with $\rho_0 = 1$)

$$\frac{\partial p}{\partial x} = -U\frac{dU}{dx} + \nu\frac{d^2U}{dx^2}$$

$$\frac{\partial p}{\partial y} = V\frac{dU}{dx} - U\frac{dV}{dx} + \nu\frac{d^2V}{dx^2} + y\left(U\frac{d^2U}{dx^2} - \left(\frac{dU}{dx}\right)^2 - \nu\frac{d^3U}{dx^3}\right).$$

Use the cross-derivative $\partial^2 p/\partial x\partial y = \partial^2 p/\partial y\partial x = 0$ to obtain the two equations (coefficients of y)

$$\frac{d}{dx}\left(V\frac{dU}{dx} - U\frac{dV}{dx} + \nu\frac{d^2V}{dx^2}\right) = \frac{d}{dx}\left(U\frac{d^2U}{dx^2} - \left(\frac{dU}{dx}\right)^2 - \nu\frac{d^3U}{dx^3}\right) = 0,$$

Integrating, we get

$$U\frac{d^2U}{dx^2} - \left(\frac{dU}{dx}\right)^2 - \nu\frac{d^3U}{dx^3} = A, \qquad V\frac{dU}{dx} - U\frac{dV}{dx} + \nu\frac{d^2V}{dx^2} = B,$$

where A and B are constants.

(b) For constant U we have $A = 0$ and $V = C\exp(Ux/\nu) - Bx/U + D$, where C and D are constants.

28.7 (a) Keep only first order in u. (b) Inserting the asymptotic expression in the equation, the leading order of the right-hand side becomes

$$s^k e^{\alpha s^2/2}\left(u''' + \alpha s u'' - 2\beta u'\right) \sim (-k\alpha^2 + 2\alpha^2 + 2\alpha\beta)s + \mathcal{O}\left(s^{-1}\right).$$

The leading term must vanish, leading to $k = 1 + 2\beta/\alpha = 1 + 4m/(1+m)$.

∗ **28.9** (a) Retain only the leading terms for $\mu \to 0$. (b) Taking $U(0) = 1$ and $\mu(0) = 4/3$, the solution is $\mu^2 = \frac{16}{9} + 12\log U$. (c) At separation $x = x_c$ we have $\mu = 0$, leading to $U_c = \exp(-4/27) = 0.862303$. (d) Solving $U(x) = U_c$ for the nine cases, one obtains $x_c = 0.138, 0.256, 0.071, 0.160, 0.077, 0.371, 0.609, 0.780, 0.531$. The deviation from the exact values is essentially as good as the full wall approximation, but that must be fortuitous.

28.11 **(a)** The roots are $\mu = -4.8866, 1.8569, 5.9806, 17.4717$. The smallest positive root determines the start value $\mu_0 = 1.8569$.

29 Subsonic flight

29.1 **(a)** 3.7 m s^{-2}, **(b)** 27 s, **(c)** 1400 m.

29.3 **(a)** From Newton's Second Law we get $MdU/dt = T = P/U$, where T is the thrust P is the thrust power. Dividing by $U = dx/dt$ we find $MdU/dx = P/U^2$. Integrating, the solution becomes $x = MU^3/3P$, which yields the runway length at take-off. Inserting the given values, one gets $P = 37$ kW, which is about 50% of the engine power. **(b)** The computed take-off time is similarly $t = MU^2/2P \approx 10$ s, which is only half the quoted value. In view of the drag having been ignored, this is not that bad.

29.5 Instead of (29.40), we have

$$\mathcal{R} = -\oint_S \rho \boldsymbol{v}\, \boldsymbol{v} \cdot d\boldsymbol{S} + \oint_S \boldsymbol{\sigma} \cdot d\boldsymbol{S}.$$

At large distances where $\rho = \rho_0 + \Delta\rho$ and $\boldsymbol{v} = \boldsymbol{U} + \Delta\boldsymbol{v}$, we get to first order

$$\mathcal{R} = -\oint_S \Delta\rho \boldsymbol{U}\boldsymbol{U} \cdot d\boldsymbol{S} - \rho_0 \oint_S \boldsymbol{U}\Delta\boldsymbol{v} \cdot d\boldsymbol{S} - \rho_0 \oint_S \Delta\boldsymbol{v}\boldsymbol{U} \cdot d\boldsymbol{S} - \oint_S \Delta p\, d\boldsymbol{S}.$$

Mass conservation becomes in the same approximation

$$\nabla \cdot (\rho \boldsymbol{v}) = \rho_0 \nabla \cdot \Delta\boldsymbol{v} + (\boldsymbol{U} \cdot \nabla)\Delta\rho = 0,$$

or after integrating

$$\rho_0 \oint_S \Delta\boldsymbol{v} \cdot d\boldsymbol{S} + \oint_S \Delta\rho \boldsymbol{U} \cdot d\boldsymbol{S} = 0.$$

In view of the uniformity of $\boldsymbol{U}$, this makes the two first terms in $\mathcal{R}$ cancel, and we arrive at the incompressible result (29.41). At great distances from the body where the velocity corrections are tiny, barotropic fluids are effectively incompressible and Bernoulli's theorem takes the usual form.

29.7 Let $\hat{\boldsymbol{e}}_U = \boldsymbol{U}/U$ be a unit vector in the direction of the asymptotic flow. Projecting on (29.42) we find

$$\mathcal{D} = \oint_S (p + \rho_0 \Delta\boldsymbol{v} \cdot \boldsymbol{U})\hat{\boldsymbol{e}}_U \cdot d\boldsymbol{S}.$$

Using that $\mathcal{L} = \mathcal{R} - \hat{\boldsymbol{e}}_U(\hat{\boldsymbol{e}}_U \cdot \mathcal{R}) = -\hat{\boldsymbol{e}}_U \times (\hat{\boldsymbol{e}}_U \times \mathcal{R})$, the lift takes the form

$$\mathcal{L} = -\rho_0 \oint_S \boldsymbol{U} \times (\Delta\boldsymbol{v} \times d\boldsymbol{S}) + \oint_S (\Delta p + \rho_0 \Delta\boldsymbol{v} \cdot \boldsymbol{u})\, \hat{\boldsymbol{e}}_U \times (\hat{\boldsymbol{e}}_U \times d\boldsymbol{S}).$$

This is the desired result. The last term evidently vanishes if S cuts the wake in a planar region orthogonal to the asymptotic velocity, so that $d\boldsymbol{S} \sim \hat{\boldsymbol{e}}_U$.

* **29.9** Replace x by complex $z = x + iy$, so that the integral is the real part of

$$I(z) = \int_0^1 \frac{1}{(1-t)\sqrt{t(1-t)}} \log \frac{t-z}{1-z}\, dt$$

for $y \to 0$ and $0 < x < 1$. Use that

$$\frac{d}{dt}\sqrt{\frac{t}{1-t}} = -\frac{1}{2(1-t)\sqrt{t(1-t)}}.$$

Perform a partial integration to get

$$I = -2\int_0^1 \sqrt{\frac{t}{1-t}}\frac{dt}{t-z} = -2\int_0^1 \left(1+\frac{z}{t-z}\right)\frac{dt}{\sqrt{t(1-t)}}$$
$$= -2\pi - 2z\int_0^1 \frac{1}{t-z}\frac{dt}{\sqrt{t(1-t)}}.$$

We must now show that the real part of the last integral vanishes for $y \to 0$ and $0 < x < 1$.
From Cauchy's theorem we have

$$\frac{1}{\sqrt{z(z-1)}} = \oint_z \frac{1}{\sqrt{t(t-1)}}\frac{1}{t-z}\frac{dt}{2\pi i}$$
$$= \left(\int_{0+i\epsilon}^{1+i\epsilon} - \int_{0-i\epsilon}^{1-i\epsilon}\right)\frac{1}{\sqrt{t(t-1)}}\frac{1}{t-z}\frac{dt}{2\pi i}$$

because there is a cut along the real axis for $0 < t < 1$. Using that $\sqrt{t(t-1+i\epsilon)} = \pm i\sqrt{t(1-t)}$, we get the desired integral,

$$\int_0^1 \frac{1}{t-z}\frac{dt}{\sqrt{t(1-t)}} = -\frac{\pi}{\sqrt{z(z-1)}}.$$

Letting $y \to 0 \pm i\epsilon$ for $0 < x < 1$, the right-hand side becomes purely imaginary, and the theorem follows.

*** 29.11** **(a)** Direct insertion confirms that the diffusion equation is satisfied. For $t \to 0$, the Gaussian becomes very sharply peaked at $y = x$, so that

$$F(x,t) \approx (4\pi t)^{-N/2} F_0(x) \int \exp\left(-\frac{(x-y)^2}{4t}\right) d^N y.$$

The integral is now standard and leads to $F(x,t) = F_0(x)$ in the limit of $t \to 0$.

(b) First write the solution as

$$F(x,t) = (4\pi t)^{-N/2} e^{-x^2/4t} \int F_0(y) e^{-x\cdot y/2t - y^2/4t}\, d^N y.$$

Assume that $F_0(y)$ is only non-vanishing for $|y| < a$. Then for $|x| < \sqrt{t}$ and $t \gg a^2$, the exponential inside the integral can be disregarded.

30 Convection

30.1 **(a)** Use the Boussinesq equations (30.1) for zero velocity. The buoyancy term varies in the horizontal direction (x) because of the temperature variation, implying that the pressure gradient also varies horizontally. Thus, the isobars are not horizontal, which they have to be for any hydrostatic solution in constant gravity (even with a spatially varying density). **(b)** In motion, the balance between the buoyancy and friction forces for the heated fluid between the plates may be estimated as

$$\alpha\Theta\rho_0 g_0 A d \sim \eta\frac{U}{d}A,$$

where A is the plate area and U is the velocity scale. Apart from numerical factors this leads to an expression of the form (30.5).

∗ 30.3 (**a**) For $\cosh\mu 1 \gg 1$ and $\tanh\mu_1 \approx 1$, the determinant (30.51) becomes (after division by $\cosh\mu_1$)

$$\det(\tau,\xi) \propto \mu_0 \sin\frac{\mu_0}{2} + (\mu_1 + \sqrt{3}\,\mu_2)\cos\frac{\mu_0}{2}.$$

The determinant vanishes for

$$\tan\frac{\mu_0}{2} = -\frac{\mu_1 + \sqrt{3}\mu_2}{\mu_0}.$$

Taking into account that the μ-values are all positive, this may be solved as

$$\mu_0 = 2\left(\pi - \arctan\frac{\mu_1 + \sqrt{3}\mu_2}{\mu_0}\right),$$

which may be cast in the form $\tau = \tau(\xi)$ given in the problem formulation.

(**b**) Straightforward numerical minimization of $\tau(\xi)$ leads to the quoted results.

31 Turbulence

31.1 From definition (31.16) we find

$$\left\langle\frac{\partial \boldsymbol{v}}{\partial t}\right\rangle = \lim_{T\to\infty}\frac{1}{T}\int_0^T \frac{\partial \boldsymbol{v}}{\partial t}(\boldsymbol{x}, t+s)\,ds = \lim_{T\to\infty}\frac{1}{T}\int_0^T \frac{\partial \boldsymbol{v}(\boldsymbol{x}, t+s)}{\partial s}\,ds$$
$$= \lim_{T\to\infty}\frac{\boldsymbol{v}(\boldsymbol{x}, t+T) - \boldsymbol{v}(\boldsymbol{x}, t)}{T}.$$

For a bounded field, the norm of the last expression is smaller than $2C/T$, which vanishes in the limit of $T \to \infty$.

31.3 (**a**) From (31.23) one finds

$$\left\langle\frac{\partial(u_i u_j)}{\partial t}\right\rangle + \langle u_i(\boldsymbol{v}\cdot\nabla)u_j + u_j(\boldsymbol{v}\cdot\nabla)u_i\rangle$$
$$+ \langle u_i(\boldsymbol{u}\cdot\nabla)v_j + u_j(\boldsymbol{u}\cdot\nabla)v_i\rangle + \langle u_i(\boldsymbol{u}\cdot\nabla)u_j + u_j(\boldsymbol{u}\cdot\nabla)u_i\rangle$$
$$= -\frac{1}{\rho_0}\langle u_i\nabla_j q + u_j\nabla_i q\rangle + \nu\left\langle u_i\nabla^2 u_j + u_j\nabla^2 u_i\right\rangle.$$

Rearranging we obtain the desired equation.

(**b**) It does not close because of the third-order fluctuation moment $\langle u_i u_j u_k\rangle$ occurring in the last term on the left-hand side. There is also a problem in the terms $\langle u_i\nabla_j q\rangle + \langle u_j\nabla_i q\rangle$ and $\langle\nabla_k u_i\nabla_k u_j\rangle$, which is not caused by lack of closure in the moments. Such terms can, in principle, be handled by calculating the more general moments $\langle u_i(\boldsymbol{x}_1, t)q(\boldsymbol{x}_2, t)\rangle$ and $\langle u_i(\boldsymbol{x}_1, t)u_j(\boldsymbol{x}_2, t)\rangle$, and afterward taking the limit of $\boldsymbol{x}_2 \to \boldsymbol{x}_1$.

31.5 Write $\Lambda_n = \Lambda(1+\delta_n)$ and expand to first order in the precision δ_n to get

$$\delta_n = -\frac{A}{\Lambda}\delta_{n-1}.$$

This shows that the approximate sequence converges rapidly for $\Lambda > A$. With the given values,

$$\Lambda_n = 1, 32.7428, 24.4746, 25.1644, 25.0985, 25.1047, 25.1041, 25.1042,$$

after which it does not change.

A Newtonian particle mechanics

A.1 (a) Defining the alternative total mass to be $M' = \sum_n m_n^2$ and the alternative "center of mass" to be $X' = \sum_n m_n^2 x_n / \sum_n m_n^2$, the global equation becomes of the same form as before, $M'(d^2X'/dt^2) = \mathcal{F}'$. (b) Since the fundamental equations (A.1) are unchanged, all the physical consequences must be unchanged and cannot depend on these definitions.

B Cartesian coordinates

B.1 In an Earth-centered Cartesian coordinate system with z-axis toward the North pole at latitude $\delta = 90°$ and x-axis toward Greenwich at longitude $\alpha = 0$ we have $x = (a+h)\cos\alpha\cos\delta$, $y = (a+h)\sin\alpha\cos\delta$, and $z = (a+h)\sin\delta$ where a is the sea-level radius of the Earth. Using the invariance of the distance the square of the distance function becomes $d^2 = (x_1-x_2)^2 + (y_1-y_2)^2 + (z_1-z_2)^2$ which may be recast into $d^2 = (a+h_1)^2 + (a+h_2)^2 - 2(a+h_1)(a+h_2)(\cos\delta_1\cos\delta_2\cos(\alpha_1 - \alpha_2) + \sin\delta_1\sin\delta_2)$.

B.3 (a) $|\boldsymbol{a}| = 7$, $|\boldsymbol{b}| = 5$. (b) $\boldsymbol{a}\cdot\boldsymbol{b} = -6$. (c) $\boldsymbol{a}\times\boldsymbol{b} = (-24, -18, -17)$.

(d)

$$\boldsymbol{a}\boldsymbol{b} = \begin{pmatrix} 6 & -8 & 0 \\ 9 & -12 & 0 \\ -18 & 24 & 0 \end{pmatrix}$$

B.5 Use that $0 \le |\boldsymbol{a} + s\boldsymbol{b}|^2 = |\boldsymbol{a}|^2 + s^2|\boldsymbol{b}|^2 + 2s\boldsymbol{a}\cdot\boldsymbol{b}$. The minimum of the right-hand side is obtained for $s = -\boldsymbol{a}\cdot\boldsymbol{b}/\boldsymbol{b}^2$, and the inequality follows. Using this result we have $(\boldsymbol{a}+\boldsymbol{b})^2 = \boldsymbol{a}^2 + \boldsymbol{b}^2 + 2\boldsymbol{a}\cdot\boldsymbol{b} \le \boldsymbol{a}^2 + \boldsymbol{b}^2 + 2|\boldsymbol{a}||\boldsymbol{b}| = (|\boldsymbol{a}| + |\boldsymbol{b}|)^2$, and the inequality follows.

B.7 Trivial.

B.9 Straightforward in index notation.

B.11 Straightforward in index notation.

B.13 The transformed tensor is, $\delta'_{ij} = \sum_{kl} a_{ik}a_{jl}\delta_{kl} = \sum_k a_{ik}a_{jk} = \delta_{ij}$.

∗ **B.15** Differentiate $(\boldsymbol{x} - \boldsymbol{y})^2 = (\boldsymbol{f}(\boldsymbol{x}) - \boldsymbol{f}(\boldsymbol{y}))^2$ with respect to $\boldsymbol{x}$ to obtain $\boldsymbol{x} - \boldsymbol{y} = (\boldsymbol{f}(\boldsymbol{x}) - \boldsymbol{f}(\boldsymbol{y}))\cdot\mathsf{A}(\boldsymbol{x})$ with $\mathsf{A}(\boldsymbol{x}) = \{a_{ij}(\boldsymbol{x})\}$ and $a_{ij} = \partial f_i/\partial x_j$. Differentiate again with respect to $\boldsymbol{y}$ to obtain $-\mathbf{1} = -\mathsf{A}(\boldsymbol{y})^\top\cdot\mathsf{A}(\boldsymbol{x})$. This means that $\mathsf{A}(\boldsymbol{x})^{-1} = \mathsf{A}(\boldsymbol{y})^\top$. The left-hand side depends only on $\boldsymbol{x}$ and the right-hand side only on $\boldsymbol{y}$ which implies that both sides are independent of $\boldsymbol{x}$ and $\boldsymbol{y}$, i.e., the matrix A is a constant, and orthogonal. Integrating $a_{ij} = \partial f_i/\partial x_j$ one gets $\boldsymbol{f}(\boldsymbol{x}) = \mathsf{A}\cdot\boldsymbol{x} + \boldsymbol{b}$.

∗ **B.17** Let $\mathsf{A}_z(\phi)$ be the transformation matrix of the simple rotation (B.29) through an angle ϕ around the z-axis. Then the three Euler angles ϕ, θ and ψ determine any rotation matrix as a product $\mathsf{A} = \mathsf{A}_z(\psi)\cdot\mathsf{A}_y(\theta)\cdot\mathsf{A}_z(\phi)$. A general reflection can be built up from a simple reflection, and an arbitrary rotation. A general translation can be built from a sum of three simple translations.

∗ **B.19** Straightforward, but cumbersome.

C Field calculus

C.1 (a) $\nabla\cdot(\boldsymbol{a}\times\boldsymbol{x}) = -(\nabla\times\boldsymbol{x})\cdot\boldsymbol{a} = 0$ because $\nabla\times\boldsymbol{x} = \mathbf{0}$.

(b) $\nabla\times(\boldsymbol{a}\times\boldsymbol{x}) = \boldsymbol{a}(\nabla\cdot\boldsymbol{x}) - \boldsymbol{a}\cdot\nabla\boldsymbol{x} = 3\boldsymbol{a} - \boldsymbol{a} = 2\boldsymbol{a}$ because $\nabla\boldsymbol{x} = \mathbf{1}$.

C.3 It is easiest to show in coordinates

$$(\nabla \times (U \times V))_i = \sum_{jk} \epsilon_{ijk} \nabla_j \left(\sum_{lm} \epsilon_{klm} U_l V_m \right)$$
$$= \sum_{jk} \epsilon_{ijk} \sum_{lm} \epsilon_{klm} \left(U_m \nabla_j V_m + V_m \nabla_j U_l \right),$$

and then making use of Problem B.8.

* **C.5** Under a general transformation $x'_i = \sum_j a_{ij} x_j$, we use the chain rule for differentiation,

$$\nabla_j = \frac{\partial}{\partial x_j} = \sum_k \frac{\partial x'_k}{\partial x_j} \frac{\partial}{\partial x'_k} = \sum_k a_{kj} \nabla'_k.$$

Multiplying by a_{ij} and summing, we get

$$\sum_j a_{ij} \nabla_j = \sum_{jk} a_{ij} a_{kj} \nabla'_k = \sum_j \delta_{ik} \nabla'_k = \nabla'_i,$$

where the orthogonality of the transformation matrix has been used.

D Curvilinear coordinates

D.1 Straightforward, but a bit cumbersome.

E Thermodynamics of ideal gases

E.1 **(a)** The total mass is $M = n M_{\rm mol} = \sum_i M_i = \sum_i n_i M_i^{\rm mol} = n \sum_i X_i M_i^{\rm mol}$. Using $V = M/\rho$ in the ideal gas law (2.26), we get the result (2.27) with $R = R_{\rm mol}/M_{\rm mol}$. **(b)** Let k_i be the degrees of freedom for each component of the mixture. The total energy becomes $\mathcal{U} = \sum_i U_i = \sum_i \frac{1}{2} k_i n_i R_{\rm mol} T = \frac{1}{2} k n R_{\rm mol} T$, where $k = \sum_i k_i n_i / n = \sum_i X_i k_i$ is the average number of degrees of freedom. The adiabatic index becomes, as usual, $\gamma = 1 + 2/k$.

References

In this book there are three kinds of references: to books, articles, and links. They have been collected in three separate lists with differently styled reference labels. Books are labeled by the author names and the year of publication. Articles are given a short label made up from the author names and two digits of the year of publication. Links are numbered by their order of appearance in the book. At the end of every reference, the citing locations are listed in italic brackets.

Books

[Acheson 1990] D. J. Acheson, *Elementary Fluid Dynamics*, Oxford University Press, 1990. *[283]*

[Anderson 1995] J. D. Anderson, Jr., *Computational Fluid Dynamics*, McGraw-Hill, 1995. *[180, 325, 326, 329]*

[Anderson 1997] J. D. Anderson, Jr., *A History of Aerodynamics*, Cambridge University Press, 1997. *[255, 296, 513, 515, 526]*

[Anderson 2001] J. D. Anderson, Jr., *Aerodynamics*, McGraw-Hill, 2001. *[513, 516, 526, 526, 531]*

[Anderson 2004] J. D. Anderson, Jr., *Modern Compressible Flow*, McGraw-Hill, 3rd edition, 2004. *[240, 453]*

[Aref 2010] H. Aref, Editor, *150 Years of Vortex Dynamics*, Springer, 2010. *[464]*

[Ball 1997] P. Ball, *The Self-Made Tapestry*, Oxford University Press, 1997. *[283, 564]*

[Batchelor 1953] G. K. Batchelor, *The Theory of Homogeneous Turbulence*, Cambridge University Press, 1953. *[565]*

[Batchelor 1996] G. K. Batchelor, *The Life and Legacy of G. I. Taylor*, Cambridge University Press, 1996. *[455]*

[Batchelor 67] G. K. Batchelor, *An Introduction to Fluid Dynamics*, Cambridge University Press, 67. *[208, 314]*

[Bodmer 2003] G. R. Bodmer, *Hydraulic Motors: Turbines and Pressure Engines*, Fredonia Books, 2003, A facsimile of the original publication from 1895. *[366]*

[Bower 2010] A. F. Bower, *Applied Mechanics of Solids*, CRC Press. Taylor & Francis Group, 2010. *[139, 171]*

[Boys 1959] C. V. Boys, *Soap Bubbles*, Dover Publications, 1959, Originally published in 1911. *[73, 74]*

[Braess 2001] D. Braess, *Finite Elements*, Cambridge University Press, 2nd edition, 2001. *[179]*

[Brennen 1995] C. E. Brennen, *Cavitation and Bubble Dynamics*, Oxford University Press, 1995. *[209]*

[Bruus 2008] H. Bruus, *Theoretical Microfluidics*, Oxford University Press, 2008. *[287]*

[Cengel and Boles 2002] Y. A. Cengel and M. A. Boles, *Thermodynamics, An Engineering Approach*, McGraw-Hill, 2002. *[627]*

[Chandrasekhar 1981] S. Chandrasekhar, *Hydrodynamic and Hydromagnetic Stability*, Dover Publications, 1981. *[283, 561]*

[Chandrasekharaiah and Debnath 1994] D. S. Chandrasekharaiah and L. Debnath, *Continuum Mechanics*, Academic Press, 1994. *[121]*

[Cohen 1985] I. B. Cohen, *The Birth of a New Physics*, Penguin Books, 1985. *[62]*

[Crowe 1994] M. J. Crowe, *A History of Vector Analysis*, Cambridge University Press, 1994. *[611]*

[da Silva 2006] V. D. da Silva, *Mechanics and Strength of Materials*, Springer, 2006. *[139]*

[Davidson 2004] P. A. Davidson, *Turbulence*, Oxford University Press, 2004. *[565]*

[Davis and Selvadurai 1994] R. O. Davis and A. P. S. Selvadurai, *Elasticity and Geomechanics*, Dover Publications, 1994. *[142]*

[de Gennes et al. 2002] P.-G. de Gennes, F. Brochard-Wyart, and D. Quéré, *Capillarity and Wetting Phenomena*, Springer, 2002. *[69, 74, 81, 81, 82]*

[Doghri 2000] I. Doghri, *Mechanics of Deformable Solids*, Springer, 2000. *[120, 134, 167, 179]*

[Douglas et al. 2001] J. F. Douglas, J. M. Gasiorek, and J. A. Swaffield, *Fluid Mechanics*, Prentice Hall, 4th edition, 2001. *[359, 361]*

[Drazin and Reid 1981] P. G. Drazin and W. H. Reid, *Hydrodynamic Stability*, Cambridge University Press, 1981. *[283]*

[Dudley 2000] R. Dudley, *The Biomechanics of Insect Flight*, Princeton University Press, 2000. *[519]*

[Dyke 1982] M. V. Dyke, *An Album of Fluid Motion*, The Parabolic Press, Stanford, CA, 1982. *[468, 564]*

[Faber 1995] T. E. Faber, *Fluid Dynamics for Physicists*, Cambridge University Press, 1995. *[240, 259]*

[Fox and McDonald 1985] R. W. Fox and A. T. McDonald, *Introduction to Fluid Mechanics*, John Wiley & Sons, Inc., 3rd edition, 1985. *[213, 359]*

[Frisch 1995] U. Frisch, *Turbulence*, Cambridge University Press, 1995. *[565, 570]*

[Graebel 2007] W. P. Graebel, *Advanced Fluid Mechanics*, Academic Press, 2007. *[264]*

[Green and Adkins 1960] A. E. Green and J. E. Adkins, *Large Elastic Deformations and Non-Linear Continuum Mechanics*, Oxford University Press, 1960. *[120, 132]*

[Green and Zerna 1992] A. E. Green and W. Zerna, *Theoretical Elasticity*, Dover Publications, 1992. *[120]*

[Griebel et al. 1998] M. Griebel, T. Dornseifer, and T. Neunhoeffer, *Numerical Simulation in Fluid Dynamics*, SIAM, 1998. *[180, 325, 326, 329]*

[Hagedorn and DasGupta 2007] P. Hagedorn and A. DasGupta, *Vibrations and Waves in Continuous Mechanical Systems*, John Wiley & Sons, Inc., England, 2007. *[403]*

[Hansen and Kawaler 1994] C. J. Hansen and S. D. Kawaler, *Stellar Interiors*, Springer, New York, 1994. *[38, 40]*

[Hunt and Vassilicos 2000] J. C. R. Hunt and J. C. Vassilicos, *Turbulence Structure and Vortex Dynamics*, Cambridge University Press, 2000. *[570]*

[Kaye and Laby 1995] G. W. C. Kaye and T. H. Laby, *Tables of Physical and Chemical Constants*, Longman Group Ltd., 16th edition, 1995, Also available online (see links). *[30, 33, 127, 376]*

[Kondepudi and Prigogine 1998] D. Kondepudi and I. Prigogine, *Modern Thermodynamics*, John Wiley & Sons, Inc., 1998. *[395, 627]*

[Lamb 1993] H. Lamb, *Hydrodynamics*, Cambridge University Press, 1993, First published 1879. *[60, 61, 62, 436]*

[Landau and Lifshitz 1980] L. D. Landau and E. M. Lifshitz, *Statistical Physics*, Volume 1, Pergamon Press, 3rd edition, 1980. *[400]*

[Landau and Lifshitz 1986] L. D. Landau and E. M. Lifshitz, *Theory of Elasticity*, Pergamon Press, 1986. *[107, 132, 139, 151, 152, 171, 557]*

[Landau and Lifshitz 1987] L. D. Landau and E. M. Lifshitz, *Fluid Mechanics*, Butterworth-Heinemann, 2nd edition, 1987. *[533, 541]*

[Lesieur 1997] M. Lesieur, *Turbulence in Fluids*, Kluwer Academic Publishers, 3rd edition, 1997. *[565]*

[Libby 1996] P. Libby, *Introduction to Turbulence*, Taylor and Francis, 1996. *[565]*

[Lide 1996] D. R. Lide, Editor, *Handbook of Chemistry and Physics*, CRC Press, 77th edition, 1996. *[33, 70, 376]*

[Lin 1955] C. C. Lin, *Hydrodynamic stability*, Cambridge University Press, 1955. *[264, 283]*

[Loeb 1961] L. B. Loeb, *Kinetic Theory of Gases*, Dover Publications, 1961. *[245]*

[Massey 1998] B. Massey, *Mechanics of Fluids*, Stanley Thornes Ltd., 1998, Revised by J. Ward-Smith. *[359]*

[Maurel and Petitjeans 2000] A. Maurel and P. Petitjeans, *Vortex Structure and Dynamics*, Springer, 2000. *[461]*

[Mei 1989] C. C. Mei, *The applied dynamics of ocean surface waves*, World Scientific, 1989. *[436]*

[Melchior 1978] P. Melchior, *The Tides of the Planet Earth*, Pergamon Press, 1978. *[62, 65, 66]*

[Milne-Thomson 1968] L. M. Milne-Thomson, *Theoretical Hydrodynamics*, Dover Publications, 5th edition, 1968. *[470]*

[Narasimhan 1993] M. N. L. Narasimhan, *Principles of Continuum Mechanics*, John Wiley & Sons, Inc., 1993. *[107]*

[Newton 1999] I. Newton, *The Principia*, University of California Press, 1999, Translated by I. B. Cohen and A. Whitman. *[352]*

[Nieuwstadt and Steketee 1995] T. M. Nieuwstadt and J. A. Steketee, *Selected Papers of J. M. Burgers*, Kluver Academic Publishers, 1995. *[473]*

[Panton 2005] R. L. Panton, *Incompressible Flow*, John Wiley & Sons, Inc., 2005. *[221]*

[Pedlosky 1987] J. Pedlosky, *Geophysical Fluid Dynamics*, Springer, 1987. *[312, 315]*

[Pope 2003] S. B. Pope, *Turbulent Flows*, Cambridge University Press, 2003. *[565]*

[Press et al. 1992] W. H. Press, S. A. Teukolsky, W. T. Vetterling, and B. P. Flannery, *Numerical Recipes in C*, Cambridge University Press, 1992. *[179, 181, 181, 326, 326, 329, 333]*

[Saffman 1992] P. G. Saffman, *Vortex Dynamics*, Cambridge University Press, 1992. *[461, 466, 470]*

[Schlichting and Gersten 2000] H. Schlichting and K. Gersten, *Boundary-Layer Theory*, Springer, 8th edition, 2000. *[295, 296, 481, 489, 501, 507]*

[Sedov 1959] L. Sedov, *Similarity and Dimensional Methods in Mechanics*, Academic Press, 1959. *[457]*

[Sedov 1975] L. Sedov, *Mécanique des Milieux Continus (I et II)*, Editions Mir, Moscow, 1975. *[139]*

[Sobey 2000] I. J. Sobey, *Introduction to Interactive Boundary Layer Theory*, Oxford University Press, 2000. *[481, 500, 501, 502, 502]*

[Sokolnikoff 1956] I. S. Sokolnikoff, *Mathematical Theory of Elasticity*, McGraw-Hill, 1956. *[151]*

[Soutas-Little 1999] R. W. Soutas-Little, *Elasticity*, Dover Publications, 1999, First edition published by Prentice-Hall in 1973. *[142]*

[Stoker 1992] J. J. Stoker, *Water Waves*, John Wiley & Sons, Inc., 1992. *[437]*

[Sychev et al. 1998] V. Sychev, A. Ruban, V. Sychev, and G. Korolev, *Asymptotic Theory of Separated Flows*, Cambridge University Press, 1998. *[502]*

[Tennekes 1997] H. Tennekes, *The Simple Science of Flight*, The MIT Press, 1997. *[519]*

[Tokaty 1994] G. A. Tokaty, *A History and Philosopy of Fluid Mechanics*, Dover Publications, 1994. *[193]*

[Townsend 1956] A. A. Townsend, *The Structure of Turbulent Shear Flow*, Cambridge University Press, 1956. *[565]*

[Tritton 1988] D. J. Tritton, *Physical Fluid Dynamics*, Oxford University Press, 1988. *[564]*

[Versteeg and Malalasekera 1995] H. K. Versteeg and W. Malalasekera, *An Introduction to Computational Fluid Dynamics*, Prentice-Hall, 1995. *[323]*

[Vogel 1988] S. Vogel, *Life's Devices*, Princeton University Press, 1988. *[83, 139]*

[Vogel 1994] S. Vogel, *Life in Moving Fluids*, Princeton University Press, 1994. *[287, 298]*

[Vogel 1998] S. Vogel, *Cats' Paws and Catapults*, Penguin Books, 1998. *[139]*

[Weinberg 1972] S. Weinberg, *Gravitation and Cosmology*, John Wiley & Sons, Inc., 1972. *[201]*

[Weinberg 2008] S. Weinberg, *Cosmology*, Oxford University Press, 2008. *[201, 203, 204]*

[White 1991] F. M. White, *Viscous Fluid Flow*, McGraw-Hill, 1991. *[272, 276, 295, 298, 481, 501, 505, 507, 510]*

[White 1999] F. M. White, *Fluid Mechanics*, McGraw-Hill, 1999. *[213, 240, 493]*

Articles

[ABS&03] A. Andersen, T. Bohr, B. Stenum, J. J. Rasmussen, and B. Lautrup. Anatomy of a bathtub vortex. *Physical Review Letters* 91, 104502–1 (2003). *[318]*

[ABS&06] A. Andersen, T. Bohr, B. Stenum, J. J. Rasmussen, and B. Lautrup. The bathtub vortex in a rotating container. *Journal of Fluid Mechanics* 556, 121–146 (2006). *[318]*

[AKO05] S. Ansumali, I. V. Karlin, and H. C. Öttinger. Thermodynamic Theory of Incompressible Hydrodynamics. *Physical Review Letters* 94, 080602.1–4 (2005). *[389]*

[ALB03] A. Andersen, B. Lautrup, and T. Bohr. Averaging method for nonlinear laminar Ekman layers. *Journal of Fluid Mechanics* 487, 81–90 (2003). *[320]*

[And05] J. D. Anderson, Jr. Ludwig Prandtl's boundary layer. *Physics Today* 58, 42–48 (2005). *[481]*

[BAB09] D. Bonn, A. Andersen, and T. Bohr. Hydraulic jumps in a channel. *Journal of Fluid Mechanics* 618, 71–87 (2009). *[447, 447]*

[Bau03] R. H. Baughman. Auxetic materials: Avoiding the shrink. *Nature* 425, 667 (2003). *[131]*

[BD01] J. M. Birch and M. H. Dickinson. Spanwise flow and the attachment of the leading-edge vortex on insect wings. *Nature* 412, 729–733 (2001). See also pages 688–689 in the same issue. *[255]*

[Bea39] J. A. Bearden. A precision determination of the viscosity of air. *Physical Review* 56, 1023–1040 (1939). *[293]*

[Ben66] D. J. Benney. Long waves in liquid films. *Journal of Mathematics and Physics (MIT)* 45, 150–155 (1966). *[267]*

[BH06a] W. Braitsch and H. Haas. Turbines for hydroelectric power. In *Advanced Materials and Technologies,* volume VIII/3C. Landolt-Börnstein (2006). *[366]*

[BH06b] J. W. M. Bush and D. L. Hu. Walking on water: Biolocomotion at the interface. *Annual Review of Fluid Mechanics* 38, 339–369 (2006). *[83]*

[BHM03] A. Juel B. Hof and T. Mullin. Scaling of the turbulence transition in a pipe. *Physical Review Letters* 91, 244502–1–4 (2003). *[270]*

[BPW97] T. Bohr, V. Putkaradze, and S. Watanabe. Averaging theory for the structure of hydraulic jumps and separation in laminar free-surface flows. *Physical Review Letters* 79, 1038–1041 (1997). *[447, 448]*

[BS91] K. Y. Billah and R. H. Scanlan. Resonance, Tacoma Narrows bridge failure, and undergraduate physics textbooks. *American Journal of Physics* 59, 118–124 (1991). *[471]*

[BT08] C. Barbarosie and A-M. Toader. Saint-Venant's principle and its connections to shape and topology optimization. *Zeitschrift für Angewandte Mathematik und Mechanik* 88, 23–32 (2008). *[142]*

[CAL95] B. Christiansen, P. Alstrøm, and M. T. Levinsen. Dissipation and ordering in capillary waves at high aspect ratios. *Journal of Fluid Mechanics* 291, 323–341 (1995). *[430]*

[CDDA&96] J. Christensen-Dalsgaard, W. Dppen, S. V Ajukov, E. R. Anderson, H. M. Antia, S. Basu, V. A. Baturin, G. Berthomieu, B. Chaboyer, S. M. Chitre, A. N. Cox, P. Demarque, J. Donatowicz, W. A. Dziembowski, M. Gabriel, D. O. Gough, D. B. Guenther, J. A. Guzik, J. W. Harvey, F. Hill, G. Houdek, C. A. Iglesias, A. G. Kosovichev, J. W. Leibacher, P. Morel, C. R. Proffitt, J. Provost, J. Reiter, E. J. Rhodes Jr., F. J. Rogers, I. W. Roxburgh, M. J. Thompson, and R. K. Ulrich. The current state of solar modeling. *Science* 272, 1286–1292 (1996). *[38, 39, 40]*

[CGM06] N. Chouaieb, A. Goriely, and J. H. Maddocks. Helices. *Proceedings of the National Academy of Sciences* 103, 9398–9403 (2006). *[173]*

[Col56] D. Coles. The law of the wake in the turbulent boundary layer. *Journal of Fluid Mechanics* 1, 191–226 (1956). *[579]*

[CW06] T. H. Colding and W. P. Minicozzi, II. Shapes of embedded minimal surfaces. *Proceedings of the National Academy of Sciences* 103, 11106–11111 (2006). *[73]*

[DG09] A. S. Dukhin and P. J. Goetz. Bulk viscosity and compressibility measurements using acoustic spectroscopy. *The Journal of Chemical Physics* 130, 124519.1–13 (2009). *[256, 385]*

[DGQC10] G. Dupeux, A. Le Goff, Davide Quéré, and C. Clanet. The spinning ball spiral. *New Journal of Physics* 12, 093004 (2010). *[225]*

[DMS05] P. Dontula, C. W. Macosko, and L. E. Scriven. Origins of concentric cylinders viscometry. *Journal of Rheology* 49, 807–818 (2005). *[277]*

[EDP84] R. Edgeworth, B. J. Dalton, and T. Parnell. The pitch drop experiment. *European Journal of Physics* 5, 198–200 (1984). *[244]*

[Egg97] J. Eggers. Nonlinear dynamics and breakup of free-surface flows. *Review of Modern Physics* 69, 865–929 (1997). *[90]*

[Ell08] G. Ellis. Patchy solutions. *Nature* 452, 158–161 (2008). *[201]*

[ES48] E. R. G. Eckert and E. Soehngen. Studies on Heat Transfer in Laminar Free Convection with the Zehnder-Mach Interferometer. Technical Report 5747 U. S. Air Force (1948). See http://www.dtic.mil. *[551]*

[EV08] J. Eggers and E. Villermaux. Physics of liquid jets. *Reports on Progress in Physics* 71, 036601 (79 pages) (2008). *[78]*

[Fal77] R. E. Falco. Coherent motions in the outer region of turbulent boundary layers. *Physics of Fluids* 20, 124–132 (1977). *[492]*

[FM71] D. Fargie and B. W. Martin. Developing laminar flow in a pipe of circular cross-section. *Proceedings of the Royal Society (London)* A321, 461–476 (1971). *[272]*

[FS31] V. M. Falkner and S.W.Skan. Some approximate solutions of the boundary layer equations. *Philosophical Magazine, Series 7* 12, 865–896 (1931). *[499]*

[GC91] Y. Gagne and B. Castaing. A universal representation without global scaling invariance of energy spectra in developed turbulence. *C. R. Acad. Sci. Paris, series II* 212, 441–445 (1991). *[567]*

[Gen85] P-G. de Gennes. Wetting: statics and dynamics. *Reviews of Modern Physics* 57, 827–863 (1985). *[81]*

[GM01] E. Gerde and M. Marder. Friction and fracture. *Nature* 413, 285–288 (2001). *[97]*

[GPG&10] S. Gekle, I. R. Peters, J. M. Gordillo, D. van der Meer, and D. Lohse. Supersonic Air Flow due to Solid-Liquid Impact. *Physical Review Letters* 104, 024501(1–4) (2010). *[237]*

[GPL05] S. Goyal, N. C. Perkins, and C. L. Lee. Nonlinear dynamics and loop formation in Kirchoff rods with implications to the mechanics of DNA and cables. *Journal of Computational Physics* 209, 371–389 (2005). *[173]*

[GSM62] H. L. Grant, R. W. Stewart, and A. Moilliet. Turbulent spectra from a tidal channel. *Journal of Fluid Mechanics* 12, 241–268 (1962). *[568]*

[Hag03] W. H. Hager. Blasius: A life in research and education. *Experiments in Fluids* 34, 566–571 (2003). *[487]*

[HCB03] D. L. Hu, B. Chan, and J. W. M. Bush. The hydrodynamics of water strider locomotion. *Nature* 424, 663–666 (2003). *[83, 83]*

[HCFJ07] V. Hunyadi, D. Chrétien, H. Flyvbjerg, and I. M. Jánosi. Why is the microtubule lattice helical. *Biology of the Cell* 99, 117–128 (2007). *[149]*

[HCG&08] L. J. Hall, V. R. Coluci, D. S Galvao, M. E. Kozlov, M. Zhang, S. O. Dantas, and R. H. Baughman. Sign change of Poisson's ratio for carbon nanotube sheets. *Science* 320, 504–507 (2008). *[131]*

[Hea82] M. R. Head. In W. Merzkirch, Editor, *Flow visualisation* volume II, pages 399–403. Hemisphere, Washington (1982). *[482, 524]*

[HST06] Hideo Hoshi, Tadahiko Shinshi, and Setsuo Takatani. Third-generation blood pumps with mechanical noncontact magnetic bearings. *Artificial Organs* 30, 324–328 (2006). *[364]*

[HZ08] N. M. Holbrook and M. A. Zwieniecki. Transporting water to the tops of trees. *Physics Today* 61, 76–77 (2008). *[78]*

[IV02] H. Irago and J. M. Viano. Saint-Venant's principle in the asymptotic analysis of elastic rods with one end fixed. *Journal of Elasticity* 66, 21–46 (2002). *[142]*

[Jea02] J. H. Jeans. The stability of a spherical nebula. *Philosphical Transactions of the Royal Society (London)* 199, 1–53 (1902). *[232]*

[Jim87] J. Jimenez. On the linear stability of the inviscid Kármán vortex street. *Journal of Fluid Mechanics* 178, 177–194 (1987). *[470]*

[Kes01] D. A. Kessler. Surface physics: A new crack at friction. *Nature* 413, 260–261 (2001). *[97]*

[Kie03] M. K. H. Kiessling. Tbe "Jeans swindle", A true story—mathematically speaking. *Advances in Applied Mathematics* 31, 132–149 (2003). *[232]*

[KSH99] K. A. Kelly, S. Singh, and R. X. Huang. Seasonal variations of sea surface height in the gulf stream region. *Journal of Physical Oceanography* 29, 313–327 (1999). *[43]*

[Lak92] R. Lakes. No contractile obligations. *Nature* 358, 713–714 (1992). *[131]*

[Lan42] H. L. Langhaar. Steady flow in the transition length of a straight tube. *Journal of Applied Mechanics* 64, A55–A58 (1942). *[265, 272]*

[LHJ07] M. A. Lombardi, Th. P. Heavner, and S. R. Jefferts. NIST primary frequency standards and the realization of the SI second. *Measure* 2, 74–89 (2007). *[11]*

[LSQ06] E. Lorenceau, T. Senden, and D. Quére. Wetting of fibers. In R. G. Weiss and P. Terech, Editors, *Molecular gels. Materials with self-assembled fibrillar networks,* chapter 7, pages 223–237. Springer (2006). *[79]*

[Lun85] T. S. Lundgren. The vortical flow above the drain-hole in a rotating vessel. *Journal of Fluid Mechanics* 155, 381–412 (1985). *[475]*

[McD54] J. E. McDonald. The Shape of Raindrops. *Scientific American* 190, 64–68 (February 1954). *[72]*

[MD97] S. Martin and R. Drucker. The effect of possible Taylor columns on the summer ice retreat in the Chukchi Sea. *Journal of Geophysical Research* 102, 10473–10482 (1997). *[314]*

[Meh85] R. D. Mehta. Aerodynamics of sports balls. *Annual Review of Fluid Mechanics* 17, 151–189 (1985). *[296]*

[MKM99] R. D. Moser, J. Kim, and N. N. Mansour. Direct numerical simulation of turbulent channel flow up to $\mathsf{Re}_\tau = 590$. *Physics of Fluids* 11, 943–945 (1999). *[575, 581, 581, 582]*

[MLJ&04] B. J. McKeon, J. Li, W. Jiang, J.C. Morrison, and A. J. Smits. Further observations on the mean velocity distribution in fully developed pipe flow. *Journal of Fluid Mechanics* 501, 135–147 (2004). *[575, 576, 578, 579, 582, 583, 584]*

[MMQ&06] I. M. Mills, P. J. Mohr, T. J. Quinn, B. N. Taylor, and E. R. Williams. Redefinition of the kilogram, ampere, kelvin and mole: a proposed approach to implementing CIPM recommendation 1(CI-2005). *Metrologia* 43, 227–246 (2006). *[3]*

[Mos64] L. Moskowitz. Estimates of the power spectrums for fully developed seas for wind speeds of 20 to 40 knots. *Journal of Geophysical Research* 69, 5161–5179 (1964). *[440]*

[MPP72] P. C. Martin, O. Parodi, and P. S. Pershan. Unified hydrodynamic theory for crystals, liquid crystals, and normal fluids. *Physical Review* A6, 2401–2420 (1972). *[107]*

[MSZ&04] B. J. McKeon, C. J. Swanson, M. V. Zagarola, R. J. Donnelly, and A. J. Smits. Friction factors for smooth pipe flow. *Journal of Fluid Mechanics* 511, 41–44 (2004). *[583, 584]*

[MY10] S. M. Mahajan and Z. Yoshida. Twisting Space-Time: Relativistic Origin of Seed Magnetic Field and Vorticity. *Physical Review Letters* 105, 095005.1–4 (2010). *[465]*

[NAT&05] R. I. Nigmatulin, I. Sh. Akhatov, A. S. Topolnikov, R. Kh. Bolotnova, and N. K. Vakhitova. Theory of supercompression of vapor bubbles and nanonscale thermonuclear fusion. *Physics of Fluids* 17, 107106+31 (2005). *[209]*

[O'M10] R. E. O'Malley. Singular perturbation theory: A viscous flow out of Göttingen. *Annual Review of Fluid Mechanics* 42, 1–17 (2010). *[481]*

[Orz71] S. A. Orzag. Accurate solution of the Orr–Sommerfeld stability equation. *Journal of Fluid Mechanics* 50, 689–703 (1971). *[264]*

[Osc95] R. J. Osczevski. The basis of wind chill. *Arctic* 48, 372–382 (1995). *[492]*

[PCC99] A. Peters, K. Y. Chung, and S. Chu. Measurement of gravitational acceleration by dropping atoms. *Nature* 400, 849–852 (1999). *[65, 65, 66, 66]*

[Pie64] W. J. Pierson. The interpretation of wave spectrums in terms of the wind profile instead of the wind measured at a constant height. *Journal of Geophysical Research* 69, 5191–5203 (1964). *[440]*

[PLJ&06] F. Pampaloni, G. Lattanzi, A. Jonás, Th. Surrey, E. Frey, and E-L. Florin. Thermal fluctuations of grafted microtubules provide evidence of a length-dependent persistence length. *Proceedings of the National Academy of Sciences* 103, 10248–10253 (2006). *[149]*

[PM64] W. J. Pierson and L. Moskowitz. A proposed spectral form for fully developed wind seas based on the similarity theory of S. A. Kitaigorodskii. *Journal of Geophysical Research* 69, 5181–5190 (1964). *[440]*

[Poh21] E. Pohlhausen. Der Wärmeaustausch zwischen festern Körpen und Flüssigkeiten mit kleiner Reibung und klainer Wẅarmeleitung. *Zeitschrift für Angewandte Mathematik und Mechanik* 1, 115–121 (1921). *[507]*

[PP77] M. S. Plesset and A. Prosperetti. Bubble dynamics and cavitation. *Annual Review of Fluid Mechanics* 9, 145–185 (1977). *[209]*

[Ray78] Lord Rayleigh. On the instability of jets. *Proceedings of the London Mathematical Society* s1-10, 4–12 (1878). *[80]*

[Ray90] Lord Rayleigh. On bells. *Philosophical Magazine* 29, 1–17 (1890). *[403]*

[Ray14] Lord Rayleigh. On the theory of long waves and bores. *Proceedings of the Royal Society (London)* A90, 234–328 (1914). *[443]*

[Ray17] Lord Rayleigh. On the pressure developed in a liquid during the collapse of a spherical cavity. *Philosophical Magazine* 34, 94–98 (1917). *[209]*

[Rey83] Osborne Reynolds. An experimental investigation of the circumstances which determine whether the motion of water shall be direct or sinuous, and of the law of resistance in parallel channels. *Philosphical Transactions of the Royal Society (London)* 174, 935–982 (1883). *[270, 274, 274, 274]*

[RJ71] D. F. Rutland and G. J. Jameson. A non-linear effect in the capillary instability of liquid jets. *Journal of Fluid Mechanics* 46, 267–71 (1971). *[78]*

[RR02] M. A. Ritter and E. B. Ringelstein. The Venturi effect and cerebrovascular ultrasound. *Cerebrovascular Diseases* 14, 98–104 (2002). *[214]*

[RZ09] S. E. Rugh and H. Zinkernagel. On the physical basis of cosmic time. *Studies in History and Philosophy of Modern Physics* 40, 1–19 (2009). *[11]*

[SBD&07] D. N. Spergel, R. Bean, O. Doré, M. R. Nolta, C. L. Bennett, J. Dunkley, G. Hinshaw, N. Jarosik, E. Komatsu, L. Page, H. V. Peiris, L. Verde, M. Halpern, R. S. Hill, A. Kogut, M. Limon, S. S. Meyer, N. Odegard, G. S. Tucker, J. L. Weiland, E. Wollack, and E. L. Wright. Three-year Wilkinson microwave anisotropy probe (WMAP) observations: implications for cosmology. *The Astrophysical Journal Supplement Series* 170, 377–408 (2007). *[201, 201, 203]*

[SBH97] V. Shtern, A. Borissov, and F. Hussain. Vortex sinks with axial flow: Solution and applications. *Physics of Fluids* 9, 2941–2959 (1997). *[474]*

[Sed46] L. I. Sedov. Propagation of strong shock waves. *Journal of Applied Mathematics and Mechanics* 10, 241–250 (1946). *[454]*

[SF02] R. M. Sadri and J. M. Floryan. Entry flow in a channel. *Computers and Fluids* 31, 133–157 (2002). *[334, 335]*

[SP45] P. A. Siple and C. F. Passel. Measurements of dry atmospheric cooling in subfreezing temperatures. *Journal of the American Philosophical Society* 89, 177–199 (1945). *[491]*

[ST02] R. B. Srygley and A. L. R. Thomas. Unconventional lift-generating mechanisms in free-flying butterflies. *Nature* 420, 660–664 (2002). *[519]*

[Str78] V. Strouhal. Über ein besondere Art der Tonerregung. *Annalen der Physik und Chemie (Leipzig)* pages 216–251 (1878). *[471]*

[SY08] Y. A. Stepanyants and G. H. Yeoh. Stationary bathtub vortices and a critical regime of liquid discharge. *Journal of Fluid Mechanics* 604, 77–98 (2008). *[475]*

[Tan79] S. Taneda. Visualization of separating stokes flows. *Journal of the Physical Society of Japan* 46, 1935–1942 (1979). *[294]*

[Tat64] T. Tate. On the magnitude of a drop of liquid formed under different circumstances. *Philosophical Magazine* 27, 176–180 (1864). *[91]*

[Tay50] G. I. Taylor. The formation of a blast wave by a very intense explosion. *Proceedings of the Royal Society (London)* A201, 159–186 (1950). *[454]*

[TBF&65] L. M. Trefethen, R. W. Bilger, P. T. Fink, R. E. Luxton, and R. I. Tanner. The bath-tub vortex in the southern hemisphere. *Nature* 207, 1084–1085 (1965). *[321]*

[TDH75] C. W. Titman, P. A. Davies, and P. M. Hilton. Taylor columns in a shear flow and Jupiter's Great Red Spot. *Nature* 255, 538–539 (1975). *[314]*

[Teg08] M. Tegmark. The second law and cosmology. In G. P. Beretta, A. F. Ghoneim, and G. N. Hatsopoulos, Editors, *Meeting the Entropy Challenge*. American Institute of Physics (2008). *[393]*

[TL03] R. Toegel and D. Lohse. Phase diagrams for sonoluminescing bubbles: A comparison between experiment and theory. *Journal of Chemical Physics* 118, 1863–1875 (2003). *[209]*

[Tyr03] M. T. Tyree. The ascent of water. *Nature* 423, 923 (2003). *[78]*

[Urs53] F. Ursell. The long-wave paradox in the theory of gravity waves. *Proceedings of the Cambridge Philosophical Society* 49, 685–694 (1953). *[421]*

[vM45] R. von Mises. On Saint-Venant's principle. *Bulletin of the American Mathematical Society* 51, 555–562 (1945). *[142]*

[VSM02] C. D. Volpe, S. Siboni, and M. Morra. Comments on some recent papers on interfacial tension and contact angles. *Langmuir* 18, 1441–1444 (2002). *[82]*

[Wer80] H. Werle. Transition et décollement: visualisations au tunnel hydrodynamique de l'ONERA. *Rech. Aerosp. (France)* 1980-5, 35–49 (1980). *[296]*

[Whe89] J. A. Wheeler. It from bit. In *Proceedings of the 3rd International Symposium on Foundations of Quantum Mechanics, Tokyo,* pages 354–368 (1989). *[17]*

[WHN09] A. G. Whittington, A. M. Hofmeister, and P. I. Nabelek. Temperature-dependent thermal diffusivity of the Earth's crust and implications for magmatism. *Nature* 458, 319–321 (2009). *[376]*

[WL05] Y. C. Wang and R. S. Lakes. Composites with inclusions of negative bulk modulus: Extreme damping and negative Poisson's Ratio. *Journal of Composite Materials* 39, 1645–1657 (2005). *[131]*

[WPB99] E. D. Wilkes, S. D. Phillips, and O. A. Basaran. Computational and experimental analysis of dynamics of drop formation. *Physics of Fluids* 11, 3577–3598 (1999). *[90]*

[WT05] J. M. Weisberg and J. H. Taylor. The relativistic binary pulsar B1913+16: Thirty years of observation and analysis. In *Binary Radio Pulsars,* volume 328 of *ASP Conference Series,* pages 25–31 (2005). *[17]*

[YFAR00] M. F. Yu, B. S. Files, S. Arepalli, and R. S. Ruoff. Tensile loading of ropes of single wall carbon nanotubes and their mechanical properties. *Physical Review Letters* 84, 5552 (2000). *[100, 127]*

[YTW06] D. Yuan, Z. Tian, and S. Wang. Mechanical and electrical design of the Three Gorges Project. In *Advanced Materials and Technologies/Hydroelectric Power,* volume VIII/3C, pages 73–128. Landolt-Börnstein (2006). *[369, 369]*

[YXB05] O. E. Yildirim, Qi Xu, and O. A. Basaran. Analysis of the drop weight method. *Physics of Fluids* 17, 062107.1–062107.13 (2005). *[91]*

[Zan98] E. D. Zanotto. Do cathedral glasses flow? *American Journal of Physics* 66, 392–395 (1998). *[58, 244]*

Links

[1] National Institute of Standards and Technology (NIST), Physical Reference Data.
http://www.nist.gov/index.html. *[3, 11, 13, 39]*

[2] G. W. C. Kaye and T. H. Laby, Tables of Physical and Chemical Constants.
http://www.kayelaby.npl.co.uk/. *[8, 245]*

[3] J. Christensen-Dalsgaard, Solar model data.
http://www.phys.au.dk/ jcd/solar_models. *[39]*

[4] U. S. Geological Survey, Geology section.
http://geology.usgs.gov. *[43]*

[5] NASA, Jupiter:Moons:Io.
http://solarsystem.nasa.gov/planets/profile.cfm?Object=Io. *[66]*

[6] J. Bush, Lecture notes on fluid mechanics.
http://web.mit.edu/1.63/www/Lec-notes/. *[73]*

[7] Extreme bubble team, World's Largest Free-floating Soap Bubble.
http://worldslargestbubble.com/index.html. *[75]*

[8] Martin Chapman, Water Properties.
http://www.lsbu.ac.uk/water/data.html. *[82]*

[9] Wikipedia, V-2.
http://en.wikipedia.org/wiki/V-2. *[239]*

[10] Wikipedia, Space Shuttle Main Engine.
http://en.wikipedia.org/wiki/SSME. *[239]*

[11] Denmark Juvenco A/S, Axial Ventilator SMP.
http://www.juvenco.dk/. *[365]*

[12] Central Intelligence Agency (CIA), The World Factbook/Rank Order/Electricity/Consumption.
https://www.cia.gov/library/publications/the-world-factbook. *[369]*

[13] Three Gorges Power Plant Animation.
http://www.youtube.com/watch?v=tjTh7A4jnbc. *[369]*

[14] China Three Gorges Project Corporation.
http://www.ctgpc.com. *[369]*

[15] US Government Energy Information Administration, International Energy Statistics.
http://www.eia.doe.gov/. *[373]*

[16] US National Weather Service, Office of Climate, Water, and Weather Services.
http://www.weather.gov/om/windchill/. *[492]*

[17] A. Filippone, Advanced Topics in Aerodynamics.
http://aerodyn.org. *[513]*

[18] R. D. Moser, J. Kim, and N. N. Mansour, DNS Data for Turbulent Channel Flow.
http://turbulence.ices.utexas.edu/MKM_1999.html. *[575]*

[19] M. V. Zagarola and A. J. Smits, Data on turbulent pipe flow.
http://gasdyn.princeton.edu/data/e247/zagarola_data.html. *[575, 576, 579, 582]*

Index